Useful Combinations of Constants

$\hbar = h/2\pi = 1.0546 \times 10^{-34}\ \text{J} \cdot \text{s} = 6.5821 \times 10^{-16}\ \text{eV} \cdot \text{s}$

$hc = 1.9864 \times 10^{-25}\ \text{J} \cdot \text{m} = 1239.8\ \text{eV} \cdot \text{nm}$

$\hbar c = 3.1615 \times 10^{-26}\ \text{J} \cdot \text{m} = 197.33\ \text{eV} \cdot \text{nm}$

$$\frac{1}{4\pi\epsilon_0} = 8.9876 \times 10^{9}\ \text{N} \cdot \text{m}^2 \cdot \text{C}^{-2}$$

Compton wavelength $\lambda_c = \dfrac{h}{m_e c} = 2.4263 \times 10^{-12}\ \text{m}$

$$\frac{e^2}{4\pi\epsilon_0} = 2.3071 \times 10^{-28}\ \text{J} \cdot \text{m} = 1.4400 \times 10^{-9}\ \text{eV} \cdot \text{m}$$

Fine structure constant $\alpha = \dfrac{e^2}{4\pi\epsilon_0 \hbar c} = 0.0072974 \approx \dfrac{1}{137}$

Bohr magneton $\mu_{\text{B}} = \dfrac{e\hbar}{2m_e} = 9.2740 \times 10^{-24}\ \text{J/T} = 5.7884 \times 10^{-5}\ \text{eV/T}$

Nuclear magneton $\mu_{\text{N}} = \dfrac{e\hbar}{2m_p} = 5.0508 \times 10^{-27}\ \text{J/T}$
$= 3.1525 \times 10^{-8}\ \text{eV/T}$

Bohr radius $a_0 = \dfrac{4\pi\epsilon_0 \hbar^2}{m_e e^2} = 5.2918 \times 10^{-11}\ \text{m}$

Hydrogen ground state $E_0 = \dfrac{e^2}{8\pi\epsilon_0 a_0} = 13.606\ \text{eV} = 2.1799 \times 10^{-18}\ \text{J}$

Rydberg constant $\text{R}_\infty = \dfrac{\alpha^2 m_e c}{2h} = 1.09737 \times 10^{7}\ \text{m}^{-1}$

Hydrogen Rydberg $R_{\text{H}} = \dfrac{\mu}{m_e} R_\infty = 1.09678 \times 10^{7}\ \text{m}^{-1}$

Gas constant $R = N_{\text{A}} k = 8.3145\ \text{J} \cdot \text{mol}^{-1} \cdot \text{K}^{-1}$

Magnetic flux quantum $\Phi_0 = \dfrac{h}{2e} = 2.0678 \times 10^{-15}\ \text{T} \cdot \text{m}^2$

Classical electron radius $= r_e = \alpha^2 a_0 = 2.8179 \times 10^{-15}\ \text{m}$

$kT = 2.5249 \times 10^{-2}\ \text{eV} \approx \dfrac{1}{40}\ \text{eV}$ at $T = 293$ K

Note: The latest values of the fundamental constants can be found at the National Institute of Standards and Technology website at http://physics.nist.gov/cuu/Constants/index.html

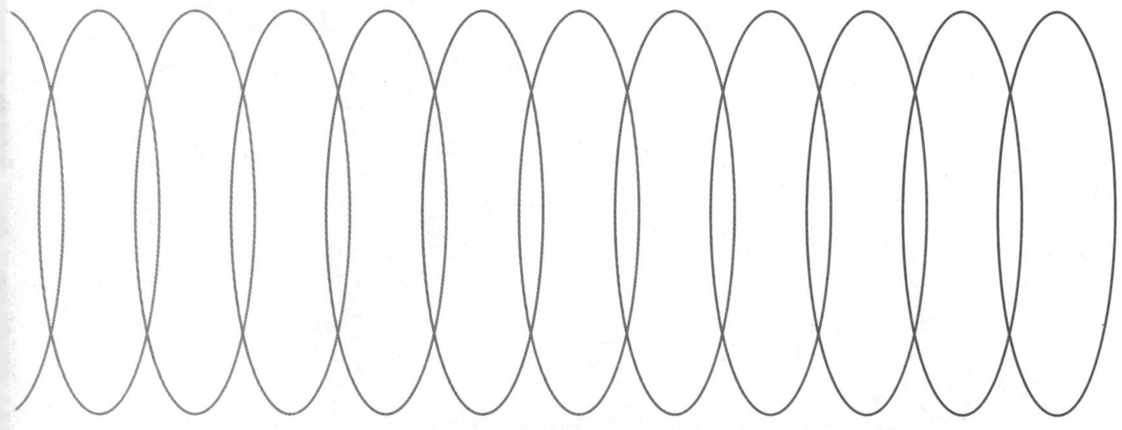

MODERN PHYSICS

For Scientists and Engineers

MODERN PHYSICS

For Scientists and Engineers

Second Edition

STEPHEN T. THORNTON
University of Virginia

ANDREW REX
University of Puget Sound

BROOKS/COLE
THOMSON LEARNING™

Australia • Canada • Mexico • Singapore • Spain
United Kingdom • United States

BROOKS/COLE

THOMSON LEARNING

Physics Editor: John Vondeling
Developmental Editor: Ed Dodd
Marketing Strategist: Pauline Mula
Production Manager: Charlene Catlett Squibb
Production Service: Graphic World Publishing Services
Art Director: Cara Castiglio
Cover Image: Randy Montoya, Sandia National Laboratories
Cover Printer: Phoenix Color
Compositor: Graphic World
Printer: Von Hoffmann Press—Jefferson City, MO

Printed in the United States of America
5 6 7 05 04 03

For more information about our products, contact us at:
Thomson Learning Academic Resource Center
1-800-423-0563

For permission to use material from this text, contact us by:
Phone: 1-800-730-2214
Fax: 1-800-730-2215
Web: http://www.thomsonrights.com

Library of Congress Catalog Card Number: 99-22847
MODERN PHYSICS, Second Edition
ISBN: 0-03-006049-4

Asia
Thomson Learning
60 Albert Street, #15-01
Albert Complex
Singapore 189969

Australia
Nelson Thomson Learning
102 Dodds Street
South Melbourne, Victoria 3205
Australia

Canada
Nelson Thomson Learning
1120 Birchmount Road
Toronto, Ontario M1K 5G4
Canada

Europe/Middle East/Africa
Thomson Learning
Berkshire House
168-173 High Holborn
London WC1 V7AA
United Kingdom

Latin America
Thomson Learning
Seneca, 53
Colonia Polanco
11560 Mexico D.F.
Mexico

Spain
Paraninfo Thomson Learning
Calle/Magallanes, 25
28015 Madrid, Spain

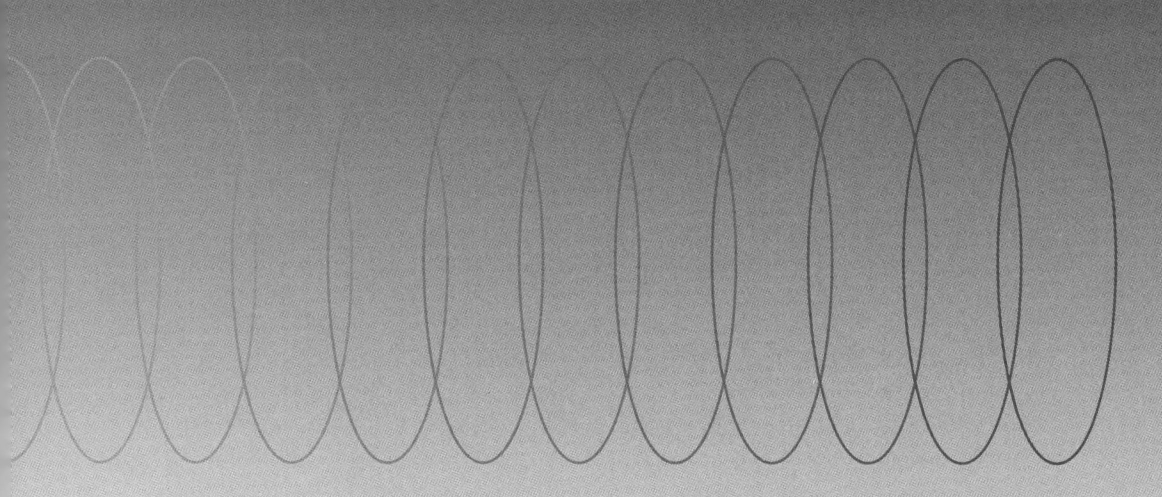

PREFACE

Our objective in writing the first edition of this book was to produce a textbook for both a one- or two-semester modern physics course for physics and engineering students. Such a course normally follows a full-year, introductory calculus-based physics course for freshmen or sophomores. When we decided to begin work on this second edition, we had the publisher send out a questionnaire to users of modern physics books. We received many replies, and we can succinctly summarize their suggestions:

1. *The topics and coverage in the first edition were done well; no major revision is necessary.*
2. *Try to make the text easier to use in a one-semester course.*
3. *Add new problems and change the number values in the old ones.*

We have taken the users' comments seriously, and the major changes in the second edition have been to transfer one of the original core chapters (general relativity) to the back of the book and to add many revised and new problems. We will present later our recommendations for a one- and two-semester modern physics course.

The first edition of our text established a trend for a contemporary approach to the exciting, thriving, and changing field of modern science. After briefly visiting the status of physics at the turn of the last century, we cover relativity and quantum theory, the basis of any study of modern physics. Almost all areas of science depend on quantum theory and the methods of experimental physics. The latter part of the book is devoted to the subfields of physics (atomic, condensed matter, nuclear, and particle) and the exciting field of cosmology. Our experience is that science and engineering majors particularly enjoy the study of modern physics after the sometimes-laborious study of classical mechanics, electricity, and magnetism. The mathematics level is not difficult for the most part, and students feel they are finally getting to the frontiers of physics. We have brought the study of modern physics alive by alluding to many current applications and challenges in physics, for example, high temperature superconductors, neutrino mass, age of the universe, gamma ray bursts, holography, and nuclear fusion. Modern physics texts need to be updated periodically to include recent advances in our understanding. Although we have emphasized modern applications, we also provide the sound theoretical basis for quantum theory that will be needed by physics majors in their upper division and graduate courses.

TEACHING SUGGESTIONS

The text has been used extensively in a one-semester course for engineering students at the University of Virginia and in a two-semester course for physics and pre-engineering students at the University of Puget Sound. Both one- and two-semester courses should cover the material through the establishment of the periodic table in Chapter 8 with few exceptions. We have eliminated the denoting of optional sections, because we believe that depends on the Instructor, but we feel Sections 2.4, 3.5, 3.9, 4.2, 6.4, 6.6, 7.2, 7.6, 8.2, and 8.3 from the first nine chapters might be optional. Our suggestions for the one- and two-semester courses are then

One-semester: Chapters 1–9 and selected other material as chosen by the Instructor. Skip the optional sections listed above.

Two-semester: Chapters 1–16 with supplementary material as desired, with possible student projects.

FEATURES

The end-of-chapter problems are a featured strength of this text. Over 200 questions and more than 700 problems provide the Instructor with ample material from which to select homework problem assignments without repeating the same problems every year. Our users suggested changing the values of numbers in many problems of the first edition, which we have done, in addition to adding over 100 new problems. We have tried to make the questions thought provoking and have given questions that have actual answers. The end-of-chapter problems have been separated by section, and general problems are included at the end to allow assimilation of the material. The easier problems are generally listed first within a section, and the more difficult ones are denoted by this special symbol ●. A few computer-based problems are given in the text, but no computer disk supplement is provided because many computer software programs are commercially available.

An *Instructor's Solutions Manual* prepared by us is available to the Instructor (by contacting your local Saunders sales representative). This manual has the *solution to every end-of-chapter problem* and has been checked by at least two physics professors. The answers to the odd-numbered problems are given at the end of the textbook itself. A *Student Solutions Manual* that contains the solutions to about 25% of the end-of-chapter problems is also available for sale to the students.

The two-color format helps to present clear illustrations and to highlight material in the text; for example, important and useful equations are highlighted in blue, and the most important part of each illustration is rendered in thick blue lines. Blue margin notes help guide the student to the important points, and the margins allow the student to make his or her own notes. The first time key words or topics are introduced they are set in **boldface**, and *italics* are used for emphasis. Although we had a large number of worked examples in the first edition, we have added additional new ones in this text. The examples have been presented in the manner in which students are expected to work the end-of-chapter problems. Problem solving does not come easy for most students, especially the problems requiring several steps (that is, not simply plugging numbers into one equation). We expect that the many text examples of varying degrees of difficulty will help the students.

Upon the advice of our users, we have deleted the essays from the first edition, but they will be placed on the textbook's website listed below. We have retained the Special Topic boxes that both students and professors find particularly interesting. We have added new ones to enliven interest in current and historical subjects, which range from the discovery of new elements, to the application of nuclear physics in discovering art forgeries, to the amazing Cavendish Laboratory in Cambridge. We include historical aspects of modern physics that some students will find interesting and that others can simply ignore. Photographs of physicists at work and of original apparatus help to enliven and humanize the material.

The most exciting, and perhaps the most useful new feature, is the establishment of a website for the book. The site can be accessed at www.brookscole.com, the University of Virginia (www.modern.physics.virginia.edu) and the University of Puget Sound (www.ups.edu/physics/faculty/rex/book.html). We will post errata, present new exciting results, add new examples and end-of-chapter problems, give links to other sites that have particularly interesting features like simulations and photos, among other things. Much of the line art from the text will be available through the website, which will allow instructors to easily retrieve the figures for use in the classroom lecture presentation.

The *Modern Physics Instructor's Resource CD-ROM* is a dynamic lecture tool available to instructors that will contain many of the text figures for easy use in the classroom lectures. It can be used in conjunction with commercial presentation software such as PowerPoint™, Persuasion™, and Podium™, or you can use the presentation tool included on the CD-ROM. The CD-ROM is for both Macintosh and Windows platforms, and also features .pdf files of the *Instructor's Solutions Manual* and the correlation guide to the Overhead Transparencies from the first edition.

ACKNOWLEDGMENTS

We acknowledge the assistance of many persons who have helped with this text. Colleagues at the University of Virginia who class tested and used the text include Julian Noble, Vittorio Celli, and Bradley Cox. We particularly acknowledge the valued help of Julian Noble in earlier stages of the first edition. We acknowledge the professional staff at Saunders College Publishing who helped make this a useful, popular, and attractive text. They include Developmental Editors Jennifer Bortel (who worked on the first edition of this text) and Edward Dodd (who worked on the second); Charlene Catlett Squibb, Production Manager, who kept the production process on track, and Cara Castiglio, Art Director, who created both the text and cover design for the text you are holding. We also want to thank the many individuals who gave us critical reviews and suggestions since the first edition. The survey respondents include Bruce Barrett, University of Arizona; Herbert L. Berk, University of Texas at Austin; Patricia C. Boeshaar, Drew University; Douglas Brothers, Benedictine College; Tom Christensen, University of Colorado at Colorado Springs; David R. Curott, University of North Alabama; A. Erwin, University of Wisconsin; Jane D. Flood, Muhlenberg College;

Robert Fuller, University of Nebraska-Lincoln; Kyle Hathcop, Union University; Lyle Hoffman, Lafayette College; C. C. Huang, University of Minnesota; Donald Isenhower, Abilene Christian University; C. Jahncke, Saint Lawrence University; Kirby W. Kemper, Florida State University; Eric Kincanon, Gonzaga University; Robert V. Krotkov, University of Massachusetts, Amherst; Michael B. Kruger, University of Missouri-Kansas City; Yuichi Kubota, University of Minnesota-Minneapolis; Juan Lin, Washington College; Vern Lindberg, Rochester Institute of Technology; Henry Nebel, Alfred University; Victor J. Newton, Fairfield University; Jae Young Park, North Carolina State University; Rodney B. Piercey, Mississippi State University; Larry Robinson, Austin College; Robert N. Rogers, San Francisco State University; Serge Rudaz, University of Minnesota-Minneapolis; Michael Sadler, Abilene Christian University; Joseph A. Schaefer, Loras College; Thor Stromberg, New Mexico State University; Edward A. Vondrak, University of Indianapolis; C. Wesley Walter, Denison University; J. J. Wilson, University of Wisconsin-Platteville; and Jerzy M. Wrobel, University of Missouri-Kansas City.

REVIEWERS

Darin Acosta, University of Florida

Roger Bengtson, University of Texas at Austin

Michael Graf, Boston College

Moo-Young Han, Duke University

Tom R. Herrmann, Eastern Oregon University

Ronald Jodoin, Rochester Institute of Technology

Joseph F. Owens, Florida State University

Robert Pompi, State University of New York at Binghamton

Thor Stromberg, New Mexico State University

We especially want to acknowledge the valuable help of Richard R. Bukrey of Loyola University of Chicago who has helped us in many ways through his enlightening reviews, careful manuscript proofing, and checking the end-of-chapter problem solutions.

We have gone to extreme measures in attempting to ensure the publication of an error-free text. The manuscript, galleys, and page proofs have been carefully checked by both authors, the editors, and numerous reviewers. While we realize that a 100% error-free text may not be humanly possible, *Modern Physics for Scientists and Engineers,* Second Edition is very close. However, should you come across an error, please feel free to bring it to our attention by contacting us, and we will see that the error is corrected in subsequent printings of this book.

Stephen T. Thornton
University of Virginia
Charlottesville, Virginia
stt@virginia.edu

Andrew Rex
University of Puget Sound
Tacoma, Washington
rex@ups.edu

CONTENTS OVERVIEW

1	One Hundred Years Ago	1
2	Special Theory of Relativity	20
3	The Experimental Basis of Quantum Theory	80
4	Structure of the Atom	118
5	Wave Properties of Matter	151
6	The Quantum Theory	185
7	The Hydrogen Atom	222
8	Many-Electron Atoms	249
9	Statistical Physics	272
10	Molecules and Solids	310
11	Semiconductor Theory and Devices	361
12	The Atomic Nucleus	392
13	Nuclear Interactions and Applications	431
14	Elementary Particles	474
15	General Relativity	506
16	Cosmology—The Beginning and the End	523
	Appendices	A-1
	Answers to Selected Problems	A-43
	Index	I-1

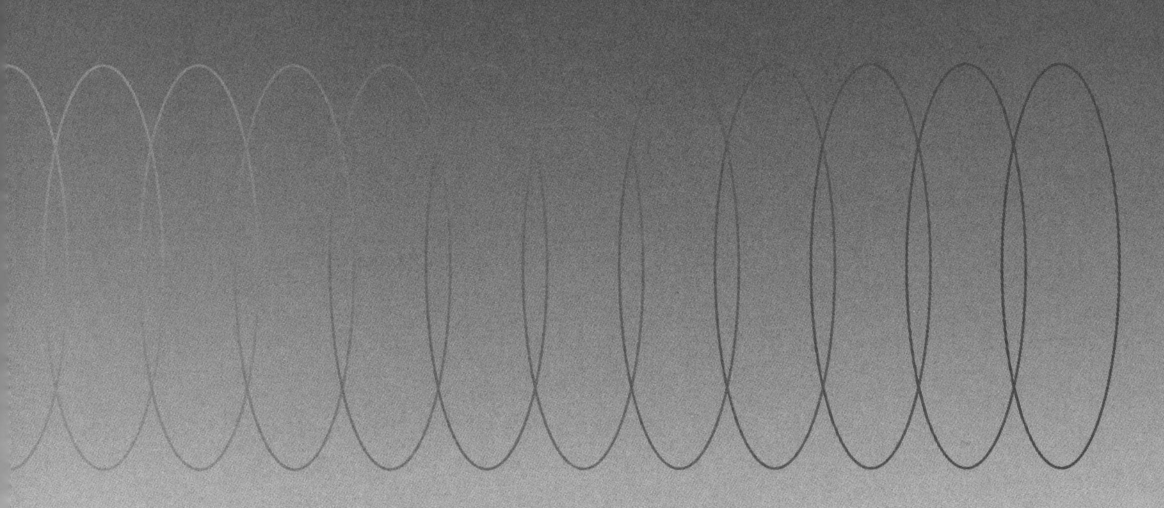

CONTENTS

CHAPTER 1
ONE HUNDRED YEARS AGO 1

1.1 CLASSICAL PHYSICS OF THE 1890S 2
Mechanics 3
Electromagnetism 5
Thermodynamics 5

1.2 THE KINETIC THEORY OF GASES 6

1.4 WAVES AND PARTICLES 9

1.4 CONSERVATION LAWS AND FUNDAMENTAL FORCES 11
Fundamental Forces 11

1.5 THE ATOMIC THEORY OF MATTER 14

1.6 OUTSTANDING PROBLEMS OF 1895 AND NEW HORIZONS 16
On the Horizon 18

SUMMARY 18

CHAPTER 2
SPECIAL THEORY OF RELATIVITY 20

2.1 HISTORICAL PERSPECTIVE 21

2.2 THE MICHELSON-MORLEY EXPERIMENT 22

2.3 EINSTEIN'S POSTULATES 27

2.4 THE LORENTZ TRANSFORMATION 30

2.5 TIME DILATION AND LENGTH CONTRACTION 33
Time Dilation 33
Length Contraction 36

2.6 ADDITION OF VELOCITIES 39

2.7 EXPERIMENTAL VERIFICATION 42
Muon Decay 43
Atomic Clock Measurement 44
Velocity Addition 45

2.8 TWIN PARADOX 46

2.9 SPACETIME 48

2.10 DOPPLER EFFECT 51

SPECIAL TOPIC:
Applications of the Doppler Effect 54

2.11 RELATIVISTIC MOMENTUM 57

2.12 RELATIVISTIC ENERGY 60
Total Energy and Rest Energy 63
Equivalence of Mass and Energy 63
Relationship of Energy and Momentum 65
Massless Particles 65

2.13 COMPUTATIONS IN MODERN PHYSICS 66
Binding Energy 68

2.14 ELECTROMAGNETISM AND RELATIVITY 70

SUMMARY 73

CHAPTER 3
THE EXPERIMENTAL BASIS OF QUANTUM THEORY 80

3.1 DISCOVERY OF THE X RAY AND THE ELECTRON 81
3.2 DETERMINATION OF ELECTRON CHARGE 84
3.3 LINE SPECTRA 87

SPECIAL TOPIC:
The Discovery of Helium 88

3.4 QUANTIZATION 91
3.5 BLACKBODY RADIATION 91
3.6 PHOTOELECTRIC EFFECT 97
Experimental Results of Photoelectric Effect 97
Classical Interpretation 99
Einstein's Theory 100
Quantum Interpretation 101
3.7 X-RAY PRODUCTION 104
3.8 COMPTON EFFECT 106
3.9 PAIR PRODUCTION AND ANNIHILATION 110
SUMMARY 113

CHAPTER 4
STRUCTURE OF THE ATOM 118

4.1 THE ATOMIC MODELS OF THOMSON AND RUTHERFORD 119
4.2 RUTHERFORD SCATTERING 122

SPECIAL TOPIC:
Lord Rutherford of Nelson 124

4.3 THE CLASSICAL ATOMIC MODEL 131
4.4 BOHR MODEL OF THE HYDROGEN ATOM 132
4.5 SUCCESSES AND FAILURES OF THE BOHR MODEL 139
4.6 CHARACTERISTIC X-RAY SPECTRA AND ATOMIC NUMBER 141
4.7 ATOMIC EXCITATION BY ELECTRONS 144
SUMMARY 147

CHAPTER 5
WAVE PROPERTIES OF MATTER 151

5.1 X-RAY SCATTERING 152

SPECIAL TOPIC:
Cavendish Laboratory 156

5.2 DE BROGLIE WAVES 159
Bohr's Quantization Condition 161
5.3 ELECTRON SCATTERING 161
5.4 WAVE MOTION 165
5.5 WAVES OR PARTICLES? 171
5.6 RELATIONSHIP BETWEEN PROBABILITY AND WAVE FUNCTION 175
5.7 UNCERTAINTY PRINCIPLE 177
SUMMARY 181

CHAPTER 6
THE QUANTUM THEORY 185

6.1 THE SCHRÖDINGER WAVE EQUATION 185
Normalization and Probability 188
Properties of Valid Wave Functions 189
Time-Independent Schrödinger Wave Equation 190
6.2 EXPECTATION VALUES 191
6.3 INFINITE SQUARE-WELL POTENTIAL 194
6.4 FINITE SQUARE-WELL POTENTIAL 198
6.5 THREE-DIMENSIONAL INFINITE-POTENTIAL WELL 200
6.6 SIMPLE HARMONIC OSCILLATOR 202
6.7 BARRIERS AND TUNNELING 208
Potential Barrier with $E > V_0$ 208
Potential Barrier with $E < V_0$ 210
Potential Well 212
Alpha-Particle Decay 213

SPECIAL TOPIC:
Scanning Probe Microscopes 214

SUMMARY 217

CHAPTER 7
THE HYDROGEN ATOM 222

7.1 APPLICATION OF THE SCHRÖDINGER EQUATION TO THE HYDROGEN ATOM 222

7.2 SOLUTION OF THE SCHRÖDINGER EQUATION FOR HYDROGEN 224
Separation of Variables 224
Relation Between the Quantum Numbers ℓ and m_ℓ 225
Solution of the Radial Equation 226
Solution of the Angular and Azimuthal Equations 227

7.3 QUANTUM NUMBERS 228
Principal Quantum Number n 229
Orbital Angular Momentum Quantum Number ℓ 229

SPECIAL TOPIC:
Rydberg Atoms 230

Magnetic Quantum Number m_ℓ 231

7.4 MAGNETIC EFFECTS ON ATOMIC SPECTRA—THE NORMAL ZEEMAN EFFECT 233

7.5 INTRINSIC SPIN 238

7.6 ENERGY LEVELS AND ELECTRON POSSIBILITIES 240
Selection Rules 241
Probability Distribution Functions 242

SUMMARY 246

CHAPTER 8
MANY-ELECTRON ATOMS 249

8.1 ATOMIC STRUCTURE AND THE PERIODIC TABLE 249
Inert Gases 255
Alkalis 255
Alkaline Earths 255
Halogens 255
Transition Metals 256
Lanthanides 256
Actinides 256

8.2 TOTAL ANGULAR MOMENTUM 257
Single-Electron Atoms 257
Many-Electron Atoms 260
LS, or Russell-Saunders, Coupling 261
jj Coupling 264

8.3 ANOMALOUS ZEEMAN EFFECT 265

SUMMARY 269

CHAPTER 9
STATISTICAL PHYSICS 272

9.1 HISTORICAL OVERVIEW 273

9.2 MAXWELL VELOCITY DISTRIBUTION 275

9.3 EQUIPARTITION THEOREM 277

9.4 MAXWELL SPEED DISTRIBUTION 280

9.5 CLASSICAL AND QUANTUM STATISTICS 284
Classical Distributions 284
Quantum Distributions 285

9.6 FERMI-DIRAC STATISTICS 288
Introduction to Fermi-Dirac Theory 288
Classical Theory of Electrical Conduction 289
Quantum Theory of Electrical Conduction 290

9.7 BOSE-EINSTEIN STATISTICS 296
Blackbody Radiation 296
Liquid Helium 298

SPECIAL TOPIC:
Superfluid ^{3}He 302

Bose Condensation in Gases 304

SUMMARY 305

CHAPTER 10
MOLECULES AND SOLIDS 310

10.1 MOLECULAR BONDING AND SPECTRA 311
Molecular Bonds 311
Rotational States 312
Vibrational States 313
Vibration and Rotation Combined 314

10.2 STIMULATED EMISSION AND LASERS 318
Scientific Applications 323
Holography 324
Other Applications 325

10.3 STRUCTURAL PROPERTIES OF SOLIDS 326

10.4 THERMAL AND MAGNETIC PROPERTIES OF SOLIDS 329
Thermal Expansion 329
Thermal Conductivity 331
Magnetic Properties 333
Diamagnetism 333
Paramagnetism 334
Ferromagnetism 335

Antiferromagnetism and Ferromagnetism 336

10.5 SUPERCONDUCTIVITY **336**
The Search for a Higher T_c 344

SPECIAL TOPIC:
Low-Temperature Methods 346

Superconducting Fullerenes 350

10.6 APPLICATIONS OF SUPERCONDUCTIVITY **351**
Josephson Junctions 351
Maglev 353
Generation and Transmission of Electricity 353
Other Scientific and Medical Applications 354

SUMMARY **356**

CHAPTER 11
SEMICONDUCTOR THEORY AND DEVICES **361**

11.1 BAND THEORY OF SOLIDS **361**
Kronig-Penney Model 364
Conductors, Insulators, and Semiconductors 366

11.2 SEMICONDUCTOR THEORY **366**

SPECIAL TOPIC:
The Quantum Hall Effect 370

Thermoelectric Effect 373

11.3 SEMICONDUCTOR DEVICES **375**
Diodes 375
Bridge Rectifiers 376
Zener Diodes 377
Light Emitting Diodes 377
Photovoltaic Cells 378
Transistors 381
Field Effect Transistors 384
MOSFETs 384
Schottky Barriers 384
Semiconductor Lasers 385
Integrated Circuits 387

SUMMARY **389**

CHAPTER 12
THE ATOMIC NUCLEUS **392**

12.1 DISCOVERY OF THE NEUTRON **393**

12.2 NUCLEAR PROPERTIES **395**
Sizes and Shapes of Nuclei 396
Intrinsic Spin 398
Intrinsic Magnetic Moment 398

12.3 THE DEUTERON **399**

12.4 NUCLEAR FORCES **400**

12.5 NUCLEAR STABILITY **402**
Nuclear Models 406

12.6 RADIOACTIVE DECAY **408**

12.7 ALPHA, BETA, AND GAMMA DECAY **412**
Alpha Decay 413
Beta Decay 415

SPECIAL TOPIC:
Neutrino Detection 418

Gamma Decay 421

12.8 RADIOACTIVE NUCLIDES **423**
Time Dating Using Lead Isotopes 424
Radioactive Carbon Dating 426

SUMMARY **427**

CHAPTER 13
NUCLEAR INTERACTIONS AND APPLICATIONS **431**

13.1 NUCLEAR REACTIONS **431**
Cross Sections 434

13.2 REACTION KINEMATICS **436**

13.3 REACTION MECHANISMS **438**
The Compound Nucleus 438
Direct Reactions 442

13.4 FISSION **442**

13.5 FISSION REACTORS **446**
Breeder Reactors 449

SPECIAL TOPIC:
Early Fission Reactors 450

13.6 FUSION **452**
Formation of Elements 452
Nuclear Fusion on Earth 454
Controlled Thermonuclear Reactions 455

13.7 SPECIAL APPLICATIONS **458**
Medicine 458
Archaeology 460
Art 460
Crime Detection 461

SPECIAL TOPIC:
How to Prove an Art Forgery 462

Agriculture 462
Mining and Oil 463

Materials 465
Industry 466
Small Power Systems 466
New Elements 467

SPECIAL TOPIC:
The Search for New Elements 468

SUMMARY 470

CHAPTER 14
ELEMENTARY PARTICLES 474

14.1 THE EARLY BEGINNINGS 475
The Positron 475
Yukawa's Meson 476

14.2 THE FUNDAMENTAL INTERACTIONS 477

14.3 CLASSIFICATION OF ELEMENTARY PARTICLES 479
Leptons 480
Hadrons 481
Particles and Lifetimes 482

14.4 CONSERVATION LAWS AND SYMMETRIES 484
Symmetries 487

14.5 QUARKS 487
Quark Description of Particles 489
Color 491
Confinement 492

14.6 THE FAMILIES OF MATTER 493

14.7 THE STANDARD MODEL AND GUTs 493
Grand Unifying Theories 495

14.8 ACCELERATORS 497
Synchrotrons 497

SPECIAL TOPIC:
Experimental Ingenuity 498

Linear Accelerators 500
Fixed-Target Accelerators 501
Colliders 501

SUMMARY 503

CHAPTER 15
GENERAL RELATIVITY 506

15.1 PRINCIPLE OF EQUIVALENCE 506

15.2 TESTS OF GENERAL RELATIVITY 510
Bending of Light 510
Gravitational Redshift 510
Perihelion Shift of Mercury 512
Light Retardation 513

15.3 GRAVITATIONAL WAVES 514

15.4 BLACK HOLES 514

SPECIAL TOPIC:
Gravitational Waves 516

15.5 FRAME DRAGGING 519

SUMMARY 520

CHAPTER 16
COSMOLOGY—THE BEGINNING AND THE END 523

16.1 EVIDENCE OF THE BIG BANG 524
Hubble's Measurements 525
Cosmic Microwave Background Radiation 527
Nucleosynthesis 528

16.2 THE BIG BANG 530

16.3 STELLAR EVOLUTION 532
Eventual Fate of Stars 534

SPECIAL TOPIC:
Planck's Time, Length, and Mass 535

16.4 ASTRONOMICAL OBJECTS 538
Active Galactic Nuclei and Quasars 538
Gamma Ray Astrophysics 540
Novae and Supernovae 540

16.5 PROBLEMS WITH THE BIG BANG 544

16.6 THE AGE OF THE UNIVERSE 547

SPECIAL TOPIC:
Future Space Telescopes 550

16.7 THE FUTURE 552
The Demise of the Sun 552
The Future of the Universe? 552
Are There Other Earths Out There? 553

SUMMARY 554

APPENDIX 1
FUNDAMENTAL CONSTANTS A-1

APPENDIX 2
CONVERSION FACTORS A-2

APPENDIX 3
MATHEMATICAL RELATIONS A-4

APPENDIX 4
PERIODIC TABLE OF THE ELEMENTS A-5

APPENDIX 5
MEAN VALUES AND DISTRIBUTIONS A-6

APPENDIX 6
PROBABILITY INTEGRALS A-8

APPENDIX 7
INTEGRALS OF THE TYPE A-10

APPENDIX 8
ATOMIC MASS TABLE A-12

APPENDIX 9
NOBEL LAUREATES IN PHYSICS A-37

ANSWERS TO ODD-NUMBERED PROBLEMS A-43

INDEX I-1

CHAPTER 1

One Hundred Years Ago

Whatever nature has in store for mankind, unpleasant as it may be, men must accept, for ignorance is never better than knowledge.

Enrico Fermi

As a science, physics has existed for hundreds of years. Sometimes it has been indistinguishable from other sciences such as chemistry and astronomy—a similarity that continues in some areas even today. In physics we can easily distinguish *classical physics,* which was mostly developed before the year 1895, from *modern physics,* which has almost exclusively occurred since 1895. The precise year is unimportant, but the time period around 1900 saw monumental changes in physics.

Queen Victoria of England had her last full decade of rule in the 1890s. During her lifetime (1819–1901) she saw remarkable changes. Bismarck was replaced as leader of Germany in 1890 by Kaiser Wilhelm II. France was still smarting from her defeat in the Franco-Prussian war of 1870–1871. England, Germany, and France were the world's leaders in science. England was at the height of its empire, and Lord Kelvin (William Thomson, 1824–1907) was perhaps its greatest scientist of the period. Germany was growing stronger, dominated by its military, and Helmholtz (1812–1894) was its most respected scientist. France was recovering from a previous neglect of science with some great individual scientists like Pasteur (1822–1895) and the Curies (Pierre, 1859–1906, and Marie, 1867–1934).

The world was abuzz with social changes, literature, intellectual development, and, of course, war. Socialism had abounded throughout Europe since the German revolution of 1848 and the works of Marx and Engels. The French Impressionists, which included Monet, Renoir, and Degas, were busy with their paintings (which were later to become highly valued). English literature was popular with the likes of Oscar Wilde, Thomas Hardy, and new authors H. G. Wells and George Bernard Shaw. The famous theme of the film *2001: A Space Odyssey* is the first section of the tone poem *Also Sprach Zarathustra* by Richard Strauss. It was written in 1896. It was just about 1900 that Sigmund Freud brought forward his revolutionary ideas on the human mind, including his unique theory of psychoanalysis. In mathematics, Sophus Lie was working on group theory in the early 1890s, and this work proved to be essential in the development of some parts of quantum physics. In 1898, the United States was involved in a short war with Spain over the liberation of Cuba, a war that resulted in the United States claiming Guam, Puerto Rico, and the Philippine Islands. Important technological inventions were the radio by Marconi around 1900, the automobile by Charles Duryea and others in the 1890s, and the airplane by the Wright brothers (and others) in 1903.

Most other things that happened during Queen Victoria's long reign (63 years, from 1837–1901) cannot be compared with the remarkable achievements that occurred in physics. The description and predictions of electromagnetism by Maxwell are responsible for the instantaneous telecommunications of today. Thermodynamics rose to become an exact science. But the birth of modern physics, with all of its good and bad applications, has had incredible ramifications. The world will never be the same.

In this chapter we will briefly review the status of physics in about 1895. We will mention Newton's laws, Maxwell's equations, and the laws of thermodynamics. These results are just as important today as they were a hundred years ago. Arguments by scientists concerning the interpretation of experimental data using wave and particle descriptions seemed resolved two hundred years ago, yet these discussions were reopened in the twentieth century. Today we look back on the evidence of a hundred years ago and wonder how anyone could have doubted the validity of the atomic view of matter. Nevertheless, this view was not fully accepted until about 1910, although the majority of physicists had accepted the atomic model much earlier. We will pay special attention to the kinetic theory of gases, which used microscopic techniques to explain macroscopic observations—itself a prominent achievement.

Physicists have always been entranced with fundamental interactions and conservation laws. Fundamental interactions like gravity, electricity, and magnetism were thought to be well understood in 1895. Since then, other fundamental forces like the nuclear and weak interactions have been added and in some cases—curious as it may seem—have even been combined. The search for the holy grail of fundamental interactions continues unabated today.

We will finish this chapter with a status report of physics before 1900. The few problems not yet understood were nevertheless extremely important. We hope the student will find this book to be interesting both for the physics learned and for the historical account.

1.1 Classical Physics of the 1890s

Scientists and engineers of the late nineteenth century were rather smug. They thought they had just about everything under control. The best scientists of the day were highly recognized and rewarded. Public lectures were frequent. Scientists had easy access to their political leaders, partly because science and engineering had been helpful to their war machines, but also because of the many useful, technical advances. Basic research was recognized as helpful because of the applications which often soon followed. Although there were only primitive automobiles and no airplanes in 1895, these were soon to follow. A few people had telephones, and plans for widespread distribution of electricity were underway.

Scientists rightly felt that given enough time and resources, they could explain just about anything. They did recognize some difficult problems. They didn't clearly understand the structure of matter—that was under intensive investigation. Nevertheless, on a macroscopic scale, they knew how to build efficient engines. Ships plied the lakes, seas, and oceans of the world. Travel between the countries of Europe was frequent and easy by train. Scientists often were born in one country, were educated in one or two others, and eventually did research in still other countries. The most recent ideas traveled quickly among the centers of research, and except for some isolated scientists, of which Einstein is the most notable example, discoveries were quickly and easily shared. Scientific journals became more accessible.

Early success of science

The classical ideas of physics include the conservation laws of energy, linear momentum, angular momentum, and charge. We state them as

Classical Conservation Laws

Conservation of Energy: The total sum of energy (in all its forms) is conserved in all interactions.

Conservation of Linear Momentum: In the absence of external force, linear momentum is conserved in all interactions (vector relation).

Conservation of Angular Momentum: In the absence of external torque, angular momentum is conserved in all interactions (vector relation).

Conservation of Charge: Electric charge is conserved in all interactions.

We might even add the Conservation of Mass to the ones above, although we know it not to be valid today. These conservation laws are reflected in the laws of mechanics, electromagnetism, and thermodynamics. Electricity and magnetism, both separate subjects for hundreds of years, had been combined by the great James Clerk Maxwell (1831–1879) in his four equations. Optics had been shown by Maxwell, among others, to be a special case of electromagnetism. Waves, which permeated especially mechanics and optics, were believed to be an important component of nature. Many natural phenomena could be explained by wave motion using the laws of physics.

Mechanics

The laws of mechanics were developed over hundreds of years by many researchers. Important contributions were made by astronomers because of the great interest in the heavenly bodies. Galileo (1564–1642) may rightfully be called the first great experimenter. His experiments and observations laid the groundwork for the important discoveries to follow *during the next two hundred years.*

Galileo, the first great experimenter

Isaac Newton (1642–1727) was certainly the greatest scientist of his time and one of the best the world has ever seen. His discoveries were in the fields of mathematics, astronomy, and physics and include gravitation, optics, motion, and forces. He also spent considerable time on alchemy and theology—two completely unrelated subjects.

Newton, the greatest scientist of his time

We owe to Newton our present understanding of motion. He understood clearly the relationships among position, displacement, velocity, and acceleration. He understood how motion was possible and that a body at rest was just a special case of a body having constant velocity. Newton was able to carefully elucidate the relationship between force and acceleration. Newton's ideas were stated in three laws that bear his name today:

Galileo is shown performing experiments of rolling balls down an inclined plane. *Courtesy of Scala/Art Resource, NY.*

Newton's laws

Newton's first law: *An object in motion with a constant velocity will continue in motion unless acted upon by some net external force.* A body at rest is just a special case of Newton's first law with zero velocity. Newton's first law is often called the *law of inertia.* Newton's first law is also used to describe inertial reference frames.

Newton's second law: *The acceleration* **a** *of a body is proportional to the net external force* **F** *and inversely proportional to the mass m of the body.* It is stated mathematically as

$$\mathbf{F} = m\mathbf{a} \tag{1.1a}$$

Issac Newton was known not only for his work on the laws of motion, but also as a founder of optics. This painting shows him performing experiments with light. *Courtesy of Bausch & Lomb Optical Co. and the AIP Niels Bohr Library.*

A more accurate statement* relates force to the time rate of change of the linear momentum **p**.

$$\mathbf{F} = \frac{d\mathbf{p}}{dt} \tag{1.1b}$$

Newton's third law: *The force exerted by body 1 on body 2 is equal and opposite to the force that body 2 exerts on body 1.* If the force on body 2 by body 1 is denoted by $\mathbf{F}_{21}$, then Newton's third law is written as

$$\mathbf{F}_{21} = -\mathbf{F}_{12} \tag{1.2}$$

It is often called the *law of action and reaction.* Not all forces obey the third law.

These three laws can be used to define the concept of force. Together with the concepts of velocity **v**, acceleration, linear momentum, rotation (angular velocity $\boldsymbol{\omega}$ and angular acceleration $\boldsymbol{\alpha}$), and angular momentum **L**, we can describe the complex motion of bodies.

*It is a remarkable fact that Newton wrote his second law not as $\mathbf{F} = m\mathbf{a}$, but as $\mathbf{F} = \frac{d}{dt}(m\mathbf{v})$, thus taking into account mass flow and change in velocity. This has applications in both fluid mechanics and rocket propulsion.

Electromagnetism

This complex subject developed over a period of several hundred years. Important contributions were made by Coulomb (1736–1806), Oersted (1777–1851), Young (1773–1829), Ampère (1775–1836), Faraday (1791–1867), Henry (1797–1878), Maxwell (1831–1879), and Hertz (1857–1894). Maxwell showed that electricity and magnetism were intimately connected and were related by a change in the inertial frame of reference. His work also led to the understanding of electromagnetic radiation, of which light and optics are special cases. Maxwell's four equations, together with the Lorentz force law, explain much of electromagnetism.

Maxwell's equations

Gauss's law for electricity $$\oint \mathbf{E} \cdot d\mathbf{A} = \frac{q}{\epsilon_0} \tag{1.3}$$

Gauss's law for magnetism $$\oint \mathbf{B} \cdot d\mathbf{A} = 0 \tag{1.4}$$

Faraday's law $$\oint \mathbf{E} \cdot d\mathbf{s} = -\frac{d\Phi_B}{dt} \tag{1.5}$$

Generalized Ampere's law $$\oint \mathbf{B} \cdot d\mathbf{s} = \mu_0\epsilon_0 \frac{d\Phi_E}{dt} + \mu_o I \tag{1.6}$$

Lorentz force law $$\mathbf{F} = q\mathbf{E} + q\mathbf{v} \times \mathbf{B} \tag{1.7}$$

Maxwell's equations indicate that charges and currents create fields, and in turn, these fields can create other fields, both electric and magnetic.

Thermodynamics

Thermodynamics deals with temperature T, heat Q, work W, and the internal energy of systems U. The understanding of the concepts used in thermodynamics—such as pressure p, volume V, temperature, thermal equilibrium, heat, entropy, and especially energy—was slow in coming. We can understand the concepts of pressure and volume as mechanical properties, but the concept of temperature has to be carefully considered. We have learned that the internal energy of a system of noninteracting point masses depends only on the temperature.

Important contributions to thermodynamics were made by Benjamin Thompson (Count Rumford, 1753–1814), Carnot (1796–1832), Joule (1818–1889), Clausius (1822–1888), and Lord Kelvin (1824–1907). The primary results of thermodynamics can be described in two laws:

Laws of thermodynamics

First law of thermodynamics: *The change in the internal energy ΔU of a system is equal to the heat Q added to the system minus the work W done by the system.*

$$\Delta U = Q - W \tag{1.8}$$

The first law of thermodynamics generalizes the conservation of energy by including heat.

Second law of thermodynamics: *It is not possible to convert heat completely into work without some other change taking place.* There are various forms of the second law that state similar, but slightly different, results. For example, it is not possible to build a perfect engine or a perfect refrigerator. It is not possible to build a perpetual motion machine. Heat does not spontaneously flow from a colder body to a hotter body without some other change taking place. The second law forbids all these from happening. The first law states the conservation of energy, but the second law says what kinds of energy processes cannot take place. For example, it is possible to completely convert work into heat, but not vice versa, without some other change taking place.

Two other "laws" of thermodynamics are sometimes expressed. One is called the "zeroth" law, and it is useful in understanding temperature. It states that *if two thermal systems are in thermodynamic equilibrium with a third system, they are in equilibrium with each other.* We can state it more simply by saying that *two systems at the same temperature as a third system have the same temperature as each other.* This concept was not explicitly stated until this century, which is why it is called *zeroth.* The "third" law of thermodynamics expresses that *it is not possible to achieve an absolute zero temperature.*

1.2 The Kinetic Theory of Gases

We understand now that gases are composed of atoms and molecules in rapid motion, bouncing off each other and the walls, but this was not so clear to the scientists of the 1700s and 1800s. The kinetic theory of gases is related to thermodynamics and to the atomic theory of matter, which we will discuss in Section 1.5. Experiments were relatively easy to perform on gases, and the Irish chemist Robert Boyle (1627–1691) showed around 1662 that the pressure times the volume of a gas was constant for a constant temperature. The relation $pV =$ constant (for constant T) is now referred to as *Boyle's law.* The French physicist Charles (1746–1823) found that $V/T =$ constant (at constant pressure), referred to as *Charles's law.* The same result was later extensively studied by Gay-Lussac (1778–1850), whose name is sometimes associated with the result. Combining these laws we obtain the ideal gas equation

Ideal gas equation

$$pV = nRT \tag{1.9}$$

where n is the number of moles and R is the ideal gas constant.

In 1811 the Italian physicist Avogadro (1776–1856) proposed that equal volumes of gases at the same temperature and pressure contained equal numbers of molecules. This hypothesis was so far ahead of its time that it was not accepted for many years. The famous English chemist Dalton opposed this idea because he apparently misunderstood the difference between atoms and molecules. Considering the rudimentary nature of the atomic theory of matter at the time, this was not surprising.

Daniel Bernoulli (1700–1782) apparently originated the kinetic theory of gases in 1738, but his results were generally ignored. Many scientists, including Newton, Laplace, Davy, Herapath, and Waterston, had contributed to the development of kinetic theory by 1850. Theoretical calculations were being compared to experiment, and by 1895 the kinetic theory of gases was widely accepted. The statistical interpretation of thermodynamics was made in the latter half of the nineteenth century by the great Scottish mathematical physicist Maxwell, the Austrian physicist Ludwig Boltzmann (1844–1906), and the American physicist Willard Gibbs (1839–1903).

In introductory physics classes, the kinetic theory of gases is usually taught by applying Newton's laws to the collisions that a molecule makes with other molecules and with the walls. A representation of a few molecules colliding is shown in Figure 1.1. In the simple model of an ideal gas, only elastic collisions are considered. By taking averages over the collisions of many molecules, the derivation of the ideal gas law, Equation (1.9), is revealed. The average kinetic energy of the molecules is interpreted as being linearly proportional to the temperature, and the internal energy U is

$$U = nN_A\langle K \rangle = \frac{3}{2}nRT \tag{1.10}$$

Statistical thermodynamics

where n is the number of moles of gas, N_A is Avogadro's number, $\langle K \rangle$ is the average kinetic energy of a molecule, and R is the ideal gas constant. This relation ignores any nontranslational contributions to the molecular energy, such as rotations and vibrations.

However, energy is not represented by only translational motion. It became clear that all *degrees of freedom,* including rotational and vibrational, also were capable of demonstrating energy. The *equipartition theorem* states that each degree of freedom of a molecule has an average energy of $kT/2$, where k is the Boltzmann constant ($k = R/N_A$). Translational kinetic energy is described in terms of three degrees of freedom because of the three directions of space, and rotations and vibrations are also excited at higher temperatures. If there are f degrees of freedom, then Equation (1.10) becomes

Equipartition theorem

$$U = \frac{f}{2}nRT \tag{1.11}$$

Internal energy

The molar ($n = 1$) heat capacity c_V at constant volume for an ideal gas is the rate of change in internal energy with respect to change in temperature and is given by

$$c_V = \frac{3}{2}R \tag{1.12}$$

Heat capacity

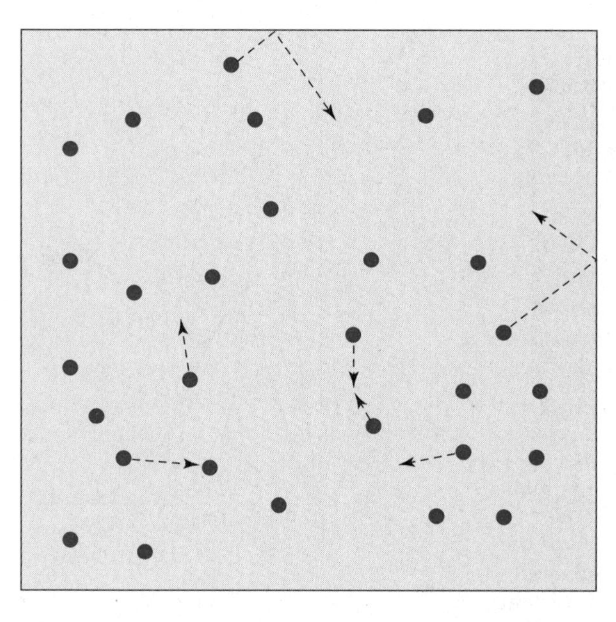

FIGURE 1.1 Molecules inside a closed container are shown colliding with the walls and with each other. The motions of a few molecules are indicated by the arrows. The number of molecules inside the container is huge.

The experimental quantity c_V/R is plotted versus temperature for hydrogen in Figure 1.2. The ratio c_V/R is equal to 3/2 for low temperatures, where only translational kinetic energy is important, but it rises to 5/2 at 300 K, where rotations occur for H_2, and finally gets close to 7/2, due to vibrations at still higher temperatures, before the molecule dissociates. Although the kinetic theory of gases fails to predict specific heats for real gases, it leads to models that can be used on a gas by gas basis. The kinetic theory is also able to provide useful information on other properties such as diffusion, speed of sound, mean free path, and collision frequency.

In the 1850s Maxwell derived a relation for the distribution of speeds of the molecules in gases. The distribution of speeds $N(v)$ is given as a function of the speed and the temperature by the equation

Maxwell's speed distribution

$$N(v) = 4\pi N\left(\frac{m}{2\pi kT}\right)^{3/2} v^2 e^{-mv^2/2kT} \tag{1.13}$$

where N is the total number of molecules, m is the mass of a molecule, and T is the temperature. This result is plotted for nitrogen in Figure 1.3 for temperatures of 300 K, 1000 K, and 4000 K. The peak of each of the distributions is the most probable speed of a gas molecule for the given temperature. In 1895 measurement was not precise enough to completely confirm Maxwell's distribution, and it was not confirmed experimentally until 1921.

Before 1895 Boltzmann had made Maxwell's calculation more rigorous, and the general relation is called the *Maxwell-Boltzmann distribution.* The distribution can be used to find the *root-mean-square* speed v_{rms},

$$v_{\text{rms}} = \sqrt{\langle v^2 \rangle} = \sqrt{\frac{3kT}{m}} \tag{1.14}$$

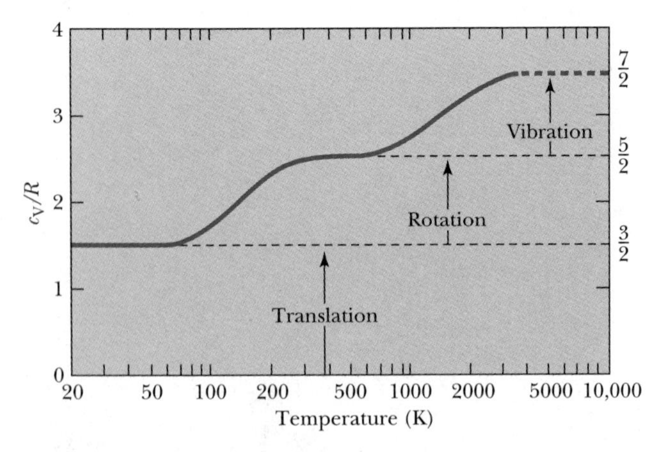

FIGURE 1.2 The molar heat capacity at constant volume (c_V) divided by R (c_V/R is dimensionless) is displayed as a function of temperature for hydrogen gas. Note that as the temperature increases, the rotational and vibrational modes become important. This experimental result is consistent with the equipartition theorem, which adds $kT/2$ of energy per molecule ($RT/2$ per mole) for each degree of freedom.

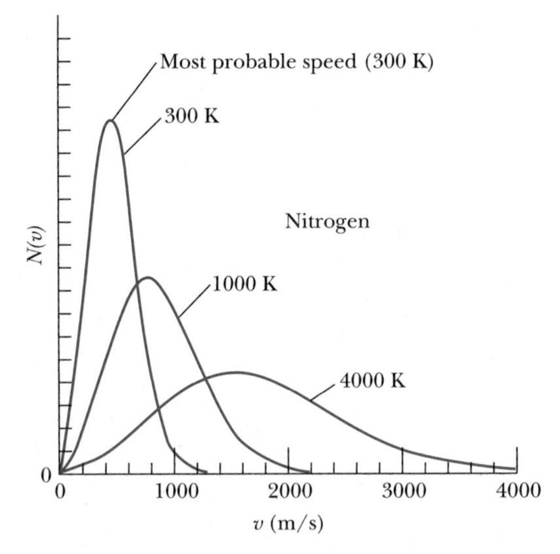

FIGURE 1.3 The Maxwell distribution of molecular speeds (for nitrogen), $N(v)$, is shown as a function of speed for three temperatures.

which shows the relationship of the energy to the temperature for an ideal gas:

$$U = nN_A\langle K\rangle = nN_A\frac{m\langle v^2\rangle}{2} = nN_A\frac{m3kT}{2m} = \frac{3}{2}nRT \tag{1.15}$$

This was the result of Equation (1.10).

1.3 Waves and Particles

In introductory physics we first learned the concepts of velocity, acceleration, force, momentum, and energy using a single particle. Bodies were first treated as if all their mass were concentrated in one small point. Eventually we realized that two- and three-dimensional bodies were necessary, and we added rotations and vibrations. However, many aspects of physics can still be treated as if the bodies are simply particles. In particular, the kinetic energy of a moving particle is one way in which energy can be transported from one place to another.

But we have found that many natural phenomena can be explained only in terms of *waves,* which are traveling disturbances that carry energy. This description includes standing waves, which are superpositions of traveling waves. Most waves, like water waves and sound waves, need an elastic medium in which to move. Curiously enough, matter is not transported in waves—energy is. Mass may oscillate, but it doesn't actually propagate along with the wave. Two examples are a cork and a boat on water. As a water wave passes, the cork gains energy as it moves up and down, and after the wave passes the cork remains. The boat also reacts to the wave, but it primarily rocks back and forth, throwing things around on the boat that are not fixed. The boat obtains considerable kinetic energy from the wave. After the wave passes, the boat eventually returns to rest.

Energy transport

Waves and particles were the subject of disagreement as early as the seventeenth century, when there were two competing theories of the nature of light. Newton supported the idea that light consisted of corpuscles (or particles). He performed extensive experiments on light for many years, and finally published his book *Opticks* in 1704. *Geometrical optics* uses straight-line, particle-like trajectories called *rays* to explain familiar phenomena like reflection and refraction. Geometrical optics was also able to explain the apparent observation of sharp shadows. The other competing theory considered light as a wave phenomenon. Its strongest proponent was the Dutch physicist Christian Huygens (1629–1695), who presented his theory in 1678 before publishing it in 1690. The wave theory could also explain reflection and refraction, but it could not explain the sharp shadows observed. Experimental physics of the 1600s and the 1700s was not able to discern between the two competing theories. Huygens's poor health and other duties kept him from working on optics much after 1678. Although Newton did not feel strongly about his corpuscular theory, the magnitude of his reputation caused the corpuscular theory to be almost universally accepted for over a hundred years and throughout most of the eighteenth century.

Nature of light: waves or particles?

Finally, in 1802, the English physician Thomas Young (1773–1829) announced the results of his two-slit interference experiment, indicating that light behaved as a wave. Even after this singular event, the corpuscular theory had its supporters. During the next few years Young and, independently, Augustin Fresnel (1788–1827) performed several experiments that clearly showed that light behaved as a wave. By 1830 most physicists believed in the wave theory—some 150 years after Newton performed his first experiments on light.

One final experiment indicated that the corpuscular theory was difficult to accept. Let c be the speed of light in vacuum and v be the speed of light in media. If light behaves as a particle, then in order to explain refraction, light must speed up when going through denser material ($v > c$). The wave theory of Huygens predicts just the opposite ($v < c$). The measurements of the speed of light in media were slowly improving, and finally, in 1850, Foucault showed that *light traveled more slowly in water than in air.* The corpuscular theory seemed incorrect. Newton would have probably been surprised that his weakly held beliefs lasted as long as they did. Now we realize that geometrical optics is correct only if the wavelength of light is much smaller than the size of the obstacles and apertures that the light encounters.

Figure 1.4 shows the "shadows" or *diffraction patterns* from light falling on sharp edges. In Part (a) the alternating black and white lines can be seen all around the razor blade's edges. Part (b) is a highly magnified photo of the diffraction from a sharp edge. The bright and dark regions can only be understood if light is a wave and not a particle. The physicists of 200–300 years ago apparently did not observe such phenomena. They were convinced that shadows were sharp, and only the particle nature of light could explain their observations.

In the 1860s Maxwell showed that electromagnetic waves consist of oscillating electric and magnetic fields. Visible light covers just a narrow range of the total electromagnetic spectrum, and all electromagnetic radiation travels at the speed of light c in free space, given by

$$c = \frac{1}{\sqrt{\mu_0 \epsilon_0}} = \lambda \nu \tag{1.16}$$

The fundamental constants μ_0 and ϵ_0 are defined in electricity and magnetism and reveal the connection to the speed of light. In 1887 the German physicist Heinrich Hertz (1857–1894) succeeded in generating and detecting electromagnetic waves having wavelengths far outside the visible range ($\lambda \approx 5$ m). The properties of these waves were just as had been predicted by Maxwell. Maxwell's

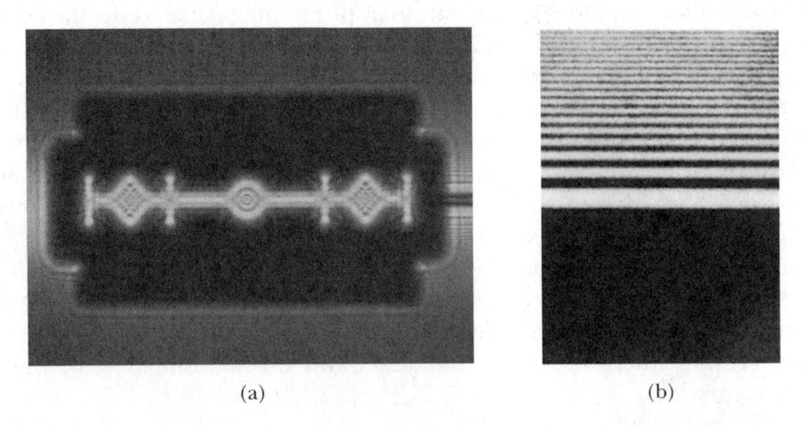

(a) (b)

FIGURE 1.4 In contradiction to what scientists thought in the seventeenth century, shadows are not sharp, but show dramatic diffraction patterns—as seen here (a) for a razor blade and (b) for a highly magnified sharp edge. *Part (a) courtesy Ken Kay/Fundamental Photographs, New York, part (b) is from The Atlas of Optical Phenomena, by Michael Cagnet, Mauric Francon, and Jean Claude Thrierr, p. 32, © 1962 Springer-Verlag, New York.*

results continue to have far-reaching effects today in our modern telecommunications system: cable TV, cellular telephones, lasers, fiber optics, and so on.

Some unresolved issues about electromagnetic waves eventually led to one of the two great modern theories, the theory of relativity (see Section 1.6). Waves play a central and essential role in the other great modern physics theory, *quantum mechanics,* which is sometimes called *wave mechanics.* Because waves play such a central role in modern physics, we will review their properties later in Chapter 5.

1.4 Conservation Laws and Fundamental Forces

Physicists love conservation laws. They are the guiding principles of physics. The application of a few laws explains a vast quantity of physical phenomena. We listed the conservation laws of classical physics in Section 1.1. They include energy, linear momentum, angular momentum, and charge. Each of these was extremely useful in introductory physics. We used linear momentum when studying collisions and studied dynamics through the conservation laws. We have seen the concept of the conservation of energy change. At first we had only the conservation of kinetic energy in a force-free region. Then we added potential energy and had the conservation of mechanical energy. In our study of thermodynamics, we added internal energy, and so on. The study of electrical circuits was made easier by the conservation of charge flow at each junction and by the conservation of energy throughout all the circuit elements.

Much of what we know about conservation laws and fundamental forces has been learned within the last hundred years. This development was underway in the nineteenth century, and the subject is so important that we review it here. In our study of modern physics we will find that mass is added to the conservation of energy, and the result is sometimes called the *conservation of mass-energy,* although the term *conservation of energy* is still sufficient and generally used. When we study elementary particles we will add the conservation of baryons and the conservation of leptons. Closely related to conservation laws are invariance principles. Some parameters are invariant in some interactions or in specific systems but not in others. Examples include time reversal, parity, and distance. We will study the Newtonian or Galilean invariance and find it lacking in our study of relativity. A new invariance principle will be needed. In our study of nuclear and elementary particles, conservation laws and invariance principles will often be used (see Figure 1.5).

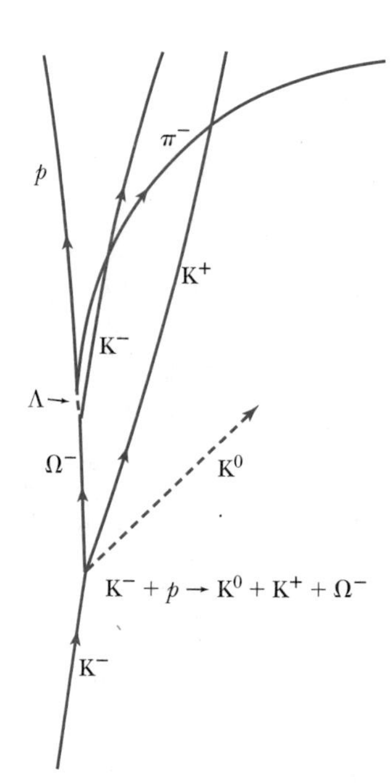

FIGURE 1.5 The conservation laws of momentum and energy are invaluable in untangling complex particle reactions like the one shown here, where a 5-GeV K^- meson interacts with a proton at rest to produce an Ω^- in a bubble chamber. The uncharged K^0 is not observed. Notice the curved paths of the charged particles in the magnetic field. Such reactions will be studied in Chapter 14.

Fundamental Forces

During our study of introductory physics, we encountered a variety of forces. We learned about force by studying the reaction of a mass at the end of a spring. The spring force can be easily calibrated. We learned about tension, friction, gravity, surface, electrical, and magnetic forces. Despite the seemingly complex array of forces, we believe presently there are only three fundamental forces. All the other forces can be derived from them. These three forces are the **gravitational, electroweak,** and **strong** forces. Some physicists refer to the electroweak interaction as separate electromagnetic and weak forces because the unification only occurs at very high energies. These physicists would quote four forces. The

TABLE 1.1
Fundamental Forces

Interaction		Relative Strength*	Range
Strong		1	Short, $\sim 10^{-15}$ m
Electroweak	Electromagnetic	10^{-2}	Long, $1/r^2$
	Weak	10^{-9}	Short, $\sim 10^{-15}$ m
Gravitational		10^{-39}	Long, $1/r^2$

*These strengths are determined by neutrons and/or protons in close proximity.

strengths and ranges of the three fundamental forces are listed in Table 1.1. Physicists sometimes use the term *interaction* when referring to the fundamental forces because it is the overall interaction among the constituents of a system that is of interest.

The gravitational force is the weakest. It is the force of mutual attraction between masses and, according to Newton, is given by

Gravitational interaction

$$\mathbf{F}_g = -G\frac{m_1 m_2}{r^2}\hat{\mathbf{r}} \tag{1.17}$$

where m_1 and m_2 are two point masses, G is the gravitational constant, r is the distance between the masses, and $\hat{\mathbf{r}}$ is a unit vector directed along the line between the two point masses (attractive force). The gravitational force is only effective on a macroscopic scale, but it has tremendous importance. It is the force that keeps the planet Earth rotating about our source of life energy—the sun. Gravity is the force that keeps us and our atmosphere anchored to the ground. Gravity is a long-range force that diminishes as $1/r^2$.

Weak interaction

The primary component of the electroweak force is *electromagnetic.* The other component is the *weak* interaction, which is responsible for beta decay in nuclei, among other processes. In the 1970s Glashow, Weinberg, and Salam predicted that the electromagnetic and weak forces were in fact facets of the same force. Their theory predicted the existence of new particles, called W and Z bosons, which were discovered in 1983. We will discuss bosons and the experiment in Chapter 14. For all practical purposes, the weak interaction is effective in the nucleus only over distances the size of 10^{-15} m. Except when dealing with very high energies, physicists mostly treat nature as if the electromagnetic and weak forces were separate. The student will sometimes see reference to the *four* fundamental forces (gravity, strong, electromagnetic, and weak).

Electromagnetic interaction

The electromagnetic force is probably the force that most affects us. It is responsible for holding atoms together, for friction, for contact forces, for tension, and for electrical and optical signals. It is responsible for all chemical and biological processes, including cellular structure and nerve processes. The list is long because the electromagnetic force is responsible for practically all nongravitational forces that we experience. The electrostatic, or Coulomb, force between two point charges q_1 and q_2, separated by a distance r, is given by

Coulomb force

$$\mathbf{F}_C = \frac{1}{4\pi\epsilon_0}\frac{q_1 q_2}{r^2}\hat{\mathbf{r}} \tag{1.18}$$

The easiest way to remember the vector direction is that like charges repel and unlike charges attract. Moving charges also create and react to magnetic fields.

Strong interaction

The third fundamental force, the strong force, is the one holding the nucleus together. It is the strongest of all the forces, but it is effective only over short distances—up to about 10^{-15} m. The strong force is so strong that it easily causes two protons to be bound together inside a nucleus even though the electrical force of repulsion over the tiny confined space is huge. The strong force is able to contain dozens of protons inside the nucleus before the electrical force of repulsion becomes strong enough to cause nuclear decay. We will study the strong force extensively in this book. We will learn that neutrons and protons are composed of *quarks,* and that the part of the strong force acting between quarks has the unusual name of *color* force.

Unification of forces

Physicists strive to combine forces into more fundamental ones. Centuries ago the forces responsible for friction, contact, and tension were all believed to be different. Today we know they are all part of the electroweak force. Two hundred years ago scientists thought the electrical and magnetic forces were independent, but after a series of experiments, physicists slowly began to see their connection. This culminated, in the 1860s, in Maxwell's work, which clearly showed they were but part of one force and at the same time explained light and other radiation. In Figure 1.6 we show a diagram indicating the unification of forces over time. Newton certainly had an inspiration when he was able to unify the planetary motions with the apple falling from the tree. We will see in Chapter 15 that Einstein was even able to link gravity with space and time.

The further unification of forces currently remains one of the most active research fields. Considerable efforts have been made to unify the electroweak and strong force through the *grand unified theories* or GUTs. There are several predictions of this theory that have not yet been verified experimentally. They include the instability of the proton, the existence of magnetic monopoles, and the existence of a massive neutrino. These experiments are among the most important ones now underway and will be discussed later. We hope to present some of

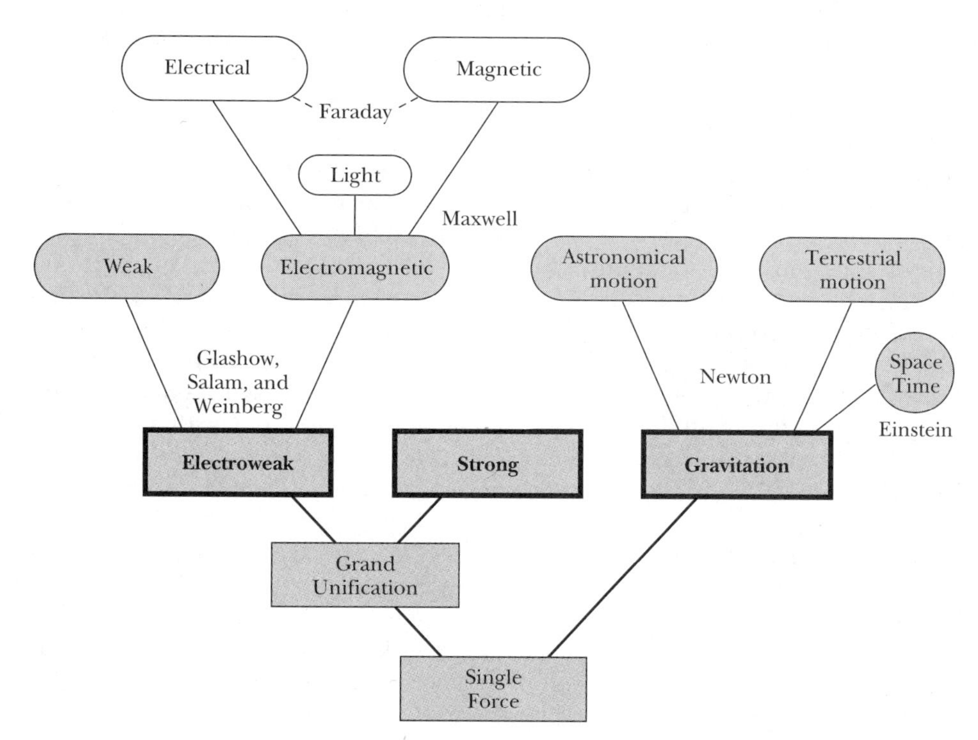

FIGURE 1.6 The three fundamental forces (shown in the heavy boxes) are themselves unifications of forces that were once believed to be fundamental. Present research is underway (see blue lines) to further unify the fundamental forces into one single force.

the excitement of present-day physics throughout this book. These experiments are the ones you will someday read about on the front pages of newspapers and in the weekly news magazines.

1.5 The Atomic Theory of Matter

It is not possible to reach college, or even high school, today without learning about the atomic theory of matter. The idea that matter is composed of tiny particles called *atoms* is taught in elementary school and expounded throughout later schooling. We are told that the Greek philosophers Democritus and Leucippus proposed the concept of atoms as early as 450 B.C. The smallest piece of matter, which could not be subdivided further, was called an *atom,* after the Greek work *atomos,* meaning "indivisible." Physicists do not discredit the early Greek philosophers for thinking that the basic entity of life consisted of atoms. For centuries, scientists were called "natural philosophers," and today the highest university degree American scientists receive is a Ph.D., which stands for Doctor of Philosophy.

Not many new ideas were proposed about atoms until the seventeenth century, when scientists started trying to understand the properties and laws of gases. The work of Boyle, Charles, and Gay-Lussac presupposed the interactions of tiny particles in gases. Many important advances were made by chemists and physical chemists. In 1799 the French chemist Proust (1754–1826) proposed the *law of definite proportions,* which states that when two or more elements combine to form a compound, the proportions by weight (or mass) of the elements are always the same. Water (H_2O) is always formed of one part hydrogen and eight parts oxygen by mass.

Dalton: Father of atomic theory

The English chemist John Dalton (1766–1844) is given most of the credit for being the originator of the modern atomic theory of matter. In 1803 he proposed that the atomic theory of matter could explain the law of definite proportions if the elements are composed of atoms. Each element has atoms that are physically and chemically characteristic. The concept of atomic weights (or masses) was the key to the atomic theory.

In 1811 the Italian physicist Avogadro proposed the existence of molecules, which consisted of individual atoms or atoms combined. He stated without proof that *all gases contain the same number of molecules in equal volumes at the same temperature and pressure.* Avogadro's ideas were ridiculed by Dalton and others who could not imagine atoms of the same element combining. If this could happen, they argued, then all the atoms of a gas would combine to form a liquid. The concept of molecules and atoms was difficult to imagine. Finally, in 1858, the Italian chemist Cannizzaro (1826–1910) explained the problem and showed how Avogadro's ideas could be used to find atomic masses. Cannizzarro was able to elucidate the difference between atomic and molecular masses. Today we think of an atom as the smallest unit of matter that can be identified with a particular element. A molecule can be a single atom or a combination of two or more atoms of either like or dissimilar elements. Molecules can consist of thousands of atoms.

Avogadro's number

The number of molecules in one gram-molecular weight of a particular element (6.02×10^{23} molecules/mol) is called Avogadro's number (N_A) in honor of Avogadro. For example, one mole of hydrogen (H_2) has a mass of 2 g and one mole of carbon has a mass of about 12 g; one mole of each substance consists of 6.02×10^{23} atoms. Avogadro's number was not even estimated until 1865, and it was finally accurately measured by Perrin, as we will discuss later.

During the mid-1800s the kinetic theory of gases was being developed, and because it was based on the concept of atoms, its successes gave validity to the atomic theory. The experimental results of specific heats, Maxwell speed distribution, and transport phenomena (see the discussion in Section 1.2) all supported the concept of the atomic theory.

In 1827 the English botanist Robert Brown (1753–1858) observed with a microscope the motion of tiny pollen grains suspended in water. The pollen appeared to dance around in random motion, while the water was still. At first the motion was ascribed to convection or organic matter, but eventually it was observed to occur for any tiny particle suspended in liquid. The explanation according to the atomic theory is that the tiny grains are constantly being bombarded by the molecules in the liquid. A satisfactory explanation was not given until the twentieth century (by Einstein).

Opposition to atomic theory

Although it may appear, according to the preceding discussion, that the atomic theory of matter was universally accepted by the end of the nineteenth century, that was not the case. Certainly, most physicists believed in it, but there was still opposition. A principal leader in the anti-atomic movement was the renowned Austrian physicist Ernst Mach. Mach was an absolute positivist, believing in the reality of nothing but our own sensations. A simplified version of his line of reasoning would be that because we have never *seen* an atom, we cannot say anything about its reality. The Nobel Prize-winning German physical chemist Wilhelm Ostwald supported Mach philosophically but also had more practical arguments on his side. In 1900 there were difficulties in understanding radioactivity, x rays, discrete spectral lines, and how atoms formed molecules and solids. Ostwald contended that we should therefore think of atoms as hypothetical constructs, useful for bookkeeping in chemical reactions.

On the other hand, there were many believers of the atomic theory. Max Planck, the originator of the quantum theory, grudgingly accepted the atomic theory of matter because his radiation law supported the existence of submicroscopic quanta. Boltzmann was convinced that atoms must exist, mainly because they were necessary in his statistical mechanics. It is said that Boltzmann committed suicide in 1905 partly because he was so despondent that so many people rejected his theory. Today we have pictures of the atom (see Figure 1.7) that

FIGURE 1.7 In this scanning tunneling microscope photo, seven xenon atoms have been arranged in a row on a nickel surface (nickel atoms indicated by the black and white stripes on the surface). The vertical scale has been greatly exaggerated here; the xenon atoms are 0.16 nm high and 0.5 nm apart. The individual atoms were actually moved across the surface and positioned. The image is magnified by about 3 million. See also the Special Topic box in Chapter 6. *Courtesy of International Business Machines.*

would undoubtedly have convinced even Mach, who died in 1916 still unconvinced of the validity of the atomic theory.

Overwhelming evidence in favor of atomic theory

Overwhelming evidence for the existence of atoms was finally presented in the first decade of the twentieth century. First, Einstein in one of his three famous papers published in 1905 (the others were about special relativity and the photoelectric effect) provided an explanation of the *Brownian motion* observed almost eighty years earlier by Robert Brown. Einstein explained the motion in terms of the molecular motion and presented theoretical calculations for the *random walk* problem. Random walk (often called the *drunkard's walk*) is a statistical process that determines how far from its initial position a tiny grain may be after many random molecular collisions. From the experimental data, Einstein was able to determine the approximate mass and sizes of atoms and molecules.

Finally, in 1908, the French physicist Jean Perrin (1870–1942) presented data, based on the kinetic theory, that agreed with Einstein's predictions. Perrin's experimental methods of observing many particles of different sizes is a classic work, for which he received the Nobel Prize for physics in 1926. Perrin's experiment utilized four different types of measurements. Each was consistent with the atomic theory and each gave a quantitative determination of Avogadro's number—the first accurate measurements that had been done. Since 1908 the atomic theory of matter has been accepted by practically everyone.

1.6 Outstanding Problems of 1895 and New Horizons

The year 1895 is about the time we can separate the periods of classical and modern physics. The thousand or so physicists living in 1895 were rightfully proud of the status of their profession. The precise experimental method was firmly established. Theories were available that could explain practically everything. In large part, scientists were busy measuring and understanding such physical parameters as specific heats, densities, compressibility, resistivity, indices of refraction, and permeabilities. The pervasive feeling was that, given enough time, everything in nature could probably be understood by applying the careful thinking and experimental techniques of physics. The field of mechanics was in particularly good shape, and its application had led to the stunning successes of the kinetic theory of gases and statistical thermodynamics.

Experiment and reasoning

In hindsight we can see now that this euphoria of success applied only to the macroscopic world. Objects of human dimensions like automobiles, steam engines, airplanes, telephones, and electric lights either existed or were soon to appear and were triumphs of science and technology. However, the atomic theory of matter was not even universally accepted, and what made up an atom was purely conjecture. The structure of matter was unknown.

Clouds on the horizon

There were certainly problems that physicists could not resolve. Only a few of the deepest thinkers seemed to be concerned with them. Lord Kelvin, in a speech in 1900 to the Royal Institution, referred to "two clouds on the horizon." These two clouds were the electromagnetic medium and the failure of classical physics to explain blackbody radiation. We mention these and other problems here. Their solutions were soon to lead to two of the greatest breakthroughs in human thought ever recorded—the theories of quantum physics and relativity.

Electromagnetic Medium. The waves that were well known and understood by physicists all had media in which the waves propagated. Water waves traveled in water, and sound waves traveled in any material. It was natural to assume that electromagnetic waves also traveled in a medium, and this medium was called the *ether.* Several experiments, the most notable of which were due to Michelson, had sought to detect the ether without success. An extremely careful experiment by Michelson and Morley in 1887 was so sensitive it should have revealed the effects of the ether. Subsequent experiments to check other possibilities were also negative. In 1895 some physicists were concerned about the elusive ether.

Electrodynamics. The other difficulty with Maxwell's electromagnetic theory had to do with the electric and magnetic fields as seen and felt by moving bodies. What appears as an electric field in one reference system may appear as a magnetic field in another system moving with respect to the first. Although the relationship between electric and magnetic fields seemed to be understood by using Maxwell's equations, the equations do not keep the same form under a Galilean transformation (see Equations [2.1] and [2.2]). Hertz and Lorentz were quite concerned. Hertz unfortunately died in 1894 at the young age of 36 and never experienced the modern physics revolution. The Dutch physicist Lorentz (1853–1928), on the other hand, proposed a radical suggestion that solved the electrodynamics problem: space was contracted along the direction of motion of the body. This was also done independently by FitzGerald in Ireland. The Lorentz-FitzGerald hypothesis, proposed in 1892, was a precursor to Einstein's theory advanced in 1905 (see Chapter 2).

Blackbody Radiation. In 1895 thermodynamics was on a strong footing; it had achieved much success. One of the interesting experiments in thermodynamics concerns an object, called a *blackbody,* that absorbs all the electromagnetic radiation incident on it. An enclosure with a small hole serves as a blackbody, because all the radiation entering the hole is absorbed. A blackbody also emits radiation, and the *emissivity* is the electromagnetic power emitted per unit area. The radiation emitted covers all frequencies, each with its own intensity. Precise measurements were carried out to determine the spectrum of blackbody radiation, such as that shown in Figure 1.8. Blackbody radiation was a fundamental issue, because the emissivity is independent of the body itself—it is characteristic of all blackbodies.

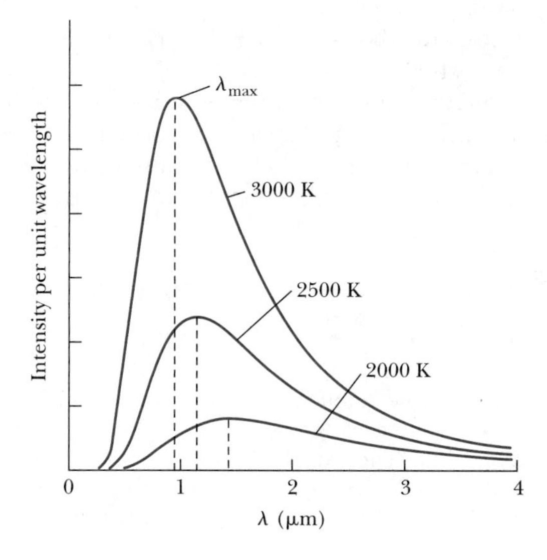

FIGURE 1.8 The blackbody spectrum, showing the emissivity of radiation emitted from a blackbody as a function of the radiation wavelength. Different curves are produced for different temperatures, but they are independent of the type of blackbody cavity. The intensity peaks at λ_{max}.

Many physicists of the period—including Kirchoff, Stefan, Boltzmann, Rubens, Pringsheim, Lummer, Wien, Lord Rayleigh, Jeans, and Planck—worked on the problem. It was possible to understand classically the spectrum both at the low-frequency end and at the high-frequency end, but no single theory could account for the entire spectrum. When the most modern theory of the day (the equipartition of energy applied to standing waves in a cavity) was applied to the problem, the result led to an *infinite* emissivity (or energy density) for high frequencies. The failure of the theory was known as the "ultraviolet catastrophe." The solution of the problem by Max Planck in 1900 was to shake the very foundations of physics.

Ultraviolet catastrophe: infinite emissivity

On the Horizon

During the years 1895–1897 there were four discoveries that were all going to require a deep understanding of the atom. The first was the discovery of x rays by the German physicist Roentgen (1845–1923) in November of 1895. Next came the accidental discovery of radioactivity by the French physicist Henri Becquerel (1852–1908), who in February 1896 placed uranium salt next to a carefully wrapped photographic plate. When the plate was developed, a silhouette of the uranium salt was evident—indicating the presence of a very penetrating ray.

Discovery of x rays

Discovery of radioactivity

Discovery of the electron

The third discovery, that of the electron, was actually due to several physicists over a period of years. Michael Faraday, as early as 1833, observed a gas discharge glow—evidence of electrons. Over the next few years, several scientists detected evidence of particles, called *cathode rays,* being emitted from charged cathodes. In 1896 Perrin proved that cathode rays were negatively charged. The discovery of the electron, however, is generally credited to the British physicist J. J. Thomson (1856–1940), who in 1897 isolated the electron (cathode ray) and measured its velocity and the ratio of its charge to mass.

The final important discovery of the period was made by the Dutch physicist Pieter Zeeman (1865–1943), who in 1896 found that spectral lines were sometimes separated into two or three lines when placed in a magnetic field. The (normal) *Zeeman effect* was quickly explained by Lorentz as being due to light emitted by the motion of electrons inside the atom. Zeeman and Lorentz showed that the frequency of the light was affected by the magnetic field according to the classical laws of electromagnetism.

Discovery of the Zeeman effect

The problems existing in 1895 and the important discoveries of 1895–1897 bring us to the subject of this book, *Modern Physics.* In 1900 Max Planck completed his radiation law, which solved the blackbody problem but required that energy be quantized. In 1905 Einstein presented his three important papers on Brownian motion, the photoelectric effect, and special relativity. The work of Planck and Einstein may have solved the problems of the nineteenth-century physicists, but they broadened the horizons of physics and have kept physicists active ever since.

Summary

Physicists of the 1890s were proud of their achievements in mechanics, electricity and magnetism, optics, and thermodynamics. They felt that almost anything in nature could be explained by the application of careful experimental methods and intellectual thought. The application of mechanics to the kinetic theory of gases and statistical thermodynamics was a great success.

The nature of light was apparently resolved in the early 1800s in favor of waves. For over 100 years the particle viewpoint had prevailed, mostly due to the weakly held belief of the great Newton. In the 1860s Maxwell showed that his electromagnetic theory predicted a much wider frequency range of electromagnetic radiation than the visible optical phenomena. Finally, again in the twentieth century, the question of waves versus particles was to reappear.

Physicists love, cherish, and are constantly trying to test conservation laws and invariance principles. The conservation laws of energy, momentum, angular momentum, and charge are well established. The three fundamental forces are gravitation, electroweak, and strong. Over the years many forces have been unified into these three. Physicists are actively pursuing attempts to unify these three forces into only two or even just one single fundamental force.

The atomic theory of matter assumes atoms are the smallest unit of matter that is identified with a characteristic element. Molecules are composed of atoms, and these atoms can be from different elements. The kinetic theory of gases assumes the atomic theory is correct, and the development of the two theories proceeded together. The atomic theory of matter was not fully accepted until around 1910, by which time Einstein had explained Brownian motion and Perrin had published overwhelming experimental evidence.

The year 1895 saw several outstanding problems that seemed to worry only a few physicists. These problems included the inability to detect an electromagnetic medium, the difficulty in understanding the electrodynamics of moving bodies, and blackbody radiation. Four important discoveries during the period 1895–1897 were to signal the atomic age: x rays, radioactivity, the electron, and the splitting of spectral lines (Zeeman effect). The understanding of these problems and discoveries (among others) is the object of this book on modern physics.

CHAPTER 2

Special Theory of Relativity

It was found that there was no displacement of the interference fringes, so that the result of the experiment was negative and would, therefore, show that there is still a difficulty in the theory itself. . . .

Albert Michelson, Light Waves and Their Uses, 1907

In introductory physics we learned that Newton's laws of motion must be measured relative to some reference frame. A reference frame is called an **inertial frame** if Newton's laws are valid in that frame. For example, if a body subject to no net external force moves in a straight line with constant velocity, then the coordinate system establishing this fact is an inertial frame. If Newton's laws are valid in one reference frame, then they are also valid in any reference frame moving at a uniform velocity relative to the first system. This is known as the **Newtonian principle of relativity** or **Galilean invariance.** Newton showed that it was not possible to determine absolute motion in space by any experiment so he decided to use relative motion. In addition, the Newtonian concepts of time and space are completely separable. Consider two inertial reference frames K and K′ that move along their x and x' axes, respectively, with uniform relative velocity v as shown in Figure 2.1. We show system K′ moving to the right with velocity v with

Galilean invariance

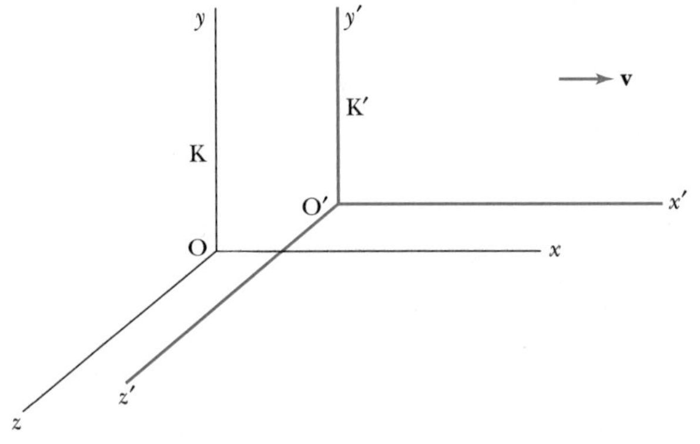

y
y′
v
K′
K
O′
x′
O
x
z′
z

FIGURE 2.1 Two inertial systems are moving with relative speed v along their x axes. We show the system K at rest and the system K′ moving with speed v relative to the system K.

respect to system K, which is fixed or stationary somewhere, such as on Earth. One result of the relativity theory is that there are no fixed, absolute frames of reference. We use the term *fixed* to refer to a system being fixed on a particular object, such as a planet, star, or spaceship that itself is moving in space. The transformation of the coordinates of a point in one system to the other system is given by

$$\begin{aligned} x' &= x - vt \\ y' &= y \\ z' &= z \end{aligned} \tag{2.1}$$

Galilean transformation

Similarly, the inverse transformation is given by

$$\begin{aligned} x &= x' + vt \\ y &= y' \\ z &= z' \end{aligned} \tag{2.2}$$

where we have set $t = t'$ because Newton considered time to be absolute. Equations (2.1) and (2.2) are known as the *Galilean transformation.* Newton's laws of motion can be shown to be invariant under a Galilean transformation; that is, they have the same form in both systems K and K′.

In the late 19th century, Albert Einstein was concerned that although Newton's laws of motion had the same form under a Galilean transformation, Maxwell's equations did not. Einstein believed so strongly in Maxwell's equations that he showed there was a significant problem in our understanding of the Newtonian principle of relativity. In 1905 he published ideas (while employed as a patent officer in the Swiss patent office) that rocked the very foundations of physics and science. He proposed that space and time are not separate and that Newton's laws are only an approximation. This special theory of relativity and its ramifications are the subject of this chapter. We begin by presenting the experimental situation historically—showing why a problem existed and what was done to try to rectify the situation. Then we discuss Einstein's two postulates on which the special theory is based. The interrelation of space and time is discussed, and several amazing and remarkable predictions based on the new theory are shown.

As the concepts of relativity became more used through the twentieth century in everyday research and development, it became essential to understand the transformation of momentum, force, and energy. We will study relativistic dynamics and the relationship between mass and energy, which leads to one of the most famous equations in physics and a new conservation law of mass-energy. Finally, we will return to electromagnetism to investigate the effects of relativity. We will learn that Maxwell's equations don't require change, and electric and magnetic effects are relative, depending on the observer.

2.1 Historical Perspective

As we discussed in the previous chapter, scientists and engineers of the late nineteenth century thought they had just about everything under control. Europe and North America had undergone the Industrial Revolution, and it appeared that the laws of physical science were known and well understood. Scientists and engineers were confident they could build a better device, no matter whether it was a faster steamship or a better photographic camera. The confidence of physicists would be shattered by the end of the 1800s.

In no area of physics was this sense of confidence stronger than in optics. In the first decade of the 1800s, Thomas Young performed his famous experiments on the interference of light. Later, Fresnel published his calculations showing the detailed understanding of interference, diffraction, and polarization. Because all known waves (other than light) required a medium in which to propagate (water waves have water, sound waves have, for example, air, etc.), it naturally was assumed that light also required a medium, even though light was able apparently to travel in vacuum through outer space. This medium was called the *luminiferous ether* or just *ether* for short, and it had some amazing properties. The ether had to have such a low density that planets could pass through it, seemingly for eternity, with no apparent loss of orbit position. Its elasticity must be strong enough to pass waves of incredibly large speeds!

The need for ether

The *tour de force* of optics was accomplished in the 1860s when James Clerk Maxwell published his electromagnetic theory of light. Among other things, his theory could even *predict the speed of light in different media,* depending only on the electric and magnetic properties of matter. In vacuum, the speed of light is given by $v = c = 1/\sqrt{\mu_0\epsilon_0}$, where μ_0 and ϵ_0 are the permeability and permittivity of free space, respectively. The properties of the ether must be consistent with this theory, and the feeling was that to be able to discern its various properties required only an experiment that was sensitive enough. The concept of ether was by then well accepted.

The laws of physics were so strongly held after Maxwell presented his electromagnetic theory that difficulties with any facet of physics were quickly jumped upon. Experience had shown that such problems often led to a new, deeper understanding of nature. Physicists, like other scientists and engineers, don't like to leave loose ends around. They prefer to tie things up in neat packages using as few basic laws as is absolutely necessary. Newton's three laws, Maxwell's four equations, the Lorentz force law, and the two (or four) laws of thermodynamics are good evidence of this desire.

Maxwell's equations predict the velocity of light in a vacuum to be c. If we have a flashbulb go off in the moving system K′, an observer in system K′ measures the speed of the light pulse to be c. However, if we make use of Equation (2.1) to find the relation between speeds, we find the speed measured in system K to be $c + v$, where v is the relative speed of the two systems. However, Maxwell's equations don't differentiate between these two systems. Physicists of the late 19th century proposed that there must be one preferred inertial reference frame in which the ether was stationary and that in this system the speed of light was c. In the other systems, the speed of light would indeed be affected by the relative speed of the reference system. Because the speed of light was known to be so enormous, 3×10^8 m/s, no experiment had as yet been able to discern an effect due to the relative speed v. The ether frame would in fact be an absolute standard, from which other measurements could be made. Scientists set out to find the effects of the ether. The Earth is orbiting around the sun at a high orbital speed, about 10^{-4} that of c, so an obvious experiment is to try to find the effects of the Earth's motion through the ether. Even though we don't know how fast the sun might be moving through the ether, certainly at various times during the solar year, the Earth must be passing rapidly through the ether.

2.2 The Michelson-Morley Experiment

Probably the most significant American physics experiment of the 1800s was performed by Albert Michelson (1852–1931), who was born in Prussia but came

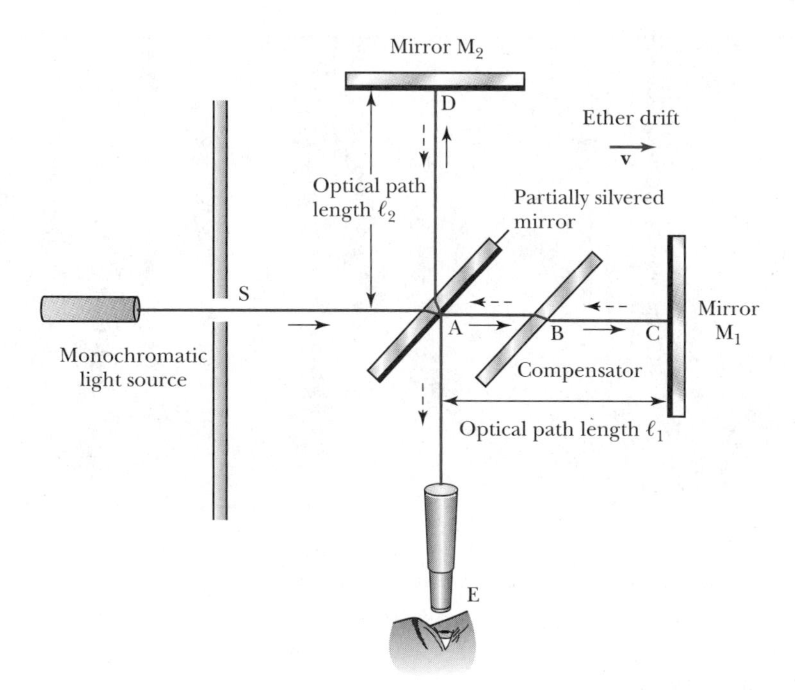

FIGURE 2.2 A schematic diagram of Michelson's interferometer experiment. Light of a single wavelength is partially reflected and partially transmitted by the glass at A. The light is subsequently reflected by mirrors at C and D and after reflection or transmission again at A enters the telescope at E. Interference fringes will appear at E.

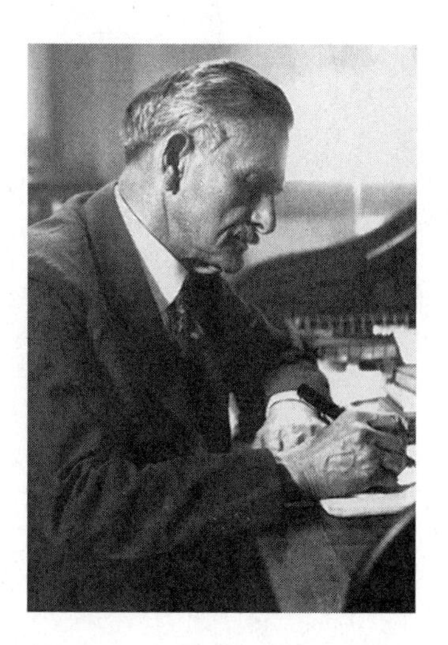

Albert A. Michelson (1852–1931) shown at his desk at the University of Chicago in 1927. Micheson had appointments at four American universities: at the Naval Academy in 1873; at the Case School of Applied Science, Cleveland, in 1883; at Clark University, Worcester, Massachusetts, in 1890; and at the University of Chicago in 1892 until his retirement in 1929. He spent his retirement years in Pasadena, California where he continued to measure the speed of light at Mount Wilson. *AIP/Niels Bohr Library.*

to the United States when he was two years old. Michelson, who was the first U.S. citizen to receive the Nobel Prize for physics (1907), was an ingenious scientist who built an extremely precise device, called an *interferometer.* During his lifetime he continued to improve on his own best measurement of the speed of light. Michelson decided that he could use his interferometer to detect the difference in the speed of light passing through the ether in different directions. The basic technique is shown in Figure 2.2. Initially, it is assumed that one of the interferometer arms (AC) is parallel to the motion of the Earth through the ether. Light leaves the source S and passes through the glass plate at A. Because the back of A is partially silvered, part of the light is reflected, eventually going to the mirror at D, and part of the light travels through A on to the mirror at C. The light is reflected at the mirrors C and D and comes back to the partially silvered mirror A, where part of the light from each path passes on to the telescope and eye at E. The compensator is added at B to make sure both light paths pass through equal thicknesses of glass. Interference fringes can be found by using a bright light source like sodium, and the apparatus is adjusted for maximum intensity of the light at E. The fringe pattern should shift if the apparatus is rotated through 90° such that arm AD becomes parallel to the motion of the Earth through the ether and arm AC is perpendicular to the motion.

We let the optical path lengths of AC and AD be denoted by ℓ_1 and ℓ_2, respectively. The observed interference pattern consists of alternating bright and dark bands, corresponding to constructive and destructive interference, respectively, similar to that shown in Figure 2.3. For constructive interference, the difference between the two path lengths (to and from the mirrors) is given by some

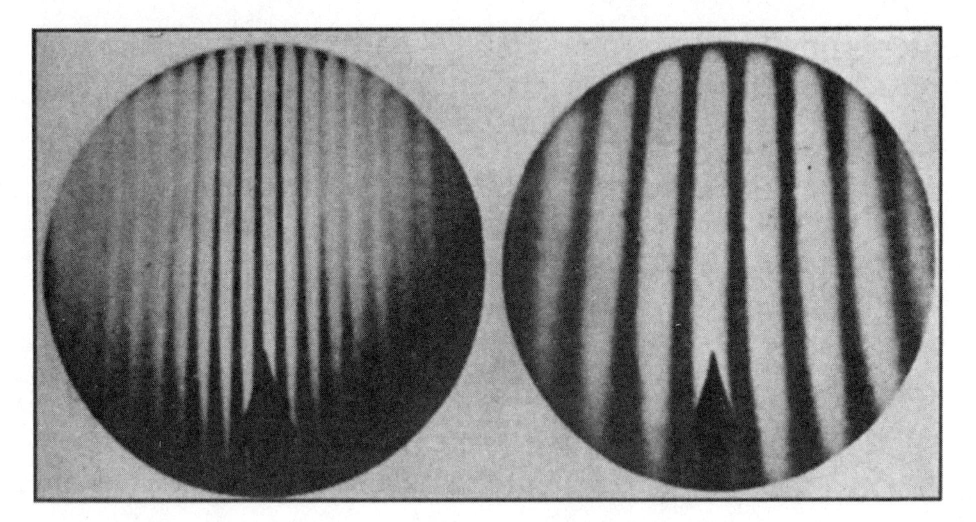

FIGURE 2.3 Inteference fringes as they would appear in the eyepiece of the Michelson-Morley experiment. *From L. S. Swenson, Jr., Invention and Discovery,* ***43*** *(Fall, 1987).*

number of wavelengths, $2(\ell_1 - \ell_2) = n\lambda$, where λ is the wavelength of the light and n is an integer.

The expected shift in the interference pattern can be calculated by determining the time difference between the two paths. When the light travels from A to C, the velocity of light according to the Galilean transformation is $c + v$ because the ether carries the light along with it. On the return journey from C to A the velocity is $c - v$ because the light travels opposite to the path of the ether. The total time for the round trip journey to mirror M_1 is t_1,

$$t_1 = \frac{\ell_1}{c+v} + \frac{\ell_1}{c-v} = \frac{2\,c\ell_1}{c^2 - v^2} = \frac{2\ell_1}{c}\left(\frac{1}{1 - v^2/c^2}\right)$$

Now imagine what happens to the light that is reflected from mirror M_2. If the light is pointed directly at point D, the ether will carry the light with it and the light misses the mirror, much as the wind can affect the flight of an arrow. If a swimmer (who can swim with speed v_2 in still water) wants to swim across a swiftly moving river (speed v_1), the swimmer must start moving upriver, so that when the current carries her downstream, she will move directly across the river. Careful reasoning shows that the swimmer's velocity is $\sqrt{v_2^2 - v_1^2}$ throughout her journey (Problem 4). Thus the time t_2 for the light to pass to mirror M_2 at D and back is given by

$$t_2 = \frac{2\ell_2}{\sqrt{c^2 - v^2}} = \frac{2\ell_2}{c} \frac{1}{\sqrt{1 - v^2/c^2}}$$

The time difference between the two journeys Δt is

$$\Delta t = t_2 - t_1 = \frac{2}{c}\left(\frac{\ell_2}{\sqrt{1 - v^2/c^2}} - \frac{\ell_1}{1 - v^2/c^2}\right) \tag{2.3}$$

We now rotate the apparatus by 90° so that the ether passes along the length ℓ_2 toward the mirror M_2. We denote the new quantities by primes and carry out an analysis similar to that just done. The time difference $\Delta t'$ is now

$$\Delta t' = t_2{}' - t_1{}' = \frac{2}{c}\left(\frac{\ell_2}{1 - v^2/c^2} - \frac{\ell_1}{\sqrt{1 - v^2/c^2}}\right) \tag{2.4}$$

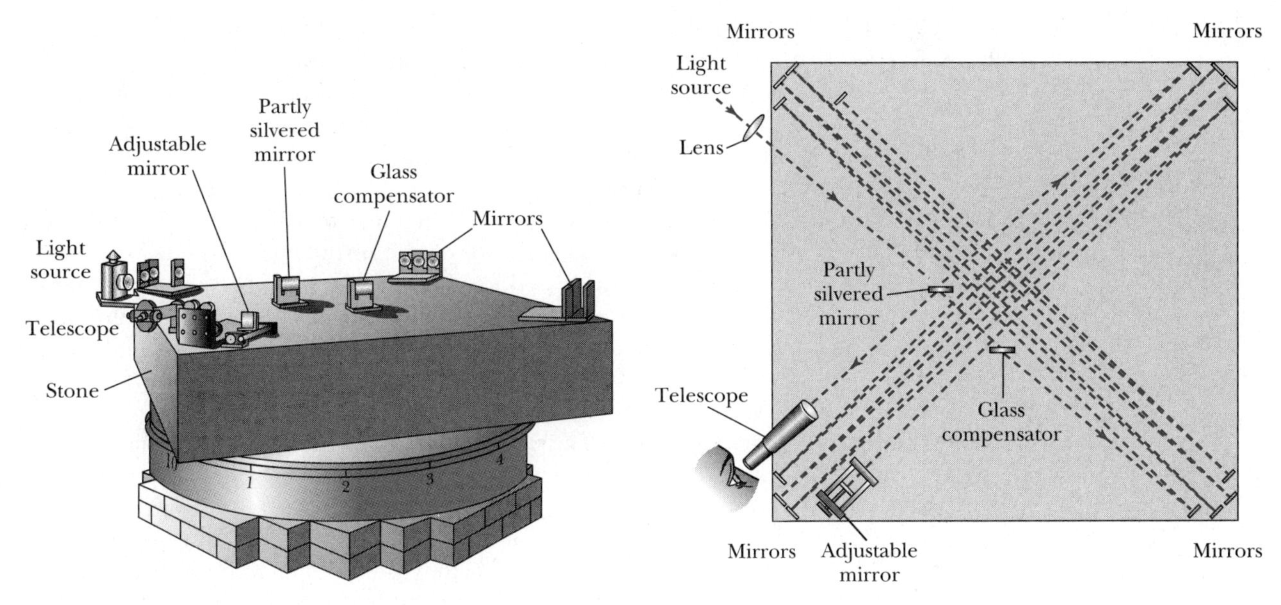

FIGURE 2.4 An adaption of the Michelson and Morley 1887 experiment taken from their publication [A. A. Michelson and E. M. Morley, *Philosophical Magazine* **190,** 449 (1887)]. The top is a perspective view of the apparatus. To reduce vibration, the experiment was done on a massive soapstone, 1.5 m square and 0.3 m thick. This stone was placed on a wooden float that rested on mercury inside the annular piece shown underneath the stone. The entire apparatus rested on a brick pier. The incoming light is focused by the lens and is both transmitted and reflected by the partly silvered mirror. The adjustable mirror allows fine adjustments in the interference fringes. The stone was slowly and uniformly rotated on the mercury to look for the interference effects of the ether.

Michelson looked for a shift in the interference pattern when his apparatus was rotated by 90°. The time difference is

$$\Delta t' - \Delta t = \frac{2}{c}\left(\frac{\ell_1 + \ell_2}{1 - v^2/c^2} - \frac{\ell_1 + \ell_2}{\sqrt{1 - v^2/c^2}}\right)$$

Because we know $c \gg v$, we can use the binomial expansion* to expand the terms involving v^2/c^2, keeping only the lowest terms.

$$\Delta t' - \Delta t = \frac{2}{c}(\ell_1 + \ell_2)\left[\left(1 + \frac{v^2}{c^2} + \cdots\right) - \left(1 + \frac{v^2}{2c^2} + \cdots\right)\right] \approx \frac{v^2(\ell_1 + \ell_2)}{c^3} \quad (2.5)$$

Michelson in Europe

Michelson left his position at the U.S. Naval Academy in 1880 and took his interferometer to Europe for postgraduate studies with some of Europe's best physicists, particularly Hermann Helmholtz in Berlin. After a few false starts he finally was able to perform a measurement in Potsdam, near Berlin, in 1881. In order to use Equation (2.5) for an estimate of the expected time difference, the value of the Earth's orbital speed around the sun, 3×10^4 m/s, was used. Michelson's apparatus had $\ell_1 \approx \ell_2 \approx \ell = 1.2$ m. Thus Equation (2.5) predicts a time difference of 8×10^{-17} s. This is an exceedingly small time, but for a visible wavelength of 6×10^{-7} m, the period of one wavelength amounts to

*Special cases of the binomial expansion include:

$$(1 \pm x)^{1/2} = 1 \pm \frac{1}{2}x - \frac{1}{8}x^2 \pm \frac{1}{16}x^3 - \cdots$$

$$(1 \pm x)^{-1} = 1 \mp x + x^2 \mp x^3 + \cdots$$

$T = 1/\nu = \lambda/c = 2 \times 10^{-15}$ s. Thus the time period of 8×10^{-17} s represents 0.04 fringes in the interference pattern. Michelson reasoned that he could detect a shift of at least half this value, but found none. Although disappointed, Michelson concluded that the hypothesis of the stationary ether must be incorrect.

The result of Michelson's experiment was so surprising that he was asked to repeat it by several well-known physicists. In 1882 Michelson accepted a position at the then-new Case School of Applied Science in Cleveland. Together with Edward Morley (1838–1923), a professor of chemistry at nearby Western Reserve College who had become interested in Michelson's work, they put together the more sophisticated experiment shown in Figure 2.4. The new experiment had an optical path length of 11 m, created by reflecting the light for eight round trips. The new apparatus was mounted on soapstone that floated on mercury to eliminate vibrations and was so effective that Michelson and Morley believed they could detect a fraction of a fringe shift as small as 0.005. With their new apparatus they expected a shift as large as 0.4 of a fringe. They reported in 1887 a *null result*—no effect whatsoever! It is this famous experiment that has become known as the Michelson-Morley experiment.

Null result of Michelson-Morley experiment

The measurement so shattered a widely held belief that many suggestions were made to explain it. What if the Earth just happened to have a zero motion through the ether at the time of the experiment? Michelson and Morley repeated their experiment during night and day and for different seasons throughout the year. It is unlikely that at least sometime during these many experiments, the Earth would not be moving through the ether. Michelson and Morley even took their experiment to a mountaintop to see if the effects of the ether might be different.

Ether drag

Of the many possible explanations of the null ether measurement, the one taken most seriously was the *ether drag* hypothesis. Suppose the Earth somehow dragged the ether with it as it rotates on its own axis and revolves around the sun. Of the several experimental measurements at variance with this hypothesis we mention that of *stellar aberration* noted by the British astronomer James Bradley in 1728. Bradley noticed that the apparent position of the stars seems to rotate in a circular motion with a period of one year. The angular diameter of this circular motion with respect to the Earth is 41 seconds of arc. This effect can be understood by an analogy. From the viewpoint of a person sitting in a car during a rainstorm, the raindrops appear to fall vertically when the car is at rest but appear to be slanted toward the windshield when the car is moving forward. The same effect occurs for light coming from stars directly above the Earth's orbital plane. If the telescope and star are at rest with respect to the ether, the light enters the telescope as shown in Figure 2.5a. However, because the Earth is

Stellar aberration

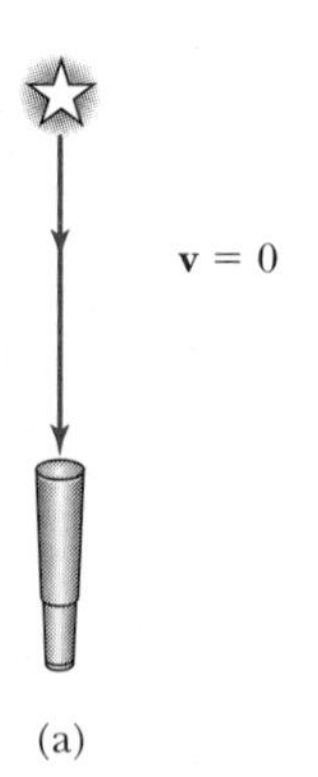

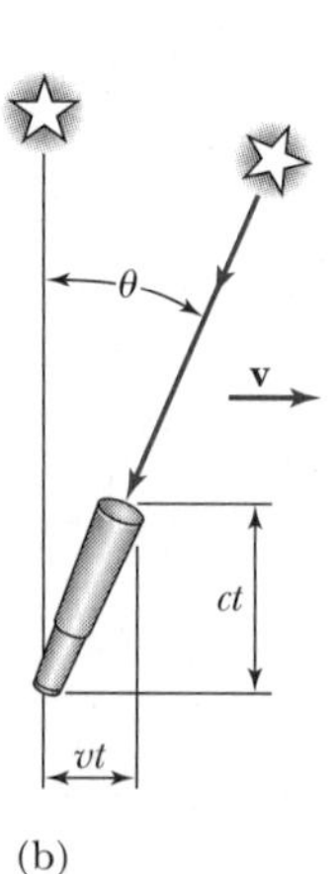

FIGURE 2.5 The effect of stellar aberration. (a) If a telescope is at rest, light from a distant star will pass directly into the telescope. (b) However, if the telescope is traveling at speed v (because it is fixed on the Earth, which has a motion about the sun), it must be slanted slightly to allow the star light to enter the telescope. This leads to an apparent circular motion of the star as seen by the telescope, as the motion of the Earth about the sun changes throughout the solar year.

moving in its orbital motion, the apparent position of the star is at an angle θ as shown in Figure 2.5b. The telescope must actually be slanted at an angle θ in order to observe the light from the overhead star. During a time t the starlight moves a vertical distance ct while the telescope moves a horizontal distance vt, so that the tangent of the angle θ is

$$\tan \theta = \frac{vt}{ct} = \frac{v}{c}$$

The orbital speed of the Earth is about 3×10^4 m/s, therefore, the angle θ is 10^{-4} rad or 20.6 seconds of arc, with a total opening as the Earth rotates, of $2\theta =$ 41 s—in agreement with Bradley's observation. The aberration reverses itself every 6 months as the Earth orbits about the sun, in effect giving a circular motion to the star's position. This observation is in disagreement with the hypothesis of the Earth dragging the ether. If the ether were dragged with the Earth, there would be no need to tilt the telescope! The experimental observation of stellar aberration together with the null result of the Michelson and Morley experiment is enough evidence to refute the suggestions that the ether exists. Many other experimental observations have now been made that also confirm this conclusion.

The inability to detect the ether was a serious blow to reconciling the invariant form of the electromagnetic equations of Maxwell. There seems to be no single reference inertial system in which the speed of light is actually c. H. A. Lorentz and G. F. FitzGerald suggested apparently independently that the results of the Michelson-Morley experiment could be understood if length is contracted by the factor $\sqrt{1 - v^2/c^2}$ in the direction of motion, where v is the speed in the direction of travel. For this situation, the length ℓ_1, in the direction of motion, will be contracted by the factor $\sqrt{1 - v^2/c^2}$, whereas the length ℓ_2, perpendicular to v, will not. The result in Equation (2.3) is that t_1 will have the extra factor $\sqrt{1 - v^2/c^2}$, making Δt precisely zero as determined experimentally by Michelson. This contraction postulate, which became known as the *Lorentz-FitzGerald contraction,* was not proven from first principles using Maxwell's equations, and its true significance was not understood for several years until after Einstein presented his explanation. An obvious problem with the Lorentz-FitzGerald contraction is that it is an ad hoc assumption that cannot directly be tested. Any measuring device would be shortened presumably by the same factor.

Albert Einstein, shown here sailing on Long Island Sound, was born in Germany and studied in Munich and Zurich, Switzerland. During his seven years in the Swiss Patent Office in Berne, he did some of his best work. His fame quickly led to appointments at Zurich, Prague, back to Zurich, and then to Berlin in 1914. In 1933, after Hitler came to power, Einstein left for the Institute for Advanced Study at Princeton University where he became a U.S. citizen in 1940 and remained until his death in 1955. Einstein's total work in physics can only be rivaled by that of Isaac Newton. *AIP/ Emilio Segrè Visual Archives.*

2.3 Einstein's Postulates

At the turn of the twentieth century, physicists were very concerned. The Michelson-Morley experiment had laid to rest the idea of finding a preferred inertial system for Maxwell's equations, yet the Galilean transformation, which worked for the laws of mechanics, was invalid for Maxwell's equations. This quandary represented a turning point for physics.

Albert Einstein (1879–1955) was only two years old when Michelson reported his first null measurement for the existence of the ether. Einstein said that he began thinking at the age of 16 about the form of Maxwell's equations in moving inertial systems, and in 1905, when he was 26 years old, he published his startling proposal* about the principle of relativity, which Einstein believed to be

*In one issue of the German journal *Annalen der Physik* **17,** No. 4 (1905). Einstein published three remarkable papers. The first, on the quantum properties of light, explained the photoelectric effect; the second, on the statistical properties of molecules, included an explanation of Brownian motion; and the third was on special relativity. All three papers contained predictions that were subsequently confirmed experimentally.

fundamental. Working without the benefit of discussions with colleagues outside his small circle of friends, Einstein was apparently unaware of the interest concerning the null result of Michelson and Morley.* Einstein instead looked at the problem in a more formal manner and believed that Maxwell's equations must be valid in all inertial frames. With piercing insight and genius, Einstein was able to bring together seemingly inconsistent results concerning the laws of mechanics and electromagnetism with two postulates (as he himself called them). These postulates are

Einstein's two postulates

1. **The Principle of Relativity:** The laws of physics are the same in all inertial systems. There is no way to detect absolute motion, and no preferred inertial system exists.
2. **The constancy of the speed of light:** Observers in all inertial systems measure the same value for the speed of light in a vacuum.

The first postulate indicates that the laws of physics will be the same in all coordinate systems moving with uniform relative motion to each other. Einstein showed that postulate 2 actually follows from the first one. Einstein returned to the Principle of Relativity as espoused by Newton, but Newton's principle only referred to the laws of mechanics. Einstein expanded it to include all laws of physics—including those of electromagnetism. In fact, Einstein was convinced by the laws of electromagnetism that the Principle of Relativity must be valid. We can now modify our previous definition of *inertial frames of reference* to be those frames of reference in which *all the laws of physics* are valid.

Inertial frames of reference

Einstein's solution requires us to take a careful look at time. Return to the two systems of Figure 2.1 and remember that we had previously assumed that $t = t'$. We assumed that events occurring in system K′ and in system K could be easily synchronized. Einstein realized that each system must have its own observers with their own clocks and meter sticks. *An event in a given system must be specified by stating both its space and time coordinates.* Consider the flashing of two light bulbs fixed in system K as shown in Figure 2.6a. Mary, in system K′ (the **M**oving system) is beside Frank, who is in system K (the **F**ixed system), when the lightbulbs flash. As seen in Figure 2.6b the light pulses travel the same distance in system K and arrive at Frank *simultaneously.* Frank sees the two flashes at the same time. However, the two light pulses do not reach Mary simultaneously, because system K′ is moving to the right, and she has moved closer to the lightbulb on the right by the time the flash reaches her. The light flash coming from the left will reach her at some later time. We can conclude that

Simultaneity

Two events that are simultaneous in one reference frame (K) are not necessarily simultaneous in another reference frame (K′) moving with respect to the first frame.

We must be very careful when comparing the same event in two systems moving with respect to one another. Time comparison can be accomplished by sending light signals from one observer to another, but this information can travel only as fast as the finite speed of light. It is best if each system has its own

*The question of whether Einstein knew of Michelson and Morley's null result before he produced his special theory of relativity is somewhat uncertain. For example, see J. Stachel, "Einstein and Ether Drift Experiments," *Physics Today,* May, 1987, p. 45.

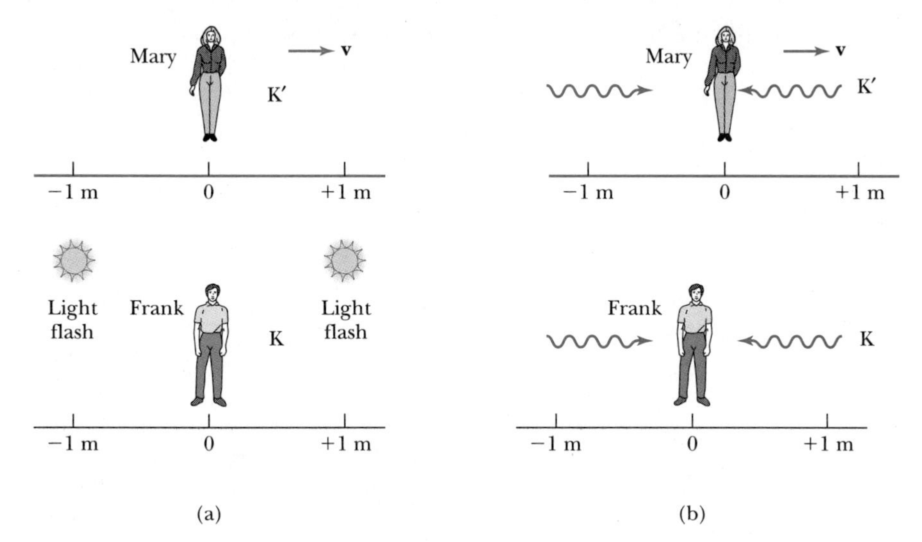

FIGURE 2.6 The problem of simultaneity. Flashbulbs positioned in system K at one meter on either side of Frank go off simultaneously in (a). Frank indeed sees both flashes simultaneously in (b). However, Mary, at rest in system K′ moving to the right with speed v, does not see the flashes simultaneously despite the fact that Mary was alongside Frank when the flashbulbs went off. During the finite time it took light to travel the one meter, Mary had moved slightly, as shown in exaggerated form in (b).

observers with clocks that are synchronized. How can we do this? We can place observers with clocks throughout a given system. If, when we bring all the clocks together at one spot at rest, all the clocks agree, then the clocks are said to be **synchronized.** However, we have to move the clocks relative to each other in order to reposition them, and that might affect the synchronization. A better way would be to flash a bulb halfway between each pair of clocks at rest and make sure the pulses arrive simultaneously at each clock. This will require many measurements, but it is a safe way of synchronizing the clocks. We can determine the time of an event occurring far away from us by having a colleague at the event with a clock fixed at rest measure the time of the particular event and send us the results, for example, by telephone or even by mail. If we need to check our clocks we can always send light signals to each other over known distances at some predetermined time.

Synchronization of clocks

In the next section we will derive the correct transformation, called the **Lorentz transformation,** that makes the laws of physics invariant between inertial frames of reference. We use the coordinate systems described by Figure 2.1. At $t = t' = 0$, the origins of the two coordinate systems are coincident and the system K′ is traveling along the x and x' axis. For this special case, the Lorentz transformation equations are

$$
\begin{aligned}
x' &= \frac{x - vt}{\sqrt{1 - v^2/c^2}} \\
y' &= y \\
z' &= z \\
t' &= \frac{t - \dfrac{vx}{c^2}}{\sqrt{1 - v^2/c^2}}
\end{aligned}
\tag{2.6}
$$

Lorentz transformation equations

We commonly use the symbols β and γ to represent

$$\beta = \frac{v}{c} \tag{2.7}$$

$$\gamma = \frac{1}{\sqrt{1 - v^2/c^2}} \tag{2.8}$$

which allows the Lorentz transformation equations to be rewritten in compact form as

$$\begin{aligned} x' &= \gamma(x - \beta ct) \\ y' &= y \\ z' &= z \\ t' &= \gamma(t - \beta x/c) \end{aligned} \tag{2.6}$$

Note that $\gamma \geq 1$ ($\gamma = 1$ when $v = 0$).

2.4 The Lorentz Transformation

In this section we want to use Einstein's two postulates to find a transformation between inertial frames of reference that will cause all the physical laws, including Newton's laws of mechanics and Maxwell's electrodynamic equations, to have the same form. We use the fixed system K and moving system K′ of Figure 2.1. At $t = t' = 0$ the origins and axes of both systems are coincident and system K′ is moving to the right along the x axis. A flashbulb goes off at the origins when $t = t' = 0$. According to postulate 2, the speed of light will be c in both systems, and the wavefronts observed in both systems must be spherical and described by

$$x^2 + y^2 + z^2 = c^2t^2 \tag{2.9a}$$

$$x'^2 + y'^2 + z'^2 = c^2t'^2 \tag{2.9b}$$

These two equations are inconsistent with a Galilean transformation because a wavefront can only be spherical in one system when the second is moving at speed v with respect to the first. The Lorentz transformation *requires* both systems to have a spherical wavefront centered on the system's origin. Another clear break with Galilean and Newtonian physics is that we do not assume that $t = t'$. Each system must have its own clocks and meter sticks. Because the systems move only along their x axes, observers in both systems agree by direct observation that

$$\begin{aligned} y' &= y \\ z' &= z \end{aligned}$$

We know that the Galilean transformation $x' = x - vt$ is incorrect, but what is the correct transformation? We require a linear transformation so that each event in system K corresponds to one, and only one, event in system K′. The simplest linear transformation is of the form

$$x' = \gamma(x - vt) \tag{2.10}$$

We will see if such a transformation suffices. The parameter γ cannot depend on x or t because the transformation must be linear. The parameter γ must be close to 1 for $v \ll c$ in order for Newton's laws of mechanics to be valid for most of our measurements.

We can use similar arguments from the standpoint of an observer stationed in system K′ to obtain an equation similar to Equation (2.10).

$$x = \gamma'(x' + vt') \tag{2.11}$$

Because postulate 1 requires that the laws of physics be the same in both reference systems, we demand that $\gamma' = \gamma$. Notice that the only difference between Equations (2.10) and (2.11) other than the primed and unprimed quantities being switched is that $v \to -v$, which is reasonable because according to the observer in each system, the other observer is moving either forward or backward.

According to postulate 2 the speed of light must be c in both systems. Therefore in each system the wavefront of the flashbulb light pulse along the respective x axes must be described by $x = ct$ and $x' = ct'$, which we substitute into Equations (2.10) and (2.11) to obtain

$$ct' = \gamma(ct - vt) \tag{2.12a}$$

and

$$ct = \gamma(ct' + vt') \tag{2.12b}$$

We divide each of these equations by c.

$$t' = \gamma t\left(1 - \frac{v}{c}\right) \tag{2.13}$$

and

$$t = \gamma t'\left(1 + \frac{v}{c}\right) \tag{2.14}$$

We substitute the value of t from Equation (2.14) into Equation (2.13).

$$t' = \gamma^2 t'\left(1 - \frac{v}{c}\right)\left(1 + \frac{v}{c}\right) \tag{2.15}$$

We solve this equation for γ^2 and obtain

$$\gamma^2 = \frac{1}{1 - v^2/c^2}$$

or

$$\gamma = \frac{1}{\sqrt{1 - v^2/c^2}} \tag{2.16}$$

In order to find a transformation for time, we substitute x' from Equation (2.10) into Equation (2.11) and obtain

$$x = \gamma^2(x - vt) + \gamma vt'$$

We solve this equation for t'.

$$t' = \frac{x}{\gamma v}(1 - \gamma^2) + \gamma t$$

If we use the value of γ in Equation (2.16) and β from Equation (2.7), a little algebra shows that $(1 - \gamma^2)/\gamma v = -\gamma\beta^2/v$ which we substitute in the previous equation to finally obtain

$$t' = \frac{t - \dfrac{vx}{c^2}}{\sqrt{1 - \beta^2}}$$

We are now able to write the complete Lorentz transformation equations as

$$\begin{aligned} x' &= \frac{x - vt}{\sqrt{1 - \beta^2}} \\ y' &= y \\ z' &= z \\ t' &= \frac{t - \dfrac{vx}{c^2}}{\sqrt{1 - \beta^2}} \end{aligned} \tag{2.17}$$

The inverse transformation equations are obtained by replacing v by $-v$ as discussed previously and by exchanging the primed and unprimed quantities.

Inverse Lorentz transformation equations

$$\begin{aligned} x &= \frac{x' + vt'}{\sqrt{1 - \beta^2}} \\ y &= y' \\ z &= z' \\ t &= \frac{t' + \dfrac{vx'}{c^2}}{\sqrt{1 - \beta^2}} \end{aligned} \tag{2.18}$$

Notice that Equations (2.17) and (2.18) both reduce to the Galilean transformation when $v \ll c$. It is only for speeds that approach the speed of light that the Lorentz transformation equations become significantly different from the Galilean equations. In our studies of mechanics we normally do not consider such high speeds, and our previous results probably require no corrections. The laws of mechanics credited to Newton are still valid over the region of their applicability. Even for a speed as high as the Earth orbiting about the sun, 30 km/s, the value of γ is 1.000000005. We show a plot of the relativistic parameter γ versus speed in Figure 2.7. As a rule of thumb, we consider using the relativistic equations when $v/c > 0.1$ ($\gamma \approx 1.005$).

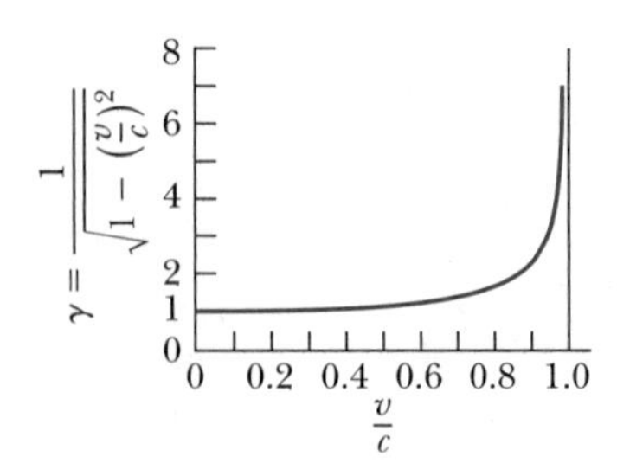

FIGURE 2.7 A plot of the relativistic parameter γ as a function of speed v/c, showing that γ becomes large quickly as v approaches c.

Finally, we should perhaps briefly discuss the implications of the Lorentz transformation. The linear transformation equations ensure that a single event in one system is described by a single event in another inertial system. However, space and time are not separate. In order to express the position of x in system K′, we must use both x' and t'. We also have found that the Lorentz transformation does not allow a speed greater than c; the transformation equations become imaginary in this case. We show later in this chapter that no object of nonzero mass can have a speed greater than c.

2.5 Time Dilation and Length Contraction

Time Dilation

Consider again our two systems K and K′ with system K fixed and system K′ moving along the x axis with velocity v as shown in Figure 2.8a. Frank lights a sparkler at position x_1 in system K. A clock placed beside the sparkler indicates the time to be t_1 when the sparkler is lit and t_2 when the sparkler goes out (Figure 2.8b). The sparkler burns for time T_0, where $T_0 = t_2 - t_1$. The time difference between two events occurring at the same position in a system as measured by a clock at rest in the system is called the **proper time.** We use the subscript zero on the time difference T_0 to denote the proper time.

Proper time

Now what is the time as determined by Mary who is passing by (but at rest in her own system K′) All the clocks in both systems have been synchronized when the systems are at rest with respect to one another. The two events (sparkler lit and then going out) do not occur at the same place according to Mary. She is beside the sparkler when it is lit, but she has moved far away from the sparkler when it goes out (Figure 2.8b). Her friend Melinda, also at rest in system K′, is beside the sparkler when it goes out. Mary and Melinda measure the two times for the sparkler to be lit and to go out in system K′ as times t_1' and t_2'. The Lorentz transformation relates these times to those measured in system K as

$$t_2' - t_1' = \frac{(t_2 - t_1) - (v/c^2)(x_2 - x_1)}{\sqrt{1 - v^2/c^2}}$$

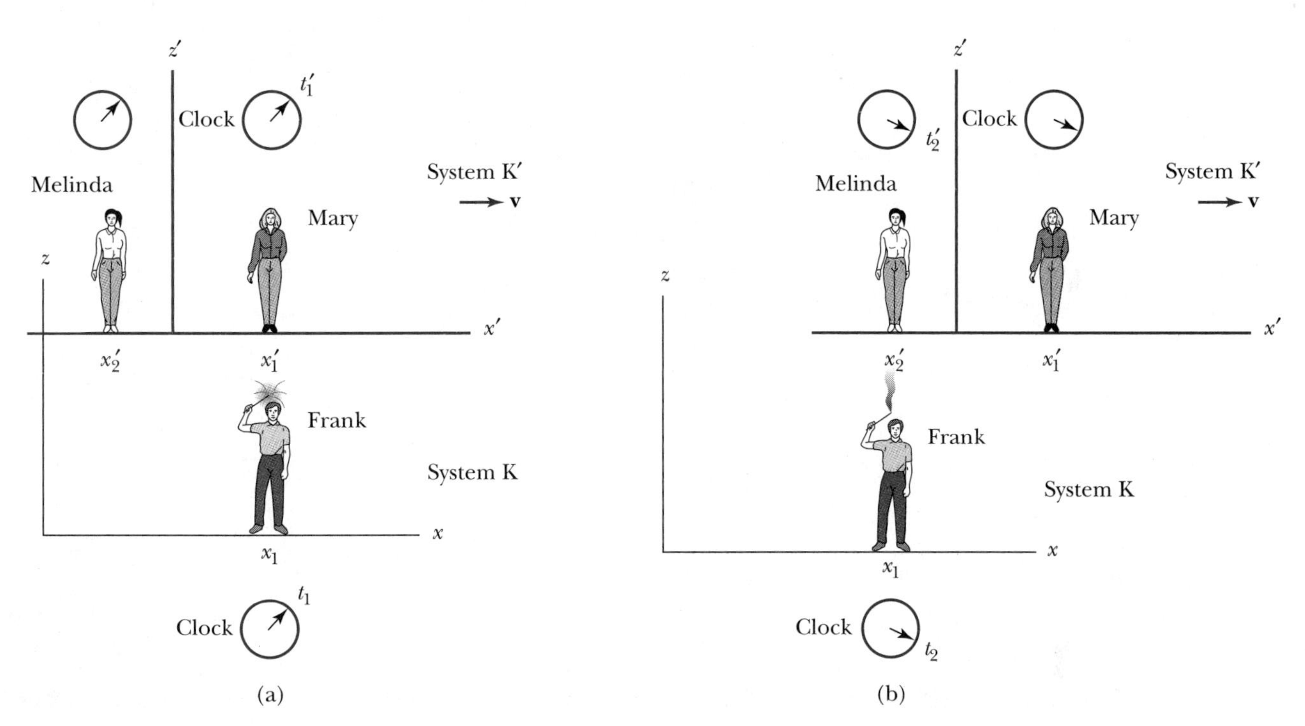

FIGURE 2.8 Frank measures the proper time for the time interval that a sparkler stays lit. His clock is at the same position in system K when the sparkler is lit in (a) and when it goes out in (b). Mary, in the moving system K′, is beside the sparkler at position x_1' when it is lit in (a), but by the time it goes out in (b), she has moved away. Melinda, at position x_2', measures the time in system K′ when the sparkler goes out in (b).

In system K the clock is fixed at x_1, so $x_2 - x_1 = 0$; that is, the two events occur at the same position. The time $t_2 - t_1$ is the proper time T_0, and we denote the time difference $t_2' - t_1' = T'$ as measured in the moving system K′.

Time dilation

$$T' = \frac{T_0}{\sqrt{1 - v^2/c^2}} = \gamma T_0 \tag{2.19}$$

Thus the time interval measured in the moving system K′ is greater than the time interval measured in system K where the sparkler is at rest. This effect is known as **time dilation** and is a direct result of Einstein's two postulates. The time measured by Mary and Melinda in their system K′ for the time difference was greater than T_0 by the factor γ, $(\gamma > 1)$. The two events, sparkler being lit and then going out, did not occur at the same position ($x_2' \neq x_1'$) in system K′ (see Figure 2.8b). This result occurs because of the absence of simultaneity. The events do not occur at the same space and time coordinates in the two systems. It requires three clocks to perform the measurement: one in system K and two in system K′.

Moving clocks run slow

The time dilation result is often interpreted by saying that *movng clocks run slow* by the factor γ^{-1}, and sometimes this is a useful way to remember the effect. The moving clock in this case can be any kind of clock. It can be the time sand takes to pass through an hourglass, the time a sparkler stays lit, the time between heartbeats, the time between ticks of a clock, or the time of a class lecture. In all cases, the actual time interval on a moving clock is greater than the proper time as measured on a clock at rest. The proper time is always the smallest possible time interval between two events.

Each person will claim the clock in the other (moving) system is running slow. If Mary had a sparkler in her system K′ at rest, Frank (fixed in system K) would also measure a longer time interval on his clocks in system K because the sparkler would be moving with respect to his system.

Example 2.1

Show that Frank in the fixed system will also determine the time dilation result by having the sparkler be at rest in the system K′.

Solution: In this case the sparkler is lit by Mary in the moving system K′. The time interval over which the sparkler is lit is given by $T_0' = t_2' - t_1'$, and the sparkler is placed at the position $x_1' = x_2'$ so that $x_2' - x_1' = 0$. In this case T_0' is the proper time. We use the Lorentz transformation from Equation (2.18) to determine the time difference $T = t_2 - t_1$ as measured by the clocks of Frank and his colleagues.

$$T = t_2 - t_1 = \frac{(t_2' - t_1') + (v/c^2)(x_2' - x_1')}{\sqrt{1 - v^2/c^2}}$$

$$= \frac{T_0'}{\sqrt{1 - v^2/c^2}} = \gamma T_0'$$

The time interval is still smaller in the system where the sparkler is at rest.

The preceding results naturally seem a little strange to us. In relativity we often carry out thought (or *gedanken* from the German word) experiments, since the actual experiments are somewhat difficult. Consider the following *gedanken*

***Gedanken* experiments**

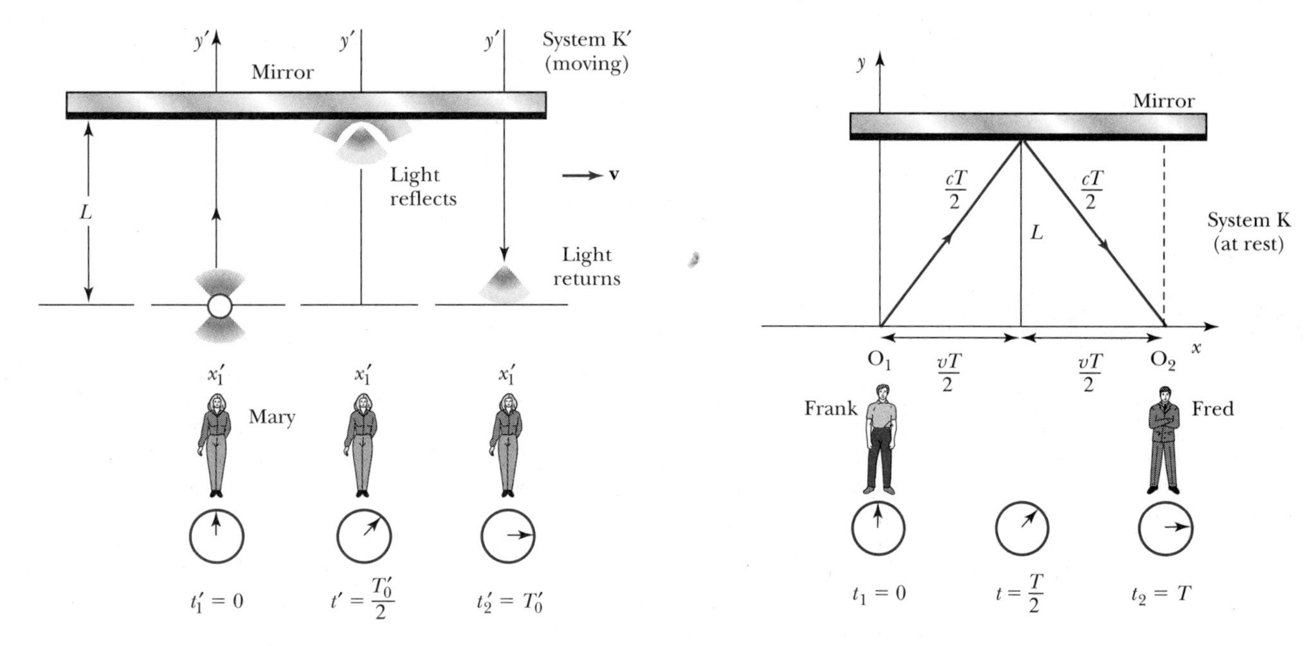

FIGURE 2.9 Mary, in system K′, flashes a light along her y' axis and measures the proper time $T_0' = 2L/c$ for the light to return. In system K Frank will see the light travel partially down his x axis, because system K′ is moving. In system K Fred times the arrival of the light. The time interval T that Frank and Fred measure is related to the proper time by $T = \gamma T_0'$.

experiment. Mary in the moving system K′ flashes a light at her origin along her y' axis. The light travels a distance L, reflects off a mirror, and returns. Mary (see Figure 2.9) says that the total time for the journey is $T_0' = t_2' - t_1' = 2L/c$, and this is indeed the proper time, because the clock in K′ beside Mary is at rest.

What do Frank and his observers in system K see? Let T be the round-trip time interval measured in system K for the light to return to the x axis. The light is flashed when the origins are coincident, as Mary passes by Frank with relative velocity v. When the light reaches the mirror in the system K′ at time $T/2$, the system K′ will have moved a distance $vT/2$ down the x axis. When the light is reflected back to the x axis, Frank will not even see the light return, because it will return a distance vT away, where another observer Fred is positioned. Because observers Frank and Fred have previously synchronized their clocks, we can still measure the total elapsed time for the light to be reflected from the mirror and return. According to observers in the K system, the total distance the light travels (as shown in Figure 2.9) is $2\sqrt{(vT/2)^2 + L^2}$, and from postulate 2 the light must travel at the speed of light, so the total time interval T measured in system K is

$$T = \frac{\text{distance}}{\text{speed}} = \frac{2\sqrt{(vT/2)^2 + L^2}}{c}$$

As can be determined from above, $L = cT_0'/2$, so we have

$$T = \frac{2\sqrt{(vT/2)^2 + (cT_0'/2)^2}}{c}$$

which reduces to

$$T = \frac{T'_0}{\sqrt{1 - v^2/c^2}} = \gamma T'_0$$

which is consistent with the earlier result. In this case $T > T'_0$. The proper time is always the shortest time interval, and we find that the clock in Mary's system K′ is running slow.

Example 2.2

It is the year 2050 and the United Nations Space Federation has finally perfected the storage of antiprotons for use as fuel in a spaceship. (Antiprotons are the antiparticles of protons. We will discuss antiprotons in Chapter 3.) Having already flown to all the planets on our solar system, preparations are underway for a manned spacecraft visit to the solar system of Alpha Centauri, some 4 lightyears away. Provisions are placed on board to allow a trip of 16 years' total duration. How fast must the spacecraft travel if the provisions are to last? Neglect the acceleration, turn around, and visiting times, because they are negligible compared to the actual travel time.

Solution: A lightyear is a convenient way to measure large distances. It is the distance light travels in one year and is denoted by ly:

$$1 \text{ ly} = \left(3 \times 10^8 \frac{\text{m}}{\text{s}}\right)(1 \text{ year})\left(365 \frac{\text{days}}{\text{year}}\right)\left(24 \frac{\text{h}}{\text{day}}\right)\left(3600 \frac{\text{s}}{\text{h}}\right)$$
$$= 9.5 \times 10^{15} \text{ m}$$

Note that a lightyear is the speed of light, c, multiplied by the time of one year. The dimension of a lightyear works out to be length. For example, $4 \text{ ly} = c(4 \text{ y}) = 38 \times 10^{15}$ m. The time interval as measured by the astronauts on the spacecraft can be no longer than 16 years, because that is how long the provisions will last. However, from Earth we realize that the spacecraft will be moving at a high relative speed v to us, and that according to our clock in the stationary system K, the trip will last $T = 2L/v$, where L is the distance to the star.

Because provisions on board the spaceship will only last for 16 years, we let the proper time T'_0 in system K′ be 16 years.

$$T'_0 = 16 \text{ years} = \text{proper time}$$

Using the time dilation result, we have

$$T = \frac{2L}{v} = \frac{T'_0}{\sqrt{1 - v^2/c^2}}$$

We solve this equation for v. Inserting the numbers gives

$$\frac{2(4 \text{ ly})(9.5 \times 10^{15} \text{ m/ly})}{v} = \frac{16 \text{ y}}{\sqrt{1 - v^2/c^2}}$$

The solution to the equation is $v = 0.447c = 1.34 \times 10^8$ m/s. The time interval as measured on Earth will be $\gamma T'_0 = 17.9$ y. Notice that the astronauts will only age 16 years (their clocks run slow), while their friends remaining on Earth will age 17.9 years. Can this really be true? We shall discuss this question again in Section 2.8.

Length Contraction

Now let's consider what might happen to the length of objects in relativity. Let an observer in each system K and K′ have a meter stick at rest in his or her own respective system. Each observer lays the stick down along his or her respective x axis, putting the left end at x_ℓ (or x'_ℓ) and the right end at x_r (or x'_r). Thus, Frank in system K measures his stick to be $L_0 = x_r - x_\ell$. Similarly, in system K′, Mary measures her stick at rest to be $L'_0 = x'_r - x'_\ell = L_0$. Every observer measures a meter stick at rest in his or her own system to have the same length, namely one meter. The length as measured at rest is called the **proper length.**

Proper length

Let system K be at rest, and system K′ move along the x axis with speed v. Frank, who is at rest in system K, measures the length of the stick moving in K′. The difficulty is to measure the ends of the stick simultaneously. We insist that

Frank measure the ends of the stick at the same time so that $t = t_r = t_\ell$. The events denoted by (x, t) are (x_ℓ, t) and (x_r, t). We use Equation (2.17) and find

$$x'_r - x'_\ell = \frac{(x_r - x_\ell) - v(t_r - t_\ell)}{\sqrt{1 - v^2/c^2}}$$

The meter stick is at rest in system K′, so the length $x'_r - x'_\ell$ must be the proper length L'_0. Denote the length measured by Frank as $L = x_r - x_\ell$. The times t_r and t_ℓ are identical, as we insisted, so $t_r - t_\ell = 0$. Notice that the times of measurement by Mary in her system, t'_ℓ and t'_r, are not identical. It makes no difference when Mary makes the measurements in her own system, because the stick is at rest. It does, however, make a big difference when Frank makes the measurements, because the stick is moving with speed v with respect to him. The measurements must be done simultaneously! With these results, the previous equation becomes

$$L'_0 = \frac{L}{\sqrt{1 - v^2/c^2}} = \gamma L$$

or, because $L'_0 = L_0$

$$L = L_0\sqrt{1 - v^2/c^2} = L_0/\gamma \tag{2.20}$$

Length contraction

Notice that $L_0 > L$, so the moving meter stick shrinks according to Frank. This effect is known as **length** or **space contraction** and is characteristic of relative motion. This effect is also sometimes called the *Lorentz-FitzGerald contraction* because they originally suggested the contraction as a way to solve the electrodynamics problem. This effect is also reciprocal. Each observer will say that the other moving stick is shorter. There is no length contraction perpendicular to the relative motion, however, because $y' = y$ and $z' = z$. Observers in both systems can check the length of the other meter stick placed perpendicular to the motion as the meter sticks pass each other. They will agree that both meter sticks are one meter long.

We can also perform another *gedanken* experiment to arrive at the same result. This time we lay the meter stick along the x' axis in the moving system K′. The two systems K and K′ are aligned at $t = t' = 0$. A mirror is placed at the end of the meter stick, and a flashbulb goes off at the origin at $t = t' = 0$, sending a light pulse down the x' axis, where it is reflected and returned. Mary sees the stick at rest in system K′ and measures the proper length L_0 (which should of course be one meter). Mary uses the same clock fixed at $x' = 0$ for the time measurements (Figure 2.10a). The stick is moving at speed v with respect to Frank in the fixed system K. The clocks at $x = x' = 0$ both read zero when the origins are aligned just when the flashbulb goes off. Notice the situation shown in system K (Figure 2.10b), where by the time the light reaches the mirror, the entire stick has moved a distance vt_1. By the time the light has been reflected back to the front of the stick again, the stick has moved a total distance vt_2. We leave to Problem 18 the solution of this in terms of length contraction.

The effect of length contraction along the direction of travel may strongly affect the appearances of two- and three-dimensional objects. We see such objects when the light reaches our eyes, not when the light actually leaves the object. Thus, if the objects are moving rapidly, we will not see them as they appear at rest. Figure 2.11 shows the appearance of several such objects as they move. In this computer simulation, the objects are drawn as if the observer were 5 units in front of the near plane of the boxes. The boxes are moving to the right

Shapes of objects at high speeds

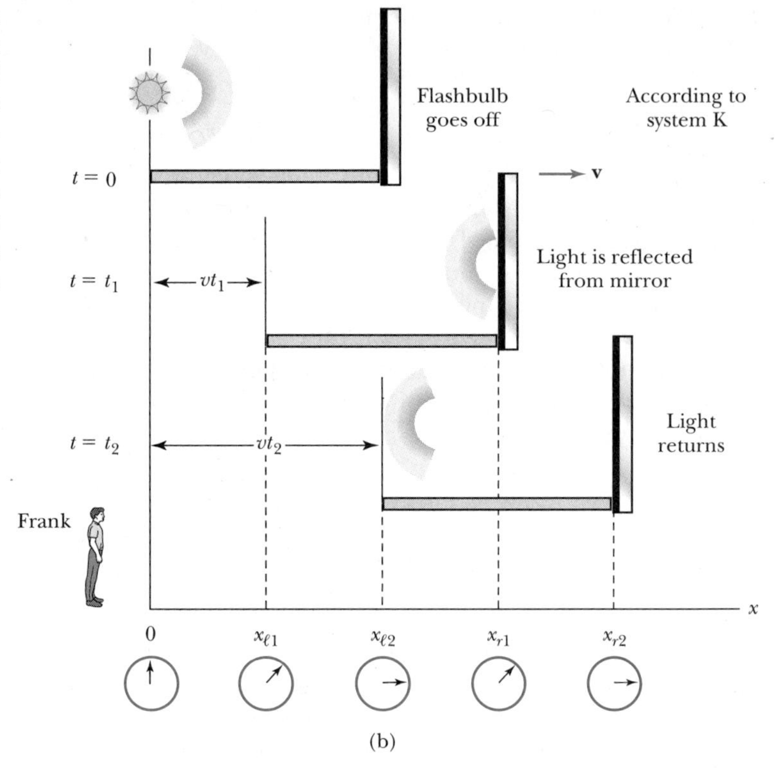

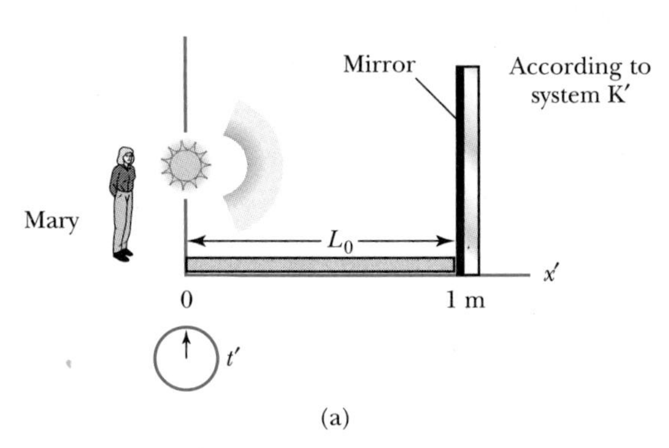

FIGURE 2.10 (a) Mary, in system K′, flashes a light down her x' axis along a stick at rest of length L_0, which is the proper length. The time interval for the light to travel down the stick and back is $2L_0/c$. (b) Frank, in system K, sees the stick moving, and the mirror has moved a distance vt_1 by the time the light is reflected. By the time the light returns back to the beginning of the stick, the stick has moved a total distance of vt_2. The times can be compared to show that the moving stick has been length contracted by $L = L_0\sqrt{1 - v^2/c^2}$.

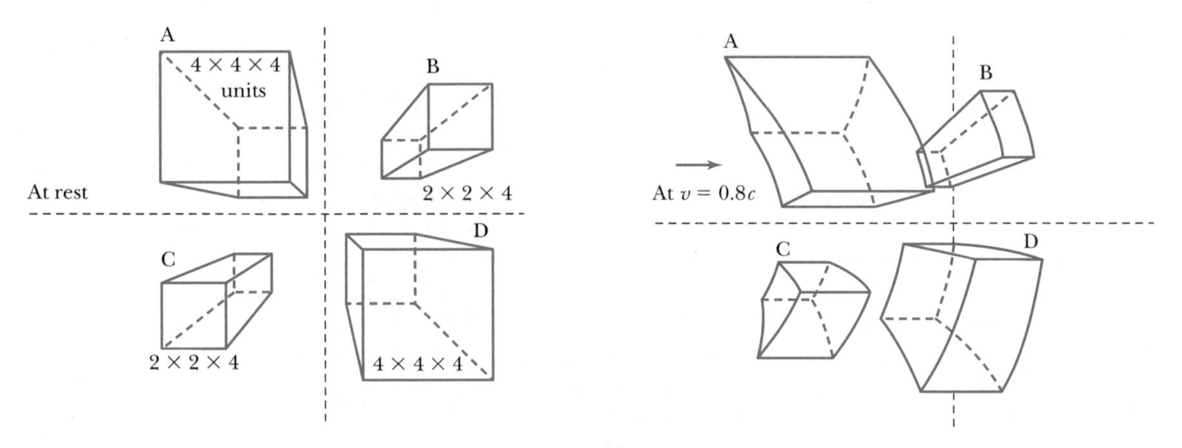

FIGURE 2.11 In this computer simulation, the rectangular boxes are drawn as if the observer were 5 units in front of the near plane of the boxes and directly in front of the origin. The boxes are moving to the right at a speed of $v = 0.8c$. The horizontal lines are only length contracted, but notice that the vertical lines become hyperbolas. The objects appear to be slightly rotated in space. The objects that are further away from the origin appear earlier because they are photographed at an earlier time, because the light takes longer to reach the camera (or our eyes). *Reprinted with permission from G. D. Scott and M. R. Viner, Am. J. Phys., 33, 534 (1965). © 1965 American Association of Physics.*

FIGURE 2.12 The array of rectangular bars is seen from above at rest in the figure on the left. In the right figure the bars are moving to the right with $v = 0.9c$. The bars appear to contract and rotate. *Quoted from P.-K. Hsiung and R.H.P. Dunn, see Science News* **137,** *232 (1990).*

at a speed of $v = 0.8c$. Only the horizontal lines are length contracted, but notice that the vertical lines become hyperbolas. The objects appear earlier because they are photographed at an earlier time, because the light takes longer to reach the camera (or our eyes). We show in Figure 2.12 a row of bars moving to the right with speed $v = 0.9c$. The result is quite surprising.

Example 2.3

Consider the solution of Example 2.2 from the standpoint of length contraction.

Solution: The astronauts have only enough provisions for a trip lasting 16 years. Thus they expect to travel for 8 years each way. If the star Alpha-Centauri is 4 lightyears away, it appears that they may need to travel at a velocity of $0.5c$ in order to make the trip. We want to consider this example as if the astronauts are at rest. Alpha-Centauri will appear to be moving towards them, and the distance to the star is length contracted. The distance of 4 ly is the proper length, and the distance measured by the astronauts will be less. The contracted distance according to the astronauts in motion is $(4\text{ ly})\sqrt{1 - v^2/c^2}$. The velocity they need to make this journey is the contracted distance divided by 8 years.

$$v = \frac{\text{distance}}{\text{time}} = \frac{(4\text{ ly})\sqrt{1 - v^2/c^2}}{8\text{ y}}$$

Dividing by c gives

$$\beta = \frac{v}{c} = \frac{(4\text{ y})(c)\sqrt{1 - v^2/c^2}}{c(8\text{ y})}$$

$$\beta = \frac{\sqrt{1 - \beta^2}}{2}$$

which gives

$$\beta = 0.447$$

$$v = 0.447c$$

which is just what we found in the previous example. The effects of time dilation and length contraction give identical results.

2.6 Addition of Velocities

A spaceship launched from a space station quickly reaches its cruising speed of $0.6c$ with respect to the space station when a band of asteroids is observed straight ahead of the ship. Mary, the commander, reacts quickly and orders her crew to blast away the asteroids with the ship's proton gun to avoid a catastrophic collision. Frank, the admiral on the space station, listens with apprehension to the communications because he fears the asteroids may eventually destroy his space station as well. Will the high-energy protons of speed $0.99c$ be able to successfully blast away the asteroids and save both the spaceship and space station? If $0.99c$ is the speed of the protons with respect to the spaceship, what speed will Frank measure?

We will use the letter u to denote speeds of objects as measured in various coordinate systems. In this case, Frank (in the fixed, stationary system K on the space station) will measure the speed of the protons to be u, whereas Mary, the commander of the spaceship (the moving system K′), will measure $u' = 0.99c$. We reserve the letter v to express the speed of the coordinate systems with respect to each other. The speed of the spaceship with respect to the space station is $v = 0.6c$.

Newtonian mechanics teaches us that to find the speed of the protons with respect to the space station, we simply add the speed of the spaceship with respect to the space station ($0.6c$) to the speed of the protons with respect to the spaceship ($0.99c$) to determine the simple result $u = v + u' = 0.6c + 0.99c = 1.56c$. However, this result is not in agreement with the results of the Lorentz transformation. We use Equation (2.18), letting x be along the direction of motion of the spaceship (and high-speed protons), and take the differentials, with the results

$$
\begin{aligned}
dx &= \gamma(dx' + v\,dt') \\
dy &= dy' \\
dz &= dz' \\
dt &= \gamma[dt' + (v/c^2)\,dx']
\end{aligned}
\tag{2.21}
$$

Velocities are defined by $u_x = dx/dt$, $u_y = dy/dt$, $u'_x = dx'/dt'$, and so on. Therefore we determine u_x by

Relativistic velocity addition

$$u_x = \frac{dx}{dt} = \frac{\gamma(dx' + v\,dt')}{\gamma[dt' + (v/c^2)\,dx']} = \frac{u'_x + v}{1 + (v/c^2)u'_x} \tag{2.22a}$$

Similarly, u_y and u_z are determined to be

$$u_y = \frac{u'_y}{\gamma[1 + (v/c^2)u'_x]} \tag{2.22b}$$

$$u_z = \frac{u'_z}{\gamma[1 + (v/c^2)u'_x]} \tag{2.22c}$$

Equations (2.22) are referred to as the **relativistic velocity additions** and the **Lorentz velocity transformation.** Notice that although the relative motion of the systems K and K′ is only along the x-direction, the velocities along y and z are affected as well. This contrasts with the Lorentz transformation equations, where $y = y'$ and $z = z'$, but the difference in velocities is simply ascribed to the transformation of time, which depends on v and x'. Thus, the transformations for u_y and u_z depend on v and u'_x. The inverse transformations for u'_x, u'_y, and u'_z can be determined by simply switching primes and unprimes and changing v to $-v$. The results are

$$
\begin{aligned}
u'_x &= \frac{u_x - v}{1 - (v/c^2)u_x} \\
u'_y &= \frac{u_y}{\gamma[1 - (v/c^2)u_x]} \\
u'_z &= \frac{u_z}{\gamma[1 - (v/c^2)u_x]}
\end{aligned}
\tag{2.23}
$$

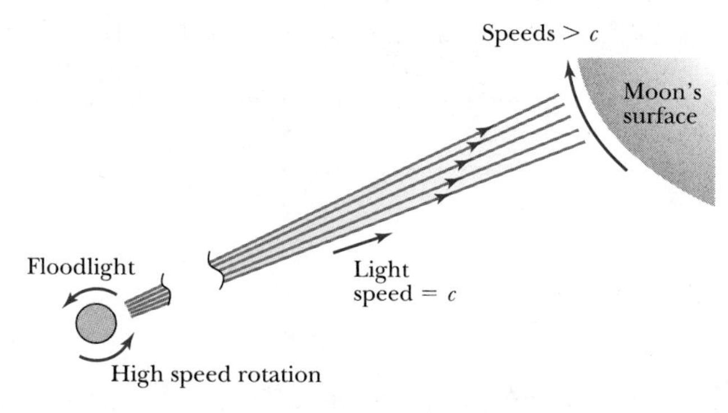

FIGURE 2.13 A floodlight revolving at high speeds can sweep a light beam across the surface of the moon at speeds exceeding c, but the velocity of the light still does not exceed c.

Note that we found the velocity transformation equations for the situation corresponding to the inverse Lorentz transformation, Equations (2.18), before finding the velocity transformation for Equations (2.17).

What is the correct result for the speed of the protons with respect to the space station? We have $u'_x = 0.99c$ and $v = 0.6c$, so Equation (2.22a) gives us the result

$$u_x = \frac{0.99c + 0.6c}{1 + \dfrac{(0.6c)(0.99c)}{c^2}} = 0.997c$$

where we have assumed we know the speeds to three significant figures. Therefore, the result is a speed only slightly less than c. The Lorentz transformation does not allow a material object to have a speed greater than c. Only massless particles, such as light, can have speed c. If the crew of the spaceship spots the asteroids far enough in advance, their reaction times should allow them to shoot down the uncharacteristically swiftly moving asteroids and save both the spaceship and the space station.

Although no particle with mass can carry energy faster than c, we can imagine a signal being processed faster than c. Consider the following *gedanken* experiment. A giant floodlight placed on a space station above the Earth revolves at 100 Hz, as shown in Figure 2.13. Light spreads out in the radial direction from the floodlight at speeds of c. On the surface of the moon, the light beam sweeps across at speeds far exceeding c (Problem 36). However, the light itself did not reach the moon at speeds faster than c. There is no energy associated with the beam of light sweeping across the moon's surface. The energy (and linear momentum) is only along the radial direction from the space station to the moon.

Example 2.4

The commander of the spaceship just discussed is holding target practice for junior officers by shooting protons at small asteroids and space debris off to the side as the spaceship passes by. What speed will an observer in the space station measure for these protons?

Solution: We use the coordinate systems and speeds of the spaceship and proton gun as described previously. Let the direction of the protons now be perpendicular to the direction of the spaceship—along the y' direction. We now use Equation (2.22b) to determine u_y and $u'_y = 0.99c$, $u'_x = 0$, and $v = 0.6c$.

$$\gamma = \frac{1}{\sqrt{1 - v^2/c^2}} = \frac{1}{\sqrt{1 - 0.6^2}} = 1.25$$

$$u_y(\text{protons}) = \frac{0.99c}{1.25[1 + (0.6c)(0)/c^2]} = 0.792c$$

$$u_x(\text{protons}) = \frac{0 + 0.6c}{[1 + (0.6c)(0)/c^2]} = 0.6c$$

$$u(\text{protons}) = \sqrt{u_x^2 + u_y^2} = \sqrt{(0.792c)^2 + (0.6c)^2} = 0.994c$$

where we have again assumed we know the velocities to three significant figures. Mary and her junior officers only observe the protons moving perpendicular to their motion. But because there are both u_x and u_y components, Frank (on the space station) sees the protons moving at an angle with respect to both his x and his y directions.

Example 2.5

By the early 1800s, experiments had shown that light tends to slow down when passing through liquids. A. J. Fresnel suggested in 1818 that there would be a partial drag of light by the medium through which the light was passing. Fresnel's suggestion explained the problem of stellar aberration if the Earth was at rest in the ether. In a famous experiment in 1851, H. L. Fizeau measured the "ether" drag coefficient for light passing in opposite directions through flowing water. Let a moving system K′ be at rest in the flowing water and let v be the speed of the flowing water with respect to a fixed observer in K (see Figure 2.14). The speed of light in the water at rest (that is, in system K′) is u', and the speed of light as measured in K is u. If the index of refraction of the water is n, Fizeau found experimentally that

$$u = u' + \left(1 - \frac{1}{n^2}\right)v \tag{2.24}$$

which was in agreement with Fresnel's prediction. This result was considered an affirmation of the ether concept. The factor $1 - 1/n^2$ became known as *Fresnel's drag coefficient.* Show that this result can be explained rather easily using relativistic velocity addition *without the ether concept.*

Solution: The velocity u' is c/n, so Equation (2.22a) gives

$$u = \frac{u' + v}{1 + u'v/c^2} = \frac{c/n + v}{1 + v/nc} = \frac{c}{n}\,\frac{1 + \dfrac{nv}{c}}{1 + \dfrac{v}{nc}}$$

Since $v \ll c$, we can expand the denominator $(1 + x)^{-1} = 1 - x + \cdots$ keeping only the lowest term in $x = v/c$. The above equation becomes

$$u = \frac{c}{n}\left(1 + \frac{nv}{c}\right)\left(1 - \frac{v}{nc} + \cdots\right) = \frac{c}{n}\left(1 + \frac{nv}{c} - \frac{v}{nc} + \cdots\right)$$

$$= \frac{c}{n} + v - \frac{v}{n^2} = u' + \left(1 - \frac{1}{n^2}\right)v$$

which is in agreement with Fizeau's experimental result and Fresnel's prediction, Equation (2.24). This relativistic calculation is another stunning success of the special theory of relativity. There is no need to consider the existence of the ether.

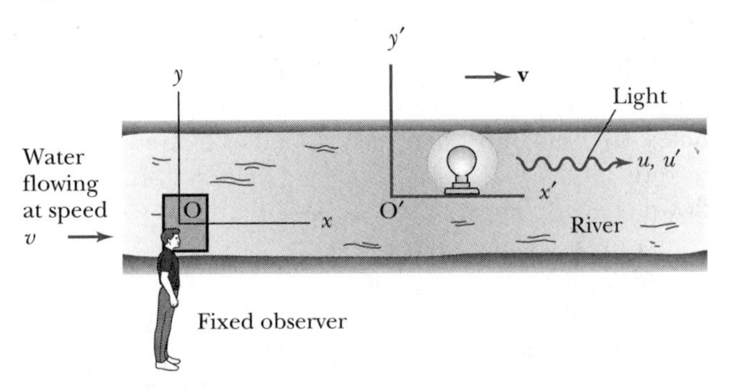

FIGURE 2.14 A stationary system K is fixed on shore, and a moving system K′ floats down the river at speed v. Light emanating from a source under water in system K′ has speed u, u' in system K, K′.

2.7 Experimental Verification

We have used the special theory of relativity to describe some unusual phenomena. The special theory has also made some startling predictions concerning length contraction, time dilation, and velocity addition. In this section we will discuss only a few of the many experiments that have been done to confirm the special theory of relativity.

Muon Decay

When high-energy particles called *cosmic rays* enter the Earth's atmosphere from outer space, they interact with particles in the upper atmosphere, creating additional particles in a *cosmic shower.* Many of the particles in the shower are *π-mesons* (pions), which decay into other unstable particles called *muons.* The properties of muons will be described later when we discuss nuclear and particle physics. Because muons are unstable, they decay according to the radioactive decay law

$$N = N_0 e^{-\left(\frac{\ln(2)t}{t_{1/2}}\right)} = N_0 e^{-\left(\frac{0.693t}{t_{1/2}}\right)}$$

Radioactive decay law

where N_0 and N are the number of muons at times $t = 0$ and $t = t$, respectively, and $t_{1/2}$ is the half-life of the muons. In the time period $t_{1/2}$ half of the muons will decay to other particles. The half-life of muons (1.52×10^{-6} s) is long enough that many of them survive the trip through the atmosphere to the Earth's surface.

We perform an experiment by placing a muon detector on top of a mountain 2000 m high and counting the number of muons traveling at a speed near $v = 0.98c$ (see Figure 2.15a). Suppose we count 10^3 muons during a given time period t_0. We then move our muon detector to sea level (see Figure 2.15b), and we determine experimentally that approximately 540 muons survive the trip without decaying. We ignore any other interactions that may remove muons.

Classically, muons traveling at a speed of $0.98c$ cover the 2000-m path in 6.8×10^{-6} s, and according to the radioactive decay law, only 45 muons should survive the trip. There is obviously something wrong with the classical calculation, because a factor of 12 more muons survive than the classical calculation predicts.

We need to consider this problem relativistically. Because the muons are moving at a speed of $0.98c$ with respect to us on Earth, the effects of time dilation will be dramatic. In the muon rest frame, the time period for the muons to travel 2000 m (on a clock fixed with respect to the mountain) is calculated from

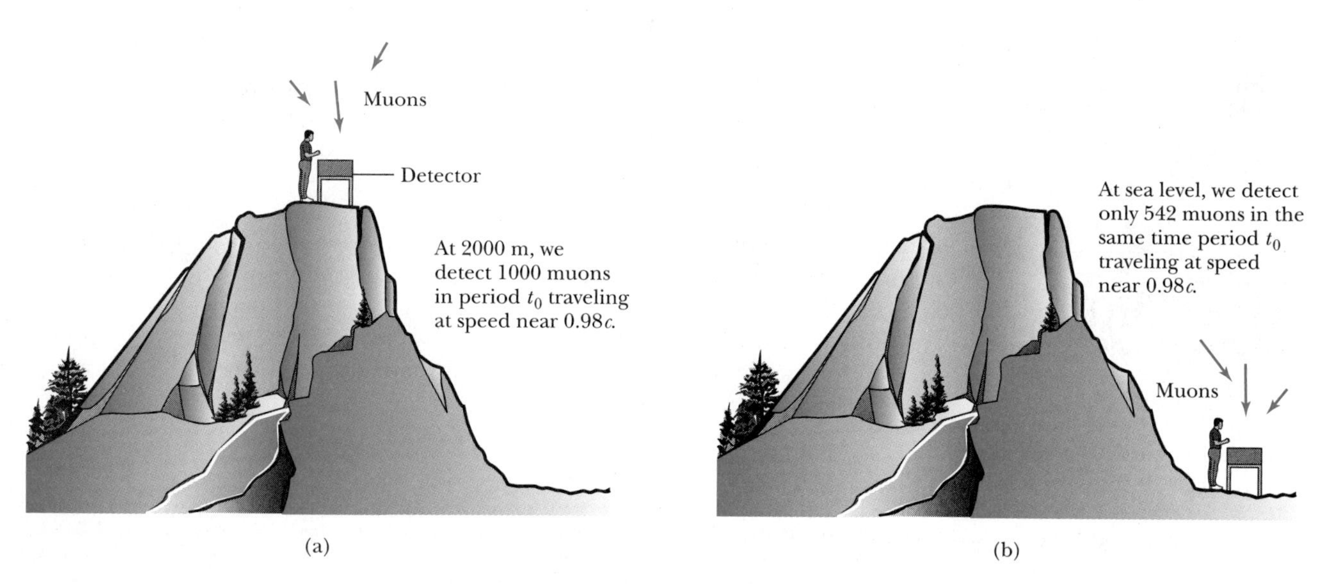

FIGURE 2.15 The number of muons detected with speeds near $0.98c$ is much different on top of a mountain (a) than at sea level (b) because of the muon's decay. The experimental result agrees with our time dilation equation.

Physicists study cosmic rays by sending balloons carrying instruments into the upper atmosphere. Arthur Compton is shown here on Stagg Field of the University of Chicago in 1939 with some students when they released several such balloons. *UPI/Corbis-Bettmann.*

Equation (2.19) to be $(6.8/5) \times 10^{-6}$ s, because $\gamma = 5$ for $v = 0.98c$. For the time $t = 1.36 \times 10^{-6}$ s, the radioactive decay law predicts that 538 muons will survive the trip, in agreement with the observations. An experiment similar to this was performed by B. Rossi and D. B. Hall* in 1941 on the top of Mount Washington in New Hampshire.

It is useful to examine the muon decay problem from the perspective of an observer traveling with the muon. This observer would not measure the distance from the top of the 2000-m mountain to sea level to be 2000 m. Rather, this observer would say that the distance is contracted and would be only 2000 m/5 = 400 m. The time to travel the 400-m distance would be $400 \text{ m}/0.98c = 1.36 \times 10^{-6}$ s according to a clock at rest with a muon. Using the radioactive decay law, an observer traveling with the muons would still predict 538 muons to survive. Therefore, we obtain the identical result whether we consider time dilation or space contraction, and both are in agreement with experiment, thus confirming the special theory.

Atomic Clock Measurement

An extremely accurate measurement of time can be made using a well-defined transition in the ^{133}Cs atom that has a frequency of 9,192,631,770 Hz. In 1972 two American physicists, J. C. Hafele and Richard E. Keating, used four cesium beam atomic clocks to test the time dilation effect. They flew with one clock eastward on regularly scheduled commercial jet airplanes around the world and with another one westward around the world. The two other clocks stayed fixed in the Earth's system at the U.S. Naval Observatory in Washington, D.C. (see Figure 2.16).

The trip eastward took 65.4 hours with 41.2 flight hours, whereas the westward trip, taken a week later, took 80.3 hours with 48.6 flight hours. The comparison with the special theory of relativity is complicated by the rotation of the Earth and by a gravitational effect arising from the general theory of relativity. The actual predictions and observations for the time differences† are

Travel	Predicted	Observed
Eastward	-40 ± 23 ns	-59 ± 10 ns
Westward	275 ± 21 ns	273 ± 7 ns

A negative time indicates that the time on the moving clock is less than the reference clock. The moving clocks lost time (ran slower) during the eastward trip, but gained time (ran faster) during the westward trip. This occurs because of the rotation of the Earth, indicating that the flying clocks ticked faster or slower than the reference clocks on Earth. The special theory of relativity is verified within the experimental uncertainties.

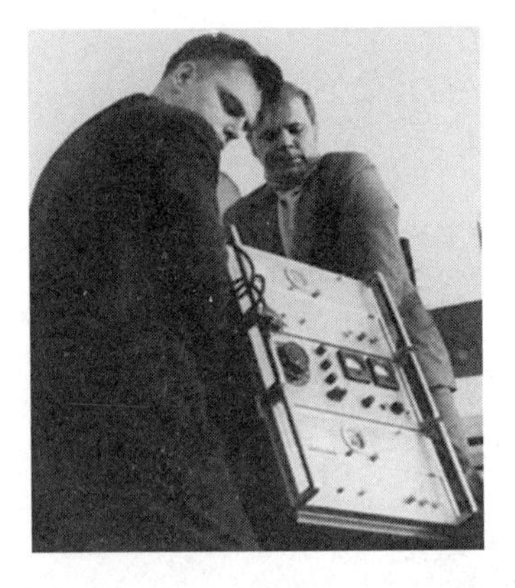

Joseph Hafele and Richard Keating are shown unloading one of their atomic clocks and the associated electronics from an airplane in Tel Aviv, Israel, during a stopover in November, 1971, on their round-the-world trip to test special relativity. *AP/Wide World Photos.*

*B. Rossi and D. B. Hall, *Physical Review* **50,** 223 (1941). An excellent, though now dated, film recreating this experiment (*Time Dilation—An Experiment with μ-mesons* by D. H. Frisch and J. H. Smith) is available from the Education Development Center, Newton, Mass. See also D. H. Frisch and J. H. Smith, *American Journal of Physics* **31,** 342 (1963).

†See J. C. Hafele and R. E. Keating, *Science* **177,** 166–170 (1972).

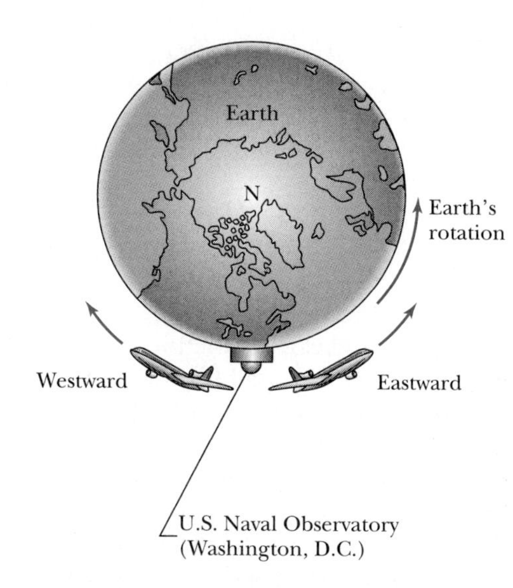

FIGURE 2.16 Two airplanes take off from Washington, D.C., where the U.S. Naval Observatory is located. The airplanes travel east and west around the Earth as the Earth rotates. Atomic clocks on the airplanes are compared with similar clocks kept at the observatory to show the effects of time dilation are correct.

Example 2.6

Assuming a jet airplane travels 300 m/s and the circumference of the Earth is about 4×10^7 m, calculate the time dilation effect expected for a round-the-world trip exclusive of the Earth's rotation and gravitational correction.

Solution: A clock fixed on Earth will measure a flight time T_0 of

$$T_0 = \frac{4 \times 10^7 \text{ m}}{300 \text{ m/s}} = 1.33 \times 10^5 \text{ s}$$

which is about 37 hours. Because a clock in the airplane will run slowly, an observer on Earth will say the time measured on the airplane is $T = T_0\sqrt{1-\beta^2}$ where $\beta = \dfrac{300 \text{ m/s}}{c} = 10^{-6}$. The time difference is

$$\Delta T = T_0 - T = T_0(1 - \sqrt{1-\beta^2})$$

Because β is such a small quantity, we can use a power series expansion of the square root, keeping only the lowest term in β^2.

$$\Delta T = T_0[1 - (1 - \beta^2/2 + \cdots)] = \frac{\beta^2 T_0}{2}$$

$$= \frac{1}{2}(10^{-6})^2(1.33 \times 10^5 \text{ s}) = 6.65 \times 10^{-8} \text{ s} = 66.5 \text{ ns}$$

Velocity Addition

An interesting test of the velocity addition relations was made by T. Alväger and colleagues* at the CERN nuclear and particle physics research facility on the border of Switzerland and France. They used a beam of almost 20-GeV (10^9-eV) protons to impact on a target to produce neutral pions (π^0) of more than 6 GeV. The $\pi^0(\beta \approx 0.99975)$ have a very short half-life and soon decay into two γ rays. In the rest frame of the π^0 the two γ rays go off in opposite directions. The experimenters measured the velocity of the γ rays going in the forward direction in the laboratory (actually 6°, but we will assume 0° for purposes of calculation, because there is little difference). The Galilean addition of velocities would require the velocity of the γ rays to be $u = 0.99975c + c = 1.99975c$, because the velocity of γ rays is already c. However, the relativistic velocity addition, with $v = 0.99975c$ being the velocity of the π^0 rest frame with respect to the laboratory

Pion decay experiment

*See T. Alväger, F. J. M. Farley, J. Kjellman, and I. Wallin, *Physics Letters* **12,** 260 (1964). See also article by J. M. Bailey, *Arkiv Fysik* **31,** 145 (1966).

and $u' = c$ being the velocity of the γ rays in the rest frame of the π^0, predicts the velocity u of the γ rays measured in the laboratory to be, according to Equation (2.22a),

$$u = \frac{c + 0.99975c}{1 + \dfrac{(0.99975c)(c)}{c^2}} = c$$

The experimental measurement was accomplished by measuring the time taken for the γ rays to travel between two detectors placed about 30 m apart and was in excellent agreement with the relativistic prediction, but not the Galilean one. We again have conclusive evidence of the need for the special theory of relativity.

Although we have mentioned only three rather interesting experiments here, there have been literally thousands of cases examined by physicists performing experiments with nuclear and particle accelerators that conclusively verify the correctness of the concepts discussed here. Measurements done in recent years have confirmed that time dilation effects of special relativity to 40 parts in one million by using lasers to repeat experiments similar in principle to that of Michelson-Morley. Quantum electrodynamics (QED) includes special relativity in its framework, and QED has been tested to 12 decimal places. Physicists continue to search for violations, including a challenge to prove that the speed of light on Earth is the same east-to-west as it is west-to-east to within 50 m/s.*

Relativity violation challenge

2.8 Twin Paradox

Probably no subject in relativity has received more attention than the twin (or clock) paradox. Almost from the time of publication of Einstein's famous paper in 1905, this subject has received considerable attention, and many variations exist. Let's summarize the paradox. Suppose twins, Mary and Frank, choose different career paths. Mary (the **M**oving twin) becomes an astronaut and Frank (the **F**ixed twin) a stockbroker. At age 30, Mary sets out on a spaceship to study a nearby star system. Mary will have to travel at very high speeds to reach the star and return during her life span. According to Frank, Mary's biological clock will tick more slowly than his own, so he will say that Mary will return from her trip younger than he. The paradox is that Mary claims that it is Frank who is moving rapidly with respect to her, so that when she returns, Frank will be the younger. To further complicate the paradox, one could argue that because nature cannot allow both possibilities, it must be true that symmetry prevails and that the twins will still be the same age. Which is the correct solution?

Who is the younger twin?

The correct answer is that Mary will return from her space journey as the younger twin. According to Frank, Mary's spaceship takes off from Earth and quickly reaches its travel speed of $0.8c$. She travels the distance 8 ly to the star system, slows down and turns around quickly, and returns to Earth at the same speed. The accelerations (positive and negative) take negligible times compared to the travel times between Earth and the star system. According to Frank, Mary's travel time to the star is 10 years ($8 \text{ ly}/0.8c = 10 \text{ y}$) and the return is also 10 years, for a total travel time of 20 years, so that Frank will be $30 + 10 + 10 \text{ y} = 50$ years old when Mary returns. However, since Mary's clock is ticking more slowly, her travel time to the star is only $10\sqrt{1 - 0.8^2} \text{ y} = 6$ years. Frank calculates that Mary will only be $30 + 6 + 6 \text{ y} = 42$ years old when she returns with respect to his own clock at rest.

*See *Science,* **250,** 1208 (1990).

TABLE 2.1
Twin Paradox Analysis*

Item	Measured by Frank (remains on Earth)	Measured by Mary (traveling astronaut)
Time of total trip	$T = 2L/v$	$T' = 2L/\gamma v$
Total number of signals sent	$\nu T = 2\nu L/v$	$\nu T' = 2\nu L/\gamma v$
Frequency of signals received at beginning of trip ν'	$\nu\sqrt{\dfrac{1-\beta}{1+\beta}}$	$\nu\sqrt{\dfrac{1-\beta}{1+\beta}}$
Time of detecting Mary's turn-around	$t_1 = L/v + L/c$	$t_1' = L/\gamma v$
Number of signals received at the rate ν'	$\nu' t_1 = \dfrac{\nu L}{v}\sqrt{1-\beta^2}$	$\nu' t_1' = \dfrac{\nu L}{v}(1-\beta)$
Time for remainder of trip	$t_2 = L/v - L/c$	$t_2' = L/\gamma v$
Frequency of signals received at end of trip ν''	$\nu\sqrt{\dfrac{1+\beta}{1-\beta}}$	$\nu\sqrt{\dfrac{1+\beta}{1-\beta}}$
Number of signals received at rate ν''	$\nu'' t_2 = \dfrac{\nu L}{v}\sqrt{1-\beta^2}$	$\nu'' t_2' = \dfrac{\nu L}{v}(1+\beta)$
Total number of signals received	$2\nu L/\gamma v$	$2\nu L/v$
Conclusion as to other twin's measure of time taken	$T' = 2L/\gamma v$	$T = 2L/v$

*After A. French, *Special Relativity*. New York: W. W. Norton, 1968, p. 158.

The important fact here is that Frank's clock is in an inertial system* during the entire trip; however, Mary's clock is not. As long as Mary is traveling at constant speed away from Frank, both of them can argue that the other twin is aging less rapidly. However, when Mary slows down to turn around, she leaves her original inertial system and eventually returns on a completely different inertial system. The result we previously mentioned of Mary's claim that Frank is younger is only valid if she had remained in her initial inertial system. There is no doubt also as to who is in the inertial system. Frank feels no acceleration during Mary's entire trip, but Mary will definitely feel an acceleration during her reversal time, just as we do when we step hard on the brakes of a car. The acceleration at the beginning and the deceleration at the end of her trip present no problem, because the fixed and moving clocks can be compared if Mary were just passing by Frank each way. It is the acceleration at the star system that is the key. If we obey the two postulates of special relativity, there is no paradox. The instantaneous rate of Mary's clock is determined by her instantaneous speed, but she must account for the acceleration effect when she turns around. The result of a careful analysis of Mary's entire trip using special relativity, including acceleration, will be in agreement with Frank's assessment that Mary is younger. Mary returns to Earth as a rich and famous woman, because her stockbroker brother has invested her salary during the 20-year period (for which she only worked 12 years!).

Mary is both younger and rich.

Following A. P. French's excellent book *Special Relativity* we present Table 2.1 analyzing the twin paradox. Both Mary and Frank send out signals at a frequency ν (as measured by their own clock). We indicate in the table the various

*The rotating and orbiting Earth is only an approximate inertial system.

journey timemarks and signals received during the trip. We make a column for the twin Frank who stayed at home and a column for the traveling twin astronaut Mary who went on the trip. Let the total time of the trip as measured on Earth be T. The speed of Mary's spaceship is v (as measured on Earth), which gives a relativistic parameter of γ. The distance Mary's spaceship goes before turning around (as measured on earth) is L. Much of this table is best analyzed by using spacetime (see the next section) and the Doppler effect (see Section 2.10).

2.9 Spacetime

Spacetime (Minkowski) diagrams

Worldline

It is convenient sometimes to represent events on a **spacetime** diagram like that shown in Figure 2.17. For convenience we use only one spatial coordinate x and specify position in this one dimension. We use ct instead of time so that both coordinates will have dimensions of length. Spacetime diagrams were first used by H. Minkowski in 1908 and are often called *Minkowski diagrams.* We have learned in relativity that we must denote both space and time in order to specify an event. This is the origin of the term *fourth dimension* for time. The events for A and B in Figure 2.17 are denoted by the respective coordinates (x_A, ct_A) and (x_B, ct_B), respectively. The line connecting events A and B is the path from A to B and is called a **worldline.** A spaceship launched from $x = 0$, $ct = 0$ with velocity v will have a particular worldline as shown in Figure 2.18. It will be a straight line in the spacetime diagram with a slope c/v. A light signal sent out from the origin will have an angle of 45° from both x and ct. Any real motion in the spacetime diagram cannot have a slope less than one (angle $<45°$), because that motion would have a speed greater than c. The Lorentz transformation does not allow such a speed.

Let us consider two events that occur at the same time ($ct = 0$), but at different positions, x_1 and x_2. We denote the events (x, ct) as $(x_1, 0)$ and $(x_2, 0)$, and we show them in Figure 2.19 in an inertial system with an origin fixed at $x = 0$ and $ct = 0$. How can we be certain that the two events happen simultaneously if they occur at different positions? We must devise a method that will allow us to determine experimentally that the events occurred simultaneously. Let us place clocks at positions x_1 and x_2 and place a flashbulb at position x_3 halfway between x_1 and x_2. The two clocks have been previously synchronized and keep identical time. At time $t = 0$, the flashbulb explodes and sends out light signals from position x_3. The light signals proceed along their worldlines as shown in

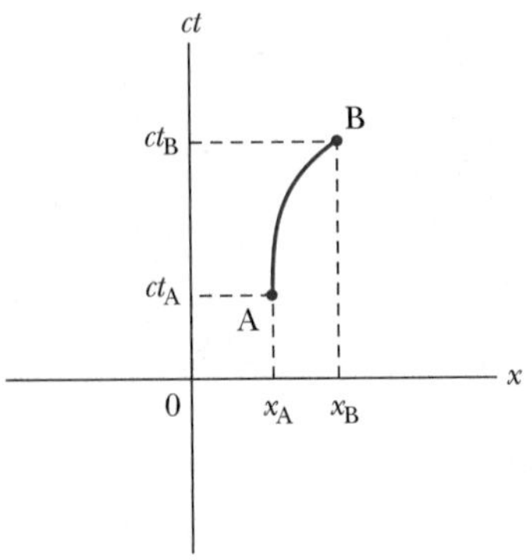

FIGURE 2.17 A spacetime diagram is used to specify events. The worldline denoting the path from event A to event B is shown.

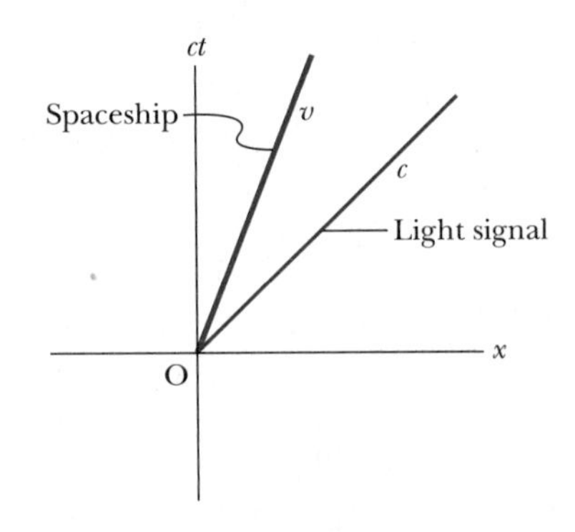

FIGURE 2.18 A light signal has the slope of 45° on a spacetime diagram. A spaceship moving along the x axis with speed v is a straight line on the spacetime diagram with a slope c/v.

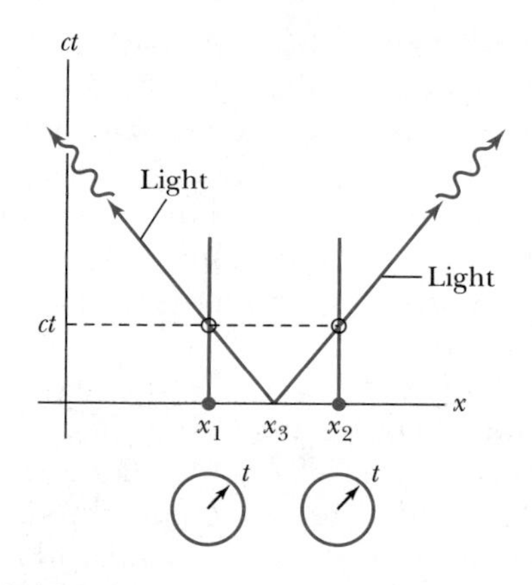

FIGURE 2.19 Clocks positioned at x_1 and x_2 can be synchronized by sending a light signal from a position x_3 halfway between. The light signals intercept the worldlines of x_1 and x_2 at the same time t.

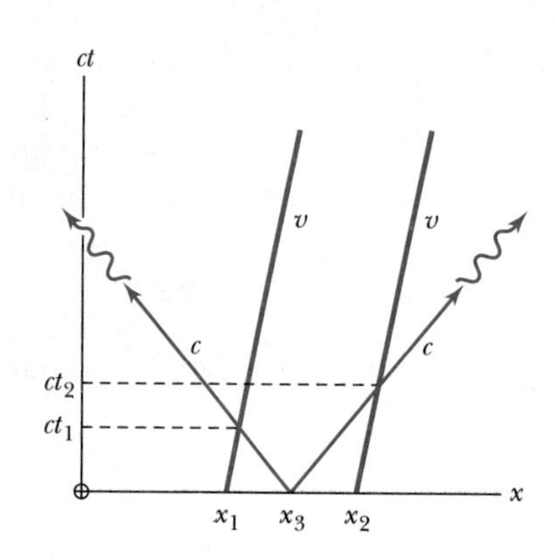

FIGURE 2.20 If the positions x_1 $(=x_1')$ and x_2 $(=x_2')$ of the previous figure are on a moving system K′ when the flashbulb goes off, the times will not appear simultaneously in system K, because the wordlines for x_1' and x_2' are slanted.

Figure 2.19. They arrive at both positions, x_1 and x_2, at identical times t as shown on the spacetime diagram. By using such techniques we can be sure that events occur simultaneously in our inertial reference system.

But what about other reference systems? We realize that the two events will not be simultaneous in a reference system K′ moving at speed v with respect to our (x, ct) system. Because the two events have different spatial coordinates, x_1 and x_2, the Lorentz transformation will preclude them from occurring at the same time t' simultaneously in the moving coordinate systems. We can see this by supposing that events 1, 2, and 3 take place on a spaceship moving with velocity v. The worldlines for x_1 and x_2 are the two slanted lines in Figure 2.20. However, when the flashbulb goes off, the light signals from x_3 still proceed at 45° in the (x, ct) reference system. The light signals intersect the worldlines from positions x_1 and x_2 at different times, so we do not see the events as being simultaneous in the moving system. Spacetime diagrams can be useful in showing such phenomena.

Anything that happened earlier in time than $t = 0$ is called the *past* and anything that occurs after $t = 0$ is called the *future*. The spacetime diagram in Figure 2.21a shows both the past and the future. Notice that only the events within the shaded area below $t = 0$ can affect the present. Events outside this area cannot affect the present because of the limitation $v \leq c$; this region is

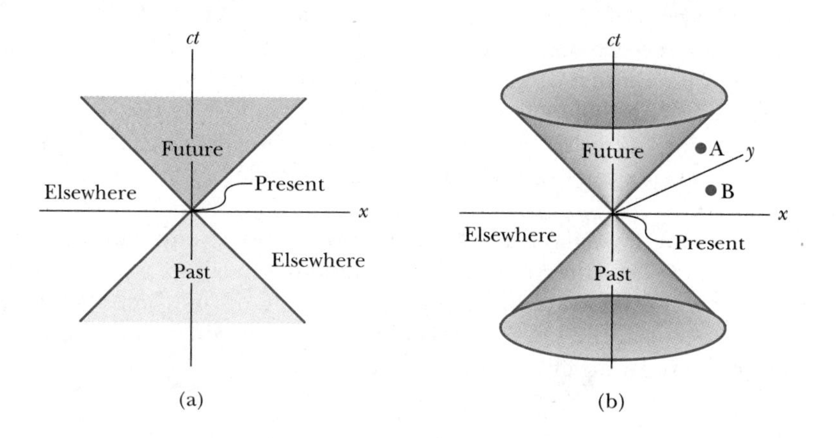

FIGURE 2.21 (a) The spacetime diagram can be used to show the past, present, and future. Only causal events are placed inside the shaded area. Events outside the shaded area below $t = 0$ cannot affect the present. (b) If we add an additional spatial coordinate y, a space cone can be drawn. The present cannot affect event A, but event B can.

called *elsewhere.* Similarly, the present cannot affect any events occurring outside the shaded area above $t = 0$, again because of the limitation of the speed of light. If we add another spatial coordinate y to our spacetime coordinates, we will have a cone as shown in Figure 2.21b, which we refer to as the **light cone.** All causal events must be within the light cone. In Figure 2.21b, anything occurring at present ($x = 0$, $ct = 0$) cannot possibly affect an event at position A; however, the event B can easily affect event A because A would be within the range of light signals emanating from B.

Light cone

Invariant quantities have the same value in all inertial frames. They serve a special role in physics, because their values do not change from one system to another. For example, the speed of light c is invariant. We are used to defining distances by $d^2 = x^2 + y^2 + z^2$, and in Galilean geometry, we obtain the same result for d^2 in any inertial frame of reference. Is there a quantity, similar to d^2, that is also invariant in the special theory? If we refer to Equations (2.9), we have similar equations in both systems K and K′. Let us look more carefully at the quantity s^2 defined by

$$s^2 = x^2 - (ct)^2 \tag{2.25}$$

and also

$$s'^2 = x'^2 - (ct')^2$$

If we use the Lorentz transformation for x and t, we find that $s^2 = s'^2$, so the quantity s^2 is an invariant quantity. This relationship can be extended to include the two other spatial coordinates, y and z, so that

Another invariant quantity

$$s^2 = x^2 + y^2 + z^2 - (ct)^2 \tag{2.26}$$

Because this is similar to the length squared of a position vector in cartesian coordinates, it is useful to think of time as the fourth dimension of spacetime. We have no need here to use the four-dimensional approach, but it is also useful to link momentum and energy together as well as space and time. For simplicity, we will continue also to use only the single spatial coordinate x.

If we consider two events, we can determine the quantity Δs^2 where

$$\Delta s^2 = \Delta x^2 - c^2\Delta t^2 \tag{2.27}$$

between the two events, and we find that it is invariant in any inertial frame. The quantity Δs is known as the **spacetime interval** between two events. There are three possibilities for the invariant quantity Δs^2.

Spacetime intervals

1. **$\Delta s^2 = 0$:** In this case $\Delta x^2 = c^2\Delta t^2$, and the two events can only be connected by a light signal. The events are said to be **lightlike.** (Lightlike)
2. **$\Delta s^2 > 0$:** Here we must have $\Delta x^2 > c^2\Delta t^2$, and no signal can travel fast enough to connect the two events. The events are not causally connected and are said to have a **spacelike** separation. In this case we can always find an inertial frame traveling at a velocity less than c in which the two events can occur simultaneously in time but at different places in space. (Spacelike)
3. **$\Delta s^2 < 0$:** Here we have $\Delta x^2 < c^2\Delta t^2$, and the two events can be causally connected. The interval is said to be **timelike.** In this case we can find an inertial frame traveling at a velocity less than c in which the two events occur at the same position in space but at different times. The two events can never occur simultaneously. (Timelike)

Example 2.7

Draw the spacetime diagram for the motion of the twins discussed in Section 2.8. Draw light signals being emitted from each twin at annual intervals and count the number of light signals received by each twin from the other.

Solution: We show in Figure 2.22 the spacetime diagram. Mary's trip has a slope $c/0.8c = 1.25$ on the outbound trip and -1.25 on the return trip. During the trip to the star system, Mary does not receive the second annual light signal from Frank until she reaches the star system. This occurs because the light signal takes considerable time to catch up with Mary. However, during the return trip Mary receives Frank's light signals at a rapid rate, receiving the last one (number 20) just as she returns. Because Mary's clock is running slow, we see the light signals being sent less often on the spacetime diagram in the fixed system. Mary sends out her sixth annual light signal when she arrives at the star system. However, this signal does not reach Frank until the 18th year! During the remaining two years, however, Frank receives Mary's signals at a rapid rate, finally receiving all 12 of them. Frank receives the last 6 signals during a time period of only 2 years.

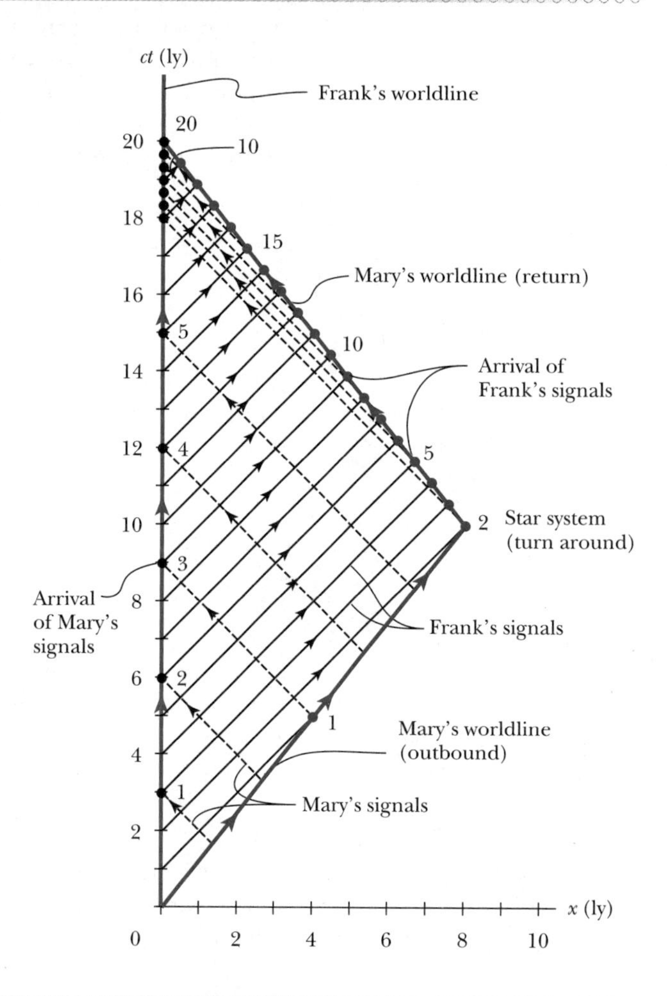

FIGURE 2.22 The spacetime diagram for Mary's trip to the star system and back. Notice that Frank's worldline is a vertical line at $x = 0$, and Mary's two worldlines do indeed have the correct slope given by c/v. The black dashed lines represent light signals sent at annual intervals from Mary to Frank. Frank's annual signals to Mary are solid blue. The solid dots denote the time when the light signals arrive.

2.10 Doppler Effect

You probably have already studied the Doppler effect of sound in introductory physics. It is represented by an increased frequency of sound as a source such as a train (with whistle blowing) approaches a receiver (our ear drum), and a decrease in pitch as the source recedes. A change in sound also occurs when the source is fixed and the receiver is moving. The change in frequency of the sound wave depends on whether the source or receiver is moving. On first thought it seems that the Doppler effect in sound violates the Principle of Relativity, until we realize that there is in fact a special frame for sound waves. Sound waves depend on media such as air, water, or a steel plate in order to propagate. For light, however, there is no such medium. It is only relative motion of the source and receiver that is relevant, and we expect some differences between the relativistic Doppler effect for light waves and the normal Doppler effect for sound. It is not possible for a source of light to travel faster than light in a vacuum, but it is possible for a source of sound to travel faster than the speed of sound.

Consider a source of light (for example, a star) and a receiver (an astronomer) approaching one another with a relative velocity v. First we consider the receiver fixed (Figure 2.23) in system K and the light source in system K′ moving toward the receiver with velocity v. The source emits n waves during the time interval T. Because the speed of light is always c and the source is moving with velocity v, the total distance between the front and rear of the wave train emitted during the time interval T is

$$\text{length of wave train} = cT - vT$$

Because there are n waves emitted during this time period, the wavelength must be

$$\lambda = \frac{cT - vT}{n}$$

and the frequency, $\nu = c/\lambda$, is

$$\nu = \frac{cn}{cT - vT} \tag{2.28}$$

In its rest frame, the source emits n waves of frequency ν_0 during the proper time T'_0.

$$n = \nu_0 T'_0 \tag{2.29}$$

The proper time interval T'_0 measured on the clock at rest in the moving system is related to the time interval T measured on a clock fixed by the receiver in system K by

$$T'_0 = \frac{T}{\gamma} \tag{2.30}$$

where γ is the relativistic factor of Equation (2.16). The clock moving with the source measures the proper time because it is present with both the beginning and end of the wave.

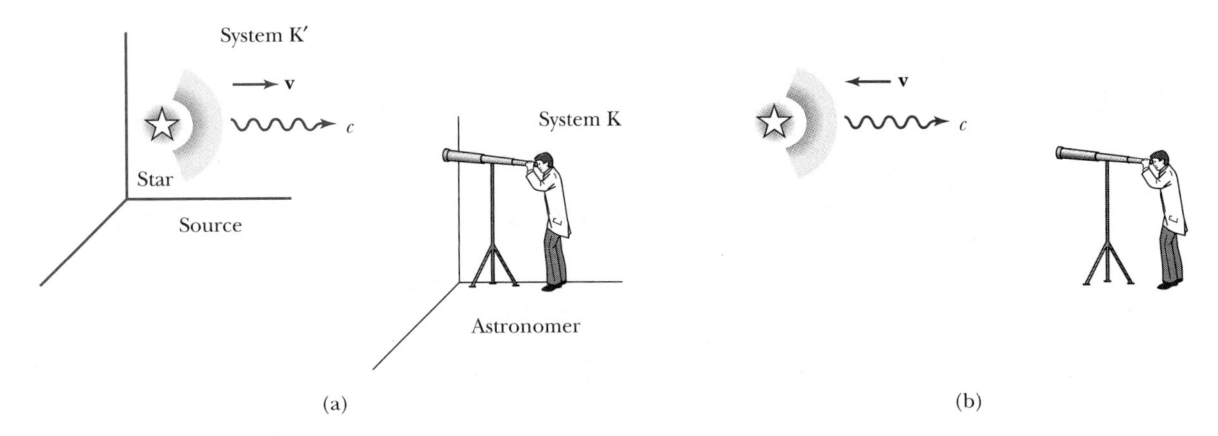

FIGURE 2.23 In (a) the source (star) is approaching the receiver (astronomer) with velocity v while it emits starlight signals with speed c. In (b) the source and receiver are receding with velocity v. The Doppler effect for light is different than that for sound, because of relativity and no medium to carry the light waves.

We substitute the proper time T'_0 from Equation (2.30) into Equation (2.29) to determine the number of waves n. Then n is substituted into Equation (2.28) to determine the frequency.

$$\nu = \frac{c\nu_0 T/\gamma}{cT - vT}$$

$$= \frac{1}{1 - v/c}\frac{\nu_0}{\gamma} = \frac{\sqrt{1 - v^2/c^2}}{1 - v/c}\nu_0$$

where we have inserted the equation for γ. If we use $\beta = v/c$, we can write the previous equation as

$$\nu = \frac{\sqrt{1+\beta}}{\sqrt{1-\beta}}\nu_0 \qquad \text{Source and receiver approaching} \qquad (2.31)$$

It is straightforward to show that Equation (2.31) is also valid when the source is fixed and the receiver approaches it with velocity v. It is the relative velocity v, of course, that is important (Problem 41).

But what happens if the source and receiver are receding from each other with velocity v? The derivation is similar to the one just done, except that the distance between the beginning and end of the wave train becomes

$$\text{Length of wave train} = cT + vT$$

because the source and receiver are receding rather than approaching. This change in sign is propagated throughout the derivation, with the final result (Problem 50),

$$\nu = \frac{\sqrt{1-\beta}}{\sqrt{1+\beta}}\nu_0 \qquad \text{Source and receiver receding} \qquad (2.32)$$

Equations (2.31) and (2.32) can be combined into one equation if we agree to use a + sign for β ($+v/c$) when the source and receiver are approaching each other and a − sign for β ($-v/c$) when they are receding. The final equation becomes

$$\nu = \frac{\sqrt{1+\beta}}{\sqrt{1-\beta}}\nu_0 \qquad \text{Relativistic Doppler effect} \qquad (2.33)$$

The Doppler effect is useful in many areas of science including astronomy, atomic physics, and nuclear physics. One of its many applications includes an effective radar system for locating airplane position and speed (see Special Topic).

Elements absorb and emit characteristic frequencies of light due to the existence of particular atomic levels. We will learn more about these later. Scientists have observed these characteristic frequencies in starlight and have observed shifts in the frequencies. One reason for these shifts is the Doppler effect, and the frequency changes are used to determine the speed of the emitting object with respect to us. This is the source of the well-known *redshifts* of starlight caused by objects moving away from us. These data have been used to ascertain that the

Redshifts

SPECIAL TOPIC:

APPLICATIONS OF THE DOPPLER EFFECT

The Doppler effect is not just a curious result of relativity. It has many interesting applications, which will be discussed in various places in this text. Perhaps the best known is in astronomy, where the Doppler shifts of well known atomic transition frequencies determine the relative velocities of astronomical objects with respect to us. Such measurements continue to be used today to find the distances of such unusual objects as quasars (objects having incredibly large masses that produce tremendous amounts of radiation, see Chapter 16). The Doppler effect has been used to discover other mundane effects in astronomy, for example the rate of rotation of Venus and the fact that Venus rotates in the opposite direction of Earth—the sun rises in the west on Venus. This was determined by observing light reflected from both sides of Venus—on one side it is blueshifted and on the other side it is redshifted as shown in Figure A. The same technique has been used to determine the rate of rotation of stars.

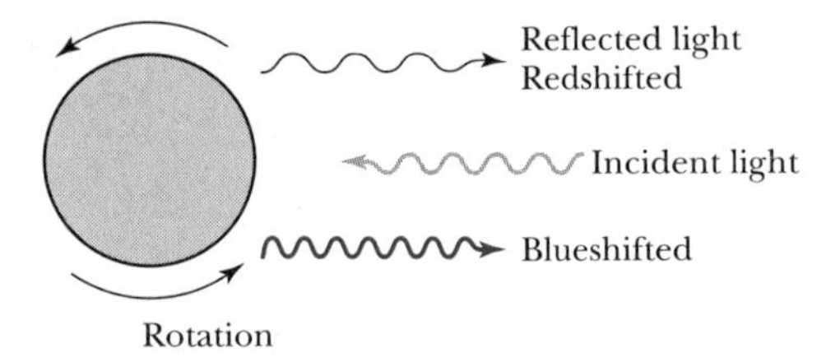

FIGURE A

Radar

Perhaps the Doppler effect is nowhere more important than it is in radar. When an electromagnetic radar signal reflects off of a moving target, the so-called *echo* signal will be shifted in frequency by the Doppler effect. Very small frequency shifts can be determined by examining the beat frequency of the echo signal with a reference signal. The frequency shift is proportional to the radial component of the target's velocity. Navigation radar is quite complex, and ingenious techniques have been devised to determine the target position and velocity using multiple radar beams. By using pulsed Doppler radar it is possible to separate moving targets from stationary targets, called clutter.

Doppler radar is also extensively used in meteorology. Vertical motion of airdrafts, sizes and motion of raindrops, motion of thunderstorms, and detailed patterns of wind distribution have all been studied with Doppler radar.

X rays and gamma rays emitted from moving atoms and nuclei will have their frequencies shifted by the Doppler effect. Such phenomena tend to broaden radiation frequencies emitted by stationary atoms and nuclei and add to the natural spectral widths observed.

Laser Cooling

In order to perform fundamental measurements in atomic physics, it is useful to limit the effects of thermal motion and to isolate single atoms. A method taking advantage of the Doppler effect can slow down

---

universe is expanding. The farther away the stars, the higher the redshift, which led Harlow Shapley and Edwin Hubble to conclude the origin of the universe started with a "Big Bang."*

So far in this section we have only considered the source and receiver to be directly approaching or receding. Of course, it is also possible for the two to be moving at an angle with respect to one another, as is shown in Figure 2.24. We omit the derivation here† but present the results. The angles θ and θ' are the

*Excellent references are "The Cosmic Distance Scale" by Paul Hodge, *American Scientist,* **72,** 474 (1984) and "Origins" by S. Weinberg, *Science,* **230,** 15 (1985). This subject will be discussed in Chapter 16.

†See Robert Resnick, *Introduction to Special Relativity.* New York: John Wiley and Sons, 1968.

even neutral atoms and eventually isolate them. Atoms emitted from an oven will have a spread of velocities around some average value. If these atoms form a beam as shown in Figure B, a laser beam impinging on the atoms from the right can slow them down by transferring momentum.

Atoms have characteristic energy levels that allow them to absorb and emit radiation of specific frequencies. Atoms moving with respect to the laser beam (of speed c) will "see" the laser frequency shifted because of the Doppler effect. For example, atoms moving toward the laser beam will encounter a laser with high frequency, and atoms moving away from the laser beam will encounter a laser with low frequency. Even atoms moving in the same direction within the beam of atoms will see slightly different frequencies depending on the velocities of the various atoms. Now, if the frequency of the laser beam is tuned to the precise frequency seen by the faster atoms so that those atoms can be excited by absorbing the radiation, those faster atoms will be slowed down by absorbing the momentum of the laser radiation. The slower atoms will "see" a laser beam that has been Doppler shifted to a lower frequency than is needed to absorb the radiation, and these atoms are not as likely to absorb the laser radiation. The net effect is that the atoms as a whole are *slowed down* and their *velocity spread is reduced.*

As the atoms slow down, they see that the Doppler shifted frequencies of the laser change, and the atoms no longer absorb the laser radiation. They continue

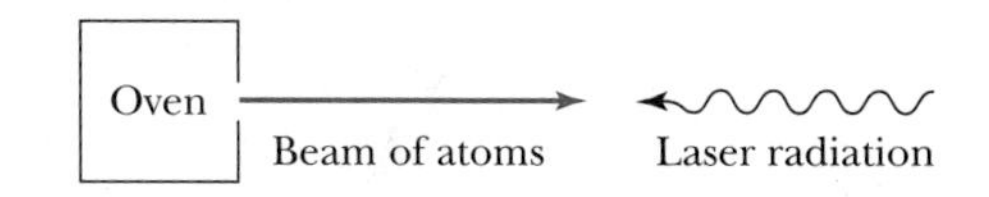

FIGURE B

with the same lower velocity and velocity spread. There are two methods to make the deceleration process continue. In one, the frequency of the laser beam is increased to keep the radiation consistent with the Doppler shifted frequency needed to excite the atoms. In the other method, the laser frequency is kept constant, but sophisticated magnetic fields are used to vary the frequencies needed to excite the atoms by changing the excited atomic energy levels. By using six intersecting laser beams coming in at different angles, an "optical molasses" has been created in which the atoms are essentially isolated and the average velocity is zero. In a remarkable series of experiments by various researchers improving on the basic technique, atoms have been cooled to temperatures below 0.2 μK (2×10^{-7} K). The 1997 Nobel Prize in Physics was awarded to Steven Chu, Claude Cohen-Tannoudji, and William Phillips for this technique. Important, fundamental physics measurements have been performed utilizing these methods including making more accurate clocks, manipulating DNA strands, and ultraprecise atom interferometers. See Steven Chu, "Laser Trapping of Neutral Particles," *Scientific American,* **266,** 70 (Feb. 1992).

---

angles the light signals make with the x axes in the K and K′ systems. They are related by

$$\nu \cos \theta = \frac{\nu_0(\cos \theta' + \beta)}{\sqrt{1 - \beta^2}} \tag{2.34}$$

and

$$\nu \sin \theta = \nu_0 \sin \theta' \tag{2.35}$$

The generalized Doppler shift equation becomes

$$\nu = \frac{1 + \beta \cos \theta'}{\sqrt{1 - \beta^2}} \nu_0 \tag{2.36}$$

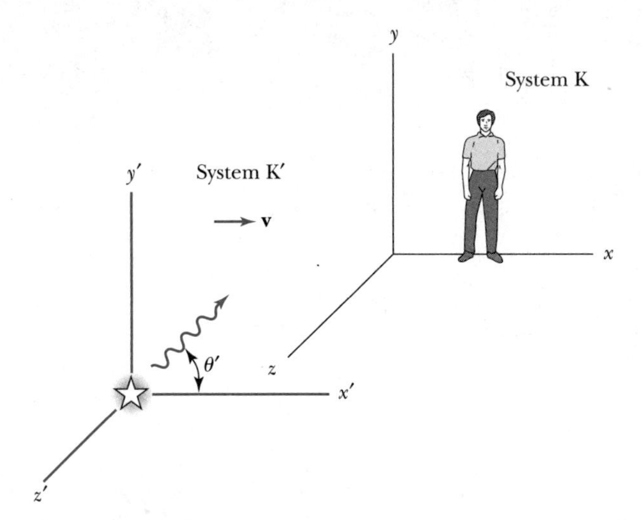

FIGURE 2.24 The light signals in system K′ are emitted at an angle θ' from the x' axis and remain in the $x'-y'$ plane.

Note that Equation (2.36) gives Equation (2.31) when $\theta' = 0$ (source and receiver approaching) and gives Equation (2.32) when $\theta' = 180°$ (source and receiver receding). This situation is known as the *longitudinal Doppler effect.*

When $\theta = 90°$ the emission is purely transverse to the direction of motion, and we have the *transverse Doppler effect,* which is purely a relativistic effect that does not occur classically. The transverse Doppler effect is directly due to time dilation and has been verified experimentally. Equations (2.34) through (2.36) can also be used to understand stellar aberration.

Example 2.8

Analyze the light signals sent out by Frank and Mary by using the relativistic Doppler effect.

Solution: First, we analyze the frequency of the light signals that Mary receives from Frank. During the outbound trip the source (Frank) and receiver (Mary) are receding so that $\beta = -0.8$. Using Equation (2.33) we have

$$\nu = \frac{\sqrt{1 + (-0.8)}}{\sqrt{1 - (-0.8)}}\,\nu_0 = \frac{\nu_0}{3}$$

Because Frank sends out signals annually, Mary will receive the signals only every three years. Therefore during the six-year trip in Mary's system to the star system, she will only receive 2 signals.

During the return trip, $\beta = 0.8$ and Equation (2.33) gives

$$\nu = \frac{\sqrt{1 + 0.8}}{\sqrt{1 - 0.8}}\,\nu_0 = 3\nu_0$$

so that Mary receives 3 signals each year for a total of 18 signals during the return trip. Mary receives a total of 20 annual light signals from Frank, and she concludes that Frank has aged 20 years during her trip.

Now let's analyze the light signals that Mary sends Frank. During the outbound trip the frequency at which Frank receives signals from Mary will also be $\nu_0/3$. During the ten years that it takes Mary to reach the star system on his clock, he will receive $\frac{10}{3}$ signals—3 signals plus $\frac{1}{3}$ of the time to the next one. Frank continues to receive signals from Mary for another eight years, because that is how long it takes the sixth signal she sent him to reach Earth. Therefore, for the first 18 years of her journey, according to his own clock he receives $\frac{18}{3} = 6$ signals. Frank has no way of knowing that Mary has turned around and is coming back until he starts receiving signals at frequency $3\nu_0$. During Mary's return trip Frank will receive signals at the frequency $3\nu_0$ or 3 per year. However, in his system, Mary returns two years after he has received her sixth signal and turned around to come back. During this two year period he will receive 6 more signals, so he concludes she has only aged a total of 12 years.

Notice that this analysis is in total agreement with the spacetime diagram of Figure 2.22 and is somewhat easier to obtain. Although geometrical constructions like spacetime diagrams are sometimes very useful, we will normally find that an analytical calculation will be easier.

2.11 Relativistic Momentum

Newton's second law, $\mathbf{F} = d\mathbf{p}/dt$, keeps its same form under a Galilean transformation, but we might not expect it to do so under a Lorentz transformation. There may be other similar transformation difficulties with the conservation laws of linear momentum and energy. We need to take a careful look at our previous definition of linear momentum to see if it is still valid at high speeds. According to Newton's second law, for example, an acceleration of a particle already moving at very high speeds could lead to a velocity greater than the speed of light. That would cause a problem with the Lorentz transformation, so we expect that Newton's second law might somehow be modified at high speeds.

Because we believe the conservation of linear momentum is a fundamental result, we begin by considering a collision that has no external forces and no accelerations. Frank (**F**ixed or stationary system) is at rest in system K holding a ball of mass m. Mary (**M**oving system) holds a similar ball in system K′ that is moving in the x direction with velocity v with respect to system K as shown in Figure 2.25a. Frank throws his ball along his y axis, and Mary throws her ball with exactly the same speed along her negative y' axis. The two balls collide in a perfectly elastic collision, and each twin measures the speed of his or her own ball to be u_0 both before and after the collision.

We show the collision according to both observers in Figure 2.25. Consider the conservation of momentum according to Frank as seen in system K. The velocity of the ball thrown by Frank has components in his own system K of

$$u_{Fx} = 0 \tag{2.37}$$
$$u_{Fy} = u_0$$

Using the definition of momentum from introductory physics, $\mathbf{p} = m\mathbf{v}$, the momentum of the ball thrown by Frank is entirely in the y direction:

$$p_{Fy} = mu_0 \tag{2.38}$$

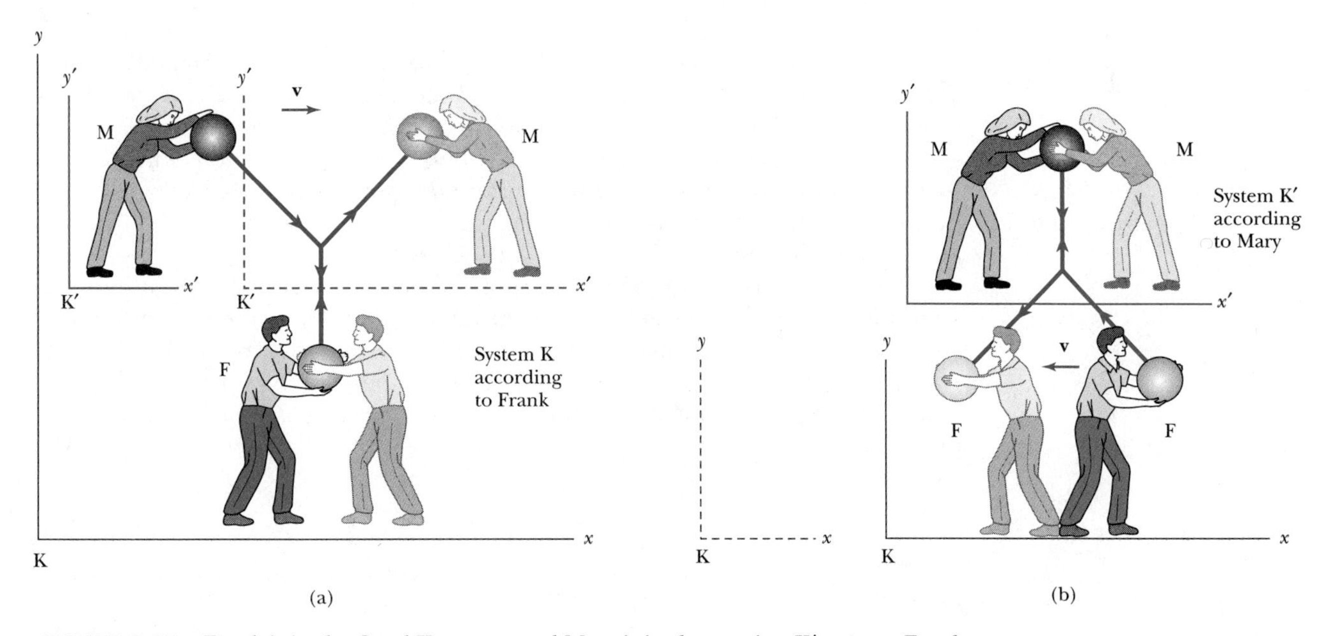

FIGURE 2.25 Frank is in the fixed K system, and Mary is in the moving K′ system. Frank throws his ball along his $+y$ axis, and Mary throws her ball along her $-y'$ axis. The balls collide. The event is shown in Frank's system in (a) and in Mary's system in (b).

Because the collision is perfectly elastic, the ball returns to Frank with speed u_0 along the negative y axis. The change of momentum of his ball as observed by Frank in system K is

$$\Delta p_F = \Delta p_{Fy} = -2mu_0 \tag{2.39}$$

In order to confirm the conservation of linear momentum, we need to determine the change in the momentum of Mary's ball *as measured by Frank*. We will let the primed speeds be measured by Mary and the unprimed speeds be measured by Frank (except that u_0 is always the speed of the ball as measured by the twin in his or her own system). Mary measures the initial velocity of her own ball to be $u'_{Mx} = 0$ and $u'_{My} = -u_0$, because she throws it along her own negative y' axis. In order to determine the velocity of Mary's ball as measured by Frank we need to use the velocity transformation equations of Equation (2.22). If we insert the appropriate values for the speeds just discussed, we obtain

$$\begin{aligned} u_{Mx} &= v \\ u_{My} &= -u_0\sqrt{1 - v^2/c^2} \end{aligned} \tag{2.40}$$

Before the collision, the momentum of Mary's ball as measured by Frank becomes

$$\begin{aligned} &\text{Before} \quad p_{Mx} = mv \\ &\text{Before} \quad p_{My} = -mu_0\sqrt{1 - v^2/c^2} \end{aligned} \tag{2.41}$$

For a perfectly elastic collision, the momentum after the collision is

$$\begin{aligned} &\text{After} \quad p_{Mx} = mv \\ &\text{After} \quad p_{My} = +mu_0\sqrt{1 - v^2/c^2} \end{aligned} \tag{2.42}$$

The change in momentum of Mary's ball according to Frank is

$$\Delta p_M = \Delta p_{My} = 2mu_0\sqrt{1 - v^2/c^2} \tag{2.43}$$

The conservation of linear momentum requires the total change in momentum of the collision, $\Delta p_F + \Delta p_M$, to be zero. The addition of Equations (2.39) and (2.43) clearly does not give zero. *Linear momentum is not conserved if we use the conventions for momentum from introductory classical physics even if we use the velocity transformation equations from the special theory of relativity.* There is no problem with the x direction, but there is a problem with the y direction along the direction the ball is thrown in each system.

Difficulty with classical linear momentum

Rather than abandon the conservation of linear momentum, let us look for a modification of the definition of linear momentum that preserves both it and Newton's second law. Similar to what we did in deriving the Lorentz transformation, we assume the simplest, most reasonable change that may preserve the conservation of momentum. We assume that the classical form of momentum $m\mathbf{u}$ is multiplied by a factor that may depend on velocity. Let the factor be $\Gamma(u)$. Our trial definition for linear momentum now becomes

$$\mathbf{p} = \Gamma(u)\, m\mathbf{u} \tag{2.44}$$

In Example 2.9 we show that momentum is conserved in the collision just described for the value of $\Gamma(u)$ given by

$$\Gamma(u) = \frac{1}{\sqrt{1 - u^2/c^2}} \tag{2.45}$$

Notice that the *form* of Equation (2.45) is the same as that found earlier for the Lorentz transformation. We even give $\Gamma(u)$ the same symbol: $\Gamma(u) = \gamma$. However, this γ is different; it contains the speed of the particle u, whereas the Lorentz transformation contains the relative speed v between the two inertial reference frames. This distinction should be kept in mind because it can cause confusion. Because the usage is so common by physicists, we will use γ for both purposes. However, when there is any chance of confusion, we will write out $1/\sqrt{1 - u^2/c^2}$ and use $\gamma = 1/\sqrt{1 - v^2/c^2}$ for the Lorentz transformation. We will also write out $1/\sqrt{1 - u^2/c^2}$ often in order to avoid confusion.

We can make a plausible determination for the correct form of the momentum if we use the proper time discussed previously to determine the velocity. The momentum becomes

$$\mathbf{p} = m\frac{d\mathbf{r}}{d\tau} = m\frac{d\mathbf{r}}{dt}\frac{dt}{d\tau} \tag{2.46}$$

we retain the velocity $\mathbf{u} = d\mathbf{r}/dt$ as used classically. All observers do not agree as to the value of $d\mathbf{r}/dt$, but they do agree as to the value of $d\mathbf{r}/d\tau$, where $d\tau$ is the proper time measured in the moving system K′. The value of $dt/d\tau$ $(= \gamma)$ is obtained from Equation (2.30), where the speed u is used in the relation for γ to represent the relative speed of the moving (Mary's) frame and the fixed (Frank's) frame.

The definition of the **relativistic momentum** becomes, from Equation (2.46),

$$\mathbf{p} = m\frac{d\mathbf{r}}{dt}\gamma$$

$$\mathbf{p} = \gamma m\mathbf{u} \quad \text{Relativistic momentum} \tag{2.47}$$

Relativistic momentum

where

$$\gamma = \frac{1}{\sqrt{1 - u^2/c^2}} \tag{2.48}$$

This result for the relativistic momentum reduces to the classical result for small values of u/c. The classical momentum expression is good to an accuracy of 1% as long as $u < 0.14c$.

Some physicists like to refer to the mass in Equation (2.47) as the *rest mass* m_0 and call the term $m = \gamma m_0$ the *relativistic mass.* In this manner the classical form of momentum, $m\mathbf{u}$, is retained. The mass is then imagined to increase at high speeds. Most research physicists prefer to keep the concept of mass as an invariant, intrinsic property of an object. We prefer this latter approach and will exclusively use the term *mass* to mean *rest mass.* Although we may use the terms *mass* and *rest mass* synonymously, we will not use the term *relativistic mass.* The use of relativistic mass too often leads the student into mistakenly inserting the term into classical expressions where it does not apply.

Example 2.9

Show that linear momentum is conserved for the collision just discussed and shown in Figure 2.25.

Solution: Using the relativistic momentum we can modify the expressions obtained for the momentum of the balls thrown by Frank and Mary. From Equation (2.38), the momentum of the ball thrown by Frank becomes

$$p_{Fy} = \gamma m u_0 = \frac{m u_0}{\sqrt{1 - u_0{}^2/c^2}}$$

For an elastic collision, the magnitude of the momentum for this ball before and after the collision is the same. After the collision, the momentum will be the negative of this value, so the change in momentum becomes, from Equation (2.39),

$$\Delta p_F = \Delta p_{Fy} = 2\gamma m u_0 = \frac{2mu_0}{\sqrt{1 - u_0^2/c^2}} \tag{2.49}$$

Now we consider the momentum of Mary's ball as measured by Frank. Even with the addition of the γ factor for the momentum in the x direction, we still have $\Delta p_{Mx} = 0$. We must look more carefully at Δp_{My}. First, we find the speed of the ball thrown by Mary as measured by Frank. We use Equations (2.40) to determine

$$u_M = \sqrt{u_{Mx}^2 + u_{My}^2} = \sqrt{v^2 + u_0^2(1 - v^2/c^2)} \tag{2.50}$$

The relativistic factor γ for the momentum for this situation is

$$\gamma = \frac{1}{\sqrt{1 - u_M^2/c^2}}$$

The value of p_{My} is now found by modifying Equation (2.41) with this value of γ.

$$p_{My} = -\gamma m u_0 \sqrt{1 - v^2/c^2} = \frac{-mu_0\sqrt{1 - v^2/c^2}}{\sqrt{1 - u_M^2/c^2}}$$

Inserting the value of u_M from Equation (2.50) gives

$$p_{My} = \frac{-mu_0\sqrt{1 - v^2/c^2}}{\sqrt{(1 - u_0^2/c^2)(1 - v^2/c^2)}} = \frac{-mu_0}{\sqrt{1 - u_0^2/c^2}} \tag{2.51}$$

The momentum after the collision will still be the negative of this value, so the change in momentum becomes

$$\Delta p_M = \Delta p_{My} = \frac{-2mu_0}{\sqrt{1 - u_0^2/c^2}} \tag{2.52}$$

The change in the momentum of the two balls as measured by Frank is given by the sum of Equations (2.49) and (2.52)

$$\Delta p = \Delta p_F + \Delta p_M = 0$$

Thus Frank indeed finds that momentum is conserved. Mary should also determine that linear momentum is conserved (see Problem 62).

2.12 Relativistic Energy

We now turn to the concepts of energy and force. When forming the new theories of relativity and quantum physics, physicists resisted changing the well-accepted ideas of classical physics unless absolutely necessary. In this same spirit we also choose to keep intact as many definitions from classical physics as possible and let experiment dictate when we are incorrect. When we studied motion in introductory physics, we learned that the concept of force is in practice defined by its use in Newton's laws. We retain here the classical definition of force, which is best represented by Newton's second law. In the previous section we studied the concept of momentum and found a relativistic expression for momentum in Equation (2.47). Therefore, we modify Newton's second law to include our new definition of linear momentum, and force becomes

Relativistic force

$$\mathbf{F} = \frac{d\mathbf{p}}{dt} = \frac{d}{dt}(\gamma m\mathbf{u}) = \frac{d}{dt}\left(\frac{m\mathbf{u}}{\sqrt{1 - u^2/c^2}}\right) \tag{2.53}$$

Aspects of this force will be examined in the problems (see Problems 55–58).

In introductory physics we introduced kinetic energy as the work done on a particle by a net (unbalanced) force. We retain the same definitions of kinetic energy and work. The work W_{12} done by a force $\mathbf{F}$ to move a particle from position 1 to position 2 along a path $\mathbf{s}$ is defined to be

$$W_{12} = \int_1^2 \mathbf{F} \cdot d\mathbf{s} = K_2 - K_1 \tag{2.54}$$

where K_1 is defined to be the kinetic energy of the particle at position 1.

For simplicity, let the particle start from rest under the influence of the force $\mathbf{F}$ and calculate the final kinetic energy K after the work is done. The force is related to the dynamic quantities by Equation (2.53). The work W and kinetic energy K are

$$W = K = \int \frac{d}{dt}(\gamma m\mathbf{u}) \cdot \mathbf{u}dt \tag{2.55}$$

where the differential path $d\mathbf{s}$ is given by $\mathbf{u}dt$. Because the mass is invariant, it can be brought outside the integral. The relativistic factor γ depends on u and cannot be brought outside the integral. Equation (2.55) becomes

$$K = m\int dt\,\frac{d}{dt}(\gamma\mathbf{u}) \cdot \mathbf{u} = m\int u\,d(\gamma u)$$

The limits of integration are from an initial value of 0 to a final value of γu.

$$K = m\int_0^{\gamma u} u\,d(\gamma u) \tag{2.56}$$

The integral in Equation (2.56) is straightforward if done by the method of integration by parts. The result, called the *relativistic kinetic energy*, is

$$K = \gamma mc^2 - mc^2 = mc^2\left(\frac{1}{\sqrt{1 - u^2/c^2}} - 1\right) = mc^2(\gamma - 1) \tag{2.57}$$

Relativistic kinetic energy

Equation (2.57) does not seem to resemble the classical result for kinetic energy, $K = \frac{1}{2}mu^2$. However, we expect it to reduce to the classical result for low speeds. Let's see if it does. For speeds $u \ll c$, we can expand γ in a power series as follows:

$$K = mc^2(1 - u^2/c^2)^{-1/2} - mc^2$$

$$= mc^2\left(1 + \frac{1}{2}\frac{u^2}{c^2} + \cdots\right) - mc^2$$

where we have neglected all terms of power $(u/c)^4$ and greater, because $u \ll c$. This gives for the relativistic kinetic energy at low speeds.

$$K = mc^2 + \frac{1}{2}mu^2 - mc^2 = \frac{1}{2}mu^2 \tag{2.58}$$

which is the expected classical result.

One of the most common mistakes that students make when first studying relativity is to use either $\frac{1}{2}mu^2$ or $\frac{1}{2}\gamma mu^2$ for the relativistic kinetic energy. It is important to *use only Equation (2.57) for the relativistic kinetic energy.* Although Equation (2.57) looks much different than the classical result, it is the only correct one, and *neither* $\frac{1}{2}mu^2$ *nor* $\frac{1}{2}\gamma mu^2$ *is a correct relativistic result.*

Equation (2.57) is particularly useful when dealing with particles accelerated to high speeds. For example, the fastest speeds produced in the United States have been in the three-kilometer-long electron accelerator at the Stanford Linear Accelerator Laboratory. This accelerator produces electrons with a kinetic energy of 8×10^{-9} J (50 GeV) or 50×10^9 eV. The electrons have velocities so close to the speed of light, that the tiny difference from c is difficult to measure directly. The speed of the electrons is inferred from the relativistic kinetic energy of Equation (2.57) and is given by $0.99999999995c$. Such calculations are difficult to do with calculators because of significant-figure limitations. As a result, we use kinetic energy or momentum to express the motion of a particle moving almost at the speed of light.

An aerial view of the 3-km long Stanford Linear Accelerator Center. An interstate highway crosses over the device. The experiments are performed in buildings both above and below ground in the foreground. *Courtesy of Stanford Linear Accelerator and Department of Energy.*

Use kinetic energy or momentum to express motion.

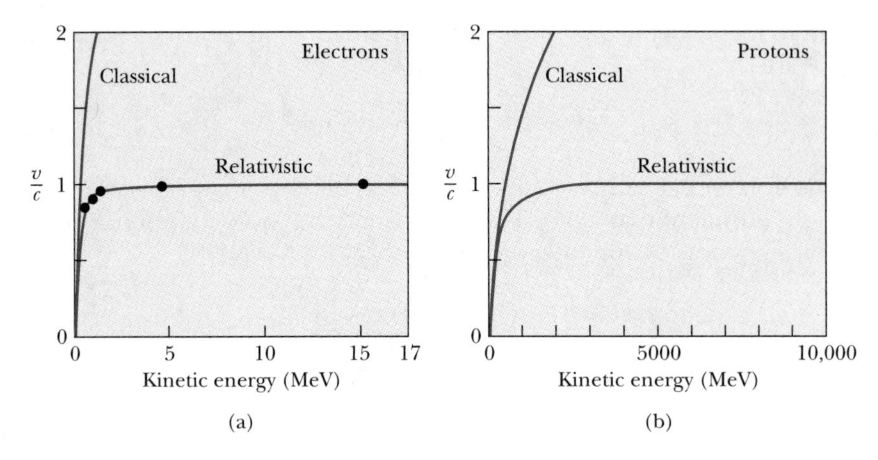

FIGURE 2.26 The velocities (v/c) of (a) electrons and (b) protons are shown versus kinetic energy for both classsical (incorrect) and relativistic calculations. *Adapted with permission from data by W. Bertozzi, American Journal of Physics, **32**, 551(1964). © 1964 American Association of Physics.*

The classical and relativistic speeds for both electrons and protons are shown in Figure 2.26 as a function of their kinetic energy. Physicists have found that experimentally it does not matter how much energy we give a massive body. Its speed can never quite reach c. Let's see if this result is consistent with our relativistic energy calculations. If we examine Equation (2.57), we see that when $u \longrightarrow c$, the kinetic energy $K \longrightarrow \infty$. Because there is not an infinite amount of energy available, we agree that no massive particle can have the speed of light—in agreement with the experimental result.

Example 2.10

Electrons in a television set are accelerated by a potential difference of 25,000 volts before striking the screen. Calculate the velocity of the electrons and determine the error in using the classical kinetic energy result.

Solution: The work done to accelerate an electron across a potential difference V is given by qV, where q is the charge of the particle. The work done to accelerate the electron from rest is the final kinetic energy K of the electron. The kinetic energy is given by

$$K = W = qV = (1.6 \times 10^{-19}\ \text{C})(25 \times 10^{3}\ \text{V})$$
$$= 4.0 \times 10^{-15}\ \text{J}$$

In order to determine the correct speed of the electrons, we must use the relativistically correct kinetic energy given by Equation (2.57). We will first determine γ; from that, we can determine the speed u. From Equation (2.57) we have

$$K = (\gamma - 1)mc^2 \tag{2.59}$$

From this equation, γ is found to be

$$\gamma = 1 + \frac{K}{mc^2} \tag{2.60}$$

The quantity mc^2 for the electron is determined to be

$$mc^2\ (\text{electron}) = (9.11 \times 10^{-31}\ \text{kg})(3 \times 10^{8}\ \text{m/s})^2$$
$$= 8.19 \times 10^{-14}\ \text{J}$$

The relativistic factor is now found from Equation (2.60) to be

$$\gamma = 1 + \frac{4.0 \times 10^{-15}\ \text{J}}{8.19 \times 10^{-14}\ \text{J}} = 1.049$$

Equation (2.8) can be rearranged to determine β^2 as a function of γ^2, where $\beta = u/c$.

$$\beta^2 = \frac{\gamma^2 - 1}{\gamma^2} \tag{2.61}$$

$$\beta^2 = \frac{(1.049)^2 - 1}{(1.049)^2} = 0.091$$

The value of β is 0.30, and the correct velocity, $u = \beta c$, is 0.90×10^8 m/s.

We determine the error in using the classical result by calculating the velocity using the nonrelativistic expression. The nonrelativistic expression is $K = \frac{1}{2}mu^2$, and the velocity is given by

$$\text{Nonrelativistic} \qquad u = \sqrt{\frac{2K}{m}}$$
$$= \sqrt{\frac{2(4.0 \times 10^{-15}\ \text{J})}{9.11 \times 10^{-31}\ \text{kg}}}$$
$$= 0.94 \times 10^{8}\ \text{m/s (nonrelativistic)}$$

The (incorrect) classical speed is about 4% greater than the (correct) relativistic speed. Such an error is significant enough to be important in designing electronic equipment and in making test measurements. Relativistic calculations are particularly important for electrons, because they have such a small mass, and therefore are easily accelerated to speeds very close to c.

Total Energy and Rest Energy

We rewrite Equation (2.57) in the form

$$\gamma mc^2 = \frac{mc^2}{\sqrt{1 - u^2/c^2}} = K + mc^2 \tag{2.62}$$

The term mc^2 is called the **rest energy** and is denoted by E_0.

$$E_0 = mc^2 \quad \text{Rest energy} \tag{2.63}$$

Rest energy

This leaves the sum of the kinetic energy and rest energy to be interpreted as the **total energy** of the particle. The total energy is denoted by E and is given by

$$E = \gamma mc^2 = \frac{mc^2}{\sqrt{1 - u^2/c^2}} = \frac{E_0}{\sqrt{1 - u^2/c^2}} = K + E_0 \tag{2.64}$$

Total energy

Equivalence of Mass and Energy

These last few equations suggest the equivalence of mass and energy, a concept attributed to Einstein. The result that energy $= mc^2$ is one of the most famous equations in physics. Even when a particle has no velocity, and thus no kinetic energy, we still believe that the particle has energy through its mass, $E_0 = mc^2$. Nuclear reactions are certain proof that mass and energy are equivalent. The concept of motion as being described by *kinetic energy* is preserved in relativistic dynamics, but a particle with no motion still has energy through its mass.

We must modify two of the conservation laws that we learned in classical physics. Mass and energy are no longer two separate conservation laws. We must combine them into one law of the **conservation of mass-energy.** We will see ample proof during the remainder of this book of the validity of this basic conservation law.

Conservation of mass-energy

Even though we often say "energy is turned into mass" or "mass is converted into energy" or "mass and energy are interchangeable," we must understand that what we mean is that mass and energy are actually *equivalent.* Mass is another form of energy, and we use the terms mass-energy and energy interchangeably. A similar occurrence happened in the late 18th century when it became clear that heat was another form of energy. The 19th-century experiments of James Joule showed that heat loss or gain was related to work. The conservation of energy was thereafter on firm footing, even though we have had to modify it since.

Consider two blocks of wood, each of mass m and having kinetic energy K, moving toward each other as shown in Figure 2.27. A spring placed between them is compressed and locks in place as they collide. Let's examine the conservation of mass-energy. The energy before the collision is

$$\text{Mass-energy before:} \qquad E = 2mc^2 + 2K \tag{2.65}$$

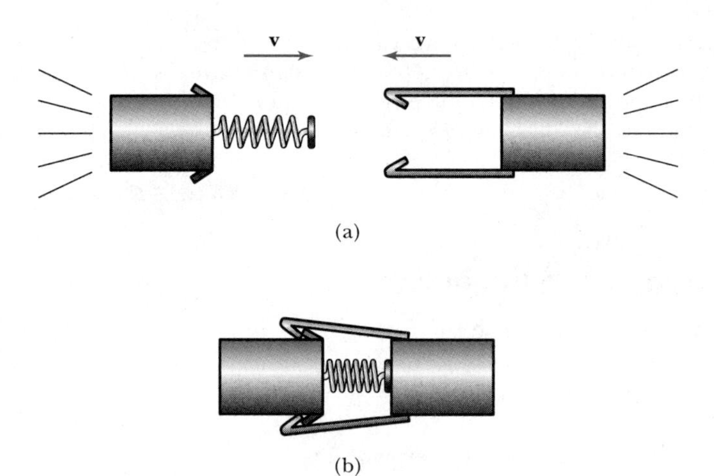

FIGURE 2.27 (a) Two blocks of wood, one with a spring attached and both having mass m, move with equal speeds v and kinetic energies K toward a head-on collision. (b) The two blocks collide, compressing the spring, which locks in place. The system now has increased mass, $M = 2m + 2K/c^2$, with the kinetic energy being converted into the potential energy of the spring.

and the energy after the collision is

$$\text{Mass-energy after:} \qquad E = Mc^2 \tag{2.66}$$

where M is the (rest) mass of the system. Because energy is conserved, we have $E = 2mc^2 + 2K = Mc^2$, and the new mass M is greater than the individual masses $2m$. The kinetic energy went into compressing the spring, so the spring has increased potential energy. Kinetic energy has been converted into mass, the result being that the potential energy of the spring has caused the system to have more mass. We find the difference in mass ΔM by setting the previous two equations for energy equal and solving for $\Delta M = M - 2m$.

$$\Delta M = M - 2m = \frac{2K}{c^2} \tag{2.67}$$

Linear momentum is conserved in this head-on collision.

The fractional mass increase in this case is quite small and is given by $f = \Delta M/2m$. If we use Equation (2.67), we have

$$f = \frac{M - 2m}{2m} = \frac{2K/c^2}{2m} = \frac{K}{mc^2} \tag{2.68}$$

For typical masses and kinetic energies of blocks of wood, this fractional increase in mass is too small to measure. For example, if we have blocks of wood of mass 0.1 kg moving rapidly at 10 m/s, Equation (2.68) gives

$$f = \frac{\frac{1}{2}mv^2}{mc^2} = \frac{1}{2}\frac{v^2}{c^2} = \frac{1}{2}\frac{(10 \text{ m/s})^2}{(3 \times 10^8 \text{ m/s})^2} = 6 \times 10^{-16}$$

where we have used the nonrelativistic expression for kinetic energy because the speed is so low. This absurdly small answer shows us that questions of mass increase are inappropriate for macroscopic objects like blocks of wood and automobiles crashing into one another. Such small increases cannot even be measured, but in the next section, we will look at the collision of two high-energy protons, in which considerable energy is available to create additional mass. Mass-energy relations will be essential in such reactions.

Relationship of Energy and Momentum

Physicists believe that linear momentum is a more fundamental concept than kinetic energy. There is no conservation of kinetic energy, whereas the conservation of linear momentum is inviolate as far as we know. A more fundamental result for the total energy in Equation (2.64) might include momentum rather than kinetic energy. Let's proceed to find a useful result. We begin with Equation (2.47) for the relativistic momentum written in magnitude form only.

$$p = \gamma m u = \frac{mu}{\sqrt{1 - u^2/c^2}}$$

We square this result, multiply by c^2, and rearrange the result.

$$\begin{aligned} p^2c^2 &= \gamma^2 m^2 u^2 c^2 \\ &= \gamma^2 m^2 c^4 \left(\frac{u^2}{c^2}\right) = \gamma^2 m^2 c^4 \beta^2 \end{aligned}$$

We use Equation (2.61) for β^2 and find

$$\begin{aligned} p^2c^2 &= \gamma^2 m^2 c^4 \left(1 - \frac{1}{\gamma^2}\right) \\ &= \gamma^2 m^2 c^4 - m^2 c^4 \end{aligned}$$

The first term on the right-hand side is just E^2, and the second term is ${E_0}^2$. The last equation becomes

$$p^2c^2 = E^2 - {E_0}^2$$

We rearrange this last equation to find the result we are seeking, a relation between energy and momentum.

$$E^2 = p^2c^2 + {E_0}^2 \tag{2.69}$$

Momentum-energy relation

or

$$E^2 = p^2c^2 + m^2c^4 \tag{2.70}$$

Equation (2.69) is an extremely useful result to relate the total energy of a particle with its momentum. The quantities $E^2 - p^2c^2$ and m are invariant quantities. Note that when a particle's velocity is zero and it has no momentum, Equation (2.69) correctly gives E_0 as the particle's total energy.

Massless Particles

Equation (2.69) can also be used to determine the total energy for particles having no mass. For example, Equation (2.69) predicts that the total energy of a photon is

$$E = pc \qquad \text{Photon} \tag{2.71}$$

The energy of a photon is completely due to its motion and not at all to its rest energy.

We can show that the previous relativistic equations correctly predict that the speed of a photon must be the speed of light c. We use Equations (2.64) and (2.71) for the total energy of a photon and set the two equations equal.

$$E = \gamma mc^2 = pc$$

If we insert the value of the relativistic momentum from Equation (2.47), we have

$$\gamma mc^2 = \gamma muc$$

Massless particles must travel at the speed of light

The fact that $u = c$ follows directly from this equation after careful consideration of letting $m \longrightarrow 0$ and realizing that $\gamma \longrightarrow \infty$.

$$u = c \qquad \text{Massless particle} \tag{2.72}$$

2.13 Computations in Modern Physics

We have been taught in introductory physics that the international system of units is preferable when doing calculations in science and engineering. This is generally true, but in modern physics, a somewhat different, more convenient set of units is often used. The smallness of the quantities used in modern physics necessitates some practical changes. We will introduce those new units in this section and indicate their practicality through several examples. Recall that the work done in accelerating a charge through a potential difference is given by $W = qV$. For a proton, with the charge $e = 1.602 \times 10^{-19}$ C, being accelerated across a potential difference of 1 V, the work done is

$$W = (1.602 \times 10^{-19}\ \text{C})(1\ \text{V}) = 1.602 \times 10^{-19}\ \text{J}$$

In modern physics calculations, the amount of charge being considered is almost always some multiple of the electron charge. Atoms and nuclei all have an exact multiple of the electron charge (or neutral). For example, some charges are proton ($+e$), electron ($-e$), neutron (0), pion (0, $\pm e$), a singly ionized carbon atom ($+e$). The work done to accelerate the proton across a potential difference of 1 V could also be written as

$$W = (1\ e)(1\ \text{V}) = 1\ \text{eV}$$

Use eV for energy

where e stands for the electron charge. Thus eV, pronounced "electron volt," is also a unit of energy. It is related to the SI unit joule by the two previous equations.

$$1\ \text{eV} = 1.602 \times 10^{-19}\ \text{J} \tag{2.73}$$

The eV was used in introductory physics, and we have already used it several times in this chapter. In modern physics, the eV unit is used more often than the SI unit J. The term eV is often used with the SI prefixes where applicable. For example, in atomic and solid state physics, eV itself is mostly used, whereas in nuclear physics MeV (10^6 eV, *mega*-electron volt) and GeV (10^9 eV, *giga*-electron volt) are predominant, and in particle physics GeV and TeV (10^{12} eV, *tera*-electron volt) are used. When we speak of a particle having a certain energy, the common usage is to refer to the kinetic energy. A 6-GeV proton has a *kinetic* energy of 6 GeV, not a *total* energy of 6 GeV. Because the rest energy of a proton is about 1 GeV, its total energy would be about 7 GeV.

Like the SI unit for energy, the SI unit for mass, kilogram, is a very large unit of mass in modern physics calculations. For example, the mass of a proton is only 1.6726×10^{-27} kg. There are two other mass units commonly used in modern physics. First, the rest energy E_0 is given by Equation (2.63) as mc^2. The rest energy of the proton is given by

$$E_0\ (\text{proton}) = (1.67 \times 10^{-27}\ \text{kg})(3 \times 10^8\ \text{m/s})^2 = 1.50 \times 10^{-10}\ \text{J}$$

$$= 1.50 \times 10^{-10}\ \text{J}\ \frac{1\ \text{eV}}{1.602 \times 10^{-19}\ \text{J}} = 9.38 \times 10^8\ \text{eV}$$

The rest energies of the elementary particles are usually quoted in MeV or GeV (to several significant figures, the rest energy of the proton is 938.27 MeV). Because $E_0 = mc^2$, the mass is often quoted in units of MeV/c^2; for example, the mass of the proton is given by 938.27 MeV/c^2. We will find that the unit of MeV/c^2 is quite useful. The masses of several elementary particles are given on the inside of the front book cover. Although we will not do so, the mass is often quoted in units of eV by research physicists.

Use MeV/c^2 for mass

The other commonly used mass unit is the (unified) *atomic mass unit.* It is based on the definition that the mass of the neutral carbon-12 (^{12}C) atom is exactly 12 u, where u is one atomic mass unit.* We obtain the conversion between kilogram and atomic mass units u by comparing the mass of one ^{12}C atom.

Atomic mass units

$$\text{mass}\ (^{12}\text{C atom}) = \frac{12\ \text{g/mole}}{6.02 \times 10^{23}\ \text{atoms/mole}}$$

$$= 1.99 \times 10^{-23}\ \text{g/atom}$$

$$\text{mass}\ (^{12}\text{C atom}) = 1.99 \times 10^{-26}\ \text{kg} = 12\ \text{u/atom} \quad (2.74)$$

Therefore, the conversion is (when properly done to 6 significant figures)

$$1\ \text{u} = 1.66054 \times 10^{-27}\ \text{kg} \quad (2.75)$$

$$1\ \text{u} = 931.494\ \text{MeV}/c^2 \quad (2.76)$$

We have added the conversion from atomic mass units to MeV/c^2 for completeness.

From Equations (2.69) and (2.71) we see that a convenient unit of momentum is energy divided by the speed of light, or eV/c. We will use the unit eV/c for momentum when appropriate. Remember also that we often quote $\beta(= v/c)$ for velocity, so that c itself is an appropriate unit of velocity.

Use eV/c for momentum

Example 2.11

A 2-GeV proton hits another 2-GeV proton in a head-on collision. (a) Calculate v, β, p, K, and E for each of the protons. (b) What happens to the kinetic energy?

Solution: (a) By the convention just discussed, a 2-GeV proton has a kinetic energy of 2 GeV. We use Equation (2.64) to determine the total energy.

$$E = K + E_0 = 2\ \text{GeV} + 938\ \text{MeV} = 2.938\ \text{GeV}$$

where we have used 938 MeV for the proton's rest energy.

The momentum is determined from Equation (2.69).

$$p^2c^2 = E^2 - E_0{}^2 = (2.938\ \text{GeV})^2 - (0.938\ \text{GeV})^2$$

$$= 7.75\ \text{GeV}^2$$

*To avoid confusion between velocity and atomic mass unit, we will henceforth use v for velocity when the possibility exists for confusing the mass unit u with the velocity variable u.

The momentum is calculated to be

$$p = \sqrt{7.75\ (\text{GeV}/c)^2} = 2.78\ \text{GeV}/c$$

In order to find β we need the relativistic factor γ. There are several ways to determine γ; one is to compare the rest energy with the total energy. From Equation (2.64) we have

$$E = \gamma E_0 = \frac{E_0}{\sqrt{1 - u^2/c^2}}$$

$$\gamma = \frac{E}{E_0} = \frac{2.938\ \text{GeV}}{0.938\ \text{GeV}} = 3.13$$

We use Equation (2.61) to determine β.

$$\beta = \sqrt{\frac{\gamma^2 - 1}{\gamma^2}} = \sqrt{\frac{3.13^2 - 1}{3.13^2}} = 0.948$$

The velocity of a 2-GeV proton is $0.95c$ or 2.8×10^8 m/s.

(b) When the two protons collide head-on, the situation is very similar to the case when the two blocks of wood collided head-on with one important exception. The time for the two protons to interact is less than 10^{-20} s. If the two protons did momentarily stop at rest, then the two-proton system would have its rest mass increased by an amount given by Equation (2.67), $2K/c^2$ or 4 GeV/c^2. The result would be a highly excited system. In fact, the collision between the protons happens very quickly, and many possibilities exist. The two protons may either remain or disappear, and new additional particles may be created. Two of the possibilities are

$$p + p \rightarrow p + p + p + \bar{p} \tag{2.77}$$

$$p + p \rightarrow \pi^+ + d \tag{2.78}$$

where the symbols are p (proton), $\bar{p}$ (antiproton), π (pion), and d (deuteron). We will study more about the possibilities later when we study nuclear and particle physics. Whatever happens must be consistent with the conservation laws of charge, energy, and momentum, as well as with other conservation laws to be learned. Such experiments are routinely done in particle physics. In the analysis of these experiments, the equivalence of mass and energy is taken for granted.

Binding Energy

The equivalence of mass and energy becomes apparent when we study the binding energy of systems like atoms and nuclei that are formed from individual particles. For example, the hydrogen atom is formed by a proton and electron bound together by the electrical (coulomb) force. A deuteron is a proton and neutron bound together by the nuclear force. The potential energy associated with the force keeping the system together is called the **binding energy** E_B. The binding energy is the work required to pull the particles out of the bound system into separate, free particles at rest. The conservation of energy is written as

$$M_{\text{bound system}}c^2 + E_B = \sum_i m_i c^2 \tag{2.79}$$

where the m_i are the masses of the free particles. The binding energy then becomes *the difference between the rest energy of the individual particles and the rest energy of the combined, bound system.*

$$E_B = \sum_i m_i c^2 - M_{\text{bound system}}c^2 \tag{2.80}$$

For the case of two final particles having masses m_1 and m_2, we have

$$E_B = (m_1 + m_2 - M_{\text{bound system}})\,c^2 = \Delta M c^2$$

where ΔM is the difference between the final and initial masses.

When two particles (for example, a proton and neutron) are bound together to form a composite (like a deuteron), part of the rest energy of the individual particles is lost, resulting in the binding energy of the system. The rest energy of the combined system must be reduced by this amount. The deuteron is a good example. The rest energies of the particles are

Proton	$E_0 = 1.007276\ c^2\ \text{u} = 938.27$ MeV
Neutron	$E_0 = 1.008665\ c^2\ \text{u} = 939.57$ MeV
Deuteron	$E_0 = 2.01355\ c^2\ \text{u} = 1875.61$ MeV

The binding energy E_B is determined from Equation (2.80) to be,

$$E_B\ (\text{deuteron}) = 938.27\ \text{MeV} + 939.57\ \text{MeV} - 1875.61\ \text{MeV} = 2.23\ \text{MeV}$$

For the hydrogen atom, the binding energy of the proton and electron is 13.6 eV, which is so much smaller than the rest energy of about 1 GeV that it can be neglected when making mass determinations. The deuteron binding energy, however, represents a much larger fraction of the rest energies and is extremely important. The binding energies of heavy nuclei like uranium can be more than 1000 MeV, and even that much energy is not large enough to keep uranium from slowly decaying to lighter nuclei. The coulomb repulsion between the many protons in heavy nuclei is mostly responsible for their instability.

Example 2.12

What is the minimum kinetic energy the protons must have in the head-on collision of Equation (2.78), $p + p \rightarrow \pi^+ + d$, in order to produce the positively charged pion and deuteron? The mass of π^+ is 139.6 MeV/c^2.

Solution: For the minimum kinetic energy required, we need just enough energy to produce the rest energies of the final particles. We let the final kinetic energies of the pion and deuteron be zero. Because the collision is head on, the momentum will be zero before and after the collision, so the pion and deuteron will truly be at rest with no kinetic energy. Conservation of energy requires

$$m_pc^2 + K + m_pc^2 + K = m_dc^2 + m_{\pi^+}c^2$$

The rest energies of the proton and deuteron were given in this section, so we solve the previous equation for the kinetic energy.

$$\begin{aligned} K &= \tfrac{1}{2}(m_dc^2 + m_{\pi^+}c^2 - 2m_pc^2) \\ &= \tfrac{1}{2}[1875.6\ \text{MeV} + 139.6\ \text{MeV} - 2(938.3\ \text{MeV})] \\ &= 69\ \text{MeV} \end{aligned}$$

Nuclear experiments like this are normally done with fixed targets, not head-on collisions, and much more energy than 69 MeV is required, because linear momentum must also be conserved.

Example 2.13

The atomic mass of the ^{4}He atom is 4.002603 u. Find the binding energy of the ^{4}He nucleus.

Solution: We use Equation (2.80) to find the binding energy.

$$E_B\ (^4\text{He}) = 2m_pc^2 + 2m_nc^2 - M_{^4\text{He}}c^2$$

Later we will learn to deal with atomic masses in cases like this, but for now we will subtract the two electron masses from the atomic mass of ^{4}He to obtain the mass of the ^{4}He nucleus.

$$\begin{aligned} M_{^4\text{He}}\ (\text{nucleus}) &= 4.002603\ \text{u} - 2(0.000549\ \text{u}) \\ &= 4.001505\ \text{u} \end{aligned}$$

The mass of the electron is given on the inside of the front cover, along with the masses of the proton and neutron. We determine the binding energy of the ^{4}He nucleus to be

$$\begin{aligned} E_B\ (^4\text{He}) &= [2(1.007276\ \text{u}) + 2(1.008665\ \text{u}) - \\ &\qquad 4.001505\ \text{u}]\,c^2 \\ &= 0.0304\ c^2\ \text{u} \end{aligned}$$

$$E_B\ (^4\text{He}) = (0.0304\ c^2\ \text{u})\frac{931.5\ \text{MeV}}{c^2\ \text{u}} = 28.3\ \text{MeV}$$

The binding energy of the ^{4}He nucleus is very large, being almost 1% of its rest energy.

Example 2.14

The molecular binding energy is called the *dissociation energy*. It is the energy required to separate the atoms in a molecule. The dissociation energy of the NaCl molecule is 4.24 eV. Determine the fractional mass increase of the Na and Cl atoms separate from NaCl. What is the mass increase for a mole of NaCl?

Solution: E_B/c^2, the binding energy divided by c^2, is the mass difference between the molecule and separate atoms. The mass of NaCl is 58.44 u. The fractional mass increase is

$$f = \frac{\Delta M}{M} = \frac{E_B/c^2}{M} = \frac{4.24\ \text{eV}/c^2}{58.44\ \text{u}} \frac{c^2\ \text{u}}{931\ \text{MeV}} \frac{1\ \text{MeV}}{10^6\ \text{eV}}$$
$$= 7.8 \times 10^{-11}$$

One mole of NaCl has a mass of 58.44 g, so the mass decrease for a mole of NaCl is $f \times 58.44$ g or only 4.6×10^{-9} g. Such small masses cannot be directly measured, which is why nonconservation of mass was not observed for chemical reactions—the changes are too small.

Example 2.15

A positively charged Σ^+ particle produced in a particle physics experiment decays very quickly into a neutron and positively charged pion before either its energy or momentum can be measured. The neutron and pion are observed to move in the same direction as the Σ^+ was originally moving, with momenta of 4702 MeV/c and 169 MeV/c, respectively. What was the kinetic energy of the Σ^+ and its mass?

Solution: The decay reaction is

$$\Sigma^+ \rightarrow n + \pi^+$$

Obviously the Σ^+ has more mass than the sum of the masses of n and π^+ or the decay would not occur. We have to conserve both momentum and energy for this reaction. We use Equation (2.69) to find the total energy of the neutron and positively charged pion. The rest energies of n and π^+ are 940 MeV and 140 MeV, respectively. The total energies of E_n and E_{π^+} are, from $E = \sqrt{p^2c^2 + E_0{}^2}$,

$$E_n = \sqrt{(4702\ \text{MeV})^2 + (940\ \text{MeV})^2} = 4795\ \text{MeV}$$
$$E_{\pi^+} = \sqrt{(169\ \text{MeV})^2 + (140\ \text{MeV})^2} = 219\ \text{MeV}$$

The sum of these energies gives the total energy of the reaction, 4795 MeV + 219 MeV = 5014 MeV, both before and after the decay of Σ^+. In order to use Equation (2.69) to determine the rest energy of Σ^+, we need to know the momentum. We can determine the Σ^+ momentum from the conservation of momentum. Because all the momenta are along the same direction, we must have

$$p_{\Sigma^+} = p_n + p_{\pi^+} = 4702\ \text{MeV}/c + 169\ \text{MeV}/c$$
$$= 4871\ \text{MeV}/c$$

This must be the momentum of the Σ^+ before decaying, so now we can find the rest energy of Σ^+ from Equation (2.69).

$$E_0{}^2(\Sigma^+) = E^2 - p^2c^2 = (5014\ \text{MeV})^2 - (4871\ \text{MeV})^2$$
$$= (1189\ \text{MeV})^2$$

The rest energy of the Σ^+ is 1189 MeV, and its mass is 1189 MeV/c^2.

We find the kinetic energy of Σ^+ from Equation (2.64).

$$K = E - E_0 = 5014\ \text{MeV} - 1189\ \text{MeV} = 3825\ \text{MeV}$$

2.14 Electromagnetism and Relativity

We have been mostly concerned with the kinematic and dynamic aspects of the special theory of relativity from strictly the mechanics aspects. However, recall that Einstein first approached relativity through electricity and magnetism. He was convinced that Maxwell's equations were invariant (have the same form) in all inertial frames. Einstein wrote in 1952,

Einstein's conviction about electromagnetism

What led me more or less directly to the special theory of relativity was the conviction that the electromagnetic force acting on a body in motion in a magnetic field was nothing else but an electric field.

Einstein was convinced that magnetic fields were nothing other than electric fields observed in another inertial frame. That is the key to electromagnetism and relativity.

Maxwell's equations and the Lorentz force law are invariant in different inertial frames. In fact, with the proper Lorentz transformations of the electric and magnetic fields (from relativity theory) together with Coulomb's law (force between stationary charges), Maxwell's equations can be obtained. We certainly do not intend to accomplish that fairly difficult mathematical task here, nor do we intend to obtain the Lorentz transformation of the electric and magnetic fields. These subjects are studied in more advanced physics classes. However, we will show qualitatively that the magnetic force that one observer sees is simply an electric force according to an observer in another inertial frame. The electric field arises from charges, whereas the magnetic field arises from *moving* charges.

Magnetism and electricity are relative

In the late 1800s electricity and magnetism were well understood. James Clerk Maxwell had presented his famous four equations in 1864. Maxwell predicted that all electromagnetic waves travel at the speed of light, and he combined electricity, magnetism, and optics all into one successful theory. This classical theory has withstood the onslaught of time and experimental tests.* There were, however, some troubling aspects of the theory when it was observed from different Galilean frames of reference. In 1895 H. A. Lorentz "patched up" the difficulties with the Galilean transformation by developing a new transformation that now bears his name, the Lorentz transformation. However, Lorentz did not understand the full implication of what he had done. It was left to Einstein, who in 1905 published a paper titled "On the Electrodynamics of Moving Bodies," to fully merge relativity and electromagnetism. Einstein did not even mention the famous Michelson-Morley experiment in this classic 1905 paper, which we take as the origin of the special theory of relativity, and the Michelson-Morley experiment apparently seems to have played little role in his thinking. Einstein's belief that *Maxwell's equations describe electromagnetism in any inertial frame* was the key that led Einstein to the Lorentz transformations. Maxwell's assertion that all electromagnetic waves travel at the speed of light and Einstein's postulate that the speed of light is invariant in all inertial frames seem intimately connected.

Einstein believed in Maxwell's equations

We now proceed to discuss qualitatively the relative aspects of electric and magnetic fields and their forces. Consider a positive test charge q_0 moving to the right with speed v outside a neutral, conducting wire as shown in Figure 2.28a in the frame of the inertial system K, where the positive charges are at rest and the negative electrons in the wire have speed v to the right. The conducting wire is long and has the same number of positive ions and conducting electrons. For simplicity, we have taken the electrons and the charge q_0 to have the same speed, but the argument can be generalized.

What is the force on the positive test charge q_0 outside the wire? The total force is given by the Lorentz force,

$$\mathbf{F} = q_0(\mathbf{E} + \mathbf{v} \times \mathbf{B}) \tag{2.81}$$

Lorentz force

and can be due to an electric field, a magnetic field, or both. Because the total charge inside the wire is zero, the electric force on the test charge q_0 in Figure 2.28a is also zero. But we learned in introductory physics that the moving electrons in the wire (current) produce a magnetic field **B** at the position of q_0

*The meshing of electricity and magnetism together with quantum mechanics, called the *theory of quantum electrodynamics* (QED), is one of the most successful theories in physics.

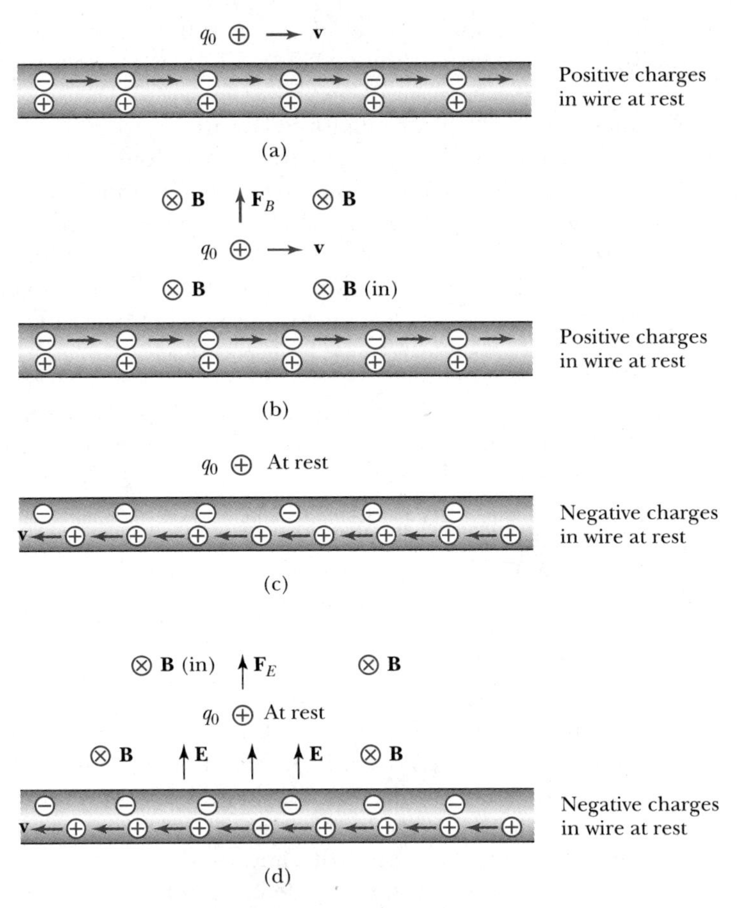

FIGURE 2.28 (a) A positive charge is placed outside a neutral, conducting wire. The figure is shown in the system where the positive charges in the wire are at rest. Note that the charge q_0 has the same velocity as the electrons. (b) The moving electrons produce a magnetic field, which causes a force $\mathbf{F}_B$ on q_0. (c) The same situation as in (a) but is shown in a system in which the electrons are at rest. (d) Now there is an abundance of positive charges due to length contraction, and the resulting electric field repels q_0. There is also a magnetic field, but this causes no force on q_0, which is at rest in this system.

that is into the page (Figure 2.28b). The moving charge q_0 *will be repelled upwards by the magnetic force* ($q_0\mathbf{v} \times \mathbf{B}$) due to the magnetic field of the wire.

Let's now see what happens in a different inertial frame K′ moving at speed v to the right with the test charge (see Figure 2.28c). Both the test charge q_0 and the negative charges in the conducting wire are at rest in system K′. In this system an observer at the test charge q_0 observes the same density of negative ions in the wire as before. However, in system K′ the positive ions are now moving to the left with speed v. Due to length contraction, the positive ions will appear to be closer together to a stationary observer in K′. Because the positive charges appear to be closer together, there is a higher density of positive charges than of negative charges in the conducting wire. The result is an electric field as shown in Figure 2.28d. The test charge q_0 will now be *repelled* in the presence of the electric field. What about the magnetic field now? The moving charges in Figure 2.28c also produce a magnetic field that is into the page, but this time the charge q_0 is at rest with respect to the magnetic field, so charge q_0 feels no magnetic force.

What appears as a magnetic force in one inertial frame (Figure 2.28b) appears as an electric force in another (Figure 2.28d). Electric and magnetic fields are *relative* to the coordinate system in which they are observed. The Lorentz contraction of the moving charges accounts for the difference. This example can be extended to two conducting wires with electrons moving and a similar result will be obtained (see Problem 84). It is this experiment, on the force between two parallel, conducting wires, in which current is defined. Because charge is defined using current, the experiment is also the basis of the definition of the electric charge.

We have come full round in our discussion of the special theory of relativity. The laws of electromagnetism represented by Maxwell's equations have a special place in physics. The equations themselves are invariant in different inertial systems; only the interpretations as electric and magnetic fields are relative. No surprising redefinitions are needed in electromagnetism as was required for mechanics. Accordingly, we spend much less time discussing the subject. The real surprises occurred in Maxwell's time.

Summary

Efforts by Michelson and Morley proved in 1887 that either the elusive ether does not exist or there must be significant problems with our understanding of nature.

Albert Einstein solved the problem in 1905 by applying two postulates:

1. The Principle of Relativity: The laws of physics are the same in all inertial systems.
2. The constancy of the speed of light: Observers in all inertial systems measure the same value for the speed of light in vacuum.

Einstein's two postulates are used to derive the Lorentz transformation relating the space and time coordinates of events viewed from different inertial systems. If system K′ is moving at speed v along the $+x$ axis with respect to system K, the two sets of coordinates are related by

$$x' = \frac{x - vt}{\sqrt{1 - v^2/c^2}}$$
$$y' = y \tag{2.17}$$
$$z' = z$$
$$t' = \frac{t - \dfrac{vx}{c^2}}{\sqrt{1 - v^2/c^2}}$$

The inverse transformation is obtained by switching the primed and unprimed quantities and changing v to $-v$.

The time interval between two events occurring at the same position in a system as measured by a clock at rest is called the proper time T_0. The time interval T' between the same two events measured by a moving observer is related to the proper time T_0 by the time dilation effect.

$$T' = \frac{T_0}{\sqrt{1 - v^2/c^2}} \tag{2.19}$$

We say that moving clocks run slow, because the least time is always measured on clocks at rest.

The length of an object measured by an observer at rest relative to the object is called the proper length L_0. The length of the same object measured by an observer who sees the object moving at speed v is L, where

$$L = L_0\sqrt{1 - v^2/c^2} \tag{2.20}$$

This effect is known as length or space contraction, because moving objects are contracted in the direction of their motion.

If u and u' are the velocities of an object measured in systems K and K′, respectively, and v is the relative velocity between K and K′; the relativistic addition of velocities (Lorentz velocity transformation) is

$$u_x = \frac{dx}{dt} = \frac{\gamma(dx' + v\,dt')}{\gamma[dt' + (v/c^2)\,dx']} = \frac{u'_x + v}{1 + (v/c^2)u'_x}$$
$$u_y = \frac{u'_y}{\gamma[1 + (v/c^2)u'_x]} \tag{2.22}$$
$$u_z = \frac{u'_z}{\gamma[1 + (v/c^2)u'_x]}$$

where

$$\gamma = \frac{1}{\sqrt{1 - v^2/c^2}} \tag{2.8}$$

Spacetime diagrams are useful to represent events geometrically. Time may be considered to be a fourth dimension for some purposes. The spacetime interval for two events defined by $\Delta s^2 = \Delta x^2 + \Delta y^2 + \Delta z^2 - c^2\Delta t^2$ is an invariant between inertial systems.

The relativistic Doppler effect for light is given by

$$\nu = \frac{\sqrt{1+\beta}}{\sqrt{1-\beta}}\nu_0 \tag{2.33}$$

where β is positive for source and receiver approaching one another and negative when receding.

The classical form for linear momentum is replaced by the special relativity form:

$$\mathbf{p} = \gamma m\mathbf{u} = \frac{m\mathbf{u}}{\sqrt{1-u^2/c^2}} \tag{2.47}$$

The relativistic kinetic energy is given by

$$K = \gamma mc^2 - mc^2 = mc^2\left(\frac{1}{\sqrt{1-u^2/c^2}} - 1\right) \tag{2.57}$$

The *total energy* E is given by

$$E = \gamma mc^2 = \frac{mc^2}{\sqrt{1-u^2/c^2}} = \frac{E_0}{\sqrt{1-u^2/c^2}} = K + E_0 \tag{2.64}$$

where $E_0 = mc^2$. This equation denotes the equivalence of mass and energy. The laws of the conservation of mass and of energy are combined into one conservation law: the conservation of mass-energy.

Energy and momentum are related by

$$E^2 = p^2c^2 + E_0{}^2 \tag{2.69}$$

In the case of massless particles (for example, the photon), $E_0 = 0$, so $E = pc$. Massless particles must travel at the speed of light.

The electron volt, denoted by eV, is equal to 1.602×10^{-19} J. The unified atomic mass unit u is based on the mass of the ^{12}C atom.

$$1\ \text{u} = 1.66054 \times 10^{-27}\ \text{kg} = 931.494\ \text{MeV}/c^2 \qquad (2.75, 2.76)$$

Momentum is often quoted in units of eV/c, and the velocity is often given in terms of β $(= v/c)$.

The difference between the rest energy of individual particles and the rest energy of the combined, bound system is called the *binding energy*.

Maxwell's equations are invariant under transformations between any inertial reference frames. What appears as electric and magnetic fields is relative to the reference frame of the observer.

Questions

1. The motion of the Earth around the sun was used to try to determine the effects of the ether by Michelson. Can you think of a more convenient experiment with a higher speed than Michelson might have used?
2. If you wanted to set out today to find the effects of the ether, what experimental apparatus would you want to use? Would a laser be included? Why?
3. For what reasons would Michelson and Morley repeat their experiment on top of a mountain? Why would they perform the experiment in summer and winter?
4. Does the fact that Maxwell's equations do not need to be modified due to the special theory of relativity, whereas Newton's laws of motion do, mean that Maxwell's work is somehow greater or more significant than Newton's?
5. Explain why the results of the special theory of relativity don't affect the great majority of measurements we make each day.
6. Why did it take so long to discover the theory of relativity? Why didn't Newton figure it out?
7. Can you think of a way by which you can make yourself older than those born on your same birthday?
8. Will meter sticks manufactured on Earth work correctly on spaceships moving at high speed? Explain.
9. Devise a system for you and three other colleagues, at rest with you, to synchronize your clocks if your clocks are too large to move and are separated by hundreds of miles.
10. In the experiment to verify time dilation by flying the cesium clocks around the Earth, what is the order of the speed of the four clocks in a system fixed at the center of the Earth, but not rotating?
11. Can you think of an experiment to directly verify length contraction? Explain.
12. Would it be easier to perform the muon decay experiment in the space shuttle orbiting above Earth and then compare with the number of muons on Earth? Explain.
13. Can events on a spacetime diagram above $t = 0$ but not in the shaded area in Figure 2.21 affect the future? Explain.
14. Why don't we include also the spatial coordinate z when drawing the light cone?
15. What would be a suitable name for events connected by $\Delta s^2 = 0$?
16. Is the relativistic Doppler effect only valid for light waves? Can you think of another situation where it might be valid?

17. In Figure 2.18 why cannot a real worldline have a slope less than one?
18. Explain how in the twin paradox, we might arrange to compare clocks at the beginning and end of Mary's journey and not have to worry about acceleration effects.
19. Which of the following is the more massive: a relaxed or compressed spring, a charged or uncharged capacitor, a piston-cylinder when closed or open?
20. In the fission of ^{235}U, the masses of the final products are less than the mass of ^{235}U. Does this make sense? What happens to the mass?
21. In the fusion of deuterium and tritium nuclei to produce a thermonuclear reaction, where does the kinetic energy that is produced come from?

Problems

2.1 Historical Perspective

1. Show that the form of Newton's second law is invariant under the Galilean transformation.
2. Show that the definition of linear momentum, $p = mv$, is invariant under a Galilean transformation.

2.2 The Michelson-Morley Experiment

3. Show that the equation for t_2 in Section 2.2 expresses the time required for the light to travel to the mirror D and back in Figure 2.2. In this case the light is traveling perpendicular to the supposed direction of the ether. In what direction must the light travel in order to be reflected by the mirror if the light must pass through the ether?
4. A swimmer wants to swim straight across a river with current flowing with speed v_1. If the swimmer swims in still water with speed v_2, at what angle should the swimmer point upstream from the shore and at what speed will the swimmer swim across the river?
5. Show that the time difference $\Delta t'$ given by Equation (2.4) is correct when the Michelson interferometer is rotated by 90°.
6. In the 1887 experiment by Michelson and Morley, the length of each arm was 11 m. The experimental limit for the fringe shift was 0.005 fringes. If sodium light was used with the interferometer ($\lambda = 589$ nm), what upper limit did the null experiment place on the speed of the Earth through the expected ether?
7. Show that if length is contracted by the factor $\sqrt{1 - v^2/c^2}$ in the direction of motion, then the result in Equation (2.3) will have the express factor needed to make $\Delta t = 0$ as needed by Michelson and Morley.

2.3 Einstein Postulates

8. Explain why Einstein argued that the constancy of the speed of light (Postulate 2) actually follows from the Principle of Relativity (Postulate 1).
9. Prove that the constancy of the speed of light (Postulate 2) is inconsistent with the Galilean transformation.

2.4 The Lorentz Transformation

10. Use the spherical wavefronts of Equation (2.9) to derive the Lorentz transformation given in Equation (2.17). Supply all the steps.
11. Show that both Equations (2.17) and (2.18) reduce to the Galilean transformation when $v \ll c$.
12. Determine the ratio $\beta = v/c$ for the following: (a) A car traveling 100 km/h. (b) A commercial jet airliner traveling 290 m/s. (c) A supersonic airplane traveling mach 2.3 (Mach number $= v/v_{\text{sound}}$). (d) The space shuttle, traveling 27,000 km/h. (e) An electron traveling 25 cm in 2 ns. (f) A proton traveling across a nucleus (10^{-14} m) in 0.35×10^{-22} s.
13. Two events occur in an inertial system K as follows:

$$\text{Event 1: } x_1 = a,\ t_1 = 2a/c,\ y_1 = 0,\ z_1 = 0$$

$$\text{Event 2: } x_2 = 2a,\ t_2 = 3a/2c,\ y_2 = 0,\ z_2 = 0$$

In what frame K′ will these events appear to occur at the same time? Describe the motion of system K′.

14. An event occurs in system K′ at $x' = 2$ m, $y' = 3.5$ m, $z' = 3.5$ m, and $t' = 0$. System K′ and K have their axes coincident at $t = t' = 0$, and system K′ travels along the x axis of system K with a speed $0.8c$. What are the coordinates of the event in system K?
15. A light signal is sent from the origin of a system K at $t = 0$ to the point $x = 3$ m, $y = 5$ m, $z = 10$ m. (a) At what time t is the signal received? (b) Find (x', y', z', t') for the receipt of the signal in a frame K′ that is moving along the x axis of K at a speed of $0.8c$. (c) From your results in (b) verify that the light traveled with a speed c as measured in the K′ frame.
16. Is there a frame K′ in which the two events described in Problem 13 occur at the same place? Explain.
17. Find the relativistic factor γ for each of the parts of Problem 12.

2.5 Time Dilation and Length Contraction

18. Show that the experiment shown in Figure 2.10 and discussed in the text leads directly to the derivation of length contraction.

19. A rocket ship carrying passengers blasts off to go from New York to Los Angeles, a distance of about 5000 km. How fast must the rocket ship go to have its length shortened by 1%?

20. Astronomers discover a planet orbiting around a star similar to our sun that is 20 light years away. How fast must a rocket ship go if the round trip is to take no longer than 40 years in time for the astronauts aboard? How much time will the trip take on Earth?

21. Particle physicists use particle track detectors to determine the lifetime of short-lived particles. A muon has a mean lifetime of 2.2 μs and makes a track 9.5 cm long before decaying into an electron and two neutrinos. What was the speed of the muon?

22. The Apollo astronauts returned from the moon under the Earth's gravitational force and reached speeds of almost 25,000 mi/h with respect to Earth. Assuming (incorrectly) they had this speed for the entire trip from the moon to Earth, what was the time difference for the trip between their clocks and clocks on Earth?

23. A clock in a spaceship is observed to run at a speed of only 3/5 of a similar clock at rest on Earth. How fast is the spaceship moving?

24. A spaceship of length 40 m at rest is observed to be 20 m long when in motion. How fast is it moving?

25. The Concorde travels 8000 km between two places in North America and Europe at an average speed of 375 m/s. What is the total difference in time between two similar atomic clocks, one on the airplane and one at rest on Earth during a one-way trip? Consider only time dilation and ignore other effects like the rotation of the Earth.

26. A mechanism on Earth used to shoot down geosynchronous satellites that house laser-based weapons is finally perfected and propels golf balls at $0.94c$. (a) How far will a detector riding with the golf ball initially measure the distance to the satellite? (Geosynchronous satellites are placed 3.58×10^4 km above the surface of the Earth.) (b) How much time will it take the golf ball to make the journey to the satellite in the Earth's frame? How much time will it take in the golf ball's frame?

27. Two events occur in an inertial system K at the same time but 4 km apart. What is the time difference measured in a system K′ moving between these two events when the distance separation of the events is measured to be 5 km?

28. Imagine that in another universe the speed of light is only 100 m/s. A person traveling along an interstate highway at 120 km/h ages how much slower than a person at rest?

29. In another universe where the speed of light is only 100 m/s, an airplane that is 40 m long at rest and flies at 300 km/hr will appear to be how long to an observer at rest?

30. Two systems K and K′ synchronize their clocks at $t = t' = 0$ when their origins are aligned as system K′ passes by system K along the x axis at relative speed $0.8c$. At time $t = 3$ ns, Frank (in system K) shoots a proton gun having proton speeds of $0.98c$ along his x axis. The protons leave the gun at $x = 1$ m and arrive at a target 120 m away. Determine the events (x, t) of the gun firing and the protons arriving as measured by observers in both systems K and K′.

2.6 Addition of Velocities

31. A spaceship is moving at a speed of $0.8c$ away from an observer at rest. A boy in the spaceship shoots a proton gun with protons having a speed of $0.7c$. What is the speed of the protons measured by the observer at rest when the gun is shot (a) away from the observer and (b) toward the observer?

32. A proton and an antiproton are moving toward each other in a head-on collision. If each has a speed of $0.8c$ with respect to the collision point, how fast are they moving with respect to each other?

33. Imagine the speed of light in another universe to be only 100 m/s. Two cars are traveling along an interstate highway in opposite directions. Person 1 is traveling 110 km/h, and person 2 is traveling 140 km/h. How fast does person 1 measure person 2 to be traveling? How fast does person 2 measure person 1 to be traveling?

34. In the Fizeau experiment described in Example 2.5, suppose that the water is flowing at a speed of 5 m/s. Find the difference in the speeds of two beams of light, one traveling in the same direction as the water and the other in the opposite direction. Use $n = 1.33$ for water.

35. Three galaxies are aligned along an axis in order A, B, C. An observer in galaxy B is in the middle and observes that galaxies A and C are moving in opposite directions away from him, both with speeds $0.60c$. What is the speed of galaxies B and C as observed by someone in galaxy A?

36. Consider the *Gedanken* experiment discussed in Section 2.6 in which a giant floodlight stationed 400 km above the Earth's surface shines its light across the moon's surface. How fast does the light flash across the moon?

2.7 Experimental Verification

37. A group of scientists decide to repeat the muon decay experiment at the Mauna Kea telescope site in Hawaii, which is 4205 m (13,796 feet) above sea level. They decided to count 10^4 muons during a certain time period. Repeat the calculation of Section 2.7 and find the classical and relativistic number of muons expected at sea level. Why did they decide to count as many as 10^4 muons instead of only 10^3?

38. Consider a reference system placed at the U.S. Naval Observatory in Washington, D.C. Two planes take off from Washington Dulles airport, one going eastward and one going westward, both carrying a cesium atomic clock. The distance around the Earth at 39° latitude (Washington, D.C.) is 31,000 km, and Washington rotates about the Earth's axis at a speed of 360 m/s. Calculate the predicted differences between the clock left at the observatory and the two clocks in the airplanes (each traveling at 300 m/s) when the airplanes return to Washington. Include the rotation of the Earth but no general relativistic effects. Compare with the predictions given in the text.

2.8 Twin Paradox

39. Derive the results in Table 2.1 for the frequencies ν' and ν''. During what time period do Frank and Mary receive these frequencies?
40. Derive the results in Table 2.1 for the time of the total trip and the total number of signals sent in the frame of both twins.

2.9 Spacetime

41. Using the Lorentz transformation, prove that $s^2 = s'^2$.
42. Prove that for a timelike interval, two events can never be considered to occur simultaneously.
43. Prove that for a spacelike interval, two events cannot occur at the same place in space.
44. Given two events, (x_1, t_1) and (x_2, t_2), use a spacetime diagram to find the speed of a frame of reference in which the two events occur simultaneously. What values may Δs^2 have in this case?
45. (a) Draw on a spacetime diagram in the fixed system a line expressing all the events in the moving system that occur at $t' = 0$ if the clocks are synchronized at $t = t' = 0$. (b) What is the slope of this line? (c) Draw lines expressing events occurring for the four times t_4', t_3', t_2', and t_1' where $t_4' < t_3' < 0 < t_2' < t_1'$. (d) How are these four lines related geometrically?
46. Consider a fixed and moving system with their clocks synchronized and their origins aligned at $t = t' = 0$. (a) Draw on a spacetime diagram in the fixed system a line expressing all the events occurring at $t' = 0$. (b) Draw on this diagram a line expressing all the events occurring at $x' = 0$. (c) Draw all the worldlines for light that pass through $t = t' = 0$. (d) Are the x' and ct' axes perpendicular? Explain.
47. Use the results of the two previous problems to show that events simultaneous in one system are not simultaneous in another system moving with respect to the first. Consider a spacetime diagram with x, ct and x', ct' axes drawn such that the origins coincide and the clocks synchronized at $t = t' = 0$. Then consider events 1 and 2 that occur simultaneously in the fixed system. Are they simultaneous in the moving system?

2.10 Doppler Effect

48. An astronaut is said to have tried to get out of a traffic violation for running a red light ($\lambda = 670$ nm) by telling the judge that the light appeared green ($\lambda = 540$ nm) to her as she passed by in her high-powered transport. How fast was the astronaut going?
49. Derive Equation (2.31) for the case where the source is fixed, but the receiver approaches it with velocity v.
50. Do the complete derivation for Equation (2.32) when the source and receiver are receding with relative velocity v.
51. A spacecraft traveling out of the solar system at a speed of $0.92c$ sends back information at a rate of 400 Hz. At what rate do we receive the information?
52. Three radio-equipped plumbing vans are broadcasting on the same frequency ν_0. Van 1 is moving east of van 2 with speed v, van 2 is fixed, and van 3 is moving west of van 2 with speed v. What is the frequency of each van as received by the others?
53. Three radio-equipped plumbing vans are broadcasting on the same frequency ν_0. Van 1 is moving north of van 2 with speed v, van 2 is fixed, and van 3 is moving west of van 2 with speed v. What is the frequency of each van as received by the others?
54. A spaceship moves radially away from Earth with acceleration 25 m/s^2. How long does it take for the sodium streetlamps ($\lambda \simeq 589$ nm) on Earth to be invisible (with a powerful telescope) to the human eye of the astronauts? The range of visible wavelengths is about 400–700 nm.

2.11 Relativistic Momentum

55. Newton's second law is given by $\mathbf{F} = d\mathbf{p}/dt$. If the force is always perpendicular to the velocity, show that $\mathbf{F} = m\gamma\mathbf{a}$, where $\mathbf{a}$ is the acceleration.
56. Using the result of the previous problem, show that the radius of a particle's circular path having charge q traveling with speed v in a magnetic field perpendicular to the particle's path is $r = p/qB$. What happens to the radius as the speed increases as in a cyclotron?
57. Newton's second law is given by $\mathbf{F} = d\mathbf{p}/dt$. If the force is always parallel to the velocity, show that $\mathbf{F} = \gamma^3 m\mathbf{a}$.
58. Find the force necessary to give a proton an acceleration of 10^{19} m/s^2 when the proton has a velocity (along the force) of (a) $0.01c$, (b) $0.1c$, (c) $0.9c$, and (d) $0.99c$.
59. A particle having a speed of $0.92c$ has a momentum of 10^{-16} kg·m/s. What is its mass?
60. A particle initially has a speed of $0.5c$. At what speed does its momentum increase by (a) 1%, (b) 10%, (c) 100%?
61. What magnetic field would be necessary to require 230-MeV protons to revolve in a circle of 15 m? (See Problem 56.)
62. Show that linear momentum is conserved in Example 2.9 as measured by Mary.

2.12 Relativistic Energy

63. Show that $1/2\, m\gamma v^2$ does not give the correct kinetic energy.

64. How much ice must melt at 0° C in order to gain 2 g of mass? Where does this mass come from? The heat of fusion for water is 334 J/g.

65. Physicists at the Stanford Linear Accelerator Center (SLAC) want to bombard 9 GeV electrons head on with 3.1 GeV positrons to create B mesons and anti-B mesons. What speeds will the electron and positron have when they collide?

66. The Tevatron accelerator at the Fermi National Accelerator Laboratory (Fermilab) outside Chicago actually consists of 5 stages to sequentially boost protons to 1 TeV (1000 GeV). What is the speed of the proton at the end of each stage: Cockcroft-Walton (750 keV), Linac (400 MeV), Booster (8 GeV), Main ring or injector (150 GeV), and finally the Tevatron itself (1 TeV)? Note: the numbers given in parentheses represent the total kinetic energy at the end of each stage.

67. Calculate the momentum, kinetic energy, and total energy of an electron traveling at a speed of (a) $0.01c$, (b) $0.1c$, and (c) $0.9c$.

68. The energy of a body is found to be twice its rest energy. How fast is it moving with respect to the observer?

69. A system is devised to exert a constant force of 8 N on an 80-kg body of mass initially at rest while moving horizontally on a frictionless table. How far does the body have to be pushed to increase its mass-energy by 25%?

70. What is the speed of a proton when its kinetic energy is equal to twice its rest energy?

71. What is the speed of an electron when its kinetic energy is (a) 10% of its rest energy, (b) equal to the rest energy, and (c) ten times the rest energy?

72. Derive the following equation:

$$\beta = \frac{v}{c} = \sqrt{1 - \left(\frac{E_0}{E_0 + K}\right)^2}$$

73. Prove that $\beta = pc/E$. This is a useful relation to find the velocity of a highly energetic particle.

74. A good rule of thumb is to use relativistic equations whenever the kinetic energies determined classically and relativistically differ by more than 1%. Find the speeds when this occurs for (a) electrons, and (b) protons.

75. How much mass-energy (in joules) is contained in a peanut weighing 0.1 ounce? What happens if you eat 10 ounces of peanuts?

76. Calculate the energy needed to accelerate a spaceship of mass 10,000 kg to a speed of $0.3c$ for intergalactic space exploration. Compare this with a projected annual energy usage on Earth of 10^{21} J.

77. Derive Equation (2.57) for the relativistic kinetic energy and show all the steps, especially the integration by parts.

78. A test automobile of mass 1000 kg moving at high speed crashes into a wall. The average temperature of the car is measured to rise by 0.5° C after the wreck. What is the change in mass of the car? Where does this change in mass come from? (Assume the average specific heat of the automobile is close to that of steel, 0.11 cal/g/°C.)

2.13 Computations in Modern Physics

79. A helium nucleus has a mass of 4.001505 u. What is its binding energy?

80. A free neutron is an unstable particle and beta decays into a proton with the emission of an electron. How much kinetic energy is available in the decay?

81. Protons of 10^{12} eV (TeV) energy are produced at the Fermilab. Calculate the speed, momentum, and total energy of the protons.

82. What is the kinetic energy of (a) an electron having a momentum of 30 GeV/c? (b) a proton having a momentum of 30 GeV/c?

83. A muon has a mass of 106 MeV/c^2. Calculate the speed, momentum, and total energy of a 200-MeV muon.

2.14 Electromagnetism and Relativity

84. Instead of one positive charge outside a conducting wire, as was discussed in Section 2.14 and shown in Figure 2.28, consider a second conducting wire parallel to the first one. Both wires have positive and negative charges, and the wires are electrically neutral. Assume that in both wires the positive charges travel to the right and negative charges to the left. (a) Consider an inertial frame moving with the negative charges of wire 1. Show that the second wire is attracted to the first wire in this frame. (b) Now consider an inertial frame moving with the positive charges of the second wire. Show that the first wire is attracted to the second. (c) Use this argument to show that electrical and magnetic forces are relative.

General Problems

85. An Ω^- particle has rest energy 1672 MeV and mean lifetime 8.2×10^{-11} s. It is created and decays in a particle track detector and leaves a track 24 mm long. What is the total energy of the Ω^- particle?

86. Show that the following form of Newton's second law satisfies the Lorentz transformation. Assume the force is parallel to the velocity

$$F = m\frac{dv}{dt}\frac{1}{\left(1 - \frac{v^2}{c^2}\right)^{3/2}}$$

87. Using the previous results listed in Table 2.1, find the time of detecting Mary's turnaround and the number of signals received at the rate ν'.

88. Using the previous results listed in Table 2.1, find the time for the remainder of the trip, the number of signals received at the rate ν'', and the total number of signals received. Determine the conclusion as to the other twin's measure of the time taken for the trip.

89. Frank and Mary are twins. Mary jumps on a spaceship and goes to Alpha-Centauri (4 lightyears away) and returns. She travels at a speed of $0.8c$ with respect to Earth and emits a radio signal every week. Frank also sends out a radio signal to Mary once a week. (a) How many signals does Mary receive from Frank before she turns around? (b) At what time does the frequency of signals Frank receives suddenly change? How many signals has he received at this time? (c) How many signals do Frank and Mary receive for the entire trip? (d) How much time does the trip take according to Frank and Mary? (e) How much time does each twin say the other twin will measure for the trip? Do they agree with (d)?

90. An electron has a total energy that is 200 times its rest energy. Determine its (a) kinetic energy, (b) speed, and (c) momentum.

91. A proton moves with a speed of $0.9c$. Find the speed of an electron that has (a) the same momentum of the proton, and (b) the same kinetic energy.

92. A high-speed K^0 meson is traveling at a speed of $0.9c$ when it decays into a π^+ and a π^- meson. What are the greatest and least speed that the mesons may have? [$M(K^0) = 498\ \text{MeV}/c^2$, $M(\pi^{\pm}) = 140\ \text{MeV}/c^2$].

CHAPTER 3

The Experimental Basis of Quantum Theory

As far as I can see, our ideas are not in contradiction to the properties of the photoelectric effect observed by Mr. Lenard.

Max Planck, 1905

As was discussed in Chapter 1, during the final decades of the 1800s scientists discovered phenomena that could not always be explained by what we now call classical physics. Many scientists, however, were not concerned with these discrepancies. The level of experimentation was such that uncertainties were large, and the results of the experiments were often slow in being reported to other investigators. But perhaps more important was the confident attitude of physical scientists that Newton's laws and Maxwell's equations contained the fundamental description of nature.

In this atmosphere it is indeed surprising that the few exceptions to the classical laws discovered during the latter part of the nineteenth century led to the fabulous thirty-year period of 1900–1930, when our understanding of the laws of physics was dramatically changed. We have already discussed in Chapter 2 the first of these new developments, the special theory of relativity, which was introduced by Einstein in 1905 and successfully explained the null result of the Michelson-Morley experiment. The other great conceptual advance of 20th-century physics, the quantum theory, began in 1900 when Max Planck introduced his explanation of blackbody radiation.

We begin this chapter by learning of Röntgen's discovery of the x ray and Thomson's discovery of the electron. Millikan later determined the electron's charge. We shall see that, although it was necessary to assume that certain physical quantities may be quantized, scientists found this idea hard to accept. We will discuss the difficulties of explaining blackbody radiation with classical physics and how Planck's proposal solved the problem. Finally, we will see that Einstein's explanation of the photoelectric effect and Compton's understanding of x-ray scattering data made the quantum hypothesis difficult to refute. After many difficult and painstaking experiments, it became clear that quantization was not only necessary, it was also the correct description of nature.

3.1 Discovery of the X Ray and the Electron

In the late 1800s scientists and engineers were familiar with the "cathode rays" that could easily be generated from one of the metal plates in an evacuated tube across which a large electric potential had been established. The origin and constitution of these cathode rays were not known. The concept of an atomic substructure of matter was widely accepted because of its use in explaining the results of chemical experiments. Therefore, it was felt that cathode rays might have something to do with atoms. It was known, for example, that cathode rays could penetrate matter. Cathode rays were of great interest and under intense investigation in the late 1800s.

In 1895 Wilhelm Röntgen (1845–1923), who had received early training as a mechanical engineer but was at the time a professor of physics at the University of Würzburg in Germany, was studying the effects of cathode rays passing through various materials. During one such experiment he noticed that a nearby phosphorescent screen was glowing vividly in the darkened room. Röntgen soon realized he was observing a new kind of ray, one that, unlike cathode rays, was unaffected by magnetic fields and was far more penetrating than cathode rays. These **x rays,** as he called them, were apparently produced by the cathode rays bombarding the glass walls of his vacuum tube. Röntgen studied their transmission through many materials and even showed that he could obtain an image of the bones in a hand when the x rays were allowed to pass through as shown in Figure 3.1. This experiment created tremendous excitement, and medical applications of x rays were quickly developed. For this discovery, Röntgen received the first Nobel Prize award for physics in 1901.

New penetrating ray: x ray

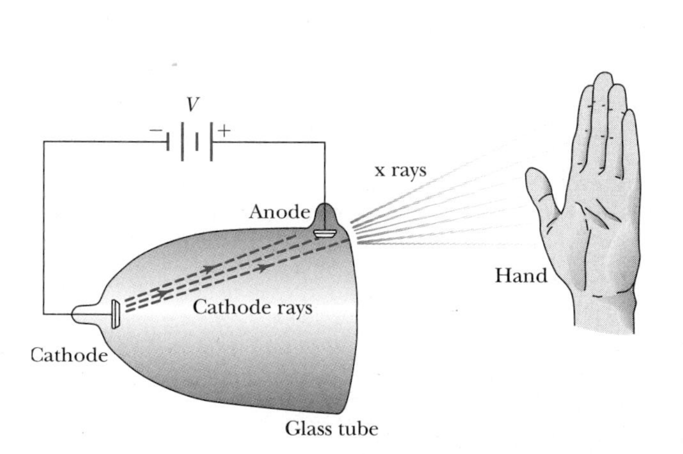

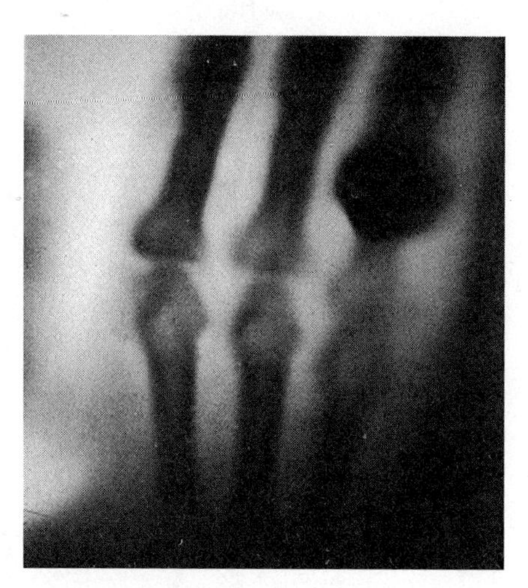

FIGURE 3.1 In Röntgen's experiment "x rays" were produced by cathode rays (electrons) hitting the glass near the anode. He studied the penetration of the x rays through several substances and even noted that if the hand was held between the glass tube and a screen, the darker shadow of the bones could be discriminated from the less dark shadow of the hand. *Photo courtesy of Deutsches Museum, München.*

Sir Joseph John Thomson, universally known as "J.J.," went to Cambridge University at age 20 and remained there for the rest of his life. Thomson's career with the Cavendish Laboratory spanned a period of over 50 years during which seven Nobel Prizes in physics were awarded. He served as Director from 1884 until 1918 when he stepped down in favor of Rutherford. *AIP Emilio Segrè Visual Archives.*

For several years before the discovery of x rays, J. J. Thomson (1856–1940), professor of experimental physics at Cambridge University, had been studying the properties of electrical discharges in gases. Thomson's apparatus was similar to that used by Röntgen and many other scientists because of its simplicity (see Figure 3.2). Thomson believed that cathode rays were particles, whereas several respected German scientists (such as H. Hertz) believed they were wave phenomena. Thomson was able to prove in 1897 that the charged particles emitted from a heated electrical cathode were in fact the same as cathode rays. The main features of Thomson's experiment are shown in the schematic apparatus of Figure 3.2. The rays from the cathode are attracted to the positive potential on aperture A (anode) and are further collimated by aperture B to travel in a straight line to impinge on the fluorescent screen in the rear of the tube, where they can be visually detected by a flash of light. A voltage across the deflection plates sets up an electrostatic field that can deflect charged particles. Previously, in a similar experiment, Hertz had observed no effect on the cathode rays due to the deflecting voltage. Thomson at first found the same result, but upon further evacuating the glass tube observed the deflection and proved that cathode rays had a negative charge. The previous experiment, in a poorer vacuum, had failed because the cathode rays had interacted with and ionized the residual gas. Thomson also studied the effects of a magnetic field upon the cathode rays by placing current coils outside the glass tube. He proved convincingly that the cathode rays acted as charged particles (electrons) in both electric and magnetic fields and received the Nobel Prize in 1906.

Thomson's method of measuring the ratio of the electron's charge to mass, e/m, is now a standard technique and generally studied as an example of charged particles passing through perpendicular electric and magnetic fields as shown schematically in Figure 3.3. With the magnetic field turned off, the electron entering the region between the plates is accelerated upward by the electric field

Measurement of electron's *e/m*

$$F_y = ma_y = qE \tag{3.1}$$

where m and q are, respectively, the mass and charge of the electron. The time for the electron to traverse the deflecting plates (length $= \ell$) is $t \approx \ell/v_0$. The exit angle θ of the electron is then given by

$$\tan\theta = \frac{v_y}{v_x} = \frac{a_y t}{v_0} = \frac{qE}{m}\frac{\ell}{v_0^2} \tag{3.2}$$

The ratio q/m can be determined if the velocity is known. By turning on the magnetic field and adjusting the strength of **B** so that no deflection of the electron

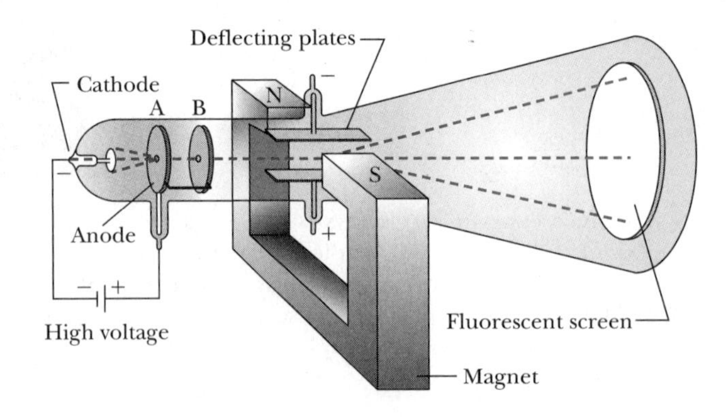

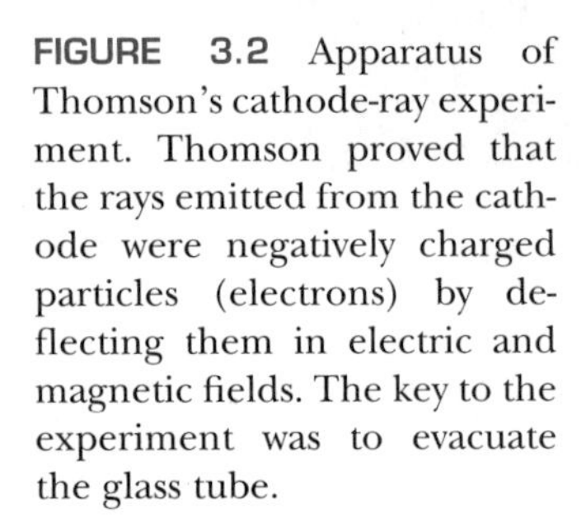

FIGURE 3.2 Apparatus of Thomson's cathode-ray experiment. Thomson proved that the rays emitted from the cathode were negatively charged particles (electrons) by deflecting them in electric and magnetic fields. The key to the experiment was to evacuate the glass tube.

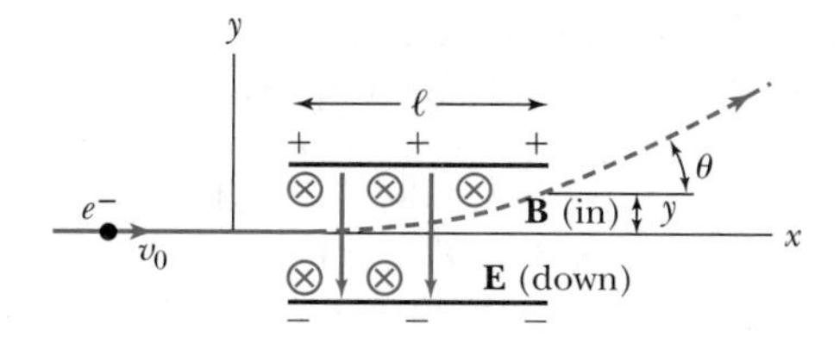

FIGURE 3.3 Thomson's method of measuring the electron's charge to mass ratio was to send electrons through a region containing a magnetic field (**B** into paper) perpendicular to an electric field (**E** down). The electrons having $v = E/B$ go through undeflected. Then, using the same energy electrons, the magnetic field is turned off and the electric field deflects the electrons, which exit at angle θ. The ratio of e/m can be determined from **B**, **E**, θ, and ℓ, where ℓ is the length of the field distance and θ is the emerging angle. See Equation (3.5).

occurs, the velocity can be determined. The condition for zero deflection is that the net force on the electron must be zero.

$$\mathbf{F} = q\mathbf{E} + q\mathbf{v} \times \mathbf{B} = 0 \tag{3.3}$$

Hence,

$$\mathbf{E} = -\mathbf{v} \times \mathbf{B}$$

or because **v** and **B** are perpendicular, the electric and magnetic field strengths are related by

$$|\mathbf{E}| = |v_x||\mathbf{B}|$$

so that

$$v_x = \frac{E}{B} = v_0 \tag{3.4}$$

If we insert this value for v_0 into Equation (3.2), we extract the ratio of q/m.

$$\frac{q}{m} = \frac{v_0{}^2 \tan\theta}{E\ell} = \frac{E\tan\theta}{B^2\ell} \tag{3.5}$$

Example 3.1

In an experiment similar to Thomson's, we use deflecting plates 5 cm in length with an electric field of 1.0×10^4 V/m. Without the magnetic field we find an angular deflection of 30°, and with a magnetic field of 8×10^{-4} T (8 gauss) we find no deflection. What is the initial velocity of the electron and its q/m?

Solution: We find the electron's velocity v_0 from Equation (3.4).

$$v_0 = \frac{E}{B} = \frac{1.0 \times 10^4\ \text{V/m}}{8.0 \times 10^{-4}\ \text{T}} = 1.25 \times 10^7\ \text{m/s}$$

Because we use all units for E and B in the international system (SI), the answer must be in meters/second.

Now we can determine q/m by using Equation (3.5):

$$\frac{q}{m} = \frac{E\tan\theta}{B^2\ell} = \frac{(1.0 \times 10^4\ \text{V/m})(\tan 30°)}{(8 \times 10^{-4}\ \text{T})^2(0.05\ \text{m})}$$

$$= 1.80 \times 10^{11}\ \text{C/kg}$$

Thomson's actual experiment, done in the manner of the previous example, obtained a result about 35% lower than the presently accepted value of 1.76×10^{11} C/kg for e/m. Thomson realized that the value of e/m (e = absolute value of electron charge) for an electron was much larger than had been anticipated and a factor of 1000 larger than any value of q/m that had been previously measured (for the hydrogen atom). He concluded that either m was small or e was large (or both), and the "carriers of the electricity" were quite penetrating as compared to atoms or molecules, which must be much larger in size.

3.2 Determination of Electron Charge

After Thomson's measurement of e/m and the confirmation of the cathode ray as the charge carrier (called *electron*), several investigators attempted to determine the electron's charge, which was poorly known in 1897. In 1911 the American physicist Robert A. Millikan (1865–1953) reported convincing evidence for an accurate determination of the electron's charge. Millikan's classic experiment began in 1907 at the University of Chicago. The experiment consisted of visual observation of the motion of uncharged and both positively and negatively charged oil drops moving under the influence of electrical and gravitational forces. The essential parts of the apparatus are shown in Figure 3.4. As the drops emerge from the nozzle, frictional forces sometimes cause them to be charged. Millikan's method consisted of balancing the upward force of the electric field between the plates against the downward force of the gravitational field.

Millikan's oil drop experiment

When an oil drop falls downward through the air, it experiences a frictional force $\mathbf{F}_f$ proportional to its velocity due to the air's viscosity:

$$\mathbf{F}_f = -b\mathbf{v} \tag{3.6}$$

The force has a minus sign because it always opposes the drop's velocity. The constant b is determined by Stokes's law and is proportional to the oil drop's radius. Millikan showed that Stokes's law for the motion of a small sphere through

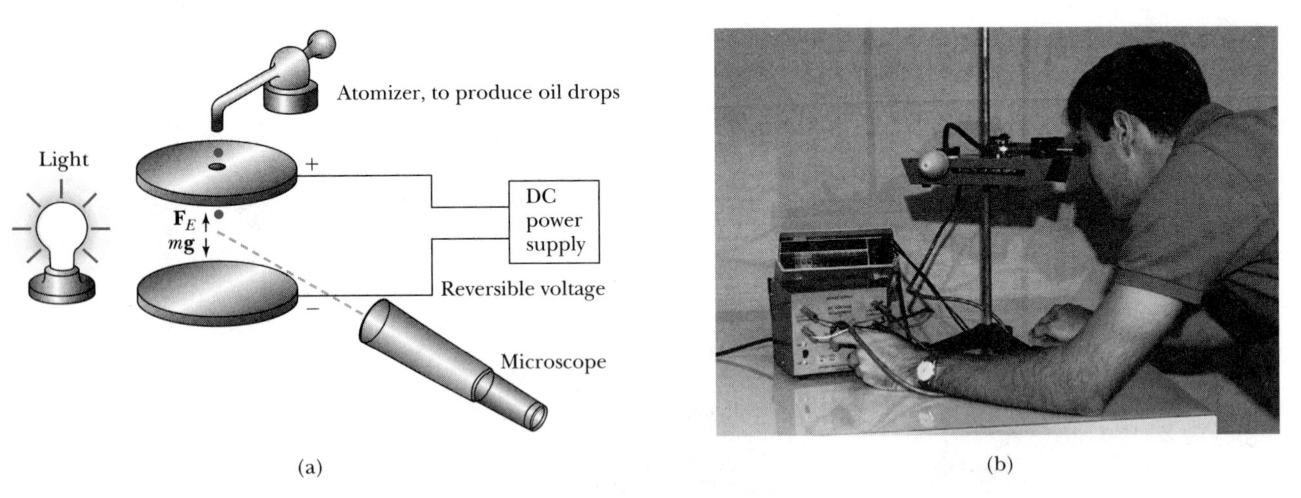

FIGURE 3.4 (a) Diagram of the Millikan oil-drop experiment to measure the charge of the electron. Some of the oil drops from the atomizer emerge charged, and the electric field (voltage) is varied to slow down or reverse the direction of the oil drops, which can have positive or negative charges. (b) A student looking through the microscope is adjusting the voltage between the plates to slow down a tiny plastic ball that serves as the oil drop.

Three great physicists (foreground), 1931: Michelson, Einstein, and Millikan. *Courtesy of California Institute of Technology.*

a resisting medium becomes incorrect for small-diameter spheres because of the atomic nature of the medium, and he found the appropriate correction. The buoyancy of the air produces an upward force on the drop, but we can neglect this effect for a first-order calculation.

To suspend the oil drop at rest between the plates, the upward electric force must equal the downward gravitational force. The frictional force is then zero because the velocity of the oil drop is zero.

$$\mathbf{F}_E = q\mathbf{E} = -m\mathbf{g} \qquad (\text{when } v = 0) \tag{3.7}$$

The magnitude of the electric field is $E = V/d$ and V is the voltage across large, flat plates separated by a small distance d. The magnitude of the electron charge q may then be extracted as

$$q = \frac{mgd}{V} \tag{3.8}$$

To calculate q we have to know the mass m of the oil drops. Millikan found he could determine m by turning off the electric field and measuring the terminal velocity of the oil drop. The radius of the oil drop is related to the terminal velocity by Stokes's law (see Problem 7). The mass of the drop can then be determined by knowing the radius r and density ρ of the type of oil used in the experiment:

$$m = \frac{4}{3}\pi r^3 \rho \tag{3.9}$$

If the power supply has a switch to reverse the polarity of the voltage and an adjustment for the voltage magnitude, the oil drop can be moved up and down in the apparatus at will. Millikan reported that in some cases he was able to observe a given oil drop for up to six hours and that the drop changed its charge several times.

Measurement of electron charge

Millikan made thousands of measurements using different oils and showed that there is a basic quantized electron charge. Millikan's value of e was very close to our presently accepted value of 1.602×10^{-19} C. Notice that we always quote a positive number for the charge e. The charge on an electron is then $-e$.

Example 3.2

For an undergraduate physics laboratory experiment we often make two changes in Millikan's procedure. First, we use plastic balls of about 1 micrometer (μm or micron) in diameter, for which we can measure the mass easily and accurately. This avoids the measurement of the oil drop's terminal velocity and the dependence on Stokes's law. The small plastic balls are still sprayed through an atomizer in liquid solution, but the liquid soon evaporates in air. The plastic balls are easily seen by a microscope. One other improvement is to bombard the region between the plates occasionally with ionizing radiation (such as x rays or α particles from radioactive sources). This radiation ionizes the air and makes it easier for the charge on a ball to change. By making many measurements we can determine whether the charges determined from Equation (3.8) are multiples of some basic charge unit.

One problem in the experiment is that occasionally one obtains fragments of broken balls or clusters of several balls. These can be eliminated by watching the flight of balls in free fall. The majority of balls will be single and fall faster than fragments, but slower than clusters. With a little experience one can select single unbroken balls.

In an actual undergraduate laboratory experiment the mass of the balls was $m = 5.7 \times 10^{-16}$ kg and the spacing eween the plates was $d = 4$ mm. Therefore q can be found from Equation (3.8).

$$q = \frac{mgd}{V} = \frac{(5.7 \times 10^{-16}\ \text{kg})(9.8\ \text{m/s}^2)(4 \times 10^{-3}\ \text{m})}{V}$$

$$q = \frac{(2.23 \times 10^{-17}\ \text{V})}{V}\text{C}$$

where V is the voltage between plates when the observed ball is stationary. Two students observed 30 balls and found the values of V shown in Table 3.1 for a stationary ball. In this experiment the voltage polarity can easily be changed, and a positive voltage represents a ball with a positive charge. Notice that charges of both signs are observed.

The values of $|q|$ are plotted on a histogram in units of $\Delta q = 0.2 \times 10^{-19}$ C. These are shown by the solid area in Figure 3.5. When 70 additional measurements from other students are added, a clear pattern of quantization develops with a charge $q = nq_0$, especially for the first three groups. The groups become increasingly smeared out for higher charges. The areas of the histogram can be separated for the various n values, and the value of q_0 found for each measurement is then averaged. For the histogram shown we find $q_0 = 1.7 \times 10^{-19}$ C for the first 30 measurements and $q_0 = 1.6 \times 10^{-19}$ C for all 100 observations.

TABLE 3.1
Student Measurements in Millikan Experiment

Particle	Voltage (V)	$q(\times 10^{-19}$ C)	Particle	Voltage	q	Particle	Voltage	q
1	−30.0	−7.43	11	−126.3	−1.77	21	−31.5	−7.08
2	+28.8	+7.74	12	−83.9	−2.66	22	−66.8	−3.34
3	−28.4	−7.85	13	−44.6	−5.00	23	+41.5	+5.37
4	+30.6	+7.29	14	−65.5	−3.40	24	−34.8	−6.41
5	−136.2	−1.64	15	−139.1	−1.60	25	−44.3	−5.03
6	−134.3	−1.66	16	−64.5	−3.46	26	−143.6	−1.55
7	+82.2	+2.71	17	−28.7	−7.77	27	+77.2	+2.89
8	+28.7	+7.77	18	−30.7	−7.26	28	−39.9	−5.59
9	−39.9	−5.59	19	+32.8	+6.80	29	−57.9	−3.85
10	+54.3	+4.11	20	−140.8	+1.58	30	+42.3	+5.27

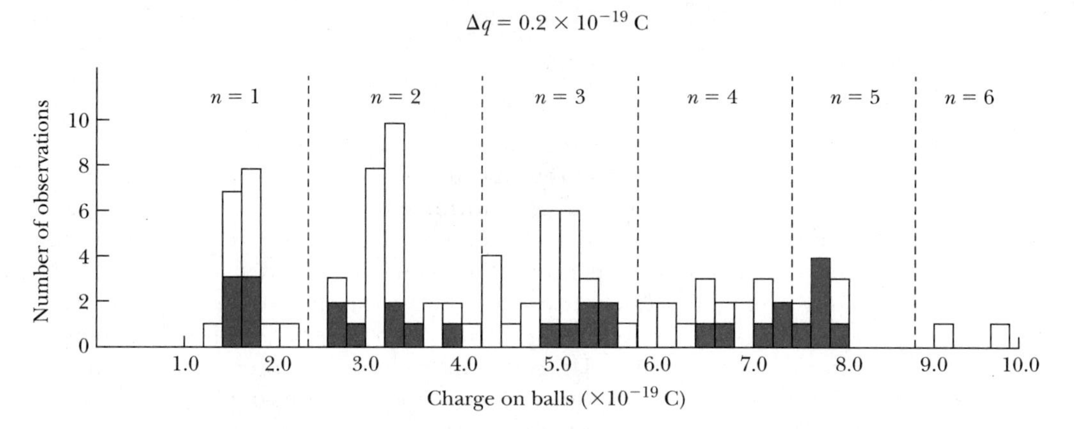

FIGURE 3.5 A histogram of the number of observations for the charge on a ball in a student Millikan experiment. The histogram is plotted for $\Delta q = 0.2 \times 10^{-19}$ C. The solid area refers to the first group's 30 measurements, and the open area to another 70 measurements. Notice the peaks, especially for the first three ($n = 1, 2, 3$) groups, indicating the electron charge quantization. When the basic charge q_0 is found from $q = nq_0$ (n = integer), $q_0 = 1.6 \times 10^{-19}$ C was determined in this experiment from all 100 observations.

3.3 Line Spectra

In the early 1800s optical spectroscopy became an important area of experimental physics, primarily because of the development of diffraction gratings. It had already been demonstrated that the chemical elements produced unique colors when burned in a flame or when excited in an electrical discharge. Prisms had been used to investigate these sources of spectra.

An example of a spectrometer used to observe optical spectra is shown in Figure 3.6. An electrical discharge excites atoms of a low-pressure gas contained in the tube. The collimated light passes through a diffraction grating with thousands of ruling lines per centimeter, and the diffracted light is separated at angle θ according to its wavelength λ. The equation expressing diffraction maxima is

$$d \sin \theta = n\lambda \tag{3.10}$$

where d is the distance between rulings, and n (an integer) is called the order number ($n = 1$ has the strongest scattered intensity). The resulting pattern of

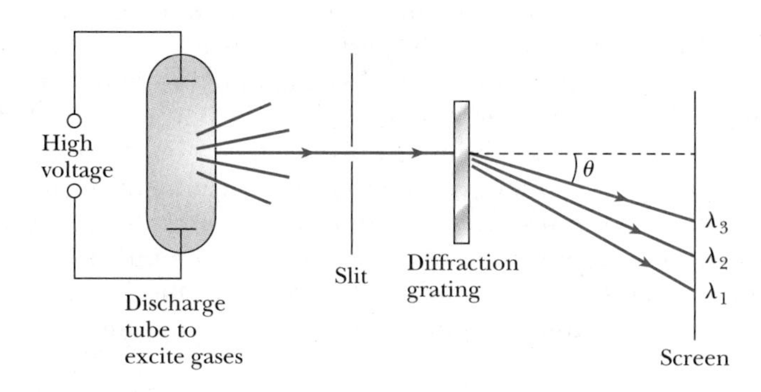

FIGURE 3.6 Schematic of an optical spectrometer. Light produced by a high-voltage discharge in the glass tube is collimated and passed through a diffraction grating, where it is deflected according to its wavelength. See Equation (3.10).

SPECIAL TOPIC:

THE DISCOVERY OF HELIUM

It might seem that the discovery of helium, the second simplest of all elements, would have occurred centuries ago. As we shall see, this is not the case, and in fact the discovery happened over a period of several years in the latter part of the 19th century as scientists were scrambling to understand unexpected results. The account here is taken from *Helium,* by William H. Keesom.*

Spectroscopes, optical devices used to measure wavelengths of light, normally consist of a slit, a collimating lens, and a prism to refract the light. Their first use in a solar eclipse was on August 18, 1868, to investigate the sun's atmosphere. Several persons traveled to the total eclipse region in India and Malaysia (including P. J. C. Janssen, G. Rayet, C. T. Haig, and J. Herschel) and all reported, either directly or indirectly, to have observed an unusual yellow line in the spectra that would later be proven due to helium. It occurred to Janssen the day of the eclipse that it must be possible to see the sun's spectrum directly without the benefit of the eclipse, and he did so with a spectroscope on the days following the eclipse. The same idea had occurred to J. N. Lockyer earlier, but he did not succeed in measuring the sun's spectrum until October, 1868, a month or so after Janssen. This method of observing the sun's atmosphere at any time was considered to be an important discovery, and Janssen and Lockyer are prominently recognized not only for the evolution of helium's discovery but for the means of studying the sun's atmosphere as well.

The actual discovery of helium was delayed by the fact that the new yellow line seen in the sun's atmosphere was very close in wavelength to two well-known yellow lines of sodium. This is apparent in the atomic line spectra of both helium and sodium seen on the inside back cover of this text. Certainly the line spectra of many elements were known by 1898, and scientists were busy cataloguing each element's characteris-

---

light bands and dark areas on the screen is called a *line spectrum.* By 1860 Bunsen and Kirchhoff realized the usefulness of the wavelengths of these line spectra in allowing identification of the chemical elements and the composition of materials. It was discovered that each element had its own characteristic wavelengths as shown on the inside back cover. The field of spectroscopy flourished because finer and more evenly ruled gratings became available, and improved experimental techniques allowed more spectral lines to be observed and catalogued. Particular interest was paid to the sun's spectrum in hopes of understanding the origin of sunlight. The helium atom was actually "discovered" by its line spectra from the sun before it was recognized on Earth (see Special Topic).

Characteristic line spectra of elements

Many scientists believed that the lines in the spectra somehow reflected the complicated internal structure of the atom, and that by carefully investigating the wavelengths for many elements, the structure of atoms and matter could be understood. That belief was eventually partially realized.

For much of the 19th century scientists attempted to find some simple underlying order for the characteristic wavelengths of line spectra. Hydrogen appeared to have an especially simple-looking spectrum, and because some chemists thought hydrogen atoms might be the constituents of heavier atoms, hydrogen was singled out for intensive study. Finally, in 1885, Johann Balmer, a Swiss schoolteacher, succeeded in obtaining a simple empirical formula that fit the wavelengths of the fourteen lines then known in the hydrogen spectrum. Four lines were in the visible region, and the remaining ultraviolet lines had

Balmer's empirical result

tic spectra. By December, 1868, Lockyer, A. Secchi, and Janssen each independently recognized that the yellow line was different than that of sodium.

Another difficulty was to prove that the new yellow line, called D_3, was not due to some other known element, especially hydrogen. For many years Lockyer thought that D_3 was related to hydrogen and he and E. Frankland performed several experiments that were not able to prove his thesis. Lockyer wrote as late as 1887 that D_3 was a form of hydrogen. However, in contradiction, Lord Kelvin reported in 1871 during his presidential address to the British Association that Frankland and Lockyer could not find the D_3 line to be related to any terrestrial (from Earth) flame. Kelvin reported that it seemed to represent a new substance, which Frankland and Lockyer proposed to call helium (from the Greek word for "sun").

It was not until 1895 that helium was finally clearly observed on Earth by Sir William Ramsay, who had received a letter reporting that W. F. Hillebrand had produced nitrogen gas by boiling uranium ores *(pitchblende)* in dilute sulphuric acid. Ramsay was skeptical of the report and proceeded to reproduce it. He was astounded, after finding a small amount of nitrogen and the expected argon gas, to see a brilliant yellow line that he compared with those from sodium, finding the wavelengths to be different. Sir William Crookes measured the wavelength and reported the following day that it was the D_3 line, proving the terrestrial existence of helium. Later in 1895 H. Kayser found the helium line in spectra taken from a gas that had evolved from a spring in Germany's Black Forest. Eventually, in 1898, helium was confirmed in the Earth's atmosphere by E. C. Baly. No one person can be credited for the discovery of helium.

The remarkable properties of *liquid* helium are discussed in Section 9.7.

*W. H. Keesom, *Helium*. Amsterdam, London, and New York: Elsevier, 1942.

---

been identified in the spectra of white stars. This series of lines, called the *Balmer series,* is shown in Figure 3.7. Balmer found that the expression

$$\lambda = 364.56 \frac{k^2}{k^2 - 4} \text{ nm} \tag{3.11}$$

(where $k = 3, 4, 5, \ldots; k > 2$) fit all the visible hydrogen lines. Wavelengths are normally given in units of nanometers* (nm). It is more convenient to take the inverse of Equation (3.11) and write Balmer's formula in the form

$$\frac{1}{\lambda} = \frac{1}{364.56 \text{ nm}} \frac{k^2 - 4}{k^2} = \frac{4}{364.56 \text{ nm}} \left(\frac{1}{2^2} - \frac{1}{k^2} \right) = R_{\text{H}} \left(\frac{1}{2^2} - \frac{1}{k^2} \right) \tag{3.12}$$

*Wavelengths were formerly listed in units of angstroms (one angstrom (Å) = 10^{-10} m), named after Ångstrom who was one of the first persons to observe and measure the wavelengths of the four visible lines of hydrogen.

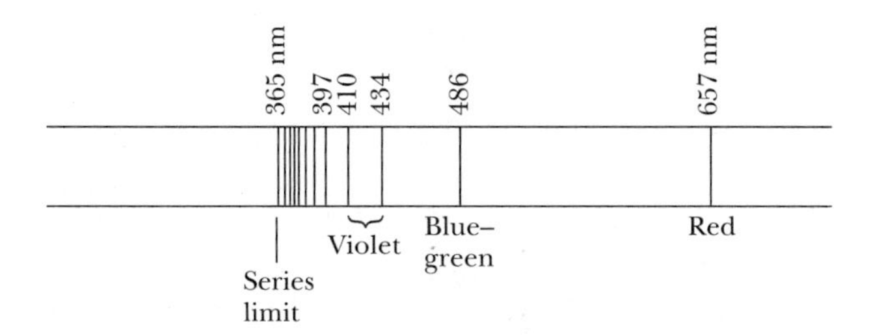

FIGURE 3.7 The Balmer series of line spectra of the hydrogen atom with wavelengths indicated in nm. The four visible lines are noted as well as the lower limit of the series.

TABLE 3.2
Hydrogen Series of Spectral Lines

Discoverer (year)	Wavelength	n	k
Lyman (1916)	Ultraviolet	1	>1
Balmer (1885)	Visible, ultraviolet	2	>2
Paschen (1908)	Infrared	3	>3
Brackett (1922)	Infrared	4	>4
Pfund (1924)	Infrared	5	>5

where R_H is called the *Rydberg constant* (for hydrogen) and has the more accurate value 1.096776×10^7 m^{-1}, and k is an integer greater than two ($k > 2$).

Efforts by Johannes Rydberg and particularly Walther Ritz eventually resulted in 1890 in a more general empirical equation for calculating the wavelengths called the *Rydberg equation.*

Rydberg equation

$$\frac{1}{\lambda} = R_H\left(\frac{1}{n^2} - \frac{1}{k^2}\right) \tag{3.13}$$

where $n = 2$ corresponds to the Balmer series and $k > n$ always. Some 20 years after Balmer's contribution, other series of the hydrogen atom's spectral lines were discovered in the early 1900s. By 1925 five series had been discovered, each having a different integer n (see Table 3.2). The understanding of the Rydberg equation (3.13) and the discrete spectrum of hydrogen were important research topics early in the 20th century.

Example 3.3

The visible lines of the Balmer series were observed first because they are most easily seen. Show that the wavelengths of spectral lines in the *Lyman* ($n = 1$) and *Paschen* ($n = 3$) series are not in the visible region. Find the wavelengths of the four visible atomic hydrogen lines. Assume the visible wavelength region is $\lambda = 400$ to 700 nm.

Solution: We use Equation (3.13) first to examine the Lyman series ($n = 1$)

$$\frac{1}{\lambda} = R_H\left(1 - \frac{1}{k^2}\right)$$

$$= 1.0968 \times 10^7\left(1 - \frac{1}{k^2}\right)\text{m}^{-1}$$

$k = 2$ $$\frac{1}{\lambda} = 1.0968 \times 10^7\left(1 - \frac{1}{4}\right)\text{m}^{-1}$$

$$\lambda = 1.216 \times 10^{-7}\text{ m} = 121.6\text{ nm} \qquad \text{(Ultraviolet)}$$

$k = 3$ $$\frac{1}{\lambda} = 1.0968 \times 10^7\left(1 - \frac{1}{9}\right)\text{m}^{-1}$$

$$\lambda = 1.026 \times 10^{-7}\text{ m} = 102.6\text{ nm} \qquad \text{(Ultraviolet)}$$

Because the wavelengths are decreasing for higher k values, all the wavelengths in the Lyman series are in the ultraviolet region and not visible by eye.

For the Balmer series ($n = 2$) we find

$k = 3$ $$\frac{1}{\lambda} = 1.0968 \times 10^7\left(\frac{1}{4} - \frac{1}{9}\right)\text{m}^{-1}$$

$$\lambda = 6.565 \times 10^{-7}\text{ m} = 656.5\text{ nm} \qquad \text{(Red)}$$

$k = 4$ $$\frac{1}{\lambda} = 1.0968 \times 10^7\left(\frac{1}{4} - \frac{1}{16}\right)\text{m}^{-1}$$

$$\lambda = 4.863 \times 10^{-7}\text{ m} = 486.3\text{ nm} \qquad \text{(Blue-green)}$$

$k = 5$ $$\frac{1}{\lambda} = 1.0968 \times 10^7\left(\frac{1}{4} - \frac{1}{25}\right)\text{m}^{-1}$$

$$\lambda = 4.342 \times 10^{-7}\text{ m} = 434.2\text{ nm} \qquad \text{(Violet)}$$

$k = 6$ $$\frac{1}{\lambda} = 1.0968 \times 10^7\left(\frac{1}{4} - \frac{1}{36}\right)\text{m}^{-1}$$

$$\lambda = 4.103 \times 10^{-7}\text{ m} = 410.3\text{ nm} \qquad \text{(Violet)}$$

$k = 7$ $$\frac{1}{\lambda} = 1.0968 \times 10^7\left(\frac{1}{4} - \frac{1}{49}\right)\text{m}^{-1}$$

$$\lambda = 3.971 \times 10^{-7}\text{ m} = 397.1\text{ nm} \qquad \text{(Ultraviolet)}$$

Therefore $k = 7$ and higher k values will be in the ultraviolet region. The four lines $k = 3$, 4, 5, and 6 of the Balmer series are visible, although the 410 nm ($k = 6$) line is difficult to see because it is barely in the visible region and is weak in intensity.

The next series, $n = 3$, named after Paschen, has wavelengths of

$$k = 4 \qquad \frac{1}{\lambda} = 1.0968 \times 10^7\left(\frac{1}{9} - \frac{1}{16}\right)\text{m}^{-1}$$

$$\lambda = 1.876 \times 10^{-6}\text{ m} = 1876\text{ nm} \qquad \text{(Infrared)}$$

$$k = 5 \qquad \frac{1}{\lambda} = 1.0968 \times 10^7\left(\frac{1}{9} - \frac{1}{25}\right)\text{m}^{-1}$$

$$\lambda = 1.282 \times 10^{-6}\text{ m} = 1282\text{ nm} \qquad \text{(Infrared)}$$

$$k = \infty \qquad \frac{1}{\lambda} = 1.0968 \times 10^7\left(\frac{1}{9} - \frac{1}{\infty}\right)\text{m}^{-1}$$

$$\lambda = 8.206 \times 10^{-7}\text{ m} = 820.6\text{ nm} \qquad \text{(Infrared)}$$

Thus the Paschen series has wavelengths entirely in the infrared region. Notice that the series limit is found for $k = \infty$. The higher series, $n \geq 4$, will all have wavelengths longer than the visible region.

3.4 Quantization

As we discussed in Chapter 1, some early Greek philosophers believed that matter must be composed of fundamental units that could not be divided further. The word "atom" means "not further divisible." Today some scientists believe, as these ancient philosophers did, that matter must eventually be indivisible. However, as we have encountered new experimental facts, our ideas about the fundamental, indivisible "building blocks" of matter have changed. More will be said about the "elementary" particles in Chapter 14.

Is matter indivisible?

Whatever the elementary units of matter may eventually turn out to be, we suppose there are some basic units of mass-energy of which matter is composed. This idea is hardly foreign to us: we have seen already that Millikan's oil drop experiment showed the quantization of electric charge. Current theories predict that charges are quantized in units (called **quarks**) of $\pm e/3$ and $\pm 2e/3$, but quarks can not be directly observed experimentally. The charges of particles that have been directly observed are quantized in units of $\pm e$.

Electric charge is quantized

In nature we see other examples of quantization. The measured atomic weights are not continuous—they have only discrete values which are close to integral multiples of a unit mass. Molecules are formed from an integral number of atoms. The water molecule is made up of exactly two atoms of hydrogen and one of oxygen. The fact that an organ pipe produces one fundamental musical note with overtones is a form of quantization arising from fitting a precise number (or fractions) of sound waves into the pipe.

Quantization occurs often in nature

The line spectra of atoms discussed in the previous section again show that characteristic wavelengths have precise values and are not distributed continuously. By the end of the 19th century, radiation spectra had been well studied. There certainly didn't appear to be any quantization effects observed in blackbody radiation spectra emitted by hot bodies. However, these radiation spectra were to have a tremendous influence on the discovery of quantum physics.

3.5 Blackbody Radiation

It has been known for many centuries that when matter is heated, it emits radiation. We can feel the heat radiation emitted by the heating element of an electric stove as it warms up. As the heating element reaches 550°C, its color becomes dark red, turning to bright red around 700°C. If the temperature were increased still further, the color would progress through orange, yellow, and finally white. We can determine experimentally that a broad spectrum of wavelengths is emitted when matter is heated. This process was of great interest to physicists of the nineteenth century. They measured the intensity of radiation being emitted as a function of material, temperature, and wavelength.

Radiation emission and absorption

All bodies simultaneously emit and absorb radiation. When a body's temperature is constant in time, the body is said to be in *thermal equilibrium* with its surroundings. In order for the temperature to be constant, the body must absorb thermal energy at the same rate as it emits it. This implies that a good thermal emitter is also a good absorber.

Physicists generally try to study the simplest or most idealized case of a problem first in order to gain the insight that is needed to analyze more complex situations. For thermal radiation the simplest case is a **blackbody,** which has the ideal property that it absorbs all the radiation falling on it and reflects none. The simplest way to construct a blackbody is to drill a small hole in the wall of a hollow container as shown in Figure 3.8. Radiation entering the hole will be reflected around inside the container and then finally absorbed. Only a small fraction of the entering rays will be re-emitted through the hole. If the blackbody is in thermal equilibrium, then it must also be an excellent emitter of radiation as well.

Blackbody radiation is unique

Blackbody radiation is theoretically interesting because of its universal character: the radiation properties of the blackbody (that is, the cavity) are independent of the particular material of which the container is made. Physicists could study the previously mentioned properties of intensity vs. wavelength (called *spectral distribution*) at fixed temperatures without having to understand the details of emission or absorption by a particular kind of atom. The question of precisely what the thermal radiation actually consisted of was also of interest, although it was assumed, for lack of evidence to the contrary (and correctly, it turned out!), to be electromagnetic radiation.

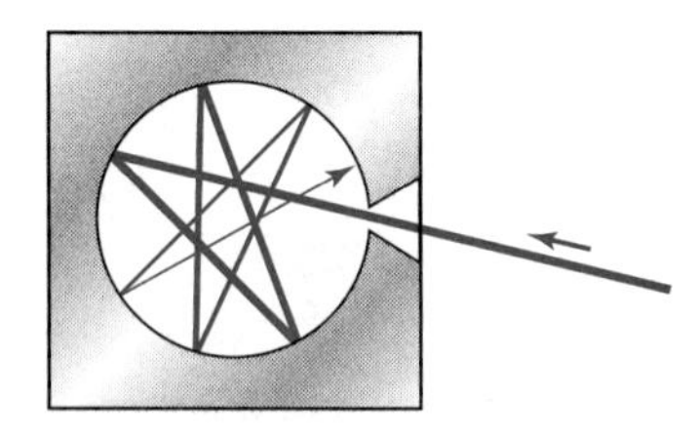

FIGURE 3.8 Blackbody radiation. Electromagnetic radiation (for example, light) entering a small hole reflects around inside the container before being eventually absorbed.

The intensity $\mathcal{I}(\lambda, T)$ is the total power radiated per unit area per unit wavelength at a given temperature. Measurements of $\mathcal{I}(\lambda, T)$ for a blackbody are displayed in Figure 3.9. Two important observations should be noted:

1. The maximum of the distribution shifts to smaller wavelengths as the temperature is increased.
2. The total power radiated increases with the temperature.

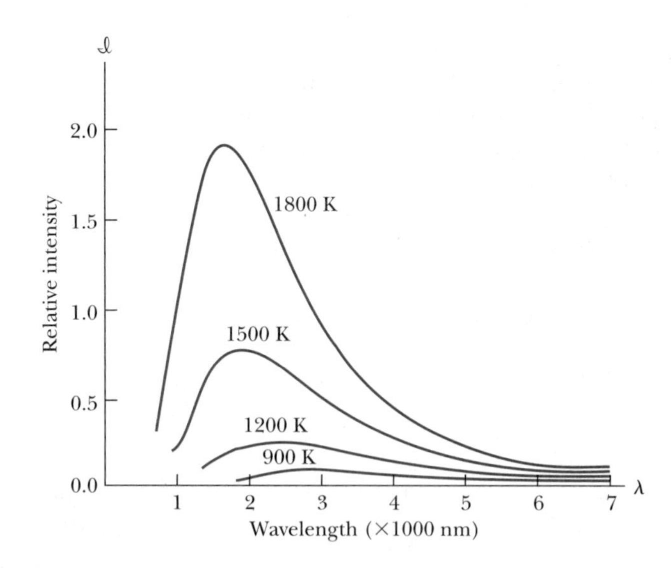

FIGURE 3.9 Spectral distribution of radiation emitted from a blackbody for different blackbody temperatures.

The first observation is commonly referred to as **Wien's displacement law,**

$$\lambda_{max} T = 2.898 \times 10^{-3} \text{ m} \cdot \text{K} \tag{3.14}$$

Wien's displacement law

where λ_{max} is the wavelength of the peak of the spectral distribution at a given temperature. Wilhelm Wien received the Nobel Prize in 1911 for his discoveries concerning radiation. We can quantify the second observation by integrating the quantity $\mathcal{I}(\lambda, T)$ over all wavelengths to find the power per unit area at T.

$$R(T) = \int_0^\infty \mathcal{I}(\lambda, T)\, d\lambda \tag{3.15}$$

Josef Stefan found empirically in 1879, and Boltzmann demonstrated theoretically several years later, that $R(T)$ is related to the temperature by

$$R(T) = \epsilon\sigma T^4 \tag{3.16}$$

Stefan-Boltzmann law

This is known as the **Stefan-Boltzmann law,** with the constant σ experimentally measured to be 5.6705×10^{-8} W/(m$^2 \cdot$ K^4). The Stefan-Boltzmann law equation can be applied to any material for which the emissivity is known. The **emissivity** ϵ ($\epsilon = 1$ for an idealized blackbody) is simply the ratio of the emissive power of an object to that of an ideal blackbody and is always less than 1. Thus, Equation (3.16) is a useful and valuable relation for practical scientific and engineering work.

Example 3.4

A furnace has walls of temperature 1600°C. What is the wavelength of maximum intensity emitted when a small door is opened?

Solution: If we assume blackbody radiation, we determine λ_{max} from Equation (3.14).

$$T = (1600 + 273)\text{K} = 1873 \text{ K}$$

$$\lambda_{max}(1873 \text{ K}) = 2.898 \times 10^{-3} \text{ m} \cdot \text{K}$$

$$\lambda_{max} = 1.55 \times 10^{-6} \text{ m} = 1550 \text{ nm}$$

Example 3.5

The wavelength of maximum intensity of the sun's radiation is observed to be near 500 nm. Assume the sun to be a blackbody and calculate (a) the sun's surface temperature, (b) the power per unit area $R(T)$ emitted from the sun's surface, and (c) the energy received by the Earth each day from the sun's radiation.

Solution: From Equation (3.14) we calculate the sun's surface temperature with $\lambda_{max} = 500$ nm.

$$(500 \text{ nm})\, T_{sun} = 2.898 \times 10^{-3} \text{ m} \cdot \text{K} \frac{10^9 \text{ nm}}{\text{m}}$$

$$T_{sun} = \frac{2.898 \times 10^6}{500} \text{ K} = 5800 \text{ K} \tag{3.17}$$

The power per unit area $R(T)$ at this temperature can be found by again assuming a blackbody:

$$R(T) = 5.67 \times 10^{-8} \frac{\text{W}}{\text{m}^2 \cdot \text{K}^4} (5800 \text{ K})^4$$

$$= 6.42 \times 10^7 \text{ W/m}^2 \tag{3.18}$$

Because this is the power per unit surface area, we need to multiply this by $4\pi r^2$, the surface area of the sun. The radius of the sun is 6.96×10^5 km.

$$\text{Surface area (sun)} = 4\pi(6.96 \times 10^8 \text{ m})^2 = 6.09 \times 10^{18} \text{ m}^2$$

Thus the total power, P_{sun}, radiated from the sun's surface is

$$P_{sun} = 6.42 \times 10^7 \frac{\text{W}}{\text{m}^2} (6.09 \times 10^{18} \text{ m}^2) = 3.91 \times 10^{26} \text{ W} \tag{3.19}$$

The fraction F of the sun's radiation received by Earth is given by the fraction of the total area over which the radiation is spread.

$$F = \frac{\pi r_e^2}{4\pi R_{es}^2}$$

where r_e = radius of Earth = 6.37×10^6 m, and R_{es} = mean Earth-sun distance = 1.49×10^{11} m. Then

$$F = \frac{\pi r_e^2}{4\pi R_{es}^2} = \frac{(6.37 \times 10^6 \text{ m})^2}{4(1.49 \times 10^{11} \text{ m})^2} = 4.57 \times 10^{-10}$$

Thus the radiation received by the Earth from the sun is

$$\begin{aligned} P_{\text{Earth}} \text{ (received)} &= 4.57 \times 10^{-10} \, (3.91 \times 10^{26} \text{ W}) \\ &= 1.79 \times 10^{17} \text{ W} \end{aligned} \tag{3.20}$$

and in one day the Earth receives

$$U_{\text{Earth}} = 1.79 \times 10^{17} \frac{\text{J}}{\text{s}} \frac{60 \text{ s}}{\text{min}} \frac{60 \text{ min}}{\text{h}} \frac{24 \text{ h}}{\text{day}} = 1.55 \times 10^{22} \text{ J} \tag{3.21}$$

The power per unit exposed area received by the Earth is

$$R_{\text{Earth}} = \frac{1.79 \times 10^{17} \text{ W}}{\pi (6.37 \times 10^6 \text{ m})^2} = 1400 \text{ W/m}^2 \tag{3.22}$$

Needless to say, this is the source of most of our energy on Earth. Measurements of the sun's radiation outside the Earth's atmosphere give a value near 1400 W/m^2, so our calculation is fairly accurate. Apparently the sun does act as a blackbody, and most of the energy received by the Earth comes primarily from the surface of the sun.

Blackbody radiation problem

Attempts to understand and derive from basic principles the shape of the blackbody spectral distribution (Figure 3.9) were unsuccessful throughout the 1890s despite the persistent effort of some of the best scientists of the day. Blackbody radiation was one of the outstanding problems of the late nineteenth century because it presented physicists with a real dilemma. The nature of the dilemma can be understood from classical electromagnetic theory, together with statistical thermodynamics. The radiation emitted from the blackbody can be expressed as a superposition of electromagnetic waves of different frequencies within the cavity. That is, radiation of a given frequency is represented by a standing wave inside the cavity. The equipartition theorem of thermodynamics assigns equal average energy kT to each possible wave configuration. For long wavelengths λ there are only few configurations whereby a standing wave can form inside the cavity. However, as the wavelength becomes shorter the number of standing wave possibilities increases, and as $\lambda \to 0$ the number of possible configurations increases without limit. This means the total energy of all configurations is infinite, because each standing wave configuration has the nonzero energy kT. This problem for small wavelengths became known as "the ultraviolet catastrophe."

Max Planck (1858–1947) spent most of his productive years as a professor at the University of Berlin (1889–1928). His theory of the *quantum of action* was slow to be accepted because of its contradiction with the heat radiation law of Wilhelm Wien. Finally, after Einstein's photoelectric effect explanation and Rutherford and Bohr's atomic model, Planck's contribution became widely acclaimed. *AIP Emilio Segrè Visual Archives.*

In the late 1890s the German theoretical physicist Max Planck (1858–1947) became interested in this problem. By this time, different empirical expressions for the blackbody spectrum had been separately fit to the data for both short wavelengths and long wavelengths, but no one had explained the whole spectrum. Planck tried various functions of wavelength and temperature until he found a single formula that fit the measurements of $\mathcal{I}(\lambda, T)$ over the entire wavelength range. He announced this result in October of 1900, and immediately his equation was compared with recent data of Rubens and Kurlbaum. The result was that Planck's formula was even more accurate over the entire spectrum than the previous empirical ones, which were only valid for either short or long wavelengths.

Planck became quite excited and started working to find a sound theoretical basis for his empirical equation. He was an expert in thermodynamics and statistical mechanics. Following Hertz's work using oscillators to confirm the existence of Maxwell's electromagnetic waves, and lacking detailed information about the atomic composition of the cavity walls, Planck assumed that the radiation in the cavity was emitted (and absorbed) by some sort of "oscillators" that were contained in the walls. Whereas we would now refer to the radiation of the electromagnetic field in the cavity, Planck referred to the radiation produced by the "oscillators," a term we will briefly continue to use. When adding up the energies of the oscillators, he assumed (for convenience) that each one had an energy that was an integral multiple of $h\nu$, where ν is the frequency of the oscillating wave. He was applying a technique invented by Boltzmann and ultimately expected to take the limit $h \to 0$, in order to include all the possibilities. However, he noticed that by keeping h finite he arrived at the equation needed for $\mathcal{I}(\lambda, T)$,

$$\mathcal{I}(\lambda, T) = \frac{2\pi c^2 h}{\lambda^5} \frac{1}{e^{hc/\lambda kT} - 1} \tag{3.23}$$

Planck's radiation law

Equation (3.23) is **Planck's radiation law.** (The derivation of Equation (3.23) is sufficiently complicated that we have omitted it here.) No matter what he tried, he could only arrive at the correct result by making two important modifications of classical theory:

1. The oscillators (of electromagnetic origin) can only have certain discrete energies determined by $E_n = nh\nu$, where n is an integer, ν is the frequency, and h is called **Planck's constant** and has the value

$$h = 6.6261 \times 10^{-34}\ \mathrm{J \cdot s} \tag{3.24}$$

Planck's constant h

2. The oscillators can absorb or emit energy in discrete multiples of the fundamental quantum of energy given by

$$\Delta E = h\nu \tag{3.25}$$

Planck himself found these results quite disturbing and spent several years trying to find a way to keep the agreement with experiment while letting $h \to 0$. Each attempt failed.

Example 3.6

Show that Planck's radiation law avoids the *ultraviolet catastrophe.*

Solution: The ultraviolet catastrophe occurs because the number of configurations (~ intensity) in the classical calculation becomes infinite as $\lambda \to 0$. If we let $\lambda \to 0$ in Equation (3.23), the value of $e^{hc/\lambda kT} \to \infty$. The exponential term dominates the λ^5 term as $\lambda \to 0$, so the denominator in Equation (3.23) is infinite, and the value of $\mathcal{I}(\lambda, T) \to 0$. Note that as the wavelength gets smaller, the frequency becomes larger ($\lambda\nu = c$), and $h\nu \gg kT$. Few oscillators will be able to obtain such large energies, partly because of the large energy necessary to take the energy step from 0 to $h\nu$. The probability of occupying the states with small wavelengths (large frequency and high energy) is vanishingly small, so the total energy of the system remains finite. The *ultraviolet catastrophe* is avoided.

Example 3.7

Show that Wien's displacement law follows from Planck's radiation law.

Solution: Wien's law, Equation (3.14), refers to the wavelength λ for which $\mathscr{I}(\lambda, T)$ is a maximum for a given temperature. Therefore, to find this maximum we let $d\mathscr{I}/d\lambda = 0$ and solve for λ.

$$\frac{d\mathscr{I}(\lambda, T)}{d\lambda} = 0 \qquad \text{for } \lambda = \lambda_{\text{max}}$$

$$2\pi c^2 h \frac{d}{d\lambda}[\lambda^{-5}(e^{hc/\lambda kT} - 1)^{-1}] = 0$$

$$-5\lambda_{\text{max}}^{-6}(e^{hc/\lambda_{\text{max}} kT} - 1)^{-1} - \lambda_{\text{max}}^{-5}(e^{hc/\lambda_{\text{max}} kT} - 1)^{-2} \cdot \left(\frac{-hc}{kT\lambda_{\text{max}}^2}\right) e^{hc/\lambda_{\text{max}} kT} = 0$$

Multiplying by $\lambda_{\text{max}}^6(e^{hc/\lambda_{\text{max}} kT} - 1)$ results in

$$-5 + \frac{hc}{\lambda_{\text{max}} kT}\left(\frac{e^{hc/\lambda_{\text{max}} kT}}{e^{hc/\lambda_{\text{max}} kT} - 1}\right) = 0$$

Let

$$x = \frac{hc}{\lambda_{\text{max}} kT}$$

then

$$-5 + \frac{xe^x}{(e^x - 1)} = 0$$

and

$$xe^x = 5(e^x - 1)$$

This is a transcendental equation and can be solved numerically (try it!) with the result, $x = 4.966$, and, therefore,

$$\frac{hc}{\lambda_{\text{max}} kT} = 4.966$$

$$\lambda_{\text{max}} T = \frac{hc}{4.966\ k} = \frac{1240\ \text{eV}\cdot\text{nm}}{4.966\left(8.617 \times 10^{-5}\dfrac{\text{eV}}{\text{K}}\right)} \frac{10^{-9}\ \text{m}}{\text{nm}}$$

and finally,

$$\lambda_{\text{max}} T = 2.898 \times 10^{-3}\ \text{m}\cdot\text{K}$$

which is Wien's displacement law.

Example 3.8

Use Planck's radiation law to derive the Stefan-Boltzmann law.

Solution: To determine $R(T)$ we integrate $\mathscr{I}(\lambda, T)$ over all wavelengths

$$R(T) = \int_0^\infty \mathscr{I}(\lambda, T)\, d\lambda$$

$$= 2\pi c^2 h \int_0^\infty \frac{1}{\lambda^5} \frac{1}{e^{hc/\lambda kT} - 1}\, d\lambda$$

Let $x = \dfrac{hc}{\lambda kT}$, then $dx = -\dfrac{hc}{kT}\dfrac{d\lambda}{\lambda^2}$. Then we have

$$R(T) = -2\pi c^2 h \int_\infty^0 \left(\frac{kT}{hc}\right)^6 x^5 \frac{1}{e^x - 1} \frac{1}{x^2}\left(\frac{hc}{kT}\right)^2 dx$$

$$= +2\pi c^2 h \left(\frac{kT}{hc}\right)^4 \int_0^\infty \frac{x^3}{e^x - 1}\, dx$$

We look up this integral in Appendix 7 and find it to be $\pi^4/15$.

$$R(T) = 2\pi c^2 h \left(\frac{kT}{hc}\right)^4 \frac{\pi^4}{15}$$

$$R(T) = \frac{2\pi^5 k^4}{15 h^3 c^2} T^4$$

Putting in the values for the constants k, h, and c results in

$$R(T) = 5.67 \times 10^{-8}\ T^4 \frac{\text{W}}{\text{m}^2\cdot\text{K}^4}$$

3.6 Photoelectric Effect

Perhaps the most compelling, and certainly the simplest, evidence for the quantization of radiation energy comes from the only acceptable explanation of the **photoelectric effect.** The photoelectric effect, discovered by Hertz in 1887 as he confirmed Maxwell's electromagnetic wave theory of light, is one of several ways in which electrons can be emitted by materials. By the early 1900s it was known that electrons are bound to matter. In metals the valence electrons are "free"—they are able to move easily from atom to atom, but are not able to leave the surface of the material. The methods known now by which electrons can be made to completely leave the material include

Methods of electron emission

1. Thermionic emission—application of heat allows electrons to gain enough energy to escape.
2. Secondary emission—the electron gains enough energy by transfer from another high-speed particle that strikes the material from outside.
3. Field emission—a strong external electric field pulls the electron out of the material.
4. Photoelectric effect—incident light (electromagnetic radiation) shining on the material transfers energy to the electrons allowing them to escape.

Photoelectrons

Work function

It is not surprising that electromagnetic radiation acts on electrons within metals giving the electrons increased kinetic energy. Because electrons in metals are weakly bound, we expect that light can give electrons enough extra kinetic energy to allow them to escape. We call the ejected electrons **photoelectrons.** The minimum extra kinetic energy that allows electrons to escape the material is called the **work function** ϕ. The work function is the minimum binding energy of the electron to the material. The work functions of alkali metals are smaller than those of other metals. We shall see why this is so in Chapter 8.

Experimental Results of Photoelectric Effect

Experiments carried out around 1900 showed that photoelectrons were produced when visible and/or ultraviolet light falls on clean metal surfaces. Photoelectricity was studied using an experimental apparatus shown schematically in Figure 3.10. Incident light falling on the **emitter** (also called the **photocathode** or **cathode**) ejects electrons. Some of the electrons travel toward the **collector**

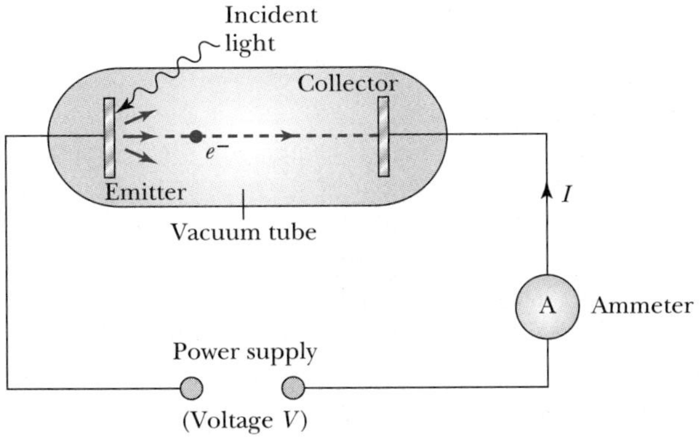

Incident light
Collector
e^-
Emitter
Vacuum tube
I
A
Ammeter
Power supply
(Voltage V)

FIGURE 3.10 Photoelectric effect. Electrons, emitted when light shines on a surface, are collected, and the photocurrent I is measured. A negative voltage, relative to that of the emitter, can be applied to the collector. When this retarding voltage is sufficiently large, the emitted electrons are repelled, and the current to the collector drops to zero.

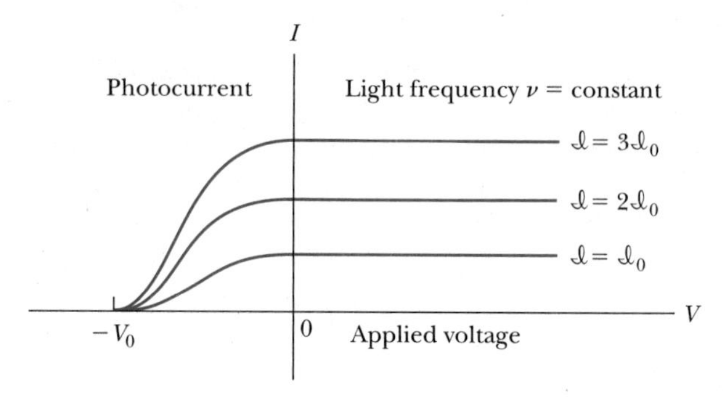

FIGURE 3.11 The photoelectric current I is shown as a function of the voltage V applied between the emitter and collector for a given frequency ν of light for three different light intensities. Notice that no current flows for a retarding potential more negative than $-V_0$ and that the photocurrent is constant for potentials near or above zero (this assumes that the emitter and collector are closely spaced or in spherical geometry to avoid loss of photoelectrons).

(also called the **anode**), where either a negative (retarding) or positive (accelerating) applied voltage V is imposed by the power supply. The current I measured in the ammeter (photocurrent) arises from the flow of photoelectrons from emitter to collector.

The pertinent experimental facts about the photoelectric effect are these:

1. The kinetic energies of the photoelectrons are independent of the light intensity. In other words, a stopping potential (applied voltage) of $-V_0$ is sufficient to stop all photoelectrons, *no matter what the light intensity,* as shown in Figure 3.11. For a given light intensity there is a maximum photocurrent, which is reached as the applied voltage increases from negative to positive values.
2. The maximum kinetic energy of the photoelectrons, for a given emitting material, depends only on the frequency of the light. In other words, for light of different frequency (Figure 3.12) a different retarding potential $-V_0$ is required to stop the most energetic photoelectrons. The value of V_0 depends on the frequency ν but not on the intensity (see Figure 3.11).
3. The smaller the work function ϕ of the emitter material, the smaller is the threshold frequency of the light that can eject photoelectrons. No photoelectrons are produced for frequencies below this threshold frequency, no matter what the intensity. Data similar to Millikan's results (discussed later) are shown in Figure 3.13, where the threshold frequencies ν_0 are measured for three different metals.
4. When the photoelectrons are produced, however, their number is proportional to the intensity of light as shown in Figure 3.14. That is, the maximum photocurrent is proportional to the light intensity.

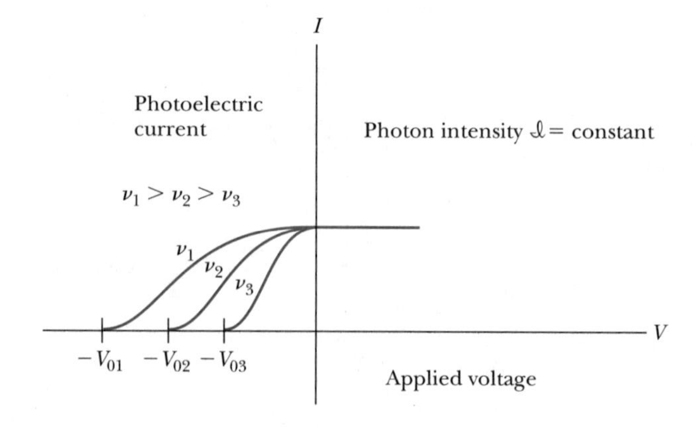

FIGURE 3.12 The photoelectric current I is shown as a function of applied voltage for three different light frequencies. The retarding potential $-V_0$ is different for each ν and is more negative for larger ν.

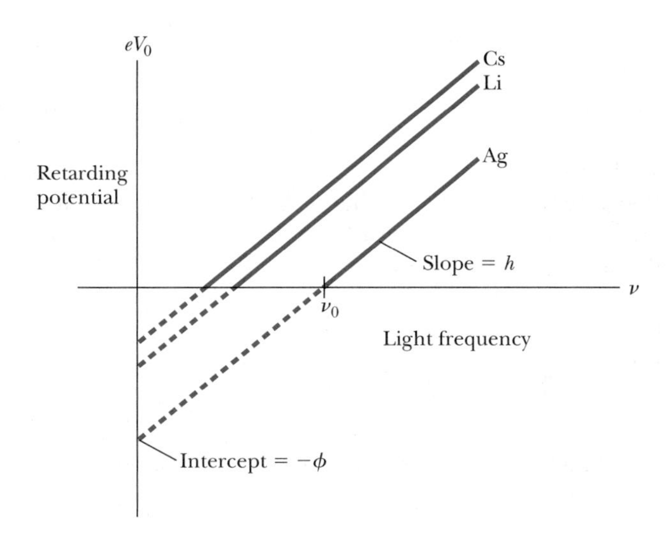

FIGURE 3.13 The retarding potential eV_0 (maximum electron kinetic energy) is plotted vs. light frequency for three different emitter materials.

5. The photoelectrons are emitted almost instantly ($\leq 3 \times 10^{-9}$ s) following illumination of the photocathode, independent of the intensity of the light.

Except for (5), these experimental facts were known in rudimentary form by 1902, primarily due to the work of Philipp Lenard, a German experimental physicist who won the Nobel Prize in 1905 for this and other research on the identification and behavior of electrons.

Classical Interpretation

As stated previously, we can understand from classical theory that electromagnetic radiation should be able to eject photoelectrons from matter. However, the classical theory predicts that the total amount of energy in a light wave increases as the light intensity increases. Therefore, classically the electrons should have more kinetic energy if the light intensity is increased. However, according to result (1) earlier and Figure 3.11, a characteristic retarding potential $-V_0$ is sufficient to stop all photoelectrons for a given light frequency ν, no matter what the intensity. Classical electromagnetic theory is unable to explain this result. Similarly, classical theory cannot explain result (2), because the maximum kinetic energy of the photoelectrons depends on the value of the light frequency ν and not on the intensity.

The existence of a threshold frequency, as shown in experimental result (3) is completely inexplicable in classical theory. Classical theory cannot predict the results shown in Figure 3.13. Classical theory does predict that the number of

Difficulties of classical theory

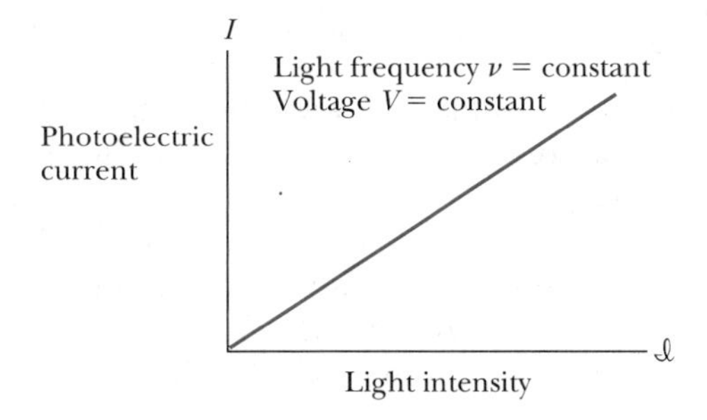

FIGURE 3.14 The photoelectric current I is a linear function of the light intensity for a constant ν and V.

photoelectrons ejected will increase with intensity in agreement with experimental result (4).

Finally, classical theory would predict that for extremely low light intensities, a long time would elapse before any one electron could obtain sufficient energy to escape. We observe, however, that the photoelectrons are ejected almost immediately. Experiments have shown that a light intensity equivalent to the illumination produced over a 1-cm² area by a 100-watt incandescent bulb at a distance of 1000 km is sufficient to produce photoelectrons within a second.

Example 3.9

Photoelectrons may be emitted from sodium ($\phi = 2.3$ eV) even for light intensities as low as 10^{-8} W/m². Calculate classically how long the light must shine in order to produce a photoelectron of kinetic energy 1 eV.

Solution: Let's assume that all of the light is absorbed in the first layer of atoms in the surface. First calculate the number of sodium atoms per unit area in a layer one atom thick.

$$\frac{\text{Avogadro's number}}{\text{Na gram molecular weight}} \times \text{Density} = \frac{\text{Number of Na atoms}}{\text{Volume}}$$

$$\frac{6.02 \times 10^{23}\ \text{atoms/mole}}{23\ \text{g/mole}} \times 0.97 \frac{\text{g}}{\text{cm}^3} =$$

$$2.54 \times 10^{22} \frac{\text{atoms}}{\text{cm}^3} = 2.54 \times 10^{28} \frac{\text{atoms}}{\text{m}^3} \tag{3.26}$$

To estimate the thickness of one layer of atoms, we assume a cubic structure.

$$\frac{1\ \text{atom}}{d^3} = 2.54 \times 10^{28} \frac{\text{atoms}}{\text{m}^3}$$

$$d = 3.40 \times 10^{-10}\ \text{m}$$

$$= \text{thickness of one layer of sodium atoms}$$

If all the light is absorbed in the first layer of atoms, the number of exposed atoms per m² is

$$2.54 \times 10^{28} \frac{\text{atoms}}{\text{m}^3} \times 3.40 \times 10^{-10}\ \text{m} = 8.64 \times 10^{18} \frac{\text{atoms}}{\text{m}^2}$$

With the intensity of 10^{-8} W/m², each atom will receive energy at the rate of

$$10^{-8} \frac{\text{W}}{\text{m}^2} \times \frac{1}{8.64 \times 10^{18}\ \text{atoms/m}^2} = 1.16 \times 10^{-27}\ \text{W}$$

$$= 1.16 \times 10^{-27} \frac{\text{J}}{\text{s}} \times \frac{1}{1.6 \times 10^{-19}\ \text{J/eV}}$$

$$= 7.25 \times 10^{-9}\ \text{eV/s}$$

We have assumed that each atom absorbs, on the average, the same energy and that a single electron in the atom absorbs all the energy. The energy needed to eject the photoelectron is 2.3 eV for the work function and 1 eV for the kinetic energy, for a total of 3.3 eV. Using the rate of energy absorption of 7.25×10^{-9} eV/s, we can calculate the time T needed to absorb 3.3 eV:

$$T = 3.3\ \text{eV} \frac{1}{7.25 \times 10^{-9}\ \text{eV/s}}$$

$$= 4.55 \times 10^{8}\ \text{s}$$

$$= 14\ \text{years}$$

The time calculated classically to eject a photoelectron is 14 years!

Einstein's Theory

Albert Einstein was intrigued by Planck's hypothesis that the electromagnetic radiation field had to be absorbed and emitted in quantized amounts. Einstein took Planck's idea one step further and suggested that the *electromagnetic radiation field itself is quantized,* and that "the energy of a light ray spreading out from a point source is not continuously distributed over an increasing space but consists of a finite number of energy quanta which are localized at points in space, which move without dividing, and which can only be produced and absorbed as complete units."* We now call these energy quanta of light **photons.** According to Einstein each photon has the energy quantum,

$$E = h\nu \tag{3.27}$$

Energy quantum

where ν is the frequency of the electromagnetic wave associated with the light and h is Planck's constant. The photon travels at the speed of light in a vacuum, and its wavelength λ is given by

$$\lambda\nu = c \tag{3.28}$$

In other words, Einstein proposed that in addition to its well-known wavelike aspect, amply exhibited in interference phenomena, light should also be considered to have a particlelike aspect. Einstein suggested that the photon (quantum of light) delivers its entire energy $h\nu$ to a single electron in the material. In order to leave the material, the struck electron must give up an amount of energy ϕ to overcome its binding in the material. The electron may lose some additional energy by interacting with other electrons on its way to the surface. Whatever energy remains will then appear as kinetic energy of the electron as it leaves the emitter. The conservation of energy requires that

$$\text{Energy before (photon)} = \text{Energy after (electron)}$$

$$h\nu = \phi + \text{K.E. (electron)} \tag{3.29}$$

Because the energies involved here are on the order of eV, we are safe in using the nonrelativistic form of the electron's kinetic energy, $\frac{1}{2}mv^2$. The electron's energy will be degraded as it passes through the emitter material, so, strictly speaking, we want to experimentally detect the maximum value of the kinetic energy.

$$h\nu = \phi + \frac{1}{2}mv_{\text{max}}^2 \tag{3.30}$$

The retarding potentials measured in the photoelectric effect are thus the opposing potentials needed to stop the most energetic electrons.

$$eV_0 = \frac{1}{2}mv_{\text{max}}^2 \tag{3.31}$$

Quantum Interpretation

We should now re-examine the experimental results of the photoelectric effect to see whether Einstein's quantum interpretation can explain all the data. The first and second experimental results are easily explained because the kinetic energy of the electrons does not depend on the light intensity at all, but only on the light frequency and the work function of the material.

$$\frac{1}{2}mv_{\text{max}}^2 = eV_0 = h\nu - \phi \tag{3.32}$$

A potential slightly more positive than $-V_0$ will not be able to repel all the electrons, and, for a close geometry of the emitter and collector, practically all the electrons will be collected when the retarding voltage is near zero. For very large

*See the English translation of A. Einstein, *Ann. Physik* **17,** 132 (1905) by A. B. Arons and M. B. Peppard, *Am. J. Phys.* **33,** 367 (1965).

positive potentials all the electrons will be collected, and the photocurrent will level off as shown in Figure 3.11. If the light intensity increases, there will be more photons per unit area, more electrons ejected, and therefore a higher photocurrent, as displayed in Figure 3.11.

If a different light frequency is used, say ν_2, then a different stopping potential is required to stop the most energetic electrons [see Equation (3.32)], $eV_{02} = h\nu_2 - \phi$. For a constant light intensity (more precisely, a constant number of photons/area/time), a different stopping potential V_0 is required for each ν, but the maximum photocurrent will not change, because the number of photoelectrons ejected is constant (see Figure 3.12). The quantum theory easily explains Figure 3.14, because the number of photons increases linearly with the light intensity, producing more photoelectrons and hence more photocurrent.

Millikan believed Einstein was wrong

Equation (3.32), proposed by Einstein in 1905, predicts that the stopping potential will be linearly proportional to the light frequency, with a slope *h, the same constant found by Planck.* The slope is independent of the metal used to construct the photocathode. The data available in 1905 were not sufficiently accurate either to prove or to disprove Einstein's theory, and the theory was received with skepticism, even by Planck himself. R. A. Millikan, then at the University of Chicago, tried to show Einstein was wrong by undertaking a series of elegant experiments that required almost ten years to complete. In 1916 he reported data confirming Einstein's prediction. From data similar to that shown in Figure 3.13, Millikan found the value of h to be in almost exact agreement with the one determined for blackbody radiation by Planck. Equation (3.32) can be rewritten

$$eV_0 = \frac{1}{2}mv_{\text{max}}^2 = h\nu - h\nu_0 \tag{3.33}$$

where $\phi = h\nu_0$ represents the negative of the y intercept. The frequency ν_0 represents the threshold frequency for the photoelectric effect (when the kinetic energy of the electron is precisely zero). Einstein's theory of the photoelectric effect was gradually accepted after 1916; finally in 1922 he received the Nobel Prize for the year 1921, primarily for his explanation of the photoelectric effect.*

Quantization of electromagnetic radiation field

We should summarize what we have learned about the quantization of the electromagnetic radiation field. First, electromagnetic radiation consists of photons which are particlelike (or corpuscular), each consisting of energy

$$E = h\nu = \frac{hc}{\lambda} \tag{3.34}$$

where ν and λ are the frequency and wavelength of the light, respectively. The total energy of a beam of light is the sum total of the energy of all the photons and for monochromatic light is an integral multiple of $h\nu$ (generally the integer is very large).

This representation of the photon picture must be true over the entire electromagnetic spectrum from radio waves to visible light, x rays, and even high-energy gamma rays. This must be true because, as we saw in Chapter 2, a photon of given frequency, observed from a moving system, can be redshifted or blueshifted by an arbitrarily large amount, depending on the system's speed and

*R. A. Millikan also received the Nobel Prize in 1923, partly for his precise study of the photoelectric effect and partly for measuring the charge of the electron. Millikan's award was the last in a series of Nobel Prizes spanning 18 years that honored the fundamental efforts to measure and understand the photoelectric effect: Lenard, Einstein, and Millikan.

direction of motion. We will examine these possibilities later. During emission or absorption of any form of electromagnetic radiation (light, x rays, gamma rays, etc.), photons must be created or absorbed. The photons have only one speed: the speed of light ($= c$ in vacuum).

Example 3.10

Light of wavelength 400 nm is incident upon lithium ($\phi =$ 2.9 eV). Calculate (a) the photon energy and (b) the stopping potential V_0.

Solution: (a) Light is normally described by wavelengths in nm, so it is useful to have an equation to calculate the energy in terms of λ.

$$E = h\nu = \frac{hc}{\lambda} = \frac{(6.626 \times 10^{-34}\ \text{J}\cdot\text{s})(2.998 \times 10^8\ \text{m/s})}{\lambda(1.602 \times 10^{-19}\ \text{J/eV})(10^{-9}\ \text{m/nm})}$$

$$E = \frac{1.240 \times 10^3\ \text{eV}\cdot\text{nm}}{\lambda} \tag{3.35}$$

For a wavelength of $\lambda = 400$ nm we have for the photon's energy:

$$E = \frac{1.240 \times 10^3\ \text{eV}\cdot\text{nm}}{400\ \text{nm}} = 3.1\ \text{eV}$$

(b) We determine the stopping potential from Equation (3.32).

$$eV_0 = h\nu - \phi = E - \phi$$
$$= 3.1\ \text{eV} - 2.9\ \text{eV} = 0.2\ \text{eV}$$
$$V_0 = 0.2\ \text{V}$$

A retarding potential of 0.2 V will stop all photoelectrons.

Example 3.11

What frequency of light is needed to produce electrons of kinetic energy 3 eV from illumination of lithium?

Solution: We determine the photon energy from Equation (3.30).

$$h\nu = \phi + \frac{1}{2}mv_{\text{max}}^2$$
$$= 2.9\ \text{eV} + 3.0\ \text{eV} = 5.9\ \text{eV}$$

The photon frequency is now found to be

$$\nu = \frac{E}{h} = \frac{(5.9\ \text{eV})(1.6 \times 10^{-19}\ \text{J/eV})}{(6.626 \times 10^{-34}\ \text{J}\cdot\text{s})}$$
$$= 1.42 \times 10^{15}\ \text{s}^{-1} = 1.42 \times 10^{15}\ \text{Hz}$$

Example 3.12

For the light intensity of Example 3.9, $\mathcal{I} = 10^{-8}\ \text{W/m}^2$, a wavelength of 350 nm is used. What is the number of photons/area/s in the light beam?

Solution: From Equation (3.35) we have

$$E_\gamma = \frac{1.24 \times 10^3\ \text{eV}\cdot\text{nm}}{350\ \text{nm}} = 3.5\ \text{eV}$$

where E_γ represents the photon's energy. Because

$$\text{Intensity } \mathcal{I} = \left[N\frac{\text{photons}}{\text{area time}}\right]\left[E_\gamma\frac{\text{energy}}{\text{photon}}\right]$$
$$= NE_\gamma\frac{\text{energy}}{\text{area time}}$$

then

$$N = \frac{\mathcal{I}}{E_\gamma} = \frac{10^{-8}\ \text{J}\cdot\text{s}^{-1}\text{m}^{-2}}{(1.6 \times 10^{-19}\ \text{J/eV})(3.5\ \text{eV/photon})}$$
$$= 1.8 \times 10^{10}\ \frac{\text{photons}}{\text{m}^2\cdot\text{s}}$$

Thus even a low-intensity light beam has a large flux of photons, and even a few photons can produce a photocurrent (albeit a very small one!).

3.7 X-Ray Production

In the photoelectric effect, a photon gives up all of its energy to an electron, which may then escape from the material to which it was bound. Can the inverse process occur? Can an electron (or any charged particle) give up its energy and create a photon? The answer is yes, but the process must be consistent with other laws of physics. Recall that photons must be created or absorbed as whole units. A photon cannot give up half its energy. Rather, it must give up all its energy. If in some physical process only part of the photon's energy were required, then a **new** photon would be created to carry away the remaining energy.

However, electrons do not act as photons. An electron may give up part or all of its kinetic energy and still be the same electron. As we now know, a photon is electromagnetic radiation. When an electron interacts with the strong electric field of the atomic nucleus and is consequently accelerated, the electron will radiate electromagnetic energy. According to classical electromagnetic theory, it would do so continuously. In the quantum picture we must think of the electron as emitting a series of photons with varying energies; this is the only way that the inverse photoelectric effect can occur. An energetic electron passing through matter will radiate photons and lose kinetic energy. The process by which photons are emitted by an electron slowing down is called **bremsstrahlung,** from the German word for "braking radiation." The process is shown schematically in Figure 3.15 where an electron (energy E_i) passing through the electric field of a nucleus slows down and produces a photon ($E = h\nu$). The final energy of the electron is then

Bremsstrahlung process

$$E_f = E_i - h\nu \tag{3.36}$$

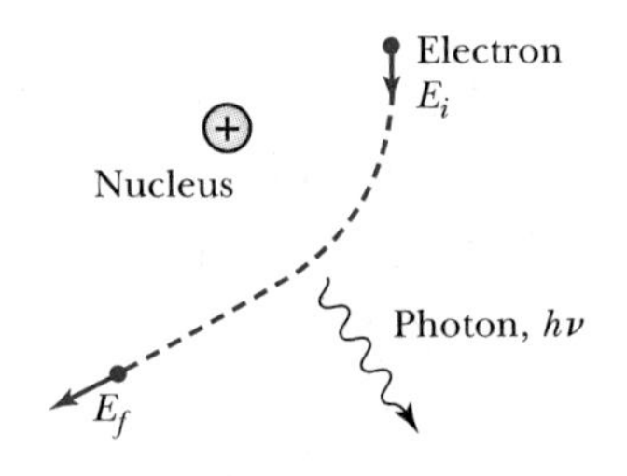

FIGURE 3.15 *Bremsstrahlung* is a process by which an electron is accelerated while under the influence of the nucleus. The accelerated electron emits a photon.

from the conservation of energy. The nucleus absorbs very little energy in order to conserve linear momentum. One or more photons may be created in this way as electrons pass through matter.

In Section 3.1 we mentioned Röntgen's discovery of x rays. The x rays are produced by the bremsstrahlung effect in apparatus shown schematically in Figure 3.16. Current passing through a filament produces copious numbers of electrons by thermionic emission. These electrons are focused by the cathode structure into a beam and are accelerated by voltages of thousands of volts until they impinge on a metal anode surface, producing x rays by bremsstrahlung (and other processes) as they stop in the anode material. Much of the electron's kinetic energy is lost by heating the anode and not by bremsstrahlung. The x-ray tube is evacuated so that the air between the filament and anode will not scatter

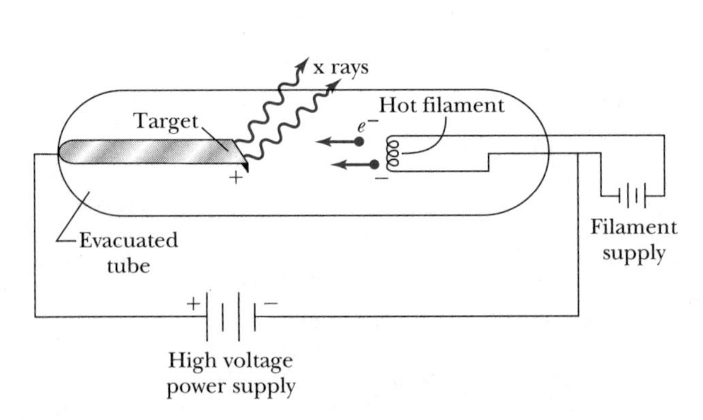

FIGURE 3.16 Schematic of x-ray tube where x rays are produced by the bremsstrahlung process of energetic electrons.

the electrons. The x rays produced pass through the sides of the tube and can be used for a large number of applications, including medical diagnosis and therapy, fundamental research in crystal and liquid structure, and, in engineering, the diagnosis of flaws in large welds and castings. X rays from a standard tube include photons of many wavelengths. By scattering x rays from crystals we can produce strongly collimated monochromatic (single wavelength) x-ray beams. Early x-ray spectra produced by x-ray tubes of accelerating potential 35 kV are shown in Figure 3.17. These particular tubes had targets of tungsten, molybdenum, and chromium. The smooth, continuous x-ray spectra are those produced by bremsstrahlung, and the sharp "characteristic x rays" are produced by atomic excitations and will be explained in Section 4.6. X-ray wavelengths typically range from 0.01 to 1 nm. However, high-energy accelerators can produce x rays with wavelengths as short as 10^{-6} nm.

Notice that in Figure 3.17 the minimum wavelength λ_{min} for all three targets is the same. The minimum wavelength λ_{min} corresponds to the maximum frequency. If the electrons are accelerated through a voltage V_0, then their kinetic energy is eV_0. The maximum photon energy therefore occurs when the electron

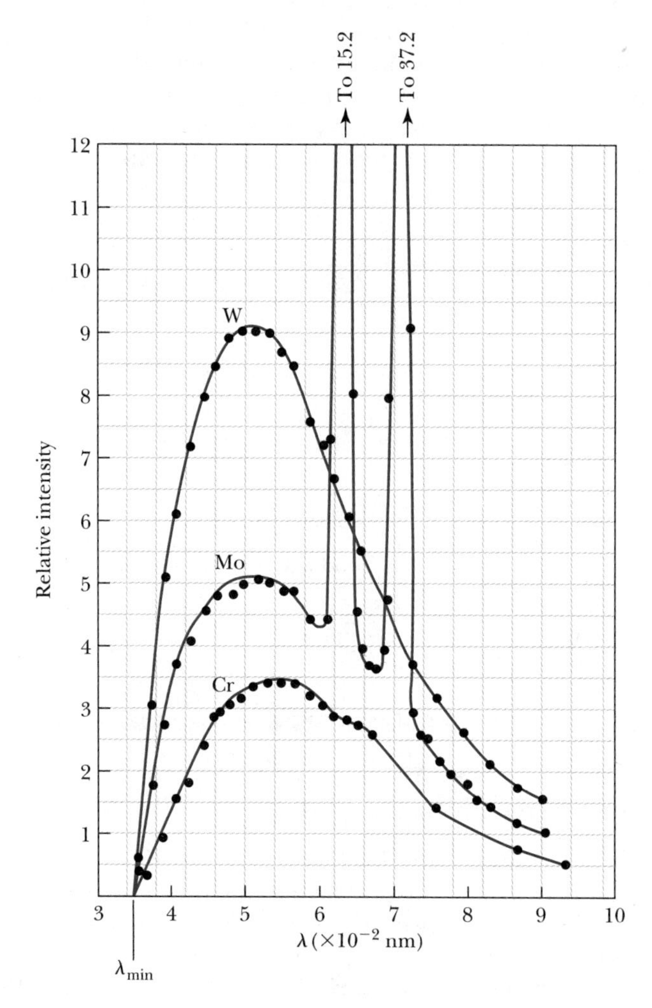

FIGURE 3.17 The relative intensity of x rays produced in an x-ray tube is shown for an accelerating voltage of 35 kV. Notice that λ_{min} is the same for all three targets. *From C. T. Ulrey, Physical Review* ***11****, 405 (1918).*

gives up all of its kinetic energy and creates one photon (this is relatively unlikely, however). This process is the **inverse photoelectric effect.** The conservation of energy requires that the electron kinetic energy equals the maximum photon energy (where we neglect the work function ϕ because it is normally so small compared to eV_0).

$$eV_0 = h\nu_{max} = \frac{hc}{\lambda_{min}}$$

or

Duane-Hunt rule

$$\lambda_{min} = \frac{hc}{e}\frac{1}{V_0} = \frac{1.24 \times 10^{-6}\ \text{V}\cdot\text{m}}{V_0} \tag{3.37}$$

The relation Equation (3.37) was first found experimentally and is known as the **Duane-Hunt rule.** Its explanation in 1915 by the quantum theory is now considered further evidence of Einstein's photon concept. The value λ_{min} depends only on the accelerating voltage and is the same for all targets.

Only the quantum hypothesis explains all of the data. Because the heavier elements have stronger nuclear electric fields, they are more effective in decelerating electrons and making them radiate. The intensity of the x rays increases with the square of the atomic number of the target. The intensity is also approximately proportional to the square of the voltage used to accelerate the electrons. This is why high voltages and tungsten anodes are so often used in x-ray machines. Tungsten also has a very high melting temperature and can withstand high electron-beam currents.

Example 3.13

If we have a tungsten anode (work function $\phi = 4.5$ eV) and electron acceleration voltage of 35 kV, why do we ignore in Equation (3.36) the initial kinetic energy of the electrons from the filament and the work functions of the filaments and anodes? What is the minimum wavelength of the x rays?

Solution: The initial kinetic energies and work functions are on the order of a few electron volts (eV), whereas the kinetic energy of the electrons due to the accelerating voltage is 35,000 eV. The error in neglecting everything but eV_0 is small.

Using the Duane-Hunt rule of Equation (3.37) we determine

$$\lambda_{min} = \frac{1.24 \times 10^{-6}\ \text{V}\cdot\text{m}}{35 \times 10^3\ \text{V}} = 3.54 \times 10^{-11}\ \text{m}$$

$$= 0.0354\ \text{nm}$$

which is in good agreement with the data of Figure 3.17.

3.8 Compton Effect

When a photon enters matter, it is likely to interact with one of the atomic electrons. Classically, the electrons will oscillate at the photon frequency because of the interaction of the electron with the electric and magnetic field of the photon and will reradiate electromagnetic radiation (photons) at this same frequency. This is called **Thomson scattering.** However, in the early 1920s Arthur Compton experimentally confirmed an earlier observation by J. A. Gray that, especially at backward-scattering angles, there appeared to be a component of the emitted radiation (called a **modified** wave) that had a longer wavelength than

Thomson scattering

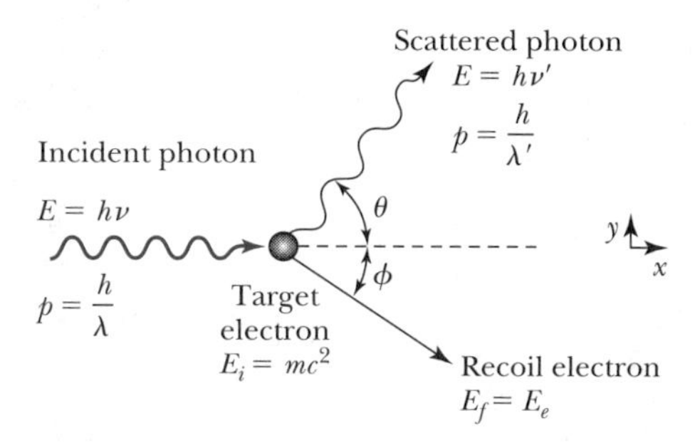

FIGURE 3.18 Compton scattering of a photon by an electron essentially at rest.

Professor Arthur Compton of the University of Chicago is shown here in 1931 looking into an ionization chamber that he designed to study cosmic rays in the atmosphere. These complex detectors had to be quite sturdy and were carefully tested and calibrated on Earth before sending up with balloons. *UPI/Corbis-Bettmann.*

the original primary (**unmodified**) wave. Classical electromagnetic theory cannot explain this modified wave. Compton then attempted to understand theoretically such a process and could only find one explanation: Einstein's photon particle concept must be correct. The scattering process is shown in Figure 3.18.

Compton proposed in 1923 that the photon is scattered from only one electron, rather than from all the electrons in the material, and that the laws of the conservation of energy and momentum apply as in any elastic collision between two particles. We recall from Chapter 2 that the momentum of a particle moving at the speed of light (photon) is given by

$$p = \frac{E}{c} = \frac{h\nu}{c} = \frac{h}{\lambda} \tag{3.38}$$

We treat the photon as a particle with a definite energy and momentum. Scattering takes place in a plane, which we take to be the xy plane in Figure 3.18. Because momentum is a vector, both x and y components must be conserved. The energy and momentum before and after the collision are given below (treated relativistically).

Compton scattering

	Initial System	**Final System**
Photon energy	$h\nu$	$h\nu'$
Photon momentum in x direction (p_x)	$\frac{h}{\lambda}$	$\frac{h}{\lambda'}\cos\theta$
Photon momentum in y direction (p_y)	0	$\frac{h}{\lambda'}\sin\theta$
Electron energy	mc^2	$E_e = mc^2 + \text{K.E.}$
Electron momentum in x direction (p_x)	0	$p_e\cos\phi$
Electron momentum in y direction (p_y)	0	$-p_e\sin\phi$

In the final system the electron's total energy is related to its momentum by

$$E_e^2 = (mc^2)^2 + p_e^2c^2 \tag{3.39}$$

We can write the conservation laws now as

$$\text{Energy:} \quad h\nu + mc^2 = h\nu' + E_e \tag{3.40a}$$

$$p_x: \quad \frac{h}{\lambda} = \frac{h}{\lambda'}\cos\theta + p_e\cos\phi \tag{3.40b}$$

$$p_y: \quad \frac{h}{\lambda'}\sin\theta = p_e\sin\phi \tag{3.40c}$$

We will relate the change in wavelength $\Delta\lambda = \lambda' - \lambda$ to the scattering angle θ of the photon. We first eliminate the recoil angle ϕ by squaring Equations (3.40b) and (3.40c) and adding them together, resulting in

$$p_e^2 = \left(\frac{h}{\lambda}\right)^2 + \left(\frac{h}{\lambda'}\right)^2 - 2\left(\frac{h}{\lambda}\right)\left(\frac{h}{\lambda'}\right)\cos\theta \tag{3.41}$$

Then we substitute E_e from Equation (3.40a) and p_e from Equation (3.41) into Equation (3.39) (setting $\lambda = c/\nu$).

$$[h(\nu - \nu') + mc^2]^2 = m^2c^4 + (h\nu)^2 + (h\nu')^2 - 2(h\nu)(h\nu')\cos\theta$$

Squaring the left-hand side and canceling terms leaves

$$mc^2(\nu - \nu') = h\nu\nu'(1 - \cos\theta)$$

Rearranging terms gives

$$\frac{h}{mc^2}(1 - \cos\theta) = \frac{\nu - \nu'}{\nu\nu'} = \frac{\dfrac{c}{\lambda} - \dfrac{c}{\lambda'}}{\dfrac{c^2}{\lambda\lambda'}} = \frac{1}{c}(\lambda' - \lambda)$$

or

Compton effect

$$\Delta\lambda = \lambda' - \lambda = \frac{h}{mc}(1 - \cos\theta) \tag{3.42}$$

which is the result Compton found in 1923 for the increase in wavelength of the scattered photon.

Compton then proceeded to check the validity of his theoretical result by performing a very careful experiment in which he scattered x rays of wavelength 0.071 nm from carbon at several angles and showed that the modified wavelength was observed in good agreement with his prediction.* A part of his data is shown in Figure 3.19 where both the modified (λ') and unmodified scattered waves (λ) are seen.

The kinetic energy and scattering angle of the recoiling electron can also be predicted. Experiments in which the recoiling electrons were detected were soon carried out, thus confirming Compton's theory completely. The process of elastic photon scattering from electrons is now called the **Compton effect.** Note that the difference in wavelength, $\Delta\lambda = \lambda' - \lambda$, only depends on the constants h, c, and m_e in addition to the scattering angle θ. The quantity $\lambda_c = h/m_ec = 2.43 \times 10^{-3}$ nm is called the **Compton wavelength** of the electron. Only for wavelengths on the same order as λ_c (or shorter) will the fractional shift $\Delta\lambda/\lambda$ be large. For visible light, for example with $\lambda = 500$ nm, the maximum $\Delta\lambda/\lambda$ is of the order of 10^{-5}, and $\Delta\lambda$ would be difficult to detect. The probability of the occurrence of the Compton effect for visible light is also quite small. However, for the x rays of wavelength 0.071 nm used by Compton, the ratio of $\Delta\lambda/\lambda$ is ~0.03 and could easily be observed. Thus, the Compton effect is important only for x rays or γ-ray photons and is small for visible light.

Compton wavelength

The physical process of the Compton effect can be described as follows. The photon elastically scatters from an essentially free electron in the material. (The

*An interesting self-account of Compton's discovery can be found in A. H. Compton, *Am. J. Phys.* **29**, 817–820 (1961).

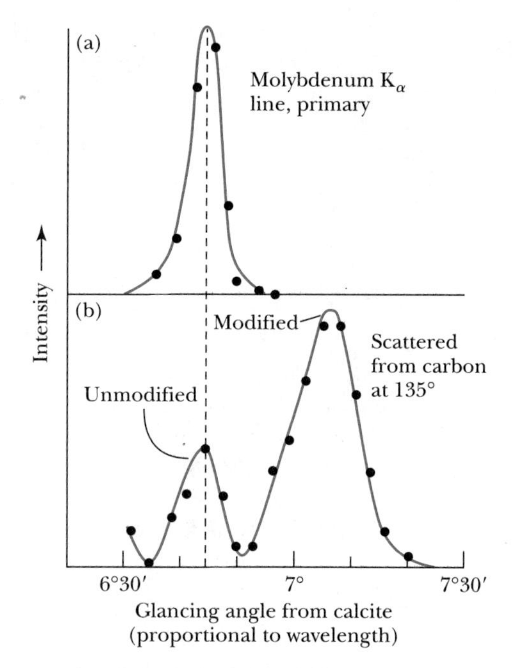

FIGURE 3.19 Compton's original data showing the primary x-ray beam from Mo unscattered in (a), and the scattered spectrum from carbon at 135° showing both the modified and unmodified wave in (b). *Adapted from Arthur H. Compton, Physical Review* **22**, *409 (1923).*

photon's energy is so much larger than the binding energy of the almost free electron that the binding energy can be neglected.) The newly created scattered photon then has a modified, longer wavelength. What happens if the photon scatters from one of the tightly bound inner electrons? Then the binding energy is not negligible, and the electron may not be able to be dislodged. The scattering would then effectively be from a much heavier system (nucleus + electrons). Then the mass in Equation (3.42) will be several thousand times larger than m_e, and $\Delta\lambda$ would be correspondingly smaller. Scattering from tightly bound electrons results in the unmodified photon scattering ($\lambda \approx \lambda'$), which also is observed in Figure 3.19. Thus, the quantum picture also explains the existence of the unmodified wavelength predicted by the classical theory (Thomson scattering) alluded to earlier.

The success of the Compton theory convincingly demonstrated the correctness of both the quantum concept and the particle nature of the photon. The use of the laws of the conservation of energy and momentum applied relativistically to pointlike scattering of the photon from the electron finally convinced the great majority of scientists of the validity of the new modern physics. Compton received the Nobel Prize in 1927.

Example 3.14

An x ray of wavelength 0.05 nm scatters from a gold target. (a) Can the x ray be Compton-scattered from an electron bound by as much as 62,000 eV? (b) What is the largest wavelength of scattered photon that can be observed? (c) What is the kinetic energy of the most energetic recoil electron and at what angle does it occur?

Solution: From Equation (3.35) the x-ray energy is

$$E_{\text{x ray}} = \frac{1.24 \times 10^3 \text{ eV} \cdot \text{nm}}{0.05 \text{ nm}} = 24{,}800 \text{ eV} = 24.8 \text{ keV}$$

Therefore, the x ray does not have enough energy to dislodge the inner electron, which is bound by 62 keV. In this case we have to use the atomic mass in Equation (3.42), which results in little change in the wavelength (Thomson scattering). Scattering may still occur from outer electrons.

The longest wavelength $\lambda' = \lambda + \Delta\lambda$ occurs when $\Delta\lambda$ is a maximum or when $\theta = 180°$.

$$\lambda' = \lambda + \frac{h}{m_e c}(1 - \cos 180°) = \lambda + \frac{2h}{m_e c}$$

$$= 0.05 \text{ nm} + 2(0.00243 \text{ nm}) = 0.055 \text{ nm}$$

The energy of the scattered photon is then a minimum and has the value

$$E'_{\text{x ray}} = \frac{1.24 \times 10^3 \text{ eV} \cdot \text{nm}}{0.055 \text{ nm}} = 2.25 \times 10^4 \text{ eV} = 22.5 \text{ keV}$$

The difference in energy of the initial and final photon must equal the kinetic energy of the electron (neglecting binding energies). The recoil electron must scatter in the forward direction at $\phi = 0°$ when the final photon is in the backward direction ($\theta = 180°$) in order to conserve momentum. The kinetic energy of the electron is then a maximum.

$$E_{\text{x ray}} = E'_{\text{x ray}} + \text{K.E. (electron)}$$

$$\text{K.E. (electron)} = E_{\text{x ray}} - E'_{\text{x ray}}$$

$$= 24.8 \text{ keV} - 22.5 \text{ keV} = 2.3 \text{ keV}$$

Because $\Delta\lambda$ does not depend on λ or λ', we can determine the wavelength (and energy) of the incident photon by merely observing the kinetic energy of the electron at forward angles (see Problem 50).

3.9 Pair Production and Annihilation

A general rule of nature is that if some process is not absolutely forbidden (by some law like conservation of energy, momentum, or charge) it will eventually occur. In the photoelectric effect, bremsstrahlung, and the Compton effect, we have studied exchanges of energy between photons and electrons. Have we covered all possible exchanges? For example, can the kinetic energy of a photon be converted into particle mass and vice versa? It would appear that if none of the conservation laws are violated, then such a process should be possible.

First, let us consider the conversion of photon energy into mass. The electron, which has a mass, $m = 0.51 \text{ MeV}/c^2$, is the lightest particle within an atom. Because an electron has negative charge, we must also create a positive charge to balance charge conservation. However, in 1932, C. D. Anderson (Nobel Prize, 1936) observed a positively charged electron (e^+) in cosmic radiation. This particle, called a **positron,** had been predicted to exist several years earlier by P. A. M. Dirac (Nobel Prize, 1933). It has the same mass as the electron but an opposite charge. Positrons are also observed when high-energy gamma rays (photons) pass through matter. Experiments show that a photon's energy can be converted entirely into an electron and a positron in the reaction

Positron

Pair production

$$\gamma \rightarrow e^+ + e^- \tag{3.43}$$

However, this process only occurs when the photon passes through matter, because energy and momentum are not conserved when the reaction takes place in isolation: the missing momentum must be supplied by interaction with a massive object such as a nucleus.

Example 3.15

Show that a photon cannot produce an electron–positron pair in free space as shown in Figure 3.20a.

Solution: Let the total energy and momentum of the electron and positron be E_-, p_- and E_+, p_+, respectively. The conservation laws are then

Energy $$h\nu = E_+ + E_- \tag{3.44a}$$

Momentum, p_x $$\frac{h\nu}{c} = p_- \cos\theta_- + p_+ \cos\theta_+ \tag{3.44b}$$

Momentum, p_y $$0 = p_- \sin\theta_- - p_+ \sin\theta_+ \tag{3.44c}$$

Equation (3.44b) can be written as

$$h\nu = p_- c\cos\theta_- + p_+ c\cos\theta_+ \tag{3.45}$$

Show that a photon cannot produce an electron–positron pair in free space as shown in Figure 3.20a.

Solution: Let the total energy and momentum of the electron and positron be E_-, p_- and E_+, p_+, respectively. The conservation laws are then

Energy $$h\nu = E_+ + E_- \tag{3.44a}$$

Momentum, p_x $$\frac{h\nu}{c} = p_- \cos\theta_- + p_+ \cos\theta_+ \tag{3.44b}$$

Momentum, p_y $$0 = p_- \sin\theta_- - p_+ \sin\theta_+ \tag{3.44c}$$

Equation (3.44b) can be written as

$$h\nu = p_- c\cos\theta_- + p_+ c\cos\theta_+ \tag{3.45}$$

If we insert $E_\pm^2 = p_\pm^2 c^2 + m^2c^4$ into Equation (3.44a), we have

$$h\nu = \sqrt{p_+^2c^2 + m^2c^4} + \sqrt{p_-^2c^2 + m^2c^4} \tag{3.46}$$

The maximum value of $h\nu$ is, from Equation (3.45),

$$h\nu_{\text{max}} = p_- c + p_+ c$$

But from Equation (3.46), we also have

$$h\nu > p_- c + p_+ c$$

Equations (3.45) and (3.46) are inconsistent and cannot simultaneously be valid. Equations (3.44), therefore, do not describe a possible reaction. The reaction displayed in Figure 3.20a is not possible, because energy and momentum are not simultaneously conserved.

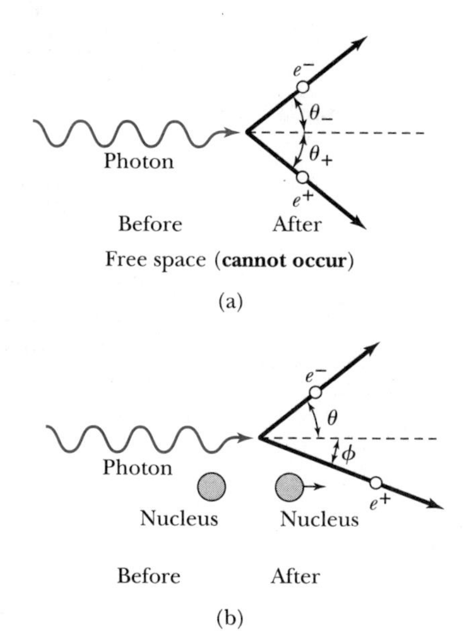

FIGURE 3.20 (a) A photon cannot decay into an electron–positron pair in free space, but (b) near a nucleus, the nucleus can absorb sufficient momentum to allow the process to proceed.

Consider the conversion of a photon into an electron and positron (called **pair production**) that takes place inside an atom where the electric field of a nucleus is large. The nucleus recoils and takes away a negligible amount of energy but a considerable amount of momentum. The conservation of energy will now be

$$h\nu = E_+ + E_- + \text{K.E. (nucleus)} \tag{3.47}$$

A diagram of the process is shown in Figure 3.20b. The photon energy must be at least equal to $2m_ec^2$ in order to create the rest masses.

$$h\nu > 2m_ec^2 = 1.02 \text{ MeV} \qquad \text{(for pair production)} \tag{3.48}$$

The probability of pair production increases dramatically both with higher photon energy and with higher atomic number Z of the nearby nucleus because of the correspondingly higher electric field which mediates the process.

The next question concerns the new particle, the positron. Why is it not commonly found in nature? We need to answer also the question posed earlier. Can mass be converted to pure kinetic energy?

Positrons are found in nature. They are detected in cosmic radiation and as products of radioactivity from a few radioactive elements. However, their lives are doomed because of their interaction with electrons. When positrons and electrons are in near proximity for even a short period of time they annihilate each other, producing photons. A positron passing through matter will quickly lose its kinetic energy through atomic collisions and with some probability will **annihilate** with an electron. After a positron slows down, it is drawn to an electron by their mutual electric attraction, and the electron and positron may then form an atomlike configuration called **positronium,** where they rotate around each other. Eventually the electron and positron come together and annihilate each other (typically in 10^{-10} s) producing electromagnetic radiation (photons). The process $e^+ + e^- \rightarrow \gamma + \gamma$ is called **pair annihilation.**

Pair annihilation

Consider a positronium "atom" in free space. It must emit at least two photons in order to conserve energy and momentum. If the positronium annihilation takes place near a nucleus, it is possible that only one photon will be created, the missing momentum being supplied by nucleus recoil as in pair production, and under certain conditions three photons may be produced. Because the emission of two photons is by far the most likely annihilation mode, let us consider this mode, displayed in Figure 3.21. The conservation laws for the process $(e^+e^-)_{\text{atom}} \rightarrow \gamma + \gamma$ will be (we neglect the atomic binding energy of about 6.8 eV)

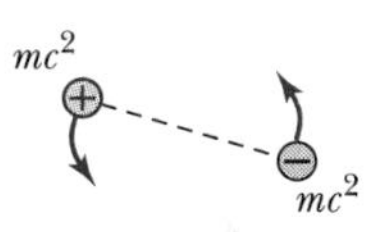

Positronium, before decay (schematic only)

(a)

hv_1

hv_2

After annihilation

(b)

FIGURE 3.21 Annihilation of positronium atom (consisting of an electron and positron), producing two photons.

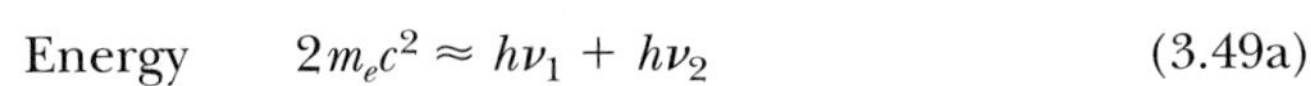

$$\text{Energy} \qquad 2m_ec^2 \approx h\nu_1 + h\nu_2 \tag{3.49a}$$

$$\text{Momentum} \qquad 0 = \frac{h\nu_1}{c} - \frac{h\nu_2}{c} \tag{3.49b}$$

where the photons obviously emerge in precisely opposite directions with equal energies, because the initial momentum is assumed to be zero (positronium at rest). Hence $\nu_1 = \nu_2 = \nu$. Thus Equation (3.49a) becomes

$$2m_ec^2 = 2h\nu$$

or

$$h\nu = m_ec^2 = 0.511 \text{ MeV} \tag{3.50}$$

In other words, the two photons from positronium annihilation will move in opposite directions, each with energy 0.511 MeV. This is exactly what is observed experimentally.

The production of two photons in opposite directions with energies of about $\frac{1}{2}$ MeV is so characteristic a signal of the presence of a positron that it has useful applications. Positron Emission Tomography (PET) scanning has become a standard diagnostic technique in medicine. A positron-emitting radioactive chemical (containing a nucleus such as ^{15}O, ^{11}C, ^{13}N, or ^{18}F) injected into the body causes two characteristic annihilation photons to be emitted from the points where the chemical has been concentrated by physiological processes. The location in the body where the photons originate is identified by measuring the directions of two gamma-ray photons of the correct energy that are detected in coincidence, as shown in Figure 3.22. Measurement of blood flow in the brain is an example of a diagnostic tool used in the evaluation of strokes, brain tumors, and other brain lesions.

PET scan

Before leaving the subject of positrons we should pursue briefly the idea of **antiparticles.** The positron is the antiparticle of the electron, having the opposite charge but the same mass.* In 1955 the antiproton was discovered by E. G. Segrè and O. Chamberlain (Nobel Prize, 1959), and by now, many antiparticles have been found. Physicists love to find symmetry in nature. We now

Antiparticles

*There are other particle properties (for example, spin) that will be described later (particularly in Chapters 7 and 14) and also need to be considered.

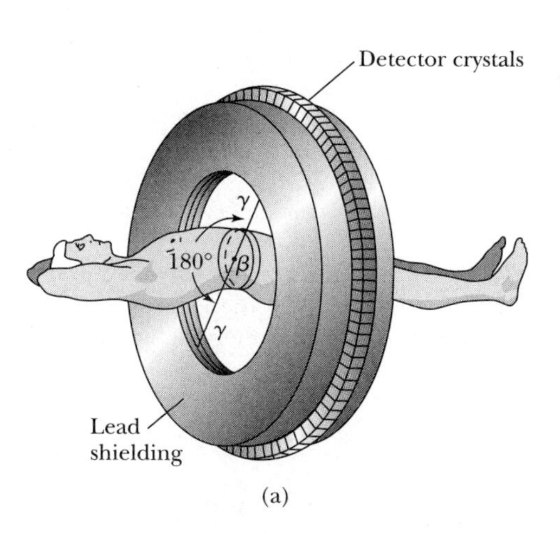

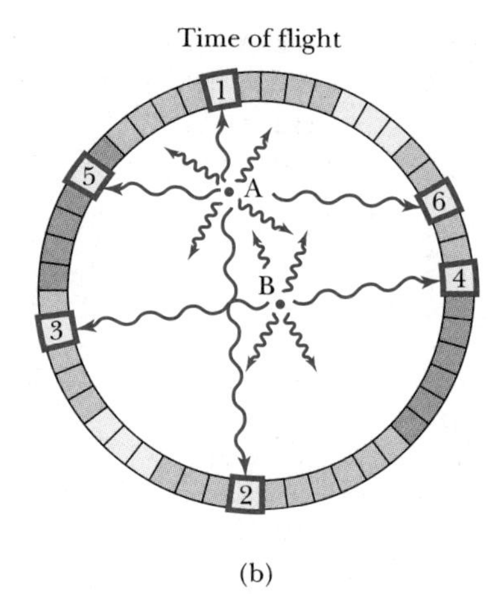

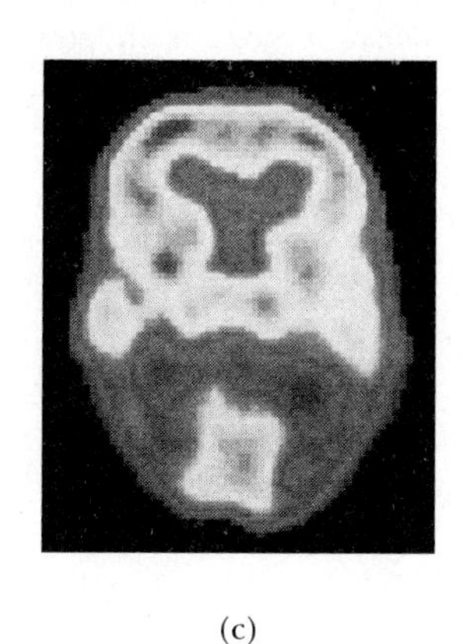

FIGURE 3.22 Positron emission tomography is a useful medical diagnostic to study the path and location of a positron-emitting radiopharmaceutical in the human body. (a) Appropriate radiopharmaceuticals are chosen to concentrate by physiological processes in the region to be examined. (b) The positron travels only a few mm before annihilation, which produces two photons that, after detection, give the positron position. (c) PET scan of a normal brain. *(a) and (b) are after G. L. Brownell, et al., Science* ***215****, 619 (1982); (c) National Institute of Health/Science Photo Library.*

believe that every particle has an antiparticle. In some cases, as for photons or neutral pi mesons, the particle and antiparticle are the same, but for most other particles (for example, the neutron and proton), particle and antiparticle are distinct.

We know that matter and antimatter cannot exist together in our world, because their ultimate fate would be annihilation. However, we may let our speculation run rampant! If we believe in symmetry, might there not be another world, perhaps in a distant galaxy, that is made of antimatter? Because galaxies are so far apart in space, annihilation would be infrequent. Modern cosmology predicts that the universe should be made up almost entirely of real particle matter and explains the obvious asymmetry this involves. However, if a large chunk of antimatter ever struck the Earth, it would tend to restore the picture of a symmetric universe. As we see from Problem 49, however, in such an event there would be no one left to receive the appropriate Nobel Prize.

Summary

In 1895 Röntgen discovered x rays, and in 1897 Thomson proved the existence of electrons and measured their charge to mass ratio. Finally, in 1911 Millikan reported an accurate determination of the electron's charge. Experimental studies resulted in the empirical Rydberg equation to calculate the wavelengths of the hydrogen atom's spectrum:

$$\frac{1}{\lambda} = R_H\left(\frac{1}{n^2} - \frac{1}{k^2}\right) \qquad k > n \tag{3.13}$$

where $R_H = 1.096776 \times 10^7\ \text{m}^{-1}$.

In order to explain blackbody radiation Planck proposed his quantum theory of radiation in 1900 to signal the era of modern physics. From Planck's theory we can derive Wien's displacement law

$$\lambda_{\text{max}} T = 2.898 \times 10^{-3}\ \text{m} \cdot \text{K} \tag{3.14}$$

and the Stefan-Boltzmann law,

$$R(T) = \epsilon\sigma T^4 \tag{3.16}$$

Planck's radiation law gives the power radiated per unit area per unit wavelength from a blackbody.

$$\mathcal{I}(\lambda, T) = \frac{2\pi c^2 h}{\lambda^5} \frac{1}{e^{hc/\lambda kT} - 1} \quad (3.23)$$

The oscillators of the electromagnetic radiation field can only change energy by quantized amounts given by $\Delta E = h\nu$, where $h = 6.6261 \times 10^{-34}$ J · s is called *Planck's constant.*

Classical theory could not explain the photoelectric effect, but in 1905, Einstein proposed that the electromagnetic radiation field itself is quantized. We call these particlelike quanta of light *photons,* and they each have energy $E = h\nu$ and momentum $p = h/\lambda$. The photoelectric effect is easily explained by the photons each interacting with only one electron. The conservation of energy gives

$$h\nu = \phi + \frac{1}{2} mv^2_{\text{max}} \quad (3.30)$$

where ϕ is the work function of the emitter. The retarding potential required to stop all electrons depends only on the photon's frequency

$$eV_0 = \frac{1}{2} mv^2_{\text{max}} = h\nu - h\nu_0 \quad (3.33)$$

where $\phi = h\nu_0$. Millikan proved experimentally in 1916 that Einstein's theory was correct.

Bremsstrahlung radiation (x rays) is emitted when charged particles (for example, electrons) pass through matter and are accelerated by the nuclear field. These x rays have a minimum wavelength

$$\lambda_{\text{min}} = \frac{hc}{eV_0} \quad (3.37)$$

where electrons accelerated by a voltage of V_0 impinge on a target.

In the Compton effect a photon scatters from an electron with a new photon created, and the electron recoils. For an incident and exit photon of wavelength λ and λ', respectively, the change in wavelength is

$$\Delta\lambda = \lambda' - \lambda = \frac{h}{mc}(1 - \cos\theta) \quad (3.42)$$

when the exit photon emerges at angle θ to the original photon direction. The Compton wavelength of the electron is $\lambda_c = h/m_e c = 2.43 \times 10^{-3}$ nm. The success of the Compton theory in 1923 convincingly demonstrated the particlelike nature of the photon.

Finally, photon energy can be converted into mass in pair production

$$\gamma \rightarrow e^+ + e^- \quad (3.43)$$

where e^+ is the positron, the antiparticle of the electron. Similarly a particle and antiparticle annihilate catastrophically in the process called *pair annihilation.*

$$e^+ + e^- \rightarrow \gamma + \gamma$$

Questions

1. How did the ionization of gas by cathode rays prevent H. Hertz from discovering the true character of electrons?
2. Why do television tubes generally deflect electrons with magnetic fields rather than with electric fields, as is done in cathode-ray-tube oscilloscopes?
3. In Thomson's *e/m* experiment, does it matter whether the electron passing through interacts first with the electric field or with the magnetic field? Explain.
4. Women in the late 1890s were terrified about the possible misuse of the new Röntgen x rays. Why do you think this fear occurred? Why was it no problem?
5. In Example 3.2, why would you be concerned about observing a cluster of several balls in the Millikan electron charge experiment?
6. In Figure 3.5 why does the histogram start smearing out for balls with multiple electron charges?
7. How is it possible for the plastic balls in Example 3.2 to have both positive and negative charges? What is happening?
8. Why do you suppose Millikan tried several different kinds of oil, as well as H_2O and Hg, for his oil drop experiment?
9. In the experiment of Example 3.2, how could you explain an experimental value of $q = 0.8 \times 10^{-19}$C?
10. Why do you suppose scientists worked so hard to develop better diffraction gratings?
11. Why was helium discovered in the sun's spectrum before being observed on Earth? Why was hydrogen observed on Earth first?
12. Do you believe there is any relation between the wavelengths of the Paschen (1908) and Pfund (1924) series and the respective dates they were discovered? Explain.
13. It is said that no two snowflakes look exactly alike, but we know that snowflakes have a quite regular, although complex, crystal structure. Discuss how this could be due to quantized behavior.
14. Why do we say that the elementary units of matter or "building blocks" must be some basic unit of mass-energy rather than of only mass?
15. Why is a red-hot object cooler than a white-hot one of the same material?
16. Why did scientists choose to study blackbody radiation from something as complicated as a hollow container rather than the radiation from something simple like a solid cylindrical thin disk (like a dime)?

17. Why does the sun apparently act as a blackbody?
18. In a typical photoelectric effect experiment, consider replacing the metal photocathode by a gas. What difference would you expect?
19. Why is it important to produce x-ray tubes with high accelerating voltages that are also able to withstand electron currents?
20. For a given beam current and target thickness, why would you expect a tungsten target to produce a higher x-ray intensity than targets of molybdenum or chromium?
21. List all possible known interactions between photons and electrons discussed in this chapter. Can you think of any more?
22. What do you believe to be an optimum lifetime for a positron-emitting radioactive nuclide used in brain tumor diagnostics? Explain.

Problems

3.1 Discovery of the X Ray and the Electron

1. Design an apparatus that will produce the correct magnetic field needed in Figure 3.2.
2. For an electric field of 2×10^5 V/m, what is the strength of the magnetic field needed to pass an electron of speed 2×10^6 m/s with no deflection? Draw **v**, **E**, and **B** directions for this to occur.
3. Across what potential difference does an electron have to be accelerated in order to reach the speed $v = 2 \times 10^7$ m/s? Should you use relativistic calculations?
4. An electron entering Thomson's e/m apparatus (Figure 3.2) has an initial velocity (in horizontal direction only) of 0.5×10^7 m/s. Lying around the lab is a permanent horseshoe magnet of strength 1.3×10^{-2} T, which you would like to use. What electric field will you need in order to produce zero deflection of the electrons as they travel through the apparatus? When the magnetic field is turned off, but the same electric field remains, how large a deflection will occur if the region of nonzero **E** and **B** fields is 2 cm long?

3.2 Determination of Electron Charge

5. Consider the following possible forces on an oil drop in Millikan's experiment: gravitational, electrical, frictional, and buoyant. Draw a diagram indicating the forces on the oil drop (a) when the electric field is turned off and the droplet is falling freely, and (b) when the electric field causes the droplet to rise.
6. Neglecting the buoyancy force on an oil droplet, show that the terminal speed of the droplet is $v_t = mg/f$, where f is the coefficient of friction when the droplet is in free fall. (Remember that the frictional force $\mathbf{F}_f$ is given by $\mathbf{F}_f = -f\mathbf{v}$ where velocity is a vector).
7. Stokes's law relates the coefficient of friction f to the radius r of the oil drop and the viscosity η of the medium the droplet is passing through: $f = 6\pi\eta r$. Show that the radius of the oil drop is given in terms of the terminal velocity v_t (see previous problem), η, g, and the density of the oil ρ by $r = 3\sqrt{\eta v_t/2g\rho}$.
8. In a Millikan oil drop experiment the terminal velocity of the droplet is observed to be 1.3 mm/s. The density of the oil is $\rho = 900$ kg/m^3 and the viscosity of air is $\eta = 1.82 \times 10^{-5}$ kg/m · s. Using the results of the two previous problems, calculate (a) the droplet radius, (b) the mass of the droplet, and (c) the coefficient of friction.

3.3 Line Spectra

9. What is the series limit (that is, the smallest wavelength) for the Lyman series? For the Balmer series?
10. Light from a slit passes through a transmission diffraction grating of 400 lines/mm, which is located 2.5 m from a screen. What are the distances on the screen (from the unscattered slit image) of the three brightest visible (first order) hydrogen lines?
11. A transmission diffraction grating of 420 lines/mm is used to study the light intensity of different orders (n). A screen is located 2.5 m from the grating. What is the separation on the screen between the three brightest red lines for a hydrogen source?

3.4 Quantization

12. Quarks have charges $\pm e/3$ and $\pm 2e/3$. What combination of three quarks could yield (a) a proton, (b) a neutron?

3.5 Blackbody Radiation

13. Calculate λ_{max} for blackbody radiation for (a) liquid helium (4.2 K), (b) room temperature (293 K), and (c) a steel furnace (2500 K).
14. Calculate the temperature of a blackbody if the spectral distribution peaks at (a) gamma rays, $\lambda = 10^{-14}$ m, (b) x rays, 1 nm, (c) red light, 670 nm, (d) broadcast television waves, 1 m, and (e) AM radio waves, 204 m.
15. A blackbody's temperature is increased from 900 K to 1900 K. By what factor does the total power radiated per unit area increase?
16. For a blackbody at a given temperature T, what is the long-wavelength limit ($\lambda \gg hc/kT$) of Planck's radiation law? (This is the Rayleigh-Jeans result known to Planck in 1895).

17. A tungsten filament of a typical incandescent light bulb operates at a temperature near 3000 K. At what wavelength is the intensity a maximum?

18. Use a computer to calculate Planck's radiation law for a temperature of 3000 K, which is the temperature of a typical tungsten filament in an incandescent light bulb. Plot the intensity versus wavelength. (a) How much of the power is in the visible region (400–700 nm) compared with the ultraviolet and infrared? (b) What is the ratio of the intensity at 400 nm and 700 nm to the maximum?

19. Show that the ultraviolet catastrophe is avoided for short wavelengths ($\lambda \to 0$) with Planck's radiation law by calculating the limiting intensity $\mathcal{I}(\lambda, T)$ as $\lambda \to 0$.

20. Estimate the power radiated by (a) a basketball at 20°C, (b) the human body (assume a temperature of 37°C).

21. At what wavelength is the radiation emitted by the human body a maximum? Assume a temperature of 37°C.

22. If we have waves in a one-dimensional box, such that the wave displacement $\Psi(x, t) = 0$ for $x = 0$ and $x = L$, where L is the length of the box, and

$$\frac{1}{c^2}\frac{\partial^2 \Psi}{\partial t^2} - \frac{\partial^2 \Psi}{\partial x^2} = 0 \qquad \text{(wave equation)}$$

show that the solutions are of the form

$$\Psi(x, t) = a(t)\sin\left(\frac{n\pi x}{L}\right) \qquad (n = 1, 2, 3, \ldots)$$

and $a(t)$ satisfies the (harmonic-oscillator) equation

$$\frac{d^2 a(t)}{dt^2} + \Omega_n{}^2 a(t) = 0$$

where $\Omega_n = \dfrac{n\pi c}{L}$ is the angular frequency, $2\pi\nu$.

23. If the angular frequencies of waves in a three-dimensional box of sides L generalize to

$$\Omega = \frac{\pi c}{L}(n_x^2 + n_y^2 + n_z^2)^{1/2}$$

where all n are integers, show that the number of distinct states in the frequency interval ν to $\nu + d\nu (\nu = \Omega/2\pi)$ is given by (where ν is large)

$$dN = 4\pi \frac{L^3}{c^3}\nu^2 d\nu.$$

24. Let the energy density in the frequency interval ν to $\nu + d\nu$ within a blackbody at temperature T be $dU(\nu, T)$. Show that the power emitted through a small hole of area ΔA in the container is

$$\frac{c}{4}\, dU(\nu, T)\, \Delta A$$

25. Derive the Planck radiation law emitted by a blackbody. Remember that light has two directions of polarization and treat the waves as an ensemble of harmonic oscillators.

3.6 Photoelectric Effect

26. An FM radio station of frequency 107.7 MHz puts out a signal of 50,000 W. How many photons/s are emitted?

27. How many photons/s are contained in a beam of electromagnetic radiation of total power 150 W if the source is (a) an AM radio station of 1100 kHz, (b) 8-nm x rays, and (c) 4-MeV gamma rays?

28. What is the threshold frequency for the photoelectric effect on lithium ($\phi = 2.9$ eV)? What is the stopping potential if the wavelength of the incident light is 400 nm?

29. What is the maximum wavelength of incident light that can produce photoelectrons from silver ($\phi = 4.7$ eV)? What will be the maximum kinetic energy of the photoelectrons if the wavelength is halved?

30. A 2-mW laser ($\lambda = 530$ nm) shines on a cesium photocathode ($\phi = 1.9$ eV). Assuming an efficiency of 10^{-5} for producing photoelectrons (that is, one photoelectron produced for every 10^5 incident photons), what is the photoelectric current?

31. An experimenter finds that no photoelectrons are emitted from tungsten unless the wavelength of light is less than 230 nm. Her experiment will require photoelectrons of maximum kinetic energy 2.0 eV. What frequency light should be used to illuminate the tungsten?

32. The human eye is sensitive to a pulse of light containing as few as 100 photons. For yellow light of wavelength 580 nm how much energy is contained in the pulse?

33. In a photoelectric experiment it is found that a stopping potential of 1.0 V is needed to stop all the electrons when incident light of wavelength 260 nm is used and 2.3 V is needed for light of wavelength 207 nm. From these data determine Planck's constant and the work function of the metal.

34. What is the limit of energies and frequencies for visible light of wavelengths 400–700 nm?

3.7 X-Ray Production

35. What is the minimum x-ray wavelength produced for an x-ray machine operated at 30 kV?

36. The Stanford Linear Accelerator can accelerate electrons to 50 GeV (50×10^9 eV). What is the minimum wavelength photons it can produce by bremsstrahlung? Is this photon still called an x ray?

37. A television tube operates at 20,000 V. What is λ_{min} for the continuous x-ray spectrum produced when the electrons hit the phosphor?

3.8 Compton Effect

38. Calculate the maximum $\Delta\lambda/\lambda$ of Compton scattering for green light ($\lambda = 530$ nm). Could this be easily observed?

39. A photon having 40 keV scatters from a free electron at rest. What is the maximum energy that the electron can obtain?

40. If a 6 keV photon scatters from a free proton at rest, what is the change in the photon's wavelength if the photon recoils at 90°?

41. Is it possible to have a scattering similar to Compton scattering from a proton in H_2 gas? What would be the Compton wavelength for a proton? What energy photon would have this wavelength?

42. An instrument has resolution $\Delta\lambda/\lambda = 0.4\%$. What wavelength incident photons should be used in order to resolve the modified and unmodified scattered photons for scattering angles of (a) 30°, (b) 90°, and (c) 170°?

43. Derive the relation for the recoil kinetic energy of the electron and its recoil angle ϕ in Compton scattering. Show that

$$\text{K.E. (electron)} = \frac{\Delta\lambda/\lambda}{1 + \dfrac{\Delta\lambda}{\lambda}}\, h\nu$$

$$\cot\phi = \left(1 + \frac{h\nu}{mc^2}\right)\tan\frac{\theta}{2}$$

44. A gamma ray of 700 keV energy Compton scatters from an electron. Find the energy of the photon scattered at 110°, the energy of the scattered electron, and the recoil angle of the electron.

45. A photon of wavelength 2 nm Compton scatters from an electron at an angle of 90°. What is the modified wavelength and the percentage change, $\Delta\lambda/\lambda$?

3.9 Pair Production and Annihilation

46. How much photon energy would be required to produce a proton–antiproton pair? Where could such a high-energy photon come from?

47. What is the minimum photon energy needed to create an e^-–e^+ pair when a photon collides (a) with a free electron at rest and (b) with a free proton at rest?

General Problems

48. What wavelength photons are needed to produce 30 keV electrons in a Compton scattering?

49. The gravitational energy of the Earth is approximately $\frac{1}{2}(GM_E^2/R_E)$ where M_E is the mass of the Earth. This is approximately the energy needed to blow the Earth into small fragments (the size of asteroids). How large would an antimatter meteorite the density of nickel-iron ($\rho \sim 5 \times 10^3$ kg/m^3) have to be in order to blow up the Earth when it strikes? Compute the energy involved in the particle–antiparticle annihilation and compare it with the total energy in all the nuclear arsenals of the world (~2000 megaton (MT), where 1 MT = 4.2×10^{15} J).

50. Show that the maximum kinetic energy of the recoil electron in Compton scattering is given by

$$\text{K.E.}_{\text{max}}\text{ (electron)} = h\nu \frac{\dfrac{2h\nu}{mc^2}}{1 + \dfrac{2h\nu}{mc^2}}$$

At what angles θ and ϕ does this occur? If we detect a scattered electron at $\phi = 0°$ of 100 keV, what energy photon was scattered?

51. Using the Wien displacement law, make a log–log plot of λ_{max} (from 10^{-8} m to 10^{-2} m) versus temperature (from 10^0 K to 10^5 K). Mark on the plot the regions of visible, ultraviolet, infrared, and microwave wavelengths. Put the following points on the line: sun (5800 K), furnace (1900 K), room temperature (300 K), and the background radiation of the universe (2.7 K). Discuss the electromagnetic radiation that is emitted from each of these sources. Does it make sense?

52. (a) What is the maximum possible energy for a Compton-backscattered x ray ($\theta = 180°$)? Express your answer in terms of λ, the wavelength of the incoming photon. (b) Evaluate numerically if the incoming photon's energy is 100 keV.

CHAPTER 4

Structure of the Atom

In the present first part of the paper the mechanism of the binding of electrons by a positive nucleus is discussed in relation to Planck's theory. It will be shown that it is possible from the point of view taken to account in a simple way for the law of the line spectrum of hydrogen.

Niels Bohr, 1913

By the end of the 19th century most physicists and chemists (with a few notable exceptions) believed in an atomic theory of matter, even though no one had ever directly observed an atom. Moreover, there was good reason to believe that atoms themselves were composite structures and not the featureless hard spheres the Greek philosophers had imagined them to be. It may seem paradoxical that the scientists of the 19th century were able to determine that objects far too small for them to see possessed an internal structure. Because of the similarities between their situation and that of present-day "elementary particle" physics, we shall explain how they were able to do this.

First of all, there seemed to be too many kinds of atoms, each belonging to a distinct chemical element. The original Greek idea was that there were four types of atoms—earth, air, water, and fire—which combined together to make the various kinds of matter we observe. But the development of chemistry made plain that there were at least seventy kinds of atoms, far too many for them all to be the ultimate elementary constituents of matter.

Second, it was found experimentally that atoms and electromagnetic phenomena were intimately related. For example, molecules can be dissociated into their component atoms by electrolysis. Some kinds of atoms form magnetic materials, and others form electrical conductors and insulators. All kinds of atoms emit light (which was known to be electromagnetic in nature) when they are heated, as well as when an electrical discharge passes through them. The visible light emitted by free or nearly free atoms of the chemical elements is not a continuum of frequencies, but rather a discrete set of characteristic colors, so substances can be analyzed according to their chemical composition using their flame spectra. The existence of **characteristic spectra** pointed to an internal structure distinguishing the elements.

Third, there was the problem of **valence**—why certain elements combine with some elements but not with others, and when they do combine, they do so in varying proportions determined by the valences of the atoms. The laws of valence suggested that the forces between atoms were very specific in nature, a characteristic that hinted at an internal atomic structure.

Finally, there were the discoveries of radioactivity, of x rays, and of the electron, all of which were at variance with earlier ideas of indivisible and elementary atoms.

Because there were these tantalizing indirect hints that the atom had a structure, the most exciting frontier of physical science in the early part of the 20th century developed into an investigation of the atom and its internal composition. The subject of this chapter is the beginning of quantum physics and its relation to the first cohesive theories of atomic structure. Although we now have a more complete theoretical framework with which to understand the early experiments than was available to the scientists themselves, it is worth repeating some of their reasoning, both for its historical interest and because it illustrates how science progresses by trying to extend well-established ideas into unknown terrain.

In this chapter we will discuss the atomic models of Thomson and Rutherford and learn how Rutherford discerned the correct structure of the atom by performing alpha-particle scattering experiments. We will see that Bohr presented a model of the hydrogen atom based on the new quantum concept that correctly produced the Rydberg equation, and we will study the successes and failures of Bohr's theory. We will also learn the origin of characteristic x-ray spectra and the concept of atomic number. Finally, we will show that electron scattering (the Franck-Hertz experiment) also confirmed the quantized structure of the atom.

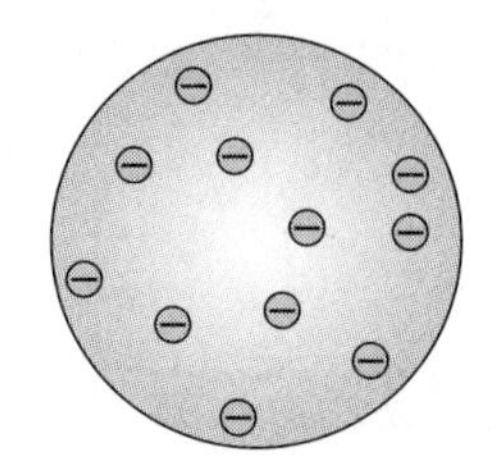

FIGURE 4.1 Schematic of J. J. Thomson's model of the atom (later proven to be incorrect). The electrons are imbedded in a homogeneous positively charged mass much like "raisins in plum pudding." The electric force on the electrons is zero, so the electrons do not move around rapidly. The oscillations of the electrons give rise to electromagnetic radiation.

4.1 The Atomic Models of Thomson and Rutherford

In the years immediately following J. J. Thomson's discovery of the electron, Thomson and others tried to unravel the mystery of the atomic structure. In just a few years it was learned that electrons were much less massive than atoms and that the number of electrons was equal to about half the number representing the atomic mass. The central question was, "How were the electrons arranged and where were the positive charges that made the atom electrically neutral?" (Remember that protons had not been discovered yet.) Thomson proposed a model wherein the positive charges were spread uniformly throughout a sphere the size of the atom, with electrons embedded in the uniform background. His model, which has been likened to "raisins in plum pudding," is shown schematically in Figure 4.1. The arrangement of charges had to be in stable equilibrium. In Thomson's view, when the atom was heated, the electrons could vibrate about their equilibrium positions, thus producing electromagnetic radiation. The emission frequencies of this radiation would fall in the range of visible light if the sphere of positive charges was of diameter $\sim 10^{-10}$ m, which was known to be the approximate size of an atom. Nevertheless, even though he tried for several

Thomson's model of the atom

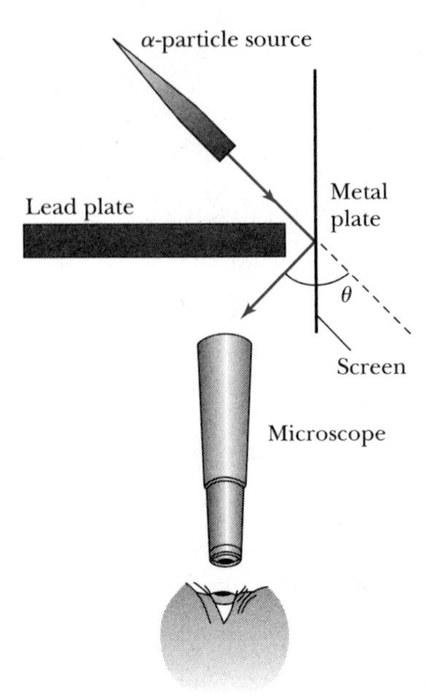

FIGURE 4.2 Schematic diagram of apparatus used by Geiger and Marsden to observe scattering of α particles past 90°. "A small fraction of the α particles falling upon a metal plate have their directions changed to such an extent that they emerge again at the side of incidence." *From Geiger and Marsden, Proceedings of Royal Society (London)* **82,** *495 (1909).*

years, Thomson was unable to calculate the light spectrum of hydrogen using his model.

In order to make further progress in deciphering atomic structure, a new approach was needed. This was supplied by Ernest Rutherford, who was already famous for his Nobel Prize–winning work (1908) on radioactivity. Because atoms are so small it was not possible to directly see or observe their internal structure. Scientists needed some other means and attempted to scatter other small objects from atoms in order to understand atomic structure. Rutherford, assisted by Hans Geiger, conceived a new technique for investigating the structure of matter by scattering energetic α particles* (emitted by radioactive sources) from atoms. Together with a young student, Ernest Marsden, Geiger showed in 1909 while working in Rutherford's lab that surprisingly many α particles were scattered from thin gold-leaf targets at backward angles greater than 90° (see Figure 4.2).

For several years Rutherford had pondered the structure of the atom. He was well aware of Thomson's model because he had worked for Thomson at the Cavendish Laboratory as a research student from 1895–1898, after receiving his undergraduate education in his native New Zealand. Although he greatly respected Thomson, Rutherford could see that Thomson's model agreed neither with spectroscopy nor with Geiger's latest experiment with alpha particles.

The experiments of Geiger and Marsden were instrumental in the development of Rutherford's model. The problem can be understood by a simple thought experiment with a .22-caliber rifle that is fired into a thin black box. If the box contains a homogeneous material such as wood or water (as in Thomson's "plum pudding" model), the bullet will pass through the box with little or no deviation in its path. However, if the box contains a few massive steel ball bearings, then occasionally a bullet will be deflected backward.

Example 4.1

Geiger and Marsden (1909) observed backward-scattered ($\theta \geq 90°$) α particles when a beam of energetic α particles was directed at a gold foil as thin as 6×10^{-7} m. Assuming an α particle scatters from an electron in the foil, what is the maximum scattering angle?

Solution: The collision must obey the laws of conservation of momentum and energy. Assume the incident α particle has mass M_α and velocity v_α, while the mass of the electron is m_e. The maximum momentum transfer occurs when the α particle hits the electron (at rest) head-on, as shown in Figure 4.3.

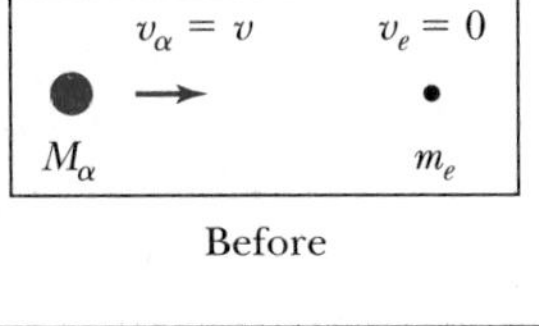

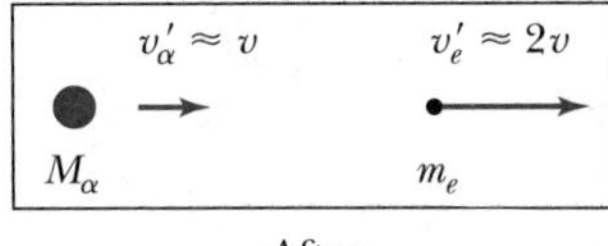

FIGURE 4.3 Schematic diagram (before and after) of an α particle of velocity $v = v_\alpha$ and mass M_α making a head-on collision with an electron initially at rest. Because the α particle is so much more massive than the electron, the α particle's velocity is hardly reduced.

Conservation of momentum (nonrelativistically) gives

$$M_\alpha \mathbf{v}_\alpha = M_\alpha \mathbf{v}'_\alpha + m_e \mathbf{v}'_e$$

Because the α particle is so much more massive than the electron ($M_\alpha / m_e \approx 4 \times 1837 = 7348$), the α particle's velocity is

*Rutherford had already demonstrated that the α particle is an ionized helium atom.

hardly affected and $v'_\alpha \approx v_\alpha$. In an elastic collision with such unequal masses, $v'_e \approx 2v_\alpha$ in order to conserve both energy and linear momentum (see Problem 3). Thus the maximum momentum change of the α particle is simply

$$\Delta \mathbf{p}_\alpha = M_\alpha \mathbf{v}_\alpha - M_\alpha \mathbf{v}'_\alpha = m_e \mathbf{v}'_e$$

or

$$\Delta p_{\text{max}} = 2m_e v_\alpha$$

Although this maximum momentum change is along the direction of motion, let's determine an upper limit for the angular deviation θ by letting Δp_{max} be perpendicular to the direction of motion as shown in Figure 4.4. (This value of θ is larger than can actually be observed because we know that the Δp_α we calculated was for a head-on collision, and the Δp_α for a glancing collision would be smaller.) Thus

$$\theta_{\text{max}} = \frac{\Delta p_\alpha}{p_\alpha} = \frac{2m_e v_\alpha}{M_\alpha v_\alpha} = \frac{2m_e}{M_\alpha} = 2.7 \times 10^{-4} \text{ rad} = 0.016°$$

Thus it is impossible for an α particle to be deflected through a large angle by a single encounter with an electron.

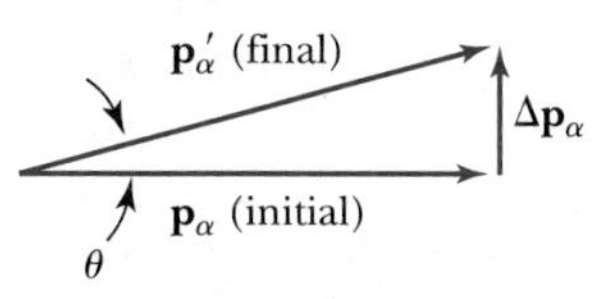

FIGURE 4.4 Vector diagram illustrating the change in momentum $\Delta \mathbf{p}_\alpha$ of the α particle after scattering from the electron.

Multiple scattering from electrons

What would happen if an α particle were scattered by *many* electrons in the target? Multiple scattering is possible, and a calculation for random multiple scattering from N targets results in an average scattering angle $\langle\theta\rangle_{\text{total}} \approx \sqrt{N}\,\theta$. The α particle is as likely to scatter on one side of its direction as the other side for each collision. We can estimate the number of atoms across the thin gold layer of 6×10^{-7} m used by Geiger and Marsden.

$$\frac{\text{number of molecules}}{\text{cm}^3} = \text{Avogadro's no.}\left(\frac{\text{molecules}}{\text{mole}}\right)$$

$$\times \frac{1}{\text{gram-molecular weight}}\left(\frac{\text{mole}}{\text{g}}\right) \times \text{Density}\left(\frac{\text{g}}{\text{cm}^3}\right)$$

$$= 6.02 \times 10^{23}\,\frac{\text{molecules}}{\text{mole}}\,\frac{1 \text{ mole}}{197 \text{ g}}\,\frac{19.3 \text{ g}}{\text{cm}^3} \qquad \text{(for gold)}$$

$$= 5.9 \times 10^{22}\,\frac{\text{molecules}}{\text{cm}^3} = 5.9 \times 10^{28}\,\frac{\text{atoms}}{\text{m}^3}$$

If there are 5.9×10^{28} atoms/m^3, then each atom occupies $(5.9 \times 10^{28})^{-1}$ m^3 of space. Assuming the atoms are equidistant, the distance d between centers is $d = (5.9 \times 10^{28})^{-1/3}$ m $= 2.6 \times 10^{-10}$ m. In the foil, then, there are

$$N = \frac{6 \times 10^{-7} \text{ m}}{2.6 \times 10^{-10} \text{ m}} = 2300 \text{ atoms}$$

along the α-particle's path. If we assume the α particle interacts with one electron from each atom, then

$$\langle\theta\rangle_{\text{total}} = \sqrt{2300}(0.016°) = 0.8°$$

where we have used the result for θ_{max} from Example 4.1. Even if the alpha particle scattered from all 79 electrons in gold, $\langle\theta\rangle_{\text{total}} = 6.8°$.

Rutherford reported* in 1911 that the experimental results were not consistent with α-particle scattering from the atomic structure proposed by Thomson and that "it seems reasonable to suppose that the deflection through

*E. Rutherford, *Philosophical Magazine* **21,** 669(1911).

a large angle is due to a single atomic encounter." Rutherford proposed that an atom consisted mostly of empty space with a central charge, either positive or negative. Rutherford wrote in 1911 that "Considering the evidence as a whole, it seems simplest to suppose that the atom contains a central charge distributed through a very small volume, and that the large single deflections are due to the central charge as a whole, and not to its constituents." Rutherford worked out the scattering expected for the α particles as a function of angle, thickness of material, velocity, and charge. Geiger and Marsden immediately began an experimental investigation of Rutherford's ideas and reported* in 1913, "we have completely verified the theory given by Prof. Rutherford." In that same year, Rutherford coined the use of the word **nucleus** for the central charged core and definitely decided that the core (containing most of the mass) was *positively* charged, surrounded by the negative electrons.

Rutherford's atomic model

The nucleus

The popular conception of an atom today, often depicted as in Figure 4.5, is due to Lord Rutherford. An extremely small positively charged core provides a Coulomb attraction for the negatively charged electrons flying at high speeds around the nucleus; this is the so-called "solar system," or "planetary" model. We now know that the nucleus is composed of positively charged protons and neutral neutrons, each having approximately the same mass, and the electrons do not execute prescribed orbital paths.

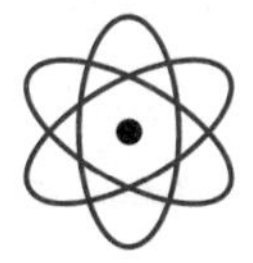

FIGURE 4.5 Solar or planetary model of the atom. Rutherford proposed that there is a massive, central core with a highly electric positive charge. According to Bohr, the electrons orbit around this nucleus. Although this is a useful pictorial, we now know this schematic is too simplistic.

4.2 Rutherford Scattering

Rutherford's "discovery of the nucleus" laid the foundation for many of today's atomic and nuclear scattering experiments. By means of scattering experiments similar in concept to those of Rutherford and his assistants, scientists have elucidated the electron structure of the atom, the internal structure of the nucleus, and even the internal structures of the nuclear constituents, protons and neutrons. Rutherford's calculations and procedures are well worth studying in some detail because of their applicability to many areas of physical and biological science.

Basic scattering experiments

In order to study matter at such a microscopic scale as atomic sizes, we perform scattering experiments. The material to be studied is bombarded with rapidly moving particles (like the 5- to 8-MeV α particles used by Geiger and Marsden) in a well-defined and collimated beam. Although the present discussion is limited to charged-particle beams, the general procedure applies to neutral particles such as neutrons; only the interaction between the beam particles and the target material is different.

Rutherford or Coulomb scattering

The scattering of charged particles by matter is called **Coulomb** or **Rutherford scattering** when it takes place at low energies, where only the Coulomb force is important. At higher beam energies other forces (for example, nuclear interactions) may become important. A typical scattering experiment is diagrammed in Figure 4.6. A charged particle of mass m, charge $Z_1 e$ and velocity v_0 is incident on the target material or scatterer of charge $Z_2 e$. The distance b is called the classical *impact parameter;* it is the closest distance of approach between the beam particle and scatterer if the projectile had continued in a straight line. The angle θ between the incident beam direction and the direction of the deflected particle is called the *scattering angle*. Normally detectors are positioned at one or more scattering angles to count the particles scattered into the small cones of solid angle subtended by the detectors (see Figure 4.7). Depending on

Scattering angle

*Hans Geiger and Ernest Marsden, *Philosophical Magazine* **25,** 604(1913).

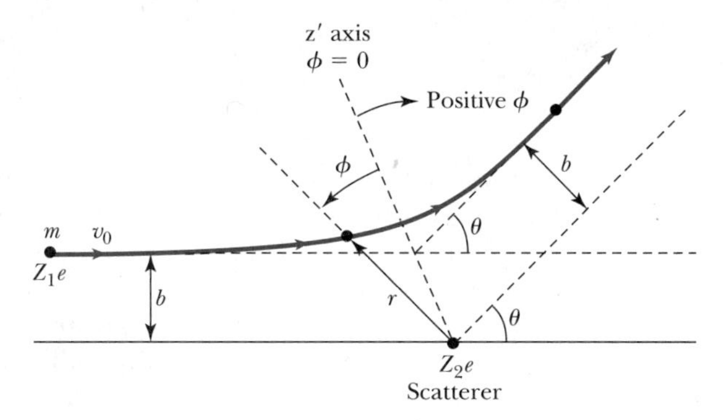

FIGURE 4.6 Representation of Coulomb or Rutherford scattering. The projectile of mass m and charge Z_1e scatters from a particle of charge Z_2e at rest. The parameters r and ϕ, which describe the projectile's orbit, are defined as shown. The angle $\phi = 0$ represents the position of closest approach. The impact parameter b and scattering angle θ are also displayed.

the functional form of the interaction between the particle and the scatterer, there will be a particular relationship between the impact parameter b and the scattering angle θ. In the case of Coulomb scattering between a positively charged α particle and a positively charged nucleus, the trajectories will resemble those in Figure 4.7. When the impact parameter is small, the distance of closest approach r_{min} is small and the Coulomb force is large. Therefore, the scattering angle can be large and the particle can be repelled backward. Conversely, for large impact parameters the particles never get close together, so the Coulomb force is small and the scattering angle is also small.

An important relationship for any interaction is that between b and θ. We wish to find this dependence for the Coulomb force. We will make the same assumptions as Rutherford:

Scattering assumptions

1. The scatterer is so massive that it does not significantly recoil; therefore the initial and final kinetic energies of the α particle are practically equal.
2. The target is so thin that only a single scattering occurs.
3. The bombarding particle and target scatterer are so small that they may be treated as point masses and charges.
4. Only the Coulomb force is effective.

Assumption 1 means that $K \equiv K.E._{initial} \approx K.E._{final}$ for the α particle. For central forces like the Coulomb force, the angular momentum, mv_0b, where v_0 is the initial velocity of the particle, is also conserved (see Problem 48). This means that the trajectory of the scattered particle lies in a plane.

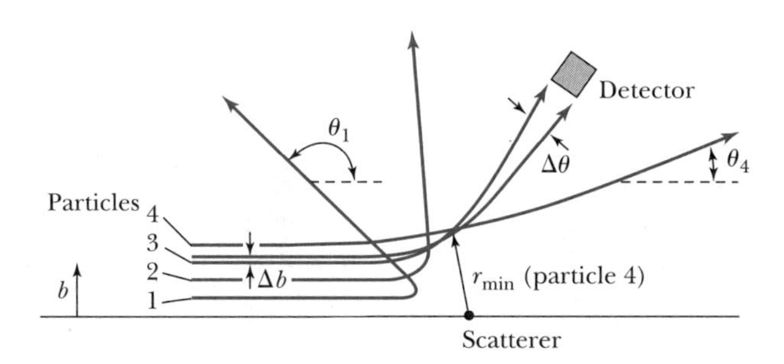

FIGURE 4.7 The relationship between the impact parameter b and scattering angle θ. Particles with small impact parameters approach the nucleus most closely (r_{min}) and scatter to the largest angles. Particles within the range of impact parameters Δb will be scattered within $\Delta\theta$.

SPECIAL TOPIC:

LORD RUTHERFORD OF NELSON

We choose to highlight the life of Ernest Rutherford not only because of his extraordinary talents as a physicist, but also because of the many contributions he made to others throughout the world. He nurtured many young physicists in the laboratories he headed, including several who later made outstanding scientific contributions themselves. They include O. Hahn and F. Soddy at McGill University in Canada; H. Geiger, E. Marsden, N. Bohr, and H. G. J. Moseley at Manchester University in England; and J. Chadwick, P. M. S. Blackett, J. D. Cockcroft, E. T. S. Walton, M. Oliphant, M. Goldhaber, and others at Cambridge University in England. The account here is from *Rutherford** and from *The World of the Atom*†.

Ernest Rutherford was born of Scottish parents on August 30, 1871 near the town of Nelson, New Zealand. Rutherford, one of eleven children, made excellent marks in school and received a scholarship to Nelson College where his interest in science blossomed. At eighteen he obtained another scholarship to attend Canterbury College of the University of New Zealand in Christchurch, where in 1894 he received both a bachelor's and a master's degree. It was there that he began significant research on the magnetization of iron and constructed a Hertz oscillator capable of generating high-frequency electric currents. His magnetic detector was able to receive electromagnetic waves over a distance of 60 feet through walls, quite a feat at the time.

In 1895 Rutherford won a competition to bring able men to British Universities from abroad. He was digging potatoes when his mother told him of his selection and he is reported to have said with a laugh as he threw the spade away, "That's the last potato I'll dig." Rutherford was the first research student to arrive to work for the famous Professor J. J. Thomson of the Cavendish Laboratory under a new program just initiated by Cambridge University. Rutherford's primary research continued to be on the transmission and detection of "wireless waves," for which he obtained a considerable reputation. Thomson was quite impressed by Rutherford and encouraged him to publish his results and make presentations at scientific meetings. He presented his work on the magnetic detector before the Royal Society in June, 1896, when he was only 24. Besides his own work on electromagnetic radiation, Rutherford began investigations with Thomson on the effects of x-ray radiation from uranium in various gases. This research continued into 1898 when Rutherford applied for and received a chaired appointment as Professor of Physics at McGill University in Montreal, Canada.

Barely 27 years of age, he arrived in Montreal in September, 1898, to take up his new duties in the Macdonald Physics Laboratories. This lab was endowed by a generous benefactor, Sir William Macdonald, who besides endowing the laboratory and Rutherford's chair, could also be called upon to provide specialized equipment when needed. Rutherford's studies on radioactivity continued, and he had frequent correspondence and occasional visits with scientists from abroad including, of course, Thomson, but also Poynting, FitzGerald, and Crookes. Early in 1900 he published a paper in the *Philosophical Magazine* in which he named *alpha, beta,* and *gamma* as the three types of radiation from thorium and uranium. In 1900 Rutherford returned briefly to New Zealand where he married Mary Newton of Christchurch and visited with his parents. Rutherford received several offers of appointment from American universities, but he longed to return to England, which he considered to be the leading center of physics. In 1901 he wrote Thomson inquiring of any possibilities.

Rutherford attracted the aid of a young research chemist, Frederick Soddy of Oxford, who had arrived at McGill in 1900 only after a position he had sought at Toronto had been filled. Soddy obtained a demonstratorship in the chemistry lab, but his radioactivity research with Rutherford progressed so well that he joined Rutherford's lab fulltime in 1901. Rutherford and Soddy discovered, in 1902, that the elements, heretofore considered immute, actually decayed to other elements. They found radiation coming from a substance they called thorium X, which was a result of thorium decay. They found that thorium X not only decayed exponentially, but that a new supply of it was also produced exponentially from thorium. During the next few years Rutherford investigated α particles and the radioactive decay chains of radium, thorium, and uranium.

*A. S. Eve, *Rutherford.* New York: Macmillan, 1939.

†H. A. Boorse, and L. Motz eds. *The World of the Atom.* New York: Basic Books, 1966.

In 1907 Rutherford returned to England as Professor of Physics at the University of Manchester where he did his greatest work. He left McGill University with deep regret, but with high hopes for the future. His first success was the proof that α particles were indeed helium ions. In 1908 Rutherford received word that he had won the Nobel Prize for the work he and Soddy had done, but Rutherford was startled and amused to learn it was in chemistry, not physics. He remarked at a banquet in the Stockholm Royal Palace that the quickest transformation (note the pun) he had observed was his own from a physicist to a chemist.

It was during the next few years that Rutherford carried out his research into the nature of the atom that culminated with his discovery of the nucleus. Rutherford was already an international figure in great demand as a lecturer while at McGill, but now his stature increased even more. Research students flocked to Manchester to work with Rutherford, and laboratory space was in great demand. In 1912 Rutherford wrote to a colleague, "Bohr, a Dane, has pulled out of Cambridge and turned up here to get some experience in radioactive work." That momentous trip resulted in the "Rutherford-Bohr atom." The work at Manchester is too voluminous to describe, but the research of Hans Geiger, Ernest Marsden, and Henry Moseley was to have dramatic consequences. Curiously enough, it was the experiment of Geiger and Marsden, not his own, that led Rutherford to the nuclear model of the atom. Rutherford coined the word *proton* to describe the fast hydrogen nuclei produced when he bombarded hydrogen and nitrogen with fast α particles. In this case, Rutherford performed most of the research himself in 1919, being aided by an assistant only in observing scintillations.

World War I broke up the family of research students working at Manchester, and in 1919 Rutherford accepted the Cavendish Professorship at Cambridge, the post just vacated by J. J. Thomson, who remained as Master of Trinity College. Thomson continued to do physics research in the lab and eventually outlived Rutherford. Being the successor at Cambridge to Maxwell, Rayleigh, and Thomson was no small feat, and Rutherford's initial research at Cambridge with James Chadwick as his collaborator produced many interesting results as well as the predictions, in 1920, of the existence of a mass 2 isotope of hydrogen and of a neutral particle with about the proton's mass. These were discovered 12 years later by Urey and Chadwick, respectively. Earlier in 1914 Rutherford had been knighted, and in 1931 he was made a Baron, choosing the town of Nelson near his boyhood home to become "Lord Rutherford of Nelson." Rutherford was greatly admired by all who came in contact with him. He was the greatest experimental physicist of his day, yet he was said to have "never made an enemy and never lost a friend." He was still doing important research until the time of his sudden death in 1937 of hernia complications at the age of 66.

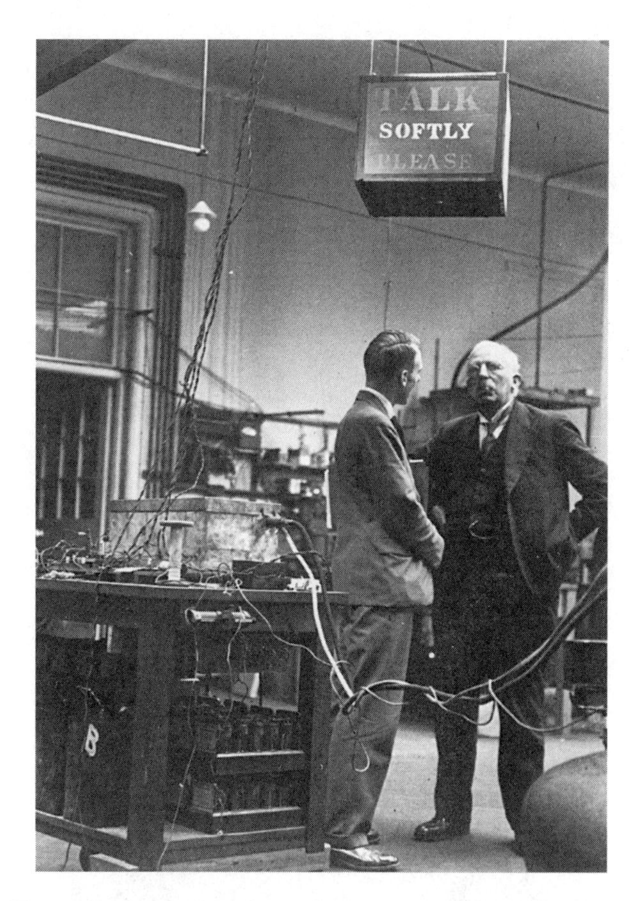

Ernest Rutherford is shown here with J. A. Ratcliffe in the Cavendish Laboratory in 1936. The sign above Rutherford (on the right) reads "TALK SOFTLY PLEASE," because the detectors being used were very sensitive to vibrations and noise. Rutherford, whose deep booming voice disturbed the detectors more than anyone else, didn't seem to think the warning applied to him and was in a loud conversation when this photo was taken. *AIP Emilio Segrè Visual Archives.*

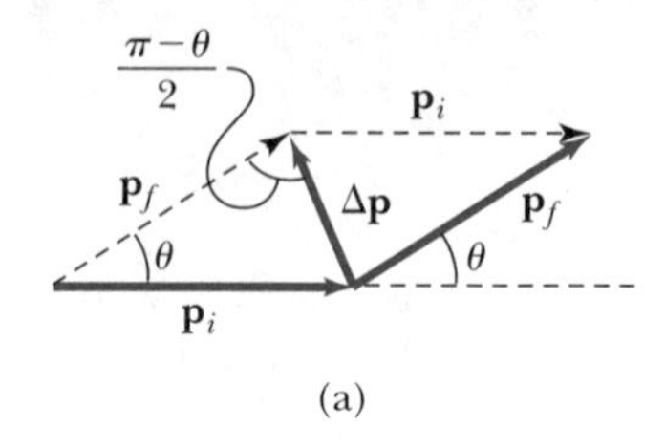

(a)

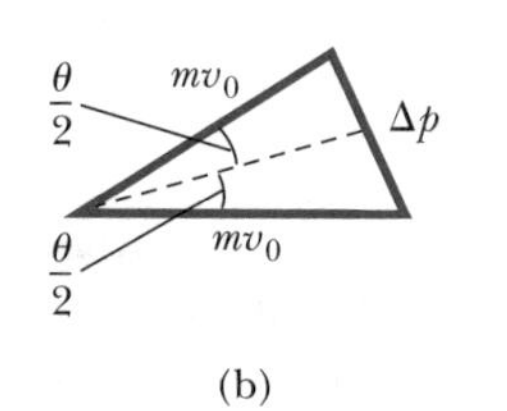

(b)

FIGURE 4.8 (a) The scattering angle θ and momentum change $\Delta\mathbf{p}$ are determined from the initial and final values of the α particle's momentum. (b) Because $\mathbf{p}_f$, $\mathbf{p}_i$, and $\Delta\mathbf{p}$ almost form an exact isosceles triangle, we determine the magnitude of $\Delta\mathbf{p}$ by bisecting the angle θ and finding the length of the triangle leg opposite the angle $\theta/2$.

We define the instantaneous position of the particle by the angle ϕ and the distance r from the force center, where $\phi = 0$ (which defines the z' axis) when the distance r is a minimum, as is shown in Figure 4.6. The change in momentum must be equal to the impulse.

$$\Delta\mathbf{p} = \int \mathbf{F}_{\Delta p}\,dt \tag{4.1}$$

where $\mathbf{F}_{\Delta p}$ is the force along the direction of $\Delta\mathbf{p}$. The massive scatterer absorbs this (small) momentum change without gaining any appreciable kinetic energy (no recoil). Using the diagram of Figure 4.8,

$$\Delta\mathbf{p} = \mathbf{p}_f - \mathbf{p}_i \tag{4.2}$$

where the subscripts i and f indicate the initial and final values of the projectile's momentum, respectively. Because $p_f \approx p_i = mv_0$, the triangle between $\mathbf{p}_f$, $\mathbf{p}_i$, and $\Delta\mathbf{p}$ is isosceles. We redraw the triangle in Figure 4.8b, indicating the bisector of angle θ. The magnitude Δp of $\Delta\mathbf{p}$ is now

$$\Delta p = 2mv_0 \sin\frac{\theta}{2} \tag{4.3}$$

The direction of $\Delta\mathbf{p}$ is the z' axis (where $\phi = 0$), so we need the component of $\mathbf{F}$ along z' in Equation (4.1). The Coulomb force $\mathbf{F}$ is along the instantaneous direction of the position vector $\mathbf{r}$ (unit vector $\mathbf{e}_r$)

$$\mathbf{F} = \frac{1}{4\pi\epsilon_0}\frac{Z_1 Z_2 e^2}{r^2}\mathbf{e}_r = F\mathbf{e}_r \tag{4.4}$$

and

$$F_{\Delta p} = F\cos\phi \tag{4.5}$$

where $F_{\Delta p}$ is the component of the force $\mathbf{F}$ along the direction of $\Delta\mathbf{p}$ that we need.

Substituting the magnitudes from Equations (4.3) and (4.5) into the components of Equation (4.1) along the z' axis ($\phi = 0$) gives

$$\Delta p = 2mv_0 \sin\frac{\theta}{2} = \int F\cos\phi\,dt$$

$$= \frac{Z_1 Z_2 e^2}{4\pi\epsilon_0}\int \frac{\cos\phi}{r^2}\,dt$$

The instantaneous angular momentum must be conserved, so

$$mr^2\frac{d\phi}{dt} = mv_0 b$$

and

$$r^2 = \frac{v_0 b}{\dfrac{d\phi}{dt}}$$

Therefore,

$$2mv_0 \sin\frac{\theta}{2} = \frac{Z_1 Z_2 e^2}{4\pi\epsilon_0}\int \frac{\cos\phi}{v_0 b}\frac{d\phi}{dt}\,dt$$

$$= \frac{Z_1 Z_2 e^2}{4\pi\epsilon_0 v_0 b}\int_{\phi_i}^{\phi_f} \cos\phi\,d\phi$$

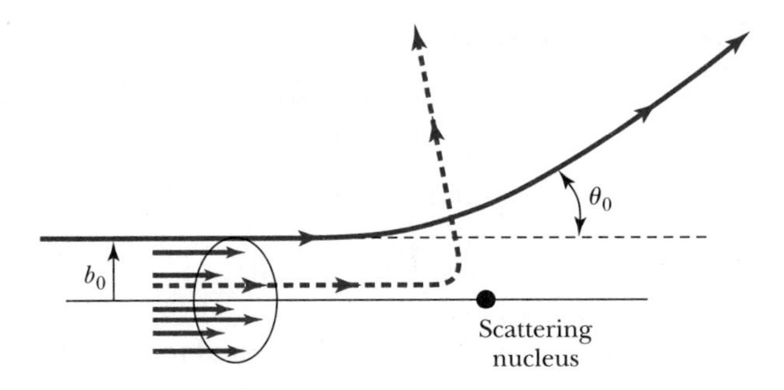

FIGURE 4.9 All particles with impact parameters less than b_0 will scatter at angles *greater* than θ_0.

We let the initial angle ϕ_i be on the negative side and the final angle ϕ_f be on the positive side of the z' axis ($\phi = 0$, see Figure 4.6). Then $\phi_i = -\phi_f$, $-\phi_i + \phi_f + \theta = \pi$, so $\phi_i = -(\pi - \theta)/2$ and $\phi_f = +(\pi - \theta)/2$.

$$\frac{8\pi\epsilon_0 m v_0{}^2 b}{Z_1 Z_2 e^2} \sin\frac{\theta}{2} = \int_{-\left(\frac{\pi-\theta}{2}\right)}^{+\left(\frac{\pi-\theta}{2}\right)} \cos\phi \, d\phi = 2\cos\frac{\theta}{2}$$

We now solve this equation for the impact parameter b.

$$b = \frac{Z_1 Z_2 e^2}{4\pi\epsilon_0 m v_0{}^2} \cot\frac{\theta}{2}$$

or with $K = \frac{1}{2} m v_0{}^2$

$$b = \frac{Z_1 Z_2 e^2}{8\pi\epsilon_0 K} \cot\frac{\theta}{2} \tag{4.6}$$

Relation between b and θ

where $K = m v_0{}^2/2$ is the kinetic energy of the bombarding particle. This is the fundamental relationship between the impact parameter b and scattering angle θ that we have been seeking for the Coulomb force.

In an experiment we are not able to select individual impact parameters b. When we put a detector at a particular angle θ, we cover a finite $\Delta\theta$ which corresponds to a range of impact parameters Δb. The bombarding particles are incident at varied impact parameters all around the scatterer as shown in Figure 4.9. All the particles with impact parameters less than b_0 will be scattered at angles greater than θ_0. Any particle with impact parameter inside the area of the circle of area $\pi b_0{}^2$ (radius b_0) will be similarly scattered. For the case of Coulomb scattering, we denote the cross section by the symbol σ, where

$$\sigma = \pi b^2 \tag{4.7}$$

is the cross section for scattering through an angle θ or more. The cross section σ is related to the *probability* for a particle being scattered by a nucleus. If we have a target foil of thickness t with n atoms/volume, the number of target nuclei per unit area is nt. Because we assumed a thin target of area A and all nuclei are exposed as shown in Figure 4.10, the number of target nuclei is simply ntA. The value of n is the density ρ (g/cm^3) times Avogadro's number N_A (molecules/mole) times the number of atoms/molecule N_M divided by the gram-molecular weight M_g (g/mole).

$$n = \frac{\rho\left(\frac{\text{g}}{\text{cm}^3}\right) N_A \left(\frac{\text{molecules}}{\text{mole}}\right) N_M \left(\frac{\text{atoms}}{\text{molecule}}\right)}{M_g\left(\frac{\text{g}}{\text{mole}}\right)} = \frac{\rho N_A N_M}{M_g} \frac{\text{atoms}}{\text{cm}^3} \tag{4.8}$$

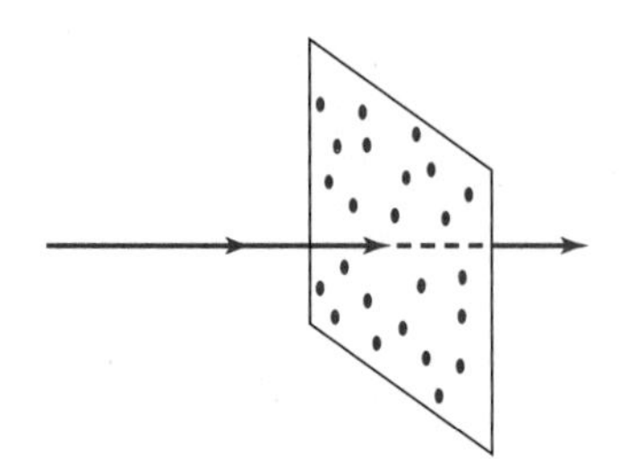

FIGURE 4.10 The target is assumed to be so thin that all nuclei are exposed to the bombarding particles. No nucleus is hidden behind another.

The number of scattering nuclei per unit area is nt.

$$nt = \frac{\rho N_A N_M t}{M_g} \frac{\text{atoms}}{\text{cm}^2} \tag{4.9}$$

If we have a foil of area A the number of target nuclei N_s is

$$N_s = ntA = \frac{\rho N_A N_M t A}{M_g} \text{ atoms} \tag{4.10}$$

The probability of the particle being scattered is equal to the total target area exposed for all the nuclei divided by the total target area A. If σ is the cross section for each nucleus, then $ntA\sigma$ is the total area exposed by the target nuclei, and the fraction of incident particles scattered by an angle of θ or greater is

$$f = \frac{\text{Target area exposed by scatterers}}{\text{Total target area}} = \frac{ntA\sigma}{A}$$

$$= nt\sigma = nt\pi b^2 \tag{4.11}$$

$$f = \pi nt\left(\frac{Z_1 Z_2 e^2}{8\pi\epsilon_0 K}\right)^2 \cot^2\frac{\theta}{2} \tag{4.12}$$

Example 4.2

Find the fraction of 7.7-MeV α particles that is deflected at an angle of 90° or more from a gold foil of 10^{-6} m thickness.

Solution: To use Equation (4.12) we first need to calculate n, the number of atoms/cm³. The density of gold is 19.3 g/cm³ and the atomic weight is 197 u. We use Equation (4.8) to determine n.

$$n = \frac{\left(19.3\ \frac{\text{g}}{\text{cm}^3}\right)\left(6.02\times 10^{23}\ \frac{\text{molecules}}{\text{mole}}\right)\left(1\ \frac{\text{atom}}{\text{molecule}}\right)}{197\ \text{g/mole}}$$

$$= 5.90\times 10^{22}\ \frac{\text{atoms}}{\text{cm}^3} = 5.90\times 10^{28}\ \frac{\text{atoms}}{\text{m}^3}$$

We insert this value of n into Equation (4.12) and find

$$f = \pi\left(5.90\times 10^{28}\ \frac{\text{atoms}}{\text{m}^3}\right)(10^{-6}\ \text{m}) \times$$

$$\left[\frac{(79)(2)(1.6\times 10^{-19}\ \text{C})^2(9\times 10^9\ \text{N}\cdot\text{m}^2/\text{C}^2)}{(2)(7.7\ \text{MeV})(1.6\times 10^{-13}\ \text{J/MeV})}\right]^2(\cot 45°)^2$$

$$= 4\times 10^{-5}$$

One α particle in 25,000 would be deflected by 90° or greater.

In an actual experiment, however, a detector is positioned only over a range of angle θ, θ to $\theta + \Delta\theta$, as shown in Figure 4.11. Thus we need to find the number of particles scattered between θ and $\theta + d\theta$ that corresponds to incident particles with impact parameters between b and $b + db$ as displayed in Figure 4.12. The fraction of the incident particles scattered between θ and $\theta + d\theta$ is $df/d\theta$. The derivative of Equation (4.12) is

$$df = -\pi nt\left(\frac{Z_1 Z_2 e^2}{8\pi\epsilon_0 K}\right)^2 \cot\frac{\theta}{2}\csc^2\frac{\theta}{2}\,d\theta$$

If the total number of incident particles is N_i, the number of particles scattered into the ring about $d\theta$ is $N_i|df|$. The area dA into which the particles scatter

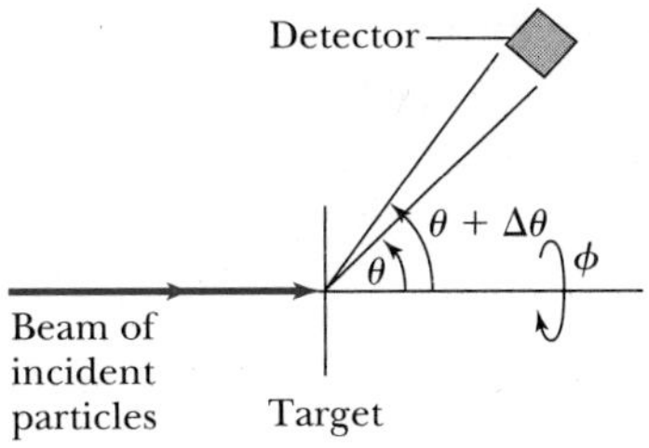

Detector
$\theta + \Delta\theta$
θ
ϕ
Beam of incident particles
Target

FIGURE 4.11 In most experiments, the detectors cover only a small angular range, from θ to $\theta + \Delta\theta$, and measurements are made for different θ. The detector also usually covers a small angular range in ϕ (angle around beam direction). Because there is usually symmetry about the beam axis, the ϕ angle is not normally varied.

is $(r\,d\theta)(2\pi r \sin\theta) = 2\pi r^2 \sin\theta\, d\theta$. Therefore, the number of particles scattered per unit area, $N(\theta)$, into the ring at scattering angle θ is

$$N(\theta) = \frac{N_i |df|}{dA} = \frac{N_i \pi n t \left(\dfrac{Z_1 Z_2 e^2}{8\pi\epsilon_0 K}\right)^2}{2\pi r^2 \sin\theta\, d\theta} \cot\frac{\theta}{2} \csc^2\frac{\theta}{2}\, d\theta$$

$$N(\theta) = \frac{N_i n t}{16}\left(\frac{e^2}{4\pi\epsilon_0}\right)^2 \frac{Z_1^{\,2} Z_2^{\,2}}{r^2 K^2 \sin^4(\theta/2)} \tag{4.13}$$

Rutherford scattering equation

This is the famous **Rutherford scattering equation.** The important points are

1. The scattering is proportional to the square of the atomic number of both the incident particle (Z_1) and the target scatterer (Z_2).
2. The number of scattered particles is inversely proportional to the square of the kinetic energy K of the incident particle.
3. The scattering is inversely proportional to the fourth power of $\sin\frac{\theta}{2}$, where θ is the scattering angle.
4. The scattering is proportional to the target thickness for thin targets.

These very specific predictions by Rutherford in 1911 were confirmed experimentally by Geiger and Marsden in 1913. The angular dependence is particularly characteristic and can be verified in a well-equipped undergraduate physics laboratory, as we see from some actual data shown in Figure 4.13.

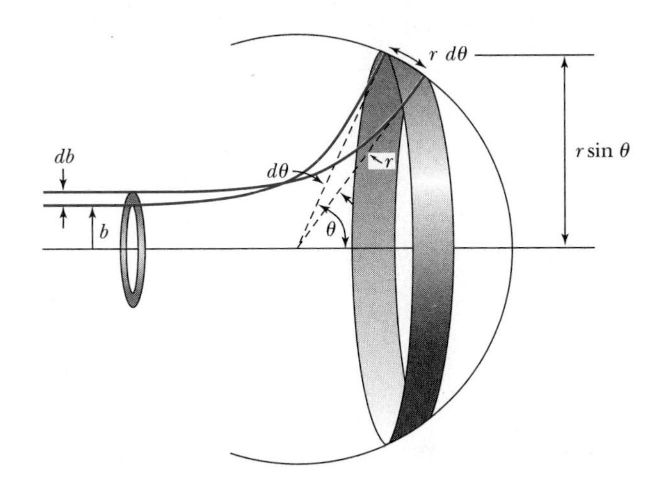

FIGURE 4.12 Particles over the range of impact parameters from b to $b + db$ will scatter into the angular range θ to $\theta + d\theta$ (with db positive, $d\theta$ will be negative).

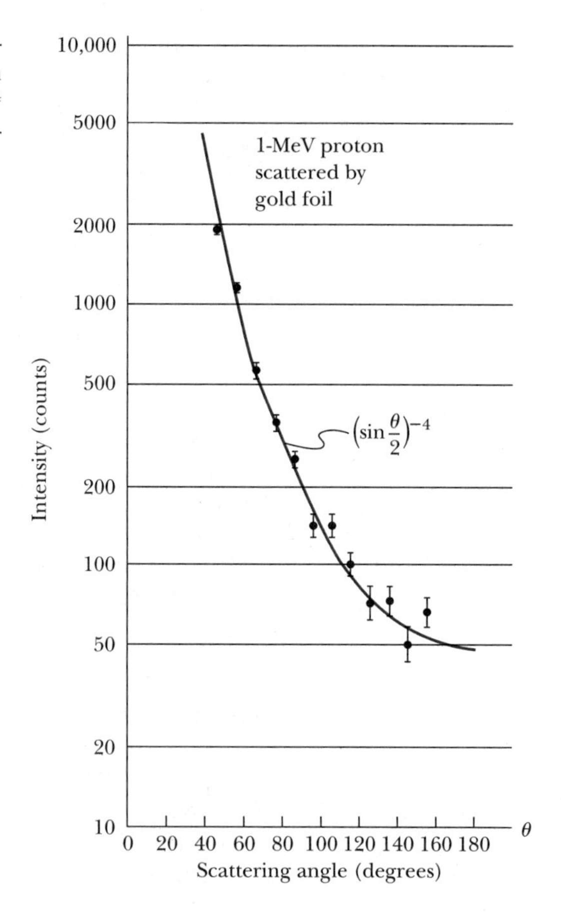

FIGURE 4.13 Results of undergraduate laboratory experiment of scattering 1-MeV protons from a gold target. The solid line shows the $(\sin\theta/2)^{-4}$ angular dependence of the data, verifying Rutherford's calculation.

Example 4.3

Calculate the fraction per mm^2 area of 7.7-MeV α particles scattered at 45° from a gold foil of thickness 2.1×10^{-8} m at a distance of 1 cm from the target.

Solution: We use Equation (4.13) to determine the fraction per unit area $N(\theta)/N_i$. We calculated $n = 5.90 \times 10^{28}$ atoms/m^3 in Example 4.2.

This is the theoretical basis for the experiment performed by Geiger and Marsden in 1913 to check the validity of Rutherford's calculation. Our calculated result agrees with their experimental result of 3.7×10^{-7} mm^{-2} when the experimental uncertainty is taken into account.

$$\frac{N(\theta)}{N_i} = \frac{\left(5.90 \times 10^{28}\,\frac{\text{atoms}}{\text{m}^3}\right)(2.1 \times 10^{-7}\text{ m})\left((1.6 \times 10^{-19}\text{ C})^2\left(9 \times 10^9\,\frac{\text{N}\cdot\text{m}^2}{\text{C}^2}\right)\right)^2}{16}$$

$$\times \frac{(2)^2(79)^2}{(10^{-2}\text{ m})^2\left(7.7\text{ MeV} \times \frac{10^6\text{ eV}}{\text{MeV}} \times \frac{1.6 \times 10^{-19}\text{ J}}{\text{eV}}\right)^2}\,\frac{1}{\sin^4(45°/2)}$$

$$\frac{N(\theta)}{N_i} = 3.2 \times 10^{-1}\text{ m}^{-2} = 3.2 \times 10^{-7}\text{ mm}^{-2}$$

For a given kinetic energy K and impact parameter b there is a distance of closest approach between a bombarding particle and target scatterer of like charge. The minimum separation occurs for a head-on collision. The particle turns around and scatters backward at 180°. At the instant the particle turns around, the entire kinetic energy has been converted into Coulomb potential energy.

$$K = \frac{(Z_1 e)(Z_2 e)}{4\pi\epsilon_0 r} \tag{4.14}$$

We solve this equation to determine r_{min}.

$$r_{\text{min}} = \frac{Z_1 Z_2 e^2}{4\pi\epsilon_0 K} \tag{4.15}$$

Distance of closest approach

For α particles of 7.7-MeV kinetic energy scattering on aluminum ($Z_2 = 13$) or gold ($Z_2 = 79$), the values of r_{min} are 5×10^{-15} m (Al) and 3×10^{-14} m (Au). We now know that nuclear radii vary from 1 to 10×10^{-15} m. Thus when α particles scatter from aluminum, an α particle may approach the nucleus quite closely and may even be affected by the nuclear force as we shall see in Chapter 12. In this case we might expect some deviation from the Rutherford scattering equation.

4.3 The Classical Atomic Model

After Rutherford presented his calculations of charged-particle scattering in 1911, it was generally conceded that the atom consisted of a small, massive, positively charged "nucleus" surrounded by moving electrons. Thomson's "plum-pudding" model was definitively excluded by the data. Actually, Thomson had previously considered a planetary model resembling the solar system (where the planets move in elliptical orbits about the sun) but had rejected it because, although both gravitational and Coulomb forces vary inversely with the square of the distance, the planets *attract* one another while orbiting around the sun, whereas the electrons would *repel* one another. Thomson considered this to be a fatal flaw from his knowledge of planetary theory.

Let us examine the simplest atom, hydrogen. We will assume circular electron orbits for simplicity rather than the more general elliptical ones. The force of attraction on the electron due to the nucleus (charge $= +e$) is

$$\mathbf{F}_e = \frac{-1}{4\pi\epsilon_0}\frac{e^2}{r^2}\,\mathbf{e}_r \tag{4.16}$$

where the negative sign indicates the force is attractive and $\mathbf{e}_r$ is a unit vector in the direction from the nucleus to the electron. This electrostatic force provides the centripetal force needed for the electron to move in a circular orbit at constant speed. Its radial acceleration is

$$a_r = \frac{v^2}{r} \tag{4.17}$$

where v is the tangential velocity of the electron. Newton's second law now gives

$$\frac{1}{4\pi\epsilon_0}\frac{e^2}{r^2} = \frac{mv^2}{r} \tag{4.18}$$

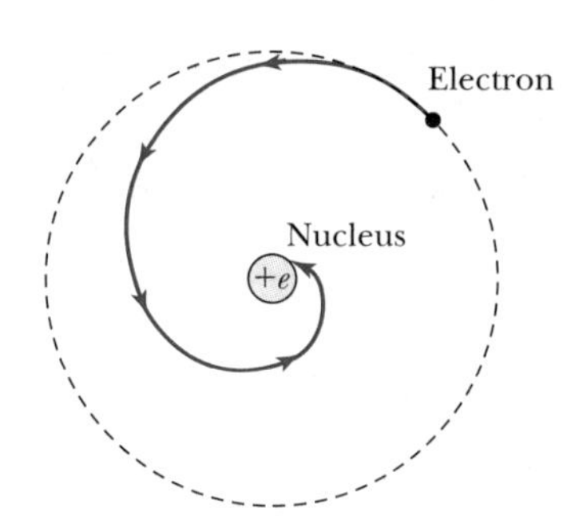

FIGURE 4.14 The electromagnetic radiation of an orbiting electron in the planetary model of the atom will cause the electron to spiral inward until it crashes into the nucleus.

and

$$v = \frac{e}{\sqrt{4\pi\epsilon_0 mr}} \tag{4.19}$$

(where we are using m without a subscript to be the electron's mass. When it is not clear which particle m refers to, we write the electron mass as m_e.) The size of an atom was thought to be about 10^{-10} m, so by letting $r = 0.5 \times 10^{-10}$ m, we can use Equation (4.19) to estimate the electron's velocity.

$$v \approx \frac{(1.6 \times 10^{-19}\text{ C})(9 \times 10^9\text{ N}\cdot\text{m}^2/\text{C}^2)^{1/2}}{(9.11 \times 10^{-31}\text{ kg})^{1/2}(0.5 \times 10^{-10}\text{ m})^{1/2}}$$

$$\approx 2.2 \times 10^6\text{ m/s} < 0.01c$$

This justifies a nonrelativistic treatment.

The kinetic energy of the system is due to the electron, $K = mv^2/2$. The nucleus is so massive compared to the electron ($m_{\text{proton}} = 1836\, m$) that the nucleus may be considered to be at rest. The potential energy V is simply $-e^2/4\pi\epsilon_0 r$, so the total mechanical energy is

$$E = K + V = \frac{1}{2}mv^2 - \frac{e^2}{4\pi\epsilon_0 r} \tag{4.20}$$

If we substitute for v from Equation (4.19), we have

$$E = \frac{e^2}{8\pi\epsilon_0 r} - \frac{e^2}{4\pi\epsilon_0 r} = \frac{-e^2}{8\pi\epsilon_0 r} \tag{4.21}$$

The total energy is negative, indicating a bound, attractive system.

Thus far, the classical atomic model seems plausible. The problem arises when we consider that the electron is accelerating due to its circular motion about the nucleus. We know from classical electromagnetic theory that an accelerated electric charge continuously radiates energy in the form of electromagnetic radiation. If the electron is radiating energy, then the total energy E of the system, Equation (4.21), must decrease continuously. In order for this to happen, the radius r must decrease. The electron will continuously radiate energy as the electron orbit becomes smaller and smaller until the electron crashes into the nucleus! This process, displayed in Figure 4.14, would occur in about 10^{-9} s (see Problem 18).

Planetary model is doomed.

Thus the classical theories of Newton and Maxwell, which had served Rutherford so well in his analysis of α-particle scattering and had thereby enabled him to discover the nucleus, also led to the failure of the planetary model of the atom. Physics had reached a decisive turning point like that encountered almost two decades earlier with Planck's revolutionary hypothesis of the quantum behavior of radiation. In the early 1910s, however, the answer would not be long in coming, as we shall see in the next section.

4.4 Bohr Model of the Hydrogen Atom

Shortly after receiving his Ph.D. from the University of Copenhagen in 1911, the 26-year-old Danish physicist Niels Bohr traveled to Cambridge University to work with J. J. Thomson. Upon meeting Ernest Rutherford early in 1912, Bohr decided to spend a few months at the University of Manchester to gain more experience in experimental physics. Upon arriving in Manchester in March, 1912, he became so involved in the mysteries of the new Rutherford model of the atom and the many experiments going on that he decided to continue his study on the structure of the atom that he had begun earlier in Cambridge under Thomson. Upon leaving Manchester for a position at the University of Copenhagen in the summer of 1912, Bohr still had many questions about atomic structure. Like several others, he believed that a fundamental length about the size of an atom (10^{-10} m) was needed for an atomic model. This fundamental length might somehow be connected to Planck's new constant h. The pieces finally came together during the fall and winter of 1912 when Bohr learned of new precise measurements of the hydrogen spectrum and of the empirical formulas describing them. He set out to find a fundamental basis from which to derive the Balmer formula Equation (3.12), the Rydberg equation (3.13), and Ritz's combination principles (see Problem 19).

Niels Bohr (1885–1962) was more than just a discoverer of modern physics theories. Born in Denmark, he was the son of a university professor and began high school at about the time Planck announced his results. After his education in Denmark, Bohr traveled to England in 1911 where he worked first with J. J. Thomson and later with Ernest Rutherford. Except for a year in 1913–1914 when he returned to Denmark, Bohr remained in England working mostly with Rutherford until 1916. Bohr nurtured many young theoretical physicists in his Institute of Theoretical Physics (now called the Niels Bohr Institute) formed in Copenhagen in 1921, the year before Bohr won the Nobel Prize. Many of the world's greatest physicists spent time in Copenhagen under the tutelage of Bohr, an outstanding teacher and developer of physicists. *American Institute of Physics.*

Bohr was well acquainted with Planck's work on the quantum nature of radiation. Like Einstein, Bohr believed that quantum principles should govern more phenomena than just the blackbody spectrum. He was impressed by Einstein's application of the quantum theory to the photoelectric effect and to the specific heat of solids (see Chapter 9).

In 1913, following several discussions with Rutherford during 1912 and 1913, Bohr published the paper* "On the Constitution of Atoms and Molecules." He subsequently published several other papers refining and restating his "assumptions" and their predicted results. We will generally follow Bohr's papers in our discussion.

Bohr assumed that electrons moved around a massive, positively charged nucleus. We will assume for simplicity (as did Bohr at first) that the electron orbits are circular rather than elliptical and that the nuclear mass is so much greater than the electron's mass that it may be taken to be infinite. The electron has charge $-e$ and mass m and revolves around a nucleus of charge $+e$ in a circle of radius a. The size of the nucleus is small compared to the radius a.

Bohr's model may best be summarized by the following "*general assumptions*" of his 1915 paper:

Bohr's general assumptions

A. Certain "stationary states" exist in atoms, which differ from the classical stable states in that the orbiting electrons do not continuously radiate electromagnetic energy. The stationary states are states of definite total energy.

B. The emission or absorption of electromagnetic radiation can occur only in conjunction with a transition between two stationary states. The frequency ν of the emitted or absorbed radiation is proportional to the difference in energy of the two stationary states (1 and 2)

$$E = E_1 - E_2 = h\nu$$

where h is Planck's constant.

C. The dynamical equilibrium of the system in the stationary states is governed by classical laws of physics, but these laws do not apply to transitions between stationary states.

D. The mean value K of the kinetic energy of the electron–nucleus system is given by $K = nh\nu_{\text{orb}}/2$, where ν_{orb} is the frequency of rotation. For a circular orbit, Bohr pointed out that this assumption is equivalent to the angular momentum of the system in a stationary state being an integral multiple of $h/2\pi$. (This combination of constants occurs so often that we give it a separate symbol, $\hbar \equiv h/2\pi$, pronounced "h bar".)

*Niels Bohr, *Philosophical Magazine* **26,** 1(1913) and **30,** 394(1915).

These four assumptions were all that Bohr needed to derive the Rydberg equation. Bohr believed that Assumptions A and C were self-evident because atoms were stable: atoms exist and do not continuously radiate energy (therefore Assumption A). It also seemed that the classical laws of physics could not explain the observed behavior of the atom (therefore Assumption C).

Bohr later stated (1915) that Assumption B "appears to be necessary in order to account for experimental facts." However, Assumption D was the hardest for Bohr's critics to accept. It is central to the derivation of the binding energy of the hydrogen atom in terms of fundamental constants; hence Bohr restated and defended it in several ways in his papers. We have emphasized here the quantization of angular momentum aspect of Assumption D. This leads to a particularly simple derivation of the Rydberg equation.

Bohr's four assumptions were chosen to keep as much as possible of classical physics by introducing just those new ideas that were needed to describe the atom. Bohr's recognition that something new was needed and his attempt to tie this to Planck's quantum hypothesis represented an advance in understanding perhaps even greater than Einstein's theory of the photoelectric effect.

Let us now proceed to derive the Rydberg equation using Bohr's assumptions. The total energy (potential plus kinetic) of a hydrogen atom was derived previously, in Equation (4.21). For circular motion, the magnitude of the angular momentum L of the electron is

$$L = |\mathbf{r} \times \mathbf{p}| = mvr$$

Assumption D states this should equal $n\hbar$:

$$L = mvr = n\hbar \tag{4.22a}$$

Principal quantum number

where n is an integer called the **principal quantum number.** We solve the previous equation for the velocity and obtain

$$v = \frac{n\hbar}{mr} \tag{4.22b}$$

Equation (4.19) yields an independent relation between v and r. If we determine v^2 from Equations (4.19) and (4.22b) and set them equal, we find

$$v^2 = \frac{e^2}{4\pi\epsilon_0 mr} = \frac{n^2\hbar^2}{m^2r^2} \tag{4.23}$$

From Equation (4.23) we see that only certain values of r are allowed.

$$r_n = \frac{4\pi\epsilon_0 n^2\hbar^2}{me^2} \equiv n^2 a_0 \tag{4.24}$$

Bohr radius

where the **Bohr radius** a_0 is given by

$$\begin{aligned} a_0 &= \frac{4\pi\epsilon_0\hbar^2}{me^2} \\ &= \frac{(1.055 \times 10^{-34}\ \mathrm{J\cdot s})^2}{\left(8.99 \times 10^9 \dfrac{\mathrm{N\cdot m^2}}{\mathrm{C^2}}\right)(9.11 \times 10^{-31}\ \mathrm{kg})(1.6 \times 10^{-19}\ \mathrm{C})^2} \\ &= 0.53 \times 10^{-10}\ \mathrm{m} \end{aligned}$$

Notice that the smallest diameter of the hydrogen atom is $2r_1 = 2a_0 \approx 10^{-10}$ m, the known size of the hydrogen atom! Bohr had found the fundamental length

a_0 that he sought in terms of the fundamental constants ϵ_0, h, e, and m. This fundamental length is determined for the value $n = 1$. Note from Equation (4.24) that the atomic radius is now quantized. The quantization of various physical values arises because of the principal quantum number n. The value $n = 1$ gives the radius of the hydrogen atom in its lowest energy state (called the "ground" state). The values of $n > 1$ determine other possible radii where the hydrogen atom is in an "excited" state.

The energies of the stationary states can now be determined from Equations (4.21) and (4.24).

$$E_n = -\frac{e^2}{8\pi\epsilon_0 r_n} = -\frac{e^2}{8\pi\epsilon_0 a_0 n^2} \equiv -\frac{E_0}{n^2} \tag{4.25}$$

The lowest-energy state ($n = 1$) is $E_1 = -E_0$ where

$$E_0 = \frac{e^2}{8\pi\epsilon_0 a_0} = \frac{e^2}{(8\pi\epsilon_0)}\frac{me^2}{4\pi\epsilon_0\hbar^2} = \frac{me^4}{2\hbar^2(4\pi\epsilon_0)^2} = 13.6 \text{ eV} \tag{4.26}$$

This is the experimentally measured ionization energy of the hydrogen atom. Bohr's assumptions C and D imply that the atom can exist only in "stationary states" with definite, quantized energies E_n, displayed in the **energy-level diagram** of Figure 4.15. Emission of a quantum of light occurs when the atom is in an excited state (quantum number $n = n_u$) and decays to a lower energy state ($n = n_\ell$). A transition between two energy levels is schematically illustrated in Figure 4.15. According to Assumption B we have

$$h\nu = E_u - E_\ell \tag{4.27}$$

where ν is the frequency of the emitted light quantum (photon). Because $\lambda\nu = c$, we have

$$\frac{1}{\lambda} = \frac{\nu}{c} = \frac{E_u - E_\ell}{hc}$$

$$= \frac{-E_0}{hc}\left(\frac{1}{n_u^2} - \frac{1}{n_\ell^2}\right) = \frac{E_0}{hc}\left(\frac{1}{n_\ell^2} - \frac{1}{n_u^2}\right) \tag{4.28}$$

where

$$\frac{E_0}{hc} = \frac{me^4}{4\pi c\hbar^3(4\pi\epsilon_0)^2} \equiv R_\infty \tag{4.29}$$

This constant R_∞ is called the **Rydberg constant** (for an infinite nuclear mass). Equation (4.28) becomes

$$\frac{1}{\lambda} = R_\infty\left(\frac{1}{n_\ell^2} - \frac{1}{n_u^2}\right) \tag{4.30}$$

which is similar to the Rydberg equation (3.13). The value of $R_\infty = 1.097373 \times 10^7 \text{ m}^{-1}$ calculated from Equation (4.29) agrees well with the experimental values given in Chapter 3, and we will obtain an even more accurate result in the next section.

Bohr's model predicts the frequencies (and wavelengths) of all possible transitions in atomic hydrogen. Several of the series are shown in Figure 4.16. The Lyman series represents transitions to the lowest state with $n_\ell = 1$; the Balmer series results from downward transitions to the stationary state $n_\ell = 2$, and the Paschen series from transitions to $n_\ell = 3$. As mentioned in Section 3.3, not all of these series were known experimentally in 1913, but it was clear that Bohr had successfully accounted for the known spectral lines of hydrogen.

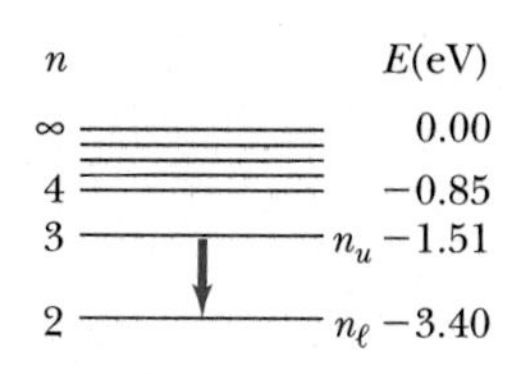

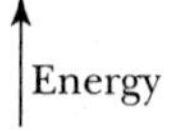

1 −13.6

FIGURE 4.15 The energy-level diagram of the hydrogen atom. The principal quantum numbers n are shown on the left, with the energy of each level indicated on the right. The ground-state energy is −13.6 eV; negative total energy indicates a bound, attractive system. When an atom is in an excited state (for example, $n_u = 3$) and decays to a lower stationary state (for example, $n_\ell = 2$), the hydrogen atom must emit the energy difference in the form of electromagnetic radiation, that is, a photon emerges.

Bohr predicts new hydrogen wavelengths.

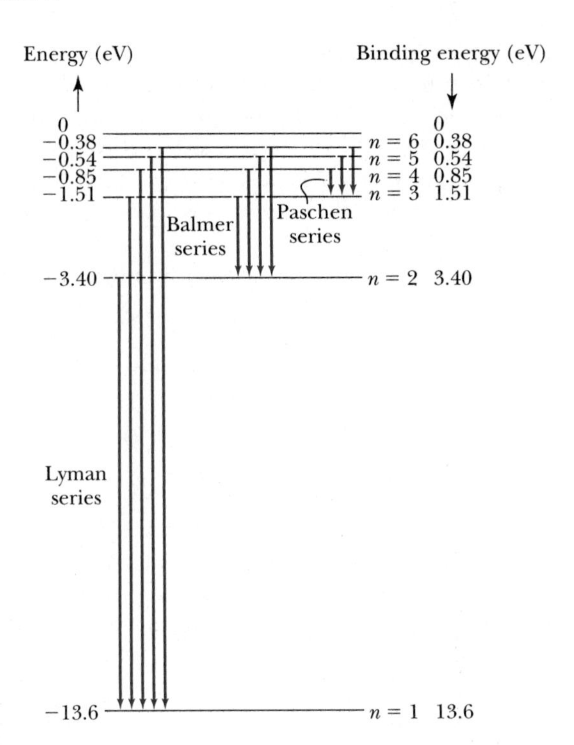

FIGURE 4.16 Transitions between many of the stationary states in the hydrogen atom are indicated. Transitions (ultraviolet) to the $n = 1$ state from the higher lying states are called the *Lyman series.* The transitions to the $n = 2$ state (Balmer series) were discovered first because they are in the visible wavelength range. The Paschen series (transitions to $n = 3$) are in the infrared. The energies of each state as well as the binding energies are denoted.

The frequencies of the photons in the emission spectrum of an element are directly proportional to the differences in energy of the stationary states. When we pass white light (composed of all visible photon frequencies) through atomic hydrogen gas, we find that certain frequencies are absent. This pattern of dark lines is called an **absorption spectrum.** The missing frequencies are *precisely* the ones observed in the corresponding **emission spectrum.** In absorption, certain photons of light are absorbed, giving up energy to the atom and enabling the electron to move from a lower (ℓ) to a higher (u) stationary state. Equations (4.27) and (4.30) describe the frequencies and wavelengths of the absorbed photons. The atom will remain in the excited state for only a very short time (on the order of 10^{-10} s) before emitting a photon and returning to a lower stationary state. Thus, at ordinary temperatures practically all hydrogen atoms exist in the lowest possible energy state, $n = 1$, and only the absorption spectra of the Lyman series are normally observed. However, these lines are not in the visible region.

Absorption and emission spectra

We can determine the electron's velocity in the Bohr model from Equations (4.22b) and (4.23).

$$v_n = \frac{n\hbar}{mr_n} = \frac{n\hbar}{mn^2 a_0} = \frac{1}{n}\frac{\hbar}{ma_0} \tag{4.31}$$

or

$$v_n = \frac{1}{n}\frac{e^2}{4\pi\epsilon_0\hbar}$$

The value of $v_1 = \hbar/ma_0 = 2.2 \times 10^6$ m/s, which is less than 1% of the speed of light. We define the dimensionless quantity ratio of v_1 to c by

Fine structure constant

$$\alpha \equiv \frac{v_1}{c} = \frac{\hbar}{ma_0 c} = \frac{e^2}{4\pi\epsilon_0\hbar c} \approx \frac{1}{137} \tag{4.32}$$

This ratio is called the **fine structure constant.** It appears often in atomic structure calculations.

Example 4.4

Atomic hydrogen in its lowest energy state absorbs a photon, raising the electron to an $n = 3$ state. If we assume the lifetime of the excited state is 10^{-10} s, and if we make the rudimentary assumption that the electron orbits around the proton, how many revolutions does the electron make in the excited state before returning to a lower energy state?

Solution: The velocity of the electron will be

$$v = \frac{\text{Circumference}}{\text{Period}} = \frac{2\pi r}{T}$$

Therefore

$$T = \frac{2\pi r_3}{v_3} = \frac{(2\pi)(9a_0)(3)(4\pi\epsilon_0\hbar)}{e^2}$$

$$T = \frac{(54\pi)(0.53 \times 10^{-10}\ \text{m})(1.055 \times 10^{-34}\ \text{J}\cdot\text{s})}{(1.6 \times 10^{-19}\ \text{C})^2\left(9 \times 10^9 \dfrac{\text{N}\cdot\text{m}^2}{\text{C}^2}\right)}$$

$$= 4.1 \times 10^{-15}\ \text{s}$$

$$\text{Number of revolutions} = \frac{10^{-10}\ \text{s}}{4.1 \times 10^{-15}\ \text{s}} = 2.4 \times 10^4$$

The electron revolves many times in the excited state before decaying to a lower energy state.

A note of caution

We must insert a word of caution at this point. Bohr's model represented a significant step forward in understanding the structure of the atom. Although it had many successes, we know now that, in principle, it is wrong. We will discuss some of its successes and failures in the next section and will discuss the correct quantum theory in Chapter 6. For example, contrary to the previous example, the electron does not exactly revolve around the nucleus in orbits. The correct explanation uses a wave picture and a probability description of the electron's position. Nevertheless, the simple picture given by Bohr is useful in our first attempt in understanding the structure of the atom.

The Correspondence Principle. Early in the 1900s physicists had trouble relating well-known and well-understood classical physics results with the new quantum ones. Sometimes completely different results were valid in their own domains. For example, there were two radiation laws: One used classical electrodynamics to determine the properties of radiation from an accelerated charge, but there was another one due to Bohr and his model. Physicists were proposing various kinds of correspondence principles to relate the new modern results with the old classical ones that had worked so well in their own domain. In his 1913 paper Bohr proposed perhaps the best *correspondence principle* to guide physicists in developing new theories. This principle was refined several times over the next few years.

Correspondence principle

Bohr's correspondence principle: In the limits where classical and quantum theories should agree, the quantum theory must reduce to the classical result.

Let us examine the predictions of the two radiation laws. The frequency of the radiation produced by the atomic electrons in the Bohr model of the hydrogen atom should agree with that predicted by classical electrodynamics in a region where the finite size of Planck's constant is unimportant—for large quantum numbers n where quantization effects are minimized. To see how this works we recall that classically the frequency of the radiation emitted is equal to the orbital frequency ν_{orb} of the electron around the nucleus:

$$\nu_{\text{classical}} = \nu_{\text{orb}} = \frac{\omega}{2\pi} = \frac{1}{2\pi}\frac{v}{r} \tag{4.33a}$$

If we substitute for v from Equation (4.19), we find

$$\nu_{\text{classical}} = \frac{1}{2\pi}\left(\frac{e^2}{4\pi\epsilon_0 m r^3}\right)^{1/2} \tag{4.33b}$$

We make the connection to the Bohr model by inserting the orbital radius r from Equation (4.24) into Equation (4.33b). We know then the classical frequency in terms of fundamental constants and the principal quantum number n.

$$\nu_{\text{classical}} = \frac{me^4}{4\epsilon_0{}^2 h^3}\,\frac{1}{n^3} \tag{4.34}$$

In the Bohr model, the nearest we can come to continuous radiation is a cascade of transitions from a level with principal quantum number $n+1$ to the next lowest and so on:

$$n + 1 \longrightarrow n \longrightarrow n - 1 \longrightarrow \cdots$$

The frequency of the transition from $n + 1 \longrightarrow n$ is

$$\nu_{\text{Bohr}} = \frac{E_0}{h}\left(\frac{1}{n^2} - \frac{1}{(n+1)^2}\right)$$

$$= \frac{E_0}{h}\left(\frac{n^2 + 2n + 1 - n^2}{n^2(n+1)^2}\right) = \frac{E_0}{h}\left(\frac{2n+1}{n^2(n+1)^2}\right)$$

which for large n becomes

$$\nu_{\text{Bohr}} \approx \frac{2nE_0}{hn^4} = \frac{2E_0}{hn^3}$$

If we substitute E_0 from Equation (4.26), the result is

Equivalence of Bohr and classical frequencies

$$\nu_{\text{Bohr}} = \frac{me^4}{4\epsilon_0{}^2 h^3}\,\frac{1}{n^3} = \nu_{\text{classical}} \tag{4.35}$$

so the frequencies of the radiated energy agree between classical theory and the Bohr model for large values of the quantum number n. Bohr's correspondence principle is thus verified for large orbits, where classical and quantum physics should agree.

By 1915, as Bohr's model gained widespread acceptance, the critics of the quantum concept were finding it harder to gain an audience. Bohr had demonstrated the necessity of Planck's quantum constant in understanding atomic structure, and Einstein's conception of the photoelectric effect was generally accepted as well. The *assumption* of quantized angular momentum $L_n = n\hbar$ led to the quantization of other quantities r, v, and E. We collect the following three equations here for easy reference.

Orbital radius
$$r_n = \frac{4\pi\epsilon_0\hbar^2}{me^2}\,n^2 \tag{4.24}$$

Velocity
$$v_n = \frac{n\hbar}{mr_n} \tag{4.22b}$$

Energy
$$E_n = \frac{-e^2}{8\pi\epsilon_0 a_0}\,\frac{1}{n^2} \tag{4.25}$$

4.5 Successes and Failures of the Bohr Model

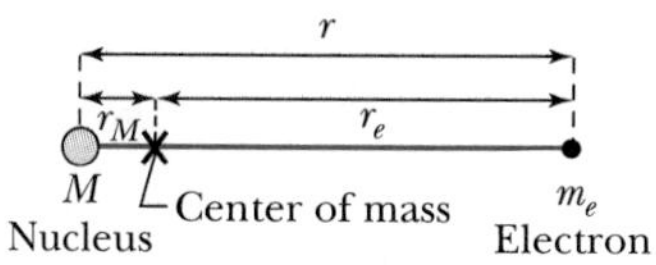

Wavelength measurements for the atomic spectrum of hydrogen are very precise and exhibit a small disagreement with the Bohr model results just presented. These disagreements can be corrected by looking more carefully at our original assumptions, one of which was to assume an infinite nuclear mass. The electron and hydrogen nucleus actually revolve about their mutual center of mass as shown in Figure 4.17. A straightforward analysis derived from classical mechanics shows that the only change required in the results of Section 4.4 is to replace the electron mass m_e by its reduced mass μ_e where

FIGURE 4.17 Because the nucleus does not actually have an infinite mass, the electron and nucleus rotate about a common center of mass that is located very near the nucleus.

Reduced mass correction

$$\mu_e = \frac{m_e M}{m_e + M} = \frac{m_e}{1 + \dfrac{m_e}{M}} \tag{4.36}$$

and M is the mass of the nucleus. The correction for the hydrogen atom is $\mu_e = 0.999456\ m_e$, only 5 parts in 10,000, but this difference can easily be measured experimentally. The Rydberg constant for infinite nuclear mass, R_∞, defined in Equation (4.29), should be replaced by R, where

$$R = \frac{\mu_e}{m_e} R_\infty = \frac{1}{1 + \dfrac{m_e}{M}} R_\infty = \frac{\mu_e e^4}{4\pi c \hbar^3 (4\pi\epsilon_0)^2} \tag{4.37}$$

The Rydberg constant for hydrogen is $R_{\text{H}} = 1.096776 \times 10^7\ \text{m}^{-1}$.

Example 4.5

Calculate the wavelength for the $n_u = 3 \rightarrow n_\ell = 2$ transition (called the H_α *line*) for the atoms of hydrogen, deuterium, and tritium.

Solution: The following masses are obtained by subtracting the electron mass from the atomic masses given in Appendix 8.

Proton	= 1.007276 u
Deuteron	= 2.013553 u
Triton (tritium nucleus)	= 3.015500 u

The electron mass is $m_e = 0.0005485799$ u. The Rydberg constants are

$$R_{\text{H}} = \frac{1}{1 + \dfrac{0.0005486}{1.00728}} R_\infty = 0.99946\ R_\infty \qquad \text{Hydrogen}$$

$$R_{\text{D}} = \frac{1}{1 + \dfrac{0.0005486}{2.01355}} R_\infty = 0.99973\ R_\infty \qquad \text{Deuterium}$$

$$R_{\text{T}} = \frac{1}{1 + \dfrac{0.0005486}{3.01550}} R_\infty = 0.99982\ R_\infty \qquad \text{Tritium}$$

The calculated wavelength for the H_α line is

$$\frac{1}{\lambda} = R\left(\frac{1}{2^2} - \frac{1}{3^2}\right) = 0.13889\ R$$

$$\lambda(\text{H}_\alpha, \text{hydrogen}) = 656.47\ \text{nm}$$

$$\lambda(\text{H}_\alpha, \text{deuterium}) = 656.29\ \text{nm}$$

$$\lambda(\text{H}_\alpha, \text{tritium}) = 656.23\ \text{nm}$$

Deuterium was discovered when two closely spaced spectral lines of hydrogen near 656.4 nm were observed in 1932. These proved to be the H_α lines of atomic hydrogen and deuterium.

The Bohr model may be applied to any single-electron atom (hydrogen-like) even if the nuclear charge is greater than 1 proton charge ($+e$), for example He^+ and Li^{++}. The only change needed is in the calculation of the Coulomb force, where e^2 is replaced by Ze^2 to account for the nuclear charge of $+Ze$. Bohr applied his model to the case of singly ionized helium, He^+. The Rydberg equation (4.30) now becomes

$$\frac{1}{\lambda} = Z^2 R\left(\frac{1}{n_\ell^2} - \frac{1}{n_u^2}\right) \tag{4.38}$$

where the Rydberg constant is given by Equation (4.37). We emphasize that Equation (4.38) is only valid for single-electron atoms (H, He^+, Li^{++}, etc.) and does not apply to any other atoms (for example He, Li, Li^+). Charged atoms like He^+ and Li^{++} are called *ions.*

In his original paper of 1913, Bohr predicted the spectral lines of He^+ although they had not yet been identified in the lab. He showed that certain lines (generally ascribed to hydrogen) that had been observed by Pickering in stellar spectra, and by Fowler in vacuum tubes containing both hydrogen and helium, could be identified with singly ionized helium. Bohr showed that the wavelengths predicted for He^+ with $n_\ell = 4$ are almost identical to those of H for $n_\ell = 2$, except that He^+ has *additional lines* between those of H (see Problem 33). The correct explanation of this fact by Bohr gave credibility to his model.

Example 4.6

Calculate the shortest wavelength that can be emitted by the Li^{++} ion.

Solution: This occurs when the electron changes from the highest state (unbound, $n_u = \infty$) to the lowest state ($n_\ell = 1$).

$$\frac{1}{\lambda} = (3)^2 R\left(\frac{1}{1^2} - \frac{1}{\infty}\right) = 9R$$

$$\lambda = \frac{1}{9R} = 10.1 \text{ nm}$$

When we let $n_u = \infty$, we have what is known as the *series limit,* which is the shortest wavelength possibly emitted for each of the named series.

As the level of precision increased in optical spectrographs, it was observed that each of the lines, originally believed to be single, actually could be resolved into two or more lines. Sommerfeld adapted the special theory of relativity (assuming some of the electron orbits were elliptical) to Bohr's hypotheses and was able to account for some of the "splitting" of spectral lines. Subsequently it has been found that other factors (especially the electron's *spin,* or *intrinsic angular momentum*) also affect the fine structure of spectral lines.

It was soon observed that external magnetic fields (the Zeeman effect) and external electric fields (the Stark effect) applied to the radiating atoms affected the spectral lines, splitting and broadening them. Although classical electromagnetic theory could quantitatively explain the (normal) Zeeman effect (see Chapter 7), it was unable to account for the Stark effect; for this the quantum model of Bohr and Sommerfeld was necessary.

Although the Bohr model was a great step forward in the application of the new quantum theory to understanding the tiny atom, it soon became apparent that the model had its limitations:

Limitations of Bohr model

1. It could be successfully applied only to single-electron atoms (H, He^+, Li^{++}, etc.).
2. It was not able to account for the intensities or the fine structure of the spectral lines.
3. Bohr's model could not explain the binding of atoms into molecules.

We discuss in Chapter 7 the full quantum mechanical theory of the hydrogen atom which accounts for all of these phenomena. The Bohr model was an ad hoc theory to explain the hydrogen spectral lines. Although it was useful in the beginnings of quantum physics, we now know that the Bohr model does not correctly describe atoms. Despite its flaws, however, Bohr's model should not be denigrated. It was the first step from a purely classical description of the atom to the correct quantum explanation. As usually happens in such tremendous changes of understanding, Bohr's model simply did not go far enough—he retained too many classical concepts. Einstein, many years later, noted* that Bohr's achievement "appeared to me like a miracle and appears as a miracle even today."

4.6 Characteristic X-Ray Spectra and Atomic Number

By 1913 when Bohr's model was published, little progress had been made in understanding the structure of many-electron atoms. It was believed that the general characteristics of the Bohr-Rutherford atom would prevail. We discussed the production of x rays from the bombardment of various materials by electrons in Section 3.7. It was known that an x-ray tube with an anode made from a given element of the periodic table produced a continuous spectrum of bremsstrahlung x rays upon which are superimposed several peaks whose frequencies are characteristic to that element (see Figure 3.17).

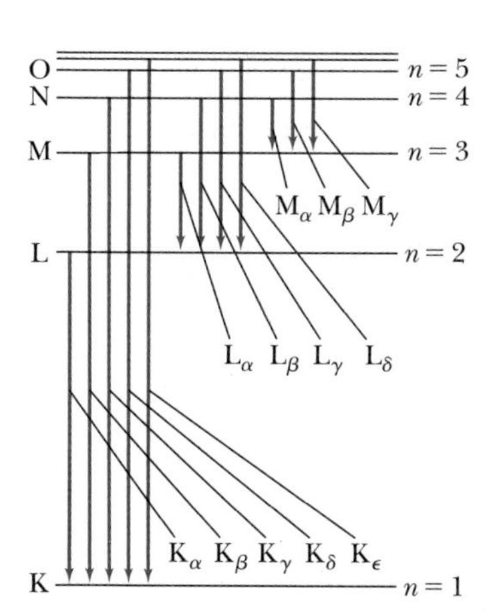

FIGURE 4.18 Historically, the stationary states were also given letter identifications: K shell ($n = 1$), L shell ($n = 2$), M shell ($n = 3$), etc. The x rays emitted when an atom changes energy states are given different names depending on the initial and final states. The Greek letter subscripts indicate the value of Δn and the roman letters the value of n for the final state.

We can now understand these **characteristic x-ray** wavelengths by adopting Bohr's *electron shell* hypothesis. Bohr's model suggests that an electron shell based on the radius r_n can be associated with each of the principal quantum numbers n. Electrons with lower values of n are more tightly bound to the nucleus than those with higher values. The radii of the electron orbits increase as n^2 [Equation (4.24)]. An energy is associated with each value of n. We may assume that when we add electrons to a fully ionized many-electron atom, the inner shells (low values of n) are filled before the outer shells, because the former have lower energies. We have not yet discussed how many electrons each shell contains or even why electrons tend to form shells, rather than a featureless glob, for example. Historically, the shells were given letter names: the $n = 1$ shell was called the K shell, $n = 2$ was the L shell, and so on. The shell structure of an atom is indicated in Figure 4.18. In heavy atoms with many electrons, we may suppose that several shells contain electrons. What happens when a high-energy electron in an x-ray tube collides with one of the K-shell electrons (we shall call these *K electrons*) in a target atom? If enough energy can be transferred to the

*P. A. Schillp, ed. *Albert Einstein, Philosopher-Scientist,* La Salle, IL: The Open Court, 1949.

K electron to dislodge it from the atom, the atom will be left with a vacancy in its K shell. The atom is most stable in its lowest energy state or *ground state,* so it is likely that an electron from one of the higher shells will change its state and fill the inner-shell vacancy at lower energy, emitting radiation as the electron changes its state. When this occurs in a heavy atom we call the electromagnetic radiation emitted an *x ray,* and it has the energy

Ground state

$$E \text{ (x ray)} = E_u - E_\ell \tag{4.39}$$

The process is precisely analogous to what happens in an excited hydrogen atom. The photon produced when the electron falls from the L shell into the K shell is called a K_α x ray; when it falls from the M shell into the K shell, the photon is called a K_β x ray. This scheme of x-ray identification is diagrammed in Figure 4.18. The relative positions of the energy levels of the various shells differ for each element, so the characteristic x-ray energies of the elements are simply the energy differences between the shells. The two strong peaks in the molybdenum spectrum of Figure 3.17 are the K_α and K_β x rays.

This simple description of the electron shells, which will be modified later by the full quantum mechanical treatment, was not understood by early 1913. The experimental field of x-ray detection was beginning to flourish (see Section 3.3), and the precise identification of the wavelengths of characteristic x rays was possible. In 1913 H. G. J. Moseley, working in Rutherford's Manchester laboratory, was engaged in cataloguing the characteristic x-ray spectra of a series of elements. He concentrated on the K- and L-shell x rays produced in an x-ray tube. In 1913 physicists in Rutherford's Manchester lab had already fully accepted the concept of the atomic number, although there was no firm experimental evidence for doing so. Most of the European physicists still believed that **atomic weight** A was the important factor, and the periodic table of elements was so structured. The **atomic number** is the number of protons in the nucleus and is denoted by Z. The makeup of the nucleus was unknown at the time, so Z was related to the positive charge of the nucleus.

Significance of atomic number

Moseley compared the frequencies of the characteristic x rays with the then-supposed atomic number of the elements and found empirically an amazing linear result when he plotted the atomic number Z versus the square root of the measured frequency as shown in Figure 4.19:

$$\nu_{K_\alpha} = \frac{3cR}{4}(Z-1)^2 \tag{4.40}$$

This result holds for the K_α x rays, and a similar result was found for the L shell. The data shown in Figure 4.19 are known as a *Moseley plot.* Moseley began his work in 1913 in Manchester and, after moving to Oxford late in 1913, completed the investigation in early 1914. Although it is clear that Bohr and Moseley discussed physics and even corresponded after Bohr left for Copenhagen, Moseley does not mention Bohr's model in his 1914 paper. Thus, it is not known whether Bohr's ideas had any influence on Moseley's work.

Using Bohr's model we can easily understand Moseley's empirical result, Equation (4.40). If a vacancy occurs in the K shell, there is still one electron remaining in the K shell. (We will learn in Chapter 8 that at most, two electrons can occupy the K shell.) An electron in the L shell will feel an effective charge of $(Z-1)e$ due to $+Ze$ from the nucleus and $-e$ from the remaining K-shell electron, because the L-shell orbit is well outside the K-shell orbit. An application of Gauss's law with a Gaussian sphere slightly larger than the K shell is useful in understanding this effect. The other electrons outside the K shell hardly affect the

L-shell electron. The x ray produced when a transition occurs from the $n = 2$ to the $n = 1$ shell has the wavelength, from Equation (4.38), of

$$\frac{1}{\lambda_{K_\alpha}} = R(Z-1)^2\left(\frac{1}{1^2} - \frac{1}{2^2}\right) = \frac{3}{4}R(Z-1)^2 \tag{4.41}$$

or

$$\nu_{K_\alpha} = \frac{c}{\lambda_{K_\alpha}} = \frac{3cR}{4}(Z-1)^2 \tag{4.42}$$

Henry G. J. Moseley (1887–1915), shown here working in 1910 in the Balliol-Trinity laboratory of Oxford University, was a brilliant young experimental physicist with varied interests. Unfortunately, he was killed in action at the young age of 27 during the English expedition to the Dardanelles. Moseley volunteered and insisted on combat duty in World War I, despite the attempts of Rutherford and others to keep him out of action. *University of Oxford, Museum of the History of Science/Courtesy AIP Niels Bohr Library.*

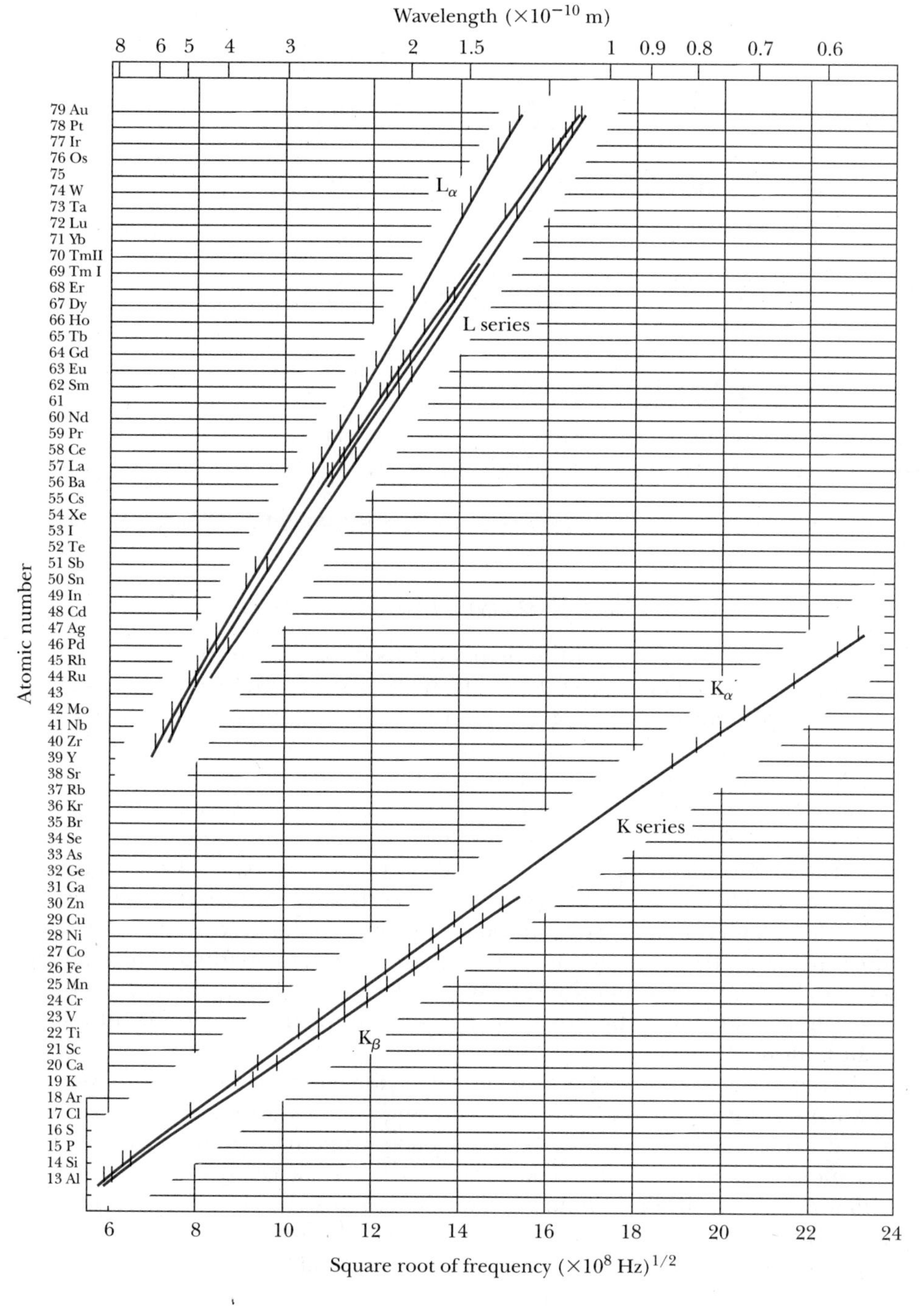

FIGURE 4.19 The original data of Moseley indicating the relationship between the atomic number Z and the characteristic x-ray frequencies. Notice the missing entries for elements $Z = 43$, 61, and 75, which had not yet been identified. There are also a few errors in the atomic number designations for the elements. © *From H. G. J. Moseley, Philosophical Magazine (6), **27**, 703 (1914).*

which is precisely the equation Moseley found describing the K_α-shell x rays. Moseley correctly concluded that the atomic number Z was the determining factor in the ordering of the periodic table, and this reordering was more consistent with chemical properties than one based on atomic weight. It put potassium ($Z = 19$, $A = 39.10$) after argon ($Z = 18$, $A = 39.95$) by atomic number rather than the reverse by atomic weight. Moseley concluded that the atomic number of an element should be identified with the number of positive units of electricity in the nucleus (that is, the number of protons). He tabulated all the atomic numbers between Al ($Z = 13$) and Au ($Z = 79$) and pointed out there were still three elements ($Z = 43$, 61, and 75) yet to be discovered! The element promethium ($Z = 61$) was not finally discovered until about 1940.

Missing atomic numbers

Example 4.7

Moseley found experimentally that the equation describing the frequency of the L_α spectral line was

$$\nu_{L_\alpha} = \frac{5}{36} cR(Z - 7.4)^2$$

How can the Bohr model explain this result?

Solution: The L_α x ray results from a transition from the M shell ($n_u = 3$) to the L shell ($n_\ell = 2$). There may be several electrons in the L shell and two electrons in the K shell that shield the nuclear charge $+Ze$ from the M-shell electron making the transition to the L shell. Let's assume the effective charge that the electron sees is $+Z_{\text{eff}}e$. Then using Equation (4.38) we have

$$\nu_{L_\alpha} = \frac{c}{\lambda_{L_\alpha}} = cR\, Z^2{}_{\text{eff}}\left(\frac{1}{2^2} - \frac{1}{3^2}\right)$$

$$\nu_{L_\alpha} = \frac{5cR\, Z^2_{\text{eff}}}{36}$$

According to Moseley's data the effective charge Z_{eff} must be $Z - 7.4$.

4.7 Atomic Excitation by Electrons

All the evidence for the quantum theory discussed so far has involved quanta of electromagnetic radiation (photons). In particular, the Bohr model explained experimental optical spectra of certain atoms. Spectroscopic experiments were typically performed by exciting the elements in some manner (for example, in a high-voltage discharge tube) and then examining the emission spectra.

The German physicists James Franck and Gustav Hertz decided to study electron bombardment of gaseous vapors in order to study the phenomenon of ionization. They explicitly set out in 1914 to study the possibility of transferring a part of an electron's kinetic energy to an atom. If such measurements were possible, they would provide a distinctive new technique for studying atomic structure. Moreover, the experiment would demonstrate that quantization would apply to an electron's mechanical energy as well as to the electromagnetic energy of photons, and would thereby provide evidence for the universality of energy quantization.

An experimental arrangement similar to that used by Franck and Hertz is shown in Figure 4.20. This particular arrangement is one actually used in a typical undergraduate physics laboratory experiment. Electrons are emitted thermionically from a hot cathode (filament), then accelerated by an electric field with its intensity determined by a variable (0 to 45 V) power supply, and pass through a grid consisting of wire mesh. After passing through the grid, the electrons feel the effects of a decelerating voltage (typically 1.5 V) between grid and

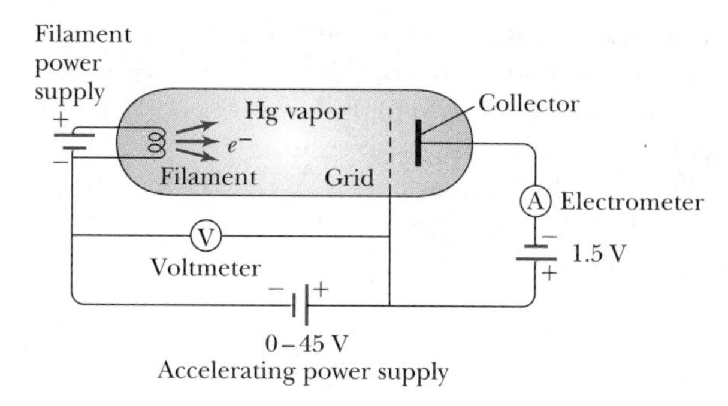

FIGURE 4.20 Schematic diagram of apparatus used in an undergraduate physics laboratory for the Franck-Hertz experiment. The hot filament produces electrons, which are accelerated through the mercury vapor toward the grid. A decelerating voltage between grid and collector prevents the electrons from registering in the electrometer unless the electron has a certain minimum energy.

James Franck (1882–1964), shown here on the left with Gustav Hertz in Tübingen, Germany in 1926, came to America in 1935 to avoid Nazi persecution and became an important American scientist who trained many experimental physicists. Gustav Hertz (1887–1975), the nephew of Heinrich Hertz who discovered electromagnetic waves, worked in German universities and industrial labs before going to the Soviet Union in 1945. They received the Nobel Prize for physics in 1925. *AIP Emilio Segrè Visual Archives.*

anode (collector). If the electrons have greater than 1.5 eV after passing through the grid they will have enough energy to reach the collector and be registered as current in an extremely sensitive ammeter (called an *electrometer*). A voltmeter measures the accelerating voltage V. The experiment consists of measuring the current I in the electrometer as a function of V.

The accelerating electrons pass through a region containing mercury (Hg) vapor (a monoatomic gas). Is there any way that energy can be transferred between an electron and a Hg atom? Franck and Hertz found that as long as the accelerating voltage V was below about 5 V (that is, the maximum kinetic energy of the electrons was below 5 eV), the electrons apparently did not lose energy. The electron current registered in the electrometer continued to increase as V increased. However, as the accelerating voltage increased above 5 V, there was a sudden drop in the current (see Figure 4.21, which was constructed using data taken by students performing this experiment). As the accelerating voltage continues to increase above 5 V, the current increases again, but suddenly drops again above 10 V. Franck and Hertz first interpreted this behavior of the current

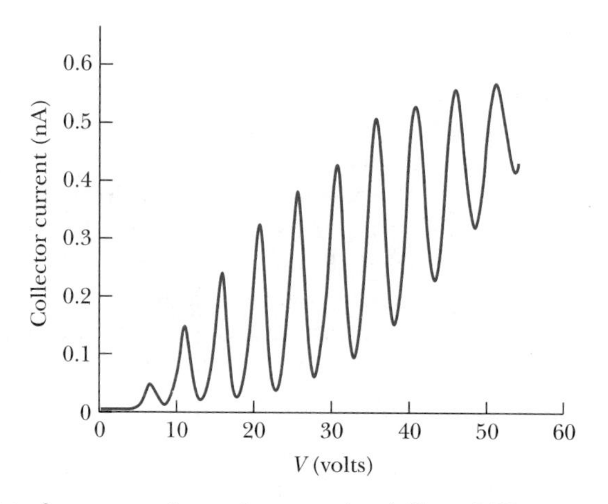

FIGURE 4.21 Data from an undergraduate student's Franck-Hertz experiment using apparatus similar to that shown in Figure 4.20. The energy difference between peaks is about 5 V, but the first peak is not at 5 V because of the work function differences of the metals used for the filament and grid.

with voltage as the onset of ionization of the Hg atom; that is, an atomic electron is given enough energy to remove it from the Hg, leaving the atom ionized. They later realized that the Hg atom was actually being excited to its first excited state.

We can explain the experimental results of Franck and Hertz quite easily within the context of Bohr's picture of *quantized atomic energy levels.* The first quantized state existing above the ground state (that is, the first excited state) in Hg is at an excitation energy of 4.88 eV. As long as the accelerating electron's kinetic energy is below 4.88 eV, no energy can be transferred to Hg because not enough energy is available to excite an electron to the next energy level in Hg. The Hg atom is so much more massive than the electron that almost no kinetic energy is transferred to the recoil of the Hg atom; the collision is *elastic.* The electron can only bounce off the Hg atom and continue along a new path with about the same kinetic energy. If the electron gains at least 4.88 eV of kinetic energy from the accelerating potential, it can transfer 4.88 eV to an electron in Hg, promoting it to the first excited state. This is an *inelastic* collision. An electron that has lost energy in an inelastic collision then has too little energy (after it passes the grid) to reach the collector. Above 4.88 V, the current dramatically drops because the inelastically scattered electrons no longer reach the collector.

When the accelerating voltage is increased to 7 or 8 V, even electrons that have already made an inelastic collision have enough remaining energy to reach the collector. Once again the current increases with V. However, when the accelerating voltage approaches 10 V, the electrons have enough energy to excite two Hg atoms in successive inelastic collisions, losing 4.88 eV in each (2×4.88 eV = 9.76 eV). The current drops sharply again. As we see in Figure 4.21, even with simple apparatus it is possible to observe several successive excitations as the accelerating voltage is increased. Notice that the energy differences between peaks are typically 4.9 eV. The first peak does not occur at 4.9 eV because of the difference in the work functions between the dissimilar metals used as cathode and anode. There are other highly excited states in Hg that could also be excited in an inelastic collision, but the probability of exciting them is much smaller than that for the first excited state, and so they are more difficult to observe.

The Franck-Hertz experiment convincingly proved the quantization of atomic electron energy levels. The bombarding electron's kinetic energy can change only by certain discrete amounts determined by the atomic energy levels of the mercury atom.

The observation of atomic excitation by collisions with electrons is a vivid example of the quantum theory. Franck and Hertz even carefully observed radiation emitted from the Hg vapor region. They observed no radiation emitted when the electron's kinetic energy was below about 5 V, but as soon as the current dropped, indicating excitation of Hg, an emission line of wavelength 254 nm (ultraviolet) was observed. Franck and Hertz set $E = 4.88 \text{ eV} = h\nu = (hc)/\lambda$ and showed that the value of h determined from $\lambda = 254$ nm was in good agreement with values of Planck's constant determined by other means.

Summary

Rutherford proposed a model of the atom consisting of a very massive, very compact (relative to the size of the atom), positively charged nucleus surrounded by electrons. His assistants, Geiger and Marsden, performed scattering experiments with energetic alpha particles and showed that the number of backward-scattered α particles could be accounted for only if the model were correct. The relation between the impact parameter b and scattering angle θ for Coulomb scattering is

$$b = \frac{Z_1 Z_2 e^2}{8\pi\epsilon_0 K} \cot\frac{\theta}{2} \tag{4.6}$$

Rutherford's equation for the number of particles scattered at angle θ is

$$N(\theta) = \frac{N_i n t}{16}\left(\frac{e^2}{4\pi\epsilon_0}\right)^2 \frac{Z_1^2 Z_2^2}{r^2 K^2 \sin^4(\theta/2)} \tag{4.13}$$

where the dependence on charges Z_1 and Z_2, the kinetic energy K, the target thickness t, and the scattering angle θ were verified experimentally. The classical planetary atomic model predicts the rapid demise of the atom because of electromagnetic radiation.

Niels Bohr was able to derive the empirical Rydberg formula for the wavelengths of the optical spectrum of hydrogen using more fundamental principles. His "general assumptions" led to the quantization of various physical parameters of the hydrogen atom, including the radius, $r_n = n^2 a_0$, where $a_0 = 0.53 \times 10^{-10}$ m, and the energy, $E_n = -E_0/n^2$, where $E_0 = 13.6$ eV. The Rydberg equation

$$\frac{1}{\lambda} = R\left(\frac{1}{n_\ell^2} - \frac{1}{n_u^2}\right)$$

gives the wavelengths, where n_ℓ and n_u are the quantum numbers for the lower and upper stationary states, respectively. The Bohr model could explain the optical spectra of hydrogenlike atoms such as He^+ and Li^{++}, but could not account for the characteristics of many-electron atoms. This indicated that the model was incomplete and only approximate. Bohr's correspondence principle relates quantum theories to classical ones in the limit of large quantum numbers.

By examining the characteristic x-ray spectra of the chemical elements, Moseley proved the fundamental significance of the atomic number. We can derive the empirical Moseley relation

$$\nu_{K_\alpha} = \frac{3cR}{4}(Z-1)^2 \tag{4.40}$$

from the structure of the atom proposed by Rutherford, together with Bohr's model of hydrogenlike energy levels.

Another way of studying atomic structure is by using electron scattering rather than photon or optical methods. Franck and Hertz were able to confirm the quantized structure of the atom and determine a value of Planck's constant h in good agreement with other methods.

Questions

1. Thomson himself was perhaps the biggest critic of the model referred to as "plum pudding." He tried for years to make it work. What experimental data could he not predict? Why couldn't he make the planetary model of Rutherford-Bohr work?
2. Does it seem fortuitous that most of the successful physicists who helped unravel the secrets of atomic structure (Thomson, Rutherford, Bohr, Geiger, and Moseley) worked either together or in close proximity in England? Why do you suppose we don't hear of names from other European countries or from the United States?
3. Could the scattering of α particles past 90° be due to scattering from electrons collected together (say 100 e^-) in one place over a volume of diameter 10^{-15} m? Explain.
4. In an intense electron bombardment of the hydrogen atom, significant electromagnetic radiation is produced in all directions upon decay. Which emission line would you expect to be most intense? Why?
5. Why are peaks due to higher-lying excited states in the Franck-Hertz experiment not more observable?
6. As the voltage increases above 5 V in the Franck-Hertz experiment, why doesn't the current suddenly jump back up to the value it had below 5 V?
7. Using Hg in the Franck-Hertz experiment, approximately what range of voltages would you expect for the first peak? Explain.
8. When are photons likely to be emitted in the Franck-Hertz experiment?
9. Is an electron most strongly bound in an H, He^+, or Li^{++} atom? Explain.

10. Why do we refer to atoms as being in the "ground" state or "at rest"? What does an "excited" state mean?
11. What lines would be missing for hydrogen in an absorption spectrum? What wavelengths are missing for hydrogen in an emission spectrum?
12. Why can the Bohr model not be applied to the He atom? What difficulties do you think Bohr had in modifying his model for He?
13. Describe how the hydrogen atom might absorb a photon of energy less than 13.6 eV. Describe a process by which a 9.8-eV photon might be absorbed. What about a 15.2-eV photon?

Problems

4.1 The Atomic Models of Thomson and Rutherford

1. In Thomson's "plum pudding" model, devise an atomic composition for carbon that consists of a pudding of charge $+6e$ along with six electrons. Try to configure a system where the charged particles move only about points in stable equilibrium.
2. How large an error in the velocity do we make by treating the velocity of a 7.7-MeV α particle nonrelativistically?
3. In Example 4.1, show that the electron's velocity must be $v_e' \approx 2v_\alpha$ in order to conserve energy and linear momentum.
4. Thomson worked out many of the calculations for multiple scattering. If we find an average scattering angle of 1° for α scattering, what would be the probability that from multiple scattering the α particle could scatter by as much as 80°? For large angles θ, the probability that the angle is larger than θ is proportional to $\exp[-(\theta/\langle\theta\rangle)^2]$. Geiger and Marsden found that about 1 in 8000 α particles were deflected past 90°. Can multiple scattering explain the experimental results of Geiger and Marsden?

4.2 Rutherford Scattering

5. Calculate the impact parameter for scattering a 7.7-MeV α particle from gold at an angle of (a) 1° and (b) 90°.
6. A beam of 8-MeV α particles scatters from a thin gold foil. What is the ratio of the number of α particles scattered to angles greater than 1° to the number scattered to angles greater than 2°?
7. For aluminum ($Z = 13$) and gold ($Z = 79$) targets, what is the ratio of α-particle scattering at any angle for equal numbers of scattering nuclei per unit area?
8. What fraction of 5-MeV α particles will be scattered through angles greater than 6° from a gold foil ($Z = 79$, density = 19.3 g/cm^3) of thickness 10^{-8} m?
9. In an experiment done by scattering 5.5 MeV α particles on a thin gold foil, students find that 10,000 α particles are scattered at an angle greater than 50°. (a) How many of these α particles will be scattered greater than 90°? (b) How many will be scattered between 70° and 80°?
10. Students want to construct a scattering experiment using a very powerful source of 5.5-MeV α particles to scatter from a gold foil. They want to be able to count 1 particle/s at 50°, but their detector is limited to a maximum count rate of 2000 particles/s. Their detector subtends a small angle. Will their experiment work without modifying the detector if the other angle they want to measure is 6°?
11. The nuclear radii of aluminum and gold are approximately $r = 3.6$ fm and 7.0 fm, respectively. The radii of protons and alpha particles are 1.3 fm and 2.6 fm, respectively. (a) What energy α particles would be needed in head-on collisions for the nuclear surfaces to just touch? (This is about where the nuclear force becomes effective.) (b) What energy protons would be needed? In both (a) and (b) perform the calculation for both aluminum and gold.
12. Consider the scattering of an alpha particle from the positively charged part of the Thomson "plum pudding" model. Let the kinetic energy of the α particle be K (nonrelativistic) and let the atomic radius be R. (a) Assuming that the maximum transverse Coulomb force acts on the α particle for a time $\Delta t = 2R/v$ (where v is the initial speed of the α particle), show that the largest scattering angle we can expect from a single atom is

$$\theta = \frac{2Z_2e^2}{4\pi\epsilon_0 KR}$$

(b) Evaluate θ for an 8-MeV α particle scattering from a gold atom of radius ~0.1 nm.
13. Using the results of the previous problem, (a) find the average scattering angle of a 10-MeV α particle from a gold atom ($R \approx 10^{-10}$ m) for the positively charged part of the Thomson model. (b) How does this compare with the scattering from the electrons?
14. The radius of a hydrogen nucleus is believed to be about $r = 1.2 \times 10^{-15}$ m. (a) If an electron moves around the nucleus at that radius, what would be its speed according to the planetary model? (b) What would the total mechanical energy be? (c) Are these reasonable?

15. Assume that the nucleus is composed of electrons and that the protons are outside. (a) If the size of an atom were about 10^{-10} m, what would be the speed of a proton? (b) What would be the total mechanical energy? (c) What is wrong with this model?

4.3 The Classical Atomic Model

16. Calculate the speed and radial acceleration for an electron in the hydrogen atom. Do the same for the Li^{++} ion.

17. What is the total mechanical energy for a ground-state electron in H, He^{+}, and Li^{++} atoms? For which atom is the electron most strongly bound? Why?

18. Calculate the time, according to classical laws, it would take the electron of the hydrogen atom to radiate its energy and crash into the nucleus. (*Hint:* The radiated power P is given by $(1/4\pi\epsilon_0)(2Q^2/3c^3)$ $(d^2\mathbf{r}/dt^2)^2$ where Q is the charge, c the speed of light, and $\mathbf{r}$ the position vector of the electron from the center of the atom).

4.4 Bohr Model of the Hydrogen Atom

19. The Ritz combination rules expressed relationships between observed frequencies of the emission optical spectra. Explain one of the more important ones:

$$\nu(K_\alpha) + \nu(L_\alpha) = \nu(K_\beta)$$

where K_α and K_β refer to the Lyman series and L_α to the Balmer series of hydrogen (Figure 4.18).

20. Calculate the angular momentum in $kg \cdot m^2/s$ for the lowest electron orbit in the hydrogen atom.

21. Use the known values of ϵ_0, h, m, and e and calculate the following to five significant figures: hc (in eV · nm), $e^2/4\pi\epsilon_0$ (in eV · nm), mc^2 (in keV), a_0 (in nm), and E_0 (in eV).

22. What is the speed (ratio of v/c) of the electron in the first three Bohr orbits of the H atom?

23. A hydrogen atom in an excited state absorbs a photon of wavelength 434 nm. What were the initial and final states of the hydrogen atom?

24. A hydrogen atom in an excited state emits a photon of wavelength 95 nm. What are the initial and final states of the hydrogen atom?

25. What is the calculated binding energy of the electron in the ground state of (a) deuterium? (b) He^{+}, and (c) Be^{+++}?

26. Describe the visible absorption spectra for (a) hydrogen atom and (b) ionized helium atom, He^{+}.

27. A hydrogen atom exists in an excited state for typically 10^{-8} s. How many revolutions would an electron make in an $n = 3$ state before decaying?

28. Electromagnetic radiation of wavelength 100 nm is incident upon the ground-state hydrogen atom at rest. What is the highest state to which hydrogen can be excited?

29. A muonic atom consists of a muon ($m = 106$ MeV/c^2) in place of an electron. For the muon in a hydrogen atom, what is (a) the smallest radius and (b) the binding energy of the muon in the ground state? (c) Calculate the series limit of the wavelength for the first three series.

30. Positronium is an atom composed of an electron and a positron ($m = m_e$, $Q = +e$). Calculate the distance between the particles and the energy of the lowest energy state of positronium. (*Hint:* what is the reduced mass of the two particles? See Problem 49.)

31. (a) Find the Bohr radius of the positronium atom described in the previous problem. (b) Find the wavelength for the transition from $n_u = 2$ to $n_\ell = 1$ for positronium.

32. What is the difference in the various Bohr radii r_n for the hydrogen atom: (a) between r_1 and r_2, (b) between r_5 and r_2, (b) between r_5 and r_6, and (c) between r_{10} and r_{11}?

4.5 Successes and Failures of the Bohr Model

33. Compare the Balmer series of hydrogen with the series where $n_\ell = 4$ for the ionized helium atom He^{+}. What is the difference between the L_α and L_β line of hydrogen and the $n_u = 6$ and 8 of He^{+}? Is there a member of the Balmer series very similar to all values where $n_\ell = 4$ in He^{+}?

34. Calculate the Rydberg constant for helium, potassium, and uranium. Compare each of them with R_∞ and determine the percentage difference.

35. In 1896 Pickering found lines from the star ζ-Puppis that had not been observed on Earth. Bohr showed in 1913 that the lines were due to He^{+}. Show that an equation giving these wavelengths is

$$\frac{1}{\lambda} = R\left(\frac{1}{n_\ell^{\,2}} - \frac{1}{n_u^{\,2}}\right)$$

What value should the Rydberg constant R have in this case?

4.6 Characteristic X-Ray Spectra and Atomic Number

36. What wavelengths for the L_α lines did Moseley predict for the missing $Z = 43$, 61, and 75 elements (see Example 4.7)?

37. If the resolution of a spectrograph is $\Delta\lambda = 10^{-12}$ m, would it be able to separate the K_α lines for platinum and gold? Explain.

38. Determine the correct equation to describe the K_β frequencies measured by Moseley. Compare that with Moseley's equation for K_α frequencies. Does the result agree with the data in Figure 4.19? Explain.

39. Calculate the K_α and K_β wavelengths for He and Li.

40. (a) Calculate the ratio of K_α wavelengths for uranium and carbon. (b) Calculate the ratio of L_α wavelengths for tungsten and calcium.

4.7 Atomic Excitation by Electrons

41. If an electron of 40 eV had a head-on collision with a Hg atom at rest, what would be the kinetic energy of the recoiling Hg atom? Assume an elastic collision.
42. In the Franck-Hertz experiment, explain why the small potential difference between the grid and collector plate is useful. Redraw the data of Figure 4.21 the way you believe the data would be without this small retarding potential.
43. Calculate the value of Planck's constant determined by Franck and Hertz when they observed the 254-nm ultraviolet radiation using Hg vapor.
44. Consider an element having excited states at 3.6 eV and 4.6 eV used as a gas in the Franck-Hertz experiment. Assume the work functions of the materials involved cancel out. List all the possible peaks that *might* be observed with electron scattering up to an accelerating voltage of 18 V.

General Problems

45. The redshift measurements of spectra from magnesium and iron are important in understanding galaxies very far away. What are the K_α and L_α wavelengths for magnesium and iron?
46. In the early 1960s the strange optical emission lines from starlike objects that also were producing tremendous radio signals confused scientists. Finally, in 1963 Maarten Schmidt of the Mount Palomar observatory discovered that the optical spectra were just those of hydrogen, but redshifted because of the tremendous velocity of the object with respect to Earth. The object was moving away from Earth at a speed of 50,000 km/s! Compare the wavelengths of the normal and redshifted spectral lines for the K_α and K_β lines of the hydrogen atom.
47. A beam of 8-MeV α particles scatters from a gold foil of thickness 0.4 μm. (a) What fraction of the α particles is scattered between 1° and 2°? (b) What is the ratio of α particles scattered through angles greater than 1° to the number scattered through angles greater than 10°; greater than 90°?
48. In Rutherford scattering we noted that angular momentum is conserved. The angular momentum of the incident α particle relative to the target nucleus is mv_0b where m is the mass, v_0 is the initial velocity of the α particle, and b is the impact parameter. Start with $\mathbf{L} = \mathbf{r} \times \mathbf{p}$ and show that angular momentum is conserved, and the magnitude is given by mv_0b along the entire path of the α particle's path while it is scattered by the Coulomb force from a gold nucleus.
49. The proton (mass M) and electron (mass m) in a hydrogen atom actually rotate about their common center of mass as shown in Figure 4.17. The distance $r = r_e + r_M$ is still defined to be the electron-nucleus distance. Show that Equation (4.24) is only modified by substituting for m by

$$\mu = \frac{m}{1 + m/M}$$

50. In Bohr's Assumption D, he assumed the mean value K of the kinetic energy of the electron-nucleus system to be $nh\nu_{\text{orb}}/2$ where ν_{orb} is the orbital frequency of the electron around the nucleus. Calculate ν_{orb} in the ground state in the following ways: (a) Use $\nu_{\text{classical}}$ in Equation (4.34). (b) Use Equation (4.33a), but first determine v and r, (c) Show that the mean value K is equal to the absolute value of the electron-nucleus system total energy and that this is 13.6 eV. (d) Use this value of K to determine ν_{orb} from the relation for K stated above.
51. Show that the quantization of angular momentum $L = n\hbar$ follows from Bohr's Assumption D that the mean value K of the kinetic energy of the electron-nucleus system is given by $K = nh\nu_{\text{orb}}/2$. Assume a circular orbit.

CHAPTER 5

Wave Properties of Matter

I thus arrived at the following overall concept which guided my studies: for both matter and radiations, light in particular, it is necessary to introduce the corpuscle concept and the wave concept at the same time.

Louis de Broglie, 1929

Chapter 3 presented compelling evidence that light (electromagnetic radiation) must be particlelike in order to explain phenomena such as the photoelectric effect and Compton scattering. The emission and absorption of photons in atoms allow us to understand the optical spectra of hydrogen atoms.

In this chapter we will learn another surprising result: wavelike properties are also exhibited by "particles" of matter. This is the only way we can interpret certain experimental observations. We begin the chapter by discussing experiments that prove that photons, in the form of x rays, behave as waves when passing through crystals. De Broglie's suggestion that particles may also behave as waves was verified by the electron-scattering experiments of Davisson and Germer.

We then present a short review of wave phenomena, including a description of the localization of a particle in terms of a collection of waves. A major hurdle is to understand how wavelike and particlelike properties can occur in nature in the same entity. Niels Bohr's principle of complementarity convinces us that *both* wavelike and particlelike properties are needed to give a complete description of matter (electrons, protons, etc.) and radiation (photons). We shall see that certain physical observables can only be determined from probabilities by using wave functions $\Psi(x, t)$. Heisenberg's uncertainty principle plays a major role in our understanding of particlelike and wavelike behavior. This principle prohibits the precise, simultaneous knowledge of both momentum and position or of both energy and time. We will see that no experiment exhibits both wave and particle properties *simultaneously*. Although modern quantum theory is applicable primarily at the atomic level, there are many macroscopic observations of its effects.

5.1 X-Ray Scattering

Following Roentgen's discovery of x rays in 1895, intense efforts were made to determine the nature and origin of the new penetrating radiation. Charles Barkla (Nobel Prize, 1917) made many x-ray measurements at Liverpool University during the early 1900s and is given credit for discovering that each element emits x rays of characteristic wavelengths and that x rays exhibit properties of polarization.

By 1912 it became clear that x rays were a form of electromagnetic radiation and must therefore have wave properties. However, because it had proved difficult to refract or diffract x rays as easily as visible light, it was suggested that their wavelengths must be much shorter than those of visible light. Max von Laue (1879–1960), a young theoretical physicist at the University of Munich, became interested in the nature of x rays primarily because of the presence at Munich of Roentgen and the theorist Arnold Sommerfeld (1868–1951), who would later play an important role in understanding atomic structure. Wilhelm Wien (1864–1928) and Sommerfeld, among others, estimated the wavelength of an x ray to be between 10^{-10} and 10^{-11} m. Knowing the distance between atoms in a crystal to be about 10^{-10} m, von Laue made the brilliant suggestion that x rays should scatter from the atoms of crystals and, if x rays were a form of electromagnetic radiation, interference effects should be observed. From the study of optics, we know that wave properties are most easily demonstrated when the sizes of apertures or obstructions are about equal to or smaller than the wavelength of the light. Von Laue suggested that crystals may act as three-dimensional gratings, scattering the waves and producing observable interference effects.

Max von Laue (Nobel Prize, 1914) designed the experiment and convinced two experimental physicists at Munich, Walter Friedrich and Paul Knipping, to perform the measurement. A schematic diagram of the transmission Laue process is shown in Figure 5.1, along with one of Friedrich and Knipping's earliest experimental results. By rotating the crystals, the positions and intensities of

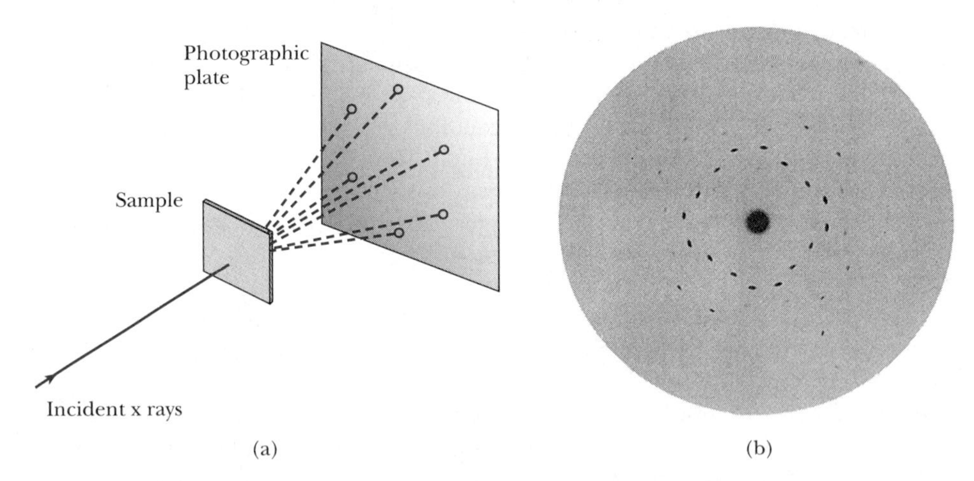

FIGURE 5.1 (a) Schematic diagram of Laue diffraction transmission method. A wide range of x-ray wavelengths scatter from a crystal sample. The x rays constructively interfere from certain planes producing dots. (b) One of the first results of Friedrich and Knipping in 1912 showing the symmetric placement of *Laue dots* of x-ray scattering from ZnS. The analysis of these results by von Laue, although complex, convincingly proved that x rays are waves. *Photo from W. Friedrich, P. Knipping, and M. Laue, Sitzungsberichte der Bayerischen Akademie der Wissenschaften (1912), 303–322 und 5 Tafeln, reprinted in Max von Laue, Gesammelte Schriften und Vorträge, Band 1, Vieweg & Sohn, Braunschweig, 1961.*

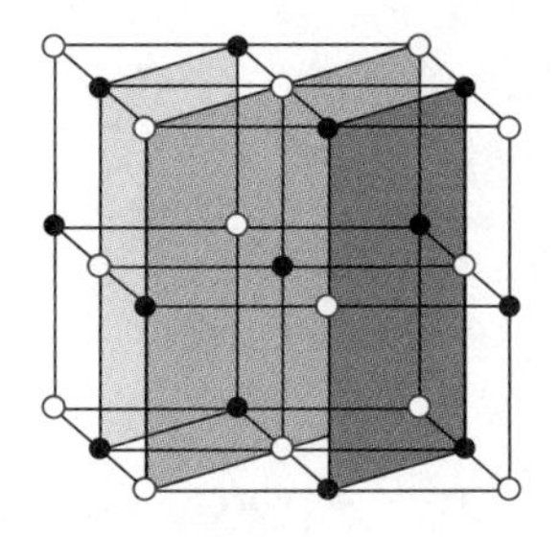
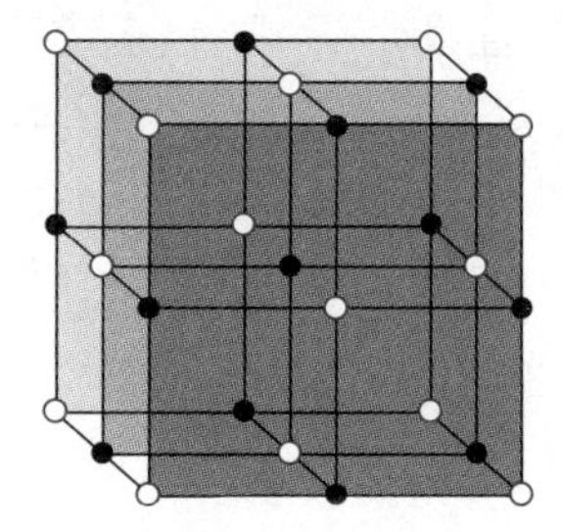

FIGURE 5.2 The crystal structure of NaCl (rock salt) showing two of the possible sets of lattice planes (Bragg planes).

the diffraction maxima were shown to change. Von Laue was able to perform the complicated analysis necessary to prove that x rays were scattered as waves from a three-dimensional crystal grating. We emphasize that von Laue was able to prove convincingly not only the wave nature of x rays but also the lattice structure of crystals.

Proof of wave nature of x rays and lattice theory of crystals

Two English physicists, William Henry Bragg and his son, William Lawrence Bragg, fully exploited the wave nature of x rays and simplified von Laue's analysis. W. L. Bragg pointed out in 1912 that each of the images surrounding the bright central spot of the Laue photographs could be interpreted as the reflection of the incident x-ray beam from a unique set of planes of atoms within the crystal. Each dot in the pattern corresponds to a different set of planes in the crystal (see Figure 5.1b).

Is x-ray scattering from atoms within crystals consistent with what we know from classical physics? From classical electromagnetic theory we know that the oscillating electric field of electromagnetic radiation polarizes an atom, causing the positively charged nucleus and negatively charged electrons to move in opposite directions. The result is an asymmetric charge distribution, or electric dipole. The electric dipole oscillates at the same frequency as the incident wave and in turn reradiates electromagnetic radiation at the same frequency, but in the form of spherical waves. These spherical waves travel throughout the matter and, in the case of crystals, may constructively or destructively interfere as the waves pass through different directions in the crystal.

If we consider x rays scattered from a simple rock salt crystal (NaCl, shown in Figure 5.2), we can, by following the Bragg simplification, determine conditions necessary for constructive interference. We will study solids in Chapter 10, but for now, note that the atoms of crystals, like NaCl, form lattice planes, called **Bragg planes.** We can see from Figure 5.3 that it is possible to have many Bragg

FIGURE 5.3 Top view of NaCl (cubic crystal), indicating possible lattice planes.

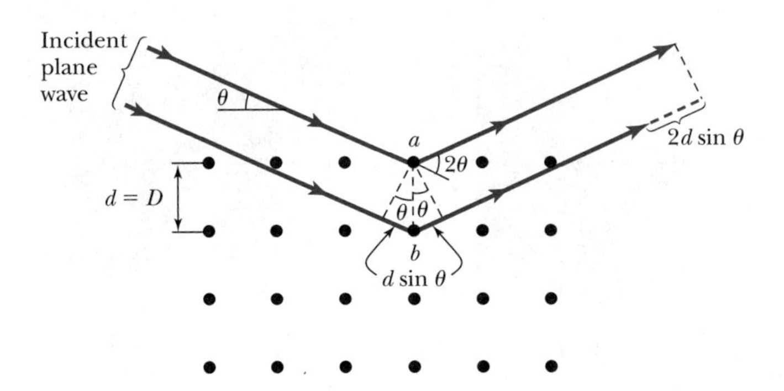

FIGURE 5.4 Schematic diagram illustrating x-ray scattering from Bragg lattice planes. The path difference of the two waves illustrated is $2d \sin \theta$. Notice that the actual scattering angle from the incident wave is 2θ.

planes in a crystal, each with different densities of atoms. Figure 5.4 shows an incident plane wave of monochromatic x rays of wavelength λ scattering from two adjacent planes. There are two conditions for constructive interference of the scattered x rays:

Conditions for constructive interference

1. The angle of incidence must equal the angle of reflection of the outgoing wave.
2. The difference in path lengths shown in Figure 5.4 ($2d \sin \theta$) must be an integral number of wavelengths.

We will not prove condition 1 but will assume it.* It is referred to as the *law of reflection* ($\theta_{\text{incidence}} = \theta_{\text{reflection}}$), although the effect is actually due to diffraction and interference. Condition 2 will be met if

Bragg's law

$$n\lambda = 2d \sin \theta \qquad n = \text{integer} \tag{5.1}$$

as can be seen from Figure 5.4, where D is the interatomic spacing (distance between atoms) and d is the distance between lattice planes. Equation (5.1) was first presented by W. L. Bragg in 1912 after he learned of von Laue's results. The integer n is called the *order of reflection,* following the terminology of ruled diffraction gratings in optics. Equation (5.1) is known as **Bragg's law** and is useful

*See L. R. B. Elton and D. F. Jackson, *Am. J. Phys.* **34,** 1036 (1966), for a correct proof.

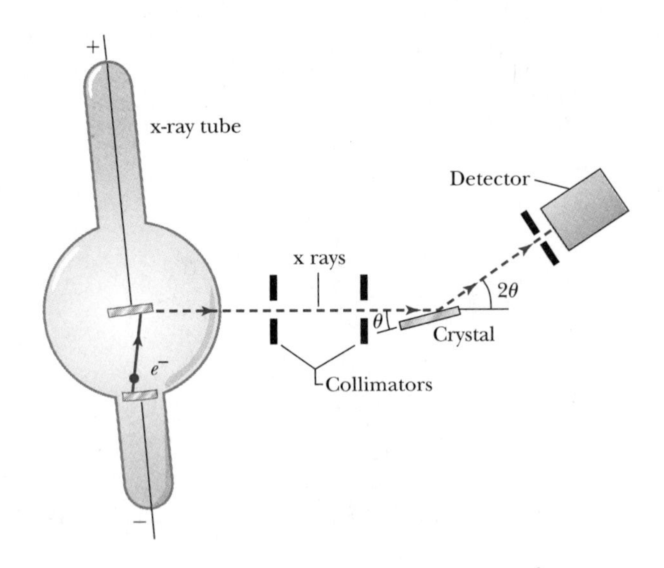

FIGURE 5.5 Schematic diagram of Bragg spectrometer. X rays are produced by electron bombardment of metal target. The x rays are collimated by lead and scatter from a crystal and are detected as a function of the angle 2θ.

for determining either the wavelength of x rays or the interplanar spacing d of the crystal if λ is already known.

W. H. Bragg and W. L. Bragg (who shared the 1915 Nobel Prize) constructed an apparatus similar to that shown in Figure 5.5, called a *Bragg spectrometer,* and scattered x rays from several crystals. By rotating the crystal and the detector, the intensity of the diffracted beam as a function of scattering angle is determined. The Braggs' studies opened up a whole new area of research that continues today.

The Bragg and Laue methods are complementary. In the Bragg method, the crystal is normally placed so that a certain set of planes produces constructive interference. The Bragg method is best for measuring x-ray wavelengths, and monochromatic x rays usually are used. The Laue method emphasizes x-ray transmission and is best for actually studying the crystal structure. Each of the symmetric Laue dots represents planes for which the two Bragg conditions are fulfilled. This technique of studying crystals is widespread, and many incident wavelengths normally are used simultaneously in order to obtain many dots. These techniques tell us almost everything we know about the structures of solids, liquids, and even complex molecules such as DNA.

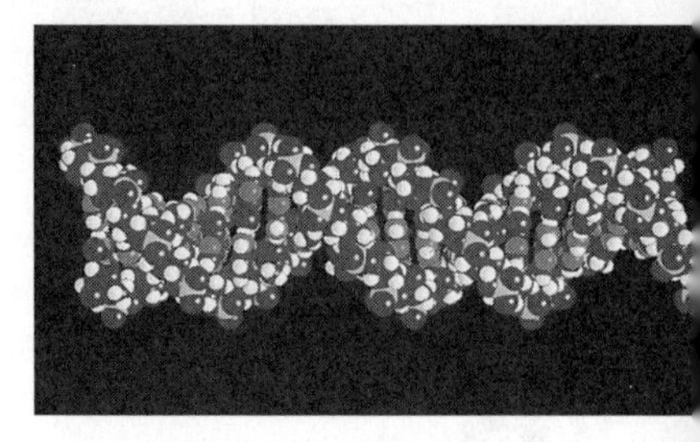

A computer graphic of the DNA double helix is shown. This complex structure was understood only after hundreds of x-ray diffraction photos were carefully studied. *Copyright Nelson Max/LLNL, Peter Arnold, Inc.*

There is still another possibility for study, although more than just a brief mention of the process is beyond the level of this text. If a single large crystal is not available, then many small crystals may be used. If these crystals are ground into a powdered form, the small crystals will then each have random orientations. Because of random orientations, when a beam of x rays passes through the powdered crystal, the dots become a series of rings. A schematic diagram of the *powder technique* is shown in Figure 5.6a along with the film arrangement to record powder photographs in Figure 5.6b. The lines indicated in part (b) are sections of rings called the *Debye-Scherrer pattern,* named after the discoverers. Figure 5.6c is a sequence of four photographs, each with an increasingly larger number of crystals, which indicates the progression from the Laue dots to the rings characteristic of the powder photographs.

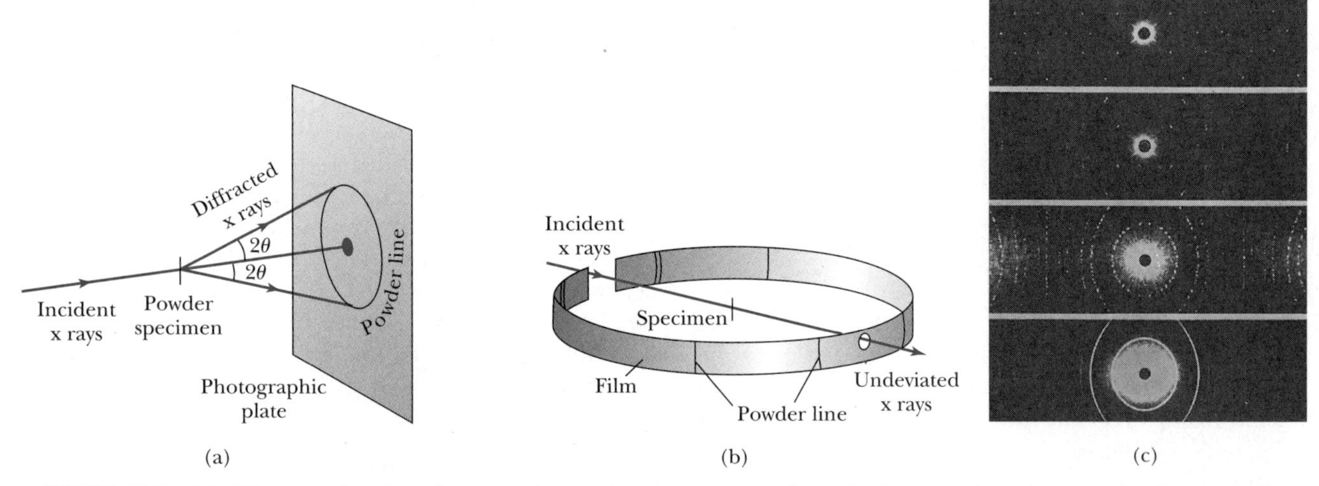

FIGURE 5.6 (a) Diagram showing the experimental arrangement of producing powder photographs from random-oriented crystals. (b) Film arrangement to record powder photographs. (c) The four photos show a progression of x-ray photographs for fluorite from a single crystal (clearly showing dots), through a few crystals, to a large number of crystals, which gives the rings the characteristic of an ideal powder photograph. *(a) and (b) are taken from N. F. M. Henry, H. Lipson, and W. A. Wooster, The Interpretation of X-ray Diffraction Photographs. London: MacMillan, 1960. (c) is from H. S. Lipson, Crystals and X-Rays. London: Wykeham Publications, 1970.*

CAVENDISH LABORATORY

Before the 1870s most of our scientific knowledge resulted from the research of persons working in their own private laboratories or in a private college room. Lord Kelvin established a laboratory at Glasgow in the 1840s, and in the 1860s efforts began at both Oxford and Cambridge to build physical laboratories. In 1871 James Clerk Maxwell was called from his Scottish home to become the first Cavendish Professor. Maxwell began planning and supervising the construction of the laboratory on Free School Lane in central Cambridge with an unexpected fervor while he gave regular lectures to students. The publication of Maxwell's treatise on electricity and magnetism in 1873 made him famous. The most important work of the day was to demonstrate the existence of Maxwell's electromagnetic waves, but they were "scooped" by Heinrich Hertz in Germany. However, the Cavendish research on waves was extensive and productive. Maxwell died in 1879, and his successor was Lord Rayleigh, a famous physicist who had not yet done his most important work. The line of Cavendish professors is impressive:

James Clerk Maxwell	1871–1879
Lord Rayleigh	1879–1884
Sir J. J. Thompson	1884–1919
Lord Rutherford	1919–1937
Sir W. Lawrence Bragg	1938–1953
Sir Nevill Mott	1954–1971
Sir Brian Pippard	1971–1982
Sir Sam Edwards	1984–1995
Richard H. Friend	1995–

Rayleigh remained only five years and then went back to his estate farm where most of his discoveries (noble gases) were made at his private laboratory. Nevertheless, in his five years at Cavendish, Rayleigh published 50 papers, setting the recognition for his Nobel Prize of 1904.

The appointment of the young J. J. Thomson at age 28 as Cavendish Professor in 1884 was the beginning of a long and fruitful era in atomic physics. The discovery of the electron in 1897, the arrival of the young Ernest Rutherford from New Zealand as a student, and the early work of C. T. R. Wilson that led to the development of the cloud chamber all helped the Cavendish Laboratory expand, prosper, and grow in stature under Thomson's leadership. Young William Lawrence Bragg started his Nobel Prize–winning research as a Cavendish student in 1912 with his father who was then a Professor of Physics at Leeds. Thomson's 35-year leadership was remarkable in many ways, particularly in the manner he stepped down in 1919 upon the opportunity of attracting Rutherford back to Cavendish to be the next Professor.

During Rutherford's 19-year reign, the Cavendish became the most renowned center of science in the world. It attracted the best students and researchers and received visitors from all over the world. Rutherford was a team leader, and he surrounded himself with a collection of young physicists whom he called "his boys." Lectures by Thomson on "Conduction of Electricity in Gases," by F. W. Aston on isotopes, by Chadwick on the discovery of the neutron, by P. M. S. Blackett on nuclear disintegration, and reports on the research of Cockcroft and Walton were on the forefront of physics. By the end of the Rutherford era in 1937, the laboratory was moving into new directions with particle accelerators and cryogenic labs. In one year, 1927–1928, the Cavendish published 53 papers.

World War II would change the face of the Cavendish forever. Physicists spread out to perform wartime research, particularly on the development of the atomic bomb and the development of radar that were to play large roles in the allied victory. William Lawrence Bragg returned as Cavendish Professor to succeed Rutherford in 1937, and the field of x-ray crystallography flourished. The years after World War II were uncertain ones, but the people at the Cavendish have had an uncanny ability to choose productive research areas. It can be said that the fields of molecular biology and radio astronomy started at the Cavendish in the late 1940s, and Bragg must be given credit for the foresight in supporting these fledgling subjects in the face of "Big Science" in the United States. Bragg's tenure as Cavendish Professor ended in 1953 just when Watson and Crick succeeded in dis-

(Special Topic text continues on p. 158)

In the upper left photo is shown the old Cavendish Laboratory on Free School Lane in Cambridge. The original building is to the left of the gate. Pictured are also the first four Cavendish Professors, James Clerk Maxwell in the upper right, Lord Rayleigh in the bottom left, and Sir J. J. Thomson (left) and Lord Rutherford in the bottom right.

covering the DNA structure. Bragg also supported J. A. Ratcliffe and Martin Ryle, who had worked on radar at the Cavendish during the war, to construct the first radio telescope. This occurred because Ryle had heard of radar operators' reports of signals coming from the stars and galaxies. This effort led to the discovery of quasars and pulsars.

When Sir Neville Mott succeeded Bragg as Cavendish Professor in 1954, the lab made a turn towards solid state physics. Mott had worked on collision theory and nuclear problems in the 1930s, but eventually turned to theoretical investigations of electronic systems. Brian Josephson did his pioneering theoretical work on the supercurrent through a tunnel barrier while a student, graduating in 1964 with his Ph.D. In 1974 the Cavendish moved to a new site in West Cambridge. Condensed matter physics now accounts for the greater part of research at the Cavendish, but the groups in radio astronomy and high energy physics are still important.

We end with a list of Nobel Prizes awarded to those who did their most important work at the Cavendish Laboratory. The asterisks (for example, Rutherford and Rayleigh) indicate examples where the Nobel Prizes were awarded primarily for work done elsewhere, but those persons are still widely associated with the Cavendish Laboratory.

Cavendish Laboratory Nobel Prizes

1904	Physics	Lord Rayleigh*	Density of gases, discovery of argon
1906	Physics	Sir J. J. Thomson	Investigations of electricity in gases
1908	Chemistry	Lord Rutherford*	Element disintegration
1915	Physics	Sir William Lawrence Bragg	X-ray analysis of crystals
1917	Physics	Charles G. Barkla	Secondary x rays
1922	Chemistry	Francis W. Aston	Isotopes discovery
1927	Physics	Charles T. R. Wilson	Cloud chamber
1928	Physics	Sir Owen W. Richardson	Thermionic emission
1935	Physics	Sir James Chadwick	Neutron discovery
1937	Physics	Sir George P. Thomson	Electron diffraction
1947	Physics	Sir Edward V. Appleton*	Upper atmosphere investigations
1948	Physics	Lord Patrick M. S. Blackett	Discoveries in nuclear physics
1951	Physics	Sir John D. Cockcroft and Ernest T. S. Walton	Nuclear transmutation
1962	Physiology or Medicine	Francis H. C. Crick and James D. Watson	DNA discoveries
1973	Physics	Brian D. Josephon	Supercurrent in tunnel barriers
1974	Physics	Sir Martin Ryle and Anthony Hewish	Radio astrophysics, pulsars
1977	Physics	Sir Nevill F. Mott	Magnetic and disordered systems
1978	Physics	P. L. Kapitsa*	Low temperature physics
1982	Chemistry	Sir Aaron Klug	Nuclei acid protein complexes

Example 5.1

X rays scattered from rock salt (NaCl) are observed to have an intense maximum at an angle of 20° from the incident direction. Assuming $n = 1$ (from the intensity), what must be the wavelength of the incident radiation?

Solution: Notice that the angle between the incident beam and scattered wave for constructive interference is always 2θ (see Figures 5.4 and 5.5). Thus $\theta = 10°$, but in order to find λ we must know d, the lattice spacing. In Section 4.1 we showed that

$$\frac{\text{Number of molecules}}{\text{Volume}} = \frac{N_A \rho}{M}$$

where N_A is Avogadro's number, ρ is the density, and M is the gram-molecular weight. For NaCl, $\rho = 2.16\ \text{g/cm}^3$ and $M = 58.5\ \text{g/mole}$.

$$\frac{N_A \rho}{M} = \frac{\left(6.02 \times 10^{23}\,\frac{\text{molecules}}{\text{mole}}\right)\left(2.16\,\frac{\text{g}}{\text{cm}^3}\right)}{58.5\,\frac{\text{g}}{\text{mole}}}$$

$$= 2.22 \times 10^{22}\,\frac{\text{molecules}}{\text{cm}^3} = 4.45 \times 10^{22}\,\frac{\text{atoms}}{\text{cm}^3}$$

$$= 4.45 \times 10^{28}\,\frac{\text{atoms}}{\text{m}^3}$$

Because NaCl has a cubic array, we take d as the distance between Na and Cl atoms, so we have a volume of d^3 per atom.

$$\frac{1}{d^3} = 4.45 \times 10^{28}\,\frac{\text{atoms}}{\text{m}^3}$$

$$d = 2.82 \times 10^{-10}\ \text{m} = 0.282\ \text{nm}$$

This technique of calculating the lattice spacing only works for a few cases because of the variety of crystal structures.

We use Equation (5.1) to find λ.

$$\lambda = \frac{2d \sin\theta}{n} = \frac{(2)(0.282\ \text{nm})(\sin 10°)}{1} = 0.098\ \text{nm}$$

which is a typical x-ray wavelength. NaCl is a useful crystal for determining x-ray wavelengths and for calibrating experimental apparatus.

5.2 De Broglie Waves

By 1920 many physicists believed that a new, more general theory was needed to replace the rudimentary Bohr model of the atom. An essential step in this development was made by a young French graduate student, Prince Louis V. de Broglie (1892–1987), who began studying the problems of the Bohr model in 1920.

De Broglie was well versed in the work of Planck, Einstein, and Bohr. He was aware of the duality of nature expressed by Einstein in which matter and energy were not independent but were in fact interchangeable. De Broglie was

After serving in World War I, Prince Louis de Broglie resumed his studies towards his doctoral degree at the University of Paris in 1924, where he reported his concept of matter waves as part of his doctoral dissertation. De Broglie spent his life in France where he enjoyed much success as an author and teacher. *AIP/Niels Bohr Library, W. F. Meggers Collection.*

particularly struck by the fact that photons (electromagnetic radiation) had both wave and corpuscular (particlelike) properties. The concept of waves is needed to understand interference and diffraction, but localized corpuscles are needed to explain phenomena like the photoelectric effect and Compton scattering. If electromagnetic radiation must have *both wave and particle properties,* then why should material particles not have both wave and particle properties as well? The symmetry of nature encourages such an idea, according to de Broglie, and no laws of physics prohibit it.

Wave properties of matter?

When de Broglie presented his new hypothesis in a doctoral thesis to the University of Paris in 1924, it aroused considerable interest. De Broglie used Einstein's special theory of relativity together with Planck's quantum theory to establish the wave properties of particles. His fundamental relationship is the prediction that

De Broglie wavelength of a particle

$$\lambda = \frac{h}{p} \tag{5.2}$$

That is, the wavelength to be associated with a particle is given by Planck's constant divided by the particle's momentum. De Broglie was guided by the concepts of phase and group velocities of waves (see Section 5.4) to arrive at Equation (5.2). Recall that for a photon

$$E = pc$$

but $E = h\nu$, so that

$$h\nu = pc = p\lambda\nu$$

$$h = p\lambda$$

and

$$\lambda = \frac{h}{p} \tag{5.3}$$

De Broglie extended this relation for photons to all particles. Particle waves were called **matter waves** by de Broglie, and the wavelength expressed in Equation (5.2) is now called the **de Broglie wavelength** of a particle.

Matter waves

Example 5.2

Calculate the de Broglie wavelength of (a) a tennis ball of mass 70 g traveling 25 m/s (about 56 mph) and (b) an electron of energy 50 eV.

Solution: (a) For the tennis ball, $m = 0.07$ kg, so

$$\lambda = \frac{h}{p} = \frac{6.63 \times 10^{-34}\ \text{J}\cdot\text{s}}{(0.07\ \text{kg})(25\ \text{m/s})} = 3.8 \times 10^{-34}\ \text{m}$$

(b) For the electron, it is more convenient to use eV units, so we rewrite the wavelength λ as

$$\lambda = \frac{h}{p} = \frac{h}{\sqrt{2mK}} = \frac{hc}{\sqrt{2(mc^2)K}}$$

$$\lambda = \frac{1240\ \text{eV}\cdot\text{nm}}{\sqrt{2(0.511 \times 10^6\ \text{eV})(50\ \text{eV})}} = 0.17\ \text{nm}$$

Note that because the kinetic energy of the electron is so small, we have used the nonrelativistic calculation. Calculations in modern physics are normally done using eV units, both because it is easier and also because eV values are more appropriate for atoms and nuclei (MeV, GeV) than are joules. The values of hc and some masses can be found inside the front cover.

How can we show whether such objects as the tennis ball or the electron in the previous example exhibit wavelike properties? The best way is to pass the objects through a slit having a width of the same dimension as the object's wavelength. We expect it to be very difficult to demonstrate interference or diffraction for the tennis ball, even if we could find a slit as narrow as 10^{-34} m (which we can't!). It is unlikely we will ever be able to demonstrate the wave properties of the tennis ball. But the de Broglie wavelength of the electron, in this case ~0.2 nm, is large enough that we should be able to demonstrate its wave properties. Because of their small mass, electrons can have a small momentum and in turn a large wavelength ($\lambda = h/p$). Electrons offer perhaps our best chance of observing effects due to matter waves.

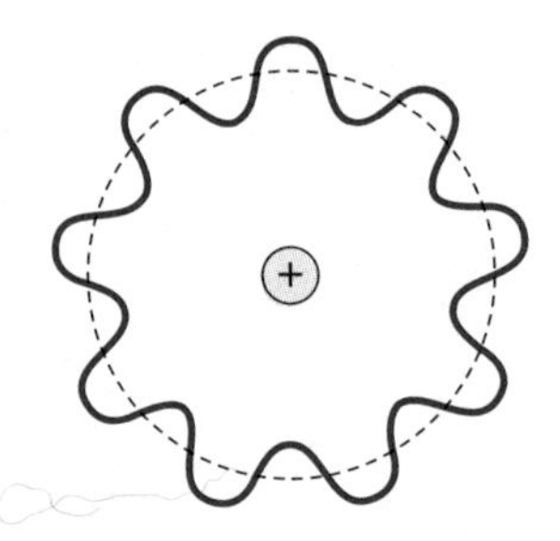

FIGURE 5.7 A schematic diagram of waves in an electron orbit around a nucleus. An integral number of wavelengths fits in the orbit. Note that the electron does not "wiggle" around the nucleus. The displacement from the dashed line represents an amplitude.

Bohr's Quantization Condition

One of Bohr's assumptions concerning his hydrogen atom model was that the angular momentum of the electron-nucleus system in a stationary state is an integral multiple of $h/2\pi$. Let's now see if we can predict this result using de Broglie's result. Represent the electron as a standing wave in an orbit around the proton. This standing wave will have nodes and be an integral number of wavelengths. We show an example of this in Figure 5.7. In order for it to be a correct standing wave, we must have

$$n\lambda = 2\pi r$$

where r is the radius of the orbit. Now we use the de Broglie relation for the wavelength and obtain

$$2\pi r = n\lambda = n\frac{h}{p}$$

The angular momentum of the electron in this orbit is $L = rp$, so we have, using the above relation,

$$L = rp = \frac{nh}{2\pi} = n\hbar$$

We have arrived at Bohr's quantization assumption by simply applying de Broglie's wavelength for an electron in a standing wave. This result hardly seems fortuitous, but though firm experimental proof was still lacking, it was soon to come.

5.3 Electron Scattering

In 1925 C. Davisson and L. H. Germer worked at Bell Telephone Laboratories investigating the properties of metallic surfaces by scattering electrons from various materials when a liquid air bottle exploded near their apparatus. Because the nickel target they were currently bombarding was at a high temperature when the accident occurred, the subsequent breakage of their vacuum system caused significant oxidation of the nickel. The target had been specially prepared and was rather expensive, so they tried to repair it by, among other procedures, prolonged heating at various high temperatures in hydrogen and under vacuum to deoxidize it.

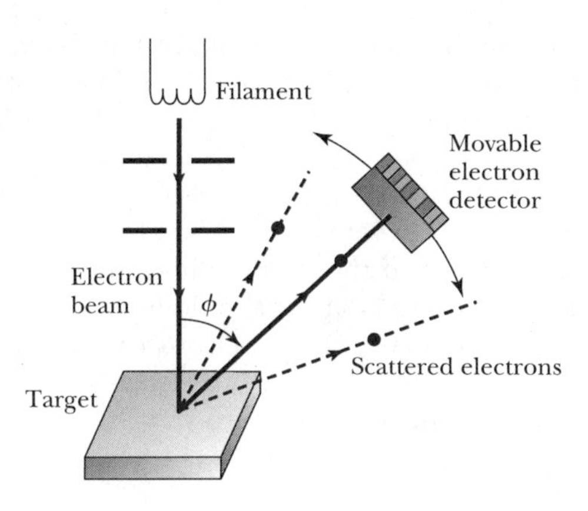

FIGURE 5.8 Schematic diagram of Davisson-Germer experiment. Electrons are produced by the hot filament, accelerated, and focused onto the target. Electrons are scattered at an angle ϕ into a detector, which is movable. The distribution of electrons is measured as a function of ϕ. The entire apparatus is located in a vacuum.

A simple diagram of the Davisson-Germer apparatus is shown in Figure 5.8. Upon putting the refurbished target back in place and continuing the experiments, Davisson and Germer found a striking change in the way electrons were scattering from the nickel surface. They had previously seen a smooth variation of intensity with scattering angle, but the new data showed large increases for certain energies at a given scattering angle. Davisson and Germer were so puzzled by their new data, that after a few days, they cut open the tube in order to examine the nickel target. They found that the polycrystalline structure of the nickel had been modified by the high temperature. The many small crystals of the original target had been changed into a few large crystals as a result of the heat treatment. Davisson surmised it was this new crystal structure of nickel—the arrangement of atoms in the crystals, not the structure of the atoms—that had caused the new intensity distributions. Some 1928 experimental results of Davisson and Germer for 54-eV electrons scattered from nickel are shown in Figure 5.9. The scattered peak occurs for $\phi = 50°$.

A lucky accident?

Diffraction of electrons

The electrons were apparently being diffracted much like x rays, and Davisson, being aware of de Broglie's results, found that the Bragg law applied to their data as well. Davisson and Germer were able to vary the scattering angles for a given wavelength and vary the wavelength (by changing the electron accelerating voltage and thus the momentum) for a given angle.

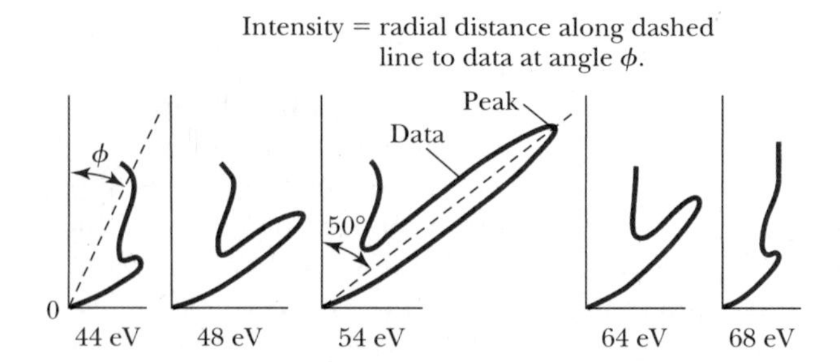

FIGURE 5.9 Davisson and Germer data for scattering of electrons from Ni. The peak $\phi = 50°$ builds dramatically as the energy of the electron nears 54 eV. *From C. J. Davisson, Journal of The Franklin Institute* ***205****, 597–623 (1928).*

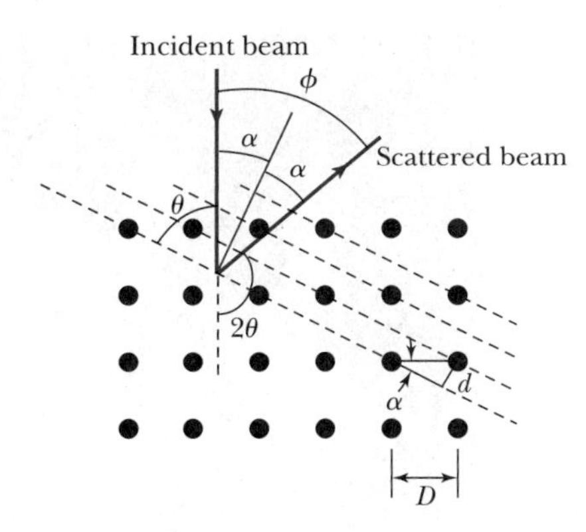

FIGURE 5.10 The scattering of electrons by lattice planes in a crystal. This figure is useful to compare the scattering relations $n\lambda = 2d \sin \theta$ and $n\lambda = D \sin \phi$ where θ and ϕ are angles shown, D = interatomic spacing, and d = lattice plane spacing.

The relationship between the incident electron beam and the nickel crystal is shown in Figure 5.10. In the Bragg Law, 2θ is the angle between the incident and exit beams. Therefore, $\phi = \pi - 2\theta = 2\alpha$. Because $\sin \theta = \cos(\phi/2) = \cos \alpha$, we have for the Bragg condition, $n\lambda = 2d \cos \alpha$. However, d is the lattice plane spacing and is related to the interatomic distance D by $d = D \sin \alpha$ so that

$$n\lambda = 2d \sin \theta = 2d \cos \alpha = 2D \sin \alpha \cos \alpha$$

$$n\lambda = D \sin 2\alpha = D \sin \phi \tag{5.4}$$

or

$$\lambda = \frac{D \sin \phi}{n} \tag{5.5}$$

For nickel the interatomic distance is $D = 0.215$ nm. If the peak found by Davisson and Germer at 50° was $n = 1$, then the electron wavelength should be

$$\lambda = (0.215 \text{ nm})(\sin 50°) = 0.165 \text{ nm}$$

Let us now compare this wavelength with that expected for a 54-eV electron. We can determine the electron's momentum nonrelativistically from the kinetic energy.

$$\frac{p^2}{2m} = \text{K.E.} = eV_0 \tag{5.6}$$

where V_0 is the voltage through which the electrons are accelerated. We find the momentum to be $p = \sqrt{(2m)(eV_0)}$. The de Broglie wavelength is now

$$\lambda = \frac{h}{p} = \frac{hc}{pc} = \frac{hc}{\sqrt{(2mc^2)(eV_0)}} = \frac{1240 \text{ eV}\cdot\text{nm}}{\sqrt{(2)(0.511 \times 10^6 \text{ eV})(eV_0)}} \tag{5.7}$$

$$= \frac{1.226 \text{ nm}\cdot\text{V}^{1/2}}{\sqrt{V_0}}$$

where the constants h, c, and m have been evaluated and V_0 is the voltage. For $V_0 = 54$ V, the wavelength is

$$\lambda = \frac{1.226 \text{ nm}\cdot\text{V}^{1/2}}{\sqrt{54 \text{ V}}} = 0.167 \text{ nm}$$

Clinton J. Davisson (1881–1958) is shown here in 1928 on the right looking at the electronic diffraction tube held by Lester H. Germer (1896–1971). They performed their work at Bell Telephone Laboratory. Davisson received the Nobel Prize in physics in 1937. *AIP Emilio Segrè Visual Archives.*

(a)

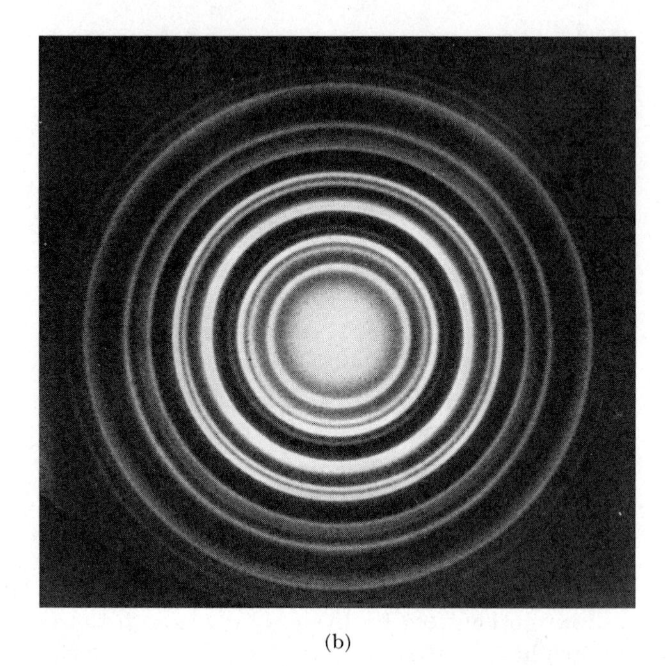

(b)

FIGURE 5.11 Examples of transmission electron diffraction photographs. (a) Produced by scattering 120-keV electrons on the quasicrystal $Al_{80}Mn_{20}$. (b) An early pattern from polycrystalline SnO_2. Notice that the dots in (a) indicate that the sample was a crystal whereas the rings in (b) indicate that a randomly oriented sample (or powder) was used. *(a) is courtesy of David Follstaedt, Sandia National Laboratory. (b) is from G. P. Thomson and W. Cochrane, Theory and Practice of Electron Diffraction. London: Macmillan, 1939.*

This value of the wavelength is in good agreement with that found earlier for the peak at 50°.

Shortly after Davisson and Germer reported their experiment, George P. Thomson (1892–1975), son of J. J. Thomson, reported seeing the effects of electron diffraction in transmission experiments. The first target was celluloid, and soon after that gold, aluminum, and platinum were used. Beautiful rings are obtained, as is shown for SnO_2 in Figure 5.11. Davisson and Thomson received the Nobel Prize in 1937 for their investigations, which clearly showed that particles exhibited wave properties. In the next few years hydrogen and helium atoms were also shown to exhibit wave diffraction. An important modern measurement technique uses diffraction of neutrons to study the crystal and molecular structure of biologically important substances. All these experiments are consistent with the de Broglie hypothesis for the wavelength of a massive particle.

Example 5.3

In introductory physics, we learned that a particle (ideal gas) in thermal equilibrium with its surroundings has a kinetic energy of $3kT/2$. Calculate the de Broglie wavelength for (a) a neutron at room temperature (300 K) and (b) a "cold" neutron at 77 K (liquid nitrogen).

Solution: We begin by finding the de Broglie wavelength of the neutron from the momentum.

$$\frac{p^2}{2m} = \text{K.E.} = \frac{3}{2}kT \tag{5.8}$$

$$p = \sqrt{3mkT}$$

$$\lambda = \frac{h}{p} = \frac{h}{\sqrt{3mkT}} = \frac{hc}{\sqrt{3(mc^2)kT}}$$

$$= \frac{1}{T^{1/2}} \frac{1240 \text{ eV} \cdot \text{nm}}{\sqrt{3(938 \times 10^6 \text{ eV})(8.62 \times 10^{-5} \text{ eV/K})}}$$

It again has been convenient to use eV units.

$$\lambda = \frac{2.52}{T^{1/2}} \text{ nm} \cdot \text{K}^{1/2} \tag{5.9}$$

$$\lambda(300 \text{ K}) = \frac{2.52 \text{ nm} \cdot \text{K}^{1/2}}{\sqrt{300 \text{ K}}} = 0.145 \text{ nm} \tag{5.9a}$$

$$\lambda(77 \text{ K}) = \frac{2.52 \text{ nm} \cdot \text{K}^{1/2}}{\sqrt{77 \text{ K}}} = 0.287 \text{ nm} \tag{5.9b}$$

These wavelengths are thus suitable for diffraction by crystals. "Supercold" neutrons, used to produce even larger wavelengths, are useful because extraneous electric and magnetic fields do not affect neutrons as much as electrons.

5.4 Wave Motion

Because particles exhibit wave behavior, it must be possible to formulate a wave description of particle motion. Our development of quantum theory will be based heavily on waves, so now we digress briefly to review the elementary physics of wave motion, which we shall ultimately apply to particles.

In elementary physics, we study waves of several kinds, including sound waves and electromagnetic waves (including light). The simplest form of wave has a sinusoidal form; at a fixed time (say, $t = 0$) its spatial variation looks like

$$\Psi(x, t)\Big|_{t=0} = A \sin\left(\frac{2\pi}{\lambda} x\right) \tag{5.10}$$

as is shown in Figure 5.12. The function $\Psi(x, t)$ represents the *instantaneous amplitude* or *displacement* of the wave as a function of position x and time t. In the case of a wave moving down a string, Ψ is the displacement of the string; in the case of a sound wave, Ψ is the displacement of the air molecules; and in the case of electromagnetic radiation, Ψ is the electric field **E** or magnetic field **B.** The maximum displacement A is normally called the **amplitude,** but a better term for a harmonic wave such as we are considering may be **harmonic amplitude.**

As time increases the position of the wave will change, so the general expression for the wave is

Wave form

$$\Psi(x, t) = A \sin\left[\frac{2\pi}{\lambda}(x - vt)\right] \tag{5.11}$$

Wavelength

Period

The position at time $t = t_0$ is also shown in Figure 5.12. The **wavelength** λ is defined to be the distance between points in the wave with the same phase, for example, positive wave crests. The **period** T is the time required for a wave to travel a distance of one wavelength λ. Because the velocity [actually phase velocity, see Equation (5.17)] of the wave is v, we have $\lambda = vT$. The frequency $\nu(= 1/T)$ of a

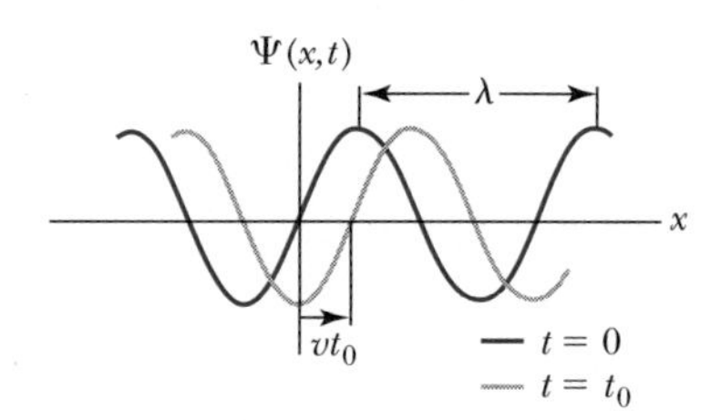

FIGURE 5.12 Wave form of a wave moving to the right at speed v shown at $t = 0$ and $t = t_0$.

harmonic wave is the number of times a crest passes a given point (a complete *cycle*) per second. A traveling wave of the type described by Equation (5.11) satisfies the wave equation,*

$$\frac{\partial^2 \Psi}{\partial x^2} = \frac{1}{v^2} \frac{\partial^2 \Psi}{\partial t^2} \tag{5.12}$$

If we use $\lambda = vT$, we can rewrite Equation (5.11).

$$\Psi(x, t) = A \sin\left[2\pi\left(\frac{x}{\lambda} - \frac{t}{T}\right)\right] \tag{5.13}$$

Wave number and angular frequency

We can write Equation (5.13) more compactly by defining the **wave number** k and **angular frequency** ω by

$$k \equiv \frac{2\pi}{\lambda} \quad \text{and} \quad \omega \equiv \frac{2\pi}{T} \tag{5.14}$$

Equation (5.13) then becomes

$$\Psi(x, t) = A \sin(kx - \omega t) \tag{5.15}$$

This is the mathematical description of a sine curve traveling in the positive x direction that has a displacement $\Psi = 0$ at $x = 0$ and $t = 0$. A similar wave traveling in the negative x direction would have the form

$$\Psi(x, t) = A \sin(kx + \omega t) \tag{5.16}$$

The **phase velocity** v_{ph} is the velocity of a point on the wave that has a given phase (for example, the crest) and is given by

Phase velocity

$$v_{ph} = \frac{\lambda}{T} = \frac{\omega}{k} \tag{5.17}$$

If the wave does not have $\Psi = 0$ at $x = 0$ and $t = 0$, we can describe the wave using a **phase constant** ϕ

$$\Psi(x, t) = A \sin(kx - \omega t + \phi) \tag{5.18}$$

For example, if $\phi = 90°$, Equation (5.18) can be written

$$\Psi(x, t) = A \cos(kx - \omega t) \tag{5.19}$$

Principle of superposition

Observation of many different kinds of waves has established the general result that when two or more waves traverse the same region, they act independently of each other. According to the **principle of superposition,** we add the displacements of all waves present. A familiar example is the superposition of two sound waves of nearly equal frequencies: The phenomenon of beats is observed. Examples of superposition are shown in Figure 5.13. The net displacement depends on the harmonic amplitude, the phase, and the frequency of each of the individual waves.

Wave packet

How might we use waves to represent a moving particle? In Figure 5.13 we see that when two waves are added together, we obtain regions of relatively large (and small) displacement. If we add many waves of different amplitudes and frequencies in particular ways, it is possible to obtain what is called a **wave packet.** The important property of the wave packet is that its net amplitude differs from

*The derivation of the wave equation is presented in most introductory physics textbooks for a wave on a string, although it is often an optional section and skipped. It would be worthwhile for the student to review its derivation now, especially the use of the partial derivatives.

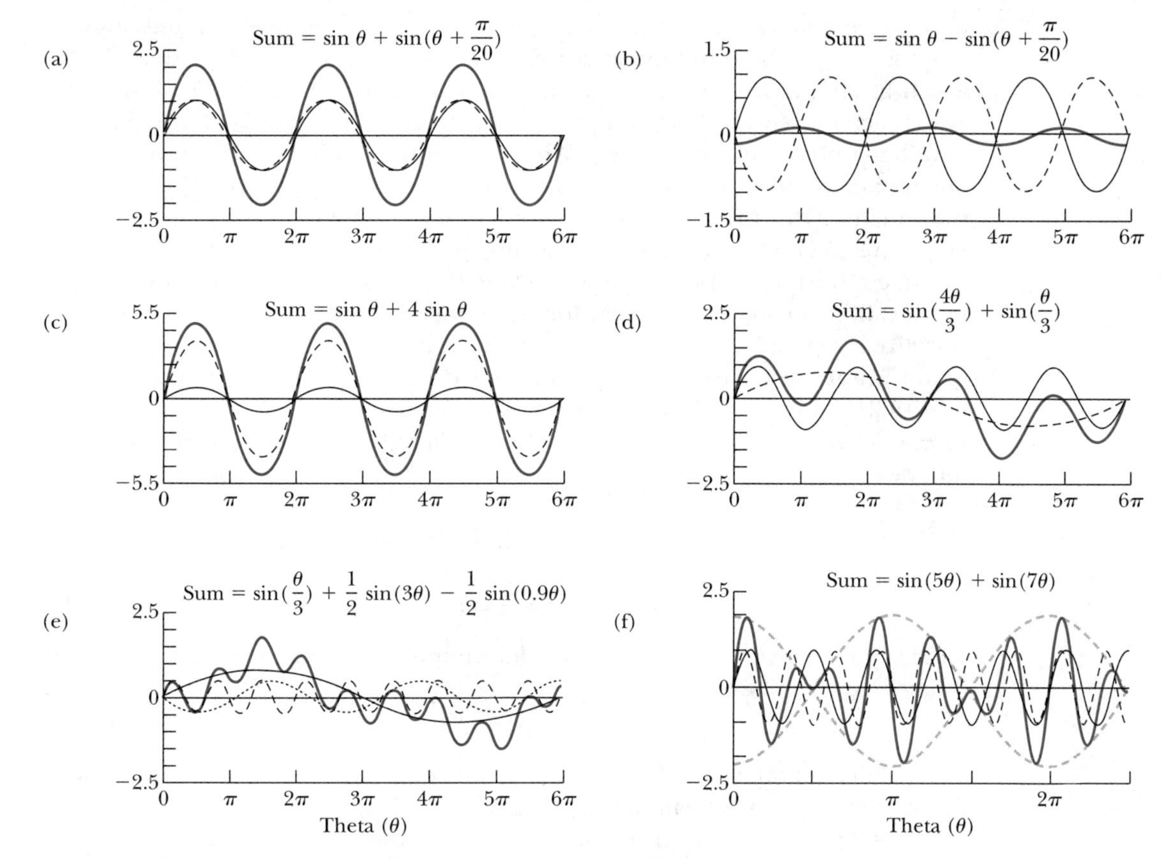

FIGURE 5.13 Superposition of waves. The heavy blue line is the resulting wave. (a) Two waves of equal frequency and amplitude that are almost in phase. The result is a larger wave. (b) As in (a) but the two waves are almost out of phase. The result is a smaller wave. (c) Superposition of two waves with the same frequency, but different amplitudes. (d) Superposition of two waves of equal amplitude but different frequencies. (e) Superposition of three waves of different amplitudes and frequencies. (f) Superposition of two waves of almost the same frequency over many wavelengths, indicating the phenomenon of beats.

zero only over a small region Δx as shown in Figure 5.14. We can localize the position of a particle in a particular region by using a wave packet description.

Let us examine in detail the superposition of two waves. Assume both waves have the same harmonic amplitude A but different wave numbers (k_1 and k_2) and angular frequencies (ω_1 and ω_2). The superposition of the two waves is the sum

$$\Psi(x, t) = \Psi_1(x, t) + \Psi_2(x, t) = A\cos(k_1 x - \omega_1 t) + A\cos(k_2 x - \omega_2 t) \tag{5.20}$$

$$= 2A\cos\left[\frac{1}{2}(k_1 - k_2)x - \frac{1}{2}(\omega_1 - \omega_2)t\right]\cos\left[\frac{1}{2}(k_1 + k_2)x - \frac{1}{2}(\omega_1 + \omega_2)t\right]$$

$$= 2A\cos\left(\frac{\Delta k}{2}x - \frac{\Delta\omega}{2}t\right)\cos(k_{\text{av}}x - \omega_{\text{av}}t) \tag{5.21}$$

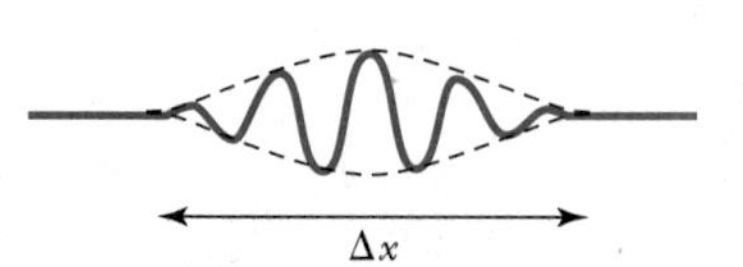

FIGURE 5.14 An idealized wave packet localized in space over a region Δx is the superposition of many waves of different amplitudes and frequencies.

where $\Delta k = k_1 - k_2$, $\Delta\omega = \omega_1 - \omega_2$, $k_{\text{av}} = (k_1 + k_2)/2$, and $\omega_{\text{av}} = (\omega_1 + \omega_2)/2$. We exhibited similar waves in Figure 5.13a–d, where the heavy solid line indicates the resulting wave. In Figure 5.13f the blue dashed line indicates an envelope which denotes the maximum displacement of the combined waves. The wave still oscillates within this envelope with the wave number k_{av} and angular frequency ω_{av}. The envelope is described by the first term of Equation (5.21), which has the wave number $\Delta k/2$ and angular frequency $\Delta\omega/2$. The individual waves each move with their own phase velocity: ω_1/k_1 and ω_2/k_2. The combined wave has a phase velocity ω_{av}/k_{av}. When combining many more than two waves, one obtains a pulse, or wave packet, which moves at the **group velocity,** as will be shown later. Only the group velocity, which describes the speed of the envelope ($u_{gr} = \Delta\omega/\Delta k$), is important when dealing with wave packets.

Phase and group velocities

In contrast to the pulse or wave packet, the combination of only two waves is not localized in space. However, for purposes of illustration, we can identify a "localized region" $\Delta x = x_2 - x_1$ where x_1 and x_2 represent two consecutive points where the envelope is zero (see Figure 5.13f). The term $\Delta k \cdot x/2$ in Equation (5.21) must be different by a phase of π for the values x_1 and x_2, because $x_2 - x_1$ represents only one half of the wavelength of the envelope confining the wave.

$$\frac{1}{2}\Delta k\, x_2 - \frac{1}{2}\Delta k\, x_1 = \pi$$

$$\Delta k(x_2 - x_1) = \Delta k\, \Delta x = 2\pi \tag{5.22}$$

Similarly, for a given value of x we can determine the time Δt over which the wave is localized and obtain

$$\Delta\omega\, \Delta t = 2\pi \tag{5.23}$$

The results of Equations (5.22) and (5.23) can be generalized to the case where there are many waves forming the wave packet. The equations, $\Delta k\, \Delta x = 2\pi$ and $\Delta\omega\, \Delta t = 2\pi$, are significant because they tell us that in order to know precisely the position of the wave packet envelope (Δx small), we must have a large range of wave numbers (Δk large). Similarly, to know precisely when the wave is at a given point (Δt small), we must have a large range of frequencies ($\Delta\omega$ large). Equation (5.23) is the origin of the bandwidth relation important in electronics. A particular circuit component must have a large bandwidth $\Delta\omega$ in order for its signal to respond in a short time (Δt).

If we are to treat particles as matter waves, we have to be able to describe the particle in terms of waves. An important aspect of a particle is its localizing in space. That is why it is so important to form the wave packet that we have been discussing. We extend Equation (5.20) by summing over many waves with possibly different wave numbers, angular frequencies, and amplitudes.

$$\Psi(x, t) = \sum_i A_i \cos(k_i x - \omega_i t) \tag{5.24}$$

Fourier series and integral

Such a result is called a **Fourier series.** When dealing with a continuous spectrum, it may be desirable to extend Equation (5.24) to the integral form called a **Fourier integral.**

$$\Psi(x, t) = \int \tilde{A}(k) \cos(kx - \omega t)\, dk \tag{5.25}$$

The amplitudes A_i and $\tilde{A}(k)$ may be functions of k. The use of Fourier series and Fourier integrals is at a more advanced level of mathematics than we want to pursue now.* We can, however, indicate their value by one important example.

*See John D. McGervey, *Introduction to Modern Physics,* Chap. 4. Orlando, FL: Academic Press, 1983.

Gaussian Wave Packet. Gaussian wave packets are often used to represent the position of particles, as illustrated in Figure 5.15, because the associated integrals are relatively easy to evaluate. At a given time t, say $t = 0$, a Gaussian wave can be expressed as

$$\Psi(x, 0) = \psi(x) = Ae^{-\Delta k^2 x^2} \cos(k_0 x) \qquad (5.26)$$

Gaussian function

where Δk expresses the range of wave numbers used to form the wave packet. The $\cos(k_0 x)$ term describes the wave oscillating inside the envelope described by the (Gaussian) exponential term $e^{-\Delta k^2 x^2}$. The intensity distribution $\mathcal{I}(k)$ for the wave numbers leading to Equation (5.26) is shown in Figure 5.15a. There is a high probability of a particular measurement of k being within one standard deviation of the mean value k_0. The function $\psi(x)$ is shown in Fig. 5.15b. For simplicity, let the constant A be one. There is a good probability of finding the particle within the values of $x = 0$ ($\psi(x) = 1$) and $x = \Delta x/2$ ($\psi(x) = \exp(-\Delta k^2 \Delta x^2/4)$). Roughly, the value of $\psi(x)$ at the position $x = \Delta x/2$ is about 0.6 (see Figure 5.15b), so we have

$$e^{-\Delta k^2 \Delta x^2/4} \approx 0.6$$

We take the logarithm of both sides and find

$$-\frac{\Delta k^2 \Delta x^2}{4} \approx -0.5 \qquad \text{or} \qquad \Delta k\, \Delta x \approx 1.4 \qquad (5.27)$$

This has been a very rough calculation, and the answer depends on the assumptions we have made. A more detailed calculation gives $\Delta k\, \Delta x = 1/2$. The important point is that with the Gaussian wave packet, we have arrived at a result similar to Equation (5.22), namely, that the product $\Delta k\ \Delta x$ is about unity. The localization of the wave packet over a small region Δx to describe a particle requires a large range of wave numbers; that is, Δk is large. Conversely, a small range of wave numbers cannot produce a wave packet localized within a small distance.

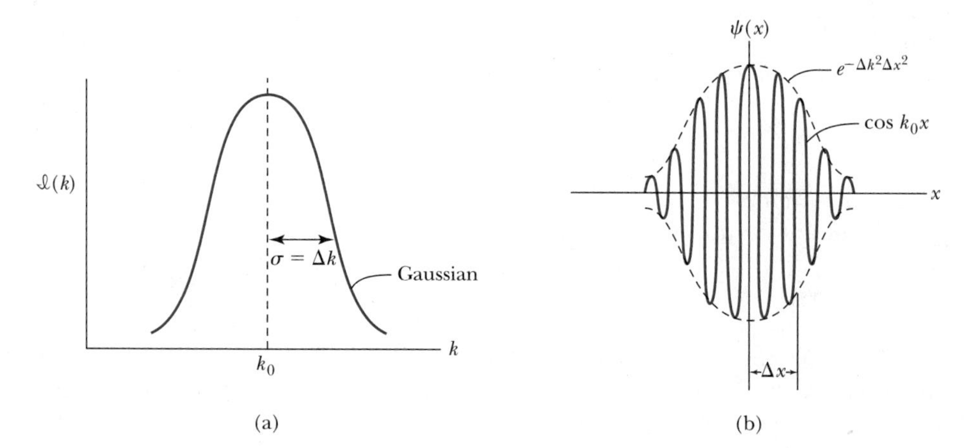

FIGURE 5.15 The form of the probability distribution or intensity $\mathcal{I}(k)$ shown in (a) is taken to have a Gaussian shape with a standard deviation of Δk (determined when the function $\exp[-(k - k_0)^2/2\sigma^2]$ has $k = k_0 \pm \Delta k$ and $\Delta k = \sigma$, the standard deviation). This $\mathcal{I}(k)$ leads to $\psi(x)$, as is shown in (b). The envelope for $\psi(x)$ is described by the $\exp(-\Delta k^2 x^2)$ term with the oscillating term $\cos(k_0 x)$ contained by the envelope. At the given time $t = 0$, the wave packet (particle) is localized to the area $x \approx 0 \pm \Delta x$ with wave numbers $k \approx k_0 \pm \Delta k$.

To complete our study of waves and the representation of particles by wave packets, we must be convinced that the superposition of waves is actually able to describe particles. We found earlier for the superposition of two waves that the group velocity, $u_{gr} = \Delta\omega/\Delta k$, represented the motion of the envelope. We can generalize this for the case of the wave packet and will find that the wave packet moves with the group velocity u_{gr} given by

Group velocity

$$u_{gr} = \frac{d\omega}{dk} \tag{5.28}$$

Because the wave packet consists of many wave numbers, we should remember to evaluate this derivative at the center of the wave packet (that is, $k = k_0$).

For a de Broglie wave, we have $E = h\nu$ and $p = h/\lambda$. We can rewrite both of these equations in terms of $\hbar$.

$$E = h\nu = \hbar(2\pi\nu) = \hbar\omega$$

$$p = \frac{h}{\lambda} = \hbar\frac{2\pi}{\lambda} = \hbar k$$

where we have used the relations $\omega = 2\pi\nu$ and $k = 2\pi/\lambda$. If we multiply the denominator and numerator in Equation (5.28) by $\hbar$, we have

$$u_{gr} = \frac{d\omega}{dk} = \frac{d(\hbar\omega)}{d(\hbar k)} = \frac{dE}{dp}$$

We use the relativistic relation $E^2 = p^2c^2 + m^2c^4$ and the derivative of E to find

$$2E\,dE = 2pc^2\,dp$$

or

$$u_{gr} = \frac{dE}{dp} = \frac{pc^2}{E} \tag{5.29}$$

This is the velocity of a particle of momentum p and total energy E. Thus, it is plausible to assume that the group velocity of the wave packet can be associated with the velocity of a particle.

The phase velocity of a wave is represented by

$$v_{ph} = \lambda\nu = \frac{\omega}{k} \tag{5.30}$$

so that $\omega = kv_{ph}$.

Then, the group velocity is related to the phase velocity by

$$u_{gr} = \frac{d\omega}{dk} = \frac{d}{dk}(v_{ph}k) = v_{ph} + k\frac{dv_{ph}}{dk} \tag{5.31}$$

Thus, the group velocity may be greater or less than the phase velocity. A medium is called *nondispersive* when the phase velocity is the same for all frequencies and $u_{gr} = v_{ph}$. An example is electromagnetic waves in vacuum. Water waves are a good example of waves in a dispersive medium. When one throws a rock in a still pond, the envelope of the waves moves more slowly than the individual waves moving outward (see Figure 5.16).

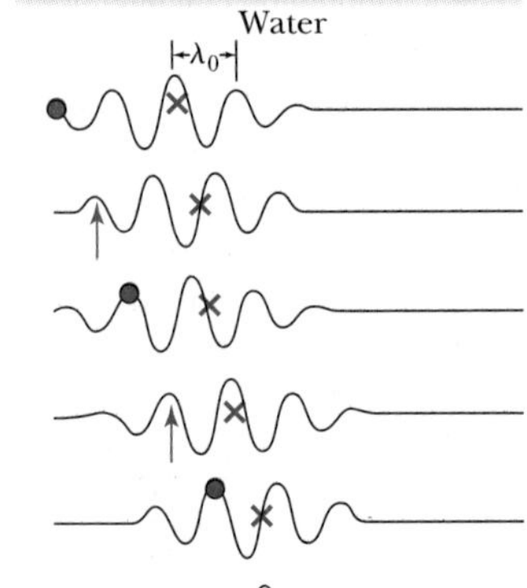

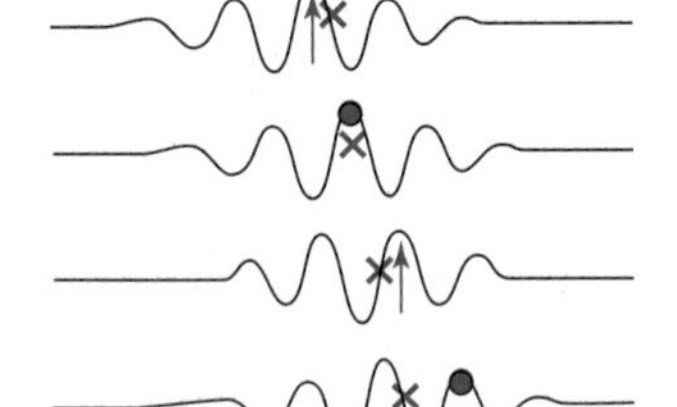

FIGURE 5.16 Progression with time of wave packet for which $u_{gr} = v_{ph}/2$. Note how the individual wave (arrow and dot alternate) moves through the wave packet (letter ×) with time.

Dispersion plays an important role in the shape of wave packets. For example, in the case of the Gaussian wave packet shown in Figure 5.15 at $t = 0$, the wave packet will spread out as time progresses. A packet that is highly localized at one time will have its waves added together in a considerably different manner at another time due to the superposition of the waves.

Example 5.4

Newton showed that deep-water waves have a phase velocity of $\sqrt{g\lambda/2\pi}$. Find the group velocity of such waves and discuss the motion.

Solution: We use Equation (5.31) to relate the group and phase velocities, but first we need to write the phase velocity v_{ph} in terms of k. If we use $\lambda = 2\pi/k$, we have

$$v_{\text{ph}} = \sqrt{\frac{g\lambda}{2\pi}} = \sqrt{\frac{g}{k}} = \sqrt{g}\,k^{-1/2}$$

Now we can take the necessary derivative for Equation (5.31).

$$u_{\text{gr}} = \sqrt{\frac{g}{k}} + k\frac{d}{dk}\left[\sqrt{g}\,k^{-1/2}\right] = \sqrt{\frac{g}{k}} + k\sqrt{g}\left[-\frac{1}{2}k^{-3/2}\right]$$

$$= \sqrt{\frac{g}{k}} - \frac{1}{2}\sqrt{\frac{g}{k}} = \frac{1}{2}\sqrt{\frac{g}{k}} = \frac{1}{2}v_{\text{ph}}$$

The group velocity is determined to be one half of the phase velocity. Such an effect can be observed by throwing a rock in a still pond. As the radial waves move out, the individual waves seem to run right through the wave crests and then disappear (see Figure 5.16).

5.5 Waves or Particles?

By this point it is not unusual to be a little confused. We have been led to believe that electromagnetic radiation behaves sometimes as waves (interference and diffraction) and other times as particles (photoelectric and Compton effects). We have been presented evidence in this chapter that particles also behave as waves (electron diffraction). Can all this really be true? If a particle is a wave, what is waving? In the preceding section we learned that, at least mathematically, we could describe particles by using wave packets. Can we represent matter as waves and particles simultaneously? And can we represent electromagnetic radiation as waves and particles simultaneously? We must answer these questions of the **wave–particle duality** before proceeding with our study of quantum theory.

Wave–particle duality

Young's Double-Slit Experiment with Light. To better understand the differences and similarities of waves and particles, we analyze Young's double-slit diffraction experiment, which is studied in detail in elementary physics courses to show the interference character of light. Figure 5.17a shows a schematic

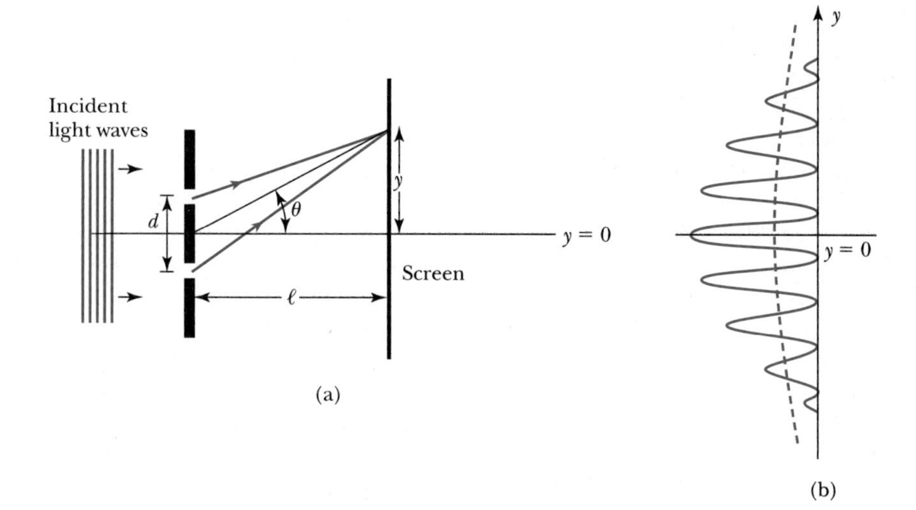

FIGURE 5.17 (a) Schematic diagram of Young's double-slit experiment. This experiment is easily performed with a laser as the light source ($\ell \gg d$, where d = slit distance). (b) The solid line indicates the interference pattern due to both slits. If either of the slits is covered, single-slit diffraction gives the result shown in the dashed curve.

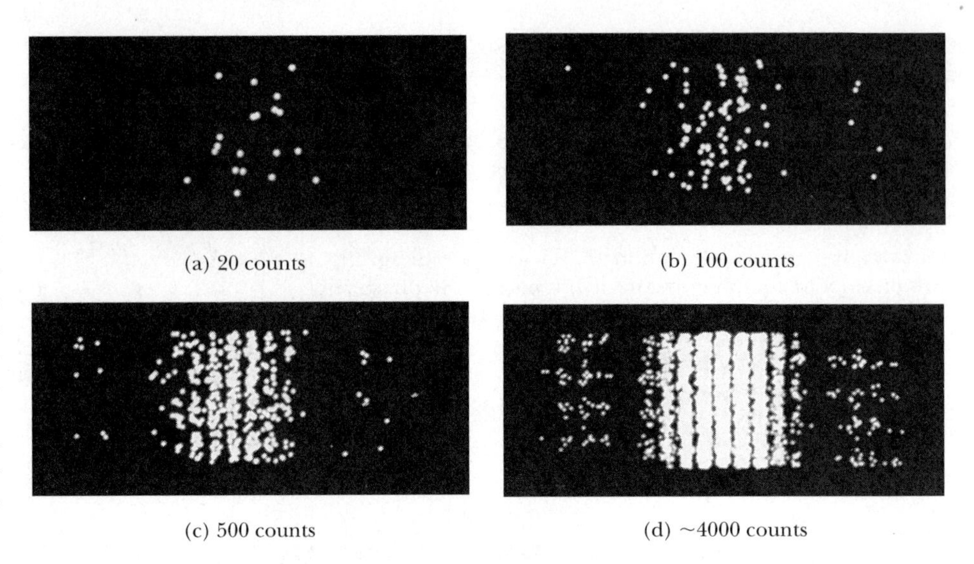

FIGURE 5.18 Computer simulation of Young's double-slit interference experiment for light or electrons. This calculation was performed for a (slit width) $= 4\lambda$, and d (slit distance) $= 20\lambda$. The four pictures are for increasing number of counts: 20, 100, 500, 4000. The interference pattern clearly has emerged for 500 counts. *Simulation and photos courtesy of Julian V. Noble.*

diagram of the experiment. This experiment is easily performed with the use of a low-power laser. With both slits open, a nice interference pattern is observed, with bands of maxima and minima. When one of the slits is covered, this interference pattern is changed, and a rather broad peak is observed. Thus, we conclude that the double-slit interference pattern is due to light passing through *both* slits—a wave phenomenon (see Figure 5.17b).

However, if the light intensity is reduced, and we observe the pattern on a screen, we learn that the light arriving on the screen produces flashes at various points, entirely representative of particle behavior! If we take pictures of the screen after varying lengths of time, the pictures will look like those shown in Figure 5.18. Eventually the interference pattern characteristic of wave behavior emerges. There is therefore no contradiction in this experiment. If we want to know the precise location of the light (photon), we must use the particle description and not the wave description.

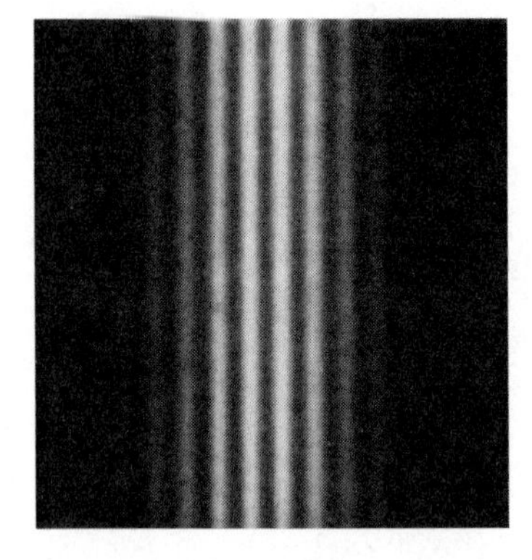

FIGURE 5.19 Demonstration of electron interference using two slits similar in concept to Young's double-slit experiment for light. This result by Claus Jönsson clearly shows that electrons exhibit wave behavior (see also Example 5.5). *Reprinted with permission from C. Jönsson, American Journal of Physics, 42, 4 (1974). © 1974 American Association of Physics.*

Electron Double-Slit Experiment. Now let us examine a similar double-slit experiment that uses electrons rather than light. If matter also behaves as waves, should not the same experimental results be obtained if we use electrons rather than light? The answer is yes, and physicists did not doubt the eventual result. This experiment is not as easy to perform as the similar one with light. The difficulty arises in constructing slits narrow enough to exhibit wave phenomena. This requires $\lambda \sim a$, where a is the slit width. For light of $\lambda = 600$ nm, slits can be produced mechanically. However, for electrons of energy 50 keV, $\lambda \approx 5 \times 10^{-3}$ nm, which is smaller than a hydrogen atom ($\sim$0.1 nm). Nevertheless, C. Jönsson* of Tübingen, Germany, succeeded in 1961 in showing double-slit interference effects for electrons (Figure 5.19) by constructing very narrow slits and using relatively large distances between the slits and the observation screen. Copper slits were made by electrolytically depositing copper on a polymer strip printed on silvered glass plates. This experiment demonstrated that precisely the same behavior occurs for both light (waves) and electrons (particles). We have seen similar behavior previously from the Debye-Scherrer rings produced by the diffraction of x rays (waves) and electrons (particles).

*C. Jönsson, *Am. J. Phys.* **42, 4** (1974), translation of *Zeitschrift f. Physik* **161,** 454 (1961).

Example 5.5

In the experiment by Jönsson, 50-keV electrons impinged on slits of width 500 nm separated by a distance of 2000 nm. The observation screen was located 350 mm beyond the slits. What was the distance between the first two maxima?

Solution: The equation specifying the orders of maxima and the angle θ from incidence is (see Figure 5.17)

$$d \sin \theta = n\lambda \tag{5.32}$$

The order $n = 0$ has $\theta = 0$, and the next maximum, $n = 1$, is

$$\sin \theta = \frac{\lambda}{d} = \frac{\lambda}{2000 \text{ nm}}$$

We have already calculated the wavelength for electrons of energy eV_0 in Equation (5.7).

$$\lambda = \frac{1.226 \text{ nm} \cdot \text{V}^{1/2}}{\sqrt{50 \times 10^3 \text{ V}}} = 5.48 \times 10^{-3} \text{ nm}$$

Because 50 keV may be too high an energy for a nonrelativistic calculation such as that done in Equation (5.7), we had better perform a relativistic calculation to be certain. We first find the momentum and insert that into $\lambda = h/p$.

$$(pc)^2 = E^2 - E_0{}^2 = (K + E_0)^2 - E_0{}^2$$
$$= (50 \times 10^3 \text{ eV} + 0.511 \times 10^6 \text{ eV})^2 -$$
$$(0.511 \times 10^6 \text{ eV})^2 = (0.231 \times 10^6 \text{ eV})^2$$

Now we can determine the wavelength.

$$\lambda = \frac{h}{p} = \frac{hc}{pc} = \frac{1240 \text{ eV} \cdot \text{nm}}{0.231 \times 10^6 \text{ eV}} = 5.36 \times 10^{-3} \text{ nm}$$

Therefore, we find the more accurate relativistic value to be somewhat less (2%) than the nonrelativistic value. Now we can determine the angle.

$$\sin \theta = \frac{5.36 \times 10^{-3} \text{ nm}}{2000 \text{ nm}} = 2.68 \times 10^{-6}$$

The distance of the first maximum along the screen is $y = \ell \tan \theta$, but for such a small angle, $\sin\theta = \tan \theta$.

$$y = \ell \tan \theta \approx \ell \sin \theta = 350 \text{ mm}(2.68 \times 10^{-6})$$
$$= 9.38 \times 10^{-4} \text{ mm} \frac{10^6 \text{ nm}}{\text{mm}} = 938 \text{ nm}$$

Such a diffraction pattern is too small to be viewed by eye. Jönsson magnified the pattern by a series of electronic lenses and then observed a fluorescent screen with a ten-power optical microscope.

Another Gedanken Experiment. If we were to cover one of the slits in the preceding Jönsson experiment, the double-slit interference pattern would be destroyed—just as it was when light was used. But our experience tells us the electron is a particle, and we know that it can go through only one of the slits. Let's devise a *gedanken* experiment, shown in Figure 5.20, to determine which slit the electron went through. We set up a light shining on the double slit and use a powerful microscope to look at the region. After the electron passes through one of the slits, light bounces off the electron; we observe the light, so we know which slit the electron came through.

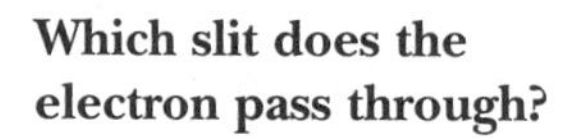
Which slit does the electron pass through?

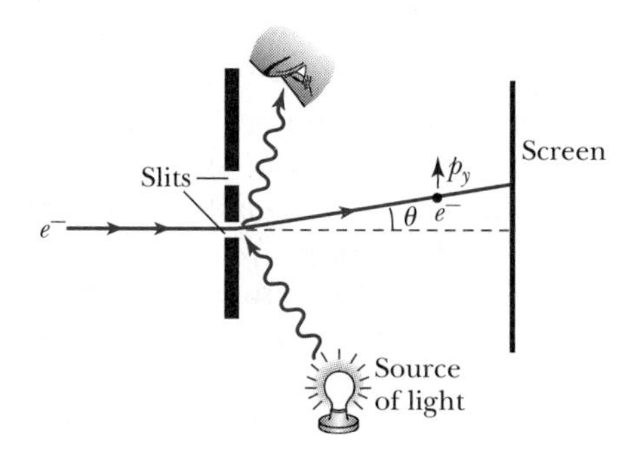

FIGURE 5.20 An attempt to measure which slit an electron passes through in the double-slit experiment. A powerful light source scatters a photon from the electron, and the scattered photon is observed. The motion of the electron is affected.

In order to do this experiment, we need to use light having wavelength narrower than the slit separation d, in order to determine which slit the electron went through. We use a subscript ph to denote variables for light (photon). Therefore, we have $\lambda_{\text{ph}} < d$. The momentum of the photon is

$$p_{\text{ph}} = \frac{h}{\lambda_{\text{ph}}} > \frac{h}{d}$$

For us to show the interference effects for the electrons passing through the slits, the electrons must also have a wavelength on the order of the slit separation d, $\lambda_{\text{el}} \sim d$. The momentum of the electrons will be on the order of

$$p_{\text{el}} = \frac{h}{\lambda_{\text{el}}} \sim \frac{h}{d}$$

The difficulty is that the momentum of the light photons used to determine which slit the electron went through is sufficiently great to strongly modify the momentum of the electron itself, thus changing the direction of the electron! The interference pattern on the screen will be changed just by requiring us to know which slit the electron went through. We will take a closer look at this experiment in Section 5.7. In trying to determine which slit the electron went through, we are examining the particlelike behavior of the electron. When we are examining the interference pattern of the electron, we are using the wavelike behavior of the electron.

Bohr resolved this dilemma by pointing out that the particlelike and wavelike aspects of nature are *complementary*. Both are needed—they just can't be observed simultaneously.

Principle of complementarity

Bohr's principle of complementarity: *It is not possible to simultaneously describe physical observables in terms of both particles and waves.*

Physical observables

Physical observables are those quantities such as position, velocity, momentum, and energy that can be experimentally measured. In any given instance we must use either the particle description or the wave description. Usually the choice is clear. The interference pattern of the double-slit experiment suggests that the light (or electron) had to go through both slits, and we must use the wave description. In our description of nature, we cannot describe phenomena by displaying both particle and wave behavior at the same time.

Solution of wave–particle duality

By the use of the principle of complementarity, we can solve the *wave–particle duality* problem, which has been plaguing us. It is not unusual for students to feel uncomfortable with this solution, because it cannot be directly proven. However, as a "principle" and not a "law," the complementarity principle does seem to describe nature, and, as such, we use it. We must pay close attention to the fact that we do not use waves and particles simultaneously to describe a particular phenomenon. Experiments dictate what actually happens in nature, and we must draw up a set of rules to describe our observations. These rules naturally lead to a probability interpretation of experimental observations. If we set up a series of small detectors along the screen in the electron double-slit experiment, we can speak of the probability of the electron being detected by one of the detectors. The interference pattern can guide us in our probability determinations. But once the electron has been registered by one of the detectors, the probability of its being seen in the other detectors is zero. Matter and radiation propagation is described by wavelike behavior, but matter and radiation interact (that is, creation/annihilation or detection) as particles.

5.6 Relationship Between Probability and Wave Function

We learned in elementary physics that the instantaneous wave intensity of electromagnetic radiation (light) is $\epsilon_0 cE^2$ where E is the electric field. Thus the probability of observing light is proportional to the square of the electric field. In the double-slit light experiment we can be assured that the electric field of the light wave is relatively large at the bright spots on the screen and small in the region of the dark places.

If Young's double-slit experiment is performed with very low intensity levels of light, individual flashes can be seen on the observing screen. We show a simulation of the experiment in Figure 5.18. After only 20 flashes (Figure 5.18a) we cannot make any prediction as to the eventual pattern. But we still know that the **probability** of observing a flash is proportional to the square of the electric field. In elementary physics we calculate this result. If the distance from the central ray along the screen we are observing in an experiment like that depicted in Figure 5.17a is denoted by y, the probability for the photon to be found between y and $y + dy$ is proportional to the intensity of the wave (E^2) times dy. For Young's double-slit experiment, the value of the electric field **E** produced by the two interfering waves is large where the flash is likely to be observed and small where it is not likely to be seen. By counting the number of flashes we relate the energy flux I (called the intensity) of the light to the number flux, N per unit area per unit time, of photons having energy $h\nu$. In the wave description, we have $\mathrm{I} = \epsilon_0 c\langle E^2\rangle$ (where E is the electric field), and in what appears to be the particle description, $\mathrm{I} = Nh\nu$. The flux of photons N, or the probability P of observing the photons, is proportional to the average value of the square of the electric field $\langle E^2\rangle$.

How can we interpret the probability of finding the electron in the wave description? First, let's remember that the localization of a wave can be accomplished by using a wave packet. We used a function $\Psi(x, t)$ to denote the superposition of many waves to describe the wave packet. We call this function $\Psi(x, t)$ the **wave function.** In the case of light, we know that the electric field **E** and magnetic field **B** satisfy a wave equation. In electrodynamics either **E** or **B** serves as the wave function Ψ. For particles (say electrons) a similar behavior occurs. In this case the wave function $\Psi(x, t)$ determines the probability, just as the flux of photons N arriving at the screen and the (electric field) **E** determined the probability in the case of light.

Wave function

The wave function: waves for matter

For matter waves having a de Broglie wavelength, it is the wave function Ψ that determines the likelihood (or probability) of finding a particle at a particular position in space at a given time.The value of the wave function Ψ has no physical significance itself, and as we will learn later, it can have a **complex** value. The quantity $|\Psi|^2$ is called the **probability density** and represents the probability of finding the particle in a given unit volume and at a given instant of time.

Probability density

In general, $\Psi(x, y, z, t)$ is a complex quantity and depends on the spatial coordinates x, y, and z as well as time t. The complex nature will be of no concern to us: we use Ψ times its complex conjugate Ψ^* when finding probabilities. We are only interested here in a single dimension y along the observing screen and for a given time t. In this case $\Psi^*\Psi\, dy = |\Psi|^2 dy$ is the probability of observing an electron in the interval between y and $y + dy$ at a given time, and we call this $\mathrm{P}(y)dy$.

$$\mathrm{P}(y)dy = |\Psi(y, t)|^2 dy \tag{5.33}$$

Normalization Because the electron has to have a probability of unity of being observed *somewhere* along the screen, we integrate the probability density over all space by integrating over y from $-\infty$ to ∞. This process is called *normalization.*

$$\int_{-\infty}^{\infty} P(y)\,dy = \int_{-\infty}^{\infty} |\Psi(y, t)|^2 dy = 1 \tag{5.34}$$

The probability interpretation of the wave function was first proposed in 1926 by Max Born (Nobel Prize, 1954), one of the founders of the quantum theory. The determination of the wave function $\Psi(x, t)$ will be discussed in much more detail in the next chapter.

The use of wave functions $\Psi(x, y, z, t)$ rather than the classical positions $x(t)$, $y(t)$, $z(t)$ represents a clean break between classical and modern physics. In order to be useful in determining values of physical observables like position, momentum, and energy, a set of rules and procedures has been developed (see Section 6.2).

Example 5.6

Consider a particle of mass m trapped in a one-dimensional box of width ℓ. Calculate the possible energies of the particle. What is the most probable location of the particle in the state with the lowest energy at a given time, say $t = 0$, so that $\Psi(x, 0) = \psi(x)$?

Solution: Let us treat the particle as a sinusoidal wave. The particle cannot physically be outside the confines of the box, so the amplitude of the wave motion must vanish at the walls and outside the box. In the language of the wave function, its probability of being outside is zero, so the wave function must vanish outside. The wave function should be continuous, and the probability distribution can have only one value at each point in the box. Several possible waves are shown in Figure 5.21. An integral number of half wavelengths $\lambda/2$ must fit into the box, so

$$\frac{n\lambda}{2} = \ell \qquad \text{or} \qquad \lambda_n = \frac{2\ell}{n} \qquad n = 1, 2, 3, \ldots \tag{5.35}$$

The possible wavelengths are quantized, and the wave shapes will have $\sin(n\pi x/\ell)$ terms. If we treat the problem nonrelativistically and assume there is no potential energy, the energy E of the particle is

$$E = \text{K.E.} = \frac{1}{2}mv^2 = \frac{p^2}{2m} = \frac{h^2}{2m\lambda^2}$$

If we insert the values for λ_n, we have

$$E_n = \frac{h^2 n^2}{2m(4\ell^2)} = n^2 \frac{h^2}{8m\ell^2} \qquad n = 1, 2, 3, \ldots \tag{5.36}$$

Therefore, the possible energies of the particle are quantized, and the lowest energy $E_1 = h^2/8m\ell^2$. Each of these possible energies is called an **energy level.**

The probability of observing the particle between x and $x + dx$ in each state is $P_n dx \propto |\psi_n(x)|^2 dx$. Notice that $E_0 = 0$ is not a possible state, because $n = 0$ corresponds to $\psi_0 = 0$. The lowest energy level is, therefore E_1 and $P_1 \propto |\psi_1(x)|^2$, shown in Figure 5.21. The most probable location for the particle in the lowest energy state is in the middle of the box.

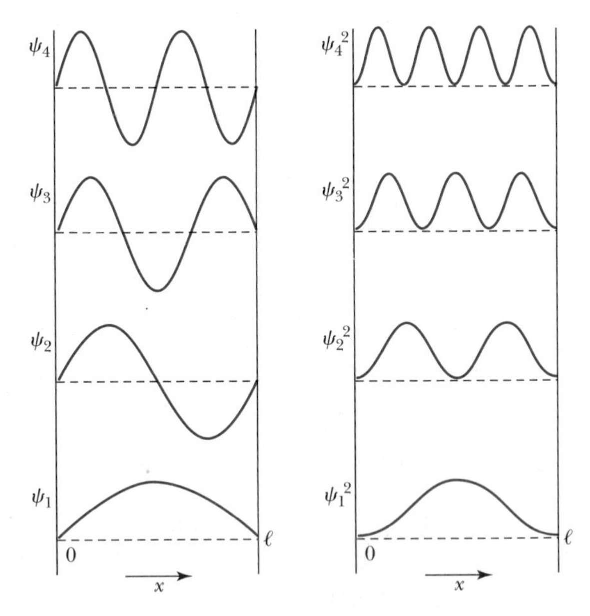

FIGURE 5.21 Possible ways of fitting waves into a one-dimensional box of length ℓ. The left side shows the wave functions for the four lowest energy values. The right side shows the corresponding probability distributions.

The previous example is an important one. It is our first application of quantum theory using waves. Notice how the quantization of energy arises from the need to fit a whole number of half-waves into the box, and how we obtained the corresponding probability densities of each of the states. We reintroduced the concept of energy levels, first discussed in the Bohr model. The procedure followed in the example is the same as finding the allowed modes of standing waves inside the box. We can use all the results that we learned about waves in elementary physics.

Werner Heisenberg (1901–1976) was born in Germany where he spent his entire career at various universities including Munich, Leipzig, and Berlin where he was appointed director of the Kaiser Wilhelm Institute in 1942, the highest scientific position in Germany. After World War II Heisenberg spent much of his effort towards supporting research and opportunities for young physicists as well as speaking out against the atom bomb. *AIP Emilio Segrè Visual Archives.*

5.7 Uncertainty Principle

In Section 5.4, when we discussed the superposition of waves, we learned that in order to localize a wave packet over a small region Δx, we had to use a large range, Δk, of wave numbers. For the case of two waves we found in Equation (5.22) that $\Delta k\,\Delta x = 2\pi$. If we examine a Gaussian wave packet closely, we would find that the product $\Delta k\,\Delta x = 1/2$. The minimum value of the product $\Delta k\,\Delta x$ is obtained when Gaussian wave packets are used.

In Section 5.4 we learned that it is impossible to measure simultaneously, with no uncertainty, the precise values of k and x for the same particle. The wave number k may be rewritten as

$$k = \frac{2\pi}{\lambda} = \frac{2\pi}{h/p} = p\frac{2\pi}{h} = \frac{p}{\hbar} \tag{5.37}$$

and

$$\Delta k = \frac{\Delta p}{\hbar} \tag{5.38}$$

so that, in the case of the Gaussian wave packet,

$$\Delta k\,\Delta x = \frac{\Delta p}{\hbar}\,\Delta x = 1/2$$

or

$$\Delta p\,\Delta x = \hbar/2 \tag{5.39}$$

for Gaussian wave packets.

The relationship in Equation (5.39) was first enunciated in 1927 by the German physicist Werner Heisenberg, who won the Nobel Prize in 1932. This uncertainty also applies in all three dimensions, so we should put a subscript on Δp to indicate the x direction Δp_x. Heisenberg's **uncertainty principle** can therefore be written

$$\Delta p_x\,\Delta x \geq \hbar/2 \tag{5.40}$$

Heisenberg uncertainty principle for p_x and x

which establishes limits on the simultaneous knowledge of the values of p_x and x.* The limits on Δp_x and Δx represent the lowest possible limits on the uncertainties in knowing the values of p_x and x, no matter how good an experimental measurement is made. It is possible to have a greater uncertainty in the values of p_x and x, but it is not possible to know them better than allowed by the uncertainty principle. The uncertainty principle does not apply to the products of Δp_z

*In some representations of the uncertainty principle, the factor $\frac{1}{2}$ is absent. Our form represents the lower limit.

and Δx or to that of Δp_y and Δz. The value of $\Delta p_z \Delta x$ can be zero. Equation (5.40) is true not only for specific waves such as water or sound, but for matter waves as well. It is a consequence of the de Broglie wavelength of matter. If we want to know the position of a particle very accurately, then we must accept a large uncertainty in the momentum of the particle. Similarly, if want to know the precise value of a particle's momentum, it is not possible to specify the particle's location very precisely. The uncertainty principle represents another sharp digression with classical physics, where it is assumed that it is possible to simultaneously specify precisely both the particle's position and momentum. Because of the small value of $\hbar$, the uncertainty principle becomes important only on the atomic level as the following example shows.

Example 5.7

Calculate the momentum uncertainty of (a) a tennis ball constrained to be in a fence enclosure of length 35 m surrounding the court, and (b) an electron within the smallest radius of a hydrogen atom.

Solution: (a) If we insert the uncertainty of the location of the tennis ball, $\Delta x = 35$ m, into Equation (5.39), we have

$$\Delta p \geq \frac{1}{2}\frac{\hbar}{\Delta x} = \frac{1.05 \times 10^{-34}\ \text{J}\cdot\text{s}}{2(35\ \text{m})} = 1.5 \times 10^{-36}\ \text{kg}\cdot\text{m/s}$$

We will have no problem specifying the momentum of the tennis ball!

(b) The diameter of the hydrogen atom in its lowest energy state (smallest radius) is $2a_0$. Let the uncertainty in x be equal to the radius, $\Delta x = a_0$ (even if we let $\Delta x = 2a_0$ the conclusions are valid).

$$\Delta x = a_0 = 0.53 \times 10^{-10}\ \text{m}$$

$$\Delta p \geq \frac{1}{2}\frac{\hbar}{\Delta x} = \frac{1.05 \times 10^{-34}\ \text{J}\cdot\text{s}}{2(0.53 \times 10^{-10}\ \text{m})}$$

$$= 0.99 \times 10^{-24}\ \text{kg}\cdot\text{m/s}$$

It is likely that the momentum of the electron is at least as big as its uncertainty, so we let $p = (\Delta p)_{\text{min}}$ and calculate the electron's minimum kinetic energy from $p^2/2m$.

$$\text{K.E.} = \frac{p^2}{2m} = \frac{(\Delta p)^2_{\text{min}}}{2m} = \frac{(0.99 \times 10^{-24}\ \text{kg}\cdot\text{m/s})^2}{(2)(9.11 \times 10^{-31}\ \text{kg})}$$

$$= 5.4 \times 10^{-19}\ \text{J}\left(\frac{1\ \text{eV}}{1.6 \times 10^{-19}\ \text{J}}\right) = 3.4\ \text{eV}$$

Remember that the binding energy of the electron in the hydrogen atom is 13.6 eV, so that on the atomic scale, the uncertainties can be large percentages of typical values themselves. Similarly, from the value of Δp determined above, we calculate

$$\Delta v = \frac{\Delta p}{m} = \frac{0.99 \times 10^{-24}\ \text{kg}\cdot\text{m/s}}{9.1 \times 10^{-31}\ \text{kg}}$$

$$= 1.1 \times 10^{6}\ \text{m/s} = 0.0036c$$

In Chapter 4 we found $v = 2.2 \times 10^6$ m/s $= 0.0073c$ for this state, so the uncertainty in the velocity is about 50%.

Equation (5.40) is not the only form of the uncertainty principle. We can find another form by using Equation (5.23) from our study of wave motion. When we superimposed two waves to form a wave packet we found $\Delta\omega\,\Delta t = 2\pi$. If we evaluate this same product using Gaussian packets, we will find

$$\Delta\omega\,\Delta t = \frac{1}{2} \tag{5.41}$$

just as we did for the product $\Delta k\,\Delta x$. A relationship like this is easy to understand. If we are to localize a wave packet in a small time Δt (instead of over an infinite time as for a single wave), we must include the frequencies of many waves in order to have them cancel everywhere but over the time interval Δt. Because $E = h\nu$, we have for each wave,

$$\Delta E = h\,\Delta\nu = h\frac{\Delta\omega}{2\pi} = \hbar\,\Delta\omega$$

therefore

$$\Delta\omega = \frac{\Delta E}{\hbar} \qquad \text{and} \qquad \Delta\omega\,\Delta t = \frac{\Delta E}{\hbar}\,\Delta t = \frac{1}{2}$$

We can therefore obtain another form of Heisenberg's uncertainty principle:

Heisenberg uncertainty principle for energy and time

$$\Delta E\,\Delta t \geq \frac{\hbar}{2} \tag{5.42}$$

There are other *conjugate variables* similar to p and x in Equation (5.40) that also form uncertainty principle relations. These variables include the angular momentum L and angle θ, for example, as well as the rotational inertia I and angular velocity ω. Similar uncertainty relations can be written for them.

We once again must emphasize that the uncertainties expressed in Equations (5.40) and (5.42) are intrinsic. They are not due to our inability to measure more precisely. No matter how well we can measure, no matter how accurate an instrument we build, and no matter how long we measure, we can never do any better than the uncertainty principle allows. Many people, including Einstein, have tried to think of situations in which it is violated, but they have not succeeded. At the 1927 Solvay conference Bohr and Einstein had several discussions about the uncertainty principle. Every morning at breakfast Einstein would present a new *gedanken* experiment that would challenge the uncertainty principle. In his careful, deliberate manner, Bohr would refute each objection. Eventually Einstein conceded—he could not provide a valid example of contradiction. These discussions continued off and on into the 1930s, because Einstein had difficulty accepting the idea that the quantum theory could give a complete description of physical phenomena. He believed the quantum theory could give a statistical description of a collection of particles but could not describe the motion of a single particle. Einstein presented several paradoxes to support his ideas. Bohr was able to analyze each paradox and present a reasonable answer. Bohr stressed his complementarity principle, which precludes a simultaneous explanation in terms of waves and particles, as well as Heisenberg's uncertainty principle, which surprisingly does allow small violations of the conservation laws of energy and momentum.

Bohr and Einstein discussions

It may seem paradoxical that energy conservation is violated in quantum physics, but no paradox is involved because the energy violation ΔE cannot be detected by any experiment. Suppose that one wishes to observe a deviation from exact energy conservation by an amount ΔE. The uncertainty principle requires that the time during which this violation takes place is on the order of $\Delta t = \hbar/2\Delta E$. To observe a time interval this short, we need a clock ticking at intervals less than Δt, that is, a train of pulses of frequency $\nu = 1/\Delta t$. But the quanta in this wave train have energies $E = h\nu = h/\Delta t = 4\pi\Delta E$, so the wave quanta have plenty of energy to disturb the system by at least the energy ΔE. Any attempt to measure ΔE must disturb the system by at least as much as the uncertainty ΔE.

FIGURE 5.22 Niels Bohr's coat of arms was designed in 1947 when he was awarded the Danish Order of the Elephant. This award was normally given only to royalty and foreign presidents. Bohr chose the Chinese yin-yang symbol because it stands for the two opposing but inseparable elements of nature. The translation of the Latin motto is "Opposites are complementary." It was hung near the king's coat of arms in the church of Frederiksborg Castle at Hillerod. *AIP Niels Bohr Library. Margarethe Bohr Collection.*

Let's return to the previous discussion of determining which slit an electron passes through in the double-slit experiment (see Figure 5.20). We again shine light on the electrons passing through the slits and look with a powerful microscope. This time we will use the uncertainty principle and make a more detailed calculation. The light photons bounce off the electron as it passes through one of the slits and scatter into the microscope where we observe them. We must be able to locate the electron's position in y to at least within $\Delta y < d/2$ (where d is the distance between the two slits) in order to know which slit the electron went through. If the position of the electron is uncertain to less than $d/2$, then according to the uncertainty principle, the electron's momentum must be uncertain to at least $\Delta p_y > \hbar/d$. Just by scattering photons off the electrons in order to know which slit the electron went through, we introduce an uncertainty in the electron's momentum. This uncertainty has been caused by the measurement itself.

Consider an electron originally moving in a particular direction; let us choose $\theta = 0$ for convenience. By scattering the photon we now have an uncertainty in the angle θ due to the "kick" given the electron by the photon in the measurement process. The uncertainty in the electron's angle due to a possible momentum change along the y axis is $\Delta\theta = \Delta p_y/p$, but because $p = h/\lambda$, we have

$$\Delta\theta = \frac{\Delta p_y}{p} = \frac{(\Delta p_y)\lambda}{h} = \frac{(\hbar)\lambda}{dh} = \frac{\lambda}{2\pi d}$$

According to Equation (5.32) the first interference maximum will be at $\sin\theta = \lambda/d$ and the first minimum at $\sin\theta = \lambda/2d$. For small angle scattering, $\sin\theta \approx \theta$, and the angle of the first minimum is $\theta_{\min} \approx \lambda/2d$. Note that the position of the first minimum is on the same order as our uncertainty in $\Delta\theta$, so the interference pattern is washed out. If we insist on identifying the electrons as particles and knowing which slit the electrons pass through, the wave characteristics of the electron disappear. We cannot simultaneously treat the electron as both a particle and a wave.

This limitation seems to be a fundamental characteristic of the laws of nature. Only the smallness of Planck's constant h keeps us from encountering this limitation in everyday life. Niels Bohr tried to turn this limitation into a philosophical principle. When he was awarded the Danish Order of the Elephant he put on his coat of arms (see Figure 5.22) the Chinese yin-yang symbol, which stands for the two opposing but inseparable elements in nature. The Latin motto on the center of the coat of arms means "Opposites are Complementary."

Example 5.8

Calculate the minimum kinetic energy of an electron that is localized within a typical nuclear radius of 6×10^{-15} m.

Solution: Let's assume the minimum electron energy is that due to the uncertainty principle with an uncertainty Δx equal to the radius ($\Delta x = \pm r$).

$$\Delta x \approx r = 6 \times 10^{-15}\ \text{m}$$

$$\Delta p \geq \frac{\hbar}{2\Delta x} = \frac{6.58 \times 10^{-16}\ \text{eV}\cdot\text{s}}{1.2 \times 10^{-14}\ \text{m}}$$

$$\geq 5.48 \times 10^{-2}\ \text{eV}\cdot\text{s/m}\ \frac{3 \times 10^{8}\ \text{m/s}}{c}$$

$$\geq 1.64 \times 10^{7}\ \text{eV}/c$$

Let's now assume that the momentum p is at least as large as the uncertainty in p.

$$p \approx \Delta p \geq 1.64 \times 10^7 \text{ eV}/c$$

Because we don't yet know the electron's energy, let's be careful and calculate it relativistically.

$$\begin{aligned} E^2 &= (pc)^2 + {E_0}^2 \\ &= \left[\left(1.64 \times 10^7 \frac{\text{eV}}{c}\right)c\right]^2 + (0.511 \text{ MeV})^2 \\ &= (16.4 \text{ MeV})^2 + (0.511 \text{ MeV})^2 \end{aligned}$$

$$\begin{aligned} E &= 16.4 \text{ MeV} \\ \text{K.E.} &= E - E_0 = 16.4 \text{ MeV} - 0.51 \text{ MeV} \\ &= 15.9 \text{ MeV} \end{aligned}$$

Note that because K.E. > E_0, a relativistic calculation was actually needed. This value of K.E. is larger than that observed for electrons emitted from nuclei in beta decay. We must conclude that electrons cannot be confined within the nucleus. Electrons emitted from the nucleus (during beta decay) must actually be created when they are emitted.

Example 5.9

An atom in an excited state normally remains in that state for a very short time ($\sim 10^{-8}$ s) before emitting a photon and returning to a lower energy state. The "lifetime" of the excited state can be regarded as an uncertainty in the time Δt associated with a measurement of the energy of the state. This, in turn, implies an "energy width," namely, the corresponding energy uncertainty ΔE. Calculate the characteristic "energy width" of such a state.

Solution: Because $\Delta E\, \Delta t \geq \hbar/2$

$$\Delta E \geq \frac{\hbar}{2\Delta t} = \frac{6.58 \times 10^{-16} \text{ eV}\cdot\text{s}}{(2)(10^{-8} \text{ s})} = 3.3 \times 10^{-8} \text{ eV}$$

This is a very small energy, but many excited energy states have such energy widths. For stable ground states, $\tau = \infty$, and $\Delta E = 0$. For excited states in the nucleus, the lifetimes can be as short as 10^{-20} s (or shorter) with energy widths of 100 keV (or more).

Summary

Max von Laue suggested the scattering of x rays from matter, thereby firmly establishing the wave nature of x rays and the lattice structure of crystals. W. H. Bragg and W. L. Bragg exploited the wave behavior of x rays by utilizing x-ray scattering to determine the spacing d between crystal planes according to Bragg's law

$$n\lambda = 2d \sin \theta \tag{5.1}$$

In an important conceptual leap, de Broglie suggested that particles might also exhibit wave properties, with a wavelength λ determined by their momentum

$$\lambda = \frac{h}{p} \quad \text{de Broglie wavelength} \tag{5.2}$$

Davisson and Germer, and G. P. Thomson independently, demonstrated the wave characteristics of particles by diffracting low-energy electrons from crystals.

Particles may be described using waves by representing them as wave packets, the superposition of many waves of different amplitudes and frequencies. The group velocity $u_{gr} = d\omega/dk$ represents the speed of the particle described by the wave packet.

Niels Bohr proposed a principle of complementarity, stating that it is not possible to describe physical behavior simultaneously in terms of both particles and waves. We must use either one form of description or the other. This principle avoids the conceptual wave–particle duality problem by precluding a simultaneous description of experiments by both wave and particle behavior.

We describe particles exhibiting wave behavior by using wave functions Ψ, which in general may be complex-valued functions of space and time. The probability of observing a particle between x and $x + dx$ at time t is $|\Psi(x, t)|^2 dx$.

Werner Heisenberg pointed out that it is not possible to know simultaneously both the exact momentum and position of a particle or to know its precise energy at a precise time. These relationships

$$\Delta p_x\, \Delta x \geq \hbar/2 \tag{5.40}$$

$$\Delta E\, \Delta t \geq \hbar/2 \tag{5.42}$$

are called *Heisenberg's uncertainty principle* and are consistent with Bohr's complementarity principle. No experiment, regardless of how clever, can measure p, x, E, and t better than the uncertainties expressed in Equations (5.40) and (5.42).

Questions

1. In 1900 did it seem clear that x rays were electromagnetic radiation? Give reasons why you think so. Was it important to perform further experiments to verify the characteristics of x rays?
2. In the early 1900s it was found that x rays were more difficult to refract or diffract than visible light. Why did this lead researchers to suppose that the wavelengths of x rays were shorter rather than longer than those of light?
3. What determines whether a given photon is an x ray? Could an x ray have a wavelength longer than ultraviolet light?
4. For a single crystal, transmission x-ray scattering will produce dots. However, if there are randomly oriented crystals, as in powder, concentric rings appear. Explain the difference qualitatively.
5. How many particles can you think of that might be shown experimentally to exhibit wavelike properties? List at least three and discuss possible experiments.
6. Why are neutrons more widely used than protons for studying crystal structure? What about using a hydrogen atom?
7. Why is it important to use "cold" neutrons for studying crystal structure? How could one obtain "cold" neutrons?
8. Are the following phenomena wave or particle behaviors? Give your reasoning (a) television picture, (b) rainbows on a rainy day, (c) football sailing through goal posts, (d) telescope observing the moon, (e) police radar.
9. The experiment by Jönsson that showed the wavelike properties of electrons passing through a double slit is considered a pedagogically interesting experiment but not a landmark experiment. Why do you suppose this is true?
10. Can you think of an experiment other than those mentioned in this chapter that might show the wavelike properties of particles? Discuss it.
11. Why doesn't the uncertainty principle restriction apply between the variables p_z and x?
12. How does the uncertainty principle apply to a known stable atomic system that apparently has an infinite lifetime? How well can we know the energy of such a system?
13. According to the uncertainty principle, can a particle having a kinetic energy of exactly zero be confined somewhere in a box of length ℓ? Explain.
14. What is similar about the conjugate variable pairs (p, x), (E, t), (L, θ), and (I, ω)?
15. What are the dimensions of the wave function $\Psi(x, t)$ that describes matter waves? Give your reasoning.
16. Soon after their discovery, Davisson and Germer were using their experimental technique to point out new crystal structures of nickel. Do you think they were justified? Explain how you think their results allowed them to make such statements.

Problems

5.1 X-Ray Scattering

1. X rays scattered from a crystal have a first-order diffraction peak at $\theta = 15°$. At what angle will the second- and third-order peaks appear?
2. X rays of wavelength 0.16 nm are scattered from NaCl. What is the angular separation between first- and second-order diffraction peaks? Assume scattering planes that are parallel to the surface.
3. Potassium chloride is a crystal with lattice spacing of 0.314 nm. The first peak for Bragg diffraction occurs at 14°. What energy x rays were diffracted? What other order peaks can be observed $(\theta \leq 90°)$?
4. A cubic crystal with interatomic spacing of 0.24 nm is used to select γ rays of energy 100 keV from a radioactive source containing a continuum of energies. If the incident beam is normal to the crystal, at what angle do the 100-keV γ rays appear?

5.2 De Broglie Waves

5. Calculate the de Broglie wavelength of a 3.0 kg rock thrown with a speed of 6 m/s into a pond. Is this wavelength similar to that of the water waves produced? Explain.
6. Calculate the de Broglie wavelength of a nitrogen molecule in the atmosphere on a hot summer day (35°C). Compare this with the diameter (less than 1 nm) of the molecule.
7. Work out Example 5.2b strictly using SI units of m, J, kg, and so on, and compare with the method of the example using eV units.
8. Assume that the total energy E of an electron greatly exceeds its rest energy. If a photon has a wavelength equal to the de Broglie wavelength of the electron, what is the photon's energy? What if $E = 2E_0$ for the electron?

9. Determine the de Broglie wavelength of a particle of mass m and kinetic energy K. Do this for both (a) a relativistic and (b) a nonrelativistic particle.
10. The Stanford Linear Accelerator can accelerate electrons to an energy of 50 GeV. What is the de Broglie wavelength of these electrons? What fraction of a proton's diameter ($d \sim 2 \times 10^{-15}$ m) can such a particle probe?
11. Find the kinetic energy of (a) photons, (b) electrons, (c) neutrons, and (d) α particles that have a de Broglie wavelength of 0.15 nm.
12. Find the de Broglie wavelength of neutrons in equilibrium at the temperatures (a) 10 K and (b) 0.1 K.
13. An electron initially at rest is accelerated across a potential difference of 3 kV. What are its wavelength, momentum, kinetic energy, and total energy?
14. What is the wavelength of an electron with kinetic energy (a) 40 eV, (b) 400 eV, (c) 4 keV, (d) 40 keV, (e) 0.4 MeV, and (f) 4 MeV? Which of these energies are most suited for study of the NaCl crystal structure?
15. Calculate the de Broglie wavelength of (a) an oxygen (O_2) molecule darting around the room at 480 m/s, (b) a bacterium of mass 1.5×10^{-15} kg moving at a speed of 10^{-6} m/s.
16. What is the de Broglie wavelength of the 1 TeV protons accelerated in the Fermi National Laboratory Tevatron accelerator?

5.3 Electron Scattering

17. In an electron-scattering experiment an intense reflected beam is found at $\phi = 32°$ for a crystal with an interatomic distance of 0.23 nm. What is the lattice spacing of the planes responsible for the scattering? Assuming first-order diffraction, what are the wavelength, momentum, kinetic energy, and total energy of the incident electrons?
18. Davisson and Germer performed their experiment with a nickel target for several energies. At what angles would they find diffraction maxima for 48-eV and 64-eV electrons?
19. A beam of 2-keV electrons incident on a crystal is refracted and observed (by transmission) on a screen 35 cm away. The radii of three concentric rings on the screen, all corresponding to first-order diffraction, are 2.1 cm, 2.3 cm, and 3.2 cm. What is the lattice-plane spacing corresponding to each of the three rings?
20. A beam of thermal neutrons (K.E. = 0.025 eV) scatters from a crystal with interatomic spacing 0.45 nm. What is the angle of the first-order Bragg peak?

5.4 Wave Motion

21. A wave, propagating along the x direction according to Equation (5.11), has a maximum displacement of 3 cm at $x = 0$ and $t = 0$. The wavespeed is 4 cm/s, and the wavelength is 7 cm. (a) What is the frequency? (b) What is the wave's amplitude at $x = 10$ cm and $t = 13$ s?
22. A wave of wavelength 4 cm has a wavespeed of 4 cm/s. What is its (a) frequency (b) period (c) wave number and (d) angular frequency?
23. Two waves are traveling simultaneously down a long slinky. They can be represented by $\Psi_1(x, t) = 0.003 \sin(6.0x - 300t)$ and $\Psi_2(x, t) = 0.003 \sin(7.0x - 250t)$. Distances are measured in meters and time in seconds. (a) Write the expression for the resulting wave. (b) What are the phase and group velocities? (c) What is Δx between two adjacent zeros of Ψ? (d) What is $\Delta k \Delta x$?
24. A wave packet describes a particle having momentum $p = mv$. Show that the group velocity is βc and the phase velocity is c/β (where $\beta = v/c$). How can the phase velocity physically be greater than c?
25. For waves in shallow water the phase velocity is about equal to the group velocity. What is the dependence of the phase velocity on the wavelength?
26. Find the group and phase velocities of 8-MeV protons and 8-MeV electrons (see Problem 24).
27. Use Equation (5.25) with $\tilde{A}(k) = A_0$ for the range of $k = k_0 - \Delta k/2$ to $k_0 + \Delta k/2$ and $\tilde{A}(k) = 0$ elsewhere to determine $\Psi(x, 0)$, that is, at $t = 0$. Sketch the envelope term, the oscillating term, and $|\Psi(x, 0)|^2$. What is approximately the width Δx over the full-width–half-maximum part of $|\Psi(x, 0)|^2$? What is the value of $\Delta k \, \Delta x$?
28. Show using Equation (5.29) that u_{gr} correctly represents the velocity of the particle both relativistically and classically.

5.5 Waves or Particles?

29. Light of intensity $\mathcal{I}_0$ passes through two sets of apparatus. One contains one slit and the other two slits. The slits have the same width. What is the ratio of the outgoing intensity amplitude for the central peak for the two-slit case compared to the single slit?
30. Design a double-slit electron-scattering experiment using 1-keV electrons that will provide the first maximum at an angle of 1°. What will be the slit separation d?
31. You want to design an experiment similar to the one done by Jönsson that does not require magnification of the interference pattern in order to be seen. Let the two slits be separated by 2000 nm. Assume that you can discriminate visually between maxima that are as little as 0.3 mm apart. You have at your disposal a lab that allows the screen to be placed 80 cm away from the slits. What energy electrons will you require? Do you think such low-energy electrons will represent a problem? Explain.

5.6 Relationship Between Probability and Wave Function

32. The wave function of a particle in a one-dimensional box of length L is $\Psi(x) = A \sin(\pi x/L)$. If we know the particle must be somewhere in the box, what must be the value of A?

33. A particle in a one-dimensional box of length L has a kinetic energy much greater than its rest energy. What is the ratio of the energy levels E_n: E_2/E_1, E_3/E_1, E_4/E_1? How do you explain this result?

34. Write down the normalized wave functions for the first three energy levels in Example 5.6. Assume there are equal probabilities of being in each state.

5.7 Uncertainty Principle

35. A neutron is confined in a deuterium nucleus (deuteron) of diameter $\approx 2 \times 10^{-15}$ m. Use the energy-level calculation of a one-dimensional box to calculate the neutron's minimum kinetic energy. What is the neutron's minimum kinetic energy according to the uncertainty principle?

36. What is the ratio uncertainty of the velocities ($\Delta v/v$) of (a) an electron and (b) a proton confined to a one-dimensional box of length 2 nm?

37. Show that the uncertainty principle can be expressed in the form $\Delta L \, \Delta\theta \geq \hbar/2$, where θ is the angle and L the angular momentum. For what uncertainty in L will the angular position of a particle be completely undetermined?

38. Some physics theories indicate that the lifetime of the proton is about 10^{36} years. What would such a prediction say about the energy of the proton?

39. What is the bandwidth $\Delta\omega$ of an amplifier for radar if it amplifies a pulse of width 2 μs?

40. Find the minimum uncertainty in the speed of a bacterium having mass 3×10^{-15} kg if we know the position of the bacterium to within 1 micron, that is, to about its own size.

41. An atom in an excited state of 4.7 eV emits a photon and ends up in the ground state. The lifetime of the excited state is 10^{-13} s. (a) What is the energy uncertainty of the emitted photon? (b) What is the spectral line width (in wavelength) of the photon?

42. An electron microscope is designed to resolve objects as small as 0.14 nm. What energy electrons must be used in this instrument?

43. Rayleigh's criterion is used to determine when two objects are barely resolved by a lens of diameter d. The angular separation must be greater than θ_R where

$$\theta_R = 1.22 \frac{\lambda}{d}$$

In order to resolve two objects 4000 nm apart at a distance of 20 cm with a lens of diameter 5 cm, what energy (a) photons or (b) electrons should be used? Is this consistent with the uncertainty principle?

44. Calculate the de Broglie wavelength of a 5.5-MeV α particle emitted from an ^{241}Am nucleus. Could this particle exist inside the ^{241}Am nucleus (diameter $\approx 1.6 \times 10^{-14}$ m)?

45. Show that the minimum energy of a simple harmonic oscillator is $\hbar\omega/2$. What is the minimum energy in joules for a mass of 2 g oscillating on a spring with a spring constant 8 N/m?

General Problems

46. Consider a wave packet having the product of $\Delta p \, \Delta x = \hbar$ at a time $t = 0$. What will be the width of such a wave packet after the time $m(\Delta x)^2/\hbar$?

47. Analyze the Gaussian wave packet carefully and show that $\Delta k \, \Delta x = 1/2$. You must justify the assumptions you make concerning uncertainties in k and x. Take the Gaussian form given in Equation (5.26). (*Hint:* the linear "spread" of the wave packet Δx is given by one standard deviation, at which point the probability amplitude ($|\Psi|^2$) has fallen to one half its peak value.)

48. Most of the particles known to physicists are unstable. For example the lifetime of the neutral pion, π^0, is about 10^{-16} s. Its mass is 135 MeV/c^2. What is the energy width of the π^0 in its ground state?

49. The range of the nuclear strong force is believed to be about 1.2×10^{-15} m. The particle that "mediates" the strong force (similar to the photon mediating the electromagnetic force) is the pion. Assume that the pion moves at the speed of light in the nucleus, and calculate the time Δt it takes to travel between nucleons. Assume that the distance between nucleons is also about 1.2×10^{-15} m. Use this time Δt to calculate the energy ΔE for which energy conservation is violated during the time Δt. This ΔE has been used to estimate the mass of the pion. What value do you determine for the mass? Compare this value with the measured value of 135 MeV/c^2 for the neutral pion.

50. The planes of atoms in a cubic crystal lie parallel to the surface, 0.8 nm apart. X rays having wavelength 0.5 nm are directed at an angle θ to the surface. (a) For what values of θ will there be a strong reflection? (b) What energy electrons could give the same result?

51. Aliens visiting Earth are fascinated by baseball. They are so advanced that they have learned how to vary $\hbar$ to make sure that a pitcher cannot throw a strike with any confidence. Assume the width of the strike zone is 0.38 m, the speed of the baseball is 35 m/s, the mass of the baseball is 145 g, and the ball travels a distance of 18 m. What value of $\hbar$ is required? (*Hint:* there are two uncertainties here: the width of the strike zone and the transverse momentum of the pitched ball.)

CHAPTER 6

The Quantum Theory

In wave mechanics there are no impenetrable barriers, and, as the British physicist R. H. Fowler put it after my lecture on that subject at the Royal Society of London . . . "Anyone at present in this room has a finite chance of leaving it without opening the door—or, of course, without being thrown out the window."

George Gamow

During the early 1920s physicists strove to correct the deficiencies of Bohr's atomic model. The hydrogen atom was the subject of intensive investigation. The origination of the quantum theory, called **quantum mechanics,** is generally credited to Werner Heisenberg and Erwin Schrödinger, whose answers were clothed in very different mathematical formulations. Heisenberg (with his colleagues Max Born and Pascual Jordan) presented the *matrix formulation* of quantum mechanics in 1925. The mathematical tools necessary to introduce matrix mechanics are not intrinsically difficult, but would require too lengthy an exposition for us to study them here. The other solution, proposed in 1926 by Schrödinger, is called *wave mechanics;* its mathematical framework is similar to the classical wave descriptions we have already studied in elementary physics. Paul Dirac and Schrödinger himself (among others) later showed that the matrix and wave mechanics formulations give identical results and differ only in their mathematical form. We shall study only the formalism of Schrödinger here.

In this chapter we will determine wave functions for some simple potentials and use these wave functions to predict the values of physical observables such as position and energy. We will learn that particles are able to tunnel through potential barriers to exist in places that are not allowed by classical physics. Several applications of tunneling will be discussed.

6.1 The Schrödinger Wave Equation

After the Austrian physicist Erwin Schrödinger (Nobel Prize, 1933) learned of de Broglie's wave theory for particles, it was suggested to him while presenting a seminar in Berlin that particles must therefore obey a wave equation.

Schrödinger then quickly found a suitable wave equation based on the relationship between geometrical optics and wave optics.

In our study of elementary physics, we learned that Newton's laws, especially the second law of motion, govern the motion of particles. We need a similar set of equations to describe the wave motion of particles; that is, we need a **wave equation** that will be dependent on the potential field that the particle experiences. We can then find the wave function Ψ (discussed in the previous chapter) that will allow us to calculate the probable values of the particle's position, energy, momentum, and so on.

We must realize that although our procedure is similar to that followed in classical physics, we will no longer be able to calculate and specify the *exact* position, energy, and momentum simultaneously. Our calculations now must be consistent with the uncertainty principle and the notion of probability. We will discuss this subject again in Section 6.2, but these notions take time and experience to get used to, and we will gain that experience in this chapter.

There are several possible paths by which we could plausibly obtain the **Schrödinger wave equation.** Because none of the methods is actually a derivation, we prefer to present the equation and indicate its usefulness. Its ultimate correctness rests on its ability to explain and describe experimental results. The Schrödinger wave equation in its **time-dependent** form for a particle of energy E moving in a potential V in one dimension is

Time-dependent Schrödinger wave equation

$$i\hbar \frac{\partial \Psi(x, t)}{\partial t} = -\frac{\hbar^2}{2m}\frac{\partial^2 \Psi(x, t)}{\partial x^2} + V\Psi(x, t) \tag{6.1}$$

Erwin Schrödinger (1887–1961) was an Austrian who worked at several European universities before fleeing Nazism in 1938 and accepting a position at the University of Dublin where he remained until his retirement in 1956. His primary work on the wave equation was performed during the period he was in Zurich from 1920–1927. Schrödinger worked in many fields including philosophy, biology, history, literature, and language. *AIP Emilio Segrè Visual Archives.*

where $i = \sqrt{-1}$ is an imaginary number and we have used partial derivatives. Both the potential V and wave function Ψ may be functions of space and time, $V(x, t)$ and $\Psi(x, t)$.

The extension for Equation (6.1) into three dimensions is fairly straightforward.

$$i\hbar \frac{\partial \Psi}{\partial t} = -\frac{\hbar^2}{2m}\left(\frac{\partial^2 \Psi}{\partial x^2} + \frac{\partial^2 \Psi}{\partial y^2} + \frac{\partial^2 \Psi}{\partial z^2}\right) + V\Psi(x, y, z, t) \tag{6.2}$$

We will restrict ourselves to the one-dimensional form until Section 6.5.

Let's compare Equation (6.1) with the classical wave equation given by

$$\frac{\partial^2 \Psi(x, t)}{\partial x^2} = \frac{1}{v^2}\frac{\partial^2 \Psi(x, t)}{\partial t^2} \tag{6.3}$$

In this equation the wave function may be as varied as the amplitude of a water wave, a guitar-string vibration, or even the electric field **E** or magnetic field **B**. Notice that the classical wave equation contains a second-order time derivative, whereas the Schrödinger wave equation contains only a first-order time derivative. This already gives us some idea that we are dealing with a somewhat new phenomenon.

Equations (6.1) and (6.2) are the starting points that we will need for this chapter. We emphasize that the time-dependent Schrödinger wave equation (6.1) has not been derived. There is no derivation because we need *new* physical principles (such as those Newton formulated in his laws). The Schrödinger wave equation is a plausible guess that describes nature. Its worth and acceptability depend on the fact that it adequately describes experimental results. In most of the remainder of this chapter, we shall apply the Schrödinger wave equation to several simple situations in order to illustrate its usefulness.

Example 6.1

The wave equation must be linear for us to use the superposition principle to form wave packets using many waves. Prove that the wave equation (6.1) is linear by showing that it is satisfied for the wave function

$$\Psi(x, t) = a\Psi_1(x, t) + b\Psi_2(x, t)$$

where a and b are constants, and Ψ_1 and Ψ_2 describe two waves each satisfying Equation (6.1).

Solution: We take the derivatives needed for Equation (6.1) and insert them in a straightforward manner.

$$\frac{\partial \Psi}{\partial t} = a\frac{\partial \Psi_1}{\partial t} + b\frac{\partial \Psi_2}{\partial t}$$

$$\frac{\partial \Psi}{\partial x} = a\frac{\partial \Psi_1}{\partial x} + b\frac{\partial \Psi_2}{\partial x}$$

$$\frac{\partial^2 \Psi}{\partial x^2} = a\frac{\partial^2 \Psi_1}{\partial x^2} + b\frac{\partial^2 \Psi_2}{\partial x^2}$$

We insert these derivatives into Equation (6.1) to yield

$$i\hbar\left(a\frac{\partial \Psi_1}{\partial t} + b\frac{\partial \Psi_2}{\partial t}\right) = -\frac{\hbar^2}{2m}\left(a\frac{\partial^2 \Psi_1}{\partial x^2} + b\frac{\partial^2 \Psi_2}{\partial x^2}\right) + V(a\Psi_1 + b\Psi_2)$$

Rearrangement of this equation gives

$$a\left[i\hbar\frac{\partial \Psi_1}{\partial t} + \frac{\hbar^2}{2m}\frac{\partial^2 \Psi_1}{\partial x^2} - V\Psi_1\right] = -b\left[i\hbar\frac{\partial \Psi_2}{\partial t} + \frac{\hbar^2}{2m}\frac{\partial^2 \Psi_2}{\partial x^2} - V\Psi_2\right]$$

Because Ψ_1 and Ψ_2 each satisfy Equation (6.1), the quantities in brackets are identically zero, and Ψ is therefore also a solution. The Schrödinger wave equation cannot include any nonlinear terms in the wave functions.

In Section 5.4 we discussed wave motion and the formation of wave packets from waves. We discussed a wave of wave number k and angular frequency ω moving in the $+x$ direction.

$$\Psi(x, t) = A\sin(kx - \omega t + \phi) \tag{5.18}$$

Equation (5.18) is not the most general form of a wave function, which may include both sines and cosines. Our wave function is also not restricted to being real. Only the physically measurable quantities must be real, and Equation (6.1) already has an imaginary number. A more general form of a wave function is

$$\Psi(x, t) = Ae^{i(kx-\omega t)} = A[\cos(kx - \omega t) + i\sin(kx - \omega t)] \tag{6.4}$$

which also describes a wave moving in the $+x$ direction. In general the amplitude A may also be complex.

Example 6.2

Show that $Ae^{i(kx-\omega t)}$ satisfies the time-dependent Schrödinger wave equation.

Solution: We first take the appropriate derivatives needed for Equation (6.1).

$$\frac{\partial \Psi}{\partial t} = -i\omega Ae^{i(kx-\omega t)} = -i\omega\Psi$$

$$\frac{\partial \Psi}{\partial x} = ik\Psi$$

$$\frac{\partial^2 \Psi}{\partial x^2} = i^2k^2\Psi = -k^2\Psi$$

Inserting these results into Equation (6.1) yields

$$i\hbar(-i\omega\Psi) = -\frac{\hbar^2}{2m}(-k^2\Psi) + V\Psi$$

$$\left(\hbar\omega - \frac{\hbar^2k^2}{2m} - V\right)\Psi = 0$$

If we use $E = h\nu = \hbar\omega$ and $p = \hbar k$, we obtain

$$\left(E - \frac{p^2}{2m} - V\right)\Psi = 0$$

which is zero in our nonrelativistic formulation. Thus $e^{i(kx-\omega t)}$ appears to be an acceptable solution at this point.

We showed in Example 6.2 that $e^{i(kx-\omega t)}$ represents an acceptable solution to the Schrödinger wave equation. It is not true that all functions of $\sin(kx - \omega t)$ and $\cos(kx - \omega t)$ are solutions. We show this in the following example.

Example 6.3

Determine whether $\Psi(x, t) = A\sin(kx - \omega t)$ is an acceptable solution to the time-dependent Schrödinger wave equation.

Solution: We take the necessary derivatives needed for Equation (6.1).

$$\frac{\partial \Psi}{\partial t} = -\omega A\cos(kx - \omega t)$$

$$\frac{\partial \Psi}{\partial x} = kA\cos(kx - \omega t)$$

$$\frac{\partial^2 \Psi}{\partial x^2} = -k^2 A\sin(kx - \omega t) = -k^2\Psi$$

After we insert these relations into Equation (6.1), we have

$$-i\hbar\omega\cos(kx - \omega t) = \left(\frac{\hbar^2 k^2}{2m} + V\right)\Psi$$

$$= \left(\frac{\hbar^2 k^2}{2m} + V\right)A\sin(kx - \omega t) \qquad \textbf{(not true)} \qquad (6.5)$$

This equation is generally not satisfied for all x and t, and $A\sin(kx - \omega t)$ is, therefore, not an acceptable wave function. This function is, however, a solution to the classical wave equation (Equation [6.3]).

Normalization and Probability

We begin by reviewing the probability interpretation of the wave function that we discussed in Section 5.6. The probability $P(x)\,dx$ of a particle being between x and $x + dx$ was given in Equation (5.33).

$$P(x)\,dx = \Psi^*(x, t)\Psi(x, t)\,dx \qquad (6.6)$$

The probability of the particle being between x_1 and x_2 is given by

Probability

$$P = \int_{x_1}^{x_2} \Psi^*\Psi\, dx \qquad (6.7)$$

The wave function must also be normalized so that the probability of the particle being somewhere on the x axis is one.

Normalization

$$\int_{-\infty}^{\infty} \Psi^*(x, t)\Psi(x, t)\,dx = 1 \qquad (6.8)$$

Example 6.4

Consider a wave packet formed by using the wave function $Ae^{-\alpha|x|}$, where A is a constant to be determined by normalization. Normalize this wave function and find the probabilities of the particle being between 0 and $1/\alpha$, and between $1/\alpha$ and $2/\alpha$.

Solution: This wave function is sketched in Figure 6.1. We use Equation (6.8) to normalize Ψ.

$$\int_{-\infty}^{\infty} A^2 e^{-2\alpha|x|}\, dx = 1$$

Because the wave function is symmetric about $x = 0$, we can integrate from 0 to ∞, multiply by 2, and drop the absolute value signs on $|x|$.

$$2\int_0^{\infty} A^2 e^{-2\alpha x}\, dx = 1 = \frac{2A^2}{-2\alpha}\, e^{-2\alpha x}\Big|_0^{\infty}$$

$$1 = \frac{-A^2}{\alpha}(0 - 1) = \frac{A^2}{\alpha}$$

The coefficient $A = \sqrt{\alpha}$, and the wave function Ψ is

$$\Psi = \sqrt{\alpha}e^{-\alpha|x|}$$

We use Equation (6.7) to find the probability of the particle being between 0 and $1/\alpha$, where we again drop the absolute signs on $|x|$ because x is positive.

$$P = \int_0^{1/\alpha} \alpha e^{-2\alpha x}\, dx$$

The integration is similar to the previous one.

$$P = \frac{\alpha}{-2\alpha} e^{-2\alpha x} \bigg|_0^{1/\alpha} = -\frac{1}{2}(e^{-2} - 1) = 0.432$$

The probability of the particle being between $1/\alpha$ and $2/\alpha$ is

$$P = \int_{1/\alpha}^{2/\alpha} \alpha e^{-2\alpha x}\, dx$$

$$P = \frac{\alpha}{-2\alpha} e^{-2\alpha x} \bigg|_{1/\alpha}^{2/\alpha} = -\frac{1}{2}(e^{-4} - e^{-2}) = 0.059$$

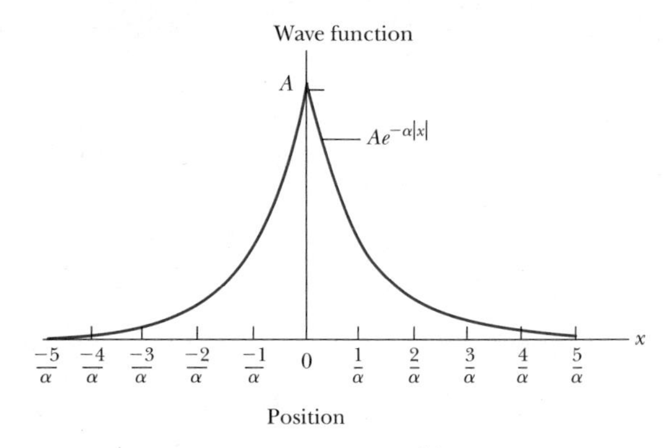

FIGURE 6.1 The wave function $Ae^{-\alpha|x|}$ is plotted as a function of x. Note that the wave function is symmetric about $x = 0$.

The wave function $e^{i(kx-\omega t)}$ represents a particle under zero net force (constant V) moving along the x axis. There is a problem with this wave function, because if we try to normalize it, we obtain an infinite result for the integral. This occurs because there is a finite probability for the particle to be anywhere along the x axis. Over the entire x axis, these finite probabilities add up, when integrated, to infinity. The only other possibility is a zero probability, but that is not an interesting physical result. Because this wave function has precise k and ω values, it represents a particle having a definite energy and momentum. According to the uncertainty principle, because $\Delta E = 0$ and $\Delta p = 0$, we must have $\Delta t = \infty$ and $\Delta x = \infty$. We cannot know where the particle is at any time. We still can use such wave functions if we restrict the particle to certain positions in space, such as in a box or in an atom. We can also form wave packets from such functions in order to localize the particle.

Properties of Valid Wave Functions

Besides the Schrödinger wave equation, there are certain properties (sometimes called **boundary conditions**) that an acceptable wave function Ψ must also satisfy. These are

1. Ψ must be finite everywhere in order to avoid infinite probabilities.
2. Ψ must be single valued in order to avoid multiple values of the probability.
3. Ψ and $\partial\Psi/\partial x$ must be continuous for finite potentials. This is required because the second-order derivative term in the wave equation must be single valued. (There are exceptions to this rule when V is infinite.)
4. In order to normalize the wave functions, Ψ must approach zero as x approaches $\pm\infty$.

Solutions for Ψ that do not satisfy these properties do not generally correspond to physically realizable circumstances.

Time-Independent Schrödinger Wave Equation

In many cases (and in most of the cases discussed here), the potential will not depend explicitly on time. The dependence on time and position can then be separated in the Schrödinger wave equation. Let

$$\Psi(x, t) = \psi(x) f(t) \tag{6.9}$$

We insert this $\Psi(x, t)$ into Equation (6.1) and obtain

$$i\hbar\psi(x)\frac{\partial f(t)}{\partial t} = -\frac{\hbar^2 f(t)}{2m}\frac{\partial^2\psi(x)}{\partial x^2} + V(x)\psi(x)f(t)$$

We divide by $\psi(x)f(t)$ to yield

$$i\hbar\frac{1}{f(t)}\frac{df(t)}{dt} = -\frac{\hbar^2}{2m}\frac{1}{\psi(x)}\frac{d^2\psi(x)}{dx^2} + V(x) \tag{6.10}$$

The left side of Equation (6.10) depends only on time, the right side only on spatial coordinates. We have changed the partial derivatives to total derivatives, because each side depends only on one variable. It follows that each side must be equal to a constant (which we label B) because one variable may change independently of the other. We integrate the left side of Equation (6.10) in an effort to determine the value of B.

$$i\hbar\frac{1}{f}\frac{df}{dt} = B$$

$$i\hbar\int\frac{df}{f} = \int B\,dt$$

We integrate both sides and find

$$i\hbar \ln f = Bt$$

$$\ln f = \frac{Bt}{i\hbar}$$

From this equation we determine f to be

$$f(t) = e^{Bt/i\hbar} = e^{-iBt/\hbar} \tag{6.11}$$

If we compare this function for $f(t)$ to the free-particle wave function that has the time dependence $e^{-i\omega t}$, we see that $B = \hbar\omega = E$. This is a general result.

We now have, from Equation (6.10),

$$i\hbar\frac{1}{f(t)}\frac{df(t)}{dt} = E \tag{6.12}$$

Time-independent Schrödinger wave equation

$$-\frac{\hbar^2}{2m}\frac{d^2\psi(x)}{dx^2} + V(x)\psi(x) = E\psi(x) \tag{6.13}$$

Equation (6.13) is known as the **time-independent Schrödinger wave equation.**

Equation (6.11) can be rewritten as

$$f(t) = e^{-i\omega t} \tag{6.14}$$

and the wave function $\Psi(x, t)$ becomes

$$\Psi(x, t) = \psi(x)e^{-i\omega t} \tag{6.15}$$

We will restrict our attention for the present to time-independent potentials in one space dimension. Many important and useful results can be obtained from this nonrelativistic and one-dimensional form of quantum mechanics, because usually only the spatial part of the wave function $\psi(x)$ is needed. Therefore, we need only use Equation (6.13), the time-independent form of the Schrödinger wave equation.

Let's examine the probability density $\Psi^*\Psi$ discussed in Section 5.6. For the case of Equation (6.15), where the potential does not depend on time, we have

$$\Psi^*\Psi = \psi^2(x)\,(e^{i\omega t}e^{-i\omega t})$$

$$\Psi^*\Psi = \psi^2(x) \tag{6.16}$$

The probability distributions are constant in time. We have seen in introductory physics certain phenomena called *standing waves* (for example, oscillations of strings fixed at both ends). Such standing waves can be formed from traveling waves moving in opposite directions. In quantum mechanics, we say the system is in a **stationary state.**

Stationary states

Comparison of Classical and Quantum Mechanics. It is worthwhile to look briefly at the similarities and differences between classical and quantum mechanics. Newton's second law ($\mathbf{F} = d\mathbf{p}/dt$) and Schrödinger's wave equation are both differential equations. They are both postulated to explain certain observed behavior, and experiments show they are successful. Actually it is possible to derive Newton's second law from the Schrödinger wave equation, so there is no doubt which is more fundamental. Newton's laws may *seem* more fundamental—because they describe the precise values of the system's parameters, whereas the wave equation only produces wave functions which give probabilities—but by now we know from the uncertainty principle that it is not possible to simultaneously know precise values of both position and momentum and of both energy and time. Classical mechanics only appears to be more precise because it deals with macroscopic values. The underlying uncertainties in macroscopic measurements are just too small to be significant.

An interesting parallel between classical mechanics and wave mechanics can be made by considering ray optics and wave optics. For many years in the 1700s, scientists argued which of the optics formulations was the more fundamental; Newton favored ray optics. Finally, it was shown that wave optics was the more fundamental. Ray optics is a good approximation as long as the wavelength of the radiation is much smaller than the dimensions of the apertures and obstacles it passes. Rays of light are characteristic of particlelike behavior: a narrow beam of light is formed of corpuscles. However, in order to describe interference phenomena, wave optics is required. For macroscopic objects, the de Broglie wavelength is so small that wave behavior is not apparent. However, at the atomic level, wave descriptions and quantum mechanics supplant classical mechanics. As far as we know now, there is only one correct theory: that of quantum mechanics. Classical mechanics is a good macroscopic approximation and is correct in the limit of large quantum numbers.

Ray and wave optics

6.2 Expectation Values

In order to be useful, the wave equation formalism must be able to determine values of the measurable quantities like position, momentum, and energy. We shall discuss in this section how the wave function is able to provide this information. We will do this here in only one dimension, but the extension to three dimensions is straightforward. We will also evaluate the values of the physical quantities for a given time t, because in general the whole system, including the values of the physical quantities, evolves with time.

Consider a measurement of the position x of a particular system (for example, the position of a particle in a box). If we make three measurements of the position, we are likely to obtain three different results. Nevertheless, if our method of measurement is inherently *accurate,* then there is some physical significance to the *average* of our measured values of x. Moreover, the *precision* of our result improves as more measurements are made. In quantum theory we use wave functions to calculate the expected result of the average of many measurements of a given quantity. We call this result the **expectation value;** the expectation value of x is denoted by $\langle x \rangle$. Any measurable quantity for which we can calculate the expectation value is called a **physical observable.** The expectation values of physical observables (for example, position, linear momentum, angular momentum, and energy) must be real, because the experimental results of measurements are real.

Expectation value

Physical observables

Let's first review the method of determining average values. Consider a particle that is constrained to move along the x axis. If we make many measurements of the particle along the x axis, we may find the particle N_1 times at x_1, N_2 times at x_2, N_i times at x_i, and so forth. The average value of x, denoted by $\bar{x}$ [or $(x)_{\text{av}}$], is then

$$\bar{x} = \frac{N_1 x_1 + N_2 x_2 + N_3 x_3 + N_4 x_4 + \cdots}{N_1 + N_2 + N_3 + N_4 + \cdots} = \frac{\sum_i N_i x_i}{\sum_i N_i}$$

We can change from discrete to continuous variables by using the probability $P(x, t)$ of observing the particle at a particular x. The previous equation then becomes

$$\bar{x} = \frac{\int_{-\infty}^{\infty} xP(x)\,dx}{\int_{-\infty}^{\infty} P(x)\,dx} \tag{6.17}$$

In quantum mechanics we must use the probability distribution given in Equation (6.6), $P(x)\,dx = \Psi^*(x, t)\Psi(x, t)\,dx$, to determine the average or expectation value. The procedure for finding the expectation value $\langle x \rangle$ is similar to that followed in Equation (6.17):

$$\langle x \rangle = \frac{\int_{-\infty}^{\infty} x\Psi^*(x, t)\Psi(x, t)\,dx}{\int_{-\infty}^{\infty} \Psi^*(x, t)\Psi(x, t)\,dx} \tag{6.18}$$

If the wave function is normalized, the denominator becomes 1. The expectation value is then given by

$$\langle x \rangle = \int_{-\infty}^{\infty} x\Psi^*(x, t)\Psi(x, t)\,dx \tag{6.19}$$

If the wave function has not been normalized, then Equation (6.18) should be used.

The same general procedure can be used to find the expectation value of any function $g(x)$ for a normalized wave function $\Psi(x, t)$.

$$\langle g(x) \rangle = \int_{-\infty}^{\infty} \Psi^*(x, t)g(x)\Psi(x, t)\,dx \tag{6.20}$$

We emphasize again that the wave function can only provide us with the expectation value of a given function $g(x)$ that can be written as a function of x. It cannot give us the value of each individual measurement. When we say the wave function provides a complete description of the system, we mean that the expectation values of the physical observables can be determined.

Any knowledge we might have of the simultaneous values of the position x and momentum p must be consistent with the uncertainty principle. To find the expectation value of p, we first need to represent p in terms of x and t. As an example, let's consider once more the wave function of the free particle, $\Psi(x, t) = e^{i(kx-\omega t)}$. If we take the derivative of $\Psi(x, t)$ with respect to x, we have

$$\frac{\partial \Psi}{\partial x} = \frac{\partial}{\partial x}[e^{i(kx-\omega t)}] = ike^{i(kx-\omega t)} = ik\Psi$$

But because $k = p/\hbar$, this becomes

$$\frac{\partial \Psi}{\partial x} = i\frac{p}{\hbar}\Psi$$

After rearrangement, this yields

$$p[\Psi(x, t)] = -i\hbar\frac{\partial \Psi(x, t)}{\partial x}$$

An **operator** is a mathematical operation that transforms one function into another. For example, an operator, denoted by $\hat{A}$, transforms the function $f(x)$ by $\hat{A}f(x) = g(x)$. In the previous wave function equation, the quantity $-i\hbar(\partial/\partial x)$ is operating on the function $\Psi(x, t)$ and is called the *momentum operator* $\hat{p}$, where the ^ sign over the letter p indicates an operator.

$$\hat{p} = -i\hbar\frac{\partial}{\partial x} \tag{6.21}$$

Momentum operator

The existence of the momentum operator is not unique. As it happens, *each* of the physical observables has an associated operator that is used to find that observable's expectation value. In order to compute the expectation value of some physical observable A, the operator $\hat{A}$ must be placed between Ψ^* and Ψ so that it *operates* on $\Psi(x, t)$ in the order shown:

$$\langle A \rangle = \int_{-\infty}^{\infty} \Psi^*(x, t)\hat{A}\Psi(x, t)\,dx \tag{6.22}$$

Thus, the expectation value of the momentum becomes

$$\langle p \rangle = -i\hbar\int_{-\infty}^{\infty} \Psi^*(x, t)\frac{\partial \Psi(x, t)}{\partial x}\,dx \tag{6.23}$$

The position x is its own operator. Operators for observables that are functions of both x and p can be constructed from x and $\hat{p}$.

Now let's take the time derivative of the free-particle wave function.

$$\frac{\partial \Psi}{\partial t} = \frac{\partial}{\partial t}[e^{i(kx-\omega t)}] = -i\omega e^{i(kx-\omega t)} = -i\omega\Psi$$

We substitute $\omega = E/\hbar$, and then rearrange to find

$$E[\Psi(x, t)] = i\hbar\frac{\partial \Psi(x, t)}{\partial t} \tag{6.24}$$

We call the quantity operating on $\Psi(x, t)$ the *energy operator.*

Energy operator

$$\hat{E} = i\hbar \frac{\partial}{\partial t} \tag{6.25}$$

It is used to find the expectation value $\langle E \rangle$ of the energy.

$$\langle E \rangle = i\hbar \int_{-\infty}^{\infty} \Psi^*(x, t) \frac{\partial \Psi(x, t)}{\partial t}\, dx \tag{6.26}$$

Although we have found the momentum and energy operators only for the free-particle wave functions, they are general results. We shall have occasion later to use these operators.

Example 6.5

Use the momentum and energy operators with the conservation of energy to determine the Schrödinger wave equation.

Solution: We begin by setting the energy E equal to the sum of the kinetic and potential energies. Because our treatment is entirely nonrelativistic, we can write the energy as

$$E = K + V = \frac{p^2}{2m} + V \tag{6.27}$$

We allow the operators of both sides of this equation to act on the wave function. The left side gives

$$\hat{E}\,\Psi = i\hbar \frac{\partial \Psi}{\partial t} \tag{6.28}$$

The application of the operators on the right side of Equation (6.27) on Ψ gives

$$\left(\frac{1}{2m}(\hat{p})^2 + V\right)\Psi = \frac{1}{2m}\left(-i\hbar \frac{\partial}{\partial x}\right)^2 \Psi + V\Psi$$
$$= -\frac{\hbar^2}{2m}\frac{\partial^2 \Psi}{\partial x^2} + V\Psi$$

Now we set the previous equation equal to Equation (6.28) and obtain

$$i\hbar \frac{\partial \Psi}{\partial t} = -\frac{\hbar^2}{2m}\frac{\partial^2 \Psi}{\partial x^2} + V\Psi \tag{6.29}$$

which is the time-dependent Schrödinger wave equation, Equation (6.1). It should be noted that this example is not a determination of the Schrödinger wave equation, but rather a check of the consistency of the definitions.

6.3 Infinite Square-Well Potential

We have thus far established the time-independent Schrödinger wave equation and have discussed how the wave functions can be used to determine the physical observables. Now we would like to find the wave function for several possible potentials and see what we can learn about the behavior of a system having those potentials. In the process of doing this we will find that some observables, including energy, have quantized values. We begin by exploring the simplest such system—that of a particle trapped in a box with infinitely hard walls that the particle cannot penetrate.

The potential, called an *infinite square well,* is shown in Figure 6.2 and is given by

$$V(x) = \begin{cases} \infty, & x \le 0,\ x \ge L \\ 0, & 0 < x < L \end{cases} \tag{6.30}$$

The particle is constrained to move only between $x = 0$ and $x = L$, where the particle experiences no forces. Although it is a simple potential, we will see that it is useful because so many physical situations can be approximated by it. We will learn also that requiring the wave function to satisfy certain boundary conditions leads to energy quantization. We will use this fact to explore energy levels of simple atomic and nuclear systems.

As we stated previously, most of the situations we encounter will allow us to use the time-independent Schrödinger wave equation. Such is the case here. If we insert $V = \infty$ in Equation (6.13), we see that the only possible solution for the wave function is $\psi(x) = 0$. Therefore, there is zero probability for the particle to be located at $x \leq 0$ or $x \geq L$. Because the kinetic energy of the particle must be finite, the particle can never penetrate into the region of infinite potential. However, when $V = 0$, Equation (6.13) becomes, after rearranging,

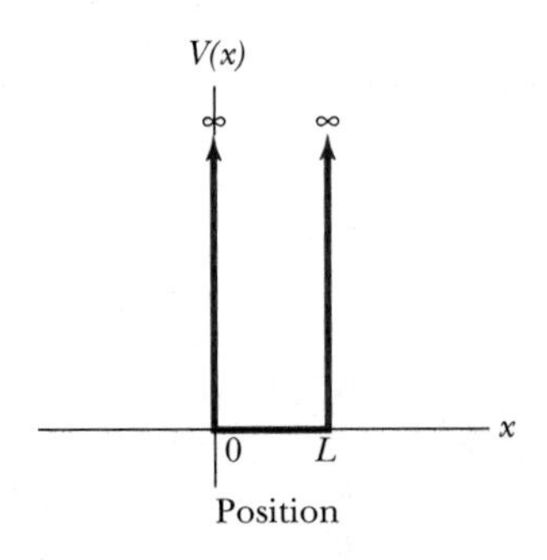

FIGURE 6.2 Infinite square well potential. Potential $V = \infty$ everywhere but $0 \leq x \leq L$, where $V = 0$.

$$\frac{d^2\psi}{dx^2} = -\frac{2mE}{\hbar^2}\psi = -k^2\psi$$

where we have used Equation (6.13) with $V = 0$ and let the wave number $k = \sqrt{2mE/\hbar^2}$. A suitable solution to this equation that satisfies the properties given in Section 6.1 is

$$\psi(x) = A \sin kx + B \cos kx \tag{6.31}$$

The wave function must be continuous, which means that $\psi(x) = 0$ at both $x = 0$ and $x = L$ as already discussed. The proposed solution in Equation (6.31) therefore must have $B = 0$ in order to have $\psi(x = 0) = 0$. In order for $\psi(x = L) = 0$, then $A \sin(kL) = 0$ and because $A = 0$ leads to a trivial solution, we must have

$$kL = n\pi \tag{6.32}$$

where n is a positive integer. The value $n = 0$ leads to $\psi = 0$, a physically uninteresting solution, and negative values of n do not give different physical solutions than the positive values. The wave function is now

$$\psi_n(x) = A \sin\left(\frac{n\pi x}{L}\right) \qquad n = 1,2,3, \ldots \tag{6.33}$$

The property that $d\psi/dx$ be continuous is not satisfied in this case, because of the infinite step value of the potential at $x = 0$ and $x = L$, but we were warned of this particular situation, and it creates no problem. We normalize our wave function over the total distance $-\infty < x < \infty$.

$$\int_{-\infty}^{\infty} \psi_n^*(x)\psi_n(x)\, dx = 1$$

Substitution of the wave function yields

$$A^2 \int_0^L \sin^2\left(\frac{n\pi x}{L}\right) dx = 1$$

This is a straightforward integral (with the help of integral tables, see Appendix 2) and gives $L/2$, so that

$$A^2 \frac{L}{2} = 1$$

and

$$A = \sqrt{\frac{2}{L}}$$

The normalized wave function becomes

$$\psi_n(x) = \sqrt{\frac{2}{L}} \sin\left(\frac{n\pi x}{L}\right) \qquad n = 1,2,3,\ldots \tag{6.34}$$

These wave functions are identical to the ones obtained for a vibrating string with its ends fixed that we studied in elementary physics. The application of the boundary conditions here corresponds to fitting standing waves into the box. It is not a surprise to obtain standing waves in this case, because we are considering time-independent solutions. Because $k_n = n\pi/L$ from Equation (6.32), we have

$$k_n = \frac{n\pi}{L} = \sqrt{\frac{2mE_n}{\hbar^2}}$$

Notice the subscript n on k_n and E_n denoting that they depend on the integer n and have multiple values. The previous equation is solved for E_n to yield

Quantized energy levels

$$E_n = n^2 \frac{\pi^2\hbar^2}{2mL^2} \qquad n = 1,2,3,\ldots \tag{6.35}$$

The possible energies E_n of the particle, called **energy levels,** are quantized. The integer n is called a **quantum number.** Notice that the results for the quantized energy levels in Equation (6.35) are identical to those obtained in Example 5.6, when we treated a particle in a one-dimensional box as a wave. The quantization of the energy occurs in a natural way from the application of the boundary conditions (standing waves) to possible solutions of the wave equation. Each wave function $\psi_n(x)$ has associated with it a unique energy E_n. In Figure 6.3 we show the wave function ψ_n, probability density $|\psi_n|^2$, and energy E_n for the lowest three values of n (1, 2, 3).

The lowest energy level given by $n = 1$, is called the *ground state,* and its energy is given by

$$E_1 = \frac{\pi^2\hbar^2}{2mL^2}$$

Note that the lowest energy cannot be zero because we have ruled out the possibility of $n = 0$ ($\psi_0 = 0$). Classically, the particle can have zero or any positive energy. If we calculate E_n for a macroscopic object in a box (for example, a tennis ball in a tennis court), we will obtain a certain number for E_1. Adjacent energy levels would be so close together that we could not measure their differences. Actual macroscopic objects must have very large values of n.

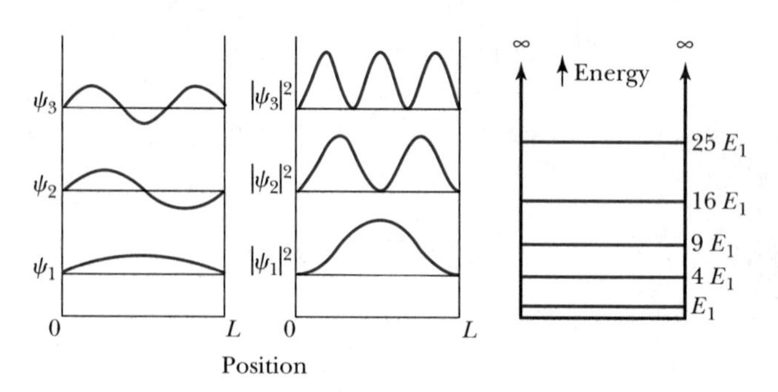

FIGURE 6.3 Wave functions ψ_n, probability densities $|\psi_n|^2$, and energy levels E_n for the lowest quantum numbers for the infinite square-well potential.

Classically, the particle has equal probability of being anywhere inside the box. The classical probability function (see Section 6.2) is $P(x) = 1/L$ (for $0 < x < L$, zero elsewhere) in order for the probability to be 1 for the particle to be in the box. According to Bohr's correspondence principle (see Section 4.4), we should obtain the same probability in the region where the classical and quantum results should agree, that is, for large n. The quantum probability is $(2/L)\sin^2(k_n x)$. For large values of n, there will be many oscillations within the box. The average value of $\sin^2\theta$ over one complete cycle is 1/2. The average value of $\sin^2\theta$ over many oscillations is also 1/2. Therefore, the quantum probability is also $1/L$ in agreement with the classical result.

Example 6.6

Show that the wave function $\Psi_n(x, t)$ for a particle in an infinite square well corresponds to a standing wave in the box.

Solution: We have just found the wave function $\psi_n(x)$ in Equation (6.34). According to Equation (6.14), we can obtain $\Psi_n(x, t)$ by multiplying the wave function $\psi_n(x)$ by $e^{-i\omega_n t}$.

$$\Psi_n(x, t) = \sqrt{\frac{2}{L}}\sin(k_n x)e^{-i\omega_n t}$$

We can write $\sin(k_n x)$ as

$$\sin(k_n x) = \frac{e^{ik_n x} - e^{-ik_n x}}{2i}$$

so that the wave function* becomes

$$\Psi_n(x, t) = \sqrt{\frac{2}{L}}\,\frac{e^{i(k_n x - \omega_n t)} - e^{-i(k_n x + \omega_n t)}}{2i}$$

This is the equation of a standing wave for a vibrating string, for example. It is the superposition of a wave traveling to the right with a wave traveling to the left. They interfere to produce a standing wave of angular frequency ω_n.

*The imaginary number i should be of little concern, because the probability values are determined by a product of $\psi^*\psi$, which gives a real number.

Example 6.7

Determine the expectation values for x, x^2, p, and p^2 of a particle in an infinite square well for the first excited state.

Solution: The first excited state corresponds to $n = 2$, because $n = 1$ corresponds to the lowest energy state or the ground state. The wave function for this case, according to Equation (6.34), is

$$\psi_2(x) = \sqrt{\frac{2}{L}}\sin\left(\frac{2\pi x}{L}\right)$$

The expectation value $\langle x\rangle_{n=2}$ is

$$\langle x\rangle_{n=2} = \frac{2}{L}\int_0^L x\sin^2\left(\frac{2\pi x}{L}\right)dx = L/2$$

We evaluate all these integrations by looking up the integral in Appendix 3. As we expect, the average position of the particle is in the middle of the box ($x = L/2$), even though the actual probability of the particle being there is zero (see $|\psi_2|^2$ in Figure 6.3).

The expectation value $\langle x^2\rangle_{n=2}$ of the square of the position is given by

$$\langle x^2\rangle_{n=2} = \frac{2}{L}\int_0^L x^2\sin^2\left(\frac{2\pi x}{L}\right)dx = 0.32L^2$$

The value of $\sqrt{\langle x^2\rangle_{n=2}}$ is $0.57L$, larger than $\langle x\rangle_{n=2} = 0.5L$. Does this seem reasonable? (*Hint:* Look again at the shape of the wave function in Figure 6.3.)

The expectation value $\langle p\rangle_{n=2}$ is determined by using Equation (6.23).

$$\langle p\rangle_{n=2} = (-i\hbar)\frac{2}{L}\int_0^L \sin\left(\frac{2\pi x}{L}\right)\left(\frac{d}{dx}\sin\left(\frac{2\pi x}{L}\right)\right)dx$$

which reduces to

$$\langle p\rangle_{n=2} = -\frac{4i\hbar}{L^2}\int_0^L \sin\left(\frac{2\pi x}{L}\right)\cos\left(\frac{2\pi x}{L}\right)dx = 0$$

Because the particle is moving left as often as right in the box, the average momentum is zero.

The expectation value $\langle p^2 \rangle_{n=2}$ is given by

$$\langle p^2 \rangle_{n=2} = \frac{2}{L}\int_0^L \sin\left(\frac{2\pi x}{L}\right)\left(-i\hbar\frac{d}{dx}\right)\left(-i\hbar\frac{d}{dx}\right)\sin\left(\frac{2\pi x}{L}\right)dx$$

$$= (-i\hbar)^2\frac{2}{L}\int_0^L \sin\left(\frac{2\pi x}{L}\right)\left(\frac{2\pi}{L}\frac{d}{dx}\right)\cos\left(\frac{2\pi x}{L}\right)dx$$

$$= -(-\hbar^2)\frac{8\pi^2}{L^3}\int_0^L \sin\left(\frac{2\pi x}{L}\right)\sin\left(\frac{2\pi x}{L}\right)dx$$

$$= \frac{4\pi^2\hbar^2}{L^2}$$

This value can be compared with E_2 (Equation (6.35)),

$$E_2 = \frac{4\pi^2\hbar^2}{2mL^2} = \frac{\langle p^2 \rangle_{n=2}}{2m}$$

which is correct, because nonrelativistically we have $E = p^2/2m + V$ and $V = 0$.

Example 6.8

A typical diameter of a nucleus is about 10^{-14} m. Use the infinite square well potential to calculate the transition energy from the first excited state to the ground state for a proton confined to the nucleus.

Solution: The energy of the ground state, from Equation (6.35), is

$$E_1 = \frac{\pi^2\hbar^2c^2}{2mc^2L^2} = \frac{1}{mc^2}\frac{\pi^2(197.3\ \text{eV}\cdot\text{nm})^2}{2(10^{-5}\ \text{nm})^2}$$

$$= \frac{1}{mc^2}(1.92 \times 10^{15}\ \text{eV}^2)$$

The mass of the proton is 938.3 MeV/c^2 which gives

$$E_1 = \frac{1.92 \times 10^{15}\ \text{eV}^2}{938.3 \times 10^6\ \text{eV}} = 2.0\ \text{MeV}$$

The first excited state energy is found from $E_2 = 4E_1 = 8$ MeV, and the transition energy is $\Delta E = E_2 - E_1 = 6$ MeV. This value is a reasonable one for protons in the nucleus. If we had done a similar calculation for an electron in the nucleus, we would find energies on the order of 10^4 MeV, much larger than the rest energy of the electron. A correct relativistic treatment is necessary, and it would give electron energies of significantly less than 10^4 MeV but still much larger than those electrons actually observed being emitted from the nucleus in β decay. Such reasoning indicates that electrons do not exist inside the nucleus.

6.4 Finite Square-Well Potential

We gained some experience in the last section in dealing with the time-independent Schrödinger wave equation. Now we want to look at a more realistic potential—one that is not infinite. The finite square-well potential is similar to the infinite one, but we let the potential be V_0 rather than infinite in the region $x \leq 0$ and $x \geq L$.

$$V(x) = \begin{cases} V_0 & x \leq 0 \quad \text{region I} \\ 0 & 0 < x < L \quad \text{region II} \\ V_0 & x \geq L \quad \text{region III} \end{cases} \tag{6.36}$$

The three regions of the potential are shown in Figure 6.4. We will consider a particle of energy $E < V_0$ that classically is bound inside the well. We will find that quantum mechanics *allows the particle to be outside the well.* We set the potential $V = V_0$ in the time-independent Schrödinger Equation (6.13) for regions I and III outside the square well. This gives

$$-\frac{\hbar^2}{2m}\frac{1}{\psi}\frac{d^2\psi}{dx^2} = E - V_0 \qquad \text{regions I, III} \tag{6.37}$$

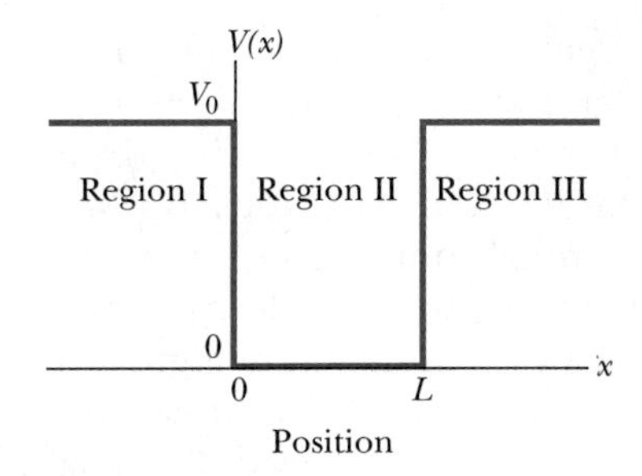

FIGURE 6.4 A finite square-well potential has the value V_0 everywhere except $0 < x < L$, where $V = 0$. The three regions I, II, and III are indicated.

We rewrite this using $\alpha^2 = 2m(V_0 - E)/\hbar^2$, a positive constant.

$$\frac{d^2\psi}{dx^2} = \alpha^2\psi$$

The solution to this differential equation has exponentials of the form $e^{\alpha x}$ and $e^{-\alpha x}$. In the region $x > L$, we can reject the positive exponential term, because it would become infinite as $x \to \infty$. Similarly, the negative exponential can be rejected for $x < 0$. The wave functions become

$$\psi_{\text{I}}(x) = Ae^{\alpha x} \qquad \text{region I, } x < 0 \tag{6.38}$$

$$\psi_{\text{III}}(x) = Be^{-\alpha x} \qquad \text{region III, } x > L \tag{6.39}$$

Inside the square well, where the potential V is zero, the wave equation becomes

$$\frac{d^2\psi}{dx^2} = -k^2\psi$$

where $k = \sqrt{(2mE)/\hbar^2}$. Instead of a sinusoidal solution, we can write it as

$$\psi_{\text{II}} = Ce^{ikx} + De^{-ikx} \qquad \text{region II, } 0 < x < L \tag{6.40}$$

We now want to satisfy the properties of Section 6.1. We have already made sure that all but properties 2 and 3 have been satisfied. The wave functions are finite throughout the x region, even at infinity. In order for the wave functions to be single valued, we must have $\psi_{\text{I}} = \psi_{\text{II}}$ at $x = 0$ and $\psi_{\text{II}} = \psi_{\text{III}}$ at $x = L$. Both ψ and $\partial\psi/\partial x$ must be continuous at $x = 0$ and $x = L$. We will not perform these tedious procedures here, but the results for the wave functions are presented graphically in Figure 6.5.

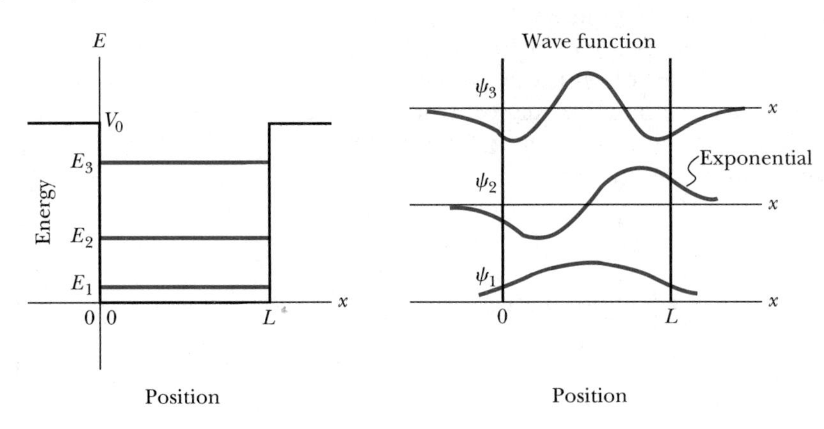

FIGURE 6.5 The energy levels E_n and wave functions ψ_n for the lowest quantum numbers for the finite square-well potential. Notice that ψ extends past $x < 0$ and $x > L$, where classically the particle is forbidden. *From R. Eisberg and R. Resnick, Quantum Physics of Atoms, Molecules, Solids, Nuclei, and Particles, 2nd ed. New York: Wiley, 1985.*

The application of the boundary conditions leads to quantized energy values E_n and to particular wave functions $\psi_n(x)$. One remarkable result is that the particle has a finite probability of being outside the square well, as is indicated by Figure 6.5. Notice that the wave functions join smoothly at the edges of the well and decrease exponentially outside the well.

What other differences can we easily discern between the infinite and finite square well? For example, by examination of Figures 6.5 and 6.3, we can see that the de Broglie wavelength is larger for the finite square well because the waves extend past the square well. This in turn leads to a smaller momentum and lower energy levels. The number of energy levels will, of course, be limited because of the potential height V_0 (see Figure 6.5). When $E > V_0$ the particle is unbound, a situation that will be discussed in Section 6.7.

The occurrence of the particle outside the square well is clearly prohibited classically, but it occurs naturally in quantum mechanics. Note that because of the exponential decrease of the wave functions ψ_{I} and ψ_{III}, the probability of the particle penetrating a distance greater than $\delta x \approx 1/\alpha$ begins to decrease markedly.

$$\delta x \approx \frac{1}{\alpha} = \frac{\hbar}{\sqrt{2m(V_0 - E)}} \tag{6.41}$$

However, we will later find values of δx as large as $10/\alpha$ and $20/\alpha$ for electrons tunneling through semiconductors (Example 6.11) and for nuclear alpha decay (Example 6.13), respectively. The fraction of particles that successfully tunnel through in these cases is exceedingly small, but the results are quite important.

It should not be surprising to find that the penetration distance that violates classical physics is proportional to Planck's constant $\hbar$. This result is also consistent with the uncertainty principle, because, in order for the particle to be in the barrier region, the uncertainty ΔE of the energy must be very large. According to the uncertainty principle ($\Delta E\,\Delta t \geq \hbar/2$), this can only occur for a very short period of time Δt.

6.5 Three-Dimensional Infinite-Potential Well

In order to use quantum theory to solve the atomic physics problems that we shall face in Chapters 7 and 8, it is necessary to extend the Schrödinger equation to three dimensions. This is easily accomplished with the operator notation already developed in Section 6.2. After obtaining the three-dimensional equation, we shall use it to study the problem of a three-dimensional infinite-potential well.

We anticipate that there will be time-independent solutions, so we shall start with the time-independent Schrödinger wave equation. The wave function ψ must be a function of all three spatial coordinates, that is, $\psi = \psi(x, y, z)$. We could just directly modify Equation (6.13) to three dimensions, but we prefer to use a simple method to arrive at the Schrödinger equation. We begin with the conservation of energy.

$$E = K + V = \frac{p^2}{2m} + V$$

We multiply this equation times the wave function ψ which gives

$$\frac{p^2}{2m}\psi + V\psi = E\psi \tag{6.42}$$

We now use Equation (6.22) to express p^2 as an operator to act on ψ. But because $p^2 = p_x^2 + p_y^2 + p_z^2$, we must apply the momentum operator in all three dimensions.

$$\hat{p}_x\psi = -i\hbar\frac{\partial\psi}{\partial x}$$

$$\hat{p}_y\psi = -i\hbar\frac{\partial\psi}{\partial y}$$

$$\hat{p}_z\psi = -i\hbar\frac{\partial\psi}{\partial z}$$

The application of $\hat{p}^2$ in Equation (6.42) gives

$$-\frac{\hbar^2}{2m}\left(\frac{\partial^2\psi}{\partial x^2} + \frac{\partial^2\psi}{\partial y^2} + \frac{\partial^2\psi}{\partial z^2}\right) + V\psi = E\psi \tag{6.43}$$

Time-independent Schrödinger wave equation in three dimensions

This is the time-independent Schrödinger wave equation in three dimensions.

You may recognize the expression in parentheses as the Laplacian operator in mathematics. It is usually written with the shorthand notation

$$\nabla^2 = \frac{\partial^2}{\partial x^2} + \frac{\partial^2}{\partial y^2} + \frac{\partial^2}{\partial z^2} \tag{6.44}$$

With this notation, we can write

$$-\frac{\hbar^2}{2m}\nabla^2\psi + V\psi = E\psi \tag{6.45}$$

Example 6.9

Consider a free particle inside a box with lengths L_1, L_2, and L_3 along the x, y, and z axes, respectively, as shown in Figure 6.6. The particle is constrained to be inside the box. Find the wave functions and energies. Then find the ground energy and wave function and the first excited state energy for a cube of sides L.

Solution: Inside the box $V = 0$, so the wave equation we must solve is

$$-\frac{\hbar^2}{2m}\nabla^2\psi = E\psi \tag{6.46}$$

We employ some of the same strategies to solve this problem as we used for the one-dimensional case. First, because we are considering the walls of the box to be absolutely closed, they are infinite potential barriers, and the wave function ψ must be zero at the walls and outside. We expect to see standing waves similar to Equation (6.31).

But how should we write the wave function so as to properly include the x, y, and z dependence of the wave function? In this case the mathematics will follow from the physics. The particle is free within the box. Therefore, the x-, y-, and z-dependent parts of the wave function must be independent of each other. It is therefore reasonable to try a wave function of the form

$$\psi(x, y, z) = A\sin(k_1x)\sin(k_2y)\sin(k_3z) \tag{6.47}$$

where A is a normalization constant. The quantities k_i ($i = 1, 2, 3$) are determined by applying the appropriate boundary conditions. For example, the condition that $\psi = 0$ at $x = L_1$ requires that $k_1L_1 = n_1\pi$ or $k_1 = n_1\pi/L_1$. The values for the k_i are

$$k_1 = \frac{n_1\pi}{L_1} \qquad k_2 = \frac{n_2\pi}{L_2} \qquad k_3 = \frac{n_3\pi}{L_3} \tag{6.48}$$

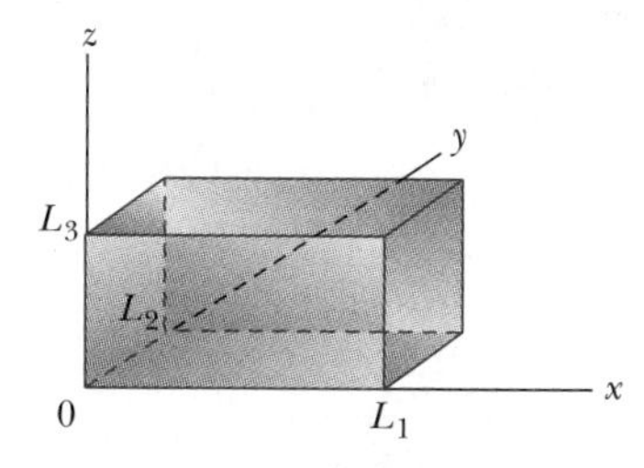

FIGURE 6.6 A three-dimensional box that contains a free particle. The potential is infinite outside the box, so the particle is constrained to be inside the box.

where n_1, n_2, and n_3 are integers. Not surprisingly, we have found that in three dimensions, it is necessary to use *three* quantum numbers to describe the physical state.

To find the energies, we simply substitute the wave function into the Schrödinger equation and solve for E. If we do this in Equation (6.43), we find

$$E = \frac{\pi^2\hbar^2}{2m}\left(\frac{n_1^2}{L_1^2} + \frac{n_2^2}{L_2^2} + \frac{n_3^2}{L_3^2}\right) \tag{6.49}$$

The allowed energy values also depend on the values of the three quantum numbers n_1, n_2, and n_3.

Let us now consider the special case of the *cubical* box, with $L_1 = L_2 = L_3 \equiv L$. The energy values of Equation (6.49) can be expressed

$$E = \frac{\pi^2\hbar^2}{2mL^2}(n_1^2 + n_2^2 + n_3^2) \tag{6.50}$$

For the ground state we have $n_1 = n_2 = n_3 = 1$, so the ground state energy is

$$E_{gs} = \frac{3\pi^2\hbar^2}{2mL^2} \tag{6.51}$$

and the ground state wave function is

$$\psi_{gs} = A \sin\left(\frac{\pi x}{L}\right)\sin\left(\frac{\pi y}{L}\right)\sin\left(\frac{\pi z}{L}\right) \tag{6.52}$$

What is the energy of the first excited state? Higher values of the quantum numbers n_i correspond to higher energies, therefore, it is logical to try something like $n_1 = 2$, $n_2 = 1$, and $n_3 = 1$. But we could just as well assign quantum numbers $n_1 = 1$, $n_2 = 2$, $n_3 = 1$ to the first excited state, or $n_1 = 1$, $n_2 = 1$, $n_3 = 2$. In each of these cases the total energy is

$$E_{1st} = \frac{\pi^2\hbar^2}{2mL^2}(2^2 + 1^2 + 1^2) = \frac{3\pi^2\hbar^2}{mL^2}$$

Degenerate state

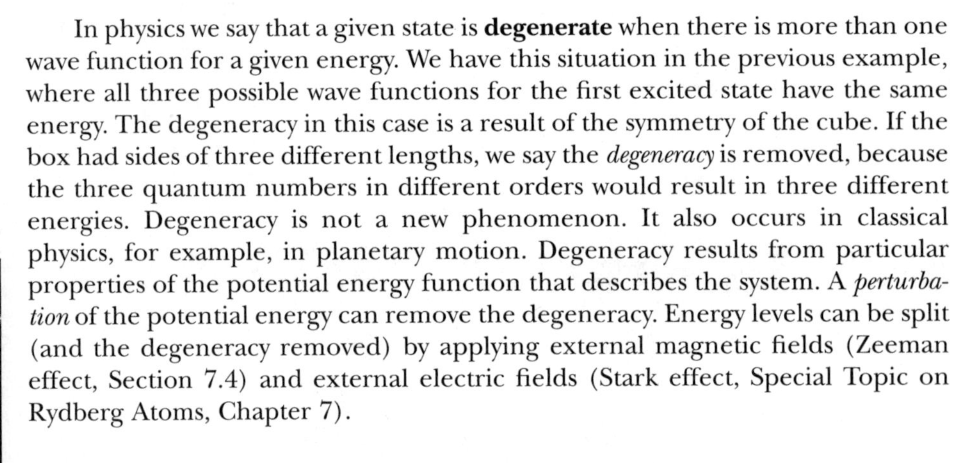

In physics we say that a given state is **degenerate** when there is more than one wave function for a given energy. We have this situation in the previous example, where all three possible wave functions for the first excited state have the same energy. The degeneracy in this case is a result of the symmetry of the cube. If the box had sides of three different lengths, we say the *degeneracy* is removed, because the three quantum numbers in different orders would result in three different energies. Degeneracy is not a new phenomenon. It also occurs in classical physics, for example, in planetary motion. Degeneracy results from particular properties of the potential energy function that describes the system. A *perturbation* of the potential energy can remove the degeneracy. Energy levels can be split (and the degeneracy removed) by applying external magnetic fields (Zeeman effect, Section 7.4) and external electric fields (Stark effect, Special Topic on Rydberg Atoms, Chapter 7).

6.6 Simple Harmonic Oscillator

Because of their wide occurrence in nature, we now want to examine simple harmonic oscillators. We have already studied in introductory physics the case of a mass oscillating in one dimension on the end of a spring. Consider a spring having spring constant κ* that is in equilibrium at $x = x_0$. The restoring force (see Figure 6.7a) along the x direction is $F = -\kappa(x - x_0)$, and the potential energy stored in the spring is $V = \kappa(x - x_0)^2/2$ (see Figure 6.7b). The resulting motion is called *simple harmonic motion* (abbreviated as SHM), and the equations describing it are well known.

Besides springs and pendula (small oscillations), many phenomena in nature can be approximated by SHM, for example, diatomic molecules and atoms in a lattice. Systems can also be approximated by SHM in a general way. As an

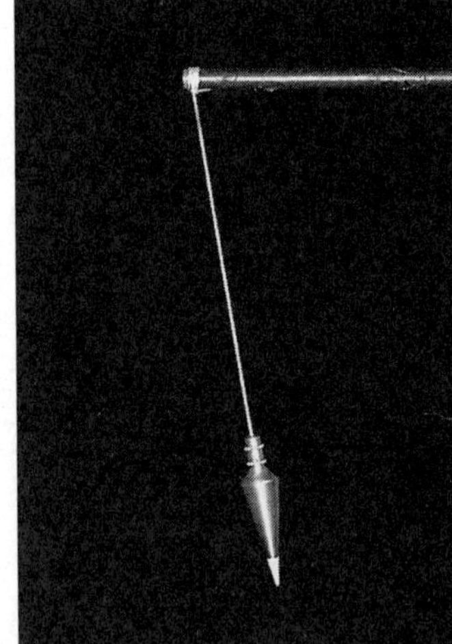

Many things in nature, including this pendulum, respond naturally in simple harmonic motion. Other examples include leaves blowing in the wind and atoms vibrating in molecules. *Leonard Lessin/Peter Arnold, Inc.*

*We let κ be the spring constant in this section rather than the normal k to avoid confusion with the wave number. It is important to note the context in which variables like k and κ are used, because either might be used as wave number or spring constant.

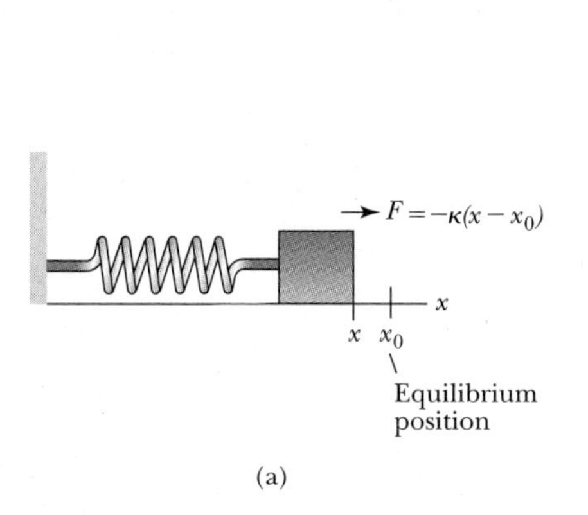

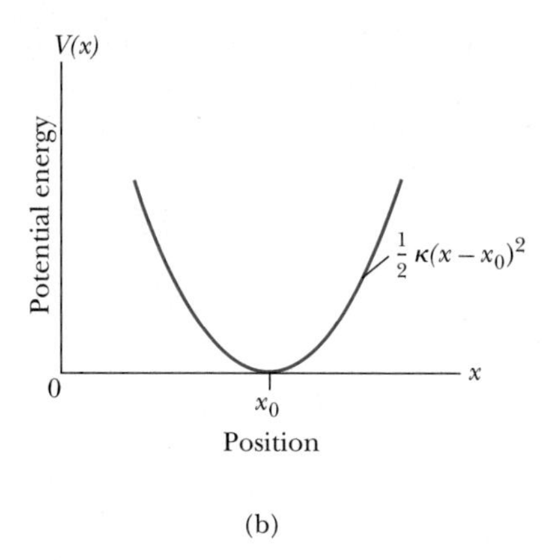

FIGURE 6.7 (a) The restoring force for a spring having a spring constant κ is $F = -\kappa(x - x_0)$. (b) The potential energy has the form $\kappa(x - x_0)^2/2$.

example, consider a lattice in which the force on the atoms depends on the distance x from some equilibrium position x_0. If we expand the potential in a Taylor series in terms of the distance $(x - x_0)$ from equilibrium, we obtain

$$V(x) = V_0 + V_1(x - x_0) + \frac{1}{2}V_2(x - x_0)^2 + \cdots \tag{6.53}$$

where V_0, V_1, and V_2 are constants, and we have kept only the three lowest terms of the series, because $(x - x_0) \approx 0$ for small excursions from the equilibrium position x_0. At $x = x_0$ we have equilibrium (e.g., a minimum of the potential), so $(dV/dx) = 0$ at $x = x_0$. This requires that $V_1 = 0$, and if we redefine the zero of potential energy to require $V_0 = 0$, then the lowest term of the potential $V(x)$ is

$$V(x) = \frac{1}{2}V_2(x - x_0)^2$$

This is the origin of the $V = \kappa x^2/2$ potential energy term that occurs so often. Near the equilibrium position many potentials may be approximated by a parabolic form as displayed in Figure 6.8.

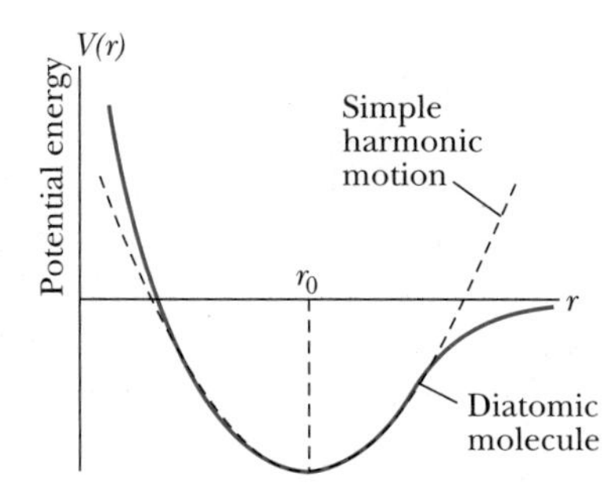

FIGURE 6.8 Many potentials in nature can be approximated near their equilibrium position by the simple harmonic potential (black dashed curve). Such is the case here for the potential energy $V(r)$ of a diatomic molecule near its equilibrium position r_0 (blue curve).

We want to study the quantum description of simple harmonic motion by inserting a potential $\kappa x^2/2$ (we let $x_0 = 0$, see Figure 6.9a) into the time-independent Schrödinger wave equation (6.13).

$$\frac{d^2\psi}{dx^2} = -\frac{2m}{\hbar^2}\left(E - \frac{\kappa x^2}{2}\right)\psi = \left(-\frac{2mE}{\hbar^2} + \frac{m\kappa x^2}{\hbar^2}\right)\psi \tag{6.54}$$

If we let

$$\alpha^2 = \frac{m\kappa}{\hbar^2} \tag{6.55a}$$

and

$$\beta = \frac{2mE}{\hbar^2} \tag{6.55b}$$

then

$$\frac{d^2\psi}{dx^2} = (\alpha^2 x^2 - \beta)\psi \tag{6.56}$$

Before discussing the solution of Equation (6.56), let us first examine what we can learn about the problem qualitatively. Because the particle is confined to the potential well, centered at $x = 0$, it has zero probability of being at $x = \infty$. Therefore, $\psi(x) \longrightarrow 0$ as $x \longrightarrow \infty$.

What is the lowest energy level possible for the harmonic oscillator? Is $E = 0$ possible? If $E = 0$, then $x = 0$ and $V = 0$ in order to allow $E \geq V$. But if E and V are zero, then K.E. $= 0$, and the momentum $p = 0$. Having both $x = 0$ and $p = 0$ (that is, both x and p are known exactly) simultaneously violates the uncertainty principle. Therefore, the minimum energy E cannot be zero. In fact, the energy levels must all be positive, because $E > V \geq 0$. The state having the lowest energy, denoted here by E_0, will have an energy like that shown in Figure 6.9a, and the wave function ψ_0 for that state will most likely be a simple wave fitting inside the region defined by the potential (see Figure 6.9b). Let $E_0 = V_0 = \kappa a^2/2$. The distances $\pm a$ define the classical limits of the particle, but we know from the previous section that the particle has a small probability of being outside the potential well dimensions of $\pm a$. Therefore, the wave function will not be zero at

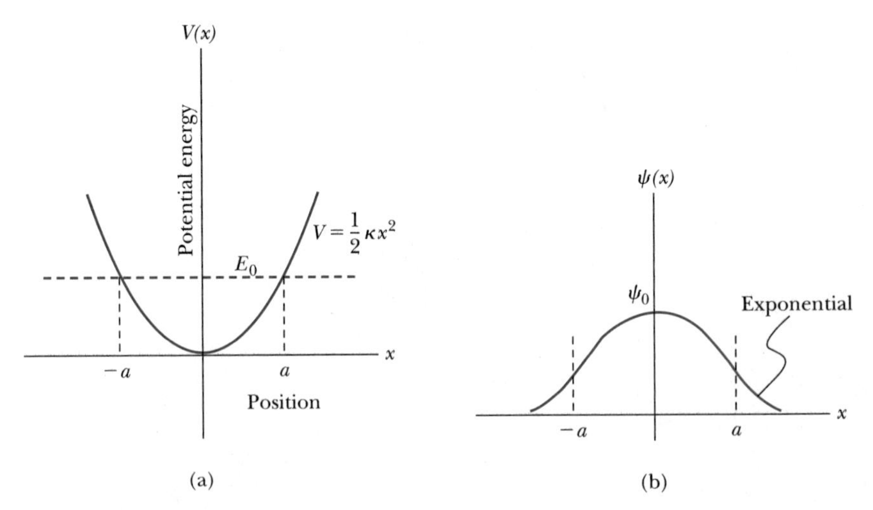

FIGURE 6.9 (a) The potential $V = \kappa x^2/2$ for a simple harmonic oscillator. The classical turning points $\pm a$ are determined for the ground state when the lowest energy E_0 is equal to the potential energy. (b) Notice that the wave function $\psi_0(x)$ for the ground state is symmetric and decays exponentially outside $\pm a$ where $V > E_0$.

$x = \pm a$, but will have a finite value that decreases rapidly to zero on the other side of the barrier. Thus a plausible guess for the lowest-order wave function ψ_0 is like that shown in Figure 6.9b. We find the minimum energy E_0, called the *zero-point energy,* in the next example.

Example 6.10

Estimate the minimum energy of the simple harmonic oscillator allowed by the uncertainty principle.

Solution: In introductory physics we learned that the average kinetic energy is equal to the average potential energy for simple harmonic oscillators over the range of motion (from $-x$ to $+x$), and both the average potential and kinetic energies are equal to one half the total energy.

$$\frac{1}{2}E = \frac{1}{2}\kappa\langle x^2\rangle_{\text{av}} = \frac{1}{2m}\langle p^2\rangle_{\text{av}}$$

The mean value of x is zero, but the mean value of $\langle x^2\rangle_{\text{av}}$ is the mean square deviation $(\Delta x)^2$. Similarly, $\langle p^2\rangle_{\text{av}} = (\Delta p)^2$. From the previous equation, we therefore have the energy $E = \kappa(\Delta x)^2 = (\Delta p)^2/m$ and, as a result, we must have $\Delta x = \Delta p/\sqrt{m\kappa}$. From the uncertainty principle $\Delta p \Delta x \geq \hbar/2$, and the minimum value of $\Delta x = \hbar/2\Delta p$. Now we have for the lowest energy E_0,

$$E_0 = \kappa(\Delta x)^2 = \kappa\left(\frac{\Delta p}{\sqrt{mk}}\right)\left(\frac{\hbar}{2\Delta p}\right)$$

$$E_0 = \frac{\hbar}{2}\sqrt{\frac{\kappa}{m}} = \frac{\hbar\omega}{2}$$

Our estimate for the zero-point energy of the harmonic oscillator is $\hbar\omega/2$. This agrees with the zero-point energy found by more rigorous means.

The zero-point energy is not just a curious oddity. For example, the zero-point energy for ^{4}He is large enough to prevent liquid ^{4}He from freezing at atmospheric pressure, no matter how cold the system, even near 0 K.

The wave function solutions ψ_n for Equation (6.56) are

$$\psi_n = H_n(x)e^{-\alpha x^2/2} \tag{6.57}$$

where $H_n(x)$ are polynomials of order n, where n is an integer ≥ 0. The functions $H_n(x)$ are related by a constant to the *Hermite polynomial functions* tabulated in many quantum mechanics books. The first few values of ψ_n and $|\psi_n|^2$ are shown in Figure 6.10. In contrast to the particle in a box, where the oscillatory wave function is a sinusoidal curve, in this case the oscillatory behavior is due to the polynomial, which dominates at small x, and the exponential tail is provided by the Gaussian function, which dominates at large x.

The energy levels are given by

$$E_n = \left(n + \frac{1}{2}\right)\hbar\sqrt{\kappa/m} = \left(n + \frac{1}{2}\right)\hbar\omega \tag{6.58}$$

where $\omega^2 = \kappa/m$, and ω is the classical angular frequency. From Equation (6.58) we see that the *zero-point energy* E_0 is

$$E_0 = \frac{1}{2}\hbar\omega \tag{6.59}$$

Notice that this result for E_0 is precisely the value found in Example 6.10 by using the uncertainty principle. The uncertainty principle is solely responsible for the minimum energy of the simple harmonic oscillator. In Section 5.7 we mentioned that the minimum value (that is, the equality sign) of the uncertainty principle is found for Gaussian wave packets. We note here that the wave

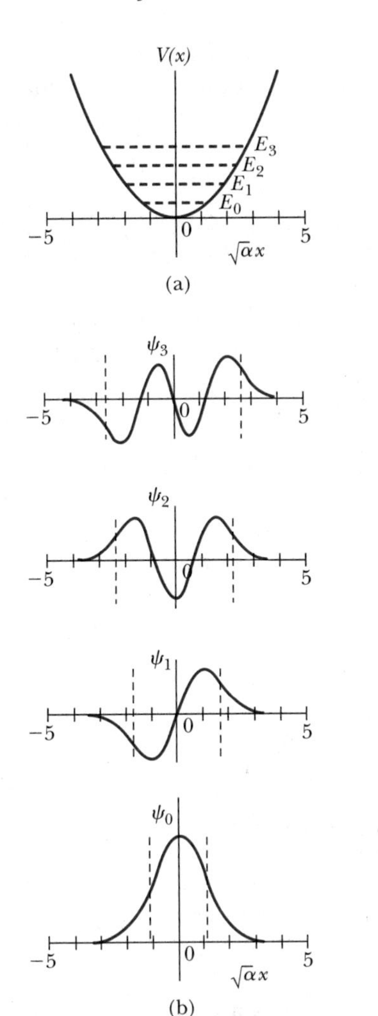

Wave functions

$$\psi_3(x) = \left(\frac{\alpha}{\pi}\right)^{1/4} \frac{1}{\sqrt{3}} (\sqrt{\alpha}x)(2\alpha x^2 - 3)e^{-\alpha x^2/2}$$

$$\psi_2(x) = \left(\frac{\alpha}{\pi}\right)^{1/4} \frac{1}{\sqrt{2}} (2\alpha x^2 - 1)e^{-\alpha x^2/2}$$

$$\psi_1(x) = \left(\frac{\alpha}{\pi}\right)^{1/4} \sqrt{2\alpha}\, x e^{-\alpha x^2/2}$$

$$\psi_0(x) = \left(\frac{\alpha}{\pi}\right)^{1/4} e^{-\alpha x^2/2}$$

$|\psi_3|^2$ $|\psi_2|^2$ $|\psi_1|^2$ $|\psi_0|^2$ −5 0 5 $\sqrt{\alpha}x$ (c)

FIGURE 6.10 Results for simple harmonic oscillator potential. (a) The energy levels for the lowest four energy states are shown with the corresponding wave functions listed. (b) The wave functions for the four lowest energy states are displayed. Notice that even quantum numbers have symmetric $\psi_n(x)$, and the odd quantum numbers have antisymmetric $\psi_n(x)$. (c) The probability densities $|\psi_n|^2$ for the lowest four energy states are displayed.

functions for the simple harmonic oscillators are of just the Gaussian form (see Figure 6.10). The minimum energy E_0 allowed by the uncertainty principle, sometimes called the *Heisenberg limit,* is found for the ground state of the simple harmonic oscillator.

Finally, let us compare the motion as described by classical and quantum theory. Classically, we recall the motion of the mass at the end of a spring. The speed is greatest as it passes through its equilibrium position. The speed is lowest (zero) at the two ends (compressed or extended positions of the spring), as the mass stops and reverses direction. Classically, the probability of finding the mass is greatest at the ends of motion and smallest at the center (that is, proportional to the amount of time the mass spends at each position). The classical probability is shown by the black dashed line in Figure 6.11.

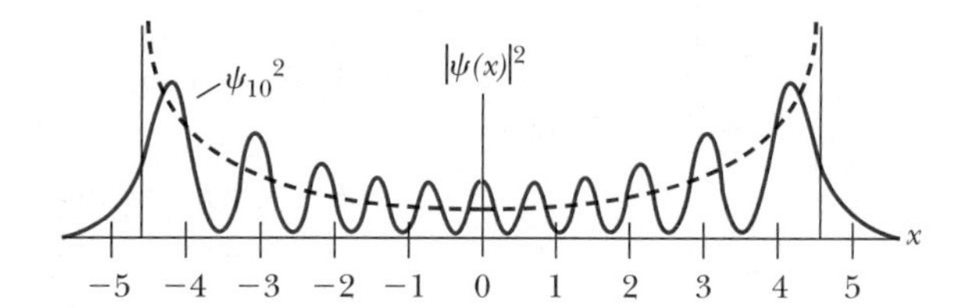

FIGURE 6.11 The probability distribution $|\psi_{10}|^2$ for the $n = 10$ state is compared with the classical probability (dashed line). As n increases, the two probability distributions become more similar.

The quantum theory probability density for the lowest energy state ($\psi_0{}^2$, see Figure 6.10) is completely contrary to the classical one. The largest probability is for the particle to be at the center. We are not surprised to see such a marked difference between classical (see Section 4.4) and quantum predictions. However, from the correspondence principle we would expect the classical and quantum probabilities to be similar as the quantum number n becomes very large. In Figure 6.11 we show $\psi_n{}^2$ for the case of $n = 10$, and we see that the average probabilities become similar. As n continues to increase, the peaks and valleys of the quantum probabilities are hardly observable, and the average value mirrors the classical result.

Example 6.11

Normalize the ground state wave function ψ_0 for the simple harmonic oscillator and find the expectation values $\langle x \rangle$ and $\langle x^2 \rangle$.

Solution: Let's assume that all we know about the wave function ψ_0 is the form given in Equation (6.57). $H_0(x)$ has no dependence on x, so we take it to be a constant A. The ground state wave function is then

$$\psi_0(x) = Ae^{-\alpha x^2/2}$$

We must normalize in order to determine A.

$$\int_{-\infty}^{\infty} \psi_0^*(x)\psi_0(x)\,dx = 1$$

$$A^2 \int_{-\infty}^{\infty} e^{-\alpha x^2} dx = 1$$

$$2A^2 \int_0^{\infty} e^{-\alpha x^2} dx = 1$$

We determine this integral with the help of integral tables (see Appendix 6), with the result:

$$2A^2\left(\frac{1}{2}\sqrt{\frac{\pi}{\alpha}}\right) = 1$$

$$A^2 = \sqrt{\frac{\alpha}{\pi}}$$

$$A = \left(\frac{\alpha}{\pi}\right)^{1/4}$$

This gives for the ground state wave function,

$$\psi_0(x) = \left(\frac{\alpha}{\pi}\right)^{1/4} e^{-\alpha x^2/2} \tag{6.60}$$

This is precisely the wave function given in Figure 6.10 and is of the Gaussian form.

The expectation value of x is

$$\langle x \rangle = \int_{-\infty}^{\infty} \psi_0^*(x)\,x\,\psi_0(x)\,dx$$

$$= \sqrt{\frac{\alpha}{\pi}} \int_{-\infty}^{\infty} xe^{-\alpha x^2} dx$$

The value of $\langle x \rangle$ must be zero, because we are integrating an odd function of x over symmetric limits from $-\infty$ to $+\infty$ (see Appendix 6). Both classical and quantum mechanics predict the average value of x to be zero because of the symmetric nature of the potential, $\kappa x^2/2$.

The expectation value $\langle x^2 \rangle$, however, should be positive.

$$\langle x^2 \rangle = \int_{-\infty}^{\infty} \psi_0^*(x)\,x^2\psi_0(x)\,dx = \sqrt{\frac{\alpha}{\pi}} \int_{-\infty}^{\infty} x^2 e^{-\alpha x^2} dx$$

$$= 2\sqrt{\frac{\alpha}{\pi}} \int_0^{\infty} x^2 e^{-\alpha x^2} dx$$

This integral can be found in a table of integrals (see Appendix 6), and the result is

$$\langle x^2 \rangle = 2\sqrt{\frac{\alpha}{\pi}}\left(\frac{\sqrt{\pi}}{4\alpha^{3/2}}\right) = \frac{1}{2\alpha}$$

Inserting the value of the constant α from Equation (6.55a) gives

$$\langle x^2 \rangle = \frac{\hbar}{2\sqrt{m\kappa}}$$

Because $\omega = \sqrt{\kappa/m}$, we have

$$\langle x^2 \rangle = \frac{\hbar}{2m\omega} \tag{6.61}$$

In Example 6.10 we argued that

$$(x^2)_{\text{av}} = (\Delta x)^2 = \frac{E_0}{\kappa}$$

and showed that $E_0 = \hbar\omega/2$, the minimum energy allowed by the uncertainty principle. We can now see that these results are consistent, because

$$\langle x^2 \rangle = (x^2)_{\text{av}} = \frac{E_0}{\kappa} = \frac{\hbar\omega}{2\kappa} = \frac{\hbar\omega}{2m\omega^2} = \frac{\hbar}{2m\omega}$$

as we determined in Equation (6.61).

6.7 Barriers and Tunneling

Potential Barrier with $E > V_0$

Consider a particle of energy E approaching a potential barrier of height V_0 for $0 < x < L$. The potential elsewhere is zero. First, let us consider the case where the particle's energy is $E > V_0$ as shown in Figure 6.12. Classically we know the particle would pass the barrier, moving with reduced velocity in the region of $V_0(mv^2/2 = E - V_0$, rather than $mv^2/2 = E)$. On the other side of the barrier, where $V = 0$, the particle will have its original velocity again. According to quantum mechanics, the particle will behave differently because of its wavelike character. In regions I and III (where $V = 0$) the wave numbers are

$$k_{\text{I}} = k_{\text{III}} = \frac{\sqrt{2mE}}{\hbar} \qquad \text{where } V = 0 \tag{6.62a}$$

In the barrier region, however, we have

$$k_{\text{II}} = \frac{\sqrt{2m(E - V_0)}}{\hbar} \qquad \text{where } V = V_0 \tag{6.62b}$$

We can consider an analogy with optics. When light in air penetrates another medium (for example, glass), the wavelength changes because of the index of refraction. Some of the light will be reflected, and some will be transmitted into the medium. Because we must consider the wave behavior of particles interacting with potential barriers, we might expect similar behavior. The wave function will consist of an incident wave, a reflected wave, and a transmitted wave (see Figure 6.13). These wave functions can be determined by solving the Schrödinger wave equation, subject to appropriate boundary conditions. The difference from classical wave theories is that the wave function allows us to compute only probabilities.

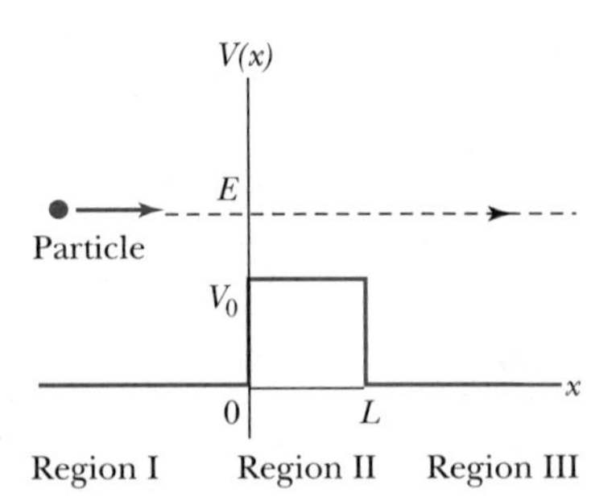

FIGURE 6.12 A particle having energy E approaches a potential barrier of width L and height V_0 with $E > V_0$. The one-dimensional space is divided into three regions as shown.

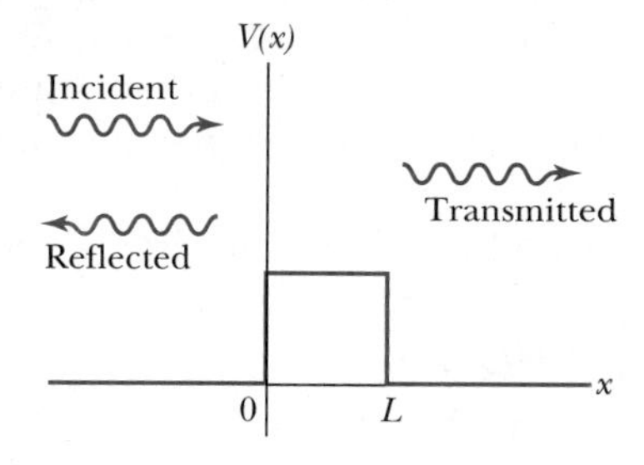

FIGURE 6.13 The incident particle in Figure 6.12 can be either transmitted or reflected.

Classical mechanics allows *no* reflection if $E > V_0$ and *total* reflection for $E < V_0$. Quantum mechanics predicts *almost total* transmission for $E \gg V_0$ and *almost complete* reflection for $E \ll V_0$. In the regime where E is comparable to V_0, unusual nonclassical phenomena may appear.

The potentials and the Schrödinger equation for the three regions are as follows:

$$\text{Region I } (x < 0) \qquad V = 0 \qquad \frac{d^2\psi_{\text{I}}}{dx^2} + \frac{2m}{\hbar^2} E\psi_{\text{I}} = 0$$

$$\text{Region II } (0 < x < L) \qquad V = V_0 \qquad \frac{d^2\psi_{\text{II}}}{dx^2} + \frac{2m}{\hbar^2} (E - V_0)\psi_{\text{II}} = 0$$

$$\text{Region III } (x > L) \qquad V = 0 \qquad \frac{d^2\psi_{\text{III}}}{dx^2} + \frac{2m}{\hbar^2} E\psi_{\text{III}} = 0$$

The wave functions obtained for these equations are

$$\text{Region I } (x < 0) \qquad \psi_{\text{I}} = Ae^{ik_{\text{I}}x} + Be^{-ik_{\text{I}}x} \tag{6.63a}$$

$$\text{Region II } (0 < x < L) \qquad \psi_{\text{II}} = Ce^{ik_{\text{II}}x} + De^{-ik_{\text{II}}x} \tag{6.63b}$$

$$\text{Region III } (x > L) \qquad \psi_{\text{III}} = Fe^{ik_{\text{I}}x} + Ge^{-ik_{\text{I}}x} \tag{6.63c}$$

We assume that we have incident particles coming from the left moving along the $+x$ direction. In this case the term $Ae^{ik_{\text{I}}x}$ in region I represents the incident particles. The term $Be^{-ik_{\text{I}}x}$ represents the reflected particles moving in the $-x$ direction. In region III there are no particles initially moving along the $-x$ direction, so the only particles present must be those transmitted through the barrier. Thus $G = 0$, and the only term in region III is $Fe^{ik_{\text{I}}x}$. We summarize these wave functions:

$$\text{Incident wave} \qquad \psi_{\text{I}}(\text{incident}) = Ae^{ik_{\text{I}}x} \tag{6.64a}$$

$$\text{Reflected wave} \qquad \psi_{\text{I}}(\text{reflected}) = Be^{-ik_{\text{I}}x} \tag{6.64b}$$

$$\text{Transmitted wave} \qquad \psi_{\text{III}}(\text{transmitted}) = Fe^{ik_{\text{I}}x} \tag{6.64c}$$

The probability of particles being reflected or transmitted is determined by the ratio of the appropriate $\psi^*\psi$. They are

$$R = \frac{|\psi_{\text{I}}(\text{reflected})|^2}{|\psi_{\text{I}}(\text{incident})|^2} = \frac{B^*B}{A^*A} \tag{6.65}$$

Probability of reflection

$$T = \frac{|\psi_{\text{III}}(\text{transmitted})|^2}{|\psi_{\text{I}}(\text{incident})|^2} = \frac{F^*F}{A^*A} \tag{6.66}$$

Probability of transmission

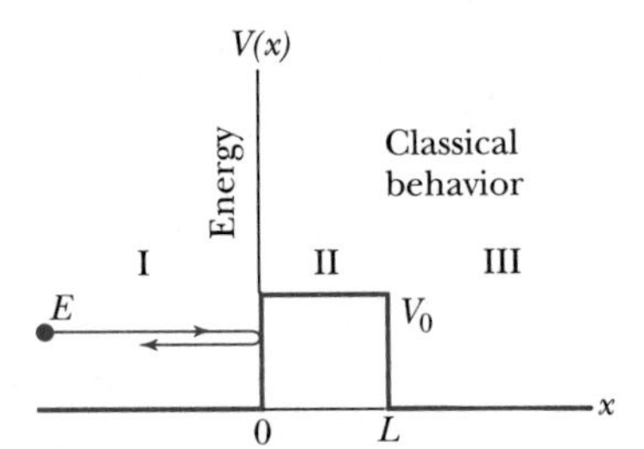

FIGURE 6.14 A particle having energy E approaches a potential barrier of height V_0 with $E < V_0$. Classically, the particle will be reflected.

where R and T are reflection and transmission probabilities, respectively. Because the particles must be either reflected or transmitted, we must have $R + T = 1$; the probability of the wave either being reflected or transmitted has to be unity.

The values of R and T are found by applying the properties(boundary conditions) of Section 6.1 as $x \longrightarrow \pm\infty$, $x = 0$, and $x = L$. These conditions will result in relationships between the coefficients A, B, C, D, and F. We will not go through that tedious math here, but the result for the transmission probability is

$$T = \left(1 + \frac{V_0{}^2 \sin^2(k_{\text{II}} L)}{4E(E - V_0)}\right)^{-1} \tag{6.67}$$

Notice that there is a situation when the transmission probability is one. This occurs when $k_{\text{II}} L = n\pi$, where n is an integer. It is possible for particles moving along the $+x$ direction to be reflected both at $x = 0$ and $x = L$. Their path difference back toward the $-x$ direction is $2L$. When $2L$ equals an integral number of the wavelengths inside the potential barrier, the reflected wave functions are completely out of phase and will completely cancel.

Potential Barrier with $E < V_0$

Now we consider the situation where classically the particle does not have enough energy to surmount the potential barrier, $E < V_0$. We show the situation in Figure 6.14. In the classical situation, the particle cannot penetrate the barrier because its kinetic energy (K.E. $= E - V_0$) would be negative. The particle is reflected at $x = 0$ and returns. The quantum mechanical result, however, is one of the most remarkable features of modern physics, and there is ample experimental proof of its existence. There is a small, but finite, probability that the particle can penetrate the barrier and even emerge on the other side. Such a surprising result requires a careful inspection of the wave functions. Fortunately, there are only a few changes to the equations already presented, and they occur in region II. The wave function in region II becomes $\psi_{\text{II}} = Ce^{\kappa x} + De^{-\kappa x}$ where $\kappa = \sqrt{(2m(V_0 - E))}/\hbar$ is a positive, real number, because $V_0 > E$. The application of the boundary conditions will again relate the coefficients of the wave functions.

The equations for the reflection and transmission probabilities of Equations (6.65) and (6.66) are unchanged, but the results will be modified by changing $ik_{\text{II}} \longrightarrow \kappa$. Quantum mechanics allows the particle to actually be on the other side of the potential barrier despite the fact that all the incident particles came in from the left moving along the $+x$ direction (see Figure 6.15). This effect is called **tunneling**. The result for the transmission probability in this case is

Tunneling

$$T = \left(1 + \frac{V_0{}^2 \sinh^2(\kappa L)}{4E(V_0 - E)}\right)^{-1} \tag{6.68}$$

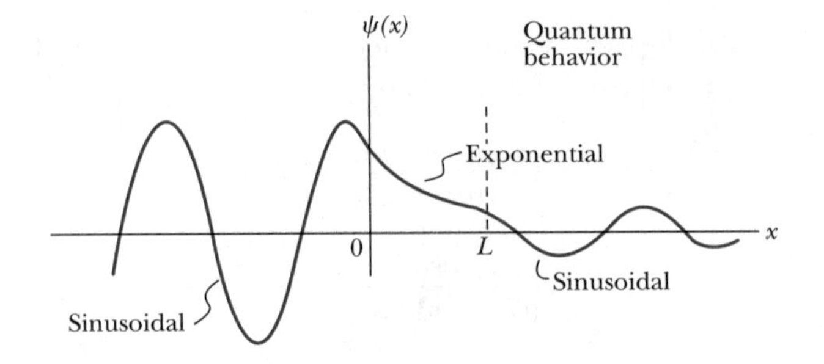

FIGURE 6.15 According to quantum mechanics, the particle approaching the potential barrier of Figure 6.14 may actually pass into the barrier and has a small probability of tunneling through the barrier and emerging at $x = L$. The particle may also be reflected at each boundary.

Note that the sine term in Equation (6.67) has been replaced by the hyperbolic sine term (sinh). When $\kappa L \gg 1$, the transmission probability equation reduces to

$$T = 16\frac{E}{V_0}\left(1 - \frac{E}{V_0}\right)e^{-2\kappa L} \tag{6.69}$$

The probability of penetration is dominated by the exponentially decreasing term (although note that for a finite thickness L, the coefficient C in Equation (6.63b) is not zero). The exponential factor in Equation (6.69) depends linearly on the barrier width but only on the square root of the potential barrier height ($\kappa \sim \sqrt{V_0 - E}$). Thus, the width of the barrier is more effective than the potential height in preventing tunneling. It comes as no surprise that tunneling is observed only at the smallest distances on the atomic scale.

A simple argument based on the uncertainty principle explains tunneling. Inside the barrier region (where $0 < x < L$), the wave function ψ_{II} is dominated by the $e^{-\kappa x}$ term, and $|\psi_{\text{II}}|^2 \sim e^{-2\kappa x}$, so that over the interval $\Delta x = \kappa^{-1}$, the probability density of observing the particle has decreased markedly ($e^{-2} = 0.14$). Because $\Delta p\, \Delta x \geq \hbar$, then $\Delta p \geq \hbar/\Delta x = \hbar\kappa$. The minimum kinetic energy in this interval must be

$$\text{K.E.}_{\text{min}} = \frac{(\Delta p)^2}{2m} = \frac{\hbar^2\kappa^2}{2m} = V_0 - E$$

where we have substituted for κ in the last step. The violation allowed by the uncertainty principle (K.E.$_{\text{min}}$) is precisely equal to the *negative* kinetic energy required! The particle is allowed by quantum mechanics and the uncertainty principle to penetrate into a classically forbidden region.

Let us return briefly to our analogy with wave optics. If light passing through a glass prism reflects from an internal surface with an angle greater than the critical angle, total internal reflection occurs as seen in Figure 6.16a. However, the electromagnetic field is not totally zero just outside the prism. If we bring another prism up very close to the first one, experiment shows that the electromagnetic wave (light) appears in the second prism (see Figure 6.16b). The situation is analogous to the tunneling described here. This effect was observed by Newton and can be demonstrated with two prisms and a laser. The intensity of the second light beam decreases exponentially as the distance between the two prisms increases.*

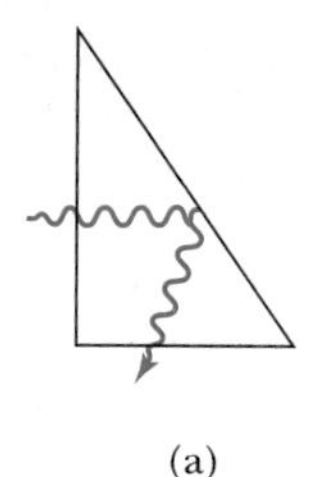

(a)

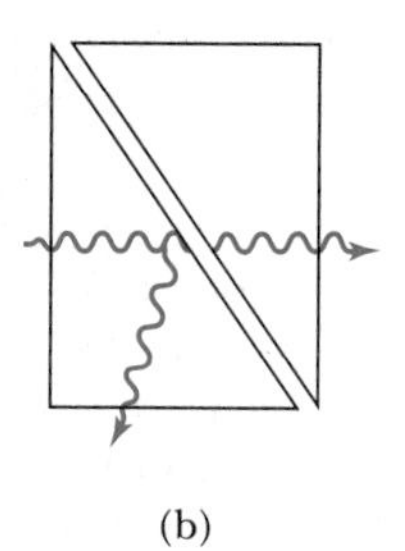

(b)

FIGURE 6.16 (a) A light wave will be totally reflected inside a prism if the reflection angle is greater than the critical angle. (b) If a second prism is brought close to the first, there is a small probability for the wave to pass through the air gap and emerge in the second prism.

Tunneling occurs for light waves

Example 6.12

In a particular semiconductor device, electrons accelerated through a potential of 5 V attempt to tunnel through a barrier of width 0.8 nm and height 10 V. What fraction of the electrons are able to tunnel through the barrier if the potential is zero outside the barrier?

Solution: We use either Equation (6.68) or (6.69) to calculate the tunneling probability, depending on the value of κL. The energy E of the electrons is $K = 5$ eV. The potential barrier has $V_0 = 10$ eV and is zero outside the barrier.

We find the value of κ by using the mass of the electron and the appropriate energies.

$$\kappa = \frac{\sqrt{2m(V_0 - E)}}{\hbar}$$

$$= \frac{\sqrt{2(0.511 \times 10^6\ \text{eV}/c^2)(10\ \text{eV} - 5\ \text{eV})}}{6.58 \times 10^{-16}\ \text{eV}\cdot\text{s}}$$

$$= \frac{3.43 \times 10^{18}\ \text{s}^{-1}}{c} = \frac{3.43 \times 10^{18}\ \text{s}^{-1}}{3 \times 10^8\ \text{m/s}} = 1.15 \times 10^{10}\ \text{m}^{-1}$$

*See D. D. Coon, *Am. J. Phys.* **34**, 240 (1966).

The value of $\kappa L = (1.15 \times 10^{10}\ \text{m}^{-1})(0.8 \times 10^{-9}\ \text{m}) = 9.2$, which might be considered to be much greater than 1, so Equation (6.69) could be used. Let's calculate the transmission probability using both equations. The approximate Equation (6.69) gives

$$T = 16\left(\frac{5\ \text{eV}}{10\ \text{eV}}\right)\left(1 - \frac{5\ \text{eV}}{10\ \text{eV}}\right)e^{-18.4} = 4.1 \times 10^{-8}$$

The more accurate Equation (6.68) gives

$$T = \left(\frac{1 + (10\ \text{eV})^2 \sinh^2(9.2)}{4(5\ \text{eV})(5\ \text{eV})}\right)^{-1} = 4.1 \times 10^{-8}$$

The approximate equation, valid when $\kappa L \gg 1$, works well in this case.

Example 6.13

Consider a particle of kinetic energy E approaching the step function of Figure 6.17 from the left, where the potential barrier steps from 0 to V_0 at $x = 0$. Find the penetration distance Δx, where the probability of the particle penetrating into the barrier drops to $1/e$. Calculate the penetration distance for a 5-eV electron approaching a step barrier of 10 eV.

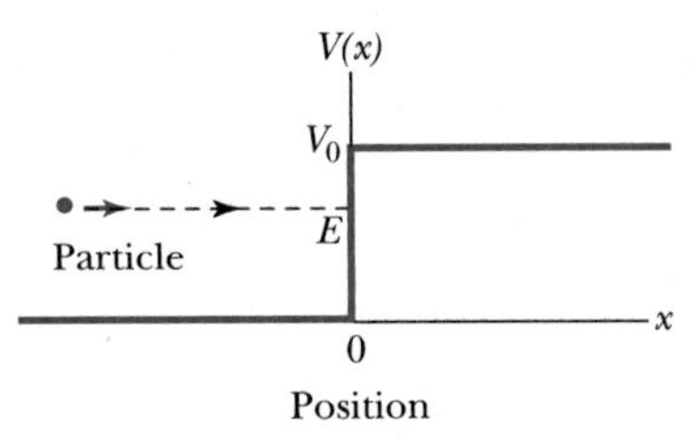

FIGURE 6.17 A particle of energy E approaches a potential barrier from the left. The step potential is $V = 0$ for $x < 0$ and $V = V_0$ for $x > 0$.

Solution: We can use the results of this section to find the wave functions in the two regions $x < 0$ and $x > 0$.

$$\psi_{\text{I}} = Ae^{ikx} + Be^{-ikx} \qquad x < 0$$

$$\psi_{\text{II}} = Ce^{\kappa x} + De^{-\kappa x} \qquad x > 0$$

where

$$k = \frac{\sqrt{2mE}}{\hbar}$$

$$\kappa = \frac{\sqrt{2m(V_0 - E)}}{\hbar}$$

Because the wave function ψ_{II} must go to zero when $x \rightarrow \infty$, the coefficient $C = 0$, so we have

$$\psi_{\text{II}} = De^{-\kappa x} \qquad x > 0$$

The probability distribution for $x > 0$ is $|\psi_{\text{II}}|^2$. The probability has dropped to e^{-1} for the penetration distance Δx, so we have

$$e^{-1} = \frac{\psi_{\text{II}}{}^2(x = \Delta x)}{\psi_{\text{II}}{}^2(x = 0)} = e^{-2\kappa\Delta x}$$

From this equation we have $1 = 2\kappa\Delta x$, and the penetration distance becomes

$$\Delta x = \frac{1}{2\kappa} = \frac{\hbar}{2\sqrt{2m(V_0 - E)}}$$

This is the result we needed.

Now we find the penetration distance for the 5-eV electron.

$$\Delta x = \frac{\hbar c}{2\sqrt{2mc^2(V_0 - E)}}$$

$$= \frac{197.3\ \text{eV} \cdot \text{nm}}{2\sqrt{2(0.511 \times 10^6\ \text{eV})(10\ \text{eV} - 5\ \text{eV})}} = 0.044\ \text{nm}$$

Electrons do not penetrate very far into the classically forbidden region.

Potential Well

Consider a particle of energy $E > 0$ passing through the potential well region (Figure 6.18), rather than a potential barrier. Let $V = -V_0$ in the region $0 < x < L$ and zero elsewhere. Classically, the particle would speed up passing the well region, because $mv^2/2 = E + V_0$. According to quantum mechanics, reflection and transmission may occur, but the wavelength inside the potential well is smaller than outside. When the width of the potential well is precisely equal to

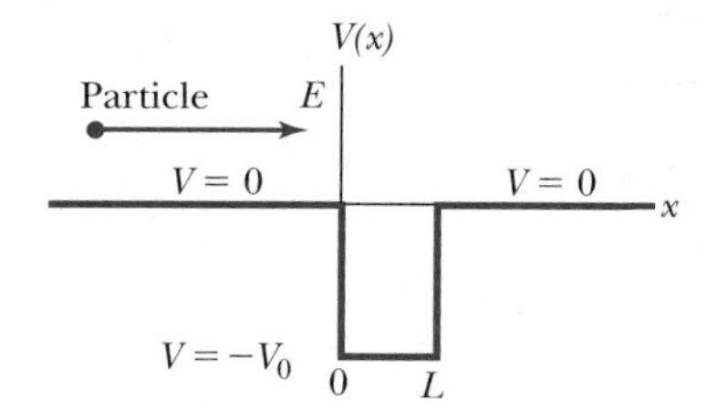

FIGURE 6.18 A particle of energy E approaches a potential well from the left. The potential is $V = 0$ everywhere except between $0 < x < L$, where $V = -V_0$.

half-integral or integral units of the wavelength, the reflected waves may be out of phase or in phase with the original wave, and cancellations or resonances may occur. The reflection/cancellation effects can lead to almost pure transmission or pure reflection for certain wavelengths. For example, at the second boundary ($x = L$) for a wave passing to the right, the wave may reflect and be out of phase with the incident wave. The effect would be a cancellation inside the well.

Alpha-Particle Decay

The phenomenon of tunneling explains the alpha-particle decay of heavy, radioactive nuclei. Many nuclei heavier than lead are natural emitters of alpha particles, but their emission rates vary over a factor of 10^{13}, whereas their energies tend to range only from 4 to 8 MeV. Inside the nucleus, an alpha particle feels the strong, short-range attractive nuclear force as well as the repulsive Coulomb force. The shape of the potential well is shown in Figure 6.19. The nuclear force dominates inside the nuclear radius r_N, and the potential can be approximated by a square well. However, outside the nucleus, the Coulomb force dominates. The so-called Coulomb potential energy barrier of Figure 6.19 can be several times the typical kinetic energy $E(\sim 5\text{ MeV})$ of an alpha particle.

The alpha particle therefore is trapped inside the nucleus. Classically, it does not have enough energy to surmount the Coulomb potential barrier. According to quantum mechanics, however, the alpha particle can "tunnel" through the barrier. The widely varying rates of alpha emission from radioactive nuclei can be explained by small changes in the potential barrier (both height and width). A small change in the barrier can manifest itself greatly in the transmission probability, because of the exponential behavior in $e^{-2\kappa L}$.

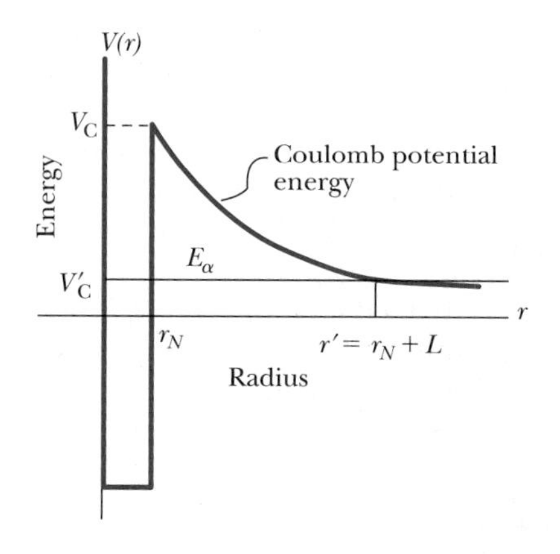

FIGURE 6.19 An α particle of energy E_α is trapped inside a heavy nucleus by the large nuclear potential. Classically, it can never escape, but quantum mechanically it may tunnel through and escape.

SCANNING PROBE MICROSCOPES

Scanning probe microscopes, consisting at present of two types, **scanning tunneling microscopes** (STM) and **atomic force microscopes** (AFM), have revolutionized the imaging of atomic surfaces. Gerd Binnig and Heinrich Rohrer (Nobel Prize, 1986) invented the STM in the early 1980s at the IBM Research Laboratory in Zürich, Switzerland. Later in 1985 while Binnig was on leave at Stanford University and IBM's Almaden Research Center, he thought up the concept of the AFM which he developed with Christoph Gerber of IBM and Calvin Quate of Stanford.

In the most common form of the STM a constant bias voltage of appropriate polarity is applied between the atoms of a tip and the sample to be examined (see Figure A). Electrons tunnel across this gap, and the sensitivity of the tunneling current to the gap distance is the key to the STM capability. The tunneling current can be as small as a few pA (10^{-12} A), and a change in the tunneling gap of only 0.4 nm can cause a factor of 10^4 in the tunneling current. In order to keep the tunneling current constant, a feedback system based on the current causes the tip to be moved up and down tracing out the contours of the sample atoms. The path of the tip is shown by the solid black line in Figure A. There are variations to this method.

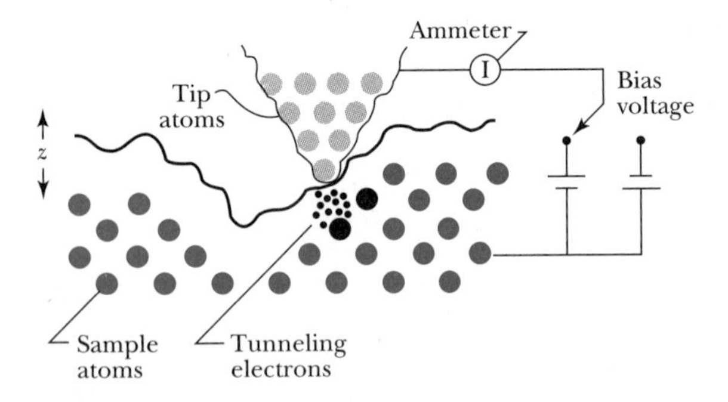

FIGURE A Highly schematic diagram of the scanning tunneling microscope process. Electrons, represented in the figure as small dots, tunnel across the gap between the atoms of the tip and sample. A feedback system that keeps the tunneling current constant causes the tip to move up and down tracing out the contours of the sample atoms.

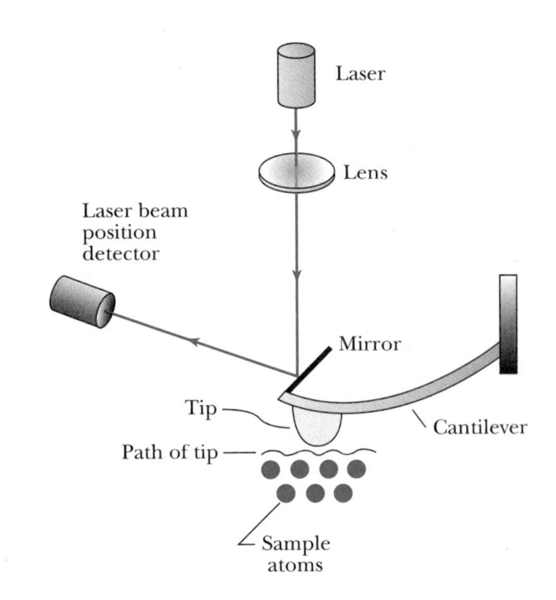

FIGURE B Highly schematic diagram of the atomic force microscope. A feedback signal from the detection of the laser beam reflecting off the mirror that is mounted on the cantilever provides a signal to move the sample atoms up or down to keep the cantilever force constant. The movement of the sample atoms traces out the contours of the sample atoms.

The AFM depends on the interatomic forces between the tip and sample atoms as shown in Figure B. In some systems, the sample atoms are scanned horizontally while the sample is moved up and down to keep the force between the tip and sample atoms constant. The interatomic forces cause the very sensitive cantilever to bend. By reflecting a laser off the end of the cantilever arm into an optical sensor, the feedback signal from this sensor controls the sample height, giving the topography of the atomic surface. The tip is scanned over the surface for a constant cantilever deflection and a constant interatomic force between tip and atom.

The interaction between tip and sample is much like that of a record player stylus moving across the record, but is about a million times more sensitive. The optical feedback system prevents the tip from actually damaging or distorting the sample atoms. Cantilevers having spring constants as small as 0.1 N/m have been microfabricated from silicon and silicon compounds. The cantilever lateral dimensions are

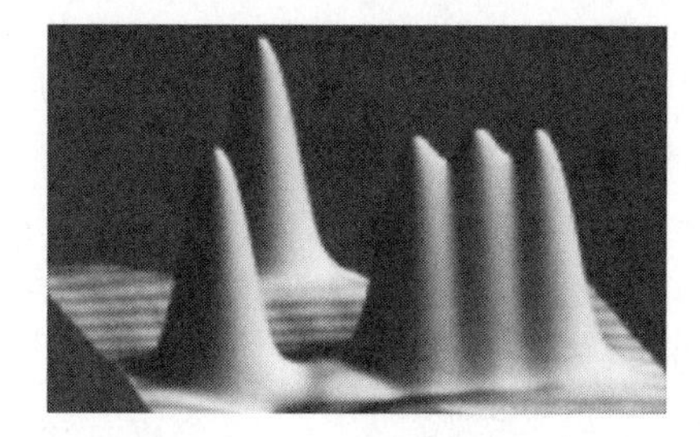
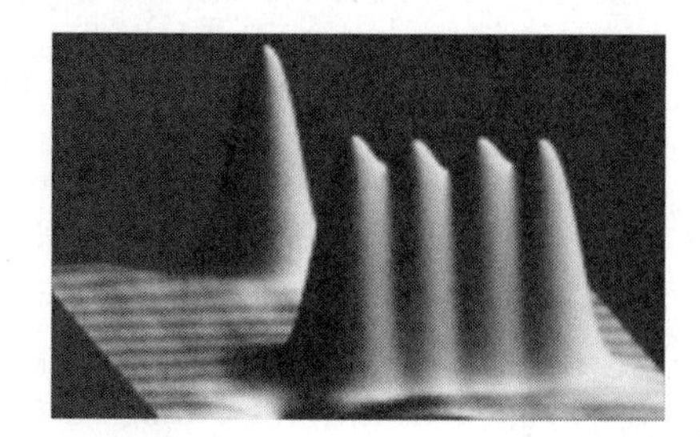

FIGURE C These three photos, taken with a STM, show xenon atoms placed on a nickel surface. The xenon atoms are 0.16 nm high and adjacent xenon atoms are 0.5 nm apart (the vertical scale has been exaggerated). The small force between the STM tip and an atom is enough to drag one xenon atom at a time across the nickel. The nickel atoms are represented by the black and white stripes on the horizontal surface. See also Figure 1.7. The image is magnified about 5 million times. *Courtesy of International Business Machines.*

on the order of 100 μm with thicknesses of about 1 μm. In comparison the spring constant of a piece of household aluminum foil 4 mm by 1 mm is about 1 N/m. The tapered tips may have an end dimension of only 50 nm. The tracking forces felt by the cantilever can be as small as 10^{-9} N. Daniel Rugar and Paul Hansma have written an excellent description of the AFM.*

The primary disadvantage of the STM compared to the AFM is that a conducting surface is required for the STM. This limits the applications of the STM, because either conductors must be scanned or a thin conductive metal coating must be placed on the sample. Because the AFM works for both insulators and conductors, it can be used for ceramics, polymers, optical surfaces, and biological structures. We showed in Figure 1.7 a photo indicating individual atoms taken using an STM. Those atoms can be individually moved as shown in Figure C. It is not only individual atomic images that make the STM and AFM so useful. They are perhaps more useful in showing gross features such as the flatness of materials, grain structures, the breakup of thin films, magnetic bit shapes, integrated circuit topography, lubricant thicknesses, inspection of optical disk stampers (see Figure D), and measurement of linewidths on integrated circuit masks. Biological applications of AFM include the imaging of amino acids, DNA, proteins and even leaf sections from a perennial cranberry vine. The AFM has been used to observe the polymerization of blood clotting protein fibrin. Real-time imaging of biological samples offers incredible possibilities, for example, the attachment of the AIDS virus onto cell membranes. Both STM and AFM instruments are now commercially available, and new variations are being developed.

*D. Rugar and P. Hansma, *Physics Today* **43**, 23 (Oct. 1990).

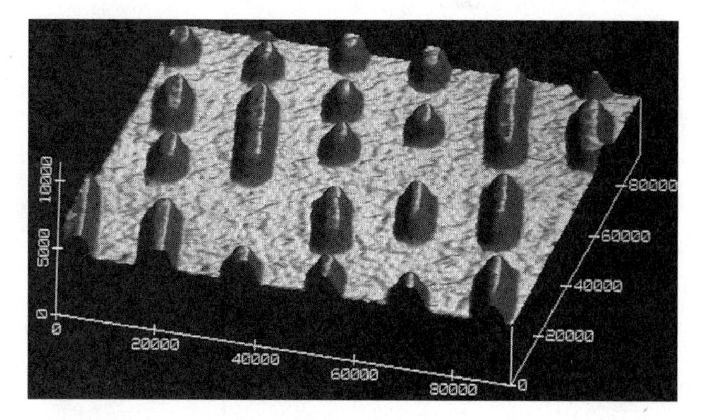

FIGURE D An atomic force microscope scan of a stamper used to mold compact disks. The numbers given are in nm. The bumps on this metallic mold stamp out 60-nm-deep holes in tracks that are 1.6 μm apart in the optical disks. *Photo courtesy of Digital Instruments/Veeco Metrology Group, Santa Barbara, CA.*

Example 6.14

Consider the α-particle emission from a ^{238}U nucleus, which emits a 4.2-MeV α particle. We shall represent the potentials as displayed in Figure 6.19. The α particle is contained inside the nuclear radius of $r_N \approx 7 \times 10^{-15}$ m. Find the barrier height and the distance the α particle must tunnel, and, using a square-top potential, calculate the tunneling probability.

Solution: We shall calculate the barrier height ($V_C(r = r_N)$ in Figure 6.19) by calculating the Coulomb potential between an α particle and the remainder of the uranium nucleus for a separation of the nuclear radius, 7×10^{-15} m.

$$V_C = \frac{Z_1 Z_2 e^2}{4\pi\epsilon_0 r_N}$$

$$= \frac{2(90)(1.6 \times 10^{-19}\text{ C})^2(9 \times 10^9\text{ N}\cdot\text{m}^2/\text{C}^2)}{7 \times 10^{-15}\text{ m}} \times \left(\frac{10^{-6}\text{ MeV}}{1.6 \times 10^{-19}\text{ J}}\right)$$

$$= 37\text{ MeV}$$

We determine the distance r' through which the α particle must tunnel by setting K.E. = $V_C(r = r')$ at that distance (see Figure 6.19). Because the K.E. = 4.2 MeV, we have

$$4.2\text{ MeV} = \frac{Z_1 Z_2 e^2}{4\pi\epsilon_0 r'}$$

We solve this equation for r' which yields

$$r' = \frac{37\text{ MeV}}{4.2\text{ MeV}} r_N = 6.2 \times 10^{-14}\text{ m} = 62\text{ fm}$$

where we have used the result above for V_C and r_N.

We make a simple, but rough, approximation of a square-top potential where $V = 37$ MeV for 7 fm $< r <$ 62 fm. We then find

$$\kappa = \frac{\sqrt{2m(V - E)}}{\hbar}$$

$$= \frac{\sqrt{2(3727\text{ MeV}/c^2)(37\text{ MeV} - 4.2\text{ MeV})}}{6.58 \times 10^{-22}\text{ MeV}\cdot\text{s}}$$

$$= 2.5 \times 10^{15}\text{ m}^{-1}$$

where the mass of the alpha particle is 3727 MeV/c^2. The barrier width L is the difference between r' and r_N.

$$L = r' - r_N$$

$$= 62\text{ fm} - 7\text{ fm} = 55\text{ fm}$$

The value of $\kappa L = (2.5 \times 10^{15}\text{ m}^{-1})(55 \times 10^{-15}\text{ m}) = 138$. Because $kL \gg 1$, we use Equation (6.69) to calculate the tunneling probability.

$$T = 16\frac{4.2\text{ MeV}}{37\text{ MeV}}\left(1 - \frac{4.2\text{ MeV}}{38\text{ MeV}}\right)e^{-275}$$

$$= 1.6e^{-275} = 6 \times 10^{-120}$$

which is an extremely small number.

Our assumption of a square-top potential of the full height and full width is very unrealistic. A closer approximation to the potential shown in Figure 6.19 would be a square-top potential of only half the maximum Coulomb potential (18 MeV rather than 37 MeV) and a barrier width of only half L (28 fm rather than 55 fm). If we use 18 MeV in the calculation of κ we obtain $1.7 \times 10^{15}\text{ m}^{-1}$. The tunneling probability now becomes

$$T = 16\frac{4.2\text{ MeV}}{18\text{ MeV}}\left(1 - \frac{4.2\text{ MeV}}{18\text{ MeV}}\right) \times \exp[-2(1.7 \times 10^{15}\text{ m}^{-1})(2.8 \times 10^{-14}\text{ m})]$$

$$= 2.9\, e^{-95} = 1.6 \times 10^{-41}$$

This still seems like a very low probability, but let us see if we can determine how long it takes the α particle to tunnel out. If the α particle has a kinetic energy of 4.2 MeV, its velocity is determined nonrelativistically by

$$\text{K.E.} = \frac{1}{2}mv^2$$

$$v = \sqrt{\frac{2\,\text{K.E.}}{m}} = \sqrt{\frac{2(4.2\text{ MeV})}{3727\text{ MeV}/c^2}} = 0.047c = 1.4 \times 10^7\text{ m/s}$$

The diameter of the nucleus is about 1.4×10^{-14} m, so it takes the α particle $1.4 \times 10^{-14}\text{ m}/(1.4 \times 10^7\text{ m/s}) \approx 10^{-21}$ s to cross. The α particle must make many traverses back and forth across the nucleus before it can escape. According to our probability it must make about 10^{41} attempts, so we estimate the α particle may tunnel through in about 10^{20} s. The half-life of a ^{238}U nucleus is 4.5×10^9 yr or about 10^{17} s. Our rough estimate does not seem all that bad.

Tunnel Diode. An extremely useful application of tunneling is that of a tunnel diode, which is a special kind of semiconductor. In a tunnel diode, electrons may pass from one region through a junction into another region. We can depict the behavior by considering a potential barrier over the region of the junction, which may be only 10 nm wide. Both positive and negative bias voltages may be applied to change the barrier height to allow the electrons to tunnel either way through the barrier. In a normal semiconductor junction, the electrons (and holes) diffuse through, a relatively slow process. In a tunnel diode, the electrons tunnel through quite rapidly when the tunneling probability is relatively high. Because the applied bias voltage can be changed rapidly, a tunnel diode is an extremely fast device and, as such, has important uses in switching circuits and high-frequency oscillators.

Summary

Werner Heisenberg and Erwin Schrödinger developed modern quantum theory in the 1920s. The time-dependent Schrödinger wave equation for the wave function $\Psi(x, t)$ is expressed as

$$i\hbar \frac{\partial \Psi(x, t)}{\partial t} = -\frac{\hbar^2}{2m}\frac{\partial^2 \Psi(x, t)}{\partial x^2} + V\Psi(x, t) \qquad (6.1)$$

The time-independent form for the spatial dependence (in one dimension) of $\psi(x)$, where $\Psi(x, t) = \psi(x)e^{-iEt/\hbar}$, is

$$-\frac{\hbar^2}{2m}\frac{d^2\psi(x)}{dx^2} + V(x)\psi(x) = E\psi(x) \qquad (6.13)$$

Certain properties of Ψ and $\partial\Psi/\partial x$ lead to quantized behavior. The wave function $\Psi(x, t)$ must be finite, single valued, and continuous; $\partial\Psi/\partial x$ must be continuous. The wave function must be normalized in order to use it to determine probabilities.

Average values of the physical observables are determined by calculating the expectation values using the wave functions. The expectation value of a function $g(x)$ is found from

$$\langle g(x)\rangle = \int_{-\infty}^{\infty} \Psi^*(x, t)g(x)\Psi(x, t)\,dx \qquad (6.20)$$

To find the expectation values of the momentum and energy, we need to know the appropriate operators. In these two cases the operators are

$$\hat{p} = -i\hbar\frac{\partial}{\partial x} \qquad (6.21)$$

$$\hat{E} = i\hbar\frac{\partial}{\partial t} \qquad (6.25)$$

and the expectation values $\langle p\rangle$ and $\langle E\rangle$ are

$$\langle p\rangle = -i\hbar\int_{-\infty}^{\infty} \Psi^*(x, t)\frac{\partial \Psi(x, t)}{\partial x}\,dx \qquad (6.23)$$

$$\langle E\rangle = i\hbar\int_{-\infty}^{\infty} \Psi^*(x, t)\frac{\partial \Psi(x, t)}{\partial t}\,dx \qquad (6.26)$$

The infinite square-well potential is a particularly simple application of the Schrödinger wave equation, and it leads to quantized energy levels and quantum numbers. The three-dimensional infinite square-well potential leads to the concept of degeneracy, different physical states with the same energy.

The simple harmonic oscillator, where the potential is $V(x) = \kappa x^2/2$, is an important application of the Schrödinger wave equation because it approximates many complex systems in nature but is exactly soluble. The energy levels of the simple harmonic oscillator are $E_n = (n + 1/2)\hbar\omega$, where $n = 0$ represents the ground state energy $E_0 = \hbar\omega/2$. The fact that the minimum energy is not zero—that the oscillator exhibits zero-point motion—is a consequence of the uncertainty principle.

Finite potentials lead to the possibility of a particle entering a region that is classically forbidden, where $V_0 > E$ (negative kinetic energy). This quantum process is called *tunneling* and is studied by considering various potential barrier shapes. Important examples of quantum tunneling are alpha decay and tunnel diodes. Tunneling is consistent with the uncertainty principle and occurs only for short distances.

Questions

1. Why can we use the nonrelativistic form of the kinetic energy in treating the structure of the hydrogen atom?
2. How do you reconcile the fact that the probability density for the ground state of the quantum harmonic oscillator (Figure 6.10c) has its peak at the center and its minima at its ends, while the classical harmonic oscillator's probability density (Figure 6.11) has a minimum at the center and peaks at each end? If you do this experiment with an actual mass and spring, what experimental result for its position distribution would you expect to obtain? Why?
3. Notice for the finite square-well potential that the wave function ψ is not zero outside the well despite the fact that $E < V_0$. Is it possible classically for a particle to be in a region where $E < V_0$? Explain this result.
4. In a given tunnel diode the p-n junction width is fixed. How can we change the time response of the tunnel diode most easily? Explain.
5. A particle in a box has a first excited state that is 3 eV above its ground state. What does that tell you about the box?
6. Does the wavelength of a particle change after it tunnels through a barrier as shown in Figure 6.15? Explain.
7. Can a particle be observed while it is tunneling through a barrier? What would its wavelength, momentum, and kinetic energy be while it tunnels through the barrier?
8. Is it easier for an electron or a proton of the same energy to tunnel through a given potential barrier? Explain.
9. Can a wave packet be formed from a superposition of wave functions of the type $e^{i(kx - \omega t)}$? Can it be normalized?
10. Given a particular potential V and wave function Ψ, how could you prove that the given Ψ is correct? Could you determine an appropriate E if the potential is independent of time?
11. Compare the infinite square-well potential with the finite one. Where is the Schrödinger wave equation the same? Where is it different?
12. Tunneling can occur for an electron trying to pass through a very thin tunnel diode. Can a baseball tunnel through a very thin window? Explain.
13. For the three-dimensional cubical box, the ground state is given by $n_1 = n_2 = n_3 = 1$. Why is it not possible to have one $n_i = 1$ and the other two equal to zero?

Problems

6.1 The Schrödinger Wave Equation

1. Try to normalize the wave function $e^{i(kx - \omega t)}$. Why can't it be done over all space? Explain why it is not possible.
2. (a) In what direction does a wave of the form $A\sin(kx - \omega t)$ move? (b) What about $B\sin(kx + \omega t)$? (c) Is $e^{i(kx-\omega t)}$ a real number? (d) In what direction is the wave in (c) moving? Explain.
3. Show directly that the trial wave function $\Psi(x, t) = Ae^{i(kx-\omega t)}$ satisfies Equation (6.1).
4. Normalize the wave function $Ae^{i(kx-\omega t)}$ in the region $x = 0$ to a.
5. Normalize the wave function $Are^{-r/\alpha}$ from $r = 0$ to ∞ where α and A are constants.
6. Property (2) specifies that Ψ must be continuous in order to avoid discontinuous probability values. Why can't we have such probabilities?
7. Consider the wave function $Ae^{-\alpha|x|}$ that we used in Example 6.4. (a) How does this wave function satisfy the boundary conditions of Section 6.1? (b) What can we conclude about this wave function? (c) If the wave function is unacceptable as is, how could it be fixed?

6.2 Expectation Values

8. A set of measurements has given the following result for the measurement of x (in some units of length): 3.4, 3.9, 5.2, 4.7, 4.1, 3.8, 3.9, 4.7, 4.1, 4.5, 3.8, 4.5, 4.8, 3.9, and 4.4. Find the average value of x, called $\bar{x}$ or $\langle x \rangle$, and average value of x^2, called $\langle x^2 \rangle$. Show that the standard deviation of x, given by

$$\sigma = \sqrt{\frac{\sum(x_i - \bar{x})^2}{N}}$$

where x_i is the individual measurement and N is the number of measurements, is also given by $\sigma = \sqrt{\langle x^2 \rangle - \langle x \rangle^2}$. Find the value of σ for the set of data.

9. If the potential V is independent of time, show that the expectation value for x is independent of time.
10. A wave function $\psi = A(e^{ix} + e^{-ix})$ in the region $-\pi < x < \pi$ and zero elsewhere. Normalize the wave function and find the probability of the particle being between $x = 0$ and $\pi/8$.
11. A wave function has the value $A \sin x$ between 0 and π, but zero elsewhere. Normalize the wave function and find the probability that the particle is between $x = 0$ and $\pi/4$.

6.3 Infinite Square-Well Potential

12. Find an equation for the difference between adjacent energy levels ($\Delta E_n = E_{n+1} - E_n$) for the infinite square-well potential. Calculate ΔE_1, ΔE_8, and ΔE_{800}.
13. Determine the average value of $\psi_n^2(x)$ inside the well for the infinite square-well potential for $n = 1, 5, 20$, and 100. Compare these averages with the classical probability of detecting the particle inside the box.
14. An electron moves with a speed $v = 10^{-4}c$ inside a one-dimensional box ($V = 0$) of length 48.5 nm. The potential is infinite elsewhere. The particle may not escape the box. What approximate quantum number does the electron have?
15. For the infinite square-well potential, find the probability that a particle in its ground state is in each third of the one-dimensional box: $0 \leq x \leq L/3$, $L/3 \leq x \leq 2L/3$, $2L/3 \leq x \leq L$.
16. Repeat the previous problem using the first excited state.
17. Repeat Example 6.8 for an electron inside the nucleus. Assume nonrelativistic equations and find the transition energy for an electron. (See Example 6.8 for an interpretation of the result.)
18. What is the minimum energy of (a) a proton and (b) an α particle trapped in a one-dimensional region the size of a uranium nucleus (radius $= 7 \times 10^{-15}$ m)?
19. An electron is trapped in an infinite square-well potential of width 0.5 nm. If the electron is initially in the $n = 4$ state, what are the various photon energies that can be emitted as the electron jumps to the ground state?

6.4 Finite Square-Well Potential

20. Consider a finite square-well potential well of width 3×10^{-15} m that contains a particle of mass 2 GeV/c^2. How deep does this potential well need to be to contain three energy levels? (Except for the energy levels, this situation approximates a deuteron.)
21. Compare the results of the infinite and finite square-well potentials. (a) Are the wavelengths longer or shorter for the finite square well compared with the infinite well? (b) Using physical arguments, do you expect the energies (for a given quantum number n) to be (i) larger or (ii) smaller for the finite square well than for the infinite square well? (c) Why will there be a finite number of bound energy states for the finite potential?
22. Apply the boundary conditions to the finite square-well potential at $x = 0$ to find the relationships between the coefficients A, C, and D and the ratio C/D.
23. Apply the boundary conditions to the finite square-well potential at $x = L$ to find the relationship between the coefficients B, C, and D and the ratio C/D.

6.5 Three-Dimensional Infinite-Potential Well

24. Find the energies of the second, third, fourth, and fifth levels for the three-dimensional cubical box. Which are degenerate?
25. Write the possible (unnormalized) wave functions for each of the first four excited energy levels for the cubical box.
26. Find the normalization constant A for the ground state wave function for the cubical box in Equation (6.52).
27. Complete the derivation of Equation (6.49) by substituting the wave function given in Equation (6.47) into Equation (6.46). What is the origin of the three quantum numbers?

6.6 Simple Harmonic Oscillator

28. In Figure 6.9 we showed a plausible guess for the wave function ψ_0 for the lowest energy level E_0 of the simple harmonic oscillator. Explain the shape of the wave function and explain why it is a maximum at $x = 0$ and not zero when $E = V_0$.
29. What is the energy level difference between adjacent levels $\Delta E_n = E_{n+1} - E_n$ for the simple harmonic oscillator? What is ΔE_0, ΔE_2, and ΔE_{50}? How many possible energy levels are there?
30. The wave function for the first excited state ψ_1 for the simple harmonic oscillator is $\psi_1 = Axe^{-\alpha x^2/2}$. Normalize the wave function to find the value of the constant A. Determine $\langle x \rangle$, $\langle x^2 \rangle$, and $\Delta x = \sqrt{\langle x^2 \rangle - \langle x \rangle^2}$.
31. An atom of mass 3.32×10^{-26} kg oscillates in one dimension at a frequency of 10^{13} Hz. What are its effective force constant and quantized energy levels?
32. One possible solution for the wave function ψ_n for the simple harmonic oscillator is

$$\psi_n = A(2\alpha x^2 - 1)e^{-\alpha x^2/2}$$

where A is a constant. What is the value of the energy level E_n?
33. What would you expect for $\langle p \rangle$ and $\langle p^2 \rangle$ for the ground state of the simple harmonic oscillator? (*Hint:* Use symmetry and energy arguments.)
34. Show that the energy of a simple harmonic oscillator in the $n = 1$ state is $3\hbar\omega/2$ by substituting the wave function $\psi_1 = Axe^{-\alpha x^2/2}$ directly into the Schrödinger equation.

35. A H_2 molecule can be approximated by a simple harmonic oscillator having spring constant $k = 1.1 \times 10^3$ N/m. Find (a) the energy levels, and (b) the possible wavelengths of photons emitted when the H_2 molecule decays from the third excited state eventually to the ground state.

6.7 Barriers and Tunneling

36. The creation of elements in the early universe and in stars involves protons tunneling into nuclei. Find the probability of the proton tunneling through ^{12}C. Let the proton and carbon be inside a star of temperature 12,000 K.

37. Compare the wavelength of a particle when it passes a barrier of height (a) $+V_0$ (see Figure 6.12) and (b) $-V_0$ where $E > |V_0|$ (see Figure 6.18). Calculate the momentum and kinetic energy for both cases.

38. (a) Calculate the transmission probability of an α particle of energy $E = 5$ MeV through a Coulomb barrier of a heavy nucleus that is approximated by a square barrier with $V_0 = 15$ MeV and barrier width $L = 1.3 \times 10^{-14}$ m. Calculate the probability (b) by doubling the potential barrier height and (c) by using the original barrier height but doubling the barrier width. Compare all three probabilities.

39. Consider a particle of energy E trapped inside the potential well shown in the accompanying figure. Sketch a possible wave function inside and outside the potential well. Explain your sketch.

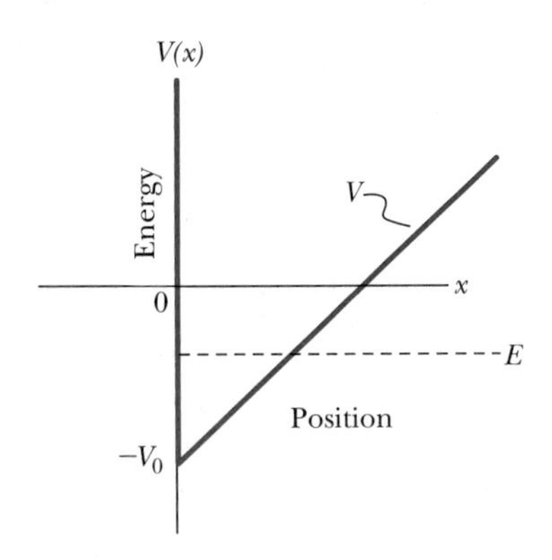

A potential well is infinite for $x < 0$ but increases linearly from $V = -V_0$ at $x = 0$.

40. When a particle of energy E approaches a potential barrier of height V_0, where $E \gg V_0$, show that the reflection coefficient is about $((V_0 \sin(kL))/2E)^2$.

41. Let 11.0-eV electrons approach a potential barrier of height 3.8 eV. (a) For what barrier thickness is there no reflection? (b) For what barrier thickness is the reflection a maximum?

42. A 1-eV electron has a 10^{-4} probability of tunneling through a 2.5-eV potential barrier. What is the probability of a 1-eV proton tunneling through the same barrier?

General Problems

43. A particle of mass m is trapped in a three-dimensional rectangular potential well with sides of length L, $L/\sqrt{2}$, and $2L$. Inside the box $V = 0$, outside $V = \infty$. Assume

$$\psi = A \sin(k_1 x) \sin(k_2 y) \sin(k_3 z)$$

inside the well. Substitute this ψ into the Schrödinger equation and apply appropriate boundary conditions to find the allowed energy levels. Find the energy of the ground and first four excited levels. Which of these levels are degenerate?

44. For a region where the potential $V = 0$, the wave function is given by $\sqrt{2/\alpha}\,\sin(3\pi x/\alpha)$. Calculate the energy of this system.

45. Consider the semi-infinite-well potential in which $V = \infty$ for $x \le 0$, $V = 0$ for $0 < x < L$, and $V = V_0$ for $x > L$. (a) Show that possible wave functions are $A \sin k_n x$ inside the well and $Be^{-\kappa_n x}$ for $x > L$, where $k_n = \sqrt{2mE_n}/\hbar$ and $\kappa_n = \sqrt{2m(V_0 - E)}/\hbar$. (b) Show that the application of the boundary conditions gives $\kappa \tan(kL) = -k$.

46. Assume that $V_0 = \hbar^2/2mL^2$ and show that the ground state energy of a particle in the semi-infinite well of the previous problem is given by $0.04\hbar^2/2mL^2$.

47. Prove that there are a limited number of bound solutions for the semi-infinite well.

48. Use the semi-infinite-well potential to model a deuteron, a nucleus consisting of a neutron and a proton. Let the well width be 3.5×10^{-15} m and $V_0 - E = 2.2$ MeV. Determine the energy E. How many excited states are there, and what are their energies?

49. The wave function for the $n = 2$ state of a simple harmonic oscillator is $A(1 - 2\alpha x^2)e^{-\alpha x^2/2}$. (a) Show that its energy level is $5\,\hbar\omega/2$ by substituting the wave function into the Schrödinger equation. (b) Find $\langle x \rangle$ and $\langle x^2 \rangle$.

50. A particle is trapped inside an infinite square-well potential between $x = 0$ and $x = L$. Its wave function is a superposition of the ground state and first excited state. Its wave function is given by

$$\Psi(x) = \frac{1}{2}\Psi_1(x) + \frac{\sqrt{3}}{2}\Psi_2(x)$$

Show that the wave function is normalized.

51. The *Morse potential* is a good approximation for a real potential to describe diatomic molecules. It is given by $V(r) = D(1 - e^{-a(r-r_e)})^2$ where D is the molecular dissociation energy, and r_e is the equilibrium distance between the atoms. For small vibrations, $r - r_e$ is small, and $V(r)$ can be expanded in a Taylor series to reduce to a simple harmonic potential. Find the lowest term of $V(r)$ in this expansion and show that it is quadratic in $(r - r_e)$.

52. Show that the vibrational energy levels E_v for the *Morse potential* of the previous problem are given by

$$E_v = \hbar\omega\left(n + \frac{1}{2}\right) - \frac{\hbar^2\omega^2}{4D}\left(n + \frac{1}{2}\right)^2$$

where
$$\omega = a\sqrt{\frac{2D}{m_r}}$$

and n is the vibrational quantum number, m_r is the reduced mass, and $E_v \ll D$. Find the three lowest energy levels for KCl where $D = 4.42$ eV, and $a = 7.8\ \text{nm}^{-1}$.

53. Consider a particle of mass m trapped inside a two-dimensional square box of sides L aligned along the x and y axes. Show that the wave function and energy levels are given by

$$\psi(x, y) = \frac{2}{L}\sin\frac{n_x\pi x}{L}\sin\frac{n_y\pi y}{L}$$

$$E = \frac{\hbar^2\pi^2}{2mL^2}(n_x^2 + n_y^2)$$

Plot the first six energy levels and give their quantum numbers.

CHAPTER 7

The Hydrogen Atom

The atom of modern physics can be symbolized only through a partial differential equation in an abstract space of many dimensions. All its qualities are inferential; no material properties can be directly attributed to it. That is to say, any picture of the atom that our imagination is able to invent is for that very reason defective. An understanding of the atomic world in that primary sensuous fashion . . . is impossible.

Werner Heisenberg

In Chapter 6 we studied the Schrödinger equation and its application to several model systems. We now have the tools to apply quantum theory to real physical systems, which we will do in the next few chapters. Our first major subject is atomic physics, and we begin by applying the Schrödinger equation to the hydrogen atom. We will learn that additional quantum numbers are needed in order to explain experimental results. A couple of the sections in this chapter (sections 7.2 and 7.6) are advanced topics and may be skipped without losing continuity.

7.1 Application of the Schrödinger Equation to the Hydrogen Atom

The hydrogen atom is the first system we shall consider that requires the full complexity of the three-dimensional Schrödinger equation. To a good approximation the potential energy of the electron–proton system is electrostatic,

$$V(r) = -\frac{e^2}{4\pi\epsilon_0 r} \tag{7.1}$$

We rewrite the three-dimensional time-independent Schrödinger Equation (6.43) as

$$-\frac{\hbar^2}{2m}\frac{1}{\psi(x, y, z)}\left[\frac{\partial^2\psi(x, y, z)}{\partial x^2} + \frac{\partial^2\psi(x, y, z)}{\partial y^2} + \frac{\partial^2\psi(x, y, z)}{\partial z^2}\right] = E - V(r) \tag{7.2}$$

As was discussed in Chapter 4, the correct mass value m to be used is the reduced mass μ of the proton and electron. We can also study other hydrogenlike (called *hydrogenic*) atoms such as He^+ or Li^{++} by inserting the appropriate μ and by replacing e^2 in Equation (7.1) with Ze^2, where Z is the atomic number.

We note that the potential $V(r)$ in Equation (7.2) depends only on the distance r between the proton and electron. To take advantage of this radial symmetry, we transform to spherical polar coordinates. The transformation is given in Figure 7.1, where the relationships between the cartesian coordinates x, y, z and the spherical polar coordinates r, θ, ϕ are shown. The transformation of Equation (7.2) into spherical polar coordinates is straightforward, but tedious. After inserting the Coulomb potential into the transformed Schrödinger equation, we have

$$\frac{1}{r^2}\frac{\partial}{\partial r}\left(r^2\frac{\partial\psi}{\partial r}\right)+\frac{1}{r^2\sin\theta}\frac{\partial}{\partial\theta}\left(\sin\theta\frac{\partial\psi}{\partial\theta}\right)+\frac{1}{r^2\sin^2\theta}\frac{\partial^2\psi}{\partial\phi^2}+\frac{2\mu}{\hbar^2}(E-V)\psi=0 \tag{7.3}$$

Schrödinger equation in spherical coordinates

The wave function ψ is now a function of r, θ, ϕ $[\psi(r, \theta, \phi)]$, but we will write it simply as ψ for brevity. In the terminology of partial differential equations, Equation (7.3) is separable, meaning a solution may be found as a product of three functions, each depending on only one of the coordinates r, θ, ϕ. (This is exactly analogous to our separating the time-dependent part of the Schrödinger equation solution as $e^{-iEt/\hbar}$.) Let us try a solution of the form

$$\psi(r, \theta, \phi) = R(r)f(\theta)g(\phi) \tag{7.4}$$

Trial solution

This substitution allows us to separate the partial differential in Equation (7.3) into three separate differential equations, each depending on one coordinate: r, θ, or ϕ.

From Chapter 6 we have a good idea what to expect the results will look like. For each of the three differential equations we must apply appropriate boundary conditions on the functions $R(r)$, $f(\theta)$, and $g(\phi)$. This will lead

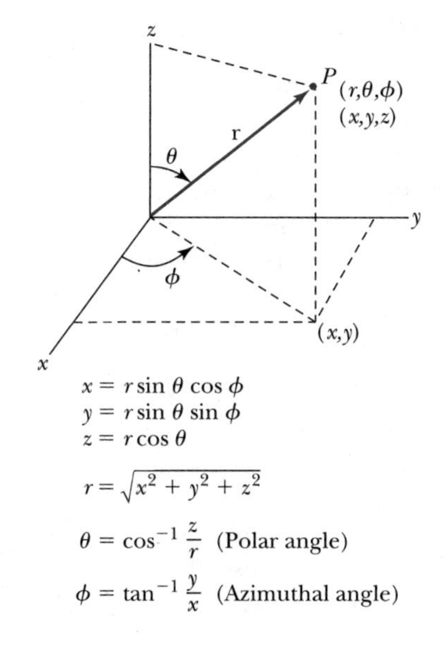

FIGURE 7.1 Relationship between spherical polar coordinates (r, θ, ϕ) and cartesian coordinates (x, y, z).

to three quantum numbers, one for each of the three separate differential equations (or one quantum number for each dimension of motion available—recall that in the previous chapter we obtained one quantum number for one-dimensional motion).

7.2 Solution of the Schrödinger Equation for Hydrogen

The first step will be to substitute the trial solution, Equation (7.4), into Equation (7.3). Then we can separate the resulting equation into three equations: one for $R(r)$, one for $f(\theta)$, and one for $g(\phi)$. The solutions to those equations will then allow us to understand the structure of the hydrogen atom, in the ground state and excited states as well.

Separation of Variables

Starting with Equation (7.4), we find the necessary derivatives to be

$$\frac{\partial \psi}{\partial r} = fg\frac{\partial R}{\partial r} \qquad \frac{\partial \psi}{\partial \theta} = Rg\frac{\partial f}{\partial \theta} \qquad \frac{\partial^2 \psi}{\partial \phi^2} = Rf\frac{\partial^2 g}{\partial \phi^2} \tag{7.5}$$

We substitute these results into the Schrödinger equation (7.3) and find

$$\frac{fg}{r^2}\frac{\partial}{\partial r}\left(r^2\frac{\partial R}{\partial r}\right) + \frac{Rg}{r^2 \sin\theta}\frac{\partial}{\partial \theta}\left(\sin\theta\frac{\partial f}{\partial \theta}\right) + \frac{Rf}{r^2\sin^2\theta}\frac{\partial^2 g}{\partial \phi^2} + \frac{2\mu}{\hbar^2}(E - V)Rfg = 0 \tag{7.6}$$

Next we multiply both sides of Equation (7.6) by $r^2\sin^2\theta/Rfg$ and rearrange to have

$$-\frac{\sin^2\theta}{R}\frac{\partial}{\partial r}\left(r^2\frac{\partial R}{\partial r}\right) - \frac{2\mu}{\hbar^2}r^2\sin^2\theta(E - V) - \frac{\sin\theta}{f}\frac{\partial}{\partial \theta}\left(\sin\theta\frac{\partial f}{\partial \theta}\right) = \frac{1}{g}\frac{\partial^2 g}{\partial \phi^2} \tag{7.7}$$

Look closely at Equation (7.7). Notice that only the variables r and θ (and their functions R and f) appear on the left side, whereas only ϕ and its function g appear on the right side. We have achieved a separation of variables, completely isolating ϕ. What does this mean? The left side of the equation cannot change as ϕ changes, because it does not contain ϕ or any function depending on ϕ. Similarly, the right side cannot change with either r or θ. The only way for this to be true is for each side of Equation (7.7) to be equal to a constant. For reasons that will be clear later, we choose to let this constant have the value $-m_\ell^2$. If we set the constant $-m_\ell^2$ equal to the right side of Equation (7.7), we have

$$\frac{1}{g}\frac{\partial^2 g}{\partial \phi^2} = -m_\ell^2$$

or, after rearranging,

Azimuthal equation

$$\frac{d^2 g}{d\phi^2} = -m_\ell^2 g \tag{7.8}$$

Notice that because ϕ is the only variable, we have replaced the partial derivative with the ordinary derivative. Because the angle ϕ in spherical coordinates corresponds to the azimuth angle in astronomy, Equation (7.8) is traditionally referred to as the **azimuthal equation.** This is just the equation of a harmonic oscillator that we have studied in introductory physics, and the solutions for $g(\phi)$ will take the form of sines and cosines.

An important restriction on the values of the quantum number m_ℓ can be obtained if we consider solutions to Equation (7.8) for the function $e^{im_\ell\phi}$. One may easily verify by direct substitution that $e^{im_\ell\phi}$ satisfies Equation (7.8) for any value of m_ℓ. However, in order to have a physically valid solution for any value of ϕ, it is necessary that the solution be single valued, that is $g(\phi) = g(\phi + 2\pi)$. This means, for example, that $g(\phi = 0) = g(\phi = 2\pi)$, which requires that $e^0 = e^{2\pi i m_\ell}$. The only way for this to be true is for m_ℓ to be zero or an integer (either positive or negative). The quantum number m_ℓ is therefore restricted to be zero or a positive or negative integer. If the sign on the right-hand side of Equation (7.8) were positive rather than negative, the solution is not physically realized, because it can't be normalized and is not single valued in ϕ. We shall defer further discussion of solutions for Equation (7.8) until later. For now it is sufficient to realize that readily obtainable solutions exist.

Now we set the left side of Equation (7.7) equal to the constant $-m_\ell^2$ and rearrange to have

$$\frac{1}{R}\frac{\partial}{\partial r}\left(r^2\frac{\partial R}{\partial r}\right) + \frac{2\mu r^2}{\hbar^2}(E - V) = \frac{m_\ell^2}{\sin^2\theta} - \frac{1}{f\sin\theta}\frac{\partial}{\partial\theta}\left(\sin\theta\frac{\partial f}{\partial\theta}\right) \tag{7.9}$$

Notice that we have again achieved a successful separation of variables, with everything depending on r on the left side and everything depending on θ on the right side. We can set each side of Equation (7.9) equal to a constant, which this time we call $\ell(\ell + 1)$. Doing so with each side of the equation in succession yields (after more rearrangement) the two equations,

$$\frac{1}{r^2}\frac{d}{dr}\left(r^2\frac{dR}{dr}\right) + \frac{2\mu}{\hbar^2}\left(E - V - \frac{\hbar^2}{2\mu}\frac{\ell(\ell+1)}{r^2}\right)R = 0 \tag{7.10}$$

Radial equation

and

$$\frac{1}{\sin\theta}\frac{d}{d\theta}\left(\sin\theta\frac{df}{d\theta}\right) + \left(\ell(\ell+1) - \frac{m_\ell^2}{\sin^2\theta}\right)f = 0 \tag{7.11}$$

Angular equation

where, after separation, we have again replaced the partial derivatives with the ordinary ones.

The process of separation of variables is now complete. The original Schrödinger equation has been separated into three ordinary second-order differential equations [(7.8), (7.10), and (7.11)], each containing only one variable.

Relation Between the Quantum Numbers ℓ and m_ℓ

Equation (7.11), which we shall call the *angular* equation, was first solved by the famous mathematician Adrien Marie Legendre (1752–1833). It is well known in the theory of differential equations as the **associated Legendre equation.** Application of the appropriate boundary conditions to Equations (7.10) and (7.11) (this process is too tedious to present here) leads to the following restrictions on the quantum numbers ℓ and m_ℓ:

The associated Legendre equation

$$\begin{aligned} \ell &= 0, 1, 2, 3, \ldots \\ m_\ell &= -\ell, -\ell + 1, \ldots, -2, -1, 0, 1, 2, \ldots, \ell - 1, \ell \end{aligned} \tag{7.12}$$

That is, the quantum number ℓ must be zero or a positive integer, and the quantum number m_ℓ must be a positive or negative integer, or zero, subject to the restriction that $|m_\ell| \leq \ell$. The choice of $\ell(\ell+1)$ as the constant for Equation (7.9) provides in the succinct results in Equation (7.12).

Solution of the Radial Equation

The associated Laguerre equation

Equation (7.10), appropriately called the *radial* equation, is another well-known differential equation. It is known as the **associated Laguerre equation** after the French mathematician Edmond Nicolas Laguerre (1834–1886). The solutions R to this equation that satisfy the appropriate boundary conditions are called *associated Laguerre functions.* We shall consider these solutions in some detail in Section 7.6. We can obtain some idea of how the ground-state wave function looks if we assume that the ground state has the lowest possible quantum number $\ell = 0$ of the system. Our conditions in Equation (7.12) then require that $m_\ell = 0$. Notice that $\ell = 0$ greatly simplifies the radial wave Equation (7.10) to be

$$\frac{1}{r^2}\frac{d}{dr}\left(r^2\frac{dR}{dr}\right) + \frac{2\mu}{\hbar^2}(E - V)R = 0 \tag{7.13}$$

The derivative of the bracketed expression in the first term of Equation (7.13) yields two terms by using the derivative product rule. We write out both of those terms and insert the Coulomb potential energy, Equation (7.1), to find

$$\frac{d^2R}{dr^2} + \frac{2}{r}\frac{dR}{dr} + \frac{2\mu}{\hbar^2}\left(E + \frac{e^2}{4\pi\epsilon_0 r}\right)R = 0 \tag{7.14}$$

Those students with some experience in solving differential equations will recognize that an exponential solution is required. We try a solution having the form

$$R = Ae^{-r/a_0}$$

where A is a normalization constant and a_0 is a constant with dimensions of length (we shall see that it was no accident that we chose the name a_0!). It is reasonable to try to verify the trial solution by inserting it into the radial equation (7.14). The first and second derivatives are

$$\frac{dR}{dr} = -\frac{1}{a_0}R \qquad \frac{d^2R}{dr^2} = \frac{R}{a_0^2}$$

We insert these derivatives into Equation (7.14) and rearrange terms to yield

$$\left(\frac{1}{a_0^2} + \frac{2\mu}{\hbar^2}E\right) + \left(\frac{2\mu e^2}{4\pi\epsilon_0\hbar^2} - \frac{2}{a_0}\right)\frac{1}{r} = 0 \tag{7.15}$$

By the same reasoning that we applied in the separation of variables method, the only way for Equation (7.15) to be satisfied for *any* value of r is for *each* of the two bracketed expressions to be equal to zero. We set the second bracket term equal to zero and solve for a_0 to find

$$a_0 = \frac{4\pi\epsilon_0\hbar^2}{\mu e^2}$$

We see that a_0 is in fact equal to the Bohr radius [see Equation (4.24)]! Now we set the first bracket in Equation (7.15) equal to zero and solve for E to find

$$E = -\frac{\hbar^2}{2\mu a_0^2} = -E_0$$

Again this is equal to the Bohr result, with E_0 having the value 13.6 eV.

Because we are not prepared to deal with the full scope of the associated Laguerre functions in this book, we shall not consider higher energy states here.

Rather, we shall summarize some of the key results. The full solution to the radial wave equation requires (not surprisingly) the introduction of another quantum number, which we shall call n, such that n is a positive integer (but not zero). There is a further restriction that the quantum number ℓ can only take on values less than n. The consequences of this, along with a full consideration of allowed sets of the three quantum numbers n, ℓ, and m_ℓ, will be considered in Section 7.3. Let us note, however, that the predicted energy levels turn out to be

$$E_n = -\frac{E_0}{n^2}$$

in agreement with the Bohr result.

Solution of the Angular and Azimuthal Equations

We now return to the azimuthal Equation (7.8). We note that its solutions can be expressed in exponential form as $e^{im_\ell\phi}$ or $e^{-im_\ell\phi}$. But because the angular equation also contains the quantum number m_ℓ, solutions to the angular and azimuthal equations are linked. It is customary to group these solutions together into what are called the **spherical harmonics** $Y(\theta, \phi)$, defined as

$$Y(\theta, \phi) = f(\theta)g(\phi) \tag{7.16}$$

Spherical harmonics

The $f(\theta)$ part of the $Y(\theta, \phi)$ is always a polynomial function of $\sin\theta$ and $\cos\theta$ of order ℓ. See Table 7.1 for a listing of the normalized spherical harmonics up to $\ell = 3$.

TABLE 7.1
Normalized Spherical Harmonics $Y(\theta, \phi)$

ℓ	m_ℓ	$Y_{\ell m_\ell}$
0	0	$\frac{1}{2\sqrt{\pi}}$
1	0	$\frac{1}{2}\sqrt{\frac{3}{\pi}}\cos\theta$
1	± 1	$\mp\frac{1}{2}\sqrt{\frac{3}{2\pi}}\sin\theta\, e^{\pm i\phi}$
2	0	$\frac{1}{4}\sqrt{\frac{5}{\pi}}(3\cos^2\theta - 1)$
2	± 1	$\mp\frac{1}{2}\sqrt{\frac{15}{2\pi}}\sin\theta\cos\theta\, e^{\pm i\phi}$
2	± 2	$\frac{1}{4}\sqrt{\frac{15}{2\pi}}\sin^2\theta\, e^{\pm 2i\phi}$
3	0	$\frac{1}{4}\sqrt{\frac{7}{\pi}}(5\cos^3\theta - 3\cos\theta)$
3	± 1	$\mp\frac{1}{8}\sqrt{\frac{21}{\pi}}\sin\theta(5\cos^2\theta - 1)e^{\pm i\phi}$
3	± 2	$\frac{1}{4}\sqrt{\frac{105}{2\pi}}\sin^2\theta\cos\theta\, e^{\pm 2i\phi}$
3	± 3	$\mp\frac{1}{8}\sqrt{\frac{35}{\pi}}\sin^3\theta\, e^{\pm 3i\phi}$

The probability density for the electron in the hydrogen atom is given by $\psi^*\psi$, therefore, the spherical harmonics together with the radial wave function R will determine the overall shape of the probability density for the various quantum states. The total wave function $\psi(r, \theta, \phi)$ will depend on the quantum numbers n, ℓ, and m_ℓ. We can now write the wave function as

$$\psi_{n\ell m_\ell}(r, \theta, \phi) = R_{n\ell}(r) Y_{\ell m_\ell}(\theta, \phi) \tag{7.17}$$

where we indicate by the subscripts that $R(r)$ depends only on n and ℓ and $Y(\theta, \phi)$ depends only on ℓ and m_ℓ, We shall look at these wave functions again in Section 7.6.

7.3 Quantum Numbers

The three quantum numbers obtained from solving Equation (7.3) are

n **principal quantum number**
ℓ **orbital angular momentum quantum number**
m_ℓ **magnetic quantum number**

Their values are obtained by applying the boundary conditions to the wave function $\psi(r, \theta, \phi)$ as discussed in Section 6.1. The restrictions imposed by the boundary conditions are

$$\begin{aligned} n &= 1, 2, 3, 4, \ldots && \text{Integer} \\ \ell &= 0, 1, 2, 3, \ldots, n-1 && \text{Integer} \\ m_\ell &= -\ell, -\ell+1, \ldots, 0, 1, \ldots, \ell-1, \ell && \text{Integer} \end{aligned} \tag{7.18}$$

These three quantum numbers must be integers. The orbital angular momentum quantum number must be less than the principal quantum number, $\ell < n$, and the magnitude of the magnetic quantum number (which may be positive or negative) must be less than or equal to the orbital angular momentum quantum number, $|m_\ell| \le \ell$. We can summarize these conditions as

$$\begin{aligned} n &> 0 \\ \ell &< n \\ |m_\ell| &\le \ell \end{aligned} \tag{7.19}$$

The lowest value of n is 1, and for $n = 1$, we must have $\ell = 0$, $m_\ell = 0$. For $n = 2$, we may have $\ell = 0$, $m_\ell = 0$ as well as $\ell = 1$, $m_\ell = -1, 0, +1$.

Example 7.1

What are the possible quantum numbers for a $n = 4$ state in atomic hydrogen?

Solution: The possible values of ℓ are $\ell = 0, 1, 2, 3$, because $\ell_{max} = n - 1$. For each value of ℓ, m_ℓ goes from $-\ell$ to $+\ell$.

n	ℓ	m_ℓ
4	0	0
4	1	−1, 0, 1
4	2	−2, −1, 0, 1, 2
4	3	−3, −2, −1, 0, 1, 2, 3

As yet these quantum numbers have little physical meaning to us. Let us examine each of them more carefully and try to find classical analogies where possible.

Principal Quantum Number n

The *principal quantum number n* results from the solution of the radial wave function $R(r)$ in Equation (7.4). Because the radial equation includes the potential energy $V(r)$, it is not surprising to find that the boundary conditions on $R(r)$ quantize the energy E. The result for this quantized energy is

$$E_n = \frac{-\mu}{2}\left(\frac{e^2}{4\pi\epsilon_0\hbar}\right)^2\frac{1}{n^2} = -\frac{E_0}{n^2} \tag{7.20}$$

which is precisely the value found in Chapter 4 from the Bohr theory [Equations (4.25) and (4.26)]. The energy levels of the hydrogen atom depend on the principal quantum number n only. The negative value of the energy E indicates that the electron and proton are bound together.

It is perhaps surprising that the total energy of the electron does not depend on the angular momentum. However, a similar situation occurs for planetary motion, where the energy depends on the semimajor axis of the elliptical planetary orbits and not on the eccentricity of the orbits. This peculiarity occurs for the solar system and the hydrogen atom because both the gravitational and Coulomb forces are central; they also both have inverse-square-law dependences on distance.

Orbital Angular Momentum Quantum Number ℓ

The *orbital angular momentum quantum number* ℓ is associated with the $R(r)$ and $f(\theta)$ parts of the wave function. The electron–proton system has orbital angular momentum as the particles pass around each other. Classically, this orbital angular momentum is $\mathbf{L} = \mathbf{r} \times \mathbf{p}$ or

$$L = mv_{\text{orbital}}r \tag{7.21}$$

where v_{orbital} is the orbital velocity, perpendicular to the radius. The quantum number ℓ is related to the magnitude of the orbital angular momentum L by

$$L = \sqrt{\ell(\ell+1)}\hbar \tag{7.22}$$

This curious dependence of L on ℓ ($L^2 \sim \ell(\ell+1)$ rather than ℓ^2) is a wave phenomenon—it results from the application of the boundary conditions on $\psi(r, \theta, \phi)$. We will present a justification for it later in the section. The quantum result disagrees with the more elementary Bohr theory of the hydrogen atom, where $L = n\hbar$. This is most obvious in an $\ell = 0$ state, where $L = \sqrt{0(1)}\hbar = 0$. Apparently we will have to discard Bohr's semiclassical "planetary" model of electrons orbiting a nucleus.

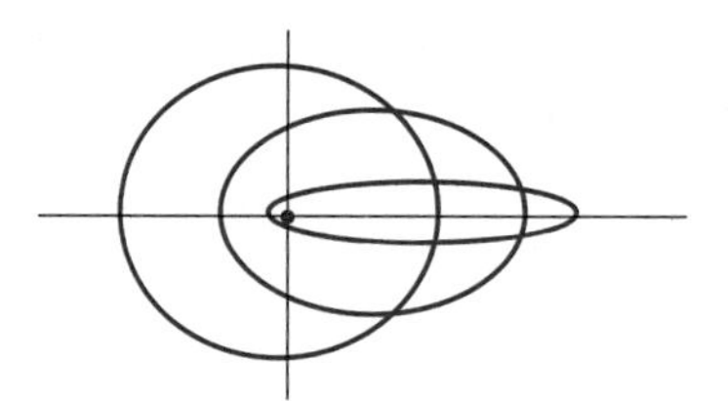

FIGURE 7.2 Various possible electron (or planetary) classical orbits. The energy depends only on the principal quantum number n and not on the angular momentum of the hydrogen atom. There is a finite probability for an $\ell = 0$ electron to be present within the nucleus. Of course, none of the planets has $\ell = 0$, and (obviously) they do not pass through the sun.

In Figure 7.2 we show several classical orbits corresponding to the same total energy. For an electron in an atom, the energy depends on n; for planetary motion, the energy depends on the semimajor axis. Do not take the elliptical orbits literally for electrons; only probability functions can describe their positions, which must be consistent with the uncertainty principle. We say that a certain energy level is **degenerate** with respect to ℓ when the energy is independent of the value of ℓ (see Section 6.5). For example, the energy for an $n = 3$ level is the same for all possible values* of ℓ ($\ell = 0, 1, 2$).

*This statement is true for single-electron atoms like hydrogen. We will learn later in Chapter 8 that for many-electron atoms (atoms with more than 1 electron) electrons with lower ℓ values lie lower in energy for a given n value.

SPECIAL TOPIC:

RYDBERG ATOMS

Rydberg atoms are highly excited atoms with their outermost electron in a high energy level, very near ionization. They are named after Johannes Rydberg who developed the empirical relation bearing his name that produces the correct wavelengths of hydrogen atoms [Equation (3.13)]. Rydberg atoms are similar to hydrogen atoms because the highly excited electron is in such a large orbit that it stays well outside the orbits of the other electrons. A Rydberg atom of atomic number Z has an electron outside a positive core of charge $+e$[Z protons and (Z − 1) electrons], just like the hydrogen atom.

Even though Rydberg atoms may have properties similar to hydrogen, they have some distinctly exotic properties. For example, they are gigantic, being as much as 100,000 times larger than normal atoms. Despite being in such a highly excited energy state, they are surprisingly long lived because the selection rules do not allow them to easily decay to lower energy levels. Their lifetime can be as long as a second, which is over a million times the lifetime of a normal excited atom. On the atomic scale, these long-lived Rydberg atoms live almost forever.

We recall from Chapter 4 that the energy levels of the hydrogen atom are given by $-E_0/n^2$ and the radius is given by $n^2 a_0$, where $E_0 = 13.6$ eV and $a_0 = 5.3 \times 10^{-11}$ m. Rydberg atoms have been observed in radio astronomy measurements from outer space with n values near 400, but those produced in the laboratory are rarely larger than 100 and are more commonly studied near 30. Note that a Rydberg atom, acting like hydrogen and having $n = 400$ would have a diameter of $10^5 \times 10^{-10}$ m or 10 μm, an incredibly large atom! A transition from $n = 401 \rightarrow 400$ results in a 4×10^{-7} eV photon emission having a wavelength near 3 m, a radio wave.

Rydberg atoms can be made in the laboratory by bombarding gaseous atoms with charged particles. A revolution in their study came about, however, from the use of tunable lasers (see Chapter 10), which allows specific states to be excited by transferring a laser photon of precise energy to an electron. The density of atoms must be kept low because a collision between Rydberg atoms and normal atoms may quickly lead to de-excitation. The reason Rydberg atoms are so easily found (relatively speaking, of course) in interstellar space is because once created, a Rydberg atom has a poor chance of colliding with another atom.

The most dramatic, and most useful, property of Rydberg atoms is due to the effect of electric fields on their energy levels, called the *Stark effect.* Because of their large values of n, Rydberg atoms are highly de-

---

It is customary for historical reasons to use letter names for the various ℓ values. These are

$$\begin{array}{lcccccc} \ell = 0 & 1 & 2 & 3 & 4 & 5 & \ldots \\ \text{letter} = s & p & d & f & g & h & \ldots \end{array}$$

These particular letter designations for the first four values resulted from empirical visual observations from early experiments: sharp, principal, diffuse, and fundamental. After $\ell = 3$, (f state), the letters generally follow alphabetical order.

Atomic states are normally referred to by the n number and ℓ letter. Thus a state with $n = 2$ and $\ell = 1$ is called a $2p$ state. Examples of other various atomic states are $1s(n = 1, \ell = 0)$, $2s(n = 2, \ell = 0)$, $4d(n = 4, \ell = 2)$, $6g(n = 6, \ell = 4)$. A state such as $2d$ is not possible, because this refers to $n = 2$ and $\ell = 2$. Our boundary conditions require $n > \ell$.

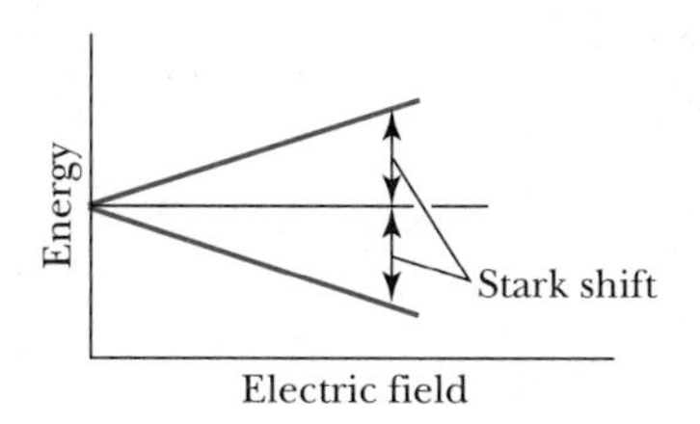

FIGURE A The thin black line represents the degenerate energy level, and the blue lines represent the maximum energy shift for a given electric field. Because of the large degeneracy, states may have many of the energies between the extremes.

generate. Remember that two states are *degenerate* when they have different quantum numbers, but have the same energy. Many states can have the same high value of n, but have different values of ℓ and m_ℓ. In highly degenerate Rydberg atoms, the Stark effect is significant because the splitting of the many energy levels varies linearly with the electric field as shown in Figure A. It requires only a weak electric field to either ionize or change the energy level of a Rydberg atom. In contrast to the electric field, a magnetic field squeezes an atom and changes its shape. This property allows the magnetic properties of Rydberg atoms to be studied under exotic situations. This area of research has received much less attention than the Stark effect.

Several applications have been proposed for Rydberg atoms besides that of fundamental atomic measurements. Detectors sensitive to electromagnetic radiation from the infrared to microwave wavelengths could be built using the very small differences between atomic energy levels in Rydberg atoms. Such long-wavelength radiation (and low frequency) is difficult to detect using normal atoms because of the larger energy differences between adjacent energy states. The use of electric fields allows the energy level differences to be fine-tuned for Rydberg atoms. Such a detector, for example, might be useful in astronomy where suitable detectors are difficult to find.

The use of Rydberg atoms has also been suggested for the separation of isotopes for uranium enrichment. A laser could be used to promote the atoms of one particular isotope, but not the others, to moderate atomic excitation, while a second laser excites the atoms of all other isotopes to highly excited states. These other atoms could then be isolated by ionization leaving the single isotope. To date, few applications* have been realized for Rydberg atoms, but they remain a useful object of experimental and theoretical inquiry.

*D. Kleppner, M. G. Littman, and M. L. Zimmerman, *Scientific American* **244**, 130 (May 1981).

Magnetic Quantum Number m_ℓ

The orbital angular momentum quantum number ℓ determines the magnitude of the angular momentum **L**, but because **L** is a vector, it also has a direction. Classically, because there is no torque in the hydrogen atom system in the absence of external fields, the angular momentum **L** is a constant of the motion and is conserved. The solution to the Schrödinger equation for $f(\theta)$ specified that ℓ must be an integer, and therefore the magnitude of **L** is quantized.

The angle ϕ is a measure of the rotation about the z axis. The solution for $g(\phi)$ specifies that m_ℓ is an integer *and* related to the z component of the angular momentum **L**.

$$L_z = m_\ell \hbar \tag{7.23}$$

The relationship of L, L_z, ℓ, and m_ℓ is displayed in Figure 7.3 for the value $\ell = 2$. The magnitude of L is fixed $(L = \sqrt{\ell(\ell+1)}\hbar = \sqrt{6}\hbar)$. Because L_z is

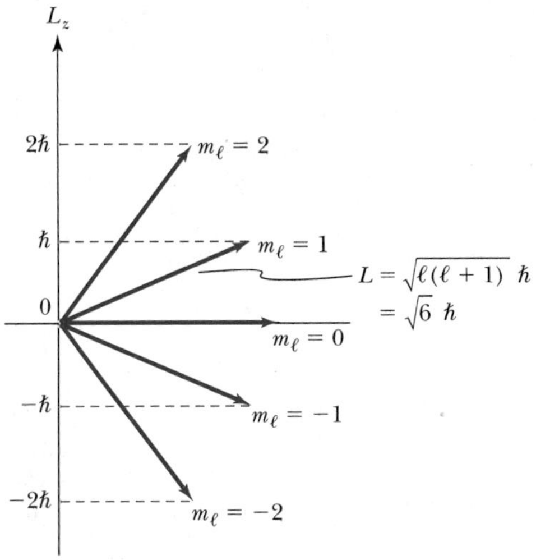

FIGURE 7.3 Schematic diagram of the relationship between **L** and L_z with the allowed values of m_ℓ.

Space quantization

quantized, only certain orientations of **L** are possible, each corresponding to a different m_ℓ (and therefore L_z). This phenomenon is called **space quantization,** because only certain orientations of **L** are allowed in space.

We can ask whether we have established a preferred direction in space by choosing the z axis. The choice of the z axis is completely arbitrary unless there is an external magnetic field to define a preferred direction in space. It is customary to choose the z axis to be along **B** if there is a magnetic field. This is why m_ℓ is called the *magnetic quantum number.*

Will the angular momentum be quantized along the x and y axes as well? The answer is that quantum theory allows **L** to be quantized along only one direction in space. Because we know the magnitude of **L**, the knowledge of a second component would imply a knowledge of the third component as well because of the relation $L^2 = L_x^2 + L_y^2 + L_z^2$. The following argument shows that this would violate the Heisenberg uncertainty principle: if all three components of **L** were known, then the direction of **L** would also be known. In this case we would have a precise knowledge of one component of the electron's position in space, because the electron's orbital motion is confined to a plane perpendicular to **L**. But confinement of the electron to that plane means that the electron's momentum component along **L** is *exactly* zero. This simultaneous knowledge of the same component of position and momentum is forbidden by the uncertainty principle.

Only the magnitude $|\mathbf{L}|$ and L_z may be specified simultaneously. The values of L_x and L_y should be consistent with $L^2 = L_x^2 + L_y^2 + L_z^2$ but cannot be specified individually. Physicists refer to the known values of L and L_z as "sharp" and the unknown L_x and L_y as "fuzzy." The angular momentum vector **L** *never* points in the z direction (see Figure 7.3) because $L = \sqrt{\ell(\ell + 1)}\hbar$ and $|\mathbf{L}| > |L_z|_{\text{max}} = \ell\hbar$. Our results from solving the Schrödinger equation of the hydrogen atom are consistent with the uncertainty principle.

The space quantization just mentioned is an experimental fact. The values of L_z range from $-\ell$ to $+\ell$ in steps of 1, for a total of $2\ell + 1$ allowed values. Because there is nothing special about the three directions x, y, and z, we expect the average of the angular momentum components squared in the three directions to be the same, $\langle L_x^2 \rangle = \langle L_y^2 \rangle = \langle L_z^2 \rangle$. The average value of $\langle L^2 \rangle$ is equal to three times the average value of the square of any one of the components, so we choose the z component, $\langle L^2 \rangle = 3\langle L_z^2 \rangle$. To find the average value of L_z^2, we just have to sum up all the squares of the quantum numbers for L_z and divide by the total number, $2\ell + 1$.

$$\langle L^2 \rangle = 3\langle L_z^2 \rangle = \frac{3}{2\ell + 1} \sum_{m_\ell = -\ell}^{\ell} m_\ell^2 \hbar^2 = \ell(\ell + 1)\hbar^2 \tag{7.24}$$

where we have used a math table for the summation result. This rather simple argument to explain the $\ell(\ell + 1)$ dependence for the expectation value of L^2 (rather than using a sophisticated quantum mechanical calculation) was originally due to Richard Feynman and simplified by P. W. Milonni.*

Example 7.2

What is the degeneracy of the $n = 3$ level? That is, how many different states are contained in the energy level, $E_3 = -E_0/9$?

Solution: The energy eigenvalues for atomic hydrogen depend only on the principal quantum number n (in the absence of a magnetic field). For each value of n, there can be n different orbital angular momentum ℓ states ($\ell = 0, 1, \ldots, n - 1$). For each value of ℓ, there are $2\ell + 1$ different magnetic quantum states ($m_\ell = -\ell, -\ell + 1, \ldots, 0, 1, \ldots, +\ell$). Therefore to find the total degeneracy for $n = 3$ we have to add up all the possibilities.

n	ℓ	m_ℓ	$2\ell + 1$
3	0	0	1
3	1	−1, 0, 1	3
3	2	−2, −1, 0, 1, 2	5
			total = 9

The $n = 3$ level is degenerate (in the absence of a magnetic field) because all nine states have the same energy but different quantum numbers. Their wave functions, however, are quite different. You may notice that, in general, the degeneracy is n^2 (see Problem 16).

7.4 Magnetic Effects on Atomic Spectra—The Normal Zeeman Effect

As early as 1896 it was shown by the Dutch physicist Pieter Zeeman that the spectral lines emitted by atoms placed in a magnetic field broaden and appear to split. Sometimes a line is split into three lines (*normal* **Zeeman effect**), but often more than three lines are found (*anomalous* Zeeman effect). The normal Zeeman effect can be understood by considering the atom to behave like a small magnet and will be discussed here. The anomalous Zeeman effect is more complicated (see Section 8.3). By the 1920s considerable **fine structure** of atomic spectral lines from hydrogen and other elements had been observed.

As a rough model, think of an electron circulating around the nucleus as a circular current loop. The current loop has a magnetic moment $\mu = IA$ where the current $I = dq/dt$ is simply the electron charge ($q = -e$) divided by the period T for the electron to make one revolution ($T = 2\pi r/v$).

$$\mu = IA = \frac{q}{T}A = \frac{(-e)\pi r^2}{2\pi r/v} = \frac{-erv}{2} = -\frac{e}{2m}L \tag{7.25}$$

where $L = mvr$ is the magnitude of the orbital angular momentum. Both the magnetic moment $\boldsymbol{\mu}$ and angular momentum $\mathbf{L}$ are vectors so that

$$\boldsymbol{\mu} = -\frac{e}{2m}\mathbf{L} \tag{7.26}$$

*P. W. Milonni, *Am. J. Phys.* **58**, 1012 (1990).

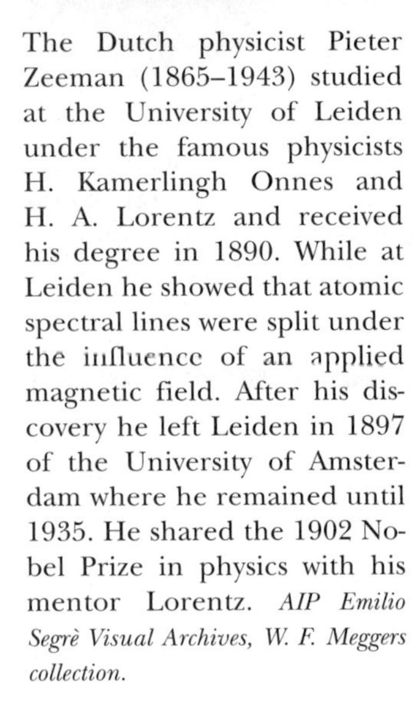

The Dutch physicist Pieter Zeeman (1865–1943) studied at the University of Leiden under the famous physicists H. Kamerlingh Onnes and H. A. Lorentz and received his degree in 1890. While at Leiden he showed that atomic spectral lines were split under the influence of an applied magnetic field. After his discovery he left Leiden in 1897 of the University of Amsterdam where he remained until 1935. He shared the 1902 Nobel Prize in physics with his mentor Lorentz. *AIP Emilio Segrè Visual Archives, W. F. Meggers collection.*

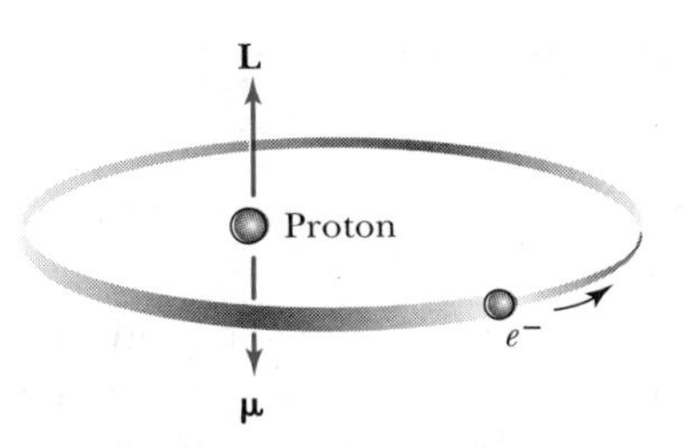

FIGURE 7.4 Representation of the orbital angular momentum **L** and magnetic moment **μ** of the hydrogen atom due to the electron orbiting the proton. The directions of **L** and **μ** are opposite because of the negative electron charge.

The relationship between **μ** and **L** is displayed in Figure 7.4.

In the absence of an external magnetic field to align them, the magnetic moments **μ** of atoms point in random directions. In classical electromagnetism, if a magnetic dipole having a magnetic moment **μ** is placed in an external magnetic field, the dipole will experience a torque $\boldsymbol{\tau} = \boldsymbol{\mu} \times \mathbf{B}$ tending to align the dipole with the magnetic field. The dipole also has a potential energy V_B in the field given by

$$V_B = -\boldsymbol{\mu} \cdot \mathbf{B} \tag{7.27}$$

If the system can change its potential energy, the magnetic moment will align itself with the external magnetic field.

Note the similarity with the case of the spinning top in a gravitational field. The gravitational field is not parallel to the angular momentum, and the force of gravity pulling down on the top results in a **precession** of the top about the field direction, not a falling down of the top. Precisely the same thing happens here with the magnetic moment. The angular momentum is aligned with the magnetic moment, and the torque between **μ** and **B** causes a precession of **μ** about the magnetic field (see Figure 7.4), not an alignment. The magnetic field establishes a preferred direction in space along which we customarily define the z axis. Then we have

$$\mu_z = \frac{e\hbar}{2m} m_\ell = -\mu_B \, m_\ell \tag{7.28}$$

Bohr magneton

where $\mu_B \equiv e\hbar/2m$ is a unit of magnetic moment called a **Bohr magneton.** Because of the quantization of L_z and the fact that $L = \sqrt{\ell(\ell+1)}\hbar > m_\ell\hbar$, we cannot have $|\boldsymbol{\mu}| = \mu_z$; the magnetic moment cannot align itself exactly in the z direction. Just like the angular momentum **L**, the magnetic moment **μ** has only certain allowed quantized orientations. Note also that in terms of the Bohr magneton, $\boldsymbol{\mu} = -\mu_B \mathbf{L}/\hbar$.

Example 7.3

Determine the precessional frequency of an atom having magnetic moment **μ** in an external magnetic field **B**. This precession is known as the *Larmor precession.*

Solution: We have already seen that the torque $\boldsymbol{\tau}$ is equal to $\boldsymbol{\mu} \times \mathbf{B}$, but from mechanics we also know that the torque is $d\mathbf{L}/dt$. The torque in Figure 7.5 is perpendicular

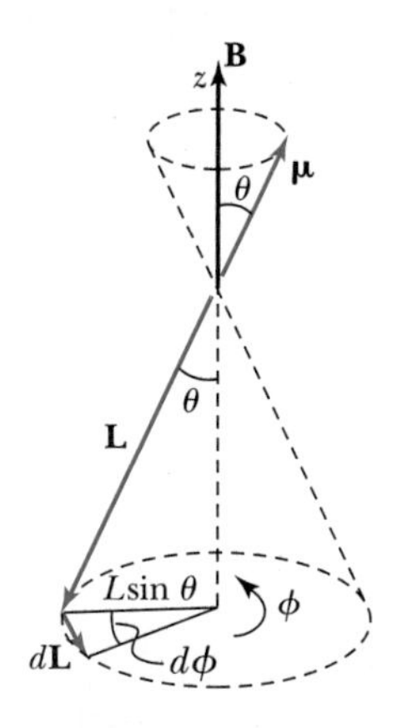

FIGURE 7.5 An atom having magnetic moment **μ** feels a torque **τ** = **μ** × **B** due to an external magnetic field **B**. This torque must also be equal to $d\mathbf{L}/dt$. The vectors **μ** and **L** are antiparallel, so the vector $d\mathbf{L}/dt$ must be perpendicular to **μ**, **B**, and **L**. As shown in the figure, $d\mathbf{L}/dt$ requires both **μ** and **L** to precess (angle ϕ) about the magnetic field **B**.

to **μ**, **L**, and **B** and is out of the page. This must also be the direction of the change in momentum $d\mathbf{L}$ as seen in Figure 7.5. Thus **L** and **μ** precess about the magnetic field. The magnitude of $d\mathbf{L}$ is given by $L\sin\theta d\phi$ (see Figure 7.5). The Larmor frequency ω_L is given by $d\phi/dt$,

$$\omega_L = \frac{d\phi}{dt} = \frac{1}{L\sin\theta}\frac{dL}{dt} \tag{7.29}$$

We now insert the magnitude of $L = 2m\mu/e$ from Equation (7.26). The value of dL/dt, the magnitude of $\boldsymbol{\mu} \times \mathbf{B}$, can be determined from Figure 7.5 to be $\mu B \sin\theta$. Equation (7.29) becomes

$$\omega_L = \left(\frac{e}{2m\mu \sin\theta}\right)\mu B \sin\theta = \frac{eB}{2m} \tag{7.30}$$

What about the energy of the orbiting electron in a magnetic field? It takes work to rotate the magnetic moment away from **B**. With **B** along the z direction, we have from Equations (7.16), (7.17), and (7.27)

$$V_B = -\mu_z B = +\mu_B m_\ell B \tag{7.31}$$

The potential energy is thus quantized according to the *magnetic quantum number* m_ℓ; each (degenerate) atomic level of given ℓ is split into $2\ell + 1$ different energy states according to the value of m_ℓ. The energy degeneracy of a given $n\ell$ level is removed by a magnetic field (see Figure 7.6a).

Example 7.4

What is the value of the Bohr magneton? Use that to calculate the energy difference between the $m_\ell = 0$ and $m_\ell = +1$ components in the $2p$ state of atomic hydrogen placed in an external field of 2 T.

Solution: We first find the Bohr magneton to be

$$\mu_B = \frac{e\hbar}{2m}$$

$$= \frac{(1.602 \times 10^{-19}\ \text{C})(1.055 \times 10^{-34}\ \text{J}\cdot\text{s})}{2(9.11 \times 10^{-31}\ \text{kg})}$$

$$= 9.27 \times 10^{-24}\ \text{J/T} \tag{7.32}$$

The international system of units has been used (T = tesla for magnetic field). The energy splitting is given by (see Figure 7.6a)

$$\Delta E = \mu_B B \Delta m_\ell \tag{7.33}$$

where $\Delta m_\ell = 1 - 0 = 1$. Hence, we have

$$\Delta E = (9.27 \times 10^{-24}\ \text{J/T})(2\ \text{T}) = 1.85 \times 10^{-23}\ \text{J}$$

$$= 1.16 \times 10^{-4}\ \text{eV}$$

An energy difference of 10^{-4} eV is easily observed by optical means.

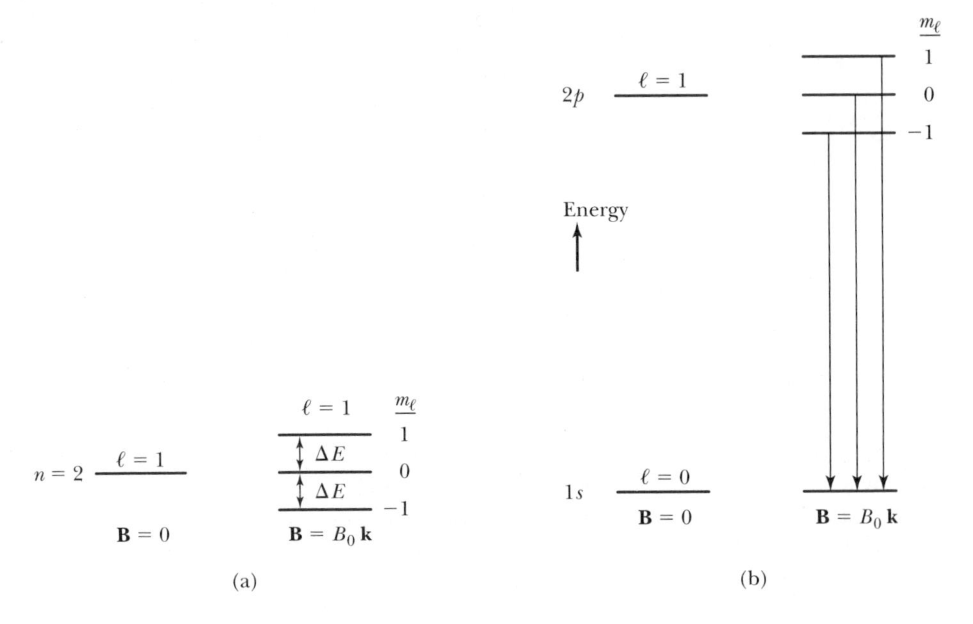

FIGURE 7.6 The normal Zeeman effect. (a) An external magnetic field removes the degeneracy of a $2p$ level and reveals the three different energy states. (b) There are now transitions with three different energies between an excited $2p$ level and the $1s$ ground state in atomic hydrogen. The energy ΔE has been grossly exaggerated along the energy scale.

The splitting of spectral lines can be partially explained by the application of external magnetic fields. This result, the normal Zeeman effect, is displayed in Figure 7.6. When a magnetic field is applied, the $2p$ level of atomic hydrogen is split into 3 different energy states with the energy difference given by Equation (7.33). A transition for an electron in the excited $2p$ level to the $1s$ ground state results in three different energy transitions as shown in Figure 7.6b. The energy differences are shown greatly exaggerated in Figure 7.6b, but as instruments were improved, such differences could be observed. The application of external magnetic fields eliminates much of the energy degeneracy, resulting in more quantized states having different energies between which electrons are

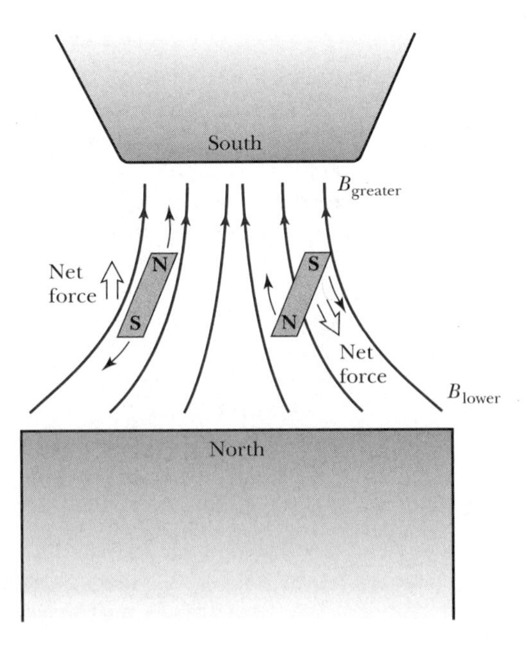

FIGURE 7.7 An inhomogeneous magnetic field is created by the smaller south pole. Two bar magnets representing atomic magnetic moments have $\boldsymbol{\mu}$ in opposite directions. Because the force on the top of the bar magnets is greater than that on the bottom, there will be a net translational force on the bar magnets (atoms).

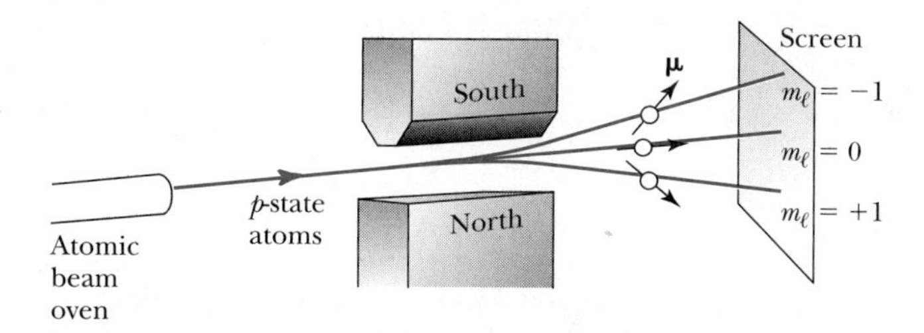

FIGURE 7.8 Schematic diagram of expected result of Stern and Gerlach experiment if atoms in a p state are used. Three patterns of atoms, due to $m_\ell = \pm 1, 0$, are expected on the screen. The magnet poles are arranged to produce a magnetic field gradient as shown in Figure 7.7. The experiment performed by Stern and Gerlach reported only two lines, not three (see Section 7.5).

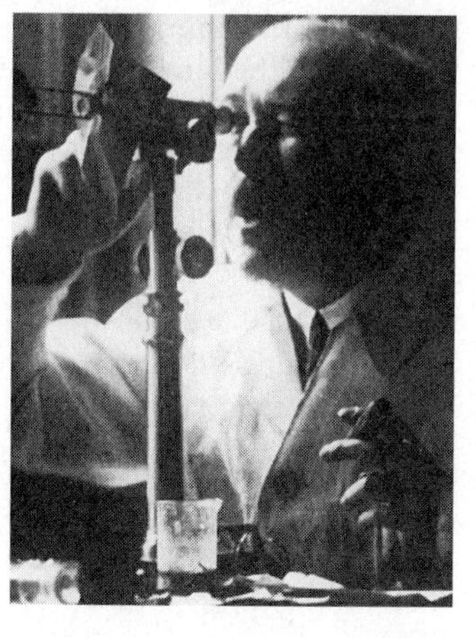

Otto Stern (1888–1969) was born in a part of Germany that is now in Poland, where he was educated and worked in several universities until he left Germany in 1933 to avoid persecution and emigrated to the United States. He was educated and trained as a theorist, but changed to experimentation when he began his molecular beam experiments in 1920 at the University of Frankfurt with Walter Gerlach. He continued his distinguished career in Hamburg and later at Carnegie Tech in Pittsburgh. He received the Nobel Prize in 1943. *AIP Emilio Segrè Visual Archives, Segrè collection.*

able to move while emitting or absorbing electromagnetic radiation. We will see in Section 7.6 that the selection rule for m_ℓ will not allow more than three different lines in the normal Zeeman effect (see Problem 29).

Efforts were begun in the 1920s to detect the effects of space quantization (m_ℓ) by measuring the energy difference ΔE as in Example 7.4. In 1922 O. Stern and W. Gerlach reported the results of an experiment that clearly showed evidence for space quantization. If an external magnetic field is inhomogeneous—for example, if it is stronger at the south pole than at the north pole—then there will be a net force on a magnet placed in the field as well as a torque. This force is represented in Figure 7.7, where the net force on $\boldsymbol{\mu}$ (direction of S to N in bar magnet) is different for different orientations of $\boldsymbol{\mu}$ in the inhomogeneous magnetic field **B**.

Now if we pass an atomic beam of particles in the $\ell = 1$ state through a magnetic field along the z direction, we have from Equation (7.31), $V_B = -\mu_z B$, and the force on the particles is $F_z = -(dV_B/dz) = \mu_z(dB/dz)$. There will be a different force on each of the three possible m_ℓ states. A schematic diagram of the Stern-Gerlach experiment is shown in Figure 7.8. The $m_\ell = +1$ state will be deflected up, the $m_\ell = -1$ state down, and the $m_\ell = 0$ state will be undeflected.

Stern and Gerlach performed their experiment with silver atoms and observed two distinct lines, not three. This was clear evidence of space quantization, although the number of m_ℓ states is always odd $(2\ell + 1)$ and should have produced an odd number of lines if the space quantization were due to the magnetic quantum number m_ℓ.

Example 7.5

In 1927 T. E. Phipps and J. B. Taylor of the University of Illinois reported an important experiment similar to the Stern-Gerlach experiment but using hydrogen atoms instead of silver. This was done because hydrogen is the simplest atom, and the separation of the atomic beam in the inhomogeneous magnetic field would allow a clearer interpretation. The atomic hydrogen beam was produced in a discharge tube having a temperature of 663 K. The highly collimated beam passed along the x direction through an inhomogeneous field (of length 3 cm) having an average value of 1240 T/m along the z direction. If the magnetic moment of the hydrogen atom is 1 Bohr magneton, what is the separation of the atomic beam?

Solution: The force can be found from the potential energy of Equation (7.31).

$$F_z = -\frac{dV}{dz} = \mu_z \frac{dB}{dz}$$

The acceleration of the hydrogen atom along the z direction is $a_z = F_z/m$. The separation of the atom along the z direction due to this acceleration is $d = a_z t^2/2$. The time that the atom spends within the inhomogeneous field is $t = \Delta x/v_x$ where Δx is the length of the inhomogeneous field, and v_x is the constant speed of the atom within the field. The separation d is therefore found from

$$d = \frac{1}{2}a_z t^2 = \frac{1}{2}\left(\frac{F_z}{m}\right)t^2 = \frac{1}{2m}\left(\mu_z \frac{dB}{dz}\right)\left(\frac{\Delta x}{v_x}\right)^2$$

We know all the values needed to determine d except the speed v_x, but we do know the temperature of the hydrogen gas. The average energy of the atoms collimated along the x direction is $\frac{1}{2}m\langle v_x^2\rangle = \frac{3}{2}kT$. We calculate $\langle v_x^2\rangle$ to be

$$v_x^2 = \frac{3kT}{m} = \frac{3(1.38 \times 10^{-23}\ \text{J/T})(663\ \text{K})}{1.67 \times 10^{-27}\ \text{kg}}$$

$$= 1.64 \times 10^7\ \text{m}^2/\text{s}^2$$

The separation d of the one atom is now determined to be

$$d = \frac{1}{2(1.67 \times 10^{-27}\ \text{kg})}\left(9.27 \times 10^{-24}\frac{\text{J}}{\text{T}}\right)\left(1240\frac{\text{T}}{\text{m}}\right)$$

$$\times \frac{(0.03\text{m})^2}{(1.64 \times 10^7\ \text{m}^2/\text{s}^2)} = 0.19 \times 10^{-3}\ \text{m}$$

Phipps and Taylor found only two distinct lines as did Stern and Gerlach for silver atoms, and the separation of the lines from the central ray with no magnetic field was 0.19 mm as we just calculated! The total separation of the two lines (one deflected up and one down) was 0.38 mm. The mystery still remained as to why there were only two lines.

7.5 Intrinsic Spin

By the early 1920s there was clearly a problem. Wolfgang Pauli was the first to suggest that a fourth quantum number (after n, ℓ, m_ℓ) assigned to the electron might explain the anomalous optical spectra. His reasoning for four quantum numbers was based on relativity, where there are four coordinates—three space and one time. The physical significance of this fourth quantum number was not made clear.

In 1925 Samuel Goudsmit and George Uhlenbeck, two young physics graduate students in Holland, proposed that *the electron must have an intrinsic angular momentum* and therefore a magnetic moment (because the electron is charged). Classically, this corresponds in the planetary model to the fact that the Earth rotates on its own axis as it orbits the sun. However, this simple classical picture runs into serious difficulties when applied to the spinning charged electron. Ehrenfest showed that the surface of the electron (or electron cloud) would have to be moving at a velocity greater than the speed of light! If such an intrinsic angular momentum exists, we must regard it as a *purely quantum-mechanical result* (see Problems 38 and 39).

Intrinsic spin quantum number

Magnetic spin quantum number

In order to explain experimental data, Goudsmit and Uhlenbeck proposed that the electron must have an **intrinsic spin quantum number** $s = 1/2$. The spinning electron reacts similarly to the orbiting electron in a magnetic field. Therefore, we should try to find quantities analogus to the angular momentum variable L, L_z, ℓ, and m_ℓ. By analogy, there will be $2s + 1 = 2(1/2) + 1 = 2$ components of the spin angular momentum vector $\mathbf{s}$. Thus the **magnetic spin quantum number** m_s has only two values, $m_s = \pm 1/2$. The electron's spin will be oriented either "up" or "down" in a magnetic field (see Figure 7.9), and the electron can never be spinning with its *magnetic moment* μ_s exactly along the z axis (the direction of the external magnetic field $\mathbf{B}$).

For each atomic state described by the three quantum numbers (n, ℓ, m_ℓ) discussed previously, there are now two distinct states, one with $m_s = +1/2$ and one with $m_s = -1/2$. These states will be degenerate in energy unless the atom is in an external magnetic field. In a magnetic field these states will have different energies due to an energy separation like that of Equation (7.33). We say the

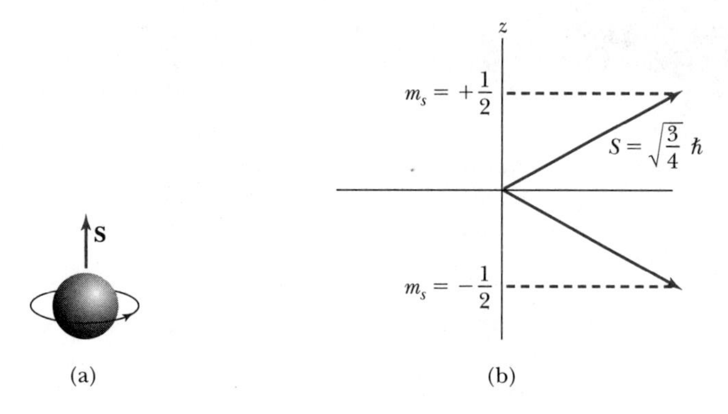

FIGURE 7.9 (a) A purely classical schematic of the intrinsic spin angular momentum, **S**, of a spinning electron. (b) The quantization of **S**, which can have only two positions in space relative to z (direction of external magnetic field). The z component of **S** is $S_z = \pm\hbar/2$.

splitting of these energy levels by the magnetic field has removed the energy degeneracy.

Intrinsic spin angular momentum vector

The **intrinsic spin angular momentum vector S** has a magnitude of $|\mathbf{S}| = \sqrt{s(s+1)}\hbar = \sqrt{3/4}\hbar$. The magnetic moment is $\boldsymbol{\mu}_s = -(e/m)\,\mathbf{S}$, or $-2\mu_\mathrm{B}\mathbf{S}/\hbar$. The fact that the coefficient of $\mathbf{S}/\hbar$ is $-2\mu_\mathrm{B}$ rather than $-\mu_\mathrm{B}$ as with the orbital angular momentum **L** is a consequence of the theory of relativity (Dirac equation), and we will not pursue the matter further here. This numerical factor relating the magnetic moment to each angular momentum vector is called the **gyromagnetic ratio.** It is designated by the letter g with the appropriate subscript (ℓ or s), so that $g_\ell = 1$ and $g_s = 2$. In terms of the gyromagnetic ratios, then,

Gyromagnetic ratio

$$\boldsymbol{\mu}_\ell = -\frac{g_\ell \mu_\mathrm{B}\mathbf{L}}{\hbar} = -\frac{\mu_\mathrm{B}\mathbf{L}}{\hbar} \tag{7.34a}$$

$$\boldsymbol{\mu}_s = -\frac{g_s \mu_\mathrm{B}\mathbf{S}}{\hbar} = -2\frac{\mu_\mathrm{B}\mathbf{S}}{\hbar} \tag{7.34b}$$

The z component of **S** is $S_z = m_s\hbar = \pm\hbar/2$.

We can now understand why the experiment of Stern and Gerlach only produced two distinct lines. If the atoms were in a state with $\ell = 0$, there would be no splitting due to m_ℓ. However, there is still space quantization due to the intrinsic spin that would be affected by the inhomogeneous magnetic field. The same arguments used previously for $\boldsymbol{\mu}_\ell$ (we now use the subscript ℓ to indicate the magnetic moment due to the orbiting electron and the subscript s to indicate the magnetic moment due to intrinsic spin) can now be applied to $\boldsymbol{\mu}_s$, and the potential energy, Equation (7.27), becomes

$$V_B = -\boldsymbol{\mu}_s \cdot \mathbf{B} = +\frac{e}{m}\,\mathbf{S}\cdot\mathbf{B} \tag{7.35}$$

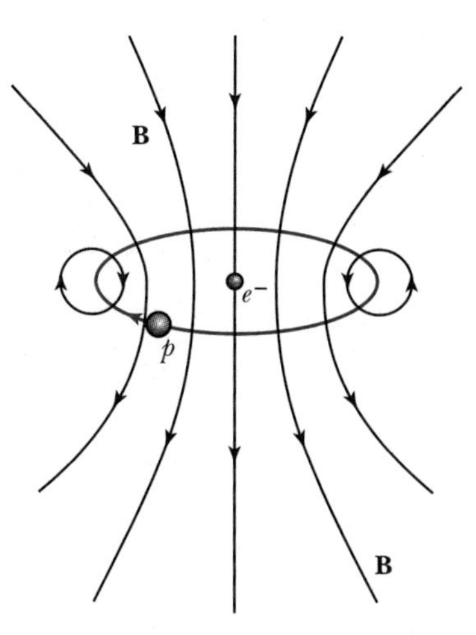

FIGURE 7.10 The hydrogen atom in the frame of reference of the electron. In this case, the orbiting proton creates a magnetic field at the position of the electron.

If we look at the hydrogen atom in the frame of the orbiting electron, we have the classical result shown in Figure 7.10. This classical picture indicates that the orbiting proton creates a magnetic field at the position of the electron. Therefore, even without an external magnetic field, the electron will feel the effects of an internal magnetic field, and Equation (7.35) predicts an energy difference depending on whether the electron's spin is up or down. Many levels are effectively split into two different states called *doublets.*

The relativistic quantum theory proposed by P. A. M. Dirac in 1928 showed that the intrinsic spin of the electron *required* a fourth quantum number as a consequence of the theory of relativity.

Example 7.6

How many distinctly different states (and therefore wave functions) exist for the 4d level of atomic hydrogen?

Solution: With the inclusion of the magnetic spin quantum number the number of states has multiplied. For the 4d level ($n = 4$, $\ell = 2$) there are $2\ell + 1 = 5$ different values of m_ℓ. For each of these m_ℓ $(-2, -1, 0, 1, 2)$, there are two m_s states $(\pm 1/2)$. Therefore there are 10 different possible individual states for a 4d level of atomic hydrogen.

Note that (in the absence of an applied magnetic field) the fourth quantum number makes the degeneracy of the nth quantum level $2n^2$.

7.6 Energy Levels and Electron Probabilities

We are now in a position to discuss a complete description of the hydrogen atom. Every possible state of the hydrogen atom has a distinct wave function that is specified completely by four quantum numbers: (n, ℓ, m_ℓ, m_s). In many cases the energy differences associated with the quantum numbers m_ℓ and m_s are insignificant (that is, the states are nearly degenerate), and we can describe the states adequately by n and ℓ alone: for example, 1s, 2p, 2s, 3d, and so on. Generally, capital letters (that is, S, P, D) are used to describe the orbital angular momentum of atomic states and lowercase letters (that is, s, p, d) to describe those for individual electrons. For hydrogen it makes little difference because each state only has a single electron, and we will use either specification.

In Figure 7.11 we show an energy-level diagram for hydrogen in the absence of an external magnetic field. The energy levels are degenerate with respect to

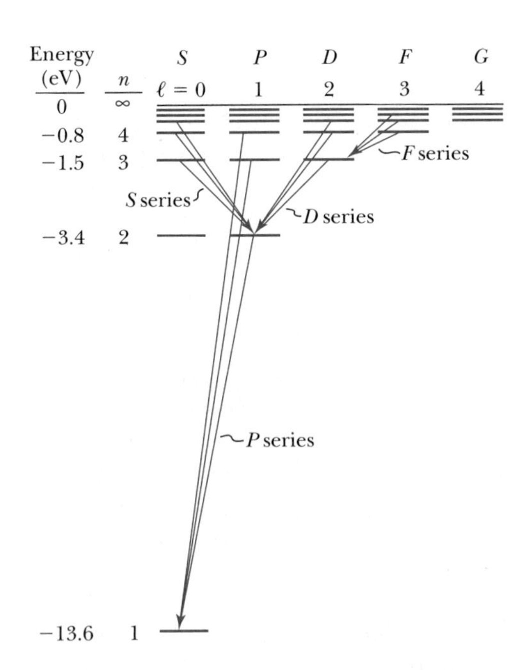

FIGURE 7.11 Energy-level diagram of hydrogen atom with no external magnetic field. Also shown are allowed photon transitions between some levels.

ℓ, m_ℓ, and m_s, but in a magnetic field this degeneracy is removed. For heavier atoms with several electrons, the degeneracy is removed—either because of internal magnetic fields within the atom or because the average potential energy due to the nucleus plus electrons is non-coulombic. In atoms with $Z > 1$ the smaller ℓ values tend to lie lower in energy for a given n (see Section 8.1). For example, in sodium or potassium, $E(4S) < E(4P) < E(4D) < E(4F)$. For hydrogen, the energy levels depend only on the principal quantum number n and are given to great accuracy by the Bohr theory.

We have previously learned that atoms emit characteristic electromagnetic radiation when they make transitions to states of lower energy. An atom in its ground state cannot emit radiation; it can absorb electromagnetic radiation, or it can gain energy through inelastic bombardment by particles, especially electrons. The atom will then have one or more of its electrons transferred to a higher energy state.

Selection Rules

Transition probabilities

We can use the wave functions obtained from the solution of the Schrödinger equation to calculate **transition probabilities** for the electron to change from one state to another. The results of such calculations show that electrons absorbing or emitting photons are much more likely to change states when $\Delta\ell = \pm 1$. Such transitions are called **allowed.** Other transitions, with $\Delta\ell \neq \pm 1$, are theoretically possible but occur with much smaller probabilities and are called **forbidden transitions.** There is no selection rule restricting the change Δn of the principal quantum number. The selection rule for the magnetic quantum number is $\Delta m_\ell = 0, \pm 1$. The magnetic spin quantum number m_s can (but need not) change between 1/2 and $-1/2$. We summarize the selection rules for allowed transitions:

Allowed and forbidden transitions

Selection rules

$$\begin{aligned} \Delta n &= \text{anything} \\ \Delta\ell &= \pm 1 \\ \Delta m_\ell &= 0, \pm 1 \end{aligned} \tag{7.36}$$

Some allowed transitions are diagrammed in Figure 7.11. Notice that there are no transitions shown for $3P \rightarrow 2P$, $3D \rightarrow 2S$, and $3S \rightarrow 1S$ because those transitions violate the $\Delta\ell = \pm 1$ selection rule.

If the orbital angular momentum of the atom changes by $\hbar$ when absorption or emission of radiation takes place, we must still check that all conservation laws are obeyed. What about the conservation of angular momentum? The only external effect on the atom during the absorption or emission process is that due to the photon being absorbed or emitted. If the state of the atom changes, then the photon must possess energy, linear momentum, and angular momentum. The $\Delta\ell = \pm 1$ selection rule strongly suggests that the photon carries one unit ($\hbar$) of angular momentum. By applying quantum mechanics to Maxwell's equations, it is possible to show* that electromagnetic radiation is quantized into photons having $E = h\nu$ and intrinsic angular momentum of $\hbar$. A consequence of the photon's intrinsic angular momentum is the circular polarization of an electromagnetic wave.

*See Leonard Schiff's *Quantum Mechanics* 3rd ed. New York: McGraw-Hill, 1968 for a discussion of both the semiclassical and quantum treatment of radiation.

Example 7.7

Which of the following transitions for quantum numbers (n, ℓ, m_ℓ, m_s) are allowed for the hydrogen atom, and if allowed, what is the energy involved?
(a) $(2, 0, 0, 1/2) \rightarrow (3, 1, 1, 1/2)$
(b) $(2, 0, 0, 1/2) \rightarrow (3, 0, 0, 1/2)$
(c) $(4, 2, -1, -1/2) \rightarrow (2, 1, 0, 1/2)$

Solution: We want to compare $\Delta\ell$ and Δm_ℓ with the selection rules of Equation (7.36). If allowed, the energies may be obtained from Equation (7.20) with $E_0 = 13.6$ eV.
(a) $\Delta\ell = +1$, $\Delta m_\ell = 1$; allowed.

$$\Delta E = E_3 - E_2 = -13.6 \text{ eV}\left(\frac{1}{3^2} - \frac{1}{2^2}\right)$$

$= 1.89$ eV, corresponding to absorption of a 1.89-eV photon.

(b) $\Delta\ell = 0$, $\Delta m_\ell = 0$; not allowed, because $\Delta\ell \neq \pm 1$.

(c) $\Delta\ell = -1$, $\Delta m_\ell = 1$; allowed. Notice that $\Delta n = -2$ and $\Delta m_s = +1$ does not affect whether the transition is allowed.

$$\Delta E = E_2 - E_4 = -13.6 \text{ eV}\left(\frac{1}{2^2} - \frac{1}{4^2}\right)$$

$= -2.55$ eV, corresponding to emission of a 2.55-eV photon.

Probability Distribution Functions

In the Bohr theory of the hydrogen atom, the electrons were pictured as orbiting around the nucleus in simple circular (or elliptical) orbits. The position vector **r** of the electron was well defined. In the wave picture of the atom, we must use wave functions to calculate the probability distributions* of the electrons. The "position" of the electron is therefore spread over space and is not well defined. The distributions can be found by examining the separable wave functions $R(r)$, $f(\theta)$, and $g(\phi)$. The $g(\phi)$ distribution is simplest because it leads to uniform probability—all values of ϕ are equally likely. It is easy to see why. Because the azimuthal part of the wave function is always of the form $e^{im_\ell\phi}$, the probability density $\psi^*\psi$ will contain a corresponding factor of $(e^{im_\ell\phi})^* e^{im_\ell\phi} = e^{-im_\ell\phi}e^{im_\ell\phi} = e^0 = 1$.

We may use the radial wave function $R(r)$ to calculate radial probability distributions of the electron (that is, the probability of the electron being at a given r). As was discussed in Section 5.6, the probability of finding the electron in a differential volume element dV is

$$dP = \psi^*(r, \theta, \phi)\psi(r, \theta, \phi)\, dV \tag{7.37}$$

We are interested in finding the probability $P(r)\,dr$ of the electron being between r and $r + dr$. The differential volume element in spherical polar coordinates is

$$dV = r^2 \sin\theta\, dr\, d\theta\, d\phi$$

Therefore,

$$P(r)\,dr = r^2 R^*(r)R(r)\,dr \int_0^\pi |f(\theta)|^2 \sin\theta\, d\theta \int_0^{2\pi} |g(\phi)|^2\, d\phi \tag{7.38}$$

We are integrating over θ and ϕ, because we are only interested in the radial dependence. If the integrals over $f(\theta)$ and $g(\phi)$ have already been normal-

*It may be useful at this time to review Section 5.6, where the relationships between probability and wave functions were discussed.

TABLE 7.2
Hydrogen Atom Radial Wave Functions

n	ℓ	$R_{n\ell}(r)$
1	0	$\frac{2}{a_0^{3/2}} e^{-r/a_0}$
2	0	$\left(2 - \frac{r}{a_0}\right) \frac{e^{-r/2a_0}}{(2a_0)^{3/2}}$
2	1	$\frac{r}{a_0} \frac{e^{-r/2a_0}}{\sqrt{3}(2a_0)^{3/2}}$
3	0	$\frac{1}{(a_0)^{3/2}} \frac{2}{81\sqrt{3}} \left(27 - 18\frac{r}{a_0} + 2\frac{r^2}{a_0^2}\right) e^{-r/3a_0}$
3	1	$\frac{1}{(a_0)^{3/2}} \frac{4}{81\sqrt{6}} \left(6 - \frac{r}{a_0}\right) \frac{r}{a_0} e^{-r/3a_0}$
3	2	$\frac{1}{(a_0)^{3/2}} \frac{4}{81\sqrt{30}} \frac{r^2}{a_0^2} e^{-r/3a_0}$

ized to unity, the probability of finding the electron between r and $r + dr$ reduces to

$$P(r)\,dr = r^2|R(r)|^2\,dr \tag{7.39}$$

Radial probability

The first few radial wave functions are listed in Table 7.2, where a_0 = Bohr radius $= 0.53 \times 10^{-10}$ m. The radial probability density is

$$P(r) = r^2|R(r)|^2 \tag{7.40}$$

This probability density depends only on n and ℓ. In Figure 7.12 we display both $R(r)$ and $P(r)$ for the lowest-lying states of the hydrogen atom.

Example 7.8

Find the most probable radius for the electron of a hydrogen atom in the $1s$ and $2p$ states.

Solution: To find the most probable radial value we take the derivative of the probability density $P(r)$ with respect to r and set it equal to zero.

1*s* state:

$$\frac{d}{dr}P(r) = 0 = \frac{d}{dr}\left(\frac{4e^{-2r/a_0}}{a_0^3}r^2\right)$$

$$0 = \frac{4}{a_0^3}\left(-\frac{2}{a_0}r^2 + 2r\right)e^{-2r/a_0}$$

$$\frac{2r^2}{a_0} = 2r$$

$r = a_0$ Most probable radius for $1s$ state electron (7.41)

2*p* state:

$$\frac{d}{dr}\left(\frac{e^{-r/a_0}}{3(2a_0)^3}\frac{r^4}{a_0^2}\right) = 0$$

$$\frac{1}{24a_0^5}\left(-\frac{r^4}{a_0} + 4r^3\right) = 0$$

$$\frac{r^4}{a_0} = 4r^3$$

$r = 4a_0$ Most probable radius for $2p$ state electron (7.42)

Notice that the most probable radii for the $1s$ and $2p$ states agree with the Bohr radii. This occurs only for the largest possible ℓ value for each n (see Problem 32).

FIGURE 7.12 (a) The radial wave function $R_{n\ell}(r)$ plotted as a function of radius (in units of Bohr radius a_0) for several states of the hydrogen atom. (b) The radial probability distribution $P_{n\ell}$, which gives the probability of the electron being between r and $r + dr$.

Radial wave functions ($R_{n\ell}$)

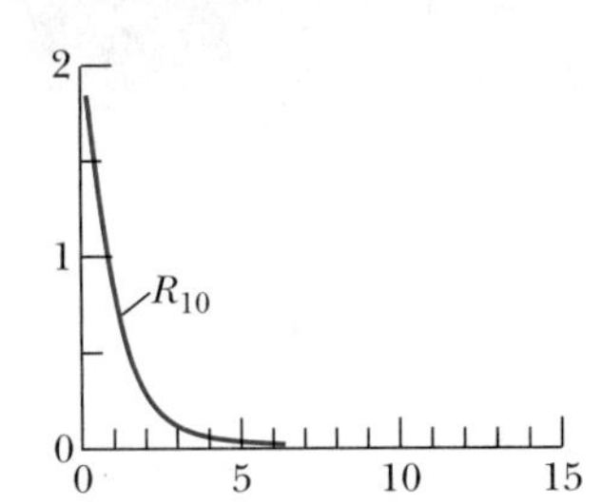

Radial probability distribution ($P_{n\ell}$)

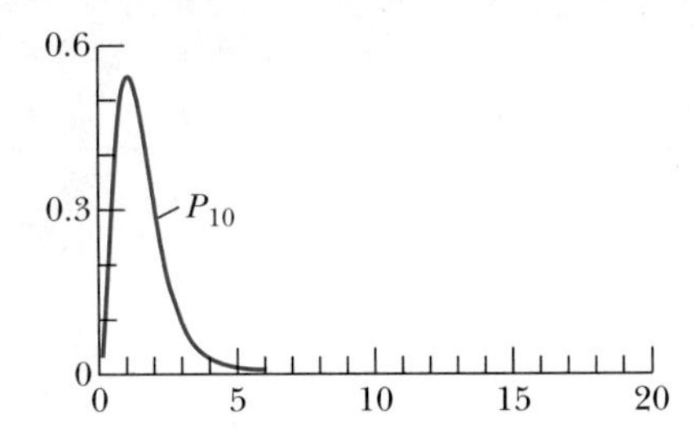

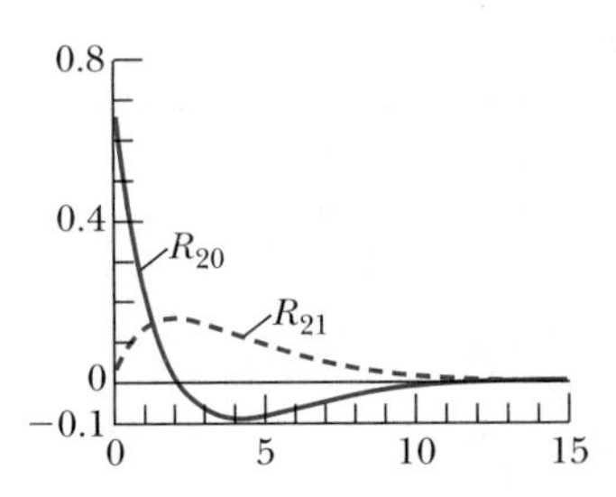

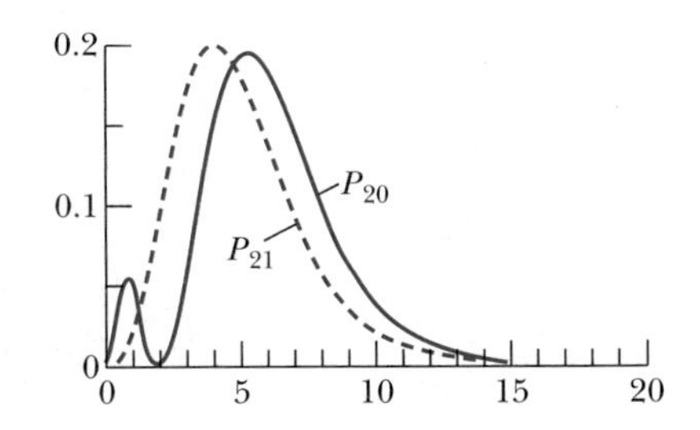

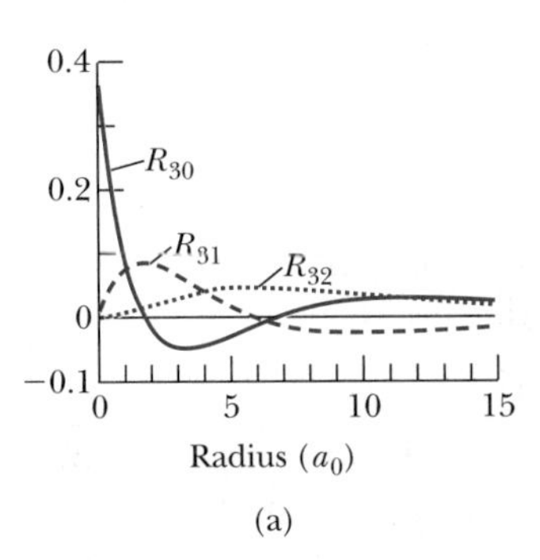

Radius (a_0)

(a)

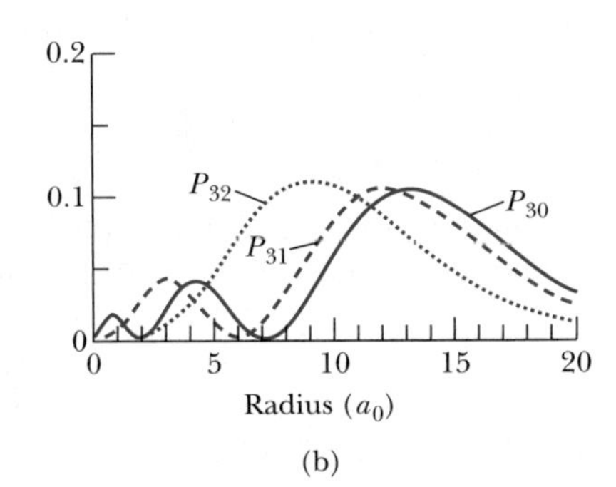

Radius (a_0)

(b)

Example 7.9

Calculate the average orbital radius of a 1s electron in the hydrogen atom.

Solution: The expectation (or average) value of r is (see Section 6.3)

$$\langle r\rangle = \int \psi^*(r, \theta, \phi)\, r\psi(r, \theta, \phi)\, dV$$
$$= \int r\mathrm{P}(r)\, dr$$

where we have again integrated over θ and ϕ.

$$\langle r\rangle = \int_0^{\infty} \frac{4}{{a_0}^3}\, e^{-2r/a_0}\, r^3 dr$$

We look up this integral in Appendix 3 and find

$$\int_0^{\infty} r^3 e^{-2r/a_0}\, dr = \frac{3{a_0}^4}{8}$$

so that

$$\langle r\rangle = \frac{4}{{a_0}^3}\,\frac{3{a_0}^4}{8} = \frac{3}{2}\, a_0 \qquad \text{For the 1}s\text{ state electron}$$

Therefore, the average electron radius in the 1s state is larger than the most probable value, the Bohr radius. We can see that this result is reasonable by examining the radial probability distribution for the 1s state displayed in Figure 7.12. The maximum (or most probable) value occurs at a_0, but the average is greater than a_0 because of the shape of the tail of the distribution.

Example 7.10

What is the probability of the electron in the $1s$ state of the hydrogen atom being at a radius greater than the Bohr radius a_0?

Solution: We simply integrate the radial probability distribution from $r = a_0$ to ∞, because $P(r)$ is already normalized (that is, it has a unit probability of being somewhere between 0 and ∞).

$$\text{Probability} = \int_{a_0}^{\infty} P(r)\,dr$$

$$= \frac{4}{a_0^3} \int_{a_0}^{\infty} e^{-2r/a_0} r^2\,dr$$

We look up the indefinite integral in Appendix 3 and evaluate to find the result

$$\text{Probability} = \frac{4}{a_0^3}\left(\frac{5}{4} a_0^3 e^{-2}\right) = 5e^{-2} = 0.68$$

The probability of the electron being outside the Bohr radius in a $1s$ state is greater than 50%. This explains why we found $\langle r \rangle_{1s} = 1.5\,a_0$. This result is consistent with the shape of the $1s$ curve in Figure 7.12b.

The probability distributions for the $\ell = 0$ state electrons are spherically symmetric, because the wave functions have no θ or ϕ dependence (see Table 7.1). For $\ell > 0$ the distributions are interesting because of the $f(\theta)$ dependence. For example, consider a p orbital. Referring to Table 7.1, we see that there are two possibilities for the angular part of the wave function. If $\ell = 1$ and $m_\ell = 0$, the Y_{10} will be a factor in the wave function, and therefore $\cos^2\theta$ will be a factor in the probability density $\psi^*\psi$. In this case the probability density will be highest near 0° and 180°, that is, near the $+z$ axis and $-z$ axis. The other possible combinations for the quantum numbers of an electron in a p orbital are $\ell = 1$ and $m_\ell = \pm 1$. Now $Y_{1\pm 1}$ will go into the wave function, and hence $\sin^2\theta$ will be a factor in the probability density $\psi^*\psi$. The probability is highest at $\theta = 90°$, that is, in the xy plane. The probability distributions seen in Figure 7.13 are consistent with this analysis.

When we look at d orbitals, the situation becomes a bit more complicated, but a similar analysis will allow us to see at what angles θ the probability density is maximized. For the $\ell = 2$, $m_\ell = 0$ state, we can see that Y_{20}^2 must have a maximum around $\theta = 0°$ and 180°. Once again, these results are shown in Figure 7.13. Similarly $Y_{2\pm 2}^2$ (corresponding to the $\ell = 2$, $m_\ell = \pm 2$ states) has a maximum in the xy plane. For the $\ell = 2$, $m_\ell = \pm 1$ states, we find a factor $\sin^2\theta\cos^2\theta$

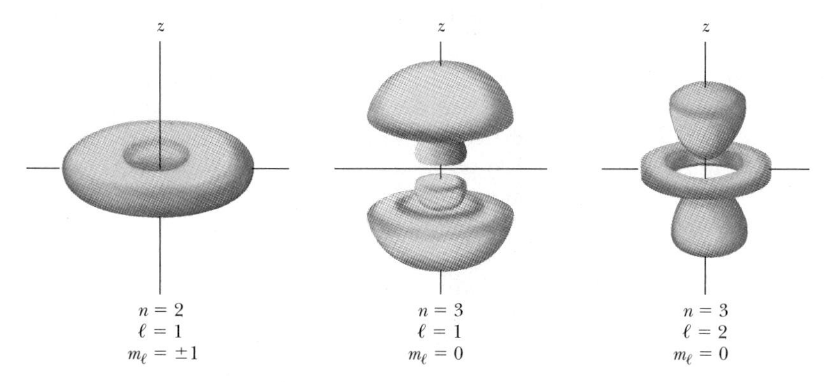

FIGURE 7.13 Pictorial representation of the probability density $|\Psi(r, \theta, \phi)|^2$ for the hydrogen atom for three different electron states. There is axial symmetry about the z axis in each case. The shaded regions indicate those parts of space with the highest probability densities.

coming from $Y^2_{2\pm1}$. For these states the probability maxima are at $\theta = 45°$ and 135°.

It is interesting to consider in which state (for a given n) the electron is closest to the origin. We can calculate $\langle r \rangle$ for the 2s and 2p states (see Problem 36) and find that the 2p average radius is smaller. However, because the $P(r)$ for the 2s state has two maxima, one with $r < a_0$, the electron in the 2s state will actually spend more time very close to the nucleus than will one in the 2p state. This effect can be seen in Figure 7.12, where the radial distribution for $P(r)$ in the 2s state extends farther out than that for 2p, but there is a secondary maximum for $P(r)$ for the 2s state near a_0.

Summary

The Schrödinger wave theory is applied to atomic physics, beginning with the hydrogen atom. The application of the boundary conditions leads to three quantum numbers:

n	principal quantum number
ℓ	orbital angular momentum quantum number
m_ℓ	magnetic quantum number

with the values and restrictions (all are integers)

$$n = 1, 2, 3, 4, \ldots \qquad n > 0$$

$$\ell = 0, 1, 2, 3, \ldots, (n-1) \qquad \ell < n$$

$$m_\ell = -\ell, -\ell+1, \ldots, 0, 1, \ldots, \ell-1, \ell \qquad |m_\ell| \le \ell$$

The energy of the electron–proton system is quantized and depends to first order only on n. The orbital angular momentum L is quantized by $L = \sqrt{\ell(\ell+1)}\hbar$ and not by $n\hbar$ as in the Bohr theory. We use letter names $s, p, d, f, g, h, \ldots$, to indicate the ℓ value for a given electron.

The z component of $\mathbf{L}$ is quantized, and $L_z = m_\ell \hbar$. This is referred to as *space quantization,* because $\mathbf{L}$ can only have certain orientations in space. In the absence of a magnetic field, the energy is degenerate with respect to ℓ and m_ℓ. In an external magnetic field each $n\ell$ level is split into $2\ell + 1$ different energy states (normal Zeeman effect).

In order to explain increasingly complex atomic spectra, a fourth quantum number was introduced by Goudsmit and Uhlenbeck. This quantum number s is related to the electron's intrinsic angular momentum, commonly referred to as *spin.* The electron spin quantum number is $s = 1/2$, and the values of the magnetic spin quantum number m_s are $\pm 1/2$. Stern and Gerlach observed in 1922 the effects of intrinsic spin, although at the time it was confused with orbital angular momentum.

The selection rules for allowed transitions for a change from one state to another are

$$\Delta n = \text{anything}$$

$$\Delta \ell = \pm 1 \tag{7.36}$$

$$\Delta m_\ell = 0, \pm 1$$

The probability of finding an electron between r and $r + dr$ is $P(r)\,dr = r^2 |R(r)|^2 dr$ where $R(r)$ is the radial wave function.

Questions

1. Do the radial wave functions depend on m_ℓ? Explain your reasons.
2. Would the radial wave functions be different for a potential $V(r)$ other than the Coulomb potential? Explain.
3. For what energy levels in the hydrogen atom will we not find $\ell = 2$ states?
4. What are the differences and similarities between atomic levels, atomic states, and atomic spectral lines? When do spectral lines occur?
5. Can the magnetic moment of an atom line up exactly with an external magnetic field? Explain.
6. What are the possible magnetic quantum numbers for an f state?
7. List all the reasons why you think a fourth quantum number (intrinsic spin) might have helped explain the complex optical spectra in the early 1920s.
8. Is it possible for the z component of the orbital magnetic moment to be zero, but not the orbital angular momentum? Explain.
9. A close examination of the spectral lines coming from starlight can be used to determine the star's magnetic field. Explain how this is possible.
10. If a hydrogen atom in the 2p excited state decays to the 1s ground state, explain how the following properties are conserved: energy, linear momentum, and angular momentum.

Problems

7.1 Application of the Schrödinger Equation to the Hydrogen Atom

1. Assume that the electron in the hydrogen atom is constrained to move only in a circle of radius a in the xy plane. Show that the separated Schrödinger equation for ϕ becomes

$$\frac{1}{a^2}\frac{d^2\psi}{d\phi^2} + \frac{2m}{\hbar^2}|E|\psi = 0$$

where ϕ is the angle describing the position on the circle. This is similar to the Bohr assumption.

2. Solve the equation in the previous problem for ψ. Find the allowed energies and angular momenta.

3. After separating the Schrödinger equation using $\psi = R(r)f(\theta)g(\phi)$, the equation for ϕ is

$$-\frac{1}{g}\frac{d^2g}{d\phi^2} = k^2$$

where k = constant. Solve for $g(\phi)$ in this equation and apply the appropriate boundary conditions. Show that k must be 0 or a positive or negative integer ($k = m_\ell$, the magnetic quantum number).

4. Using the transformation equations between cartesian coordinates and spherical polar coordinates given in Figure 7.1, transform the Schrödinger Equation (7.2) from cartesian to spherical coordinates as given in Equation (7.3).

7.2 Solution of the Schrödinger Equation for Hydrogen

5. Show that the radial wave function R_{20} for $n = 2$ and $\ell = 0$ satisfies Equation (7.14). What energy E results? Is this consistent with the Bohr model?

6. Show that the radial wave function R_{21} for $n = 2$ and $\ell = 1$ satisfies Equation (7.10). What energy results? Is this consistent with the Bohr model?

7. Show that the radial wave function R_{21} for $n = 2$ and $\ell = 1$ is normalized.

8. The normalized wave function ψ for the ground state of hydrogen is given by

$$\psi_{100}(r, \theta, \phi) = \frac{1}{\sqrt{\pi a_0^{\,3}}}e^{-r/a_0}$$

Show that the wave function is normalized over all space.

7.3 Quantum Numbers

9. List all the possible quantum numbers (n, ℓ, m_ℓ) for the $n = 6$ level in atomic hydrogen.

10. For a $3p$ state give the possible values of n, ℓ, m_ℓ, L, L_z, L_x, and L_y.

11. Write down all the wave functions for the $3p$ level of hydrogen. Identify the wave functions by their quantum numbers. Use the solutions in Tables 7.1 and 7.2.

12. Prove that $\langle L^2\rangle = \ell(\ell + 1)\hbar^2$ by actually performing the summation for Equation (7.24).

13. What is the degeneracy of the $n = 6$ shell of atomic hydrogen considering (n, ℓ, m_ℓ) and no magnetic field?

14. For a $3d$ state draw all the possible orientations of the angular momentum vector **L**. What is $L_x^{\,2} + L_y^{\,2}$ for the $m_\ell = -1$ component?

15. What is the smallest value that ℓ may have if **L** is within 3° of the z axis?

16. Prove that the degeneracy of an atomic hydrogen state having principal quantum number n is n^2 (ignore the spin quantum number).

7.4 Magnetic Effects on Atomic Spectra—The Normal Zeeman Effect

17. Calculate the possible z components of the orbital angular momentum for an electron in a $4p$ state.

18. For hydrogen atoms in a $4d$ state what is the *maximum* difference in potential energy between atoms when placed in a magnetic field of 2.5 T? Ignore intrinsic spin.

19. Show that the wavelength difference between adjacent transitions in the normal Zeeman effect is given approximately by

$$\Delta\lambda = \frac{\lambda_0^{\,2}\,\mu_B B}{hc}$$

20. For hydrogen atoms in a d state sketch the orbital angular momentum with respect to the z axis. Use units of $\hbar$ along the z axis and calculate the allowed angles of μ_ℓ with respect to the z axis.

21. For a hydrogen atom in the $6f$ state, what is the minimum angle between the orbital angular momentum vector and the z axis?

22. The red Balmer series line in hydrogen ($\lambda = 656.5$ nm) is observed to split into three different spectral lines with $\Delta\lambda = 0.04$ nm between two adjacent lines when placed in a magnetic field B. What is the value of B if $\Delta\lambda$ is due to the energy splitting between two adjacent m_ℓ states?

23. A hydrogen atom in an excited $5f$ state is in a magnetic field of 3 T. How many different energy states can the electron have in the $5f$ subshell? (Ignore the magnetic spin effects.) What is the energy of the $5f$ state in the absence of a magnetic field? What will be the energy of each state in the magnetic field?

24. The magnetic field in a Stern-Gerlach experiment varies along the vertical direction as $dB_z/dz = 20$ T/cm. The horizontal length of the magnet is 7.1 cm, and the speed of the silver atoms averages 925 m/s. The mass of the silver atoms is 1.8×10^{-25} kg. Show that the z component of its magnetic moment is 1 Bohr magneton. What is the separation of the two silver atom beams as they leave the magnet?

25. An experimenter wants to separate silver atoms in a Stern-Gerlach experiment by at least 1 cm (a large separation) as they exit the magnetic field. To heat the silver she has an oven that can reach 1000°C and needs to order a suitable magnet. What should be the magnet specifications?

7.5 Intrinsic Spin

26. In an external magnetic field can the electron spin vector **S** point in the direction of **B**? Draw a diagram with $\mathbf{B} = B_0\,\mathbf{k}$ showing **S** and S_z.

27. Using all four quantum numbers (n, ℓ, m_ℓ, m_s) write down all possible sets of quantum numbers for the $5f$ state of atomic hydrogen. What is the total degeneracy?

28. Prove that the total degeneracy for an atomic hydrogen state having principal quantum number n is $2n^2$.

7.6 Hydrogen Atom Energy Levels and Electron Probabilities

29. Show that, for transitions between any two n states of atomic hydrogen, no more than three different spectral lines can be obtained for the normal Zeeman effect.

30. Find whether the following transitions are allowed, and if they are, find the energy involved and whether the photon is absorbed or emitted for the hydrogen atom: (a) $(5, 2, 1, \frac{1}{2}) \to (5, 2, 1, -\frac{1}{2})$; (b) $(4, 3, 0, \frac{1}{2}) \to (4, 2, 1, -\frac{1}{2})$; (c) $(5, 2, -2, -\frac{1}{2}) \to (1, 0, 0, -\frac{1}{2})$; (d) $(2, 1, 1, \frac{1}{2}) \to (4, 2, 1, \frac{1}{2})$.

31. In Figure 7.12, the radial distribution function $P(r)$ for the $2s$ state of hydrogen has two maxima. Find the values of r (in terms of a_0) where these maxima occur.

32. Find the most probable radial position for the electron of the hydrogen atom in the $2s$ state. Compare this value with that found for the $2p$ state in Example 7.8.

33. Sketch the probability function as a function of r for the $2s$ state of hydrogen. At what radius is the position probability equal to zero?

34. Calculate the probability of an electron in the ground state of the hydrogen atom being inside the region of the proton (radius $\approx 1 \times 10^{-15}$ m). (*Hint:* Note that $r \ll a_0$.)

35. Calculate the probability that an electron in the ground state of the hydrogen atom can be found between $0.95a_0$ and $1.05a_0$.

36. Find the expectation value of the radial position for the electron of the hydrogen atom in the $2s$ and $2p$ states.

37. Calculate the probability of an electron in the $2s$ state of the hydrogen atom being inside the region of the proton (radius $\approx 1 \times 10^{-15}$m). Repeat for $2p$ electron. (*Hint:* Note that $r \ll a_0$).

General Problems

38. Assume the following (incorrect!) classical picture of the electron intrinsic spin. Take the electrical energy of the electron to be equal to its mass energy concentrated into a spherical shell of radius R.

$$\frac{e^2}{4\pi\epsilon_0 R} = mc^2$$

Calculate R (called the *classical electron radius*). Now let this spherical shell rotate and calculate the velocity in order to obtain the electron intrinsic spin.

$$\text{Angular momentum} = I\omega = I\frac{v}{R} = \frac{\hbar}{2}$$

where $I =$ moment of inertia of a spherical shell $= 2mR^2/3$. Is the value of v obtained in this manner consistent with the theory of relativity? Explain.

39. As in the previous problem, we want to calculate the speed of the rotating electron. Now let's assume that the diameter of the electron is equal to the Compton wavelength of an electron. Calculate v and comment on the result.

40. Consider a hydrogenlike atom such as He^+ or Li^{++} that has a single electron outside a nucleus of charge $+Ze$. (a) Rewrite the Schrödinger equation with the new Coulomb potential. (b) What change does this new potential have on the separation of variables? (c) Will the radial wave functions be affected? Explain. (d) Will the spherical harmonics be affected? Explain.

41. For the previous problem find the wave function ψ_{100}.

42. Consider a hydrogen atom in the $3p$ state. (a) At what radius is the electron probability equal to zero? (b) At what radius will the electron probability be a maximum? (c) For $m_\ell = 1$, at what angles θ will the electron probability be equal to zero? What about for $m_\ell = -1$?

43. Consider a "muonic atom," which consists of a proton and a negative μ^-. Compute the ground state energy following the methods used for the hydrogen atom.

CHAPTER 8

Many-Electron Atoms

What distinguished Mendeleev was not only genius, but a passion for the elements. They became his personal friends; he knew every quirk and detail of their behavior.

J. Bronowski

We began our study of atomic physics in the previous chapter with a study of the hydrogen atom. Now we will examine more complex atoms with multiple electrons. The highlight of this chapter is the understanding of how quantum mechanics explains the periodic table of the chemical elements. We shall also discover how even a qualitative understanding of atomic structure allows us to explain some of the physical and chemical properties of the elements.

8.1 Atomic Structure and the Periodic Table

We now have a good basis for understanding the hydrogen atom. How do we proceed to understand atoms with more than one electron? The obvious procedure is to add one more electron (helium atom) to the Schrödinger equation and solve for the wave functions. We soon run into formidable mathematical problems. Not only do we now have a nucleus with charge $+2e$ attracting two electrons, we also have the interaction of the two electrons repelling one another. The energy level obtained previously for the hydrogen atom for the first electron will be changed because of these new interactions. In general the problem of many-electron atoms cannot be solved exactly with the Schrödinger equation because of the complex potential interactions. However, with modern computers great progress has been made, and numerical calculations can be carried out with great precision. If we cannot solve exactly for the wave functions, then what can we learn about many-electron atoms?

Let us see how far we can go toward explaining experimental results without actually computing the wave functions of many-electron atoms. Physicists and chemists have been studying the properties of the elements for centuries. We

Dmitri Ivanovich Mendeleev (1834–1907), a Russian chemist, studied and was a professor at the University of St. Petersburg where he developed the periodic table. He successfully predicted the properties of several elements and discovered gallium (1875), scandium (1879), and germanium (1886). *AIP Emilio Segrè Visual Archives, W. F. Meggers collection.*

Wolfgang Pauli (1900–1958) was born in Austria, studied at Munich, and spent brief periods at Göttingen, Copenhagen, and Hamburg before accepting an appointment at Zurich in 1925 where he remained, except for brief periods at American universities including Princeton University during World War II. Pauli was a brilliant theoretical physicist who received the Nobel Prize in 1945. *Photo taken by S. A. Goudsmidt, AIP/Niels Bohr Library.*

know much about atomic sizes, chemical behavior, ionization energies, magnetic moments, and spectroscopic properties, including x-ray spectra. In 1869 the Russian chemist Dmitri Mendeleev arranged the elements into a periodic table that systematized many of their chemical properties. Mendeleev predicted several hitherto unknown elements that were ultimately discovered (most notably, germanium). The elucidation of the underlying physical basis of this (empirical) periodic table became one of the outstanding goals of physics. This goal was attained by the end of the 1920s.

The rise of quantum physics was accompanied by a vast accumulation of precise atomic spectroscopic data for optical frequencies. Attempting to understand these data, Wolfgang Pauli (Nobel Prize, 1945) proposed in 1925 his famous **exclusion principle:**

> Pauli Exclusion Principle: No two electrons in an atom may have the same set of quantum numbers (n, ℓ, m_ℓ, m_s).

This principle has far-reaching implications. Using it we can describe in a reasonable, precise fashion the organization of atomic electrons into shells and subshells. Pauli's exclusion principle applies to all particles of half-integer spin which are called *fermions* and can be generalized to include particles in the nucleus, where it is crucial to nuclear structure because neutrons and protons are both fermions.

The atomic electron structure* leading to the observed ordering of the periodic table can be understood by the application of two rules:

> **1.** The electrons in an atom tend to occupy the lowest energy levels available to them.
> **2.** Only one electron can be in a state with a given (complete) set of quantum numbers (Pauli exclusion principle). (8.1)

Let us apply these rules to the first few atoms in the periodic table. Hydrogen has quantum numbers (n, ℓ, m_ℓ, m_s) equal to $(1, 0, 0, \pm 1/2)$ in its lowest energy state (ground state). In the absence of a magnetic field, the $m_s = 1/2$ state is degenerate with the $m_s = -1/2$ state. In neutral helium the quantum numbers must be different for the two electrons, so if the quantum numbers are $(1, 0, 0, 1/2)$ for the first electron, those for the second electron must be $(1, 0, 0, -1/2)$. Direct experimental evidence shows that the two electrons in the He atom have their spins antialigned (spin angular momentum opposed) rather than aligned (spin angular momentum aligned). This supports the Pauli exclusion principle. These two electrons form a rather strong bond with their spin angular momentum antialigned. We speak of the electrons as being *paired,* and the total spin is zero.

The principal quantum number n has also been given letter codes:

$$\begin{array}{rcccc} n = 1 & 2 & 3 & 4 & \ldots \\ \text{Letter} = \text{K} & \text{L} & \text{M} & \text{N} & \ldots \end{array} \tag{8.2}$$

Because the binding energies depend mainly on n, the electrons for a given n are said to be in **shells.** We speak of the K shell, L shell, and so on (recall from Chapter 4 that this was nomenclature used to describe Moseley's x-ray results). The $n\ell$ descriptions are called **subshells.** We have $1s$, $2p$, $3d$ subshells. Both electrons in

*The use of hydrogenic quantum numbers for other atoms implies a hydrogenlike central field for the outer electrons of these atoms.

the He atom are in the K shell and $1s$ subshell (which is a shell in itself). We use a superscript to denote the number of electrons in each subshell. The hydrogen atom description is $1s^1$ or $1s$ (the superscript 1 is sometimes omitted), and the helium atom is $1s^2$.

The next atom in the table is lithium. There is no more space in the K shell because only two electrons are allowed. The next shell is the L shell ($n = 2$), and the possible subshells are $2s$ and $2p$. Rule 1 says the electrons will occupy the state with the lowest energy. Remember that semiclassically the $2s$ state (with zero angular momentum) has an orbit through the nucleus, whereas the $2p$ state has a more nearly circular orbit. An electron in the $2p$ subshell (Li) will experience a $+3e$ nuclear charge, but the positive nuclear charge will be partially screened* by the two electrons in the $1s$ shell. The effective charge that the $2p$ electron sees (or feels) will therefore be $Z_{\text{eff}} \sim +1e$. The $2s$ electron, on the other hand, spends more time than a $2p$ electron actually passing near the nucleus, hence the effective charge it experiences will be $Z_{\text{eff}} > +1e$. Therefore, an electron in the $2s$ subshell will experience a more attractive potential than a $2p$ electron and will thus lie lower in energy. The electronic structure of Li is $1s^22s^1$. The third electron has the quantum numbers $(2, 0, 0, \pm 1/2)$.

How many electrons may be in each subshell in order not to violate the Pauli exclusion principle?

	Total
For each m_ℓ: two values of m_s	2
For each ℓ: $(2\ell + 1)$ values of m_ℓ	$2(2\ell + 1)$

Thus each $n\ell$ subshell can have $2(2\ell + 1)$ electrons. The $1s$, $2s$, $3s$, $4s$ subshells may have only two electrons. The $2p$, $3p$, $4p$ subshells may have up to six. The $3d$, $4d$, $5d$ subshells may have up to ten.

We can now describe the electronic configurations of many-electron atoms. Although there are effects due to internal magnetic fields, in the absence of external magnetic fields the m_ℓ and m_s quantum numbers do not affect the atom's total energy. Thus, the different states available within the same subshell are nearly degenerate. For a qualitative understanding we need only refer to $n\ell$.

The filling of electrons in an atom generally proceeds until each subshell is full. When a subshell has its maximum number of electrons, we say it is *closed* or *filled*. Electrons with lower ℓ values spend more time inside the (inner) closed shells. Classically, we understand this result, because the lower ℓ values have more elliptical orbits than the higher ℓ values. The electrons with higher ℓ values are therefore more shielded from the nuclear charge $+Ze$, feel less Coulomb attraction, and lie higher in energy than those with lower ℓ values. For a given n the subshells fill in the order *s, p, d, f, g,* . . . . This shielding effect becomes so pronounced that the $4s$ subshell actually fills before the $3d$ subshell even though it has a larger n. This happens often as the higher-lying shells fill with electrons. Experimental evidence shows that the order of subshell filling given in Table 8.1 is generally correct. There are some important variations from this order, which produce the rare earth lanthanides and actinides. A schematic diagram of the subshell energy levels is shown in Figure 8.1.

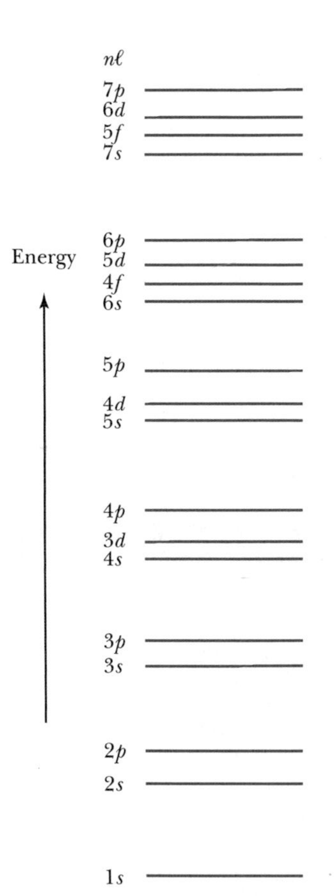

FIGURE 8.1 Approximate energy ordering of the subshells for the outermost electron in an atom. This representation assumes that the given subshell is receiving its first electron and that all lower subshells are full and all higher subshells are empty.

One nomenclature for identifying atoms is ${}_ZX$ where Z is the atomic number of the element (the number of protons), and X is the chemical symbol that

*"Screened" in this case means the electron will react to both the $+3e$ nucleus charge and $-2e$ electron charge within its own orbit.

TABLE 8.1
Order of Electron Filling in Atomic Subshells

n	ℓ	Subshell	Subshell Capacity	Total Electrons in All Subshells
1	0	$1s$	2	2
2	0	$2s$	2	4
2	1	$2p$	6	10
3	0	$3s$	2	12
3	1	$3p$	6	18
4	0	$4s$	2	20
3	2	$3d$	10	30
4	1	$4p$	6	36
5	0	$5s$	2	38
4	2	$4d$	10	48
5	1	$5p$	6	54
6	0	$6s$	2	56
4	3	$4f$	14	70
5	2	$5d$	10	80
6	1	$6p$	6	86
7	0	$7s$	2	88
5	3	$5f$	14	102
6	2	$6d$	10	112

identifies the element. The number of electrons in a neutral atom is Z. The Z is superfluous because every element has a unique Z. For example, ${}_8$O and O stand for the same element, because oxygen has $Z = 8$. In Chapter 12 we will discuss isotopes of elements in which the mass number of the element may vary because the number of neutrons in the nucleus may be different.

Example 8.1

Give the electron configuration and the $n\ell$ value of the last electrons in the subshell (called *valence* electrons) for the following atoms: ${}_{11}$Na, ${}_{18}$Ar, ${}_{20}$Ca, ${}_{35}$Br.

Solution: **${}_{11}$Na:** Sodium has 11 electrons, 10 of which are in the filled $1s^22s^22p^6$ subshells, which is called the *core* because it is filled. According to the order of filling given in Table 8.1, the extra electron must be in the $3s$ subshell with $n = 3$, $\ell = 0$. The electronic configuration is then $1s^22s^22p^63s^1$. The chemical properties of Na are determined almost exclusively by this one extra electron outside the core. The core is rather inert with the orbital and intrinsic angular momenta of the electrons paired to zero.

${}_{18}$Ar: From Table 8.1 we see that 18 electrons complete the $3p$ subshell, so the last electron has $n = 3$, $\ell = 1$, and the electronic configuration is $1s^22s^22p^63s^23p^6$. Argon has completely closed subshells and no extra (valence) electrons. This is the reason argon, one of the inert gases, is chemically inactive.

${}_{20}$Ca: After Ar the next two electrons go into the $4s$ subshell, so $n = 4$, $\ell = 0$, and the electronic configuration for calcium is $1s^22s^22p^63s^23p^64s^2$. There is a rather large energy gap between the $3p$ subshell and the $4s$ and $3d$ subshells (see Figure 8.1). The two electrons in the $4s$ subshell are situated rather precariously outside the inert core of Ar and can react strongly with other atoms.

${}_{35}$Br: One more electron past ${}_{35}$Br finishes the $4p$ subshell and makes the strongly inert gas krypton. The last electron in ${}_{35}$Br has $n = 4$ and $\ell = 1$, and the electronic configuration of the last few subshells is $3p^64s^23d^{10}4p^5$. Bromine badly needs one more electron to complete its subshell and is very active chemically with a high electron affinity—searching for that last electron to fill its $4p$ subshell.

It is now relatively easy to understand the structure of the periodic table shown in Figure 8.2. The ordering of electrons into subshells follows from the two rules of Equation (8.1). In Figure 8.2 the horizontal groupings are according to separate subshells. Atomic electron configurations are often denoted by only the last subshell, and all previous subshells are assumed to be filled. In

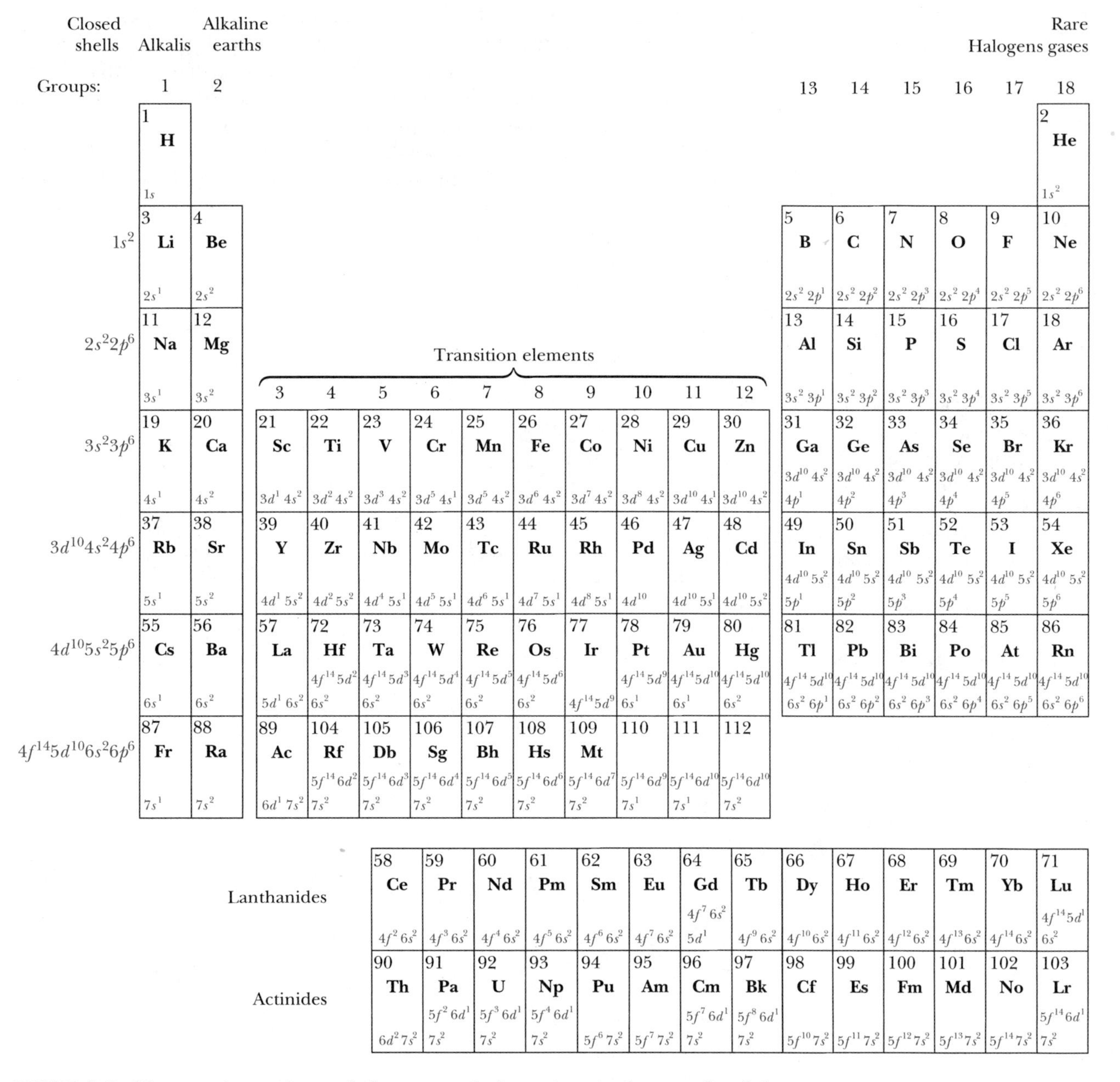

FIGURE 8.2 The atomic number and element symbol are given in the top of each box. The electron configuration for each element is specified by giving the values of the principal quantum numbers n, the angular momentum quantum numbers ℓ (s, p, d, or f), and the number of electrons outside closed shells. The configuration of some of the closed shells is given on the left.

Figure 8.2 only the last unfilled subshell configurations are shown. There are occasions when a smooth order doesn't occur. For example, ${}_{40}$Zr has $5s^2 4d^2$, that is, the $5s$ subshell is filled, but the next element, ${}_{41}$Nb, has the structure $5s^1 4d^4$. An electron has been taken from the $5s$ subshell and placed in the $4d$ subshell with an additional electron to make a total of four electrons in the $4d$ subshell. There are several such unusual cases as the atomic number increases. These details reflect the complex electron–electron interactions in a system of many particles.

Groups and periods

Let us briefly review some of the special arrangements of the periodic table. The vertical columns (or **groups**) have similar chemical and physical properties. This occurs because they have the same valence electron structure—that is, they have the same number of electrons in an ℓ orbit and can form similar chemical bonds. The horizontal rows are called **periods,** and they correspond to filling of the subshells. For example, in the fourth row the $4s$ subshell is filled first with 2 electrons, next the $3d$ subshell is filled with 10 electrons, and finally the $4p$ subshell is filled with 6 electrons. The fourth row consists of 18 elements and the filling of the $4s$, $3d$, and $4p$ subshells.

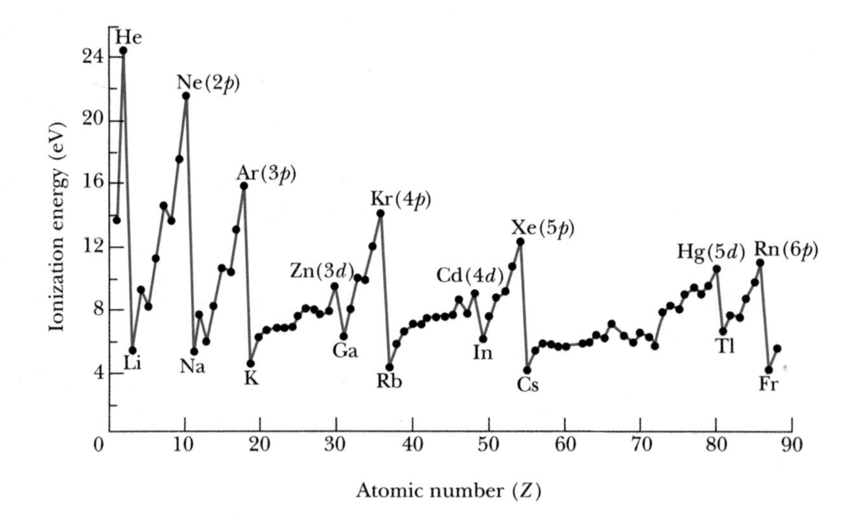

FIGURE 8.3 The ionization energies of the elements are shown versus the atomic numbers. The element symbols are shown for the peaks and valleys with the subshell closure in parentheses where appropriate. When a single electron is added to the p and d subshells, the ionization energy significantly decreases indicating the shell effects of atomic structure.

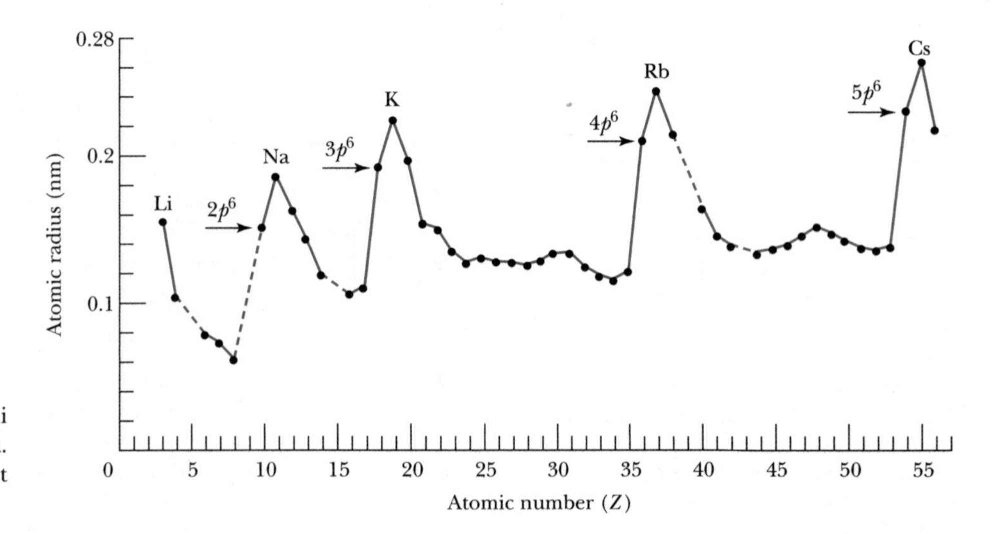

FIGURE 8.4 Atomic radii from ionic crystal atomic data. The radii are the smallest when the subshells are filled.

In order to compare some properties of elements we show the ionization energies of elements in Figure 8.3 and atomic radii in Figure 8.4. (The ionization energy is the energy required to remove the most weakly bound electron, forming a positive ion.) The electrical conductivity and resistivity also show subshell effects. Good electrical conductors need free electrons that are only weakly bound to their nuclei. In Chapter 10 we shall see similar patterns in superconducting properties. The differences according to subshells are remarkable.

Inert Gases

The last group of the periodic table is the inert gases. They are unique in that they all have closed subshells, a p subshell except for helium. They have no valence electrons, and the p subshell is tightly bound. These elements therefore are inert chemically. They do not easily form chemical bonds with other atoms. They have zero net spins, large ionization energies (Figure 8.3), and poor electrical conductivities. Their boiling points are quite low, and at room temperature they are monoatomic gases, because their atoms interact so weakly with each other.

Alkalis

Hydrogen and the alkali metals (Li, Na, K, etc.) form Group 1 of the periodic table. They have a single s electron outside an inert core. This electron can be easily removed, so the alkalis easily form positive ions with a charge $+1e$. Therefore, we say that their *valence* is $+1$. Figure 8.3 shows that the alkali metals have the lowest ionization energies. The drop in ionization energies between the inert gases and the alkalis is precipitous. The alkali metals are relatively good electrical conductors, because the valence electrons are free to move around from one atom to another.

Alkaline Earths

The alkaline earths are in Group 2 of the periodic table. These elements (Be, Mg, Ca, Sr, etc.) have two s electrons in their outer subshell, and although these subshells are filled, the s electrons can extend rather far from the nucleus and can be relatively easily removed. The alkali metals and alkaline earths have the largest atomic radii (Figure 8.4), because of their loosely bound s electrons. The ionization energies (Figure 8.3) of the alkaline earths are also low, but their electrical conductivity is high. The valence of these elements is $+2$, and they are rather active chemically.

Halogens

Immediately to the left of the inert gases, Group 17 lacks one electron from having a filled outermost subshell. These elements (F, Cl, Br, I, etc.) all have a valence of -1 and are chemically very active. They form strong ionic bonds (for example, NaCl) with the alkalis (valence $+1$) by gaining the electron given up easily by the alkali atom. In effect, a compound such as NaCl consists of a Na^+ and a Cl^- ion strongly bound by their mutual Coulomb interaction. The groups to the immediate left of the halogens have fewer electrons in the p shell. In Figure 8.4 it is apparent that the radii of the p subshell decrease as electrons are added. A more stable configuration occurs in the p subshell as it is filled, resulting in a more tightly bound atom.

Transition Metals

The three rows of elements in which the $3d$, $4d$, and $5d$ subshells are being filled are called the *transition elements* or *transition metals.* Their chemical properties are similar—primarily determined by the s electrons, rather than by the d subshell being filled. This occurs because the s electrons, with higher n values, tend to have greater radii than the d electrons. The filling of the $3d$ subshell leads to some important characteristics for elements in the middle of the period. These elements (Fe, Co, and Ni) have several d-shell electrons with unpaired spins (as dictated by Hund's rules, see Section 8.2). The spins of neighboring atoms in a crystal lattice align themselves, producing large magnetic moments and the ferromagnetic properties of these elements (see Section 10.4). As the d subshell is filled, the electron spins eventually pair off, and the magnetic moments, as well as the tendency for neighboring atoms to align spins, are reduced.

Lanthanides

The lanthanides ($_{58}$Ce to $_{71}$Lu), also called the *rare earths,* all have similar chemical properties. This occurs because they all have the outside $6s^2$ subshell completed while the smaller $4f$ subshell is being filled. The ionization energies (see Figure 8.3) are similar for the lanthanides. As occurs in the $3d$ subshell, the electrons in the $4f$ subshell often have unpaired electrons. These unpaired electrons align themselves, and because there can be so many electrons in the f subshell, large magnetic moments may occur. The large orbital angular momentum ($\ell = 3$) also helps the large ferromagnetic effects that some of the lanthanides have. The element holmium can have an extremely large internal magnetic field at low temperatures, much larger than even that of iron.

Actinides

The actinides ($_{90}$Th to $_{103}$Lr) are similar to the lanthanides in that inner subshells are being filled while the $7s^2$ subshell is complete. It is difficult to obtain reliable chemical data for these elements, because they are all radioactive. It is possible to keep significant quantities of a few actinide isotopes, which have sufficiently long half-lives. Examples of these are thorium-232, uranium-235, and uranium-238, which occur naturally, and neptunium-237 and plutonium-239, which are produced in the laboratory.

Example 8.2

Copper and silver have the two highest electrical conductivities. Explain how the electronic configurations of copper and silver can account for their very high electrical conductivities.

Solution: We need to refer to Figure 8.2 to investigate their electron configurations. We see that $_{28}$Ni has the structure $3d^8 4s^2$, but $_{29}$Cu has the structure $3d^{10} 4s^1$ and the next element, $_{30}$Zn, has $3d^{10} 4s^2$. Copper is unique in that one electron from the $4s$ subshell has changed to the $3d$ subshell. The remaining $4s$ electron is very weakly bound—in fact it is almost free.

Something similar happens to $_{47}$Ag in the next period. The elements on either side have completed the $4d^{10}$ subshell, and for $_{47}$Ag the $5s$ electron is only weakly bound. The elements $_{41}$Nb through $_{45}$Rh have an unpaired $5s$ electron, but incomplete $4d$ subshells, and so less screening—their $5s$ electrons are less free to wander than that of $_{47}$Ag.

8.2 Total Angular Momentum

If an atom has an orbital angular momentum and a spin angular momentum due to one or more of its electrons, we expect that, as is true classically, these angular momenta combine to produce a **total angular momentum.** We learned previously, in Section 7.5, that an interaction between the orbital and spin angular momenta in one-electron atoms causes splitting of energy levels into doublets, even in the absence of external magnetic fields. In this section we shall examine how the orbital and spin angular momenta combine and see how this results in energy-level splittings.

Single-Electron Atoms

We discuss initially only atoms having a single electron outside an inert core (for example, the alkalis). For an atom with orbital angular momentum **L** and spin angular momentum **S**, the total angular momentum **J** is given by

$$\mathbf{J} = \mathbf{L} + \mathbf{S} \tag{8.3}$$

Total angular momentum

Because L, L_z, S, and S_z are quantized, the total angular momentum and its z component J_z are also quantized. If j and m_j are the appropriate quantum numbers for the single electron we are considering, quantized values of J and J_z are, in analogy with the single electron of the hydrogen atom,

$$J = \sqrt{j(j+1)}\,\hbar \tag{8.4a}$$

$$J_z = m_j\hbar \tag{8.4b}$$

Because m_ℓ is integral and m_s is half-integral, m_j will always be half-integral. Just as the value of m_ℓ ranges from $-\ell$ to ℓ, the value of m_j ranges from $-j$ to j, and therefore j will be half-integral.

The quantizations of the magnitudes of **L**, **S**, and **J** are all similar.

$$L = \sqrt{\ell(\ell+1)}\,\hbar$$
$$S = \sqrt{s(s+1)}\,\hbar \tag{8.5}$$
$$J = \sqrt{j(j+1)}\,\hbar$$

The total angular momentum quantum number for the single electron can only have the values

$$j = \ell \pm s \tag{8.6}$$

which, because $s = 1/2$, can only be $\ell + 1/2$ or $\ell - 1/2$ (but j must be 1/2 if $\ell = 0$). The relationships of **J**, **L**, and **S** are shown in Figure 8.5. For an ℓ value of 1, the quantum number j is 3/2 or 1/2, depending on whether **L** and **S** are aligned or antialigned. The notation commonly used to describe these states is

$$nL_j \tag{8.7}$$

where n is the principal quantum number, j the total angular momentum quantum number, and L is an uppercase letter (S, P, D, etc.) representing the orbital angular momentum quantum number.

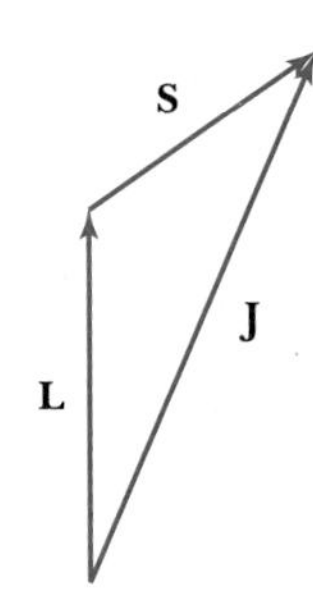

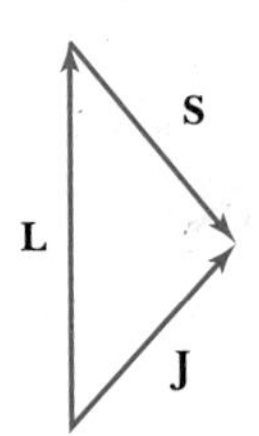

FIGURE 8.5 When forming the total angular momentum from the orbital and spin angular momenta, the addition must be done vectorially, **J** = **L** + **S**. We show schematically the addition of **L** and **S** with $\ell = 1$ and $s = 1/2$ to form vectors **J** with quantum numbers $j = 1/2$ and 3/2.

In Section 7.5 we briefly mentioned that the single electron of the hydrogen atom can feel an internal magnetic field $\mathbf{B}_{\text{internal}}$ due to the proton, which in the rest system of the electron appears to be circling it (see Figure 7.10). A careful examination of this effect shows that the spin of the electron and the orbital angular momentum can be coupled, an effect called *spin–orbit coupling.* The potential energy $V_{s\ell}$ will be equal to $-\boldsymbol{\mu}_s \cdot \mathbf{B}_{\text{internal}}$. The spin magnetic moment is proportional to $-\mathbf{S}$, and $\mathbf{B}_{\text{internal}}$ is proportional to $\mathbf{L}$, so that $V_{s\ell} \sim \mathbf{S} \cdot \mathbf{L} = SL \cos \alpha$, where α is the angle between $\mathbf{S}$ and $\mathbf{L}$. The result of this effect is to make the states with $j = \ell - 1/2$ slightly lower in energy than for $j = \ell + 1/2$, because α is smaller when $j = \ell + 1/2$. The same applies for the atom when placed in an external magnetic field. The same effect leads us to accept j and m_j as better quantum numbers than m_ℓ and m_s, even for single-electron atoms like hydrogen. We mean "better" in this case, because j and m_j are more directly related to a physical observable. A given state having a definite energy can no longer be assigned a definite L_z and S_z, but it can have a definite J_z. The wave functions will now depend on n, ℓ, j, and m_j. The spin–orbit interaction splits the $2P$ level into two states, $2P_{3/2}$ and $2P_{1/2}$, with $2P_{1/2}$ being lower in energy. There are additional relativistic effects, which will not be discussed here, that give corrections to the spin–orbit effect.

Spin–orbit coupling

In the absence of an external magnetic field, the total angular momentum is conserved in *magnitude and direction.* The effect of the internal magnetic field is to cause $\mathbf{L}$ and $\mathbf{S}$ to precess about $\mathbf{J}$. In an external magnetic field, however, $\mathbf{J}$ will precess about $\mathbf{B}_{\text{ext}}$, while $\mathbf{L}$ and $\mathbf{S}$ still precess about $\mathbf{J}$. The motion of $\mathbf{L}$ and $\mathbf{S}$ then becomes quite complicated.

Optical spectra are due to transitions between different energy levels. We have already discussed transitions for the hydrogen atom in Section 7.6 and gave the rules listed in Equation (7.36). For single-electron atoms, we now add the selection rules for Δj. The restriction of $\Delta\ell = \pm 1$ will require $\Delta j = \pm 1$ or 0. The allowed transitions for a single-electron atom are

$$\begin{aligned} \Delta n &= \text{anything} \qquad & \Delta \ell &= \pm 1 \\ \Delta m_j &= 0, \pm 1 \qquad & \Delta j &= 0, \pm 1 \end{aligned} \tag{8.8}$$

The selection rule for Δm_j follows from our results for Δm_ℓ in Equation (7.36) and from the result that $m_j = m_\ell + m_s$, where m_s is not affected.

Figure 7.10 presented an energy-level diagram for hydrogen showing many possible transitions. We show in Figure 8.6 a highly exaggerated portion of the hydrogen energy-level diagram for $n = 2$ and $n = 3$ levels showing the spin–orbit splitting. All of the states (except for the s states) are split into doublets. What appeared in Figure 7.10 to be one transition is now actually seven different

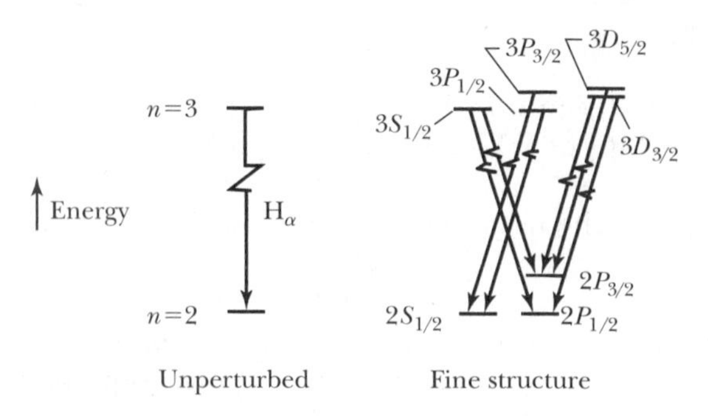

FIGURE 8.6 At the left the unperturbed H_α line is shown due to a transition between the $n = 3$ and $n = 2$ shells of the hydrogen atom. At right is shown the more detailed level structure (not to scale) of the hydrogen atom that leads to optical fine structure. The spin–orbit interaction splits each of the $\ell \neq 0$ states.

transitions. The splitting is quite small, but measurable—typically on the order of 10^{-5} eV in hydrogen. For example, the splitting between the $2P_{3/2}$ and $2P_{1/2}$ levels has been found to be 4.5×10^{-5} eV.

We show in Figure 8.7 the energy levels of a single-electron atom, sodium, compared with that of hydrogen. The single electron in sodium is $3s^1$, and the energy levels of sodium should be similar to that of $n = 3$ and above for hydrogen. However, the strong attraction of the electrons with small ℓ causes those energy levels to be considerably lower than for higher ℓ. Notice in Figure 8.7 that the $5f$ and $6f$ energy levels of sodium closely approach the hydrogen energy levels, but the $3s$ energy level of sodium is considerably lower. The transitions between the energy levels of sodium displayed in Figure 8.7 are consistent with the selection rules of Equation (8.8).

The fine splitting of the levels for different j are too small to be seen in Figure 8.7. Nevertheless these splittings are important, and they are easily detected in the optical spectra of sodium. The energy levels $3P_{3/2}$ and $3P_{1/2}$ are separated by about 2×10^{-3} eV, for example. This splits the $3p \rightarrow 3s$ ($\sim$2.1 eV) optical line into a doublet: the famous yellow sodium doublet, with $\lambda = 589.0$ nm and 589.6 nm (see also Example 8.6).

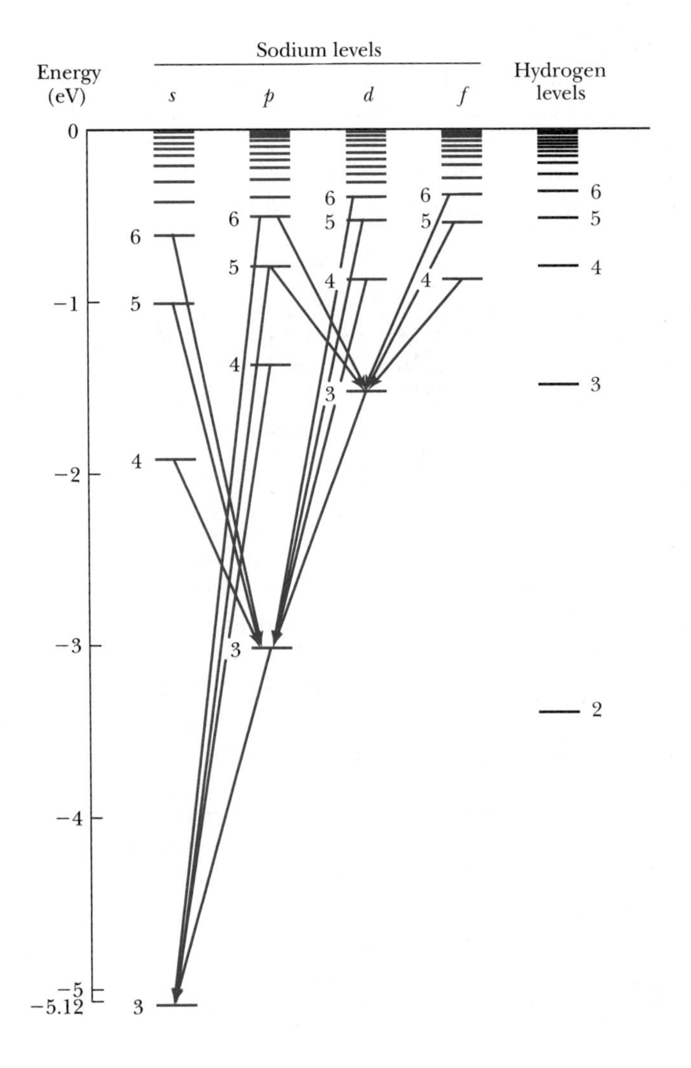

FIGURE 8.7 The energy-level diagram of sodium (single electron outside inert core) is compared to that of hydrogen. Coulomb effects cause the lower ℓ states of sodium to be lower than the corresponding levels of hydrogen. Several allowed transitions are shown for sodium.

Example 8.3

Show that an energy difference of 2×10^{-3} eV for the $3p$ subshell of sodium accounts for the 0.6-nm splitting of a spectral line at 589.3 nm.

Solution: The wavelength λ of a photon is related to the energy of a transition by

$$E = \frac{hc}{\lambda}$$

For a small splitting, we approximate by using a differential,

$$dE = \frac{-hc}{\lambda^2} d\lambda$$

Then, letting $\Delta E = dE$ and $\Delta\lambda = d\lambda$ and taking absolute values yields

$$|\Delta E| = \frac{hc}{\lambda^2}|\Delta\lambda| \quad \text{or} \quad |\Delta\lambda| = \frac{\lambda^2}{hc}|\Delta E|$$

$$|\Delta\lambda| = \frac{(589.3 \text{ nm})^2(2 \times 10^{-3} \text{ eV})}{1.240 \times 10^3 \text{ eV} \cdot \text{nm}} = 0.6 \text{ nm}$$

Many-Electron Atoms

The situation becomes formidable for more than two electrons outside an inert core. Various empirical rules (for example, Hund's rules) help in applying the quantization results to such atoms. We will consider here the case of two electrons outside a closed shell (for example, helium and the alkaline earths.)*

The order in which a given subshell is filled is governed by Hund's rules, which state that

Hund's rules

1. The total spin angular momentum S should be maximized to the extent possible without violating the Pauli exclusion principle.
2. Insofar as Rule 1 is not violated, L should also be maximized.

For example, the first five electrons to occupy a d subshell should all have the same value of m_s. This requires that each one has a different m_ℓ (because the allowed m_ℓ values are $-2, -1, 0, 1, 2$). By Rule 2 the first two electrons to occupy a d subshell should have $m_\ell = 2$ and $m_\ell = 1$ *or* $m_\ell = -2$ and $m_\ell = -1$.

Besides the spin–orbit interaction already discussed, there are now spin–spin and orbital–orbital interactions. There are also effects due to the spin of the nucleus that lead to *hyperfine* structure, but the nuclear effect is much smaller than the ones we are presently considering. For the two-electron atom, we label the electrons 1 and 2 so that we have $\mathbf{L}_1$, $\mathbf{S}_1$ and $\mathbf{L}_2$, $\mathbf{S}_2$. The total angular momentum $\mathbf{J}$ is the vector sum of the four angular momenta:

$$\mathbf{J} = \mathbf{L}_1 + \mathbf{L}_2 + \mathbf{S}_1 + \mathbf{S}_2 \tag{8.9}$$

There are two schemes, called *LS coupling* and *jj coupling*, for combining the four angular momenta to form J. The decision of which scheme to use depends on relative strengths of the various interactions. We shall see that jj coupling predominates for heavier elements.

*The reader is referred to H. G. Kuhn's *Atomic Spectra* 2nd ed. New York: Academic Press, 1969, or H. E. White's *Introduction to Atomic Spectra* New York: McGraw-Hill, 1934 for further study.

LS, or Russell-Saunders, Coupling

The LS coupling scheme is used for most atoms when the magnetic field is weak. The orbital angular momenta $\mathbf{L}_1$ and $\mathbf{L}_2$ combine to form a total orbital angular momentum $\mathbf{L}$ and similarly for $\mathbf{S}$.

$$\mathbf{L} = \mathbf{L}_1 + \mathbf{L}_2 \tag{8.10}$$

$$\mathbf{S} = \mathbf{S}_1 + \mathbf{S}_2 \tag{8.11}$$

Then $\mathbf{L}$ and $\mathbf{S}$ combine to form the total angular momentum.

$$\mathbf{J} = \mathbf{L} + \mathbf{S} \tag{8.12}$$

One of Hund's rules states that the electron spins combine to make $\mathbf{S}$ a maximum. Physically this occurs because of the mutual repulsion of the electrons, which want to be as far away from each other as possible in order to have the lowest energy. If two electrons in the same subshell have the same m_s, they must then have different m_ℓ, normally indicating different spatial distributions. Similarly, the lowest energy states normally occur with a maximum $\mathbf{L}$. We can understand this physically, because the electrons would revolve around the nucleus in the same direction if aligned, thus staying as far apart as possible. If $\mathbf{L}_1$ and $\mathbf{L}_2$ were antialigned, they would pass each other more often, tending to have a higher interaction energy.

For the case of two electrons the total spin angular momentum quantum number* may be $S = 0$ or 1 depending on whether the spins are antiparallel or parallel. For a given value of L, there are $2S + 1$ values of J, because J goes from $L - S$ to $L + S$ (for $L > S$). For $L < S$ there are fewer than $2S + 1$ possible values of J (see Examples 8.4 and 8.5). The value of $2S + 1$ is called the **multiplicity** of the state.

Multiplicity

The notation nL_j discussed before for a single-electron atom becomes

$$n^{2S+1}L_J \tag{8.13}$$

Spectroscopic symbols

This code is called *spectroscopic* or *term symbols.* For two electrons we have **singlet** states ($S = 0$) and **triplet** states ($S = 1$), which refer to the multiplicity $2S + 1$. Recall that a single-electron state (with $s = 1/2$) is a doublet, with $2s + 1 = 2$.

Singlet and triplet states

Consider two electrons: One is in the $4p$ and one is in the $4d$ subshell. We have the following possibilities ($S_1 = 1/2$, $S_2 = 1/2$, $L_1 = 1$, and $L_2 = 2$) for the atomic states shown in Table 8.2. A schematic diagram showing the relative energies of these states is shown in Figure 8.8. The spin–spin interaction breaks the unperturbed state into the singlet and triplet states. The Coulomb effect, due to the electrons, orders the highest L value for each of these states to be lowest in energy. Finally, the spin–orbit splitting causes the lowest J value to be lowest in energy ($\mathbf{L}$ and $\mathbf{S}$ antialigned).

As an example of the optical spectra obtained from two-electron atoms, we consider the energy-level diagram of magnesium in Figure 8.9. The most obvious characteristic of this figure is that we have separated the energy levels according to whether they are $S = 0$ or $S = 1$. This is because *allowed* transitions must have

*It is customary to use capital letters, L, S, and J for the angular momentum quantum numbers of many-electron atoms. This can lead to confusion, because we are accustomed to thinking, for example, $S = |\mathbf{S}|$. To avoid confusion, remember that the magnitude of an angular momentum vector is always some number times $\hbar$, while the new angular momentum quantum numbers, L, S, and J are simply integers or half-integers.

TABLE 8.2
Spectroscopic Symbols for Two Electrons: One in $4p$ and One in $4d$

S	L	J	Spectroscopic Symbol
	1	1	4^1P_1
0 (singlet)	2	2	4^1D_2
	3	3	4^1F_3
		2	4^3P_2
1 (triplet)	1	1	4^3P_1
		0	4^3P_0
		3	4^3D_3
1 (triplet)	2	2	4^3D_2
		1	4^3D_1
		4	4^3F_4
1 (triplet)	3	3	4^3F_3
		2	4^3F_2

$\Delta S = 0$, and no allowed transitions are possible between singlet and triplet states. This does not mean that it is impossible for such transitions to occur. Remember that *forbidden* transitions occur, but with much lower probability than allowed transitions.

The two electrons in magnesium outside the closed $2p$ subshell are $3s^2$. Therefore, the ground state of magnesium is $3^1S_0(S = 0,\ L = 0,\text{ and } J = 0)$. The 3^3S_1 state $(S = 1, J = 1)$ cannot exist because $m_s = 1/2$ for both electrons in order to have $S = 1$, and this is forbidden by the exclusion principle. A $^3S_1$ state is

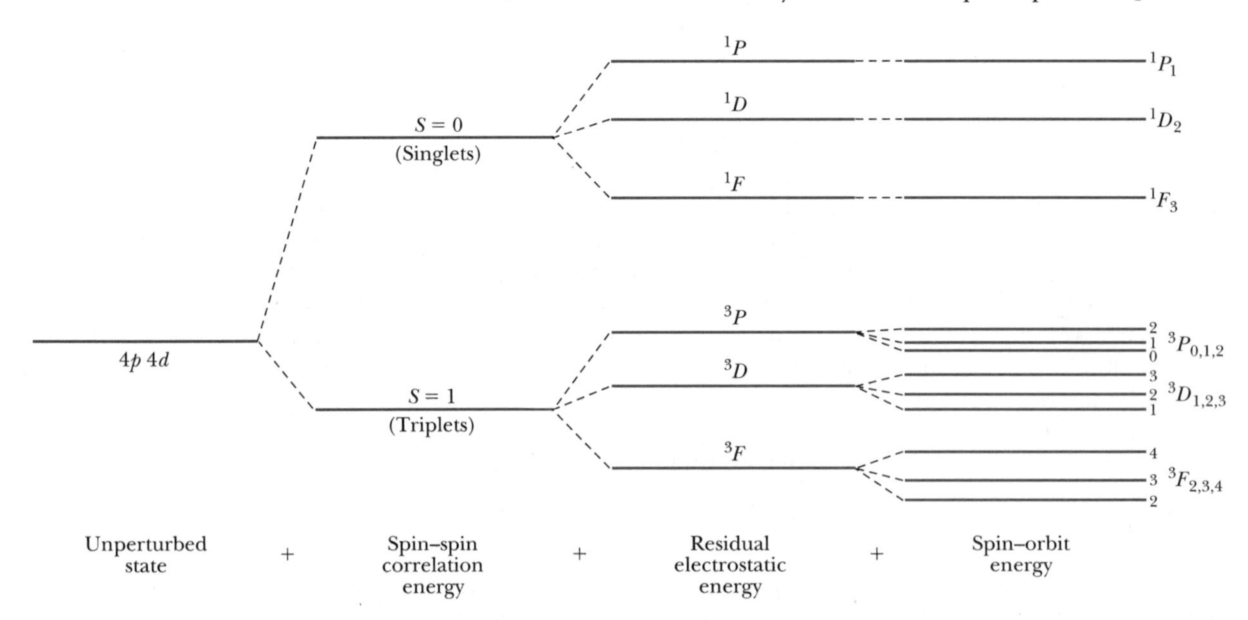

FIGURE 8.8 Schematic diagram indicating the increasing fine structure splitting due to different effects. This case is for an atom having two valence electrons, one in the $4p$ and the other in the $4d$ state. The energy is not to scale. *From R. B. Leighton, Principles of Modern Physics, New York: McGraw-Hill, 1959, p. 261.*

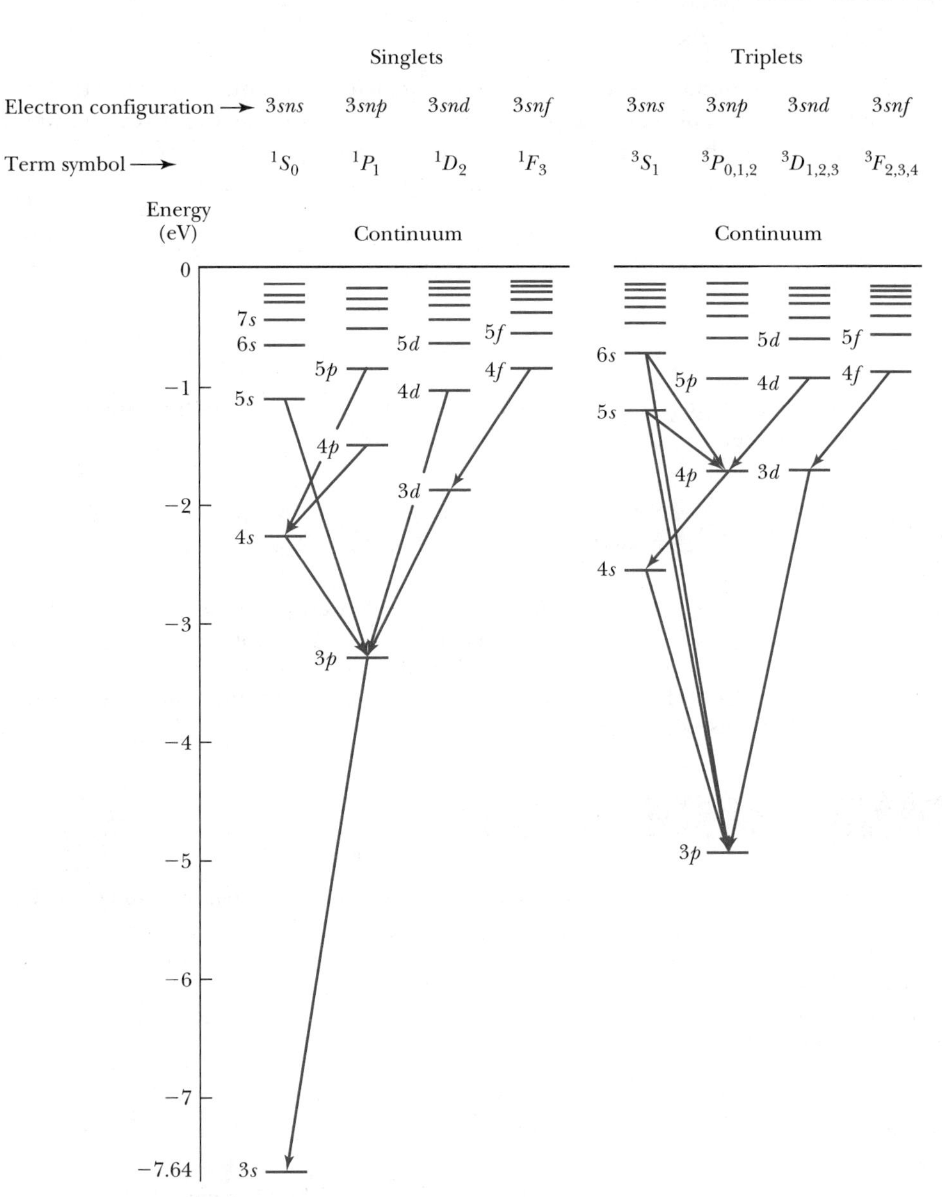

FIGURE 8.9 Energy-level diagram for magnesium with one electron in the 3*s* subshell and the other electron (two-electron atom) excited into the $n\ell$ subshell indicated. The singlet and triplet states are separated, because transitions between them are not allowed by the $\Delta S = 0$ selection rule. Several allowed transitions are indicated.

allowed if one of the electrons is in a higher *n* shell. The energy-level diagram of Figure 8.9 is generated by one electron remaining in the 3*s* subshell while the other electron is promoted to the subshell indicated on the diagram. The allowed transitions (for the *LS* coupling scheme) are

$$\begin{aligned} \Delta L &= \pm 1 \qquad \Delta S = 0 \\ \Delta J &= 0, \pm 1 \qquad (J = 0 \rightarrow J = 0 \text{ is forbidden}) \end{aligned} \tag{8.14}$$

A magnesium atom excited to the 3*s*3*p* triplet state has no lower triplet state to which it can decay. The only state lower in energy is the 3*s*3*s* ground state, which is singlet. Such an excited triplet state may exist for a relatively long time ($\gg 10^{-8}$ s) before it finally decays to the ground state as a forbidden transition. Such a 3*s*3*p* triplet state is called **metastable,** because it lives for such a long time on the atomic scale.

Metastable states

jj Coupling

This coupling scheme predominates for the heavier elements, where the nuclear charge causes the spin–orbit interactions to be as strong as the forces between the individual $\mathbf{S}_i$ and the individual $\mathbf{L}_i$. The coupling order becomes

$$\mathbf{J}_1 = \mathbf{L}_1 + \mathbf{S}_1 \tag{8.15a}$$

$$\mathbf{J}_2 = \mathbf{L}_2 + \mathbf{S}_2 \tag{8.15b}$$

and then

$$\mathbf{J} = \sum_i \mathbf{J}_i \tag{8.16}$$

The spectroscopic or term notation is also used to describe the final states in this coupling scheme.

Example 8.4

What are the total angular momentum and the spectroscopic notation for the ground state of helium?

Solution: The two electrons for helium are both $1s$ electrons. Because helium is a light atom, we will use the LS coupling scheme. We have $L_1 = 0$ and $L_2 = 0$, and therefore $L = 0$. We can have $S = 0$ or 1 for two electrons, but not in the same subshell. The spins must be antialigned and $S = 0$. Therefore $J = 0$ also. The ground state spectroscopic symbol for helium is 1^1S_0.

Example 8.5

What are the L, S, and J values of the first few excited states of helium?

Solution: The lowest excited states of helium must be $1s^12s^1$ or $1s^12p^1$—that is, one electron is promoted to either the $2s^1$ or $2p^1$ subshell. It turns out that *all* excited states of helium are single-electron states, because to excite both electrons requires more than the ionization energy. We expect the excited states of $1s^12s^1$ to be lower than those of $1s^12p^1$, because the subshell $2s^1$ is lower in energy than the $2p^1$ subshell. The possibilities are:

$1s^12s^1 \quad L = 0$

if $S = 0$, then $J = 0$

if $S = 1$, then $J = 1$

with $S = 1$ being lowest in energy. The lowest excited state is $^3S_1$ and then comes $^1S_0$.

$1s^12p^1 \quad L = 1$

if $S = 0$, then $J = 1$

if $S = 1$, then $J = 0, 1, 2$

The state $^3P_0$ has the lowest energy of these states, followed by $^3P_1$, $^3P_2$, and $^1P_1$.

The energy-level diagram for helium is shown in Figure 8.10.

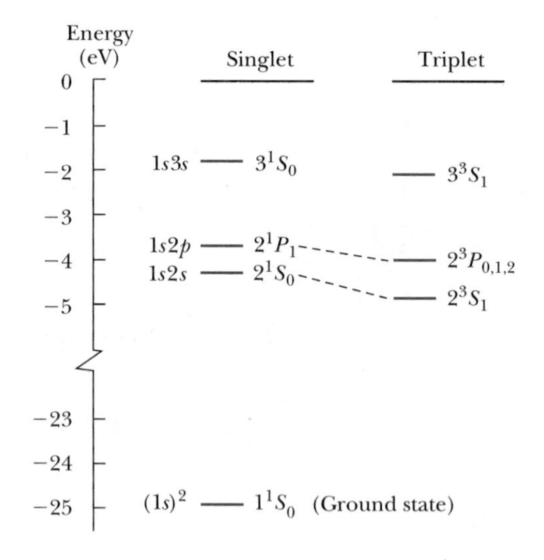

FIGURE 8.10 The low-lying atomic states of helium are shown. The ground state ($^1S_0$) is some 20 eV below the grouping of the lowest excited states. The level indicated by $^3P_{0,1,2}$ is actually three states ($^3P_0$, $^3P_1$, $^3P_2$), but the separations are too small to be indicated.

Example 8.6

If the spin–orbit splitting of the $3P_{3/2}$ and $3P_{1/2}$ states of sodium is 0.002 eV, what is the internal magnetic field causing the splitting?

Solution: The potential energy due to the spin magnetic moment is

$$V = -\boldsymbol{\mu}_s \cdot \mathbf{B} \tag{8.17}$$

By analogy with Equation (7.34), the z component of the total magnetic moment is

$$\mu_z = -g_s\left(\frac{e\hbar}{2m}\right)\frac{J_z}{\hbar} \tag{8.18}$$

where we have used the gyromagnetic ratio $g_s = 2$, because this splitting is actually due to spin. The difference in spins between the $3P_{3/2}$ and $3P_{1/2}$ states is $\hbar$ so that

$$\Delta E = g_s\left(\frac{e\hbar}{2m}\right)\frac{\hbar}{\hbar}B = \frac{e\hbar}{m}B$$

Then

$$B = \frac{m\Delta E}{e\hbar} = \frac{(9.11 \times 10^{-31}\text{ kg})(0.002\text{ eV})}{(1.6 \times 10^{-19}\text{ C})(6.58 \times 10^{-16}\text{ eV}\cdot s)}$$

$$= 17\text{ T, a large magnetic field}$$

Example 8.7

What are the possible energy states for atomic carbon?

Solution: The element carbon has two $2p$ subshell electrons outside the closed $2s^2$ subshell. Both electrons have $\ell = 1$, so we can have $L = 0$, 1, or 2 using the LS coupling scheme. The spin angular momentum is $S = 0$ or 1. We list the possible states:

S	L	J	Spectroscopic Notation	
0	0	0	$^1S_0$	
	1	1	$^1P_1$	not allowed
	2	2	$^1D_2$	
1	0	1	$^3S_1$	not allowed
	1	0, 1, 2	$^3P_{0,1,2}$	
	2	1, 2, 3	$^3D_{1,2,3}$	not allowed

The $^3S_1$ state is not allowed by the Pauli exclusion principle, because both electrons in the $2p^2$ subshell would have $m_s = +1/2$ and $m_\ell = 0$. Similarly, the $^3D_{1,\,2,\,3}$ states are not allowed, because both electrons would have $m_s = +1/2$ and $m_\ell = 1$. According to Hund's rules, the triplet states $S = 1$ will be lowest in energy, so the ground state will be one of the $^3P_{0,1,2}$ states. The spin–orbit interaction then indicates the $^3P_0$ state to be the ground state; the others are excited states.

The fact that the $^1P_1$ state is not allowed is a result of the antisymmetrization of the wave function, which we have not discussed. This rule, which requires electrons to have antisymmetric wave functions, is basically an extension of the Pauli exclusion principle for this example, and it allows the states in which the m_ℓ of the two electrons are equal to combine only with $S = 0$ states. The m_ℓ values for the electrons forming the $S = 1$ state must be unequal. This rule is the theoretical basis for Hund's rules, described previously. It precludes the ^1P, ^3S, and ^3D states from existing for $2p^2$ electrons. The states with one electron having $m_\ell = 1$, $m_s = +1/2$ and the other with $m_\ell = 0$, $m_s = -1/2$ still exist, but they can be included in the $^1D_2$ state, for example, because $m_s = 0$ and $m_L = 2, 1, 0, -1, -2$.

The low-lying excited states of carbon are then $^3P_1$, $^3P_2$, $^1D_2$, and $^1S_0$.

8.3 Anomalous Zeeman Effect

In Section 7.4 we discussed the normal Zeeman effect and showed that the splitting of an optical spectral line into three components in the presence of an external magnetic field could be understood by considering the interaction $(\boldsymbol{\mu}_\ell \cdot \mathbf{B}_{\text{ext}})$ of the orbital angular momentum magnetic moment m_ℓ and the external magnetic field. Soon after the discovery of this effect by Zeeman in 1896, it was found that often more than the three closely spaced optical lines were observed. This observation was called the *anomalous* Zeeman effect. We are now able to explain both Zeeman effects. We shall see that the anomalous effect depends on the effects of electron intrinsic spin.

The interaction that splits the energy levels in an external magnetic field $\mathbf{B}_{\text{ext}}$ is still caused by the $\boldsymbol{\mu} \cdot \mathbf{B}$ interaction. However, the magnetic moment is due not only to the orbital contribution $\boldsymbol{\mu}_\ell$, it must depend on the spin magnetic moment $\boldsymbol{\mu}_s$ as well. The $2J + 1$ degeneracy (due to m_J) for a given total angular momentum state J is removed by the effect of the external magnetic field. If the external field $\mathbf{B}_{\text{ext}}$ is small in comparison with the internal magnetic field (say $B_{\text{ext}} < 0.1$ T), then $\mathbf{L}$ and $\mathbf{S}$ (using the *LS* coupling scheme) precess about $\mathbf{J}$ while $\mathbf{J}$ precesses *slowly* about $\mathbf{B}_{\text{ext}}$.

We can see this more easily by calculating $\boldsymbol{\mu}$ in terms of $\mathbf{L}$, $\mathbf{S}$, and $\mathbf{J}$. The total magnetic moment $\boldsymbol{\mu}$ is

$$\boldsymbol{\mu} = \boldsymbol{\mu}_\ell + \boldsymbol{\mu}_s \tag{8.19}$$

$$= -\frac{e}{2m}\mathbf{L} - \frac{e}{m}\mathbf{S} \tag{8.20}$$

where $\boldsymbol{\mu}_\ell$ is obtained from Equation (7.26) and $\boldsymbol{\mu}_s$ from Section 7.5.

$$\boldsymbol{\mu} = -\frac{e}{2m}(\mathbf{L} + 2\mathbf{S}) = -\frac{e}{2m}(\mathbf{J} + \mathbf{S}) \tag{8.21}$$

The vectors $-\boldsymbol{\mu}$ and $\mathbf{J}$ are along the same direction only when $S = 0$. We show schematically in Figure 8.11 what is happening. The $\mathbf{B}_{\text{ext}}$ defines the z direction. We plot $-\boldsymbol{\mu}$ instead of $+\boldsymbol{\mu}$ in order to emphasize the relationship between $\boldsymbol{\mu}$ and $\mathbf{J}$. In a weak magnetic field the precession of $\boldsymbol{\mu}$ around $\mathbf{J}$ is much faster than the precession of $\mathbf{J}$ around $\mathbf{B}_{\text{ext}}$. Therefore, we first find the average $\boldsymbol{\mu}_{\text{av}}$ about $\mathbf{J}$ and then find the interaction energy of $\boldsymbol{\mu}_{\text{av}}$ with $\mathbf{B}_{\text{ext}}$. We leave this as an exercise for the student (see Problem 29). The result is

$$V = \frac{e\hbar B_{\text{ext}}}{2m} g m_J = \mu_B B_{\text{ext}} g m_J \tag{8.22}$$

where μ_B is the Bohr magneton and

Landé *g* factor

$$g = 1 + \frac{J(J+1) + S(S+1) - L(L+1)}{2J(J+1)} \tag{8.23}$$

is a dimensionless number called the **Landé *g* factor.** The magnetic total angular momentum numbers m_J range from $-J$ to J in integral steps. The external field $\mathbf{B}_{\text{ext}}$ splits each state J into $2J + 1$ equally spaced levels separated by $\Delta E = V$, with V determined in Equation (8.22), each level being described by a different m_J.

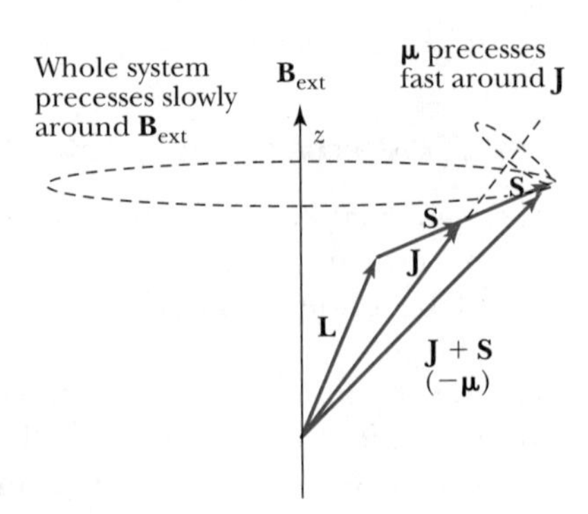

FIGURE 8.11 Relationships between $\mathbf{S}$, $\mathbf{L}$, $\mathbf{J}$, and $\boldsymbol{\mu}$ are indicated. The $\mathbf{B}_{\text{ext}}$ is the z direction. The magnetic moment $\boldsymbol{\mu}$ precesses fast around $\mathbf{J}$ as $\mathbf{J}$ precesses more slowly around the weak $\mathbf{B}_{\text{ext}}$. *After J. D. McGervey, Introduction to Modern Physics. New York: Academic Press, 1983, p. 329.*

In addition to the previous selection rules [Equation (8.14)] for photon transitions between energy levels, we must now add one for m_J:

$$\Delta m_J = \pm 1, 0 \tag{8.24}$$

but $m_{J_1} = 0 \longrightarrow m_{J_2} = 0$ is forbidden when $\Delta J = 0$.

Example 8.8

Show that the normal Zeeman effect should be observed for transitions between the $^1D_2$ and $^1P_1$ states.

Solution: Because $2S + 1 = 1$ for both states, then $S = 0$ and $J = L$. The g factor from Equation (8.23) is equal to 1 (as it always will be for $S = 0$). The $^1D_2$ state splits into five equally spaced levels, and the $^1P_1$ state splits into three (see Figure 8.12). Using the selection rules from Equations (8.14) and (8.24), there are only nine allowed transitions between the two states as labeled in Figure 8.12. The other transitions are disallowed by the selection rule for Δm_J. Even though there are nine different transitions, there are only three possible energies for emitted or absorbed photons, because transition energies labeled 1, 3, 6 are identical, as are 2, 5, 8, and also 4, 7, 9. Thus the three equally spaced transitions of the normal Zeeman effect are observed whenever $S = 0$.

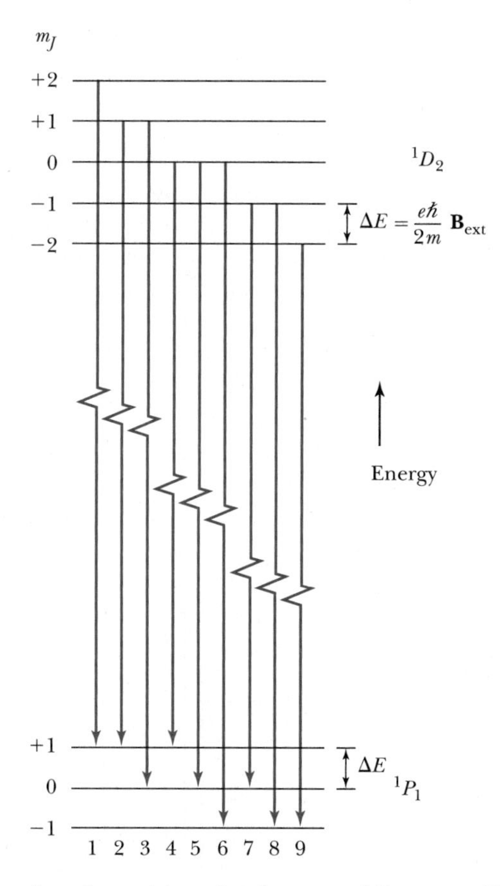

FIGURE 8.12 Examples of transitions for the normal Zeeman effect. The nine possible transitions are labeled, but there are only three distinctly different energies because the split energy levels are equally spaced (ΔE) for both the $^1D_2$ and $^1P_1$ states.

The anomalous Zeeman effect is a direct result of intrinsic spin. Let us consider transitions between the ^2P and ^2S states of sodium as shown in Figure 8.13. In a completely unperturbed state the $^2P_{3/2}$ and $^2P_{1/2}$ states are degenerate. However, the internal spin–orbit interaction splits them, with $^2P_{1/2}$ being lower in energy. The $^2S_{1/2}$ state is not split by the spin–orbit interaction because $\ell = L = 0$.

When sodium is placed in an external magnetic field, all three states are split into $2J + 1$ levels with different m_J (see Figure 8.13). The appropriate Landé g factors are

$$^2S_{1/2} \qquad g = 1 + \frac{\frac{1}{2}\left(\frac{1}{2}+1\right) + \frac{1}{2}\left(\frac{1}{2}+1\right)}{2 \cdot \frac{1}{2}\left(\frac{1}{2}+1\right)} = 2$$

$$^2P_{1/2} \qquad g = 1 + \frac{\frac{1}{2}\left(\frac{1}{2}+1\right) + \frac{1}{2}\left(\frac{1}{2}+1\right) - 1(1+1)}{2 \cdot \frac{1}{2}\left(\frac{1}{2}+1\right)} = 0.67$$

$$^2P_{3/2} \qquad g = 1 + \frac{\frac{3}{2}\left(\frac{3}{2}+1\right) + \frac{1}{2}\left(\frac{1}{2}+1\right) - 1(1+1)}{2 \cdot \frac{3}{2}\left(\frac{3}{2}+1\right)} = 1.33$$

All g factors are different, and the energy-splitting ΔE calculated using Equation (8.22) for the three states will be different. By using the selection rules, the allowed transitions are shown in Figure 8.13. There are four different energy transitions for $^2P_{1/2} \rightarrow {}^2S_{1/2}$ and six different energy transitions for $^2P_{3/2} \rightarrow {}^2S_{1/2}$.

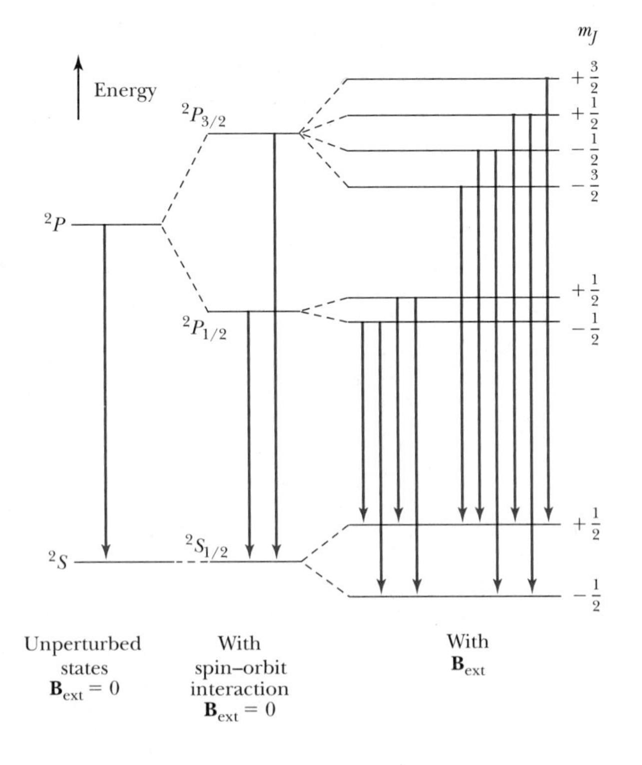

FIGURE 8.13 Schematic diagram of anomalous Zeeman effect for sodium (energy levels not to scale). With $\mathbf{B}_{ext} = 0$ for the unperturbed states, there is only one transition. With the spin–orbit interaction splitting the ^2P state into 2 states, there are two possible transitions when $\mathbf{B}_{ext} = 0$. Finally, the $\mathbf{B}_{ext}$ splits J into $2J + 1$ components, each with a different m_J. The energy splitting ΔE for each major state is different because $\Delta E = g m_J (e\hbar/2m) B_{ext}$ and the Landé g factor for $g[\Delta E(^2S_{1/2})] > g[\Delta E(^2P_{3/2})] > g[\Delta E(^2P_{1/2})]$. All allowed transitions are shown.

If the external magnetic field is increased, then **L** and **S** precess too rapidly about $\mathbf{B}_{ext}$ and our averaging procedure for $\boldsymbol{\mu}$ around $\mathbf{B}_{ext}$ breaks down. In that case, the equations developed in this section are incorrect. This occurrence, called the *Paschen-Back effect,* must be analyzed differently. We will not pursue this calculation further.*

*See H. E. White's, *Introduction to Atomic Spectra.* New York: McGraw-Hill, 1934 for more information.

Summary

The Pauli exclusion principle states that no two electrons in an atom may have the same set of quantum numbers (n, ℓ, m_ℓ, m_s). Because electrons normally occupy the lowest energy state available, the Pauli exclusion principle may be used to produce the periodic table and understand many properties of the elements.

The total angular momentum **J** is the vector sum of **L** and **S**, $\mathbf{J} = \mathbf{L} + \mathbf{S}$. The coupling of **S** and **L**, called the *spin–orbit interaction,* leads to lower energies for smaller values of J. For two or more electrons in an atom we can couple the $\mathbf{L}_i$ and $\mathbf{S}_i$ of the valence electrons by either LS or jj coupling. The spectroscopic notation for an atomic state is $n^{2S+1}L_J$.

The allowed transitions now have

$$\Delta L = \pm 1 \qquad \Delta S = 0$$
$$\Delta J = 0, \pm 1 \qquad (J = 0 \rightarrow J = 0 \text{ is forbidden}) \tag{8.14}$$

The anomalous Zeeman effect is explained by the removal of the $2J + 1$ degeneracy when an atom is placed in a weak magnetic field. Each state has a different m_J, which has the selection rule for transitions of $\Delta m_J = \pm 1, 0$ (with exceptions). The normal Zeeman effect (three spectral lines) occurs when $S = 0$.

Questions

1. Explain in terms of the electron shell configuration why it is dangerous to throw sodium into water.
2. Why are the inert gases in gaseous form at room temperature?
3. Which groups of elements have the best and which the poorest electrical conductivities? Explain.
4. Why are the elements with good electrical conductivities also generally good thermal conductors?
5. Boron, carbon, and aluminum are not part of the alkalis or alkaline earths, yet they are generally good electrical conductors. Explain.
6. The alkali metals have the lowest ionization energies (Figure 8.3), yet they have the largest atomic radii (Figure 8.4). Is this consistent? Explain.
7. List four compounds that you believe should be strongly bound. Explain why.
8. Explain why the transition metals have good thermal and electrical conductivities.
9. Why do the alkaline earths have low resistivities?
10. Why is there no spin–orbit splitting for the ground state of hydrogen?
11. Is it possible for both atoms in a hydrogen molecule to be in the $(1, 0, 0, -1/2)$ state? Explain.
12. Discuss in your own words the differences between **L** and ℓ, between m_ℓ and m_s, and between J_z and m_j.

Problems

8.1 Atomic Structure and the Periodic Table

1. A lithium atom has three electrons. Allow the electrons to interact with each other and the nucleus. Label each electron's spin and angular momentum. List all the possible interactions.
2. For all the elements through neon list the electron descriptions of these elements in their ground state using $n\ell$ notation (for example, helium is $1s^2$).
3. How many subshells are there in the following shells: L, N, and O?

4. What electron configuration would you expect ($n\ell$) for the first excited state of argon and krypton?
5. Using Table 8.1 and Figure 8.2 write down the electron configuration ($n\ell$ notation) of the following elements: potassium, vanadium, selenium zirconium, samarium, and uranium.
6. Using Figure 8.2 list all the (a) inert gases, (b) alkalis, and (c) alkaline earths.
7. The $3s$ state of Na has an energy of -5.14 eV. Determine the effective nuclear charge.
8. List the quantum numbers (n, ℓ, m_ℓ, m_s) for all the electrons in a nitrogen atom.
9. What atoms have the configuration (a) $1s^22s^22p$, (b) $1s^22s^22p^63s$, (c) $3s^23p^6$?

8.2 Total Angular Momentum

10. If the zirconium atom ground state has $S = 1$ and $L = 3$, what are the permissible values of J? Write the spectroscopic notation for these possible values of S, L, and J. Which one of these is likely to represent the ground state?
11. Using the information in Table 8.2, determine the ground state spectroscopic symbol for gallium.
12. List all the elements through calcium that you would expect not to have a spin–orbit interaction that splits the ground state energy. Explain.
13. For the hydrogen atom in the $3d$ excited state find the possible values of ℓ, m_ℓ, j, s, m_s, and m_j. Give the term notation for each possible configuration.
14. What are S, L, and J for the following states: ${}^1S_0$, ${}^2D_{5/2}$, ${}^5F_1$, ${}^3F_4$?
15. What are the possible values of J_z for the $8^2G_{7/2}$ state?
16. (a) What are the possible values of J_z for the $6^2F_{7/2}$ state? (b) Determine the minimum angle between the total angular momentum vector and the z axis for this state.
17. Explain why the spectroscopic term symbol for lithium in the ground state is ${}^2S_{1/2}$.
18. What is the spectroscopic term symbol for aluminum in its ground state? Explain.
19. The $4P$ state in potassium is split by its spin–orbit interaction into the $4P_{3/2}$ ($\lambda = 766.41$ nm) and $4P_{1/2}$ ($\lambda = 769.90$ nm) states (the wavelengths are for the transitions to the ground state). Calculate the spin–orbit energy splitting and the internal magnetic field causing the splitting.
20. Draw the energy-level diagram for the states of carbon discussed in Example 8.7. Draw lines between states that have allowed transitions and list ΔL, ΔS, and ΔJ.
21. An $n = 2$ shell (L shell) has a $2s$ state and two $2p$ states split by the spin–orbit interaction. Careful measurements of the K_α x ray ($n = 2 \rightarrow n = 1$ transition) reveal only two spectral lines. Explain.
22. What is the energy difference between a spin-up state and spin-down state for an electron in an s state if the magnetic field is 1.7 T?
23. Which of the following elements can have either (or both) singlet and triplet states and which have neither: He, Al, Ca, Sr? Explain.
24. If the minimum angle between the total angular momentum vector and the z axis is 32.3° (in a single-electron atom), what is the total angular momentum quantum number?
25. Use the Biot-Savart law to find the magnetic field in the frame of an electron circling a nucleus of charge Ze. If the velocity of the electron around the nucleus is $\mathbf{v}$ and the position vector of the proton with respect to the electron is $\mathbf{r}$, show that the magnetic field at the electron is

$$\mathbf{B} = \frac{Ze}{4\pi\epsilon_0}\frac{\mathbf{L}}{mc^2r^3}$$

where m is the electron mass and $\mathbf{L}$ is the angular momentum, $\mathbf{L} = m\mathbf{r} \times \mathbf{v}$.

26. Using the internal magnetic field of the previous problem show that the potential energy of the spin magnetic moment μ_s interacting with $\mathbf{B}_{\text{internal}}$ is given by

$$V_{s\ell} = \frac{Ze^2}{4\pi\epsilon_0}\frac{\mathbf{S}\cdot\mathbf{L}}{m^2c^2r^3}$$

There is an additional factor of 1/2 to be added from relativistic effects called the *Thomas factor.*

27. The difference between the $2P_{3/2}$ and $2P_{1/2}$ doublet in hydrogen due to the spin–orbit splitting is 4.5×10^{-5} eV. (a) Compare this with the potential energy given in the previous problem. (b) Compare this with a more complete calculation giving the potential energy term V as

$$V = -\frac{Z^4\alpha^4}{2n^3}mc^2\left(\frac{2}{2j+1} - \frac{3}{4n}\right)$$

where α is the fine-structure constant, $\alpha = 1/137$.

8.3 Anomalous Zeeman Effect

28. For which L and S values does an atom exhibit the normal Zeeman effect? Does this apply to both ground and excited states? Can an atom exhibit both the normal and anomalous Zeeman effects?
29. Derive Equations (8.22) and (8.23). First find the average value of $\boldsymbol{\mu}$ and $\mathbf{J}$. Use

$$\boldsymbol{\mu}_{\text{av}} = (\boldsymbol{\mu}\cdot\mathbf{J})\frac{\mathbf{J}}{\mathbf{J}\cdot\mathbf{J}} \quad \text{and} \quad V = -\boldsymbol{\mu}_{\text{av}}\cdot\mathbf{B}$$

(Remember $|\mathbf{J}|^2 = J(J+1)\hbar^2$.)

30. In the early 1900s the normal Zeeman effect was useful to determine the electron's e/m if Planck's constant was assumed known. Calcium is an element that exhibits the normal Zeeman effect. The difference between adjacent components of the spectral lines is observed to be 0.013 nm for $\lambda = 422.7$ nm when calcium is placed in a magnetic field of 1.5 T. From these data calculate the value of $e\hbar/m$ and compare with the accepted value today. Calculate e/m assuming the known value of $\hbar$.

31. Calculate the Landé g factor for an atom with a single (a) s electron, (b) p electron, (d) d electron.

32. An atom with the states ${}^2G_{9/2}$ and ${}^2H_{11/2}$ is placed in a weak magnetic field. Draw the energy levels and indicate the possible allowed transitions between the two states.

33. Repeat the previous problem for ${}^3P_1$ and ${}^3D_2$ states.

34. With no magnetic field, the spectral line representing the transition from the ${}^2P_{1/2}$ state to the ${}^2S_{1/2}$ state in sodium has the wavelength 589.76 nm (see Figure 8.13). This is one of the two strong yellow lines in sodium. Calculate the difference in wavelength between the shortest and longest wavelength between these two states when placed in a magnetic field of 0.5 T.

35. When sodium in the ${}^2P_{3/2}$ state is placed in a magnetic field of 0.5 T, the energy level splits into four levels (see Figure 8.13). Calculate the energy difference between these levels.

CHAPTER 9

Statistical Physics

Ludwig Boltzmann, who spent much of his life studying statistical mechanics, died in 1906 by his own hand. Paul Ehrenfest, carrying on his work, died similarly in 1933. Now it is our turn to study statistical mechanics. Perhaps it will be wise to approach the subject cautiously.

David L. Goodstein (States of Matter. Mineola, New York: Dover Press, 1985.)

Statistics are important in a number of areas of physics, one of which is atomic physics. In Section 7.6 the position of the electron in the hydrogen atom was described in terms of a probability distribution. Another important application of the concept of probability to atomic physics arises for transitions between atomic states. Like most atomic systems, the hydrogen atom may be in the ground state or in one of a number of excited states. An atom in an excited state is likely to make a transition to a lower energy state and eventually to the ground state. The question that arises logically is: just how likely are each of the allowed transitions? One may also ask: what are the relative probabilities of finding an atom in any particular state? These are not simple questions, and the answers require a knowledge of the wave functions for each of the states involved. We shall not attempt to solve these problems here. We should simply be aware of the fact that transitions between quantum states must usually be described in probabilistic terms because there is no simple causal mechanism we can use to track the electron(s) from one level to another.

There is a simpler way to begin to understand how statistics and probability theory are used in physics. Historically, the need for these mathematical tools became apparent to those studying problems in heat and thermodynamics. Accordingly, we shall begin this chapter with a review of how probability and statistics found their way into physics in the 19th century. The important results of Maxwell and Boltzmann will be outlined and used to derive some of the basic laws of the kinetic theory of gases. Then we shall look at how quantum statistics differ from classical statistics. The last two sections deal with applications of quantum statistics (Fermi-Dirac and Bose-Einstein) to a number of problems in modern physics. The use of statistics in quantum systems, particularly in solids, will be continued in more detail in Chapter 10.

9.1 Historical Overview

At the beginning of the 19th century, physicists were generally not inclined to use probability and statistics to describe physical processes. Indeed, there was a strong tendency to view the universe as a machine run by strict, unvarying, deterministic laws. This mechanistic view was due in large part to Newton and to the successful run of Newtonian physics through the 18th and 19th centuries. New mathematical methods developed by Lagrange around 1790 and Hamilton around 1840 added significantly to the computational power of Newtonian mechanics. They enabled physicists to describe extremely complicated physical systems with relatively simple second-order differential equations.

Probably the most extreme defender of the absolute power of classical mechanics (not to mention the fact that he was one of its greatest practitioners) was Pierre-Simon de Laplace (1749–1826). Laplace held that it should be possible in principle to have perfect knowledge of the physical universe. This could be accomplished by knowing exactly at one time the position and velocity of every particle of matter and knowing the physical laws governing interactions between particles. Because of the absolute immutability of Newton's laws, this knowledge could be extended indefinitely into the future and the past, even to the creation of the universe. We realize now that Heisenberg's uncertainty principle (Section 5.7) creates serious problems for Laplace's position. A famous but perhaps apocryphal story will serve to demonstrate the conviction of the defenders of the mechanistic view. Laplace once presented the principle of perfect knowledge just described to the Emperor Napoleon. After hearing it in some detail, Napoleon asked where God fit into this mechanistic system. Laplace is supposed to have answered, "I have no need of this hypothesis." That was a rather extreme statement for anyone to make in the early 19th century, but it should indicate the great sense of power felt by those true believers in the Newtonian program.

In fairness to Laplace we must point out that he did make major contributions to the theory of probability. This may seem to be ironic, but perhaps he was inspired to study probability by his understanding of the practical limits of measurement in classical mechanics. His 1812 treatise *Théorie analytique des probabilitées* was the standard reference on the subject for much of the 19th century.

The development of statistical physics in the 19th century was tied closely to the development of thermodynamics (which in turn was driven by the industrial revolution). The common view in 1800 was that heat was a material substance known as *caloric,* a fluid that could flow through bodies to effect changes in temperature. In 1798 Benjamin Thompson (Count Rumford) put forward the idea that what we call *heat* is merely the motion of individual particles in a substance. Rumford's idea was essentially correct, but it was not accepted quickly. His work was too sketchy and qualitative, although it did plant the seeds that later blossomed in the work of Maxwell and Boltzmann. In 1822 Joseph Fourier published his theory of heat, which was the first truly mathematical treatise on the subject. It was not statistical in nature, but it did serve to provide a quantitative basis for later work.

The concept of energy is central to modern thermodynamics. Of supreme importance in the history of energy is the work of James Prescott Joule (1818–1889). Joule is probably best known for his experiment demonstrating

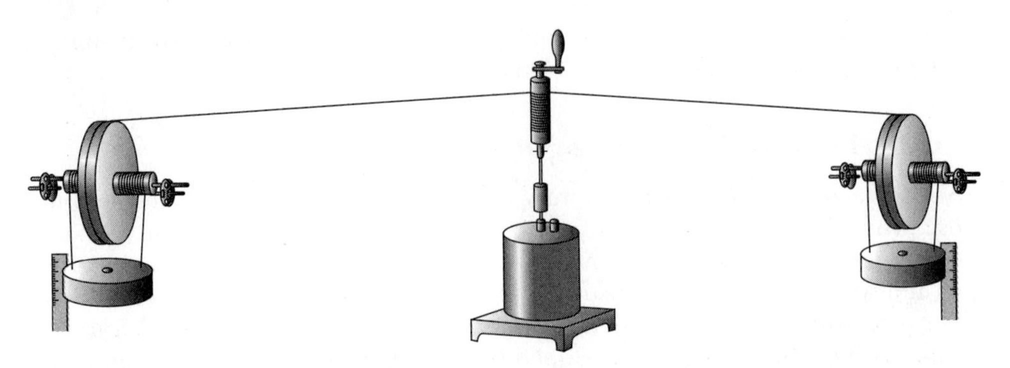

FIGURE 9.1 A schematic drawing of J. P. Joule's paddle wheel apparatus used to determine the mechanical equivalent of heat. Falling weights were used to turn paddle wheels through water (in the center drum), which raised the temperature of the water. Joule concluded that a weight of 772 lb (mass = 350.2 kg) must fall through a distance of 1 ft (30.48 cm) in order to raise the temperature of 1 lb of water (2.2 kg) by 1°F ($\frac{5}{9}$°C).

the mechanical equivalent of heat, first performed in about 1843. In that experiment (Figure 9.1) a falling weight was used to turn a paddle wheel through water. Joule showed conclusively that the energy of the falling weight was transferred to internal energy in the water.

James Clerk Maxwell did the great work of bringing the mathematical theories of probability and statistics to bear on the physical problems of thermodynamics. Using the understanding of energy gained by Joule and others, Maxwell derived expressions for the distributions of velocities and speeds of molecules in an ideal gas and showed that these distributions can be used to derive the observed macroscopic phenomena (see Sections 9.2–9.4). In spite of the great importance of this work, Maxwell is probably best known for his synthesis of electromagnetic theory (Maxwell's equations). There are some similarities between Maxwell's electromagnetic and thermodynamic theories. They are both highly mathematical (Maxwell was one of the foremost mathematicians of his day). They are also both very mechanical, in keeping with the Newtonian tradition. In his electromagnetic treatises Maxwell set up elaborate molecular vortices to explain the electric and magnetic properties of matter. In thermodynamics he believed that all relevant properties were due to the motions of individual molecules. These similarities are important, because the great and immediate success of Maxwell's electromagnetic theory played a role in winning over scientists to the statistical view of thermodynamics.

But it was not really until the early 20th century that statistical mechanics won the day. In 1905 Einstein (perhaps in his spare time after developing special relativity and his theory of the photoelectric effect!) published a theory of Brownian (random) motion, a theory that helped support the view that atoms are real. Experiments done several years later by Perrin confirmed Einstein's results. Soon after this came Bohr's successful atomic theory, and then the quantum theory we have developed in the previous chapters of this book.

We conclude this section with a brief discussion of the philosophy of statistical physics. Some people have difficulty accepting some of the uses of probability and statistics because of what they perceive as an implied indeterminism or lack of causality. It is perhaps difficult to reconcile this indeterminism with the fairly strict determinism still found in much of physics. Can the probabilistic laws governing the behavior of atoms and subatomic particles really be that different from our everyday experience of cause and effect? No less a figure than Einstein worried about this. He said, "God does not play dice," implying that there is some causal mechanism at work that is simply beyond the scope of our experiments. This philosophical discussion is an interesting and ongoing one, but it should be made clear that statistical physics is necessary regardless of the ultimate nature of physical reality. This is true for three reasons. First, even as simple a problem as determining the outcome of the toss of a coin is so complex that it is most often useful to reduce it to statistical terms. Air resistance, rotational dynamics, and restitution problems make predicting the outcome of a simple coin toss a formidable mechanics problem. Other problems, particularly in quantum theory, are even more challenging. Second, when the number of particles is large (e.g., on the order of 10^{23} particles for a modest sample of an ideal gas), it is highly impractical to study individual particles if one is more interested in the overall behavior of a system of particles (for example, the pressure, temperature, or specific heat of an ideal gas). This will be made more clear in subsequent sections of this chapter. Third, as Heisenberg showed, uncertainties are inherent in physics, and they are of significant size in atomic and subatomic systems. Therefore, it appears that statistics will always be an important part of physics.

9.2 Maxwell Velocity Distribution

As Laplace pointed out, we could, in principle, know everything about an ideal gas by knowing the position and instantaneous velocity of every molecule. This actually entails knowing six parameters per molecule, because three numbers are needed to specify the position (x, y, z) and three more to specify the velocity (v_x, v_y, v_z). Many relevant physical quantities must depend on one or more of these six parameters. In physics we sometimes think of those parameters as the components of a six-dimensional pseudospace called **phase space.**

Phase space

Maxwell focused on the three velocity components because he was most interested in the properties of ideal gases. The velocity components of the molecules of an ideal gas are clearly more important than the (random) instantaneous positions, because the energy of a gas should depend only on the velocities and not on the instantaneous positions. The crucial question for Maxwell was: what is the distribution of velocities for an ideal gas at a given temperature? Let us define (see Appendix 5) a **velocity distribution function** $f(\mathbf{v})$ such that

Velocity distribution function

$f(\mathbf{v})\,d^3\mathbf{v}$ = The probability of finding a particle with velocity between $\mathbf{v}$ and $\mathbf{v} + d^3\mathbf{v}$

where $d^3\mathbf{v} = dv_x dv_y dv_z$. Note that because $\mathbf{v}$ is a vector quantity, the preceding statement implies three separate conditions. The vector $\mathbf{v}$ has components v_x, v_y, and v_z. Therefore $f(\mathbf{v})\,d^3\mathbf{v}$ is the probability of finding a particle with v_x between v_x and $v_x + dv_x$, with v_y between v_y and $v_y + dv_y$, *and* with v_z between v_z and $v_z + dv_z$. We can think of the distribution function $f(\mathbf{v})$ as playing a role analogous to the probability density $\Psi^*\Psi$ in quantum theory.

Maxwell was able to prove that the probability distribution function is proportional to $\exp(-\frac{1}{2}mv^2/kT)$, where m is the molecular mass, v is the molecular speed, k is Boltzmann's constant, and T is the absolute temperature. Therefore, we may write

$$f(\mathbf{v})\,d^3\mathbf{v} = C\exp\left(-\frac{1}{2}\beta m v^2\right)d^3\mathbf{v} \tag{9.1}$$

where C is a proportionality factor and $\beta \equiv (kT)^{-1}$. We can easily rewrite Equation (9.1) in terms of the three velocity components, because as usual $v^2 = v_x^2 + v_y^2 + v_z^2$. Then

$$f(\mathbf{v})\,d^3\mathbf{v} = C\exp\left(-\frac{1}{2}\beta m v_x^2 - \frac{1}{2}\beta m v_y^2 - \frac{1}{2}\beta m v_z^2\right)d^3\mathbf{v} \tag{9.2}$$

Equation (9.2) can be rewritten in turn as the product of three factors, each of which contains one of the three velocity components. Let us define them as

$$\begin{aligned} g(v_x)\,dv_x &\equiv C'\exp\left(-\frac{1}{2}\beta m v_x^2\right)dv_x \\ g(v_y)\,dv_y &\equiv C'\exp\left(-\frac{1}{2}\beta m v_y^2\right)dv_y \\ g(v_z)\,dv_z &\equiv C'\exp\left(-\frac{1}{2}\beta m v_z^2\right)dv_z \end{aligned} \tag{9.3}$$

with a new constant $C' = C^{1/3}$. Equations (9.3) give us the distributions of the three velocity components.

In order to perform any useful calculations using the distributions (9.3), we will need to know the value of the constant C'. Now $g(v_x)\,dv_x$ is simply the probability that the x component of a gas molecule's velocity lies between v_x and $v_x + dv_x$. If we sum (or integrate) $g(v_x)\,dv_x$ over all possible values of v_x, the result must be one, because every molecule has a velocity component v_x somewhere in this range. This is just the same process of normalization that we have followed throughout our study of quantum theory in Chapters 6 and 7. Performing the integral (see Appendix 6) yields

$$\int_{-\infty}^{\infty} g(v_x)\,dv_x = C'\left(\frac{2\pi}{\beta m}\right)^{1/2} = 1 \tag{9.4}$$

Then

$$C' = \left(\frac{\beta m}{2\pi}\right)^{1/2}$$

and

$$g(v_x)\,dv_x = \left(\frac{\beta m}{2\pi}\right)^{1/2}\exp\left(-\frac{1}{2}\beta m v_x^2\right)dv_x \tag{9.5}$$

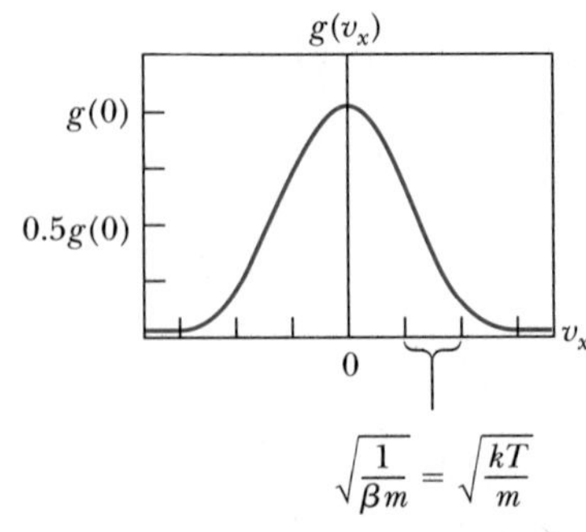

FIGURE 9.2 The Maxwell velocity distribution as a function of one velocity dimension (v_x). Notice that at $v_x = \sqrt{1/\beta m} = \sqrt{kT/m}$ we have $g(v_x) = g(0)e^{-1/2} \approx 0.607\,g(0)$.

With this distribution we can calculate the mean value of v_x (see Appendix 5):

$$\overline{v_x} = \int_{-\infty}^{\infty} v_x g(v_x)\,dv_x = C'\int_{-\infty}^{\infty} v_x \exp\left(-\frac{1}{2}\beta m v_x^2\right)dv_x = 0 \tag{9.6}$$

because v_x is an odd function (see Appendix 6). This result makes sense physically, because in a random distribution of velocities one would expect the velocity components to be distributed evenly around the peak at $v_x = 0$ (Figure 9.2).

Similarly, the mean value of v_x^2 is

$$\begin{aligned} \overline{v_x^2} &= C'\int_{-\infty}^{\infty} v_x^2\exp\left(-\frac{1}{2}\beta m v_x^2\right)dv_x \\ &= 2C'\int_0^{\infty} v_x^2\exp\left(-\frac{1}{2}\beta m v_x^2\right)dv_x \end{aligned}$$

$$\overline{v_x^2} = \left(\frac{\beta m}{2\pi}\right)^{1/2}\frac{\sqrt{\pi}}{2}\left(\frac{2}{\beta m}\right)^{3/2} = \frac{1}{\beta m} = \frac{kT}{m} \tag{9.7}$$

Of course there is nothing special about the x direction (the gas makes no distinction among x, y, and z!), so the results for the x, y, and z velocity components are identical. The three components may be used together to find the mean translational kinetic energy of a molecule:

$$\overline{K} = \frac{1}{2}\overline{mv^2} = \frac{1}{2}m\left(\overline{v_x^2} + \overline{v_y^2} + \overline{v_z^2}\right) = \frac{1}{2}m\left(\frac{3kT}{m}\right) = \frac{3}{2}kT \tag{9.8}$$

We have just confirmed one of the principal results of kinetic theory! The fact that we have done so from purely statistical considerations is good evidence of the validity of this statistical approach to thermodynamics.

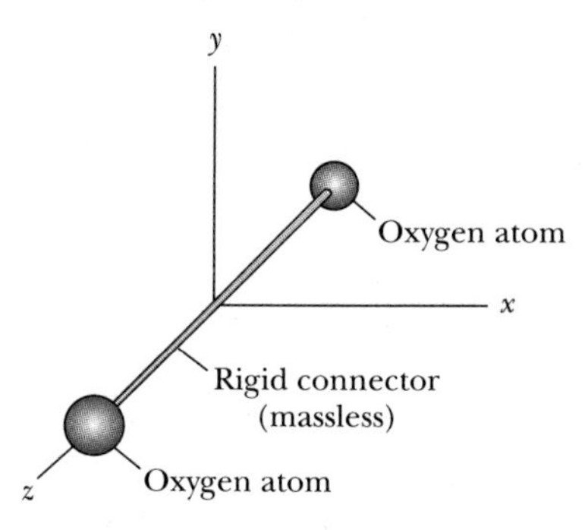

FIGURE 9.3 The rigid rotator model of the O_2 molecule, with the two oxygen atoms connected by a rigid, massless rod along the z axis.

9.3 Equipartition Theorem

The results of Section 9.2 can be extended into a rather general statement relating the internal energy of a thermodynamic system to its temperature. According to Equation (9.7)

$$\frac{1}{2}m\overline{v_x^2} = \frac{1}{2}kT \tag{9.9}$$

Similarly, there is an average energy $\frac{1}{2}kT$ associated with each of the other two velocity components, producing a net average translational kinetic energy of $\frac{3}{2}kT$ per molecule. In a monatomic gas such as helium or argon, virtually all of the gas's energy is in this form. But consider instead a diatomic gas, such as oxygen (O_2). If we think of this molecule as two oxygen atoms connected by a massless rod, it is clear that this molecule can also have *rotational* kinetic energy. The crucial question is, How much rotational energy is there and how is it related to temperature?

The answer to this question is provided by the equipartition theorem, which we state without proof.

Equipartition theorem

Equipartition theorem: *In equilibrium there is a mean energy of* $\frac{1}{2}kT$ *per molecule associated with each independent quadratic term in the molecule's energy.*

We shall see that those independent quadratic terms may be quadratic in coordinate, velocity component, angular velocity component, or anything else that when squared is proportional to energy. Each independent phase space coordinate is called a **degree of freedom** for the system.

Degree of freedom

Let us apply the equipartition theorem to the model of the oxygen molecule described earlier and shown in Figure 9.3. The molecule is free to rotate about either the x or y axis, and the corresponding rotational energies can be written in terms of rotational inertia and angular velocity components as $\frac{1}{2}I_x\omega_x^2$ and $\frac{1}{2}I_y\omega_y^2$, respectively. Each of these is quadratic in angular velocity, so the equipartition theorem tells us that we should add $2\left(\frac{1}{2}kT\right) = kT$ per molecule to the translational kinetic energy, for a total of $\frac{5}{2}kT$. Stated another way, there are five degrees of freedom (three translational and two rotational), so the energy per molecule is $5 \times \left(\frac{1}{2}kT\right) = \frac{5}{2}kT$. Why don't we include rotations about the z axis? The answer lies in quantum theory. In the quantum theory of the rigid rotator the allowed energy levels are

$$E = \frac{\hbar^2\ell(\ell + 1)}{2I}$$

where I is the rotational inertia and ℓ is a quantum number equal to zero or a positive integer. Notice that this result is consistent with the quantization of angular momentum given previously in Equation (8.5), because the kinetic energy of a rotator with angular momentum L is $L^2/(2I)$. Consider again our diatomic molecule modeled in Figure 9.3. We learned in studying atomic physics that the vast majority of the mass of an atom is confined to a nucleus that is four or five orders of magnitude smaller than the whole atom. Therefore the diatomic molecule's rotational inertia (I_z) about the axis connecting the two atoms is orders of magnitude smaller than I_x and I_y. A small value of I_z in the denominator of the energy equation given above leads to a high energy, relative to that obtained with I_x or I_y and comparable quantum numbers. Thus, when the rotational energy is relatively low and small quantum numbers are required, only rotations about the x and y axes are allowed.*

How can we check our calculation of $\frac{5}{2}kT$? This is done by measuring the heat capacity at constant volume. In a gas of N oxygen molecules the total internal energy should be

$$U = N\overline{E} = \tfrac{5}{2}NkT$$

The heat capacity at constant volume is $C_V = (\partial U/\partial T)_V$, or

$$C_V = \tfrac{5}{2}Nk$$

We will use the standard notation of uppercase C_V for the heat capacity of a general amount (such as N molecules) of a substance, and the lowercase c_V when describing the heat capacity for one mole. For one mole $N = N_A$, Avogadro's number, and

$$c_V = \tfrac{5}{2}N_A k = \tfrac{5}{2}R = 20.8\ \text{J/K} \tag{9.10}$$

where we have used the fact that $N_A k = R = 8.31$ J/K, the ideal gas constant.

The measured molar heat capacity of O_2 is very close to this value. In Table 9.1 the molar heat capacities of a number of gases are listed. For most

*A more complete description of why quantum theory overrules the equipartition theorem in this case is given by Clayton A. Gearhart, *Am. J. Phys.* **64**, 995–1000 (1996).

TABLE 9.1
Molar Heat Capacities for Selected Gases at 15°C and 1 Atmosphere

Gas	c_V (J/K)	c_V/R
Ar	12.5	1.50
He	12.5	1.50
CO	20.7	2.49
H_2	20.4	2.45
HCl	21.4	2.57
N_2	20.6	2.49
NO	20.9	2.51
O_2	21.1	2.54
Cl_2	24.8	2.98
CO_2	28.2	3.40
CS_2	40.9	4.92
H_2S	25.4	3.06
N_2O	28.5	3.42
SO_2	31.3	3.76

diatomic gases the measured heat capacities are similar to the value predicted in Equation (9.10). Note also that the monatomic gases have heat capacities very near to the value predicted for them, $\frac{3}{2}kT$. Some of the diatomic gases (principally Cl_2) do not match up as well, though. Why not? The physics of the two-atom molecule is evidently not as simple as we have pictured it. From our study of atomic physics we know that atoms are complex systems and that the quantum theory must be used to replace classical mechanics when studying atomic systems. Further, we have not even begun to discuss the nature of molecular bonding; the solid massless rod joining the two atoms is undoubtedly a crude approximation. Given all these approximations it is a wonder that the equipartition theorem serves as well as it does!

In some circumstances it is a better approximation to think of atoms connected to each other by a massless spring rather than a rigid rod. How many degrees of freedom does this add? One may be tempted to say just one, because of the potential energy $\frac{1}{2}\kappa(r - r_0)^2$, where κ is the spring's force constant, r the separation between atoms, and r_0 the equilibrium separation between atoms. But there is another degree of freedom that is associated with the vibrational velocity (dr/dt), because the vibrational kinetic energy should be $\frac{1}{2}m(dr/dt)^2$. This makes sense because in classical physics the average kinetic energy is equal to the average potential energy for a harmonic oscillator (see Problem 43). If this molecule is still free to rotate, there will now be a total of seven degrees of freedom: three translational, two rotational, and two vibrational. The resulting molar heat capacity would be $\frac{7}{2}R$.

Is this vibrational mode ever a factor? Yes it is, but high temperatures are normally required to excite the vibrational modes in diatomic gases. In fact, the heat capacities of diatomic gases are temperature dependent, indicating that the different degrees of freedom are "turned on" at different temperatures. One of the more striking examples of this is H_2 (see Figure 9.4). At temperatures just above the boiling point (about 20 K), the molar heat capacity is about $\frac{3}{2}R$. At just under 100 K the rotational mode is excited, and c_V then increases gradually until it reaches $\frac{5}{2}R$, where it remains from about 250 K up to about 1000 K. At that point the vibrational mode turns on (also gradually). We can see that c_V is

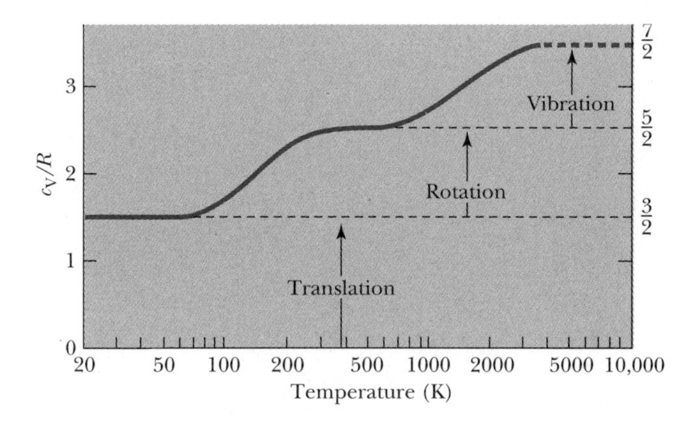

FIGURE 9.4 Molar heat capacity c_V as a function of temperature for H_2, a typical diatomic gas. The heat capacity $c_V = 3R/2$ at low temperatures, rises to $5R/2$ at higher temperatures as the rotational mode is excited, and finally approaches $7R/2$ when the molecule dissociates at very high temperatures.

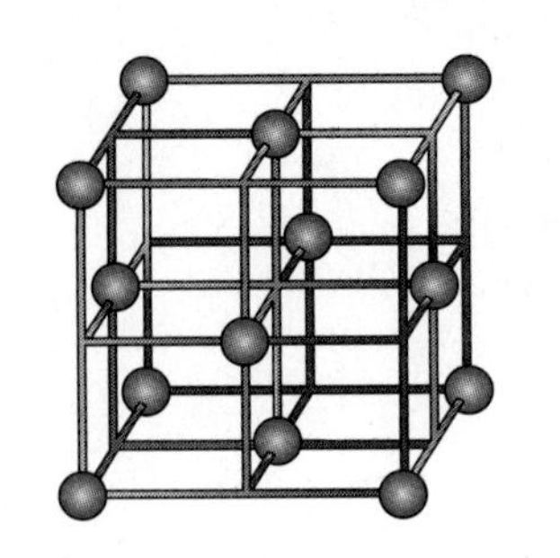

FIGURE 9.5 The lattice structure of copper, an example of the face-centered cubic lattice.

beginning to approach $\frac{7}{2}R$, but the curve terminates at this point because the molecule dissociates.

Apparently it is difficult to excite the vibrational mode in a diatomic gas. In a polyatomic gas, we may choose to think of the molecule as consisting of a number of masses (atoms) connected by springs. In classical mechanics such systems have a frequency corresponding to each "normal mode" of oscillation, with the number of normal modes increasing as the number of masses in the system increases. Therefore polyatomic systems may have a number of different vibration frequencies, each turning on at a different temperature. This can cause the vibrational spectra of polyatomic molecules to be quite complex.

Let us briefly turn our attention to the thermal molecular motion in a solid. Now the vibrational mode will be of supreme importance, for the atoms in a solid are not free to translate or rotate. As an example, we consider a sample of pure copper, the atoms of which are arranged in a face-centered cubic lattice, as shown in Figure 9.5.

How many degrees of freedom are there in this system? It is possible to think of each atom as a three-dimensional harmonic oscillator. Just as described earlier, a one-dimensional harmonic oscillator has two degrees of freedom, one coming from the kinetic energy and one from the potential energy. Therefore there must be **six** degrees of freedom for our three-dimensional oscillator, and the molar heat capacity should be $6 \times \frac{1}{2}R = 3R$. The experimental value of molar heat capacity is almost exactly $3R$ for copper near room temperature. It is observed also that, near room temperature, the molar heat capacity increases slightly with increasing temperature. As we will see in Section 9.6, this can be attributed to the conduction electrons, which have been neglected in our consideration of the equipartition theorem.

9.4 Maxwell Speed Distribution

Let us return to the Maxwell velocity distribution in the form

$$f(\mathbf{v})\, d^3\mathbf{v} = C \exp\left(-\frac{1}{2}\beta m v^2\right) d^3\mathbf{v} \tag{9.1}$$

where we know that

$$C = C'^3 = \left(\frac{\beta m}{2\pi}\right)^{3/2} \tag{9.11}$$

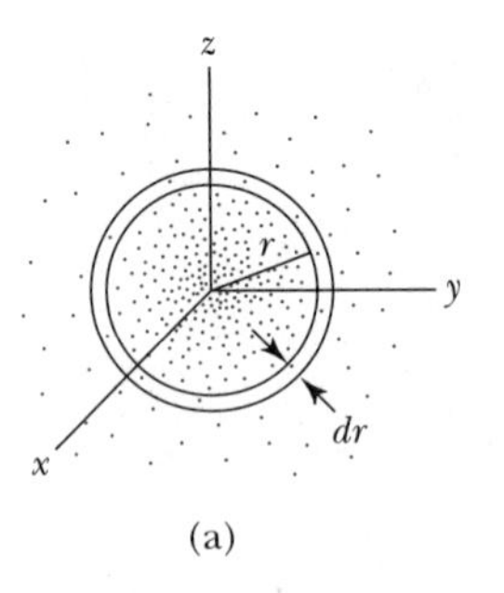

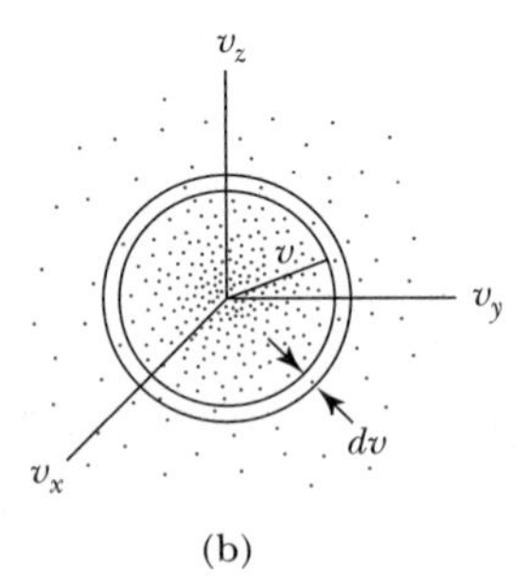

FIGURE 9.6 (a) A distribution of particles in three-dimensional space. The distribution function $f(r)$ is proportional to the number of particles in a spherical shell, between r and $r + dr$. (b) A similar distribution in three-dimensional (velocity) phase space. This shows that the speed distribution $f(v)$ must be proportional to the number of particles found in a spherical shell in phase space between v and $v + dv$.

Even though $f(\mathbf{v})$ is a function of the speed (v) and not the velocity ($\mathbf{v}$), this is still a velocity distribution because the probability equation (9.1) contains the differential velocity element. It will be useful for us to turn this velocity distribution into a speed distribution $F(v)$, where by the usual definition

$F(v)\,dv =$ the probability of finding a particle with speed between v and $v + dv$

As we shall soon see, it is *not* possible simply to assume that $f(\mathbf{v}) = F(v)$.

This is just the kind of problem for which the phase-space concept is useful. Consider the analogous problem in normal three-dimensional (x, y, z) space. Suppose there exists some distribution of particles $f(x, y, z)$, as in Figure 9.6a. A particle at the point (x, y, z) is a distance $r = (x^2 + y^2 + z^2)^{1/2}$ from the origin, and $\mathbf{r}$ is a position vector directed from the origin to the point (x, y, z). Then

$f(x, y, z)\,d^3\mathbf{r} =$ the probability of finding a particle between $\mathbf{r}$ and $\mathbf{r} + d^3\mathbf{r}$

with $d^3\mathbf{r} = dx\,dy\,dz$. Now let us change to a radial distribution $F(r)$, such that

$F(r)\,dr =$ the probability of finding a particle between r and $r + dr$

The space between r and $r + dr$ is a spherical shell. Therefore this problem is simply one of counting all of the particles in a spherical shell of radius r and thickness dr. The volume ($d^3\mathbf{r}$) of a spherical shell is $4\pi r^2 dr$. Thus we may write

$$F(r)\,dr = f(x, y, z)\,4\pi r^2\,dr \tag{9.12}$$

Returning to our problem of obtaining a speed distribution $F(v)$ from a velocity distribution $f(\mathbf{v})$, we see that all we have to do is count the number of particles in a spherical shell in phase space. Simply replace the coordinates x, y, and z with the phase space coordinates (i.e., the velocity components) v_x, v_y, and v_z (Figure 9.6b). The speed $v = (v_x^2 + v_y^2 + v_z^2)^{1/2}$ is the phase space analog of radius $r = (x^2 + y^2 + z^2)^{1/2}$. The preceding analysis indicates that the "volume" of our spherical shell in phase space is $4\pi v^2 dv$, and the desired speed distribution is $F(v)$ where

$$F(v)\,dv = f(\mathbf{v})\,4\pi v^2 dv \tag{9.13}$$

Using Equation (9.1) we obtain the Maxwell speed distribution.

$$F(v)\,dv = 4\pi C \exp\left(-\frac{1}{2}\beta m v^2\right) v^2\,dv \tag{9.14}$$

Maxwell speed distribution

A graph of a typical Maxwell speed distribution is shown in Figure 9.7. There is a qualitative similarity between this distribution and some of the radial probability distributions we encountered in Chapter 7 (see for example Figure 7.12).

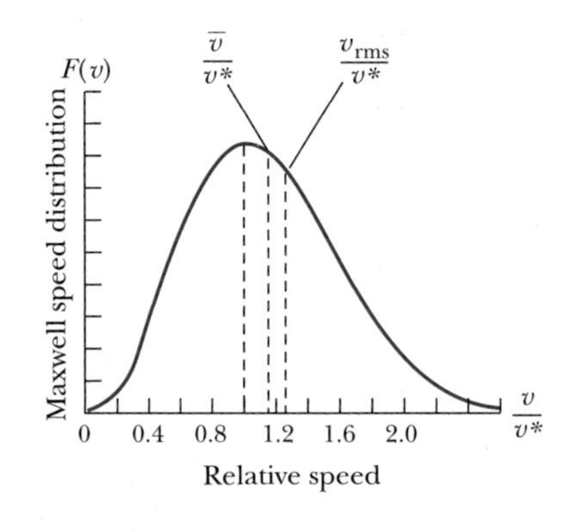

FIGURE 9.7 The Maxwell speed distribution, expressed in terms of the most probable speed v^*. Note the positions of $\bar{v}$ and v_{rms} relative to v^*.

There is a quantitative distinction, however, in that the quadratic (v^2) argument in the exponential in Equation (9.14) differs from the linear (r) power found throughout Section 7.6. Notice also that the speed distribution is qualitatively different from the velocity distribution in that it is not symmetric about its peak. It is simple to show that $F(v)$ approaches zero in the limiting cases of very high and very low speeds (Problem 8). Because it was derived from purely classical considerations, the Maxwell speed distribution gives a nonzero probability of finding a particle with a speed greater than c (Problem 5). We know this cannot be true, therefore, Equation (9.14) is only valid in the classical limit. This presents no serious problems in using the distribution because (you should convince yourself of this) the predicted probability of $v > c$ is extremely low at any reasonable temperature. Further, the other formulas we use in the kinetic theory of ideal gases (e.g., $E_k = \frac{1}{2}mv^2$) already restrict us to the classical limit.

The asymmetry of the distribution curve leads to an interesting result: the most probable speed v^*, the mean speed $\bar{v}$, and the root mean square speed v_{rms} are all slightly different from each other. The most probable speed v^* corresponds to the peak of the curve. Thus

$$\frac{d}{dv}\left[4\pi C \exp\left(-\frac{1}{2}\beta m v^2\right)v^2\right]\Bigg|_{v=v^*} = 0$$

$$\exp\left(-\frac{1}{2}\beta m v^{*2}\right)(2v^*) - \frac{1}{2}\beta m(2v^*)\exp\left(-\frac{1}{2}\beta m v^{*2}\right)v^{*2} = 0$$

Solving for v^*:

Most probable speed v^*

$$v^* = \sqrt{\frac{2}{\beta m}} = \sqrt{\frac{2kT}{m}} \tag{9.15}$$

A curious corollary of this result is that the kinetic energy of a molecule moving with the most probable speed v^* is

$$K^* = \frac{1}{2}mv^{*2} = kT \tag{9.16}$$

which is exactly two thirds of the mean kinetic energy.

Now the mean speed is

$$\bar{v} = \int_0^\infty vF(v)\,dv = 4\pi C\int_0^\infty v^3 \exp\left(-\frac{1}{2}\beta m v^2\right)dv$$

See Appendix 6 for an evaluation of this integral. The result is

$$\bar{v} = 4\pi C\left(\frac{1}{2\left(\frac{1}{2}\beta m\right)^2}\right) = 8\pi\left(\frac{\beta m}{2\pi}\right)^{3/2}\left(\frac{1}{\beta m}\right)^2$$

Mean speed $\bar{v}$

$$\bar{v} = \frac{4}{\sqrt{2\pi}}\sqrt{\frac{kT}{m}} \tag{9.17}$$

Comparing the results (9.15) and (9.17), we see that

$$\frac{\bar{v}}{v^*} = \sqrt{\frac{4}{\pi}} \approx 1.13 \tag{9.18}$$

In other words, $\bar{v}$ is about 13% greater than v^* *at any temperature.*

We define the **root mean square speed** v_{rms} to be

Root mean square speed

$$v_{\text{rms}} \equiv \left(\overline{v^2}\right)^{1/2}$$

In order to find the root mean square speed, it is first necessary to calculate $\overline{v^2}$:

$$\overline{v^2} = \int_0^\infty v^2 F(v)\,dv = 4\pi C \int_0^\infty v^4 \exp\left(-\frac{1}{2}\beta m v^2\right) dv$$

$$= 4\pi C\left[\frac{3\sqrt{\pi}}{8\left(\frac{1}{2}\beta m\right)^{5/2}}\right] = \frac{3}{\beta m} = \frac{3kT}{m} \tag{9.19}$$

Then

$$v_{\text{rms}} = \left[\overline{v^2}\right]^{1/2} = \sqrt{\frac{3kT}{m}} \tag{9.20}$$

As before, it is instructive to compare this result with the most probable speed:

$$\frac{v_{\text{rms}}}{v^*} = \sqrt{\frac{3}{2}} \approx 1.22 \tag{9.21}$$

The rms speed is about 22% greater than the most probable speed for any temperature. Notice that the result (9.19) is again in keeping with our basic law of kinetic theory, namely

$$\overline{K} = \frac{1}{2}m\overline{v^2} = \frac{3}{2}kT$$

Finally, we calculate the standard deviation of the molecular speeds (see Appendix 5)

$$\sigma_v = \left(\overline{v^2} - \overline{v}^2\right)^{1/2} = \left(\frac{3kT}{m} - \frac{8kT}{\pi m}\right)^{1/2}$$

$$= \left[3 - \left(\frac{8}{\pi}\right)\right]^{1/2}\left(\frac{kT}{m}\right)^{1/2} \approx 0.48\ v^* \tag{9.22}$$

The result expressed in Equation (9.22) indicates that σ_v increases in proportion to $\sqrt{T}$. The distribution of molecular speeds therefore widens somewhat as temperature increases.

Example 9.1

What fraction of the molecules in an ideal gas in equilibrium have speeds within $\pm 1\%$ of v^*?

Solution: Recall that $F(v)\,dv$ is the probability of finding a particle with speed between v and $v + dv$. In principle we could integrate the distribution $F(v)$ in Equation (9.14) from the limits $0.99\ v^*$ to $1.01\ v^*$.

$$P(\pm 1\%) = \int_{0.99v^*}^{1.01v^*} F(v)\,dv \tag{9.23}$$

Unfortunately, the indefinite integral cannot be done in closed form. We can obtain an approximate solution by calculating $F(v^*)$ and multiplying by $dv \approx \Delta v = 0.02\ v^*$. The product of $F(v^*)(0.02\ v^*)$ gives us our probability

$$P(\pm 1\%) \approx F(v^*)(0.02\ v^*)$$

$$\approx 4\pi C \exp\left(-\frac{1}{2}\beta m v^{*2}\right) v^{*2}(0.02\ v^*)$$

$$\approx 4\pi\left(\frac{\beta m}{2\pi}\right)^{3/2} e^{-1}(0.02)\left(\frac{2}{\beta m}\right)^{3/2}$$

$$\approx \frac{4}{\sqrt{\pi}}\, e^{-1}(0.02) \approx 0.017$$

Students with computer programming experience are encouraged to do the integration in Equation (9.23) numerically and compare the result with this approximation. A numerical integration, though still an approximation, should yield a more accurate result.

9.5 Classical and Quantum Statistics

In the previous section it was mentioned that the Maxwell speed distribution is only valid in the classical (i.e., nonrelativistic) limit, but that this fact will only prohibit the use of the Maxwell distribution at exceptionally high temperatures. A more severe restriction of the use of the Maxwell distribution comes from quantum theory. You may have already wondered why it has not been necessary to apply quantum theory to the ideal gas systems we have been studying to this point. After all, the particles involved are of molecular and atomic size. The important point is that an ideal gas is dilute. The molecules of the gas are so far apart that they can be considered not to interact with each other. When collisions between molecules do occur, they can be considered totally elastic, and therefore they have no effect on the distributions and mean values calculated in the previous sections of this chapter.

But what happens when matter is in the liquid or solid state, where the density of matter is generally several orders of magnitude higher than the density of gases? In liquids and solids the assumption of a collection of noninteracting particles may no longer be valid. If molecules, atoms, or subatomic particles are close enough together, the Pauli exclusion principle (see Chapter 8) will be an important factor. It will limit the allowed energy states of any particle subject to the Pauli principle, and therefore the distribution of energies for a system of particles will undoubtedly be affected too.

There is another fundamental difference between classical and quantum energy states. In classical physics there is no restriction on particle energies, but in quantum systems only certain energy values are allowed. This will certainly affect the overall distribution of energies.

Classical Distributions

Because energy levels are of such fundamental importance in quantum theory, it will be useful for us to rewrite the results of Section 9.4 in terms of energy rather than velocity. The Maxwell speed distribution was given by

$$F(v)\,dv = 4\pi C \exp\left(-\frac{1}{2}\beta m v^2\right) v^2 dv \tag{9.14}$$

For a monatomic gas the energy is all translational kinetic energy. Thus

$$E = \frac{1}{2} m v^2$$

$$dE = mv\,dv$$

$$dv = \frac{dE}{mv} = \frac{dE}{\sqrt{2mE}} \tag{9.24}$$

With Equation (9.24) the speed distribution (9.14) can be turned into an energy distribution (see Problem 17):

$$F(v)\,dv = F(E)\,dE \tag{9.25}$$

where

$$F(E) = \frac{8\pi C}{\sqrt{2}m^{3/2}} \exp(-\beta E)\,E^{1/2} \tag{9.26}$$

James Clerk Maxwell (left) 1831–1879) and Ludwig Boltzmann (1844–1906), two of the greatest physicists of the 19th Century. Maxwell and Boltzmann worked independently on developing the laws governing the statistical behavior of classical particles. *(Left) AIP/Niels Bohr Library, (Right) University of Vienna, courtesy AIP/Niels Bohr Library.*

There are two factors in Equation (9.26) that contain E. The factor $E^{1/2}$ can be traced back to the phase space analysis done in Section 9.4. It is a feature of this particular kind of distribution (i.e, the distribution of molecular speeds in an ideal gas, for which the energy of a molecule is $\frac{1}{2}mv^2$). The factor $\exp(-\beta E)$ is of more fundamental importance. Boltzmann showed that it is a characteristic of any classical system, regardless of how quantities other than molecular speeds may affect the energy of a given state. Thus we define the **Maxwell-Boltzmann factor** for classical systems as

$$F_{\text{MB}} = A\exp(-\beta E) \tag{9.27}$$

Maxwell-Boltzmann factor

where A is a normalization constant. The energy distribution for a classical system will have the form

$$n(E) = g(E)F_{\text{MB}} \tag{9.28}$$

where, in the usual sense, $n(E)$ is a distribution such that $n(E)\,dE$ represents the number of particles with energies between E and $E + dE$. The function $g(E)$, known as the **density of states,** is the number of states available per unit energy range. It makes sense that $n(E)$ is proportional to $g(E)$. The density of states can therefore be thought of as an essential element in all distributions, and we shall keep this in mind as we develop the quantum distributions. The factor F_{MB} tells us the relative probability that an energy state will be occupied at a given temperature.

Density of states

Quantum Distributions

Now we turn our attention from classical to quantum distribution functions. In quantum theory, particles are described by wave functions. Identical particles cannot be distinguished from one another if there is a significant overlap of their wave functions. It is this characteristic of indistinguishability that makes quantum statistics different from classical statistics.

To illustrate this point consider the following example. Suppose that we have a system of just two particles, each of which has an equal probability (0.5) of being in either of two energy states. If the particles are distinguishable (call them A and B), then the possible configurations we may measure are as follows:

State 1	State 2
A B	
A	B
B	A
	A B

There are a total of four equally likely configurations: therefore the probability of each is one fourth (0.25). If on the other hand the two particles are indistinguishable, then our probability table changes:

State 1	State 2
X X	
X	X
	X X

Now there are only three equally likely configurations, each having a probability of one third ($\approx$0.33).

It turns out that two kinds of quantum distributions are needed. This is because some particles obey the Pauli exclusion principle and others do not. As was mentioned earlier in this section, the Pauli principle has a significant impact on how energy states can be occupied and therefore on the corresponding energy distribution. It is easy to tell whether a particle will obey the Pauli principle: we determine the spin of the particle. Particles with half-integer spins obey the Pauli principle and are known collectively as **fermions;** those with zero or integer spins do not obey the Pauli principle and are known as **bosons.** Protons, neutrons, and electrons are fermions. Photons and pions are bosons. Also, atoms and molecules consisting of an even number of fermions must be bosons when considered as a whole, because their total spin will be zero or an integer. Similarly, those (relatively few) atoms and molecules made up of an odd number of fermions will be fermions.

Fermions and bosons

We state here without proof the Fermi-Dirac distribution, which is valid for fermions:

$$n(E) = g(E)F_{\mathrm{FD}} \tag{9.29}$$

where

Fermi-Dirac distribution

$$F_{\mathrm{FD}} = \frac{1}{B_1 \exp(\beta E) + 1} \tag{9.30}$$

Similarly the Bose-Einstein distribution, valid for bosons, is

$$n(E) = g(E)F_{\mathrm{BE}} \tag{9.31}$$

where

Bose-Einstein distribution

$$F_{\mathrm{BE}} = \frac{1}{B_2 \exp(\beta E) - 1} \tag{9.32}$$

In each case $B_i (i = 1 \text{ or } 2)$ is a normalization factor, and $g(E)$ is the density of states appropriate for a particular physical situation. Notice that the Fermi-Dirac and Bose-Einstein distributions look very similar; they differ only by the normalization constant and by the sign attached to the 1 in the denominator.

TABLE 9.2
Classical and Quantum Distributions

Distributors	Properties of the Distribution	Examples	Distribution Function
Maxwell-Boltzmann	Particles are identical but distinguishable	Ideal gases	$F_{MB} = A\exp(-\beta E)$
Bose-Einstein	Particles are identical and indistinguishable with integer spin	Liquid ^{4}He, photons	$F_{BE} = \dfrac{1}{B_2\exp(\beta E) - 1}$
Fermi-Dirac	Particles are identical and indistinguishable with half-integer spin	Electron gas	$F_{FD} = \dfrac{1}{B_1\exp(\beta E) + 1}$

This sign difference causes a significant difference in the properties of bosons and fermions, as will become evident in Sections 9.6 and 9.7. It is also important to see that both the Fermi-Dirac and Bose-Einstein distributions reduce to the classical Maxwell-Boltzmann distribution when $B_i\exp(\beta E)$ is much greater than one* (in that case the normalization constant $A = 1/B_i$). This means that the Maxwell-Boltzmann factor $A\exp(-\beta E)$ is much less than one (i.e., the probability that a particular energy state will be occupied is much less than one). This is consistent with our earlier use of Maxwell-Boltzmann statistics for a dilute, noninteracting system of particles. See Table 9.2 for a summary of the properties of the three distribution functions.

*This happens at high temperatures and low densities. A good rule of thumb is to compare the interparticle spacing with the average de Broglie wavelength. If the interparticle spacing is much greater than the de Broglie wavelength, then Maxwell-Boltzmann statistics are fairly accurate. Otherwise one should use the quantum statistics.

Example 9.2

In a gas of atomic hydrogen (assume that the Maxwell-Boltzmann distribution is valid), what is the relative number of atoms in the ground state and first excited state at 293 K (room temperature), 5000 K (the temperature at the surface of a star), and 10^6 K (a temperature in the interior of a star)?

Solution: The desired ratio is

$$\frac{n(E_2)}{n(E_1)} = \frac{g(E_2)}{g(E_1)}\exp[\beta(E_1 - E_2)]$$

In the ground state ($n = 1$) of hydrogen there are two possible configurations for the electron, that is, $g(E_1) = 2$. There are eight possible configurations in the first excited state (see Chapter 7), so $g(E_2) = 8$. For atomic hydrogen $E_1 - E_2 = -10.2$ eV. Therefore

$$\frac{n(E_2)}{n(E_1)} = 4\exp[\beta(-10.2\text{ eV})] = 4\exp(-10.2\text{ eV}/kT)$$

for a given temperature T. The desired numerical results are

$$\begin{aligned} \frac{n(E_2)}{n(E_1)} &= 4\exp(-404) \approx 10^{-175} && \text{for } T = 293\text{ K} \\ &= 4\exp(-23.7) \approx 2\times10^{-10} && \text{for } 5000\text{ K} \\ &= 4\exp(-0.118) \approx 3.55 && \text{for } 10^6\text{ K} \end{aligned}$$

Notice that at very high temperatures ($T \gg 10^6$ K), the exponential factor approaches 1, so the ratio $n(E_2)/n(E_1)$ approaches 4, the ratio of the densities of states. In fact, atomic hydrogen cannot exist at such high temperatures (10^6 K or greater). The electrons dissociate from the nuclei to form a state of matter known as a *plasma*.

Enrico Fermi (above) (1901–1954) and Paul Adrien Maurice Dirac (1902–1984) developed the correct quantum mechanical laws (Fermi-Dirac statistics) governing the statistics of half-integer spin particles, now known as fermions. *(Top) AIP/Niels Bohr Library, (bottom) A. B. Örtzells Tryckeri, courtesy AIP/Niels Bohr Library,*

9.6 Fermi-Dirac Statistics

Introduction to Fermi-Dirac Theory

The Fermi-Dirac distribution, as expressed in Equations (9.29) and (9.30), provides the basis for our understanding of the behavior of a collection of fermions. Let us examine the distribution in some detail before applying it to the problem of electrical conduction.

First we consider the role of the factor B_1. In principle one may compute B_1 for a particular physical situation by integrating $n(E)\,dE$ over all allowed energies. Because the parameter β $(= 1/kT)$ is contained in F_{FD}, it is clear that B_1 should be temperature dependent. It is possible to express this temperature dependence as

$$B_1 = \exp(-\beta E_F) \qquad (9.33)$$

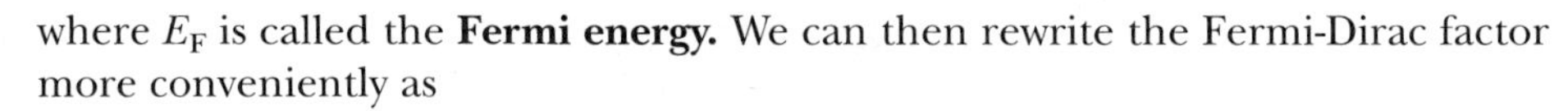

where E_F is called the **Fermi energy.** We can then rewrite the Fermi-Dirac factor more conveniently as

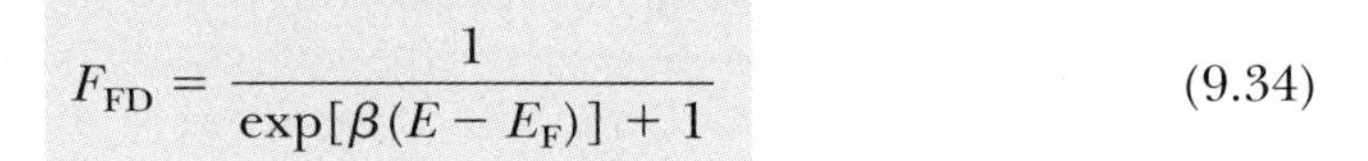

$$F_{FD} = \frac{1}{\exp[\beta(E - E_F)] + 1} \qquad (9.34)$$

Equation (9.34) shows us an important fact about the Fermi energy: when $E = E_F$, the exponential term is one, and therefore $F_{FD} = \frac{1}{2}$ (exactly). In fact it is common to define the Fermi energy as the energy at which $F_{FD} = \frac{1}{2}$.

Consider now the temperature dependence of F_{FD}, as expressed in Equation (9.34). In the limit as $T \to 0$, it is seen (Problem 28) that

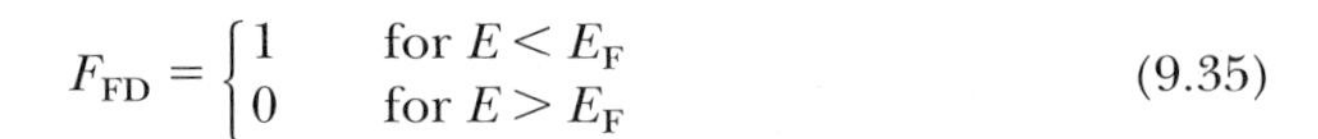

$$F_{FD} = \begin{cases} 1 & \text{for } E < E_F \\ 0 & \text{for } E > E_F \end{cases} \qquad (9.35)$$

The physical basis for Equation (9.35) is easily understood. At $T = 0$ fermions occupy the lowest energy levels available to them. They cannot all be in the lowest level, because that would violate the Pauli principle. Rather, fermions will fill all the available energy levels up to a particular energy (the Fermi energy). Near $T = 0$ there is little chance that thermal agitation will kick a fermion to an energy greater than E_F.

As the temperature increases from $T = 0$, more and more fermions may be in higher energy levels. The Fermi-Dirac factor "smears out" from the sharp step function [Figure 9.8(a)] to a smoother curve [Figure 9.8(b)]. It is sometimes useful to consider a **Fermi temperature,** defined as $T_F \equiv E_F/k$. A plot of F_{FD} at $T = T_F$ is shown in Figure 9.8(c). When $T \ll T_F$ the step function approximation for F_{FD} (Equation 9.35) is reasonably accurate. When $T \gg T_F$, F_{FD} approaches a simple decaying exponential [see Figure 9.8(d)]. This is just as one would expect, for at sufficiently high temperatures we expect Maxwell-Boltzmann statistics to be reasonably accurate [see Equation (9.27) and the subsequent discussion in Section 9.5].

Fermi temperature

We can apply the Fermi-Dirac theory to the problem of understanding electrical conduction in metals. For comparison we will first present a brief review of the classical theory. A more complete description of the classical theory can be found in most good introductory physics texts.

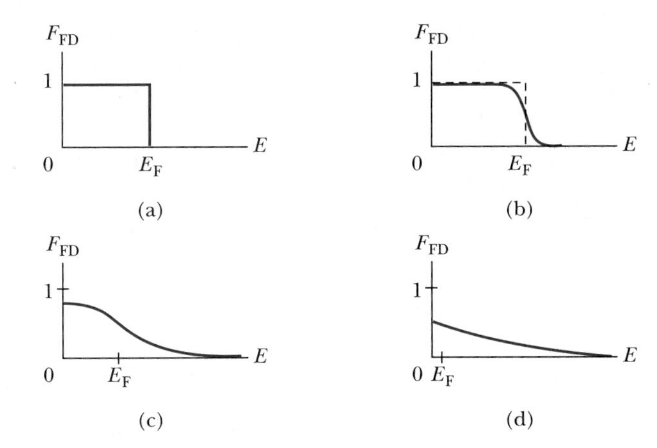

FIGURE 9.8 The Fermi-Dirac factor F_{FD} at various temperatures: (a) $T = 0$, (b) $T > 0$, (c) $T = T_F = E_F/k$, and (d) $T \gg T_F$. At $T = 0$ the Fermi-Dirac factor is a step function. As the temperature increases, the step is gradually rounded. Finally, at very high temperatures, the distribution approaches the simple decaying exponential of the Maxwell-Boltzmann distribution.

Classical Theory of Electrical Conduction

In 1900 Paul Drude developed a theory of electrical conduction in an effort to explain the observed conductivity of metals. His model assumed that the electrons in a metal existed as a gas of free particles. This is a fair assumption, because in a conductor the outermost electron(s) (there may be on average up to several per atom) are so weakly bound to an atom that they may be stripped away easily by even a very weak electric field. In the Drude model the metal is thought of as a lattice of positive ions with a gas of electrons free to flow through it. Just as in the ideal gases we considered earlier in this chapter, the electrons have a thermal kinetic energy proportional to temperature. The mean speed of an electron at room temperature can be calculated [using Equation (9.17)] to be about 10^5 m/s. But remember that the velocities of the particles (earlier molecules, now electrons) in a gas are directed randomly. Therefore there will be no net flow of electrons unless an electric field is applied to the conductor.

When an electric field is applied, the negatively charged electrons flow in the opposite direction to the field. According to Drude, their flow is severely restricted by collisions with the lattice ions. Based on several simple assumptions from classical mechanics, Drude was able to show that the current in a conductor should be linearly proportional to the applied electric field. That is consistent with Ohm's law, a well-known experimental fact. The principal success of Drude's theory was that it did predict Ohm's law.

Unfortunately, the numerical predictions of the theory were not so successful. One important prediction was that the electrical conductivity could be expressed* as

$$\sigma = \frac{ne^2\tau}{m} \tag{9.36}$$

where n is the number density of conduction electrons, e is the electron charge, τ is the average time between electron–ion collisions, and m is the electronic

*See the introductory text, R. Serway, *Physics for Scientists and Engineers,* 4th ed. Philadelphia: Saunders College Publishing, 1996, pp. 783–785.

mass. It is possible to measure n by the Hall effect (see Chapter 11). The parameter τ is not as easy to measure, but it can be estimated using transport theory. The best estimates of τ, when combined with the other parameters in Equation (9.36), produced a value of σ that is about one order of magnitude too small for most conductors. The Drude theory is therefore incorrect in this prediction.

A restatement of Equation (9.36) will show another deficiency of the classical theory. The mean time between collisions τ should be simply the mean distance ℓ traveled by an electron between collisions (called the **mean free path**) divided by the mean speed $\bar{v}$ of electrons. That is,

$$\tau = \ell/\bar{v}$$

Then the electrical conductivity can be expressed as

$$\sigma = \frac{ne^2\ell}{m\bar{v}} \tag{9.37}$$

Equation (9.17) shows that the mean speed is proportional to the square root of the absolute temperature. Hence, according to the Drude model, the conductivity should be proportional to $T^{-1/2}$. But for most conductors the conductivity is very nearly proportional to T^{-1} except at very low temperatures, where it no longer follows a simple relation. Clearly the classical model of Drude has failed to predict this important experimental fact.

Finally, there is the problem of calculating the electronic contribution to the heat capacity of a solid conductor. As was discussed in Section 9.3, the heat capacity of a solid can be almost completely accounted for by considering the six degrees of freedom in the lattice vibrations. This gives a molar heat capacity of $6 \times \frac{1}{2}R = 3R$. According to the equipartition theorem, we should add another $3 \times \frac{1}{2}R = \frac{3}{2}R$ for the heat capacity of the electron gas, giving a total of $\frac{9}{2}R$. This is not consistent with experimental results. The electronic contribution to the heat capacity depends on temperature and is typically only about $0.02R$ per mole at room temperature. Clearly a different theory is needed to account for the observed values of electrical conductivity and heat capacity as well as the temperature dependence of the conductivity.

Quantum Theory of Electrical Conduction

To obtain a better solution we will have to turn to quantum theory. It is necessary to understand how the electron energies are distributed in a conductor. Electrons are fermions, and therefore we will need to use the Fermi-Dirac distribution described earlier in this section. The real problem we face is to find $g(E)$, the number of allowed states per unit energy. The question is: what energy values should we use? Let us retain the "free electron" assumption of the Drude model and use the results obtained in Chapter 6 for a three-dimensional infinite-square-well potential (this will correspond physically to a cubical lattice of ions). The allowed energies are

$$E = \frac{h^2}{8mL^2}(n_1^2 + n_2^2 + n_3^2) \tag{9.38}$$

where L is the length of a side of the cube and n_i are the integer quantum numbers. For the time being we will not worry about the effects an elevated temperature might have on the distribution of energies. We will solve the problem at $T = 0$ first and consider the effect of temperature later. It turns out that the distribution at room temperature is very much like the $T = 0$ distribution anyway because we are in the $T \ll T_F$ regime (for example, $T_F \approx 80{,}000$ K for copper, see Table 9.4).

Equation (9.38) can be rewritten as

$$E = r^2 E_1 \tag{9.39}$$

where $r^2 \equiv n_1{}^2 + n_2{}^2 + n_3{}^2$ and $E_1 = h^2/8mL^2$. The parameter r is the "radius" of a sphere in phase space and is a dimensionless quantity—it should not be confused with a radius in Euclidean space. Note that E_1 is just a constant, not the ground-state energy (it is actually one third of the ground-state energy). We have defined r in this way in order to construct a geometric solution to the problem of counting the number of allowed quantum states per unit energy. Think of the n_i as the "coordinates" of a three-dimensional number space, as in Figure 9.9. The number of allowed states up to "radius" r (or up to energy $E = r^2E_1$) will be directly related to the spherical "volume" $\frac{4}{3}\pi r^3$. The exact number of states up to radius r is

$$N_r = (2)\left(\frac{1}{8}\right)\left(\frac{4}{3}\pi r^3\right) \tag{9.40}$$

The extra factor of 2 is due to spin degeneracy: for each set of quantum numbers there may be two electrons, one with spin up and one with spin down. The factor of $\frac{1}{8}$ is necessary because we are restricted to positive quantum numbers,

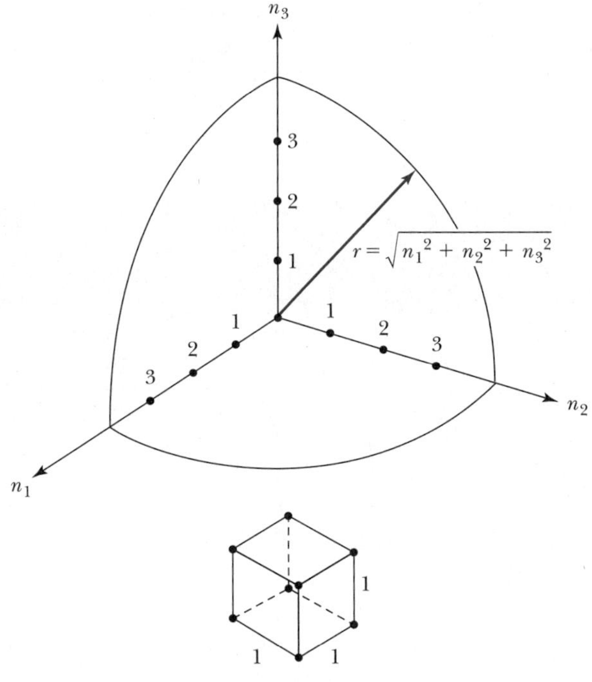

FIGURE 9.9 The three-dimensional number space used to count the number of particles within a sphere of radius $r = \sqrt{n_1{}^2 + n_2{}^2 + n_3{}^2}$. We may also think of this as counting the number of unit ($1 \times 1 \times 1$) cubes within this octant of space.

and therefore to one octant of the three-dimensional number space. Equations (9.39) and (9.40) can be used to express N_r as a function of E:

$$N_r = \frac{1}{3}\pi\left(\frac{E}{E_1}\right)^{3/2} \tag{9.41}$$

At $T = 0$ the Fermi energy is the energy of the highest occupied energy level [as we saw in Figure 9.8(a)]. If there are a total of N electrons, then

$$N = \frac{1}{3}\pi\left(\frac{E_F}{E_1}\right)^{3/2}$$

Solving for E_F:

$$E_F = E_1\left(\frac{3N}{\pi}\right)^{2/3} = \frac{h^2}{8m}\left(\frac{3N}{\pi L^3}\right)^{2/3} \tag{9.42}$$

Equation (9.42) is useful, because the ratio N/L^3 is well known for most conductors: it is simply the number density of conduction electrons, a quantity easily measured using the Hall effect (see Chapter 11).

Example 9.3

Calculate the Fermi energy and Fermi temperature for copper.

Solution: The number density of conduction electrons in copper is about $8.47 \times 10^{28}\ \text{m}^{-3}$ (see Problem 25 and Table 9.3). Using this value of N/L^3 in Equation (9.42) gives

$$E_F = \frac{(6.626 \times 10^{-34}\ \text{J}\cdot\text{s})^2}{8(9.11 \times 10^{-31}\ \text{kg})}\left(\frac{3(8.47 \times 10^{28}\ \text{m}^{-3})}{\pi}\right)^{2/3}$$

$$= 1.13 \times 10^{-18}\ \text{J} = 7.03\ \text{eV}$$

Within rounding errors, this result is equivalent to that given in Table 9.4.

$$T_F = E_F/k = \frac{7.03\ \text{eV}}{8.62 \times 10^{-5}\ \text{eV/K}} = 8.16 \times 10^4\ \text{K}$$

Fermi energies and Fermi temperatures for other common conductors are listed in Table 9.4. Note that E_F changes little between $T = 0$ and room temperature.

TABLE 9.3
Free-electron Number Densities for Selected Elements at $T = 300$ K

Element	N/V ($\times 10^{28}\ \text{m}^{-3}$)	Element	N/V ($\times 10^{28}\ \text{m}^{-3}$)
Cu	8.47	Mn (α)	16.5
Ag	5.86	Zn	13.2
Au	5.90	Cd	9.27
Be	24.7	Hg (78 K)	8.65
Mg	8.61	Al	18.1
Ca	4.61	Ga	15.4
Sr	3.55	In	11.5
Ba	3.15	Sn	14.8
Nb	5.56	Pb	13.2
Fe	17.0		

From N. W. Ashcroft and N. D. Mermin, *Solid State Physics*. Philadelphia: Saunders College Publishing, 1976.

TABLE 9.4
Fermi Energies (T = 300 K), Fermi Temperatures, and Fermi Velocities for Selected Metals.

Element	E_F (eV)	T_F ($\times 10^4$ K)	u_F ($\times 10^6$ m/s)
Li	4.74	5.51	1.29
Na	3.24	3.77	1.07
K	2.12	2.46	0.86
Rb	1.85	2.15	0.81
Cs	1.59	1.84	0.75
Cu	7.00	8.16	1.57
Ag	5.49	6.38	1.39
Au	5.53	6.42	1.40
Be	14.3	16.6	2.25
Mg	7.08	8.23	1.58
Ca	4.69	5.44	1.28
Sr	3.93	4.57	1.18
Ba	3.64	4.23	1.13
Nb	5.32	6.18	1.37
Fe	11.1	13.0	1.98
Mn	10.9	12.7	1.96
Zn	9.47	11.0	1.83
Cd	7.47	8.68	1.62
Hg	7.13	8.29	1.58
Al	11.7	13.6	2.03
Ga	10.4	12.1	1.92
In	8.63	10.0	1.74
Tl	8.15	9.46	1.69
Sn	10.2	11.8	1.90
Pb	9.47	11.0	1.83
Bi	9.90	11.5	1.87
Sb	10.9	12.7	1.96

From N. W. Ashcroft and N. D. Mermin, *Solid State Physics.* Philadelphia: Saunders College Publishing, 1976.

The density of states can be calculated by differentiating Equation (9.41) with respect to energy:

$$g(E) = \frac{dN_r}{dE} = \frac{\pi}{2} E_1{}^{-3/2} E^{1/2}$$

This result can be expressed more conveniently in terms of E_F rather than E_1. Using Equation (9.42) for E_1 we find:

$$g(E) = \frac{\pi}{2}\left(E_F{}^{-3/2}\frac{3N}{\pi}\right)E^{1/2} = \frac{3N}{2} E_F{}^{-3/2} E^{1/2} \tag{9.43}$$

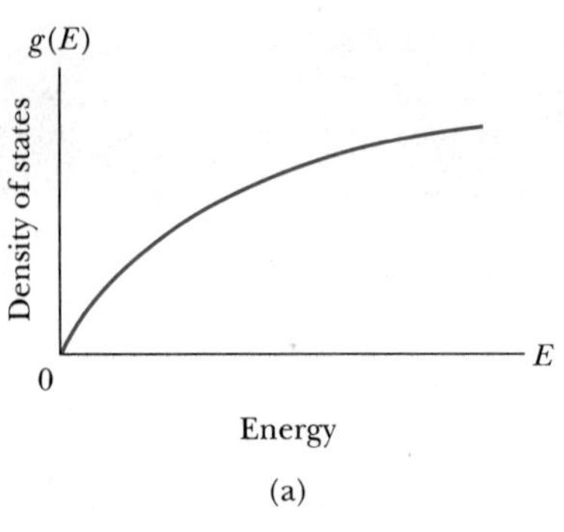

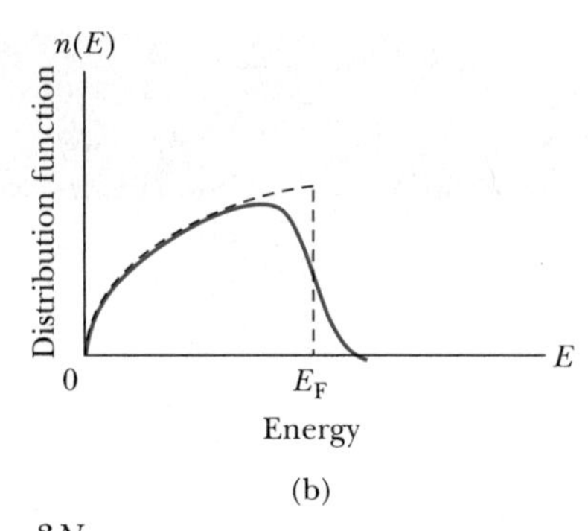

FIGURE 9.10 (a) The density of states $g(E) = \frac{3N}{2} E_F^{-3/2} E^{1/2}$ and (b) distribution function $n(E) = g(E)F_{FD}$ for an electron gas. The function $n(E)$ is shown at $T = 0$ (dashed line) and $T = 300$ K (solid line).

The distribution of electronic energies is then given by Equation (9.29). Because we are considering the $T = 0$ case, it is possible to use the step function form of the Fermi-Dirac factor (Equation 9.35). Therefore at $T = 0$ we have

$$n(E) = \begin{cases} g(E) & \text{for } E < E_F \\ 0 & \text{for } E > E_F \end{cases} \tag{9.44}$$

With the distribution function $n(E)$ the mean electronic energy can be calculated easily:

$$\begin{aligned} \bar{E} &= \frac{1}{N}\int_0^\infty E\, n(E)\, dE = \frac{1}{N}\int_0^{E_F} E\, g(E)\, dE \\ &= \frac{1}{N}\int_0^{E_F} \left(\frac{3N}{2}\right) E_F^{-3/2}\, E^{3/2} dE \\ &= \frac{3}{2} E_F^{-3/2} \int_0^{E_F} E^{3/2} dE = \frac{3}{5} E_F \end{aligned} \tag{9.45}$$

This is a reasonable result, considering the shapes of the curves in Figure 9.10.

We can now proceed to find the electronic contribution to the heat capacity of a conductor. Recall that the general expression for heat capacity at constant volume is $C_V = \partial U/\partial T$, where U is the internal energy of the system in question. By Equation (9.45)

$$U = N\bar{E} = \frac{3}{5} NE_F$$

at $T = 0$. The important question for determining the heat capacity is how U increases with temperature. Because the energy levels are filled up to E_F, we expect that only those electrons within about kT of E_F will be able to absorb thermal energy and jump to a higher state. Therefore the fraction of electrons capable of participating in this thermal process is on the order of kT/E_F. The exact number of electrons depends on temperature, because the shape of the curve $n(E)$ changes with temperature (see Figure 9.10). In general we can say that

$$U = \frac{3}{5} NE_F + \alpha N \frac{kT}{E_F} kT \tag{9.46}$$

*See C. Kittel, *Introduction to Solid State Physics*, 6th ed. New York: Wiley, 1986.

where α is a constant >1 due to the shape of the distribution curve. Therefore the electronic contribution to the heat capacity is

$$C_V = \frac{\partial U}{\partial T} = 2\alpha N k^2 \frac{T}{E_F}$$

As with ideal gases (see Section 9.3) it is common to express this result as a molar heat capacity c_V. For one mole we have $N = N_A$. Then using $N_A k = R$ (the ideal gas constant) and $E_F = kT_F$, we find that

$$c_V = 2\alpha R \frac{T}{T_F} \tag{9.47}$$

Sommerfield used the correct distribution $n(E)$ at room temperature and found a value* for α of $\pi^2/4$. With the value $T_F \approx 80{,}000$ K for copper, we obtain $c_V \approx 0.02R$, which is just what is measured experimentally! The increase in c_V with increasing temperature is also seen experimentally. The quantum theory has proved to be a success in a case where the classical theory failed. The small value of the electronic contribution to the heat capacity can be attributed to the unique nature of the Fermi-Dirac distribution, and in particular to the Pauli exclusion principle, which severely restricts the number of electrons that may participate in heat absorption.

Turning our attention to the electrical conductivity, we can make some improvement by replacing the mean speed $\bar{v}$ in Equation (9.37) with what is called a *Fermi speed* u_F, defined from $E_F = \frac{1}{2} m u_F{}^2$. We can justify this change by noting that conduction electrons in a metal are those most loosely bound to their atoms. Therefore these electrons must be at the highest energy level. At room temperature the highest energy level is close to the Fermi energy, as was discussed previously. This means that we should use

$$u_F = \sqrt{\frac{2E_F}{m}} \approx 1.6 \times 10^6 \text{ m/s}$$

for copper. Unfortunately, this is an even higher speed than $\bar{v}$, which seems undesirable because the value we calculated for σ using $\bar{v}$ was already too low. It turns out that there is also a problem with the classical value Drude used for the mean free path ℓ. He felt that because the ions were so large and occupied so much space in a solid, the mean free path could be no more than several tenths of a nanometer. But in quantum theory the ions are not hard spheres and the electrons can be thought of as waves. Therefore it is not so certain that the mean free path cannot be longer. Indeed, Einstein calculated the value of ℓ to be on the order of 40 nm in copper at room temperature. This gives a conductivity of

$$\sigma = \frac{ne^2\ell}{mu_F} \approx 6 \times 10^7 \ \Omega^{-1} \cdot \text{m}^{-1}$$

which is just right.

Finally, let us consider the temperature dependence of the electrical conductivity. We have already replaced $\bar{v}$ in Equation (9.37) with the Fermi speed u_F, which is nearly temperature independent. However, as the temperature of a conductor is increased, ionic vibrations become more severe. Thus electron–ion collisions will become more frequent, the mean free path will become smaller, and the conductivity will be reduced. According to elementary transport theory (this result is shown in many introductory physics texts) the mean free path is inversely proportional to the cross-sectional area of these ionic scatterers. Let us assume that the ions are harmonic oscillators. The energy of a harmonic oscillator is proportional to the square of its vibration amplitude. But the effective cross-sectional area should also be proportional to the square of the vibration amplitude. Therefore, the mean free path is inversely proportional to vibration energy (and temperature, because thermal energy is proportional to temperature). We may summarize this dizzying array of proportions by saying that

$$\sigma \propto \ell \propto r^{-2} \propto U^{-1} \propto T^{-1}$$

In other words, the electrical conductivity varies inversely with temperature. This is another success for the quantum theory, because electrical conductivity is observed to vary inversely with temperature for most pure metals.

Example 9.4

Use the Fermi theory to compute the electronic contribution to the molar heat capacity of (a) copper and (b) silver, each at temperature $T = 293$ K. Express the results as a function of the molar gas constant R.

Solution: (a) From the Fermi theory the molar heat capacity is $c_V = 2\alpha RT/T_F$. Recall that we can use $\alpha = \pi^2/4$ at room temperature ($T = 293$ K), so

$$c_V = 2\left(\frac{\pi^2}{4}\right)R\frac{T}{T_F} = \frac{\pi^2 RT}{2T_F}$$

Inserting numerical values (from Table 9.4, $T_F = 8.16 \times 10^4$ K for copper),

$$c_V = \frac{\pi^2(293\text{ K})}{2\,(8.16 \times 10^4\text{ K})}R = 0.0177R$$

for copper, which rounds to the value $0.02\,R$ quoted in the text.

(b) For silver, we simply replace the Fermi temperature with 6.38×10^4 K (Table 9.4):

$$c_V = \frac{\pi^2(293\text{ K})}{2(6.38 \times 10^4\text{ K})}R = 0.0227R$$

We see that the electronic contribution to the molar heat capacity is not much different for these two good conductors, and in each case it is small compared with the lattice contribution (Section 9.3).

9.7 Bose-Einstein Statistics

Like Fermi-Dirac statistics, Bose-Einstein statistics can be used to solve problems that are beyond the scope of classical physics. In this section we will concentrate on two examples: a derivation of the Planck formula for blackbody radiation and an investigation of the properties of liquid helium.

Blackbody Radiation

You may wish to review Section 3.5 on blackbody radiation. We will use the ideal blackbody described in that section: a nearly perfectly absorbing cavity that emits a spectrum of electromagnetic radiation. The problem is to find the intensity of the emitted radiation as a function of temperature and wavelength [Equation (3.23)].

$$\mathcal{I}(\lambda, T) = \frac{2\pi c^2 h}{\lambda^5}\frac{1}{e^{hc/\lambda kT} - 1} \tag{3.23}$$

Of course, in quantum theory we must begin with the assumption that the electromagnetic radiation is really a collection of photons of energy hc/λ. Recall that photons are bosons with spin 1. Our approach to this problem will be to use the Bose-Einstein distribution to find how the photons are distributed by energy, and then use the relationship $E = hc/\lambda$ to turn the energy distribution into a wavelength distribution. The desired temperature dependence should already be included in the Bose-Einstein factor [see Equation (9.32)].

As in the previous section, the key to the problem is being able to find the density of states $g(E)$. In fact, it is possible to model the photon gas just as we did the electron gas: a collection of free particles within a three-dimensional infinite-potential well. We cannot use Equation (9.38) for the energy states, however, for we are now dealing with massless particles. To solve this problem it is necessary to recast the solution to the particle in a box problem in terms of momentum states rather than energy states. For a free particle of mass m the energy is $p^2/2m$. We may rewrite Equation (9.38) in terms of momentum

$$p = \sqrt{p_x^{\,2} + p_y^{\,2} + p_z^{\,2}} = \frac{h}{2L}\sqrt{n_1^{\,2} + n_2^{\,2} + n_3^{\,2}} \tag{9.48}$$

The energy of a photon is pc, so

$$E = \frac{hc}{2L}\sqrt{n_1^{\,2} + n_2^{\,2} + n_3^{\,2}} \tag{9.49}$$

We proceed now to calculate the density of states as we did in Section 9.6. Think of the n_i as the coordinates of a number space and define $r^2 \equiv n_1^{\,2} + n_2^{\,2} + n_3^{\,2}$. The number of allowed energy states within "radius" r is

$$N_r = 2\left(\frac{1}{8}\right)\left(\frac{4}{3}\pi r^3\right) \tag{9.50}$$

where the factor $\frac{1}{8}$ again comes from the restriction to positive values of n_i. This time the factor of 2 is due to the fact that there are two possible photon polarizations. Note that energy is proportional to r, namely,

$$E = \frac{hc}{2L}r \tag{9.51}$$

Thus, we can rewrite N_r in terms of E:

$$N_r = \frac{8\pi L^3}{3h^3c^3}E^3 \tag{9.52}$$

The density of states $g(E)$ is

$$g(E) = \frac{dN_r}{dE} = \frac{8\pi L^3}{h^3c^3}E^2 \tag{9.53}$$

The energy distribution is as always the product of the density of states and the appropriate statistical factor, in this case the Bose-Einstein factor:

$$\begin{aligned} n(E) &= g(E)F_{\text{BE}} \\ &= \frac{8\pi L^3}{h^3c^3}E^2\frac{1}{e^{E/kT} - 1} \end{aligned} \tag{9.54}$$

Notice that the normalization factor B_2 in the Bose-Einstein factor has been set equal to unity. This is because we have an unnormalized collection of photons. As photons are absorbed and emitted by the walls of the cavity, the number of photons is not constant. The distribution (9.54) will serve to provide a relative number of photons at different energies, but it is impossible to normalize to a particular number of photons.*

The next step is to convert from a number distribution to an energy density distribution $u(E)$. To do this it is necessary to multiply by the factor E/L^3 (that is, energy per unit volume):

$$u(E) = \frac{E\, n(E)}{L^3} = \frac{8\pi}{h^3c^3} E^3 \frac{1}{e^{E/kT} - 1}$$

For all photons in the range E to $E + dE$,

$$u(E)\, dE = \frac{8\pi}{h^3c^3} \frac{E^3 dE}{e^{E/kT} - 1} \tag{9.55}$$

Using $E = hc/\lambda$ and $|dE| = (hc/\lambda^2)\, d\lambda$,† we find

$$u(\lambda, T)\, d\lambda = \frac{8\pi hc}{\lambda^5} \frac{d\lambda}{e^{hc/\lambda kT} - 1} \tag{9.56}$$

In the SI system multiplying by a constant factor $c/4$ is required‡ to change the energy density [$u(\lambda, T)$ is energy per unit volume per unit wavelength inside the cavity] to a spectral intensity [$\mathcal{I}(\lambda, T)$ is power per unit area per unit wavelength for radiation emitted from the cavity]:

$$\mathcal{I}(\lambda, T) = \frac{2\pi c^2 h}{\lambda^5} \frac{1}{e^{hc/\lambda kT} - 1} \tag{9.57}$$

which is identical with Equation (3.23).

Planck did not use the Bose-Einstein distribution to derive his radiation law. Nevertheless, it is an excellent example of the power of the statistical approach to see that such a fundamental law can be derived with relative ease. This problem was first worked this way by the young Indian physicist Satyendra Nath Bose in 1924. Einstein's name was added to the distribution because he helped Bose publish his work in the West and later applied the distribution to other problems. It is remarkable that Bose did this work prior to the full development of quantum mechanics, and in particular prior to the elucidation of the concept of spin in quantum theory.

Heike Kamerlingh Onnes (1853–1926), the first great low temperature physicist. Onnes was the first to liquify helium, and in turn this discovery allowed him to discover the phenomenon of superconductivity in 1911. *Burndy Library, courtesy AIP/Niels Bohr Library.*

Liquid Helium

Helium is an element with a number of remarkable properties. It has the lowest boiling point of any element (4.2 K) and has no solid phase at normal pressures. Liquid helium has been studied extensively since its discovery in 1911 by Heike Kamerlingh Onnes (1853–1926, Nobel Prize 1913). It continues to be used by many experimental physicists as a cryogenic (cooling) device. There are many effects that can only be observed at extremely low temperatures, and liquid helium can be used to cool materials to 4.2 K and lower. Later, (Chapter 10) we will ex-

*For more detail on the derivation of Equation (9.54), see F. Reif, *Fundamentals of Statistical and Thermal Physics.* New York: McGraw-Hill, 1965, pp. 339–340.

†The negative sign is dropped because it would be meaningless in a distribution in which a (positive) number representing a probability is required. Physically the negative sign means that energy increases as wavelength decreases, but that fact is not relevant here.

‡The factor c is as may be expected dimensionally, and the extra factor of $\frac{1}{4}$ comes from a geometrical analysis. For a detailed calculation see J. J. Brehm and W. J. Mullin, *Introduction to the Structure of Matter.* New York: Wiley, 1989, p. 80. See also our Chapter 3, Problem 24.

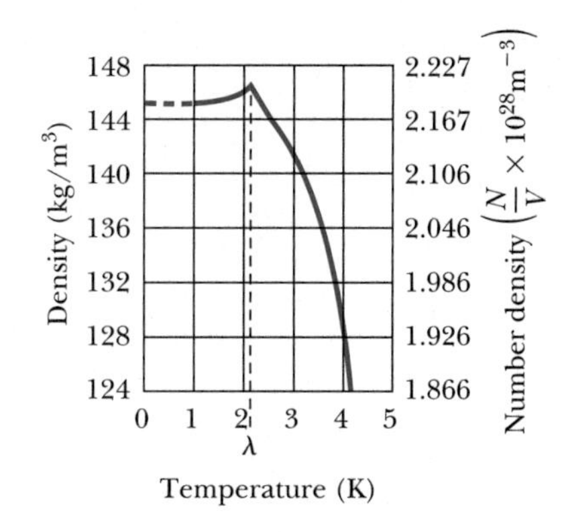

FIGURE 9.11 Density and number density vs. temperature for liquid helium. Notice the sharp drop in ρ above the lambda point, $T = 2.17$ K. *From F. London, Superfluids. New York: Dover Publications, Inc., 1964. Reprinted with permission.*

amine low-temperature methods in more detail, along with our study of superconductivity, perhaps the best known low-temperature phenomenon.

But liquid helium is quite interesting in its own right, especially at temperatures somewhat less than its boiling point. In 1924 Kamerlingh Onnes (along with J. Boks) measured the density of liquid helium as a function of temperature and obtained the curve shown in Figure 9.11. The specific heat of liquid helium as a function of temperature was measured in 1932 by W. Keesom and K. Clausius. Their results are shown in Figure 9.12.

It should be clear that something extraordinary is happening at about 2.17 K. That temperature is commonly referred to as the *critical temperature* (T_c), *transition temperature,* or simply the *lambda point,* a name derived from the shape of the curve in Figure 9.12. Indeed, visual inspection of the liquid helium is sufficient to demonstrate that something special happens at the lambda point. As the temperature is reduced from 4.2 K toward the lambda point, the liquid boils vigorously. But at 2.17 K the boiling suddenly stops and the liquid becomes quite calm. At lower temperatures there continues to be evaporation from the surface, but there is no further boiling.

Normal and superfluid phases

We call what happens a 2.17 K a transition from the **normal phase** to the **superfluid phase.** The word *superfluid* comes from yet another peculiar property of liquid helium below the lambda point. The flow rate of liquid helium through

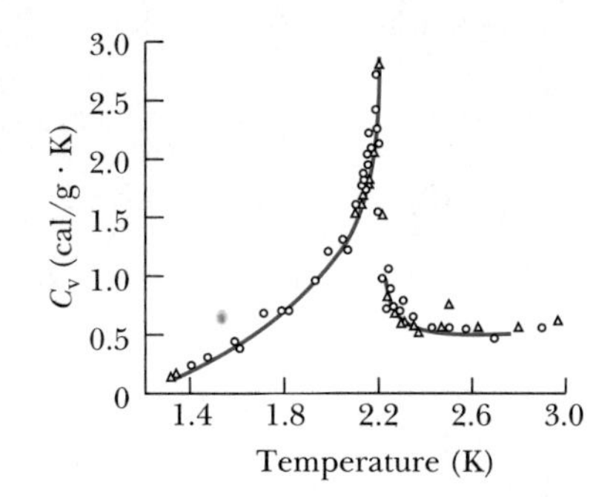

FIGURE 9.12 Specific heat of liquid helium as a function of temperature. The characteristic shape of this curve gives the transition point of 2.17 K the name "lambda point." *From F. London, Superfluids. New York: Dover Publications, Inc., 1964. Reprinted with permission.*

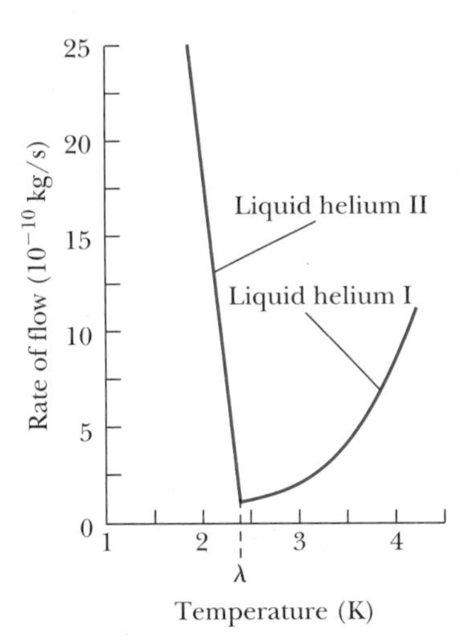

FIGURE 9.13 The rate of capillary flow in liquid helium II (superfluid) and liquid helium I (normal liquid). The capillary flow rate increases dramatically with decreasing temperature below the lambda point. *From M. Zemansky, Temperatures Very Low and Very High. New York: Dover Publications, Inc., 1964. Reprinted with permission.*

a capillary tube as a function of temperature is shown in Figure 9.13. There is a dramatic increase in the rate of flow as the temperature is reduced (Figure 9.14). Another way of describing this phenomenon is to say that the superfluid has a lower viscosity than the normal fluid. The viscosity can be so low that superfluid helium has been observed to form what is called a **creeping film** and flow up and

Creeping film

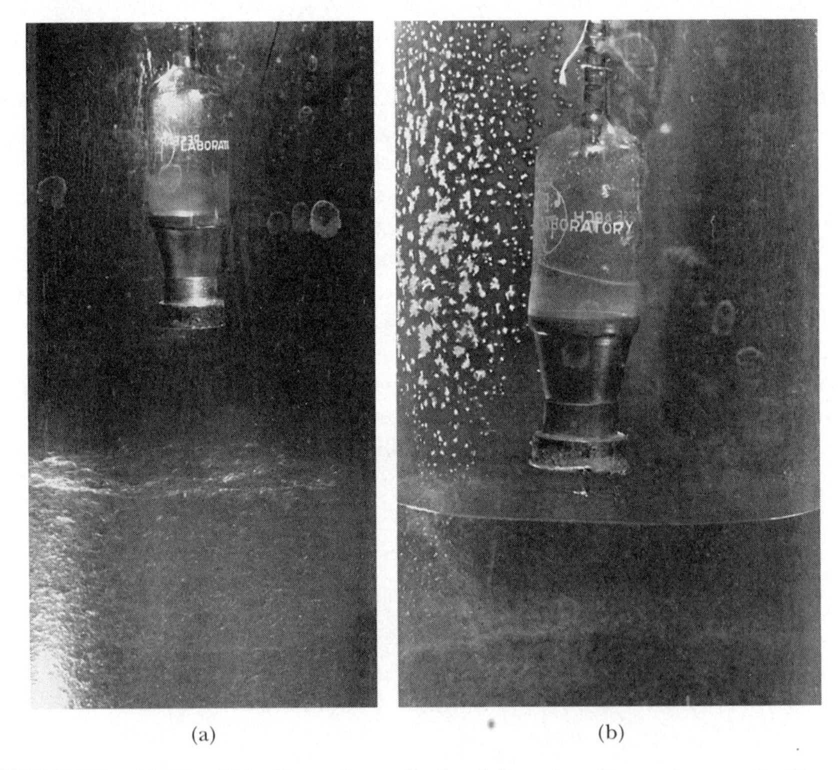

(a) (b)

FIGURE 9.14 (a) Liquid helium above the lambda point shows vigorous boiling. The suspended vessel contains holes in the bottom that are so fine that they do not allow passage of the normal fluid. (b) Liquid helium below the lambda point no longer boils, and the superfluid can pass through the small holes in the bottom of the vessel. *Reproduced by kind permission of the Department of Physics, Clarendon Laboratory, Oxford.*

over the (vertical) walls of its container. But there is a curious note to add regarding the viscosity of liquid helium. If one tries to measure the viscosity by, for example, measuring the drag on a metal plate as it is passed over the surface of the liquid, the result is just about what one would get with a normal fluid, even at temperatures somewhat below the lambda point. In other words, there appears to be a contradiction between the results of this experiment and the capillary flow experiment just described.

A solution to this conundrum is provided by a theory proposed by Fritz London in 1938. London claimed that liquid helium below the lambda point is part superfluid and part normal. The superfluid component increases as the temperature is reduced from the lambda point, approaching 100% superfluid as the temperature approaches absolute zero. It has been shown that the fraction F of helium atoms in the superfluid state follows fairly closely the relation

$$F = 1 - \left(\frac{T}{T_c}\right)^{3/2} \tag{9.58}$$

The two-fluid model can explain our viscosity paradox, if we assume that only the superfluid part participates in capillary flow, while there is always enough normal fluid present to retard the motion of a metal plate dragged across its surface.

With two protons, two neutrons, and two electrons, the helium atom is a boson and therefore subject to Bose-Einstein statistics. How does this help explain the behavior of superfluid helium? Perhaps one of the most important things about bosons is that they are not subject to the Pauli exclusion principle. There is no limit to the number of bosons that may be in the same quantum state. What we see in liquid helium is often referred to as a **Bose condensation** into a single state, in this case the superfluid state. All the particles in a Bose condensate are in the same quantum state, which is not forbidden for bosons. As we saw in the case of the Fermi electron gas, fermions must "stack up" into their energy states, no more than two per energy state. Therefore, such a condensation process is not possible with fermions. As a striking demonstration of this fact, consider the isotope ^{3}He. With one less neutron than the more common ^{4}He, this isotope is a fermion. Liquid ^{3}He undergoes a superfluid transition at 2.7 mK (a temperature almost a factor of 1000 lower than T_c in ^{4}He!). The superfluid mechanism for the fermion ^{3}He is radically different than the Bose condensation described earlier. It is more like the electron (fermion) pairing that one finds in the superconducting transition (see Section 10.5). Although they are very similar chemically (as two isotopes of the same element usually are), ^{3}He and ^{4}He are radically different when it comes to superfluidity, simply because one is a fermion and the other a boson. (See also the Special Topic box on Superfluid ^{3}He in this section.)

Bose condensation

Bose-Einstein statistics may be used to estimate T_c for the superfluid phase of ^{4}He. Rather than derive a new density of states function for bosons, let us take advantage of the density of states function we have already developed for fermions. Using that result [Equation (9.43)] and substituting for the constant E_F yields

$$g_{FD}(E) = \frac{\pi}{2}\left(\frac{h^2}{8mL^2}\right)^{-3/2} E^{1/2} \tag{9.59}$$

where the FD subscript indicates a fermion distribution. The only difference for bosons is that they do not obey the Pauli principle, and therefore the density of

SPECIAL TOPIC:

SUPERFLUID ^{3}He

Superfluidity in ^{3}He was discovered in the early 1970s. This makes it a relatively recent discovery, compared with superfluidity in ^{4}He. There are good reasons, both experimental and theoretical, for the delay. The isotope ^{3}He accounts for only 0.00013% of naturally occurring helium. This is not enough to produce significant amounts of ^{3}He for experimental purposes. Since World War II it has been possible to obtain ^{3}He through the radioactive decay of tritium, which is produced in nuclear reactors. Also recall that the superfluid transition in ^{3}He occurs at just 2.7 mK, a temperature not reached until the 1950s. Even then many physicists believed that a superfluid phase in ^{3}He was impossible because the Bose condensation responsible for the superfluid phase of ^{4}He cannot occur in ^{3}He because it is a fermion.

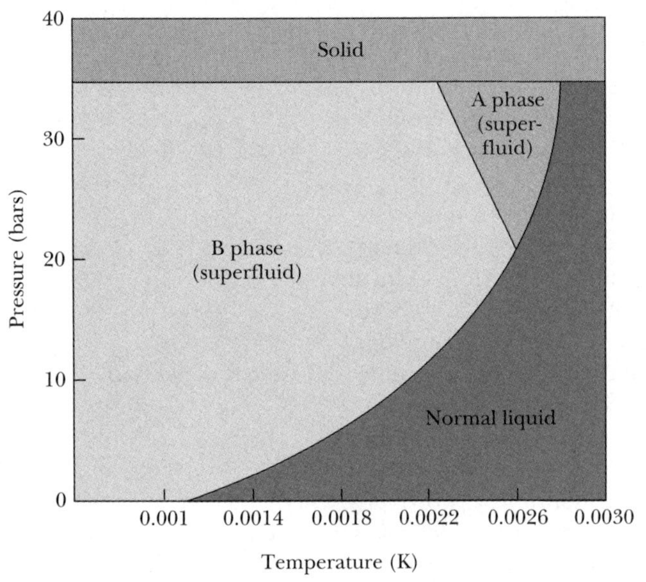

FIGURE A Phase diagram for ^{3}He with zero applied magnetic field. In this case there are two superfluid phases (A and B), with phase A existing only at very high pressures (1 bar = 10^5 Pa $\approx$ 1 atm). *From N. D. Mermin and D. M. Lee, Scientific American* ***235****, 68 (December 1976) pages 57–71, Courtesy George V. Kelvin/Scientific American.*

Superfluid ^{3}He was finally discovered in 1971 by Douglas Osheroff, Robert Richardson, and David Lee at Cornell University. As is sometimes the case in science, they made the discovery while studying something else: the magnetic properties of solid ^{3}He (^{3}He, like ^{4}He, has a solid phase at extremely high pressures). Liquid ^{3}He was present in the experiment only as a refrigeration ingredient. The researchers saw a sudden change in the rate at which the pressure in their apparatus changed, and they were ultimately able to show that a transition from normal ^{3}He to superfluid ^{3}He was responsible.

Subsequent studies showed something even more remarkable about ^{3}He: it has three distinct superfluid phases. Two of them are shown in Figure A. The third phase can be produced only in the presence of a magnetic field (Figure B). This raises an interesting point about ^{3}He: with three spin $\frac{1}{2}$ particles in its nucleus, it has a net magnetic moment. The magnetic behavior of ^{3}He, along with the fermion/boson distinction discussed in the text, is largely responsible for the different behavior of the two isotopes. It also helps us distinguish between the fermion pairing mechanism in superfluid ^{3}He and the fermion pairing mechanism in superconductors (Section 10.5). The pairs of ^{3}He

---

states for bosons should be less by a factor of 2. Dividing by 2 and rearranging we find

$$g_{\mathrm{BE}}(E) = \frac{2\pi V}{h^3}(2m)^{3/2}E^{1/2} \tag{9.60}$$

with $V \equiv L^3$. Of course, the mass m is now the mass of a helium atom.

The number distribution $n(E)$ is now

$$\begin{aligned} n(E) &= g_{\mathrm{BE}}(E)F_{\mathrm{BE}} \\ &= \frac{2\pi V}{h^3}(2m)^{3/2}E^{1/2}\frac{1}{B_2 e^{E/kT} - 1} \end{aligned} \tag{9.61}$$

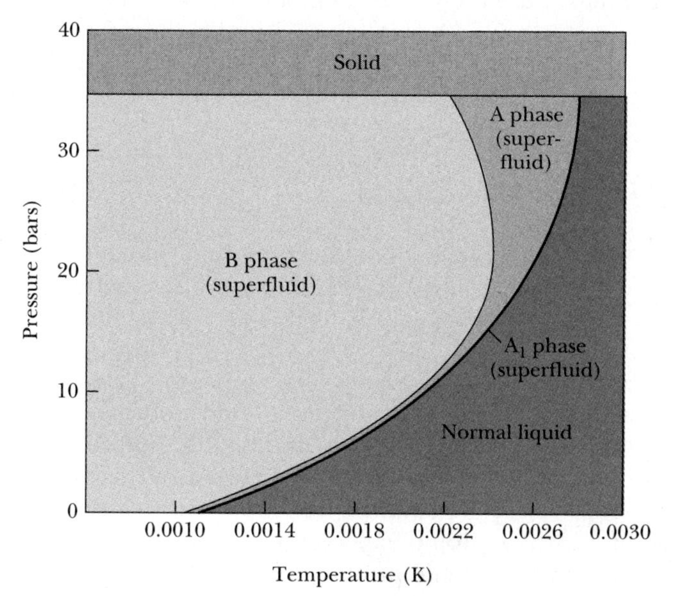

FIGURE B Phase diagram for ^{3}He in the presence of a strong magnetic field. The third superfluid phase (A_1) appears in a very narrow band on the pressure–temperature diagram. *From N. D. Mermin and D. M. Lee, Scientific American* ***235****, 68 (December 1976) pages 57–71, Courtesy George V. Kelvin/Scientific American.*

atoms form in such a way as to have a net spin (the magnetic moments of the two atoms tend to align with each other) and net angular momentum (they also tend to revolve around one another), but electron pairs in a superconductor have zero spin and zero angular momentum. However, the fact that physicists had previously considered the possibility of a superconducting state with nonzero angular momentum helped them quickly understand the nonzero angular momentum state of superfluid ^{3}He when it was observed.

How do we characterize the three superfluid phases? In the A_1 phase, which exists only in the presence of an applied magnetic field and just below the superfluid transition temperature (Figure B), the magnetic moments align with the applied field and the mutual pair revolution just described takes place in a plane parallel to the applied field. The A and B phases are more difficult to describe. Let it suffice to say that they correspond to various superpositions of the allowed fermion wave functions. Although the A and B phases do not depend on the existence of an external magnetic field, their behavior when such a field is imposed is quite interesting. The superfluid is then said to exhibit *anisotropic* behavior. That is, the properties of the superfluid are dependent upon the direction one considers with respect to the magnetic field. For example, the superfluid flow rate is quite different when measured parallel to the applied field than when it is measured in a plane perpendicular to the applied field.

Recently physicists have been interested in studying the vortex motion of superfluid ^{3}He. Upon rotation the superfluid tends to break into separate vortices—small regions of rotation with rotation patterns distinct from those in neighboring regions. The interesting thing about this discovery is that the vortex patterns are similar to those observed in superconductors. This new connection between superconductivity and superfluidity in ^{3}He continues to be the subject of careful study. In 1996 Osheroff, Richardson, and Lee were awarded the Nobel Prize in physics for their discovery of superfluid ^{3}He.

In a collection of N helium atoms the normalization condition is

$$N = \int_0^\infty n(E)\,dE$$

$$= \frac{2\pi V}{h^3}(2m)^{3/2}\int_0^\infty \frac{E^{1/2}}{B_2 e^{E/kT} - 1}\,dE \qquad (9.62)$$

With a substitution $u = E/kT$, this reduces to

$$N = \frac{2\pi V}{h^3}(2mkT)^{3/2}\int_0^\infty \frac{u^{1/2}}{B_2 e^{u} - 1}\,du \qquad (9.63)$$

We do not know the value of B_2, but we will proceed using the minimum allowed value $B_2 = 1$. Then the integral in Equation (9.63), when evaluated (as in Appendix 7), has a value of 2.315. Because we used the minimum value of B_2, this result will correspond to the maximum value of N. In other words

$$N \leq \frac{2\pi V}{h^3}(2mkT)^{3/2}(2.315) \tag{9.64}$$

Rearranging, we find

$$T \geq \frac{h^2}{2mk}\left(\frac{N}{2\pi V(2.315)}\right)^{2/3} \tag{9.65}$$

Equation (9.65) can be evaluated numerically, because N/V is simply the number density of liquid helium in the normal state ($2.11 \times 10^{28}\ \mathrm{m^{-3}}$). The result is

$$T \geq 3.06\ \mathrm{K} \tag{9.66}$$

The value 3.06 K is an estimate of T_c, because our analysis has predicted that this is the minimum temperature at which we may expect to find a normal Bose-Einstein distribution. At lower temperatures there is condensation into the superfluid state. Our result is a bit off, because we used a density of states derived for a noninteracting gas rather than a liquid. Still, we have come within one kelvin of the correct value of T_c. As you would expect, calculations using a density of states constructed especially for a liquid produce even better results.

Example 9.5

For a gas of N_2 at room temperature (293 K) and one atmosphere pressure, calculate the Maxwell-Boltzmann constant A and thereby show that Bose-Einstein statistics can be replaced by Maxwell-Boltzmann statistics in this case.

Solution: Equation (9.60) gives the density of states for this boson gas:

$$g(E) = \frac{2\pi V}{h^3}(2m)^{3/2}E^{1/2}$$

Use the Maxwell-Boltzmann factor $F_{\mathrm{MB}} = A\exp(-E/kT)$ to get the distribution $n(E) = g(E)F_{\mathrm{MB}}$. Now the normalization condition for a gas of N molecules is

$$N = \int_0^\infty n(E)\,dE$$

$$= \frac{2\pi V}{h^3}(2m)^{3/2}A\int_0^\infty E^{1/2}\exp(-E/kT)\,dE$$

The integral is a standard definite integral, which yields

$$N = \frac{2\pi V}{h^3}(2m)^{3/2}A\frac{\sqrt{\pi}}{2}(kT)^{3/2}$$

$$= \frac{V}{h^3}(2\pi mkT)^{3/2}A$$

Therefore

$$A = \frac{h^3 N}{V}(2\pi mkT)^{-3/2}$$

Under normal conditions (atmospheric pressure and room temperature) the number density of nitrogen gas is $N/V = 2.50 \times 10^{25}\ \mathrm{m^{-3}}$. Plugging this into our result for A along with the molecular mass of N_2 and $T = 293$ K yields a value $A = 1.8 \times 10^{-7}$. Because this is much less than unity, the use of Maxwell-Boltzmann statistics is justified (see the discussion at the end of Section 9.5).

Bose Condensation in Gases

For years physicists attempted to demonstrate Bose condensation in gases. They were hampered, however, by the strong Coulomb interactions among the gas particles. This kept researchers from obtaining the low temperatures and high densities needed to produce the condensate.

Finally in 1995 success was achieved by a group led by Eric Cornell, Carl Weiman, and Michael Anderson working at Boulder, Colorado. This group used a magnetic trap to cool a gas of ^{87}Rb atoms to a temperature of about 20 nK (2.0×10^{-8} K). In their magnetic trap the researchers varied the magnetic field at radio frequencies in such a way that atoms with higher speeds and further from the center are driven away. What remained was an extremely cold, dense cloud. At temperatures of about 170 nK, the rubidium cloud was observed to pass through a transition, from a gas with a normal, broad velocity distribution to an extremely narrow one. It was also observed that the fraction of atoms in the condensed state increases as the temperature was lowered further, just as in a superfluid.

Later in 1995 Bose condensation in a gas was confirmed, this time with sodium gas, by a group led by Wolfgang Ketterle at MIT. For sodium the transition to the condensed state was observed to begin at a relatively warm 2 μK. Today physicists are striving to understand the properties of these and other Bose condensates and reconcile their observations with quantum theory.

Summary

By the end of the 19th century, the work of Maxwell and Boltzmann had made it clear that statistics could be useful in describing physical processes. In particular, Maxwell's statistical distribution for the velocities of molecules in an ideal gas is $f(\mathbf{v})$ where

$$f(\mathbf{v})\,d^3\mathbf{v} = C\exp\left(-\frac{1}{2}\beta m v_x^{\,2} - \frac{1}{2}\beta m v_y^{\,2} - \frac{1}{2}\beta m v_z^{\,2}\right)d^3\mathbf{v} \tag{9.2}$$

The corresponding speed distribution $F(v)$ is

$$F(v)\,dv = 4\pi C\exp\left(-\frac{1}{2}\beta m v^2\right)v^2\,dv \tag{9.14}$$

where $C = \left(\dfrac{\beta m}{2\pi}\right)^{3/2}$ and $\beta = (kT)^{-1}$. The speed distribution can be used to predict many of the observed properties of ideal gases. Computing the mean kinetic energy of a monatomic ideal gas molecule using the Maxwell velocity distribution yields

$$\overline{K} = \frac{3}{2}kT \tag{9.8}$$

This result is in accord with the equipartition theorem, which states that in equilibrium there is a mean energy of $\frac{1}{2}kT$ associated with each independent quadratic term in a molecule's energy. The equipartition theorem can be applied to the rotational and vibrational modes of a diatomic molecule, and it can thereby be used to compute the heat capacities of diatomic gases at various temperatures.

Under conditions of higher densities and lower temperatures the particles' wave functions overlap, and the indistinguishability of the particles becomes a factor. As a result it is necessary to use quantum statistics: Fermi-Dirac statistics for fermions and Bose-Einstein statistics for bosons. These distributions differ because fermions obey the Pauli exclusion principle and bosons do not. The statistical factors associated with the classical (Maxwell-Boltzmann) and quantum distributions are given by

$$F_{\text{MB}} = A\exp(-\beta E) \tag{9.27}$$

$$F_{\text{FD}} = \frac{1}{B_1\exp(\beta E) + 1} \tag{9.30}$$

$$F_{\text{BE}} = \frac{1}{B_2\exp(\beta E) - 1} \tag{9.32}$$

where A, B_1, and B_2 are normalization factors. In each case the distribution function $n(E)$ can be expressed as the product of the density of states $g(E)$ [where $g(E)$ is defined as the number of states available per unit energy range] and the appropriate statistical factor.

Fermi-Dirac statistics are needed in order to predict the correct behavior of conduction electrons in a metal. Using the Fermi-Dirac distribution, one can find the electrical conductivity (and the correct temperature dependence thereof) as well as the electronic contribution to the specific heat of a metal.

Bose-Einstein statistics can be used to derive the Planck law for blackbody radiation

$$\mathcal{I}(\lambda, T) = \frac{2\pi c^2 h}{\lambda^5}\,\frac{1}{e^{hc/\lambda kT} - 1} \tag{9.57}$$

Bose-Einstein statistics also help us understand some of the properties of liquid helium, which undergoes a transition to the superfluid state at the temperature $T_c = 2.17$ K. The extraordinarily low viscosity of a superfluid is due to the fact that the molecules in a superfluid obey Bose-Einstein statistics.

Questions

1. How relevant is the Heisenberg uncertainty principle in frustrating Laplace's goal of determining the behavior of an essentially classical system of particles (say an ideal gas) through a knowledge of the motions of individual particles?
2. Why might the measured molar heat capacity of Cl_2 not match the prediction of the equipartition theorem as well as that of O_2?
3. In a diatomic gas why should it be more difficult to excite the vibrational mode than the rotational mode?
4. What is the physical significance of the root mean square speed in an ideal gas?
5. At room temperature and one atmosphere pressure compare v^* for N_2 and O_2. Do the same comparison for $\bar{v}$ and v_{rms}.
6. An insulated container filled with an ideal gas moves through field-free space with a constant velocity. Describe the effect this has on the Maxwell velocity distributions and Maxwell speed distribution.
7. Maxwell conceived of a device (later called *Maxwell's demon*) that could use measurements of individual molecular speeds in a gas to separate faster molecules from slower ones. This device would operate a trap door between two (initially identical) compartments, allowing only faster molecules to pass one way and slower molecules the other way, thus creating a temperature imbalance. Then the temperature difference could be used to run a heat engine, and the result would be the production of mechanical work with no energy input. Of course this would violate the laws of thermodynamics. Discuss reasons why you think Maxwell's demon cannot work.
8. If the distribution function $f(q)$ for some physical property q is even, then it follows that $\bar{q} = 0$. Does it also follow that the most probable value $q^* = 0$?
9. Which of the following act as fermions and which as bosons: hydrogen atoms, deuterium atoms, neutrinos, muons, ping-pong balls?
10. Theorists tend to believe that free quarks (with charge $\pm\frac{2}{3}e$ and $\pm\frac{1}{3}e$) do not exist, but the question is by no means decided. If free quarks are found to exist, what can you say about the distribution function they would obey?
11. Explain why you expect $\bar{E}$ to be greater than $\frac{1}{2}E_F$ [Equation (9.45)].
12. How would the behavior of metals be different if electrons were bosons rather than fermions?
13. What would happen to the Planck distribution and the behavior of liquid helium if we let $h \rightarrow 0$ in the Bose-Einstein density of states [Equation (9.53)]?
14. If only the superfluid component of liquid helium flows through a very fine capillary, is it possible to use capillary flow to separate completely the superfluid component of a sample of liquid helium from the normal component?
15. Why is $B_2 = 1$ the minimum allowed value in the integral in Equation (9.63)?

Problems

9.2 Maxwell Velocity Distribution

1. (a) Use Equation (9.5) to show that the one-dimensional rms speed is

$$v_{x\text{rms}} = \left[\overline{v_x^2}\right]^{1/2} = \left[\frac{kT}{m}\right]^{1/2}$$

(b) Show that Equation (9.5) can be rewritten as

$$g(v_x)\,dv_x = (2\pi)^{-1/2}\, v_{x\text{rms}}^{-1} \exp\left[-\frac{1}{2} v_x^2 / v_{x\text{rms}}^2\right] dv_x$$

2. The result of Problem 1 can be used to estimate the relative probabilities of various velocities. Pick a small interval $\Delta v_x = 0.002\, v_{x\text{rms}}$. For one mole of an ideal gas compute the number of molecules within the range Δv_x centered at (a) $v_x = 0.01\, v_{x\text{rms}}$, (b) $v_x = 0.20\, v_{x\text{rms}}$, (c) $v_x = v_{x\text{rms}}$, (d) $v_x = 5\, v_{x\text{rms}}$, (e) $v_x = 100\, v_{x\text{rms}}$.
3. Consider an ideal gas enclosed in a spectral tube. When a high voltage is placed across the tube, many atoms are excited and all excited atoms will emit electromagnetic radiation at characteristic frequencies. According to the Doppler effect, the frequencies observed in the laboratory depend upon the velocity of the emitting atom. The nonrelativistic Doppler shift of radiation emitted in the x direction is $\nu = \nu_0 (1 + v_x / c)$. The resulting wavelengths observed in the spectroscope are spread to higher and lower values due to the (respectively) lower and higher frequencies, corresponding to negative and positive values of v_x. We say that the spectral line has been *Doppler broadened* (this is what allows us to see the lines easily in the spectroscope, because the Heisenberg uncertainty principle does not cause significant line broadening in atomic transitions). (a) What is the

mean frequency of the radiation observed in the spectroscope? (b) To get an idea of how much the spectral line is broadened at particular temperatures, derive an expression for the standard deviation of frequencies, defined to be

$$\text{Standard deviation} = \left[\overline{(\nu - \nu_0)^2}\right]^{1/2}$$

Your result should be a function of ν_0, T, and constants. (c) Use your results from (b) to estimate the fractional line width, defined by the ratio of the standard deviation to ν_0, of hydrogen (H_2) gas at $T = 293$ K. Repeat for a gas of atomic hydrogen at the surface of a star, with $T = 5500$ K.

9.3 Equipartition Theorem

4. Consider the model of the diatomic gas oxygen (O_2) shown in Figure 9.3. (a) Assuming the atoms are point particles separated by a distance of 0.85 nm, find the rotational inertia I_x for rotation about the x axis. (b) Now compute the rotational inertia of the molecule about the z axis, assuming almost all of the mass of each atom is in the nucleus, a nearly uniform solid sphere of radius 3.0×10^{-15} m. (c) Compute the rotational energy associated with the first ($\ell = 1$) quantum level for a rotation about the x axis. (d) using the energy you computed in (c), find the quantum number ℓ needed to reach that energy level with a rotation about the z axis. Comment on the result in light of what the equipartition theorem predicts for diatomic molecules.

9.4 Maxwell Speed Distribution

5. Using the Maxwell speed distribution, (a) write an integral expression for the number of molecules in an ideal gas that would have speed $v > c$ at $T = 293$ K. (b) Explain why the numerical result of the expression you found in (a) is negligible.

6. Use a computer to explore the numerical value of the definite integral you constructed in the previous problem.

7. It is important for nuclear engineers to know the thermal properties of neutrons in a nuclear reactor. Assuming that a gas of neutrons is in thermal equilibrium, find $\bar{v}$ and v^* for neutrons at (a) 500 K and (b) 2500 K.

8. Show that the Maxwell speed distribution function $F(v)$ approaches zero by carefully taking the limit as $v \to 0$ and as $v \to \infty$.

9. Find v^* for N_2 gas in air (a) on a cold day at $T = -10°$ C and (b) on a hot day at $T = 35°$C.

10. For an ideal gas O_2 at $T = 293$ K find the two speeds v that satisfy the equation $2F(v) = F(v^*)$. Which of the two speeds you found is closer to v^*. Does this make sense?

11. For the ideal gas Ar at $T = 293$ K use a computer to show that

$$\int_0^\infty F(v)\,dv = 1$$

and thereby verify that $C = (\beta m/2\pi)^{3/2}$.

12. Consider the ideal gas H_2 at $T = 293$ K. Use a numerical integration program on a computer to find the fraction of molecules with speeds in the following ranges: (a) 0 to 1 m/s, (b) 0 to 100 m/s, (c) 0 to 1000 m/s, (d) 1000 m/s to 2000 m/s, (e) 2000 m/s to 5000 m/s. (f) 0 to 5000 m/s.

9.5 Classical and Quantum Statistics

13. Using the Maxwell-Boltzmann energy distribution Equation (9.26), (a) find the mean translational kinetic energy of an ideal gas and (b) compare your results with $\frac{1}{2}m\bar{v}^2$ and $\frac{1}{2}m\overline{v^2}$.

14. From the Maxwell-Boltzmann energy distribution find the most probable energy E^*. Plot $F(E)$ vs. E and indicate the position of E^* on your plot.

15. Inside a certain kind of star there are regions in which approximately one hydrogen atom per million is in the first excited level ($n = 2$). The other atoms can be assumed to be in the $n = 1$ level. Use this information to estimate the temperature there, assuming that Maxwell-Boltzmann statistics are valid. (Hint: In this case the density of states depends on the number of possible quantum states available on each level, which is 8 for $n = 2$ and 2 for $n = 1$.)

16. One way to decide whether Maxwell-Boltzmann statistics are valid is to compare the de Broglie wavelength λ of a typical particle with the average interparticle spacing d. If $\lambda \ll d$ then Maxwell-Boltzmann statistics are generally acceptable. (a) Using de Broglie's relation $\lambda = h/p$, show that

$$\lambda = \frac{h}{[3mkT]^{1/2}}$$

(b) Use the fact that $N/V = 1/d^3$ to show that the inequality $\lambda \ll d$ can be expressed as

$$\frac{N}{V}\frac{h^3}{[3mkT]^{3/2}} \ll 1$$

(c) Use the result of (b) to determine whether Maxwell-Boltzmann statistics are valid for Ar gas at room temperature (293 K) and for the conduction electrons in pure silver at $T = 293$ K.

17. Use Equation (9.24) to turn the Maxwell speed distribution Equation (9.14) into an energy distribution [Equations (9.25) and (9.26)].

18. Consider an atom with a magnetic moment μ and a total spin of $\frac{1}{2}$. The atom is placed in a uniform magnetic field of magnitude B at temperature T. (a)

Assuming Maxwell-Boltzmann statistics are valid at this temperature, find the ratio of atoms with spins aligned with the field to those aligned opposite the field. (b) Evaluate numerically with $B = 8$T, for $T = 77$ K, $T = 273$ K and $T = 800$ K.

9.6 Fermi-Dirac Statistics

19. The Fermi energy can be defined as the energy at which the Fermi factor $F_{FD} = 0.5$. Using this definition show that the constant B_1 in Equation (9.30) is equal to $\exp(-\beta E_F)$ and that

$$F_{FD} = \frac{1}{\exp[\beta(E - E_F)] + 1}$$

In other words, verify Equations (9.33) and (9.34).

20. At $T = 0$ what fraction of electrons have energy $E < \bar{E}$?

21. Silver has exactly one conduction electron per atom. (a) Using the density of silver (1.05×10^4 kg/m^3) and the mass of 107.87 g/mole, find the density of conduction electrons in silver. (b) At what temperature is $A = 1$ for silver (where A is the normalization constant in the Maxwell-Boltzmann distribution)? (c) At what temperature is $A = 10^{-3}$?

22. What fraction of the electrons in a good conductor have energies between $0.90E_F$ and E_F at $T = 0$?

23. Using the data in Problem 21 compute (a) E_F and (b) u_F for silver.

24. The Fermi energy for gold is 5.51 eV at $T = 293$ K. (a) Find the average energy of a conduction electron at that temperature. (b) Compute the temperature at which the average kinetic energy of an ideal gas molecule would equal the average energy you found in (a). (c) Comment on the relative temperatures in (a) and (b).

25. The density of pure copper is 8.92×10^3 kg/m^3, and its molar mass is 63.546 grams. Using the experimental value of the conduction electron density 8.47×10^{28} m^{-3} compute the number of conduction electrons per atom.

26. Aluminum has a density of 2.70×10^3 kg/m^3 at a temperature of 293 K, and its molar mass is 26.98 g. (a) Compute the number of aluminum atoms per unit volume at that temperature. (b) Use the fact that $E_F = 11.63$ eV for aluminum at 293 K to find the number density of free electrons. (c) Combine your results from (a) and (b) to estimate the number of conduction electrons per atom—the valence number for aluminum.

27. Compute the Fermi speed for (a) Sr ($E_F = 3.93$ eV) and (b) Zn ($E_F = 9.47$ eV).

28. Verify that Equation (9.35) is valid in the limit $T \to 0$.

29. Show that in general (i.e., $T \neq 0$) the energy distribution of N electrons in a conductor with Fermi energy E_F at temperature T is

$$n(E) = \frac{3N}{2} E_F^{-3/2} \frac{E^{1/2}}{\exp[\beta(E - E_F)] + 1}$$

30. Use the result of Problem 29 with copper ($E_F = 7.0$ eV) to sketch $n(E)$ at (a) $T = 0$, (b) $T = 293$ K, and (c) $T = 1800$ K.

31. Use numerical integration of the function given in Problem 29 to verify that

$$\int_0^\infty n(E)\, dE = N$$

Choose the parameters $T = 300$ K and use $E_F = 7.00$ eV for copper.

32. Use numerical integration of the function given in Problem 29 to find the fraction of conduction electrons with energies between 6.00 eV and 7.00 eV in copper at $T = 293$ K. Comment on your results.

33. In a neutron star the entire star's mass has collapsed essentially to nuclear density. For a neutron star with radius 10 km and mass 4.50×10^{30} kg, find the Fermi energy of the neutrons.

9.7 Bose-Einstein Statistics

34. Consider the problem of photons in a spherical cavity at temperature T, as described in Section 9.7. (a) For the entire collection of photons, what is the number density (number of photons per unit volume)? (b) Evaluate your result from (a) numerically at $T = 400$ K and $T = 5500$ K (the approximate temperature of the sun's surface).

35. Use numerical integration on a computer to verify the output of the definite integral in Equation (9.63).

General Problems

36. Use the method described in Appendix 6 to evaluate the integral

$$\int_0^\infty x^n \exp(-ax^2)\, dx$$

in terms of the constant a for $n = 3$, 4, and 5.

37. Assume that air is an ideal gas under a uniform gravitational field, so that the potential energy of a molecule of mass m at altitude z is mgz. Show that the distribution of molecules varies with altitude as given by the distribution function $f(z)\,dz = C_z \exp(-\beta mgz)\,dz$ and that the normalization constant $C_z = kT/mg$. This distribution is referred to as the *law of atmospheres*.

38. Use the law of atmospheres (Problem 37) to compare the air densities at sea level, Denver (altitude 1610 m), and the summit of Mt. Rainier (altitude 4390 m), assuming the same temperature 273 K in each case.

39. Consider the law of atmospheres (Problem 37). First show that the pressure difference ΔP corresponding to an altitude change Δz is approximately

$$\Delta P = -\rho g\, dz = -\frac{Nmg}{V}\, dz$$

Next assume that the temperature is constant over small altitude changes and then show that $P \approx P_0 \exp(-\beta mgz)$ where P_0 is the pressure at $z = 0$.

40. Consider a thin-walled, fixed-volume container of volume V that holds an ideal gas at constant temperature T. It can be shown by dimensional analysis that the number of particles striking the walls of the container per unit area per unit time is given by $n\bar{v}/4$, where as usual n is the particle number density. The container has a small hole of area A in its surface through which the gas can leak slowly. Assume that A is much less than the surface area of the container. (a) Assuming that the pressure inside the container is much greater than the outside pressure (so that no gas will leak from the outside back in), estimate the time it will take for the pressure inside to drop to one-half the initial value. Your answer should contain A, V, and the mean molecular speed $\bar{v}$. (b) Obtain a numerical result for a spherical container with a diameter of 40 cm containing air at 293 K, if there is a circular hole in the surface of diameter 1.0 mm.

41. For the situation described in Problem 40, show that the speed distribution of the escaping molecules is proportional to $v^3 \exp(-\frac{1}{2}\beta m v^2)$ and that the mean energy of the escaping molecules is $2kT$.

42. (a) What would the density of conduction electrons in copper need to be in order for the Maxwell-Boltzmann normalization constant to be $A = 1$ at $T = 293$ K? (b) Repeat the calculation for neutrons at the same temperature. (c) Repeat the calculation for He gas at the same temperature.

43. For a (classical) simple harmonic oscillator with fixed total energy E, find the mean value of kinetic energy $\overline{K}$ and the mean value of potential energy $\overline{V}$. Show that $\overline{K} = \overline{V} = E/2$.

CHAPTER 10

Molecules and Solids

The secret of magnetism, now explain that to me! There is no greater secret, except love and hate.

Johann Wolfgang von Goethe

In this chapter we shall study a variety of systems in which atoms have been joined together to form molecules and solids. We first consider the diatomic molecule, studying how the two atoms are joined. The resulting rotational and vibrational energy states are quantized, and we shall describe the distinctive rotational and vibrational spectra in some detail. One application of our knowledge of atomic and molecular energy levels that we shall study in detail is the use of stimulated emission in lasers. In Section 10.2 the working mechanisms of several kinds of lasers will be explained and some of their many applications, including holography, will be discussed.

The remainder of the chapter will be devoted to a study of solids. Solids can have many different crystal structures, although in some cases they lack long-range order altogether. Thermal and magnetic properties (including thermal expansion, thermal conductivity, and magnetic susceptibility) will be discussed in some detail. A knowledge of these properties is important in many uses of solids, because variations in temperature and applied magnetic fields can alter significantly the behavior of solid materials.

Superconductivity is a remarkable phenomenon observed in many solid materials at low temperatures. It is characterized by the absence of electrical resistance and the expulsion of magnetic flux from the superconductor. We shall study the development of superconducting materials, the theory of superconductivity, and the prospects for future basic research in this field. In the final section of the chapter we shall describe many of the developed and proposed applications of superconductivity. Resistanceless electrical circuits and transmission lines, magnetic levitation, and high-field magnets are just a few of the applications being studied. We shall see that the amazing properties of superconductors give rise to many more exciting possibilities for the future.

10.1 Molecular Bonding and Spectra

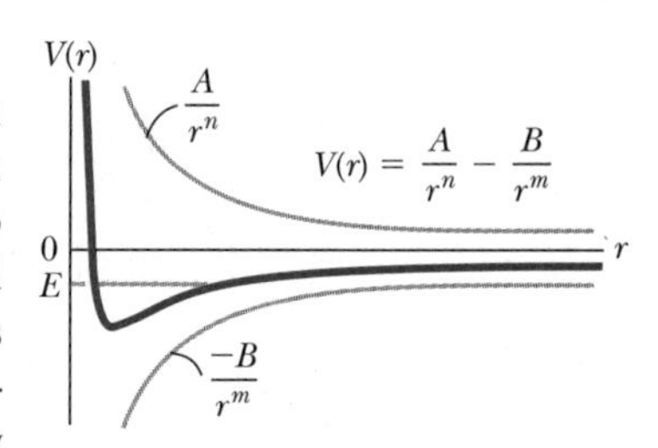

FIGURE 10.1 Attractive and repulsive potentials (lightly shaded lines) experienced by one atom in a vicinity of another atom. The sum of attractive and repulsive potentials is represented by the solid blue line. Bound states exist for a total energy $E < 0$.

How are atoms held together to form molecules? There is a lot of good physics in the answer to this apparently simple question. Clearly the attractive forces between atoms in molecules must be due to the Coulomb force, because the Coulomb force is the only one that has both the strength and long range necessary to bind atoms at the distances observed (normally on the order of 10^{-10} m). All atoms have both positive (the nucleus) and negative charges (electrons) needed to provide attractive forces, though not all atoms readily form molecules. Because every neutral atom contains both kinds of charge, the net force of one atom on another must be a combination of attractive and repulsive forces. It is the combination of attractive and repulsive forces that creates a stable molecular structure.

As an example let us consider the bonding in a diatomic molecule (the mechanisms are similar, though sometimes more complex, in multiatom molecules). We will find it useful to look at molecular binding using potential energy V. Recall that in conservative systems, force is related to potential energy by $F = -dV/dr$, where r is the distance of separation. This means that, in cases in which the magnitude of the force decreases with increasing distance (as it does with Coulomb forces), we should associate a negative slope $(dV/dr < 0)$ with repulsive forces and a positive slope $(dV/dr > 0)$ with attractive forces. An approximation of the force felt by one atom in the vicinity of another atom may then be written

$$V = \frac{A}{r^n} - \frac{B}{r^m} \tag{10.1}$$

where A and B are positive constants that depend on the types of atoms involved and n and m are small, positive numbers. You may suspect that n and m are equal to one because the forces are Coulomb, but, because of the complicated shielding effects of the various electron shells, this is not so.

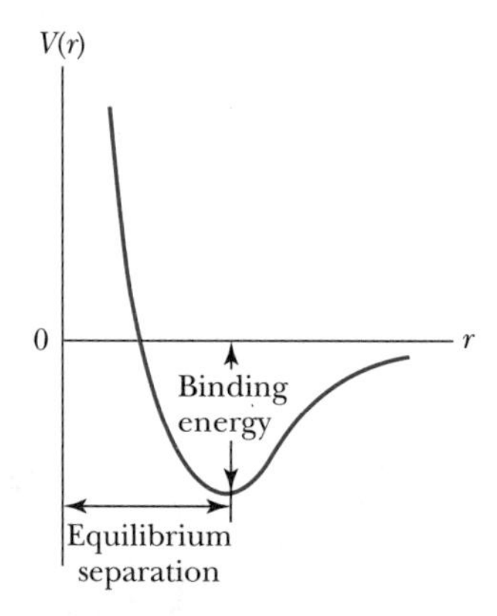

FIGURE 10.2 The net potential energy curve, showing the equilibrium separation and binding energy.

As shown in Figure 10.1, the sum of attractive and repulsive potentials produces a potential well that provides a stable equilibrium for total energy $E < 0$. The exact shape of the curve depends on the parameters A, B, n, and m from Equation (10.1). You should convince yourself that $n > m$ is required in order to produce the potential well shown in Figure 10.1. The exact level of E depends on temperature, with E at the bottom of the potential well when $T = 0$. Once a pair of atoms is joined, one would then have to supply enough energy to raise the total energy of the system to zero (effectively taking the second atom to $r = \infty$) in order to separate the molecule into two neutral atoms. These potential energy curves typically have a minimum value, and the corresponding value of r is an *equilibrium separation.*

Binding energy

The amount of energy required to separate the two atoms completely is known as the **binding energy** for that molecule. The binding energy is roughly equal to the depth of the potential well shown in Figure 10.2. The binding energy and well depth may not be exactly equal, however, because by the Heisenberg uncertainty principle, the ground-state energy cannot lie exactly at the bottom of the well.

Molecular Bonds

Ionic bond

There are different kinds of bonds. The bonding mechanism for a particular molecule depends principally on the electronic structure of the atoms involved. The simplest is the **ionic bond;** this typically occurs when the two atoms involved are easily ionized. For example, sodium, which has the electronic configuration

$1s^2 2s^2 2p^6 3s^1$, readily gives up its $3s$ electron to become Na^+, while chlorine, with electronic configuration $1s^2 2s^2 2p^6 3s^2 3p^5$, readily gains an electron to become Cl^-. Notice that both Na^+ and Cl^- have filled electronic shells. The Na^+ and Cl^- ions are electrostatically attracted to form the NaCl molecule

Covalent bond

In a **covalent bond,** the atoms are not as easily ionized. In order to form a molecule, they must somehow share their outer shell electrons. Diatomic molecules formed by the combination of two identical atoms (H_2, N_2, O_2, etc., sometimes referred to as *homopolar* molecules) tend to be covalent, because neither atom is more likely than the other to gain or lose an electron. Most larger molecules are formed principally with covalent bonds.

There are several more exotic bonding mechanisms that we shall not describe in detail here. The *van der Waals bond* is a relatively weak bond found mostly in liquids and solids at low temperatures. The *hydrogen bond* is important in holding together many organic molecules. In a *metallic bond* essentially free valence electrons may be shared by a number of atoms.

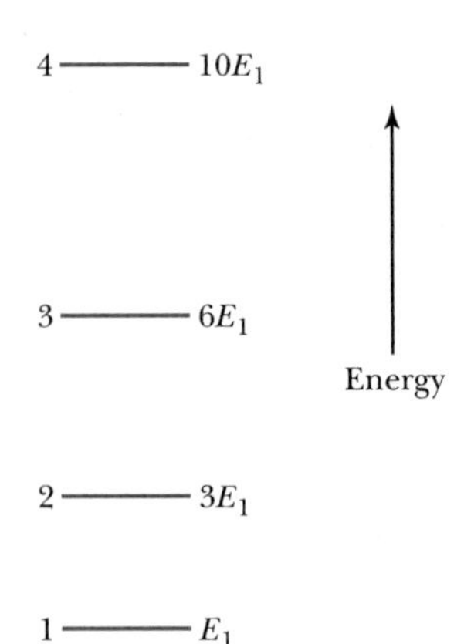

FIGURE 10.3 Energy levels for a rigid rotator. The spacing between levels increases with increasing energy. The levels are given in terms of the $\ell = 1$ energy level E_1, where $E_1 = \hbar^2/I$.

Rotational States

Regardless of the types of molecular bonds present, we can learn much about the properties of molecules by studying how molecules absorb, emit, and scatter electromagnetic radiation. This kind of study is referred to broadly as *molecular spectroscopy*. Let us begin by considering a simple two-atom molecule, such as N_2. As we learned in Chapter 9, there are different ways of modeling this molecule, depending on the physical situation. For example, one may think of the N_2 molecule as two N atoms held together with a massless, rigid rod. This is known as the *rigid rotator* model. Because these are purely rotational modes of motion, the quantum theory of angular momentum can be used to determine those energy levels. In a purely rotational system, the kinetic energy can be expressed in terms of the angular momentum L and the rotational inertia I as

$$E_{\text{rot}} = \frac{L^2}{2I}$$

The angular momentum will be quantized (see Section 7.3) in the form

$$L = \sqrt{\ell(\ell+1)}\,\hbar \tag{7.22}$$

where ℓ is the angular momentum quantum number. Then the energy levels will be given by

Energy levels of a simple rotational state

$$E_{\text{rot}} = \frac{\hbar^2\,\ell(\ell+1)}{2\,I} \tag{10.2}$$

For a rigid rotator I is constant and therefore E_{rot} varies only as a function of the quantum number ℓ, as shown in Figure 10.3.

Example 10.1

Estimate the value of E_{rot} for the lowest rotational energy state of N_2.

Solution: If we consider the nitrogen atoms to be point masses (each with mass m) separated by a distance R (see Figure 10.4), then the rotational inertia about an axis passing through the center of the molecule and perpendicular to the line joining the atoms is

$$I = m\left(\frac{R}{2}\right)^2 + m\left(\frac{R}{2}\right)^2 = \frac{mR^2}{2}$$

For nitrogen $m = 2.33 \times 10^{-26}$ kg. If we assume $R \approx 10^{-10}$ m in this order of magnitude estimate, then

$$I \approx \frac{(2 \times 10^{-26}\ \text{kg})(10^{-20}\ \text{m}^2)}{2} = 10^{-46}\ \text{kg} \cdot \text{m}^2$$

For $\ell = 1$

$$E_{\text{rot}} = \frac{2\,\hbar^2}{2\,I} = \frac{\hbar^2}{I} \approx \frac{10^{-68}\ \text{J}^2 \cdot \text{s}^2}{10^{-46}\ \text{kg} \cdot \text{m}^2} = 10^{-22}\ \text{J} \approx 10^{-3}\ \text{eV}$$

Remember, however, that this is just the energy of the **lowest** state. The energy of the $\ell = 4$ state [with $\ell(\ell + 1) = 20$] is ten times higher, or about 0.01 eV. This is still at least two orders of magnitude less than the energy associated with an electronic transition in hydrogen (the kind that produces visible photons). Therefore we can expect photons generated by transitions between adjacent rotational states to be in the *infrared* or *microwave* portion of the spectrum.

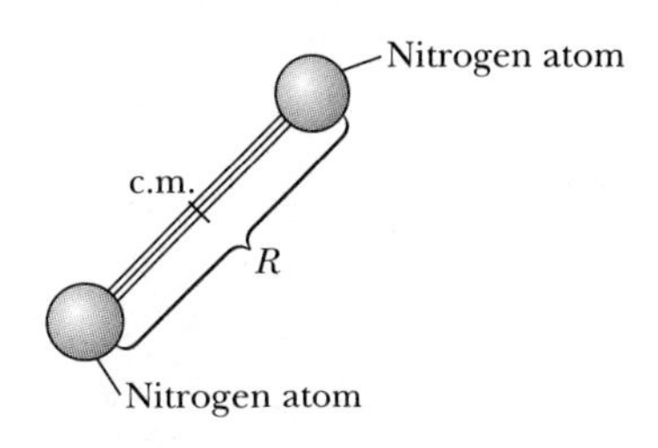

FIGURE 10.4 Schematic of a diatomic molecule (N_2) as a rigid rotator, with an equilibrium separation R between atomic centers.

Vibrational States

We must consider also the possibility that a vibrational energy mode will be excited. As we learned in Section 9.3, there will be no thermal excitation of this mode in a diatomic gas at ordinary temperatures. However, it is possible to stimulate vibrations in molecules using electromagnetic radiation. As a crude model of vibration in a diatomic molecule we again assume that the two atoms are point masses connected by a massless spring, as in Figure 10.5. Then the atoms can execute simple harmonic motion, just as in classical physics. The energy levels must be those of a quantum mechanical oscillator (see Section 6.6), namely

$$E_{\text{vibr}} = \left(n + \frac{1}{2}\right)\hbar\omega \tag{10.3}$$

where ω is the natural (classical) angular frequency of the oscillator.

One way of estimating E_{vibr} is first to estimate ω from purely classical considerations. For a classical two-particle oscillator

$$\omega = \sqrt{\frac{\kappa}{\mu}}$$

where $\mu = \dfrac{m_1 m_2}{m_1 + m_2}$ is the reduced mass of the system and κ is a force (spring) constant. Now we can estimate κ by assuming that the force holding the masses together is Coulomb. If the bond were purely ionic, then the point masses would be point charges $+e$ and $-e$. Thus we can compute the force constant κ (force per unit distance):

$$\kappa = \left|\frac{dF}{dr}\right| \approx \left|\frac{d}{dr}\left[\frac{e^2}{4\pi\epsilon_0 r^2}\right]\right| = \frac{e^2}{2\pi\epsilon_0 r^3} \tag{10.4}$$

and

$$\omega \approx \sqrt{\frac{e^2}{2\pi\epsilon_0 \mu r^3}} \tag{10.5}$$

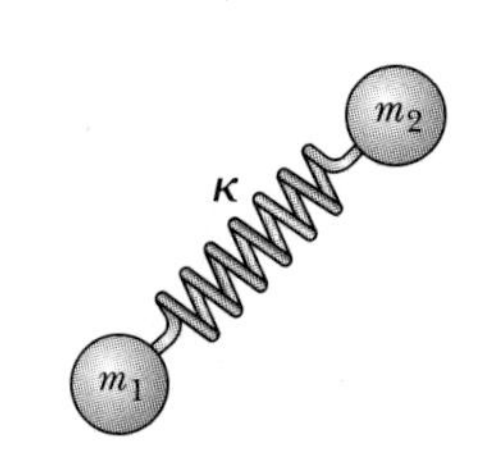

FIGURE 10.5 A model of a diatomic molecule with the two masses m_1 and m_2 connected by a massless spring with force constant κ.

TABLE 10.1
Fundamental Vibrational Frequencies and Effective Force Constants for Some Diatomic Molecules

Molecule	Frequency (Hz), $n = 0$ *to* $n = 1$	Force Constant (N/m)
HF	8.72×10^{13}	970
HCl	8.66×10^{13}	480
HBr	7.68×10^{13}	410
HI	6.69×10^{13}	320
CO	6.42×10^{13}	1860
NO	5.63×10^{13}	1530

From G. M. Barrow, *The Structure of Molecules.* New York: W. A. Benjamin, 1963.

If we use $r \approx 10^{-10}$ m, Equation (10.4) yields a force constant of $\kappa \approx 460$ N/m. Then $\omega = \sqrt{\kappa/\mu} \approx 2.0 \times 10^{14}$ rad/s (or $\nu \approx 3.2 \times 10^{13}$ Hz), and Equation (10.3) gives an energy $E_{\text{vibr}} \approx 0.2$ eV for the $n = 1$ vibrational level of N_2. Actual values of the fundamental vibrational frequencies and effective force constants are shown in Table 10.1.

Example 10.2

(a) Given that the spacing between vibrational energy levels of the HCl molecule is 0.36 eV, calculate the effective force constant.
(b) Find the classical temperature associated with this difference between vibrational energy levels in HCl.

Solution: (a) Because $\kappa = \mu\omega^2$ we will first need to find μ and ω. $\Delta E = h\nu = \hbar\omega$, so

$$\omega = \Delta E/\hbar = \frac{0.36 \text{ eV}}{6.58 \times 10^{-16} \text{ eV} \cdot \text{s}} = 5.47 \times 10^{14} \text{ rad/s}$$

Now, $\mu = m_1 m_2/(m_1 + m_2) = 1.61 \times 10^{-27}$ kg (assuming we have the chlorine-35 isotope). Then

$$\kappa = \mu\omega^2 = (1.61 \times 10^{-27} \text{ kg})(5.47 \times 10^{14} \text{ rad/s})^2 = 480 \text{ N/m}$$

which is in agreement with the value in Table 10.1.

(b) At the lowest level $n = 0$ and $E_{\text{vibr}} = \hbar\omega/2$. There are two degrees of freedom associated with a one-dimensional oscillator, one from the kinetic energy and one from the potential (see Section 9.3). Therefore we can say that

$$\Delta E = \hbar\omega = 2 \times \left(\frac{kT}{2}\right) = kT$$

(where k is Boltzmann's constant).

$$T = \frac{\Delta E}{k} = \frac{0.36 \text{ eV}}{(8.62 \times 10^{-5} \text{ eV/K})} = 4200 \text{ K}$$

This exceptionally high temperature is a good indication of why vibrational levels are not thermally excited at more ordinary temperatures.

Vibration and Rotation Combined

In a real molecular system it is possible to excite the rotational and vibrational modes of motion simultaneously. Therefore the total energy of our simple vibration-rotation system can be written

$$E = E_{\text{rot}} + E_{\text{vibr}} = \frac{\hbar^2 \ell(\ell + 1)}{2I} + \left(n + \frac{1}{2}\right)\hbar\omega \tag{10.6}$$

A diatomic molecule that has been stimulated to an excited state will, as in atomic systems, emit a photon upon decaying to a lower energy state. Generally, then, it is possible to observe a wide spectrum of emitted photons, corresponding to various rotational and vibrational transitions.

One outstanding characteristic of emission spectra can be deduced by examining Equation (10.6). Because the vibrational energies are spaced at regular intervals ($E_{\text{vibr}} = \frac{1}{2}\hbar\omega$, $\frac{3}{2}\hbar\omega$, etc.), emission features due to vibrational transitions will appear at regular intervals. This is also the case for rotational features, though for a different reason. Consider, for example, a transition from the $\ell + 1$ state to the ℓ state. The photon produced by that transition will have an energy

$$
\begin{aligned}
E_{\text{ph}} &= \frac{\hbar^2}{2I}[(\ell + 1)(\ell + 2) - \ell(\ell + 1)] \\
&= \frac{\hbar^2}{2I}[\ell^2 + 3\ell + 2 - \ell^2 - \ell] = \frac{\hbar^2}{I}(\ell + 1) \qquad (10.7)
\end{aligned}
$$

Now we see an energy-level spacing that varies with ℓ. Specifically, the higher the starting energy level, the greater the photon energy for a transition with $\Delta\ell = -1$. This is consistent with the energy-level spacings shown in Figure 10.3. Because the photon energy increases linearly with the quantum number ℓ, photon energies increase at regular intervals.

As our crude estimates have indicated, vibrational energies are typically greater than rotational energies by as much as an order of magnitude. This significant energy difference, along with the different spacing characteristic noted earlier, results in the **band spectrum** shown in Figure 10.6. We see an evenly spaced vibrational spectrum with a more closely spaced rotational spectrum superimposed on each vibrational line.

Band spectrum

In any molecular spectrum the positions and intensities of the observed bands are governed by the rules of quantum mechanics. We note two features in particular. First, the relative intensities of the bands are due to different transition probabilities. The probabilities of transitions from an initial state (for example, $n = 5$) to each possible final state ($n = 4$, 3, etc.) are not necessarily the same. The more probable transitions occur more frequently, and the result is a brighter line corresponding to that transition. Second, some transitions are forbidden by the quantum mechanical selection rule that requires $\Delta\ell = \pm 1$. This must be so, because upon emission the photon carries away its intrinsic angular momentum of one quantum unit ($\hbar$).

An interesting application of the $\Delta\ell = \pm 1$ selection rule is found in the study of **absorption spectra.** When electromagnetic radiation is incident upon a collection of a particular kind of molecule (for example, in a closed gas cell), molecules can absorb photons and make transitions to a higher vibrational state

Absorption spectra

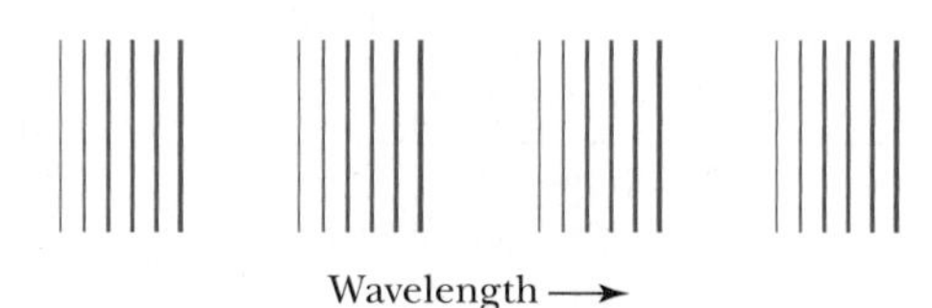

FIGURE 10.6 Typical section of the emission spectrum of a diatomic molecule. There are equally spaced groups of lines, corresponding to the equal spacings between vibrational levels. The structure within each group is due to transitions between rotational levels.

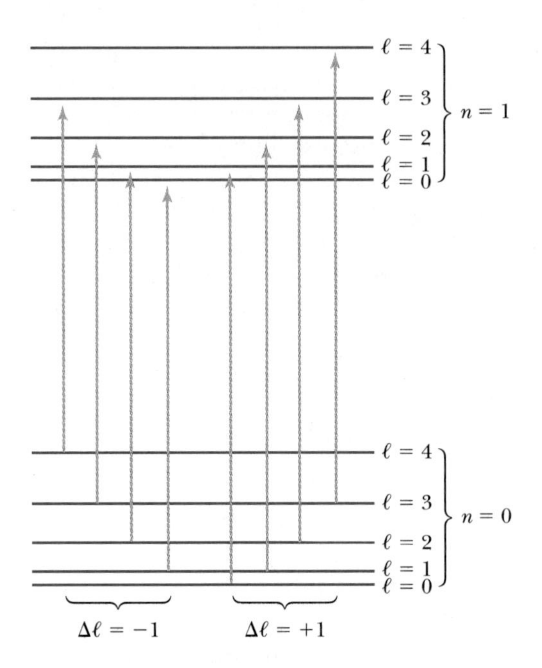

FIGURE 10.7 A schematic diagram of the absorptive transitions between adjacent vibrational states ($n = 0$ to $n = 1$) in a diatomic molecule.

only if the rotational state changes by $\Delta\ell = \pm 1$. A schematic of the allowed transitions between two vibrational states is shown in Figure 10.7. Because ΔE increases linearly with ℓ as in Equation (10.7), one expects to see absorption bands at regular intervals of energy (or frequency, which is proportional to energy). This is evident in the absorption spectrum of HCl (Figure 10.8). The regular spacing between the peaks can be used to compute I (Problem 15a). The missing peak in the center corresponds to the forbidden $\Delta\ell = 0$ transition. Therefore that central frequency ν is just $\nu = \frac{1}{2\pi}\sqrt{\kappa/\mu}$, the frequency of a vibrational transition from $n = 1$ to $n = 0$.

Physicists and chemists have developed extremely sophisticated equipment and data reduction methods for the sole purpose of studying molecular spectra. One of the most popular methods is known as Fourier Transform Infrared

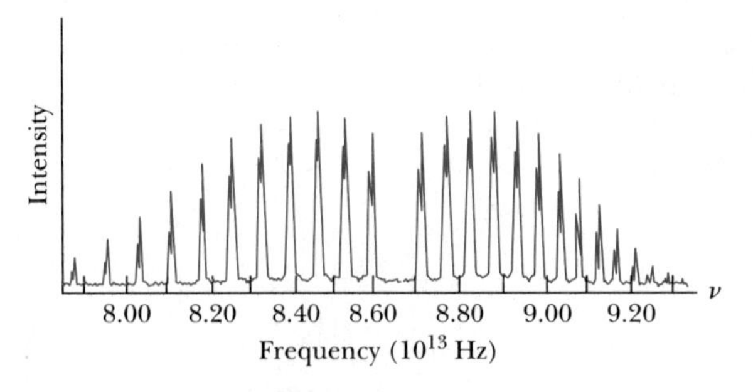

FIGURE 10.8 An absorption spectrum of a diatomic molecule, HCl. The spacing is regular, as predicted in Equation (10.7), and each line is split due to the small mass difference between the chlorine isotopes ^{35}Cl and ^{37}Cl. The missing central peak at $\nu = 8.65 \times 10^{13}$ Hz corresponds to the forbidden $\Delta\ell = 0$ transition.

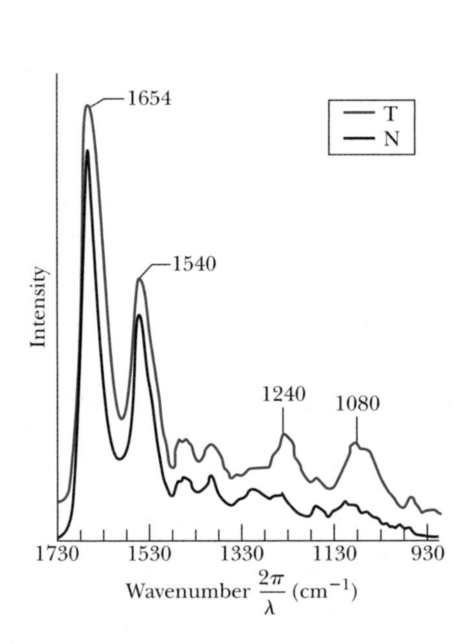

FIGURE 10.9 FTIR spectrum comparing tumor cells (T) and normal cells (N) from human lungs. Absorption in the infrared region is shown as a function of wavenumber ($2\pi/\lambda$), so the wavelengths range from 36 μm to 68 μm. The FTIR method makes peak locations and intensities distinct enough so that this could be a useful diagnostic tool. *From E. Benedetti et al., Appl. Spectrosc.* **44,** 1276–1280 (1990).

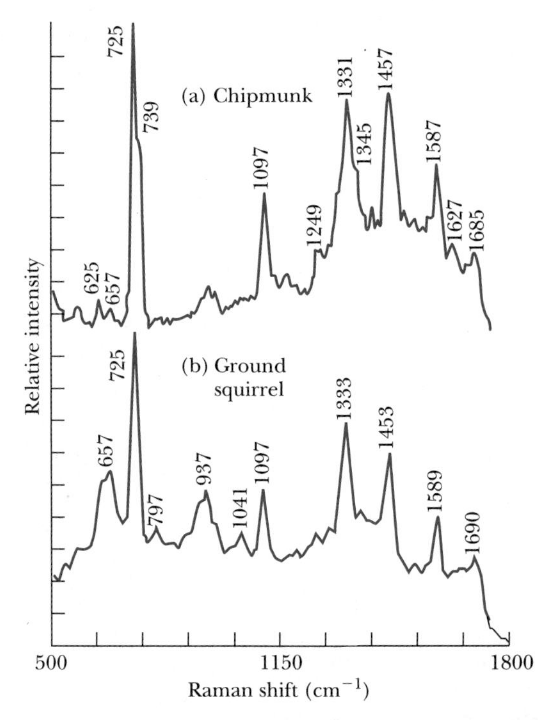

FIGURE 10.10 Raman spectra of (a) chipmunk and (b) ground squirrel eye pigments. The "Raman shift" is the difference in wavenumber, $2\pi/\lambda$, corresponding to the frequency difference in Equation (10.9). These spectra are used to study metabolic and photochemical generation of lens pigments. *From S. Nie et al., Appl. Spectrosc.* **44,** 571–575 (1990).

(FTIR) Spectroscopy. Most Fourier transform spectrometers use an interferometric system based on the Michelson interferometer in order to make precise determinations of photon wavelengths. The Fourier transform is used to analyze the spectrum. In Fourier analysis, a spectrum can be decomposed into an infinite series of sine and cosine functions. With the proper knowledge of spectral characteristics, random and instrumental noise can be greatly reduced in order to produce a "clean" spectrum, that is, one with a high signal-to-noise ratio (see Figures 10.9 and 10.10).

It is not necessary that an incoming photon's energy precisely match the transition energy ΔE. If a photon of energy greater than ΔE is absorbed by a molecule, the excess energy may be released in the form of a scattered photon of lower energy. This process is known as **Raman scattering.** It is possible to look at the spectrum of Raman scattered photons and from it learn some of the properties of the molecules being studied. In Raman scattering the angular momentum selection rule becomes $\Delta\ell = \pm 2$ because of the second photon involved.*

Raman scattering

*Of course $\Delta\ell = 0$ is also possible in a two-photon process. The $\Delta\ell = 0$ case is known as *Rayleigh scattering*. Quantum theory predicts that Rayleigh scattering becomes more likely (relative to Raman scattering) at shorter wavelengths. Rayleigh scattering of photons by the atmosphere causes the sky to be blue for the most part and orange-red near sunrise and sunset, because shorter-wavelength photons (blue) are Rayleigh-scattered more than longer-wavelength photons (red).

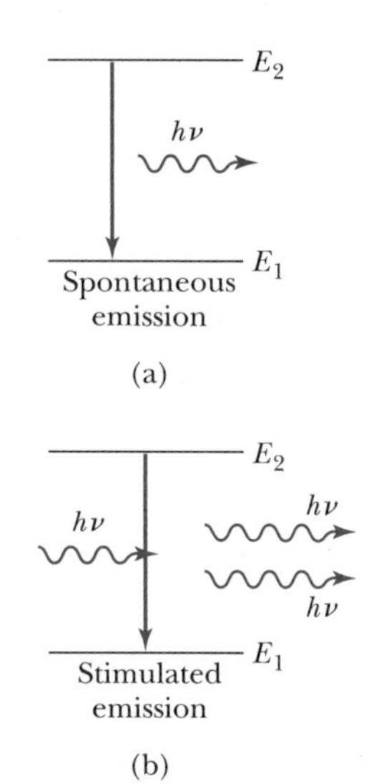

FIGURE 10.11 Schematic of (a) spontaneous emission, with $E_2 - E_1 = h\nu$ and (b) stimulated emission in a two-level system. In stimulated emission one photon triggers the emission of a second photon.

Consider a transition from a state ℓ to a state $\ell + 2$. The rotational part of the transition energy is

$$\Delta E_{\text{rot}} = \frac{\hbar^2}{2I}[(\ell + 2)(\ell + 3) - \ell(\ell + 1)]$$

$$= \frac{\hbar^2}{I}[2\ell + 3] \tag{10.8}$$

Suppose an incoming photon with energy $h\nu$ is Raman scattered and the scattered photon has energy $h\nu'$. Then the frequency of the scattered photon can be found in terms of the relevant rotational variables

$$h\nu' = h\nu - \Delta E_{\text{rot}} = h\nu - \frac{\hbar^2}{I}[2\ell + 3]$$

Thus

$$\nu' = \nu - \frac{\hbar}{2\pi I}[2\ell + 3] \tag{10.9}$$

Raman spectroscopy can be a useful tool in determining rotational properties (specifically ℓ and I) of molecules. It is also used to study polyatomic systems, in which the analysis of molecular properties is not so straightforward. Today Raman spectroscopy is extremely popular among molecular spectroscopists for the wealth of information it provides.

10.2 Stimulated Emission and Lasers

Spontaneous emission

The emission of photons by molecules as described in Section 10.1 is known as **spontaneous emission.** A molecule in an excited state will decay to a lower energy state and emit a photon spontaneously, without any stimulus from the outside (Figure 10.11a). The laws of quantum mechanics do not allow us to say when the transition will occur. Because the process is probabilistic, the best we can do is calculate the mean lifetime of an excited state or the probability that a spontaneous transition will occur within a given amount of time. As a consequence, one can expect the phase of the emitted photon to be random. We note as an aside that the mean lifetime can often be estimated from the width of the emission spectrum line. If a spectral line has a width ΔE, then Heisenberg's uncertainty principle, Equation (5.42), gives a lower bound estimate of the lifetime of $\hbar/2\Delta E$.

Stimulated emission

It is possible, however, to make the emission process occur in a more controlled way. Electromagnetic radiation (a photon) incident upon a molecule in an excited state can cause the inherently unstable system to decay to a lower state. This is called **stimulated emission.** An important feature of stimulated emission is that the photon emitted in the process tends to have the same phase and direction as the stimulating radiation. And if the incoming photon has the same energy as the emitted photon, the result will be *two* photons of the same wavelength and phase traveling in the same direction, because the incoming photon is not absorbed, but rather is used only to trigger emission of the second photon. The two photons (of the same wavelength and phase) are then said to be *coherent.* A schematic diagram of the stimulated emission process is shown in Figure 10.11b.

Stimulated emission is the fundamental physical process in the operation of the laser. A simple argument by Einstein from his 1917 paper "On the Quantum

Theory of Radiation" will help explain in a rather straightforward way why we can expect stimulated emission to occur. It is a testament to Einstein's genius that he did this work, as he did in many other areas of physics, on purely theoretical grounds, long before it was confirmed in the laboratory.

Here, in brief, is Einstein's analysis. We shall consider transitions between two molecular states with energies E_1 and E_2 (where $E_1 < E_2$). The photon associated with either emission or absorption has an energy E_{ph} and frequency ν, where $E_{ph} = h\nu = E_2 - E_1$. If stimulated emission (that is, a process in which incoming radiation causes a transition from E_2 to E_1) occurs, the *rate* of those transitions must be proportional to the number of molecules in the higher state (call this N_2) and the energy density of the incoming radiation [call this $u(\nu)$]. Therefore let us say that the rate at which stimulated transitions are made from E_2 to E_1 is $B_{21}N_2u(\nu)$, where B_{21} is a proportionality constant that depends on the quantum mechanical probability of stimulated emission. Similarly, the probability that a molecule at E_1 will *absorb* a photon can be expressed $B_{12}N_1u(\nu)$. We must also take into account the possibility that spontaneous emission will occur. The rate of spontaneous emission is independent of $u(\nu)$, however, and can be expressed simply as AN_2, where A is a constant related to the quantum mechanical probability of spontaneous emission.

Once the system has reached equilibrium with the incoming radiation, the total number of downward transitions must equal the total number of upward transitions. Then

$$B_{21}N_2u(\nu) + AN_2 = B_{12}N_1u(\nu) \tag{10.10}$$

In thermal equilibrium each of the N_i are proportional to their respective Boltzmann factors $e^{-E_i/kT}$. Therefore

$$[B_{21}u(\nu) + A]e^{-E_2/kT} = B_{12}u(\nu)e^{-E_1/kT} \tag{10.11}$$

In the classical limit $T \rightarrow \infty$ (see Section 9.5). Then $e^{-E_1/kT} = e^{-E_2/kT}$ and the energy density $u(\nu)$ becomes very large, so the factor A becomes insignificant. Thus we see that in the classical limit

$$B_{12} \approx B_{21} \equiv B \tag{10.12}$$

that is, the probability of stimulated emission is approximately equal to the probability of absorption. What this basically means is that if the transition from E_1 to E_2 (absorption) can occur, then we should also expect that stimulated emission will occur.

Finally, we can use Equation (10.11) to obtain another useful relationship. Solving Equation (10.11) for $u(\nu)$ yields

$$u(\nu) = \frac{A}{B_{12}e^{(E_2 - E_1)/kT} - B_{21}} = \frac{A}{B_{12}e^{h\nu/kT} - B_{21}}$$

or, if we use Equation (10.12),

$$u(\nu) = \frac{A/B}{e^{h\nu/kT} - 1} \tag{10.13}$$

Equation (10.13) should look familiar because it closely resembles the Planck radiation law, Equation (3.23). In fact, when the Planck law is expressed in terms of frequency instead of wavelength, it is

$$u(\nu, T) = \frac{8\pi\nu^2}{c^3} \frac{h\nu}{e^{h\nu/kT} - 1} \tag{10.14}$$

Therefore by Equations (10.13) and (10.14) it is required that

$$\frac{A}{B} = \frac{8\pi h\nu^3}{c^3} \tag{10.15}$$

In other words, the stimulated emission probability coefficient B is proportional to the spontaneous emission probability coefficient A in equilibrium. We may interpret this result to mean that in a process for which the probability of spontaneous emission is high, the probability of stimulated emission will also be high.

The laser

How then does a **laser*** actually work? The schematic drawing of a helium–neon laser (Figure 10.12), the type commonly found in most college and university physics departments today, will help you to understand the process. The main body of the laser is a closed tube, filled with about a 90%/10% ratio of helium and neon. Photons bouncing back and forth between the two mirrors are used to stimulate the transition in helium, which in turn produces more photons. Recall that photons produced by stimulated emission will be coherent, and thus the beam of photons that escapes through the partially silvered mirror will be the coherent beam we observe.

How are atoms put into the excited state in the first place? We cannot rely on the photons in the tube to do so; if we did, then there would be no net output of energy from the laser. Any photon produced by stimulated emission would have to be "used up" to excite another atom in order for the process to be continuous. A second potential difficulty is that there may be nothing to prevent spontaneous emission from atoms in the excited state. In that case the beam would not be coherent.

Three-level system

The way around both of those problems is to use a multilevel atomic system (of course this will also work with molecular systems). Consider first the **three-level system** shown in Figure 10.13. Atoms in the ground state (E_1) are *pumped* to a higher state (E_3) by some external source of energy. The atom then decays quickly to E_2. The key is that the transition from E_2 to E_1 must be forbidden, for example, by a $\Delta\ell = \pm 1$ selection rule. Then the state with energy E_2 is said to be metastable. Now a large number of atoms can exist for a relatively long time at E_2, where they are just waiting for a photon to come along and stimulate the transition to E_1. In normal operation there will be many more atoms in the ex-

*The word **laser** is an acronym for **l**ight **a**mplification by the **s**timulated **e**mission of **r**adiation. One also hears of **masers,** in which microwaves are used instead of visible light. The first working maser was made by Charles H. Townes in 1954, and the first laser by Theodore H. Maiman in 1960. As of the end of the 20th century, "phasers" exist only on television's "Star Trek."

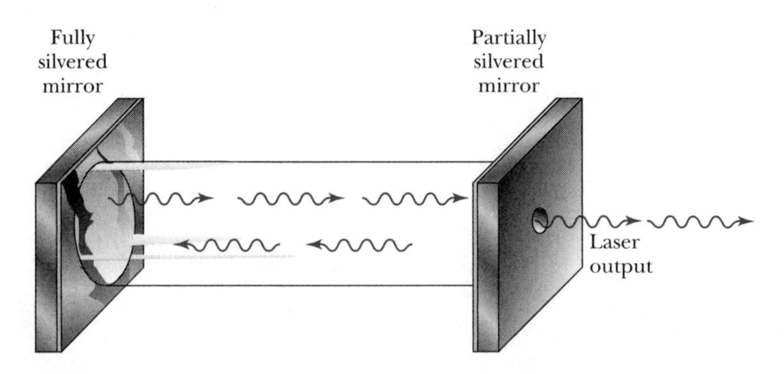

FIGURE 10.12 A schematic diagram of a He–Ne laser. The coherent photon beam escapes through the partially silvered mirror.

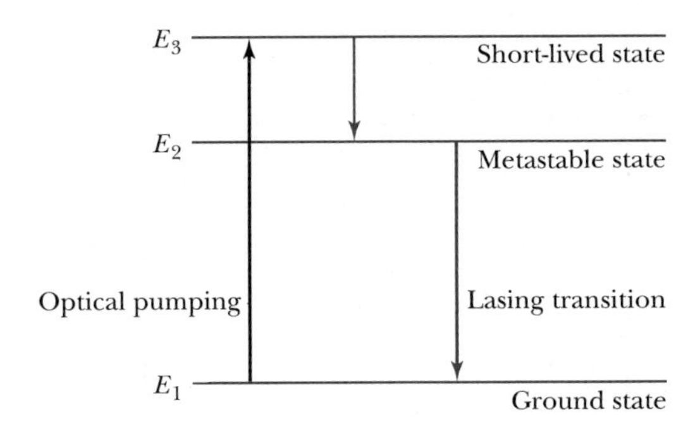

FIGURE 10.13 Transitions in a three-level laser. The lasing transition takes the system from the metastable state (E_2) to the ground state (E_1).

cited (metastable) state than in the ground state. This situation is, of course, contrary to what one finds under normal conditions. Therefore it is known as a *population inversion.* As we have described here, the population inversion is an essential feature of the operation of lasers.

There is yet another potential difficulty with the three-level system as we have described it. What happens to an atom after it has been returned to the ground state from E_2 by a stimulated emission? We would like the external power supply to return it immediately to E_3, but in practice it may take some time for this to happen, during which it is possible that a photon with energy $E_2 - E_1$ (just the energy that a photon created by stimulated emission would have) can be absorbed. That would be undesirable, for that photon would then be unavailable for stimulating another transition and for passing through the partially silvered mirror, as we expect all of them to do eventually. The result would be a much weaker beam, or perhaps none at all.

Four-level system

We can get over this last hurdle by using a **four-level system,** as shown in Figure 10.14. As before, atoms are pumped from the ground state to a higher state (now E_4), where they decay quickly to the metastable state E_3. The stimulated emission takes atoms from E_3 to E_2 The spontaneous transition from E_2 to E_1 is not forbidden, so the state E_2 will not exist long enough for a photon to be used

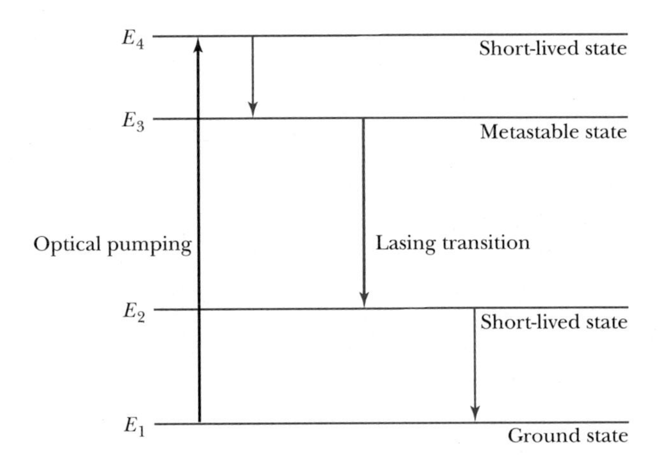

FIGURE 10.14 Transitions in a four-level laser. The lasing transition takes the system from the metastable state (E_3) to another short-lived state (E_2). The system then returns quickly to the ground state (E_1), so that photons cannot be reabsorbed in returning the system from E_2 to E_3.

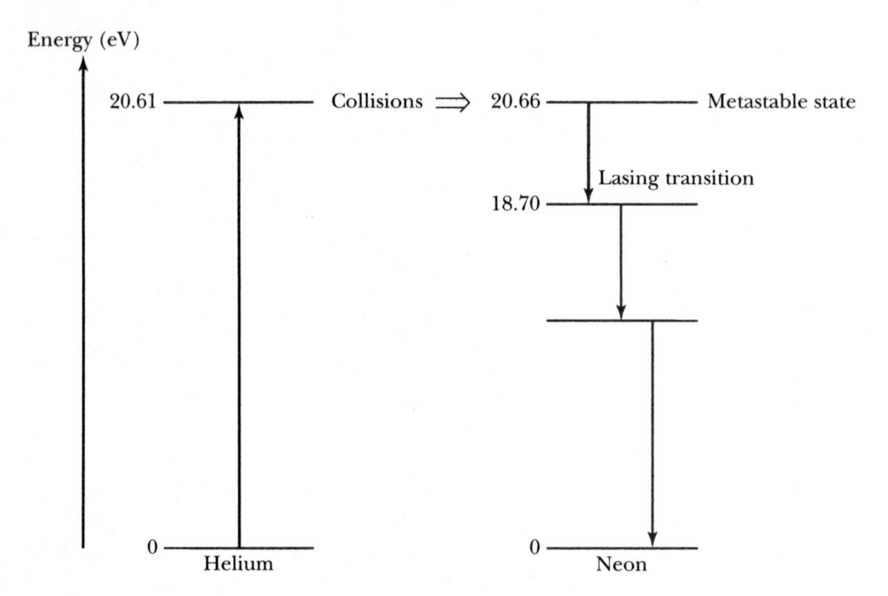

FIGURE 10.15 The energy levels and transitions in a helium–neon laser.

up in kicking the system from E_2 to E_3. There is little chance that an atom in the ground state will absorb a photon with energy $E_3 - E_2$, so the lasing process can proceed efficiently.

The red helium-neon laser uses transitions between energy levels in both helium and neon, as shown in Figure 10.15. The applied voltage excites a helium atom from its ground state to an excited level at approximately 20.61 eV. An excited helium atom will occasionally collide with a neon atom and transfer this excess energy to it, because this is very close to the energy needed to excite a neon atom from its ground state to the $2p^5 5s^1$ level (the metastable state). The lasing process then proceeds as shown in Figure 10.15, with the result being a coherent beam of red light at $\lambda = 632.8$ nm. Of course, there are other allowed transitions in neon, and some energy is invariably lost through those photons that do not participate in the lasing process. Some of these transitions are used to make green and orange helium-neon lasers, which are now available commercially.

Tunable laser

In a **tunable laser** the wavelength of the emitted radiation can be adjusted over a sometimes wide range (as wide as 200 nm). In the past tunable lasers have been made by using an organic dye as the lasing material. The organic compounds are chosen so that they have a great number of closely spaced energy levels. This is what makes stimulated emission over a range of wavelengths possible. Tuning is effected by changing the concentration of the dye and/or the length of the dye cell. For a more complete description see Eugene Hecht's *Optics.** In many applications dye lasers are now being replaced by semiconductor lasers (see Section 11.3).

Free-electron laser

Another kind of tunable laser is the **free-electron laser,** shown schematically in Figure 10.16. This laser relies on the fact that charged particles, in this case electrons, emit electromagnetic radiation when accelerated. A series of magnets called *wigglers* is used to accelerate a beam of electrons transversely as shown in the figure. The electron velocity is matched to the oscillations in the magnetic

*E. Hecht, *Optics,* 2nd ed. Reading, Massachusetts: Addison Wesley, 1988, pp. 588–590.

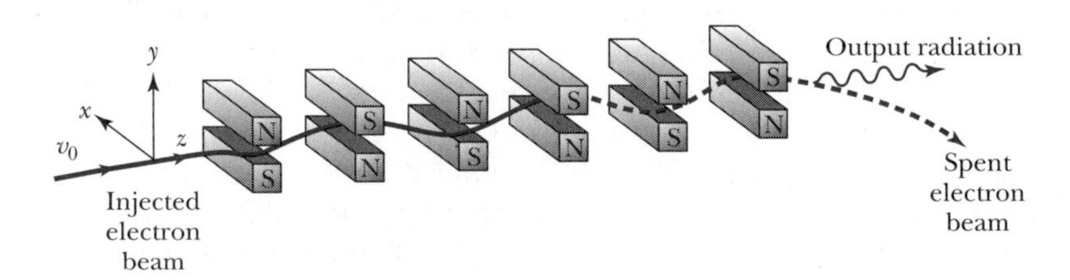

FIGURE 10.16 Schematic of the operation of a free-electron laser. The laser output is in the same direction (z in this drawing) as the injected electron beam. *From P. Sprangler and T. Coffey, Physics Today (March 1984) Fig. 3, p. 48.*

field in such a way that the emitted radiation is as coherent as in a gas-filled laser. Electron-beam lasers have been made to produce outputs in the ultraviolet, visible, and infrared ranges. The exceptional fine-tuning that is possible with these devices, along with their relatively high efficiency, make electron-beam lasers and masers useful for a number of scientific applications, including spectroscopy, accelerator technology, radar, and fusion research.

Scientific Applications

In recent years lasers have been used in a wide range of scientific applications. One obvious application is the use of the extremely coherent and nondivergent beam in making precise determinations of distances both large and small. Thanks in part to the accuracy of laser measurements, the speed of light in a vacuum has been *defined* to be $c = 299{,}792{,}458$ m/s. This definition of the speed of light has led to a redefinition of the meter as the length of the path traveled by light in a vacuum during a time interval of 1/299,792,458 of a second. On the longer end of the distance scale, reflecting mirrors placed on the moon by Apollo astronauts have enabled scientists to reflect laser light from the moon's surface and thereby determine the earth–moon distance to within 10 cm. Variations in this distance have enabled us to better understand the orbital mechanics of the solar system. Lasers have been set up to measure land shifts near geologic fault lines in California, with the hope of better understanding and perhaps predicting earthquakes.

Pulsed lasers are now used to evaporate small amounts of materials in thin-film deposition. This pulsed-laser technique can make extremely thin layers (on the order of 1 nm) with unprecedented control. Such thin-layered materials are used in basic research to study the electronic properties of different materials and in the manufacture of integrated circuits, where smaller-sized circuit elements are desirable.

A potentially important use of lasers is in fusion research. The greatest problem in achieving controlled fusion in the laboratory is the difficulty in containing enough nuclei within a confined space for a sufficient length of time in order for nuclei to fuse and thereby produce energy. In one scheme, known as *inertial confinement,* a pellet of deuterium and tritium would be induced into fusion by an intense burst of laser light coming simultaneously from many directions. Fusion will be discussed in Chapter 13.

Holography

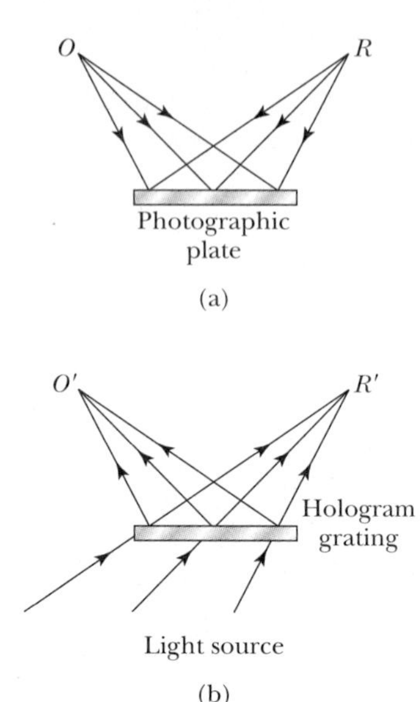

FIGURE 10.17 How a hologram is produced. (a) The production of interference fringes on a photographic plate, using interference of reference (R) and object (O) beams. (b) Illumination of the plate from the opposite side produces a real image of O at O'.

The field of **holography** is one in which lasers have already been put to good use. The basic mechanism for constructing and viewing a hologram is shown in Figure 10.17. Consider laser light emitted by a **reference source** R. Through a simple combination of mirrors and lenses, this light can be made to strike both a piece of photographic plate and an object O. Because the laser light is coherent, the image on the film will essentially be an interference pattern with the interfering beams coming from the reference source and object. After exposure this interference pattern is a hologram, and when the hologram is illuminated from the other side (as in Figure 10.17b), a real image of O is formed. The fact that not only intensity but also phase information is contained in the hologram means that a full three-dimensional image is formed. The other fascinating feature of holograms results from the fact that, if the lenses and mirrors are properly situated, light from virtually every part of the object will strike every part of the film, along with some of the reference beam. This means that each portion of the film contains enough information to reproduce practically the whole object! It is indeed possible to reconstruct a view of an object from a tiny piece of a hologram, although one generally loses some of the three-dimensional effect in that case. Figure 10.18 shows a close-up of a holographic plate and several images produced from it.

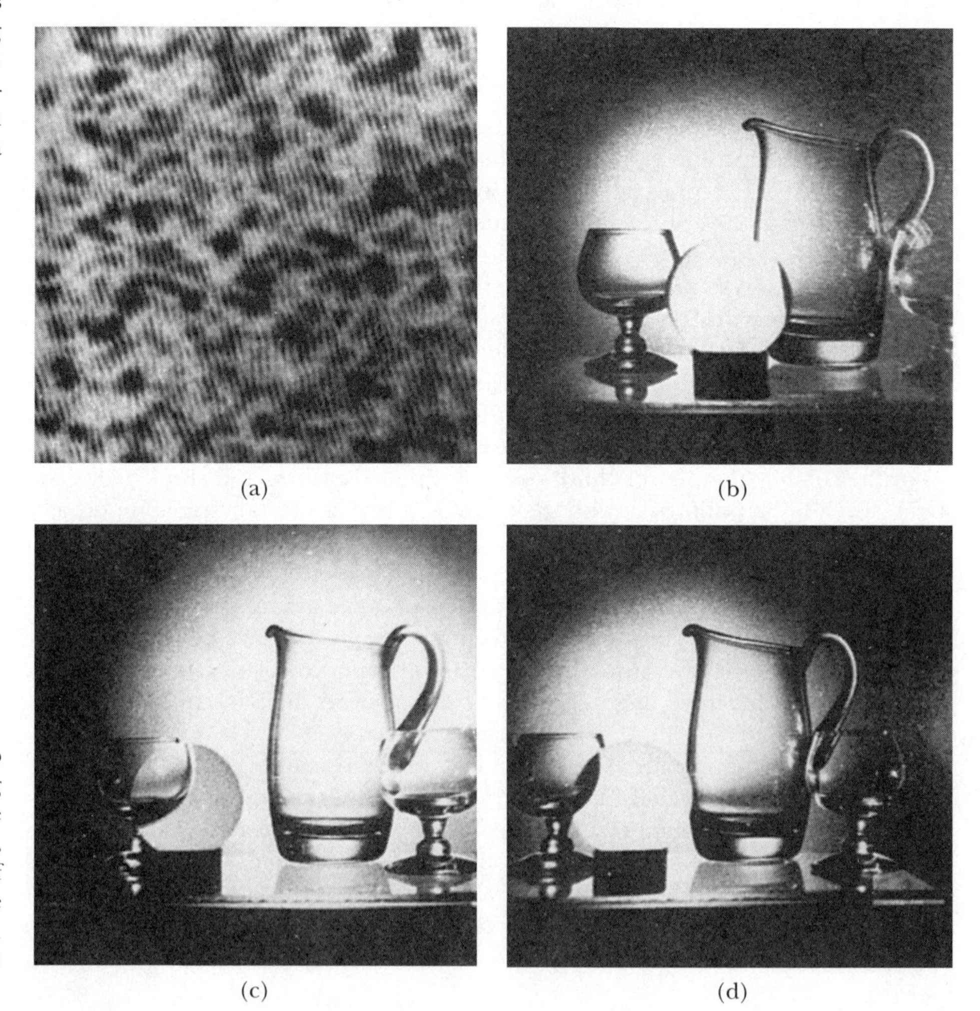

FIGURE 10.18 (a) A close-up view of a hologram, showing the interference pattern in the photographic plate. (b), (c), (d) Three different views of the image produced by the hologram in (a). *From Smith, H.M., Principles of Holography, 2nd ed. New York: Wiley, 1975.*

FIGURE 10.19 Interferometric observation of the growth of a live mushroom through a hologram. *Photo by T. Jeong.*

Transmission, reflection, and white light holograms

We have just described a **transmission hologram,** that is, one in which the reference beam is on the same side of the film as the object and the illuminating beam is on the opposite side. It is possible to make a **reflection hologram** by reversing the positions of the reference and illuminating beams, and if an overlaying series of holograms is made using several lasers of different colors, the result will be a **white light hologram,** in which the different colors contained in white light provide the colors seen in the image. This is what one commonly sees in magazine pictures and on identification cards and credit cards, because the casual observer does not have a laser handy for viewing a monochromatic hologram. Holograms have proven to be useful security tools on credit cards, checks, and so on. Holograms are exceptionally difficult to counterfeit because of the complexity of the interference pattern they contain.

An interesting scientific application of holography is interferometry. Two holograms of the same object produced at different times can be used to detect small motion or growth that could not otherwise be seen. In Figure 10.19 we see a holographic image of a mushroom made seconds earlier superimposed on the mushroom itself. The interference fringes indicate growth patterns. Holographic interferometry is now widely used in industry to scan for imperfections, for example, in machined parts and computer chips. See *The Industrial Physicist,* September 1997, pp. 37–39.

Other Applications

Lasers have been important in medicine for years. They are used in surgery to make precise incisions in different kinds of body tissue, often with not only better precision than conventional means but also reduced bleeding, because the laser tends to help the blood coagulate while cutting. Lasers are used in a number of eye operations, particularly in retinal reattachment and in the treatment of glaucoma (excessive fluid pressure) by burning holes that allow small amounts of fluid to leak out and thereby reduce pressure. There have already been thousands of cases in which laser surgery has prevented blindness.*

*If you or anyone you know ever questions the value of basic research (thinking perhaps that a more direct "practical" approach is always better), consider the following statement from Arthur Schawlow, who worked with Townes in the 1950s to lay the foundation for masers and lasers. Schawlow, who was later awarded the Nobel Prize for his work in laser spectroscopy, wrote: "When we were working on the laser concepts, I had no idea that there was such a thing as a detached retina. If we had been trying to prevent blindness, I do not think we would have concerned ourselves with amplification of stimulated emission by atoms. Research cannot always go directly toward the goals; you sometimes have to explore and hope something will come of it."

The most common application we see in everyday life is in the scanning devices used by supermarkets and other retailers. The reflection of a laser beam from the barcode of a packaged product can be analyzed by an optical scanner. When properly analyzed, the barcode is translated by a microprocessor into the appropriate product and price information. This device is the result of a substantial amount of engineering, which one must appreciate by realizing that laser scanners have a high reliability nearly independent of the angular orientation of the code and the speed with which it is swept over the beam.

There are numerous other common applications of the laser. These include fiber optic communications, compact disc players, and laser printers.

10.3 Structural Properties of Solids

The remainder of this chapter and all of Chapter 11 fall under the general heading of **condensed matter physics.** Under this broad heading fall many kinds of research.* Probably the single most important subfield of condensed matter physics is the study of the electronic properties of solids. The fundamentals of electrical conductivity were covered in Chapter 9. In the last parts of this chapter and all of Chapter 11, you will learn about the remarkable properties of superconductors and semiconductors. Other important work is devoted to understanding the fundamental properties of the structure of solids and their bulk thermal and magnetic properties. It is these fundamental properties we shall study now, along with some of the many applications in science and engineering.

Crystal structures

The structures of different solids are quite varied. Many solids exhibit a **crystal structure,** in which the atoms are arranged in extremely regular, periodic patterns. This was proved in 1912 by Max von Laue, who used the still popular method of x-ray diffraction (see Section 5.1). In an ideal crystal the same basic structural unit is repeated indefinitely throughout space. The set of points in space occupied by atomic centers is called a *lattice.* Figure 10.20 shows the principal lattice types in three dimensions. A perfect crystal is rare. Most solids are in a *polycrystalline* form, meaning that they are made up of many smaller crystals, whose size may be anything from a few atoms on a side to thousands. A solid lacking any significant lattice structure is called *amorphous* (literally "without form"). Common glass is amorphous, and therefore amorphous materials are referred to colloquially as "glasses."

Why do the atoms of a solid arrange themselves in a particular crystal lattice? A qualitative answer to this question is that as the material is cooled and allowed to change from the liquid to solid state, the atoms can each find a place relative to their neighbors that creates the minimum energy configuration. It is rather like an electron being captured by an atom "finding" its way to the ground state. This is why some solids that normally tend to form good crystals will become polycrystalline or amorphous if cooled too quickly from the liquid to the solid state. To give a more quantitative answer to the question of why solids form as they do, let us use as an example the structure of the sodium chloride crystal. The basic cubic structure of sodium chloride is shown in Figure 10.21. Sodium and chlorine easily ionize to form Na^+ and Cl^-, and we may think of the solid as

*Strictly speaking, condensed matter physics includes studies of solids and liquids. Because much more research is done on solids than on liquids, some people still refer to condensed matter physics as "solid state physics." To get an idea of just how significant the role of condensed matter physics is today, consider the fact that in a typical month the number of articles published in *Physical Review B,* the section of *Physical Review* devoted to condensed matter, is nearly equal to the number of articles published in *Physical Review A, C, D,* and *E* (devoted to the other major fields) *combined.*

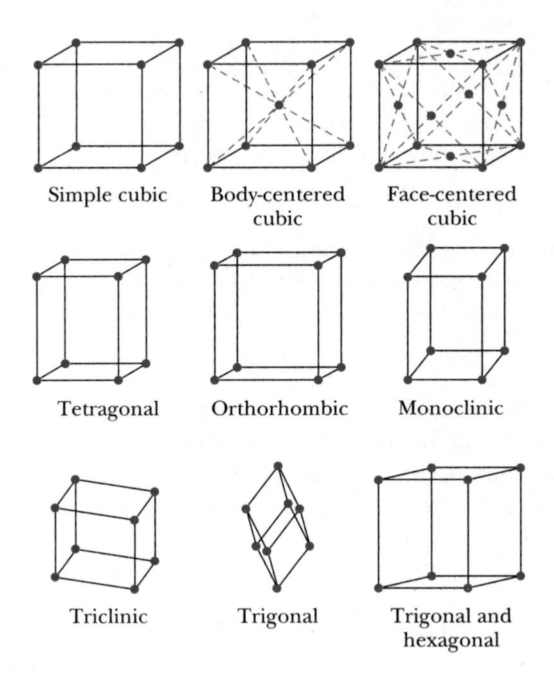

FIGURE 10.20 Some of the crystal lattices found in solids.

a collection of (spherically symmetric) Na^+ and Cl^- ions alternating indefinitely in each of the three orthogonal directions. The spatial symmetry results because there is no preferred direction for bonding the spheres together. The fact that different atoms have different symmetries (consider the shapes of p and d orbitals) should give you a hint of why the many different structures shown in Figure 10.20 exist.

In order for this to be a stable configuration, each ion must experience a net attractive potential energy, which we shall model as

$$V_{\text{att}} = -\frac{\alpha e^2}{4\pi\epsilon_0 r} \tag{10.16}$$

where r is the nearest-neighbor distance (0.282 nm in NaCl). This looks like a normal Coulomb potential energy, except for the introduction of the constant α, known as the **Madelung constant.** V_{att} is the net potential energy of an ion in a lattice due to all the other ions in the lattice. The Madelung constant therefore depends on the type of crystal lattice. In the NaCl crystal, each ion has 6 nearest neighbors of the opposite kind of charge that supply an attractive potential. Therefore, the nearest-neighbor contribution to the Madelung constant is exactly 6. The next nearest neighbors are of **like** charge to the ion we are con-

Madelung constant

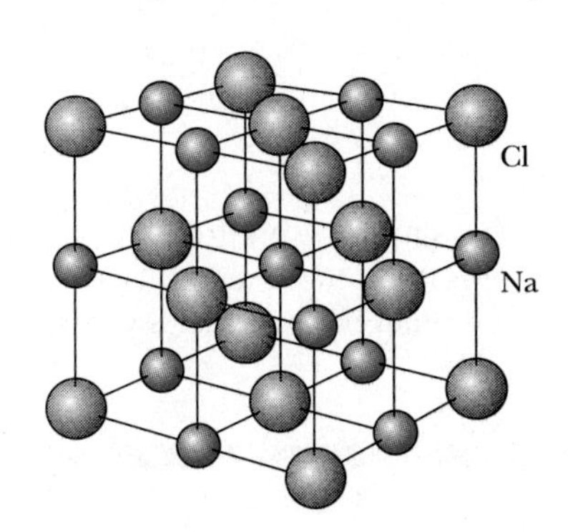

FIGURE 10.21 The NaCl crystal structure.

sidering. There are 12 of them, each located a distance $\sqrt{2}\ r$ away. Those ions contribute $-12/\sqrt{2}$ to the Madelung constant. Next there are 8 ions of the opposite charge $\sqrt{3}\ r$ away, which contribute $8/\sqrt{3}$ to the Madelung constant. Continuing this process, one has an infinite series:

$$\alpha = 6 - 12/\sqrt{2} + 8/\sqrt{3} - \cdots \approx 1.7476 \tag{10.17}$$

In addition to the attractive potential of Equation (10. 16) there is a *repulsive* potential due to the Pauli exclusion principle and the overlap of electron shells. A good theoretical model of the repulsive potential is

$$V_{\text{rep}} = \lambda e^{-r/\rho} \tag{10.18}$$

where λ and ρ are also constants of the particular lattice and compound. The exponential factor in Equation (10.18) is a common feature of "screened" potentials. Because the value of $e^{-r/\rho}$ diminishes rapidly for $r > \rho$, the constant ρ is roughly regarded as the *range* of the repulsive force. We shall see that its value can be calculated from experimental data.

The net potential energy is

$$V = V_{\text{att}} + V_{\text{rep}} = -\frac{\alpha e^2}{4\pi\epsilon_0 r} + \lambda e^{-r/\rho} \tag{10.19}$$

At the equilibrium position $(r = r_0)$, $F = -dV/dr = 0$. Thus

$$0 = \frac{\alpha e^2}{4\pi\epsilon_0 {r_0}^2} - (\lambda/\rho)e^{-r_0/\rho} \tag{10.20}$$

Therefore

$$e^{-r_0/\rho} = \frac{\rho\alpha e^2}{4\pi\epsilon_0\lambda {r_0}^2} \tag{10.20a}$$

and

$$V(r = r_0) = -\frac{\alpha e^2}{4\pi\epsilon_0 r_0}[1 - (\rho/r_0)] \tag{10.21}$$

Typically the repulsive potential is very short range, so that the ratio ρ/r_0 is much less than one (it is about 0.11 for NaCl, see Example 10.3). This is important, because Equation (10.21) indicates that ρ/r_0 must be less than 1 in order for the net potential energy at $r = r_0$ to be negative, as is required.

Example 10.3

The **dissociation energy,** that is, the energy needed to break a NaCl crystal into individual sodium and chlorine atoms, is determined experimentally to be 764.4 kJ/mol at STP (see Table 10.2). Use this value to calculate the range parameter ρ for NaCl.

Solution: First divide the experimental value by Avogadro's number (the number of ion pairs per mole) to obtain a value of 1.269×10^{-18} J/ion pair. Then (again letting the equilibrium position be $r = r_0$) we have $V(r = r_0) = -1.269 \times 10^{-18}$ J. Solving (10.21) for ρ/r_0, we obtain

$$\rho/r_0 = 1 + \frac{4\pi\epsilon_0 r_0 V(r = r_0)}{\alpha e^2}$$

$$= 1 + \frac{(0.282 \text{ nm})(-1.269 \times 10^{-18}\text{ J})}{(9 \times 10^9 \text{ N}\cdot\text{m}^2/\text{C}^2)(1.7476)(1.6 \times 10^{-19}\text{ C})^2}$$

$$= 0.112$$

Therefore $\rho = 0.112 r_0 = 0.0316$ nm, which is in agreement with the value listed in Table 10.2. Indeed, this shows that the repulsive potential is very short range.

TABLE 10.2
Properties of Salt Crystals with the NaCl Structure

	Nearest-Neighbor Separation (nm)	Repulsive Range Parameter ρ (nm)	Dissociation Energy (kJ/mol)
LiF	0.214	0.029	1014
LiCl	0.257	0.033	832.6
LiBr	0.275	0.034	794.5
LiI	0.300	0.037	743.8
NaF	0.232	0.029	897.5
NaCl	0.282	0.032	764.4
NaBr	0.299	0.033	726.7
NaI	0.324	0.035	683.2
KF	0.267	0.030	794.5
KCl	0.315	0.033	694.0
KBr	0.330	0.034	663.5
KI	0.353	0.035	627.5
RbF	0.282	0.030	759.3
RbCl	0.329	0.032	666.8
RbBr	0.345	0.034	638.8
RbI	0.367	0.035	606.6

From C. Kittel, *Introduction to Solid State Physics,* 5th ed. New York: Wiley, 1976, p. 92.

10.4 Thermal and Magnetic Properties of Solids

Thermal Expansion

One of the more ubiquitous properties of solids is **thermal expansion,** the tendency of a solid to expand as its temperature increases. A qualitative understanding can be obtained by studying Figure 10.22, the potential energy curve for an ion bound in a solid. We learned earlier in this chapter (see Figure 10.1) that this potential energy curve is characteristic of the combined attractive and repulsive forces experienced by this ion.

At $T = 0$, the ion is nearly "frozen solid" at $r = r_0$ because it has just the minimum energy possible (the zero-point energy, see Section 6.6). As the temperature of the solid is increased from zero, the average energy of each ion increases. The result is an oscillation between points $r_1(<r_0)$ and $r_2(>r_0)$. To a first approximation the motion is simple harmonic, but it is not *exactly* so. The potential energy for a simple harmonic oscillator is of the form $V = (\frac{1}{2})\kappa\,(r - r_0)^2$, a function that is symmetric about the point $r = r_0$. The potential energy shown in Figure 10.22 is not symmetric. The effect of this is that the mean lattice spacing $r_T \approx (r_2 + r_1)/2$ is slightly greater than r_0 when $T > 0$ and increases gradually with increasing temperature. The bulk effect of an increase in the mean lattice spacing is an overall expansion of the solid (Figure 10.23).

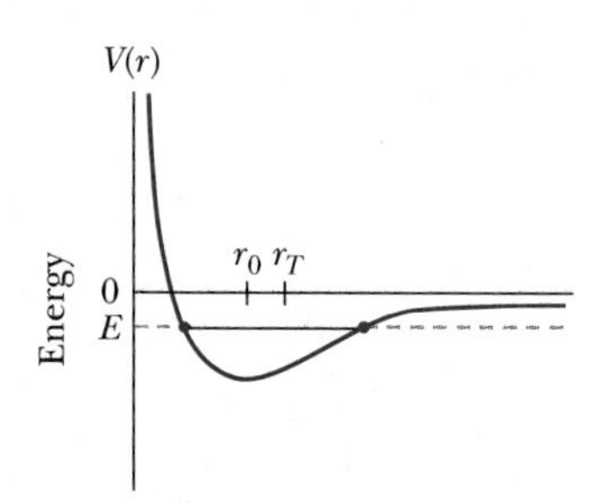

FIGURE 10.22 The asymmetry of the potential energy curve leads to thermal expansion. At $T = 0$ the mean spacing is r_0 but as T increases so does the mean spacing.

Now we shall develop a quantitative model of thermal expansion. Let $x = r - r_0$, so that we will consider small oscillations of an ion about the equilibrium position $x = 0$. A good model of the potential energy close to $x = 0$ is

$$V = ax^2 - bx^3 \tag{10.22}$$

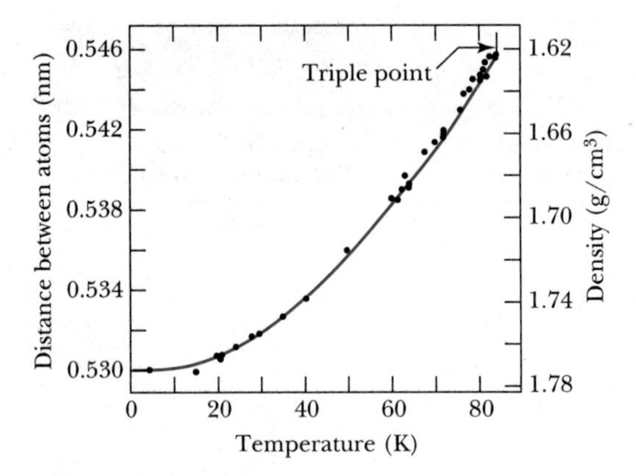

FIGURE 10.23 Thermal expansion of solid argon. The expansion is fairly linear except at very low temperatures. *From O. G. Peterson, D. N. Batchelder, and R. O. Simmons, Physical Review* ***150****, 703 (1966).*

where the x^3 term is responsible for the anharmonicity (that is, the deviation from the standard harmonic oscillator) of the oscillation. The mean displacement $\langle x \rangle$ can be calculated using the Maxwell-Boltzmann distribution function $e^{-\beta V}$:

$$\langle x \rangle = \frac{\int_{-\infty}^{\infty} x e^{-\beta V} \, dx}{\int_{-\infty}^{\infty} e^{-\beta V} \, dx} \tag{10.23}$$

where we have used the usual notation $\beta = (kT)^{-1}$. In the numerator we can use a Taylor expansion for the x^3 term because b is small:

$$x e^{-\beta V} = x e^{-\beta a x^2} e^{+\beta b x^3} = x e^{-\beta a x^2}(1 + \beta b x^3 + \cdots) \approx e^{-\beta a x^2}(x + \beta b x^4)$$

This allows us to evaluate the integral in the numerator of Equation (10.23) easily, because only the even (x^4) term survives integration from $-\infty$ to ∞ (see Appendix 6):

$$\int_{-\infty}^{\infty} e^{-\beta a x^2}(x + \beta b x^4) \, dx = \int_{-\infty}^{\infty} e^{-\beta a x^2} \beta b x^4 \, dx = \frac{3\sqrt{\pi}}{4} b a^{-5/2} \beta^{-3/2} \tag{10.24}$$

Because we are interested only in the first-order dependence on T, it is acceptable to approximate the denominator of Equation (10.23) by

$$\int_{-\infty}^{\infty} e^{-\beta V} \, dx \approx \int_{-\infty}^{\infty} e^{-\beta a x^2} \, dx = \left(\frac{\pi}{a\beta}\right)^{1/2} \tag{10.25}$$

Combining Equations (10.24) and (10.25), we obtain

$$\langle x \rangle = \frac{3b}{4a^2 \beta} = \frac{3bkT}{4a^2} \tag{10.26}$$

Therefore thermal expansion is nearly linear with temperature in the classical limit. At very low temperatures the expansion is nonlinear, because the term calculated in Equation (10.26) vanishes as $T \to 0$. This is in agreement with experiment, as is seen in Figure 10.23.

Thermal Conductivity

Another important property of solids is their **thermal conductivity.** Materials that are good electrical conductors also tend to be good thermal conductors. Why should that be? Evidently the conduction electrons in a solid, because they are free to move around, are primarily responsible for the conduction of heat. We shall see, in fact, that the quantum theory of electrical conductivity developed in Chapter 9 will be necessary in order to develop a good model for heat conduction.

The standard way to define thermal conductivity is in terms of the flow of heat along a solid rod of uniform cross-sectional area A, as is shown in Figure 10.24. It is found experimentally that the flow of heat per unit time along the rod is proportional to A and to the temperature gradient dT/dx. We define the thermal conductivity K to be the proportionality constant, so that

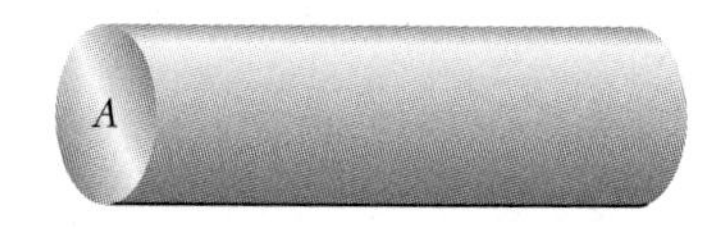

FIGURE 10.24 A uniform rod of cross-sectional area A, used to illustrate thermal conductivity.

$$\frac{dQ}{dt} = -KA\frac{dT}{dx} \tag{10.27}$$

The negative sign in Equation (10.27) is due to the fact that heat flows in a direction opposite to the thermal gradient (i.e., from hotter to colder).

In classical theory the thermal conductivity of an ideal free electron gas* is

$$K = \frac{n\bar{v}\ell c_V}{3N_A} \tag{10.28}$$

where n is the volume density of free electrons, $\bar{v}$ is the mean (thermal) speed, ℓ is the mean free path (see Chapter 9), and c_V is the molar heat capacity. Classically $c_V = (\frac{3}{2})R = (\frac{3}{2})N_A k$, so

$$K = \frac{1}{2}n\bar{v}\ell k \tag{10.29}$$

Because of their apparently close relationship, it is interesting to compare the thermal and electrical conductivities:

$$\frac{K}{\sigma} = \frac{\frac{1}{2}n\bar{v}\ell k}{(ne^2\ell/m\bar{v})} = \frac{m\bar{v}^2 k}{2e^2} \tag{10.30}$$

From classical thermodynamics the mean speed is (see Equation 9.17)

$$\bar{v} = \sqrt{\frac{8kT}{\pi m}}$$

Therefore

$$\frac{K}{\sigma} = \frac{4k^2 T}{\pi e^2} \tag{10.31}$$

*See for example F. Reif, *Statistical Physics.* New York: McGraw-Hill, 1967, pp. 331–333.

Amazingly, then, the ratio K/σ is proportional to T. It is convenient to calculate the *constant* ratio

Wiedemann-Franz law

$$L \equiv \frac{K}{\sigma T} = \frac{4k^2}{\pi e^2} \tag{10.32}$$

Lorenz number

Equation (10.32) is called the **Wiedemann-Franz Law,** and the constant L is the **Lorenz number.** The numerical value of L is about $1.0 \times 10^{-8}\ \text{W} \cdot \Omega \cdot \text{K}^{-2}$. Experiments show that $K/\sigma T$ is indeed constant, but it has a numerical value about 2.5 times higher than predicted by Equation (10.32). Some experimental values are listed in Table 10.3.

What is wrong with our analysis? The only problem is that we have used classical expressions for $\bar{v}$ and c_V. We should replace $\bar{v}$ with the Fermi speed u_F (because only electrons near the Fermi energy will be able to contribute to the conductivity) and replace $c_V = (\frac{3}{2})R$ with the quantum mechanical result [see Equation (9.47)]

$$c_V = \frac{\pi^2 RkT}{2E_F} \tag{10.33}$$

so that Equation (10.28) can be rewritten

$$K = \frac{1}{3}\frac{nu\,\ell}{N_A}\frac{\pi^2 RkT}{2E_F} \tag{10.34}$$

Note that $R = N_A k$ and $E_F = \frac{1}{2}mu_F^2$. Thus

$$K = \frac{n\ell\pi^2 k^2 T}{3mu_F} \tag{10.35}$$

TABLE 10.3
Lorenz Number $L = K/\sigma T$ in Units of $10^{-8}\ \text{W} \cdot \Omega/\text{K}^2$ at Temperatures 273 K and 373 K

Metal	273 K	373 K
Ag	2.31	2.37
Au	2.35	2.40
Cd	2.42	2.43
Cu	2.23	2.33
Ir	2.49	2.49
Mo	2.61	2.79
Pb	2.47	2.56
Pt	2.51	2.60
Sn	2.52	2.49
W	3.04	3.20
Zn	2.31	2.33

From C. Kittel, *Introduction to Solid State Physics,* 5th ed. New York: Wiley, 1976, p. 178.

The proof of the validity of Equation (10.35) will be the correct quantum version of the Wiedemann-Franz Law, that is, one in agreement with experimental results. As before, $L = K/\sigma T$, but now we use $\sigma = ne^2\ell/mu_F$:

$$L = \frac{K}{\sigma T} = \frac{n\ell\pi^2k^2}{3mu_F}\frac{mu_F}{ne^2\ell}$$

$$L = \frac{\pi^2k^2}{3e^2} \tag{10.36}$$

Quantum Lorenz number

The quantum mechanically correct Lorenz number contains the same physical constants, k^2 and e^2, as the classical one, but strangely it is higher by a factor $\pi^3/12$. This is just enough to bring the Lorenz number to a numerical value of 2.45×10^{-8} W $\cdot$ Ω $\cdot$ K^{-2}, which is in agreement with experiment!

Magnetic Properties

The study of magnetic properties of solids constitutes an important subfield of solid state physics. Solids are characterized by their intrinsic magnetic moments (or lack thereof) and their responses to applied magnetic fields. Materials with a net magnetic moment even in zero applied magnetic field are called **ferromagnets.** Ferromagnets are sometimes referred to as "permanent magnets," although as we shall see later this is somewhat of a misnomer. In a **paramagnet** there is a net magnetic moment only in the presence of an applied field. The magnetic dipoles in a paramagnet tend to align themselves with the applied field. In a **diamagnet,** on the other hand, there is a (usually weak) tendency to have an induced magnetic moment opposite to the applied field. We shall consider each of these three principal kinds of materials separately.

Ferromagnets, paramagnets, and diamagnets

A useful quantity in studying magnetic materials is the **magnetization** M, which we define as the net magnetic moment per unit volume. Then the **magnetic susceptibility** χ is defined by

Magnetization

$$\chi = \frac{\mu_0 M}{B} \tag{10.37}$$

Magnetic susceptibility

In other words, we may think of magnetic susceptibility as the induced magnetic moment per applied field, with a proportionality constant equal to the permeability constant μ_0. The magnetic susceptibility is positive for paramagnets and negative for diamagnets. Notice that χ is a dimensionless quantity (see Problem 37).

Diamagnetism

The behavior of a diamagnet may seem contrary to common sense, because in a diamagnet the magnetization opposes the applied field. However, one may think of the material as a whole responding to the applied field according to Lenz's law. There are generally no unpaired electron spins in a diamagnet, and this keeps individual spins from aligning with the field. A crude model will perhaps clarify the situation further. Consider two atoms (supposedly in the same solid), identical in every respect, except that an outer electron orbits in a clockwise direction in one and counterclockwise in the other, as shown in Figure 10.25. Ultimately we will want to think of these as two electrons in the same atom, but we

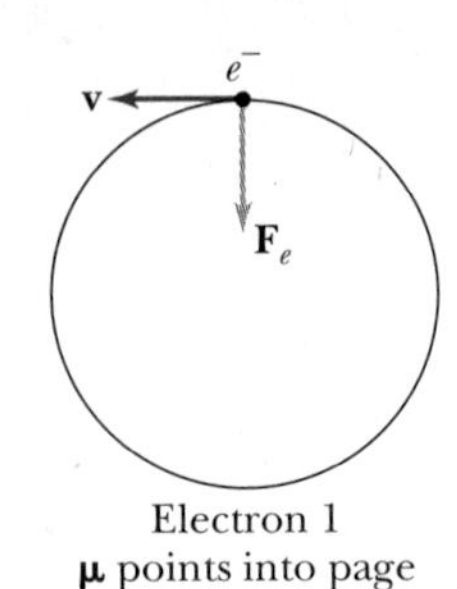

Electron 1
μ points into page

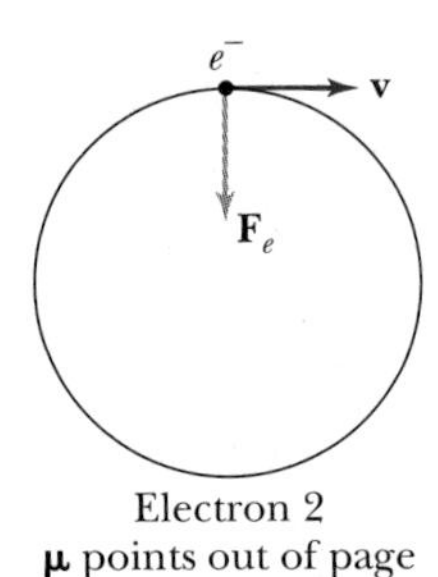

Electron 2
μ points out of page

FIGURE 10.25 Opposite magnetic moments used to illustrate diamagnetism.

separate them now to simplify the picture. Electron 1, orbiting counterclockwise, produces a magnetic moment pointing into the page, while electron 2 produces a magnetic moment pointing out of the page. Each electron initially experiences the same force, the Coulomb force of attraction F_e toward the nucleus, which points down in Figure 10.25. Now we apply a magnetic field pointing into the page to both atoms. The resulting magnetic force ($-e\mathbf{v} \times \mathbf{B}$) is up for electron 1 and down for electron 2. This magnetic force causes a reduction in the net force on electron 1 (and hence a reduction in the atom's magnetic moment) and an increase in the net force on electron 2 (and a corresponding increase in that atom's magnetic moment). Therefore, there will be an imbalance in the magnetic moments, in favor of electron 2, so there will be a net magnetic moment pointing out of the page, opposed to the applied magnetic field. The result, then, on a pair of electrons with opposite orbits (or, more importantly, a pair of electrons in a filled shell with opposing spins) is an induced magnetic moment opposite to the applied field. This accounts for the negative magnetic susceptibility in diamagnetic materials.

Paramagnetism

In a paramagnet there exist unpaired magnetic moments that can be aligned by an external field. This is the case for rare earth elements and for many transition metals. The paramagnetic susceptibility χ is strongly temperature dependent. We can determine the temperature dependence by considering a collection of N unpaired magnetic moments per unit volume. At a given temperature there will be N^+ moments aligned parallel to the applied field and N^- moments aligned antiparallel to the applied field. The energy associated with a magnetic moment is $V = -\boldsymbol{\mu} \cdot \mathbf{B}$, so $V = -\mu B$ for a parallel alignment and $V = +\mu B$ for an antiparallel alignment. In the classical limit (i.e., at ordinary temperatures) the distribution of magnetic moments will be governed by Maxwell-Boltzmann statistics, so that

$$N^+ = ANe^{\beta\mu B} \quad \text{and} \quad N^- = ANe^{-\beta\mu B} \tag{10.38}$$

where A is a normalization constant, and as usual $\beta \equiv (kT)^{-1}$. Then the net magnetic moment (per unit volume) μ_{net} is

$$\mu_{\text{net}} = \mu(N^+ - N^-)$$

$$\mu_{\text{net}} = \mu AN[e^{\beta\mu B} - e^{-\beta\mu B}] \tag{10.39}$$

Rather than directly calculating χ from μ_{net}, it is useful first to eliminate the constant A by considering $\bar{\mu}$, the mean magnetic moment per atom:

$$\bar{\mu} = \frac{\mu_{\text{net}}}{N} = \frac{\mu_{\text{net}}}{N^+ + N^-} = \frac{\mu AN(e^{\beta\mu B} - e^{-\beta\mu B})}{AN(e^{\beta\mu B} + e^{-\beta\mu B})}$$

$$\bar{\mu} = \mu\frac{e^{\beta\mu B} - e^{-\beta\mu B}}{e^{\beta\mu B} + e^{-\beta\mu B}} = \mu \tanh(\beta\mu B) \tag{10.40}$$

Because we used the Maxwell-Boltzmann distribution, this expression should only be valid for $T \gg 0$. Note that this also means $\mu B \ll kT$, in which case $\tanh(\beta\mu B) = \tanh(\mu B/kT) \approx \mu B/kT$ and $\bar{\mu} = \mu^2 B/kT$. Therefore in the classical limit (note $M = N\bar{\mu}$)

Curie law

$$\chi = \frac{\mu_0 M}{B} = \frac{\mu_0 N\bar{\mu}}{B} = \frac{\mu_0 N\mu^2}{kT} \tag{10.41}$$

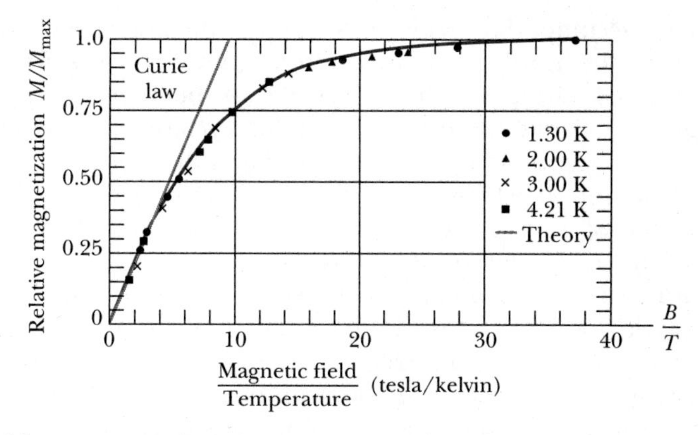

FIGURE 10.26 An actual plot of magnetization for potassium chromate sulfate, a paramagnetic salt. Note the agreement with the Curie law for low magnetic fields. *From W. E. Henry, Physical Review* **88,** *559 (1952).*

Equation (10.41) is called the **Curie law** and is often simply stated as $\chi = C/T$, where $C = \mu_0 N\mu^2/k$ is a constant (the **Curie constant**) for a given paramagnetic material. Sample magnetization curves are shown in Figure 10.26, where the utility and the limitations of the Curie law can be seen. In Figure 10.26 it is clear that the M vs. B curve is nearly linear over a wide range of magnetic fields. It is apparent that the Curie law breaks down at higher values of B, when the magnetization reaches a "saturation point" at which as many magnetic moments as possible have been aligned.

Example 10.4

Using the Bohr magneton $\mu_B = e\hbar/2m = 9.27 \times 10^{-24}$ J/T as a typical magnetic moment and $B = 0.5$ T,

(a) find T at which $\mu B = 0.1kT$.

(b) For $\mu B = 0.1kT$ compare $\tanh(\beta\mu B)$ with $\beta\mu B$ and thereby check the suitability of this value of T as a "classical" temperature.

(c) Make the same comparison in the expression $\tanh(\beta\mu B) \approx \beta\mu B$ at $T = 100$ K.

Solution: (a) We are given that $\mu B = 0.1\ kT$. We let $\mu = \mu_B$ and solve for T,

$$T = \frac{10\mu B}{k} = \frac{10(9.27 \times 10^{-24}\ \text{J/T})(0.5\ \text{T})}{1.38 \times 10^{-23}\ \text{J/K}} = 3.36\ \text{K}$$

(b) Now $\beta\mu B = \dfrac{\mu B}{kT} = 0.10$.

$$\tanh(\beta\mu B) = \tanh(0.10) = 0.10$$

to two significant digits. Therefore we conclude that the approximation is a good one for most purposes, even at this low temperature.

(c) Now $\beta\mu B = \dfrac{\mu B}{kT} = \dfrac{(9.27 \times 10^{-24}\ \text{J/T})(0.5\ \text{T})}{(1.38 \times 10^{-23}\ \text{J/K})(100\ \text{K})}$

$$= 3.36 \times 10^{-3}.$$

$$\tanh(\beta\mu B) = \tanh(3.36 \times 10^{-3}) = 3.36 \times 10^{-3}$$

so that $\beta\mu B$ and $\tanh(\beta\mu B)$ are the same to three significant figures. At these relatively high temperatures the approximation is an excellent one.

Ferromagnetism

Ferromagnetic materials are fairly rare. Of all the elements only five (Fe, Ni, Co, Gd, Dy) are ferromagnetic. A number of compounds are ferromagnetic, including some that do not contain any of these ferromagnetic elements. In the most powerful magnetic compounds (such as $Nd_2Fe_{14}B$) the magnetic field at the surface can exceed 1 T. In order to have a ferromagnet it is necessary to have not only unpaired spins, but also sufficient interaction between the magnetic moments so that, by their mutual interaction, a high degree of magnetic order is maintained. Opposing the maintenance of order is the continual randomizing effect of thermal motion, which contributes to the eventual "running down" of the magnetization of a ferromagnet. This phenomenon should not be surprising, in light of the second law of thermodynamics.

Sufficient thermal agitation at an elevated temperature can completely disrupt the magnetic order, to the extent that above a certain temperature (known as the **Curie temperature** T_C) a ferromagnet changes to a paramagnet. T_C for various ferromagnetic materials are listed in Table 10.4. At temperatures approaching T_C (from below), the magnetization of the ferromagnetic material begins to drop significantly as the thermal motion begins to disrupt the long-range magnetic order.

Curie temperature

Ferromagnetic materials are used widely in scientific research and industry. Electrical generators use permanent magnets to take advantage of Faraday's law. Electric motors use permanent magnets in the reverse process, to turn electrical current into mechanical motion. Scientific uses include the wiggler magnets described at the end of Section 10.2, low-field nuclear magnetic resonance, and Stern-Gerlach experiments (Section 7.4). In the future, one likely application will be in magnetically levitated transport systems (Section 10.6).

Antiferromagnetism and Ferrimagnetism

There are two more exotic kinds of magnetism. In **antiferromagnetic** materials, adjacent magnetic moments have opposing directions, as seen in Figure (10.27). The net effect is zero net magnetization below the ordering temperature (similar to T_C in ferromagnetic materials) called the *Neel temperature*, T_N. Above T_N, antiferromagnetic materials become paramagnetic. Negative ions seem to be most effective in providing the mechanism for antiparallel alignment, as in the antiferromagnetic materials MnO, FeO, MnS, MnF_2, and $NiCl_2$. In a **ferrimagnetic** substance a similar antiparallel alignment occurs, except that there are two different kinds of positive ions present. Thus the antiparallel moments do not precisely cancel, leaving a small net magnetization. The most common example of ferrimagnetic order is in magnetite, $FeO \cdot Fe_2O_3$.

Today magnetic materials are found in many applications outside the research laboratory. The small size and extreme stability of magnetic domains (small regions of magnetization) in many materials makes them ideal for any device that requires data storage and retrieval. Computers, electronic instruments, and audio and video tapes all take advantage of magnetic materials in this way.

TABLE 10.4 Selected Ferromagnets, with Curie Temperatures T_C

Material	T_C (K)
Fe	1043
Co	1388
Ni	627
Gd	293
Dy	85
$CrBr_3$	37
Au_2MnAl	200
Cu_2MnAl	630
Cu_2MnIn	500
EuO	77
EuS	16.5
MnAs	318
MnBi	670
$GdCl_3$	2.2

From F. Keffer, *Handbuch der Physik*, **18**, pt. 2, New York: Springer-Verlag, 1966 and P. Heller, *Rep. Progr. Phys.*, **30**, (pt. II), 731 (1967).

10.5 Superconductivity

Superconductivity is perhaps the most remarkable phenomenon ever studied in solid state physics. For some time after its discovery over 80 years ago, superconductivity was a curiosity for research scientists with little potential for practical use. In recent years, particularly with the advent of high-temperature superconductors (i.e., with temperatures exceeding 100 K), this has changed. Today superconductors are used in a wide variety of applications. Also, the prospect of even higher-temperature superconductors makes it essential that physicists continue to study and learn more about their properties. We shall describe what is known today about superconductors in this section and consider the applications in Section 10.6.

Superconductivity is a special physical state characterized by two distinctive macroscopic features. The first of these is zero resistivity. This property was discovered in solid mercury by the great Dutch physicist Heike Kamerlingh Onnes in Leiden in 1911. He made this discovery shortly after liquifying helium, itself a significant accomplishment (remember that the boiling point of helium is 4.2 K at a pressure of 1 atmosphere). Kamerling Onnes could achieve temperatures approaching 1 K by reducing the pressure of the vapor surrounding his liquid helium. Today much lower temperatures are achieved by more sophisticated means (see the Special Topic on low-temperature methods). For his work in low-temperature physics Kamerlingh Onnes was awarded the Nobel Prize for physics in 1913.

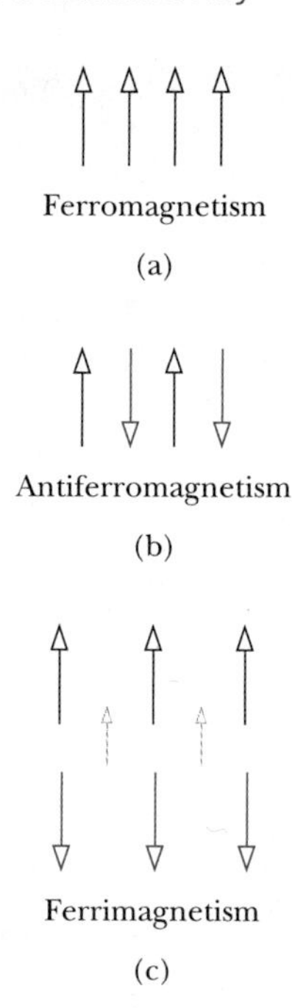

FIGURE 10.27 The alignment of spins in ferromagnets, antiferromagnets, and ferrimagnets.

Figure 10.28 illustrates how the resistivity of a superconductor differs from that of a normal conductor. In a superconductor the resistivity drops abruptly to zero at what is called the *critical* (or *transition*) *temperature*, T_c. It is important to realize that the resistivity is not merely very low; it really is zero. In the years 1956–1958 a group of British physicists led by S. C. Collins established a current in a superconducting ring and allowed it to flow with no external power source. It lasted until they were tired of watching it (about $2\frac{1}{2}$ years) with no detectable loss of current. Extrapolating from the uncertainty in their measuring instruments, they calculated that there would be current remaining after at least 100 million years (see Problem 52). By contrast, the current in a similar loop made of copper would be virtually gone in seconds.

Different superconductors have different transition temperatures, as seen in Table 10.5. Niobium is the champion of the pure elements, with $T_c = 9.25$ K. Notice that copper, silver, and gold, three of the four best conductors at room temperature, are not superconductors. This illustrates a rule of thumb that holds true with rare exceptions: the best conductors make the worst superconductors. Another interesting result is that superconducting behavior, like chemical behavior, tends to be similar within a given column of the periodic table. This

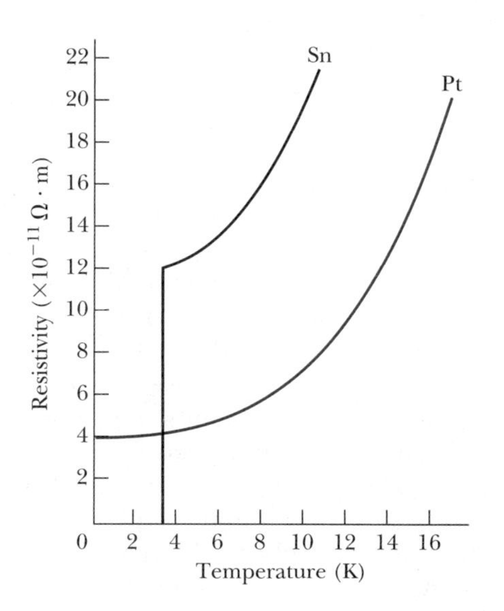

FIGURE 10.28 Resistivity of a normal conductor (platinum) and a superconductor (tin) at low temperature. The resistivity of tin drops dramatically to zero at its T_c, 3.7 K.

TABLE 10.5
Superconductivity Parameters of the Elements

Superconducting transition temperatures and critical fields

Upper number: Transition temperature in K

Lower number: Critical magnetic field at absolute zero in 10^{-4} tesla

Li	Be											B	C	N	O	F	Ne
Na	Mg											Al 1.18 105	Si	P	S	Cl	Ar
K	Ca	Sc	Ti 0.40 56	V 5.40 1408	Cr	Mn	Fe	Co	Ni	Cu	Zn 0.85 54	Ga 1.08 58	Ge	As	Se	Br	Kr
Rb	Sr	Y	Zr 0.61 47	Nb 9.25 2060	Mo 0.92 96	Tc 7.77 1410	Ru 0.49 69	Rh	Pd	Ag	Cd 0.52 28	In 3.41 282	Sn 3.72 505	Sb	Te	I	Xe
Cs	Ba	La 6.00 1046	Hf 0.13 13	Ta 4.47 829	W 0.02 1.15	Re 1.70 200	Os 0.66 70	Ir 0.11 16	Pt	Au	Hg 4.15 411	Tl 2.38 178	Pb 7.20 803	Bi	Po	At	Rn
Fr	Ra	Ac															

Ce	Pr	Nd	Pm	Sm	Eu	Gd	Tb	Dy	Ho	Er	Tm	Yb	Lu 0.1 350
Th 1.38 1.60	Pa 1.4	U	Np	Pu	Am	Cm	Bk	Cf	Es	Fm	Md	No	Lr

From B. W. Roberts, *Properties of Selected Superconductive Materials.* Supplement, NBS Technical Note 983, Washington D.C.: U.S. Government Printing Office, 1978.

should not be surprising, for it is the outermost electrons that are responsible for both chemical reactions and electrical conduction.

Meissner effect

The second important macroscopic phenomenon associated with superconductivity is the **Meissner effect,** discovered by W. Meissner and R. Ochsenfeld in 1933. Succinctly stated, the Meissner effect is the complete expulsion of magnetic flux from within a superconductor. To do this it is necessary for the superconductor to generate currents, called *screening currents.* There is just enough current generated to expel the magnetic flux one tries to impose upon it. One can therefore view the superconductor as a perfect diamagnet, with a magnetic susceptibility $\chi = -1$. In Figure 10.29 we see a demonstration of the Meissner effect in which the induced currents within the superconductor create a magnetic field that opposes the field of a cubical magnet, thereby providing sufficient force to float it against gravity. There is an equilibrium position for the magnet, where the gravitational force on it is just balanced by the magnetic force at that distance from the superconductor.

FIGURE 10.29 Levitation of a cubical magnet over superconducting $YBa_2Cu_3O_7$ cooled to 77 K (well below its T_c of 93 K), illustrating the Meissner effect. Screening currents are generated in the superconductor, which provide a magnetic field to oppose the field of the magnet and thereby suspend it. *© Richard Megna, Fundamental Photographs, NYC*

The Meissner effect only works to a certain point, however. If a particular value of magnetic field (called the **critical field** B_c) is exceeded, magnetic flux does penetrate the material, and the superconductivity is lost until the magnetic field is reduced to below B_c. When the magnetic flux penetrates, the zero resistivity property is also lost, showing that zero resistance and the Meissner effect do indeed go together in a superconductor.

Critical field

The critical field is also different for different superconductors. Superconductors with higher transition temperatures tend to have higher critical fields, although there are some exceptions. The critical field varies with temperature as shown in Figure 10.30. Just below T_c the critical field is low; that is, it takes very little magnetic field to eliminate the superconductivity. The fact that B_c for pure metals even at absolute zero is only on the order of 0.1 tesla is important, because one often encounters higher fields in laboratory situations. The critical field places a strict limit on how a particular superconductor can be used. Current-carrying wires generate magnetic fields, both inside and outside the wire. Therefore, if one wishes to use a superconducting wire to carry current without resistance, there will be a limit (known as the **critical current**) to the current that can be used. These effects severely limited the applications of superconductors for several decades.

Critical current

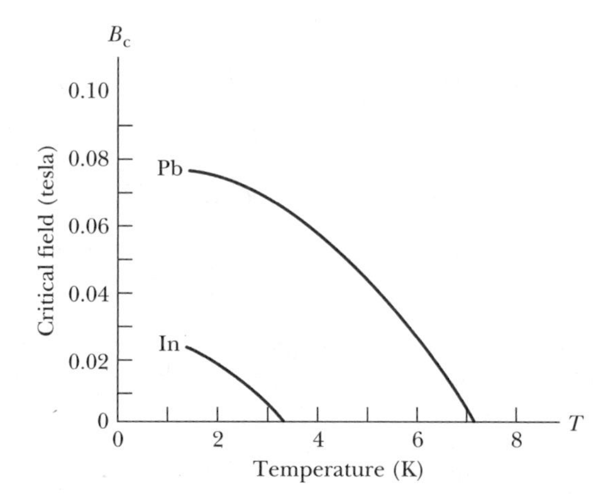

FIGURE 10.30 The temperature dependence of the critical fields of two superconductors. Note that B_c approaches zero as T approaches T_c (3.4 K for In and 7.2 K for Pb).

Type I and type II superconductors

The superconducting state we have just described is that of the pure metals Hg, Al, and many others. They are collectively known as **type I superconductors.** In **type II superconductors** (this category includes most superconducting alloys) there are two critical fields, a lower critical field B_{c1} and an upper critical field B_{c2}. Below B_{c1} and above B_{c2}, type II superconductors behave in the same manner as type I superconductors below and above B_c. Between B_{c1} and B_{c2}, however (known as the *vortex state*), there is a partial penetration of magnetic flux, as is shown in Figure 10.31, although the zero resistance property is generally not lost. The good news is that B_{c2} can sometimes be very high, hundreds of a tesla or more. The bad news is that B_{c1} is seldom more than a few *hundredths* of one tesla. This must be kept in mind when considering applications of superconducting materials that depend on either the zero resistance property or the Meissner effect.

The Meissner effect may remind you of a phenomenon from classical physics known as *Lenz's law,* which states that a changing magnetic flux generates a current in a conductor in such a way that the current produced will oppose the change in the original magnetic flux. In classical physics the current lasts only as long as the magnetic flux is changing (Faraday's law). One might think that in a superconductor the current simply persists because of the zero resistance property, but that is not what happens, and a simple experiment serves to demonstrate. One can impose a constant magnetic field on a material above its T_c, so that initially there is no current. If the material is then cooled to below T_c, the field is expelled instantly! This shows that there is really something unique happening in a superconductor, something that cannot be explained by classical physics.

Isotope effect

What makes a superconductor work? For years there were only vague guesses. In 1950 a phenomenon known as the **isotope effect** was discovered, and this eventually helped lead to a successful theory. Many superconductors follow the equation

$$M^{0.5}T_c = \text{constant} \tag{10.42}$$

where M is the mass of the particular superconducting isotope. This means that T_c is just a bit higher for lighter isotopes. For example, a mercury sample with an average mass per atom of 199.5 u has $T_c = 4.185$ K, and a mercury sample with an average mass per atom of 203.4 u has $T_c = 4.146$ K. In other elements this general trend is followed, although for some the exponent differs slightly from 0.5.

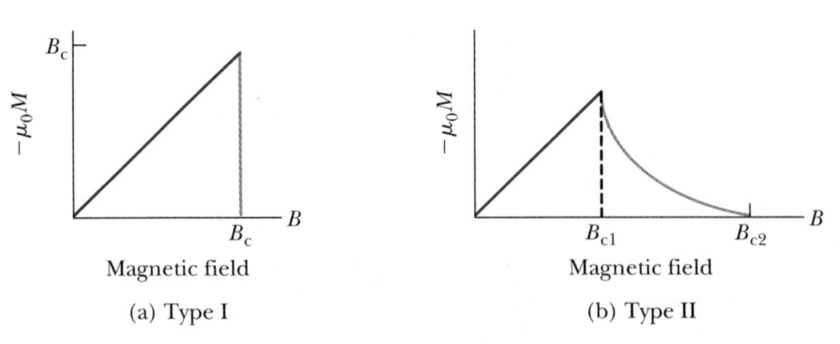

FIGURE 10.31 A comparison of the temperature dependence of critical fields for (a) type I and (b) type II superconductors. In type I superconductors below B_c, we have $-\mu_0 M = B$, corresponding to $\chi = -1$, or complete expulsion of magnetic flux. Up to B_{c1}, the type II material expels all magnetic flux, and above B_{c2} it allows complete penetration. Between B_{c1} and B_{c2} (the vortex state) there is partial flux penetration.

Left to right, John Bardeen (1908–1991), Leon Cooper (1930–), and J. Robert Schrieffer (1931–). In the 1950s these physicists developed the first successful comprehensive theory of superconductivity. Their theory was based on the pairing of electrons (Cooper pairs) through the electron-phonon interaction and correctly predicted a number of experimental results. *AIP/Niels Bohr Library.*

What the isotope effect indicates is that the lattice ions are important in the superconducting state. This is at odds with the classical model of conduction, which leads one to believe that zero resistance can result only from zero interaction between electrons and lattice ions.

A very successful theory* was developed in the mid-1950s by John Bardeen, Leon Cooper, and Robert Schrieffer (Nobel Prize, 1972) and is referred to by their initials: **BCS.** The two principal features of the BCS theory are (1) that electrons form pairs (**Cooper pairs**), which propagate throughout the lattice, and (2) that such propagation is without resistance because the electrons move in resonance with the lattice vibrations (**phonons**, so called because, like *photons* of electromagnetic radiation, they represent quanta of energy). Hence the interaction described by the BCS theory is known as the **electron–phonon interaction.**

BCS theory
Cooper pairs

Phonons

Electron–phonon interaction

How is it possible for two electrons to form a coherent pair? As a first approximation consider the crude model shown in Figure 10.32, in which two electrons propagate in tandem along a single lattice row. Each of the two electrons experiences a net attraction toward the nearest positive ion. If the electrons get too close to each other in the region between ions, they are repelled by their mutual Coulomb force. There is a kind of balance between attraction and repulsion, just as we saw in our study of molecular bonding. In this way relatively stable electron pairs can be formed. The two fermions (electrons) combine to form a boson. Then the collection of these bosons in a bulk sample condense to form the

*J. Bardeen, L. N. Cooper, and J. R. Schrieffer, *Phys. Rev.* **108,** 1175–1204 (1957).

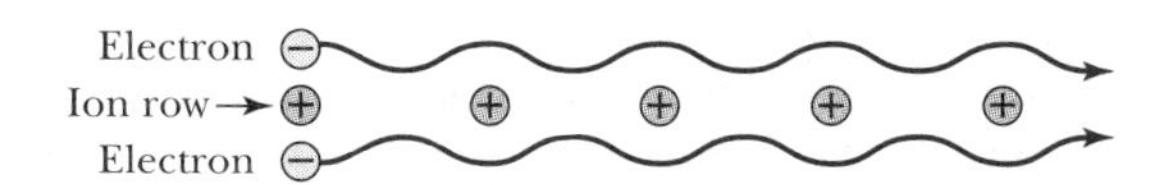

FIGURE 10.32 A schematic of electron motion around a one-dimensional lattice, showing how the attraction of the two paired electrons varies as they propagate through the lattice.

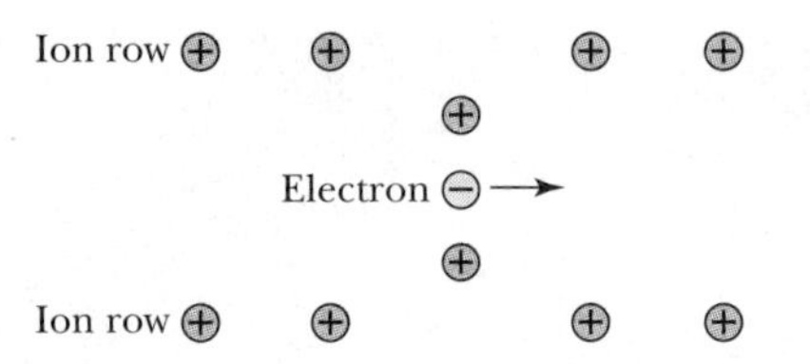

FIGURE 10.33 Propagation of a single electron between rows of a two-dimensional lattice.

superconducting state, similar to the Bose condensate superfluid helium discussed in Section 9.7.

How can the zero resistivity property be explained? Even at low temperatures there is some ionic motion (remember that a harmonic oscillator has a zero-point energy of $\frac{1}{2}\hbar\omega$). That is why one would expect some resistance, even at the lowest temperatures. An electron moving through a wire should eventually collide with an ion. In this inelastic collision, electrical energy is converted to thermal energy and the wire heats up.

But if we neglect for a moment the second electron in the pair, we can understand how a single electron can travel between adjacent rows of ions without transfer of energy (Figure 10.33). The Coulomb attraction between the electron and ions causes a deformation of the lattice, which propagates along with the electron. This propagating wave is associated with phonon transmission, and the electron-phonon resonance allows the electron (along with its pair elsewhere in the lattice) to move without resistance. Keep in mind that this model is crude. The complete BCS theory contains very sophisticated mathematics and is based solidly on the foundations of the quantum theory.

The BCS theory does, in fact, predict successfully several other observed phenomena. First, it predicts an isotope effect, with an exponent [Equation (10.42)] very close to 0.5. Second, it gives a critical field varying with temperature as

$$B_c(T) = B_c(0)\left[1 - \left(\frac{T}{T_c}\right)^2\right] \tag{10.43}$$

which is in agreement with experiment (see Figure 10.30). It also accounts for the fact that the metals with higher resistivity at room temperature tend to be better superconductors. The BCS theory can be used to show that the magnetic flux through a superconducting ring is quantized in the form

Quantum fluxoid

$$\Phi = n\Phi_0 = \frac{nh}{2e} \tag{10.44}$$

where n is an integer and $\Phi_0 = h/2e \approx 2.068 \times 10^{-15}$ Wb is known as the **quantum fluxoid.** The quantization of magnetic flux, confirmed experimentally by B. S. Deaver, Jr. and W. M. Fairbank in 1961, is the basis for the Josephson junction (see Section 10.6).

Another correct prediction of the BCS theory concerns the energy gap (E_g) between the ground state (the superconducting state) and first excited state for conduction electrons. Electrons above the ground state can no longer move with zero resistance. Basically this means that E_g is the energy needed to break a Cooper pair apart, and the effect is that the larger the energy gap, the more stable the superconductor. The BCS theory predicts that

$$E_g(0) \approx 3.54\ kT_c \tag{10.45}$$

TABLE 10.6
Energy Gaps in Superconductors, at $T = 0$

Upper number: $E_g(0)$ in 10^{-4} eV
Lower number: $E_g(0)/k_BT_c$

										Al 3.5 3.4	Si
Sc	Ti	V 15.8 3.4	Cr	Mn	Fe	Co	Ni	Cu	Zn 2.3 3.2	Ga 3.3 3.5	Ge
Y	Zr	Nb 30.3 3.80	Mo 2.7 3.4	Tc	Ru	Rh	Pd	Ag	Cd 1.4 3.2	In 10.6 3.6	Sn 11.2 3.5
La 19 2.3	Hf	Ta 13.9 3.6	W	Re	Os	Ir	Pt	Au	Hg 16.5 4.6	Tl 7.32 3.6	Pb 26.7 4.3

From N. W. Ashcroft and N. D. Mermin, *Solid State Physics*. Philadelphia: Saunders College Publishing, 1976, p. 745.

at $T = 0$ (see Table 10.6). This prediction is easily verified by studying the absorption of electromagnetic radiation by superconductors. Only photons with energy greater than or equal to E_g are absorbed (this was first observed by Michael Tinkham in 1960). At higher temperatures (just below T_c) BCS theory predicts

$$E_g(T) \approx 1.74\, E_g(0)\,(1 - T/T_c)^{1/2} \tag{10.45a}$$

Again, the agreement with experiment is striking (Figure 10.34).

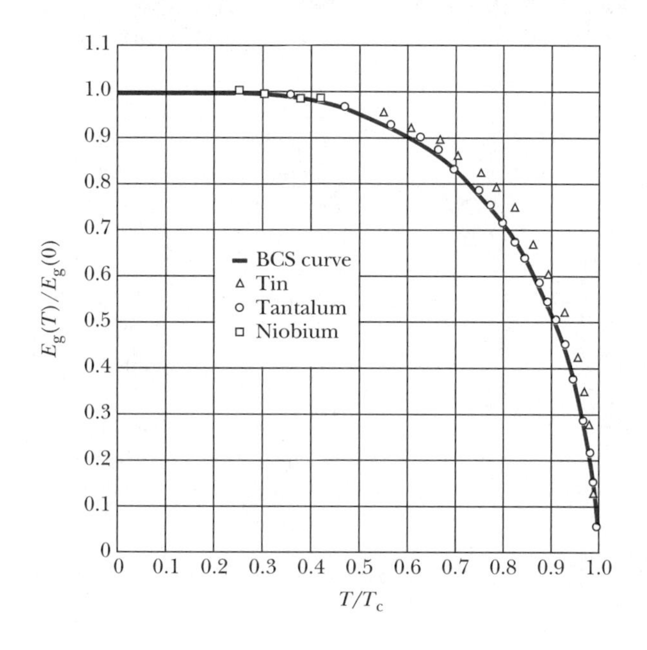

FIGURE 10.34 The superconductive energy gap as a function of temperature for various superconductors and according to the BCS theory. The agreement with theory is quite good for these pure metals. *From P. Townsend and J. Sutton, Physical Review* ***128,*** *591–595 (1962).*

Example 10.5

Estimate the energy gap E_g for niobium at $T = 0$, and find the minimum photon wavelength needed to break the Cooper pair.

Solution: From Table 10.5 we see that $T_c = 9.25$ K for Nb. Therefore we use Equation (10.45) to find

$$E_g \approx 3.54\ (1.38 \times 10^{-23}\ \mathrm{J/K})\,(9.25\ \mathrm{K}) = 4.52 \times 10^{-22}\ \mathrm{J} = 2.82\ \mathrm{meV}$$

This small energy corresponds to a photon wavelength of

$$\lambda = hc/E_g = 4.39 \times 10^{-4}\ \mathrm{m}$$

This is in the far-infrared region of the electromagnetic spectrum. Photons with this wavelength or lower have sufficient energy to break the Cooper pair in niobium. Note that this estimate of E_g is close to the experimental value of 3.05 meV.

The Search for a Higher T_c

In order for us to put recent developments into perspective, it is instructive to look at the history of transition temperature increases (Table 10.7, Figure 10.35). In 1911 Kamerlingh Onnes measured the transition temperature of mercury to be 4.2 K (just a coincidence that this is approximately the same as the boiling point of helium!), and in 1930 Meissner found 9.3 K for niobium. The next significant increase came in the 1950s with various niobium alloys. NbN, a compound used in many applications today for its stability and relatively high critical current, was found to have $T_c = 15$ K. In 1973 Berndt Matthias measured $T_c = 23.2$ K for Nb_3Ge, a record that stood until 1986. Nb_3Ge belongs to a class of compounds called **A-15,** which includes the relatively high-temperature superconductors Nb_3Sn, V_3Si, and others. This serves to illustrate that if you find one new superconductor, you have probably found several, because it is normally possible to substitute for one or more constituent atoms with another from the same column of the periodic table.

TABLE 10.7
Superconductivity Records Through the Years

Material	Type	T_c(K)	Year of Discovery
Hg	Element	4.2	1911
Pb	Element	7.2	1913
Nb	Element	9.3	1930
Nb_3Sn	Alloy	18.1	1954
$Nb_3(Al_{0.75}Ge_{0.25})$	Intermetallic	20–21	1966
Nb_3Ga	Intermetallic	20.3	1971
Nb_3Ge	Intermetallic	23.2	1973
$Ba_xLa_{5-x}Cu_5O_{5(3-y)}$	Ceramic	30–35	1986
$(La_{0.9}Ba_{0.1})_2CuO_{4-\delta}$, at 1 GPa	Ceramic	52.5	1986
$YBa_2Cu_3O_{7-\delta}$	Ceramic	90–95	1987
BiSrCaCuO	Ceramic	105–120	1988
TlBaCaCuO	Ceramic	110–125	1988
$HgBa_2Ca_2Cu_4O_{1+x}$	Ceramic	133	1993

From C. P. Poole Jr. *et al., Copper Oxide Superconductors.* New York: Wiley Interscience, 1988, p. 7.

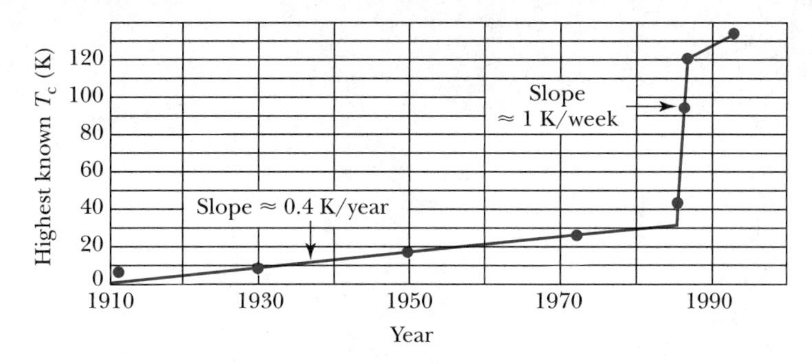

FIGURE 10.35 The highest known superconducting T_c by year from 1911 to 1993. Note the dramatic increase beginning in 1986. As of early 1999 there had been no further significant increases since 1993.

In 1986 the excitement began. Georg Bednorz and Karl Alex Müller, working at the IBM Zürich research laboratory, found that a compound containing lanthanum, barium, copper, and oxygen had a transition temperature of at least 30 K. By playing the substitution game just described, Bednorz and Müller* were soon able to achieve a T_c of 40 K in $La_{1-x}Sr_xCuO$ with $x = 0.15$. Their discovery was revolutionary, certainly because they had nearly doubled the old record for T_c, but also because the essential ingredients in the new superconductors were copper and oxygen, previously thought to be anathemas to superconductivity. For this discovery Bednorz and Müller were awarded the 1987 Nobel Prize for physics.†

But this was not the end of it. In early 1987 a group at the University of Houston led by Paul Chu more than doubled the record T_c again when they substituted yttrium for lanthanum and barium for strontium and changed the composition slightly. Chu reported a maximum T_c of about 93 K for $YBa_2Cu_3O_7$. The ratio of the three metallic elements has caused this compound to be referred to as "1-2-3." A T_c of 93 K was a fantastic advance because it surpasses 77 K, the boiling point of nitrogen. The cooling mechanism for superconductors has always been a liquified gas, such as helium. But liquid helium is very expensive to make,‡ and to keep anything at that temperature (4.2 K or lower) requires cumbersome and expensive insulation techniques. Further, helium gas itself is rare and expensive. It is found only in trace amounts in the atmosphere (it is so light that its mean molecular speed is a significant fraction of escape speed from the earth's surface; therefore it tends to leave the atmosphere!). Helium can be obtained from beneath the earth's surface. It is found at the top of deposits of natural gas and in geological formations such as dolomite. Hydrogen, with a boiling point of 20 K, is a possible substitute, but its high flammability makes it undesirable for most uses. Nitrogen, on the other hand, is plentiful, safe, and relatively easy to liquify and store. That is why 77 K had long been the goal of physicists looking for higher-temperature superconductors.

Another great thing about the (type II) copper oxide superconductors is that they have extremely high upper critical fields. For $YBa_2Cu_3O_7$ the upper

*J. G. Bednorz and K. A. Müller, Z. *Phys.* **B64,** 189 (1986).

†This was the fastest such recognition in the history of the Nobel Prize in physics.

‡A useful observation by B. S. Deaver was that in the 1970s liquid He cost about the same (for a given amount) as a fine scotch whiskey, while liquid N_2 cost about the same as milk! Today liquid He and liquid N_2 are somewhat cheaper by comparison.

SPECIAL TOPIC:

LOW-TEMPERATURE METHODS

Research in superconductivity and other low-temperature phenomena requires special laboratory equipment and techniques. There are well-established methods of achieving, maintaining, and measuring temperatures close to absolute zero. Here we shall introduce you to some of the hardware and techniques used by physicists who study materials at these very low temperatures.

Normally a very cold (cryogenic) liquid is used as a low-temperature bath. The sample being studied is immersed in the liquid, or else it is placed in good thermal contact with a material that does have contact with the bath. In designing low-temperature apparatus, one wishes to insulate the sample and bath from their room-temperature surroundings so that the sample does not absorb too much heat and thereby suffer an unwanted rise in temperature.

Heat can be transferred by any of three ways: conduction, convection, and radiation. The **dewar** flask, named for the low-temperature pioneer James Dewar (1842–1923), limits heat transfer by all three modes (see Figure A). The surfaces the liquid touches are designed to be poor conductors of heat. The cold liquid is insulated from convective transfer by being surrounded by a vacuum jacket. If the dewar is metallic, the shiny outer surface will tend to limit heat transfer by radiation. Otherwise the surface should be "silvered" that is, coated with a thin layer of reflective material.

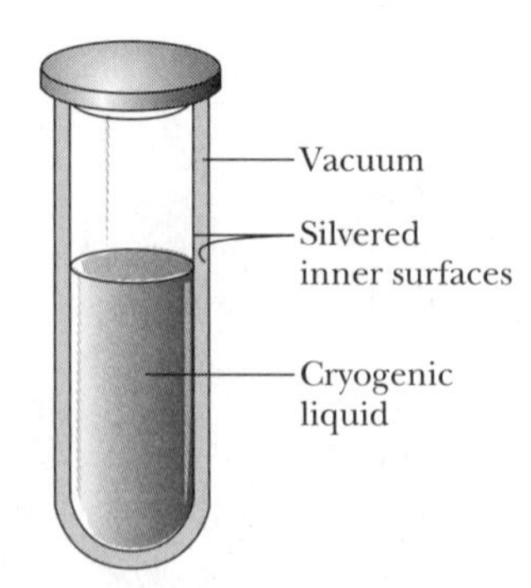

FIGURE A Schematic drawing of the dewar flask. The silvered inner surfaces reduce radiation, while the vacuum jacket reduces conduction of heat from the outside.

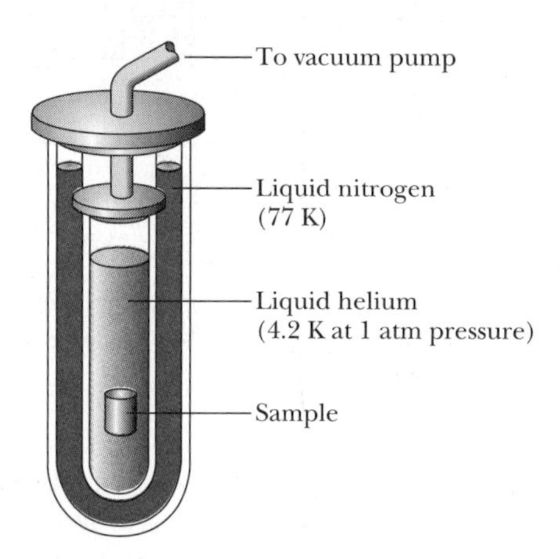

FIGURE B Schematic drawing of a double-dewar apparatus. The outer dewar is maintained at 77 K with liquid nitrogen, and the inner dewar contains liquid helium. The inner dewar may be connected to a pump if temperatures less than 4.2 K are required.

Figure B shows a schematic drawing of a double dewar apparatus typically used for very-low-temperature work. The dewar containing liquid helium, which has a temperature of 4.2 K at a pressure of 1 atm, is surrounded by a second dewar containing liquid nitrogen at 77 K. The logic of this device is that much of the heat that does enter from the surroundings will be absorbed by the liquid nitrogen, which will cause the (relatively inexpensive) liquid nitrogen to boil away, rather than the liquid helium.

One may achieve temperatures lower than 4.2 K by pumping the vapor above the helium bath. The remaining liquid and vapor then cool by adiabatic expansion. As the liquid cools and the vapor pressure decreases, it becomes increasingly difficult to maintain the lower pressure by pumping. Therefore there is a practical temperature limit of about 1 K with this method. In fact, at 0.7 K the creeping film (discussed in Section 9.7) extracts heat from the container walls, vaporizes, and gives up this heat to the bath by condensation. This presents a more serious limitation to the pumping method.

In 1926 W. F. Giauque and P. Debye (working independently) developed the idea of **adiabatic demagnetization** for achieving lower temperatures. To be precise, adiabatic demagnetization is the second step in a two-step procedure. First, a paramagnetic salt is put into a vessel containing helium gas, which in turn is connected thermally to the helium bath. This is done with the dewar between the poles of an electromagnet. When the electromagnet is switched on, the magnetic dipoles in the salt align with the field. This magnetization takes place at a fixed temperature in the helium bath, and therefore this first step is referred to as *isothermal magnetization.* The key to understanding what is about to happen is to consider the entropy of the salt. Because the alignment of dipoles corresponds to a more ordered system (at the same temperature), the salt's entropy has decreased. This is permitted by the second law of thermodynamics, because the gas surrounding the salt can transfer some heat to the helium bath and thereby generate an entropy increase that at least offsets the entropy decrease just described.

Now the helium gas in contact with the salt is pumped away, so that no further heat transfer can take place. The magnet is removed, so that the magnetic dipoles will resume a random orientation. The demagnetization of the salt takes place without heat transfer (i.e., adiabatically), and therefore it is this step that is properly called the adiabatic demagnetization. What is the result? The disordering of the magnetic dipoles surely corresponds to an entropy increase. But with no heat transfer possible, the total entropy of the salt should be constant. Therefore there must be a corresponding entropy *decrease,* and the only way this can occur is for the temperature of the salt to decrease. Figure C illustrates the entropy-temperature relationship in both steps of the process. This procedure can be repeated until (given the insulation constraints of the apparatus) no further heat can be removed from the salt. Temperatures of less than 1 K are routinely achieved by adiabatic demagnetization.

Today the most popular device for maintaining temperatures below 1 K is the ^{3}He–^{4}He dilution refrigerator, which also uses the principle of entropy exchange. This process was suggested by London in 1951, and the technology was developed in the 1960s and 1970s. ^{3}He–^{4}He dilution is an effective method of cooling below 1 K, because at 1 K ^{4}He, in the superfluid state (see Section 9.7), has extremely little entropy, while ^{3}He still has a significant amount of entropy. The entire process is too detailed to be described here. Let it suffice to say that adiabatic dilution serves the same purpose as the adiabatic demagnetization described earlier. Commercially available dilution refrigerators routinely reach temperatures on the order of 10^{-3} K.

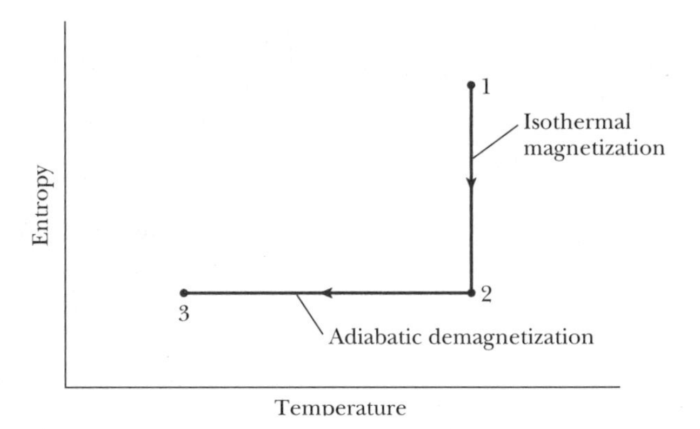

FIGURE C In isothermal magnetization a sample is taken from 1 to 2 along the vertical line of the entropy-temperature diagram, as the magnetic dipoles are aligned at a constant temperature. The path from 2 to 3 is adiabatic demagnetization. Because the total entropy is constant as the magnetic alignment vanishes, the temperature must drop.

Low-temperature physicists must also be concerned with how best to measure such low temperatures. Normally they do so by measuring the resistance of a device attached to the sample or bath. Semiconductor resistors are often favored at low temperatures (say from 20 K to less than 1 K) because the resistance of semiconductors tends to increase sharply at very low temperatures (see Section 11.2). The sharp rise in resistance with decreasing temperature makes semiconductor resistors very sensitive thermometers. The increase is typically nonlinear, however, and this puts a premium on calibration methods and standards.

Paramagnetic materials, which obey the Curie law [Equation (10.41)], also are used as temperature-measuring instruments at low temperatures. According to the Curie law, magnetic susceptibility varies inversely with temperature at low temperatures, and therefore the magnetic susceptibility is sensitive to small changes in temperature in this regime.

Since the time of Kamerlingh Onnes, physicists have strived to cool materials to lower and lower temperatures. By the late 1990s, temperatures of about 10^{-5} K had been reached in the laboratory. Note that this temperature is for a bulk sample, compared with temperatures of nK for individual nuclei. The "microkelvin" laboratory at the University of Florida expects to achieve its namesake 1 μK early in the 21st century. In the past, lower temperatures have led to unexpected discoveries (e.g., superconductivity and superfluids). Physicists today hope that with the achievement of even lower temperatures there will be new discoveries, along with new data to enhance our understanding of theoretical physics and materials science.

For more information see Pobell, F., *Matter and Methods at Low Temperatures,* New York: Springer-Verlag, 1995.

---

critical field $B_{c2} \approx 100$ T at 77 K. B_{c2} at 0 K is so high that it cannot be measured, but by extrapolation it is taken to be about 300 T. This allows us to imagine applications that use the Meissner effect or high currents. One manufacturer (HITC Superconco) advertises high-T_c superconducting material with a critical current density of 3.5×10^7 A/m^2 at a temperature of 77 K and 1.0×10^9 A/m^2 at a temperature of 4.2 K in a 1-T magnetic field. By early 1992 a research team at Oak Ridge National Laboratory had surpassed this mark in laboratory trials. Their filamentary yttrium-barium-copper oxide wire had a critical current density of 10^9 A/m^2 in zero applied magnetic field at 77 K and a critical current density of 2×10^8 A/m^2 in an applied magnetic field of 8 T at 77 K. For comparison, a normal conductor, #10 gauge copper wire, has a diameter of 2.59 mm and a recommended maximum current of 30 A (if insulated), or a maximum current density of 5.7×10^6 A/m^2. A critical current density 10^9 A/m^2 has been reported for newer mercury-based oxides at T = 110 K. More recently there have been reports of 10^{10} A/m^2 critical currents at 77 K in specially prepared yttrium-barium-copper oxide tapes.

Ceramics

The copper oxide superconductors fall into a general category of materials called **ceramics.** Unfortunately, like most ceramic materials, they are extremely brittle and therefore are not easy to mold into convenient shapes. The copper oxide superconductors are relatively easy to make compared with the A-15 compounds or with the semiconductor wafers used in computers, but making them into something useful for practical applications is a challenge and requires innovative technology. By the late 1990s, it was possible to produce long, flexible, high-T_c superconducting wire that could replace much thicker conventional wires, as illustrated in Figure 10.36. One manufacturer (Pacific Superconductors) advertises a high-T_c wire kilometers in length with a 7-cm bending radius and a critical current of at least 10 A at a temperature of 50 K in zero applied magnetic field. Their wire consists of a layer of 12 superconducting fibers soldered into a 5.7-mm-by-0.9-mm copper channel.

In early 1988 Chu and his group developed a similar compound using bismuth and aluminum, $BiAl_{1-y}CaSrCoO_{7-d}$, which has $T_c = 114$ K with $0 < y < 0.3$ and $d < 0.45$. The best that could be done by 1992 that was reproducible (in spite of some wild claims from around the world) was 125 K in the thallium-based compound $Tl_2Ba_2Ca_{n-1}Cu_nO_{2n+4}$ with $n = 3$. The interesting thing here is that there is a regular variation of T_c with n; $T_c = 80$ K for $n = 1$, $T_c = 110$ K for $n = 2$, and $T_c = 125$ K for $n = 3$. The curve shown in Figure 10.37 suggests that if we could only make this thallium-copper oxide with $n > 3$, T_c could be

FIGURE 10.36 Although almost 100 times smaller in cross-section, four strands of American Superconductor's multifilamentary HTS wire (foreground) can transmit as much electrical current as conventional copper cable. *(Courtesy of American Superconductor).*

increased further, perhaps indefinitely. In the structure (Figure 10.38) these higher values of n correspond to more stacked layers of copper and oxygen. So far it has not been possible to stack more than three. Also, thallium is discouraging because it is so toxic; it is a major component of rat poison and is dangerous to handle. A similar and slightly less toxic version that offers some hope is $Bi_2Sr_2Ca_{n-1}Cu_nO_x$, but its highest T_c is only about 110 K with $x = 1.6$.

In 1993 a higher T_c was achieved in mercury-based copper oxides. A T_c of 133 K has been reported in the compound $HgBa_2Ca_2Cu_3O_{1+x}$ with x a small positive number. It has also been shown that these mercury-based superconductors reach superconducting transition temperatures of 164 K at extremely high pressures—about 300,000 atmospheres.

The use of high-T_c superconductors is increasing in other applications for which long, flexible wires are unnecessary. For example, the wireless communications industry uses circuits with high-T_c components in filters. These new filters are less noisy and consume less power than conventional, copper-based filters. Ceramic superconductors are also finding their way into nuclear magnetic resonance applications (see Section 10.6).

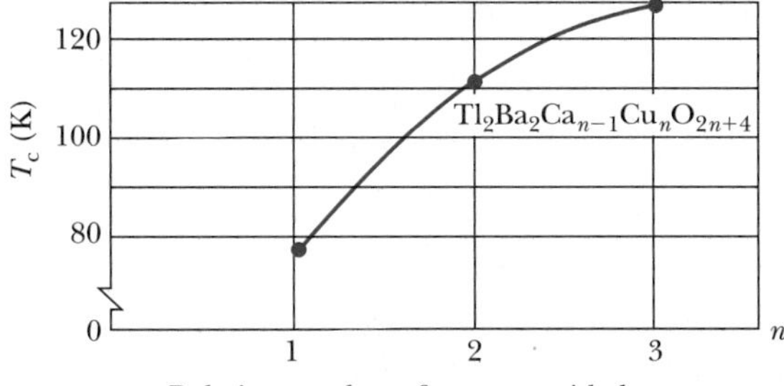

FIGURE 10.37 Superconducting transition temperatures in the thallium-based superconductor $Tl_2Ba_2Ca_{n-1}Cu_nO_{2n+4}$. The variation with n suggests that if n could be increased further, even higher transition temperatures could be achieved.

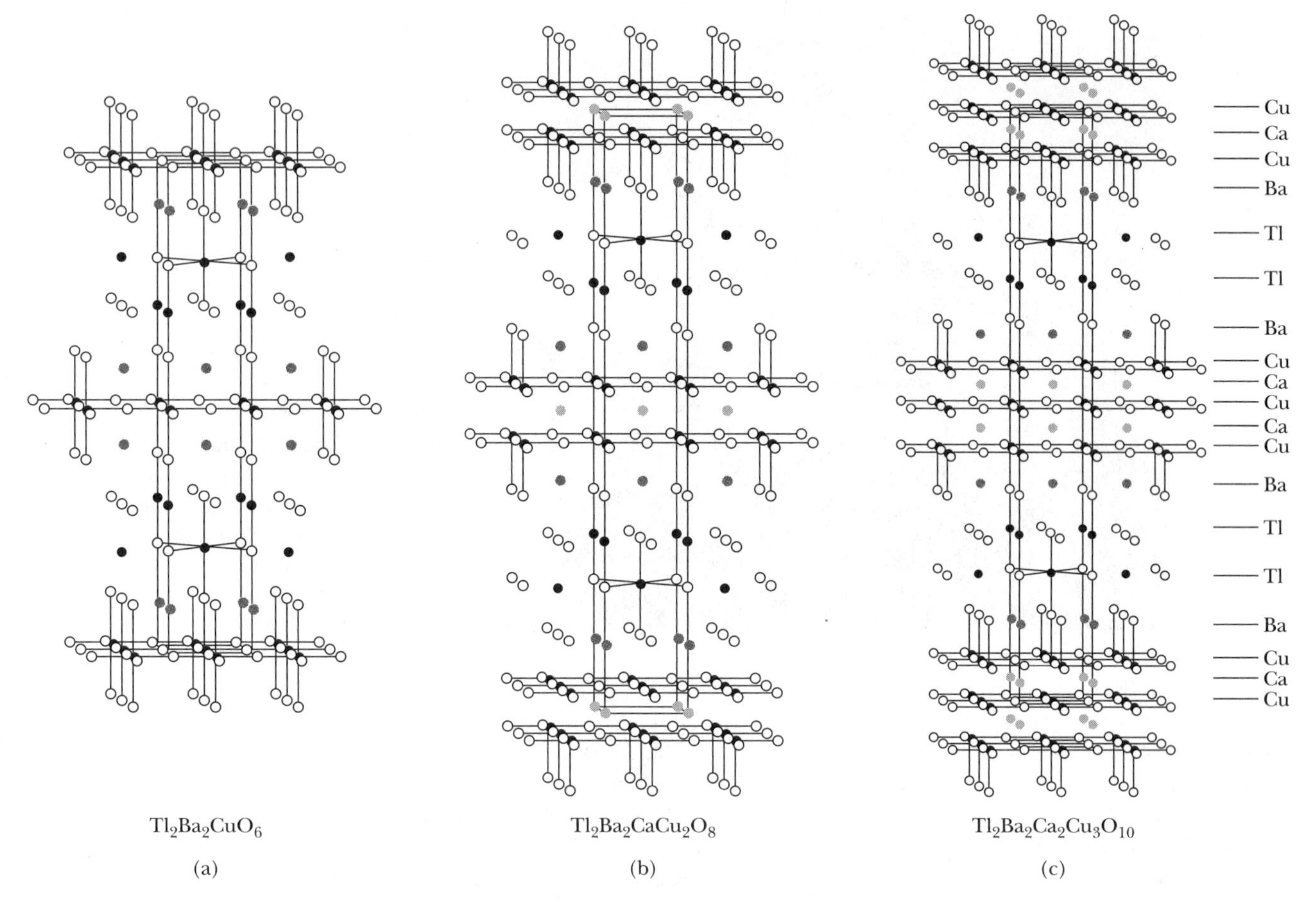

FIGURE 10.38 Model of the thallium-based superconductor. Note the stacking of the copper-oxide layers (oxygen is shown by open circles). *From C. C. Torardi et al., Science **240**, 631–634 (1988).*

There is still much work to be done, both experimentally and theoretically, before we can understand the mechanism for superconductivity in the copper oxides. Generally they exhibit no isotope effect, which suggests that there may be a new, non-BCS mechanism at work. And while the advance in the highest T_c from 23 K to 133 K was spectacular, the mechanical properties of copper oxides makes them less than desirable. Also, the greatest dream of low-temperature physicists has been to achieve a T_c not of 77 K but of 300 K (room temperature), thereby eliminating the need for cryogenic fluids. Whether this can be achieved in copper oxides or any other materials remains to be seen.

Superconducting Fullerenes

Another class of exotic superconductors was announced in 1991 by a research group at Bell Labs.* This class is based on the organic molecule C_{60}, which had been discovered just a few years earlier. The basic C_{60} molecule is called

*D. R. Hoffman, "Solid C_{60}." *Physics Today* **44,** 11 (November 1991), pp. 22–29.

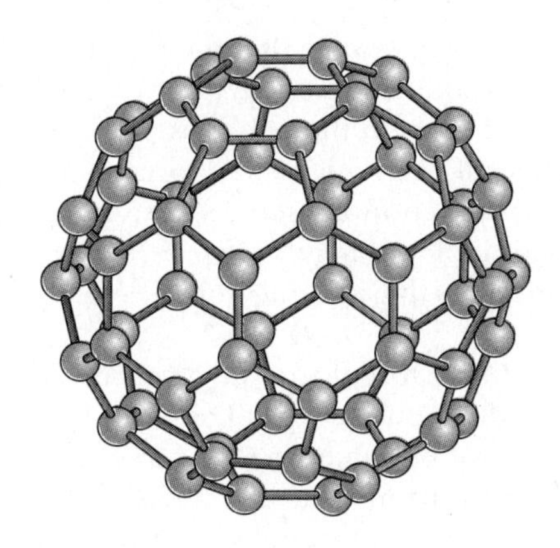

FIGURE 10.39 A schematic drawing of the C_{60} molecule. The structures of the superconducting compounds based on C_{60}, in which the atoms of the dopant are generally found on the inside of the "soccer ball" structure, were not well understood as of early 1999.

"buckminsterfullerene" (after the mathematician, architect, and general gadfly Buckminster Fuller) because of the large molecule's resemblance to Fuller's geodesic dome (Figure 10.39). The 1996 Nobel Prize in chemistry was awarded to Robert F. Curl, Harold W. Kroto, and Richard E. Smalley for their 1985 discovery of fullerenes.

While pure C_{60} is not superconducting, the addition of certain other elements can make it so. For example, when C_{60} is doped with the right amount of potassium, it forms the compound K_3C_{60} with a superconducting transition temperature of 18 K. When C_{60} is combined with thallium and rubidium, the T_c can be as high as 42.5 K. While these transition temperatures are not as high as those achieved in the copper oxides, it is still encouraging whenever a new class of superconductors with a *relatively* high T_c is found. Further, each new discovery of this sort brings with it the potential for deeper understanding of the phenomenon of superconductivity, which in turn can lead to more experimental discoveries.

10.6 Applications of Superconductivity

The remarkable properties of zero resistance and the Meissner effect make superconductors ideal for many applications. Some of these have been in use for years, while others may become feasible only if higher temperature superconductors become available. The cost of cryogenic systems and fluids must always be taken into account when one considers whether to use a superconductor.

Josephson Junctions

One of the earliest applications of superconductors was in a device known as a **Josephson junction.** In 1962 Brian Josephson predicted that electron pairs can tunnel from one superconductor through a thin layer of insulator into another superconductor. The superconductor/insulator/superconductor layer constitutes the Josephson junction. In the absence of any applied magnetic or electric field, a DC current will flow across the junction (the **DC Josephson effect**). When a voltage V is applied across the junction, the electron pair current across the junction oscillates with a frequency

$$\nu_j = \frac{2eV}{h} \tag{10.46}$$

This is the **AC Josephson effect.** Notice that the frequency of oscillation and the applied voltage are in the simple ratio $2e/h$. Because frequencies can be measured to extremely high accuracy in these ranges ($\nu_j = 483.6$ GHz for $V = 1$ mV), Equation (10.46) provides a convenient way to measure and maintain voltage standards. The Josephson junction is used for just this purpose at the National Institute for Standards and Technology (NIST, formerly the National Bureau of Standards). With this device, the accuracy of the voltage standard is approximately 1 part in 10^{10}.

It has been suggested that Josephson junctions could be used in integrated circuits, the heart of modern computers. These are superconducting devices, and therefore they would not be subject to the power losses that semiconductor-based circuits suffer. Studies throughout the 1970s showed that, given the costs of fabrication and cooling, superconducting computers were just not a competitive option. Needless to say, this view will have to be re-evaluated whenever new and higher-temperature superconductors become available.

Today Josephson junctions are used routinely in devices known as SQUIDs, Superconducting QUantum Interference Devices (Figure 10.40). The SQUID uses a pair of Josephson junctions in a current loop. It turns out that the current is extremely sensitive to the magnetic flux applied to the loop. Therefore, SQUIDs are useful in measuring very small amounts of magnetic flux. They can be used to measure the quantum fluxoid $\Phi_0 = h/2e$ to within 1 part in 10^6. As ordinary magnetometers, SQUIDS are capable of measuring magnetic fluctuations on the order of 10^{-13} T!

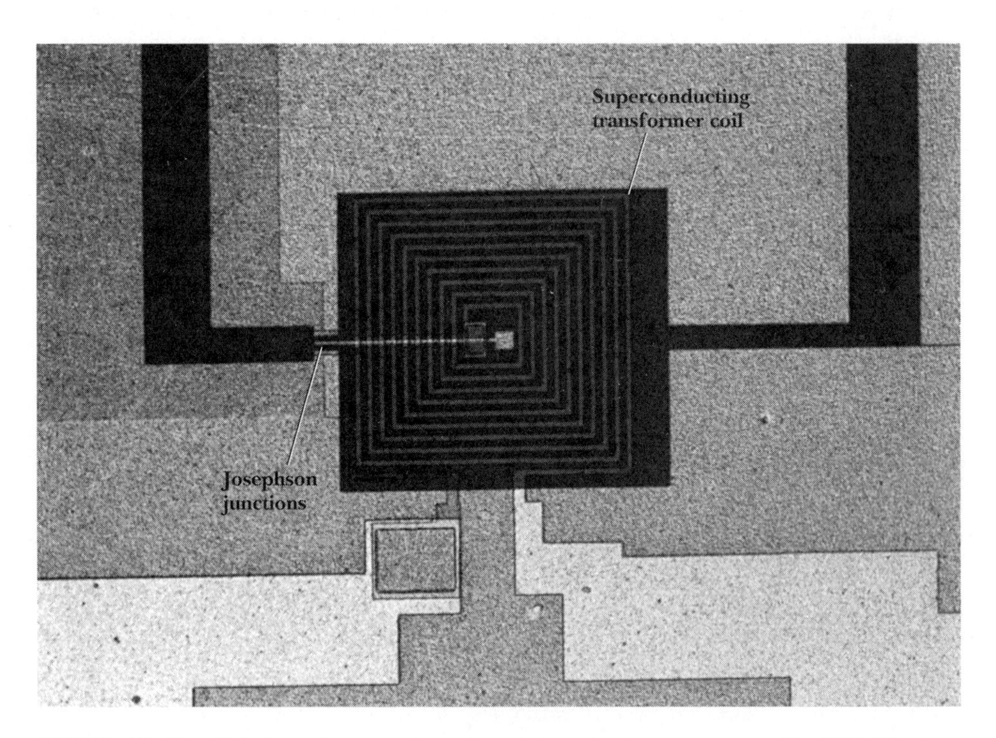

FIGURE 10.40 This is a close-up view of a yttrium–barium–copper-oxide SQUID chip, manufactured by Conductus, Inc. This highly sophisticated chip contains 15 layers of material, of which three are superconducting. *Photo courtesy Conductus, Inc.*

Example 10.6

A uniform magnetic field is perpendicular to a loop of radius 1 cm. Find the value of the magnetic field such that the magnetic flux through the loop is equal to Φ_0.

Solution: Because the magnetic field is perpendicular to the loop, $\Phi_0 = BA$, where A is the area of the loop. Therefore

$$B = \Phi_0/A = (2.068 \times 10^{-15}\ \text{T} \cdot \text{m}^2)/\pi(10^{-2}\ \text{m})^2 = 6.58 \times 10^{-12}\ \text{T}.$$

This exceptionally small value is an indication of just how small the quantum fluxoid is.

SQUIDs have been available commercially since the early 1970s. Since then many new applications for the SQUID have been found. For example, there are a number of biomedical applications, the most important of which is imaging soft tissues such as the brain. SQUIDs have also been used to examine materials for defects, search for explosives, detect the presence of bacteria, and search for oil. A survey of applications can be found in *The Industrial Physicist* **4,** 2(June 1998) pp. 20–23.

Maglev

What might be other applications of high-temperature superconductors, either at 77 K or 300 K? One suggestion has been to use the Meissner effect to levitate various transport systems, particularly trains. The idea is to make a more comfortable ride at higher speeds while reducing frictional losses. Several prototype magnetic levitation (**maglev**) trains have been built around the world. In late 1997 a speed of 550 km/h was achieved by a maglev train on a 18.4-km test track in Yamanashi prefecture, Japan. Though expensive to operate, it is hoped that with technological improvements maglev trains can be used commercially. Maglev is being studied actively in Japan, Europe, and the United States.

Generation and Transmission of Electricity

There exists a potential for significant energy savings if superconductors can be used in electrical generators and motors. Part of the savings would obviously come from cutting resistive losses, but there would be even more savings if the heavy iron cores used today could be replaced by lighter superconducting magnets. The largest generator used now can produce energy at a rate of about 1 GW, with this upper limit due to the size of the iron core. Larger generators would make for better economies of scale.

Once electrical power is generated it must be transmitted through power lines to industries and homes. Superconducting transmission lines would save significant amounts of energy (approaching 5% in some cases), again because there would be no resistive loss. Another advantage of superconducting transmission is that expensive transformers would no longer have to be used to step up voltage for transmission and down again for use. Transformers are used at both ends of conventional lines because the energy loss rate is

$$P_{\text{lost}} = I^2R = P_{\text{trans}}^2 R/V^2 \tag{10.47}$$

It is desirable to make V as large as possible for transmission, but this is not necessary if $R = 0$. Finally, the larger current densities possible in superconducting wire could make it possible to reduce the number and size of transmission lines.

Superconducting rings may be used for energy storage. Today, power plants are strained during "peak hours," when it is necessary to produce electrical energy at up to several times the average use rate. Superconducting storage rings would allow plants to produce at just the average rate, with extra energy generated during low usage hours and stored for use during peak hours. This storage option might also allow us to make better use of other forms of energy, especially solar and wind.

Other Scientific and Medical Applications

There are other ways in which superconductors are already used in scientific research. Magnets used to confine plasma in fusion research are superconducting. Magnets for large particle accelerators, including existing ones at Fermilab and CERN and the planned large hadron collider (LHC) at CERN, must be superconducting in order to produce the large fields required. This is feasible using existing superconducting technology and liquid helium. However, higher-T_c magnets can improve the operation of large particle accelerators in two ways. First, the cost of operating is much lower with liquid nitrogen than with liquid helium. Second, recent work has shown that some ceramic superconductors can produce larger magnetic fields than traditional superconductors, due to their high critical currents. With higher fields experimenters can choose between increasing the accelerator's energy or decreasing its size, because the net energy is proportional to both the radius and magnetic field strength. The technology for making magnet coils out of brittle ceramic materials is now becoming available.

Incidentally, it is not possible to make a large electromagnet simply by winding large coils of superconducting wire. Statistically infrequent events known as

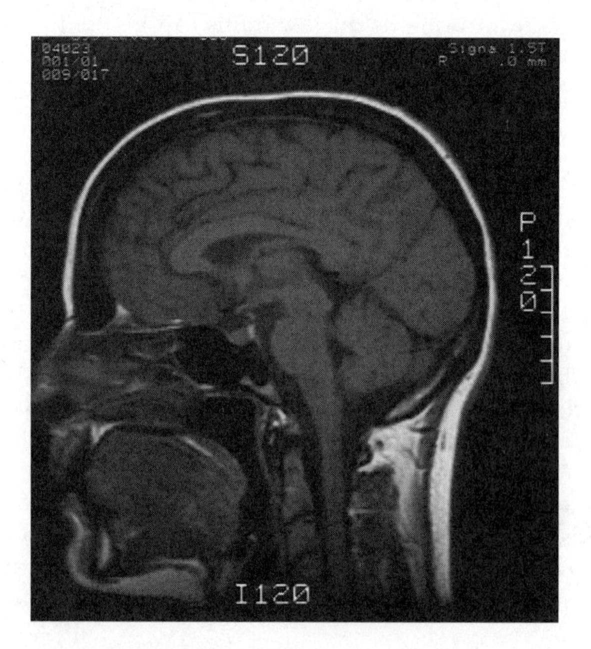

FIGURE 10.41 A picture of the soft tissues of the brain produced by magnetic resonance imaging. *Courtesy Tacoma Magnetic Imaging.*

flux jumps occasionally make the wire a normal conductor in small regions. Unfortunately this small region grows and propagates rapidly due to resistive heating. When one part of the wire undergoes resistive heating, the heat flow from warmer regions soon drives the cooler regions above T_c. For this reason, long superconducting wires must be embedded in a copper matrix. Copper is such a good conductor of heat at low temperatures that it shunts away heat much faster than the superconductor gone normal. This allows the superconductor to recover from those temporary fluctuations.

There is at least one significant medical application of superconducting magnets in wide use today. It is called *magnetic resonance imaging* (MRI).* Medical workers use MRI to get clear pictures of the body's soft tissues, allowing them to detect tumors and other disorders of the brain, muscles, organs, and connective tissues (see Figures 10.41 and 10.42). With high-field (several tesla) supercon-

*The physical process used in MRI is the well-established quantitative analysis technique called *nuclear magnetic resonance* (NMR). In MRI the data is processed by a computer with the specific goal of making a picture of the sample. It has been suggested that the word *nuclear* was dropped so that the general public would not have an irrational fear of the procedure. Of course MRI has nothing to do with gamma radiation or radioactive isotopes!

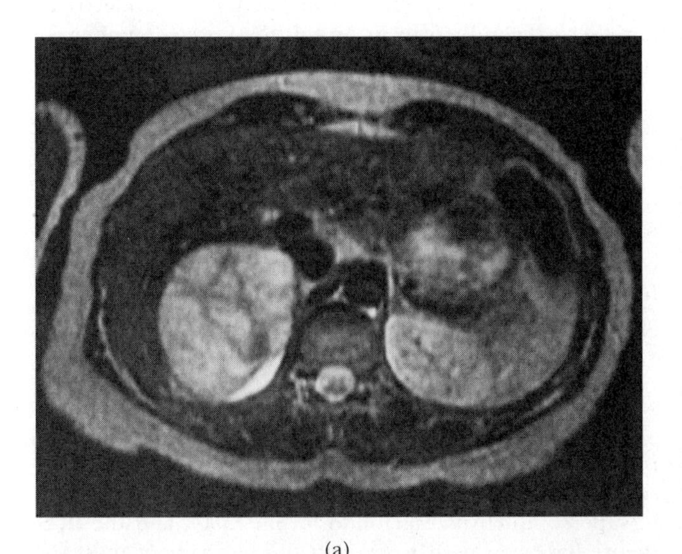

(a)

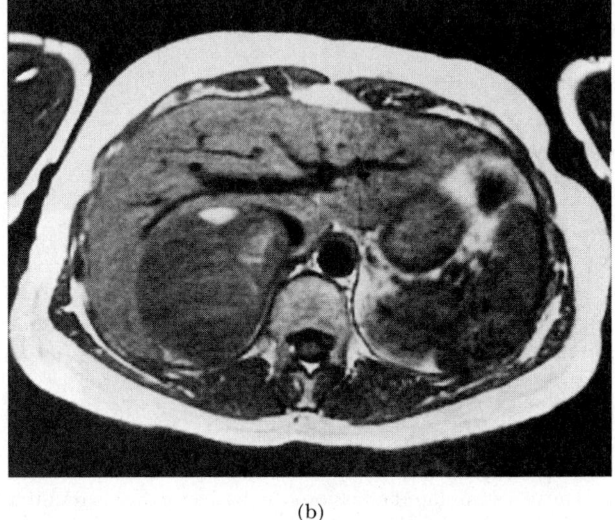

(b)

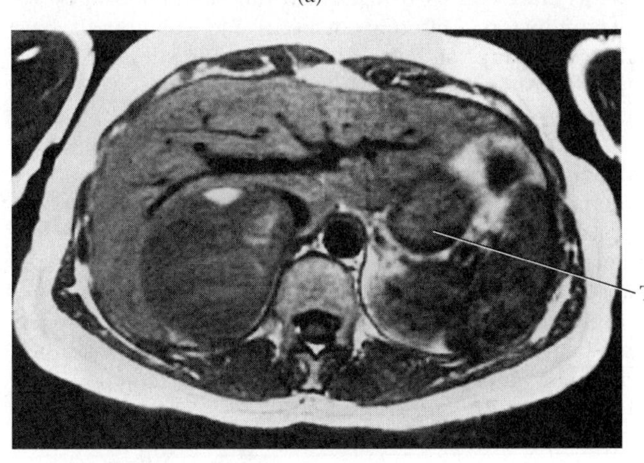

Tumor

(c)

FIGURE 10.42 These magnetic resonance images from the thoracic region of a patient are horizontal "slices" approximately 0.5 cm apart. These pictures were used to detect a malignant tumor in the liver. *Courtesy of Haleigh Kurtz.*

ducting magnets the MRI diagnosis can be made earlier than with other methods and, an important fact, without surgical intrusion. Many lives have already been saved by MRI, and many others have been vastly improved. The major drawback is that the cost of a single MRI image is on the order of $1000. Higher-temperature superconductors may eventually reduce the cost and thereby make MRI available to more people.* Just as with Schawlow and the laser, Kamerlingh Onnes could not have imagined this life-saving possibility when he first pondered superconductivity.

*For a good review see F. W. Wehrli, "The Origins and Future of Nuclear Magnetic Resonance Imaging." *Physics Today* **45,** 6 (June 1992), pp. 34–42.

Summary

The Coulomb force holds together the atoms of a molecule or solid. In a diatomic molecule, quantum theory can be used to deduce the allowed rotational and vibrational energy levels:

$$E_{\text{rot}} = \frac{\hbar^2 \ell(\ell+1)}{2I} \tag{10.2}$$

and

$$E_{\text{vibr}} = \left(n + \frac{1}{2}\right)\hbar\omega \tag{10.3}$$

Transitions between these levels are observed in both emission and absorption spectra. In Raman scattering a photon is scattered from a molecule, and the change in frequency of the scattered photon is used to deduce rotational properties of the molecule.

A photon incident upon a molecule in an excited state can cause a stimulated emission, the basic mechanism in the operation of a laser. Three-level and four-level systems can be used to control the output and efficiency of the laser. The monochromatic, intense beam has a number of applications in research and industry, including holography and biomedical uses.

Many solids exhibit regular crystal structures. Solids generally tend to expand when heated. It can be shown that the mean separation between atoms in a solid is very nearly proportional to temperature in the classical limit:

$$\langle x \rangle = \frac{3b}{4a^2\beta} = \frac{3bkT}{4a^2} \tag{10.26}$$

Good electrical conductors tend to be good thermal conductors too. Using the quantum theory of conduction, we found the correct relationship between the electrical and thermal conductivity in a solid (the Lorenz number L) to be

$$L = \frac{K}{\sigma T} = \frac{\pi^2 k^2}{3e^2} \tag{10.36}$$

A few solids, known as *ferromagnets,* have permanent magnetic moments. In most materials a magnetic moment is induced by an applied magnetic field. The magnetic susceptibility (which is positive for paramagnets and negative for diamagnets) relates the induced magnetization to the applied magnetic field:

$$\chi = \frac{\mu_0 M}{B} \tag{10.37}$$

Superconductors exhibit the complete loss of electrical resistance and expulsion of magnetic fields below their transition temperature T_c. A successful theory (the BCS theory) of superconductivity was found in the 1950s, and the BCS theory has been used to explain the behavior of most superconductors. Different elements and compounds have different values of T_c, and the search for a higher T_c has been the study of a great deal of research. In the late 1980s compounds with $T_c > 100$ K were discovered.

The search for higher T_c values continues today, with the aim of using higher-temperature superconductors in a number of applications. Low-loss energy generation and transmission, magnetic levitation systems, superconducting electronic devices, and high-field magnets are some of the applications made possible by the extraordinary properties of superconductors. Today superconductors are already used in research, industry, and medicine, and the extension of superconductors into other areas is an exciting possibility for the future.

Questions

1. Explain in your own words why the sky is blue.
2. Explain why $n > m$ is required in Equation (10.1) in order to produce the potential energy curve shown in Figure 10.1.
3. Are the force constants for diatomic molecules (Table 10.1) at all similar to those for common laboratory springs (remember your introductory college or high school lab experience)? Explain.
4. Explain how to use the rotational spectra to determine the equilibrium separation between the two nuclei in a diatomic molecule.
5. Why do the gases He, Ne, and Xe tend to be monatomic rather than diatomic?
6. Do you expect the fundamental vibrational frequency to be higher for HCl or NaCl? Explain.
7. Is it necessary that the substance used in a laser have at least three energy levels? Why?
8. Critique the following statement: "The Pauli exclusion principle is responsible for keeping solids from collapsing to zero volume."
9. The average nearest-neighbor distance between nuclei in solid NaCl is 0.282 nm, but the distance is 0.236 nm in a free NaCl molecule. How do you account for the difference?
10. What patterns do you notice within the groups of salts in Table 10.2? Explain those patterns.
11. Explain why the paramagnetic susceptibilities of rare earth elements tend to be higher than those of the transition elements.
12. Considering the electronic configurations of the five ferromagnetic elements, justify why they should be ferromagnetic. Why are some elements in the same columns as these five, with similar electronic configurations, not ferromagnetic?
13. Notice that ferromagnetic elements tend to come from the middle of the rows of rare earth elements and transition elements in the periodic table. Explain.
14. Why should elements and compounds with positive paramagnetic susceptibilities not be good candidates for superconductivity?
15. Explain similarities and differences between the Meissner effect and Lenz's law.
16. Consider the superconducting transition temperatures of the elements as shown in Table 10.5. In cases in which there is more than one superconductor in a column of the periodic table, are the transition temperatures consistent with the spirit of the isotope effect (that is, does the heavier element have a lower T_c)?
17. A superconducting ring can carry a current for an indefinite length of time. Isn't this a perpetual motion machine, which violates the first or second law of thermodynamics?
18. Consider a sample at a temperature initially above its superconducting T_c and in a magnetic field. When the sample is cooled to below T_c, currents are generated to expel the magnetic flux from the interior. What is the source of the energy for these currents?
19. Would you expect that a material with little or no crystal structure (a "glass") could exhibit superconductivity? Why or why not?
20. A superconducting wire and a copper wire with very low resistance are connected together in parallel. When a potential difference is applied, is it really true that the copper wire carries no current? Explain.

Problems

10.1 Molecular Bonding and Spectra

1. Consider again the rotational energy states of the N_2 molecule as described in Example 10.1. Find the energy involved in a transition (a) from the $\ell = 2$ to $\ell = 1$ state, and (b) from the $\ell = 20$ to $\ell = 19$ state.
2. (a) Use the data in Table 10.1 to find the approximate spacing between vibrational energy levels in NO. (b) What temperature would be needed to excite this vibration thermally?
3. Estimate the amplitude of the smallest vibration of the HCl molecule (see Example 10.2).
4. The distance between the centers of the H and Cl atoms in the HCl molecule is approximately 0.128 nm. (a) Find the angular velocity of the molecule about its center of mass when $\ell = 1$ and $\ell = 10$. (b) What is the speed of the H atom in each of the cases in (a)? (c) What value of ℓ is required in order for the H atom to have a speed of 0.1 c? (d) Estimate the classical temperature associated with the ℓ you found in (c).
5. Derive an expression for the allowed rotational levels in a homopolar diatomic molecule using the Bohr quantization rule for angular momentum. Discuss your result in comparison with the correct quantum mechanical result, Equation (10.2).
6. The wavelength of a microwave absorption line in CO corresponding to a transition from $\ell = 0$ to $\ell = 1$ is 1.30 mm. (a) Calculate the rotational inertia of the

CO molecule. (b) Show that it is impossible for this amount of energy (corresponding to a photon of wavelength 1.30 mm) to be absorbed by CO in a *vibrational* transition.

7. If the energy of a vibrational transition from the $n = 0$ state to the $n = 1$ state in CO could be absorbed in a rotational transition that begins in the ground state ($\ell = 0$), what would be the value of ℓ for the final state? Explain why such a rotational transition is impossible.

8. Show that $I = \mu R^2$ for a diatomic molecule, where R is the distance between the two atomic centers and $\mu = m_1 m_2/(m_1 + m_2)$ is the reduced mass. (Note: You should assume that the atoms are point particles.)

9. The energy of a transition from the $\ell = 2$ to the $\ell = 3$ state in CO is 1.43×10^{-3} eV. (a) Compute the rotational inertia of the CO molecule. (b) What is the average separation between the centers of the C and O atoms?

10. Consider the model of the H_2O molecule shown below. (a) Find the rotational inertia of H_2O about the dashed line. (b) Estimate the energies of the first two rotational energy levels ($\ell = 1$ and $\ell = 2$). (c) What is the wavelength of a photon required to excite a transition from $\ell = 0$ to $\ell = 1$? (Note: This is roughly the wavelength of radiation used in a microwave oven.)

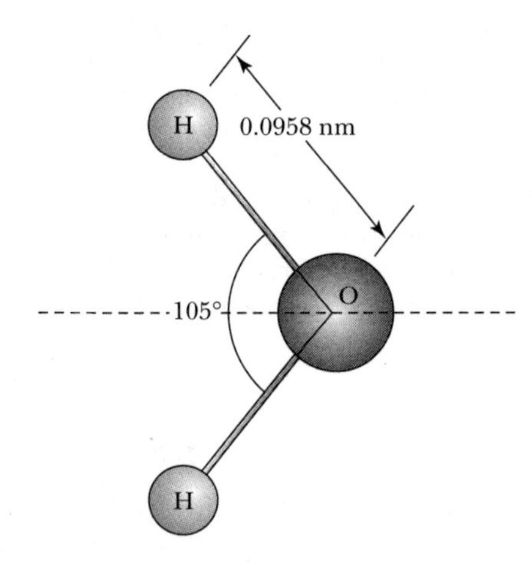

11. Consider the rotational energy of a single helium atom. Assume that the electrons are uniformly distributed throughout a sphere of radius equal to the Bohr radius and that the nucleus is a uniform solid sphere of radius 1.9×10^{-15} m. (a) Estimate the energy of the first (nonzero) rotational energy level in helium. (b) Is this state likely to be observed? Why or why not?

12. Consider the NaCl molecule, for which the rotational inertia is 1.46×10^{-45} kg · m^2. If infrared radiation with wavelength 30 μm is Raman scattered from a free NaCl molecule, what are the allowed wavelengths of the scattered radiation?

13. This problem deals with the splitting of rotational energy levels of diatomic molecules. If one atom of the molecule has more than one stable isotope, then both isotopes are normally present in a sample. Show that the fractional change $\Delta\nu/\nu$ in the observed frequency of a photon emitted in a transition between adjacent rotational states is equal to the fractional difference in the reduced mass $\Delta\mu/\mu$ for molecules containing the two different isotopes.

14. (a) At $T = 293$ K what are the relative Maxwell-Boltzmann factors for N_2 molecules in the $\ell = 0$, $\ell = 1$, and $\ell = 2$ states? Use $I \approx 10^{-46}$ kg · m^2. (b) Use your answer to (a) along with the fact that there is also a degeneracy factor of $2\ell + 1$ on the ℓth angular momentum level to find the relative populations of the $\ell = 0$, $\ell = 1$, and $\ell = 2$ states of N_2 at room temperature. (c) Explain why the $2\ell + 1$ degeneracy factor is more important for lower rotational states, but the Maxwell-Boltzmann factor dominates for higher states.

15. Using the HCl absorption spectrum shown in Figure 10.8 (a) compute the rotational inertia I of the molecule and (b) compute the force constant κ and compare with the value given in Table 10.1.

16. The equilibrium separation between the two ions in the KCl molecule is 0.267 nm. (a) Assuming that the K^+ and Cl^- ions are point particles, compute the electric dipole moment of the molecule. (b) Compute the ratio of your result in (a) to the measured electric dipole moment of 5.41×10^{-29} C · m. This ratio is known as the *fractional ionic character* of the molecular bond.

17. Find the energy of the photon required to excite the transition from the ground state to the first excited vibrational state in HF. In what part of the electromagnetic spectrum is this?

10.2 Stimulated Emission and Lasers

18. (a) How many photons are emitted each second from a 2.5-milliwatt helium-neon laser ($\lambda = 632.8$ nm)? (b) If the laser contains 0.02 mole of neon gas, what fraction of the neon atoms in the tube participate in the lasing process during each second of operation? (c) Comment on the relatively low numerical result in (b).

19. (a) For the helium–neon laser estimate the Doppler broadening (see Chapter 9, Problem 3) of the output wavelength 632.8 nm at $T = 293$ K. (b) Estimate the broadening of the same wavelength due to the Heisenberg uncertainty principle, assuming that the metastable state has a lifetime of about 1 ms.

20. Consider the problem of using laser light to measure the distance from the Earth to the moon. (a) What is the maximum uncertainty in timing the round-trip for a light pulse in order to determine the distance with an uncertainty of one meter? (b) Estimate the effect of the Earth's atmosphere on this experiment, using the fact that the speed of light in air (at sea level) is slower than the speed of light in vacuum by a factor of about (1.0003). Assume an 8-km-high atmosphere of uniform sea level density.

21. What is the minimum fraction of the lasing molecules in a three-level laser that must be in the excited state in order for the laser to operate? Answer the same question for a four-level laser.

22. The 3*s* state of neon (see Figure 10.15) is 16.6 eV above the ground state. (a) Estimate the relative populations of the ground state and the 3*s* state at $T = 293$ K. (b) Repeat for $T = 200$ K. (c) Repeat for $T = 500$ K. (d) What implications (if any) do your answers for parts (a)–(c) have for the operation of a He–Ne laser at various temperatures?

10.3 Structural Properties of Solids

23. The density of solid KCl is about 1980 kg/m^3. Compute the nearest-neighbor distance in KCl, that is, the distance between neighboring K^+ and Cl^- ions. Note that KCl has the same lattice structure as NaCl.

24. Show that the Madelung constant for a one-dimensional lattice of alternating positive and negative ions is $\alpha = 2 \ln 2$.

25. Write the first five terms of the Madelung constant for a two-dimensional lattice of alternating positive and negative ions.

26. Use Equation (10.19) to evaluate the net force ($F = -dV/dr$) on an atom in a sodium chloride lattice. Then show that the force can be expressed as

$$F = \frac{\alpha e^2}{4\pi\epsilon_0 r_0^2}\left(-\frac{r_0}{r^2} + e^{-(r-r_0)/\rho}\right)$$

27. Starting with the result of Problem 26, approximate $r \approx r_0 + \delta r$ and show that

$$F \approx \frac{\alpha e^2}{4\pi\epsilon_0 r_0^2}\left(-\frac{r_0}{(r_0+\delta r)^2} + e^{-\delta r/\rho}\right)$$

Perform a Taylor series expansion about $r = r_0$, keeping terms up to and including $(\delta r)^2$. Show that

$$F \approx K_1 \delta r + K_2 (\delta r)^2$$

where

$$K_1 = \frac{\alpha e^2}{4\pi\epsilon_0 r_0^2}(2/r_0 - 1/\rho)$$

and

$$K_2 = \frac{\alpha e^2}{4\pi\epsilon_0 r_0^2}[-3/r_0^2 + 1/(2\rho^2)].$$

28. If we consider again the linear term (δr) in the result of Problem 27, we have a harmonic oscillator. (a) Find an expression for the frequency of oscillation and evaluate using the correct values of α, r_0, and ρ for NaCl. (b) Find the photon wavelength corresponding to the frequency you computed in (a) and compare with the observed absorption wavelength for NaCl of about 61 μm.

29. Refer again to the result of Problem 27. (a) Use the fact that the average new force on an ion must be zero (if there is no overall translation of the crystal) to show that the average values of δr and $(\delta r)^2$ are related to $\overline{\delta r} = -(K_2/K_1)\overline{(\delta r)^2}$. (b) According to the equipartition theorem the mean potential energy of an oscillator is $\frac{1}{2}K_1\overline{(\delta r)^2} = \frac{1}{2}kT$. Use this to show that $\overline{\delta r} = (K_2/K_1^2)\, kT$, and thereby show that the coefficient of thermal expansion is approximately $(1/r_0)(kK_2/K_1^2)$. (c) Use the result of (b) to evaluate the coefficient of thermal expansion for NaCl and compare with the experimental value of 4×10^{-6} K^{-1}.

10.4 Thermal and Magnetic Properties of Solids

30. (a) Derive Equation (10.26). (b) Evaluate the constant $3bk/4a^2$ in Equation (10.26) for copper, given that the coefficient of linear expansion [defined as $\alpha = \Delta x/(x\Delta T)$] for copper is found experimentally to be 1.67×10^{-5} K^{-1} at $T = 293$ K.

31. (a) Explain why the parameter *a* in Equation (10.22) is essentially twice the effective force constant of a spring connecting adjacent atoms. Then use this result along with the result of Problem 30 (b) to estimate a value for the parameter *b* in Equation (10.22), the coefficient of the x^3 term in the potential energy.

32. (a) Show that the ideal gas law can be written as

$$PV = \frac{2N\overline{E}}{3}$$

where *N* is the number of particles in the sample and $\overline{E}$ is the mean energy. (b) Use the result of (a) to estimate the pressure of the conduction electrons in copper, assuming an ideal Fermi electron gas. Comment on the numerical result, noting that one atmosphere equals 1.01×10^5 Pa.

33. Show that the *bulk modulus,* defined as

$$B = -V\frac{\partial P}{\partial V}$$

(where *P* is pressure and *V* is volume) can be written as

$$B = \frac{5P}{3} = \frac{2NE_F}{3V}$$

for a Fermi electron gas with Fermi energy E_F. (*Hint:* Use the relationship between *P* and *V* given in Problem 32(a).)

34. (a) From the result of Problem 33, compute the bulk modulus of pure silver. (b) Compare your result with the experimental value of 1.01×10^{11} N/m^2.

35. Starting with Equation (10.40) derive an expression for $\bar{\mu}$ valid in the *low temperature* limit $kT \ll \mu B$.

36. (a) Plot $\bar{\mu}$ vs. $\mu B/kT$ over the range $\mu B/kT = 0$ to $\mu B/kT = 4$. (b) Compute $\bar{\mu}$ at $\mu B/kT = 5$ and compare your results with the approximate value given in Problem 35. (c) Compute $\bar{\mu}$ at $\mu B/kT = 0.10$ and compare your results with the approximate value given in Section 10.4, $\tanh(\beta\mu B) \approx \beta\mu B$.

37. Prove that magnetic susceptibility χ is a dimensionless quantity. Note that the definition in Equation (10.37) presumes SI units.

38. (a) Compute the maximum magnetization of a bulk sample of iron, assuming perfect alignment of the spins and one unpaired spin per atom. (b) Compare with the observed maximum magnetization of about 1.6×10^6 A/m. (c) Based on your results in (a) and (b), what can you say about the actual number of unpaired spins per atom of iron?

10.5 Superconductivity

39. At what temperature (expressed as a fraction of T_c) is $B_c = 0.1B_c(0)$, according to the BCS theory? Repeat for $B_c = 0.5B_c(0)$ and $B_c = 0.9B_c(0)$.

40. It is found that for a given pure metal superconductor, photons of wavelength 0.568 mm are sufficient to break the Cooper pairs at $T = 2.0$ K. Identify the superconductor.

41. Compute T_c of the mercury isotopes ^{201}Hg and ^{204}Hg.

42. Estimate the BCS prediction for the change in T_c if all of the ^{16}O atoms in $YBa_2Ca_3O_7$ (assume $T_c = 93.0$ K) could be replaced by ^{18}O. This change is *not* observed experimentally!

43. From the data given in the text, estimate T_c for $Tl_2Ba_2Ca_{n-1}O_{2n+4}$ with $n = 4$.

44. A solenoid made of superconducting wire has exactly 30 turns/cm length and a diameter of 2.5 cm. If the wire carries a current of 5.0 A, what is the magnetic field strength near the center of the solenoid? What is the magnetic flux through a cross-section of the solenoid taken near the center? To how many flux quanta does this correspond? Comment on the number of flux quanta.

45. Recall that the magnetic field at the surface of a uniform cylindrical wire of radius R carrying a current I is

$$B = \frac{\mu_0 I}{2\pi R}$$

Find the minimum possible diameter for a wire of pure niobium so that the $T = 0$ critical field would not be exceeded if the wire carried a current of 5.5 A.

46. In a normal conductor heat is generated at a rate I^2R. Therefore a current-carrying conductor must dissipate heat effectively or it can melt or overheat the device in which it is used. Consider a long cylindrical copper wire (resistivity $1.72 \times 10^{-8}\ \Omega \cdot$ m) of diameter 0.75 mm. If the wire can dissipate 100 W/m^2 along its surface, what is the maximum current this wire can carry?

47. (a) Compute the maximum current that a niobium wire 0.75 mm in diameter can carry at $T = 4.2$ K. (b) Compare your result in (a) with the copper wire of the same diameter described in Problem 46.

10.6 Applications of Superconductivity

48. What is the maximum uncertainty in the measurement of the oscillation frequency in a Josephson junction if the voltage standard of 1 mV is to be maintained within 1 part in 10^{10}? Assume a reference frequency of 483.6 GHz.

49. Find the minimum acceleration needed for a maglev train to reach a speed of 550 km/h in 3.5 km, half the length of the 7 km track in Japan. Express your answer as a fraction of g, the free-fall acceleration near the Earth. Would this acceleration be noticeable?

50. (a) Compute the escape speed of a particle from the Earth's surface. The earth's radius is 6378 km and its mass is 5.98×10^{24} kg. (b) Find the mean speed $\bar{v}$ for a helium atom at a temperature of 293 K. (c) Comment on the fact that your answer to (b) is less than the answer to (a). Why then does helium not remain in the atmosphere in significant quantities?

51. A superconducting Nb_3Sn magnet can achieve a peak magnetic field of 13.5 T in a magnet designed for use in the Large Hadron Collider. Find the maximum energy that a singly charged particle (for example, a proton or electron) can have if that field is maintained around a circular ring of circumference 27 km. (*Note:* In reality particle energies will be about 35% less, because the peak field is not maintained throughout the ring.)

General Problems

52. In the persistent current experiment described in Section 10.5, let us assume that the current persisted without detectable reduction for exactly 2.5 years. Given that the inductance of the ring was approximately 3.14×10^{-8} H and the sensitivity of the current measurement was 1 part in 10^9, (a) estimate an upper bound for the resistance of the ring, and (b) estimate how long we could be assured that at least 90% of the current will remain.

53. From thermodynamics the entropy difference per unit volume between the normal and superconducting states is

$$\frac{\Delta S}{V} = -\frac{\partial}{\partial T}\left[\frac{B^2}{2\mu_0}\right]$$

where $B^2/2\mu_0$ is the magnetic energy density needed to return a superconductor to the normal state. Use this fact to compute the entropy difference between the normal and superconducting states in 1 mole of niobium at a temperature of 6.0 K.

CHAPTER 11

Semiconductor Theory and Devices

Once experienced, the expansion of personal intellectual power made available by the computer is not easily given up.

Sheila Evans Widnall, Science, *August 12, 1983*

In the last half of the 20th century a revolution occurred in electronics. The development and widespread use of integrated circuits have changed the way we live. Automobiles, televisions, audio equipment, microwave ovens, and other home appliances now contain microcomputers to enhance their efficiency. The electronics revolution has had an even more profound effect on the work of scientists and engineers. In the early days of the space program (in the 1950s), the slide rule, pencil, and paper were still the standard tools of the physicist and the aeronautical engineer. For the most difficult calculations, they could use the best computers of their day. Those computers filled an entire room and lacked the speed and computational power of desktop personal computers available today (not to mention that the cost was orders of magnitude higher!).

The remarkable properties of semiconductor materials have made possible the advances just described. In this chapter we shall examine those properties and see how they are used in various devices. It is possible to understand the behavior of semiconductors by studying the quantum theory of solids. But as it was with superconductors, that theory is to a great extent quite advanced and therefore beyond the scope of this text. Still, we can present a good portion of the theory in descriptive fashion and thereby make evident its beauty and utility.

11.1 Band Theory of Solids

There are three categories of solids, based on their conducting properties: conductors, semiconductors, and insulators. As seen in Table 11.1, the electrical conductivity at room temperature is quite different for each of these three kinds of solids. We have already accounted for the conductivity of ordinary metals in Section 9.6.

TABLE 11.1
Electrical Resistivity and Conductivity of Selected Materials at 293 K

Material	Resistivity ($\Omega \cdot m$)	Conductivity ($\Omega^{-1} \cdot m^{-1}$)
	Metals	
Silver	1.59×10^{-8}	6.29×10^{7}
Copper	1.72×10^{-8}	5.81×10^{7}
Gold	2.44×10^{-8}	4.10×10^{7}
Aluminum	2.82×10^{-8}	3.55×10^{7}
Tungsten	5.6×10^{-8}	1.8×10^{7}
Platinum	1.1×10^{-7}	9.1×10^{6}
Lead	2.2×10^{-7}	4.5×10^{6}
	Alloys	
Constantan	4.9×10^{-7}	2.0×10^{6}
Nichrome	1.5×10^{-6}	6.7×10^{5}
	Semiconductors	
Carbon	3.5×10^{-5}	2.9×10^{4}
Germanium	0.46	2.2
Silicon	640	1.6×10^{-3}
	Insulators	
Wood	10^{8}–10^{11}	$10^{-8} - 10^{-11}$
Rubber	10^{13}	10^{-13}
Amber	5×10^{14}	2×10^{-15}
Glass	10^{10}–10^{14}	$10^{-10} - 10^{-14}$
Quartz (fused)	7.5×10^{17}	1.3×10^{-18}

From R. Serway, *Physics for Scientists and Engineers,* 3rd ed. Philadelphia: Saunders College Publishing, 1990, p. 747.

The free-electron model used in Chapter 9 does not apply to semiconductors and insulators. In these materials there are simply not enough free electrons for the material to conduct in a free-electron mode. One may be tempted to apply the free-electron Fermi-Dirac model to semiconductors and explain their relatively low conductivity at room temperature by a relative lack of free electrons. If the electrical conductivity of semiconductors is about 10 orders of magnitude lower, could it be that the charge carrier density is lower by that factor? It turns out that carrier density is only part of the story. In fact there is a very different conduction mechanism for semiconductors than for normal conductors. Striking evidence of this fact is seen in Figure 11.1. While the free-electron theory correctly predicts a linear increase in resistivity with temperature, semiconductors generally exhibit *decreasing* resistivity with increasing temperature. We will need a new theory, known as the **band theory,** in order to account for this and other properties of semiconductors.

Band theory

A simple and straightforward justification of the existence of electronic energy bands in solids was suggested by William Shockley, one of the co-inventors of the transistor (Section 11.3). We first consider what happens when two atoms of hydrogen are brought together (the argument is good for any kind of atom; we choose hydrogen because it has known wave functions). When the two atoms

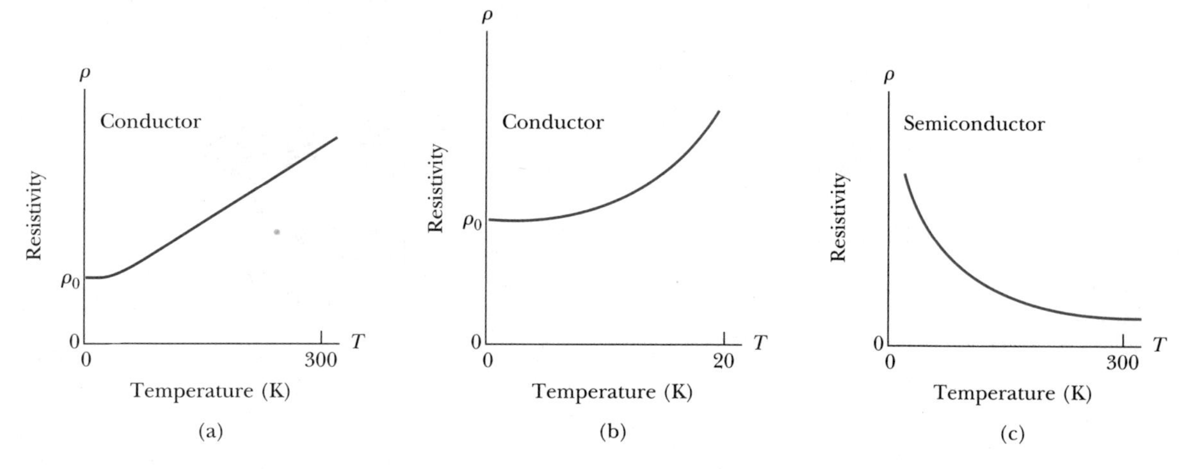

FIGURE 11.1 (a) Resistivity versus temperature for a typical conductor. Notice the linear rise in ρ with increasing temperature at all but very low temperatures. (b) Resistance versus temperature for a typical conductor at very low temperatures. Notice that the curve flattens and approaches a nonzero resistance as $T \rightarrow 0$. (c) Resistivity versus temperature for a typical semiconductor. The resistivity increases dramatically as $T \rightarrow 0$.

are far apart, the electronic wave functions can be thought of as noninteracting. As the atoms are brought closer together, the wave functions begin to overlap. But because any linear combination of wave functions is possible, there can be either a symmetric or an antisymmetric overlap (see Figure 11.2). These two situations correspond to two slightly different energies. Hence, there is a splitting of all possible energy levels (1*s*, 2*s*, etc.), as is seen in Figure 11.3a. In each case the symmetric state ($\Psi_A + \Psi_B$) has the lower energy. An electron in the symmetric state has a nonzero probability of being halfway between the two atoms (Figure 11.2b), while an electron in the antisymmetric state ($\Psi_A - \Psi_B$) has a zero probability of being at that location. This causes the binding to be slightly stronger in the symmetric case, and hence the energy of that state is lower.

When more atoms are added (as in a real solid), there is a further splitting of energy levels. With a large number of atoms the levels are split into nearly continuous **energy bands,** as seen in Figure 11.3b. There may or may not exist an **energy gap** between bands, depending on a number of factors, including the type of atom or atoms in the solid, lattice spacing, lattice structure, and temperature. We shall consider the effects of the existence and size of the energy gap later in this chapter.

Energy bands
Energy gap

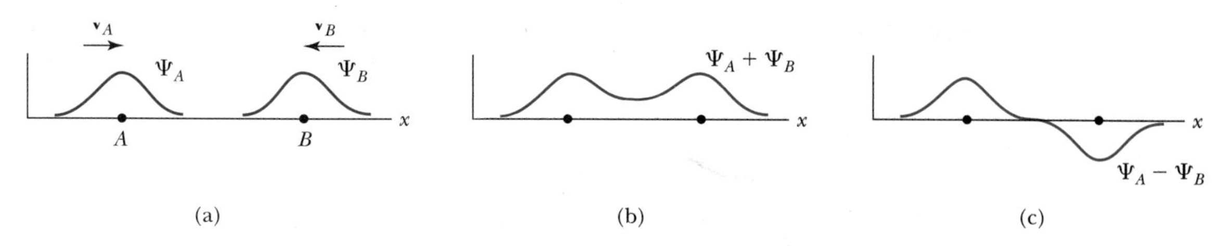

FIGURE 11.2 A rough representation of the wave functions of two approaching hydrogen atoms. When they are far apart (a) there is negligible overlap of their wave functions. In (b) and (c) the atoms are closer, and the wave functions begin to overlap. In (b) they combine with the same sign, and in (c) with opposite signs.

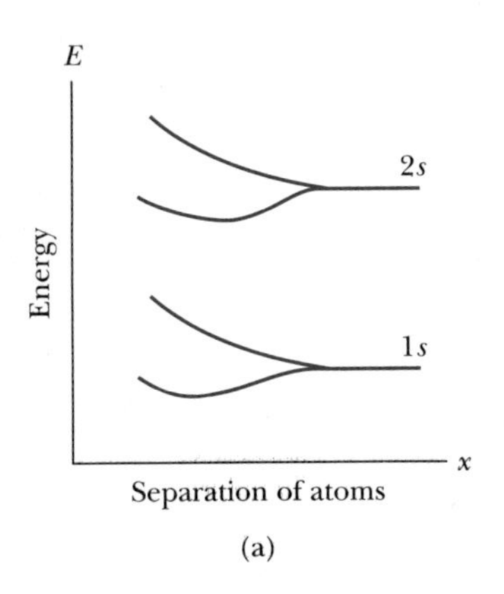

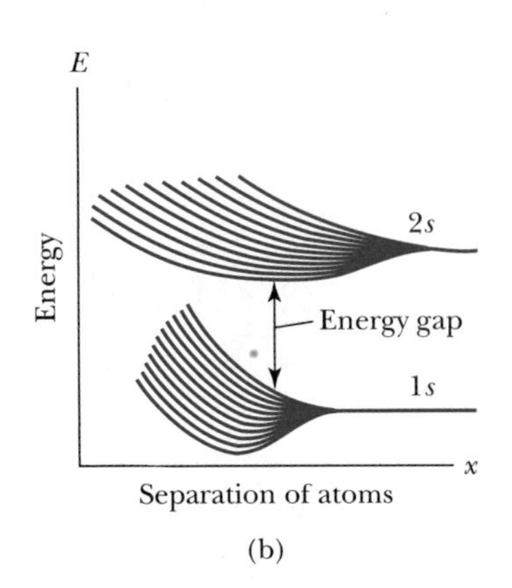

FIGURE 11.3 The 1*s* and 2*s* energy-level splittings of approaching hydrogen atoms for (a) 2 atoms, and (b) 11 atoms. Notice the splitting of each energy level into a nearly continuous band.

Kronig-Penney Model

Another more quantitative way to understand the energy gap in semiconductors is to model the interaction between the electrons and the lattice of atoms. This interaction is more important in semiconductors than in good conductors, because the much higher resistivity implies tighter binding and/or more interaction. A very useful one-dimensional model was developed in 1931 by Kronig and Penney.* They assumed that an electron experiences the potential shown in Figure 11.4, an infinite one-dimensional array of finite potential wells. The size of the wells must correspond roughly to the lattice spacing.

The Kronig-Penney method for finding the allowed energy levels for the electron is the same as we used in Chapter 6. The electrons are not free; therefore we assume that the total energy E of an electron is less than the height V_0 of each barrier/well in the Kronig-Penney potential shown in Figure 11.4. The electron is essentially free in the gap $0 < x < a$, where it has a wave function of the form

$$\psi = Ae^{ikx} + Be^{-ikx} \tag{11.1}$$

where the wave number k is given by the usual relation $k^2 = 2mE/\hbar^2$. In the barrier region $a < x < a + b$, however, the electron can tunnel through. As we saw in Chapter 6, this means that the wave function loses its sinusoidal character and becomes

$$\psi = Ce^{\kappa x} + De^{-\kappa x} \tag{11.2}$$

with $\kappa^2 = 2m(V_0 - E)/\hbar^2$.

Next, Kronig and Penney used the procedure of matching wave functions and first derivatives thereof at the various boundaries. The fact that these are

*R. de L. Kronig and W. G. Penney, *Proc. Roy. Soc. (London)*, **A130,** 499 (1931).

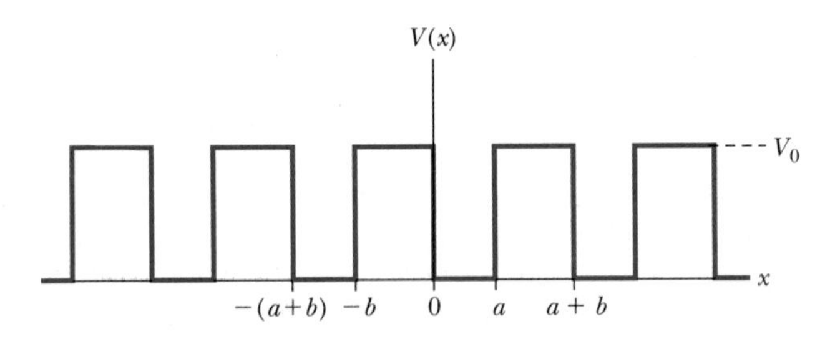

FIGURE 11.4 The Kronig-Penney square-well potential.

finite potential wells makes this a tedious problem to solve, and we shall not attempt to do so here. Application of the appropriate boundary conditions yields the important relation

$$\frac{\kappa^2 b}{2k}\sin(ka) + \cos(ka) = \cos(Ka) \tag{11.3}$$

where K is another wave number. When the left side of Equation (11.3) is plotted against the argument ka (Figure 11.5a), the relation (11.3) cannot be satisfied for all values of κ and k, because the sine and cosine functions are restricted to the range -1 to $+1$. There is no solution whenever the left side of Equation (11.3) has a value outside the range -1 to $+1$, because the right side of the equation *must* have a value in the range -1 to $+1$. Apparently there are **forbidden zones** for the wave numbers, and hence there are gaps in the allowed energies. Figure 11.5b shows how the energy gaps and allowed energy bands alternate with each other as one sweeps over increasing energies. The gaps occur regularly at $ka = n\pi$, with n an integer. Then if $k = n\pi/a = 2\pi/\lambda$, we see that $\lambda = 2a/n$. Thus, twice the lattice spacing $(2a)$ corresponds to an integer multiple of the free-particle wavelength $(n\lambda)$ and a free particle with this wavelength would be reflected by the lattice.

Forbidden zones

Before proceeding, we note some important differences between this simplified Kronig-Penney model and the single potential well studied in Chapter 6. First, for an infinite lattice the allowed energies within each band are continuous rather than discrete. In a real crystal the lattice is not infinite, but even if chains are only thousands of atoms long, the allowed energies are nearly continuous. Second, note that in a real three-dimensional crystal it is appropriate to speak of a **wave vector k.** The allowed ranges for **k** constitute what are referred to in solid state theory as **Brillouin zones.** Finally, in a real crystal the potential function is somewhat more complicated than the Kronig-Penney squares. A result of this is that the energy gaps are by no means uniform in size. The gap sizes may be changed by the introduction of impurities or imperfections of the lattice. These facts concerning the energy gaps are of paramount importance in understanding the electronic behavior of semiconductors.

Wave vector
Brillouin zones

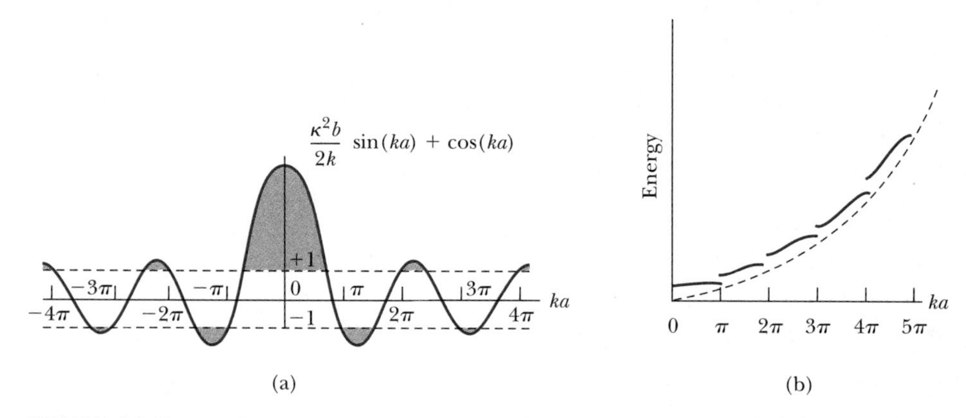

FIGURE 11.5 (a) Plot of the left side of Equation (11.3) versus ka for $\kappa^2 ba/2 = 3\pi/2$. Allowed energy values must correspond to the values of $k = \sqrt{2mE/\hbar^2}$ for which the plotted function lies between -1 and $+1$. Forbidden values are shaded in light blue. (b) The corresponding plot of energy versus ka for $\kappa^2 ba/2 = 3\pi/2$, showing the forbidden energy zones (gaps).

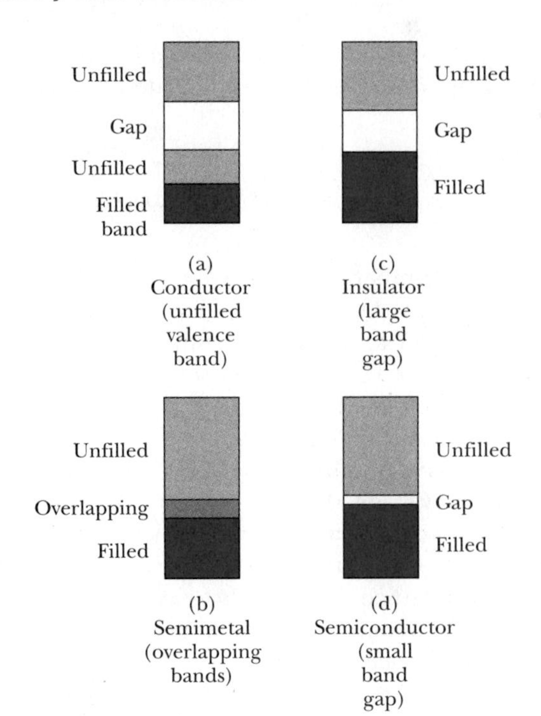

FIGURE 11.6 Possible band structures: (a) a conductor with an unfilled valence band, (b) a conductor with overlapping valence and conduction bands (a semimetal), (c) an insulator due to its large band gap, and (d) a semiconductor (due to its small band gap).

Conductors, Insulators, and Semiconductors

Now we proceed to the problem of what makes a conductor, insulator, or semiconductor. Good conductors like copper can be understood using the free-electron model, but it is also possible to make a conductor using a material with its highest band filled, in which case no electron in that band can be considered free. If this filled band overlaps with the next higher band, however (so that effectively there is no gap between these two bands, as shown in Figure 11.6b), then an applied electric field can make an electron from the filled band jump to the higher level. This allows conduction to take place, although typically with slightly higher resistance than in normal metals. Such materials are known as **semimetals.** Several widely used semimetals are arsenic, bismuth, and antimony.

Semimetals

The band structures of insulators and semiconductors resemble each other qualitatively. Normally there exists in both insulators and semiconductors a filled energy band (referred to as the **valence band**) separated from the next higher band (referred to as the **conduction band**) by an energy gap. If this gap is large (at least several electron volts), the material is an insulator. It is just too difficult for an applied field to overcome that large an energy gap. But if the energy gap is only about one electron volt, it is possible for enough electrons to be in the conduction band in order for an applied electric field to produce a modest current. We shall explain this mechanism more fully in the next section.

Valence band
Conduction band

11.2 Semiconductor Theory

At $T = 0$ we expect all of the atoms in a solid to be in the ground state. The distribution of electrons (fermions) at the various energy levels is governed by the Fermi-Dirac distribution of Equation (9.30)

$$F_{\mathrm{FD}} = \frac{1}{B_1 \exp(\beta E) + 1} \tag{9.30}$$

where B_1 is a normalization constant and $\beta = (kT)^{-1}$. As was shown in Figure 9.8, more and more atoms are found in excited states when the temperature is increased from $T = 0$.

This fact can be used to explain the anomalous temperature dependence of the resistivity of semiconductors. Only those electrons that have jumped from the valence band to the conduction band are available to participate in the conduction process in a semiconductor. As the temperature is increased, more electrons are in the conduction band and the resistivity of the semiconductor therefore decreases.

While it is not possible to use the Fermi-Dirac factor to derive an exact expression for the resistivity of a semiconductor as a function of temperature, we can make a couple of observations. First, the energy E in the exponential factor makes it clear why the band gap is so crucial. An increase in the band gap by a factor of 10 (say from 1 eV to 10 eV) will, for a given temperature, increase the value of $\exp(\beta E)$ by a factor of $\exp(9\beta)$. This generally makes the factor F_{FD} so small that the material has to be an insulator. Our second observation is that, based on this analysis, one may expect the resistance of a semiconductor to decrease exponentially with increasing temperature. This is approximately true—although not exactly, because the function F_{FD} is not a simple exponential, and because the band gap does vary somewhat with temperature (Table 11.2).

A useful empirical expression developed by Clement and Quinnell for the temperature variation of standard carbon resistors is

$$\log R + K/\log R = A + B/T \tag{11.4}$$

Clement-Quinnell equation

where A, B, and K are constants. Figure 11.7a shows the data Clement and Quinnell obtained, plotted in such a way as to test Equation (11.4). The plot of $\log R + K/\log R$ versus $1/T$ is a straight line up to $1/T \approx 0.6$, so we conclude

TABLE 11.2
Energy Gaps for Selected Semiconductor Materials at $T = 0$ K and $T = 300$ K

	E_g (eV)	
Material	**0 K**	**300 K**
Si	1.17	1.11
Ge	0.74	0.66
InSb	0.23	0.17
InAs	0.43	0.36
InP	1.42	1.27
GaP	2.32	2.25
GaAs	1.52	1.43
GaSb	0.81	0.68
CdSe	1.84	1.74
CdTe	1.61	1.44
ZnO	3.44	3.2
ZnS	3.91	3.6

From C. Kittel, *Introduction to Solid State Physics,* 6th ed. New York: John Wiley, 1986, p. 185.

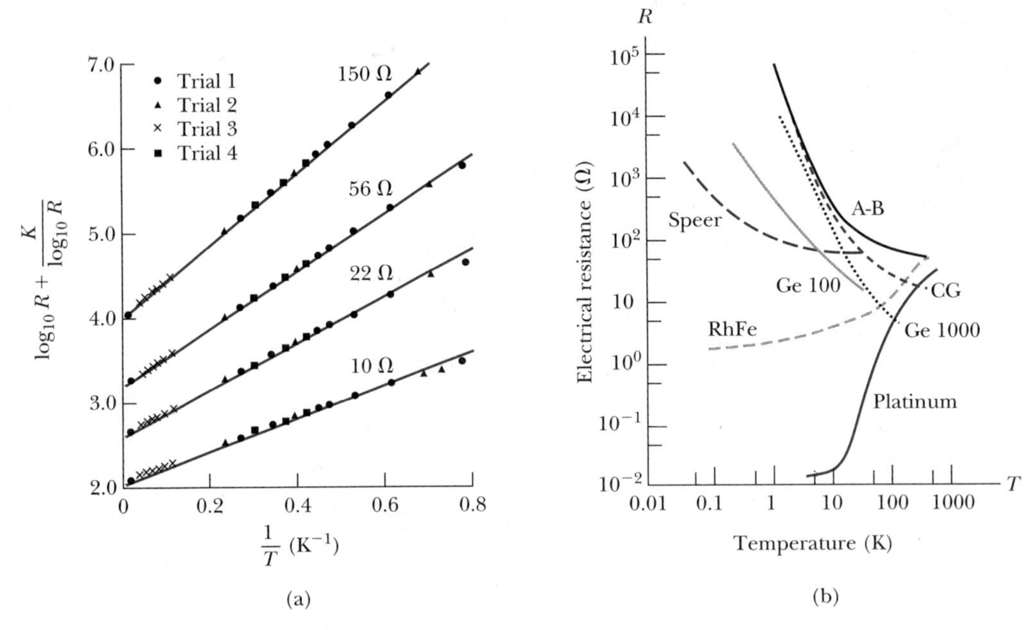

FIGURE 11.7 (a) An experimental test of the Clement-Quinnell equation, using resistance versus temperature data for four standard carbon resistors. The fit is quite good up to $1/T \approx 0.6$, corresponding to $T \geq 1.6$ K. (b) Resistance versus temperature curves for some thermometers used in research. A-B is an Allen-Bradley carbon resistor of the type used to produce the curves in (a). Speer is a carbon resistor, and CG is a carbon-glass resistor. Ge 100 and 1000 are germanium resistors. *From G. White, Experimental Techniques in Low Temperature Physics. Oxford: Oxford University Press, 1979.*

that the Clement-Quinnell equation (Equation 11.4) is good down to $T \approx 1$ K/0.6 $\approx$ 1.7 K. In Figure 11.7b we see a simple plot of resistance versus temperature for a number of different resistance thermometers. For all semiconductor materials shown, the variation of R with T is particularly rapid in the low-temperature range, from $T = 0$ up to about $T = 20$ K. For this reason carbon and other semiconductors are widely used as resistance thermometers in low-temperature physics.

Example 11.1

Assuming $B_1 = 1$ in Equation (9.30), find the relative number of electrons with energies 0.1 eV, 1 eV, and 10 eV above the valence band at room temperature (293 K). (*Note:* Setting $B_1 = 1$ is equivalent to setting the Fermi energy at the top of the valence band.)

Solution: For $E = 1$ eV, $E/kT = (1.6 \times 10^{-19}\,\text{J})/(1.38 \times 10^{-23}\,\text{J/K})(293\,\text{K}) = 39.61$. Then

$$F_{FD}(0.1\text{ eV}) \approx \frac{1}{e^{3.961} + 1} = 0.019$$

$$F_{FD}(1\text{ eV}) \approx \frac{1}{e^{39.61} + 1} = 6.27 \times 10^{-18}$$

$$F_{FD}(10\text{ eV}) \approx \frac{1}{e^{396.1} + 1} = 1.0 \times 10^{-172}$$

This example illustrates clearly how strongly the Fermi-Dirac factor depends on the size of the band gap. The number of electrons available for conduction drops off sharply as the band gap increases.

A curious (and important) phenomenon in semiconductors results from the fact that when electrons move into the conduction band, they leave behind vacancies in the valence band. We call these vacancies **holes,** and because they represent the absence of negative charges, it is useful to think of them as positive charges. While the electrons move in a direction opposite to the applied electric field, the holes move in the direction of the electric field. In an *intrinsic semiconductor* such as pure carbon or germanium, there is a balance between the number of electrons in the conduction band and the number of holes in the valence band.

Holes

In many applications it is desirable to create an *impurity semiconductor* by adding a small amount of another material (often but not necessarily another semiconductor) called a *dopant* to a semiconductor. As an example consider what happens when we add a small amount of arsenic to silicon. Each arsenic atom will in effect replace a silicon atom in the lattice. What will this do to the conductive properties of the material? Notice that silicon has four electrons ($3s^2 3p^2$) in its outermost shell (this corresponds to the valence band) and arsenic has five ($4s^2 4p^3$). Therefore, while four of arsenic's outer shell electrons will participate in covalent bonding with its nearest neighbors (just as another silicon atom would), the fifth electron is very weakly bound. In fact, it takes only about 0.05 eV to move this extra electron into the conduction band. The effect is that adding only a small amount of arsenic to silicon greatly increases the electrical conductivity.

The addition of arsenic to silicon creates what is known as an ***n*-type** semiconductor (*n* for negative), because it is the electrons close to the conduction band that will eventually carry electrical current. The new arsenic energy levels just below the conduction band are called **donor levels** (see Figure 11.8a), because an electron there is easily donated to the conduction band.

***n*-type semiconductors**

Now consider what happens when indium ($5s^2 5p^1$) is added to silicon. In this case there is one less electron than before the impurity was added. The result is an extra hole per indium atom. The existence of these holes creates extra energy levels just above the valence band, because it takes relatively little energy to move another electron into a hole. Those new indium levels are called **acceptor levels** (see Figure 11.8b), because they can easily accept an electron from the valence band. Again, the result is an increased flow of current (or, equivalently, lower electrical resistance) as the electrons move to fill holes under an applied electric field. It is always easier to think in terms of the flow of positive

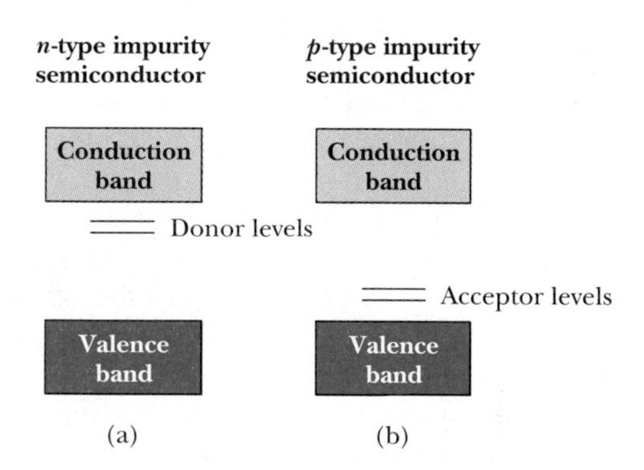

FIGURE 11.8 Energy bands of impurity semiconductors, showing (a) donor levels in an *n*-type impurity semiconductor and (b) acceptor levels in a *p*-type impurity semiconductor.

SPECIAL TOPIC:

THE QUANTUM HALL EFFECT

The quantum Hall effect has been the subject of intense investigation by semiconductor physicists since its discovery by the German physicist Klaus von Klitzing in 1980 (Nobel Prize, 1985). Schematically, the apparatus for observing the quantum Hall effect is similar to that used for the normal Hall effect. A thin strip of material is positioned perpendicular to a uniform magnetic field. A current is then made to flow along the length of the strip, and voltage leads detect the potential difference in two directions: along the direction of current flow and perpendicular to the current flow (see Figure 11.9). Both of these voltages are directly proportional to the current flowing through the strip. Therefore in order to characterize the material being studied it is customary to divide both voltages by the current to obtain resistance. The sample's ordinary electrical resistance is the voltage drop in the direction of current flow divided by the current, and the Hall resistance is the voltage in the perpendicular direction divided by the current.

In the normal Hall effect, the Hall resistance increases in direct proportion to the strength of the applied magnetic field. Von Klitzing discovered the quantum Hall effect in semiconductor materials in which the current-carrying electrons were confined to an extremely thin layer (5–10 nm thick), and when the sample was placed in a very strong magnetic field (2–10 T) and cooled to about 1 K. Under these conditions von Klitzing found that the Hall resistance, when plotted against the increasing magnetic field, showed well-defined steps at levels of h/ne^2, where h is Planck's constant, e is the electronic charge, and n is an integer (Figure A). Strictly speaking, then, it is the electrical conductance (inverse resistance) that is quantized in units of e^2/h. However, we shall stick to the convention of speaking in terms of the Hall resistance. The "base" Hall resistance h/e^2 is about 25,812.8 Ω. A second striking feature of the quantum Hall effect is that when the Hall resistance is on a plateau, the ordinary electrical resistance is practically zero (again refer to Figure A).

In 1982 another curious phenomenon was found by Daniel Tsui and Horst Störmer. They discovered that some semiconductor materials exhibited a *fractional* quantum Hall effect. To date electrical conductances have been found in fractions of $\frac{1}{3}$, $\frac{1}{5}$, $\frac{1}{7}$, $\frac{2}{3}$, $\frac{4}{5}$, and $\frac{6}{7}$ of the base value e^2/h. The fractional quantum Hall effect can be observed only in samples made of ultrapure materials.

The 1998 Nobel Prize for Physics was awarded to Tsui and Störmer, along with Robert Laughlin, who explained the fractional quantum Hall effect theoretically using a quantum fluid model.

***p*-type semiconductors**

charges (holes) in the direction of the applied field, so we call this a ***p*-type** semiconductor (p for positive).

In both p-type and n-type semiconductors, it is possible to control the electrical conductivity by selecting the desired amount of dopant. But doping is not the only way to take advantage of the properties of donor and acceptor levels. Many semiconductors are made with both donor and acceptor levels built in. This is accomplished by combining (for example) indium and antimony. The combination of those materials' electronic configurations produces a compound with large numbers of holes and electrons available for conduction. InSb and GaAs are among the most popular compounds of this type.

In introductory physics some students find it bothersome at first that the particles that actually move in a conductor, the electrons, travel opposite to the direction of the applied field and opposite to what we call "positive cur-

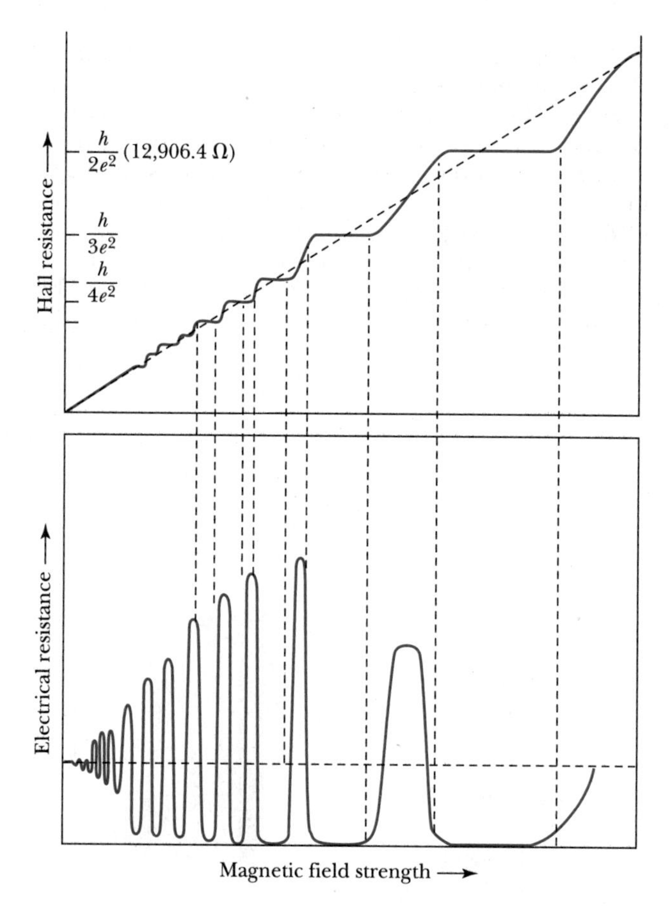

Semiconductor theorists have developed models that explain both the quantum Hall effect and the fractional quantum Hall effect. These models are too elaborate to detail here.* We can say, however, that they are based on constructing electron wave functions in *two* dimensions (remember, the electrons are basically confined to a plane). Such a two-dimensional structure is possible in FETs and in devices known as *heterojunctions,* in which electrons are confined to an interface region between two semiconductor materials.

One way that the quantum Hall effect and fractional quantum Hall effect have been applied is in the field of *metrology,* which seeks to develop measurements and standards for industrial and research purposes. The quantum Hall effects can be used to provide very accurate standards for resistance measurements. Because the base resistance h/e^2 depends only on the physical constants h and e, it should be possible to fabricate resistance standards that are relatively insensitive to how they were made, as long as they do exhibit a quantum Hall effect.

*See D. R. Yennie, *Rev. Mod. Phys.* **59,** 781–824 (1987) for many of the details.

FIGURE A The quantized steps in the Hall resistance are evident in the top graph. In the bottom graph we see the corresponding disappearance of electrical resistance. *Tomko/Scientific American.*

rent."* One may well ask to what extent the holes are real, because in both p-type and n-type semiconductors electrons are moving. It can be shown experimentally (using the Hall effect) that p-type materials really do behave as if the charge carriers were positive. The magnitude and sign of the Hall voltage allow one to calculate the density and sign of the charge carriers in a conductor or semiconductor and verify that charges in carriers in p-type and n-type materials have different signs. In fact, hole conduction is not a phenomenon limited to semiconductors. Among metallic conductors zinc is well known to exhibit positive charge carriers.

*This accident of history is usually blamed on Benjamin Franklin, who popularized the "single-fluid" model of electricity. Franklin considered electrical forces to be the result of excesses (hence "positive") and deficiencies (hence "negative") of an electrical fluid. Of course his designation is rather arbitrary, and if he had chosen the reverse, electrons would be positive.

Example 11.2

A very thin rectangular strip of zinc has been deposited on an insulating substrate. Let the length, width, and thickness of the strip be x, y, and z, respectively. The length and width are measured to be 10.0 cm and 2.0 cm. When a potential difference of 20 mV is applied as shown in Figure 11.9a, the current through the strip is 400 mA.

(a) Use the fact that the resistivity of zinc is (at room temperature) $5.92 \times 10^{-8}\ \Omega \cdot \text{m}$ to find the thickness of the strip.

(b) Now a magnetic field of 0.25 T is applied perpendicular to the strip as shown in Figure 11.9b. A Hall voltage $V_H = +0.56\ \mu\text{V}$ appears when the voltmeter leads (+ and −) are connected as shown. Determine the sign of the charge carriers and their number density.

Solution:

(a) The resistance is related to the resistivity by $R = \rho x/A$, where A is the cross-sectional area. In this case $A = yz$. Also, $R = V/I$, so $R = V/I = \rho x/yz$. Therefore

$$z = \frac{\rho x I}{yV} = \frac{(5.92 \times 10^{-8}\ \Omega \cdot \text{m})(0.10\ \text{m})(0.400\ \text{A})}{(0.02\ \text{m})(0.02\ \text{V})}$$

$$= 5.92 \times 10^{-6}\ \text{m} \tag{11.5}$$

(b) In equilibrium the magnetic force on a charge carrier is equal to the electric force due to the Hall voltage. Thus

$$eE = evB$$

$$e\frac{V_H}{y} = eB\frac{I}{neA} \tag{11.6}$$

where we have used the drift velocity $v = I/neA$ and the fact that for this geometry $E = V_H/y$. Finally, because $A = yz$, Equation (11.6) reduces to

$$n = \frac{IB}{eV_H z}$$

$$= \frac{(0.400\ \text{A})(0.25\ \text{T})}{(1.6 \times 10^{-19}\ \text{C})(5.6 \times 10^{-7}\ \text{V})(5.92 \times 10^{-6}\ \text{m})}$$

$$= 1.88 \times 10^{29}\ \text{m}^{-3} \tag{11.7}$$

This is a very high value (slightly higher, in fact, than one finds for copper). It turns out that the higher the density of conductors (electrons or holes), the thinner the strip needs to be in order to obtain a decent Hall voltage. That is why this experiment would be much easier to do with a semiconductor, most of which have a much lower value of n.

Now in order to determine the sign, notice that if the charge carriers were negative, the voltmeter as connected in Figure 11.9b would read *negative.* If instead the majority charge carriers are positive, the magnetic field will cause them to drift to the right on the strip. This is consistent with a positive voltmeter reading, and therefore we conclude that the majority charge carriers for zinc are *positive.* This is a somewhat unusual fact (though zinc is not unique—the majority carriers in Cd and Be are also positive), because for most metals the charge carriers are negative.

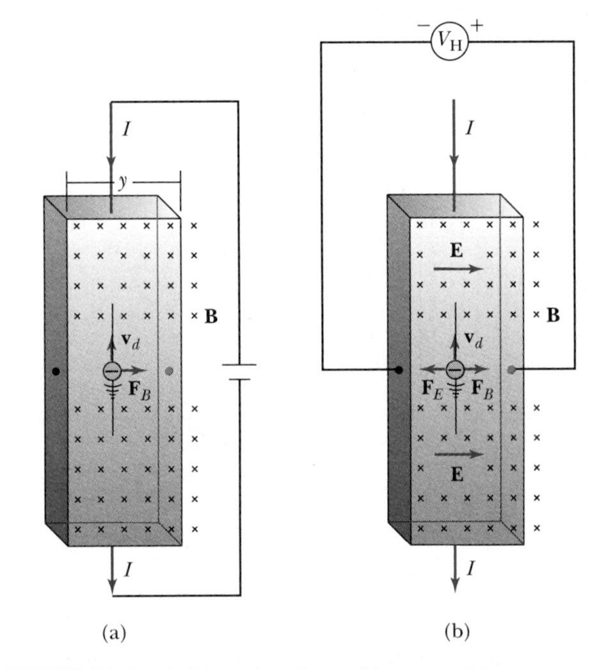

FIGURE 11.9 A thin strip of metal immersed in a magnetic field used to test the Hall effect. In (a) negative charge carriers are forced to the right. In (b) the buildup of negative charge on the right side (with a corresponding positive charge on the left) sets up the electric field as shown. This creates an electric force on the charge carriers equal and opposite to the magnetic force. The voltmeter (reading V_H) can detect the magnitude and sign of the potential difference across the strip.

Thermoelectric Effect

The **thermoelectric effect** is often used to study the properties of semiconductors. When there is a temperature gradient in a thermoelectric material, an electric field appears.* In one dimension the induced electric field E is proportional to the temperature gradient, so that

$$E = Q\frac{dT}{dx} \tag{11.8}$$

Thermoelectric power

where Q is called the **thermoelectric power.** It turns out that the direction of the induced field depends on whether the semiconductor is p-type or n-type (see Figure 11.10), so the thermoelectric effect can be used to determine the extent to which n- or p-type carriers dominate in a complex system. More detailed analysis yields information about carrier concentration and band structures.

The thermoelectric effect we have just described is sometimes called the **Seebeck effect.** It is the most commonly used thermoelectric effect, but there are two others. In a normal conductor, heat is generated at the rate of I^2R. But a temperature gradient across the conductor causes additional heat to be generated. This is known as the **Thomson effect.** Strangely enough, the Thomson effect is entirely reversible. If the direction of the current is toward the higher-temperature end of the conductor, heat is generated, but if the current flows toward the lower-temperature end, heat is absorbed from the surroundings. The **Peltier effect** occurs when heat is generated at a junction between two conductors as current passes through the junction.

*It is easier to understand why this happens in a pure metal, in which case we can assume a gas of free electrons. As in an ideal gas, the density of free electrons is greater at the colder end of the wire, and therefore the electrical potential should be higher at the warmer end and lower at the colder end. Of course the free-electron model is not valid for semiconductors; nevertheless, the conducting properties of a semiconductor are temperature dependent, as we have seen, and therefore it is reasonable to believe that semiconductors should exhibit a thermoelectric effect.

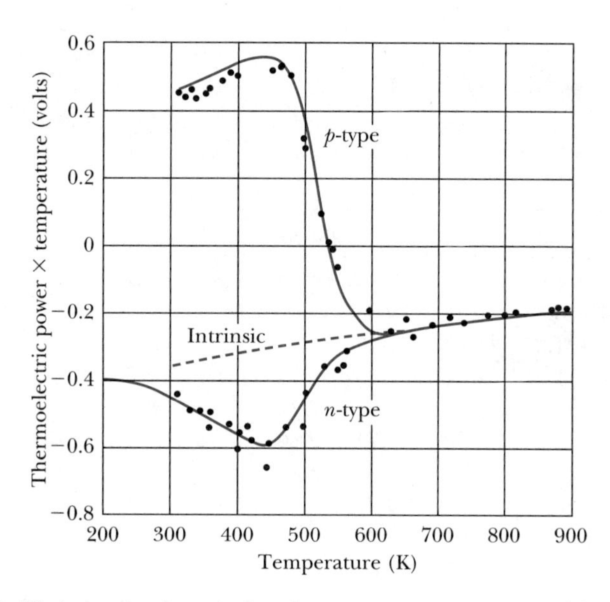

FIGURE 11.10 Variation in thermoelectric power × temperature with temperature for p-type and n-type silicon *From T.H. Geballe and G.W. Hall, Physical Review **98,** 940 (1955).*

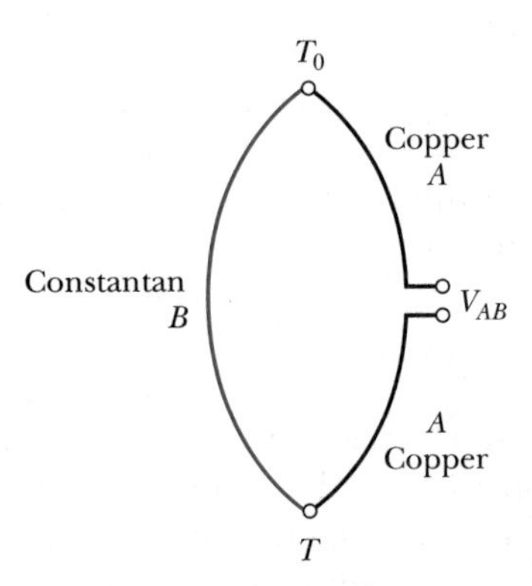

FIGURE 11.11 Schematic diagram of a thermocouple circuit. Two materials A and B (in this case, copper and constantan) are joined at their ends. One junction is held at a reference temperature T_0, and the other junction is used to measure the temperature T.

Thermocouple

The other important application of the Seebeck thermoelectric effect is in thermometry. The thermoelectric power of a given conductor varies as a function of temperature, and the variation can be quite different for two different conductors. This fact makes possible the operation of a **thermocouple** (Figure 11.11). Two conductors A and B are joined at each end. One end is held at a reference temperature T_0, and the other is placed at some unknown temperature T. The differential thermopowers of A and B cause a voltage V_{AB} to be induced between the two ends of the thermocouple. Knowing the temperature variation of each thermopower allows one to calculate the temperature difference $T - T_0$ (and hence the unknown temperature T) as a function of V_{AB}. In practice it is not necessary to measure thermopowers to use a thermocouple. Well-established voltage versus temperature tables can be found in many textbooks and handbooks for different conducting pairs over wide ranges of temperature (see Table 11.3).

TABLE 11.3
Thermocouple Voltage (in millivolts) versus Temperature (in °C) for an Iron-Constantan Thermocouple

°C	0	1	2	3	4	5	6	7	8	9
					Millivolts					
(+)0	0.00	0.05	0.10	0.15	0.20	0.25	0.30	0.35	0.40	0.45
10	0.50	0.56	0.61	0.66	0.71	0.76	0.81	0.86	0.91	0.97
20	1.02	1.07	1.12	1.17	1.22	1.28	1.33	1.38	1.43	1.48
30	1.54	1.59	1.64	1.69	1.74	1.80	1.85	1.90	1.95	2.00
40	2.06	2.11	2.16	2.22	2.27	2.32	2.37	2.42	2.48	2.53
50	2.58	2.64	2.69	2.74	2.80	2.85	2.90	2.96	3.01	3.06
60	3.11	3.17	3.22	3.27	3.33	3.38	3.43	3.49	3.54	3.60
70	3.65	3.70	3.76	3.81	3.86	3.92	3.97	4.02	4.08	4.13
80	4.19	4.24	4.29	4.35	4.40	4.46	4.51	4.56	4.62	4.67
90	4.73	4.78	4.83	4.89	4.94	5.00	5.05	5.10	5.16	5.21
100	5.27	5.32	5.38	5.43	5.48	5.54	5.59	5.65	5.70	5.76

From *CRC Handbook of Chemistry and Physics,* 55th ed. Cleveland: CRC Press, 1974, p. E-109.

11.3 Semiconductor Devices

Their unique electronic properties have made semiconductors useful in a wide range of applications in science and industry. In this section we shall describe some of the most important semiconductor devices and explain how they are used. But really we can only scratch the surface. The range of applications is wide, and semiconductor technology is still evolving at a rapid rate. Many of you who are students today will have much to do with the shape of the electronics revolution as it continues into the 21st century.

Diodes

We begin with a simple device known as a ***p-n*–junction diode,** in which *p*-type and *n*-type semiconductors are joined together, as shown in Figure 11.12. The principal characteristic of a *p-n*–junction diode is that it allows current to flow easily in one direction but hardly at all in the other direction. We call these situations **forward bias** and **reverse bias,** respectively.

***p-n*–junction diode**

Forward and reverse bias

To explain how this happens, we must first consider the situation when no external voltage is applied (the "no bias" case—Figure 11.12a). Excess electrons from the *n* side can drift through random motion to the *p* side, and their migration leaves a small net positive charge on the *n* side. This flow of electrons is shown as the electron recombination current I_r in Figure 11.12a. Equilibrium is achieved very quickly, because the potential difference set up by the charge migration (with the *n* side now at a higher potential than the *p* side) tends to prohibit further migration. There is typically a small current of electrons from the *p* side to the *n* side, due to the fact that at normal temperatures, electrons on the *p* side can be thermally excited from the valence band to an acceptor level. The thermal electron current is designated I_t in Figure 11.12a. With no external voltage source, the electron currents I_t and I_r exactly cancel each other, thus preserving the equilibrium.

In the reverse bias case, a potential difference is applied across the junction as shown in Figure 11.12b. In a normal conductor, electrons would tend to flow freely from the negative toward the positive terminal. But in the *p-n* junction, there remains the tendency for electrons to drift back from the *n* side toward the *p* side whenever an imbalance is created. This compensates for most of the electron flow. The result is only a small net flow of electrons from the *p* side to the

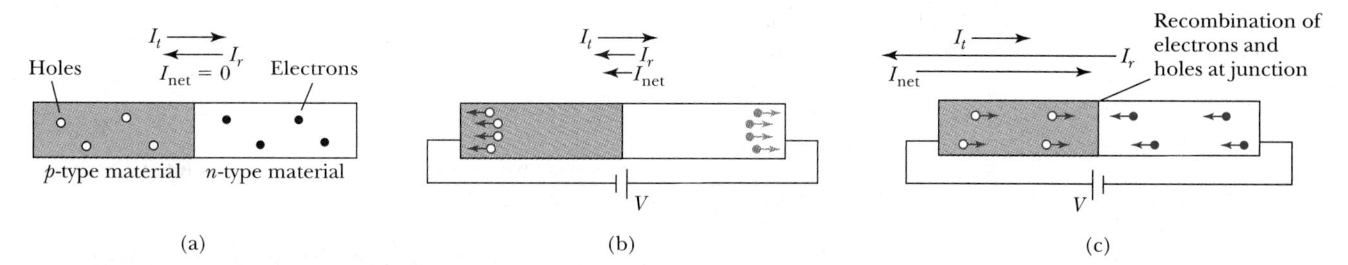

FIGURE 11.12 The operation of a *p-n*–junction diode. (a) This is the no bias case. The small thermal electron current (I_t) is offset by the electron recombination current (I_r). The net positive current (I_{net}) is zero. (b) With a DC voltage applied as shown the diode is in reverse bias. Now I_r is slightly less than I_t. Thus there is a small net flow of electrons from *p* to *n* and positive current from *n* to *p*. (c) Now the diode is in forward bias. Because current can readily flow from *p* to *n*, I_r can be much greater than I_t. (*Note:* In each case, I_t and I_r are electron (negative) currents, but I_{net} indicates positive current.)

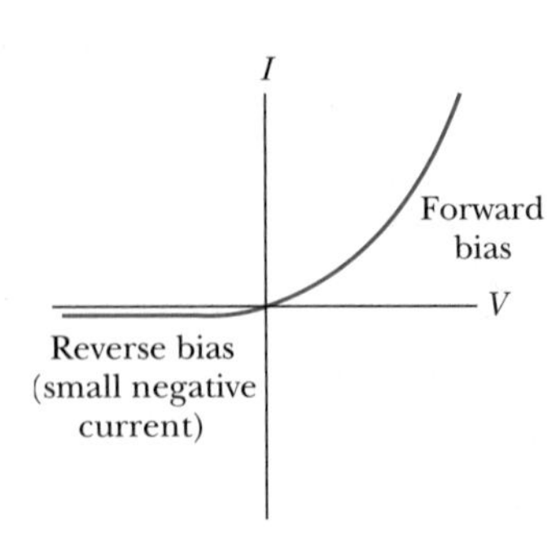

FIGURE 11.13 A typical I-V curve for a p-n–junction diode.

n side, or in other words, a small positive current from the high-potential side of the battery to the low-potential side of the battery.

In forward bias (Figure 11.12c) a potential difference is applied with the positive terminal connected to the p side and the negative terminal to the n side of the junction. Now we are pushing the electrons the way they would tend to move anyway. The only compensating factor, the thermal flow of electrons from p to n just described, is typically too small to retard the flow of electrons. The result is a steady flow of positive current from higher to lower potential, inhibited only by the natural resistance of the device. Figure 11.13 shows the behavior of the p-n–junction diode in both forward and reverse bias.

We can use the tools of statistical physics developed in Chapter 9 to model the I-V characteristics of the p-n–junction diode and thereby obtain a quantitative justification of the empirical curve shown in Figure 11.13. Let I_0 and V_0 be the current and voltage in the no-bias situation, and let N be the number of electrons on the n side in the conduction band. At room temperature, Maxwell-Boltzmann statistics are sufficient to describe the behavior of the electrons. In the no-bias case the number of electrons able to move from the n side to the p side must be proportional to $Ne^{-eV_0/kT}$, and therefore I_0 is proportional to the same factor. Under the influence of a forward bias voltage V, however, the number is proportional to $Ne^{-e(V_0-V)/kT} = Ne^{-eV_0/kT}e^{eV/kT}$. Therefore the electron current under forward bias must be $I = I_0 e^{eV/kT}$. Because there is still an additional current $-I_0$ in forward bias due to the motion of holes from the n side to the p side, the total current in forward bias is

$$I = I_0(e^{eV/kT} - 1) \tag{11.9}$$

Equation (11.9) is a rather good approximation of the I-V curve of Figure 11.13.

Example 11.3

Consider a simple p-n–junction diode. Suppose this diode carries a current of 50 mA with a forward bias voltage of 200 mV at room temperature (293 K). What is the current when a *reverse* bias of 200 mV is applied?

Solution: We could use Equation (11.9) to solve for I_0 in the forward bias case and then use that I_0 to find the current in reverse bias. But it is simpler to use the fact that the forward bias current I_f and the reverse bias current I_r are related by

$$\frac{I_r}{I_f} = \frac{I_0(e^{eV_r/kT} - 1)}{I_0(e^{eV_f/kT} - 1)}$$

Using $I_f = 50$ mA, $V_f = +200$ mV, and $V_r = -200$ mV, we find that

$$\frac{eV_f}{kT} = \frac{(1.6 \times 10^{-19}\text{ C})(0.200\text{ V})}{(1.38 \times 10^{-23}\text{ J/K})(293\text{ K})} = 7.921$$

$$I_r = I_f \frac{e^{eV_r/kT} - 1}{e^{eV_f/kT} - 1}$$

$$I_r = (50\text{ mA})\frac{e^{-7.921} - 1}{e^{+7.921} - 1} = (50\text{ mA})\frac{-1}{e^{+7.921}}$$

$$I_r = \frac{-50\text{ mA}}{2754} = -18\ \mu\text{A}$$

This is an extremely small current, showing clearly the "one-way" property of the diode.

Bridge Rectifiers

The diode is an important tool in many kinds of electrical circuits. As an example, consider the **bridge rectifier** circuit shown in Figure 11.14. The bridge rectifier is set up so that it allows current to flow in only one direction through the resistor R when an alternating current supply is placed across the bridge. The current through the resistor is then a rectified sine wave—it is of the form

$$I = I_{max}|\sin(\omega t)| \quad (11.10)$$

This is the first step in changing alternating current to direct current. The power supply design can be completed by adding capacitors and resistors in appropriate proportions. Obviously this is an important application, because direct current is needed in many devices and the current that we get from our wall sockets is alternating.

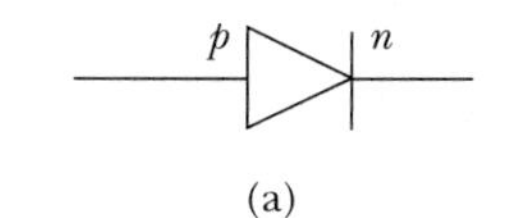

(a)

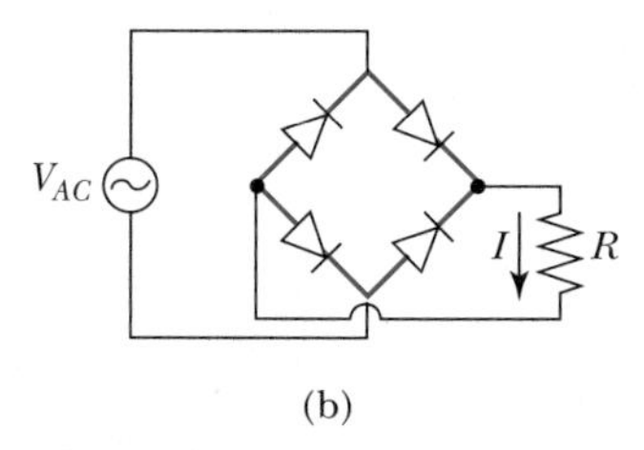

(b)

FIGURE 11.14 (a) The circuit diagram symbol for a diode, with p and n sides indicated. (b) Circuit diagram for a diode bridge rectifier.

Zener Diodes

The **Zener diode** is made to operate under reverse bias once a sufficiently high voltage has been reached. The *I-V* curve of a Zener diode is shown in Figure 11.15. Notice that under reverse bias and low voltage the current assumes a low negative value, just as in a normal *p-n*–junction diode. But when a sufficiently large reverse bias voltage is reached, the current increases at a very high rate.

Depending on which semiconductor materials are used to make the diode, the Zener *I-V* phenomenon may occur in one of two ways. In the process known as *Zener breakdown,* the applied voltage induces electrons from the valence band on the p side to move to the conduction band on the n side. Once a high reverse bias voltage is reached, large numbers of electrons can be pulled immediately into the conduction band, thus accounting for the sudden breakdown. In another process known as *avalanche multiplication,* the applied voltage is high enough to accelerate electrons to energies sufficient to ionize atoms through collision. The "avalanche" occurs when electrons released in this process are in turn accelerated and ionize other atoms. Again, this happens rather suddenly, accounting for a sharp increase in current at a particular voltage. In heavily doped material, Zener breakdown is likely to occur first.

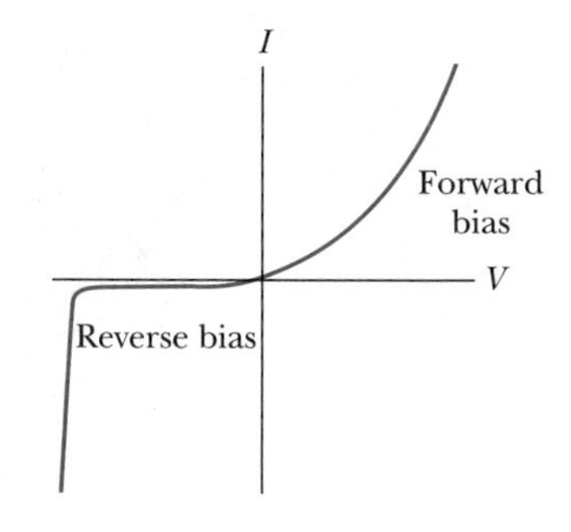

FIGURE 11.15 A typical *I-V* curve for a Zener diode.

A common application of Zener diodes is in regulated voltage sources, important components in many electronic instruments (see Figure 11.16). The idea behind the operation of this regulator circuit is that, as long as we operate on the steep part of the *I-V* curve, any change (say an increase) in the supply voltage V tends to be compensated for by a sharp increase in the current through the Zener diode. Then the voltage across the resistor increases, which in turn tends to keep the output voltage $V_Z(= V - IR)$ fairly constant.

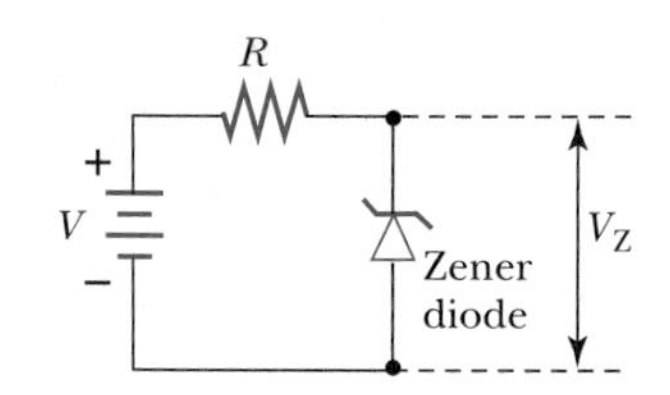

FIGURE 11.16 A Zener diode reference circuit.

Light Emitting Diodes

Another important kind of diode is the **light emitting diode (LED).** Whenever an electron makes a transition from the conduction band to the valence band (effectively recombining the electron and hole) there is a release of energy in the form of a photon (Figure 11.17). In some materials the energy levels are spaced so that the photon is in the visible part of the spectrum. In that case, the continuous flow of current through the LED results in a continuous stream of nearly monochromatic light.

LEDs are found in many places today. The visible displays of many calculators, clocks, automobile dashboards, and other instruments consist of combinations of LEDs. The LED can also be used as a principal component of a laser. Photons released from the LED can be used to stimulate the emission of other photons, as described in Section 10.2.

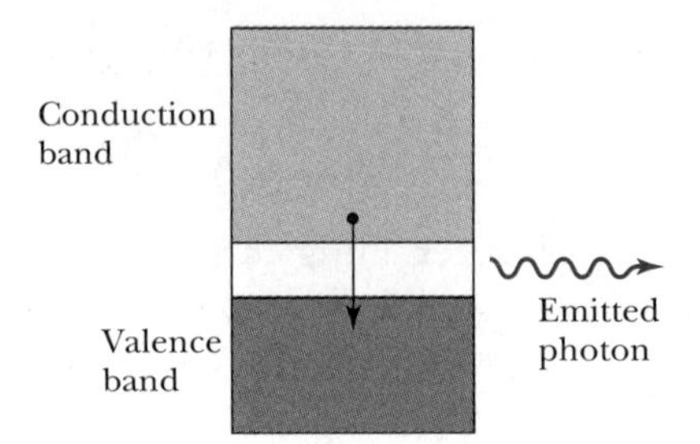

FIGURE 11.17 Schematic of a LED. A photon is released as an electron falls from the conduction band to the valence band. The band gap may be sufficiently large so that the photon will be in the visible portion of the spectrum.

Photovoltaic Cells

An exciting application closely related to the LED is the **solar cell,** also known as the **photovoltaic cell.** A good way to think of the solar cell is to consider the LED in reverse (Figure 11.18). A *p-n*–junction diode can absorb a photon of (say) solar radiation by having an electron make a transition from the valence band to the conduction band. In doing so, both a conducting electron and a hole have been created. If a circuit is connected to the *p-n*–junction, the holes and electrons will move so as to create an electric current, with positive current flowing from the *p* side to the *n* side. Even though the efficiency of most solar cells is low, their widespread use could potentially generate significant amounts of electricity. Remember that the "solar constant" (the energy per unit area of solar radiation reaching the Earth) is over 1400 W/m^2, and more than half of this makes it through the atmosphere to the Earth's surface. There has been tremendous progress in recent years toward making solar cells more efficient. This, combined with the increasing cost and decreasing availability of other energy sources, will make photovoltaic cells an important energy source in the years ahead.

The photovoltaic effect was first observed in 1839 by Alexandre-Edmond Becquerel. For over 100 years the discovery remained a curiosity for scientists with access to materials that naturally exhibit the photovoltaic effect. In the 1950s, however, researchers interested in using semiconductors in electronics

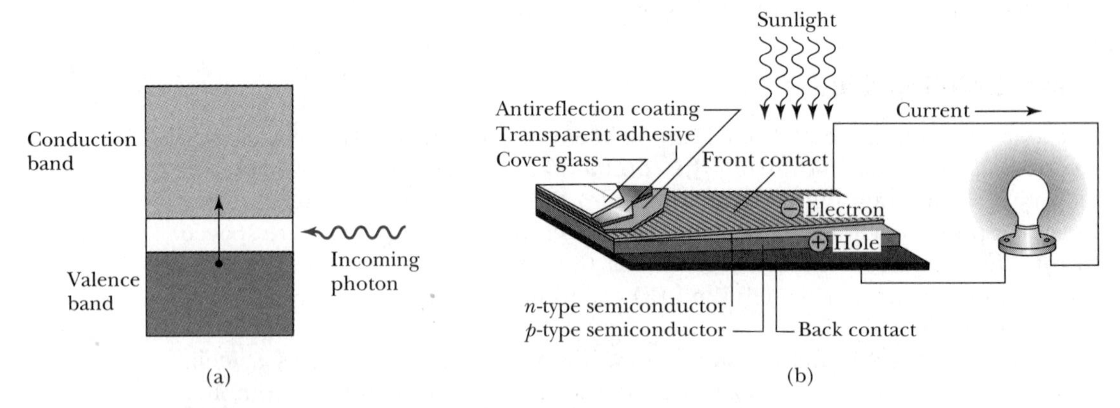

FIGURE 11.18 (a) Schematic of a photovoltaic cell. Note the similarity to Figure 11.17. (b) A schematic showing more of the working parts of a real photovoltaic cell. *From H.M. Hubbard, Science* ***244****, 297–303 (21 April 1989).*

began to develop better ways of manufacturing semiconductor devices, including photovoltaics. In 1954 a silicon-based solar cell with an efficiency of 6% was made at Bell Laboratories (efficiency is defined simply as the ratio of electrical power output to the power of incident radiation). Since then the advances in design and manufacture of semiconductor devices have been applied to solar cells. As a result, both silicon and gallium arsenide solar cells can now be made with efficiencies of over 20%. Lower manufacturing costs and higher efficiencies have reduced the cost of photovoltaic electricity by a factor of 10^2, from \$15 per kilowatt-hour in the 1950s to about \$0.15 per kilowatt-hour in the late 1990s. This cost is still about one order of magnitude higher than the cost of producing electricity using our other principal renewable energy source, hydroelectric power, and it is roughly a factor of two higher than the cost of using nonrenewable sources. Experts believe that research in the next decade or two will make solar cells a practical source of energy on a large scale by early in the 21st century.

There are two basic types of photovoltaic devices now in use: crystalline and thin film. The former has the advantage of higher efficiency, while the latter has the advantage of lower fabrication cost. The most widely used and studied crystalline photovoltaics are silicon-based. The technology exists for making large single crystals of silicon. It is in silicon-based cells that efficiencies of over 20% are reached. Unfortunately, the cost of making good single crystals of silicon is prohibitive. The cost of making cells with polycrystalline and amorphous silicon is lower, but so is the efficiency of the solar cells made with these materials.

There is now hope that thin-film devices (which are easier and more inexpensive to fabricate) can approach the efficiency of silicon devices. The prime candidate is GaAs and its alloys, including AlGaAs and InGaAs. Recently a device with an efficiency of 25% was made on a reusable GaAs substrate. These reusable single crystals of GaAs can now be made fairly inexpensively by growing them on single crystals of Ge.

Another advantage of GaAs is that it makes more efficient use of the solar spectrum than Si. Figure 11.19 shows that the response of GaAs and its alloys is almost entirely within the visible portion of the spectrum, where the output of solar radiation is highest (recall the Wien law). Recently the use of multiple layers, with each layer sensitive to a different wavelength range, has proven to be

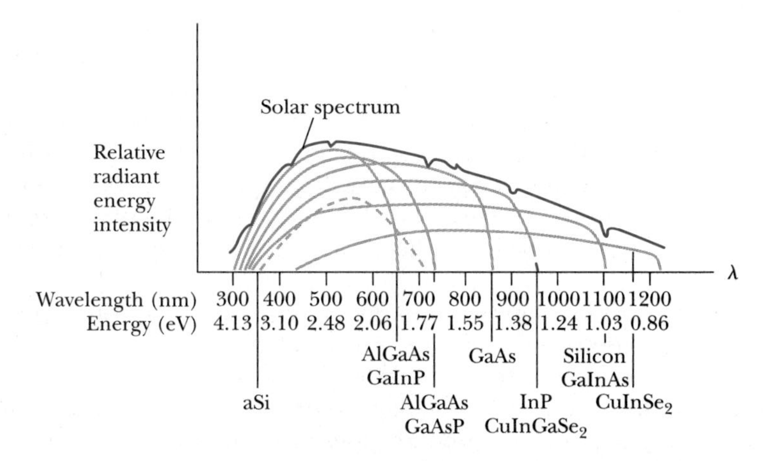

FIGURE 11.19 The response of various solar cells as a function of wavelength. Due to the variation in band gaps of the different semiconductors, solar cells will be more sensitive to some wavelengths than to others. *From H.M. Hubbard, Science* ***244****, 297–303 (21 April 1989).*

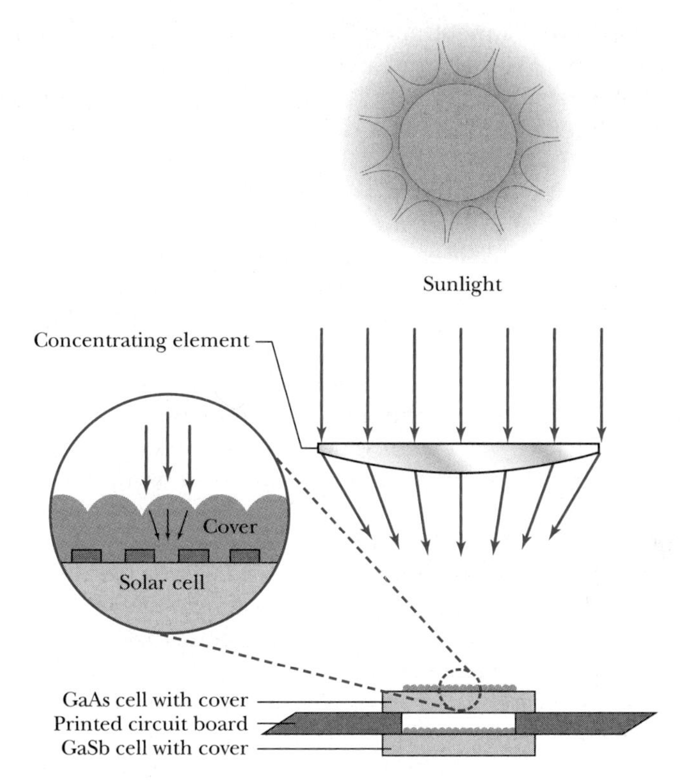

FIGURE 11.20 Boeing's 31% efficient solar cell, containing a gallium arsenide upper layer and a gallium antimonide lower layer used to capture infrared radiation. *From B.W. Henderson, Av. Wk. Space Tech* ***131,*** *61 (23 October 1989).*

very successful. Figure 11.20 shows a two-layer device created by Boeing. The upper layer semiconductor, GaAs, has a band gap of 1.42 eV. That energy corresponds to a photon wavelength of

$$\lambda = hc/E = 875 \text{ nm}$$

that is, a photon in the near infrared portion of the spectrum. The lower layer is made of GaSb, with a band gap of 0.72 eV (corresponding to $\lambda = 1730$ nm). Photons with wavelengths shorter than 875 nm will have sufficient energy to boost an electron from the valence to the conduction band in GaAs, and therefore the top layer is able to use near-IR, visible, and ultraviolet solar radiation to produce energy. Less-energetic IR photons make it through to the GaSb layer, where they can be absorbed and converted to electrical energy if their wavelength is less than 1730 nm. This device achieved an efficiency of 31% in 1989. Other multilayer systems are currently being studied, with the hope of raising the efficiency even further.

The use of solar radiation as an energy source has some intrinsic advantages and disadvantages compared with other sources. One outstanding advantage relative to fossil-fuel burning is that solar cells generate no environmental waste in their use and relatively little in their manufacture. The release of polluting particles, poisonous gases, and greenhouse gases (especially CO_2) in burning fossil fuels may make it desirable to replace fuel-burning generators with solar-cell generators even before the absolute cost per kilowatt hour of solar cells becomes much lower. Another advantage of solar cells is that they can be made in various

FIGURE 11.21 A photovoltaic center in Carrisa Plains, California. *Copyright Harald Sund, The IMAGE Bank.*

sizes. Large arrays (Figure 11.21) can be used to produce electricity for cities and factories, while smaller arrays can be put on rooftops in vacant areas to generate power for homes and small businesses. The principal disadvantage of even very efficient solar cells is that it is not always sunny. The use of solar energy at night or during cloudy periods would require large, generally inefficient storage systems. Perhaps this is where the superconducting storage rings discussed in Section 10.6 can best be put to use. (See also Problems 10 and 22.)

Transistors

Another use of semiconductor technology is in the fabrication of **transistors,** devices used to amplify voltages or currents in many kinds of circuits. The first transistor was developed in 1948 by John Bardeen, William Shockley, and William Brattain (Nobel Prize, 1956).* As an example we consider an ***n-p-n*–junction transistor,** which consists of a thin layer of *p*-type semiconductor sandwiched between two *n*-type semiconductors. The three terminals (one on each semiconducting material) are known as the **collector, emitter,** and **base.** A good way of thinking of the operation of the *n-p-n*–junction transistor is to think of two *p-n*–junction diodes back to back (Figure 11.22b). Then the emitter side consists of a diode in forward bias, and the collector side consists of a diode in reverse bias. The base would therefore be the *p* side of each diode. Explicitly showing

n-p-n–junction transistor

Collector, emitter, base

*For a history of transistor development see *Physics Today* (December 1997) pp. 34–39.

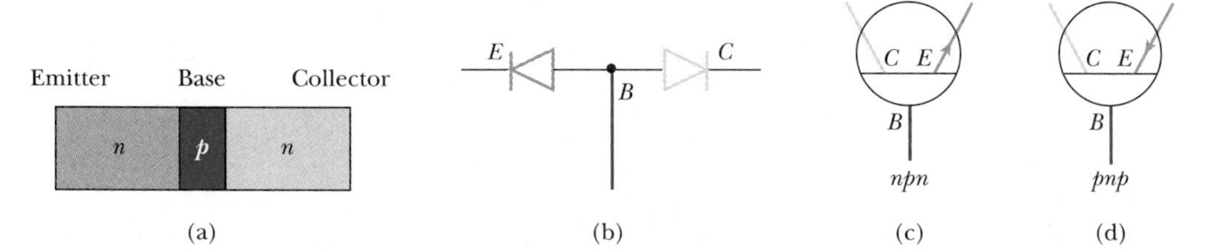

FIGURE 11.22 (a) In the *n-p-n* transistor, the base is a *p*-type material, and the emitter and collector are *n*-type. (b) The two-diode model of the *n-p-n* transistor. (c) The *n-p-n* transistor symbol used in circuit diagrams. (d) The *p-n-p* transistor symbol used in circuit diagrams.

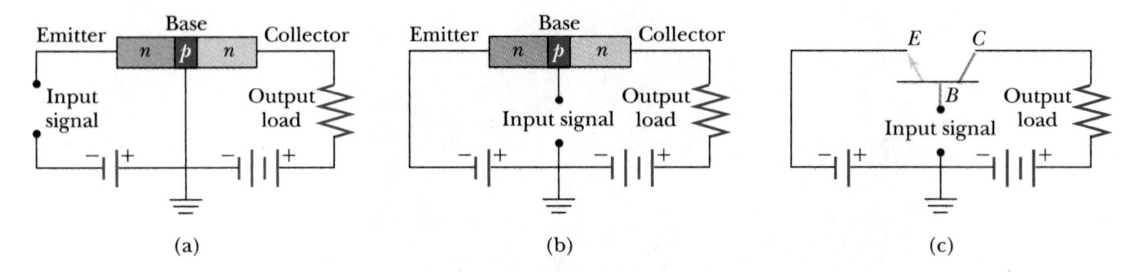

FIGURE 11.23 (a) The *n-p-n* transistor in a voltage amplifier circuit. (b) The circuit has been modified to put the input between base and ground, thus making a current amplifier. (c) The same circuit as in (b) using the transistor circuit symbol.

the *n-p-n* construction of each transistor would prove cumbersome in a complicated circuit diagram containing many transistors. A more compact notation for those cases is shown in Figure 11.22c and 11.22d.

Consider now the *n-p-n* junction in the circuit shown in Figure 11.23a. If the emitter is more heavily doped than the base, then there is a heavy flow of electrons from left to right into the base. The base is made thin enough so that virtually all of those electrons can pass through the collector and into the output portion of the circuit. As a result the output current is a very high fraction of the input current. The key now is to look at the input and output voltages. Because the base-collector combination is essentially a diode connected in reverse bias, the voltage on the output side can be made higher than the voltage on the input side. Recall that the output and input currents are comparable, so the resulting output power (current × voltage) is much higher than the input power.

In the circuit we have just described, the transistor is used as a voltage amplifier. The circuit in Figure 11.23a can be modified to serve as a current amplifier by moving the input signal to a position between the base and ground, as shown in Figure 11.23b. As we have already shown, a very small current flows in that branch of the circuit. Therefore, in this configuration the output current will be much higher than the input current. It is also possible to make a ***p-n-p*–junction transistor,** which may be understood using the same model as we used for the *n-p-n* junction, but with hole conduction taking the place of electron conduction.

p-n-p–junction transistor

As an example of an amplifier circuit consider Figure 11.24. The voltages V_{bb} and V_{cc} are fixed. The resistances R_c and R_e may in part be separate from the transistor, but they must include the intrinsic base-collector and base-emitter resistances, respectively. We wish to amplify a signal V_s. Let us assume that there is a

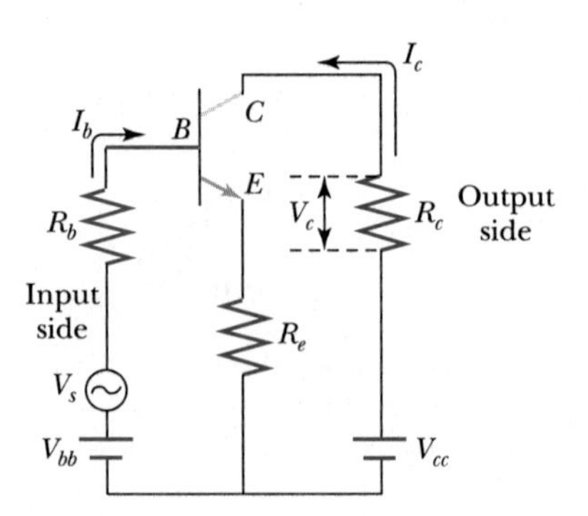

FIGURE 11.24 A transistor amplifier circuit.

current gain β, which means that $I_c = \beta I_b$. In order to calculate the voltage gain, we first apply Kirchhoff's loop rule to the left-hand loop to obtain

$$I_b = \frac{V_s + V_{bb}}{R_b + (1 + \beta)R_e} \tag{11.11}$$

Then

$$I_c = \beta I_b = \beta \frac{V_s + V_{bb}}{R_b + (1 + \beta)R_e} \tag{11.12}$$

The output voltage on the load resistor is

$$V_c = I_c R_c = \beta \frac{V_s + V_{bb}}{R_b + (1 + \beta)R_e} R_c \tag{11.13}$$

Now we can calculate the voltage gain A_v:

$$A_v \equiv \frac{V_c}{V_s} = \beta \frac{1 + V_{bb}/V_s}{R_b + (1 + \beta)R_e} R_c \tag{11.14}$$

Because the voltages V_{cc} and V_{bb} are present only to establish the proper bias on the transistor, it is possible in practice to make $V_{bb} \ll V_s$, so that

$$A_v \approx \beta \frac{R_c}{R_b + (1 + \beta)R_e} \tag{11.15}$$

With the appropriate choice of resistors, one can in theory achieve any desired voltage gain. In real circuits, however, there are severe limitations based on the characteristics of the particular transistor and power limitations in various parts of the circuit. Often, capacitors are put into the circuit to limit power surges. It is also not unusual to amplify a small signal through several (or many) stages before reaching the desired output voltage. For example, the initial output from the magnetic or optical reading device is typically on the order of microvolts, and it has to be amplified several times before it is strong enough to drive the speakers of an audio system.

Solid state transistors are not made by gluing together pieces of n-type and p-type semiconductors. It is possible to make layers of n- and p-type materials by diffusing or implanting acceptor and donor atoms at the appropriate thicknesses in a slab of pure germanium or silicon. It is this technology that has made possi-

Left to right, John Bardeen (1908–1991), William Shockley (1910–1989), and Walter Brattain (1902–1987) developed the semiconductor transistor at Bell Labs in the late 1940s. *AIP/Niels Bohr Library, Brattain Collection.*

ble the exceptional miniaturization of electronic circuits, which we shall discuss in more detail later.

Field Effect Transistors

The devices we have just described are referred to as a group by the term **bipolar transistors.** A similar device is the **field effect transistor** (FET). A schematic diagram of a FET is shown in Figure 11.25. The three terminals of the FET are known as the *drain, source,* and *gate,* and these correspond to the collector, emitter, and base, respectively, of a bipolar transistor. Comparing the circuit of Figure 11.25 with the corresponding circuit for the bipolar transistor (Figure 11.23), we see the principal difference is that the *p*-type gate is connected in reverse bias with *both* the *n*-type source and *n*-type drain, while the *p*-type base was connected in reverse bias with the *n*-type collector but a forward bias with the *n*-type emitter. The effect of this is that there is a severe depletion of charge carriers, both holes and electrons, at the *p-n*–junction of the FET, and therefore very little current can flow through that junction. For that reason the FET is said to be a high input impedance device and voltage controlled, while the bipolar transistor is current controlled. This turns out to be an advantage in the design of many kinds of circuits, because the FET draws a relatively small amount of current.

MOSFETs

In common use today is the metal oxide semiconductor FET, or **MOSFET.** In a MOSFET the gate is some kind of metal, and it is separated from the channel by a thin layer of oxide (an insulator), usually silicon dioxide. The oxide layer makes the input impedance of the MOSFET much higher than that of the standard FET. It can be made as high as 10^{15} Ω by making the oxide layer thicker. The other principal advantage of MOSFETs is that they can be made extremely small using thin-film deposition methods. This aids in the miniaturization process in the design and manufacture of integrated circuits.

Schottky Barriers

A closely related device is called the **Schottky barrier,** in which direct contact is made between a metal and a semiconductor. If the semiconductor is *n*-type, elec-

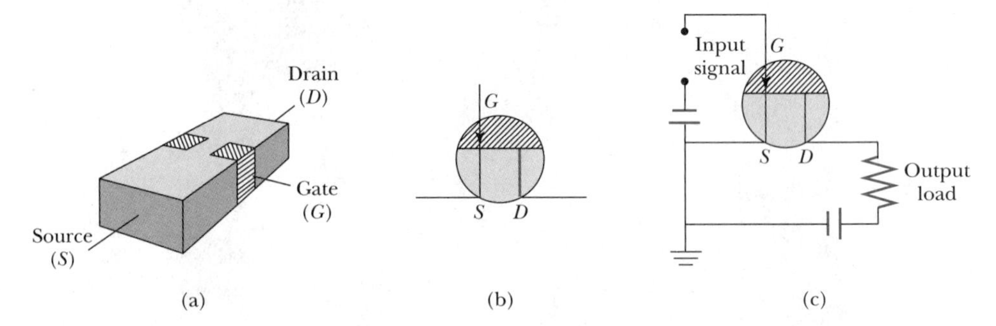

FIGURE 11.25 (a) A schematic of a FET. The two gate regions are connected internally. (b) The circuit symbol for the FET, assuming the source to drain channel is of *n*-type material and the gate is *p*-type. If the channel is *p*-type and the gate *n*-type, then the arrow is reversed. (c) An amplifier circuit containing a FET.

trons from it tend to migrate into the metal, leaving a depleted region within the semiconductor. This will happen as long as the work function of the metal is higher (or lower, in the case of a *p*-type semiconductor) than that of the semiconductor. The width of the depleted region depends on the properties of the particular metal and semiconductor being used, but it is typically on the order of microns. The *I-V* characteristics of the Schottky barrier are similar to those of the *p-n*–junction diode (Equation [11.9], Figure 11.13). When a *p*-type semiconductor is used, the behavior is similar but the depletion region now has a deficit of holes.

There is a variation of the Schottky barrier known as an **ohmic contact.** The ohmic contact also involves contact between a metal and a semiconductor, but now the work function of the semiconductor is higher than that of the metal if the semiconductor is *n*-type, and lower than that of the metal for a *p*-type. In this case the interface region will be enriched in majority carriers (whether *p* or *n*), and current can flow easily across the boundary.

In Figure 11.26 a typical Schottky-barrier gate FET (also known as a metal semiconductor FET, or MESFET) is shown. In this device the doped *n*-type GaAs serves as the channel. The contacts between the AlGe layers and the doped GaAs layer are ohmic, while the Al gate contact forms a Schottky barrier. When the gate is connected in reverse bias, the effect is the same as obtained using the silicon dioxide layer in the MOSFET (i.e., a higher input impedance). The CrPt interface serves only to prevent the Au and Al from forming an alloy.

In Figure 11.26b we see the MESFET amplifier circuit gain plotted as a function of frequency. The superior characteristics of the MESFET are due to the fact that no diffusion of impurities into any of the layers is needed. The source, gate, and drain are simply evaporated or sputtered directly onto the doped GaAs. Another benefit of this is that the channel can be made extremely narrow (a fraction of a micron), making the device smaller and its operation faster.

Semiconductor Lasers

In recent years semiconductor lasers have been used widely in scientific research and industrial applications. Like the gas lasers described in Section 10.2, semiconductor lasers operate using population inversion—an artificially high number of electrons in excited states. In a semiconductor laser, the band gap determines the energy difference between the excited state and the ground state.

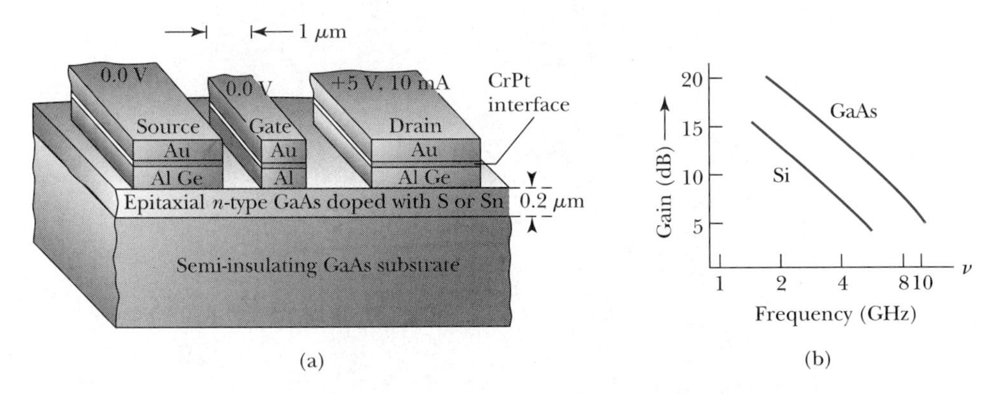

FIGURE 11.26 (a) Schematic drawing of a typical Schottky-barrier FET. (b) Gain versus frequency for two different substrate materials, Si and GaAs. *From D.A. Fraser, Physics of Semiconductor Devices. Oxford: Clarendon Press, 1979.*

Rather than the optical pumping and electrical discharge used to operate gas lasers, semiconductor lasers use *injection pumping,* where a large forward current is passed through a diode. This creates electron-hole pairs, with electrons in the conduction band and holes in the valence band. A photon is emitted when an electron falls back to the valence band to recombine with the hole. Thus the photon energy is determined by the band gap in the semiconductor.

The first semiconductor laser was made in 1962 and consisted of a GaAs *p-n*–junction with a band gap of 1.47 eV. It is straightforward to compute the photon wavelength produced in a transition from the conduction band to the valence band:

$$\lambda = \frac{hc}{\Delta E} = \frac{1240\ \text{eV} \cdot \text{nm}}{1.47\ \text{eV}} = 844\ \text{nm}$$

which is in the near infrared. Because of the band gaps found in typical semiconductors, outputs of most semiconductor lasers are in the infrared or red portions of the spectrum.

Since their development, semiconductor lasers have been used in a number of applications, most notably in fiber-optics communication. One advantage of using semiconductor lasers in this application is their small size; like other semiconductor devices, semiconductor lasers can be made quite small, with dimensions typically on the order of 10^{-4} m. They are solid state devices, so semiconductor lasers are more robust than gas-filled tubes.

The relatively small (1-eV to 2-eV) band gaps in semiconductors would seem to limit the operation of semiconductor lasers to longer wavelengths. For example, a 2-eV photon has a wavelength of 620 nm, and lower energies correspond to even longer wavelengths. Over the years scientists worked to develop semiconductor lasers with shorter wavelengths. Finally in 1996 Shuji Nakamura demonstrated a blue laser based on the semiconductor gallium nitride. It is hoped that blue lasers can replace the infrared lasers now used in compact disc players. With shorter laser wavelengths, information can be packed with higher density onto a compact disc, with less chance of error in reading the information. For more information see *Scientific American* (September 1997), p. 36.

FIGURE 11.27 Working on the ENIAC computer. *The University of Pennsylvania Archives.*

FIGURE 11.28 The Intel 4004, the first commercial microprocessor (1971). This integrated circuit measured just 3 mm by 4 mm, and with 2300 transistors it could perform 60,000 operations per second. *Courtesy Intel Corporation.*

Integrated Circuits

The most important use of all these semiconductor devices today is not in discrete components, but rather in **integrated circuits.** It is possible to fabricate a silicon-based substrate (commonly called a **chip**) containing a million or more components, including resistors, capacitors, transistors, and logic switches. This extreme miniaturization has done two things for the electronics industry. First, it has made it possible to put very sophisticated integrated circuits into even the smallest products, such as wristwatches and cameras. Second, it has enabled those integrated circuits to work much faster. Although there are still limits based on the speeds at which switches can operate,* the transit time of an electronic signal has been reduced by packing the components so closely together. Those signals travel at essentially the speed of light—0.3 μm/fs, in units relevant to the distance/time scale of integrated circuits.

Integrated circuits have had a tremendous impact on the work of scientists and engineers by enhancing the computing power available to them. The Manhattan Project (which produced the atomic bomb in 1945) and other scientific work during World War II inspired the development of large mechanical computers that could perform many computations relatively quickly. For example the ENIAC (**E**lectronic **N**umerator, **I**ntegrator, **A**nalyzer, and **C**omputer), built in 1945, could multiply 333 ten-digit numbers per second. Unfortunately it weighed 30 tons, contained over 17,000 vacuum tubes, 70,000 resistors, 10,000 capacitors, and 1500 relay switches. It was 2.5 m high and wide and 25 m long (see Figure 11.27), and it used 174 kW of power in normal operation! By 1963 the best computers (e.g., the Control Data 6600) still filled a normal-sized room (6 m × 6 m) but could perform one million operations per second. In 1971 the first true commercial integrated circuit was released: the Intel 4004 (Figure 11.28). Today, personal computers (which fit on desktops) can perform millions of operations per second while plugged into a standard 110-volt outlet, while the largest supercomputers are thousands of times faster. In late 1998, a

*By the late 1990s switch rates in bipolar semiconductor transistors exceeded 100 GHz, corresponding to switching times of less than 10 ps. See *Scientific American* (March 1994) pp. 62–67.

computer developed by IBM and Lawrence Livermore National Lab set a high-speed computing record of 3.9×10^{12} operations per second.

One indication of the rapid progress in computing is the increase in the number of transistors that can be fit on a single microchip. That number was just 2300 for the Intel 4004 chip in 1971. The number of transistors per chip grew to 134,000 with the release of the 80286 chip in 1982; 275,000 on the 386 chip in 1985, and 1.2 million on the 486 chip in 1989. The rapid growth in microchip technology continued through the 1990s. The Pentium 2 chip, produced by Intel in 1997, contained 7.5 million transistors on a single chip. In order to achieve such high densities, it is necessary to build extremely small components. In the Pentium 2, for example, circuit elements are about 3.5×10^{-7} m across in each direction. Notice that this is less than the wavelength of visible light. For that reason chip manufacturers must now use ultraviolet lithographic techniques. By 2010 manufacturers hope to use shorter ultraviolet wavelengths to produce circuit elements no more than 1×10^{-7} m across, thus producing even smaller and faster microprocessors.

Another measure of progress is summarized in Moore's law, which states that microchip capacity doubles roughly every 18 to 24 months. This "law" was reasonably accurate from the 1970s all the way through the 1990s.

Another important factor in technology is not the computing itself but the storage and retrieval of information. In the late 1990s the standard storage medium was magnetic. It is possible to store and reliably retrieve as many as 4×10^{12} bits of information per m^2 of magnetic disc or tape. This is approaching the theoretical limit, because at that density as few as 200 magnetic particles are used to store each bit of information. At higher densities one begins to have problems with reliability, due to statistical fluctuations in the magnetizations of the particles and overlap of magnetic fields from one small bit to another. If further miniaturization is to be achieved, it may come through the use of optical media. A familiar example of this is the compact disc (CD), which can be scanned by a precisely focused laser to retrieve the digitally encoded information. Optical discs can contain over 10^9 bits/cm^2 of information. Optical discs are encoded with information as they are manufactured, but that information cannot be erased and rewritten. CD/ROM devices (Figure 11.29) are used extensively in such fields as astronomy, where there is often a need to store

FIGURE 11.29 A CD/ROM device used to store large amounts of data, in this case infrared emissions from polar regions obtained by orbiting satellites. *Photograph courtesy of Alan Thorndike.*

enormous amounts of data. Magneto-optical discs, which store information magnetically and retrieve optically, became available for general use in 1991.* They are fully erasable and can contain up to 300 megabytes of information on a $3\frac{1}{2}''$ disc, compared to 1.4 megabytes on a high-density $3\frac{1}{2}''$ magnetic disc. By 1999, standard optically read CDs were available in rewritable formats for home computers.

*J. Devlin, "Magnet-optical disk drives," *Infoworld* **13,** 53–70 (Dec. 2, 1991).

Summary

The properties of semiconductors are responsible for their widespread use today in computers, electronic instruments, and other applications. In a semiconductor there exists a small band gap (≈ 1 eV) between the valence band and the conduction band. When a modest electric field is applied, sufficient numbers of electrons can overcome the band gap in order to conduct.

Normal conductors have resistivities that increase with increasing temperature, but in semiconductors the resistivity decreases with increasing temperature, due to the increased statistical probability that the conduction band will be occupied. An empirical formula relating the resistance of many semiconductors to temperature is the Clement-Quinnell equation

$$\log R + K/\log R = A + B/T \tag{11.4}$$

where A, B, and K are constants.

Some semiconductors are n-type, meaning that the majority carriers are negative electrons. Other semiconductors are p-type, meaning that the majority carriers are positive holes. The Hall effect can be used to determine the sign and magnitude of the charge carriers.

Semiconductors exhibit the thermoelectric effect, the presence of an electric field when a temperature gradient is established. The thermoelectric effect is the basis for the design of thermocouples.

Semiconductors are used to make a variety of electronic devices. p-type and n-type materials can be placed end to end to form a p-n–junction diode, which permits current to flow easily in one direction (forward bias) but hardly at all in the opposite direction (reverse bias). A Zener diode will operate in reverse bias once a sufficiently high voltage is applied.

Other useful semiconductor devices include light emitting diodes and photovoltaic (solar) cells. Solar cells are currently the focus of an intense research effort. It is hoped that more efficient and cheaper solar cells can provide a useful source of energy by converting sunlight to electrical energy. This will become increasingly important as nonrenewable sources of energy are depleted.

Semiconducting transistors are important components in computers and other electronic instruments. Transistors can be used to amplify signals (either voltage or current). Related devices, such as field effect transistors, MOSFETs, and Schottky barriers are also used extensively. The most important use of these electronic devices today is in integrated circuits. Silicon-based chips contain millions of components. Miniaturization has led to significant increases in the speed and efficiency of electronic circuits as well as to greater convenience. Perhaps the most significant application has been in the design of modern computers. Thanks to improvements in semiconductors technology, the large, complex computers of several decades ago have been supplanted by much smaller, cheaper, and easier-to-use microprocessors and personal computers in use today.

Questions

1. Why is the free-electron model not applicable to semiconductors and insulators?
2. How does a semimetal differ from a normal conductor?
3. Compare the room temperature resistivities of conductors, semiconductors, and insulators?
4. Why does the addition of impurities to a conductor increase its resistivity, while the addition of impurities to a semiconductor generally decreases its resistivity?
5. Would you expect carbon resistors to be useful as thermometers at room temperature and above? Explain.
6. Use the size of the energy gap to decide whether the following should tend to be transparent to visible light: conductors, semiconductors, and insulators.
7. Repeat Question 6 considering near infrared light with $\lambda = 1$ micron.
8. Describe the effect of high temperatures on a semiconductor diode.
9. What role do capacitors play in the conversion of alternating current to direct current?
10. Why is it appropriate to think of a photovoltaic cell as a LED operating in reverse?

Problems

11.2 Semiconductor Theory

1. For a certain resistor with a normal (i.e. room-temperature) resistance of 150 Ω, the values of the constants in Equation (11.4) are found to be $A = 4.05$, $B = 4.22$, and $K = 4.11$, in units such that R will be in ohms and T in kelvin. What is the resistance of this resistor at (a) $T = 77$ K, (b) $T = 20$ K, and (c) $T = 1$ K?
2. For a nominal 10 Ω resistor (as described in Figure 11.7) the resistance at various temperatures is as follows: $R = 10\ \Omega$ at $T = 293$ K, $R = 40\ \Omega$ at $T = 10$ K, and $R = 5800\ \Omega$ at $T = 1$ K. Determine the constants A, B, and K in Equation (11.4) for this resistor.
3. Consider the experimental arrangement of Figure 11.9, set up to observe the Hall effect. With the supplies and meters in this configuration, what will be the sign of the voltage on the voltmeter if the sample is a semiconductor in which the majority of charge carriers are holes? Explain.
4. (a) Hall effect data in an actual student laboratory was as follows: for a doped indium arsenide strip of thickness 0.15 mm, with a current of 100 mA flowing through the strip and a magnetic field of 36 mT perpendicular to the strip, the measured Hall voltage was 8.4 mV. Use these data to find the density of charge carriers. (b) Find the density of charge carriers using the complete data set shown in the table below.

$I = 100$ mA

B (mT)	V_H (mV)
17.5	4.4
27	6.4
36	8.4
31	7.4
25	6.1
47	10.9
50	11.5
59.5	13.6

5. A pure lead bar 10-cm long is maintained with one end at $T = 273$ K and the other at 283 K. The thermoelectric potential difference thus induced across the ends is 12.6 μV. Find the thermoelectric power for lead in this temperature range. (*Note:* Q varies nonlinearly with temperature, but over this narrow temperature range you may use a linear approximation.)
6. The reference junction of an iron-constantan thermocouple is maintained at 0°C, and the other side is at an unknown temperature. Find the unknown temperature if the other side is 2.43 mV higher in potential than the reference junction (see Table 11.3).
7. Assuming that the potential corresponding to *any* temperature T in Table 11.3 could be known (say through a computer model) to at least five significant digits, what maximum uncertainty would be needed in your voltmeter if you were to use an iron-constantan thermocouple to measure temperatures to within 0.01°C?
8. What kind (p-type or n-type) of semiconductor is made if pure silicon is doped with a small amount of (a) aluminum (b) selenium (c) indium?

11.3 Semiconductor Devices

9. Find the fraction of the standard solar flux reaching the Earth (about 1000 W/m²) available to a solar collector lying flat on the Earth's surface at each of the following places at noon on the winter solstice, spring equinox, and summer solstice: (a) San Diego, latitude 33°; (b) Regina, latitude 50°; (c) St. Petersburg, Russia, latitude 60°.
10. Assuming that the *average daily* solar constant at a particular place is 200 W/m², how large an array of 30% efficient solar cells is required to equal the power output of a typical power plant, about 10^9 W?
11. For the diode described in Example 11.3, find the forward bias current with $V = 250$ mV at (a) $T = 250$ K, (b) $T = 300$ K, and (c) $T = 500$ K.

General Problems

12. Normally the Fermi-Dirac factor is expressed [see Equation (9.34)] as

$$F_{FD} = \frac{1}{\exp[(E - E_F)/kT] + 1}$$

In a semiconductor or insulator, with an energy gap E_g between the valence and conduction bands, we can take E_F to be halfway between the bands, so that $E_F = E_g/2$. (a) Show that for a typical semiconductor or insulator at room temperature the Fermi-Dirac factor is approximately equal to $\exp[-E_g/2kT]$. (b) Use the result in (a) to compute the Fermi-Dirac factor for a typical insulator, with $E_g = 6.0$ eV at $T = 293$ K. (c) Repeat for a semiconductor, silicon, with $E_g = 1.1$ eV at $T = 293$ K. (d) Your result in (c) is still small, but sufficiently large to explain why there will be conduction. Explain.
13. Consider what happens when silicon is doped with arsenic. Suppose that the extra, weakly bound electron from arsenic moves in the first Bohr orbit around a silicon atom. The first Bohr orbit has a radius of

$$a_0 = \frac{4\pi\epsilon_0\hbar^2}{me^2}$$

Due to the effects of screening, it is necessary to replace ϵ_0 with the electric permittivity $\epsilon = \kappa\epsilon_0$, where κ is the dielectric constant ($\kappa = 11.7$ for silicon). (a) Compute the effective Bohr radius for this electron. (b) Compare your result with the lattice spacing in silicon, about 0.235 nm and comment on the result.

14. Following the same procedure as in Problem 13, find the binding energy E_0 for the first Bohr orbit of the extra electron in silicon. Comment on the result.

15. (a) Using the thermocouple data in Table 11.3, use a computer to perform a linear least-squares fit of voltage versus temperature for the range 0°C to 50°C and for the range 50°C to 100°C. Over which range is the fit better? (b) Repeat your analysis of (a), this time using a second-order least-squares fit (voltage will be a function of T and T^2). Compare your results with those you obtained in (a).

16. A certain diode has a reverse bias current of 1.75 μA. Now this diode is connected in forward bias in the circuit shown, in series with a resistor and with a constant voltage source of 6.0 volts. (a) Find the value of resistance R such that a current of 80 mA will flow at room temperature (293 K). (b) Under the condition described in (a), find the voltage drop across the resistor.

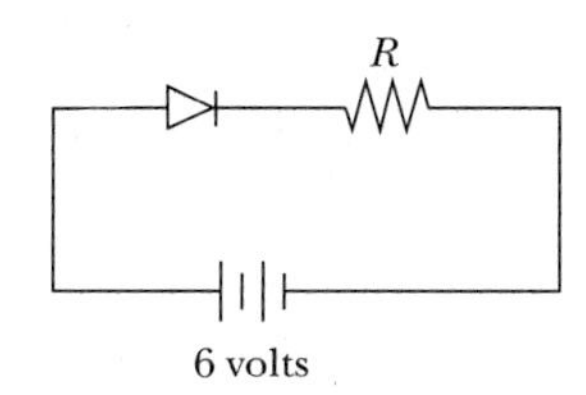

17. Assuming a temperature of 293 K, for what value V of the bias voltage in Equation (11.9) does (a) $I = 8I_0$? (b) $I = -0.8I_0$?

18. A light-emitting diode made of the semiconductor GaAsP gives off red light ($\lambda = 650$ nm). Determine the energy gap for this semiconductor.

19. How large an energy gap is required for a blue semiconductor laser with $\lambda = 460$ nm?

20. (a) Find the length of each side of a square computer chip with 7.5 million transistors, if each transistor occupies a square of side 0.35 μm. (b) Find the number of transistors on a chip of the same size you found in (a) if the transistor size can be reduced to 0.10 μm on a side.

21. Early research on semiconductor materials focused on silicon and germanium, which have band gaps of 1.1 eV and 0.67 eV, respectively. (a) Using the result of Problem 12(a) compute the Fermi-Dirac factor for silicon and germanium at $T = 0$°C and $T = 75$°C (this represents a fair temperature range for semiconductors in use). Based on your computed value for F_{FD} for germanium at the higher temperature, explain why silicon is preferred in most applications.

22. Suppose the average solar flux reaching the United States is 200 W/m^2. This average is taken over a whole year and takes into account seasonal effects and weather (clouds), and assumes fixed solar cells. (a) Find the total energy produced in one year by a 1 m^2 cell producing energy with an efficiency of 15%. (b) How much area would have to be covered with these solar cells in order to supply the United States with all its electricity, if the yearly electrical energy consumption in the United States is about 2.4×10^{12} kW · h? (c) Real solar arrays require about 2.5 times as much area as you found in (b) in order to avoid one array of cells from shading another. What fraction of the land area of the United States (about 9×10^6 km^2) would have to be covered with solar cells to meet the nation's energy requirements?

23. In this problem you will examine the temperature dependence of the forward/reverse bias modes of a diode. (a) For a *p-n*–junction diode compute the ratio of forward bias current to reverse bias current with an applied voltage of 1.50 volts at each of the following temperatures: 77 K, 273 K, 340 K, 500 K. (b) Comment on the results with respect to possible applications.

24. Pure silicon is used as a photon detector. An incoming photon can strike the surface and excite electrons from the valence band to the conduction band, where they can be counted. (a) Compute the number of electrons you would expect to count if a silicon detector is struck with a 1.04-MeV gamma ray produced in the decay of a ^{136}Cs nucleus. (b) Explain why the counting of electrons should be more precise if the detector is cooled well below room temperature.

CHAPTER 12

The Atomic Nucleus

It is said that Cockcroft and Walton were interested in raising the voltage of their equipment, its reliability, and so on, more and more, as so often happens when you are involved with technical problems, and that eventually Rutherford lost patience and said, "If you don't put a scintillation screen in and look for α particles by the end of the week, I'll sack the lot of you." And they went and found them [the first nuclear transmutations].

Sir Rudolf Peierls in Nuclear Physics in Retrospect, *ed. Roger Stuewer*

Ernest Rutherford can rightly be called the "Father of the Nucleus." As we learned earlier, he proposed a model of atomic structure that placed the heavy, positively charged nucleus at the center and the much lighter electrons at the periphery.

Around 1900, early investigators (including Becquerel, Rutherford, and Marie and Pierre Curie) found that radioactive emissions from atoms were comprised of three types of radiation, called α(alpha), β(beta), and γ(gamma). Alpha radiation is the least penetrating—it can be stopped by a piece of paper. Beta rays are more penetrating—they are also common, appearing in the radioactive emissions of many nuclei. Gamma radiation is the most penetrating—it can pass through the human hand, for example. Many early experiments established that α rays were doubly charged positive particles, β rays were probably electrons, and γ rays were electrically neutral. Rutherford proved in a series of experiments that alpha particles were the nuclei of helium atoms.

We begin this chapter by discussing the discovery of the constituents of the nucleus, one of science's most interesting series of experiments. We then study the properties of the nucleus and its constituents, the neutrons and protons. We will discuss nuclear forces and explain why some nuclei are stable and others are unstable, that is, radioactive. The study of nuclear radioactivity has many important applications, which we shall discuss in Chapter 13.

12.1 Discovery of the Neutron

Although Rutherford proposed the atomic structure with the massive nucleus at the center in 1912, it was not until 20 years later that scientists knew which particles comprised the nucleus. This study is still active today (see Chapter 14), because when physicists strive to find the essence of the fundamental nuclear particles, they continue to find even more particles.

Even though Rutherford and others thought that a neutral particle like the neutron might exist, the existence of neutrons was not proven experimentally until 1932, when James Chadwick found them by using one of the first particle accelerators. In the early 1900s the nucleus had been erroneously assumed to consist of protons and electrons. However, there are several reasons why electrons cannot exist within the nucleus. We will only mention a few.

Electrons can't exist within the nucleus

1. *Nuclear size* We showed previously, in Example 5.8, that in order to confine an electron in a space as small as a nucleus, the uncertainty principle puts a lower limit on its kinetic energy that is much larger than any kinetic energy observed for an electron emitted from nuclei.
2. *Nuclear spin* Protons and electrons have spin 1/2. If a deuteron (mass number $A = 2$ and atomic number $Z = 1$) consists of protons and electrons, the deuteron must contain 2 protons and 1 electron, in order to have $A = 2$ and $Z = 1$. However, 3 fermions must have a half-integral spin, whereas the nuclear spin of the deuteron has been measured to be 1.
3. *Nuclear magnetic moment* The magnetic moment of an electron is over 1000 times larger than that of a proton. However, the measured nuclear magnetic moments are of the same order of magnitude as the proton's. In the proton and electron model, we would expect the nuclear magnetic moment to be on the same order as that of the electron, so it is difficult to understand how an electron can exist within the nucleus.

Example 12.1

What is the minimum kinetic energy of a proton in a medium-sized nucleus having a diameter of 8×10^{-15} m?

Solution: We will determine this answer in a similar manner to that in Example 5.8, where we found the minimum kinetic energy for an electron in a nucleus. First, we use the uncertainty principle to find the uncertainty Δp.

$$\Delta p\,\Delta x \geq \frac{\hbar}{2}$$

$$\Delta p \geq \frac{\hbar}{2\,\Delta x} = \frac{6.58 \times 10^{-16}\ \text{eV}\cdot\text{s}}{2(8 \times 10^{-15}\ \text{m})}$$

$$\Delta p \geq 0.041\ \text{eV}\cdot\text{s/m}$$

The momentum p must be at least as large as Δp, hence we have for $p_{\text{min}}c$:

$$p_{\text{min}}c = (0.041\ \text{eV}\cdot\text{s/m})(3 \times 10^8\ \text{m/s}) = 12\ \text{MeV}$$

Because this energy is only about 1% of the proton's rest energy, we can treat the problem nonrelativistically. The kinetic energy of a proton in this nucleus must be at least as large as

$$\text{K.E.} = \frac{(p_{\text{min}})^2}{2m} = \frac{(p_{\text{min}}c)^2}{2mc^2} = \frac{(12\ \text{MeV})^2}{2(938\ \text{MeV})} = 0.08\ \text{MeV}$$

which is an entirely reasonable experimental value. The result for a neutron would be similar.

We have seen there is strong experimental and theoretical evidence that electrons are not bound within the nucleus. Although all this evidence was not available to Rutherford in 1920, he proposed that a neutral particle, called a *neutron,* might exist. The three arguments just given against the existence of electrons within a nucleus do not apply to a nucleus made of neutrons and protons.

The discovery of the neutron is a classic experimental investigation. In 1930 the German physicists W. Bothe and H. Becker were using a radioactive polonium source that emitted α particles. Bothe and Becker found that when these α particles bombarded beryllium, a very penetrating radiation was produced. Irene Curie and F. Joliot subsequently showed in 1932 that the radiation could even penetrate several centimeters of lead (see Figure 12.1). This radiation could not be charged particles, because charged particles of the energies available then could not penetrate even a short distance through lead. It was naturally assumed that electromagnetic radiation (photons) was produced in the α bombardment of beryllium. Photons are called *gamma rays* when they emanate from the nucleus. Gamma rays produced in the nucleus have energies on the order of MeV (as compared to the order of keV x-ray photons produced by transitions in an atom).

Curie and Joliot performed several measurements to study the effects of this new penetrating radiation (produced by α + Be) on various materials. When the radiation passed through paraffin (which contains hydrogen), they found that protons with energies up to 5.7 MeV were ejected. The simplest assumption was that the radiation—assumed to be gamma rays—scattered by the Compton process from the hydrogen nuclei and knocked the protons out of the paraffin. However, the hypothesis of Compton scattering requires gamma-ray energies of at least 50 MeV to produce 5.7-MeV protons (see Problem 1). Energies as large as 50 MeV were unprecedented at that time. No known reaction could produce such high-energy gamma rays.

In 1932 James Chadwick proposed that the new radiation produced by α + Be consisted of neutrons, the hypothetical electrically neutral particles with the mass of the proton. Such neutrons would pass through material rather easily. The neutron has only a very small electromagnetic interaction via its magnetic moment. The nuclear force is very short-range, and a neutron having

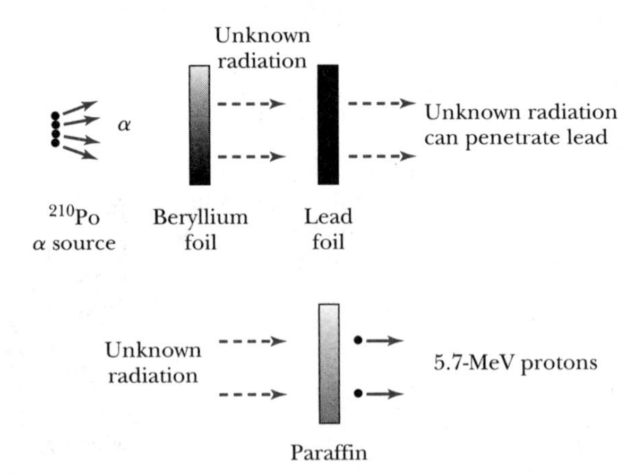

FIGURE 12.1 Schematic diagram of events leading to neutron discovery. A polonium α-particle source emits α particles that produce unknown radiation when incident on beryllium. This unknown radiation is so penetrating that it can pass through a lead foil, which indicates it may be gamma rays. When the unknown radiation is incident on parafin, however, 5.7-MeV protons are produced. Only gamma rays with energy above 50 MeV are able to do this, and they are unlikely to be produced by a nucleus. Chadwick suggested the radiation was a neutral particle of about the same mass as a proton.

MeV energies may have only a 10^{-6} probability of interacting with a nucleus. Chadwick correctly surmised that if neutrons of about 5.7 MeV kinetic energy were produced in the α + Be reaction, the neutrons could then collide elastically with the protons in paraffin, thereby accounting for the 5.7-MeV protons.

12.2 Nuclear Properties

The primary constituents of nuclei are the proton and neutron, and the nuclear mass is roughly the sum of its constituent proton and neutron masses (the slight difference being the binding energy of the nucleus). The nuclear charge is $+e$ times the number (Z) of protons ($e = 1.6 \times 10^{-19}$ C). Helium has $Z = 2$, oxygen has $Z = 8$, and uranium has $Z = 92$.

The simplest form of hydrogen has a single proton for a nucleus. However, we know that several forms of hydrogen exist. Deuterium—sometimes called "heavy hydrogen"—has a neutron as well as a proton in its nucleus. Another isotope of hydrogen is called *tritium*—it has two neutrons and one proton. The nuclei of the deuterium and tritium atoms are called *deuterons* and *tritons,* respectively.

The atomic (and nuclear) mass number A is the total integral number of protons and neutrons in a nucleus. Atoms with the same Z, but different A, are called *isotopes*. For example, deuterium ($A = 2$) and tritium ($A = 3$) are both isotopes of regular hydrogen ($A = 1$). The atomic mass M is the mass of the entire atom (including electrons), measured, for example, with a *mass spectrograph*.

We will designate an atomic nucleus by the symbol

Nucleus designations

$${}^{A}_{Z}X_{N}$$

where Z = atomic number (number of protons)
N = neutron number (number of neutrons)
A = mass number ($Z + N$)
X = chemical element symbol

Each nuclear species with a given Z and A is called a *nuclide*. As we discussed in Chapter 8, each Z characterizes a chemical element, symbol X; for example, Al for aluminum ($Z = 13$), Ca for calcium ($Z = 20$). In the last few decades the naming of the newly discovered elements has become a prestigious affair, often surrounded by controversy. Although it is superfluous, we will sometimes include Z to help us remember the number of protons when dealing with elements with which we are not so familiar. When we write the nuclidic symbol for the better-known elements, Z is often omitted. The nuclidic symbol A is always shown, but N is often omitted, because $A = N + Z$. Thus, ${}^{16}_{8}O_8$, ${}^{16}_{8}O$, and ${}^{16}O$ all represent the most abundant isotope of oxygen with $Z = 8$, $N = 8$, $A = 16$. Other stable isotopes of oxygen are ${}^{17}O$ and ${}^{18}O$, which differ from ${}^{16}O$ only in having more neutrons. Nuclides with the same neutron number are called *isotones* (for example, ${}^{14}_{6}C$, ${}^{15}_{7}N$, ${}^{16}_{8}O$ and ${}^{17}_{9}F$). Nuclides with the same value of A are called *isobars* (for example, ${}^{16}_{6}C$, ${}^{16}_{7}N$, ${}^{16}_{8}O$, and ${}^{16}_{9}F$).

The chemical properties of an atom are determined by its electron configuration. Because the numbers of electrons and protons are equal in a neutral atom, the chemical properties are essentially determined by Z. The dependence of the chemical properties on N is negligible.

Atomic masses are measured in atomic mass units, which are denoted by the symbol u. Atomic mass units are defined in terms of the mass of the isotope ${}^{12}C$ whose atomic mass is defined to be exactly 12 u. The atomic masses of many

TABLE 12.1
Some Nucleon and Electron Properties

Particle	Symbol	Rest Energy (MeV)	Charge	Mass (u)	Spin	Magnetic Moment
Proton	p	938.272	$+e$	1.0072765	1/2	$2.79\ \mu_N$
Neutron	n	939.566	0	1.0086649	1/2	$-1.91\ \mu_N$
Electron	e	0.51100	$-e$	5.4858×10^{-4}	1/2	$-1.00116\ \mu_B$

nuclides are given in Appendix 8. The reason we present atomic masses rather than nuclear masses will be explained later. As we have seen in Chapter 2, the atomic mass unit works out to be

Atomic mass unit

$$1\ \text{u} = 1.66054 \times 10^{-27}\ \text{kg} = 931.49\ \text{MeV}/c^2 \tag{12.1}$$

The masses of the proton and neutron are given in Table 12.1. The fact that neutrons and protons have almost the same mass is no accident. As we shall see in Chapter 14, both neutrons and protons, collectively called **nucleons,** are constructed of other particles called *quarks*. Neutrons are slightly more massive than protons.

Sizes and Shapes of Nuclei

The size of the nucleus has been determined in a variety of ways. Rutherford concluded from the alpha-particle scattering experiments of his assistants Geiger and Marsden that the range of the nuclear force must be less than about 10^{-14} m, because deviations from Coulomb's law at that distance could be inferred from their data (see Section 4.2).

Let us assume that nuclei are spheres of radius R. Particles, such as electrons, protons, neutrons, and alphas, scatter when projected close to the nucleus. It is not immediately obvious whether the maximum interaction distance measured in such collisions refers to the nuclear size *(matter radius)*, or whether the nuclear force extends beyond the nuclear matter *(force radius)*. Electrons do not respond to the nuclear force but scatter from the electromagnetic field of the nucleus. Thus, electron scattering measures the nuclear *charge radius.*

The nuclear force between nucleons is the strongest of the three known forces (nuclear, gravitational, and electroweak) at short distances. As such, the nuclear force is often called the **strong** force, and we shall use the terms *nuclear* and *strong* force interchangeably. Because neutrons interact only with the nuclear force, the scattering of neutrons determines the nuclear force radius. Through many measurements using beams of different particles, physicists have found that

$$\text{Nuclear force radius} \approx \text{Mass radius} \approx \text{Charge radius}$$

The nuclear radius may be approximated from a spherical charge distribution to be

Nuclear radius

$$R = r_0 A^{1/3} \tag{12.2}$$

where $r_0 \approx 1.2 \times 10^{-15}$ m. Various measurements for r_0 range between 1.0 and 1.5×10^{-15} m. Because the nucleus is so small, we use the **femtometer,** abbrevi-

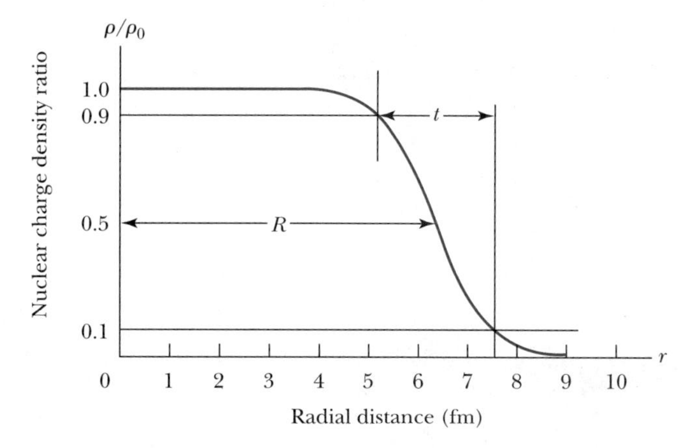

FIGURE 12.2 The shape of the Fermi distribution for the nuclear charge density given by Equation (12.2), where $\rho_0(r)$ is the central charge density, R is the distance at which the nuclear density has dropped to 50% of its central value, and $t = 4.4a$ is the surface thickness, measured from 90% to 10% of the central density.

ated fm, with 1 fm $= 10^{-15}$ m. However, the term **fermi,** named after Enrico Fermi, one of the founders of nuclear physics, is more often used for 10^{-15} m and has the same abbreviation, fm. For example, 2.4 fm is read as 2.4 fermi.

Robert Hofstadter (Nobel Prize, 1961) and his colleagues at Stanford University in the 1950s performed the first precision electron-scattering measurements of the nuclear charge distribution using electron energies from 100 to 500 MeV. In order to probe the actual shape of most nuclei, we need a particle having a short wavelength. The de Broglie wavelength of a 500-MeV electron is about 2.5 fm, and by now, measurements have been made with much shorter wavelengths using higher energy electrons. These measurements are approximately described for all but the lightest nuclei by the Fermi distribution for the nuclear charge density $\rho(r)$ of the following form (see Figure 12.2):

$$\rho(r) = \frac{\rho_0}{1 + e^{(r-R)/a}} \tag{12.3}$$

The shape of this distribution is shown in Figure 12.2 where ρ_0 is the central nuclear density, R is the distance at which the nuclear density has dropped to 50% of its central value, and $t = 4.4a$ is the surface thickness, measured from 90% to 10% of the central density.

Robert Hofstadter (1915–1990) was educated at the College of the City of New York and Princeton University. Hofstadter used high-energy electrons at Stanford University to measure the charge distribution inside the nucleus, thereby determining the radius of the nucleus. He also studied the structure of the neutron and proton, proving that they were similar particles, only differing in their charge properties. *Courtesy of Stanford University.*

Example 12.2

What is the nuclear radius of ^{40}Ca? What energy electrons and protons are required to probe the size of ^{40}Ca if one wants to "see" at least half the radius?

Solution: We first determine the radius of ^{40}Ca using Equation (12.2). We determine

$$R = 1.2 \text{ fm } (40)^{1/3} = 4.1 \text{ fm}$$

In order to distinguish a distance at least half the radius, we need a de Broglie wavelength of 2 fm. If we use the relation for the de Broglie wavelength, $\lambda = h/p$; we have $p = h/\lambda$. The total energy of the probing particle is

$$E^2 = (mc^2)^2 + (pc)^2 = (mc^2)^2 + \frac{h^2c^2}{\lambda^2}$$

For a wavelength of 2 fm, the last term becomes

$$\frac{h^2c^2}{\lambda^2} = \left(\frac{1240 \text{ MeV} \cdot \text{fm}}{2 \text{ fm}}\right)^2 = 3.844 \times 10^5 \text{ MeV}^2$$

Now if we insert the rest energy of the appropriate particle, we can determine the total energy and required kinetic energy of either the electron or proton.

electron energy

$$E^2 = (0.511\ \text{MeV})^2 + 3.844 \times 10^5\ \text{MeV}^2$$

$$E = 620\ \text{MeV}$$

$$K = E - mc^2 = 620\ \text{MeV} - 0.5\ \text{MeV} = 620\ \text{MeV}$$

proton energy

$$E^2 = (938.3\ \text{MeV})^2 + 3.844 \times 10^5\ \text{MeV}^2$$

$$E = 1125\ \text{MeV}$$

$$K = E - mc^2 = 1125\ \text{MeV} - 938\ \text{MeV} = 187\ \text{MeV}$$

Example 12.3

What is the ratio of the radii of ^{238}U and ^{4}He?

Solution: We use Equation (12.2) to determine the ratio of the two radii.

$$R(^{238}\text{U}) = (1.2\ \text{fm})(238)^{1/3} = 7.4\ \text{fm}$$

$$R(^{4}\text{He}) = (1.2\ \text{fm})(4)^{1/3} = 1.9\ \text{fm}$$

The ratio is

$$\frac{R(^{238}\text{U})}{R(^{4}\text{He})} = \frac{7.4\ \text{fm}}{1.9\ \text{fm}} = 3.9$$

Even though ^{238}U has 60 times the number of nucleons of ^{4}He, its radius is only 4 times greater.

If we approximate the nuclear shape as a sphere, we have $V = 4\pi R^3/3$, or, by using Equation (12.2) for R,

$$V = \frac{4}{3}\pi {r_0}^3 A \tag{12.4}$$

The nuclear mass density (mass/volume) can be determined from $(A\ \text{u})/V$ to be 2.3×10^{17} kg/m^3. The nucleus is about 10^{14} times denser than ordinary matter!

Intrinsic Spin

The neutron and proton are fermions with spin quantum numbers $s = 1/2$. The spin quantization rules are those we have already learned for the electron (see Chapter 7).

Intrinsic Magnetic Moment

The proton's intrinsic magnetic moment points in the same direction as its intrinsic spin angular momentum because the proton's charge is positive. This is contrasted with the electron, where the spin and magnetic moment point in opposite directions (see Figure 7.4). Nuclear magnetic moments are measured in units of the nuclear magneton μ_N, which is defined, by analogy to the Bohr magneton for electrons, by the relation

$$\mu_N = \frac{e\hbar}{2m_p} \tag{12.5}$$

Note that the proton mass m_p (rather than the electron mass) appears in μ_N, which makes the nuclear magneton some 1800 times smaller than the Bohr magneton.

The proton magnetic moment is measured to be $\mu_p = 2.79\mu_N$. This contrasts strongly with the magnetic moment of the electron, $\mu_e = -1.00116\mu_B$. What is even more surprising is that the neutron, which is electrically neutral, also has a magnetic moment, $\mu_n = -1.91\mu_N$. The negative sign indicates that the magnetic moment points opposite to the neutron spin. The large deviation from unity of the proton's magnetic moment and the fact that the neutron even has a magnetic moment indicate that nucleons are more complicated structurally than electrons. The *nonzero* neutron magnetic moment implies that the neutron has negative and positive internal charge components at different radii, hence a complex internal *charge distribution.*

12.3 The Deuteron

After the proton, the next simplest nucleus is the deuteron, the nucleus of ^{2}H. A deuteron consists of one proton and one neutron and allows our first look at the nuclear force. First, let us determine how strongly the neutron and proton are bound together in a deuteron. The deuteron mass is 2.013553 u, and the mass of a deuterium atom is 2.014102 u. The difference in masses is 2.014102 u − 2.013553 u = 0.000549 u, which is just the mass of an electron. The electron binding energy (13.6 eV for hydrogen) is so small that it can be neglected for our purposes. The deuteron nucleus is bound by an energy B_d, which represents mass-energy. The mass of a deuteron is then

$$m_d = m_p + m_n - B_d/c^2 \tag{12.6}$$

The deuteron mass is less than the sum of the masses of a neutron and proton by just the nuclear binding energy B_d. If we add an electron mass to each side of Equation (12.6), we have

$$m_d + m_e = m_p + m_n + m_e - B_d/c^2 \tag{12.7}$$

But $m_d + m_e$ is the atomic deuterium mass $M(^2\text{H})$ and $m_p + m_e$ is the atomic hydrogen mass (if we neglect the small amount of electron binding energy). Thus Equation (12.7) becomes

$$M(^2\text{H}) = m_n + M(^1\text{H}) - B_d/c^2 \tag{12.8}$$

and we can use *atomic masses.* Because the electron masses cancel in almost all nuclear-mass difference calculations like Equation (12.8), we routinely use atomic masses (see Appendix 8) rather than nuclear masses.* Note that we use uppercase M for atomic masses and lowercase m for nuclear and particle masses. The binding energy B_d of the deuteron is now easily determined:

$$m_n = 1.008665\ \text{u} \qquad \text{Neutron mass}$$

$$M(^1\text{H}) = 1.007825\ \text{u} \qquad \text{Atomic hydrogen mass}$$

$$M(^2\text{H}) = 2.014102\ \text{u} \qquad \text{Atomic deuterium mass}$$

$$B_d/c^2 = m_n + M(^1\text{H}) - M(^2\text{H}) = 0.002388\ \text{u}$$

*What is quoted in many reference tables of atomic masses is the **mass excess** Δ, given by $M - A$. This avoids having to quote the masses to so many significant figures, because M and A are almost equal. We have listed in Appendix 8 the actual atomic mass M rather than Δ in order to make the calculations more transparent to the student.

We convert this mass-difference to energy using $\text{u} = 931.5\ \text{MeV}/c^2$.

$$B_d = 0.002388\ c^2 \cdot \text{u}\left(\frac{931.5\ \text{MeV}}{c^2 \cdot \text{u}}\right) = 2.224\ \text{MeV} \tag{12.9}$$

Our neglect of the atomic electron binding energy of 13.6 eV is justified, because the nuclear binding energy of 2.2 MeV is almost one million times greater. Even for heavier nuclei we normally neglect the electron binding energies. Although they become considerably larger, the nuclear binding energies increase as well. The electron binding energies cancel to a great extent in an equation like Equation (12.8) for heavy masses anyway.

The binding energy of any nucleus ${}^A_ZX$ is the energy required to separate the nucleus into free neutrons and protons. It can be determined using atomic masses, $M({}^1\text{H})$, and $M({}^A_ZX)$:

Nuclear binding energies

$$B({}^A_ZX) = [Nm_n + ZM({}^1\text{H}) - M({}^A_ZX)]c^2 \tag{12.10}$$

Experimental Determination of Nuclear Binding Energies. We can check our result for the 2.22-MeV binding energy of the deuteron by using a nuclear reaction. We scatter gamma rays (photons) from deuterium gas and look for the breakup of a deuteron into a neutron and proton:

$$\gamma + d \longrightarrow n + p \tag{12.11}$$

This type of nuclear reaction is called *photodisintegration* or a *photonuclear reaction,* because a photon causes the target nucleus to change form (see Figure 12.3). The mass-energy relation corresponding to Equation (12.11) is

$$h\nu + M({}^2\text{H})c^2 = m_nc^2 + M({}^1\text{H})c^2 + K_n + K_p \tag{12.12}$$

where $h\nu$ is the incident photon energy and K_n and K_p are the neutron and proton kinetic energies, respectively. If we want to find the minimum energy required for the photodisintegration, we let $K_n = K_p = 0$. We then find

$$h\nu_{\text{min}} = m_nc^2 + M({}^1\text{H})c^2 - M({}^2\text{H})c^2 = B_d \tag{12.13}$$

Equation (12.13) is not quite correct, because momentum must also be conserved in the reaction (K_n and K_p can't both be zero). The precise relation is

$$h\nu_{\text{min}} = B_d\left(1 + \frac{B_d}{2M({}^2\text{H})c^2}\right) \tag{12.14}$$

This value of $h\nu_{\text{min}}$ is almost exactly B_d, the deuteron binding energy, because the second term is so small. Experiment shows that a photon of energy less than 2.22 MeV cannot dissociate a deuteron.

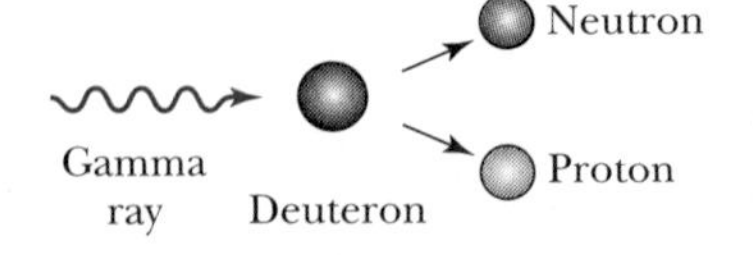

FIGURE 12.3 A gamma ray of energy greater than 2.22 MeV is able to dissociate a deuteron into a neutron and proton. This photodisintegration effect confirms that the binding energy of the deuteron is 2.22 MeV.

Deuteron Spin and Magnetic Moment. Another striking property of a deuteron is its nuclear spin quantum number of 1. This indicates the neutron and proton spins are aligned parallel to each other. The nuclear magnetic moment of a deuteron is $0.86\mu_N$, which is close to the sum of the values for the free proton and neutron: $2.79\mu_N - 1.91\mu_N = 0.88\mu_N$. This supports our hypothesis of parallel spins.

12.4 Nuclear Forces

Many techniques are used to study nuclear forces. The most straightforward are based on scattering experiments. We examine the nuclear force by first studying

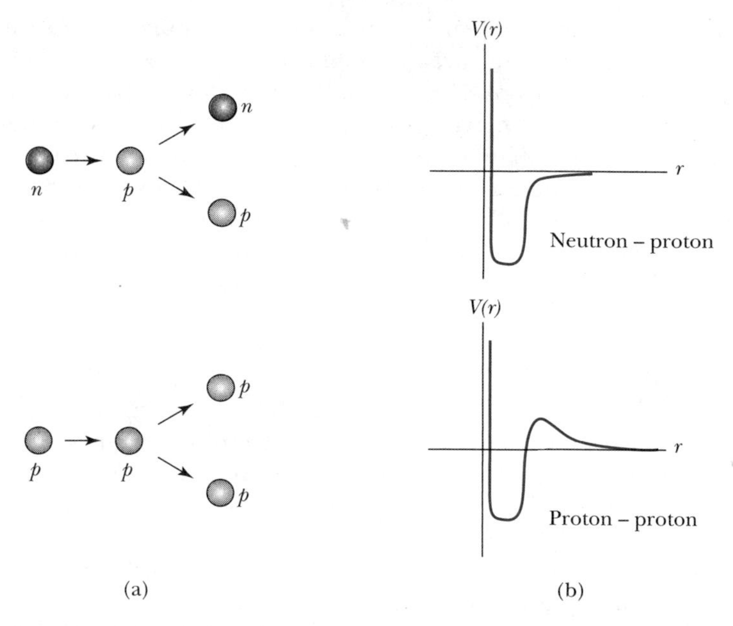

FIGURE 12.4 (a) A detailed study of neutron + proton and proton + proton scattering reveals (b) the shape of the potential describing the interaction. The proton–proton interaction includes the Coulomb effect (not to scale).

the simplest systems. We looked at the deuteron in the previous section. In scattering neutrons from protons, a deuteron is sometimes formed in the nuclear reaction,

$$n + p \rightarrow d + \gamma \tag{12.15}$$

We can also study the angular distribution of neutrons elastically scattered by protons, as shown in Figure 12.4a. Neutron + proton and proton + proton elastic scattering reveals that the nuclear potential is shaped roughly as shown in Figure 12.4b. The internucleon potential has a "hard core" that prevents the nucleons from approaching each other much closer than about 0.4 fm. A proton has a charge radius up to about 1 fm. We believe the neutron is roughly the same size. Two nucleons within about 2 fm of each other feel an attractive nuclear force. Outside about 3 fm the nuclear force is essentially zero. We call the nuclear force *short range* because it falls to zero so abruptly with interparticle separation. Because the nuclear force is short range, nucleons mostly interact with their nearest-neighbor nucleons. The nuclear force is said to be *saturable,* because the interior nucleons are completely surrounded by other nucleons with which they interact. However, nucleons on the nuclear surface are not so completely bound, and their nuclear force is not saturated. Of course, we are speaking classically about phenomena that must be described quantum mechanically.

Nuclear forces are short range

The only difference between the *np* and *pp* potentials shown in Figure 12.4b is the Coulomb potential shown for $r \geq 3$ fm for the *pp* force. Inside 3 fm the nuclear force clearly dominates, but outside 3 fm only the Coulomb force is effective. The depth of the nucleon–nucleon potential is about 40 MeV, with the *np* potential being slightly greater because of the absence of the Coulomb force.

The nuclear force is known to be spin dependent because the bound state of the deuteron has the neutron and proton spins aligned, but there is no bound state with the spins antialigned (that is, coupled to total spin 0).

The neutron–neutron system is more difficult to study because free neutrons (not bound in a nucleus) are not stable, and we do not know how to construct a target of free neutrons. However, indirect evidence, together with analyses of experiments where moving neutrons scatter from each other (as in simultaneous nuclear bomb explosions), indicates the *nn* potential is similar to the *np* potential. The nuclear potential between two nucleons seems independent of their charges. We call this *charge independence of nuclear forces.* For many purposes, the neutron and proton can be considered different charge states of the same particle. This is why we use the term *nucleon* to refer to either neutrons or protons.

Nuclear forces are charge independent

12.5 Nuclear Stability

In Equation (12.10) we presented a method to determine the binding energies of nuclides in terms of atomic masses. If B is positive, then the nuclide is said to be stable against dissociating into free neutrons and protons. We need to generalize Equation (12.10), however, because a nucleus containing A nucleons is said to be *stable* if its mass is smaller than that of any other possible combination of A nucleons. The binding energy of a nucleus ${}^A_Z X$ against dissociation into any other possible combination of nucleons, for example nuclei R and S, is

$$B = [M(R) + M(S) - M({}^A_Z X)]c^2 \tag{12.16}$$

In particular, the energy required to remove one proton (or neutron) from a nuclide is called the proton (or neutron) *separation energy,* and Equation (12.16) is useful for finding this energy. Even if B is negative for a particular dissociation, there may be other reasons why the nucleus is stable.

Example 12.4

Show that the nuclide ^{8}Be has a positive binding energy but is unstable with respect to decay into two alpha particles.

Solution: The binding energy of ^{8}Be is determined by Equation (12.10).

$$B(^8\text{Be}) = [4m_n + 4M(^1\text{H}) - M(^8\text{Be})]c^2$$

We look up the atomic masses of ^{1}H and ^{8}Be in Appendix 8 and calculate the binding energy to be

$$B(^8\text{Be}) = [4(1.008665\text{ u}) + 4(1.007825\text{ u}) - 8.005305\text{ u}]c^2\left(\frac{931.5\text{ MeV}}{c^2\cdot\text{u}}\right) = 56.5\text{ MeV}$$

Now we calculate the binding energy of the decay of ^{8}Be into two α particles, $^8\text{Be} \rightarrow 2\alpha$, by using Equation (12.16)

$$B(^8\text{Be} \rightarrow 2\alpha) = [2M(^4\text{He}) - M(^8\text{Be})]c^2$$
$$= [2(4.002603\text{ u}) - 8.005305\text{ u}]c^2\left(\frac{931.5\text{ MeV}}{c^2\cdot\text{u}}\right)$$
$$= -0.093\text{ MeV}$$

Because the latter B is negative, ^{8}Be is *unstable* against decay to two alpha particles. From the standpoint of energy, there is no reason why a ^{8}Be nucleus will not decay into two alpha particles. Sometimes a nuclide may be stable even if another combination of A nucleons has a lower mass, because some conservation law, like spin angular momentum, prevents the radioactive decay, but in this case we find experimentally that ^{8}Be does spontaneously decay into two alpha particles. The instability of ^{8}Be is also responsible for stars consisting of mostly hydrogen and helium. Because of the instability of ^{8}Be, it is difficult for helium nuclei to join together to make heavier nuclei.

In Figure 12.5 we exhibit all known stable nuclei as well as many known unstable nuclei that are long-lived enough to be observed (nowadays this includes nuclides that decay in a millisecond or less). There are several important facts we

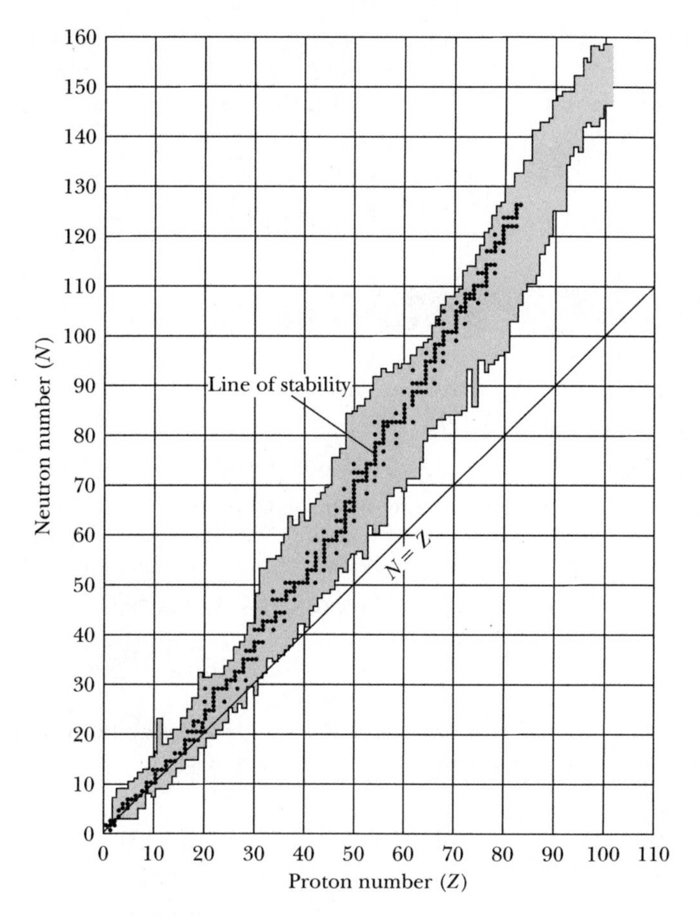

FIGURE 12.5 A plot of the known nuclides with neutron number N versus proton number Z. The solid points represent stable nuclides, and the shaded area represents unstable nuclei. A smooth line through the solid points would represent the *line of stability*.

can extract from Figure 12.5. First, it appears that for $A \leq 40$, nature prefers the number of protons and neutrons in the nucleus to be about the same, $Z \sim N$. However, for $A \geq 40$, there is a decided preference for $N > Z$. We can understand this difference in the following way. As we noted earlier, the strength of the nuclear force is independent of whether the particles are *nn*, *np*, or *pp*. Equal numbers of neutrons and protons may give the most attractive average internucleon nuclear force, but the Coulomb force must be considered as well. As the number of protons increases, the Coulomb force between all the protons becomes stronger and stronger until it eventually affects the binding significantly.

The electrostatic energy required to contain a charge Ze evenly spread throughout a sphere of radius R can be calculated by determining the work required to bring the charge inside the sphere from infinity (see Problem 51) and is determined to be

$$\Delta E_{\text{Coul}} = \frac{3}{5} \frac{(Ze)^2}{4\pi\epsilon_0 R} \tag{12.17}$$

For a single proton, Equation (12.17) gives for the self-energy,

$$\Delta E_{\text{Coul}} = \frac{3}{5} \frac{e^2}{4\pi\epsilon_0 R}$$

This term represents the work done to assemble the proton itself, and we do not want to include it in the electrostatic self-energy of a nucleus composed of Z protons. Therefore we must subtract Z such terms from the total given in Equation (12.17) to give us the total Coulomb repulsion energy in a nucleus:

$$\Delta E_{\text{Coul}} = \frac{3}{5}\frac{Z(Z-1)e^2}{4\pi\epsilon_0 R} \tag{12.18}$$

Example 12.5

Show that Equation (12.18) can be written as

$$\Delta E_{\text{Coul}} = 0.72\ [Z(Z-1)]\ A^{-1/3}\ \text{MeV} \tag{12.19}$$

and use this equation to calculate the total Coulomb energy of $^{238}_{92}\text{U}$.

Solution: We use Equation (12.2) for R (with $r_0 = 1.2$ fm) and insert into Equation (12.18) to find the energy first in joules and then in MeV.

$$\Delta E_{\text{Coul}} = \frac{3}{5}[Z(Z-1)](1.6\times 10^{-19}\ \text{C})^2 \times (9\times 10^9\ \text{N}\cdot\text{m}^2/\text{C}^2)\frac{1}{1.2\times 10^{-15}\ \text{m}\cdot A^{-1/3}}$$

$$= 1.15\times 10^{-13}[Z(Z-1)]A^{-1/3}\ \text{J}\ \frac{\text{MeV}}{1.6\times 10^{-13}\ \text{J}}$$

$$= 0.72\ [Z(Z-1)]\ A^{-1/3}\ \text{MeV}$$

Now we insert $Z = 92$ and $A = 238$ into Equation (12.19) to find

$$E_{\text{Coul}} = 0.72\ (92)(91)(238)^{-1/3} = 973\ \text{MeV}$$

Is this large or small? If we look up the masses in Appendix 8, we calculate that the total binding energy of $^{238}_{92}\text{U}$ with respect to dissociation into its component nucleons is

$$B_{\text{tot}} = [146(1.008665\ \text{u}) + 92(1.007825\ \text{u}) - 238.050783\ \text{u}]c^2\left(\frac{931.5\ \text{MeV}}{c^2\cdot\text{u}}\right) = 1802\ \text{MeV}$$

The total Coulomb energy is a significant fraction of the binding energy of a large nucleus.

We conclude that for heavy nuclei, the nucleus will have a preference for fewer protons than neutrons because of the large Coulomb repulsion energy. In fact, Figure 12.5 reveals there are no stable nuclei with $Z > 83$ because of the increasingly larger Coulomb force. The heaviest known stable nucleus is $^{209}_{83}\text{Bi}$. All nuclei with $Z > 83$ and $A > 209$ will eventually decay spontaneously into some combination of A nucleons with lighter masses. Adding one proton to a heavy nucleus adds a constant amount of nuclear binding energy, but the repulsive Coulomb energy increases as $\Delta E_{\text{Coul}} \sim (Z+1)^2 - Z^2 \approx 2Z$. Because the Coulomb force is long-range, the proton interacts electromagnetically with all the protons already in the nucleus. And because this repulsive energy increases with Z, nuclei with higher Z eventually become unstable. The neutrons dilute the Coulomb repulsion slightly because they intersperse among the protons, causing the protons to be slightly farther apart.

Another interesting fact discernable from Figure 12.5 is that most stable nuclides have *both* even Z and even N (called "even-even" nuclides). Only four stable nuclides, all light nuclei, have odd Z and odd N (called "odd-odd" nuclides). These nuclides are ^2_1H, ^6_3Li, $^{10}_5\text{B}$, and $^{14}_7\text{N}$. All the other stable nuclides are odd-even or even-odd, that is, with either an odd number of Z or N. Nature apparently prefers nuclei with even numbers of protons and neutrons.

We can understand this empirical observation in terms of the Pauli exclusion principle. Neutrons and protons are distinguishable fermions, hence they separately obey the exclusion principle. Only two neutrons (or protons) may coexist in each spatial orbital (quantum state), one with spin "up" and the other with spin "down." Each nuclear energy level is thus able to hold two particles whose spins are paired to 0. This configuration of opposite spins is particularly stable because placing the same number of particles in any other arrangement will produce a (less stable) state of higher energy. Therein lies the preference for even N and Z.

Niels Bohr, Carl F. von Weizsäcker and others were able to explain many nuclear phenomena in the 1930s by treating the nucleus as a collection of interacting particles in a liquid drop. This model of the nucleus is known as the **liquid drop model.** In the preceding discussion we understood qualitatively the line of stability curve displayed in Figure 12.5. In much the same way von Weizsäcker proposed in 1935 his semi-empirical mass formula based on the liquid drop model.

Written in terms of the total binding energy, the semi-empirical mass formula is

$$B\left({}^{A}_{Z}\mathrm{X}\right) = a_V A - a_A A^{2/3} - \frac{3}{5}\frac{Z(Z-1)e^2}{4\pi\epsilon_0 r} - a_S\frac{(N-Z)^2}{A} + \delta \tag{12.20}$$

von Weizsäcker semi-empirical mass formula

The volume term (a_V) indicates that the binding energy is approximately the sum of all the interactions between the nucleons. Because the nuclear force is short range and each nucleon interacts only with its nearest neighbors, this interaction is proportional to A, the total number of nucleons.

The second term, called the *surface effect,* is simply a correction to the first term (similar to surface tension), because the nucleons on the nuclear surface are not completely surrounded by other nucleons. The surface nucleons do not have saturated interactions, and a correction should be made proportional to the liquid drop surface, $4\pi R^2$. Because $R \sim A^{1/3}$, the correction is proportional to $A^{2/3}$.

The third term is the Coulomb energy discussed and presented in Equations (12.17) and (12.18).

The fourth term is due to the symmetry energy, also previously discussed. In the absence of Coulomb forces, the nucleus prefers to have $N = Z$. This term has a quantum mechanical origin, depending on the exclusion principle. Notice that the sign of the fourth term is independent of the sign of $N - Z$.

The last term is due to the pairing energy and reflects the fact that the nucleus is more stable for even-even nuclides. We can determine this term empirically.

One set of values given by Enrico Fermi* for the parameters of Equation (12.20) is

$$\begin{aligned} a_V &= 14\ \text{MeV} \quad &&\text{Volume} \\ a_A &= 13\ \text{MeV} \quad &&\text{Surface} \\ a_S &= 19\ \text{MeV} \quad &&\text{Symmetry} \end{aligned}$$

$$\text{pairing } \delta = \begin{cases} +\Delta & \text{for even-even nuclei} \\ 0 & \text{for odd-}A\text{ (even-odd, odd-even) nuclei} \\ -\Delta & \text{for odd-odd nuclei} \end{cases}$$

where $\Delta = 33\ \text{MeV} \cdot A^{-3/4}$

*See B. Cohen, *Concepts of Nuclear Physics.* New York: McGraw-Hill, 1971.

The entire table of stable isotopes can be understood by applying the ideas in the von Weizsäcker semi-empirical mass formula. No nuclide heavier than $^{238}_{92}$U has been found in nature. Such nuclides, if they ever existed, must have decayed so quickly that quantities sufficient to measure no longer exist. Many of the nuclides between $^{209}_{83}$Bi and $^{238}_{92}$U are still found in nature, either because their decay rates are slow enough that they have not sufficiently decayed since their formation in the interior of stars or because they are produced continuously by the radioactive decay of another nuclide.

By calculating the binding energy of each known nucleus, and dividing by its mass number, we obtain the plot shown in Figure 12.6. We see that the average binding energy per nucleon peaks near $A = 56$ and slowly decreases for the heaviest nuclei. For the lighter nuclei the curve increases rapidly from hydrogen until all the nucleons are surrounded by other nucleons. This curve demonstrates the saturation effect of nuclear forces. After the very light nuclei ($A < 20$), the curve is reasonably flat at 8 MeV/nucleon. There are sharp peaks for the even-even nuclides ^{4}He, ^{12}C, and ^{16}O, which are particularly tightly bound.

Example 12.6

Calculate the binding energy per nucleon for $^{20}_{10}$Ne, $^{56}_{26}$Fe, and $^{238}_{92}$U.

Solution: We first find the binding energy of each of these nuclides using Equation (12.10) and then divide by the mass number to obtain the binding energy per nucleon.

$$B(^{20}_{10}\text{Ne}) = [10m_n + 10M(^1\text{H}) - M(^{20}_{10}\text{Ne})]c^2$$

$$= [10(1.008665\text{ u}) + 10(1.007825\text{ u}) - 19.992440\text{ u}]c^2\left(\frac{931.5\text{ MeV}}{c^2\cdot\text{u}}\right)$$

$$= 160.6\text{ MeV}$$

$$\frac{B(^{20}_{10}\text{Ne})}{20\text{ nucleons}} = 8.03\text{ MeV/nucleon}$$

$$B(^{56}_{26}) = [30m_n + 26M(^1\text{H}) - M(^{56}_{26}\text{Fe})]c^2$$

$$= [30(1.008665\text{ u}) + 26(1.007825\text{ u}) - 55.934942\text{ u}]c^2\left(\frac{931.5\text{ MeV}}{c^2\cdot\text{u}}\right)$$

$$= 492\text{ MeV}$$

$$\frac{B(^{56}_{26}\text{Fe})}{56\text{ nucleons}} = 8.79\text{ MeV/nucleon}$$

$$B(^{238}_{92}\text{U}) = [146m_n + 92M(^1\text{H}) - M(^{238}_{92}\text{U})]c^2$$

$$= [146(1.008665\text{ u}) + 92(1.007825\text{ u}) - 238.050783\text{ u}]c^2\left(\frac{931.5\text{ MeV}}{c^2\cdot\text{u}}\right)$$

$$= 1802\text{ MeV}$$

$$\frac{B(^{238}_{92}\text{U})}{238\text{ nucleons}} = 7.57\text{ MeV/nucleon}$$

All three nuclides have a binding energy per nucleon near 8 MeV, with ^{56}Fe having the largest binding energy per nucleon.

Nuclear Models

Physicists do not fully understand the nuclear force or how nucleons interact inside the nucleus. Current research stresses the constituent quarks (see Chapter 14) that make up the nucleons. Because of this lack of knowledge, physicists have used a multitude of models to explain nuclear behavior. These models have been more or less successful in explaining various nuclear properties.

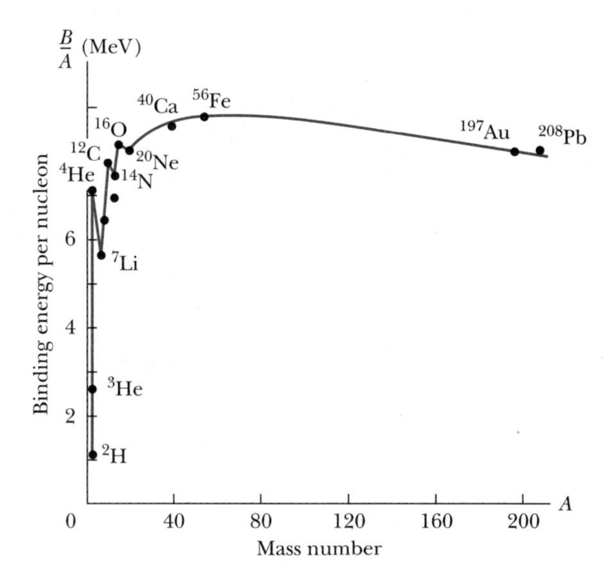

FIGURE 12.6 The binding energy per nucleon versus the mass number A. Notice the subpeaks at ^{4}He, ^{12}C, and ^{16}O.

The models generally fall into two categories:

1. *Independent Particle Models,* in which the nucleons move nearly independently in a common nuclear potential. The shell model has been the most successful of these.
2. *Strong Interaction Models,* in which the nucleons are strongly coupled together. The liquid drop model already discussed is characteristic of these models and has been quite successful in explaining nuclear masses as well as nuclear fission (see Chapter 13).

Space does not permit us a full discussion of each of the many models. We have already discussed the liquid drop model in this section, so we now present the simplest of the independent-particle models. We show in Figure 12.7a representation of the nuclear potential felt by the neutron and proton. Because of the Coulomb interaction, the shape and depth of the proton potential is somewhat different than that of the neutron. For example, typical depths are about 43 MeV for neutrons but only 37 MeV for protons. Energy levels, which represent states that can be filled by the nucleons, are shown inside the potential. Note that nuclei have a Fermi energy level, just as do atoms, which is the highest energy level filled in the nucleus. A typical Fermi energy level has a depth of about 8 MeV. In the ground state of a nucleus, all the energy levels below the Fermi level are filled, but when a nucleus becomes excited, one or more of the nucleons is raised to one of the previously unoccupied levels above the Fermi level.

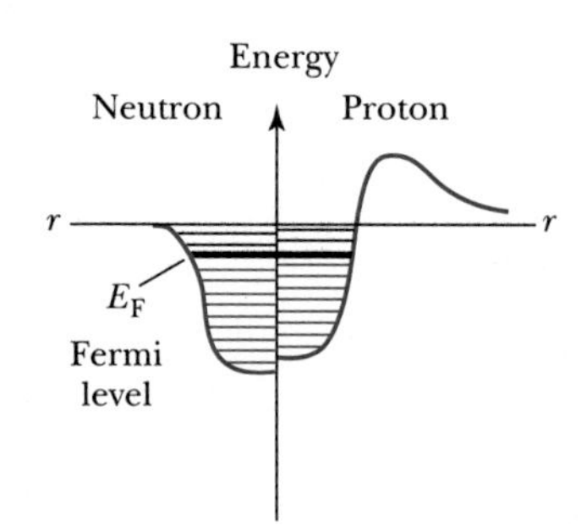

FIGURE 12.7 Diagram of nuclear potential wells as felt by neutrons and protons. Neutrons are more strongly bound than protons because of the Coulomb potential. All levels below the Fermi energy E_F are filled.

Nuclei are formed by a collection of nucleons. The nucleons sort themselves into the lowest possible energy levels. In Figure 12.8 we exhibit energy-level diagrams for several possible nuclides between ^{12}C and ^{16}O. These energy-level diagrams assume the zero of the energy scale to be the bottom of the nuclear potential, so we can deal with positive energy values. Both ^{12}C and ^{16}O are particularly stable because they are even-even. We show schematically the neutron energy levels slightly lower than the proton levels because of the additional

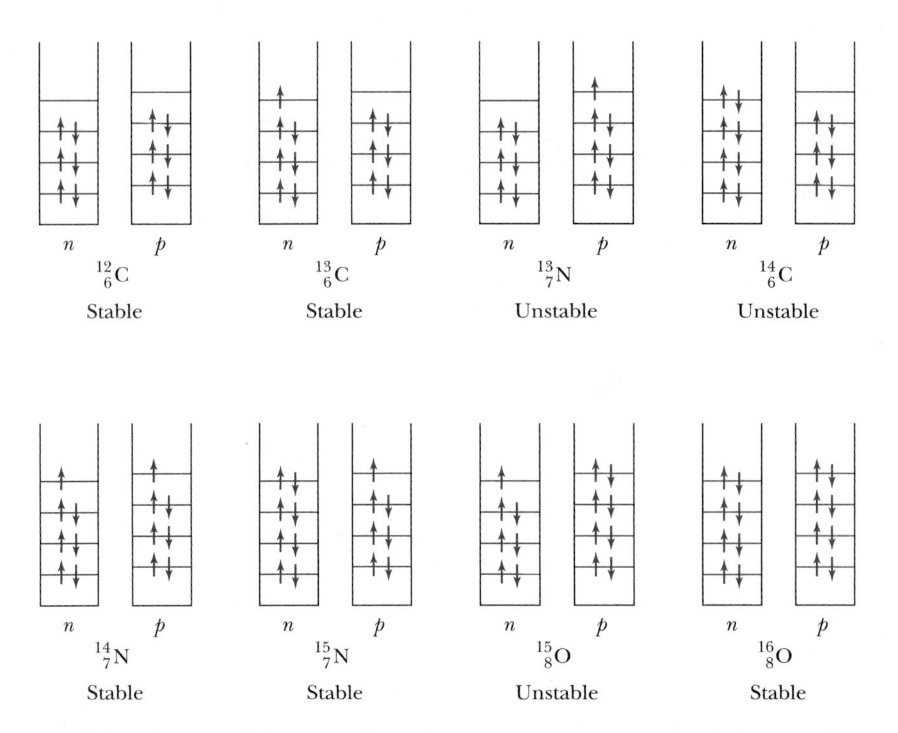

FIGURE 12.8 Schematic diagram of proton and neutron energy levels for several nuclei between ^{12}C and ^{16}O. The nuclei ^{12}C and ^{16}O are particularly stable, but the effects of $N \approx Z$ and the spin-pairing effects are important in this region.

Coulomb repulsion of the protons. If we add one proton to ^{12}C to make $^{13}_{7}$N, we find it is unstable, whereas if we add a neutron—to make ^{13}C—we find it to be stable. Even when we add another neutron to produce ^{14}C, we find it is barely unstable. In this mass region nature prefers the number of neutrons and protons to be about equal ($N \approx Z$), but it doesn't want $Z > N$. This helps explain why ^{13}C is stable, but not ^{13}N; ^{14}C has too many more neutrons (8) than protons (6) in this mass region to be stable. When we add a proton to ^{13}C we make ^{14}N, one of the few stable odd-odd nuclides. If we next add a proton to ^{14}N we obtain the unstable ^{15}O. However, if we add a neutron to ^{14}N we find stable ^{15}N, again indicating neutron energy levels to be lower in energy than the corresponding proton ones. Finally, if we add one more proton to ^{15}N, we pair the extra proton and make the extra stable ^{16}O. However, if we add a neutron to ^{15}N, we have the very unstable nuclide ^{16}N. Only ^{14}N and ^{15}N are stable isotopes of nitrogen. The shell model of nuclei takes advantage of the pairing effect and places only two neutrons or two protons in each shell (or energy level). The ordering of the energy levels is established by angular momentum rules, which couple the nucleon spins in a prescribed manner similar to that already discussed for *jj* coupling in Chapter 8.

12.6 Radioactive Decay

We learned in the previous section that many nuclei are unstable and can decay spontaneously to some other combination of A nucleons that has a lower mass. These decays take different forms. The simplest is that of a gamma ray, which represents the nucleus changing from an excited state to a lower-energy state (no change in N or Z). Other modes of decay include emission of α particles, β particles, protons, neutrons, and fission. We will discuss α, β, and γ decay in the next section and defer the discussion of fission to the next chapter. In this section we will learn the nature of the radioactive decay law.

The general form of the law of radioactivity is the same for all decays because it is a statistical process. Given a sample of radioactive material we measure the distintegrations or decays per unit time, which we define as **activity.** If we have N unstable atoms of a material, the activity R is given by

$$\text{Activity} = -\frac{dN}{dt} = R \tag{12.21}$$

where we insert the minus sign to make R positive (dN/dt is negative because the total number N is decreasing with time). The SI unit of activity is the becquerel (1 Bq = 1 decay/s). More commonly used over the past few decades is the curie (Ci) which is 3.7×10^{10} decays/s. In keeping with the worldwide trend to use SI units, here we will use primarily the becquerel for activity. A typical radioactive source used in a student laboratory experiment (for example, ^{226}Ra or ^{210}Po α emitters), the ^{241}Am α emitter used in a smoke alarm, or the radium used in a luminous watch may contain material having an activity of only about 10^4 Bq (a few μCi), and these are exempt from federal licensing requirements.

Marie Curie (1867–1934) (on the right) and Irene Joliot-Curie (1897–1956) are the most famous mother–daughter pair in science. Marie won two Nobel Prizes, one in physics with her husband Pierre in 1903 and another in chemistry in 1911, for her work in radiation phenomena and the discovery of the elements radium and polonium. Irene, who began working in her mother's lab as a teenager, won a Nobel Prize in chemistry in 1935 with her husband Frederic Joliot for their production of new radioactive elements. *AIP Emilio Segrè Visual Archives, William G. Myers collection.*

We observe experimentally that the activity of a given sample falls off exponentially with time. If $N(t)$ is the number of radioactive nuclei in a sample at time t, and λ (called the *decay constant*) is the probability per unit time that any given nucleus will decay, then the activity R is

$$R = \lambda N(t) \tag{12.22}$$

The number dN of nuclei decaying during the time interval dt is

$$dN(t) = -R\,dt = -\lambda N(t)\,dt \tag{12.23}$$

where we have used Equations (12.21) and (12.22). If we rearrange and integrate this equation, we have

$$\int \frac{dN}{N} = -\int \lambda\,dt$$

$$\ln N = -\lambda t + \text{constant}$$

$$N(t) = e^{-\lambda t + \text{constant}}$$

If we let $N(t = 0) \equiv N_0$, the previous equation becomes

Radioactive decay law

$$N(t) = N_0 e^{-\lambda t} \tag{12.24}$$

This is the radioactive decay law and it applies to all decays. The exponential decay rate is consistent with experimental observation. The activity R is

$$R = \lambda N(t) = \lambda N_0 e^{-\lambda t} = R_0 e^{-\lambda t} \tag{12.25}$$

where R_0 is the initial activity at $t = 0$. The activity of a radioactive sample also falls off exponentially.

It is more common to refer to the half-life $t_{1/2}$ or the mean lifetime τ rather than its decay constant. The half-life is the time it takes one half of the radioactive nuclei to decay.

$$N(t_{1/2}) = \frac{N_0}{2} = N_0 e^{-\lambda t_{1/2}} \tag{12.26}$$

$$\ln\left(\frac{1}{2}\right) = \ln\left(e^{-\lambda t_{1/2}}\right) = -\lambda t_{1/2}$$

The half-life is determined to be

The half-life

$$t_{1/2} = \frac{-\ln(1/2)}{\lambda} = \frac{\ln(2)}{\lambda} = \frac{0.693}{\lambda} \tag{12.27}$$

The mean (or average) lifetime τ is calculated to be (see Problem 23)

The mean lifetime

$$\tau = \frac{1}{\lambda} = \frac{t_{1/2}}{\ln(2)} \tag{12.28}$$

The number of radioactive nuclei as a function of time is displayed in Figure 12.9. Because the decay of a radioactive nucleus is a statistical process, it will take a very large sample to give such a smooth curve as that shown in Figure 12.9. If we have a radioactive sample (for example, 1 mg) with a half-life of 1 h, we know that about 50% of the 1-mg sample will decay in 1 h. What fraction decays in 2 h? It is not 100%. During the second hour period the probability is still 50% for each remaining nucleus to decay. The total probability that a given nucleus did *not* decay is 0.50(0.50) = 0.25 or 25%. The probability of decay is 75%, and this is the fraction of the original nuclei expected to be gone in 2 hours.

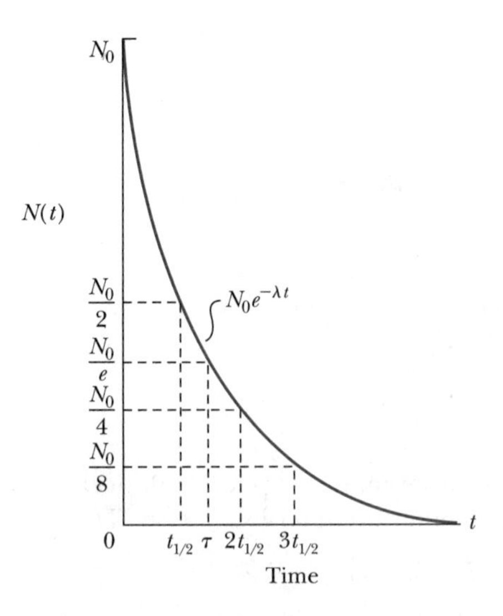

FIGURE 12.9 The number of remaining nuclei $N(t)$ as a function of time t for a sample of radioactive nuclides. The half-life $t_{1/2}$ and mean lifetime τ are indicated for the exponential radioactive decay law.

Example 12.7

A sample of ^{210}Po which α decays with $t_{1/2} = 138$ days is observed by a student to have 2000 disintegrations/s (2000 Bq).
(a) What is the activity in μCi for this source?
(b) What is the mass of the ^{210}Po sample?

Solution: (a) First, we find the number of Ci that 2000 decays/s represents. The fraction of one Ci is given by

$$\frac{2000 \text{ decays/s}}{3.7 \times 10^{10} \text{ decays/s}}(1 \text{ Ci}) = 0.054 \times 10^{-6} \text{ Ci} = 0.054 \ \mu\text{Ci}$$

(b) Because we know the activity (2000 decays/s), we can use Equations (12.22) and (12.28) to find the number of radioactive nuclei. From this we can determine the mass of the ^{210}Po sample.

$$N = \frac{R}{\lambda} = \frac{(R)(t_{1/2})}{\ln(2)} = \frac{2000 \text{ decays/s}}{\ln(2)}(138 \text{ d})\frac{24 \text{ h}}{1 \text{ d}}\frac{3600 \text{ s}}{1 \text{ h}}$$

$$= 3.44 \times 10^{10} \text{ nuclei}$$

We use Avogadro's number to determine the mass from the number of atoms (nuclei).

$$\text{Mass} = 3.44 \times 10^{10} \text{ atoms} \frac{1 \text{ mol}}{6.02 \times 10^{23} \text{ atoms}} \frac{210 \text{ g}}{1 \text{ mol}}$$

$$= 1.2 \times 10^{-11} \text{ g}$$

This is an extremely small amount of mass!

Example 12.8

A sample of ^{18}F is used internally as a medical diagnostic tool to look for the effects of the positron decay ($t_{1/2} =$ 110 min). How long does it take for 99% of the ^{18}F to decay?

Solution: We use the radioactive decay law, Equation (12.24), to determine the time needed.

$$N = N_0 e^{-\lambda t} = N_0 e^{-\ln(2)t/t_{1/2}}$$

If we want 99% of the initial sample to decay, then only 1% will be left, and $N/N_0 = 1\%$. We then have

$$\frac{N}{N_0} = 0.01 = e^{-\ln(2)t/t_{1/2}}$$

If we take the natural logarithm, we have

$$\ln(0.01) = -\ln(2)\frac{t}{t_{1/2}}$$

$$t = -\frac{\ln(0.01)}{\ln(2)}t_{1/2} = -\frac{-4.61}{0.693}110 \text{ min}$$

$$= 731 \text{ min} = 12.2 \text{ h}$$

Example 12.9

What is the alpha activity of a 10-kg sample of ^{235}U that is used in a nuclear reactor?

Solution: We find in a table of radioactive nuclides (or Appendix 8) that ^{235}U has a half-life for emitting α particles of $t_{1/2} = 7.04 \times 10^8$ y. We use Equation (12.22) to find the activity. First, we find the number of radioactive atoms by using Avogadro's number and the gram-molecular weight.

$$N = M\frac{N_A}{M(^{235}\text{U})}$$

$$= (10 \text{ kg})\left(\frac{10^3 \text{ g}}{1 \text{ kg}}\right)\left(\frac{6.02 \times 10^{23} \text{ atoms/mol}}{235 \text{ g/mol}}\right)$$

$$= 2.56 \times 10^{25} \text{ atoms} = 2.56 \times 10^{25} \text{ nuclei}$$

The activity is

$$R = \lambda N = \frac{\ln(2) \cdot N}{t_{1/2}} = \frac{\ln(2) \cdot (2.56 \times 10^{25} \text{ nuclei})}{7.04 \times 10^8 \text{ y}}$$

$$= 2.52 \times 10^{16} \text{ decays/y} = 8.0 \times 10^8 \text{ Bq}$$

12.7 Alpha, Beta, and Gamma Decay

The three common decay modes of nuclei (α, β, and γ) were all observed by the early 20th century. When a nucleus decays, all the conservation laws must be observed: mass-energy, linear momentum, angular momentum, and electric charge. To these laws we must add another one for radioactive decay, called the **conservation of nucleons.** It states that *the total number of nucleons (A, the mass number) must be conserved in a low-energy nuclear reaction (say, less than 100 MeV) or decay.* Neutrons may be converted into protons, and vice versa, but the total number of nucleons must remain constant. At higher energies enough rest energy may be available to create nucleons, but other conservation laws to be discussed in Chapter 14 still apply.

The conservation of nucleons

Radioactive decay may occur for a nucleus when some other combination of the A nucleons has a lower mass. Let the radioactive nucleus ${}^A_ZX$ be called the parent and have the mass $M({}^A_ZX)$. There can be two or more products produced in the decay. In the case of two products let the mass of the lighter one be M_y and the mass of the heavier one (normally called the *daughter*) be M_D. The conservation of energy is

$$M({}^A_ZX) = M_D + M_y + Q/c^2 \tag{12.29a}$$

where Q is the energy released and is equal to the total kinetic energy of the reaction products. It is given in terms of the masses by

Disintegration energy

$$Q = [M({}^A_ZX) - M_D - M_y]c^2 \tag{12.29b}$$

Note that the **disintegration energy** Q is the negative of the binding energy B [see Equation (12.16)]. The binding energy normally refers to stable nuclei, whereas Q is normally used with unstable nuclei. If $B > 0$, a nuclide is bound and stable; if $Q > 0$, a nuclide is unbound, unstable, and may decay. If we look at the naturally abundant radioactive nuclei, we find that decays emitting nucleons do not occur, because the masses are such that $Q < 0$. The reason for this is that if nucleon decay were possible, it would have taken place too quickly for such radioactive nuclei to be naturally abundant.

Example 12.10

Show that ${}^{230}_{92}\text{U}$ does not decay by emitting a neutron or proton.

Solution: The decays in question are

(a) ${}^{230}_{92}\text{U} \rightarrow n + {}^{229}_{92}\text{U}$ and (b) ${}^{230}_{92}\text{U} \rightarrow p + {}^{229}_{91}\text{Pa}$.

We can determine whether the decays occur by looking up the atomic masses in Appendix 8.

(a) $M({}^{230}_{92}\text{U}) = 230.033927$ u; $m_n = 1.008665$ u; $M({}^{229}_{92}\text{U}) = 229.033496$ u.

We use Equation (12.29b) to determine Q.

$$Q = [230.033927 \text{ u} - 229.033496 \text{ u} - 1.008665 \text{ u}]c^2 \left(\frac{931.5 \text{ MeV}}{c^2 \cdot \text{u}}\right) = -7.7 \text{ MeV}$$

Because $Q < 0$, neutron decay is not allowed.

(b) $m({}^1\text{H}) = 1.007825$ u; $M({}^{229}_{91}\text{Pa}) = 229.032089$ u.

$$Q = [230.033927 \text{ u} - 229.032089 \text{ u} - 1.007825 \text{ u}]c^2 \times \left(\frac{931.5 \text{ MeV}}{c^2 \cdot \text{u}}\right) = -5.6 \text{ MeV}$$

Because $Q < 0$, proton decay is not allowed.

In both cases the decay is not allowed, because the mass of the products is greater than that of the parent. The nucleus ${}^{230}\text{U}$ is stable against nucleon emission.

Alpha Decay

Is it possible for a collection of nucleons inside the nucleus to decay? The nucleus ^{4}He is particularly stable. Its binding energy is 28.3 MeV. The combination of two neutrons and two protons is particularly strong because of the pairing effects previously discussed. If the last two protons and two neutrons in a nucleus are bound by less than 28.3 MeV, then the emission of an alpha particle, called *alpha decay*, is energetically possible. For alpha decay, Equation (12.29) becomes

$$^{A}_{Z}X \rightarrow {}^{A-4}_{Z-2}D + \alpha \qquad (12.30)$$

Alpha decay

$$Q = [M(^{A}_{Z}X) - M(^{A-4}_{Z-2}D) - M(^4\text{He})]c^2 \qquad (12.31)$$

If $Q > 0$, then alpha decay is possible.

Consider the nucleus $^{230}_{92}$U that we studied in Example 12.10. The alpha-decay reaction is given by

$$^{230}_{92}\text{U} \rightarrow \alpha + {}^{226}_{90}\text{Th}$$

We look up the appropriate masses to find

$$M(^{230}_{92}\text{U}) = 230.033927 \text{ u}; \; M(^4\text{He}) = 4.002603 \text{ u}; \; M(^{226}_{90}\text{Th}) = 226.024891 \text{ u}$$

If we insert the masses into Equation (12.31), we find

$$\begin{aligned} Q &= [M(^{230}\text{U}) - M(^{226}\text{Th}) - M(^4\text{He})]c^2 \\ &= [230.033927 \text{ u} - 226.024891 \text{ u} - 4.002603 \text{ u}]c^2\left(\frac{931.5 \text{ MeV}}{c^2 \cdot \text{u}}\right) \\ &= 6.0 \text{ MeV} \end{aligned}$$

Alpha decay is allowed, because $Q > 0$. The mass of the products is less than the mass of the decaying nuclide. Many of the nuclei above $A = 150$ in fact are susceptible to alpha decay. These heavy nuclei have increasingly stronger Coulomb repulsion as protons are added. The expulsion of two protons (along with two neutrons) in the form of an alpha particle may decrease this Coulomb energy and make the resulting nucleus more stable.

We might wonder why any nuclei exist with $A > 150$. First, it is believed that nuclei are not necessarily made up of a collection of alpha particles. In order for alpha decay to occur, two neutrons and two protons form together within the nucleus prior to decay. Second, the alpha particle, even when formed, has great difficulty in overcoming the nuclear attraction from the remaining nucleons to escape. Consider the potential energy diagram shown in Figure 12.10. The barrier height V_B for alpha particles is normally greater than 20 MeV. The kinetic energies of alpha particles emitted from nuclei range from 4 to 10 MeV. It is classically impossible for the alpha particles to escape, because the potential energy barrier is greater than the kinetic energy. If we project 5-MeV α particles onto a heavy nucleus we find that the alpha particle is repelled by the Coulomb force (see Figure 12.09) and doesn't get close enough to feel the attraction of the short-range nuclear force. It is virtually impossible classically for the alpha particle to reach the nucleus. How then can the alpha particle ever surmount the barrier if it is trapped inside the potential barrier? As we discussed in Chapter 6, the alpha particles are able to tunnel through the barrier. This is a pure quantum mechanical effect, and there is a small, but finite, chance for the alpha particle to appear on the other side of the barrier. The probability depends critically on

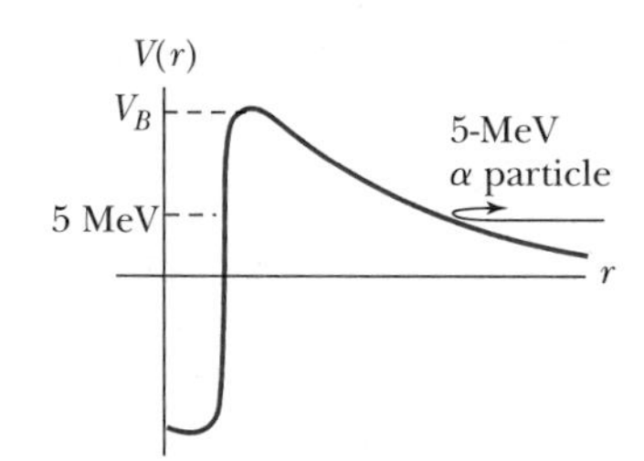

FIGURE 12.10 The potential energy barrier for an alpha particle is shown. The Coulomb barrier V_B is much greater than the typical alpha-particle energies produced by radioactive sources. Classically a 5-MeV particle inside the nucleus or scattered from outside cannot penetrate the barrier.

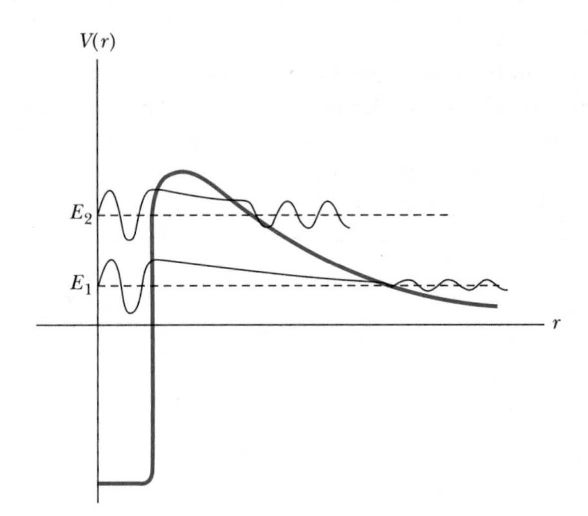

FIGURE 12.11 Quantum theory allows an alpha particle to tunnel through the barrier. A higher-energy alpha E_2 has a much higher probability (shorter lifetime) than a lower-energy alpha particle E_1. The waves shown are schematics only.

the barrier height and width. A higher-energy alpha particle, E_2 in Figure 12.11, has a much higher probability (shorter lifetime) than a lower-energy, E_1, alpha particle to tunnel through the barrier. In Figure 12.12 we compare the lifetimes of various alpha emitters with the kinetic energies of the alpha particles. We see that there is a strong correlation between lower energies and greater difficulty of escaping (longer lifetimes).

In Example 6.13 we discussed the α decay of ^{238}U. It might be worthwhile looking at that example again. Because of the low probability of tunneling, we showed that the α particle must make about 10^{41} traverses back and forth across the nucleus before it can escape. Because only two products occur in alpha decay, we can calculate the kinetic energy of the α particle from the disintegration energy Q. Assume the parent nucleus is initially at rest so that the total momentum is zero. As shown in Figure 12.13, the final momenta of the daughter p_D and alpha particle p_α are equal and opposite. By using the conservation of energy

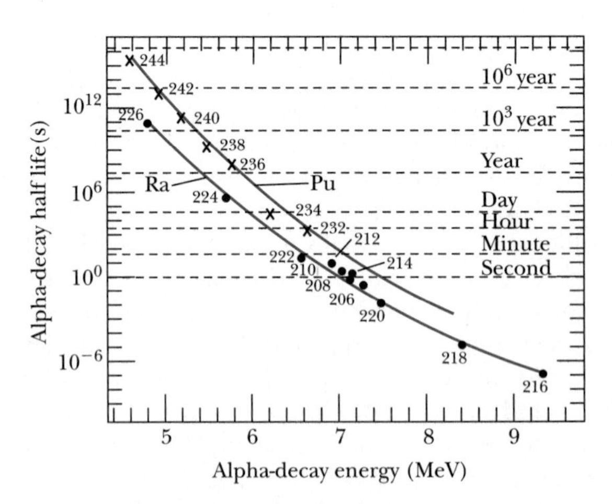

FIGURE 12.12 The half-lives for several radioactive alpha emitters of radium and plutonium isotopes are plotted versus their alpha energy. The two curves show that the higher-energy alphas result from nuclei having a much shorter lifetime where the alpha particles have a higher probability of tunneling through the Coulomb barrier.

and conservation of linear momentum, we can determine a unique energy for the alpha particle.

$$Q = K_\alpha + K_D$$

$$p_\alpha = p_D$$

$$K_\alpha = Q - K_D = Q - \frac{p_D{}^2}{2M_D} = Q - \frac{p_\alpha{}^2}{2M_D}$$

$$K_\alpha = Q - \frac{2M_\alpha K_\alpha}{2M_D} = Q - \frac{M_\alpha}{M_D} K_\alpha$$

$$K_\alpha\left(1 + \frac{M_\alpha}{M_D}\right) = Q$$

$$K_\alpha = \frac{M_D}{M_D + M_\alpha} Q \approx \left(\frac{A-4}{A}\right) Q \qquad (12.32)$$

Because the parent mass A is normally over 200, the alpha particle takes most of the kinetic energy.

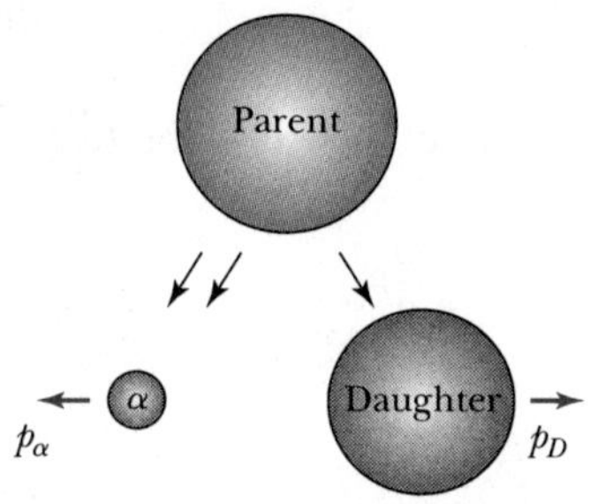

FIGURE 12.13 When a radioactive parent at rest alpha decays to an alpha particle and a heavy daughter, conservation of mass-energy and momentum still must occur. The momentum of the alpha and daughter are equal and opposite.

Beta Decay

Radioactive decay occurs because some nuclides are not stable. In Figure 12.5 we showed a plot of stable nuclei. The centroid position of these stable nuclei describe what is called the *line of stability*. Note that in alpha decay the parent nucleus reverts to a daughter nucleus that is down 2 units in N (Figure 12.5) and to the left 2 units in Z. In many cases alpha decay leaves the daughter nucleus farther from the line of stability than the parent. One method that unstable nuclei may move toward the line of stability is by beta decay. The simplest example of beta decay is the decay of a free neutron.

$$n \rightarrow p + \beta^- \qquad (12.33)$$

As we discussed in Section 12.1, electrons cannot exist within the nucleus, so when beta decay occurs for a nuclide, the beta particle, denoted by β^- (we now know it is an electron), is created at the time of the decay. We showed in Figure 12.8 that ^{14}C is unstable—it has an excess of neutrons. As a result ^{14}C beta decays to form ^{14}N, a stable nucleus.

$$^{14}_{6}\text{C} \rightarrow {}^{14}_{7}\text{N} + \beta^- \qquad (12.34)$$

This decay produces two products, like α decay, and we expect to measure a monoenergetic electron spectrum (see Problem 34). However, the electron energy spectrum from the beta decay of ^{14}C (see Figure 12.14) shows a continuous

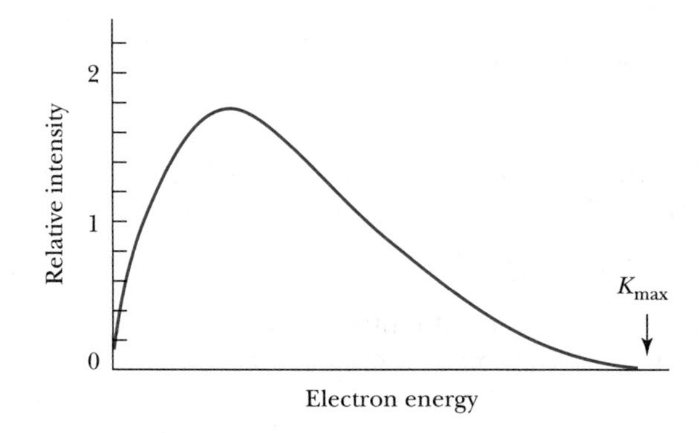

FIGURE 12.14 The relative intensity of electrons as a function of kinetic energy is shown for ^{14}C decay. If there were only two products in beta decay, the electron energy would be monoenergetic and the energy equal to K_{max}.

energy spectrum up to a maximum energy. This experimental result was a major puzzle for many years. In addition to the strange energy spectrum, there was a problem with spin conservation. In neutron decay, the spin 1/2 neutron cannot decay to two spin 1/2 particles, a proton and electron. Also ^{14}C has spin 0, ^{14}N has spin 1, and the electron has spin 1/2. We cannot add spin 1/2 and 1 to obtain a spin of 0. Both the electron energy spectrum and the spin angular momentum conservation posed major difficulties with our understanding of beta decay.

Difficulties with beta decay

The neutrino

The correct explanation was proposed in 1930 by Wolfgang Pauli, who suggested that a third particle—later called a **neutrino**—must also be produced in beta decay. The neutrino, with the symbol ν, has spin quantum number 1/2, charge 0, and carries away the additional energy missing in Figure 12.14. In Figure 12.14 an occasional electron is detected with the kinetic energy K_{max} required to conserve energy, but in the great majority of cases the electron's kinetic energy is less than K_{max}. This means the neutrino has little or no mass, and its energy may be all kinetic, like the photon. The photon cannot be the missing particle because it has spin 1. Pauli's suggestion seemed to explain the difficulties, and all circumstantial evidence supported the neutrino hypothesis. However, the detection of the elusive neutrino was difficult, and its existence was not proven experimentally until 1956 by C. Cowan and F. Reines (see the Special Topic box on Neutrino Detection). Reines received the Nobel Prize in 1995.

Frederick Reines (left) and Clyde Cowan discovered the neutrino in 1956, almost three decades after it had been postulated by Pauli. See the Special Topic box on Neutrino Detection. *Courtesy of Frederick Reines.*

Neutrinos interact so weakly with matter that they pass right through the Earth with little chance of being absorbed. They have no charge and do not interact *electromagnetically*. They are not affected by the *strong* force of the nucleus. The mass of the neutrino is still under intensive experimental investigation, and recent evidence indicates it has a small mass. We now believe that beta decay is the reflection of a special kind of force, called simply the *weak* interaction. Neutrinos are created (or absorbed) in weak processes, including β decay. The electromagnetic and weak forces are two manifestations of the *electroweak* force and will be discussed in Chapter 14.

β^- Decay. We now know there are *anti*neutrinos $\bar{\nu}$ as well as neutrinos. The beta decay of a free neutron and of ^{14}C is now correctly written as

$$n \rightarrow p + \beta^- + \bar{\nu} \qquad \beta^- \text{ decay} \tag{12.35}$$

$$^{14}\text{C} \rightarrow {}^{14}\text{N} + \beta^- + \bar{\nu} \qquad \beta^- \text{ decay} \tag{12.36}$$

where it is the antineutrino $\bar{\nu}$ actually produced in β^- decay.

In the general beta decay of the parent nuclide ^A_ZX to the daughter $_{Z+1}^{\ \ A}D$, the reaction is

β^- decay

$$^A_ZX \rightarrow {}_{Z+1}^{\ \ A}D + \beta^- + \bar{\nu} \qquad \beta^- \text{ decay} \tag{12.37}$$

The disintegration energy Q is given by

$$Q = [M(^A_ZX) - M(_{Z+1}^{\ \ A}D)]c^2 \qquad \beta^- \text{ decay} \tag{12.38}$$

In order for β^- decay to occur, we must have $Q > 0$. When using Equation (12.38) we must be careful to use atomic masses because the Z of the decaying nuclide changes and the number of electron masses has been accounted for in Equation (12.38). In β^- decay, the nucleus A is constant, but Z changes to $Z + 1$, so that in Figure 12.5, it is the unstable nuclei to the left of the line of stability that are moving closer (one number down and one number right) to the line of stability.

β^+ Decay. One might ask what happens for unstable nuclides with too many protons, that is, for nuclei to the right of the line of stability in Figure 12.5. Nature does allow such a transition, and a positive electron, called a *positron* (e^+), is produced. The positron is the antiparticle of the electron. In beta decay the electron and positron are normally referred to as β^- and β^+, respectively. Using β^- and β^+ helps remind us that the electron and positron are created in the nucleus during beta decay. Current experimental evidence indicates that a free proton does not decay ($t_{1/2} > 10^{32}$ y), but a proton bound within the nucleus may transmutate, if energy is taken from the nucleus and the result is a more stable nucleus. A nucleus with an excess of protons, for example, ^{14}O is a good candidate. It is unstable and will decay by emitting a positron to become the stable ^{14}N. The reaction is

$$^{14}\text{O} \rightarrow {}^{14}\text{N} + \beta^+ + \nu \qquad \beta^+ \text{ decay} \tag{12.39}$$

Nuclei near ^{14}N with an excess neutron (^{14}C) or proton (^{14}O) will decay to ^{14}N by the appropriate beta decay.

The general β^+ decay is written as

$$^{A}_{Z}X \rightarrow {}_{Z-1}^{A}D + \beta^+ + \nu \qquad \beta^+ \text{ decay} \tag{12.40}$$

β^+ decay

A careful analysis of the atomic masses shows that the disintegration energy Q for positron decay is

$$Q = [M(^{A}_{Z}X) - M(_{Z-1}^{A}D) - 2m_e]c^2 \qquad \beta^+ \text{ decay} \tag{12.41}$$

In this case the electron masses do not cancel when atomic masses are used. The mass of the positron is equal to that of an electron.

Note that in Figure 12.5 the unstable nuclei to the right of the line of stability move one number up and one number to the left when they β^+ decay, again moving closer to the line of stability.

Electron Capture. There is one other important possibility for beta decay. Because they are tightly bound and their (classical) orbits are highly elliptical, the inner K-shell and L-shell electrons spend a reasonable amount of time passing through the nucleus, thereby increasing the possibility of atomic electron capture (EC). A proton in the nucleus absorbs the e^-, producing a neutron and neutrino. The reaction for a proton is

$$p + e^- \rightarrow n + \nu \tag{12.42}$$

The general reaction is written as

$$^{A}_{Z}X + e^- \rightarrow {}_{Z-1}^{A}D + \nu \qquad \text{Electron capture} \tag{12.43}$$

Electron capture

When one of the inner atomic electrons is captured, another electron will take its place, producing a series of characteristic atomic x-ray spectra. This is a signature of electron capture, because these x rays are produced in the absence of any other kind of radiation. X rays can also be produced in other kinds of nuclear decay, because the decay products (for example, an α particle) may knock out electrons.

Electron capture has the same effect as positron decay, a proton is converted to a neutron. Electron capture occurs more frequently for higher-Z nuclides because the inner atomic electron shells are more tightly bound and there is a

NEUTRINO DETECTION

Imagine particles that have little probability of being stopped while passing completely through the Earth. They have no charge and very little mass yet they are everywhere throughout the universe, perhaps as many as 10^9 in every m^3 of "empty" space. Millions of them are now passing through your eyeballs, but you have no chance of seeing them. Such is the neutrino, and no wonder it took over a quarter of a century to experimentally prove its existence after it was postulated by Pauli in 1930 to solve the problem of nonconservation of energy in beta decay.

Frederick Reines (1918– ; Nobel Prize, 1995) and Clyde Cowan, working at Los Alamos, decided in 1951 to observe the inverse beta process,

$$\bar{\nu} + p \rightarrow n + \beta^+ \tag{B12.1}$$

At first they thought they would detect (anti) neutrinos resulting from an atomic bomb blast, but they eventually decided on neutrinos (we use the term generically to mean either neutrinos or antineutrinos) coming from a nuclear reactor. They took their

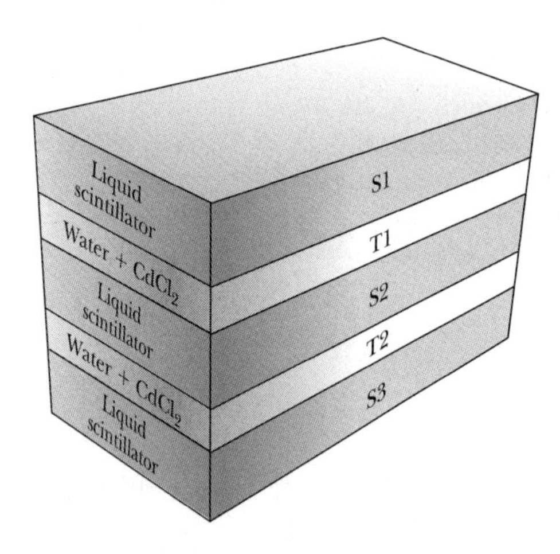

FIGURE A Schematic diagram of Cowan and Reines neutrino detector used at the Savannah River reactor. Neutrinos scatter from protons in the water. The height of the detector is 2 m.

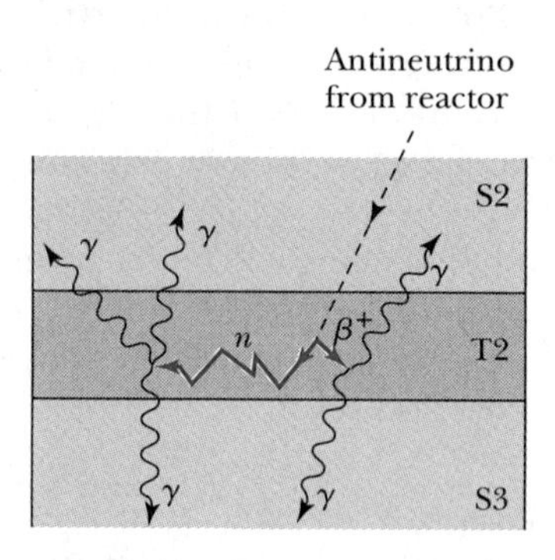

FIGURE B An antineutrino from the reactor scatters from a proton to create a neutron and β^+ that eventually produced γ rays that were detected in coincidence in the two scintillators S2 and S3.

first apparatus to a Hanford, Washington, reactor in 1953, but were not convinced they had detected neutrinos. Their experiment was called *Project Poltergeist* because, even though most physicists believed the neutrino existed, some thought it may prove not to be observable. In 1956 they took a larger detector to the Savannah reactor, where the neutrino rate from the beta decay of fission fragments was $10^{13}\ \bar{\nu}_e/\text{cm}^2/\text{s}$. There a well-shielded area 12 m below ground helped reduce cosmic ray background. A schematic of their Savannah detector is shown in Figure A. It consists of three liquid scintillator detectors (S1, S2, and S3), each viewed by 110 photomultiplier tubes. Between each scintillator was placed a target (T1 and T2) containing $CdCl_2$ dissolved in water.

The detection scheme is shown in Figure B. An antineutrino from the reactor interacts with a proton in target T2, for example, producing a neutron and positron. The positron quickly slows down, annihilates with an electron, and produces two 0.5-MeV gamma rays, $e^+ + e^- \rightarrow 2\gamma$. The γ rays pass through the water target and are detected in coincidence in the scintillators S2 and S3. The neutron undergoes several collisions, slows down, and is eventually captured by ^{114}Cd, which has a very large neutron capture cross section. Several gamma rays may be produced by the neutron capture, and they are detected in one or both of the scintillators. The signature of the antineutrino detection is a delayed coincidence between the gamma ray resulting from β^+ and those from the neutron.

Meanwhile Raymond Davis, Jr., of Brookhaven National Laboratory, had designed an experiment using 4000 liters of CCl_4 to detect neutrinos. The reaction in this case is

$$\nu + {}^{37}\text{Cl} \longrightarrow \beta^- + {}^{37}\text{Ar} \tag{B12.2}$$

Helium gas was bubbled through the solution to remove the ^{37}Ar that was later detected as it decayed back to ^{37}Cl. No reactions were observed when this apparatus was placed next to a reactor. That is because a reactor produces *anti*neutrinos, whereas reaction (B12.2) requires neutrinos. This is further evidence that Cowan and Reines were detecting antineutrinos.

In the 1960s Davis constructed a large 100,000-gallon detector (Figure C) containing perchloroethylene cleaning fluid (C_2Cl_4) placed 1500 m below ground in the Homestake Gold Mine in South Dakota. Such a depth shields against most cosmic ray background. This detector has been used for more than 25 years to detect the rate of neutrinos coming from our sun (solar neutrinos). That rate is only about one half of the rate expected, and the concern over this result has led to several more neutrino detectors being constructed throughout the world (see Chapter 14) as the field of neutrino physics has flourished.

FIGURE C The 100,000 gallon tank of cleaning fluid in the Homestake Gold Mine. Note the size of the tank relative to the two men—one on the catwalk and the other (lower left) below the tank. *The University of Pennsylvania Archives.*

In recent years the number of neutrinos released from the sun has been calculated using theoretical models of the sun. At least five neutrino experiments have found the neutrino flux from the sun to be low by about a factor of 2. This is known as the "Solar Neutrino Problem" or SNP. Experiments already in operation include the Homestake mine, very expensive gallium detectors under the Caucasus Mountains near Baksan in Russia and in the Gran Sasso tunnel in Italy, and ultrapure water detectors deep in a mine in Kamioka, Japan, where it was shown indirectly in 1998 that neutrinos have mass. When absorbed by gallium, neutrinos induce a reaction that produces radioactive germanium. The water detectors operate by an anti neutrino being absorbed by a proton, producing a neutron and β^+; the fast moving positron produces light, which is detected by photomultiplier tubes.

Several other experiments are in the construction and planning stages. A large detector in Ontario, the Sudbury Neutrino Observatory, uses heavy water and is perhaps the most sophisticated detector yet constructed. It began operation in 1998. Several experiments are under construction for use in the Gran Sasso tunnel. Neutrinos have also been detected in a huge mass of iron deep in the Soudan mine in northern Minnesota. A unique experiment proposes to detect the arrival and type of neutrinos in the Soudan mine that have traveled 700 km through the Earth from the Fermilab accelerator near Chicago.

The detection of neutrinos has become a huge effort because of the importance of neutrinos in the solar neutrino problem, proton decay, and mass of the universe.

greater probability of an electron being absorbed. The disintegration energy Q for electron capture is

$$Q = [M(^A_Z X) - M(_{Z-1}^{\ A} D)]c^2 \qquad \text{Electron capture} \qquad (12.44)$$

Because $Q > 0$ for β^+ or EC to occur, there will be some cases where EC is possible, but not β^+ decay, because of the difference between Equations (12.41) and (12.44).

Example 12.11

Show that the relations expressed for the disintegration energy Q in Equations (12.38), (12.41), and (12.44) are correct.

Solution: *β^- decay* We begin with β^- decay and write the energy equation for the reaction in Equation (12.43). In all the cases here, we neglect any neutrino mass.

$$M_{\text{nucl}}(^A_Z X) = M_{\text{nucl}}(_{Z+1}^{\ A} D) + m_e + Q/c^2$$

where we use M_{nucl} to indicate the nuclear mass. In order to change to atomic masses we add Zm_e to each side above

$$M_{\text{nucl}}(^A_Z X) + Zm_e = M_{\text{nucl}}(_{Z+1}^{\ A} D) + (Z+1)m_e + Q/c^2$$

We neglect the difference in atomic binding energies on the two sides of the equation and write this equation now in terms of atomic masses.

$$M(^A_Z X) = M(_{Z+1}^{\ A} D) + Q/c^2$$

We solve this equation for Q to determine Equation (12.38).

$$Q = [M(^A_Z X) - M(_{Z+1}^{\ A} D)]c^2 \qquad \beta^- \text{ decay} \qquad (12.38)$$

β^+ decay We write the mass-energy equation for the reaction in Equation (12.40) and follow a similar procedure to that above.

$$M_{\text{nucl}}(^A_Z X) = M_{\text{nucl}}(_{Z-1}^{\ A} D) + m_e + Q/c^2$$

We again add Zm_e to each side and have

$$M_{\text{nucl}}(^A_Z X) + Zm_e = M_{\text{nucl}}(_{Z-1}^{\ A} D) + (Z+1)m_e + Q/c^2$$

In this case we only need $(Z-1)m_e$ for the daughter atomic mass, which gives us a remaining mass of $2m_e$.

$$M(^A_Z X) = M(_{Z-1}^{\ A} D) + 2m_e + Q/c^2$$

We solve this equation for Q to determine Equation (12.41).

$$Q = [M(^A_Z X) - M(_{Z-1}^{\ A} D) - 2m_e]c^2 \qquad \beta^+ \text{ decay} \qquad (12.41)$$

Electron capture The mass-energy equation for the electron capture reaction of Equation (12.43) is

$$M_{\text{nucl}}(^A_Z X) + m_e = M_{\text{nucl}}(_{Z-1}^{\ A} D) + Q/c^2$$

We add $(Z-1)m_e$ to each side above and obtain

$$M_{\text{nucl}}(^A_Z X) + Zm_e = M_{\text{nucl}}(_{Z-1}^{\ A} D) + (Z-1)m_e + Q/c^2$$

In this case we have just the right number of electron masses to change to atomic masses.

$$M(^A_Z X) = M(_{Z-1}^{\ A} D) + Q/c^2$$

We solve this equation for Q to find Equation (12.44).

$$Q = [M(^A_Z X) - M(_{Z-1}^{\ A} D)]c^2 \qquad \text{Electron capture} \qquad (12.44)$$

Example 12.12

Show that ^{55}Fe may undergo electron capture, but not β^+ decay.

Solution: The two possible reactions are

$$^{55}_{26}\text{Fe} + e^- \rightarrow {}^{55}_{25}\text{Mn} + \nu \qquad \text{Electron capture}$$

$$^{55}_{26}\text{Fe} \rightarrow {}^{55}_{25}\text{Mn} + \beta^+ + \nu \qquad \beta^+ \text{ decay}$$

We need to obtain all the masses needed to determine the disintegration energy Q of Equations (12.41) and (12.44): $M(^{55}_{26}\text{Fe}) = 54.938298$ u, $M(^{55}_{25}\text{Mn}) = 54.938050$ u, and $m_e = 0.000549$ u.

Positron decay

$$Q = [54.938298 \text{ u} - 54.938050 \text{ u} - 2(0.000549 \text{ u})c^2 \times \left(\frac{931.5 \text{ MeV}}{c^2 \cdot \text{u}}\right)$$

$$= -0.79 \text{ MeV} \qquad \text{Positron decay not allowed}$$

Electron capture

$$Q = [54.938298 \text{ u} - 54.938050 \text{ u}]c^2\left(\frac{931.5 \text{ MeV}}{c^2 \cdot \text{u}}\right)$$

$$= 0.23 \text{ MeV} \qquad \text{Electron capture is allowed}$$

We find experimentally that only electron capture is allowed, with $t_{1/2} = 2.7$ years.

Example 12.13

Find whether alpha decay or any of the beta decays are allowed for $^{226}_{89}\text{Ac}$.

Solution: For each of the possible four reactions, we first write the reaction, list the disintegration energy Q equation, look up the appropriate masses, and calculate the disintegration energy Q. If $Q > 0$, the decay is allowed.

Alpha decay $\quad ^{226}_{89}\text{Ac} \rightarrow {}^{222}_{87}\text{Fr} + \alpha$

$$Q = [M(^{226}_{89}\text{Ac}) - M(^{222}_{87}\text{Fr}) - M(^{4}\text{He})]c^2$$

$$= 5.50 \text{ MeV} \qquad \text{Alpha decay is allowed}$$

β^- decay $\quad ^{226}_{89}\text{Ac} \rightarrow {}^{226}_{90}\text{Th} + \beta^- + \bar{\nu}$

$$Q = [M(^{226}_{89}\text{Ac}) - M(^{226}_{90}\text{Th})]c^2$$

$$= 1.12 \text{ MeV} \qquad \beta^- \text{ decay is allowed}$$

β^+ decay $\quad ^{226}_{89}\text{Ac} \rightarrow {}^{226}_{88}\text{Ra} + \beta^+ + \nu$

$$Q = [M(^{226}_{89}\text{Ac}) - M(^{226}_{88}\text{Ra}) - 2m_e]c^2$$

$$= -0.38 \text{ MeV} \qquad \beta^+ \text{ decay is not allowed}$$

Electron capture $\quad ^{226}_{89}\text{Ac} + e^- \rightarrow {}^{226}_{88}\text{Ra} + \nu$

$$Q = [M(^{226}_{89}\text{Ac}) - M(^{226}_{88}\text{Ra})]c^2$$

$$= 0.64 \text{ MeV} \qquad \text{Electron capture is allowed}$$

We find that α decay, β^- decay, and electron capture are all possible from the same nucleus. Experiment shows that alpha decay occurs only 0.006% of the time for ^{226}Ac, β^- decay 83%, and electron capture 17%.

Gamma Decay

The nuclide ^{230}U can alpha decay to the ground state or any of the low-lying excited states of ^{226}Th (see Figure 12.15). Similarly, experimental evidence indicates that ^{226}Ac beta decays primarily to the ground and two excited states of ^{226}Th. The energies calculated by the disintegration energy Q for alpha and beta decay give the appropriate transitions to the ground state. If the decay proceeds to an excited state of energy E_x (for example, from the ground state of ^{230}U to the $E_x = 0.072$ MeV excited state of ^{226}Th), rather than to the ground state, then the disintegration energy Q for the transition to the excited state can be determined with respect to the transition to the ground state. If we label the disintegration energy Q to the ground state as Q_0, the Q for a transition to the excited state E_x is given by

$$Q = Q_0 - E_x \tag{12.45}$$

For example, the disintegration value Q for the α decay transition from ^{230}U to the excited state at $E_x = 0.072$ MeV of ^{226}Th (see Figure 12.15) is given in terms of $Q_0 = 5.992$ MeV and $E_x = 0.072$ MeV by $Q = Q_0 - E_x = 5.992 \text{ MeV} - 0.072 \text{ MeV} = 5.920$ MeV.

A nucleus has excited states in much the same way atoms do. The excitation energies, however, tend to be much larger, many keV or even MeV, as a result of the stronger nuclear interaction. One of the possibilities for the nucleus to rid itself of this extra energy is to emit a photon (gamma ray) and undergo a transition to some lower energy state. The gamma-ray energy $h\nu$ is given by the difference of the higher energy state $E_>$ and the lower one $E_<$.

$$h\nu = E_> - E_< \tag{12.46}$$

In order to conserve momentum, the nucleus normally must absorb some of this energy difference. However, for a nucleus initially at rest, Equation (12.46) is a very good approximation.

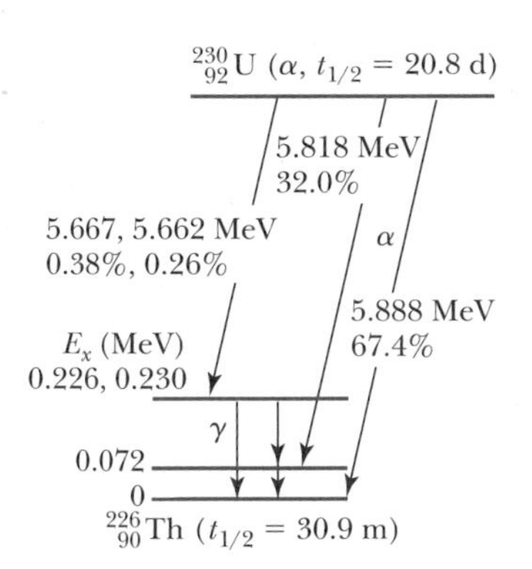

FIGURE 12.15 The relevant energy levels for the alpha decay of $^{230}_{92}\text{U}$. The alpha-particle energies and their percentage of occurrence are shown for several decays for $^{230}_{92}\text{U} \rightarrow {}^{226}_{90}\text{Th}$. The decay percentages may not add up to 100% because less likely decays to higher lying excited states are not shown. *From R. B. Firestone and V. S. Shirley, eds., Table of Isotopes, 8th ed., New York: John Wiley & Sons, Inc., 1996.*

Example 12.14

Consider the γ decay from the 0.072-MeV excited state to the ground state of ^{226}Th at rest shown in Figure 12.15. Find an exact expression for the gamma-ray energy by including both the conservation of momentum and energy. Determine the error obtained by using the approximate value in Equation (12.46).

Solution: We denote the final momentum of ^{226}Th by p. Because the decaying nucleus is initially at rest, the total linear momentum is zero, and the linear momentum p of the daughter nucleus must be equal and opposite to the momentum of the gamma ray, h/λ.

$$p = \frac{h}{\lambda} = \frac{h\nu}{c}$$

The conservation of energy gives

$$h\nu + \frac{p^2}{2M} = E_> - E_<$$

where M is the mass of ^{226}Th. Substituting for the momentum p in the latter equation, we have

$$h\nu + \frac{(h\nu)^2}{2Mc^2} = E_> - E_< \qquad (12.47)$$

We could solve Equation (12.47) for $h\nu$ by using the quadratic equation, but that would be tedious. Let us determine whether we can use an approximation. Rewrite Equation (12.47) as

$$h\nu\left(1 + \frac{h\nu}{2Mc^2}\right) = E_> - E_<$$

$$h\nu = \frac{E_> - E_<}{1 + \dfrac{h\nu}{2Mc^2}}$$

If the value $h\nu/2Mc^2 = x$ is very small, then we can use the binomial expansion $(1 + x)^{-1} = 1 - x + x^2 - \cdots$. We determine x to be

$$x = \frac{h\nu}{2Mc^2} = \frac{0.072 \text{ MeV}}{2(226 \text{ u} \cdot c^2)\left(\dfrac{931.5 \text{ MeV}}{c^2 \cdot \text{u}}\right)} = 1.7 \times 10^{-7}$$

so the error in using the approximate Equation (12.46), which amounts to letting $(1 + x)^{-1} \approx 1$, amounts to an error of only about 10^{-7}, hardly worth worrying about. We can safely use Equation (12.46) where $h\nu \approx E_> - E_< = 0.072$ MeV.

The decay of an excited state of $^AX^*$ to its ground state is denoted by

Gamma decay

$$^AX^* \longrightarrow {^AX} + \gamma \qquad (12.48)$$

A transition between two nuclear excited states $E_>$ and $E_<$ is denoted by

$$^AX^*(E_>) \longrightarrow {^AX^*}(E_<) + \gamma \qquad (12.49)$$

The excited state energies or *levels* are characteristic of each nuclide, and a careful study of the gamma energies is usually sufficient to identify a particular nuclide. The gamma rays are normally emitted soon after the nucleus is created in an excited state. Coincidence measurements between the β- and/or α-decay products and the subsequent γ rays are a powerful technique in experimental nuclear physics to study properties of nuclear excited states.

We can contrast nuclear spectroscopy with that of the optical spectra of atoms. The solution of the Schrödinger equation for the Coulomb force of the atom is possible, but no such success has yet been obtained for the nuclear force. Although nuclear models are used to predict certain characteristics of excited nuclear levels, no encompassing theory exists.

There are sophisticated, quantum mechanical selection rules that determine what kind and how fast a gamma-ray transition occurs between nuclear states. Sometimes these selection rules prohibit a certain transition, and the excited state may live for a long time—even years! These states are called **isomers** or **isomeric states** and are denoted by a small m for *metastable* next to the mass number A. An example is the spin 9 state of $^{210\text{m}}_{83}$Bi at 0.271 MeV excitation energy

Isomers

which has $t_{1/2} = 3 \times 10^6$ y (alpha decay) and apparently does not gamma decay. The lower energy states have spins 0 and 1, and such a large spin difference transition is prohibitive for gamma decay. The ground state of $^{210}_{83}$Bi has a half-life $t_{1/2} = 5$ days (99+% β^- decay, 10^{-4}% α decay). Another isomer is $^{93m}_{41}$Nb ($E_x = 0.03$ MeV, $t_{1/2} = 13.6$ y), which eventually does gamma decay to the ground state; in this case we say the decay is prohibited, but that just means the probability of its occurring is very small. Most isomeric states, of course, have much shorter lifetimes. An extremely useful isomer for clinical work in medicine is ^{99m}Tc (gamma emitter, $t_{1/2} = 6$ h), which will be discussed further in the next chapter.

12.8 Radioactive Nuclides

Most scientists now believe that the universe was created in a tremendous explosion (the *Big Bang*) about 13 billion years ago (13×10^9 y). Neutrons and protons fused together to form deuterons and light nuclei within the first few minutes. The heavy elements were formed much later by nuclear reactions within stars. We will discuss these processes in Chapters 13 and 16.

We say that the unstable nuclei found in nature exhibit *natural radioactivity*. Those radioactive nuclides made in the laboratory (for example, with accelerators or reactors) exhibit *artificial radioactivity* and include all known nuclides heavier than ^{238}U. There are many natural radioactive nuclides left on Earth with lifetimes long enough to be observed. Only those with half-lives longer than a few billion years could have existed since primordial times; most of them are heavy elements, but there are several with $A < 150$ that are listed in Table 12.2. Some nuclides with long half-lives have been produced as a result of the decay of another radioactive nucleus.

In addition to nuclear fission, heavy radioactive nuclides can change their mass number only by alpha decay ($^AX \rightarrow {}^{A-4}D$) but can change their charge number Z by either alpha or beta decay. As a result, there are only four paths that the heavy naturally occurring radioactive nuclides may take as they decay to stable end products. The four paths have mass numbers expressed by either $4n$, $4n + 1$, $4n + 2$, or $4n + 3$ (n = integer), because only alpha decay can change

TABLE 12.2
Some Naturally Occurring Radioactive Nuclides

Nuclide	$t_{1/2}$ (y)	Natural Abundance
$^{40}_{19}$K	1.28×10^9	0.01%
$^{87}_{37}$Rb	4.8×10^{10}	27.8%
$^{113}_{48}$Cd	9×10^{15}	12.2%
$^{115}_{49}$In	4.4×10^{14}	95.7%
$^{128}_{52}$Te	7.7×10^{24}	31.7%
$^{130}_{52}$Te	2.7×10^{21}	33.8%
$^{138}_{57}$La	1.1×10^{11}	0.09%
$^{144}_{60}$Nd	2.3×10^{15}	23.8%
$^{147}_{62}$Sm	1.1×10^{11}	15.0%
$^{148}_{62}$Sm	7×10^{15}	11.3%

TABLE 12.3
The Four Radioactive Series

Mass Numbers	Series Name	Parent	$t_{1/2}$ (y)	End Product
$4n$	Thorium	$^{232}_{90}Th$	1.40×10^{10}	$^{208}_{82}Pb$
$4n+1$	Neptunium	$^{237}_{93}Np$	2.14×10^{6}	$^{209}_{83}Bi$
$4n+2$	Uranium	$^{238}_{92}U$	4.47×10^{9}	$^{206}_{82}Pb$
$4n+3$	Actinium	$^{235}_{92}U$	7.04×10^{8}	$^{207}_{82}Pb$

the mass number. These four series are listed in Table 12.3. All of these radioactive series occur in nature except that of neptunium. The member of the neptunium series with the longest half-life, ^{237}Np ($t_{1/2} = 2.14 \times 10^6$ y), has a lifetime so much less than the age of our solar system that virtually all the members have already decayed.

The sequence of one of the radioactive series, ^{232}Th, is shown in Figure 12.16. Note that ^{212}Bi can decay by either alpha or beta decay; this is called *branching*. The subsequent decay is usually a beta or alpha decay, respectively, to eventually reach the same end product. Normally one path of the branch is heavily favored, but $^{212}_{83}$ Bi is singled out in Figure 12.16 because it has a 36% probability of alpha decay and a 64% probability of beta decay. As shown in Figure 12.16 the effect of the successive alpha and beta decays is to bring the nuclide closer to the line of stability until a stable nuclide is finally reached.

Time Dating Using Lead Isotopes

Because the isotope ^{204}Pb is not radioactive and no other nuclide decays to it, its abundance is presumably constant. The stable isotopes ^{206}Pb and ^{207}Pb, however, are at the end of the radioactive chains of ^{238}U and ^{235}U, respectively. However, because ^{235}U has a relatively short half-life (0.70×10^9 y) compared to the age of the Earth, most of the ^{235}U has already decayed to ^{207}Pb, and the ratio of $^{207}Pb/^{204}Pb$ has been relatively constant over the past two billion years. Since the half-life of ^{238}U is so long, the ratio of $^{206}Pb/^{204}Pb$ is still increasing. A plot of the abundance ratio of $^{206}Pb/^{204}Pb$ versus $^{207}Pb/^{204}Pb$ can be a sensitive indi-

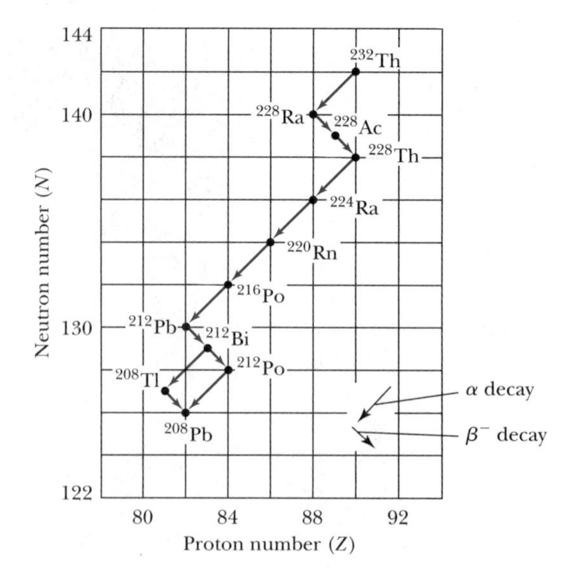

FIGURE 12.16 The predominant path for the decay chain of $^{232}_{90}Th$. The mass number can only change by alpha decay, but both alpha and beta decay can change the atomic number Z. A branch is shown for ^{212}Bi. Not all branching is shown.

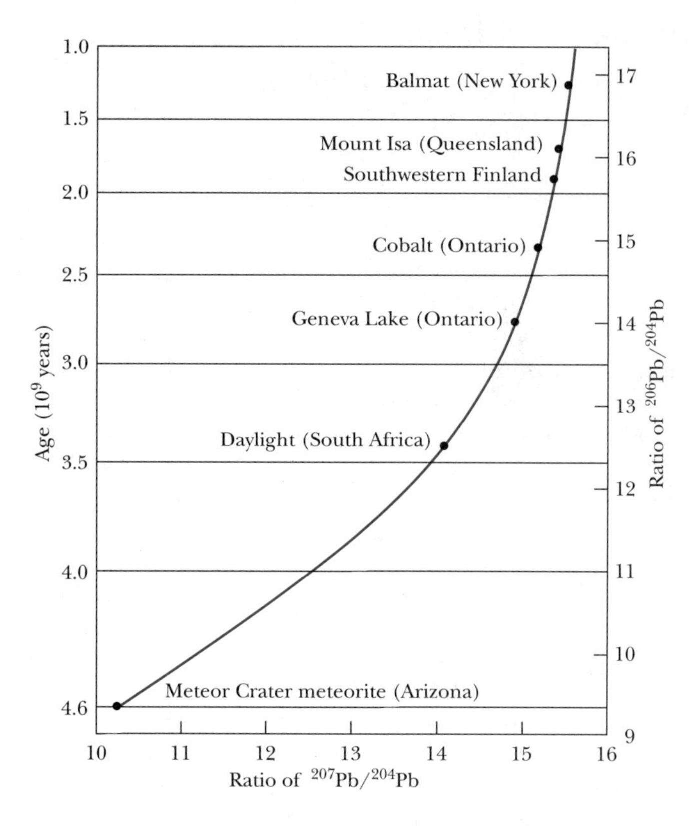

FIGURE 12.17 The growth curve for lead ores from various deposits is shown. The age of the specimens can be obtained from the abundance ratio of $^{207}Pb/^{204}Pb$. When extrapolated backward in time, the growth curve gives an age of 4.6 billion years for the specimens from the Meteor Crater in Arizona. *From S. Moorbath, Scientific American* ***236****, 92 (1977).*

cator of the age of lead ores as shown in Figure 12.17. Such techniques have been used to show that meteorites (Meteor Crater in Arizona) believed to be left over from the formation of the solar system are about 4.5 billion years old. This is in agreement with dating measurements made on moon rocks returned to Earth. Although no 4.5-billion-year-old terrestrial rocks have yet been found on Earth, indirect evidence based on radioactive dating techniques leads us to believe the Earth was formed about 4.5 billion years ago. Rocks as old as 3.8 billion years have been found in Greenland.

Example 12.15

Assume that all the ^{206}Pb found in a given sample of uranium ore resulted from decay of ^{238}U and that the ratio of $^{206}Pb/^{238}U$ is 0.6. How old is the ore?

Solution: Let N_0 be the original number of ^{238}U nuclei that existed. The ^{238}U nuclei eventually decay to ^{206}Pb, and the longest time in the radioactive decay chain $^{238}U \rightarrow {}^{206}Pb$ is the half-life of ^{238}U, $t_{1/2} = 4.47 \times 10^9$ y. The numbers of nuclei for ^{238}U and ^{206}Pb are then

$$N(^{238}U) = N_0 e^{-\lambda t}$$

$$N(^{206}Pb) = N_0 - N(^{238}U) = N_0(1 - e^{-\lambda t})$$

The abundance ratio is

$$R' = \frac{N(^{206}Pb)}{N(^{238}U)} = \frac{1 - e^{-\lambda t}}{e^{-\lambda t}} = e^{\lambda t} - 1 \qquad (12.50)$$

We can solve Equation (12.50) for t, because we know experimentally the ratio R' and the decay constant λ for ^{238}U. The result is

$$t = \frac{1}{\lambda}\ln(R' + 1) = \frac{t_{1/2}}{\ln(2)}\ln(R' + 1)$$

$$= \frac{4.47 \times 10^9 \text{ y}}{\ln(2)}\ln(1.6) = 3.03 \times 10^9 \text{ y}$$

The sample of uranium ore is more than 3 billion years old.

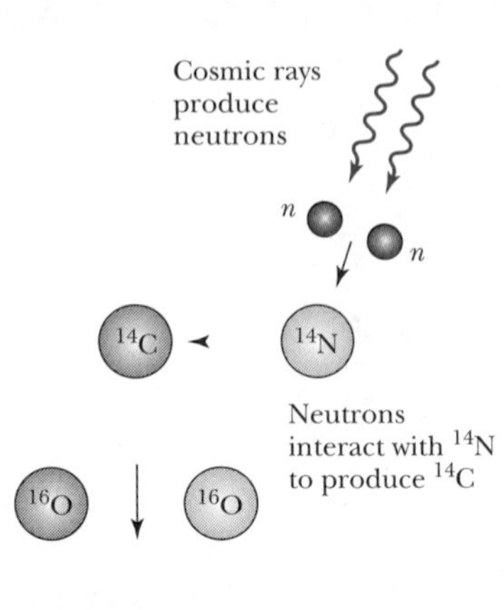

FIGURE 12.18 Cosmic rays produce neutrons, which produce ^{14}C after reacting with ^{14}N in the atmosphere. The ^{14}C nuclei enter into living organisms in the form of CO_2. After the living organism dies, the ratio of ^{14}C/^{12}C decreases according to the decay rate of ^{14}C.

Radioactive Carbon Dating

Radioactive ^{14}C is produced in our atmosphere by the bombardment of ^{14}N by neutrons produced by cosmic rays.

$$n + {}^{14}\text{N} \rightarrow {}^{14}\text{C} + p \tag{12.51}$$

A natural equilibrium of ^{14}C to ^{12}C exists in molecules of CO_2 in the atmosphere which all living organisms take in (see Figure 12.18). However, when the living organisms die, their intake of ^{14}C ceases, and the ratio of ^{14}C/^{12}C ($=R$) decreases as ^{14}C decays. In order to use ^{14}C for dating, corrections must be made for the changes in composition of the Earth's atmosphere and for variations in the flux of cosmic rays due to changes in the Earth's magnetic field. Evidence early in the 1990s indicated that the period just before 9000 years ago had a higher ^{14}C/^{12}C ratio by a factor of about 1.5 than it does today. At the time of death, the initial decay rate is 14 decays/min per gram of carbon ($R = 1.2 \times 10^{-12}$). This method has been calibrated with tree-ring counting to ages of about 10,000 years. Because the half-life of ^{14}C is 5730 years, it is convenient to determine the age of objects over a wide range up to perhaps 45,000 years. Indeed, quite sophisticated mass spectrometry techniques using accelerators to accelerate carbon ions have enabled scientists to test samples as small as 10^5 atoms (about 10^{-17} g). Much older dates can now be obtained than by using the previous techniques of actually measuring the ^{14}C decay rate, which required samples of up to 10 g. Willard Libby received the 1960 Nobel Prize in chemistry for this ingenious technique.

Other particularly useful radioisotopes for dating purposes include ^{10}Be, ^{26}Al, ^{36}Cl, and ^{129}I. The half-life of ^{10}Be is 1.5 million years and may be useful in studying the evolution both of humans and of the ice ages. The isotope ^{36}Cl is particularly well suited for dating and tracing groundwater movement and for determining the suitability of radioactive waste depositories. The dating of ^{10}Be and ^{26}Al in marine sediments has confirmed their extraterrestrial origin, possibly from comets.

Example 12.16

A bone suspected to have originated during the period of the Roman emperors was found in Great Britain. Accelerator techniques gave its ^{14}C/^{12}C ratio as 1.1×10^{-12}. Is the bone old enough to have Roman origins?

Solution: The initial ratio of ^{14}C/^{12}C at the time of death was $R_0 = 1.2 \times 10^{-12}$. The number of ^{14}C atoms decays as $e^{-\lambda t}$.

$$N({}^{14}\text{C}) = N_0 e^{-\lambda t}$$

The ratio of ions is given by

$$R = \frac{N({}^{14}\text{C})}{N({}^{12}\text{C})} = \frac{N_0({}^{14}\text{C})e^{-\lambda t}}{N({}^{12}\text{C})} = R_0 e^{-\lambda t}$$

where R_0 is the original ratio. We can solve this equation for t.

$$e^{-\lambda t} = \frac{R}{R_0}$$

$$t = \frac{-\ln(R/R_0)}{\lambda} = -t_{1/2}\frac{\ln(R/R_0)}{\ln(2)}$$

$$= -(5730\text{ y})\left(\frac{-0.087}{0.693}\right) = 719\text{ y}$$

where we have inserted the known values of $t_{1/2}$, R_0, and R to find the age of the bone. The bone does not date from Roman times, but from the Middle Ages.

This Egyptian sarcophagus was dated by ^{14}C techniques in 1952 to be approximately 3940 years old with an uncertainty of 230 years. *Courtesy of The University Museum, University of Pennsylvania.*

Summary

The discovery of the neutron in 1932 solved several outstanding problems, including the understanding of the nuclear constituents and the origin of very penetrating radiation.

A nuclide ${}^A_Z X$ has mass M and is composed of Z protons and N neutrons. Its mass number is $A = Z + N$. Masses are measured in terms of atomic mass units u. The radius of a nucleus is $R = r_0 A^{1/3}$, where $r_0 \approx 1.2 \times 10^{-15}$ m = 1.2 fm. Electron scattering is useful to measure the size and shapes of nuclei. The properties of the nucleons are

Property	Neutron	Proton
mass(u)	1.008665	1.007276
charge (e)	0	+1
spin ($\hbar$)	1/2	1/2
magnetic moment ($e\hbar/2m_p$)	−1.91	+2.79

The study of the deuteron and nucleon–nucleon scattering indicates that the nuclear force is attractive and much stronger than the Coulomb force. However, it is effective only over a short range (up to about 3 fm). The nuclear force is charge independent and has a hard core.

A nuclide is stable if its mass is smaller than any other possible combination of the A nucleons. Stable nuclides tend to have $N \approx Z$ for small A and $N > Z$ for medium and large A. The total binding energy for a nuclide is

$$B({}^A_Z X) = [Nm_n + ZM({}^1\text{H}) - M({}^A_Z X)]c^2 \quad (12.10)$$

There are no stable nuclei with $Z > 83$ or $A > 209$. Nuclei tend to be more stable with an even number of protons and/or neutrons. Nuclei near ^{56}Fe have the highest binding energies per nucleon, and the average binding energy per nucleon for most nuclei is about 8 MeV.

The radioactive decay law is $N = N_0 e^{-\lambda t}$, where λ is the decay constant and the half-life $t_{1/2} = 0.693/\lambda$. The activity $R = \lambda N$. A becqueral (Bq) is 1 decay/s. Radioactive decay occurs when the disintegration energy $Q > 0$. The four kinds of alpha and beta decay are

Alpha decay $\quad {}^A_Z X \to {}^{A-4}_{Z-2}D + \alpha \quad (12.30)$

$$Q = [M({}^A_Z X) - M({}^{A-4}_{Z-2}D) - M({}^4\text{He})]c^2 \quad (12.31)$$

β^- decay $\quad {}^A_Z X \to {}_{Z+1}^{A}D + \beta^- + \bar{\nu} \quad (12.37)$

$$Q = [M({}^A_Z X) - M({}_{Z+1}^{A}D)]c^2 \quad (12.38)$$

β^+ decay $\quad {}^A_Z X \to {}_{Z-1}^{A}D + \beta^+ + \nu \quad (12.40)$

$$Q = [M({}^A_Z X) - M({}_{Z-1}^{A}D) - 2m_e]c^2 \quad (12.41)$$

Electron capture $\quad {}^A_Z X + e^- \to {}_{Z-1}^{A}D + \nu \quad (12.43)$

$$Q = [M({}^A_Z X) - M({}_{Z-1}^{A}D)]c^2 \quad (12.44)$$

Questions

1. Explain why neutrons are not prohibited from being in the nucleus for the same three reasons that electrons were excluded as discussed in the text.
2. Why does the atomic number Z determine the chemical properties of a nuclide? What differences in chemical properties would you expect for ^{16}O, ^{17}O, and ^{18}O, which are all stable? What about ^{15}O, which is unstable?
3. Explain how the nuclear charge radius could be different from the nuclear mass or nuclear force radii.
4. Do you believe it is easier to measure atomic masses or nuclear masses? Explain how you could experimentally measure both for ^{2}H. What about for ^{40}Ca?
5. Do you think it is significant that the nucleus has a hard core? What would happen if it did not?
6. Is the nuclear potential shown in Figure 12.4 consistent with a short-range nuclear force? Explain.
7. How likely is it that there are further, yet undetected, stable nuclides that are not shown in Figure 12.5? Explain.
8. Why does the binding energy curve of Figure 12.6 rise so fast for the light nuclei, but fall off so slowly for the heavy nuclei?
9. Hundreds of nuclides are known to decay by alpha emission. Why is decay by ^{3}He emission never (or rarely) observed?
10. If we find eventually that the neutrino does have zero mass, explain why it must be distinct from a photon, because both must travel at the speed of light.
11. Why is electron capture more probable than β^+ decay for the very heavy radioactive elements?
12. Why do unstable nuclei below the line of stability in Figure 12.5 undergo β^+ decay, whereas unstable nuclei above the line of stability undergo β^- decay?

Problems

12.1 Discovery of the Neutron

1. Assume that gamma rays are Compton scattered by hydrogen. What energy gamma rays are needed to produce 5.7-MeV protons?
2. Are the nuclear spins of the following nuclei (a) integral or (b) half-integral: ^{3}He, ^{6}Li, ^{7}Li, ^{18}F, ^{19}F?

12.2 Nuclear Properties

3. What are the number of protons, number of neutrons, mass number, atomic number, charge, and atomic mass for the following nuclei: $^3_2$He, $^4_2$He, $^{17}_{8}$O, $^{42}_{20}$Ca, $^{210}_{82}$Pb, and $^{235}_{92}$U?
4. What are the number of neutrons and protons for the following nuclides: ^{6}Li, ^{13}C, ^{40}K, and ^{64}Cu?
5. List from Appendix 8 all the isotopes, isobars, and isotones of ^{40}Ca.
6. What is the ratio of density of the nucleus to that of water?
7. What is the ratio of the magnetic moment of the proton to that of the electron?
8. Calculate the density and mass of the nuclide ^{56}Fe in SI units.

12.3 The Deuteron

9. What is the ratio of the electron binding energy in deuterium to the rest energy of the deuteron? Is it reasonable to ignore electronic binding energies when doing nuclear calculations?
10. Consider the photodisintegration of a deuteron at rest. Using both the conservation of energy and momentum, what is the minimum photon energy required? What percentage error is made by neglecting the conservation of momentum?

12.4 Nuclear Forces

11. Compute the gravitational and Coulomb force between two protons in ^{3}He. Assume the distance between the protons is equal to the nuclear radius. If the average nuclear potential is an attractive 40 MeV effective over a distance of 3 fm, compare the nuclear force with both the gravitational and Coulomb forces.
12. Consider two protons in the ^{27}Al nucleus located 2 fm apart. How strong does the nuclear force need to be to overcome the Coulomb force?

12.5 Nuclear Stability

13. (a) Show that the equation giving the binding energy of the last neutron in a nucleus ^A_ZX is

$$B = [M(^{A-1}_{\ \ Z}X) + m_n - M(^A_ZX)]c^2$$

(b) Calculate the binding energy of the most loosely bound neutron of ^{6}Li, ^{17}O, and ^{207}Pb.

14. (a) Show that the equation giving the binding energy of the last proton in a nucleus ^A_ZX is

$$B = [M(^{A-1}_{Z-1}Y) + M(^1\text{H}) - M(^A_ZX)]c^2$$

(b) Calculate the binding energy of the most loosely bound proton of ^{8}Be, ^{15}O, and ^{32}S.

15. What is the energy released when three α particles combine to form ^{12}C?

16. Estimate the nuclear spins of the ^{3}He and ^{4}He nuclei. Explain your reasoning.
17. The energy required to remove an inner K-shell electron from a silver atom is 25.6 keV. Compare this electron binding energy (the most tightly bound electron) with the binding energy of the most loosely bound proton of $^{107}_{47}$Ag.
18. Compare the total Coulomb repulsion between protons for ^{4}He, ^{40}Ca, and ^{208}Pb. Assume the protons in ^{4}He are, on the average, a distance equal to the nuclear radius apart and use Equation (12.23) for the Coulomb repulsion of the larger nuclei.
19. Continue adding neutrons into ^{16}O as in Figure 12.8. We find that ^{17}O and ^{18}O are stable, but not ^{19}O. Explain.
20. Explain why $^{42}_{20}$Ca is stable, but not $^{41}_{20}$Ca. If $^{42}_{20}$Ca is stable, why is $^{42}_{22}$Ti unstable?

12.6 Radioactive Decay

21. A radioactive sample of ^{60}Co ($t_{1/2} = 5.271$ y) has a β^- activity of 2.4×10^7 Bq. How many grams of ^{60}Co are present?
22. An unknown radioactive sample is observed to decrease in activity by a factor of five in a one-hour period. What is its half-life?
23. Show that the mean (or average) lifetime of a radioactive sample is $\tau = 1/\lambda = t_{1/2}/\ln(2)$.
24. For the reactor described in Example 12.9, compute the alpha activity of the ^{238}U present in the fuel cell. Assume that the uranium has been enriched to 4% ^{235}U which is typical for a commercial power reactor fuel cell.
25. Potassium is a useful element in the human body and is present at about the 0.3% level of body weight. Calculate the $^{40}_{19}$K activity in a 70-kg person (^{40}K has 0.012% natural abundance.)
26. The nuclide $^{18}_{9}$F (β^+ emitter, $t_{1/2} = 109.8$ min) is a useful radioactive tracer for human consumption. An amount of ^{18}F having an activity of 10^7 Bq is administered to a patient. What is the activity 48 hours later?
27. Tritium ($t_{1/2} = 12.33$ y) is mostly produced for military purposes. If we had stopped producing tritium in 1960 with a 2000-kg stockpile, how much would be left in the year 2010?
28. If we have the same mass of the following nuclides, rank the activities of the following material: ^{3}H (tritium), ^{222}Rn (radon gas), or ^{239}Pu (alpha source for power generation).

12.7 Alpha, Beta, and Gamma Decay

29. Using atomic masses show that nucleon decay does not occur for $^{52}_{26}$Fe, although this nuclide is highly unstable. If one could produce $^{40}_{26}$Fe, do you believe it might nucleon decay? Explain.
30. In Example 12.10 we showed that $^{230}_{92}$U does not decay by nucleon emission. What are the neutron and proton separation energies?
31. Show directly using masses that protons do not undergo any of the beta decays.
32. Calculate whether $^{144}_{62}$Sm and $^{147}_{62}$Sm may alpha decay. The natural abundance of ^{144}Sm is 3.1% and that of ^{147}Sm is 15.0%. How can this be explained?
33. How much kinetic energy does the daughter have when $^{241}_{95}$Am undergoes α decay from rest?
34. Show from the conservation of energy and momentum that, if Equation (12.34) correctly describes the β^- decay of ^{14}C initially at rest, the electron energy spectrum must be monoenergetic and not like that shown in Figure 12.14.
35. Find which of the α and β decays are allowed for $^{80}_{35}$Br.
36. Find which of the α and β decays are allowed for $^{227}_{89}$Ac.
37. Show that α, β^-, β^+, and EC decay are all possible for the nucleus $^{230}_{91}$Pa.
38. List all the possible energies of γ decay for ^{230}U based on Figure 12.15.
39. Calculate the partial pressure of helium gas for a 10^{-6}-m^3 volume of $^{222}_{86}$Rn gas after 6 days. The radon gas was originally placed in an evacuated container at 1 atm, and the temperature remains constant at 0°C. What is the partial pressure of the radon gas after 6 days? ($t_{1/2} = 3.82$ d for ^{222}Rn).
40. The nuclide ^{60}Co decays by β^-. Yet ^{60}Co is often used, especially in medical applications, as a source of γ rays. Explain how this is possible.
41. Explain why β^- predominates over β^+ decay in the natural radioactive decay chains of the heavy elements.
42. Give reasons why ^{14}O can β^+ decay, but the proton can not. (*Hint:* Use masses to prove this.)

12.8 Radioactive Nuclides

43. Two rocks are found that have different ratios R' of ^{238}U to ^{206}Pb: $R' = 0.76$ and 3.1. What are the ages of the two rocks? Did they likely have the same origin?
44. Using Table 12.2 list some radioactive nuclides that may be useful for dating the age of the Earth.
45. If scientists are only able to determine the ratio of R' in Equation (12.50) to ± 0.01, what is the minimum time possible for dating?
46. If the age of the Earth is 4.5 billion years, what should be the ratio of ^{206}Pb/^{238}U in a uranium-bearing rock as old as the Earth?
47. Using only Z and A values, calculate the number of α and β particles produced from the decay of $^{235}_{92}$U to its stable end product $^{207}_{82}$Pb.
48. The Earth is about 4.5 billion years old. If ^{235}U is 0.72% abundant today, how abundant was it when the Earth formed? (*Hint:* Look in Appendix 8 for useful information.)

49. Consider 100 g of ^{252}Fm, which decays in a sequence of five α decays to eventually reach ^{232}Th. (a) How much of the original sample is ^{252}Fm and how much is ^{248}Cf after one day? (b) After one month? (c) Explain why it is mostly curium after 5 years. (d) What isotope is it mostly after 100 years? (e) Approximate how long it will take to be mostly thorium.

50. Note in Table 12.2 that the half-lives of two abundant isotopes of tellurium are more than 10^{21} years. (a) What is the decay rate per unit mass of ^{128}Te, which decays by emitting two β^-, in units of $s^{-1} \cdot kg^{-1}$? (b) How much mass of a natural sample of tellurium would it take to measure a decay rate of 10 β^-/s for ^{128}Te?

General Problems

51. Show that the total Coulomb self-energy of a sphere of radius R containing a charge Ze evenly spread throughout the sphere is given by

$$\Delta E_{\text{Coul}} = \frac{3}{5}\frac{(Ze)^2}{4\pi\epsilon_0 R} \tag{12.17}$$

It might be useful to calculate the work done to bring a charge from infinity in to a spherical shell of radius r and then integrate the spherical shell from 0 to R.

52. Using the uncertainty principle $\Delta p\, \Delta x \geq \hbar/2$, calculate the minimum kinetic energy of a nucleon known to be confined in ^{2}H. What is the wavelength of this nucleon? Is this reasonable?

53. The nucleus $^{180}_{73}$Ta is unusual because it has both odd Z and odd N, yet it is barely unstable with a half-life of 8 h. It has an isomeric state at excitation energy 0.075 MeV that experimental measurements indicate has a half-life greater than 10^{15} y. For many years it was believed that the long-lived state was the ground state and may be stable. All the stable odd Z and N nuclei are smaller than ^{16}O. How can you explain this? Why are all heavy elements with both odd Z and N unstable? The spins of the ^{180}Ta ground state and isomeric states are believed to be 1^+ and 9^-, respectively. Explain how it might be possible that it was believed for so long that ^{180}Ta was stable.

54. The only stable isotope of holmium is $^{165}_{67}$Ho. Explain this. Can ^{165}Ho α decay? Is it likely to α decay? Explain.

55. The nuclide $^{226}_{88}$Ra decays to gaseous $^{222}_{86}$Rn with a $t_{1/2}$ = 1600 y. The nuclide $^{222}_{86}$Rn in turn α decays with a shorter lifetime, $t_{1/2}$ = 3.82 d. If the radium was originally placed in an evacuated closed container, the amount of radon gas builds up and can be measured. It is found that radon gas builds up to a constant value, and that as much ^{222}Rn is being produced as decays. This process is called *secular equilibrium*, and the activities are equal. (a) Show that radon builds up at the rate $dN_2/dt = \lambda_1 N_1 - \lambda_2 N_2$ where λ_1, N_1 and λ_2, N_2 are the decay constants and number of nuclei present for radium and radon, respectively. (b) Because the decay of ^{226}Ra is so slow, assume that N_1 = constant and show that

$$N_2 = \frac{\lambda_1}{\lambda_2} N_1 (1 - e^{-\lambda_2 t})$$

(c) Show that after a long time secular equilibrium is reached.

56. Rudolf Mössbauer discovered in 1957 that transitions from an excited nuclear state occur with negligible nuclear recoil when the nucleus is embedded in a large crystal lattice, because the entire lattice absorbs the recoil. A transition like that in ^{191}Ir from the 129-keV excited state to the ground state has a lifetime of 1.9×10^{-10} s. (a) Determine the energy width of the decay. Similarly, if the photon is absorbed by ^{191}Ir embedded in a crystal, the recoil is negligible. However, even a slight motion of the absorber will lead to a Doppler shift sufficient to destroy the resonance absorption. Calculate the speed necessary to shift the energy absorption by 5Γ, where Γ is the nuclear decay width.

57. Radon gas in the form of ^{222}Rn is a health hazard, because it is a gas that occurs as a result of one of the naturally occurring radioactive decay chains. It tends to collect in basements and can be inhaled by humans. (a) Which decay chain produces this isotope of radon? (b) Show that ^{222}Rn produces five more disintegrations before a stable isotope is reached. (c) Choose one of the paths of the decay chain from ^{222}Rn to the stable isotope and sum the half-lives. Approximate the number of days it would take for more than half these decays to occur for a given amount of radon.

CHAPTER 13

Nuclear Interactions and Applications

During a Christmas snow in Sweden in 1938, Lise Meitner and her nephew Otto Frisch performed calculations and concluded that Otto Hahn's experiment of bombarding uranium with neutrons must have produced fission. Upon hearing Frisch's explanation upon returning to Copenhagen, Niels Bohr exclaimed, "Oh, what idiots we all have been. This is just as it must be!"

***Adapted from C.P. Snow's* The Physicists**

Early investigators studied the atomic nucleus in the first decades of the 20th century by using naturally radioactive emitters as sources of high-speed particles to probe the nucleus. The invention of various kinds of particle accelerators in the 1930s enabled investigators to control the intensity and energy of these high-speed particles.

We will begin by studying nuclear reactions and mechanisms. Then we will describe nuclear power generation by fission and fusion reactors. Several interesting applications of nuclear phenomena will be discussed.

13.1 Nuclear Reactions

Rutherford produced the first nuclear reaction in a laboratory experiment in 1919. He used 7.7-MeV alpha particles from the decay of $^{214}_{84}\text{Po}$ to bombard a nitrogen target, and he observed protons being emitted. Although Rutherford was not certain of the exact nuclear reaction taking place, he convinced himself that he had broken down the nuclear structure. We now know that the reaction was

$$\alpha + {}^{14}_{7}\text{N} \rightarrow p + {}^{17}_{8}\text{O} \tag{13.1}$$

The first nucleus written is the projectile (α or ^{4}He nucleus), and the second is the target (^{14}N), normally at rest. These two nuclei interact and undergo a transmutation to one or more final particles. The detected particle is normally listed first after the arrow (p or ^{1}H), and the residual nucleus listed last (^{17}O). A shorthand way of writing this reaction is

$$^{14}\mathrm{N}(\alpha, p)^{17}\mathrm{O} \tag{13.2}$$

with several variations.

A general reaction like $x + X \rightarrow y + Y$ is written as

Nuclear reaction shorthand

$$X(x, y)Y \tag{13.3}$$

Normally p, d, t, α are used as symbols for the nuclei ^{1}H, ^{2}H, ^{3}H, and ^{4}He when they are the projectile or detected particle.

The study of nuclear reactions received a tremendous boost from three important technological advances:

1. The high-voltage multiplier circuit, developed in 1932 by the British physicists J. D. Cockcroft and E. T. S. Walton.
2. The first electrostatic generator (Van de Graaff accelerator), developed in 1931 by R. Van de Graaff in the United States (see Figure 13.1).
3. The first cyclotron, built in 1932 at the University of California in Berkeley by E. O. Lawrence (Nobel Prize, 1939) and M. S. Livingston (see Figure 13.2).

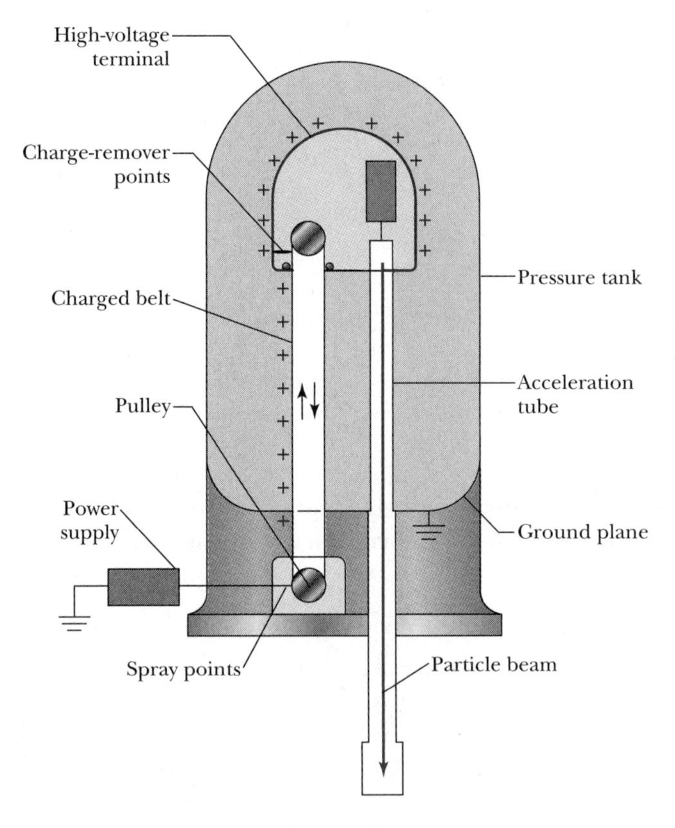

FIGURE 13.1 Schematic diagram of a Van de Graaff electrostatic accelerator. Charge is transferred to an insulated moving belt by spray points and moved to a high-voltage terminal where the charge is removed. The charge builds up on the terminal, creating a large electrostatic potential. Ionized particles are accelerated to ground potential in an acceleration tube, where they become a particle beam that can be steered to a target.

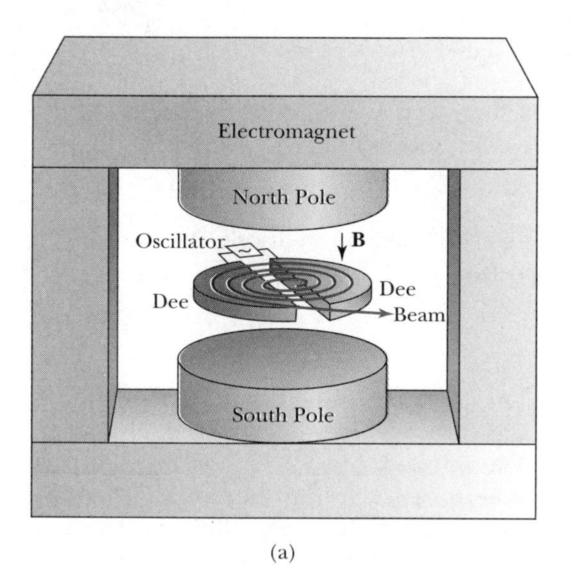

(a)

(b)

FIGURE 13.2 (a) The components of a simple cyclotron are shown. Particles are injected into the middle between the dees and move in circular paths perpendicular to the magnetic field. Each time they pass between the two dees they are accelerated by a RF voltage. Eventually the radius of the particle's path becomes large and the particles are extracted as a particle beam. (b) E. O. Lawrence (1901–1958) is shown on the right working on the 60-inch-diameter cyclotron at the University of California at Berkeley in the 1930s. *AIP/Niels Bohr Library.*

These three accelerators and the ones that followed them allowed nuclear reactions to be studied in the laboratory using accelerated particles. Accelerator technology continues today to allow physicists to delve even deeper into the understanding of nature's forces (see Chapter 14).

We have already mentioned several nuclear reactions in the previous chapter. Among them are the production of neutrons in the $^{9}\mathrm{Be}(\alpha, n)^{12}\mathrm{C}$ reaction. Nuclear photodisintegration is the initiation of a nuclear reaction by a photon, as in the $^{2}\mathrm{H}(\gamma, n)^{1}\mathrm{H}$ reaction to measure the binding energy of the deuteron. Neutron or proton radioactive capture occurs when the nucleon is absorbed by the target nucleus, with energy and momentum conserved by γ-ray emission. Examples are $^{9}\mathrm{Be}(p, \gamma)^{10}\mathrm{B}$ and $^{16}\mathrm{O}(n, \gamma)^{17}\mathrm{O}$.

If we want to draw attention to the fact that one of the residual particles is left in an excited state, we put a superscript asterisk on the nucleus's symbol: $^{12}\mathrm{C}(^{12}\mathrm{C}, \alpha)^{20}\mathrm{Ne}^*$. If we want to indicate the explicit excited state that a nucleus is in, we can so indicate by writing the energy in parentheses after the nucleus: $^{12}\mathrm{C}(^{12}\mathrm{C}, \alpha)^{20}\mathrm{Ne}(4.247\ \mathrm{MeV})$. The asterisk would be redundant in this case.

Entrance and exit channels

The projectile and target are said to be in the *entrance channel* of a nuclear reaction; for example, $^{12}\mathrm{C}$ and $^{12}\mathrm{C}$ represent a beam of carbon ions incident on a carbon target. The reaction products are in the *exit channel,* for example, α and $^{20}\mathrm{Ne}$. If the conservation laws allow, there may be many exit channels for any given entrance channel. For example, $^{12}\mathrm{C}(^{12}\mathrm{C}, n)^{23}\mathrm{Mg}$, $^{12}\mathrm{C}(^{12}\mathrm{C}, p)^{23}\mathrm{Na}$, $^{12}\mathrm{C}(^{12}\mathrm{C}, {}^{3}\mathrm{He})^{21}\mathrm{Ne}$, and $^{12}\mathrm{C}(^{12}\mathrm{C}, {}^{6}\mathrm{Li})^{18}\mathrm{F}$ all have different exit channels for the same entrance channel.

Elastic and inelastic scattering

In *elastic scattering* the entrance and exit channels are identical, and the particles in the exit channels are not in excited states. In *inelastic scattering* the entrance and exit channels are also identical, but one or more of the reaction products in the exit channel are left in an excited state as in the $^{20}Ne(\alpha, \alpha)^{20}Ne^*$ reaction.

We have listed the preceding reactions as if two particles are always in the exit channel. That is certainly not the case. Reactions like

$$\alpha + {}^{20}_{10}Ne \rightarrow n + p + {}^{22}_{11}Na \tag{13.4}$$

can also occur. If both the outgoing neutron and proton are detected in coincidence, the previous reaction can be written as $^{20}Ne(\alpha, np)^{22}Na$.

Cross Sections

The properties of the nucleus have mostly been studied by detecting one or more particles in the exit channel in a nuclear reaction. The probability of a particular nuclear reaction occurring is determined by measuring the *cross section* σ. By experimentally measuring the number of particles produced in a given nuclear reaction, the cross section is determined, which in turn allows physicists to learn about the nuclear force that caused the nuclear reaction. We were first introduced to the concept of the cross section in Section 4.2, when we discussed how Rutherford and his colleagues "discovered" the nucleus in their early experiments. We follow the same procedure as in Section 4.2 and define n = number of target atoms/volume, t = target thickness, A = area of the target, ρ = density, N_A = Avogadro's number, N_M = atoms/molecule, and M_g = gram-molecular weight. The number of target nuclei N_S is

$$N_S = ntA = \frac{\rho N_A N_M t A}{M_g} \tag{4.10}$$

where the parameters are discussed in Section 4.2. The probability of the particle being scattered is proportional to the product of the cross section times the total number of target nuclei N_S. We normalize this product to obtain the probability of scattering by dividing by the total target area A.

Probability of scattering

$$\text{Probability of scattering} = \frac{N_S \sigma}{A} = \frac{ntA\sigma}{A} = nt\sigma \tag{13.5}$$

The product nt is the number of target nuclei exposed per unit area, and it is related to the probability of scattering by multiplying by the cross section.

We measure the cross section by counting the number of detected particles as a function of the number of incoming particles. If the cross sections are measured as a function of the scattering angle θ (between the incoming beam of particles and the detected particle), we call them *differential cross sections* $\sigma(\theta)$. The geometry is shown in Figure 13.3. Differential cross sections are determined

Differential and total cross sections

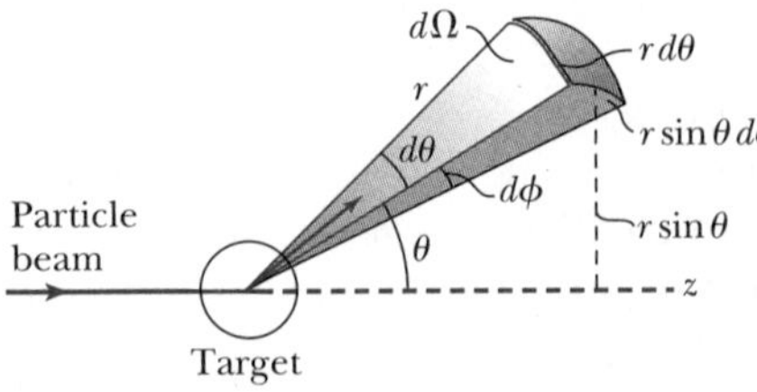

$d\Omega$, r, $r\,d\theta$, $r\sin\theta\,d\phi$, $d\theta$, $d\phi$, Particle beam, θ, $r\sin\theta$, z, Target

$$d\Omega = \frac{(r\sin\theta)(r\,d\theta)(d\phi)}{r^2} = \sin\theta\,d\theta\,d\phi$$

FIGURE 13.3 A particle from a beam interacts with a target, sending a reaction particle into a differential solid angle $d\Omega$. The solid angle is defined by the angles θ and ϕ in spherical coordinates: $d\Omega = \sin\theta\,d\theta\,d\phi$.

by the number of particles scattered into a small solid angle $d\Omega$, measured in units of steradian (sr), surrounding the scattering angle θ. In spherical coordinates, with θ being the angle measured from the incident beam direction, the solid angle $d\Omega = \sin\theta\, d\theta\, d\phi$. The differential cross section can also be written as $d\sigma/d\Omega$, which expresses more clearly the number of particles scattered into the differential solid angle $d\Omega$: $\sigma(\theta) = d\sigma/d\Omega$. If the cross sections are integrated over the entire range of scattering angles, we call them *total cross sections* σ_T.

$$\sigma_T = \int \sigma(\theta)\, d\Omega = \int_0^{2\pi} d\phi \int_0^{\pi} \frac{d\sigma}{d\Omega} \sin\theta\, d\theta$$

In general, the cross sections depend on the incident kinetic energy and other properties including the spins of the particles. Total cross sections are traditionally measured in units of barns (b) with 1 barn $\equiv 10^{-28}$ m^2 = 100 fm^2, so that a barn is about the cross-sectional area of an $A = 100$ nucleus. The units of differential cross section are barns/steradian, or b/sr.

Example 13.1

J. L. Black and colleagues* measured the total cross section for the $^{12}C(\alpha, n)^{15}O$ reaction at an incident energy of $E_\alpha = 14.6$ MeV and determined it to be $\sigma_T = 25$ mb. If a 1-μA α-particle beam ($^4He^{++}$) is incident on a 4-mm^2 carbon target of thickness 1 μm (density = 1.9 g/cm^3) for one hour, how many neutrons are produced?

Solution: To find the probability of scattering, we first need to determine n, the number of nuclei/volume, which we established in Equation (4.8).

$$n = \frac{\rho N_A N_M}{M_g} \frac{\text{atoms}}{\text{cm}^3} \tag{4.8}$$

Some of the values needed are

$$\rho = 1.9\ \text{g/cm}^3 \qquad N_A = 6.02 \times 10^{23}\ \text{molecules/mol}$$

$$N_M = 1\ \text{atom/molecule} \qquad M_g = 12\ \text{g/mol}$$

If we substitute these values in the equation for n, we have

$$n = \frac{1.9\ \text{g}}{\text{cm}^3} \cdot \frac{6.02 \times 10^{23}\ \text{molecules}}{\text{mol}} \frac{1\ \text{atom}}{\text{molecule}} \frac{\text{mol}}{12\ \text{g}}$$

$$= 9.53 \times 10^{22} \frac{\text{atoms}}{\text{cm}^3}$$

The probability of scattering can now be determined from Equation (13.5).

$$P = nt\sigma = (9.53 \times 10^{22}\ \text{nuclei/cm}^3)(10^{-6}\ \text{m}) \times (25 \times 10^{-31}\ \text{m}^2)(10^6\ \text{cm}^3/\text{m}^3)$$

$$= 2.4 \times 10^{-7}$$

The number of incident α particles on the target N_I can be determined by the beam current and length of time the beam is on the target.

$$N_I = \left(10^{-6}\ \frac{\mu\text{C}}{\text{s}}\right)(1\ \text{h})\left(3600\ \frac{\text{s}}{\text{h}}\right)\left(10^{-6}\ \frac{\text{C}}{\mu\text{C}}\right) \times \frac{1\ \text{alpha}}{2(1.6 \times 10^{-19}\ \text{C})}$$

$$= 1.1 \times 10^{10}\ \text{alphas}$$

Note that we have taken the charge of the incident α particles to be $+2e$.

The ratio of detected neutrons to incident alpha particles (N_n/N_I) is the probability of scattering. We therefore have

$$N_n = N_I(\text{Probability}) = N_I nt\sigma$$

$$= (1.1 \times 10^{10}\ \text{alphas})(2.4 \times 10^{-7}\ \text{neutrons/alpha})$$

$$= 2.6 \times 10^3\ \text{neutrons}$$

Nuclear Physics* **115, 683 (1968).

Example 13.2

E. M. Bernstein and colleagues* measured the differential cross section $\sigma(\theta)$ at the same energy $E_\alpha = 14.6$ MeV and at the same scattering angle for the $^{12}C(\alpha, n)^{15}O$ and $^{12}C(\alpha, p)^{15}N$ reactions. They found differential cross sections of 3 mb/sr and 0.2 mb/sr, respectively, for the neutron and proton production at the same scattering angle θ. How much more likely is it that a neutron is produced than a proton?

Solution: The probability of scattering is simply $nt\sigma$, so let P_n and P_p be the neutron and the proton probability, respectively. We also denote the neutron and proton differential cross section at scattering angle θ by $\sigma_n(\theta)$ and $\sigma_p(\theta)$, respectively.

$$\frac{P_n}{P_p} = \frac{nt\sigma_n(\theta)}{nt\sigma_p(\theta)} = \frac{\sigma_n(\theta)}{\sigma_p(\theta)} = \frac{3 \text{ mb/sr}}{0.2 \text{ mb/sr}} = 15$$

For the particular scattering angle θ, this ratio is larger than might be expected from the fact that neutrons and protons have similar nuclear interactions. Other factors, such as overcoming the Coulomb barrier and the existence of resonances, are important and will be discussed in Section 13.3.

13.2 Reaction Kinematics

In this chapter we are discussing low-energy nuclear reactions, in which the kinetic energies are typically much lower than the rest energies. Therefore we can use nonrelativistic kinematics in many cases. Consider the reaction $X(x, y)Y$ depicted in Figure 13.4, where the momentum and kinetic energy of the projectile x are $\mathbf{p}_x$ and K_x. The target X is assumed at rest, so $\mathbf{p}_X = 0$ and $K_X = 0$.

The conservation of energy for the reaction is

$$M_x c^2 + K_x + M_X c^2 = M_y c^2 + K_y + M_Y c^2 + K_Y \tag{13.6}$$

If we rearrange this equation and put all the masses on one side and all the kinetic energies on the other side, we find a quantity similar to the disintegration energy of Chapter 12:

Q value

$$Q = M_x c^2 + M_X c^2 - (M_y c^2 + M_Y c^2) = K_y + K_Y - K_x \tag{13.7}$$

The difference between the final and initial kinetic energies is precisely the difference between the initial and final mass energies. We call this difference the **Q value** or energy release. Nuclear masses are listed in Equation (13.7), but as we learned in Chapter 12, we may use atomic masses. When using the tabulated atomic masses from Appendix 8, Equation (13.7) determines the ground-state Q value, that is, all the nuclei are in their lowest energy state. If one or more of the nuclei are left in an excited state E^*, that energy must be added to the mass energy listed in Equation (13.7) in order to calculate the reaction Q value.

We don't need to worry about this particular point if we use the kinetic energies in Equation (13.7) to determine the Q value. The electron masses cancel, and the electron binding energies are so small that we neglect their differences when using masses to determine the Q value. Energy is released in a nuclear reaction when $Q > 0$, and we call this an **exoergic** (or *exothermic*) reaction. When $Q < 0$, kinetic energy is converted to mass energy and we call the reaction **endoergic** (or *endothermic*). In an elastic collision, $X(x, x)X$, we must have $Q = 0$. In an inelastic collision, $X(x, x)X^*$, we must have $Q < 0$.

Endoergic and exoergic reactions

Physical Review* **C3, 427 (1971).

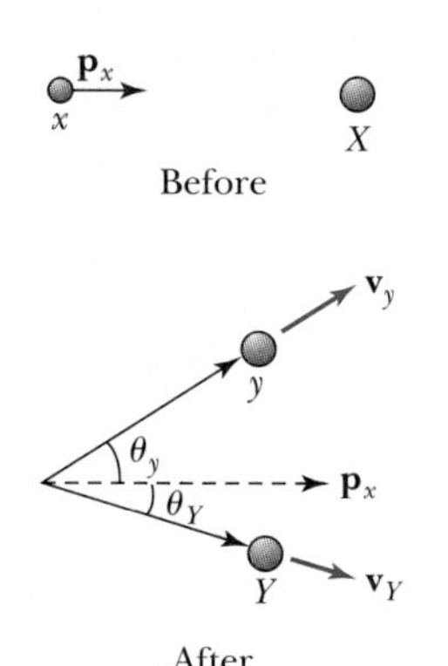

FIGURE 13.4 Two-body nuclear reaction, $X(x, y)Y$, as observed in the laboratory system showing the before and after.

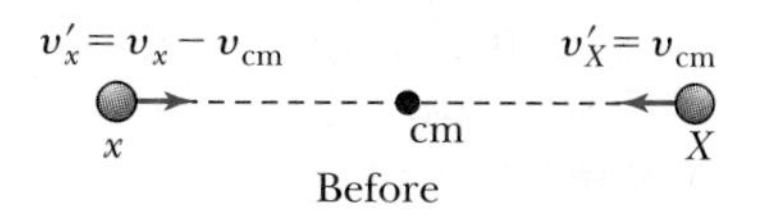

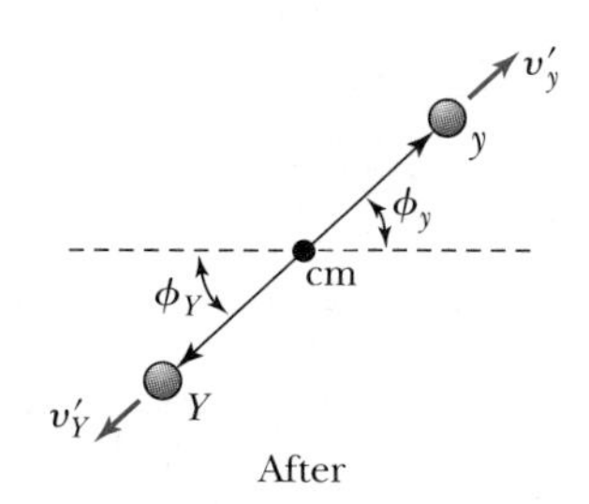

FIGURE 13.5 The same two-body nuclear reaction seen in Figure 13.4, but as observed in the center-of-mass system in which the center of mass (cm) is at rest.

Example 13.3

Calculate the ground-state Q value for the reaction $^{14}\text{N}(\alpha, p)^{17}\text{O}$ in which Rutherford first observed a nuclear reaction. The kinetic energy of the α particles was 7.7 MeV. What was the sum of the kinetic energies of the exit channel?

Solution: We can determine the Q value from the atomic masses listed in Appendix 8. They are

$$M(^4\text{He}) = 4.002603 \text{ u} \qquad M(^1\text{H}) = 1.007825 \text{ u}$$

$$M(^{14}\text{N}) = 14.003074 \text{ u} \qquad M(^{17}\text{O}) = 16.999132 \text{ u}$$

Notice that we must use the atomic masses for ^{1}H and ^{4}He in order to have the complete cancellation of the electron masses. The ground state Q value is calculated from Equation (13.7).

$$\frac{Q}{c^2} = M(^4\text{He}) + M(^{14}\text{N}) - [M(^1\text{H}) + M(^{17}\text{O})]$$

$$= 4.002603 \text{ u} + 14.003074 \text{ u} - (1.007825 \text{ u} + 16.999132 \text{ u}) = -0.001280 \text{ u}$$

$$Q = (-0.001280\ c^2 \cdot \text{u})\left(\frac{931.5 \text{ MeV}}{c^2 \cdot \text{u}}\right) = -1.192 \text{ MeV}$$

The reaction is endoergic. We calculate the kinetic energies of the products also by using Equation (13.7).

$$Q = K(p) + K(^{17}\text{O}) - K(\alpha)$$

$$K(p) + K(^{17}\text{O}) = Q + K(\alpha)$$

$$= -1.192 \text{ MeV} + 7.7 \text{ MeV} = 6.5 \text{ MeV}$$

The final reaction products share 6.5 MeV of energy.

An endoergic ($Q < 0$) reaction will not occur unless there is enough kinetic energy K_x to supply the needed nuclear rearrangement energy. Because linear momentum must also be conserved, the energy must be available in the center-of-mass (cm) system where v_{cm} is zero. The total momentum is zero in the cm system (see Figure 13.5). The minimum kinetic energy needed to initiate a nuclear reaction is called the **threshold energy** K_{th}, and for this energy the particles y and Y will be at rest in the cm system. Of course, particle y will still be moving in the laboratory system.

Let us examine conservation of energy in the center-of-mass system. The speed of the center of mass is given by

$$v_{cm} = v_x\left(\frac{M_x}{M_x + M_X}\right) \tag{13.8}$$

As shown in Figure 13.5 the speeds of x and X in the center-of-mass system are $v'_x = v_x - v_{cm}$ (to the right) and $v'_X = v_{cm}$ (to the left), respectively, where the primes indicate the quantities in the cm system. At threshold we must have $v'_y = v'_Y = 0$ in the center-of-mass system. The conservation of energy in the center-of-mass system is

$$\frac{1}{2}M_x(v_x - v_{\rm cm})^2 + \frac{1}{2}M_X v_{\rm cm}^2 + M_x c^2 + M_X c^2 = M_y c^2 + M_Y c^2 + \frac{1}{2}M_y v_y'^2 + \frac{1}{2}M_Y v_Y'^2 \qquad (13.9)$$

However, because $v'_y = v'_Y = 0$ at threshold, we have, upon rearranging terms,

$$\frac{1}{2}M_x\left(v_x - \frac{M_x v_x}{M_x + M_X}\right)^2 + \frac{1}{2}M_X\left(\frac{M_x^2 v_x^2}{(M_x + M_X)^2}\right) = -Q$$

This equation reduces to

$$\frac{1}{2}v_x^2\left(\frac{M_x M_X}{M_x + M_X}\right) = -Q$$

But the threshold energy is defined by $K_{\rm th} = \frac{1}{2}M_x v_x^2$, so we have

Threshold energy

$$K_{\rm th} = -Q\left(\frac{M_x + M_X}{M_X}\right) \qquad (13.10)$$

Equation (13.10) is the threshold energy calculated nonrelativistically for an endoergic reaction. The Q value determined in Example 13.3 for the $^{14}N(\alpha, p)^{17}O$ reaction was -1.191 MeV. The threshold kinetic energy is 1.191 MeV times $(4 + 14)/14$, which gives 1.531 MeV. The reaction discussed in Example 13.3 will not take place if the incident α particle has less than 1.531 MeV kinetic energy.

13.3 Reaction Mechanisms

As we discussed in Chapter 12, physicists use models to describe the nucleus because they have been unable to completely understand the nuclear force. The primary technique for studying nuclei and nuclear forces has been to perform scattering reactions. Physicists measure cross sections, which are proportional to the reaction probabilities, which in turn depend on details of nuclear structure and the strengths and ranges of the interaction.

Just as we discussed different nuclear models in the previous chapter, there are also different types of nuclear reactions—each occurring over varying bombarding energies. Reaction mechanisms for projectiles like electrons are much different than those initiated by alpha particles, for example. Often, however, they measure similar properties. For heavy charged particles (protons and alpha particles) at low energies, $E < 10$ MeV, the scattering is dominated by the Coulomb force, and the **compound nucleus** reaction mechanism (discussed later) is appropriate. **Direct reactions** dominate for bombarding energies in the range 10–100 MeV. At energies above about 140 MeV, a new particle called a *pion* emerges (see Chapter 14); the pion interacts strongly through the nuclear force. The energy region above 200 MeV to a few GeV is the realm of *medium-energy physics.* Experiments with protons, pions, and electrons dominate this region, which has some overlap with *high-energy* or *elementary particle physics* (studied in Chapter 14).

The Compound Nucleus

In 1936 Niels Bohr proposed that nuclear reactions take place through formation and decay of a *compound nucleus.* The compound nucleus is a composite

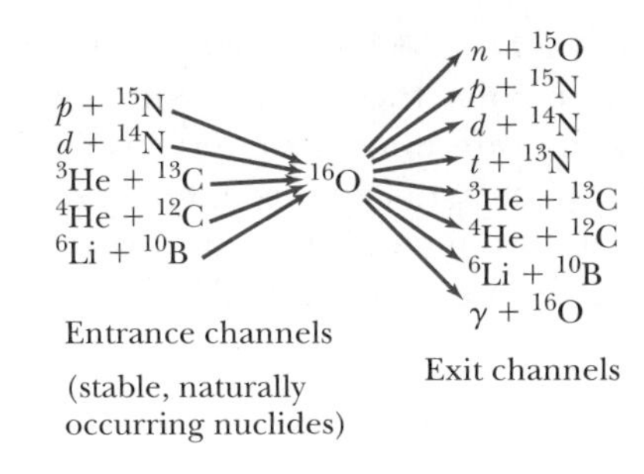

FIGURE 13.6 Several different entrance-channel two-body reactions may form the nucleus ${}^{16}O$. Similarly, the excited ${}^{16}O$ nucleus may decay to one of many exit channels that the conservation laws allow.

of the projectile and target nuclei, usually in a high state of excitation. In the ${}^{12}C(\alpha, n){}^{15}O$ reaction discussed in Examples 13.1 and 13.2, there is enough energy available from just the masses to leave ${}^{16}O^*$ in an excited state of 7.2 MeV when an α particle and ${}^{12}C$ join to produce ${}^{16}O^*$. The kinetic energy *available in the center-of-mass,* K'_{cm} (see Problem 18),

$$K'_{cm} = \frac{M_X}{M_x + M_X} K_{lab} \tag{13.11}$$

is available to excite the compound nucleus to even higher excitation energies than that from just the masses.

Once formed, the compound nucleus may exist for a relatively long time compared to the time taken by the bombarding particle to cross the nucleus. This latter time is sometimes referred to as the *nuclear time scale* t_N, and, for a 5-MeV proton ($v \sim 0.1c$) crossing a typical nuclear diameter of about 9 fm, is about $t_N \sim 3 \times 10^{-22}$ s. Compound nuclei may live as long as 10^{-15} s or $10^6 t_N$. Bohr's hypothesis was that this is such a long time that the nucleus "forgets" how it was formed, and the excitation energy is shared by all the nucleons in the nucleus. When the compound nucleus finally does decay from its highly excited state, it decays into all the possible exit channels according to statistical rules consistent with the conservation laws. Examples of the formation and decay of ${}^{16}O$ are shown in Figure 13.6. In Figure 13.7 we show the excitation energy in ${}^{16}O^*$ equivalent to the various entrance channel rest energies.

Example 13.4

In Example 13.2 we noted that the ${}^{12}C(\alpha, n){}^{15}O$ reaction cross section was much larger than that of the ${}^{12}C(\alpha, p){}^{15}N$ at $E_\alpha = 14.6$ MeV. To what energy is ${}^{16}O^*$ excited in this reaction?

Solution: We can determine the available kinetic energy in the center of mass from Equation (13.11).

$$K'_{cm} = \frac{12}{4 + 12}(14.6 \text{ MeV}) = 11.0 \text{ MeV}$$

The excitation energy $E({}^{16}O^*)$ due to just the masses when an α particle and ${}^{12}C$ form to make ${}^{16}O$ is determined by

$$M({}^4He)c^2 + M({}^{12}C)c^2 = M({}^{16}O)c^2 + E({}^{16}O^*) \tag{13.12}$$

The appropriate masses are $M({}^4He) = 4.002603$ u, $M({}^{12}C) = 12.0$ u, and $M({}^{16}O) = 15.994915$ u. The excitation energy of ${}^{16}O^*$ becomes

$$\begin{aligned} E({}^{16}O^*) &= (4.002603 \text{ u} + 12.0 \text{ u} - 15.994915 \text{ u})c^2 \\ &= 0.007688\ c^2 \cdot \text{u} \\ &= (0.007688\ c^2 \cdot \text{u})\left(\frac{931.5 \text{ MeV}}{c^2 \cdot \text{u}}\right) = 7.16 \text{ MeV} \end{aligned}$$

This energy, 7.2 MeV, is indicated for $\alpha + {}^{12}C$ in Figure 13.7. The total excitation energy in ${}^{16}O^*$ is the sum of the available K'_{cm} and 7.2 MeV.

$$E({}^{16}O^*) = 11.0 \text{ MeV} + 7.2 \text{ MeV} = 18.2 \text{ MeV}$$

We mentioned earlier that nuclei have discrete energy levels, much as atoms do, and in Figure 13.7 we exhibit the low-lying states of ^{16}O. The lowest state, called the *ground state,* is the reference; its energy is 0.0 MeV. The first excited state is 6.05 MeV, with the next state close by at 6.13 MeV. The ^{16}O ground state is a particularly strongly bound nuclear state. In order to excite ^{16}O, it takes an anomalously large amount of energy (6.05 MeV).

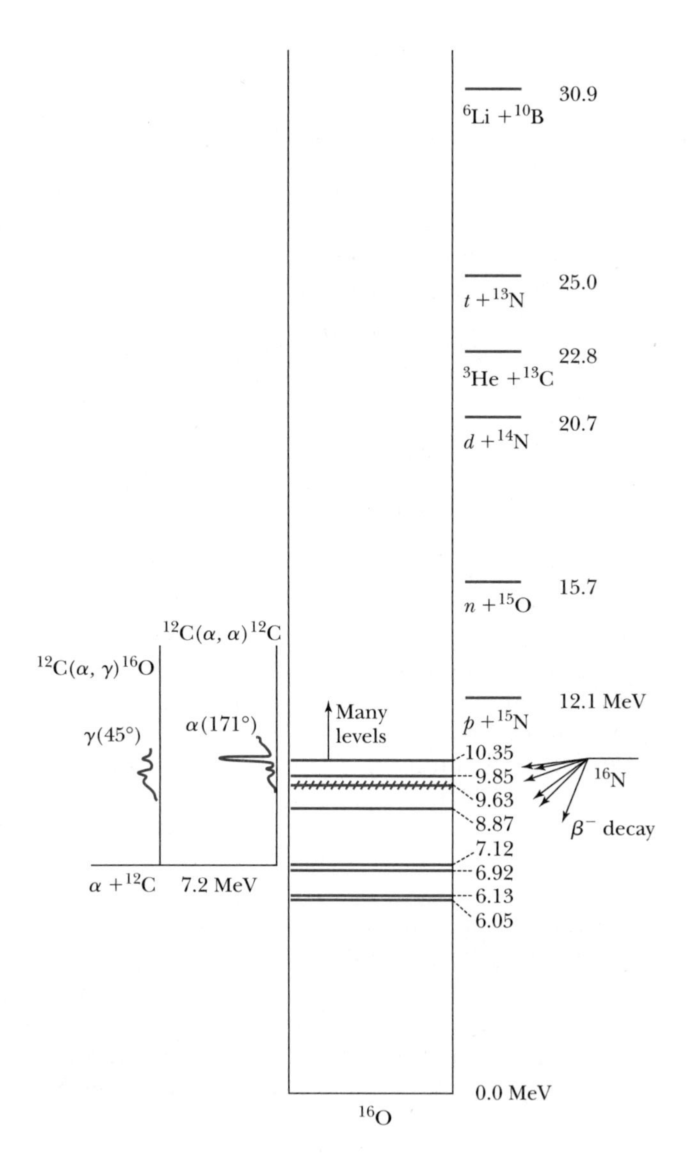

FIGURE 13.7 Energy-level diagram of the nucleus ^{16}O. The first excited state is at 6.05 MeV. If $d + ^{14}N$ could join together at rest to form ^{16}O, the excitation energy of ^{16}O would be 20.7 MeV. The nucleus ^{16}O is much more tightly bound than any of the two-body systems shown, with $\alpha + ^{12}C$ being the lowest at 7.2 MeV in ^{16}O. On the left are shown excitation functions for $\alpha + ^{12}C$ for the outgoing channels of γ at 45° and α at 171°. Note that resonances are observed in some reactions at bombarding energies appropriate for a particular resonance in ^{16}O; see, for example, 10.35 MeV. Note that ^{16}N can beta decay to levels in ^{16}O with excitation energies below about 10 MeV. The cross-hatched level at 9.63 MeV is quite broad.

Nuclear physicists study nuclear excited states by varying the projectile bombarding energy K_x and measuring the cross section at each energy, generally at fixed angles for the outgoing particles. This is called an *excitation function.* Excitation functions for the ground states of $^{12}C(\alpha, \alpha)^{12}C$ and $^{12}C(\alpha, \gamma)^{16}O$ are shown on the left in Figure 13.7. Notice that there are peaks and sudden changes in the smooth curves. Such sharp peaks in the excitation function of the reacting particles are called **resonances,** and they represent a quantum state of the compound nucleus being formed; in this case an excited state of ^{16}O is formed when α and ^{12}C interact. This is confirmed in the excitation functions of Figure 13.7 where the bumps coincide with particular energy levels in $^{16}O^*$.

Resonances

Now we can understand why $^{12}C(\alpha, n)^{15}O$ had such a large cross section in Example 13.2. The reaction populated a resonance near $E(^{16}O^*) = 18.2$ MeV as determined in Example 13.4. The quantum numbers of the $^{12}C(\alpha, n)^{15}O$ exit channel select this energy level in $^{16}O^*$ to be populated, whereas the quantum numbers for the $^{12}C(\alpha, p)^{15}N$ exit channel apparently do not.

The uncertainty principle may be used to relate the energy width of a particular nuclear state (called Γ) to its lifetime (called τ). The relationship is

$$\Gamma\tau \geq \frac{\hbar}{2} \tag{13.13}$$

If the width of a certain nuclear state is measured to be Γ by an excitation function, Equation (13.13) can be used to determine its lifetime. Ground states for stable nuclei, for example ^{16}O, have an infinite lifetime and therefore have zero energy width.

Neutron Activation. Because neutrons have zero net charge, they more easily interact with nuclei at low energies than do charged particles, because of the Coulomb barrier. If the nuclide $^{113}_{48}Cd$ interacts with a neutron, the compound nucleus $^{114}_{48}Cd^*$ may be formed. We call this process *neutron activation.* One common mode for decay of this compound nucleus is the emission of a γ ray. The reaction $^{113}Cd(n, \gamma)^{114}Cd$, an example of *neutron radioactive capture,* produces γ-ray energies that depend on the energy-level structure of ^{114}Cd. The γ-ray energies and intensities are characteristic of the ^{114}Cd nucleus and no other. They are like a unique fingerprint that lets us determine the compound nucleus to be ^{114}Cd and thus indicates the original presence of ^{113}Cd. The general technique, called *neutron activation analysis,* is a powerful practical method for identifying elements without damaging the sample. We will discuss it further in Section 13.7.

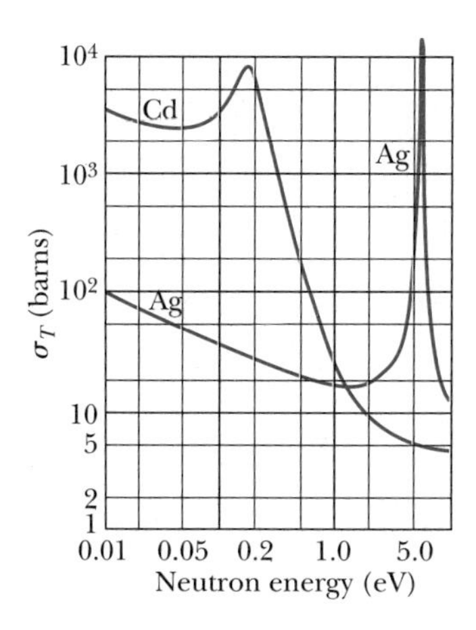

FIGURE 13.8 Total neutron capture cross section at low energies for silver and cadmium. The peaks indicate the presence of resonances in the compound nucleus. Note the $1/v$ dependence for the cross section of Ag at low energies. Cadmium's high cross section makes it an excellent material to control neutron flux in nuclear reactors.

The neutron capture reaction (n, γ) often has a large cross section, which can be as large as thousands of barns at a resonance. As neutrons pass through matter, they lose energy by having many elastic and inelastic collisions and eventually reach a (thermal) kinetic energy $3kT/2$, when they attain equilibrium with their surroundings. The average neutron capture cross section (at energies up to about 100 keV) varies empirically as $1/v$, where v is the neutron's velocity (see the cross section for Ag in Figure 13.8). The $1/v$ dependence can be explained in terms of the time the neutron spends near the nucleus. This time is $2r/v$ ($2r =$ nuclear diameter). The longer the neutron is within the range of the nuclear force, the higher the probability it will be captured, because the average capture probability per unit time is nearly independent of incident energy.

Direct Reactions

As the energy of the bombarding particle rises, the excitation energy of the compound nucleus becomes high. The compound nuclear excited states become broad (large Γ), and the number of nuclear states becomes very large. As the lifetimes of the states decrease, they approach that of the nuclear time scale. In addition, the compound nuclear states overlap, and the idea of a compound nucleus loses its utility for analyzing reactions. If a compound nucleus state is formed, it is very likely to emit one or more nucleons (particularly neutrons) to rid itself of the extra excitation energy. The nuclear force is much stronger than the Coulomb force, so nucleon decay will almost always occur before γ decay when it is allowed by the conservation laws.

For higher bombarding energies, the bombarding particle spends much less time within the range of the nuclear force. Simply *stripping* one or more nucleons off the projectile or *picking up* one or more nucleons from the target become more probable. It is also possible for the projectile to *knock out* energetic nucleons from the target nucleus. These are called **direct reactions.**

The chief advantage of direct reactions is that the final residual nucleus may be left in any one of many low-lying excited states. By using different direct reactions, the nuclear excited states can be studied in a variety of ways to learn more about nuclear structure. Many different reactions may be used, for example, to study ^{16}O energy levels; $^{15}\text{N}(^3\text{He}, d)^{16}\text{O}$, $^{13}\text{C}(\alpha, n)^{16}\text{O}$, $^{12}\text{C}(^6\text{Li}, d)^{16}\text{O}$, $^{18}\text{O}(p, t)^{16}\text{O}$, and $^{17}\text{O}(d, t)^{16}\text{O}$ are just a few of many. All the states shown for ^{16}O in Figure 13.7 are populated in the $^{14}\text{N}(^3\text{He}, p)^{16}\text{O}$ reaction.

13.4 Fission

In Chapter 12 we learned that nuclei near $A = 56$ have the highest average binding energy per nucleon. Some nuclei with $A > 100$ are able to alpha decay, and many nuclei with $A > 220$ are unstable with respect to fission. In fission a nucleus separates into two somewhat equal-sized *fission fragments.* We can determine which nuclei are able to fission by using Equation (12.20) for the binding energy.

Fission occurs for heavy nuclei because of the increased Coulomb forces between the protons. We can understand fission by using the semi-empirical mass formula based on the liquid drop model (Chapter 12). For a spherical nucleus of $A \sim 240$, the attractive short-range nuclear forces (volume term) more than offset the Coulomb repulsive term. However, as a nucleus becomes deformed (nonspherical), the surface energy is increased, and the effect of the short-range nuclear interactions is reduced. Nucleons on the surface are not surrounded by other nucleons, and the unsaturated nuclear force reduces the overall nuclear attraction. For a certain deformation, a critical energy is reached, and the *fission barrier* is overcome. This is understood in terms of the liquid drop model where the Coulomb force pushes a deformed drop further apart. A careful examination of the semi-empirical mass formula reveals that *spontaneous fission* occurs for nuclei with $Z^2/A \geq 49$ ($Z \sim 115$, $A \sim 270$). The term Z^2/A comes from the surface energy term ($A^{2/3}$) and the Coulomb energy term ($\sim Z^2/A^{1/3}$). Spontaneous fission can also occur for $Z^2/A < 49$ by the process of tunneling through the Coulomb barrier, but the half-lives are much increased. The naturally occurring nuclide with the highest spontaneous fission rate is ^{238}U, which has a half-life for fission of 8.2×10^{15} y. This is to be compared with its alpha decay half-life of 4.5×10^9 y. Thus ^{238}U is over a million times more likely to alpha decay than to fission.

Spontaneous fission

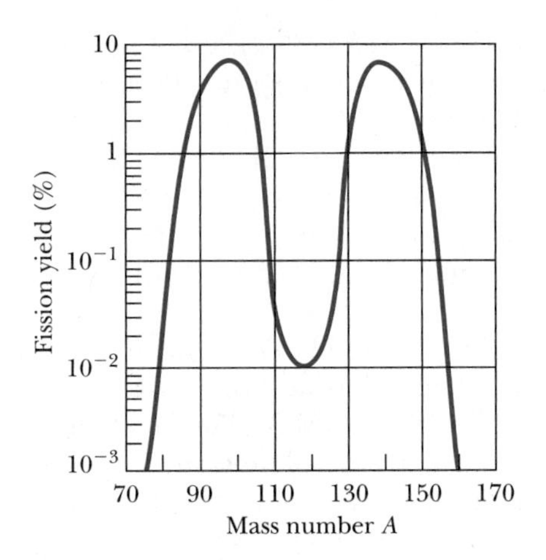

FIGURE 13.9 The percentage distribution of fission fragment yields from the thermal neutron induced fission of ^{235}U. Notice asymmetric fission is much more probable than symmetric fission (equal masses). *From R. D. Evans, The Atomic Nucleus. New York: McGraw-Hill, 1955.*

Induced Fission. The previous discussion has been for nuclei in their naturally occurring ground states. Fission may also be *induced* by a nuclear reaction. A neutron absorbed by a heavy nucleus forms a highly excited compound nucleus that may quickly fission. An example of induced fission is

Induced fission

$$n + {}^{235}_{92}\text{U} \longrightarrow {}^{236}_{92}\text{U}^* \longrightarrow {}^{99}_{40}\text{Zr} + {}^{134}_{52}\text{Te} + 3n \qquad (13.14)$$

The fission products have a ratio of N/Z much too high to be stable for their A value. Normally 2 or 3 neutrons will be emitted during fission. There are many possibilities for the Z and A of the fission products, as is shown in Figure 13.9. Symmetric fission (equal Z products) is possible, but the most probable fission is asymmetric, as shown by the two peaks in Figure 13.9, one mass being larger than the other.

Example 13.5

Calculate the ground-state Q value of the induced fission reaction in Equation (13.14) if the neutron is thermal. A neutron is said to be *thermal* when it is in thermal equilibrium with its environment; it then has an average kinetic energy given by $\frac{3}{2}kT$.

Solution: Because the kinetic energy of a thermal neutron is so small, its kinetic energy can be neglected; even for a temperature of 10^6 K, the thermal energy is only 130 eV. The Q value is given by Equation (13.7). We look up the atomic masses in Appendix 8 and determine the Q value to be

$$Q = [M(^{235}\text{U}) + m(n) - [M(^{99}\text{Zr}) + M(^{134}\text{Te}) + 3m(n)]]c^2$$

$$= [235.0439 \text{ u} - 98.9165 \text{ u} - 133.9115 \text{ u} - 2(1.0087 \text{ u})]c^2 = 0.1985\ c^2 \cdot \text{u}$$

$$= (0.1985\ c^2 \cdot \text{u})\left(\frac{931.5 \text{ MeV}}{c^2 \cdot \text{u}}\right) = 185 \text{ MeV}$$

Even if the fission is induced by a thermal neutron of negligible kinetic energy on the nuclear scale, a tremendous amount of energy is released.

Thermal Neutron Fission. The result of Example 13.5 can be understood by examining the binding-energy curve of Figure 12.6. The binding energy per nucleon of ^{236}U is about 7.6 MeV. The binding energy per nucleon of the fission fragments, however, is about 8.4 MeV. Therefore because of the difference in binding energies, the energy released per nucleon is 0.8 MeV. The total energy released is 236 nucleons times 0.8 MeV/nucleon or 190 MeV. This rough calculation of the binding-energy difference agrees with the Q-value determination of Example 13.5.

We can use the liquid drop model to understand how fission can be induced so easily in heavy nuclei. When ^{235}U absorbs a thermal neutron to form ^{236}U*, the excitation energy of ^{236}U* is 6.5 MeV. This nucleus, in a highly excited and unstable state, is agitated and becomes deformed as shown in Figure 13.10. Finally, it becomes so deformed that the Coulomb force overcomes the nuclear force, and the nucleus separates—much like a liquid drop.

Experiment shows that ^{235}U and ^{239}Pu fission easily after absorbing thermal neutrons (very-low energy neutrons). The nuclide ^{238}U needs a neutron of at least 1 MeV kinetic energy to easily fission. If lower-energy neutrons are absorbed by ^{238}U, the resulting ^{239}U* is more likely to decay by emitting a γ ray. As we discussed previously, the cross section for heavy nuclei to absorb a low-energy neutron varies as $1/v$, so the cross section is larger for lower-energy neutrons.

Example 13.6

Calculate the excitation energy of the compound nuclei produced when ^{235}U and ^{238}U absorb thermal neutrons.

Solution: The two reactions are

$$n + {}^{235}\mathrm{U} \longrightarrow {}^{236}\mathrm{U}^*$$

$$n + {}^{238}\mathrm{U} \longrightarrow {}^{239}\mathrm{U}^*$$

As we did in Equation (13.12) of Example 13.4, we find the excitation energy from the atomic masses. A thermal neutron has a negligible kinetic energy (about 0.03 eV).

$$\begin{aligned} E({}^{236}\mathrm{U}^*) &= [m(n) + M({}^{235}\mathrm{U}) - M({}^{236}\mathrm{U})]c^2 \\ &= [1.0087\ \mathrm{u} + 235.0439\ \mathrm{u} - 236.0456\ \mathrm{u}]c^2 \\ &= (0.0070\ c^2 \cdot \mathrm{u})\left(\frac{931.5\ \mathrm{MeV}}{c^2 \cdot \mathrm{u}}\right) = 6.5\ \mathrm{MeV} \end{aligned}$$

$$\begin{aligned} E({}^{239}\mathrm{U}^*) &= [m(n) + M({}^{238}\mathrm{U}) - M({}^{239}\mathrm{U})]c^2 \\ &= [1.0087\ \mathrm{u} + 238.0508\ \mathrm{u} - 239.0543\ \mathrm{u}]c^2 \\ &= (0.0052\ c^2 \cdot \mathrm{u})\left(\frac{931.5\ \mathrm{MeV}}{c^2 \cdot \mathrm{u}}\right) = 4.8\ \mathrm{MeV} \end{aligned}$$

Thus ^{236}U* has almost 2 MeV more excitation energy than ^{239}U* when both are produced by thermal neutron absorption. This helps explain why ^{235}U more easily undergoes thermal neutron fission.

Fission fragments are highly unstable because they are so neutron-rich. This occurs because heavy nuclei deviate further and further away from the $N = Z$ line of Figure 13.11. After the fission, the resulting fission fragments are relatively further away from the line of stability, also seen in Figure 13.11. *Prompt* neutrons are emitted simultaneously with the fissioning process. Even after prompt neutrons are released, the fission fragments undergo beta decay, releasing more energy. Most of the ~200 MeV released in fission goes to the kinetic

Prompt neutrons

energy of the fission products, but the neutrons, beta particles, neutrinos, and gamma rays carry away 30–40 MeV of the kinetic energy.

Chain Reactions. Because several neutrons are produced in fission, these neutrons may subsequently produce another fission. This is the basis of the *self-sustaining chain reaction.* If slightly more than one neutron, on the average, results in another fission, the chain reaction becomes *critical.* Because a sufficient amount of mass is required to increase the chances of a neutron being absorbed, a *critical mass* of fissionable material must be present. If less than one neutron, on the average, produces another fission, the reaction is said to be *subcritical.* If more than one neutron, on the average, produces another fission, the reaction is said to be *supercritical.* An atomic bomb is an extreme example of a supercritical fission chain reaction.

A critical-mass fission reaction may be controlled by absorbing neutrons. In a nuclear reactor the average number of neutrons producing another fission is very nearly one (or slightly greater). It only takes a few microseconds for a neutron to be absorbed and cause another fission. If 1.01 neutrons were captured on the average within 5 μs of each fission, then within 50 ms the *rate* would increase by a factor of 10,000 or $(1.01)^{10,000} = 1.6 \times 10^{43}$ fissions. An energy of 3.2×10^{45} MeV would be produced within 50 ms. Obviously, the control of the reactor would be impossible. Fortunately, not all the neutrons are *prompt.* Some of the neutrons are *delayed* by several seconds and are emitted by daughter nuclides resulting from the (slow) beta decay of the fission fragments. It is these delayed neutrons that allow a nuclear reactor to be controlled by movement of *control rods* that absorb more or fewer neutrons to keep the chain reaction sustained, but under control.

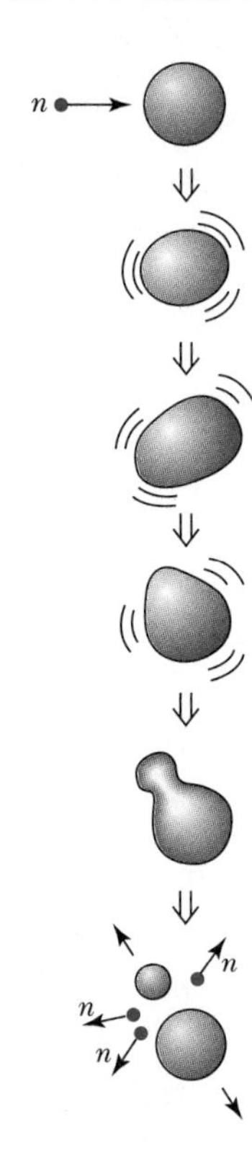

FIGURE 13.10 After absorbing a neutron, a large nucleus becomes excited and unstable. It becomes deformed, and eventually the Coulomb force causes it to fission and produce two fragments of unequal mass and, in addition, three neutrons.

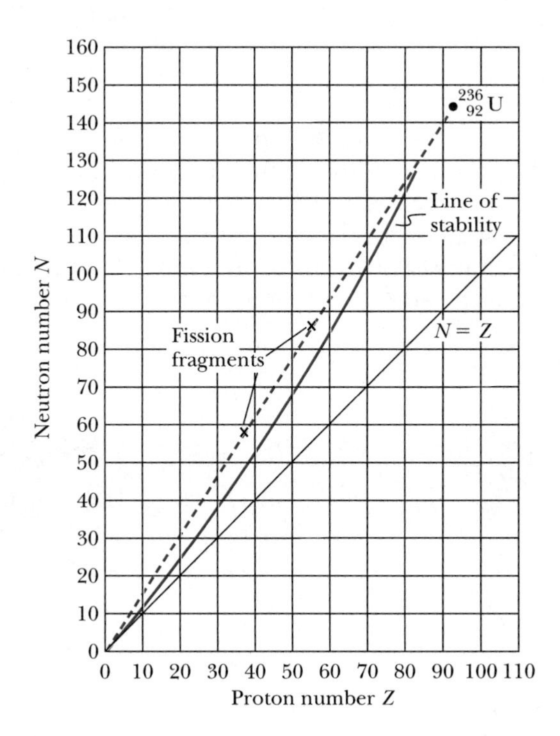

FIGURE 13.11 When ^{236}U fissions, its fragments are very neutron-rich and the nuclides are far away from the line of stability. As a result, neutrons are produced in the fission process. The fission fragments beta decay to return to the line of stability.

13.5 Fission Reactors

Because so much energy is released in nuclear fission, it is a useful energy source for commercial power production. The energy content of several fuels are shown in Table 13.1, and the fuel requirements for a 1000-megawatt (MWe, the e indicating electrical power) power plant are shown in Table 13.2.

There are several important components necessary for a controlled nuclear reactor:

Reactor components

1. Fissionable fuel.
2. Moderator to slow down neutrons.
3. Control rods for safety and to control criticality of reactor.
4. Reflector to surround moderator and fuel in order to contain neutrons and thereby improve efficiency.
5. Reactor vessel and radiation shield.
6. Energy transfer systems if commercial power is desired.

We have already learned that ^{235}U fissions with thermal neutrons. Because only $2\frac{1}{2}$ neutrons, on the average, result from each fission, it is important not to lose neutrons. There are two main effects that "poison" reactors: (1) Neutrons may be absorbed without producing fission (for example, by neutron radiative capture $^{235}\text{U}(n, \gamma)^{236}\text{U}$), and (2) neutrons may escape from the fuel zone.

In order to produce a critical mass of ^{235}U, it is necessary to process natural uranium ore to enrich the ^{235}U content (0.7%) from the more abundant ^{238}U (99%). Such enrichment is difficult because chemical techniques cannot be used. Giant uranium processing plants enrich uranium (in the form of UF_6) by a gaseous diffusion process in the United States. The molecules of ^{235}U diffuse slightly more easily than ^{238}U, and after about 1500 stages of diffusion, the more valuable ^{235}U is obtained in enriched form. A gas ultracentrifuge process is used in Europe to enrich uranium at lower production costs. Laser enrichment techniques are also being developed.

Fission neutrons may have 1–2 MeV of kinetic energy, and it helps to slow them down in order to increase the chance of producing another fission, because the cross section increases as $1/v$ at low energies. A *moderator* is used to elastically scatter the high-energy neutrons and thus reduce their energies. A neutron loses the most energy in a single collision with a light stationary particle. Hydrogen (in water), carbon (graphite), boron, and beryllium are all good moderators.

TABLE 13.1
Energy Content of Fuels

Material	Amount	Energy (J)
Coal	1 kg	3×10^7
Oil	1 barrel (0.16 m^3)	6×10^9
Natural gas	1 ft^3 (0.028 m^3)	10^6
Wood	1 kg	10^7
Gasoline	1 gallon (0.0038 m^3)	10^{10}
Uranium (fission)	1 kg	10^{14}
Deuterium (fusion)	1 kg	2×10^{14}

TABLE 13.2
Daily Fuel Requirements for 1000-MWe Power Plant

Material	Amount	
Coal	8×10^6 kg	(1 trainload/day)
Oil	40,000 barrels (6400 m^3)	(1 tanker/week)
Natural gas	2.5×10^6 ft^3 (7.1×10^4 m^3)	
Uranium	3 kg	

The graphite reactor, built in only 11 months in 1943 at Oak Ridge National Laboratory, operated until 1963 producing for many years most of the world's supply of radioisotopes for medicine, agriculture, and industry. It is the world's oldest nuclear reactor still in existence today. Note that elements could be loaded inside the reactor through the holes in the reactor side. *Courtesy of Oak Ridge National Laboratory, Oak Ridge, Tennessee.*

The simplest method to reduce the loss of neutrons escaping from the fissionable fuel is to make the fuel zone larger. The fuel elements are normally placed in regular arrays within the moderator as shown in Figure 13.12. Fission neutrons from one fuel cell will be moderated before entering an adjacent fuel cell to produce another fission. The fuel must be clumped in cells because of large $^{238}U(n, \gamma)$ cross sections at 7 and 21 eV, which would cause the neutrons to be absorbed as they slow down. The neutrons should be moderated outside the fuel cells.

The reaction rates of nuclear reactors are controlled by control rods. Cadmium is an excellent control material to absorb neutrons because of its extremely large (n, γ) cross section (see Figure 13.8). The delayed neutrons produced in fission allow the mechanical movement of the rods to control the fission reaction. A "failsafe" system automatically drops the control rods into the reactor in an emergency shutdown. If the fuel and moderator are surrounded by a material with a very low neutron capture cross section, there is a reasonable chance that after one or even many scatterings, the neutron will be backscattered or "reflected" back into the fuel area. Water is often used both as moderator and reflector.

Commercial reactors

Finally, the reactor must be contained within a secure vessel with adequate radiation shielding to protect personnel. If commercial power is desired, there must be a method to transfer energy. The most common method is to pass hot water heated by the reactor through some form of heat exchanger. In *boiling water reactors* (BWRs) the moderating water turns into steam, which drives a turbine producing electricity as shown in Figure 13.13. In *pressurized water reactors* (PWRs) the moderating water is under high pressure and circulates from the reactor to an external heat exchanger where it produces steam, which drives a turbine.

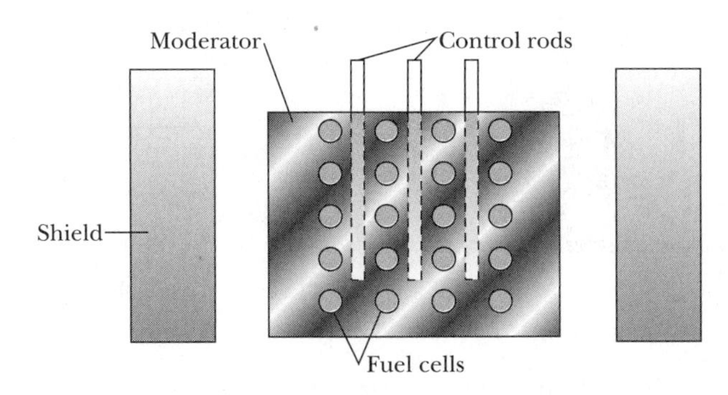

FIGURE 13.12 Cross-section schematic of an idealized nuclear reactor.

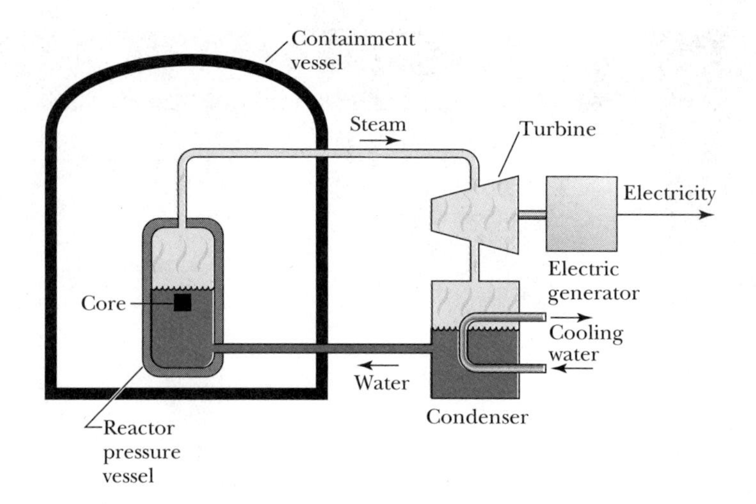

FIGURE 13.13 Schematic diagram of a boiling water nuclear reactor. Note that steam directly heated by the reactor core drives the turbine.

This two-step process is shown in Figure 13.14. Boiling water reactors are inherently simpler than pressurized water reactors. However, the possibility that the steam driving the turbine may become radioactive is greater with the BWR. The two-step process of the PWR helps to isolate the power generation system from possible radioactive contamination.

Reactors are designed and operated for different purposes. We have been describing primarily *power* reactors used for commercial electricity production. There were over 400 power reactors in operation worldwide in 1998 that produced 17% of the world's electricity. More than 100 of these reactors were in the United States, and over 50 were in France where power reactors produced 77%

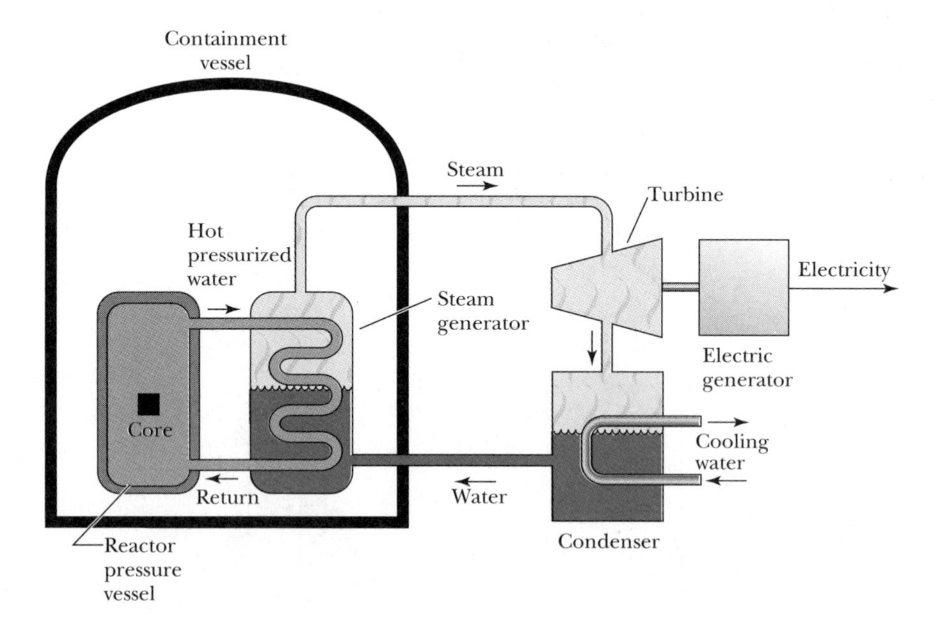

FIGURE 13.14 Schematic diagram of a pressurized water nuclear reactor. Note that the steam driving the turbine is one step removed from the water in the reactor core.

of the electricity. Sweden (50%), Russia, and South Korea (36%) are other countries heavily dependent on nuclear power for electricity, whereas in the United States a little over 20% comes from reactors. The amount of electricity from power reactors has been increasing for several years throughout the world and is expected to continue doing so for a couple of decades before decreasing. This decrease will occur, especially in the United States, because existing nuclear reactors are not being replaced as they age due to the public's fear of nuclear radiation.

Research reactors are operated to produce high neutron fluxes for research such as neutron-scattering experiments. Smaller reactors, on the order of 100 MW, are operating in some colder countries like Russia for *heat production* to warm both homes and businesses. Other reactors are designed to produce *radioisotopes* for industrial and medical purposes. Several *training* reactors are located on college campuses.

Reactor Problems. Problems certainly exist with nuclear power plants. The danger of a serious accident in which radioactive elements are released into the atmosphere or groundwater is of great concern to the general public. A more serious problem is the safe disposal of the radioactive wastes produced in the fissioning process, because some fission fragments have half-lives of thousands of years. The proliferation of fissionable fuel, even from used fuel elements, to countries capable of atomic weapon production is of great concern. Thermal pollution both in the atmosphere and in lakes and rivers used for cooling may be a significant ecological problem and is under close scrutiny.

Two widely publicized accidents—one at Three Mile Island in Pennsylvania in 1979, the other at Chernobyl in Ukraine in 1986—have significantly dampened the general public's support. The number of deaths as a result of the Chernobyl accident has been officially reported to be 32, but it has been estimated that the toll could reach 10,000 from cancer and other radiation-related diseases. Despite these two accidents, the safety record of nuclear power is generally accepted as being better than most other methods of power generation except natural gas, of which there is a limited supply. Burning of fossil fuels causes air pollution and greenhouse-effect gases.

Breeder Reactors

A promising choice for a more advanced kind of reactor is the *breeder* reactor, which produces more fissionable fuel than it consumes. When ^{238}U undergoes the (n, γ) reaction, the resulting ^{239}U beta decays to ^{239}Np and subsequently to ^{239}Pu in the chain

$$\begin{aligned} n + {}^{238}_{92}\mathrm{U} \longrightarrow {}^{239}_{92}\mathrm{U}^* \longrightarrow \gamma + {}^{239}_{92}\mathrm{U} & \\ \longrightarrow \beta^- + {}^{239}_{93}\mathrm{Np} + \bar{\nu} & \\ \longrightarrow \beta^- + {}^{239}_{94}\mathrm{Pu} + \bar{\nu} & \qquad (13.15) \end{aligned}$$

The resulting ^{239}Pu nucleus has a half-life of 24,100 years and is an α emitter. The intermediate beta decays take place in a matter of days. The plutonium is easily separated from uranium by chemical means, and ^{239}Pu fissions easily with both thermal and fast neutrons. Because ^{239}Pu fission produces an average of 2.7 neutrons (compared with 2.3 for ^{235}U), it is a highly desirable fuel for both reactors and bombs. The critical mass can be made smaller for ^{239}Pu because

EARLY FISSION REACTORS

In the late 1930s European scientists were busy studying the properties of the heaviest nuclei. Several groups, including Enrico Fermi and his collaborators in Italy, Curie and Savitch in France, and O. Hahn and F. Strassman in Germany, were bombarding uranium with neutrons. In 1938 Hahn and Strassman observed that barium ($Z = 56$) had been formed from their neutron bombardment of uranium ($Z = 92$). The Austrian-born Lise Meitner was a well-known physicist who had spent almost her entire career in Germany before fleeing to the Nobel Institute in Stockholm in 1938 where she joined her nephew O. R. Frisch. Meitner received word from Hahn of his experimental results during the turmoil leading up to World War II. Meitner and Frisch were the first to report in 1939 the correct analysis of the Hahn and Strassman experiment as being due to the fission of uranium.

By 1939 Fermi had fled Italy, again due to the persecution of some scientists, and was at Columbia University in New York. Bohr brought word of the Hahn and Strassman experiment, and the resulting excitement caused a flurry of experiments in the United States as well as at nuclear laboratories throughout the world. Fermi was one of the best nuclear experimental physicists of the period, but he also had a strong grasp of theoretical physics. The Hungarian-born Leo Szilard, working with Fermi, encouraged Einstein to write President Franklin Roosevelt and encourage the effort that was to lead to the Manhattan Project of World War II.

Bohr and John Wheeler showed theoretically that the nuclide ^{235}U, and not the more abundant ^{238}U, underwent fission more readily. The fission of ^{235}U was known to occur with slow neutrons, so a process to slow down the neutrons produced in nuclear fission was needed in order to produce the chain reaction. The effort toward producing the first controlled chain reaction moved to the University of Chicago, where, in a former squash court under the stands of the football stadium (Stagg Field), the team led by Fermi succeeded on December 2, 1942. The race toward producing the atomic bomb developed quickly. Eventually, in 1955, the first electricity for public use was generated by a nuclear reactor in Idaho. The *Nautilus*, the world's first nuclear submarine, was launched in 1954, and the world's first large-scale nuclear reactor began producing electricity in England in 1956. The nuclear age was underway.

However, there is some evidence that a natural fission reactor occurred in the Republic of Gabon on the west coast of Africa almost two billion years ago. The present natural abundance of ^{235}U is 0.7% and of ^{238}U 99.3%. It is believed that the relative abundances were about equal when the Earth was formed, but more of the ^{235}U has decayed because of its shorter half-life. However, two billion years ago the natural abundance of ^{235}U was 3.7%, and calculations indicate a critical mass of ^{235}U could have been present in natural uranium ore.

The groundwater or possibly water flooding served as a moderator to slow down the fission neutrons. When the natural reactor became too hot, the water probably boiled and the neutrons were not slowed down, thus reducing the probability of causing another fission. This served as an automatic control much as the water moderator might do in a reactor today. The isotopic abundances of the fission products found in the uranium ore in Gabon (see Figure A) lead to the conclusion that for several hundred thousand years a natural fission reactor may have existed.

Scientists believe that at least six reactor zones exist in the Oklo mine of Gabon and that conditions probably existed elsewhere for similar natural fission reactors. Such an event is not possible today because of the low percentage of ^{235}U in natural uranium ores.

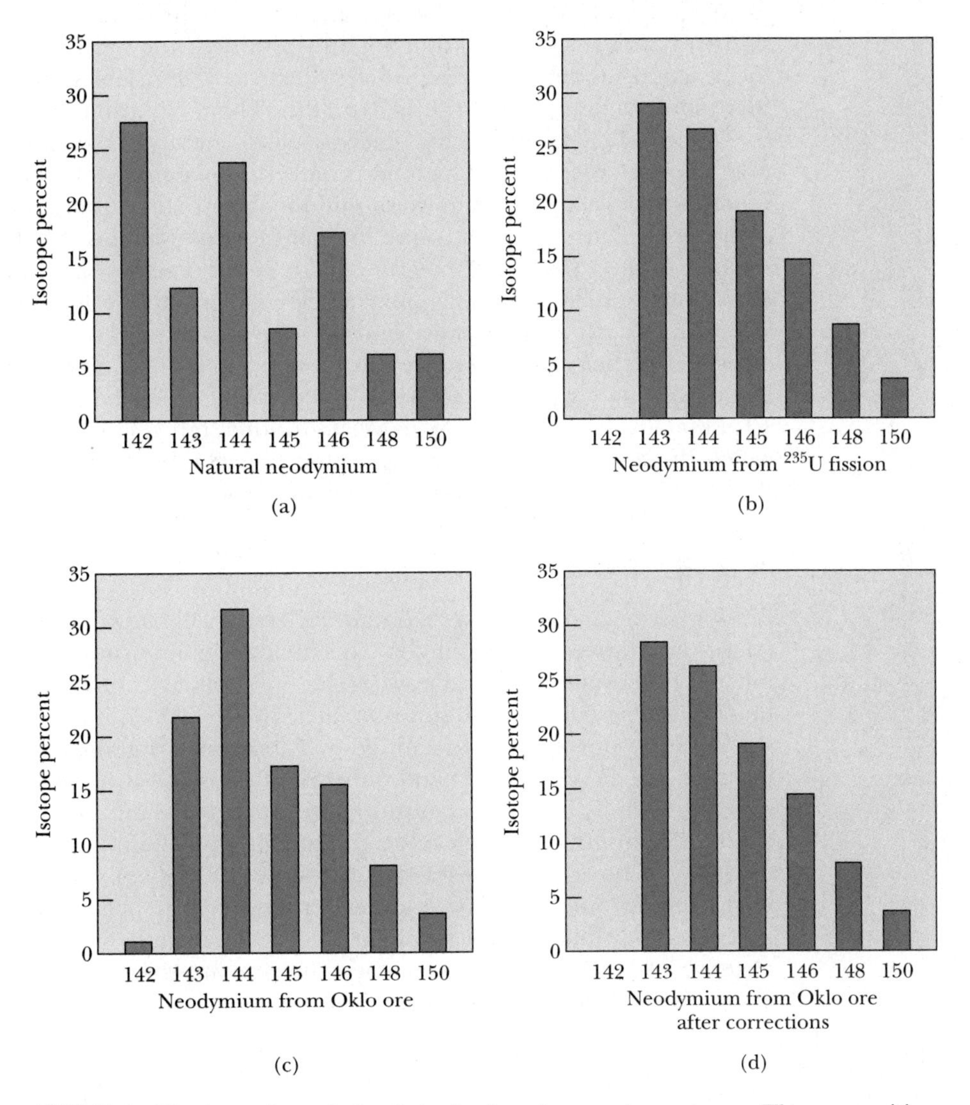

FIGURE A The isotopic analysis of neodymium from various sources. The composition of neodymium from $n + {}^{235}U$ fission in (b) is characteristically different from that of natural neodymium in (a). Corrections have to be made on the composition of neodymium from the Oklo mines in (c) due to the natural neodymium present and to changes due to neutron absorption. The resulting analysis shown in (d) is very similar to (b); this shows a fission reaction may have taken place in the ore. *From George A. Cowan, Scientific American* ***235****, 36 (July 1976); Courtesy of Jerome Kuhl.*

1.7 of the neutrons may be lost on the average, rather than 1.3 for fissioning of ^{235}U.

Fast breeder reactors have been built that convert ^{238}U to ^{239}Pu. The reactors are designed to use fast neutrons. If the loss of neutrons is kept small, we occasionally have a fission event that not only produces another fission event to keep the reaction sustained but also converts ^{238}U to ^{239}Pu. The amount of ^{239}Pu will then slowly build up (or, we "breed" ^{239}Pu). This ^{239}Pu can eventually be used as fuel for another power reactor. There is 99.3% natural ^{238}U and only 0.7% natural ^{235}U, so breeder reactors hold promise as an almost unlimited supply of fissionable material. To be fair, we should mention that plutonium is highly toxic, and there is concern about its use in unauthorized weapons production.

The United States had experimental breeder reactors operational as early as 1951, but in 1984 the government decided to postpone indefinitely the construction of a prototype power plant because of safety concerns and new projections of electricity usage. France has moved much more quickly in fast breeder technology and started a 233-MWe power plant (the Phénix) in operation in 1974 that was still in operation in 1999. Another, much larger plant (1200-MWe) called the Superphénix began operation in 1985, but was shut down in 1998. Both Japan and Russia operate fast breeder reactors for power production.

13.6 Fusion

Except for nuclear fission and geothermal power, all known terrestrial sources of energy are derived from sunlight. This includes combustion of wood, coal, gas, oil, and both water and wind power. The only primary source in widespread use that is *not* derived from the sun is nuclear power (which incidentally is also the sun's energy source). Energy emitted by stars arises from *nuclear fusion* reactions, in which the enormous heat and pressure of the star's core cause light nuclei to fuse together. This process contrasts with nuclear fission, in which large nuclei divide. The origin of fusion can be understood by examining Figure 12.6. Nuclei near $A = 56$ have the highest binding energy per nucleon. When ^{236}U* fissions, it divides into nuclei having a larger binding energy per nucleon, thereby releasing energy. Similarly, if two light nuclei fuse together, they also form a nucleus with a larger binding energy per nucleon and energy is released. The most energy is released if two isotopes of hydrogen, ^{2}H and ^{3}H, fuse together in the reaction,

$$^2\text{H} + {}^3\text{H} \rightarrow n + {}^4\text{He} \qquad Q = 17.6 \text{ MeV} \tag{13.16}$$

About 3.5 MeV per nucleon is released because of the strong binding of ^{4}He. Less than 1 MeV per nucleon is released in fission (200 MeV/236 nucleons = 0.85 MeV/nucleon). The lower-mass side of Figure 12.6 is much steeper than the higher-mass side, and this leads to the nuclear fusion process being a prolific source of energy.

Formation of Elements

When the primordial "Big Bang" occurred 10–15 billion years ago, the light elements of hydrogen and helium were formed in the first few minutes. It was millions of years later before the heavier elements were formed in stars through nuclear fusion. These extremely interesting phenomena will be examined further in Chapter 16, but now we want to study two of the main cycles for producing energy in stars.

Proton–proton chain

The first is the **proton–proton chain,** which includes a series of reactions that eventually converts four protons into an alpha particle. As stars form due to gravitational attraction of interstellar matter, the heat produced by the attraction is enough to cause protons to overcome their Coulomb repulsion and fuse by the following reaction:

$$^{1}\text{H} + {}^{1}\text{H} \rightarrow {}^{2}\text{H} + \beta^{+} + \nu \qquad Q = 0.42 \text{ MeV} \tag{13.17}$$

This reaction produces ^{2}H and is a special kind of weak interaction beta decay process. It is extremely slow, because only 1 collision in about 10^{26} produces a reaction. This is good, because otherwise the sun would explode! The deuterons are then able to combine with ^{1}H to produce ^{3}He:

$$^{2}\text{H} + {}^{1}\text{H} \rightarrow {}^{3}\text{He} + \gamma \qquad Q = 5.49 \text{ MeV} \tag{13.18}$$

The ^{3}He are then able to produce ^{4}He:

$$^{3}\text{He} + {}^{3}\text{He} \rightarrow {}^{4}\text{He} + {}^{1}\text{H} + {}^{1}\text{H} \qquad Q = 12.86 \text{ MeV} \tag{13.19}$$

Note that two each of reactions (13.17) and (13.18) must occur to produce (13.19). A total of six ^{1}H are required to produce ^{4}He and two ^{1}H. This process consumes four protons. The total Q value, or kinetic energy produced, for the six ^{1}H to produce ^{4}He is 24.7 MeV. An additional 2 MeV is derived from the annihilation of the two positrons for a total of 26.7 MeV.

The proton–proton chain is particularly slow because reaction (13.17) limits the entire process. As the reaction proceeds, however, the star's temperature increases, and eventually ^{12}C nuclei are formed, for example, by a process that converts three ^{4}He into ^{12}C.

Carbon cycle

Another cycle due to carbon is also able to produce ^{4}He. The series of reactions responsible for the **carbon** or **CNO** cycle are

$$\begin{aligned}
&^{1}\text{H} + {}^{12}\text{C} \rightarrow {}^{13}\text{N} + \gamma \\
&\qquad \hookrightarrow {}^{13}\text{C} + \beta^{+} + \nu \qquad t_{1/2} = 9.96 \text{ min} \\
&^{1}\text{H} + {}^{13}\text{C} \rightarrow {}^{14}\text{N} + \gamma \\
&^{1}\text{H} + {}^{14}\text{N} \rightarrow {}^{15}\text{O} + \gamma \\
&\qquad \hookrightarrow {}^{15}\text{N} + \beta^{+} + \nu \qquad t_{1/2} = 2.04 \text{ min} \\
&^{1}\text{H} + {}^{15}\text{N} \rightarrow {}^{12}\text{C} + {}^{4}\text{He}
\end{aligned} \tag{13.20}$$

Notice that four ^{1}H and one ^{12}C nuclei are required to produce ^{4}He and ^{12}C; the ^{12}C nucleus merely serves as a catalyst. We believe the proton–proton chain is probably responsible for most of our sun's energy, but the carbon cycle is a much more rapid fusion reaction. It requires higher temperatures (perhaps 20×10^{6} K) than are present in the sun, because of the higher Coulomb barrier of ^{12}C relative to ^{1}H for the protons.

A hydrostatic equilibrium exists in the sun between the gravitational attraction tending to contract a star and a gas pressure pushing out due to all the particles. As the lighter nuclides are "burned up" to produce the heavier nuclides, the gravitational attraction succeeds in contracting the star's mass into a smaller volume and the temperature increases. A higher temperature allows the higher-Z nuclides to fuse. This process continues in a star until a large part of the star's mass is converted to iron. The star then collapses under its own gravitational attraction to become, depending on its mass, a white dwarf star, neutron star, or black hole; or it may even undergo a supernova explosion (see Chapter 16).

Nuclear Fusion on Earth

Many scientists believe that controlled nuclear fusion ultimately represents our best source of terrestrial energy. Among the several possible fusion reactions, three of the simplest involve the three isotopes of hydrogen.

$$^2\text{H} + {}^2\text{H} \rightarrow n + {}^3\text{He} \qquad Q = 3.3 \text{ MeV} \tag{13.21}$$

$$^2\text{H} + {}^2\text{H} \rightarrow p + {}^3\text{H} \qquad Q = 4.0 \text{ MeV} \tag{13.22}$$

$$^2\text{H} + {}^3\text{H} \rightarrow n + {}^4\text{He} \qquad Q = 17.6 \text{ MeV} \tag{13.23}$$

Deuterium exists in vast quantities in seawater. If we estimate there are 10^{21} liters of water on Earth, the natural abundance of deuterium (0.015%) gives 10^{43} deuterons. These deuterons, when fused together in reaction (13.21), would produce over 10^{30} J of energy, enough to support present world energy consumption for a few billion years.

There are three main conditions necessary for controlled nuclear fusion:

1. The temperature must be hot enough to allow the ions, for example, deuterium and tritium, to overcome the Coulomb barrier and fuse their nuclei together. This requires a temperature of 100–200 million K.
2. The ions have to be confined together in close proximity to allow the ions to fuse. A suitable ion density is $2\text{–}3 \times 10^{20}$ ions/m^3.
3. The ions must be held together in close proximity at high temperature long enough to avoid plasma cooling. A suitable time is 1–2 s.

The suitable values given above assume magnetic confinement, which will be discussed soon. In order to reach the breakeven point of producing as much energy as input to produce fusion, the product of the plasma density n and the containment time τ must be

Lawson criteria

$$n\tau \geq 3 \times 10^{20} \text{ s/m}^3 \tag{13.24}$$

at sufficiently high temperature to initiate fusion. This relation is called the *Lawson criterion* after the British physicist J. D. Lawson who first derived it in 1957. A triple product of $n\tau T$ called the *Fusion Product* is sometimes used (where T is the ion temperature).

Fusion Product

$$n\tau T \geq 6 \times 10^{28} \text{ s} \cdot \text{K/m}^3 \quad \text{or} \quad 6 \times 10^{21} \text{ s} \cdot \text{keV/m}^3 \tag{13.25}$$

The factor Q is used to represent the ratio of the power produced in the fusion reaction to the power required to produce the fusion (heat). This Q factor is not to be confused with the Q value. Breakeven is for $Q = 1$, and ignition occurs for $Q = \infty$. For controlled fusion produced in the laboratory, temperatures on the order of 20 keV are satisfactory. For uncontrolled fusion ($Q = \infty$), as in the hydrogen bomb or "H-bomb," high temperatures and densities are achieved over a very brief time by using an atomic bomb as a trigger. Unfortunately, there are enough thermonuclear warheads (H-bombs) to destroy most of the life on Earth. Some scientists have predicted that even a limited nuclear war could produce so much dust in the Earth's atmosphere that sunlight would be partially blocked. This might lead to a "nuclear autumn" (or even "winter"), with a drop in temperature of up to 30°C that could last for several months. Fusion bombs do not produce nearly as severe radiation effects as do fission bombs, because the primary products (n, p, and ^{4}He) are not dangerously radioactive.

Example 13.7

Calculate the ignition temperature needed for the reaction (13.23).

Solution: Enough energy is needed to overcome the Coulomb barrier. The charges of the participants in reaction (13.23) are both $+e$. We will use 3 fm as the distance where the nuclear force first becomes effective. The particles must approach each other to at least this distance. The Coulomb potential energy that must be overcome is

$$V = \frac{q_1 q_2}{4\pi\epsilon_0 r}$$

$$= \frac{(9 \times 10^9\ \mathrm{N \cdot m^2/C^2})(1.6 \times 10^{-19}\ \mathrm{C})^2}{3 \times 10^{-15}\ \mathrm{m}} = 7.7 \times 10^{-14}\ \mathrm{J}$$

The thermal energy required is equal to $\frac{3}{2}kT$, so we have for the ignition temperature,

$$T = \frac{2V}{3k} = \frac{2(7.7 \times 10^{-14}\ \mathrm{J})}{3(1.38 \times 10^{-23}\ \mathrm{J/K})} = 3.7 \times 10^9\ \mathrm{K}$$

A temperature of almost 4 billion K (or °C) is needed to ignite the D + T reaction. This is an overestimate for several reasons. First, a deuteron and triton are extended objects, and their nucleons will probably feel an attraction before the centers of the nuclei are 3 fm apart. Second, the protons in D + T will tend to repel each other as the two nuclides approach, compared to the behavior of the neutrons. A more appropriate distance to use in the Coulomb potential energy could be a distance as great as 5 fm, which would result in a lower temperature. Third (and most important), the distribution of energies for a plasma (ionized particles) in thermal equilibrium at temperature T follows a statistical process. We have used the mean energy, $\frac{3}{2}kT$. However, far out on the tail of the distribution (see Figure 9.7), there are many particles with energies several times greater than $\frac{3}{2}kT$. It only takes a few particles out of the total of 10^{20}–10^{22} particles/m^3 to initiate the reaction. Fourth, we have assumed that each particle needs $\frac{3}{2}kT$ energy. If the collision were head-on, then each of the D and T ions would need only half this energy. More accurate ignition temperature estimates for the D + T fusion reaction are in the range of 100–200 million K, which seems a reasonable correction to our original estimate.

Controlled Thermonuclear Reactions

A controlled thermonuclear reaction of nuclear fusion in the laboratory is one of the primary goals of science and engineering. This effort should continue for several more decades. Because of the large amount of energy produced and the relatively small Coulomb barrier, the first fusion reaction will most likely be the D + T reaction, Equation (13.23). The tritium will be derived from two possible reactions:

$$n + {}^6\mathrm{Li} \rightarrow {}^3\mathrm{H} + {}^4\mathrm{He} \tag{13.26}$$

$$n + {}^7\mathrm{Li} \rightarrow {}^3\mathrm{H} + {}^4\mathrm{He} + n \tag{13.27}$$

The lithium is required to generate the tritium and is also used as the heat transfer medium and a neutron radiation shield.

For $Q = 1$ (breakeven), a product of $n\tau T$ of a few times $10^{21}\ \mathrm{keV \cdot s \cdot m^{-3}}$ will be required for a commercial reactor using D + T. The problem of controlled fusion involves significant scientific and engineering difficulties. The two major methods of controlled thermonuclear reactions are magnetic and inertial confinement. We will describe the main ideas of each.

Magnetic Confinement of Plasma. The primary effort of research laboratories around the world for several years has been a device called the **tokamak,** first developed in the USSR in the 1960s. Several other magnetic confinement

Tokamak

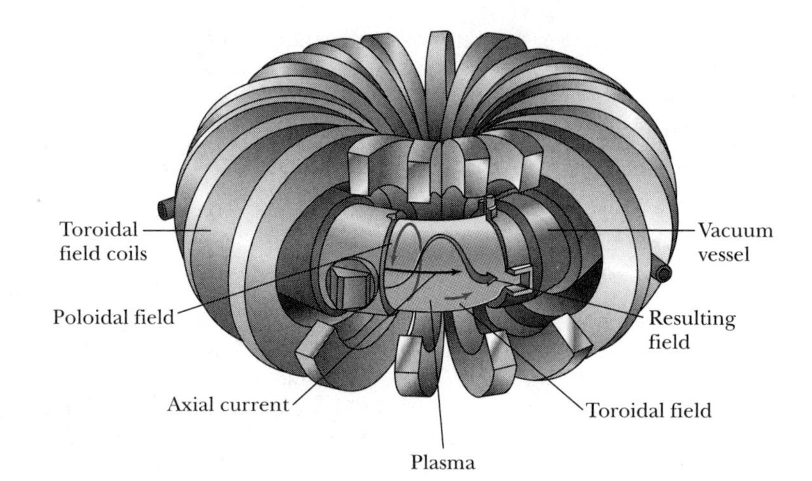

FIGURE 13.15 Diagram of a toroidal fusion device, the tokamak. The most important magnetic fields are the toroidal and polodial fields to contain the plasma inside the vacuum vessel.

schemes have been tried. A schematic diagram of a tokamak is shown in Figure 13.15. As many as six separate magnetic fields may be used to contain and heat the plasma.

A cross section of a typical magnetic containment vessel for a tokamak that might eventually be used as a commercial device is shown in Figure 13.16. The center is the location of the hot plasma where the D + T reaction takes place. The plasma is surrounded by vacuum to keep out impurities, which would poison the reaction. A wall, which is subjected to intense radiation, surrounds the plasma and vacuum region. The plasma must be kept from touching the first wall (which is subjected to the most hostile environment of any device yet designed). The next layer is a lithium blanket, which absorbs neutrons to breed more tritium. Next comes the radiation shield to prevent radiation from reaching the magnets, which may be superconducting in a commercial reactor. The lithium blanket and radiation shield are not present in existing test machines.

The heating of the plasma to sufficiently high temperatures is initially accomplished by the resistive heating from the electric current flowing in the plasma. Because this is insufficient to attain the high ignition temperature, there are two other schemes to add additional heat: (1) injection of high-energy (40–120 keV) neutral (so they pass through the magnetic field) fuel atoms that interact with the plasma, and (2) radio-frequency (RF) induction heating of the plasma (similar to a microwave oven).

The best results to date have come from the Princeton University Tokamak Fusion Test Reactor (TFTR) in the United States (shut down in 1997) and the Joint European Torus (JET) in England. Both fusion reactors have used tritium

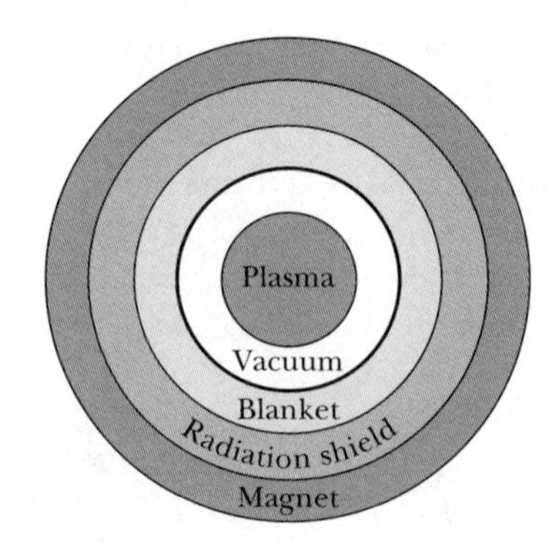

FIGURE 13.16 Schematic cross section of a typical magnetic confinement vessel (a tokamak) that might be used for commercial power production. Heat absorbed in the blanket is used to produce electricity.

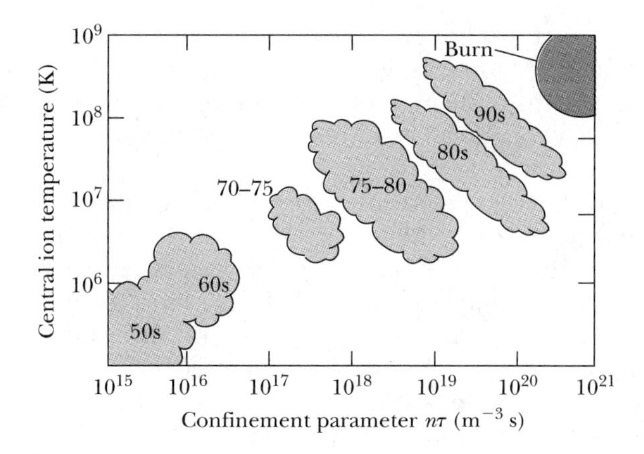

FIGURE 13.17 Plot of plasma ion temperature versus the Lawson confinement parameter $n\tau$. Note the gradual progress made since the 1950s.

as fuel. In 1997 JET reached a power output of 2.1×10^7 J and had a Q-value of 0.65, the highest to date. We show in Figure 13.17 a plot of plasma ion temperature versus the confinement parameter $n\tau$. During the past few decades significant progress has been made in reaching the elusive $Q = 1$ and higher needed for power production.

Plans are now uncertain regarding the next large magnetic fusion reactor, but the International Thermonuclear Experimental Reactor (ITER) has been in planning for several years by its four partners: Europe, Japan, Russia, and the United States. The ITER was expected to assess the engineering requirements of a full-scale fusion reactor and might have taken the form shown in Figure 13.18.

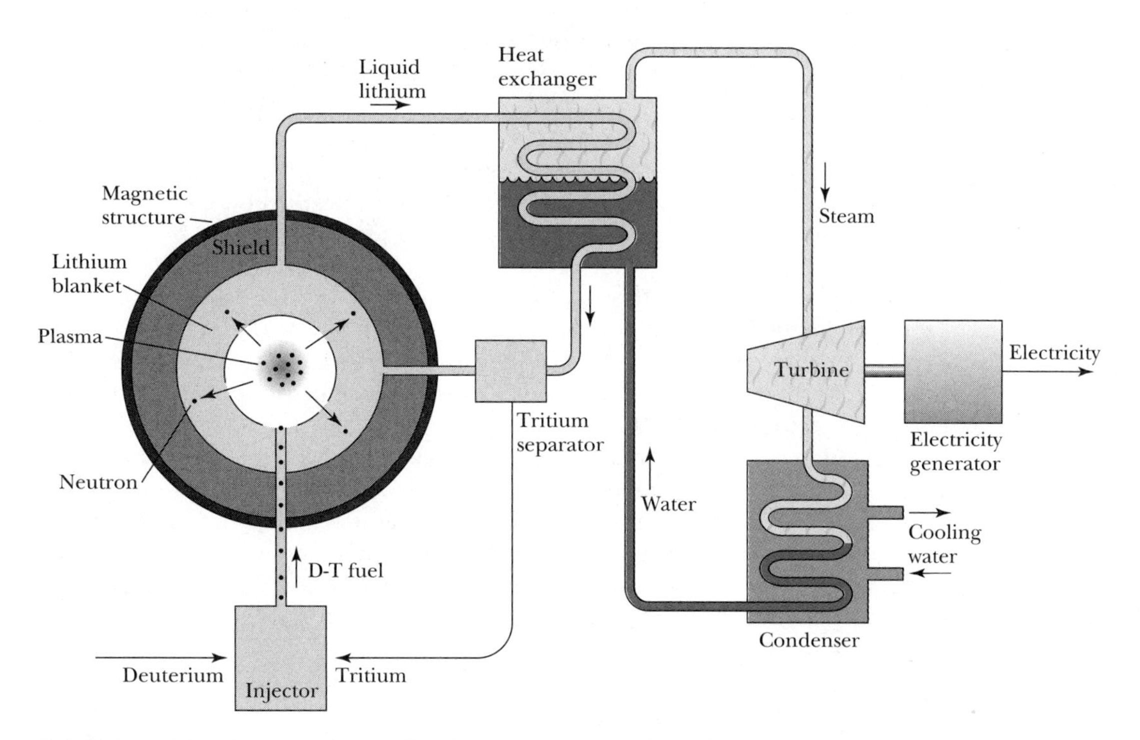

FIGURE 13.18 Diagram of a possible fusion reactor configuration using magnetic confinement. The tritium is obtained from neutron reactions with lithium, and the tritium must be separated before going back to the fusion reactor as fuel. As in fission reactors the heat produced by fusion is used to run a turbine, which in turn generates electrical power.

The high cost of about $10 billion has the four partners considering smaller alternative possibilities. A commercial power station is not envisioned until the middle of the 21st century. Meanwhile, a new device at Princeton Plasma Physics Lab will generate data on plasma confinement and stability.

Inertial Confinement. The concept of inertial confinement fusion is to use an intense high-powered beam of heavy ions or light (laser) called a *driver* to implode a pea-sized target (a few mm in diameter) composed of D + T to a density and temperature high enough to cause fusion ignition. Some inertial confinement results are included in Figure 13.17. Several institutions in the United States are doing research and development in laser fusion including Lawrence Livermore National Laboratory and the University of Rochester. The National Ignition Facility at Livermore will use 192 lasers to create a thermonuclear burn for research purposes. Sandia National Laboratories has used a device called a *Z-pinch* that uses a huge jolt of current to create a powerful magnetic field that squeezes ions into implosion and heats the plasma (see book cover). They have proposed an upgrade that may be a serious contender in the fusion race.

Z-pinch

13.7 Special Applications

In this chapter and the previous one we have mentioned several applications of nuclear science. There are many applications of particle beams, radioactive nuclides, and nuclear effects. Most of these applications depend on a specific isotope of a radioactive element called a **radioisotope.** The usefulness of a given radioisotope may depend on the specific decay particle it produces, for example, an α particle, β ray, γ ray, or fission fragment, and even on the half-life of the radioisotope. Radioisotopes are produced for useful purposes by different methods:

Radioisotopes

1. By particle accelerators as reaction products.
2. In nuclear reactors as fission fragments or decay products.
3. In nuclear reactors using neutron activation.

An important accelerator-produced radioisotope is ^{67}Ga, which is useful in diagnostic medicine. An important fission fragment is ^{99}Mo, which beta decays to ^{99m}Tc, probably the most useful radioisotope in medicine. Neutron activation is used in a whole host of applications from crime detection to the production of nuclides heavier than ^{238}U.

One other important area of applications is the search for a very small concentration of a particular element, called a *trace* element. By irradiating an object suspected of containing a trace element, a radioactive nucleus may be produced that can be inspected for its particular decay, and even for the γ rays produced by excited states of the trace element. Such techniques have been invaluable in detecting minute quantities of trace elements.

Medicine

Radioisotopes are useful in medical research, diagnostics, and treatment. They have been used to study virtually every organ and tissue in the body. The β^- emitters ^{3}H, ^{14}C, and ^{32}P are all widely used in medical research. Over 1100 radioisotopes are available for clinical use. By far the most widely used is ^{99m}Tc, an isomer of technetium with a $t_{1/2} = 6$ h that produces a 140-keV gamma ray when it decays to the ground state of ^{99}Tc. A patient is given an intravenous injection of ^{99m}Tc, for example, TcO_4^-. The ^{99m}Tc is trapped by cancerous cells, for example, the thyroid or salivary glands, the choroid plexus of the brain, or the gastric mucosa. A few minutes after the injection, the patient is scanned by an array of NaI γ-ray detectors (see Chapter 14) that are able to pinpoint the ^{99m}Tc activity. This is only one of several applications for ^{99m}Tc.

The nuclide ^{99}Mo is produced as a fission product in a nuclear reactor or as a *byproduct material* in a nuclear reactor formed by neutron activation of the stable isotope ^{98}Mo in the reaction $^{98}\text{Mo}(n, \gamma)^{99}\text{Mo}$. The latter method is particularly easy because the ^{98}Mo can be inserted encapsulated into the reactor. After retrieval, some of the material has been converted to ^{99}Mo. After separating out the radioisotope ^{99}Mo, it is shipped typically once a week by commercial companies producing radioisotopes to hospitals all over the country as a "^{99m}Tc generator" of about 10^{11} Bq. The ^{99}Mo beta decays to form the useful ^{99m}Tc.

$$^{99}_{42}\text{Mo} \rightarrow \beta^- + {}^{99m}_{43}\text{Tc} \qquad t_{1/2} = 66 \text{ h} \tag{13.28}$$

$$\hookrightarrow \gamma + {}^{99}_{43}\text{Tc} \qquad t_{1/2} = 6 \text{ h}, E_\gamma = 140 \text{ keV}$$

The radionuclide ^{131}I (iodine) ($t_{1/2} = 8$ d) became the first radionuclide used in nuclear medicine soon after its discovery by Glenn Seaborg in 1937. It has had widespread use for both diagnostic and therapeutic applications in nuclear medicine. Its diagnostic use is similar to ^{123}I, and it has been used therapeutically to treat hyperthyroidism and thyroid cancer.

The radionuclide ^{123}I ($t_{1/2} = 13$ h) is useful for a thyroid function test. ^{123}I is produced by an accelerator in a reaction such as

$$^{121}_{51}\text{Sb}(\alpha, 2n)^{123}_{53}\text{I} \tag{13.29}$$

After being administered to the patient, ^{123}I decays by electron capture to excited states of $^{123}_{52}$Te.

$$^{123}_{53}\text{I} + e^- \rightarrow {}^{123}_{52}\text{Te}^* \tag{13.30}$$

$$\hookrightarrow \gamma + {}^{123}_{52}\text{Te}$$

About 98% of the time a 159-keV γ ray is produced, which is detected by an array of NaI detectors.

The radionuclide $^{67}_{31}$Ga is useful as a tumor-localizing agent for Hodgkin's disease. The nuclide ^{67}Ga electron captures to $^{67}_{30}$Zn*, which then gamma decays with energies of 93 keV, 185 keV, 300 keV, and 394 keV, among others.

Other useful radioisotopes are ^{153}Gd for detection of bone mineral loss or osteoporosis in elderly people; ^{192}Ir for industrial radiography of welds in steel, oil well rigs, and pipelines; ^{241}Am for oil exploration and smoke alarms; and ^{85}Kr for leak testing of sealed electrical components such as transistors.

Tomography

Radioisotopes are also used in **tomography,** a technique for displaying images of practically any part of the body used to look for abnormal physical shapes or for testing functional characteristics of organs. By using detectors (either surrounding the body or rotating around the body) together with computers, three-dimensional images of the body can be obtained. Tomography now includes various techniques—including Single-Photon Emission Computed Tomography (SPECT), Positron Emission Tomography (PET, see Section 3.9), and Magnetic Resonance Imaging (MRI, see Section 10.6), among others. In the first two techniques the radiation (normally γ rays) is detected externally to place the location of the radionuclide inside the body. The use of very-short-lived *radiopharmaceuti-*

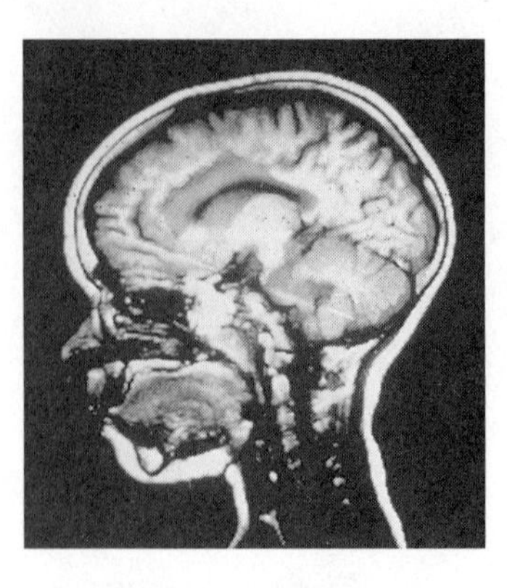

FIGURE 13.19 Photo of brain scan using PET and MRI combined. *Courtesy of C. Pelizzari, University of Chicago Radiation Oncology.*

cals, compounds produced with the radioisotopes, have resulted in about 50 accelerators placed in medical centers around the country to produce the necessary radioisotopes. A PET scan used in conjunction with a noninvasive MRI of the brain is shown in Figure 13.19.

Archaeology

The use of ^{14}C in radioactive dating has already been discussed in Chapter 12. Investigators can now measure a large number of trace elements in many ancient specimens and then compare the results with the concentrations of components having the same origin. Materials may include glass, metal, pottery, stone, minerals, paper, and fabric. This technique has been used to examine pottery fragments excavated from the ancient Agora in Athens.

Probably the greatest controversy of the New World archaeology is the question of who were the first settlers and when did they come to the Americas. Many experts believe they crossed a land bridge over the Bering Strait from Siberia to Alaska. In 1932, archaeologists found distinctive fluted spear points at a site near Clovis, New Mexico, that indicate, from ^{14}C radioactive dating, humans had a settlement there 12,000 years ago. Several claims have surfaced in the past few years, especially from South America, that dispute this earliest finding, but no conclusive proof has been confirmed. The controversy rages on.

The Chauvet Cave, discovered in France in 1995, is one of the most important archaeology finds in decades. More than 300 paintings and engravings and many traces of human activity, including hearths, flintstones, and footprints, were found. These works are believed, from ^{14}C radioactive dating, to be from the Paleolithic era, some 32,000 years ago.

Art

Neutron activation

Neutron activation is a nondestructive technique that is becoming more widely used to examine oil paintings. A thermal neutron beam from a nuclear reactor is spread broadly and evenly over the painting. Several elements within the painting become radioactive. X-ray films sensitive to beta emissions from the radioactive nuclei are subsequently placed next to the painting for varying lengths of time. For example, this method, called an *autoradiograph,* has been used to examine Van Dyck's *Saint Rosalie Interceding for the Plague-Stricken of Palermo,* from the New York Metropolitan Museum of Art collection. A remarkable find shows an over-painted self-portrait of Van Dyck himself (see Figure 13.20). Art historians are able to see modern repairs of the painting as well as the underdrawings of the original figures in the painting.

Neutron autoradiography may also be used to establish art fraud, for example, to inspect the painter's signature. Neutron activation may also be used to compare trace elements of the various paints. The composition can be quite characteristic for a given painter because of his geographical location and particular methods of producing colors.

Charged particle beams have also been used in ways similar to neutron activation because of different sensitivities for some elements. For example, they have been used to study ancient coins, particularly to measure their gold or silver content. A large fraction of copper in coins covered by a layer of gold or silver speaks of counterfeiting or even difficult times for a monarch's reign.

(a)

(b)

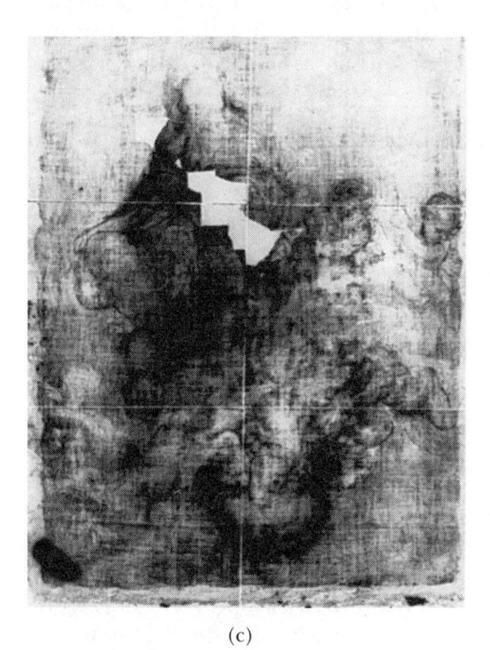

(c)

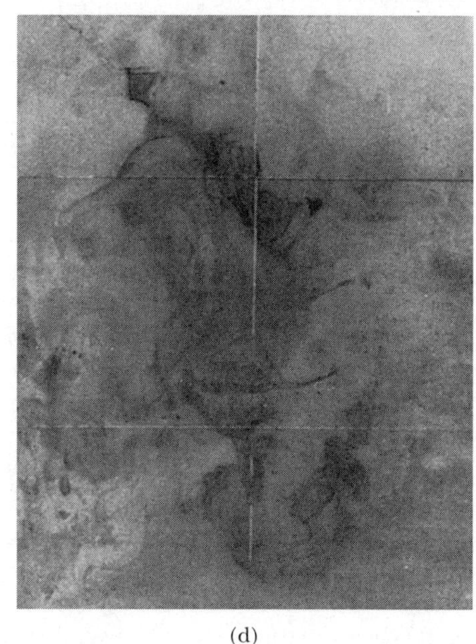

(d)

FIGURE 13.20 (a) Photo of Van Dyck's painting *Saint Rosalie Interceding for the Plague-Stricken of Palermo,* in the New York Metropolitan Museum of Art. (b) An x-ray barely shows the image of a man upside down in the bottom. (c) A few hours after the painting was irradiated with neutrons, the presence of manganese shows up. (d) Four days later, the phosphorus shows up in a charcoal sketch of Van Dyck himself (upside down) that he had painted over. *Courtesy of the Metropolitan Museum of Art, Purchase 1871. (71.41).*

Crime Detection

Neutron activation analysis is also useful to search for particular elements indicative of crime. This sophisticated technique requires access to a nuclear reactor, and as a result, few police departments, other than national agencies, have this capability on a day-to-day basis.

The examination of gunshots by measuring trace amounts of barium and antimony from the gunpowder has proven to be 100 to 1000 times more sensitive than looking for the residue itself. Amounts as small as 0.005 μg of barium and

SPECIAL TOPIC:

HOW TO PROVE AN ART FORGERY

FIGURE A The painting of the Van Meegeren forgery *Christ and His Disciples at Emmaus.* It was sold as a genuine Vermeer in 1937, but a radioactive measurement in 1968 of a portion of the lead showed the painting was not genuine. *Courtesy of Museum Boijmans Van Beuningen, Rotterdam.*

In 1937 the artist Han Van Meegeren sold* a painting titled *Christ and His Disciples at Emmaus* (see Figure A) that he claimed had been painted by the famous 17th-century Dutch artist Vermeer. The newly found painting was hailed as Vermeer's greatest work and found a place of honor in a Rotterdam museum. Over the next few years Van Meegeren discovered other Vermeer paintings, some of which had similarities to known Vermeers (see Figure B). After World War II, one of these newly found paintings was found in the collection of the Nazi leader Hermann Göring. Because it was illegal to sell national treasures to the Nazis, Van Meegeren was arrested.

Van Meegeren soon claimed innocence, stating that his "discovered" Vermeers were actually fakes, painted by himself in order to impress the critics, who had been less impressed with paintings already credited to him. Many people still thought the new Vermeers were real and that Van Meegeren was just

*This account has been taken from *Secrets of the Past,* by Bernard Keisch, U.S. Atomic Energy Commission Office of Information Services, 1972.

0.001 μg of antimony may be detected by (n, γ) techniques. Specialists are able to ascertain firing distances up to 2 m and whether a hole or slit in some material like cloth, flesh, wood, leather, and so on, was caused by a bullet. Sensitive trace elements include barium, antimony, lead (from the primer and the bullet), and copper (from jacketed bullets, the cartridge case, and the primer case).

Scientists are also able to detect toxic elements in hair by neutron activation analysis. Human head hair grows about 10 cm/y. Small amounts of arsenic and mercury may be detected, and the time of poisoning may even be determined. A famous study by several Scottish scientists examined samples of Napoleon's hair, taken during the last five years of his life and just after his death, for arsenic content. The beta emitter ^{76}As ($t_{1/2}$ = 26.4 h) was studied after being activated in a nuclear reactor. Analysis of strands of hair as short as 1 mm (about 3 days' growth) showed that Napoleon had arsenic concentrations as high as 40 times normal. Speculation indicates that Napoleon may have been poisoned, but a more reasonable explanation is that his physicians were simply trying to cure him of his many illnesses.

Was Napolean poisoned?

Agriculture

Many third-world countries lack adequate facilities for food storage and refrigeration. As a result over 30% of all foodstuffs may be wasted by spoilage. If this

trying to save himself. As part of Van Meegeren's trial in 1946 he actually painted another supposed Vermeer, and after the charge against him was changed from collaboration to forgery, he was convicted and died in prison within a year. The evidence against the famous *Emmaus* was not conclusive, and up until 1968, many people still believed it was an original Vermeer.

The radioactive nucleus ^{226}Ra decays through ^{210}Pb on its way to its final stable product ^{206}Pb. ^{210}Pb has a half-life of 22 y, and in centuries-old samples of lead ore, the activities of ^{210}Pb and ^{226}Ra will be in equilibrium. However, when lead ore is smelted, the ^{226}Ra is eliminated. Therefore, in new lead samples there will be very small ratios of ^{226}Ra/^{210}Pb compared to old samples of lead. In 1968 this technique of lead dating was performed on several Van Meegeren paintings including the *Emmaus*. The small ratios of ^{226}Ra/^{210}Pb proved the *Emmaus* had been painted in the 20th century and could not have been an original Vermeer.

FIGURE B (a) A genuine Vermeer painting titled *A Woman Writing.* (b) A similar Van Meegeren forgery. The controversy has probably increased the value of Vermeer's art. *(a) Courtesy of National Gallery of Art, (b) Courtesy of Rijksmuseum Amsterdam.*

could be prevented, literally millions of humans could be helped from hunger. Studies in India have shown that food irradiated by ^{60}Co gamma rays will last much longer: onions (100% longer), potatoes (80%), Bombay duck (factor of 8), shrimp (factor of 6). Over 30 countries allow the use of irradiation to preserve more than 35 types of food. However, consumer resistance to the irradiation of food still exists. Irradiation of medical supplies and spices is done routinely, but efforts to kill the salmonella bacteria occurring in poultry and *E. coli* bacteria in beef have been widely debated in the United States.

Food irradiation

Radioisotopes are also useful in agricultural research. The β^+ emitter ^{11}C has been used in New Zealand to study photosynthesis in plants. Similarly, the beta emitter ^{13}N was used to investigate the uptake of nitrogen by plant roots as well as the movement of nitrogen to the growth points within plants.

Mining and Oil

Radioactive sources have long been useful in the petroleum industry, for example, by placing them in the oil in pipelines to signal a change in the product being shipped, or just to log the arrival time. Geologists and petroleum engineers use radioactive sources routinely to search for oil and gas. A source and detector are inserted down an exploratory drill hole to examine the material at different depths. Neutron sources called PuBe (plutonium and beryllium) or AmBe

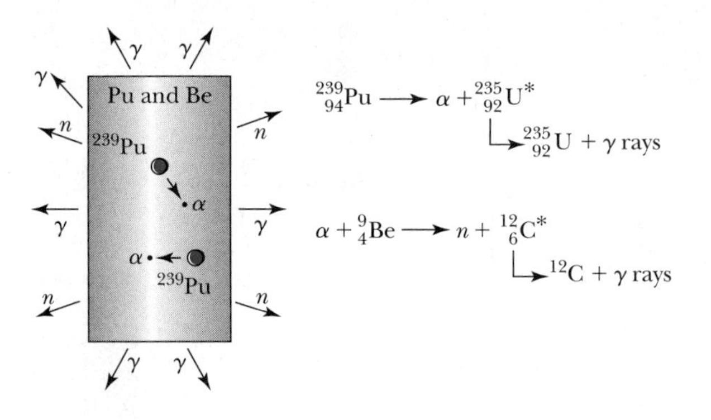

FIGURE 13.21 A plutonium-beryllium (PuBe) source consists of a can of beryllium powder mixed with ^{239}Pu, which α decays. The alpha interacts with the ^{9}Be and produces a neutron. Copious numbers of gamma rays also result, mostly from the decay of ^{235}U*.

(americium and beryllium) are particularly useful as shown in Figure 13.21. Small Van de Graaff accelerators producing 14-MeV neutrons have been inserted into the casing that has already been placed inside the borehole. Casing diameters are typically between 14 and 25 cm, leaving enough room for the spectroscopic tool. The neutrons activate nuclei in the material surrounding the borehole, and these nuclei produce gamma decays characteristic of the particular element. NaI and, more recently, cooled germanium detectors pick up characteristic elemental decays when they pass through different geological formations. Especially of interest are the oil- and gas-bearing regions (see Figure 13.22).

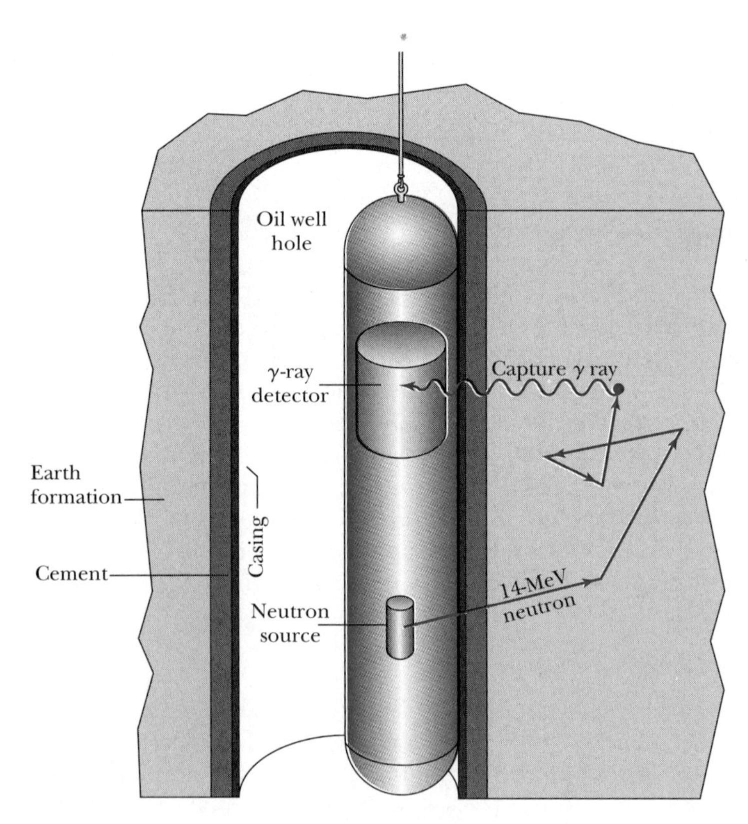

FIGURE 13.22 Example of a system used to search for oil and gas in a well boring. The (n, γ) reaction is used to examine for particular elements indicative of oil and gas.

Materials

The problem of radiation damage of an electronic device on a large single chip of silicon has long been recognized. For example, an alpha-particle decay from a contaminant uranium or thorium nucleus could cause "soft" computer errors as it ionizes the silicon. Cosmic ray particles can create similar problems, especially in satellites and space probes. Scientists have discovered, however, that fast-neutron irradiation of bulk computer memory components can decrease the soft-error rate by a factor of 10. Apparently the neutrons reduce the intrinsic resistivity in the silicon substrate so that the extraneous ionization caused later is much less likely to reset a bit.

Phosphorus-doped silicon may also be produced with fast-neutron irradiation. Silicon consists of 3.1% of the isotope ^{30}Si which undergoes the reaction

$$n + {}^{30}_{14}\text{Si} \rightarrow {}^{31}\text{Si} \qquad (13.31)$$

$$\hookrightarrow \beta^- + {}^{31}_{15}\text{P} \qquad t_{1/2} = 2.62 \text{ h}$$

Because the ^{31}Si was evenly distributed throughout the silicon, the phosphorus doping will also be much more uniformly distributed than could have been achieved by diffusive doping techniques. Silicon treated in this manner (called *neutron-transmutation-doped silicon*) is better able to handle high power levels in rectifiers, among several other uses.

Neutrons have long been used to study properties of materials, primarily because the neutron's wave nature allows it to easily probe atomic dimensions. Such effects include crystal structures, magnetic properties, interatomic and lattice forces, alloys, structures and dynamics of liquids, superconductors, phase transitions, and voids in materials and oil shales. Neutrons are particularly useful because they have no charge and do not ionize the material as do charged particles and photons. They penetrate matter easily and introduce uniform lattice distortions or impurities. Because they have a magnetic dipole moment, neutrons can probe bulk magnetization and spin phenomena. Thermal neutrons have just the right momenta and energies to probe vibrational states, their acoustic modes, and the underlying interatomic forces in solid lattices.

Example 13.8

Neutrons are used to study structures of solids and their properties. What energy (and temperature) neutrons are needed if the atomic structures are of the size of 0.06 nm?

Solution: The required wavelength needed for the neutrons is 0.06 nm. From this we determine the neutron momentum from the de Broglie wavelength relation.

$$p = \frac{h}{\lambda} = \frac{6.63 \times 10^{-34}\ \text{J}\cdot\text{s}}{0.06 \times 10^{-9}\ \text{m}} = 1.1 \times 10^{-23}\ \text{kg}\cdot\text{m/s}$$

Because we expect the kinetic energy to be low, we use a nonrelativistic relation to determine the kinetic energy.

$$K = \frac{p^2}{2m} = \frac{(1.1 \times 10^{-23}\ \text{kg}\cdot\text{m/s})^2}{2(1.67 \times 10^{-27}\ \text{kg})} = 3.6 \times 10^{-20}\ \text{J}$$

$$= 0.23\ \text{eV}$$

It was certainly adequate to use the nonrelativistic relation. In thermal equilibrium, the temperature is found from $K = \frac{3}{2}kT$.

$$T = \frac{2K}{3k} = \frac{2(3.6 \times 10^{-20}\ \text{J})}{3(1.38 \times 10^{-23}\ \text{J/K})} = 1740\ \text{K}$$

Such an energy is easily obtained by thermalizing neutrons from a nuclear reactor.

Industry

Radioactive sources (for example, ^{60}Co gamma emitters) are useful to inspect welds or castings. A small source is placed on one side of the material, and photographic film is placed on the other side. The developed film may reveal flaws, especially voids (see Figure 13.23).

Negative beta sources are used as thickness gauges for paper, plastic, sheet metal, and so on. The β^- pass through the material, and the detected count rate on the other side goes up or down depending on whether the thickness is too thin or thick, respectively. Normally, a feedback system makes the appropriate corrections automatically.

Small amounts of radioisotopes can be used as tracers to find leaks in pipes. Wear on bearings and sliding surfaces (for example, piston rings in internal combustion engines) can be studied by using slightly radioactive parts or by using neutron activation to make the finished product radioactive. Numerous other applications exist in industry.

Small Power Systems

Alpha-emitting radioactive sources like ^{241}Am or ^{238}Pu have been used as power sources in heart pacemakers. ^{241}Am sources of alpha particles are also used in smoke detectors as current generators. The scattering of the alpha particles by the smoke particles reduces the current flowing to a sensitive solid state device, which results in an alarm.

Spacecraft power

Spacecraft have been powered by radioisotope generators (RTGs) since the early 1960s. These devices use the heat produced by the α decay of ^{238}Pu to produce electricity in a thermocouple circuit. About two dozen U.S. spacecraft have used RTGs including Apollo, Pioneer 10 and 11, Viking 1 and 2, Voyager 1 and 2, Galileo, Ulysses, and more recently, Cassini launched in 1997 on its way to visit Saturn in 2004. Current RTGs supply almost 300 watts each, and usually more than one unit is placed on a spacecraft (Cassini has three). All past and present operational RTGs have exceeded their original design requirements both in

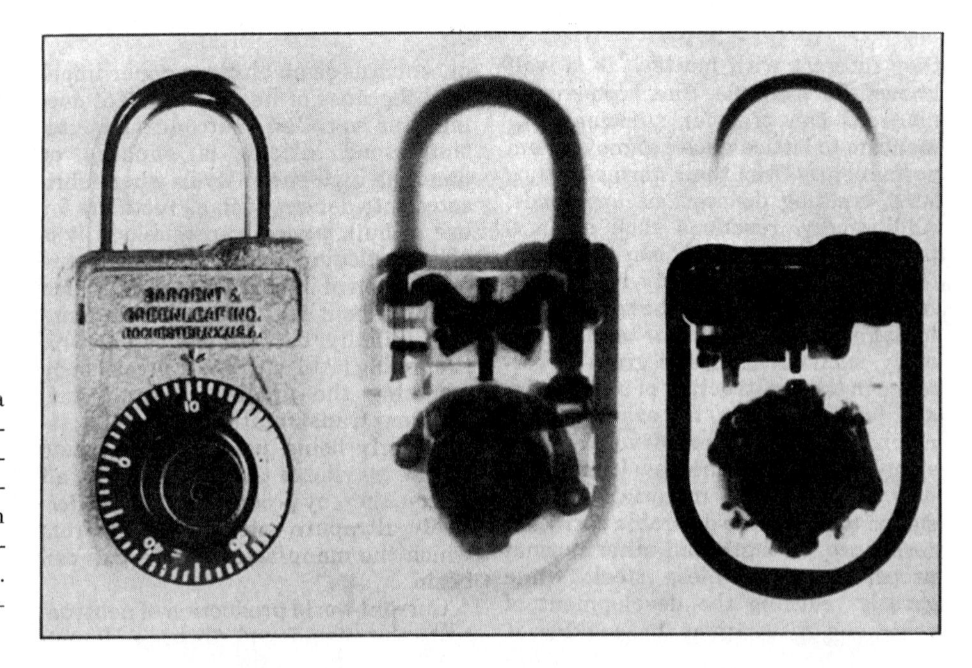

FIGURE 13.23 Photos of a combination lock made by regular photo (left), neutron radiograph (center), and x radiograph (right). The neutron image shows greater detail, especially for light elements. *Courtesy of the United States Department of Energy.*

power output and longevity. Voyagers 1 and 2 were both launched in 1977 and their RTGs are expected to be operational through at least 2020, a period of over 40 years! Voyager 1 is the most remote human emissary in space; in 1998 it was 10^{10} km from Earth, and from there its signals took 10 hours to travel to Earth.

Considerable research and development have been done in the United States on a small nuclear reactor to be used in space. The latest program was terminated in 1994, because there was no clear mission that would justify construction. Possible uses of a space reactor include outer planetary exploration, manned science outposts on the moon, and astronaut visits to Mars. Space nuclear power has always been highly controversial. The United States has only put one reactor in space (1965), but the former Soviet Union is reported to have placed almost three dozen, with some spectacular mishaps. The most famous one spread its radioactive debris across northwestern Canada in 1978.

An artist's rendering of Voyager 2 as it neared Saturn in 1981 when six new moons were discovered. The three radioisotope thermoelectric generators (RTGs) shown on the arm coming down and to the front provided 430 watts of electrical power. The RTGs are still operating more than 20 years after launch. *Courtesy of NASA.*

New Elements

No elements with atomic number greater than $Z = 92$ (uranium) are found in nature because of their short half-lives. However, with the use of reactors and especially accelerators, scientists have been able to produce 16 of these new **transuranic** elements up to $Z = 109$. Over 150 new isotopes heavier than uranium have been discovered. Neptunium ($Z = 93$) and plutonium ($Z = 94$) were discovered in Berkeley in 1940. Glenn Seaborg (Nobel Prize in Chemistry, 1951) was able to reclassify the periodic table and place the actinide series under the lanthanide series. Seaborg, a nuclear chemist, used his knowledge of chemistry and physics to predict the chemical properties of all the elements with $Z = 97$ to 103.

Physicists have reasons to suspect that superheavy elements with atomic numbers of 110–120 with 184 neutrons may be stable. These elements should have particularly interesting chemical and physical properties. Research was reported in 1999 that one nuclide of atomic number 114 that lived for 30 s had been produced at the Joint Institute for Nuclear Research in Russia.

Certain isotopes including ^{252}Cf (californium) have been produced in sufficient quantities in high-flux reactors to be generally available for special applications. ^{252}Cf, discovered in 1953, has $t_{1/2} = 2.6$ y with both alpha emission (97%) and spontaneous fission (3%). Its high spontaneous fission rate makes it particularly useful as a source of fission fragments as well as neutrons. It is extremely useful in treating malignant tumors. A tiny amount of ^{252}Cf may be placed adjacent to the tumor, so that the body tissues surrounding the malignant tumor are not nearly as damaged as they can be in normal γ-ray radiotherapy projected from outside the body. The potential use of ^{252}Cf as a neutron source includes inspection of welds in aluminum aircraft, corrosion effects in operating aircraft, oil well logging, fuel-rod scans, fission waste monitors, copper and nickel ore analyzers, determining sulfur content in coal or oil, moisture monitors, and cement analyzers.

Glenn T. Seaborg (1912–1999) is shown here in 1951 at the University of California at Berkeley with the apparatus that he used to chemically separate out the newly created transuranic elements. Seaborg, a nuclear chemist, had both a distinguished research and public service career, having served as Chairman of the U.S. Atomic Energy Commission. *Courtesy of Lawrence Berkeley National Laboratory.*

THE SEARCH FOR NEW ELEMENTS

Francium ($Z = 87$) was the last element to be discovered in nature. It was found in the decay products of uranium by Marguerite Perey working at the Curie Laboratory in 1939. Even before this, the first synthesized element, technetium ($Z = 43$), was artificially created in a nuclear reaction in 1937 at the University of California, Berkeley, by Carlo Perrier and Emilio Segrè. After bombarding molybdenum with deuterons, the discoverers took the products back to Italy where technetium was identified. This discovery was made possible by the invention of the cyclotron at Berkeley by E. O. Lawrence for whom the Lawrence Berkeley National Lab (LBNL) is named. The lab is still generally referred to as "Berkeley," and it sits on a hill overlooking the campus and San Francisco Bay.

During the next three decades the Berkeley researchers, except for relocation to the University of Chicago during World War II, dominated the discovery of new elements. Astatine ($Z = 85$) was discovered by Dale Corson and colleagues in 1940 at Berkeley by bombarding bismuth with α particles. Also in 1940 at Berkeley, Edwin McMillan and Philip Abelson discovered the first *transuranium* element, which is an element with more protons than uranium ($Z > 92$). They found neptunium ($Z = 93$) by bombarding uranium with neutrons. Similarly in 1940 Glenn Seaborg, Arthur Wahl, and Joseph Kennedy discovered plutonium ($Z = 94$) by bombarding uranium with deuterons which produced a heavy isotope of neptunium, which beta decayed to plutonium. McMillan and Seaborg received the 1951 Nobel Prize in Chemistry. Although plutonium was discovered in secret in 1940, it was not announced publicly until 1946, because of atomic bomb production in World War II. Plutonium has been often used as an energy source for deep space probes to produce electricity.

Americium ($Z = 95$) was discovered at Berkeley in 1944 by Seaborg and others by successive neutron capture reactions by plutonium in a nuclear reactor. Americium is used in smoke detectors. Using a different reaction, the next element, curium ($Z = 96$), was discovered by Seaborg, Ralph James, and Albert Ghiorso at Berkeley in 1944 by bombarding a tiny amount of plutonium with α particles.

Three elements (promethium, einsteinium, and fermium) were discovered during the development of nuclear weapons. Promethium ($Z = 61$) was the last of the "lighter elements" to be discovered. It was found at Oak Ridge, Tennessee, in 1947 by J. A. Marinsky, Lawrence Glendenin, and Charles Coryell by chemical identification of residues from a nuclear reactor. Earlier claims of the discovery of promethium in 1924 and 1941 were not substantiated. Both einsteinium and fermium (elements 99 and 100, respectively) were discovered in 1952 by a group of scientists led by Ghiorso from Argonne National Lab, Los Alamos National Lab, and Berkeley. They isolated the new elements from the radioactive debris from the first large hydrogen bomb test in the Pacific Ocean. The fermium isotope produced in the blast had been produced by 17 successive neutron captures in uranium, followed by beta decay.

After World War II several more new elements were discovered at Berkeley by various groups of scientists led by Seaborg and Ghiorso. These include berkelium (97) in 1949, californium (98) in 1950, mendelevium (101) in 1955, lawrencium (103) in 1961, rutherfordium (104) in 1969, dubnium (105) in 1970, and seaborgium (106) in 1974. All these new elements were synthesized by bombardment of transuranic elements by particles from cyclotrons. In the 1960s a new group of scientists doing particle accelerator experiments at the Joint Institute for Nuclear Research in Dubna, Russia, came on the scene. They also claimed the discovery of elements 103, 104, 105, and 106. There is some controversy about the discovery precedence that led to a long delay in the naming of the new elements. The naming of elements 101 to 106 was not settled (if it even is now!) until 1997 by acknowledging the efforts of both labs.

An international team working at the Nobel Institute for Physics in Stockholm claimed the discovery of nobelium (102) in 1957 by bombarding curium with carbon ions. The Berkeley group announced in 1958 that they could not reproduce the results, and the Dubna group soon agreed with the Berkeley findings. Both Berkeley and Dubna were later able to make authenticated discoveries of element 102.

Since 1980 a team led by Peter Armbruster working at the Institute for Heavy Ion Research (referred to by its initials in German, GSI) at Darmstadt, Germany, has dominated the search for discovery of new elements. This group bombarded targets with accelerated heavy ions, and in fairly quick succession discovered bohrium (107) in 1981, hassium (108) in 1984, and meitnerium (109) in 1982. Fusing an accelerated iron nucleus with bismuth, for example, produced this last element. There was no controversy over which team had first discovered elements 107, 108, and 109.

It would be ten more years before another element would be discovered in 1994, and all three labs, Berkeley, Dubna, and GSI, claim the discovery of element 110. All three labs produced the new element in accelerator experiments. Normally the lab that discovers an element has the right to name it. However, with the controversy surrounding precedence of several discoveries, it took many years to agree on the names of elements 101 through 109. It is expected that the agreement reached in 1997 will prevail, and we have used those names. There is general consensus that all three labs will agree on the name for element 110, but no agreement has yet been reached.

Later in 1994 GSI discovered element 111 and then in 1996 the same lab discovered element 112. Names for elements 110 to 112 are expected soon, as well as the discovery of elements 113 and 114. Element 114 is expected to be particularly long-lived, because 114 is one of the "magic numbers" of the shell model. The labs at Berkeley, Dubna, and GSI, together with the Argonne National Laboratory, were all in a race to discover element 114, and it appeared early in 1999 that Dubna had won. The single nuclide produced lived for about 30 s, so it appears that other nearby elements may be long-lived as well. Years in the future, the element 126 is also expected to be produced, but theory is less uncertain about it. The future is bright for the super-heavy element discoverers. It is also interesting to look at the names for the new elements discussed here. See if you can figure out the reasoning behind the names of the elements.

Summary

The construction of accelerators in the 1930s heralded a new era for physicists, allowing them to study the nuclear force by performing nuclear reactions. The energy released in a nuclear reaction $X(x, y)Y$, called the Q-value, can be determined from the atomic masses.

$$Q = (M_x + M_X - M_y - M_Y)c^2 \tag{13.7}$$

Different nuclear reaction mechanisms include compound nucleus, Coulomb excitation, and direct reactions, among others. Nuclei have excited states, which may appear as resonances in a compound nucleus reaction. The lifetimes τ of nuclear states are related to their widths Γ by the uncertainty principle, $\Gamma\tau \approx \hbar/2$.

Heavy nuclei may fission into two nearly equal fission fragments because of the increasingly large Coulomb force. Spontaneous fission occurs for nuclei with $Z^2/A \geq 49$, and fission can be induced with a nuclear reaction.

The nuclide ^{235}U fissions with the absorption of a slow neutron. A self-sustaining chain reaction is possible, because fission produces neutrons which can cause another fission.

Nuclear reactors may be built in different ways for special purposes. A breeder reactor produces more fissionable fuel than it consumes.

Nuclear fusion is an efficient energy source, and isotopes of hydrogen, ^{2}H and ^{3}H, appear to be the most useful. Although fusion is the source of our sun's energy, it has not yet been controlled on Earth.

Applications of nuclear science are plentiful and include medicine, archaeology, art, crime detection, agriculture, mining, oil, material studies and production, industry, and small power systems.

Questions

1. Rutherford was able to initiate nuclear reactions with α particles before 1920. Why wasn't he able to initiate nuclear reactions with protons?
2. In Example 13.2 we learned that the $^{12}\text{C}(\alpha, n)^{15}\text{O}$ cross section is much larger than the $^{12}\text{C}(\alpha, p)^{15}\text{N}$ reaction for $E_\alpha = 14.6$ MeV. We believe this is evidence of a resonance in ^{16}O. If it is a resonance, why aren't both neutron and proton exit channels populated strongly? Why do we conclude the difference must be due to quantum numbers in the exit channel? Can the Coulomb barrier in the exit channels make a difference?
3. Why do the lifetimes of nuclear excited states decrease for higher excitation energies?
4. Why is the density of nuclear excited states larger for higher excitation energies?
5. Both deuterons and alpha particles can cause direct reactions by stripping. Which are more effective? Explain.
6. Discuss the changes of the cross section for neutron-induced and proton-induced reactions as the initial kinetic energy is decreased from 50 MeV. Ignore resonances.
7. Think about how a chain reaction could be controlled without delayed neutrons. Is it possible? What would be the difficulties?
8. Think carefully about the fissioning process. Does it seem peculiar that symmetric fission is not the most probable? Does the distribution shown in Figure 13.9 seem reasonable? Explain.
9. Why it is useful to slow down neutrons produced by fission in a nuclear reactor?
10. Why is fission fuel placed in 4-m-long rods placed parallel but separated, rather than in one lump of mass?
11. Discuss how each of the following sources of energy is ultimately derived from the sun: wood, coal, gas, oil, water, and wind.
12. Why does a star's temperature increase as fusion proceeds? Why does it require higher temperatures for the carbon cycle than for the proton–proton chain?
13. The fusion process continues until a star consists of nuclei near ^{56}Fe. Explain why this occurs.
14. The first wall of a magnetic fusion containment vessel has been said to contain the most hostile environment yet designed by man. Justify this statement.
15. Neutron-activation analysis is much more widely used than charged-particle activation. Why do you suppose that is true?
16. Explain in your own words the origin of the names of elements 97 through 103; that is, who or what the elements were named after and the reasons for doing so.
17. Explain in your own words the origin of the names of elements 104 through 109; that is, who or what the elements were named after and the reasons for doing so.

Problems

13.1 Nuclear Reactions

1. Write down the precise nuclide identification for the missing element x for the following reactions. (a) $^{16}O(d, x)^{14}N$, (b) $^{7}Li(x, n)^{7}Be$, (c) $^{13}C(\alpha, n)x$, (d) $x(d, p)^{74}Ge$, (e) $^{107}Ag(^{3}He, d)x$, and (f) $^{162}Dy(x, d)^{163}Ho$.
2. For each of the reactions listed in Problem 1, write down one other possible exit channel.
3. The cross section for a 2-MeV neutron being absorbed by a ^{238}U nucleus and producing a fission is 0.6 barn. For a pure ^{238}U sample of thickness 3 cm, what is the probability of a neutron producing fission? ($\rho = 19\ g/cm^3$ for uranium).
4. List at least three entrance channels using stable nuclei that can produce the exit channel $d + {}^{20}Ne$.
5. The cross section for neutrons of energy 10 eV being captured by silver ($\rho = 10.5\ g/cm^3$) is 17 barns. What is the probability of a neutron being captured as it passes through a layer of silver 2 mm thick?
6. In order to measure the cross section of the $^{12}C(\alpha, p)^{15}N$ reaction of Example 13.2, a detector subtending a solid angle of 3×10^{-3} sr is used at the scattering angle θ. A 0.2-μA beam of 14.6-MeV α particles is incident on a 100 μg/cm² thick ^{12}C target for one hour. If the differential cross section is 0.2 mb/sr, how many protons are detected at the angle θ?
7. Write down the complete reaction for an ^{16}O target for the following reactions. List which products are stable. (a) (n, α), (b) (d, n), (c) (γ, p), (d) (α, p), (e) $(d, {}^{3}He)$, (f) $({}^{7}Li, p)$.

13.2 Reaction Kinematics

8. Calculate the ground-state Q-values for the following reactions. Are the reactions endothermic or exothermic? (a) $^{16}O(d, \alpha)^{14}N$, (b) $^{12}C(^{12}C, d)^{22}Na$, (c) $^{23}Na(p, {}^{12}C)^{12}C$.
9. For the endothermic reactions of the previous problem, calculate the threshold kinetic energy.
10. A state in $^{16}O^*$ at an excitation energy of 9.63 MeV has a broad width $\Gamma = 510$ keV. It is indicated in Figure 13.7 by hatch marks. In the excitation functions for $^{12}C(\alpha, \alpha)^{12}C$ and $^{12}C(\alpha, \gamma)^{16}O$ shown in Figure 13.7, broad peaks reflect this state. At what laboratory bombarding energy K_α will the resonance be observed (the actual peak may be shifted slightly due to interference effects)? What is the approximate lifetime of this excited state?
11. In a certain nuclear reaction initiated by 5.5-MeV α particles, the outgoing particles are measured to have kinetic energies of 1.1 MeV and 6.4 MeV. (a) What is the Q-value of the reaction? (b) If exactly the same reaction were initiated by 10-MeV α particles, what is the Q-value? (The outgoing energies will change).
12. Calculate the Q-value and threshold energy for the $^{20}Ne(\alpha, {}^{12}C)^{12}C$ reaction. What will be the kinetic energies of the ^{12}C nuclides if the alpha particle initially has 45 MeV kinetic energy in the lab?
13. The threshold kinetic energy is calculated nonrelativistically in Equation (13.10). For the reaction $A(a, b)B$ show that the threshold kinetic energy calculated relativistically is

$$K_{th} = -\frac{Q(m_a + m_A + m_b + m_B)}{2m_A}$$

14. A slow neutron is absorbed by ^{10}B in the reaction $^{10}B(n, \gamma)^{11}B$. Approximate the energy of the γ ray.
15. In a PuBe source, plutonium produces α particles of about 5 MeV. These α particles interact with beryllium by the $^{9}Be\ (\alpha, n)^{12}C$ reaction. How much kinetic energy do the reaction products have?
16. Calculate the ground-state Q-value and the threshold kinetic energy for the reactions (a) $^{16}O(\alpha, p)^{19}F$ and (b) $^{12}C(d, {}^{3}He)^{11}B$.
17. ^{60}Co is produced by neutron activation of ^{59}Co placed in a nuclear reactor where the neutron flux is 10^{18} neutrons/m² · s. The cross section is 20 b, and the sample of ^{59}Co has mass 40 mg. (a) If the ^{59}Co is left in the reactor for one week, how many ^{60}Co nuclei are produced? (b) What would be the activity of the ^{60}Co? (density of cobalt = 8.9 g/cm³) (c) Describe a procedure for producing 10^{14} Bq of ^{60}Co for medical use.

13.3 Reaction Mechanisms

18. Consider the reaction $X(x, y)Y$ depicted in Figure 13.4, where the target X is at rest. The energy of the center of mass is given by

$$K_{cm} = \frac{1}{2}(M_x + M_X)v_{cm}^2$$

where v_{cm} is the speed of the center of mass given by Equation (13.8). Show that the energy available in the center-of-mass system K'_{cm} is given by Equation (13.11)

$$K'_{cm} = K_{lab} - K_{cm} = \frac{M_X}{M_x + M_X}K_{lab}$$

where $K_{lab} = M_x v_x^2/2$.

19. A 7.7-MeV alpha particle initiates the $^{14}N(\alpha, p)^{17}O$ reaction in air. What is the excitation energy of the compound nucleus ^{18}F?
20. A 14-MeV neutron is captured by a ^{208}Pb nucleus. At what excitation energy is the resulting ^{209}Pb? What decay mechanism would you expect for this highly excited ^{209}Pb nucleus?

21. Make an estimate for the Coulomb barrier that the alpha particle must overcome for the reaction $^{14}N(\alpha, p)^{17}O$ in Example 13.3. Also make an estimate for the proton kinetic energy at a forward scattering angle. Will the proton have enough energy to tunnel out of the nucleus?

22. The ground state of ^{17}Ne is unstable. Its half-life has been measured to be 109 ms. What is the energy width of the state? List two possible decay mechanisms.

23. The first excited state of ^{17}F is at 0.495 MeV. Can the $p + {}^{16}O$ reaction populate this state? Give your reasons.

24. ^{239}Pu absorbs a thermal neutron and the resulting nucleus gamma decays to the ground state. (a) What is the energy of the gamma ray? (b) What would be the energy of the gamma ray if a 1-MeV neutron is absorbed by ^{239}Pu at rest?

25. List as many nuclear reactions as you can that use deuterons and alpha particles for projectiles with stable targets that will populate ^{22}Ne as the final state in direct reactions.

13.4 Fission

26. Calculate how much energy is released when ^{239}Pu absorbs a thermal neutron and fissions in the reaction

$$n + {}^{239}_{94}Pu \rightarrow {}^{240}_{94}Pu^* \rightarrow {}^{95}_{40}Zr + {}^{142}_{54}Xe + 3n$$

Note: The mass of $^{142}_{54}Xe$ is 141.29700 u.

27. A sample of shale contains 0.04% ^{238}U by weight. Calculate the number of spontaneous fissions in one day in a 10^6 kg pile of the shale by determining (a) the mass of ^{238}U present, (b) the number of ^{238}U atoms, (c) the fission activity, and finally (d) the number of fissions. The density of uranium is 19 g/cm^3. The spontaneous fission activity rate is 0.69 fissions/kg · s.

28. Calculate the percentage abundance of ^{235}U and ^{238}U two billion years ago if the abundance today is 0.7% and 99.3%, respectively. The increased percentage of ^{235}U probably allowed natural nuclear reactors to occur. Explain why such a reaction could not occur today.

29. Use Figure 13.9 to write down at least three common sets of fission fragments for the fission products of ^{236}U.

13.5 Fission Reactors

30. A fission reactor operates at 1000-MWe level. Assume all this energy comes from the 200 MeV released by fission caused by thermal neutron absorption by ^{235}U. At what daily rate is the mass of ^{235}U used? (In practice, of course, the energy conversion is not 100% efficient, nor is all the ^{235}U in a fuel cell used.)

31. Calculate the energy released in kilowatt hours from the fission of 1 kg of ^{235}U. Compare this with the energy released from the combustion of 1 kg of coal. The heat of combustion of coal is about 29,000 Btu per kg.

13.6 Fusion

32. Neutrons in equilibrium with their surroundings at temperature T are called thermal neutrons and have an average kinetic energy $\frac{3}{2}kT$. Calculate the thermal neutron energy for (a) room temperature (300 K), and (b) the sun (15×10^6 K).

33. Determine the ground-state Q-values for each of the reactions in the carbon cycle and show that the overall energy released is the same as for the proton–proton chain (26.7 MeV).

34. Assume that two-thirds of the Earth's surface is covered with water to an average depth of 3 km. Calculate how many nuclei of deuterium exist (^2H is 0.015% abundant). Estimate using reaction (13.22) how many joules of energy this represents through fusion.

35. The ignition temperature of fusion reactions is referred to both in temperature and kinetic energy. (a) Explain why this is done. (b) What is the relation between the two? (c) At what temperature is the energy 6 keV?

36. The following reactions may be useful in producing energy for fusion reactions. Find their Q-values. $^4He(^3He, \gamma)^7Be$, $^2H(d, p)^3H$, $^2H(p, \gamma)^3He$, $^{12}C(p, \gamma)^{13}N$, $^3He(^3He, pp)^4He$, $^7Li(p, \alpha)^4He$, $^3H(d, n)^4He$, $^3He(d, p)^4He$.

37. Determine how hot the environment must be in order for the first reaction of the CNO cycle to occur. (*Hint:* First find the threshold kinetic energy for the proton and the Coulomb barrier. After determining the kinetic energy, then determine the temperature.)

38. One of the possibilities for producing energy in a star after the hydrogen has burned to helium is $3\alpha \rightarrow {}^{12}C$ (that is, three alpha particles react to form ^{12}C). How much energy is released in this process?

39. For a thermal neutron (300 K) what is its (a) energy, (b) speed, and (c) de Broglie wavelength?

13.7 Special Applications

40. In order to determine the wear of an automobile engine, a steel compression ring is placed in a nuclear reactor where it becomes neutron activated due to the formation of $^{59}Fe(t_{1/2} = 44.5\text{ d}, \beta^-)$. The activity of the ring when placed in the engine is 4×10^5 Bq. Over the next 60 days, the car is driven 100,000 km on a test track. The engine oil is extracted and the activity rate of the oil is measured to be 512 β^-/min. What fraction of the ring was worn off during the test?

41. (a) Why does a ^{99m}Tc generator need to be shipped once a week to hospitals? (b) What is the activity of a 10^{11} Bq ^{99m}Tc generator source 9 days after it was produced? (c) If the activity is 0.9×10^{11} Bq on Monday morning when it arrives, what will be the activity on Friday morning, the last day of the working week?

42. The Los Angeles County police want to use neutron activation analysis to look for a tiny residue of barium in gunpowder. The suspected residue is placed in a nuclear reactor, where it is activated by the neutron flux. Natural barium contains 71.7% ^{138}Ba. The β^- emitter ^{139}Ba is produced in the ^{138}Ba(n, γ) ^{139}Ba reaction. The half-life of ^{139}Ba is 83.1 min. ^{139}Ba beta decays to ^{139}La, 72% going to the ground state and 27% going to the first excited state at 0.166 MeV. Scientists think they need a count rate for the 166-keV γ ray (decay to the ground state) of at least 1000 Bq 30 min. after the residue is removed from the reactor in order to make a positive identification of barium. (a) How many ^{139}Ba nuclei must be present at the end of the activation? (Remember the decay and fraction going to the first excited state.) (b) How many grams of ^{139}Ba must be produced? If the original amount of barium was 0.01 μg, what fraction of the ^{138}Ba was activated?

43. A 4×10^5 Bq ^{241}Am alpha source is used in a smoke alarm. The device is arranged so that 15% of the decay alphas are detected. (a) What current is detected? (b) If the introduction of smoke causes a 10% change in the intensity of the alpha particles, what sensitivity must the electronic circuit have to cause an alarm?

44. Consider a power source for a spacecraft consisting of ^{210}Po, which emits a 5.3-MeV alpha particle, $t_{1/2} =$ 138 d. (a) How many kg of ^{210}Po are needed to initially produce a power source of 5 kW? (b) If the power source must produce 7 kW after 2 years in space, how much ^{210}Po is needed?

45. A hospital has a 3×10^{14} Bq ^{60}Co source for cancer therapy. What is the rate of γ rays incident on a patient of area 0.3 m^2 located 4 m from the source? ^{60}Co emits a 1.1- and 1.3-MeV γ ray for each disintegration.

46. Rework Example 13.8 if the neutron is to probe the diameter of a ^{238}U nucleus. Could neutrons from a nuclear reactor be used? Explain.

47. Assume that a 10-kg sample of ^{239}Pu is used to produce electrical power from its α decay. If your device is 60% efficient in producing electrical power, how much power can be produced?

General Problems

48. Compare the following: (a) total atomic binding energy of 1 kg of hydrogen atoms, (b) nuclear binding energy of 1 kg of deuterons, (c) annihilation energy of 0.5 kg of protons with 0.5 kg of antiprotons.

49. One method used to determine unknown atomic masses consists of precisely measuring the kinetic energies of the particles involved in a nuclear reaction and using known atomic masses. The mass of ^{34}Si is determined by the ^{30}Si$(^{18}$O$, ^{14}$O$)^{34}$Si reaction initiated by 100-MeV ^{18}O particles. The outgoing particles have 86.63 MeV energy, which can be determined only by measuring the ^{14}O energy and using the conservation of momentum and energy. (a) What is the Q-value of the reaction? (b) What is the mass of ^{34}Si assuming the other three masses involved are known (see Appendix 8)?

50. ^{90}Sr is one of the most deadly products of nuclear fission. Assume that 4% of the fission fragment yield from a ^{235}U atomic bomb is ^{90}Sr. In a nuclear interchange on the planet Inhospitable, 1000 atomic bombs, each corresponding to the fission of 100 kg of ^{235}U, are detonated. (a) How many atoms of ^{90}Sr are released? (b) Assuming the ^{90}Sr is spread evenly over the planet of diameter 12,000 km, what is the resulting activity for each m^2? The half-life of ^{90}Sr is 28.8 y.

51. Assume a temperature of 2×10^8 K in a controlled thermonuclear reactor. (a) Calculate the most probable energy of deuterons at this temperature. (b) Use the Maxwell-Boltzmann distribution from Chapter 9 to determine the fraction of deuterons having an energy that is 2, 5, and 10 times the most probable energy.

52. A PuBe source has a neutron activity of 10^5 Bq. The neutrons are produced by the ^{9}Be$(\alpha, n)^{12}$C reaction with an effective cross section of 90 mb and thickness of 3 cm. (a) What is the probability of an incident alpha particle interacting with a ^{9}Be nucleus? (b) What must the incident alpha particle rate be on ^{9}Be? (c) What must the amount of mass be of the ^{239}Pu producing the alpha particles?

53. A typical person of mass 70 kg contains 0.35% potassium, by weight. Of the potassium, 0.012% is ^{40}K, an unstable nucleus which β^- decays (89.3%) and EC (10.7%) with a $t_{1/2} = 1.28 \times 10^9$ y. What is the ^{40}K activity due to β^- decay in a typical person's body?

CHAPTER 14

Elementary Particles

The electron is not as simple as it looks.

Sir William Lawrence Bragg

The understanding of basic facts about nature is one of science's foremost goals. What are the basic building blocks of matter? What are the forces that hold matter together? How did the universe begin? Will the universe end, and if so, how and when?

We try to use the ideas, concepts, and laws of physics to answer these questions. The ancient Greek philosophers, among them Aristotle, supposed things were made of earth, fire, wind, and water. Democritus coined the word *atom,* believing that things were made of some very small entities and empty space. In this book we have shown that matter is made of molecules and atoms, and electromagnetic forces are responsible for holding atoms together. But, at a smaller level, an atom is made of electrons and a nucleus. The electromagnetic force is responsible for attracting the electrons to the nucleus, but the strong (nuclear) force is responsible for keeping neutrons and protons together in the nucleus. Previously we have thought of electrons, neutrons, and protons as **elementary particles,** because we believe they are basic building blocks of matter. However, in this chapter we learn that the term *elementary particle* is used very loosely to refer to hundreds of particles, most of which are unstable. They can be produced by adding energy, for example, to the three particles already mentioned. Some of these particles are indeed very crucial to our understanding of matter and of the forces that hold matter together. We will learn in this chapter that neutrons and protons are made up of even more fundamental particles called *quarks.* These cannot be observed outside the nucleus, yet we believe they must exist. How far can this division into basic building blocks continue (see Figure 14.1)?

Elementary particles

We will find in this chapter, and especially in Chapter 16 on cosmology, that there is an intertwined connection between the oldest science, astronomy, and that of particle physics, in which matter is studied in the tiniest dimensions. The question of how the universe started and how it may end is in fact buried in the mysteries of elementary particle physics. For this reason many physicists believe that particle physics is at the forefront of physics, and indeed of all science. The mysteries are deep and profound, but finding the answers gives great satisfaction. Although the subject is complex and requires the meshing of relativity, quantum mechanics, electrodynamics, and gravitation, we hope to convey to the student some of the excitement and flavor of particle physics. This excitement is even evident in the tools that experimental physicists use. In the last section of this chapter we discuss some of the accelerators used by physicists, especially nuclear and particle physicists.

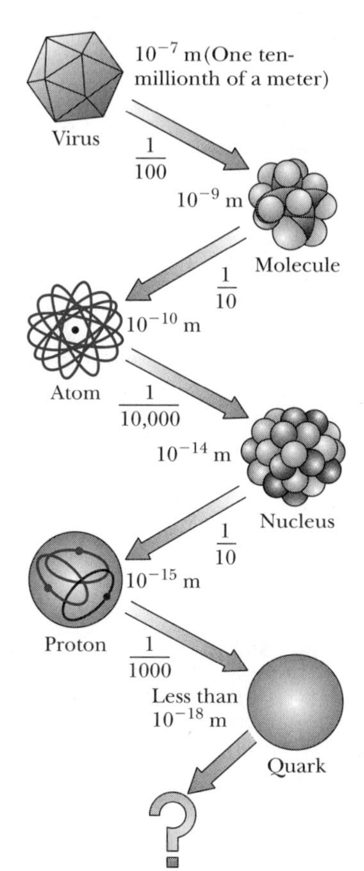

FIGURE 14.1 Starting from a virus, the structure of matter can be divided into smaller and smaller entities down to the pointlike quark and to whatever lies beyond. *Courtesy of Universities Research Association.*

14.1 The Early Beginnings

In 1930 the known elementary particles were the proton, electron, and photon. The electron had been identified by Thomson in 1897, and Einstein's work on the photoelectric effect can be said to define the photon (originally called a *quantum*) in 1905. The proton is the nucleus of the hydrogen atom. Despite the rapid progress of physics in the first couple of decades of the 20th century, no more elementary particles were discovered until 1932, when Chadwick proved the existence of the neutron, and the positron was identified in cosmic rays by Carl Anderson.

The Positron

Paul A. M. Dirac (1902–1984), a British theoretical physicist, received his bachelor's degree in electrical engineering and doctoral degree in mathematics. His training and brilliant insight into nature allowed him to make many contributions to physics, which included his own general form of quantum mechanics, as well as showing that Heisenberg's matrix mechanics and Schrödinger's wave mechanics were special cases of his own general theory. Dirac also generalized the concept of "action-at-a-distance." When discussing gravitation and electromagnetism, the concept of fields is useful to understand the forces on objects placed in the external fields. Dirac developed the early form of quantum electrodynamics in which the absorption and emission of photons is a quantum process of the radiation field itself. His quantum electrodynamics (QED) theory was later generalized by Richard Feynman, Julian Schwinger, and Sin-itiro Tomonaga into the most accurately tested theory of physics today.

Feynman presented a particularly simple graphical technique to describe particle reactions. For example, when two electrons approach each other, according to the quantum theory of fields, they exchange a series of photons. These photons are called *virtual,* because they cannot be directly observed. The action of the electromagnetic field (for example, the Coulomb force) can be interpreted as the exchange of photons. In this case we say that the photons are the *carriers* or *mediators* of the electromagnetic force. The procedure is represented schematically in a spacetime diagram like that shown in Figure 14.2, called a **Feynman diagram,** in which a photon is exchanged between two electrons. Feynman diagrams have been generalized to represent quite complicated interactions.

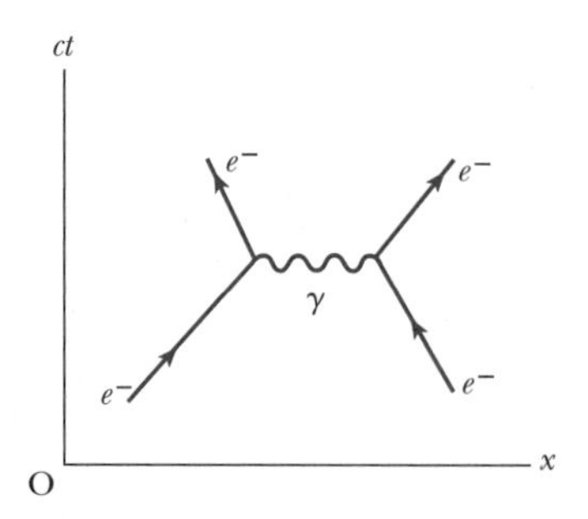

FIGURE 14.2 Example of a Feynman spacetime diagram. Electrons interact through mediation of a photon. The axes are normally omitted.

Perhaps Dirac's greatest success was his 1928 relativistic theory of the electron, for which he received the Nobel Prize in 1933. When Dirac used his considerable mathematical skills to combine quantum mechanics with relativity, he found that his wave equation had negative, as well as positive, energy solutions. These negative energy solutions were eventually interpreted in terms of positively charged particles, or positive electrons, called *positrons* and denoted e^+.

Antiparticles

Dirac's theory and refinements by others opened up the possibility of **antiparticles,** which have the same mass and lifetime as their associated particles, and the same magnitude, but with opposite sign, for such physical quantities as electric charge and various quantum numbers. All particles, even neutral ones (with some notable exceptions like the neutral pion), have antiparticles (see Section 14.3).

Cosmic rays

Cosmic rays are highly energetic particles, mostly protons, that cross interstellar space and enter the Earth's atmosphere, where they interact with particles creating cosmic "showers" of many distinct particles. Cosmic rays contain the highest particle energies observed (up to 10^{21} eV), although they normally are in the GeV (10^9 eV) range. The Austrian physicist V. F. Hess (1883–1964; Nobel Prize, 1936) discovered cosmic rays in 1912 by detecting them at high altitudes during balloon flights. In 1932 the American physicist Carl D. Anderson (1905–1991; Nobel Prize, 1936) discovered positrons by observing the paths of cosmic ray showers passing through a cloud chamber placed in a magnetic field.

The ultimate fate of positrons (antielectrons) is annihilation with electrons. After a positron slows down by passing through matter, it is attracted by the Coulomb force to an electron where it annihilates through the reaction,

$$e^+ + e^- \rightarrow 2\gamma \tag{14.1}$$

We have already discussed this procedure in Chapter 3. The characteristic of positron annihilation is the emission of two back-to-back 0.511-MeV gamma rays with the kinetic energy coming from the rest energies of the electron and positron. The kinetic energies of the charged particles are usually very small when they meet and are neglected.

Yukawa's Meson

The idea of the photon being the mediator of the electromagnetic force had been discussed by several physicists besides Dirac, including H. Bethe, E. Fermi, C. Møller, and G. Breit. The Japanese physicist Hideki Yukawa (1907–1981; Nobel Prize, 1949) had the idea of developing a quantum field theory for the force between nucleons analogous to that for the electromagnetic force. If that is the case, what is the carrier or mediator of the nuclear strong force analogous to the photon in the electromagnetic force? What particle is exchanged between nucleons to keep the nucleons strongly attracted?

In this case the mediating particle is called a **meson,** named after the Greek word *meso* which means "middle." The attraction of nucleons depends on the exchange of a virtual meson, much like the exchange of a virtual photon in the Coulomb attraction of an electron and proton. The energy ΔE required to create the meson is $\Delta E = m_\pi c^2$ where m_π is the meson mass. The Feynman diagram indicating the meson exchange between a neutron and a proton is displayed in Figure 14.3. Quantum theory shows that if the range of the force is small (10^{-15} m), the mass of the mediating particle must be large. We can give a simple argument for this based on the Heisenberg uncertainty principle. During the time Δt that the meson is created, energy conservation is violated because of the

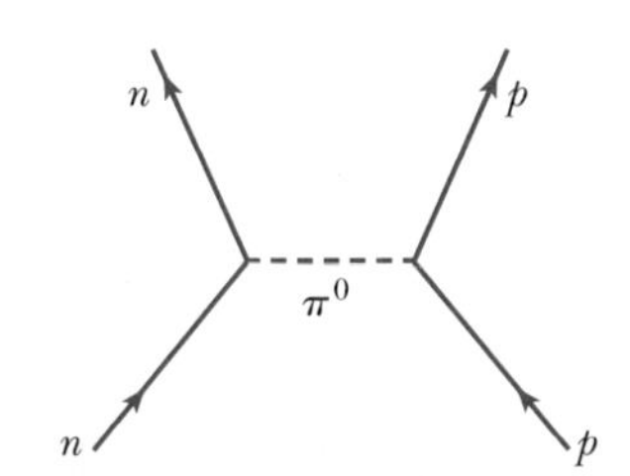

FIGURE 14.3 A Feynman diagram indicating the exchange of a pion between a neutron and a proton.

excess energy ΔE required to create the meson. This energy ΔE and the time Δt are related by the uncertainty principle, $\Delta E \Delta t \approx \hbar$, so Δt is given by

$$\Delta t \approx \frac{\hbar}{\Delta E} = \frac{\hbar}{m_\pi c^2} \tag{14.2}$$

The fastest speed possible for the meson is c, so the distance the meson travels in time Δt is $c\,\Delta t$. This distance $c\,\Delta t$ must be about the range of the nuclear force R_N, so we have $R_N = c\,\Delta t = \hbar/m_\pi c$. We can now solve Equation (14.2) for the meson mass,

$$m_\pi c^2 = \frac{\hbar c}{R_N}$$

Yukawa thought the mean range of the nuclear force was about 2 fm, so he found

$$m_\pi c^2 = \frac{1.973 \times 10^2 \text{ eV} \cdot \text{nm}}{2 \times 10^{-15} \text{ m}} \approx 100 \text{ MeV} \tag{14.3}$$

In 1935 Yukawa predicted a particle mass of about 200 electron masses for the virtual meson. There were no accelerators at the time that could create such a high-energy particle, but they could perhaps be observed in cosmic rays.

In 1938 Carl Anderson and his collaborators observed a particle (later called the **muon** or *mu-meson* or μ-meson)* in cosmic radiation that at first was believed to be Yukawa's particle. It proved to have a mass of 106 MeV/c^2, but subsequent experiments showed that it did not interact strongly with the nucleus. This new particle could not be the propagator of the strong force, and we now know that the muon is not even a member of the meson family (see Section 14.3); it interacts only through the weak and electromagnetic interactions.

Muon

Yukawa's meson, called a **pion** (or *pi-meson* or π-meson), was finally identified in 1947 by C. P. Powell (1903–1969) and G. P. Occhialini (1907–1993). Charged pions $\pi^\pm$ have masses of 140 MeV/c^2, and a neutral pion π^0 was later discovered that has a mass of 135 MeV/c^2. Yukawa said the spin of the meson must be zero (or integral), which the pion has, but the muon has spin $\frac{1}{2}$. We will learn later that neutrons, protons, and pions are made of quarks, and that the nuclear interaction actually takes place between the quarks. It is preferable to speak of the "strong" force rather than the "nuclear" force, because the interaction is at a level more basic than the nucleus.

Pion

14.2 The Fundamental Interactions

In Chapter 1 we discussed the fundamental forces in nature responsible for all interactions. Those forces include the gravitational, electroweak, and strong forces. For all practical purposes, there are really four fundamental forces, because the electroweak interaction is a unification of the electromagnetic and weak interactions, just as the electromagnetic interaction is a unification of electric and magnetic interactions. However, the electromagnetic and weak forces act independently except at particle energies available only in cosmic rays, produced by accelerators, or in the early stages of the creation of the universe. We will discuss this further in Chapter 16.

*Although the muon was initially called a μ-meson, the muon is not a meson. The particle should be called a muon, not a μ-meson.

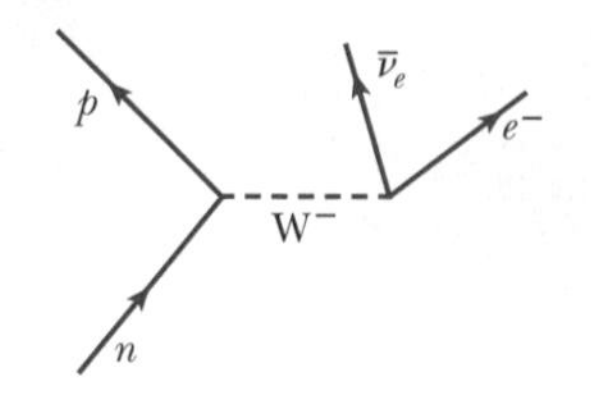

FIGURE 14.4 A Feynman diagram indicating the beta decay of a neutron. Note that the W^- mediates this beta decay.

We have learned that the fundamental forces act through the exchange or mediation of particles according to the quantum theory of fields. The exchanged particle in the electromagnetic interaction is the photon. All particles having either electric charge or a magnetic moment (and also the photon) interact with the electromagnetic interaction. The electromagnetic interaction has very long range.

In the 1960s Sheldon Glashow, Steven Weinberg, and Abdus Salam (Nobel Prize, 1979) predicted that new particles called W (for weak) are responsible for the weak interaction. This theory, called the *electroweak* theory, unified the electromagnetic and weak interactions much as Maxwell had unified electricity and magnetism into the electromagnetic theory a hundred years earlier. The details of the electroweak theory are too complex to be discussed here, but we know the W particle must be very massive, because the weak interaction is very short-range (remember the uncertainty principle argument in Section 14.1, $R \sim \hbar/mc$). The W is a boson (integral spin). There are three versions of it: $W^{\pm}$ (mass 80.4 GeV/c^2), and Z^0 (mass 91.2 GeV/c^2), all with spin 1. In 1983 these massive bosons were discovered at CERN by a group led by C. Rubbia (Nobel Prize, 1984, see Special Topic), confirming the theoretical predictions and the unification of the interactions. A Feynman diagram of the neutron beta decay is displayed in Figure 14.4, showing the W^- as the carrier of the interaction.

W bosons

We previously mentioned (Section 14.1) that Yukawa's pion is responsible for the nuclear force. Now we know there are other mesons that interact with the strong force. Later, in Section 14.5, we will learn that the nucleons and mesons are part of a general group of particles formed from even more fundamental particles called **quarks.** The particle that mediates the strong interaction between quarks is called a **gluon** (for the "glue" that holds the quarks together); it is massless and has spin 1, just like the photon. Thus, at a more fundamental level, we believe it is the gluons that are responsible for the strong force. No one has observed an isolated quark or a gluon, because they stay hidden within particles, confined by the strong force. Experimental evidence (to be presented in Section 14.5) convinces us that quarks and gluons exist. Particles that interact with the strong force are called **hadrons** and include the neutron, proton, and mesons, among many other elementary particles.

Gluons

Hadrons

It has been suggested that the particle responsible for the gravitational interaction be called a **graviton.** It must be massless, travel at the speed of light, have spin 2, and interact with all particles that have mass. The graviton has never been observed because of its extremely weak interaction with laboratory objects. Very sensitive measurements are underway that hope to prove the existence of gravitons by detecting gravitational waves from supernova explosions.

Gravitons

The four interactions are shown schematically in Figure 14.5, and the mediators are listed in Table 14.1. Note that all the mediating particles are bosons with integral spin.

TABLE 14.1
The Fundamental Interactions

Interaction	Relative Strength	Range	Mediating Particle
Strong	1	10^{-15} m	Gluons
Electroweak:			
Electromagnetic	10^{-2}	∞	Photon
Weak	10^{-5}	10^{-18} m	$W^{\pm}$, Z^0 bosons
Gravitation	10^{-39}	∞	Graviton

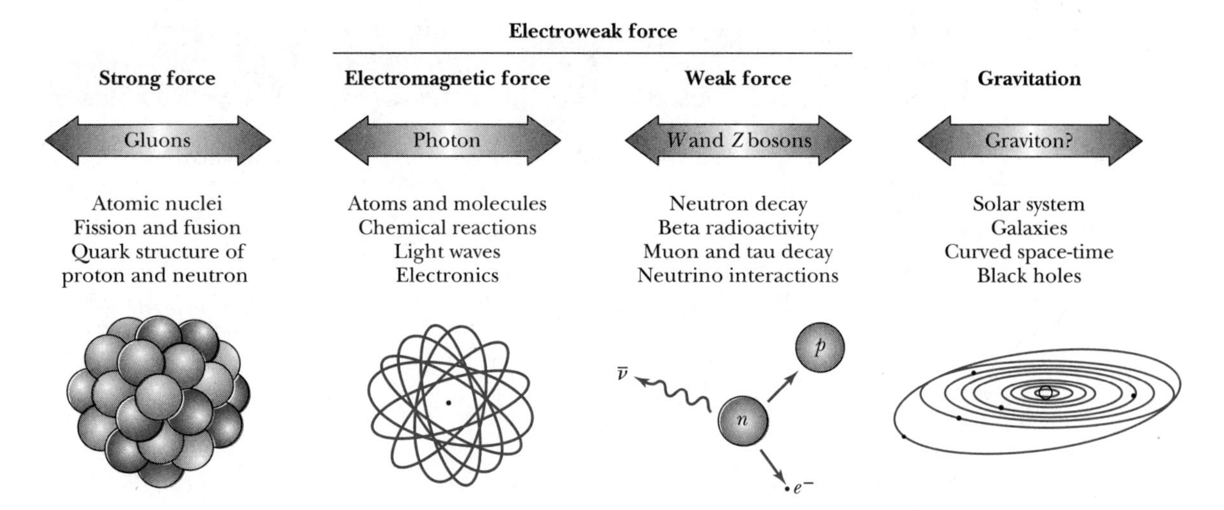

FIGURE 14.5 Some manifestations of the fundamental forces of nature. The mediating particles are shown as well as the areas in which the forces are effective. *Courtesy Universities Research Association.*

Example 14.1

Using the mass of the W^- particle, estimate the range of the weak interaction responsible for the neutron beta decay.

Solution: This calculation is much like the one we did in Section 14.1, when we discussed Yukawa's meson. In that case we knew the nuclear force range, but in this case, the force range is unknown. We begin with Equation (14.2), obtained by applying the uncertainty principle to the W^- mass (80.4 GeV/c^2) responsible for the beta decay.

$$\Delta t = \frac{\hbar}{\Delta E} = \frac{\hbar}{m_{W^-}c^2} \tag{14.4}$$

We again assume the mediating particle travels at the fastest speed possible, so that $c\,\Delta t$ is the range of the force, R_W.

$$R_W = c\,\Delta t = \frac{\hbar c}{m_{W^-}c^2} \tag{14.5}$$

$$= \frac{1.973 \times 10^2 \text{ eV} \cdot \text{nm}}{(80.4 \text{ GeV}/c^2)(c^2)} = 2.45 \times 10^{-18} \text{ m}$$

In this case, it may not be true that the W^- travels at the speed of light, because it is such a massive particle. The calculation here is therefore an upper limit. Note in this case the violation of the conservation of energy is massive, so the violation must occur over a very short period of time. We can calculate Δt from Equation (14.4) to be

$$\Delta t = \frac{1.055 \times 10^{-34} \text{ J} \cdot \text{s}}{80.4 \text{ GeV}} \left(\frac{1 \text{ GeV}}{1.6 \times 10^{-10} \text{ J}}\right) \approx 10^{-26} \text{ s}$$

The lifetime of the neutron is much longer than the time it takes for the decay process itself.

14.3 Classification of Elementary Particles

As the number of elementary particles discovered continued to increase in the 1950s and 1960s, various schemes were tried to make reasonable order of what some physicists referred to as a "zoo" of elementary particles. Since the 1960s experimentalists have needed accelerators with ever-increasing energies in order to test elementary particle theories. Eventually, an understanding of the zoo occurred due to many contributions. In this section we will present a compact picture of the organization of elementary particles that represents some of the most important advances in particle physics during the past 25 years.

TABLE 14.2
Boson Properties:* Gauge, Higgs, and Graviton

Boson	Mass	Spin	Electric Charge	Comments
Gauge:				
Photon	0	1	0	Stable
W^+, W^-	80.41 GeV/c^2	1	1, −1	Γ = 2.06 GeV, decays observed
Z^0	91.19 GeV/c^2	1	0	Γ = 2.49 GeV, decays observed
Gluon	0	1	0	Bound in hadrons, not free
Higgs H^0	>77.5 GeV/c^2	0	0	Not observed
$H^\pm$	>54.5 GeV/c^2	0	±1	Not observed
Graviton	0	2	0	Stable, not observed

*1998 Review of Particle Physics, C. Caso et al. (Particle Data Group), *European Physical Journal* **C3**, 1 (1998).

Fermions and bosons

We discussed in Chapter 9 that particles with half-integral spin are called *fermions* and those with integral spin are called *bosons.* This is a particularly useful way to divide elementary particles because *all stable matter in the universe appears to be composed at some level of constituent fermions.* We have already discussed some bosons in the previous section. Photons, gluons, $W^\pm$, and the Z^0 are called **gauge bosons** and are responsible for the strong and electroweak interactions. Gravitons are also bosons, having spin 2. Fermions exert attractive or repulsive forces on each other by exchanging gauge bosons which are the force carriers.

Gauge bosons

There is one other boson that has been predicted, but not yet detected, that seems to be necessary in quantum field theory to explain why the $W^\pm$ and Z^0 have such large masses, yet the photon has no mass. This missing boson is called the **Higgs particle** (or **Higgs boson**) after Peter Higgs, who first proposed it. We don't know whether the Higgs particle is an elementary boson or a composite particle. It may also be responsible for a new fundamental force. The properties of the gauge and Higgs bosons, as well as the graviton, are given in Table 14.2. Another set of bosons is the meson family, of which the pion is a member.

Higgs boson

We shall spend the remainder of this section on the fermions, which come in two varieties (leptons and quarks) and the mesons, which are bosons made also from quarks. Leptons include charged particles (electrons and muons) as well as uncharged particles (neutrinos). Quarks make up hadrons. We will discuss leptons and hadrons in turn.

Leptons

The leptons are perhaps the simplest of the elementary particles. They appear to be pointlike and seem to be truly elementary. Thus far there has been no plausible suggestion they are formed from some more fundamental particles. There are only six leptons (displayed in Table 14.3), plus their six antiparticles. We have already discussed the electron and muon. Each of the charged particles has an associated neutrino, named after its charged partner (for example, muon neutrino ν_μ). The electron and all the neutrinos are stable. The muon decays into an electron, and the tau can decay into an electron, a muon, or even hadrons (which is most probable). The muon decay (by the weak interaction) is

$$\mu \longrightarrow e + \nu_\mu + \overline{\nu}_e \tag{14.6}$$

TABLE 14.3
The Leptons*

Particle Name	Symbol	Anti-particle	Mass (MeV/c^2)	Mean Lifetime (s)	Main Decay Modes
Electron	e^-	e^+	0.511	Stable	
e-neutrino	ν_e	$\bar{\nu}_e$	$< 7 \times 10^{-6}$		
Muon	μ^-	μ^+	105.7	2.2×10^{-6}	$e^- \bar{\nu}_e \nu_\mu$
μ-neutrino	ν_μ	$\bar{\nu}_\mu$	< 0.17		
Tau	τ^-	τ^+	1777	2.9×10^{-13}	$\mu^- \bar{\nu}_\mu \nu_\tau$, $e^- \bar{\nu}_e \nu_\tau$
τ-neutrino	ν_τ	$\bar{\nu}_\tau$	< 18		

*See Footnote, Table 14.2.

We are already familiar with the electron neutrino that occurs in the beta decay of the neutron (Chapter 12). Neutrinos have zero charge. Their masses are known to be very small. The precise mass of neutrinos has a strong bearing on current cosmological theories of the universe because of the gravitational attraction of mass. All leptons have spin $\frac{1}{2}$, and all three neutrinos have been identified experimentally. Neutrinos are particularly difficult to detect, because they have no charge and little mass, and their probability of being stopped while going through a mass like our planet Earth is exceedingly small. Leptons do not experience the strong force.

Hadrons

As was mentioned previously, hadrons are particles that interact through the strong force. There are two classes of hadrons: **mesons** and **baryons.** Mesons are particles with integral spin having masses greater than that of the muon (106 MeV/c^2; note that the muon is a lepton and not a meson). All baryons have masses at least as large as the proton and have half-integral spins.

Mesons and baryons

Mesons. We have already discussed the pion in Section 14.1; the pion is a meson that can either have charge or be neutral. Mesons are also bosons because of their integral spin. The meson family is rather large and consists of many variations, distinguished according to their composition of quarks. We list in Table 14.4 only a few of the mesons. In addition to the pion there is also a K meson which exists in both charged ($K^\pm$) and neutral forms (K^0). The K^- meson is the antiparticle of the K^+, and their common decay mode is into muons or pions. The K^0 meson is particularly interesting because it has two decay lifetimes: K^0_S has a shorter mean lifetime of 9×10^{-11} s and decays to $\pi^+\pi^-$ or $2\pi^0$, whereas K^0_L has a longer mean lifetime of 5×10^{-8} s and has many decay modes, including the two-pion mode (only 0.3% of all K^0_L decays), which violates the combined conservation laws of charge and parity (see Section 14.4). All mesons are unstable and are not abundant in nature. They are routinely produced in cosmic radiation and in nuclear and particle physics experiments. The π^0 is its own antiparticle.

Baryons. The neutron and proton are the best-known baryons, and the baryons are a prolific group. Baryons having nonzero strangeness numbers (a new quantum number to be discussed in the next section) are called hyperons. The proton is the only stable baryon, but some theories predict that it is also unstable with a lifetime greater than 10^{30} years. Ongoing experiments have set a lower limit for

TABLE 14.4
The Hadrons

Particle Name	Symbol	Anti-particle	Mass (MeV/c^2)	Mean Lifetime (s)	Main Decay Modes	Spin	Baryon Number B	Strangeness Number S	Charm Number C
Mesons									
Pion	π^-	π^+	140	2.6×10^{-8}	$\mu^+ \nu_\mu$	0	0	0	0
	π^0	Self	135	8.4×10^{-17}	2γ	0	0	0	0
Kaon	K^+	K^-	494	1.2×10^{-8}	$\mu^+\nu_\mu$, $\pi^+\pi^0$	0	0	1	0
	K_S^0	$\overline{K}_S^0$	498	8.9×10^{-11}	$\pi^+\pi^-$, $2\pi^0$	0	0	1	0
	K_L^0	$\overline{K}_L^0$	498	5.2×10^{-8}	$\pi^\pm e^\mp \nu$, $3\pi^0$, $\pi^\pm \mu^\mp \nu$, $\pi^+\pi^-\pi^0$	0	0	1	0
Eta	η^0	Self	547	5×10^{-19}	2γ, $3\pi^0$ $\pi^+\pi^-\pi^0$	0	0	0	0
Charmed D's	D^+	D^-	1869	1.1×10^{-12}	e^+, $K^\pm$, K^0, $\overline{K}^0$ + anything	0	0	0	1
	D^0	$\overline{D}^0$	1864	4.2×10^{-13}	Same as D^+	0	0	0	1
	D_S^+	$\overline{D}_S^-$	1969	4.7×10^{-13}	Various	0	0	1	1
Bottom B's	B^+	B^-	5279	1.6×10^{-12}	Various	0	0	0	0
	B^0	$\overline{B}^0$	5279	1.6×10^{-12}	Various	0	0	0	0
J/Psi	J/ψ	Self	3097	10^{-20}	Various	1	0	0	0
Upsilon	Υ	Self	9460	10^{-20}	Various	1	0	0	0
Baryons									
Proton	p	$\overline{p}$	938.3	Stable (?)		$\frac{1}{2}$	1	0	0
Neutron	n	$\overline{n}$	939.6	887	$pe^-\overline{\nu}_e$	$\frac{1}{2}$	1	0	0
Lambda	Λ	$\overline{\Lambda}$	1116	2.6×10^{-10}	$p\pi^-$, $n\pi^0$	$\frac{1}{2}$	1	−1	0
Sigmas	Σ^+	$\overline{\Sigma}^-$	1189	8.0×10^{-11}	$p\pi^0$, $n\pi^+$	$\frac{1}{2}$	1	−1	0
	Σ^0	$\overline{\Sigma}^0$	1193	7.4×10^{-20}	$\Lambda\gamma$	$\frac{1}{2}$	1	−1	0
	Σ^-	$\overline{\Sigma}^+$	1197	1.5×10^{-10}	$n\pi^-$	$\frac{1}{2}$	1	−1	0
Xi	Ξ^0	$\overline{\Xi}^0$	1315	2.9×10^{-10}	$\Lambda\pi^0$	$\frac{1}{2}$	1	−2	0
	Ξ^-	Ξ^+	1321	1.6×10^{-10}	$\Lambda\pi^-$	$\frac{1}{2}$	1	−2	0
Omega	Ω^-	Ω^+	1672	0.82×10^{-10}	ΛK^-, $\Xi^0\,\pi^-$	$\frac{3}{2}$	1	−3	0
Charmed lambda	Λ_C^+	$\overline{\Lambda}_C^-$	2285	2.1×10^{-13}	Various	$\frac{1}{2}$	1	0	1

the proton lifetime to be at least 10^{32} years (see Section 14.4). The longest-lived baryons are listed in Table 14.4 along with the mesons. They include the lambda (Λ), sigma ($\Sigma^\pm$, Σ^0), Xi (Ξ^0, Ξ^-), and omega (Ω^-). All baryons, except the proton, eventually decay into protons.

Particles and Lifetimes

The lifetimes of particles are also indications of their force interactions. Particles decaying through the strong interaction are usually the shortest-lived, normally decaying in less than 10^{-20} s. The decays caused by the electromagnetic interaction generally are of the order of about 10^{-16} s, and the weak interaction decays are even slower, longer than 10^{-10} s. There are several important exceptions to these general statements; for example, some nuclear beta decays take a long time, including the beta decay of the free neutron, which has a lifetime of about 15 min.

The length of a particle's lifetime has sometimes been used to define what we mean by a *particle*. The argument is that an object cannot rightfully be called a particle if its short lifetime prevents a direct observation. There is no single definition of a particle. An experimental physicist might say that a particle is an object having a well-defined charge and mass that behaves like a point (particle) while being accelerated or being detected.

A theoretical physicist might define a particle as an object having a complete set of numbers for charge, spin, mass, lifetime, and various other quantum numbers like charm, strangeness, and isospin (some of which we have not yet discussed). Some particles are observed only as a resonance, and although there is a precise definition of a resonance, its complexity prevents us from pursuing it here. We previously discussed compound nucleus resonances in Chapter 13; the resonances we are discussing here are wavelike phenomena and occur in particle scattering. The first excited state of the nucleon at 1232 MeV is a good example of a resonance called the delta [$\Delta(1232)$]; its lifetime is about 10^{-23} s and is too short for the delta to be directly observed. It decays to a nucleon and a pion and is easily observed in pion and electron scattering from a proton (see Figure 14.6). Objects having lifetimes as long as 10^{-10} to 10^{-14} s are normally re-

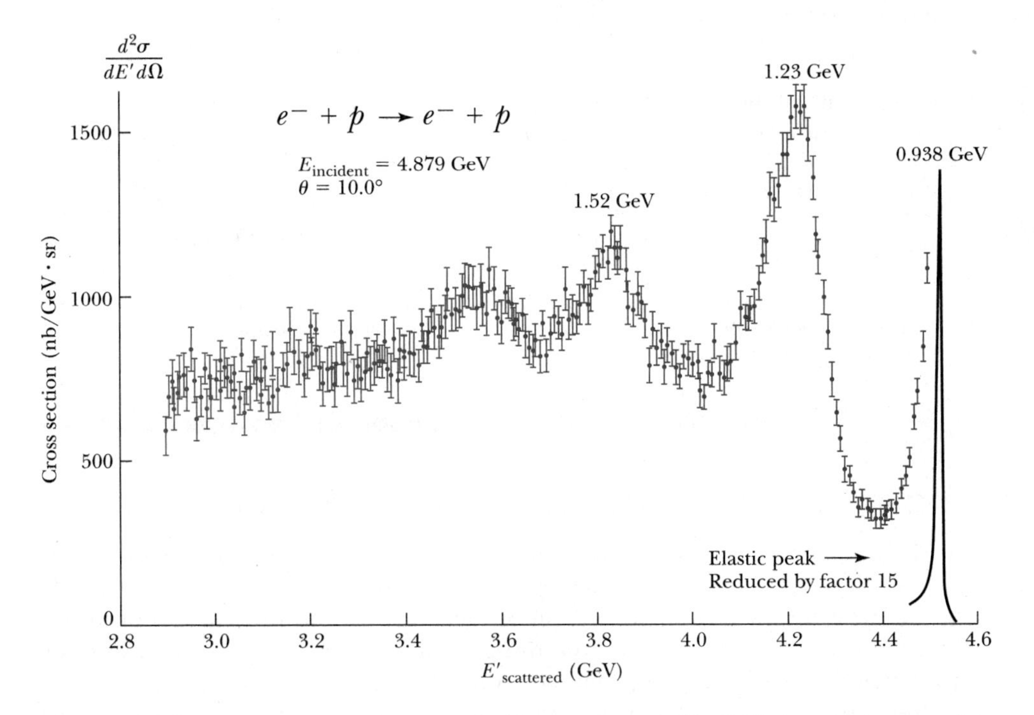

FIGURE 14.6 The spectrum of scattered electrons from 4.879 GeV electrons scattered inelastically from protons taken at DESY showing the presence of the first excited state of the proton, called the $\Delta(1232)$, at an invariant mass energy of 1.23 GeV (the ground state has 0.938 GeV). The particle is short-lived and quickly decays. Peaks due to other higher-lying resonances are also seen at 1.52 GeV and multiple resonances near 1.68 GeV. *After W. Bartel et al., Phys. Lett.* ***28B****, 148 (1968).*

garded as particles because these times are long enough for the particles to be detected and induce other reactions. The lambda, sigma, xi, and omega baryons are examples of particles.

Fundamental and composite particles

We call certain particles *fundamental;* other particles are made of them, and we believe leptons, quarks, and gauge bosons are fundamental particles. Although the Z^0 and W bosons have very short lifetimes, they are regarded as particles, so a definition of particles dependent only on lifetimes is too restrictive. Other particles are *composites,* made from the fundamental particles.

Example 14.2

The J/ψ particle was discovered at both SLAC and at Brookhaven in 1974. The SLAC data from the head-on collision of e^+ and e^- are shown in Figure 14.7. The width of the state is $\Gamma = 0.063$ MeV. Discuss the evidence that this resonance represents an elementary particle and determine the particle's lifetime.

Solution: The data shown in Figure 14.7 represent a clear experimental example of a resonance. The peak near 3.1 GeV was also observed in other outgoing reaction channels, indicating that the resonant peak occurs in the "compound" system and not in the outgoing reaction channel.

The width of the resonance observed in Figure 14.7 is on the order of 5 MeV and is limited by experimental resolution, namely, the beam momentum. With better experimental energy resolution, the resonance would become much sharper. As stated, the actual energy width of the resonant state has been established to be 0.063 MeV and was determined by other means. We can determine the lifetime of the J/ψ particle by using the uncertainty principle,

$$\Delta t = \frac{\hbar}{\Gamma} = \frac{1.055 \times 10^{-34}\,\text{J} \cdot \text{s}}{0.063\,\text{MeV}} \left(\frac{1\,\text{MeV}}{1.6 \times 10^{-13}\,\text{J}} \right) \approx 10^{-20}\,\text{s}$$

The lifetime of 10^{-20} s is very long compared to the characteristic time taken by the e^+ and e^-, traveling practically at the speed of light, to interact over the distance of a few fermi.

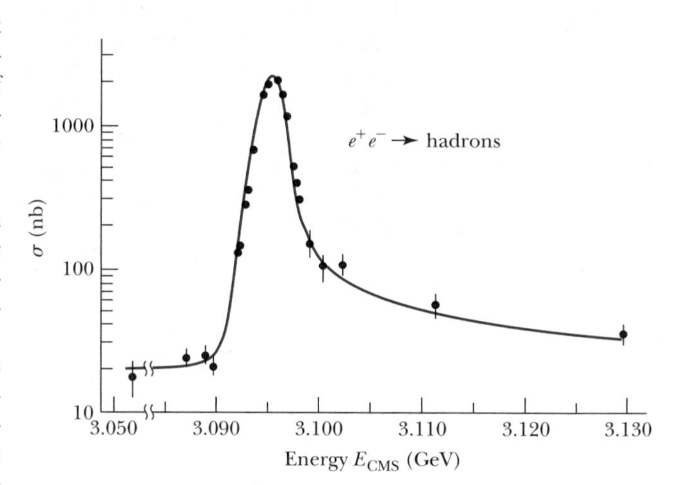

FIGURE 14.7 The experimental results of Burton Richter and his group at SLAC showing the observation of the J/ψ resonance near a mass 3.1 GeV/c^2. The Nobel prize-winning experiment was done with e^+ and e^- colliding head-on in the SPEAR storage ring at SLAC. The energies are in the center-of-mass system (CMS). *Data from Augustin et al., Phys. Rev. Lett. 33, 1406 (1974). Figure after D. H. Perkins, Introduction to High Energy Physics. Reading, MA: Addison Wesley, 1982, p. 205.*

14.4 Conservation Laws and Symmetries

As the number of elementary particles continued to increase, physicists were often perplexed by the occurrence of some reactions and decays. Physicists like to have clear rules or laws that determine whether a certain process can occur or not. We believe that in nature everything occurs that is not forbidden. Certain conservation laws are already familiar from our study of classical physics. These include energy, charge, linear momentum, and angular momentum. These are absolute conservation laws: they are always obeyed. In this section we will introduce additional conservation laws that are helpful in understanding the many possibilities of elementary particle interactions. As we shall see, some of these laws will be absolute, but others may be valid for only one or two of the fundamental interactions.

Baryon Conservation. In lower energy nuclear reactions, the number of nucleons is always conserved. We now know empirically that this is part of a more general conservation law for baryon number. We assign a new quantum number called *baryon number* that has the value $B = +1$ for baryons, -1 for antibaryons, and 0 for all other particles. *The conservation of baryon number requires the same total baryon number before the reaction as afterward.*

Let us examine this conservation law in a few examples. In the neutron decay, $n \rightarrow p + e + \bar{\nu}_e$, the baryon number is +1 on the left and the right sides, so baryon number is conserved. The antiproton was discovered in the reaction $p + p \rightarrow p + p + p + \bar{p}$. Note that as many as four particles must be produced in the reaction in order to create one antiproton; the conservation of baryons requires it: $B = 1 + 1 = 2$ on the left and $B = 1 + 1 + 1 - 1 = 2$ on the right. No fewer than three protons must be produced in order to create the one antiproton.

Example 14.3

Examine the conservation of baryon number in the reaction producing the Ω^- particle shown in the bubble chamber photograph of Figure 14.8.

Solution: This should be a straightforward application of the baryon conservation law. The reaction is

$$\text{K}^- + p \rightarrow \text{K}^0 + \text{K}^+ + \Omega^- \qquad (14.7)$$

The kaons all have $B = 0$, so on the left side we have $B = 0 + 1 = 1$. The Ω^- has $B = +1$, so on the right side we have $B = 0 + 0 + 1 = 1$. This was the experiment that first identified the Ω^- particle, some three years after Gell-Mann predicted its existence.

Lepton Conservation. The leptons are all fundamental particles, and *there is a conservation of leptons for* ***each*** *of the three kinds (families) of leptons; the number of leptons from each family is the same both before and after a reaction.* We let $L_e = +1$ for the electron and the electron neutrino, $L_e = -1$ for their antiparticles, and $L_e = 0$ for all other particles. We assign the quantum numbers L_μ for the muon and its neutrino and L_τ for the tau and its neutrino in a similar way. We now have three additional conservation laws, one each for L_e, L_μ, and L_τ, that must be obeyed in every reaction and decay.

Let us examine a few reactions to study the effects of these conservation laws. First, we look again at the neutron beta decay, $n \rightarrow p + e + \bar{\nu}_e$. The values of L_μ and L_τ are zero on both sides. On the left side we have $L_e = 0$, and on the right side we have $L_e = 0 + 1 - 1 - 0$, so L_e is also conserved. It now becomes clear why the antineutrino, rather than the neutrino, is produced in beta decay; the antiparticles play an important role in the weak interactions. The decay of the pion is another good example. The π^- first decays to μ^-, which in turn decays into e^-. However, we must also have neutrinos in order to conserve lepton number. The decays are

$$\pi^- \rightarrow \mu^- + \bar{\nu}_\mu$$

$$\mu^- \rightarrow e^- + \bar{\nu}_e + \nu_\mu$$

The first reaction has $L_\mu = 0$ on the left and $L_\mu = +1 - 1 = 0$ on the right. For the second reaction, we must conserve both L_e and L_μ. On the left side we have $L_e = 0$ and $L_\mu = 1$, and on the right side we have $L_e = 1 - 1 + 0 = 0$ and $L_\mu = 0 + 0 + 1 = 1$, so both L_e and L_μ are conserved.

At the present time there is no general theoretical foundation for the baryon and three lepton conservation laws. They are empirical results and should be tested. Some current theories predict that the proton is unstable with a lifetime greater than 10^{30} years. Because the proton is the lightest baryon, if it decays, the conservation of baryons would be violated. It has been predicted that the proton would decay into leptons. Several experimental groups have recently attempted measurements of the proton decay. Through 1998, no proton decays had been observed, and the proton lifetime had been pushed to greater than 10^{32} years. These experiments are of continuing interest to physicists.

Strangeness. In the early 1950s physicists had considerable difficulty understanding the myriad of observed reactions and decays. For example, the behavior of the K mesons seemed very odd. There is no conservation law for the production of mesons, but it appeared that K mesons, as well as the Λ and Σ baryons, were always produced in pairs in the reaction with protons studied most often, namely the $p + p$ reaction. In addition, the very fast decay of the π^0 meson into two photons (10^{-16} s) is the preferred mode of decay. One would expect the K^0 meson to also decay into two photons very quickly, but it does not. The long and short decay lifetimes of the K^0 are 10^{-8} and 10^{-10} s, respectively.

This strange behavior was described by assigning a new quantum number called *strangeness* to certain particles. Strangeness is conserved in the strong and electromagnetic interactions, but not in the weak interaction. The values of the strangeness quantum number S are listed in Table 14.4. The kaons have $S = +1$, lambda and sigmas have $S = -1$, the xi has $S = -2$, and the omega has $S = -3$. Their antiparticles all have the opposite sign for the S quantum number. When the strange particles are produced by the $p + p$ strong interaction, they must be produced in pairs to conserve strangeness. Although π^0 can decay into two photons by the strong interaction, it is not possible for K^0 to decay at all by the strong interaction. The K^0 is the lightest $S \neq 0$ particle, and there is no other strange particle to which it can decay. It can only decay by the weak interaction, which violates strangeness conservation. Because the typical decay times of the weak interaction are on the order of 10^{-10} s, this explains the longer decay time for K^0. Only $\Delta S = \pm 1$ violations are allowed by the weak interaction.

Example 14.4

Explain the various lifetimes of the hyperons shown in Table 14.4.

Solution: All the lifetimes for the hyperons are on the order of 10^{-10} s except for the Σ^0 which has a lifetime of 7×10^{-20} s. Note that the decay is $\Sigma^0 \rightarrow \Lambda + \gamma$, which has $S = -1$ on the left side and $S = -1 + 0 = -1$ on the right side. The Σ^0 is able to decay by the strong interaction to another strange particle with the same value of strangeness.

All the other hyperons in Table 14.4, however, have a decay that violates strangeness. Both Σ^+ and Σ^- decay to nucleons, which violates strangeness ($|\Delta S| = 1$). The Ξ^0 and Ξ^- both decay to Λ, which violates strangeness, because the left side has $S = -2$, whereas the right side has $S = -1$. Similarly, the Ω^- decays to either Ξ^0 or Λ; both violate strangeness, because $S = -3$ for Ω^-. These decay times are all on the order of 10^{-10} s, which is characteristic of the weak interaction.

The decay of Ω^- into the Λ is particularly interesting. At first glance it appears to violate strangeness by $|\Delta S| = 2$, but note that the reaction is actually $\Omega^- \rightarrow \Lambda + K^-$, so $S = -3$ on the left side, and $S = -1 - 1 = -2$ on the right side; thus, $\Delta S = 1$, because both Λ and K^- have $S = -1$.

One more quantity, called **hypercharge,** has also become widely used. The hypercharge quantum number Y is defined by $Y = S + B$. Hypercharge is the sum of the strangeness and baryon quantum numbers and is conserved in strong interactions. Because hypercharge is not independent of strangeness, the conservation laws of hypercharge and strangeness are also related. The hypercharge and strangeness conservation laws hold for the strong and electromagnetic interactions, but are violated for the weak interaction.

Val Fitch (1923–), Nobel Prize 1980), together with James Cronin, discovered a decay of the K meson that violated a widely held belief in the symmetry of charge conjugation and parity. Fitch and Cronin, both born in the United States, were professors at Princeton University at the time of their surprising experimental discovery. *Photo by Clem Fiori. Courtesy AIP Emilio Segrè Visual Archives.*

Symmetries

Symmetry of equations describing a system under some operation is a useful aid to physicists in understanding particle reactions. Familiar symmetry operations, for example, are translation or rotation of a system in space. Because symmetries lead directly to conservation laws, it is important that we discuss three important symmetry operators called *parity, charge conjugation,* and *time reversal.*

The conservation of parity P describes the inversion symmetry of space, that is $x \longrightarrow -x$, $y \longrightarrow -y$, and $z \longrightarrow -z$. Inversion, if valid, does not change the laws of physics. The conservation of parity is valid for the strong and electromagnetic interactions, but T. D. Lee and C. N. Yang (Nobel Prize for Physics, 1957) pointed out in the early 1950s that there was no experimental evidence for parity conservation in the weak interaction. They suggested, as a test, an experiment involving nuclear beta decay. Evidence was subsequently found by C. S. Wu for the nonconservation of parity in the beta decay of ^{60}Co, which verified the suspicions of Lee and Yang.

Charge conjugation C reverses the sign of the particle's charge and magnetic moment. It has the effect of interchanging every particle with its antiparticle. Charge conjugation is also not conserved in the weak interactions, but it is valid for the strong and electromagnetic interactions. Even though both C and P are violated for the weak interaction, for several years after Lee and Yang's work, it was believed that when both charge conjugation and parity operations are performed (called CP), conservation was still valid. We can also understand this from the general theoretical result that when all three operations are performed (CPT), where T is the time reversal symmetry, conservation holds. We believe that nature will proceed in the same manner both forward and backward in time for microscopic systems. We speak of the *invariance* of the symmetry operators, such as T, CP, and CPT.

This was the situation in 1964 when V. Fitch and J. Cronin (Nobel Prize, 1980) with their colleagues found that the K_L^0 meson decayed 0.3% of the time into two pions, rather than into three pions. Even though the decay into two pions is rare, it violates CP conservation and has tremendous ramifications for time reversal symmetry because of the strong belief in CPT invariance. The complete understanding and consequences of this result are still in some doubt today.

The symmetries discussed here are just a small part of those used in particle physics, which relies heavily on group theory, the branch of mathematics that utilizes symmetry. We will return to symmetry in Section 14.7.

14.5 Quarks

We are finally prepared to discuss quarks and how quarks form the many baryons that have been discovered experimentally. In 1961 Murray Gell-Mann and Yuval Ne'eman independently proposed a classification system called the **eightfold way** that separated the known particles into multiplets based on charge, hypercharge,

Eightfold way

and another quantum number called *isospin*, which we have not previously discussed. Isospin is a method of classifying different charged particles that have similar mass and interaction properties. The neutron and proton are members of an isospin multiplet we call the *nucleon*. In this case the isospin quantum number (I) has the value 1/2, with the proton having the substate value +1/2 ("spin up") and the neutron having −1/2 ("spin down"). Isospin is conserved in strong interactions, but not in electromagnetic interactions. After the eightfold way was worked out, it was noticed that some members of the multiplets were missing. Because of physicists' strong belief in symmetry, experimentalists set to work to find them, a task made easier because many of the particles' properties were predicted. The Ω^- was detected in 1964 at Brookhaven National Laboratory (see Figure 14.8 and Example 14.2) in this manner and confirmed the usefulness of the eightfold way.

It soon became clear, however, as other particles were discovered, that the eightfold way was not the final answer. In 1963 Gell-Mann and, independently, George Zweig proposed that hadrons were formed from fractionally charged particles called *quarks* (the term was coined by Gell-Mann from a line in James Joyce's book *Finnegans Wake* that says "three quarks for Muster Mark"; Gell-Mann is well known for his many other interests besides producing innovative theories). This quark theory was unusually successful in describing properties of the

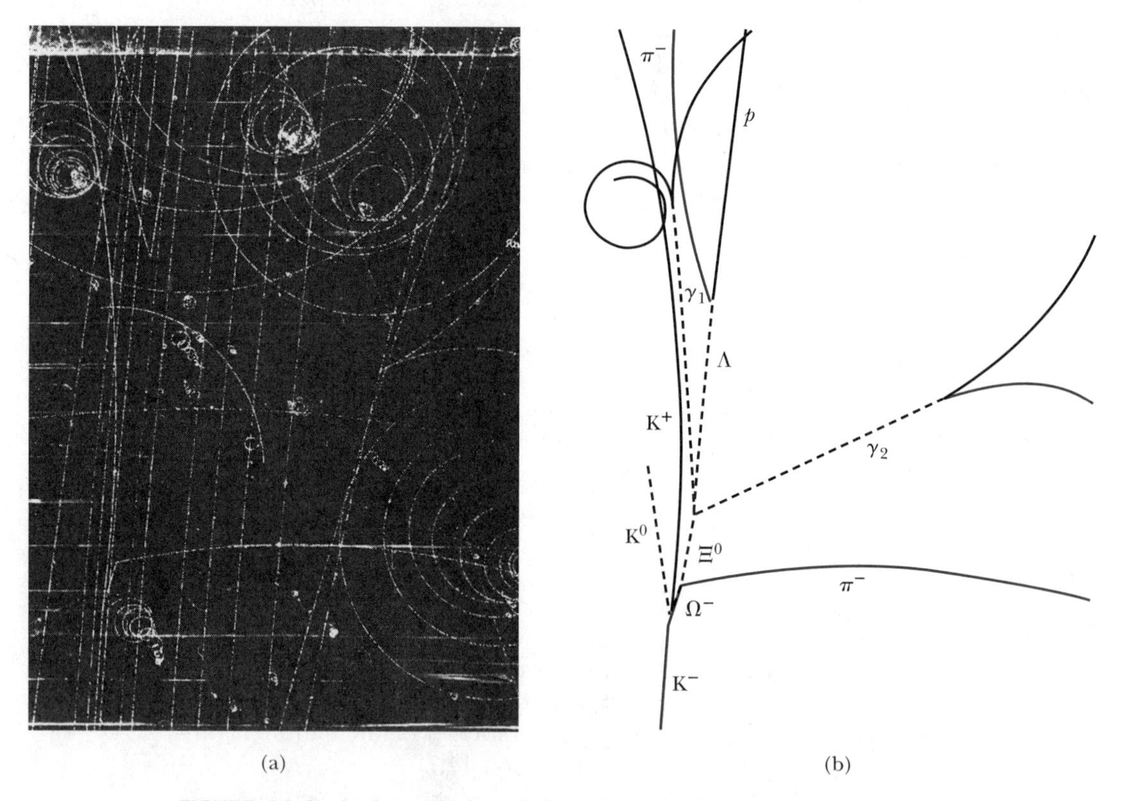

FIGURE 14.8 A photograph and the schematic diagram of a reaction from a liquid hydrogen bubble chamber at Brookhaven National Lab showing the production of Ω^- by the interaction of a K^- meson with a proton (a hydrogen nucleus in the bubble chamber). The neutral particles (dashed lines) leave no tracks in the bubble chamber. The Ω^- decays, producing the Ξ^0 and a π^-. Notice the variety of particles produced in the reaction and the eventual decays of short-lived particles. *Courtesy of Brookhaven National Laboratory.*

particles and in understanding particle reactions and decay. Three quarks were proposed, namely the *up* (u), *down* (d), and strange (s) (originally called "sideways"), with the charges $+2e/3$, $-e/3$, and $-e/3$, respectively. The strange quark has the strangeness value of -1, whereas the other two quarks have $S = 0$. Quarks are believed to be essentially pointlike, just like leptons.

With these three quarks, all the known hadrons could be specified by some combination of quarks and antiquarks. But there were still at least two problems: (1) It would be expected from symmetry that there would be an even number of quarks, not just three. (2) We discussed in the last section how the strangeness quantum number was able to explain the lifetimes of some of the known particles. For example, the differences in lifetimes of the hyperons were discussed in Example 14.4 and were explained in terms of the violation of strangeness conservation. Similarly, it was again believed that a new quantum number might be able to explain some additional discrepancies in the lifetimes of some of the known particles, and a fourth quark called the *charmed* quark (c) was proposed in 1970. A new quantum number called **charm C** was introduced so that the new quark would have $C = +1$ while its antiquark would have -1. Particles without the charmed quark have $C = 0$. Charm is similar to strangeness in that it is conserved in the strong and electromagnetic interactions, but not in the weak interactions. This behavior was sufficient to explain the particle lifetime difficulties. Another justification for charm, based on symmetry, was simply the fact that there were four known leptons at the time (the tau and its neutrino had not yet been discovered), and it seemed reasonable that there should also be four quarks. However fallacious this argument may seem, physicists have a very strong belief in the symmetry of nature. We now believe there are *six leptons and six quarks* and that both leptons and quarks come in pairs.

Charmed quark proposed

Experimentalists continued in their never-ending search for new elementary particles. In 1974 groups working independently and led by Burton Richter at SLAC and by Samuel Ting from MIT at Brookhaven National Laboratory found evidence in particle-scattering experiments for a heavy meson that required the existence of the charmed quark and its antiquark (see Example 14.2). Both Richter and Ting received the Nobel Prize in 1976 for the discovery of this new particle, named the *J/ψ particle* (the teams assigned different names). Other particles soon followed that had even higher masses and also needed the properties of charm.

Discovery of charmed quark

In 1977 physicists led by Leon Lederman at Fermi National Laboratory discovered the massive Υ (Upsilon) meson, which contains the **bottom** quark (b). It was also soon verified by scanning resonances from e^+e^- scattering, where three narrow resonances near 10 GeV of the Υ (Upsilon) meson require the existence of the b quark. There are strong theoretical arguments for believing that quarks come in pairs, and finally in 1995 the **top** quark (t) was discovered at the Fermi National Laboratory after a long search. These last two quarks are also called *truth* and *beauty*. Quantum numbers called **bottomness B** and **topness T** are assigned to these quarks.

Top and bottom quarks

Quark Description of Particles

We can now present the given quark properties and see how they are used to make up the hadrons. In Table 14.5 we give the name, symbol, mass, charge, and the quantum numbers for strangeness, charm, bottomness, and topness. The spin of all quarks (and antiquarks) is $\frac{1}{2}$.

TABLE 14.5
Quark Properties*

Quark Name	Symbol	Mass (GeV/c^2)	Charge	Baryon Number	Strangeness S	Charm C	Bottomness B	Topness T
Up	u	0.0015 to 0.005	$2e/3$	$\frac{1}{3}$	0	0	0	0
Down	d	0.003 to 0.009	$-e/3$	$\frac{1}{3}$	0	0	0	0
Strange	s	0.060 to 0.170	$-e/3$	$\frac{1}{3}$	-1	0	0	0
Charmed	c	1.1 to 1.5	$2e/3$	$\frac{1}{3}$	0	1	0	0
Bottom	b	4.1 to 4.4	$-e/3$	$\frac{1}{3}$	0	0	-1	0
Top	t	~174	$2e/3$	$\frac{1}{3}$	0	0	0	1

Antiquarks, $\bar{u}$, $\bar{d}$, $\bar{s}$, $\bar{c}$, $\bar{b}$, and $\bar{t}$, have opposite signs for charge, baryon number, S, C, B, and T.

*The u, d, and s quark masses are estimates of so-called "current-quark masses." See Footnote, Table 14.2.

A meson consists of a quark–antiquark pair, which gives the required baryon number of 0. Baryons consist of three quarks. We present the quark content of several mesons and baryons in Table 14.6. The structure is quite simple. For example, a π^- consists of $\bar{u}d$, which gives a charge of $(-2e/3) + (-e/3) = -e$, and the two spins couple to give 0 $(-\frac{1}{2} + \frac{1}{2} = 0)$. A proton is uud, which gives a charge of $(2e/3) + (2e/3) + (-e/3) = +e$; its baryon number is $\frac{1}{3} + \frac{1}{3} + \frac{1}{3} = 1$; and two of the quarks' spins couple to zero, leaving a spin $\frac{1}{2}$ for the proton $(\frac{1}{2} + \frac{1}{2} - \frac{1}{2} = \frac{1}{2})$.

Example 14.5

Check the neutron beta decay for the quark composition of each particle and check that the charge and baryon numbers are correct.

Solution: The neutron beta decay is given by

$$n \rightarrow p + e + \bar{\nu}_e$$

The corresponding quark composition for each of the particles is given by

$$\text{Neutron beta decay:} \qquad udd \rightarrow uud$$

where we don't have to worry about the quark composition of the electron and its antineutrino, because they are leptons, not hadrons. We have already checked that the quark composition of the proton gives a charge $+e$, so let us check that udd gives a zero charge for the neutron: $(2e/3) + (-e/3) + (-e/3) = 0$. The baryon number on both sides is 1, because there are three quarks. In neutron beta decay a *down* quark becomes an *up* quark.

What about the quark composition of the Ω^- that has a strangeness of $S = -3$? We look in Table 14.6 and find that its quark composition is sss. According to the properties in Table 14.5 its charge must be $3(-e/3) = -e$, and its spin is due to three quark spins aligned, $3(1/2) = 3/2$. Both of these values are correct. There is no other possibility for a stable omega (lifetime $\sim 10^{-10}$ s) in agreement with Table 14.4.

Color

There is one difficulty that perhaps you have noticed. Because the quarks have spin $\frac{1}{2}$, they are all fermions. According to the Pauli exclusion principle, no two fermions can exist in the same state. Yet we have three strange quarks in the Ω^-. This is not possible unless there is some other quantum number that distinguishes each of these same quarks in one particle. This problem is circumvented by establishing a new quantum number called **color.** A theory called **quantum chromodynamics (QCD)** is based on this concept. There are three colors for quarks, which we shall, for simplicity, call red (R), green (G), and blue (B). This color has absolutely nothing to do with the visual colors that we see. It is merely an attempt to distinguish this new property, which in some ways is analogous to the behavior of colored light. We can then call the corresponding colors for antiquarks: antired ($\overline{\text{R}}$), antigreen ($\overline{\text{G}}$), and antiblue ($\overline{\text{B}}$). *Color* is the "charge" of the strong nuclear force, in analogy to *electric charge* for electromagnetism.

TABLE 14.6 Quark Composition of Selected Hadrons

Particle	Quark Composition
Mesons	
π^+	$u\bar{d}$
π^-	$\bar{u}d$
K^+	$u\bar{s}$
K^0	$d\bar{s}$
D^+	$c\bar{d}$
D^0	$c\bar{u}$
Baryons	
p	uud
n	udd
Λ	uds
Σ^+	uus
Σ^0	uds
Ξ^0	uss
Ξ^-	dss
Ω^-	sss
Λ_C^+	udc

The two theories, quantum electrodynamics and quantum chromodynamics, are similar in structure; color is often called *color charge* and the force between quarks is sometimes referred to as *color force.* Back in Section 14.2 we mentioned that *gluons* are the particles that hold the quarks together. We show a Feynman diagram of two quarks interacting in Figure 14.9. A red quark comes in from the left and interacts with a blue quark coming in from the right. They exchange a gluon, changing the blue quark into a red one and the red quark into a blue one.

The rules for combining color are the following:

1. A color and its anticolor cancel out. We call this *colorless* (or white).
2. All three colors or all three anticolors in combination also cancel out and give colorless.
3. All hadrons are colorless.

In Figure 14.9 the gluon itself must have the color B$\overline{\text{R}}$ in order for the diagram to work. Quarks change color when they emit or absorb a gluon, and quarks of the same color repel, whereas quarks of different color attract.

Now let us see whether this works. Mesons are composed of a quark and antiquark of the same color, so the colors cancel to give colorless. The attraction of color to anticolor explains the stability of mesons. The baryons are composed of three quarks that must each have a different color. This avoids the difficulties of the Pauli exclusion principle. The three different colors in a baryon also cause stability and explain why the colors must all be different.

To finish the story we should mention that the six different kinds of quarks are referred to as *flavor.* There are six flavors of quarks (u, d, s, c, b, t). Each flavor has three colors. Finally, how many different gluons are possible? Using the three colors red, blue, and green, there are nine possible combinations for a gluon. They are B$\overline{\text{B}}$, B$\overline{\text{R}}$, B$\overline{\text{G}}$, R$\overline{\text{B}}$, R$\overline{\text{R}}$, R$\overline{\text{G}}$, G$\overline{\text{B}}$, G$\overline{\text{R}}$, and G$\overline{\text{G}}$. Note in Figure 14.10 that the gluon is B$\overline{\text{R}}$ and not BR. The combination B$\overline{\text{B}}$ + R$\overline{\text{R}}$ + G$\overline{\text{G}}$ does not have any net color change and cannot be independent. Therefore, there are only eight independent gluons. Gluons can interact with each other, because each gluon carries a color charge. Note that in this case, gluons, as the mediator of the strong force, are much different from photons, the mediator of the electromagnetic force.

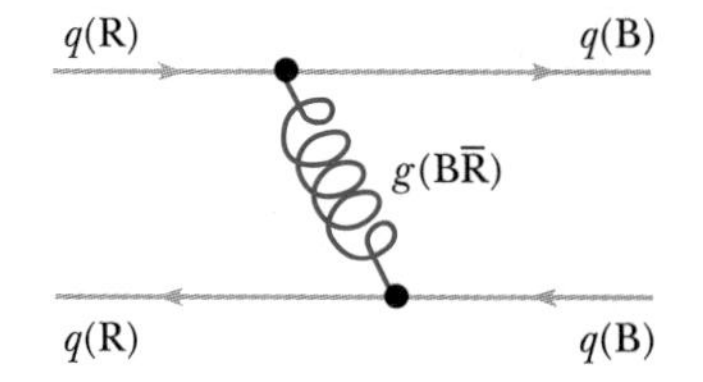

FIGURE 14.9 A Feynman diagram showing the exchange of a gluon (B$\overline{\text{R}}$) between a quark having color R and a quark having color B. The colors of the quarks are changed in the interaction.

To date no one has ever clearly observed a free quark. However, in 1967, Jerome Friedman, Henry Kendall, and Richard Taylor (Nobel Prize, 1990) performed experiments at the Stanford Linear Accelerator (SLAC) by scattering

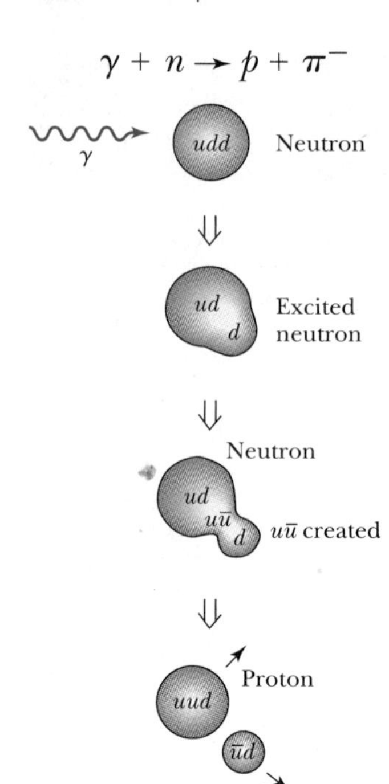

FIGURE 14.10 When a high-energy γ ray is scattered from a neutron, a free quark cannot escape because of confinement. For high enough energies, an antiquark-quark pair is created (for example, $u\bar{u}$), and a pion and proton are the final particles.

20-GeV electrons deep into protons and were surprised by the significant number of scattered electrons. Eventually this scattering was interpreted as evidence for pointlike quarks inside the proton. Notice the similarity between this experiment and that of Rutherford almost 50 years earlier. Both concluded there was something hard inside the object. Yet it was several years after 1967 before the idea of quarks was widely accepted by the general physics community.

Confinement

Physicists now believe that free quarks cannot be observed; they can only exist within hadrons. This is called **confinement.** We can explain what happens with a simple diagram like that shown in Figure 14.10, which shows three quarks confined within a neutron. At low energies the three quarks are easily contained and are free to move around. If an incident photon scatters from the neutron, one of the quarks may become so energetic that it tries to escape. Quarks have the property, however, that the color force transmitted by the exchanged gluons increases as the quarks get further apart. As one of the quarks moves away, the restoring force increases. If there is so much energy that color force can't confine the quarks within the neutron, then this extra energy will be able to create a quark–antiquark pair, and a meson is created. In the case shown in Figure 14.10, a proton and π^- are the final result. Apparently a single quark will not escape when a photon or electron interacts at high energy with a baryon like a neutron or proton. With enough energy several mesons may be produced, as long as all the conservation laws and quark rules are observed. The production of the delta

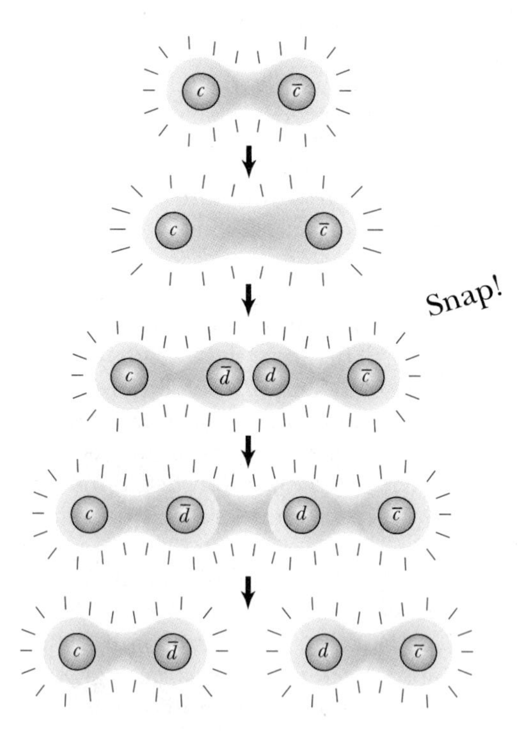

FIGURE 14.11 Quarks cannot be pulled away from a hadron. As the color-force field increases to pull out the quark, the increasing energy eventually goes into creating a quark pair. An analogy might be a spring that is pulled apart. More and more energy goes into the potential energy of the spring that eventually snaps. The energy in this case creates the new quark pair, and the system relaxes.

resonance in Section 14.3 is an example of this. High-energy electrons scattered from the proton produce the delta resonance at 1232 MeV, which subsequently decays to a $p + \pi^0$ or $n + \pi^+$. In both cases a quark–antiquark pair is produced.

Consider a hadron in which one of the quarks is being pulled away. The color-force field "stretches" between the quarks as shown in Figure 14.11. As the quark is pulled away, more and more energy is added to the field to keep it from stretching, much like the force on a spring that is stretched. In the quark case, however, the color-force field eventually "snaps" (like a spring breaking) into two new quarks, rather than pulling the quark out of the hadron. The energy is converted into the mass of new quarks, and the system relaxes back to the unstretched state (see Figure 14.11).

Quark masses cannot be measured directly, because they are confined within hadrons and are not observed as physical particles. The values of quark masses depend on the computational scheme used to determine them within a particular model. There is no one definition of quark masses that is the obvious choice. Therefore one must keep in mind that the quark masses tabulated in data listings like in Table 14.5 are determined from one particular fit of a phenomenological potential used and make sense only in the context of a particular quark model. The masses given here for *u*-, *d*-, and *s*-quarks are the so-called *current-quark masses*. These are only estimates and are somewhat controversial. The quark masses remain under active investigation.

14.6 The Families of Matter

Now that we have had a brief review of the particle classifications and have learned how the hadrons are made from the quarks, we should summarize. We presently believe that the leptons and quarks are fundamental particles. These fundamental particles can be divided into three simple families or generations as shown in Figure 14.12. Each generation consists of two leptons and two quarks. The two leptons are a charged lepton and its associated neutrino. The quarks are combined by twos or threes to make up the hadrons.

The vast majority of mass in the universe is made from the components in the first generation (electrons and *u* and *d* quarks). The second generation consists of the muon, its neutrino, and the charmed and strange quarks. The members of this generation are found in certain astrophysical objects of high energy and in cosmic rays, and are produced in high-energy accelerators. The third generation consists of the tau and its neutrino and two more quarks, the bottom (or beauty) and top (or truth). The members of this third generation existed in the early moments of the creation of the universe and can be created with very-high-energy accelerators. Note that in each group, the mass increases with the generation number, with the third generation having the greatest mass.

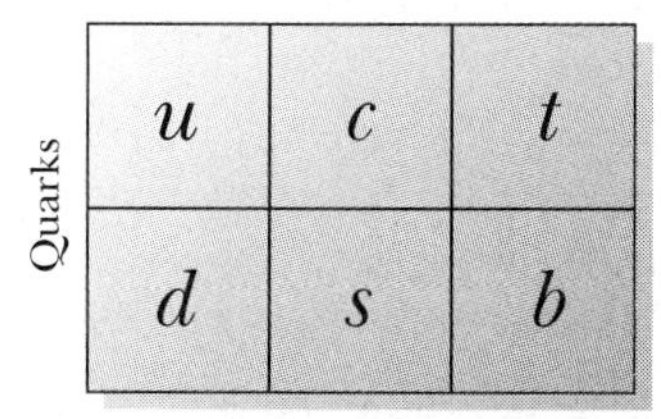

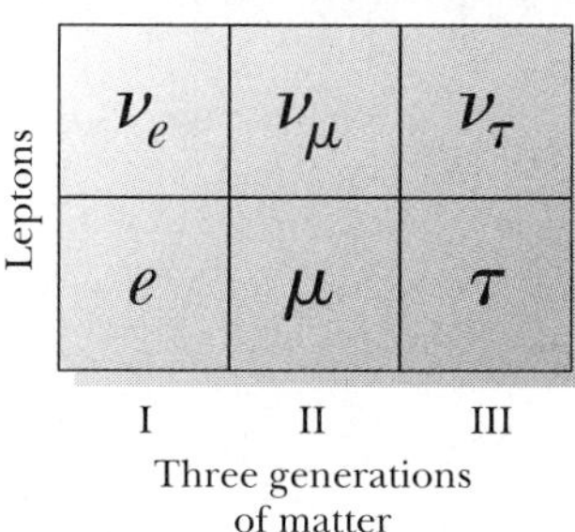

FIGURE 14.12 The three generations (or families) of matter. Note that both quarks and leptons exist in 3 distinct sets. One of each charge type of quark and lepton make up a generation. All visible matter in the universe is made from the first generation; second and third generation particles are unstable and decay into first generation particles.

14.7 The Standard Model and GUTs

The quest for unifying all the fundamental forces of nature into one theory has been an ultimate goal of some physicists for over a century. Maxwell unified the electric and magnetic forces into the electromagnetic theory in the 1860s. Glashow, Weinberg, and Salam unified the electromagnetic and weak interactions into the electroweak interaction in the 1960s. But the unsuccessful attempts are legend. Dirac badly wanted the proton to be the particle in the negative energy states of the electron in his relativistic electrodynamics theory of 1928. His theory would then encompass all the particles then known (proton, electron, and photon). Heisenberg was purported to have discovered a unified theory in the 1950s, but the details were never forthcoming. This rumor prompted Pauli's famous work of art shown in Figure 14.13 and his quote that "the details remain to be sketched in."

Standard Model

The most widely accepted theory of elementary particle physics at present is the **Standard Model.** It is a combination of the electroweak theory and quantum chromodynamics (QCD). Its details are too complicated to present here, but its general ingredients have already been presented in this chapter, and we summarize them now. All matter consists of particular *fermions,* which exert attractive or repulsive forces on each other by exchanging the force-mediating particles called *gauge bosons.* The two varieties of fermions are called *leptons* and *quarks,* and they are shown divided into the three generations in Figure 14.12.

Leptons are essentially pointlike and are fundamental elementary particles. There are three leptons with mass and three others with little or no mass (the *neutrinos*). Quarks and antiquarks make up the *hadrons* (*mesons* and *baryons*). Quarks may also be pointlike ($<10^{-18}$ m) and are confined together, never being in a free state. There are six flavors of quarks (up, down, strange, charmed, bottom, and top) and there are three colors (green, red, and blue) for each flavor. Rules for combining the colored quarks allow us to represent all known hadrons.

Bosons mediate the four fundamental forces of nature: gluons are responsible for the strong interaction, photons for the electromagnetic interaction, $W^{\pm}$ and Z^0 for the weak interaction, and the as yet unobserved graviton for the gravitational interaction. In our study of nuclear physics we discussed the pion as being the mediator of the strong force. At a more fundamental level, we can now say that the gluon is responsible. The gluon is responsible for the attraction between the antiquark and quark that make up the pion, and the gluon is responsible for the attraction between the quarks that make up the nucleons.

Although the Standard Model has been successful in particle physics, it doesn't answer nearly all the questions. For example, it is not by itself able to predict the particle masses. It is believed that the *Higgs boson* may be the key to unlocking this "black box" mystery of particle masses. Why are there only three generations or families of fundamental particles? Do quarks and/or leptons actually consist of more fundamental particles? All of our past experience would lead us to suspect this to be true. The understanding of the nuclear components was

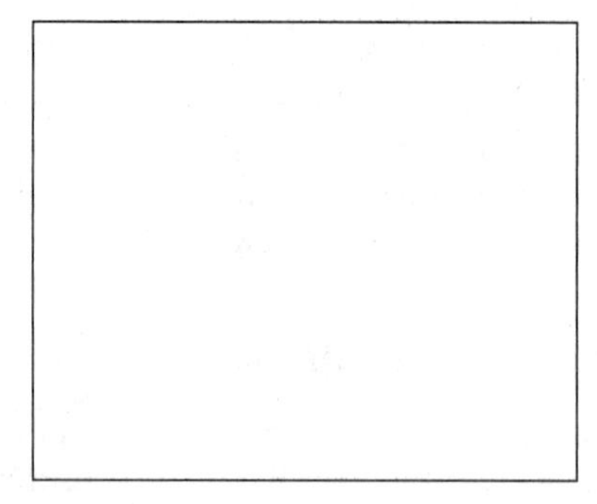

FIGURE 14.13 Wolfgang Pauli was a skeptic of elementary particle theories that purported to explain everything. Once, in the 1950s when a rumor of such a theory by Werner Heisenberg was circulating, Pauli noted that the details of Heisenberg's theory remained to be sketched in. He drew the figure shown and announced, "Below is the proof that I am as great an artist as Rembrandt; the details remain to be sketched in." *From O. W. Greenberg, Am. Scien.* ***76,*** *361 (1988).*

completed in 1932 with Chadwick's discovery of the neutron. The next level of understanding, that of the quarks, did not come for another 35 years. If there is another level, we may take even longer to reach it, because experiments are becoming more difficult and expensive to perform.

Physicists continue to search for answers. The Z^0 is useful because it interacts directly with both leptons and quarks. The Z^0 decays into a variety of particle–antiparticle pairs, each of which may allow physicists to unravel one more clue. At this time considerable data are being amassed on the Z^0.

Matter–Antimatter. According to existing theories of the creation of our universe in the Big Bang, matter and antimatter should have been created in exactly equal quantities. It appears that matter dominates over antimatter in our universe, and the reason for this has concerned physicists and cosmologists for years. Fortunately, we do have an opportunity to understand this dilemma. The tiny violation of *CP* symmetry in the kaon decay tilts the scales in terms of matter over antimatter. However, estimates indicate that this *CP* violation is too small by at least ten orders of magnitude to explain the predominance of matter.

In the last few years, physicists have proposed that experiments with B mesons might reveal considerably more about *CP* violation than kaons possibly could. B mesons are similar to kaons, but have a strange quark replaced by the much more massive bottom quark. Calculations indicate that *CP* violation might be 100 times greater for B mesons than for kaons. A careful study of *CP* violation in the decay of the B meson and the anti-B meson would be extremely interesting.

Several laboratories are racing to complete accelerator upgrades and/or new detector systems that will allow them to detect *CP* violation in B meson decays. Much higher intensities of B mesons are required. Most electron accelerators, like those at Stanford and Cornell in the United States, KEK in Japan, and DESY in Germany, will produce the upsilon (Υ) particle in head-on collisions of electrons and positrons at the resonance energy of 10.58 GeV, which in turn decays into a B meson and anti-B meson. Plans are also underway to use the proton accelerators at Fermilab and CERN to study B decay asymmetries at much higher energies. The next few years promise to be a rich harvest of B meson experiments.

Grand Unifying Theories

Einstein is known to have worked during the last 30 years of his life on a *unified field theory* that would encompass some of the four fundamental forces. Neither he nor anyone else has completely succeeded. There have been several attempts in recent years toward a Grand Unified Theory (GUT) to combine the weak, electromagnetic, and strong interactions. The theory of Howard Georgi and Sheldon Glashow of the 1970s has been one of the most successful, but the simplest version of their model predicted a proton lifetime too short by orders of magnitude. The various GUT theories make several predictions, including

1. The proton is unstable with a lifetime of 10^{29} to 10^{31} years. Current experimental measurements have shown the lifetime to be greater than 10^{32} years.
2. Neutrinos may have a small, but finite, mass.
3. Massive magnetic monopoles may exist. There is presently no confirmed experimental evidence for magnetic monopoles.
4. The proton and electron electric charges should have the same magnitude.

The unification of the strong and electroweak interactions is an important part of cosmological attempts to understand the origin of the universe. This will be examined more thoroughly in Chapter 16.

Superstrings. In recent years there has been a tremendous amount of effort by theorists in the *theory of superstrings.* In this revolutionary theory the universe is described by as many as ten dimensions rather than the four (x, y, z, t) that we have discussed. In superstring theory elementary particles do not exist as points, but rather as tiny, wiggling loops that are only 10^{-35} m in length. Particles of 10^{19} GeV (the Planck energy) would be required to probe a size as small as 10^{-35} m. Superstring theory seems so far removed from the real world that it would appear to be difficult to bridge the wide gap between theory and experiment. However, at the present time, superstring theory is the most promising approach to unifying the four fundamental forces, including gravity. Superstring theories are beginning to make predictions, and the masses of quarks and leptons are presently being calculated. The predictions for the bottom and the top quarks are close to experiment. The theories also predict the proton to be unstable, but its lifetime is greater than the present experimental limits. Neutrinos also apparently have mass. Superstring theory may be able to account for the dark matter in the universe (dark matter has been proposed as an explanation for the "missing mass" in the universe, see Chapter 16).

Supersymmetry. Supersymmetry is a necessary ingredient in many of the theories trying to unify the forces of nature. It is contained in the superstring theory just discussed. The details of supersymmetry are beyond the level of this book, but its basic premise is to transform particles or fields with one spin into other particles or fields whose spins differ by $\hbar/2$. It therefore relates the spin $\frac{1}{2}$ fermions (quarks and leptons) with the spin 1 gauge bosons (photon, gluon, $W^{\pm}$, Z^0). It even includes the spin 2 particles of the gravitational field, the gravitons. Supersymmetry starts with the GUTs and then adds gravity, or it can begin with gravity and add the other interactions.

At some fundamental level, matter is believed to be composed of spin $\frac{1}{2}$ particles (quarks and leptons). The gauge bosons mediate the color and electroweak interactions. At some very high energies, supersymmetry is valid. However, if supersymmetry is relevant at all to nature, it must be broken spontaneously, and we refer to *spontaneous symmetry breaking.* An example is shown in Figure 14.14, where we show the coupling strengths of the four interactions on a logarithmic scale as a function of energy (or mass). At some high energy there is unification that is spontaneously broken at lower energies. There is a richness of physics in the evolution of the state of the universe as various symmetries are broken. The violation of *CP* in kaon decay is a good example.

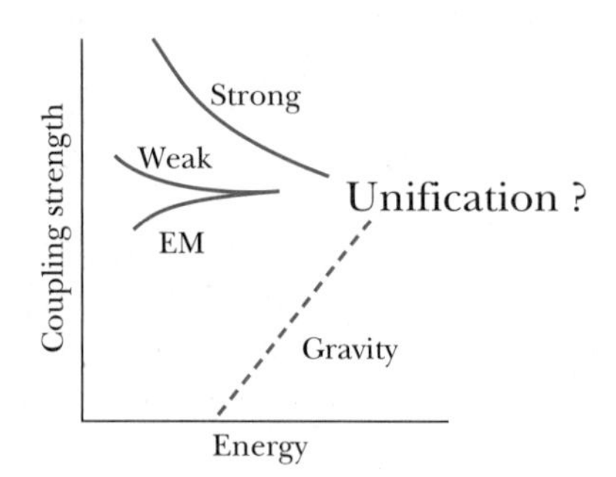

FIGURE 14.14 The coupling strength of the four fundamental interactions is shown plotted vs. energy. Both are log scales. Unification may occur at some very high energy, but at lower energies symmetry breaking separates the four interactions.

The inclusion of Einstein's general theory of relativity into the ideas that explain the other interactions of particles is a success of supersymmetry, although much remains to be done. An important prediction of supersymmetry concerns the possibility of many additional particles with masses that should be detectable. There might be a doubling of the number of elementary particles, with each current particle having a *superpartner* differing in spin by $\hbar/2$. Presently, none of the known leptons, quarks, or gauge bosons can be identified with a superpartner of any other.

The number of particles that have been predicted from a variety of different theories include the fanciful names of sleptons, squarks, axions, winos, photinos, zinos, gluinos, and preons. Only experiment will be able to wade through the vast number of unifying theories to weed out the good from the bad. The next few years will be very interesting in this exciting field of physics.

14.8 Accelerators

Although cosmic rays contain high-energy particles, the small intensity of those particles allows physicists to perform only limited experiments. Particle physics was not able to develop fully until particle accelerators were constructed with high enough energies to create particles of masses of about 1 GeV/c^2 or greater. Some early accelerators were briefly mentioned in Chapter 13.

The precursor of most modern accelerators is the cyclotron shown in Figure 13.2. The pioneering work of E. O. Lawrence (1901–1958, Nobel Prize, 1939) and M. S. Livingston led to the development of the cyclotron in the 1930s. Cyclotrons rely on charged particles moving in a circular orbit perpendicular to a magnetic field. The radius of curvature is given by $R = p/(qB)$ where p is the particle's momentum and q is its charge. The orbital frequency is given by

$$\nu = \frac{\omega}{2\pi} = \frac{v}{2\pi R} = \frac{mv}{2\pi mR} = \frac{p}{R}\frac{1}{2\pi m} = \frac{qB}{2\pi m} \tag{14.8}$$

where m is the mass of the particle. In a cyclotron, the gap between the two dees (in the shape of semicircles) contains a radio frequency voltage that accelerates the particle every time it passes through the gap. The charged particle travels thousands of orbits gaining two bursts of energy during each orbit.

There are three main types of accelerators used presently in particle physics experiments: synchrotrons, linear accelerators, and colliders. We will discuss each of these in turn.

Synchrotrons

Eventually at higher energies the cyclotron technique becomes limited by relativistic effects. The equation for the radius of curvature doesn't change as long as the relativistic momentum is used, but the orbital frequency becomes

$$\nu = \frac{qB}{2\pi m}\sqrt{1 - \frac{v^2}{c^2}} \tag{14.9}$$

In *synchrocyclotrons* the frequency of the RF voltage between the dees is adjusted to match this changing frequency. In *synchrotrons* the magnetic field is changed to keep the radius ($R = p/qB$) constant and match the frequency. In both cases the particles are accelerated in pulses or bunches, because the variation of RF voltage or magnetic field can only match one particular particle momentum. The Bevatron at Berkeley was an early (1954) high-energy proton synchro-

EXPERIMENTAL INGENUITY

Some people think of physicists as long-haired old men holed up in offices with indecipherable scribbling on a blackboard. Other people think of laboratories crammed with wires, pipes, and weird looking apparatus on which people are crawling, peering, adjusting, and taking measurements. Although there might be a tiny bit of truth to these stereotypes, physicists do tend to be either theorists or experimentalists. The best known scientists are perhaps the ones who propose radical new breakthroughs in thought that propel us forward in great jumps in our understanding of nature. Certainly Newton, Maxwell, and Einstein are in this group. But these great leaps in intellect generally require much experimental measurement to understand nature and to test the theories that have been proposed.

The field of experimental particle physics began around 1950, but we might argue that it started with the advent of particle accelerators around 1930. Experiments did not actually begin until many years later when energies reached about 1 GeV. Since 1955 there have been four Nobel Prizes awarded for theoretical work in elementary particle physics (1957, 1965, 1969, and 1979, see Appendix 9), but there have been eleven awarded for experimental work (1959, 1960, 1961, 1968, 1976, 1980, 1984, 1988, 1990, 1992, and 1995). Several of these experiments resulted in an increased understanding of physics, but some of them were for pure experimental development, which eventually led to experiments performed by others. It is this latter category that we want to highlight here because of the experimenters' special ingenuity.

We begin with the invention of the bubble chamber by physicist Donald Glaser at the University of California at Berkeley in the early 1950s. Bubble chambers, having much greater densities than cloud chambers, allowed much higher energy particles to be detected. Glaser tried several liquids that could be superheated, just short of boiling, and eventually settled on liquid hydrogen and xenon. He developed an ingenious method of photographing the tracks of the particles passing through the liquid. His efforts revolutionized experimental particle physics, and bubble chambers remained the mainstay of high-energy particle detectors for many years.

Though Glaser was the inventor of the bubble chamber, his colleague at Berkeley, Luis Alvarez, was its developer, and although nominated to receive the Nobel Prize with Glaser, Alvarez did not receive the Prize until 1968, eight years after Glaser. Alvarez and his colleagues used a large bubble chamber to discover the first of the new "resonance" particles. Although his Nobel Prize citation mentioned his development of the bubble chamber and measurements

Donald Glaser (1926–) is shown examining a detector at Lawrence Berkeley Laboratory around 1960. The bubble chamber was mostly developed while Glaser was a professor at the University of Michigan during the period 1950–1959. *Courtesy of the Lawrence Radiation Laboratory, University of California, Berkeley.*

Luis Alvarez (1911–1988), shown here with his bubble chamber, was born in California and spent almost his entire career at the University of California, Berkeley. Alvarez was also well known for his work on analyzing the film tape of John Kennedy's assassination, his searches for burial chambers in pyramids, and research with his son Walter on the demise of dinosaurs by the debris caused by an asteroid colliding with Earth. *Courtesy of Lawrence Berkeley Laboratory.*

with it, Alvarez undoubtedly was rewarded for his many contributions to science, which include forty patents comprising a wide range of projects including radar, color television, and a golf training device. He was particularly well known for his efforts in looking for unknown tombs beneath the Egyptian pyramids.

Carlo Rubbia and Simon van der Meer represent an unusual pair of Nobel Prize winners whose success depended on each other. Rubbia was an aggressive leader who made convincing arguments in the late 1970s to reconfigure an existing CERN accelerator to scatter protons and antiprotons head-on. He conducted the successful experiment to discover the W and Z particles in the early 1980s. Rubbia managed a huge team of hundreds of technicians, engineers, and scientists who built and operated the detector.

In contrast to the flamboyant Rubbia, the shy and unassuming van der Meer was responsible for inventing the stochastic cooling of the beam particles that allowed the accumulation of sufficient numbers of antiprotons to make Rubbia's experiment feasible. Although a surprise winner of the Nobel Prize with Rubbia, the selection of the Dutch engineer turned physicist was highly welcomed, because of his tremendous ingenuity in the very complex method of stochastic cooling.

The research labs of the world are full of scientists and engineers developing clever, new ways of building and operating equipment that will lead to important advances in the future.

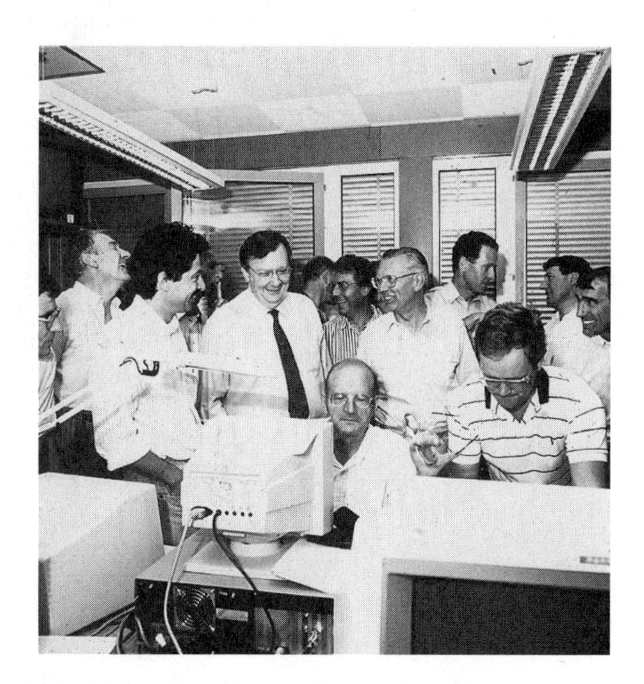

Carlo Rubbia (wearing the tie) and his colleagues await the start up of the LEP accelerator at CERN in 1989. *Courtesy of CERN.*

tron. Its energy was 6.4 GeV, just enough for E. G. Segrè (1905–1989) and O. Chamberlain (1920–) to produce antiprotons in 1955, for which they received the Nobel Prize in 1959.

Most modern proton accelerators are synchrotrons. The magnetic fields do not span the entire circle but rather encircle a closed pipe at a fixed radius. The proton synchroton (called the Tevatron) at the Fermi National Lab is currently the world's highest-energy proton accelerator at almost 1000 GeV (or 1 TeV).

Synchroton Radiation. A difficulty with cyclic accelerators is that when charged particles are accelerated, they radiate electromagnetic energy called *synchrotron radiation.* This problem is particularly severe when electrons, moving very close to the speed of light, move in curved paths. If the radius of curvature is small, electrons can radiate as much energy as they gain. The Large Electron–Positron (LEP) ring at the European Center for Nuclear Research (CERN) has a radius of 4.3 km in order to limit synchrotron radiation losses.

Light sources

Physicists have learned to take advantage of these synchrotron radiation losses and now build special electron accelerators (called **light sources**) that produce copious amounts of photon radiation for both basic and applied research in physics, chemistry, materials science, metallurgy, biology, and medicine. For example, the radiation is used in x-ray lithography to produce miniaturized computer chips with higher speed and greater capacity. Synchrotron radiation accelerators are being constructed at an increasing rate, even by industry for their own particular purposes. There are dozens of accelerators throughout the world used by thousands of users, including universities, government laboratories, and industries.

Linear Accelerators

The *linear accelerators* or **linacs** typically have straight electric-field-free regions between gaps of RF voltage boosts (see Figure 14.15). Because the particles gain speed with each boost, and the voltage boost is on for a fixed period of time, the distance between gaps becomes increasingly larger as the particles accelerate. Linacs are sometimes used as preacceleration devices for large circular accelerators. The longest linear accelerator is the 3-km-long Stanford Linear Accelerator (SLAC), which accelerates electrons up to 50 GeV. The SLAC linear accelerator gains its energy from a more complicated traveling wave, which accelerates the electron continuously along its 3-km path. It operated originally at 20 GeV, but a very successful upgrade later allowed it to operate at 50 GeV.

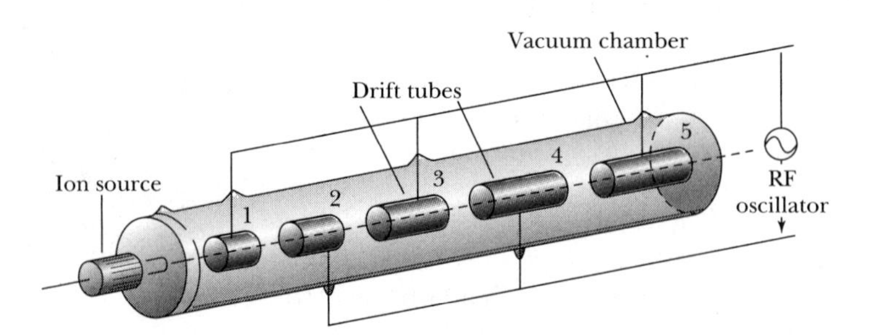

FIGURE 14.15 A schematic diagram of one type of linear accelerator. Each succeeding drift tube has to be longer because of the increasing speed of the particle. An RF voltage accelerates the charged particle in the region between the drift tubes. *Adapted from A. Arya, Elementary Modern Physics. Reading, MA: Addison-Wesley, 1974.*

Fixed-Target Accelerators

Most of the accelerators discussed so far have fixed targets. The accelerated particles are directed on a fixed target, where the reaction takes place. Because of conservation of momentum, the energy of the beam's particle is not fully available to cause reactions and create new particles. If we consider the reaction, $m_1 + m_2 \rightarrow$ anything, with the bombarding particle of mass m_1 having kinetic energy K on the fixed target of mass m_2, the amount of energy available in the center of mass system is

$$E_{cm} = \sqrt{(m_1c^2 + m_2c^2)^2 + 2m_2c^2K} \tag{14.10}$$

In Chapter 13 we found the threshold kinetic energy required to initiate a nuclear reaction with a fixed target. If we use relativistic relations, the threshold kinetic energy is

$$K_{th} = (-Q)\,\frac{\text{Total masses involved in reaction}}{2m_2} \tag{14.11}$$

See the following example for an indication of how little energy is available for reaction studies with fixed-target accelerators.

Example 14.6

(a) How much energy was available in the center of mass for the experiment of Segre and Chamberlain, who used 6.4-GeV protons on a fixed proton target to produce antiprotons in the reaction given here?

$$p + p \rightarrow p + p + p + \bar{p}$$

(b) How much beam energy was necessary to produce the antiprotons? (c) How much energy is available for a similar reaction with 1-TeV protons from the Tevatron on a fixed proton target?

Solution: (a) We can use Equation (14.10) to calculate the energy available in the center of mass. The rest energy of the proton is 938 MeV.

$$E_{cm} = \sqrt{[2(0.938\text{ GeV})]^2 + 2(0.938\text{ GeV})(6.4\text{ GeV})}$$
$$= 3.94\text{ GeV}$$

The total mass of the reaction products is $4(0.938\text{ GeV}/c^2) = 3.75\text{ GeV}/c^2$, so the Bevatron was constructed with enough energy to create the four particles in the final state.
(b) The threshold kinetic energy can be found by using Equation (14.11). The Q-value was defined in Equation (13.7) and is calculated to be

$$Q = (\text{Initial mass energies}) - (\text{Final mass energies})$$
$$= 2m_pc^2 - 4m_pc^2 = -2m_pc^2$$
$$= -2(0.938\text{ GeV}) = -1.88\text{ GeV}$$

If we insert this value into Equation (14.11), we obtain

$$K_{th} = (-Q)\,\frac{6m_p}{2m_p} = (1.88\text{ GeV})(3) = 5.6\text{ GeV}$$

The kinetic energy of 6.4 GeV was clearly enough to initiate the reaction, which is consistent with our result in (a).
(c) Now we insert the kinetic energy of 1 TeV into Equation (14.10) to determine

$$E_{cm} = \sqrt{[2(0.938\text{ GeV})]^2 + 2(0.938\text{ GeV})(1000\text{ GeV})}$$
$$= 43\text{ GeV}$$

Because of the conservation of momentum requirement, there is a tremendous reduction in the available energy for reactions for the 1000-GeV proton beam from the Tevatron.

Colliders

Because of the limited energy available for reactions like that found in part (c) of the preceding example, physicists decided they had to resort to colliding beam experiments where the particles meet head-on. If the colliding particles have equal masses and kinetic energies, the total momentum is zero and all the energy is available for the reaction and the creation of new particles. By

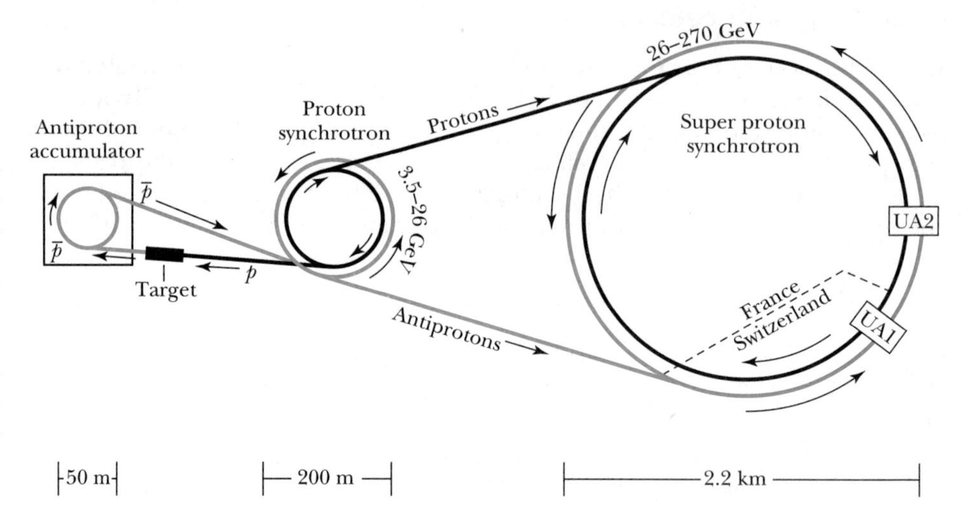

FIGURE 14.16 A schematic diagram of the Super Proton Synchroton at CERN, which straddles France and Switzerland, indicating how the protons and antiprotons are created, are accelerated, and finally interact in two detector regions at UA1 and UA2. The protons originate from the synchrotron accelerator system and produce antiprotons in the target in front of the antiproton accumulator. *After A. Kernan, Am. Scien.* ***74,*** *21 (1986).*

the 1960s physicists had gained enough experience in accelerator technology to build **colliders,** a new concept for accelerators. Colliding-beam accelerators usually have to utilize storage rings in order to collect enough particles to ensure sufficient intensity for the expected reaction. The CERN Large Electron–Positron (LEP) storage ring has electrons and positrons colliding as does the SLAC Stanford Linear Collider (SLC). The CERN Super Proton Synchrotron (SPS) uses an ingenious technique shown in Figure 14.16 to cause 270-GeV protons and antiprotons to collide head-on. The Tevatron has 1 TeV protons and antiprotons colliding. These machines require fantastic precision in order to cause particles moving at essentially the speed of light to interact head-on.

The LEP collider at CERN is presently the world's largest accelerator. It is contained in a tunnel 27 km in circumference located 100 m beneath ground straddling the border of France and Switzerland. Electrons and positrons collide head on at an energy of 180 GeV. Presently under development at CERN is the world's highest energy accelerator called the Large Hadron Collider (LHC) where 7-TeV protons will collide head-on with 7-TeV protons for a total energy of 14 TeV. Many countries, including the United States, are partners in this project, which should be operational by 2005. The LHC particle beams will be accelerated in the same tunnel as the LEP.

Example 14.7

How much energy would a fixed-target proton accelerator require to match the energy available in the Tevatron at Fermilab for a $p + \bar{p}$ reaction?

Solution: The energy available in the Tevatron is the sum of the colliding beams or 2 TeV. If we insert 2 TeV (2000 GeV) into Equation (14.15), we can solve for K, the kinetic energy.

$$2000 \text{ GeV} = \sqrt{[2(0.938 \text{ GeV})]^2 + 2(0.938 \text{ GeV})K}$$

The second term on the right-hand side will have to be much larger than the first term, so we can neglect the first term. We then solve for K.

$$K = \frac{(2000 \text{ GeV})^2}{2(0.938 \text{ GeV})} = 2.1 \times 10^6 \text{ GeV} \approx 10^{15} \text{ eV}$$

Such a fixed-target accelerator can currently not be constructed.

Summary

The beginnings of elementary particle physics occurred in the 1930s with the discoveries of the positron, the neutron, and the muon. The development of accelerators and particle detectors during the 1930s and the following decades has given physicists the tools needed to delve deeper into the structure of the nucleus and finally the elementary particles. Accelerators have included cyclotrons, synchrotrons, linear accelerators, and variations including storage rings and colliders.

The fundamental interactions are the gravitational, electroweak, and strong. For all practical purposes, except at very high energies, the electroweak is two interactions: electromagnetic and weak. Gluons are the mediators of the strong force, photons mediate the electromagnetic interaction, the $W^{\pm}$ and Z^0 bosons mediate the weak interaction, and an as yet unobserved particle, the graviton, mediates the gravitational interaction.

Elementary particles are classified in various ways. For example, fermions and bosons have half-integral and integral spins, respectively. Stable particles appear to be composed at some level of constituent fermions. We believe leptons are truly elementary, pointlike particles: electrons, muons, taus, and their respective neutrinos. Hadrons are particles that interact through the strong force and consist of two classes: mesons (integral spin) and baryons (half-integral spin). Mesons are unstable. The only stable free baryon is the proton, but some theories predict it to also be unstable. Neutrons and protons are stable when bound in nuclei. Baryons heavier than the nucleons are called *hyperons.*

The building blocks of matter can be described by three simple families shown in Figure 14.12. Each family consists of two leptons and two quarks. The two leptons are a charged lepton and its associated neutrino. Quarks are combined by twos (mesons) and threes (baryons) to make the hadrons.

The conservation of baryon number and three separate laws for the conservation of leptons appear to be universally valid. Symmetry breaking seems to be spontaneous and widespread. For example, there are other conservation laws that appear to be valid for one or more of the interactions, but not for all of them. These include strangeness, charge conjugation, and parity.

A breakthrough in the understanding of elementary particle structure began in the 1960s with the introduction of quarks. Quarks have fractional electric charges and only exist as constituents of hadrons. There are six quarks: up, down, strange, charmed, bottom, and top. Quantum chromodynamics (QCD) theory establishes how quarks are combined to form elementary particles. A property called *color* is required to understand how quarks and antiquarks combine.

The standard model combines the electroweak and QCD theories and has been quite successful in explaining elementary particle physics. For example, another particle (or group of particles) called the *Higgs boson* is expected, but has not yet been identified. Grand Unifying Theories (GUTs) have taken another step in unifying the fundamental interactions, but some of their predictions have already been negated. Research on GUTs and new theories called *superstrings* continues.

Questions

1. What are the characteristics of the following conservation laws: mass-energy, electric charge, linear momentum, angular momentum, baryon number, lepton number? Explain how they are related to fundamental laws of nature.
2. Why are storage rings useful for high-energy accelerators?
3. Why are colliding beams useful for particle physics experiments?
4. Can a baryon be produced when an antibaryon interacts with a meson? Explain.
5. What kinds of neutrinos are produced in the reaction $\mu^+ + e^- \to 2\nu$? Explain.
6. What mediating particles are exchanged between two positrons? Between two quarks?
7. Which families of particles seem to be truly elementary?
8. Does it appear that the total baryon number of the universe is zero? What would that mean?
9. Explain how a magnet may be used to distinguish a range of energies for protons. How can a monoenergetic beam of protons be obtained?
10. Why is it a problem when a particle gets out of phase with the frequency of a pulsed accelerator?

Problems

14.1 The Early Beginnings

1. What are the frequencies of two photons produced when a proton and antiproton annihilate each other at rest?
2. The mass of the charged pion is 140 MeV/c^2. Determine the range of the nuclear force if the pion is the mediator.

14.2 The Fundamental Interactions

3. The strong interaction must interact within the time it takes a high-energy nucleon to cross the nucleus. Using an appropriate speed and distance, estimate the time for the strong interaction to occur.

14.3 Classification of Elementary Particles

4. Supply the missing neutrinos in the following reactions or decays.
(a) $\mu^+ \rightarrow e^+ + ?$
(b) $? + p \rightarrow n + e^+$
(c) $\pi^- \rightarrow \mu^- + ?$
(d) $\mathrm{K}^- \rightarrow \mu^- + ?$
(e) $? + n \rightarrow p + \mu^-$
5. Calculate the approximate baryon number of the Earth.

14.4 Conservation Laws and Symmetries

6. Consider the two following reactions:

$$\mathrm{K}^0 + p \rightarrow \Sigma^+ + \pi^0$$

$$\overline{\mathrm{K}}^0 + p \rightarrow \Sigma^+ + \pi^0$$

Do both of these reactions obey the conservation rules? Does this explain why K^0 is not its own antiparticle?
7. Explain why each of the following reactions is forbidden.
(a) $p + p \rightarrow p + p + n$
(b) $p + p \rightarrow p + \pi^+ + \gamma$
8. Explain why each of the following reactions is forbidden.
(a) $p + \overline{p} \rightarrow \mu^+ + e^-$
(b) $\gamma + p \rightarrow n + \pi^0$
(c) $\gamma \rightarrow \pi^+ + \pi^-$
9. Determine the energy of a γ ray produced in the decay of Σ^0 at rest into $\Lambda + \gamma$.
10. A π^0 of kinetic energy 600 MeV decays in flight into 2 γ rays of equal energies. Determine the angles of the γ rays from the incident π^0 direction.
11. Complete the following reactions.
(a) $\mu^- + p \rightarrow n + ?$
(b) $n + p \rightarrow \Sigma^0 + n + ?$

14.5 Quarks

12. Show that electric charge, baryon number, and strangeness for the neutron, Σ^+, and Λ_C^+ are all equal to the sum of their quark configurations.
13. Show that electric charge, baryon number, and strangeness for the π^+, K^+, and D^0 are all equal to the sum of their quark configurations.
14. Determine the quark composition of the D^0 and D^+ charmed mesons.
15. Determine the quark composition of the B^+, B^- and B^0 mesons.
16. What kind of particle would you expect to be made of the $\bar{c}u$ configuration?

14.7 The Standard Model and GUTs

17. Assume the half-life of the proton is 10^{33} y. How many decays per year would you expect in a tank of water containing 100,000 gallons of water? Assume the bound protons can decay.
18. Some GUT theories allow the proton to be unstable. What conservation laws are broken in the following proton decays?
(a) $p \rightarrow \pi^+ + \overline{\nu}_e$
(b) $p \rightarrow \mu^+ + \pi^0$
(c) $p \rightarrow e^+ + \mathrm{K}^0 + \nu_e$
19. Assume that half of the mass of a 50-kg person consists of protons. If the half-life of the proton is 10^{32} years, calculate the number of proton decays per day from the body.
20. A Σ^+ particle of kinetic energy 3 GeV is produced inside a bubble chamber. Ignoring energy losses, what is the mean distance the Σ^+ travels in the detector?

14.8 Accelerators

21. Calculate the speed of the 7-TeV protons to be accelerated in the LEP collider at CERN.
22. A cyclotron is used to accelerate protons to 60 MeV. If these protons are elastically scattered from deuterons (^{2}H) and tritons (^{3}H), what are the maximum energies of the ^{2}H and ^{3}H?
23. Calculate the minimum kinetic energy of a proton to be scattered from a fixed proton target in order to produce an antiproton.
24. A magnetic field of 1 T is used to accelerate 10-MeV protons in a cyclotron. (a) What is the radius of the magnet? (b) What is the cyclotron frequency?
25. Show that for higher particle energies the simple cyclotron frequency in Equation (14.8) becomes limited by relativistic effects. Show that Equation (14.9) is the correct orbital frequency:

$$\nu = \frac{qB}{2\pi m}\sqrt{1 - \frac{v^2}{c^2}}$$

26. Show that Equation (14.10) is the correct relativistic result for the amount of energy available in the center of mass system when the reaction $m_1 + m_2 \rightarrow$ anything occurs with a bombarding particle of mass m_1

f mass m_2.
nit to

m, de-
owing
Discuss

beam

lculated
10-MeV
a 1-GeV

bout half
s around
Gradient
ement.

$\rightarrow p + \Lambda +$
equired for
a stationary
l-on?
ys is not al-

0

n

33. Determine which of the following decays or reactions is not allowed and explain why.
(a) $\overline{p} + p \rightarrow \overline{\Lambda} + \Lambda$
(b) $n \rightarrow p + e^- + \nu_e$
(c) $\Xi^0 \rightarrow n + \pi^0$
(d) $\pi^- + p \rightarrow \pi^- + \Sigma^+$
(e) $\pi^+ \rightarrow \mu^+ + \nu_\mu$

34. Determine which of the following decays or reactions is not allowed and explain why.
(a) $p \rightarrow e^+ + \pi^0$
(b) $\pi^- + p \rightarrow \Lambda + K^0$
(c) $p + p \rightarrow p + p + \pi^0$
(d) $n \rightarrow p + e^- + \overline{\nu}_e$

35. Determine which of the following decays or reactions is not allowed and explain why.
(a) $\Xi^- \rightarrow \Lambda + \pi^-$
(b) $\Lambda \rightarrow p + \pi^0$
(c) $p + p \rightarrow p + \pi^+ + \overline{\Lambda} + \overline{K}^0$
(d) $\Omega^+ \rightarrow \Xi^+ + \pi^0$

36. What is the relation between the de Broglie wavelength λ, mass m, and kinetic energy K for (a) a low-energy proton, and (b) a very-high-energy proton?

37. Consider the 7 TeV protons that will be accelerated in the LHC collider at CERN. (a) Use the results of Problem 26 or 27 to find the available center of mass energy if these protons collide with other protons in a fixed-target experiment. (b) Compare your results in (a) with the center of mass energy available in a colliding-beam experiment.

CHAPTER 15

General Relativity

There is nothing in the world except empty, curved space. Matter, charge, electromagnetism, and other fields are only manifestations of the curvature.

John Archibald Wheeler

Despite the great usefulness of the *special* theory, it was the introduction of his *general* theory of relativity in 1916 that made Einstein a celebrity. Although it is more complicated than the special theory, Einstein proposed several experiments to test his general theory. A well publicized journey in 1919 to South America and Africa by scientists to test the gravitational bending of light in a solar eclipse resulted in agreement between experiment and prediction, sealing Einstein's fame. In this chapter we will discuss this and other remarkable experiments that support Einstein's general theory. We also will mention some other effects that are still the subject of present-day research.

15.1 Principle of Equivalence

In the special theory of relativity we encountered inertial frames of reference moving at uniform relative velocities. A more general situation has systems moving in nonuniform motion with respect to one another. Linearly accelerated systems or rotating frames are examples of such motion. It was rather easy to accept the idea that uniform motion is relative, for the special theory can be used to understand the physics in any of the relative inertial systems. But what about nonuniform motion? Is it also relative? Einstein believed so, and after several years of thinking about the problem, he presented his *general theory of relativity* in 1916.

We can imagine another *gedanken* experiment to help us understand. Imagine an astronaut sitting in a confined space on a rocket placed on Earth ready to blast off for a mission to Mars (see Figure 15.1a). The astronaut is strapped into a chair that is mounted on a weighing scale that indicates a mass M. The

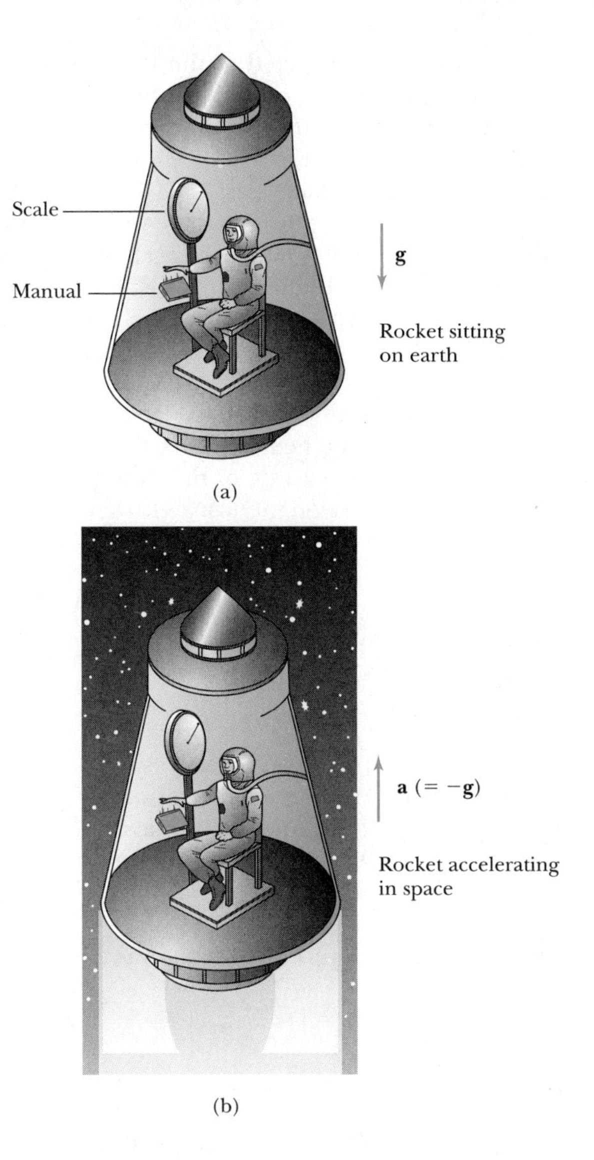

FIGURE 15.1 An astronaut sits in a confined space inside a rocket with no windows. The astronaut sits on a chair mounted on an ordinary weighing scale and drops a safety manual to the floor. In (a) the rocket is sitting at rest on Earth, and in (b) the rocket is accelerating upward with acceleration equal to $-\mathbf{g}$. No experiment the astronaut can do can distinguish between gravity and acceleration (forget the noise!).

astronaut drops a safety manual that falls to the floor. Now contrast this situation with the one shown in Figure 15.1b. The rocket is now on its way to Mars and is far enough away from Earth that the gravitational attraction is negligible. However, the rocket is still accelerating in order to gain speed for the trip to Mars. If the acceleration at this point has exactly the same magnitude as g on Earth, then the weighing scale will indicate precisely the same mass M that it did on Earth, and the safety manual, when dropped, will still fall with the same acceleration. The question is: How can the astronaut tell whether the rocket is sitting on Earth or accelerating in space? Einstein provided the answer in the **Principle of Equivalence:**

Principle of Equivalence

> *There is no experiment that can be done in a small confined space that can detect the difference between a uniform gravitational field and an equivalent uniform acceleration.*

Astronauts experience this equivalence to a certain degree when they are in free fall, rotating around the Earth in the space shuttle orbiter. Does the fact they feel weightless mean they are not attracted to the Earth by gravity? No. In fact, they are "falling around the Earth." The inside of the orbiter is almost an inertial reference system, and the astronauts can do experiments inside the orbiter on a small particle that will obey Newton's laws. However, according to an inertial reference system fixed in the distant stars, the shuttle will be in an accelerated reference system. If the astronauts look out the window at the Earth and the distant stars, they can figure out what is happening. However, if we cover the windows and do not let them see outside their tiny capsule, they might not be able to understand precisely why they are weightless. In fact, the astronauts can figure out they are in a gravitational field because weightlessness actually occurs only at the center of mass of the orbiter. If they put a drop of water in a corner of the orbiter, it will not have a precisely spherical shape because the gravitational field of the Earth is not uniform. The drop becomes very slightly bulged on both sides closest and farthest from the Earth because of the change in the Earth's gravitational field. In this manner the astronauts can tell they are in a gravitational field. The forces are called *tidal,* because the same forces are responsible for the ocean tides.

Tidal forces

Let us look briefly at gravitational and inertial forces. According to Newton's second law, a body accelerates in reaction to a force according to its *inertial mass* m_I,

$$\mathbf{F} = m_I \mathbf{a}$$

Inertial mass m_I measures how strongly an object resists a change in its motion.

The law of gravitation states that a *gravitational mass* reacts to a similar force in a field $\mathbf{g}$ that depends on position and other masses. Gravitational mass m_G measures how strongly an object is attracted to other masses.

$$\mathbf{F} = m_G \mathbf{g}$$

For the same force we set the two preceding equations equal.

$$\mathbf{a} = \left(\frac{m_G}{m_I}\right)\mathbf{g} \tag{15.1}$$

Equivalence of gravitational and inertial masses

According to the principle of equivalence, the inertial and gravitational masses are equal.

The principle of equivalence has been proposed in various forms for hundreds of years—including versions by Galileo, Newton, and Einstein. We encountered it in introductory physics when we discussed the equivalence of inertial and gravitational masses. Galileo and Newton both performed tests of the mass equivalence using a pendulum and found a result good to 10^{-3}. Pendulums of equal lengths would have periods proportional to the ratio $\sqrt{m_I/m_G}$ (see Problem 1). During a couple of decades around 1900, Eötvös performed a remarkable series of experiments showing the two masses are equivalent to 1 part in 10^8. More recent experiments show they are equivalent to 1 part in 10^{11}. The equivalence of inertial and gravitational masses helps in understanding Einstein's assertion that inertial and gravitational forces are equivalent. Einstein believed that if no experiment can distinguish inertial and gravitational forces, then they must be the same thing.

The principle of equivalence is the key to Einstein's general theory of relativity. The advanced mathematics needed prevents us from delving too far into the theory. We will only present some results and several predictions of the theory. For a few years after Einstein presented his general theory, there was a flurry of activity to test several predictions that Einstein had made. After this initial period, when no experiments seemed to contradict his theory, the subject was practically ignored until the 1960s. Experiments had not conclusively proven Einstein was correct, and there were competing theories (these other versions will not be discussed here). Since the 1960s, partly because of space probes, many experiments, including those in astrophysics, have confirmed Einstein's general theory. Few scientists now doubt it, and the general theory has opened new fields of study in astrophysics and cosmology that include research in mathematics, particle physics, nuclear physics, and astronomy. We will return to these matters again in the next chapter when we discuss cosmology and the origin of the universe.

There is one simple, but perhaps surprising, prediction of the equivalence principle. Consider a rocket accelerating through a region of space where gravitational force is negligible. A hole masked over a window allows a burst of light from a distant star to enter the spacecraft. Consider in Figure 15.2 what the

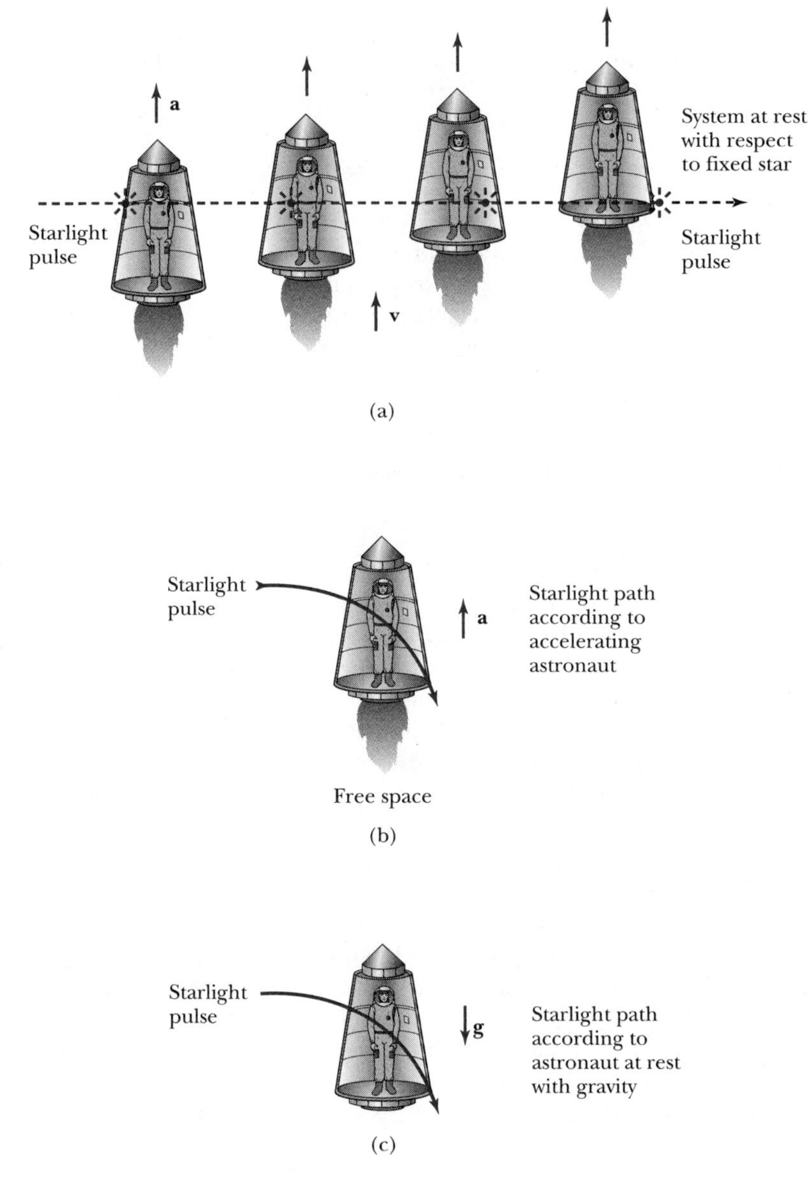

FIGURE 15.2 Starlight enters a small hole in a spacecraft while the rocket is accelerating. (a) The burst of starlight will hit a spot on the opposite wall at a point lower than where the light came in (greatly exaggerated here). (b) According to the astronaut inside, the light pulse curved downward and must have been affected by the acceleration. (c) According to the equivalence principle, the same thing must happen on the Earth because of gravity.

astronaut inside sees. By the time the horizontally moving light pulse hits the opposite wall, the rocket has moved up considerably (Figure 15.2a). In Figure 15.2b we show what happens to the light pulse *according to the astronaut inside:* the light pulse *curves downward* as it travels through the spacecraft. This effect occurs because the rocket is accelerating vertically while the light pulse traverses the spacecraft. Now consider the same thing happening on the rocket placed on Earth in a gravitational field. According to the equivalence principle, exactly the same thing happens: *the light pulse is attracted by the gravitational field and curves downward* as shown in Figure 15.2c. Of course, we have greatly exaggerated the effect in Figure 15.2.

Light is affected by a gravitational field

If we think about it a little, this should not really come as a great surprise. We showed in Section 2.12 that energy and mass are equivalent. A light pulse has energy, and it therefore can act as if it had a mass. We can think of the bending of the light pulse as simply the gravitational attraction of light. The effect is small and has little effect on Earth. A beam of light sent across a distance as wide as the United States and under a gravitational attraction of g would be deflected less than 1 mm. However, we will show in the next section that the bending of light has been observed experimentally and that other gravitational effects of light have even been observed on Earth.

15.2 Tests of General Relativity

Many tests of general relativity have been performed since Einstein announced his theory in 1916. Einstein himself proposed three tests, one of which was performed by 1919. Because of extreme experimental difficulties, only a few experimental tests were done until the 1960s. A constant stream of experiments has been done in recent years.

Bending of Light

As we have already discussed and as was predicted by Einstein, light should be bent while passing through a strong gravitational field. Starlight passing close to the sun will be bent away from its normal direction while passing close to the sun, as is shown in Figure 15.3. A total eclipse of the sun was to occur in May 1919, and preparations were quickly made following the end of World War I to mount observations in both South America and Africa, both good places to observe the eclipse. Einstein's general theory predicted a deflection of 1.75 seconds of arc, and the two measurements found 1.98 ± 0.16 and 1.61 ± 0.40 seconds. Einstein became an international celebrity because of the experiment. Many experiments, using both starlight and radio waves from quasars, have obtained agreement with increasingly good accuracy.

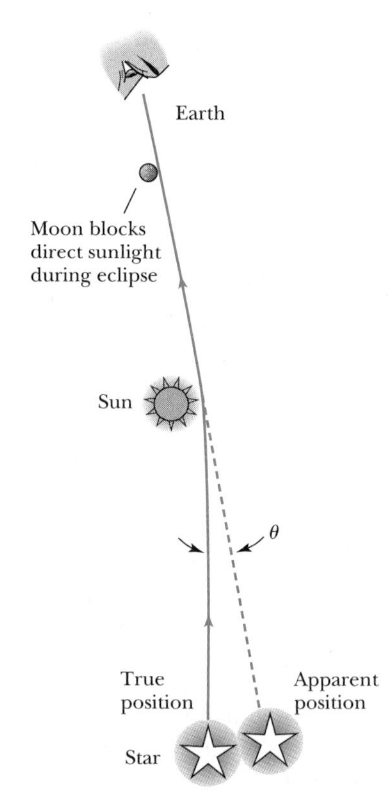

FIGURE 15.3 Starlight passing close to the sun will be bent due to gravitational attraction. The effect is that the position of stars are not always where they appear.

Gravitational Redshift

The second test of general relativity also relies on the gravitational attraction on light. Imagine a light pulse being emitted from the surface of the Earth that travels vertically upward. The gravitational attraction of the Earth cannot slow down light, but it can do work on the light pulse to lower its energy. This is similar to a rock being thrown straight up. As it goes up, its gravitational potential energy increases while its kinetic energy decreases. A similar thing happens to a light pulse. A light pulse's energy depends on its frequency ν through Planck's constant, $E = h\nu$. As the light pulse travels up vertically, it loses kinetic energy and its frequency decreases. Its wavelength increases, and visible light will be shifted toward the red end of the visible spectrum. We say that the light is *redshifted,* and the general phenomenon is referred to as the **gravitational redshift.**

We will perform the calculation close to the surface of the Earth where g is constant. The energy lost when traveling up a distance H is mgH. If ν is the frequency of the light at the bottom, and ν' the frequency at the top, energy conservation gives

$$h\nu = h\nu' + mgH \tag{15.2}$$

We can substitute an effective mass of light by letting $m = E/c^2 = h\nu/c^2$ to obtain

$$h\nu = h\nu' + h\nu\frac{gH}{c^2}$$

If we cancel Planck's constant h and let $\Delta\nu = \nu - \nu'$, we have

$$\frac{\Delta\nu}{\nu} = \frac{gH}{c^2} \tag{15.3}$$

Measurements comparing frequency differences are very sensitive, and such an experiment was performed by Pound and Rebka in 1960 at the Harvard tower using gamma rays from radioactive ^{57}Co ($^{57}Co \rightarrow {}^{57}Fe \rightarrow$ gamma ray). They sent the γ rays down the tower, so the γ rays gained energy, increasing their frequency. In this case, a *blueshift* occurs. Pound and Rebka used the Mössbauer effect* to obtain the needed sensitivity of $\Delta\nu/\nu \approx 10^{-15}$.

The general relationship for Equation (15.3) will not depend on the gravitational potential energy mgH but rather on $GMm(1/r_1 - 1/r_2)$. The ratio of the difference in frequency is

$$\frac{\Delta\nu}{\nu} = -\frac{GM}{c^2}\left(\frac{1}{r_1} - \frac{1}{r_2}\right) \tag{15.4}$$

where clocks measuring the frequency are placed at distances r_1 and r_2 from the center of the gravitational field. When $r_1 < r_2$, $\Delta\nu$ is negative. A signal leaving r_1 will appear to have a lower frequency when arriving at r_2. This means that an atom emitting light signals in a strong gravitational field (r_1 is small near the mass center) will have its wavelength redshifted when the light arrives at r_2 far away from the mass center.

Gravitational redshifts from stars are difficult to measure, because of the larger *Doppler redshift* due to the star moving away from us at high speed; a few stars are observed to be moving toward us. Several experiments using light from the sun and from white dwarfs have confirmed the existence of the gravitational redshift. A very accurate experiment was done by comparing the frequency of an atomic clock flown on a Scout D rocket to an altitude of 10,000 km with the frequency on a similar clock on the ground. The measurement agreed with Einstein's general relativity theory to within 0.02%.

These instruments were used on the island of Sobral (off the coast of Brazil) to observe the eclipse of May 29, 1919. The mirrors mounted in front of the telescopes reflected the light into the two horizontally mounted telescopes. The results of Eddington and Dyson confirmed Einstein's predictions. *From Arthur Eddington's Space, Time, and Gravitation, Cambridge University Press, 1921.*

*The energy of photons emitted or absorbed by nuclei is shifted due to the recoil of the nucleus. The Doppler effect can offset somewhat the change in frequency of the photon. In 1958 R. L. Mössbauer (Nobel Prize, 1961) performed experiments in which the nucleus was embedded in a solid crystal, thereby producing essentially a *recoil-free* emission or absorption of photons. The result is a tremendous increase in sensitivity in the resonance emission or absorption of photons, a situation that allows the measurement of extremely small differences in energies (or photon frequencies).

Example 15.1

In the test of Hafele and Keating of flying atomic clocks around the Earth (see Chapter 2), the gravitational redshift had to be considered. Calculate the effect and compare it with the special relativity time dilation found in Example 2.6.

Solution: The ratio $\Delta\nu/\nu$ will be equal to the time difference $\Delta T/T$ measured by the clocks on Earth and on the jet airplanes. Because the height of the airplanes is not so high above the surface of the Earth compared with the Earth's radius, we use the simpler of the two equations for the gravitational redshift, Equation (15.3). We have

$$\frac{\Delta T}{T} = \frac{gH}{c^2}$$

The value of H for the clock on the surface of the Earth is r_e, and the value for the flying airplane is $r_e + A$, where A is the altitude of the airplane, about 33,000 feet (10,000 m). The value of H is the difference of height of the two clocks in the Earth's gravitational field g, and $H = A$. We neglect the change of g over this altitude, and $\Delta T/T$ becomes

$$\frac{\Delta T}{T} = \frac{(9.8\ \text{m/s}^2)(10{,}000\ \text{m})}{(3 \times 10^8\ \text{m/s})^2} = 1.09 \times 10^{-12}$$

The eastward and westward airplane trips took about $T = 45$ hours flying time. The difference in the two clocks due to the gravitational redshift is

$$\Delta T = (1.09 \times 10^{-12})\ (45\ \text{h}) \left(\frac{3600\ \text{s}}{1\ \text{h}}\right) = 177 \times 10^{-9}\ \text{s}$$

$$= 177\ \text{ns}$$

This *gravitational time dilation* effect of 177 ns is larger than the 66-ns *special relativity time dilation* effect that we calculated in Example 2.6. There was a third correction due to the rotation of the Earth.

Perihelion Shift of Mercury

The orbits of the planets are ellipses, and the closest point on their orbits to the sun is called the *perihelion*. It has been known for hundreds of years that Mercury's orbit precesses about the sun as shown in Figure 15.4. The point of perihelion precesses very slowly, but the rate of precession has been accurately measured. Most of the effect we observe is due to the Earth's rotation itself, but after that is subtracted there is still a perihelion shift of 575 seconds of arc *per century*. Most of this shift is due to the gravitational perturbation caused by the other planets, but in 1859 Le Verrier announced that after even these corrections were accounted for, there was still a remaining perihelion shift amounting to 43 seconds of arc per century. This was a great mystery to scientists at the time, and the answer had to await some 60 years, until Einstein worked out his general theory of relativity. Einstein showed that general relativity also predicted a perihelion shift, and his calculation showed that the shift was just the missing 43 seconds needed. Einstein was overjoyed by this calculation—together with the observed deflection of starlight by the sun in 1919, it gave great credence to his theory.

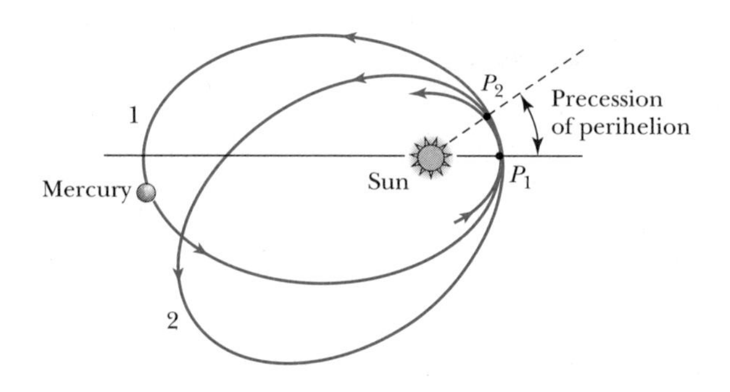

FIGURE 15.4 The orbit of Mercury slowly precesses about the sun. Points P_1 and P_2 are the perihelion for orbits 1 and 2. The effects are exaggerated here.

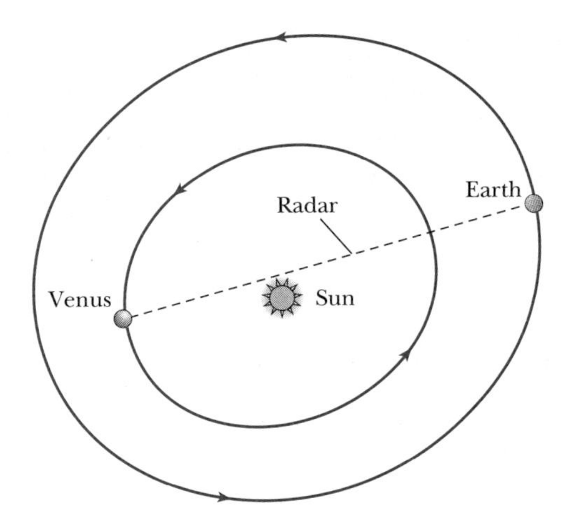

FIGURE 15.5 The superior conjunction position of Venus when it is exactly on the other side of the sun from the Earth. Radar waves sent between the two planets will be delayed slightly due to the gravitational attraction of the sun.

Light Retardation

Light retardation is similar to the deflection of light by gravitational fields. Although a complete description is too complex for us to discuss here, light is retarded, not just deflected, when passing through a strong gravitational field. As a result there is a time delay for a light pulse traveling close to the sun. The time difference between similar paths, with one being close to the sun, can be compared with the general relativity prediction.

Irwin Shapiro showed that such an effect could be measured by sending a radar wave to Venus where it was reflected back to Earth. The position of Venus has to be in the "superior conjunction" position on the other side of the sun from the Earth as shown in Figure 15.5. In this path, the light signal passed close to the sun and experienced a time delay. Shapiro's measurements of January 25, 1970, of a time delay of about 200 μs was in excellent agreement with the general theory (Figure 15.6).

Several experiments have since been done with spacecraft to measure this effect. The experiment to land part of the Viking spacecraft on Mars in 1976 produced the most notable measurement. It produced agreement with the general theory to within the experimental uncertainty of about 0.1%.

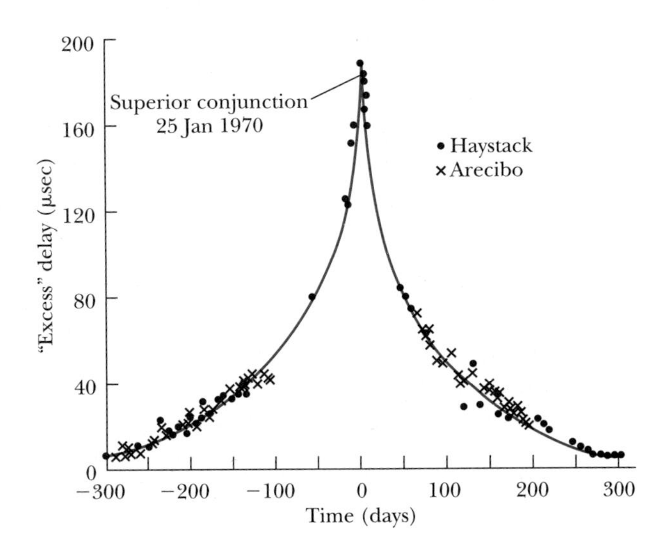

FIGURE 15.6 Shapiro's 1970 data of the time-delay measurements of the radar waves between Earth and Venus. The solid line is the general relativity prediction. *From Shapiro et al., Physical Review Letters,* ***26,*** *1132 (1971).*

More recently, the Hipparcos satellite launched by the European Space Agency in 1989 spent four years fixing the position of 120,000 stars to within 1 millisecond of arc and another million stars to within 2.5 milliseconds of arc. The experimenters had to account for significant light bending of the starlight not only when it passed close to the sun (1 second of arc), but also when the light passed at right angles to the solar direction (4 milliseconds of arc). The entire data set had to be reduced within a general relativistic framework. This remarkable set of data and analysis is consistent with the general relativity theory.

15.3 Gravitational Waves

The four experiments discussed in the previous section represent tests that have confirmed the validity of general relativity. There are other phenomena predicted by general relativity that are currently under significant investigation. One of these is gravitational waves, and we will discuss them in this section. Gravitational waves are not too hard to imagine. We know that electromagnetic waves are produced by oscillating charges. When a charge accelerates, the electric field surrounding the charge redistributes itself. This change in the electric field produces an electromagnetic wave, which is easily detected. In much the same way, an accelerated mass should also produce gravitational waves.

Einstein's general theory of relativity is also a modern theory of gravitation. Einstein showed that his general theory had wave solutions, but because the expected wave amplitudes are so small, they were not taken seriously until 1968 when Joseph Weber announced that he had detected gravitational radiation (waves) from space (extraterrestrial). Subsequent investigations by other experimenters have not confirmed Weber's results, but his announcement spurred a new field of gravitational astronomy. There are several sources of gravitational radiation that could possibly be detected on Earth. These include the collapse of two neutron stars rotating around each other, a neutron star falling into a black hole, and the gravitational collapse of a star to form a black hole.

Astronomers and physicists believe with some conviction that they have identified a system exhibiting gravitational radiation, although no direct detection of gravitational waves has occurred. Russel Hulse and Joseph Taylor (Nobel Prize, 1993) discovered in 1974 a binary system consisting of a pulsar (rapidly spinning neutron star) and another star rotating around each other with a period of about 8 h. The unidentified second star is also believed to be a neutron star. If the system (called PSR 1913 + 16) is radiating gravitational waves, it loses energy, and the two stars come closer together, spiraling faster and faster around each other. As the gravitational radiation increases, the stars will finally lose enough energy to crash into one another. The predicted and observed decrease in the orbital period are in good agreement with the production of gravitational waves.

15.4 Black Holes

While stars are burning, the heat produced by the thermonuclear reactions pushes out the star's matter and balances the force of gravity. When the star's fuel is used up, there is no heat left to counteract the tremendous force of gravity, which becomes dominant. The star's remaining mass collapses into an incredibly dense ball, much smaller than the burning star. The amount of mass left in the star determines what it becomes. A star the size of our sun will become a *white dwarf.* Stars somewhat larger than the sun, but having a mass less than about three solar masses, will become a *neutron star.* We will discuss white dwarfs and neutron stars further in Section 16.3. For stars having greater than three solar masses, an astounding phenomenon occurs. The local gravitational force is so strong that nothing can ever leave, not even light! Oppenheimer and Snyder predicted in 1939 that the gravitational collapse of a star could produce such a body, now called a **black hole.** Because nothing can escape a black hole, it is very hard to detect.

Let's discuss the case of a spherical object as the star's fuel is used up, and the gravitational attraction starts to contract the mass. The object gets smaller and smaller until the point is reached that the gravitational force is so strong that nothing inside the object can escape. For a spherical body, the radius of the tiny ball was determined by Karl Schwarzschild in 1916 using Einstein's general theory of relativity. He showed that this radius, $r = r_S$, now called the *Schwarzschild radius,* is given by

$$r_S = \frac{2GM}{c^2} \quad \text{The Schwarzschild radius} \tag{15.5}$$

Schwarzschild radius

This expression is for a nonrotating spherical mass. The body continues contracting past the Schwarzschild radius, but nothing happening inside the collapsing body can affect the spacetime region outside r_S. Any light ray or particle emitted inside the black hole is kept from going outside by the gravitational force.

An observer outside the Schwarzschild radius cannot even tell when the radius passes through r_S. The light leaving the body right before the body crossed over to $r < r_S$ is so strongly attracted to the body that it takes progressively longer and longer to reach the observer. This is merely an optical effect, because physically the apparent luminosity of the body decreases rapidly with time, and the light is strongly redshifted.

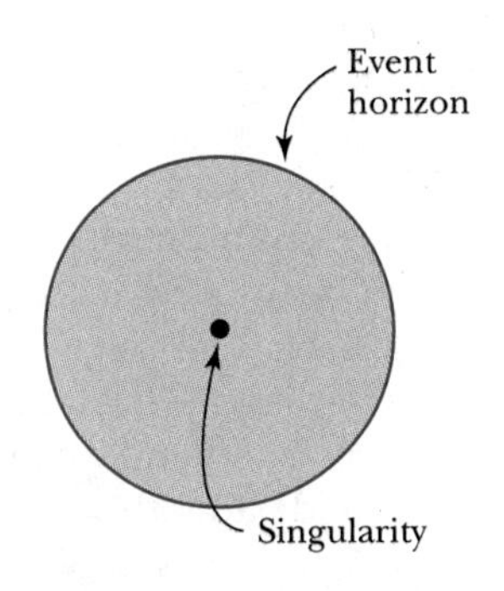

FIGURE 15.7 At the center of a black hole is a *singularity.* Within a certain distance, called the *event horizon* of the singularity, the gravitational pull is so strong that nothing, including light, can escape. The event horizon is not a physical boundary, but rather a limit beyond which nothing can leave. The event horizon describes the size of the black hole.

The boundary region of a black hole is called the *event horizon* (see Figure 15.7). The point at the center of a black hole is called the *singularity.* For a spherical black hole, the event horizon is located at r_S. When people talk about the size of a black hole, they are referring to the size of the event horizon. Even the term *event horizon* makes sense, because it is the region where outside observers can see nothing happening inside the black hole. If we insert the gravitational constant and the speed of light into the Schwarzschild radius we obtain

$$r_S = \frac{2GM}{c^2} = (1.5 \times 10^{-27} \text{ m/kg})\ M \tag{15.6}$$

This allows us to calculate the radius of a black hole for a given mass, as seen in the following example.

Example 15.2

Calculate the Schwarzschild radius for the sun and Earth.

Solution: If we substitute the mass of the sun and Earth into Equation (15.6), we obtain

$$r_S\ (\text{sun}) = (1.5 \times 10^{-27} \text{ m/kg})\ (2 \times 10^{30} \text{ kg}) = 3 \text{ km}$$

$$r_S\ (\text{Earth}) = (1.5 \times 10^{-27} \text{ m})\ (6 \times 10^{24} \text{ kg}) = 9 \text{ mm}$$

If the Earth were a black hole, the dark spot would be very tiny indeed! Of course neither the Earth nor the sun have enough mass to become a black hole.

SPECIAL TOPIC:

GRAVITATIONAL WAVES

The general theory of relativity predicts that gravitational wave radiation will be produced when mass is accelerated, much as electromagnetic radiation is produced when electric charge accelerates. Gravitational waves carry energy and momentum, travel at the speed of light, and are characterized by frequency and wavelength. Gravitational waves interact with all forms of matter. Physicists believe that a particle called the *graviton* mediates the gravitational force, similar in principle to the photon that mediates the electromagnetic force.

It seems unlikely that gravitational waves produced in the laboratory could be detected, but possibly waves from astronomical objects such as black holes, neutron stars, or supernovae (see Chapter 16) may have enough accelerating mass to produce detectable radiation. It might even be possible to detect gravitational radiation coming from the first fraction of a second after the Big Bang. Gravitational waves might affect a mass shown in Figure A. A circular mass would be distorted, and such motion might be represented by a spring connected inside a ring.

During the 1960s Joseph Weber of the University of Maryland formulated creative experiments in gravitational wave detection, and finally in 1969 he electrified the scientific world by announcing that he had detected gravitational waves. Weber's announcement prompted many other investigators, both experimental and theoretical, to consider these phenomena. On the theoretical side, there seems to be no doubt of the existence of gravitational waves. Only a few years after Weber's announcement, other experimenters had built gravitational-wave antennas, several more sensitive than Weber's, in an attempt to replicate Weber's observations. None were successful, and to date, no direct confirmation of the elusive gravitational waves has been reported, although scientists are more convinced than ever that the waves eventually will be detected. Indirect evidence of the orbital decay of a neutron star has been attributed to gravitational waves.

The detection of gravitational waves requires detectors having extremely low noise values. In Weber's original experiment, he used a long, massive aluminum

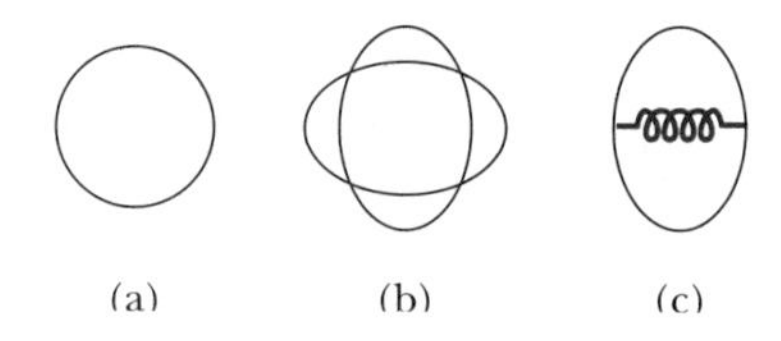

FIGURE A A gravitational wave acting on a spherical mass (a) may cause the mass to change shape (b), which can be represented by a spring placed across the diameter of a circular ring (c). The oscillating spring causes the ring to change shape.

The problem of detecting a black hole is particularly difficult. If light cannot leave a black hole, how can it be detected? There are several indirect means. During the formation of a black hole, as the star collapses the gravitational redshifts should be large. The signals will abruptly cease when the black hole is formed. Observation of such signals would be pure luck. Stephen Hawking has applied quantum theory to the problem of a black hole and has devised a method by which energy evaporates from the black hole much like a thermal energy spectrum. Pieces of the black hole could evaporate, but the process is slow and would be very difficult to detect.

Black holes may be detected indirectly by their gravitational influence on their surroundings. Consider a binary system of a black hole and a companion star. The strong gravitational force of the black hole pulls gaseous matter off the star, but the gas does not go directly into the black hole. Because the companion star has some rotational motion, the gas moves into an orbit around the black hole (see Figure 15.8). This rotating mass collects into an **accretion disk** sur-

bar that was isolated from vibration in a vacuum environment. A gravitational wave passing through would cause a slight strain, or elongation, of the bar. Because detection of an amplitude as small as 10^{-18} m might be expected from a supernova in our galaxy, Weber had to devise ingenious techniques using low temperatures and resonance to cause ringing of the waves in the bar in order to reduce the effects of thermal noise. During the past 20 years, many improvements have been made in the resonant-bar method, and these experiments continue.

The other method of possibly detecting gravitational waves takes advantage of optical interference and is shown in Figure B. Four test masses, two on each arm, react to a gravitational wave passing through, changing the path lengths L_1 and L_2 slightly. This basic Michelson interferometer device takes advantage of the Fabry-Perot method by placing mirrors on the masses, so that the light bounces back and forth between masses (2 km or 4 km) many times before interfering. This greatly increases the sensitivity. Detectors placed in the states of Washington and Louisiana (over 1000 km apart) will aid in looking for correlations in the gravitational waves. This ambitious project, called LIGO for Laser Interferometer Gravitational-Wave Observatory, has been under development for several years by groups from Cal Tech and MIT, and data searches are expected to begin in 2000. The researchers hope their success will rival that of radio astronomy over the past 60 years, possibly detecting additional objects besides pulsars, quasars, galactic nuclei, and tremendous jets. These discoveries occurred by simply observing electromagnetic waves of different frequencies other than visible light. Consider the possibility of actually being able to observe waves of a completely different character, that is, gravitational waves as opposed to electromagnetic waves.

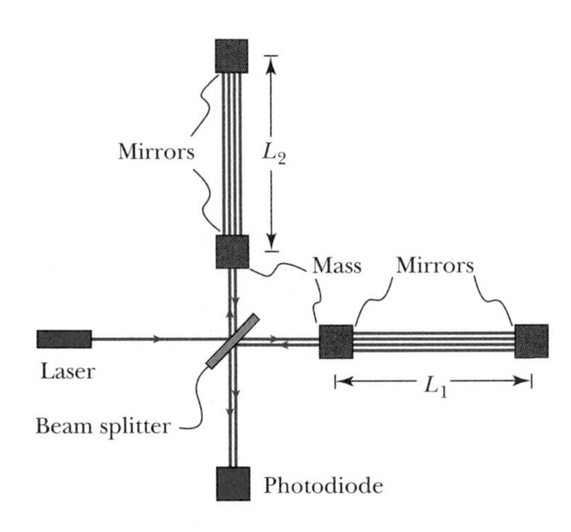

FIGURE B A schematic diagram of the LIGO device. A gravitational wave passing through large masses suspended on thin wires causes the distance $L_1 - L_2$ to change. The interference pattern observed in the photodiode should be capable of recognizing the gravitational wave.

rounding the black hole. As the gas moves toward the inside of the disk, it revolves at speeds approaching the speed of light, and the internal friction between layers of gas heats up the gas to high temperatures. This superheated gas emits x rays that can be observed, because the accretion disk is not inside the black hole's event horizon. The strongest galactic x-ray sources are binary systems, and only black holes and neutron stars are believed capable of producing the immense x-ray emissions observed. Experiments indicate that many of the x-ray sources are due to neutron stars, but remember that general relativity predicts that neutron stars with masses greater than three solar masses cannot exist.

A signature of a black hole is copious amounts of x rays emitted from a binary system where one of the companions is unseen and has a mass larger than three solar masses. For larger black holes much of the accretion disk's energy is emitted as electromagnetic radiation before the matter is absorbed by the black hole. Another process called *advection* is also possible. If the black hole accumulates mass at a slower rate, there is not as much friction in the accretion disk, and

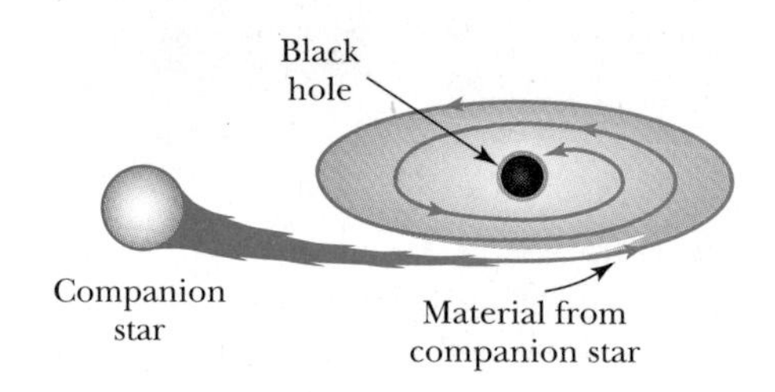

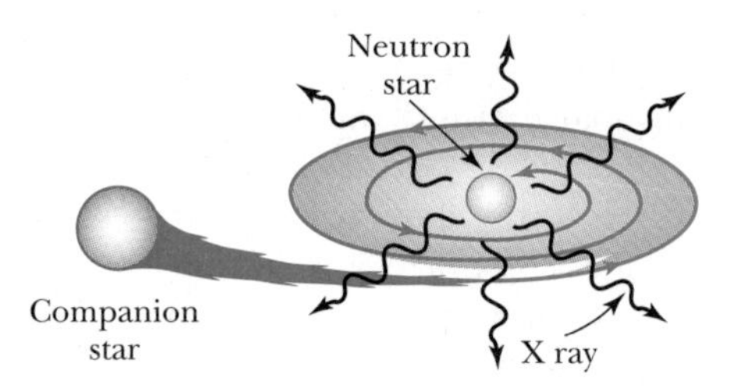

FIGURE 15.8 Matter from a companion star is attracted by the strong gravitational pull of (a) a black hole or (b) a neutron star. The matter spirals inward forming a disk before being pulled into the black hole or neutron star. Friction within the disk creates tremendous temperatures and copious numbers of x rays. This occurs for both neutron stars and black holes. However, for some black holes, the advection model predicts much fewer x rays, because the black hole accumulates matter at a slower rate. In this case most of the energy disappears into the black hole instead of producing x rays. It is believed that the Milky Way's own black hole may be of the advection type resulting in its being only one-thousandth as bright as expected. *Adapted from drawing of Ramesh Narayan and Michael Garcia, Harvard-Smithsonian Center for Astrophysics.*

the energy is absorbed by the black hole. Thus it is possible for black holes, even in a binary system, to only dimly emit radiation.

To date it is likely that astrophysicists have discovered a number of black holes. The earliest and best-known candidate is Cygnus X-1, which was found in 1965 to be an x-ray source. The X-1 designation indicates it was the first x-ray emitter found in the constellation Cygnus. In the 1970s it became known that Cygnus X-1 and an observable star were a binary system. Many observations and analyses have been made on this system in the intervening decades. It appears to be a classic system having an accretion disk whose gaseous molecules get hot and emit x rays. Continuing evidence mounted that the mass of the unobserved object must be at least five solar masses, which is convincing evidence to most scientists that Cygnus X-1 is indeed a black hole.

A convincing case can also be made that the unusual giant galaxy M87 has a black hole near its center. A disk of hot gas can be seen in the lower left of Figure 15.9, taken with the Hubble Space Telescope. The observations of the rotating gas indicate the mass of the object, while the size of the disk indicates the approximate volume of the central object. These results yield an object with a central density so high that only a black hole is possible. The photo also dramatically indicates a plasma jet shooting mass up and away from the central object.

With the tremendous production of the Hubble Space Telescope and the output of new telescopes coming on line, and with better detection systems and

FIGURE 15.9 There is convincing evidence of a black hole in the center of the giant galaxy M87 seen in the lower left of this Hubble Space Telescope photograph. The highly energetic jet emanating upward and to the right is composed of fast moving charged particles and has broken into knots as small as 10 light years across. *Courtesy of H. Ford (JHU, ST ScI, R. Harms, Z. Tsvantanov and NASA. Materials created with support to AURA/ST ScI from NASA contract NAS5-26555 is reproduced here with permission.*

computer analysis, more and more candidates for black holes appear each year. The scientific community generally accepts the existence of these strange creatures that can only be inferred indirectly, because we can't actually see them.

15.5 Frame Dragging

Soon after Einstein proposed his general theory of relativity in 1915, the Austrian physicists Josef Lense and Hans Thirring used it to propose in 1918 that a rotating body's gravitational force can literally drag space and time around with it as the body rotates. This effect, sometimes called the Lense-Thirring effect, is normally referred to as **frame dragging.** All celestial bodies that rotate can modify the spacetime curvature, and the larger the gravitational force, the greater the effect. It is one of the last of the general theory's predictions to be confirmed.

Because of the strong gravitational force near black holes, researchers were first able to confirm frame dragging in 1997 by noticing that x-ray emissions from several black holes varied in intensity. This variation was repetitious and could be explained if the object's orbit were precessing. As the matter orbits the black hole, the spacetime that is being dragged around the black hole drags the matter around with it as shown in Figure 15.10. If we could observe the black hole binary system we might see the accretion disk wobble like a top out of balance. Astrophysicists were first able to observe the spacetime distortion of black holes by observing x rays with the NASA's Rossi X-ray Timing Explorer spacecraft.

Because this effect should occur for all rotating bodies, it has been a goal of astrophysicists to detect the effect around Earth. The Gravity Probe B satellite, scheduled to be launched in 2000, will contain precision gyroscopes inside a

FIGURE 15.10 An artist's conception of a spinning black hole (invisible at the center) twisting spacetime as it turns. The lines curving to the outside represent spacetime. Excess energy shoots off in jets of hot gas along the directions of the rotation axes. The white ball in the center represents the event horizon. *Artwork used by permission of artist Joe Bergeron.*

liquid helium bath. It will be able to detect the induced precession caused by frame dragging with great accuracy.

In 1998 Italian and American researchers were able to use two Earth-orbiting Laser Geodynamics Satellites, LAGEOS I and LAGEOS II, to detect the effects of frame dragging near Earth. The LAGEOS satellites are passive, dedicated to laser ranging where Earth-based lasers are reflected off the satellites. The researchers were able to detect that the plane of the satellites shifted about two meters per year in the direction of the Earth's rotation, in agreement with the prediction of general relativity. This relativistic effect is about ten million times smaller than classical Newtonian disturbances and required painstaking analysis of data that required four years to acquire. The Gravity Probe B satellite will be able to reduce the uncertainty of this effect by more than a factor of ten.

Summary

Einstein's general theory of relativity is a new theory of gravity that replaces Newton's force laws by a system based on curvature of spacetime. Einstein's principle of equivalence states that there is no experiment that can be done in a small confined space that can tell the difference between a uniform gravitational field and an equivalent uniform acceleration. The principle of equivalence leads to the bending of light in a gravitational field.

There are several tests of general relativity. The bending of light was first measured in 1919 and has been repeated in several experiments since. Light can gain and lose energy when passing through gravitational fields. When light loses energy, the frequency of visible light is decreased, and we say it is *redshifted.* To first order, in small regions where g is approximately constant, the gravitational redshift is given by

$$\frac{\Delta \nu}{\nu} = \frac{gH}{c^2} \tag{15.3}$$

Many experiments confirm the existence of gravitational redshifts. The general theory also predicts a noticeable effect on the perihelion of Mercury's orbit. Such a calculation accounted for an anomaly in astronomical measurements known since 1859. The deflection of light by gravitational fields has been detected by light passing close to the sun between Venus and the Earth.

One prediction made by general relativity that has not yet been confirmed is the existence of gravitational waves, which is currently of great interest. Black holes are objects that have collapsed under their own gravitational attraction. Nothing can escape a black hole, not even light. The size of a nonrotating spherical black hole is given by the Schwarzschild radius.

$$r_S = \frac{2GM}{c^2} \tag{15.5}$$

Scientists believe that the existence of black holes has been confirmed, despite the difficulty of their detection.

More recently the existence of frame dragging has been confirmed for black holes and perhaps for the Earth as well.

Questions

1. Laser light from Earth is received for an experiment by an Earth satellite. Is the light redshifted or blueshifted? What happens to the light if it is reflected back to Earth?
2. Explain why true weightlessness occurs only at the center of mass of the space shuttle orbiter as it rotates around the Earth.
3. Why does a drop of water become bulged in the space shuttle due to the Earth's gravitational field? Draw the water drop showing the direction to the Earth's center.
4. Devise a way for the occupants of a spaceship to know whether they are being pulled into a black hole. What can they do if they determine they are within the Schwarzschild radius?
5. Astronauts riding in the space shuttle are said to be in "zero-free" or "micro" gravity. Explain why this is not really so. Is the net force on them zero?
6. In the experiment discussed in Chapter 2 of the atomic clocks flown around Earth, what was the effect regarding general relativity?
7. In 1919 when the gravitational deflection of light was measured, why did the scientists travel to Africa and South America?
8. How likely is it for a black hole to collide with Earth? Would we have much warning?
9. Why is it difficult to experimentally test models of general relativity?

Problems

15.1 Principle of Equivalence

1. Devise an experiment like the one Newton performed to test the equivalence of inertial and gravitational masses. Using different masses on pendula of equal length, show that the period depends on the ratio of $\sqrt{m_I/m_G}$.

15.2 Tests of General Relativity

2. Controllers want to communicate with a satellite in orbit 400 km above the Earth. If they use a signal of frequency 100 MHz, what is the gravitational redshift? Assume g is constant.
3. For clocks near the surface of the Earth, show that Equation (15.4) reduces to Equation (15.3) for $\Delta\nu/\nu$.
4. Repeat Example 15.1 using the more accurate Equation (15.4) for the gravitational redshift. Compare with the result of Example 15.1.
5. In Shapiro's experiment on the time delay for the superior conjunction of Venus and Earth, how long did it take for the radar signals to travel the round trip to Venus? What percentage change was he looking for?
6. Consider a light beam that travels a distance equivalent to the Earth's circumference (40,000 km) under the influence of a constant gravitational field ($g = 9.8\ \text{m/s}^2$). How far down would the light beam drop over this distance due to the gravitational attraction? (Of course a light beam cannot circle around the Earth like this).
7. Radiation is emitted from the sun over a wide range of wavelengths. Calculate the gravitational redshift of light emitted from the sun that is in the center of the visible range (550 nm).
8. Calculate the gravitational redshift of radiation emitted from a neutron star having a mass of about 0.5×10^{31} kg and a radius of about 10 km for the middle of the visible range (550 nm).
9. In the experiment of Pound and Rebka, a 14.4-keV gamma ray fell through a distance of 22.5 m near the Earth's surface. What is the change in frequency and the percentage frequency change due to the gravitational redshift?

15.4 Black Holes

10. What is the value of the Schwarzschild radius for the moon? ($m_{\text{moon}} = 7.35 \times 10^{22}$ kg)
11. Calculate the Schwarzschild radius for Jupiter. ($m_{\text{Jupiter}} = 1.9 \times 10^{27}$ kg)
12. Stephen Hawking has predicted the temperature of a black hole to be $T = hc^3/8\pi kGM$, where h is Planck's constant and k is Boltzmann's constant. Calculate the temperature of a black hole with the mass of the sun.
13. Calculate the mass and Schwarzschild radius of a black hole at room temperature. (See Problem 12.) How many solar masses is this?

General Problems

14. Derive Equation (15.4).
15. One of the communication frequencies that the space shuttle orbiter uses is 294 MHz. Find the gravitational frequency change with respect to Earth when the orbiter is at an altitude of 300 km. Assume g is constant.
16. Weightlessness occurs only at the center of mass of the space shuttle orbiter as it rotates 300 km above the Earth. Calculate the effective g that an astronaut in the orbiter would feel who is 3 m closer to the Earth than the center of mass. You may choose to ignore relativistic effects.

17. Find the mass of a particle whose Compton wavelength is πr_S where r_S is the Schwarzschild radius. This mass is called the *Planck mass,* and the energy required to create the mass is called the *Planck* energy, $E_P = m_{Pl}c^2$. Determine the values of both the Planck mass and energy.

18. The length scale on which the quantized nature of gravity should first become evident is called the *Planck length.* (a) Determine it using dimensional analysis using the fundamental constants G, h, and c. (b) Determine it by finding the de Broglie wavelength of the Planck mass of the previous problem. Are the values close to the value of 10^{-35} m?

19. Use the fundamental constants G, h, and c and dimensional analysis to determine a time constant called the *Planck time.* How much time would it take light to travel the *Planck length* discovered in the previous problem? Are these two times consistent?

20. A communications satellite is at a geosynchronous orbit position (35,870 km above the Earth's surface) and communicates with the Earth at a frequency of 2×10^9 Hz. What is the frequency change due to gravity?

CHAPTER 16

Cosmology—The Beginning and the End

I too can see the stars on a desert night, and feel them. But do I see less or more? The vastness of the heavens stretches my imagination—stuck on this carousel my little eye can catch one-million-year-old light. A vast pattern—of which I am a part—perhaps my stuff was belched from some forgotten star, as one is belching there.

Richard Feynman

Most physicists and astronomers believe our universe evolved from a primordial event called the *Big Bang*. In this chapter we will present some of the experimental evidence supporting this belief. Our knowledge of cosmology is ever-growing, and we will also present some of the unexplained evidence and conflicting theories. What students perhaps find most surprising is how much cosmology and astrophysics depend on other fields of physics. The formation of stars depends on quantum physics. The understanding of nucleosynthesis of elements depends on exact measurements of nuclear physics cross sections to determine how certain elements may have been formed in the early stages of the universe. As we will see in Section 16.2 on the Big Bang, elementary particle physics and cosmology have continued to merge so much that the fields now overlap considerably. The fundamental forces are subjects of extreme interest to both cosmology and elementary particle physics. This field of physics is changing perhaps faster than any other. Strongly held theories of today can be disproved by the experimental observations of tomorrow. New Earth- and space-based telescopes will undoubtedly provide evidence to modify our present understanding. New accelerator experiments may help increase our knowledge about the origin of the universe.

Not only will we attempt in this chapter to understand the origins of our universe, we will examine what the demise of life on Earth will be like. Our sun will undoubtedly eventually burn out, but present evidence seems to indicate our universe will expand forever.

16.1 Evidence of the Big Bang

In the last few decades, several cosmological theories have attempted to explain the origin of our universe—what was missing was observational data. In the 1950s and early 1960s there were two rival theories of cosmology, but data were insufficient to discriminate between them. Both theories accounted for the known expansion of the universe, which was demonstrated conclusively by the redshift data of Edwin Hubble in 1929. One theory, known as the **steady-state theory,** held that matter is being continuously created; as the universe expands, the density of the universe remains constant. The other theory, the Big Bang model, considers the universe to be created in a primeval fireball of incredible density and high temperature. The Big Bang model will be presented in Section 16.2.

Before presenting the Big Bang model, we will present three important pieces of evidence that have led most scientists to accept it as the most likely model for the origin of the universe. The pieces of evidence include

The evidence

1. The observations between 1929 and 1952 by Edwin P. Hubble (Figure 16.1), using the giant telescopes of Mount Wilson and Mount Palomar, that the galaxies of the universe are moving away from each other at high speeds. The universe is apparently expanding from some primordial event.
2. The observation in 1964 by Arno Penzias and Robert Wilson, two Bell Laboratory scientists, that a cosmic microwave background radiation permeates the universe. This background radiation has been attributed to the Big Bang.
3. There is good agreement between predictions of the primordial nucleosynthesis of the elements and the known abundance of elements in the universe. This applies to the light elements that were produced in the early stages of the Big Bang.

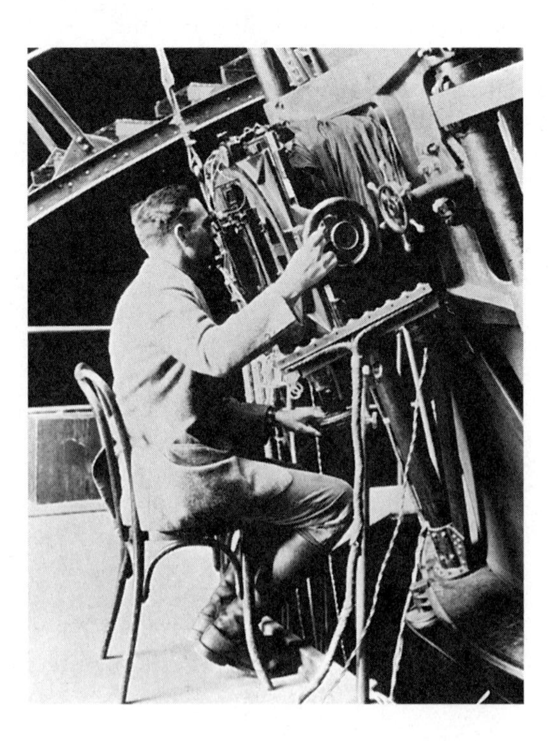

FIGURE 16.1 Edwin Hubble is shown taking photographs with the 100-inch telescope in 1924 at the Mount Wilson Observatory. *Courtesy of the Huntington Library, San Marino, CA.*

Although there are still controversies concerning some of the predictions and results of the Big Bang model, its main ideas are widely accepted. It has been modified somewhat, as we shall see in Section 16.5, where we will discuss some of the difficulties. We examine here the three pieces of evidence in some detail.

Hubble's Measurements

Edwin Hubble arrived at the Mount Wilson observatory in California in 1919 after service in World War I. His career spanned four decades of brilliant observation and understanding of our universe. He began by showing in the mid-1920s that there were indeed galaxies other than our own. This pioneering work conflicted with other published data of the time, but Hubble carefully and painstakingly presented an overwhelming case.

The recessional velocity of astronomical objects is inferred from the shift toward lower frequencies (redshift) of certain spectral lines emitted by very distant objects. We derived the redshift relation in Chapter 2 (Equation 2.32). Using data like those shown in Figure 16.2, Hubble was able to determine the recessional velocities of many objects. Other astronomers, particularly V. M. Slipher of Lowell Observatory in 1912, had reported that certain nebulae appeared to be receding at high radial velocities from us, but it was Hubble who put the tech-

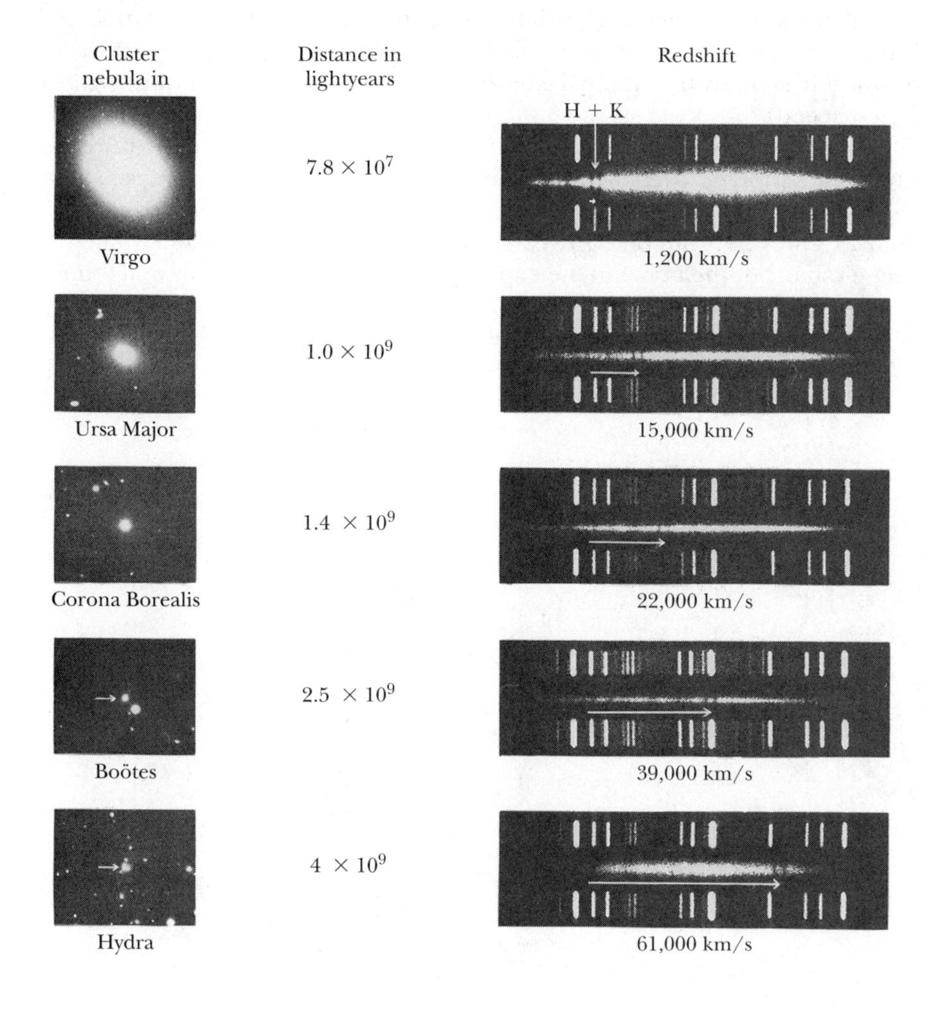

FIGURE 16.2 Redshift data for various galaxies are shown with their distance in lightyears. The spectrum of each galaxy is the wide, hazy band placed in the middle between the laboratory comparison spectra. The K and H lines of calcium are shown redshifted by the arrow and move to the right for higher velocities. These early data convincingly showed that the universe was expanding. *Courtesy of The Observatories of the Carnegie Institution of Washington.*

nique on firm footing. Galaxies receding from us with speed v are related to the distance R from Earth by the relation known as **Hubble's law,**

Hubble's law and constant

$$v = HR \tag{16.1}$$

The parameter H is called **Hubble's constant,** and it is related to a scale factor a that is proportional to the distance between galaxies by

$$H = \frac{1}{a}\frac{da}{dt} \tag{16.2}$$

Because the universe has been expanding, Hubble's constant is not actually constant, but changes slowly over long periods of time. Its value today is sometimes denoted by H_0.

In order to determine whether Hubble's law is valid, it is necessary to know the distance R to objects for which the redshift has been measured. Hubble developed very sophisticated techniques that used the brightness of stars and galaxies to determine the distances R. Hubble was able to do this with some certainty for stars out to distances of 10 million light years and, with some additional assumptions, for galaxies out to distances of 500 million light years.

Together with his gifted colleague Milton Humason, Hubble examined, over a period of many years, hundreds of stellar objects to determine their redshifts. By 1929 he was already able to show that 49 galaxies (which he called *extragalactic nebulae*) fit the velocity–distance relationship. Hubble's results required painstaking measurements of brightness and redshifts. By 1935 Hubble and Humason had catalogued the redshifts of 100 additional galaxies; these data unequivocably showed that the galaxies farthest away from us were moving at the highest speeds.

Hubble showed that Equation (16.1) is valid. The distance measurements have been corrected over the intervening years, but the linearity between v and R remains valid (see Figure 16.3). Today we believe that Hubble's constant H is about 21 km/s per million lightyears. There is still some controversy about the precise value, but the belief in the expansion of the universe seems well founded.

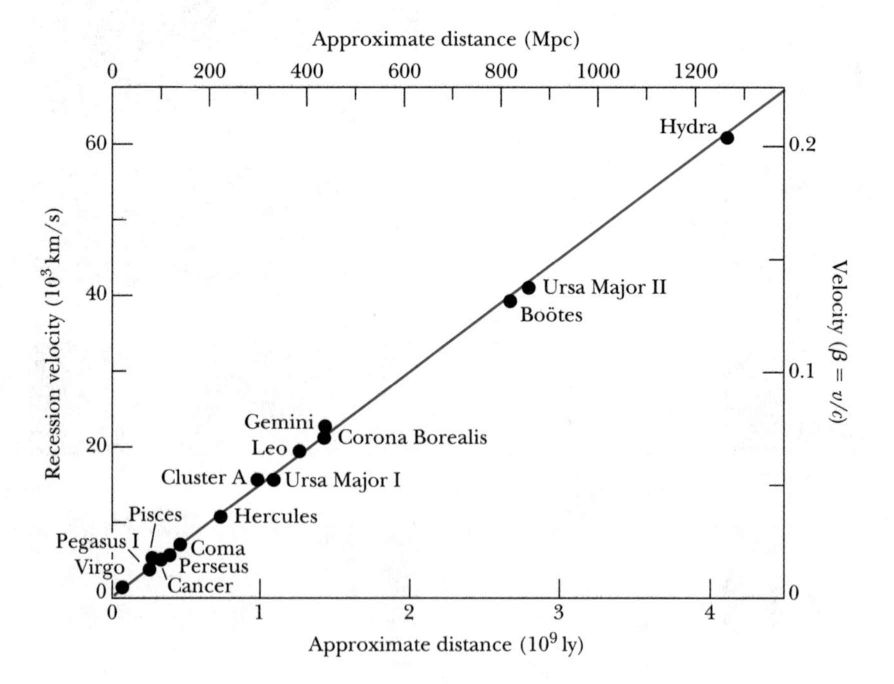

FIGURE 16.3 More recent analyses of 15 clusters of galaxies for the recession velocity as a function of distance. The solid line represents a Hubble constant of 15 km/s per million lightyears (or 49 km · s^{-1} · Mpc^{-1}). Various analyses of the distances give different values of H. *From G. Siegfried Kutter, The Universe and Life. Boston: Jones and Bartlett, 1987. p. 50.*

FIGURE 16.4 A representation of how galaxies are receding with respect to each other. As the balloon blows up, each dot is further away from the other dots. The expansion of the universe has been somewhat homogeneous.

It is not necessary for Earth to be at the center of the universe in order to observe the expansion. We show in Figure 16.4 a balloon with dots. Notice that as the balloon blows up, *all* the dots move further apart from each other. The surface of the balloon is two-dimensional; a three-dimensional example often quoted is raisins in bread dough. As the bread bakes, it rises and expands in three dimensions, and the raisins separate. The raisins all move further apart, with the ones on the outsides moving faster. Something similar happens in the expansion of the universe, and the galaxies separate. We will discuss in Section 16.6 the possibility of using Hubble's constant to determine the age of the universe.

Example 16.1

Astronomers use a measure of distance called the *parsec,* abbreviated pc, more often than the lightyear, another measure of distance. A star that is one parsec away has a parallax of one second of arc relative to the Earth's orbit about the sun. (Parallax is the apparent displacement of a body due to its observation from two different positions.) One parsec is about 3.26 lightyears. Determine Hubble's constant in km/s per Mpc.

Solution: One Mpc is a megaparsec or 10^6 pc. We use the value just given for Hubble's constant, 21 km/s per million lightyears, and make the conversion to Mpc.

$$\frac{21 \text{ km/s}}{\text{million ly}} \frac{3.26 \text{ ly}}{\text{pc}} = 68 \frac{\text{km/s}}{\text{Mpc}}$$

The range quoted presently by astronomers for Hubble's constant is 58–73 $\text{km} \cdot \text{s}^{-1} \cdot \text{Mpc}^{-1}$.

Cosmic Microwave Background Radiation

In 1964 Arno Penzias and Robert Wilson were studying radio emissions in the microwave wavelength region. They kept picking up a very annoying characteristic signal in their specially designed low-noise antenna. When repeated attempts to eliminate the source of the noise were not successful, Penzias and Wilson were at a loss to understand their result. In a conversation with another colleague they learned of calculations by P. James E. Peebles and his group at Princeton that a remnant background of radiation from the Big Bang had been predicted. Another Princeton physicist, Robert Dicke, was mounting an experiment to measure it. It was this remnant background radiation that Penzias and Wilson were observing.

Because of the rapid expansion of the universe, the blackbody radiation characteristic of 3000 K several billion years ago would be Doppler-shifted to a peak near 3 K today. The calculated redshift based on the extremely high velocity of that part of the universe with respect to Earth today is about a factor of 1000. The blackbody radiation spectrum is shown in Figure 16.5. Subsequent satellite measurements have mapped out the complete blackbody spectrum to amazing accuracies and shown it to be nearly, but not completely, isotropic. We will return to this subject later. The observation of the cosmic microwave background radiation by Penzias and Wilson (Nobel Prize, 1978) was a triumph of the Big Bang model, and the result was very difficult to explain with the steady-state model. After the initial observation it became widely known that George Gamow, Ralph Alpher, and Robert Herman had performed calculations in the late 1940s and early 1950s that had predicted the Big Bang remnant radiation would appear in the range of 5–7 K.

Nucleosynthesis

A few minutes after the Big Bang, the universe had cooled enough so that neutrons and protons could undergo thermonuclear fusion and form light elements. The cosmic microwave radiation background provides a view back to about 700,000 years after the creation of the universe, but the formation of elements began after only a few minutes. Therefore, the nucleosynthesis of elements provides a much more stringent test of the Big Bang model. By measuring the present relative abundances of the elements, physicists are able to work backward and test the conditions of the universe that may have existed as early as a fraction of a second, when neutrons and protons were first joined to produce nuclei.

We examined the proton–proton nucleosynthesis cycle in Chapter 13. In reactions like this, the light elements ^{2}H, ^{3}He, ^{4}He, and ^{7}Li are formed, and predictions of the Big Bang model can be compared with current observations. Heavier elements are formed in stars, but the vast majority of the presently known mass in the universe is composed of hydrogen and helium. The other

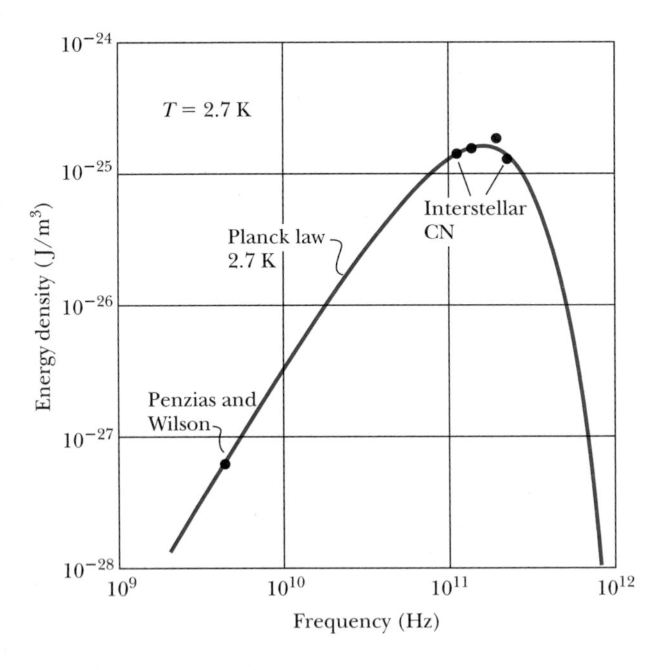

FIGURE 16.5 Calculated blackbody radiation distribution is shown as a function of frequency; the datum point that Penzias and Wilson measured is noted. Radiation due to the cyanogen (CN) radical in interstellar space was observed in 1937, but it was incorrectly thought at the time that the radiation was due to collisions of electrons with CN.

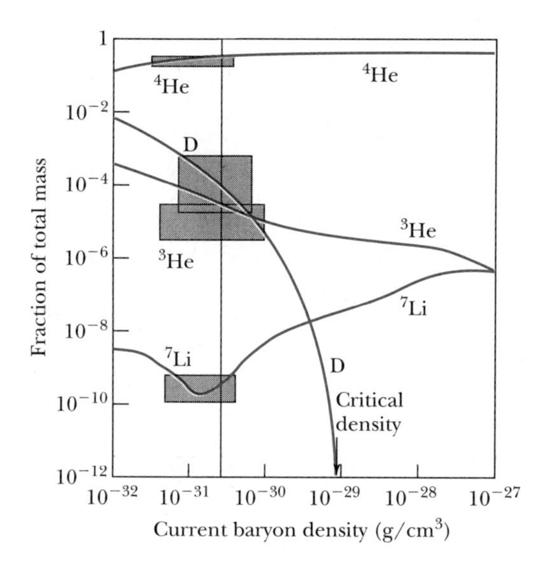

FIGURE 16.6 The fractional mass abundances of several light elements are displayed against possible mass densities. The boxes represent experimental observations, and the solid curves are calculations of the standard model of the Big Bang. The best agreement between observation and calculation occurs for a density of 3×10^{-31} g/cm^3. These data strongly support the standard Big Bang model, but do not agree with the critical density, $\sim 9 \times 10^{-30}$ g/cm^3 (see Example 16.8). *From R. A. Malaney and W. A. Fowler, American Scientist* ***76****, 472(1988).*

elements exist in only minute quantities. The standard model of elementary particle physics, together with nuclear physics, has been used in complex and difficult calculations to compare with current experimental observations of the elemental abundances. The observations and Big Bang predictions are in remarkable agreement, as shown in Figure 16.6, for a density of about 3×10^{-31} g/cm^3. One variation of the Big Bang model that we will discuss in Section 16.5 proposes an inflationary universe, and that model suggests a density of about 9×10^{-30} g/cm^3. We will discuss this further in Section 16.5.

Approximately 25% of the known mass of the universe is composed of ^{4}He, the remainder being almost all free protons. (The reason we keep writing the *known* mass of the universe will become clear in Section 16.5 when we discuss the *missing mass* problem.) The synthesis of ^{4}He depended critically on the ratio of neutrons to protons and the density of matter during early stages of the universe. The ^{4}He abundance has been used to suggest that at most four families of elementary particles are possible. The suggestion and data are in good agreement with the current belief from elementary particle physics data that only three families exist (see Figure 16.7).

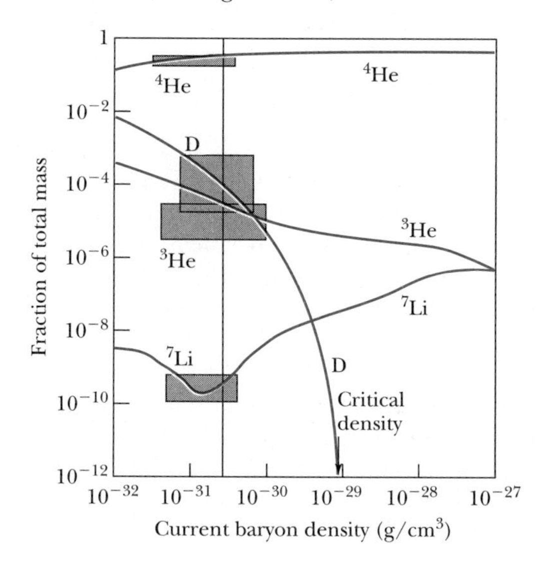

FIGURE 16.7 The predicted ^{4}He mass fraction is shown as a function of the density of nucleons for three different possible family groups of particles. These results indicate a maximum of four different families having neutrinos. Evidence from elementary particle physics experiments indicates that there are only three families (three neutrinos), which is consistent with this figure. *Courtesy of Slim Films/Scientific American.*

Example 16.2

If the current mass of the universe consists of 75% protons and 25% ^{4}He, what is the current ratio of protons to neutrons?

Solution: The mass of a proton is about 1 u, and the mass of ^{4}He is about 4 u. If the ratio of mass of free protons to ^{4}He is 3:1 (75% protons, 25% ^{4}He), there must be 12 free protons for every ^{4}He. Therefore out of a total of 16 nucleons (12 free protons and 4 nucleons in ^{4}He), only 2 are neutrons, and the remaining 14 are protons (two are in ^{4}He). The ratio of protons to neutrons is about 7. This is in reasonable agreement with the fraction calculated from temperature considerations made for the time when protons and neutrons formed deuterons. When the temperature dropped below 10^9 K, photons no longer had enough energy to dissociate the deuteron (2.2-MeV binding energy). After that time, free neutrons decayed (they are unstable with a half life of 10.4 minutes), and deuterons combined to form helium.

16.2 The Big Bang

Since the 1960s the Big Bang has generally been accepted as the event that began our universe. We can use our knowledge of nuclear and elementary particle physics (Chapters 12–14) and general relativity (Chapter 15) to understand the formation of our universe since the Big Bang. In this section we define the time $t = 0$ as the beginning of the universe, or the Big Bang, and describe the major steps in the intervening ~13 billion years until today. The temperature of the universe is shown as a function of time in Figure 16.8.

The unknown

$t = 0 \rightarrow 10^{-43}$ s. In this era we have no theories that can tell us with certainty what happened, because the known laws of physics do not apply. In the beginning, the universe most likely had infinite mass density and zero spacetime curvature. This condition is known as a *cosmological singularity*. The radius of the visible universe by the time 10^{-43} s was probably less than 10^{-52} m. Our present theories of gravity and quantum physics were not valid. The four forces of strong, electromagnetic, weak, and gravity were all unified into one force. Only a theory of gravitation including quantum physics, or quantum gravity, has any hope of explaining what happened. The applicable theory has been dubbed **The Theory of Everything.** There are no viable candidates for this theory at the present time. The temperature of matter was probably greater than 10^{30} K. We don't know if any of the particles known today could have even existed.

Gravity separates

$t = 10^{-43}$ s $\rightarrow 10^{-35}$ s. As strange as it seems, we believe that we have some understanding of what happened in this era. By 10^{-35} s the universe has expanded somewhat, and the temperature is slightly reduced, maybe to 10^{28} K. Gravity is established as the first separate force. The Grand Unified Theories (GUTs) try to explain this era, but there is no widely accepted theory as yet.

Quark-electron soup

$t = 10^{-35}$ s $\rightarrow 10^{-13}$ s. The strong force has now broken off and become a separate force. The fundamental particles (quarks and leptons) have formed as well as their antiparticles. The small universe can be described as a hot, quark-electron soup. The universe continues to cool, perhaps to 10^{16} K by 10^{-13} s. No present accelerator has enough particle energy to explore this region. At present only some cosmic rays have comparable energies.

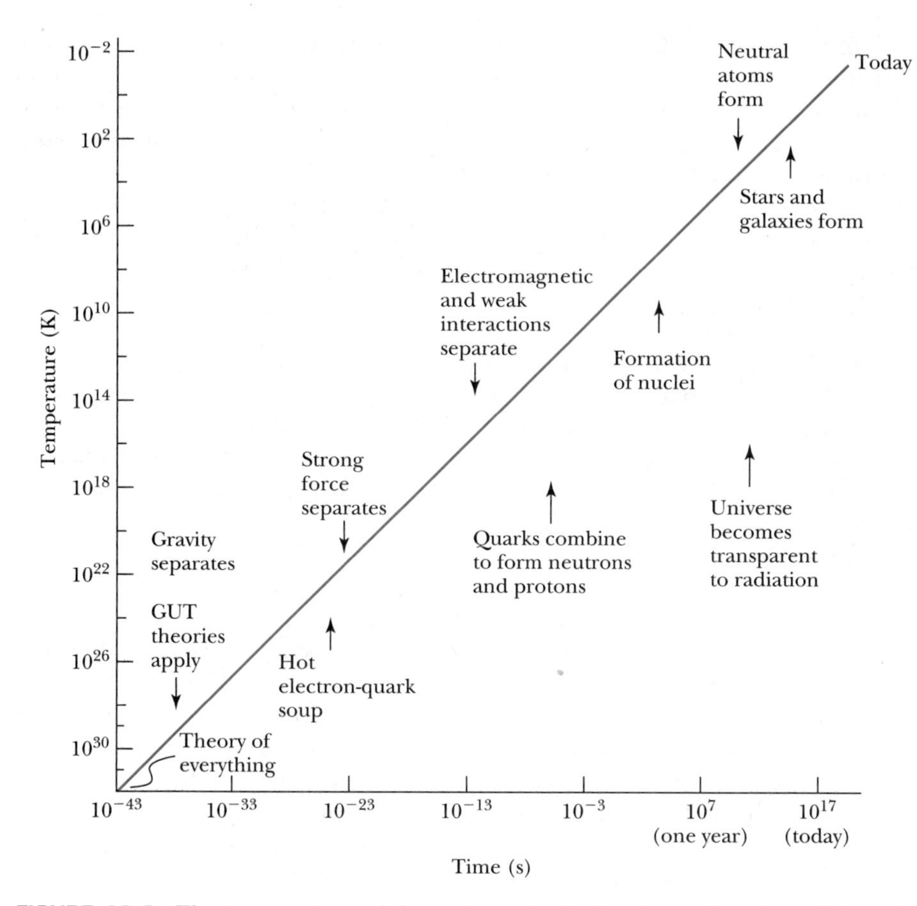

FIGURE 16.8 The temperature of the universe is displayed as a function of time since the Big Bang. Note that both scales are logarithmic. The comments are placed vertically above the times at which the phenomena occurred, except for the Theory of Everything which might be applicable for the time $< 10^{-43}$ s.

***t* = 10^{-13} s → 10^{-3} s.** During this period the quarks bind together to form neutrons and protons. Only the two lightest quarks, the up and down, are very effective in the thermodynamics of the universe, because the more massive quarks have such low velocities at this time that they cannot transport much energy. There are more protons than neutrons, because the protons are slightly less massive. At energies above 100 GeV (10^{15} K) the electromagnetic and weak interaction are unified into the electroweak interaction. However, below 100 GeV, the $W^{\pm}$ and Z^0 bosons behave like massive particles and the photon is massless, so the electromagnetic and weak interactions have broken their symmetry and act separately. We say that "symmetry is broken" below 100 GeV. At the end of this period the temperature has dropped to about 10^{11} K. The four forces of today are now distinct. By the end of this period the universe mostly consists of a soup of photons, electrons, neutrinos, protons, and neutrons as well as various antiparticles (for example, positrons and antiprotons).

Neutrons and protons form

Electromagnetic and weak forces separate

***t* = 10^{-3} s → 3 min.** By the end of this period the universe has cooled enough (to 10^9 K) that deuterons do not immediately fly apart when formed, because the cooler photons do not have enough energy to dissociate the deuterons. This is the beginning of nucleosynthesis—the formation of nuclei.

Deuterons form

Light nuclei begin forming

$t = 3$ min $\rightarrow$ 500,000 y. Helium and other light atomic nuclei form by nucleosynthesis (for example, the proton–proton cycle). Neutron decay occurs, so there are many more protons than neutrons. The universe continues to cool, and the temperature is as low as 10^4 K at the end of this era. The universe consists primarily of photons, protons, helium nuclei, and electrons. Atoms are not able to form because the intense electromagnetic radiation ionizes them almost as soon as they are formed. Photons interact freely with charged particles through the electromagnetic interaction and are absorbed, emitted, and scattered by matter.

Matter-dominated universe

$t = 500,000$ y $\rightarrow$ Present. During this period the universe has finally cooled enough that electromagnetic radiation (photons) has decoupled from matter. Until about 700,000 years, the universe was "radiation dominated," meaning that most of the energy was in the form of photons, which were continually being absorbed and emitted by ions. At about 3000 K the temperature is low enough that protons can combine with electrons to form neutral hydrogen atoms. At this point the scattering of photons from neutral hydrogen (as opposed to free protons and electrons) drops dramatically, and electromagnetic radiation is free to pass throughout the universe. From this time on, the universe is "matter dominated," with more energy being in the form of matter than radiation. Because photons are now free to pass throughout the universe, a blackbody radiation of temperature 3000 K should persist forever. Remember from Section 16.1 that this radiation characteristic of 3000 K is redshifted with respect to us, and we measure the vast majority of photons in the universe today to be due to the 3 K background. This is the cosmic microwave background radiation discussed in the previous section. Atoms are now able to form, and matter begins to clump together to form molecules, gas clouds, stars and eventually galaxies. The rest is history!

Example 16.3

Nucleosynthesis begins around the time 10^{-3} s, when protons and neutrons can finally remain together in the deuteron without flying apart from the interaction of radiation. Calculate the ratio of protons to neutrons when the temperature of the universe was about 10^{10} K.

Solution: The ratio of protons to neutrons is purely a statistical distribution based on the available energy and the masses of the proton and neutron. The ratio is determined by the thermodynamics and the difference in masses. Because the proton has lower mass, we expect more protons to exist. Let $\Delta m = m_n - m_p = 939.566\ \text{MeV}/c^2 - 938.272\ \text{MeV}/c^2 = 1.294\ \text{MeV}/c^2$. The ratio of protons to neutrons is calculated to be

$$\frac{\text{Number of protons}}{\text{Number of neutrons}} = \frac{e^{-m_p c^2/kT}}{e^{-m_n c^2/kT}}$$

$$= e^{\Delta mc^2/kT} = \exp\left(\frac{\Delta mc^2}{kT}\right) \qquad (16.3)$$

$$= \exp\left(\frac{1.294 \times 10^6\ \text{eV}}{(8.6 \times 10^{-5}\ \text{eV/K})(10^{10}\ \text{K})}\right) = 4.5$$

As the temperature continued to decrease, the ratio of protons to neutrons continued to increase due both to the decay of the neutron and to the factor of T in the exponential. However, the ratio of protons to neutrons bound in nuclei eventually stabilized due to the production of helium.

16.3 Stellar Evolution

Some 700,000 years after the Big Bang, matter in the form of electrons, protons, and ^{4}He drifted throughout the universe much like gas particles in a room. Even-

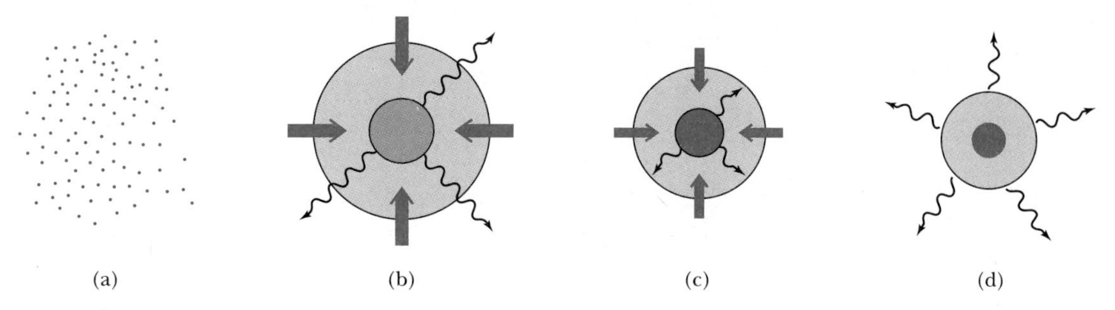

FIGURE 16.9 Stars form by interstellar gas and clouds condensing by gravitational attraction as in (a). As the matter contracts in (b), a core forms that heats up and radiates energy. Eventually, as in (c), the outer region becomes so dense that the radiation from the hot inner core can no longer escape. The collapse slows but the matter continues to heat and eventually a protostar forms of high density and temperature. For a star of about one solar mass (our sun), the contraction eventually heats up enough to sustain nuclear fusion. The radiation pressure produced by nuclear fusion balances the gravitational contraction, and the star stabilizes into a main sequence star as in (d). The star will burn for 10 billion years converting its hydrogen into helium.

tually, as the temperature continued to decrease, gravitational forces managed to bring some of the matter together into massive gaseous clouds. From this beginning, stars were formed. As the protons were attracted together by their gravitational interaction, their kinetic energy rose. This process continued as the interior temperature of the infant star kept increasing. The interior of the gaseous cloud had a higher density and temperature than the outside. The cloud continued to contract until finally the temperature reached about 10^7 K, and the nuclear fusion process began. It may take a million or more years for the contraction of the cloud to be able to produce fusion and the star to be born. The schematic formation of a star is shown in Figure 16.9.

We have previously discussed the nuclear fusion process as the energy source of stars. The proton–proton chain releases energy, which is observed as radiation. The eventual result of this fusion process is ^{4}He, which collects at the center of the star. Other processes form ^{12}C and heavier masses if the temperature in the star is high enough.

Of course, we know most about our own star, the sun. We can measure the surface temperature of stars by measuring the color of the radiated light, but it is difficult to know the interior temperature. We believe the surface temperature of the sun to be about 5500 K and the interior temperature to be as high as 6×10^6 K. A star the size of our sun may burn for 10^{10} y, but a larger star may use up its fusion fuel much faster. The light presently received by Earth was most likely produced in the interior of the sun more than 10^5 years ago and underwent many scatterings until it was emitted from the surface of the sun.

Example 16.4

Estimate the mean temperature of the sun by assuming its protons behave as a gas.

Solution: As gas clouds collected together through the gravitational interaction, the decrease in the gravitational potential energy was accompanied by an increase in the kinetic energy of the particles. We can estimate the mean temperature of the sun by setting the change in the kinetic energy equal to the negative of the change in potential energy, $\Delta(\text{K.E.}) = -\Delta(\text{P.E.})$. In other words, the total energy,

K.E. + P.E., is constant. We assume the sun is a uniform sphere of mass M and radius R. Its self-potential energy can be calculated to be (see Problem 26)

$$\text{P.E.} = -\frac{3}{5}\frac{GM^2}{R}$$

$$= -\frac{3}{5}\frac{(6.67 \times 10^{-11}\ \text{N}\cdot\text{m}^2/\text{kg}^2)(1.99 \times 10^{30}\ \text{kg})^2}{6.96 \times 10^8\ \text{m}}$$

$$= -2.28 \times 10^{41}\ \text{J}$$

The kinetic energy of the particles within the sun is then 2.28×10^{41} J. If we assume the sun is made entirely of protons, the number of protons in the sun is $N = M/m_p$, where M is the mass of the sun. We write the kinetic energy of the sun as N protons each having speed v.

$$\frac{1}{2}Nm_pv^2 = \frac{1}{2}\frac{M}{m_p}m_pv^2 = 2.28 \times 10^{41}\ \text{J}$$

From this relation we can determine the proton velocity.

$$v^2 = \frac{2(2.28 \times 10^{41}\ \text{J})}{M} = \frac{2(2.28 \times 10^{41}\ \text{J})}{1.99 \times 10^{30}\ \text{kg}}$$

$$= 2.3 \times 10^{11}\ \text{m}^2/\text{s}^2$$

$$v = 4.8 \times 10^5\ \text{m/s}$$

Using the kinetic theory of gases relation for the velocity of a gas particle at temperature T, we can determine the sun's mean temperature.

$$v = \sqrt{\frac{3kT}{m}}$$

$$T = \frac{mv^2}{3k}$$

If we assume the particle is a proton, we determine the temperature to be

$$T = \frac{(1.67 \times 10^{-27}\ \text{kg})(2.3 \times 10^{11}\ \text{m}^2/\text{s}^2)}{3(1.38 \times 10^{-23}\ \text{J/K})} = 9 \times 10^6\ \text{K}$$

This is a reasonable result for the mean temperature of the sun.

In one of the cycles of the proton–proton chain in the sun, $^3\text{He} + {}^4\text{He} \rightarrow {}^7\text{Be} + \gamma$, and the ^{7}Be undergoes electron capture to produce ^{7}Li and a neutrino of 0.862 MeV. These solar neutrinos have been detected in a large tank of perchloroethylene (C_2Cl_4) by the reaction

$$\nu + {}^{37}\text{Cl} \rightarrow {}^{37}\text{Ar} + e^- \qquad (16.4)$$

Solar neutrino problem

This long-running experiment by Raymond Davis, Jr. of Brookhaven National Laboratory in the Homestake Gold Mine in South Dakota began in 1967. In a 100,000-gallon tank of C_2Cl_4 (cleaning fluid) only one ^{37}Ar per day is expected. Over a 20-year period of Davis's experiment, the measured rate of neutrinos was only about one third of the expected rate. This discrepancy, called the *solar neutrino problem,* has been examined in some detail. We discussed this problem in the Special Topic box on *Neutrino Detection* in Chapter 12.

In 1990 John Bahcall and Hans Bethe proposed a possible solution (a variation of an earlier theory by Mikheyev, Smirnov, and Wolfenstein). They proposed that electron neutrinos can be converted into muon neutrinos on their way out of the sun. If at least one of the neutrinos has mass, then the discrepancy between the number of solar neutrinos as solar output and the number predicted on the basis of extensively correlated solar models can be eliminated. This result is possible if the mass difference between the electron and muon neutrinos is 0.001 eV. The problem of the solar neutrino production is one of the most interesting in astrophysics today.

Eventual Fate of Stars

The final stages of stellar evolution begin when the hydrogen fuel is exhausted. At this point the gravitational attraction continues; the density and temperature increase. Eventually, the temperature becomes hot enough that the helium nuclei begin to fuse. Heavier elements are eventually created in nuclear fusion processes that are well understood. The fusion process continues until nuclei

SPECIAL TOPIC:

PLANCK'S TIME, LENGTH, AND MASS

We have recognized several fundamental constants in our study of physics. They include the gravitational constant G, Planck's constant h, and the speed of light c. The gravitational force depends on G, and several properties of a particle depend on h and c. These include the energy, $E = mc^2$, the wavelength as a function of a particle's momentum, $\lambda = h/p$, the Compton wavelength of a particle, $\lambda_C = h/mc$, and the energy of a photon, $E = h\nu$.

It is interesting to use dimensional analysis and to determine some characteristic values of time, length, and mass using the fundamental constants G, c, and h. Of course, the actual values depend on the system of units used, in our case SI, but what significance might these characteristic values have? The values of the fundamental constants and their dimensions in terms of length (L), time (T), and mass (M) are

Gravitational constant $G = 6.6726 \times 10^{-11}\ \text{N} \cdot \text{m}^2/\text{kg}^2$
Dimensions $M^{-1}L^3T^{-2}$

Speed of light $c = 2.9979 \times 10^8$ m/s
Dimensions LT^{-1}

Planck's constant $h = 6.6261 \times 10^{-34}$ J · s
Dimensions ML^2T^{-1}

If we use dimensional analysis, we find that we obtain mass by using the combination of $\sqrt{hc/G}$. We call this **Planck's mass** m_P and determine its value to be

$$m_P = \sqrt{\frac{hc}{G}} = 5.46 \times 10^{-8}\ \text{kg}$$

The physical significance of Planck's mass is that general relativity does not allow a black hole to be created from a mass smaller than Planck's mass.

We determine a characteristic length, called **Planck's length** λ_P, similarly by using dimensional analysis to determine

$$\lambda_P = \sqrt{\frac{Gh}{c^3}} = 4.05 \times 10^{-35}\ \text{m}$$

The time that it takes the speed of light to travel across Planck's length is called **Planck's time,** t_P, but it could also be determined from dimensional analysis. Its value is given by

$$t_P = \frac{\lambda_P}{c} = \sqrt{\frac{Gh}{c^5}} = 1.35 \times 10^{-43}\ \text{s}$$

In particular, physicists do not understand what happened before Planck's time in the creation of the universe. In essence, we can consider Planck's time to be the moment of creation of the universe, because the laws of physics as we know them now are not valid before Planck's time of 10^{-43} s.

---

near the iron region are produced, where elements have the highest binding energy per nucleon. The nuclear fusion process can no longer continue, and the reactions stop. The star's eventual fate depends on its mass.

For stars somewhat more massive than the sun, the gravitational attraction of the mass continues until the density of the star is incredibly large. Let us look at this process in some detail. Let there be N nucleons, each of mass m, in the star. The gravitational self-potential energy of a uniform sphere of mass Nm and radius R is (see Problem 26)

$$\text{P.E.} = U_{\text{grav}} = -\frac{3}{5}\frac{G(Nm)^2}{R}$$

where we have used U for potential energy rather than V to avoid confusion with the volume V. We determine the gravitational pressure by determining the force per unit area.

$$P_{\text{grav}} = \frac{F}{A} = -\frac{1}{4\pi R^2}\frac{dU_{\text{grav}}}{dR} = \frac{3G}{5}\frac{(Nm)^2}{4\pi R^4}$$

We write the gravitational pressure in terms of the volume by using $V = \frac{4}{3}\pi R^3$ and obtain

$$P_{\text{grav}} = 0.3G\frac{(Nm)^2}{V^{4/3}} \tag{16.5}$$

Matter is kept from total collapse by the outward electron pressure. This occurs because the Pauli exclusion principle effectively keeps two electrons from occupying the same state. However, for a sufficiently massive star, gravity will eventually force the electrons to interact with the protons through the reaction

$$e^- + p \rightarrow n + \nu_e \tag{16.6}$$

Neutron star

The result, called a **neutron star,** is composed mostly of neutrons.

However, neutrons also obey the exclusion principle, and outward pressure similar to that of the electrons will also be caused by the neutrons. Using the techniques in Chapter 9, the pressure of an electron gas is shown to be

$$P_e = \frac{2f\pi^2}{3}\frac{\hbar^2}{2m_e}\left(\frac{N_e}{V}\right)^{5/3}$$

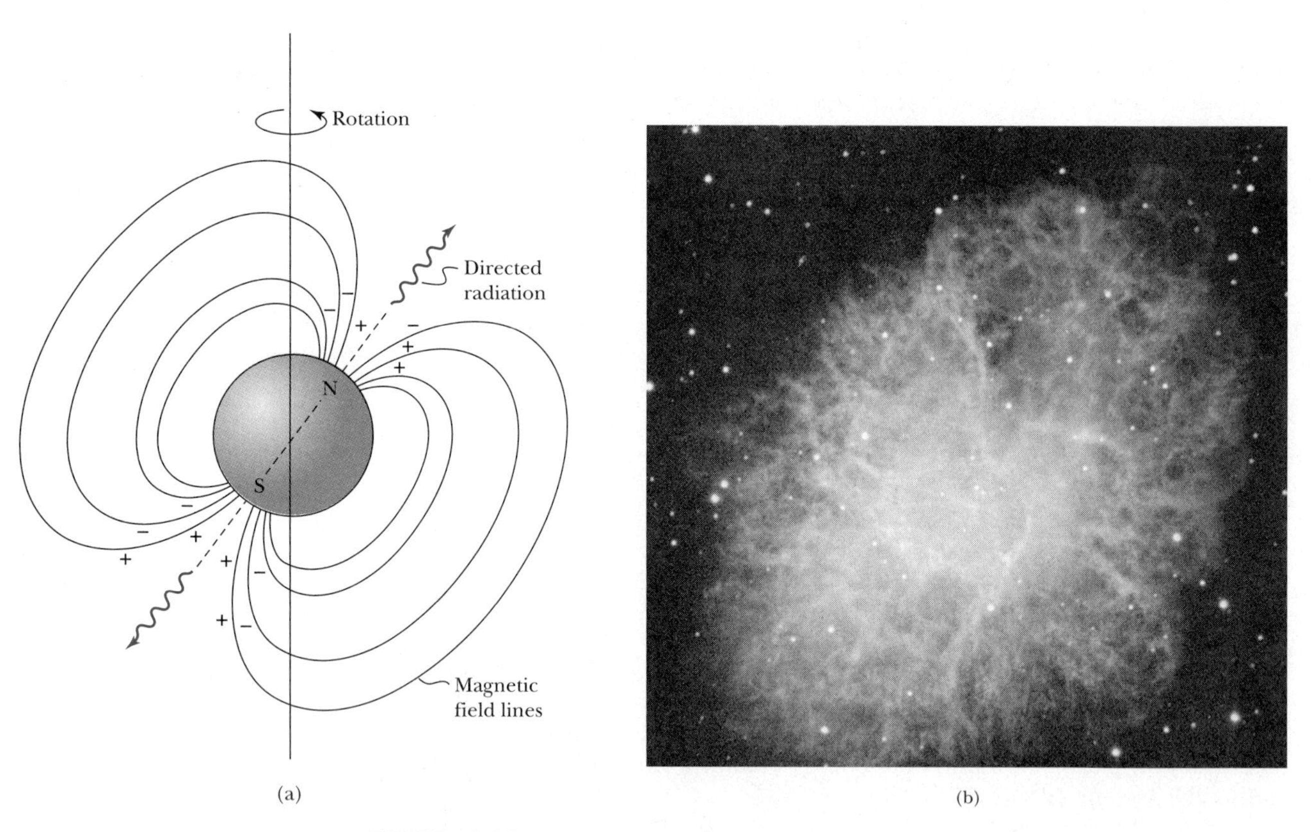

FIGURE 16.10 (a) A **pulsar** is a rapidly rotating neutron star. Large magnetic fields most likely accelerate charged particles near the pole regions producing copious amounts of electromagnetic radiation along the magnetic polar regions as shown in the figure. As the star rotates in space, this highly directional radiation appears as a pulsed radiation source to an observer located along one of the polar directions. Pulsars have periods ranging mostly from $\frac{1}{30}$ second to 3 minutes. (b) Photo of the **Crab Nebula,** which contains a pulsar whose supernova explosion in 1054 could be observed in daylight for three weeks. *Courtesy of Max Planck Institute for Astronomy and Calar Alto Observatory; K. Meisenheimer and A. Quetz.*

where f is the fraction of the average energy between 0 and E_F. For a Fermi gas, the average energy is about $\frac{3}{5}E_F$ (see Chapter 9), so we can take $f \approx 3/5$. We can use the same relation for neutrons by replacing N_e by N and m_e by m. The neutron pressure becomes

$$P_n = \frac{3.9\hbar^2(N/V)^{5/3}}{2m} \tag{16.7}$$

In equilibrium, the gravitational pressure will be balanced by the outward pressure of the neutrons due to the exclusion principle. We set Equations (16.5) and (16.7) equal.

$$0.3G\frac{(Nm)^2}{V^{4/3}} = \frac{3.9\hbar^2(N/V)^{5/3}}{2m}$$

We solve this equation for the cube root of the volume (see Problem 10).

$$V^{1/3} = \frac{6.5\hbar^2}{N^{1/3}\, m^3 G} \tag{16.8}$$

We can use this result to find the size of neutron stars as in the following example.

Example 16.5

Determine the radius of a neutron star with a mass of two solar masses.

Solution: We use Equation (16.8) to determine the volume and from that we can determine the radius. First, we find the number of neutrons N.

$$N = \frac{2N_{\text{sun}}}{m_{\text{neutron}}} = \frac{2(1.99 \times 10^{30}\text{ kg})}{1.675 \times 10^{-27}\text{ kg}} = 2.4 \times 10^{57}\text{ neutrons}$$

We now determine the cube root of the volume to be

$$V^{1/3} = \frac{6.5(1.06 \times 10^{-34}\text{ J}\cdot\text{s})^2}{(2.4 \times 10^{57})^{1/3}(1.675 \times 10^{-27}\text{ kg})^3(6.67 \times 10^{-11}\text{ J}\cdot\text{m/kg}^2)} = 1.7 \times 10^4\text{ m}$$

The radius R is calculated to be, from $V = \frac{4}{3}\pi R^3$,

$$R = \left(\frac{3}{4\pi}\right)^{1/3} V^{1/3} = \left(\frac{3}{4\pi}\right)^{1/3}(1.7 \times 10^4\text{ m}) = 11\text{ km}$$

The radius of a neutron star twice as massive as our sun is only 11 km! It is interesting to compare the density of a neutron star with that of a typical nucleus and with a nucleon (see Problem 11).

Neutron stars have densities as high as nuclei. They were first predicted in the 1930s but were not clearly observed until 1967, when a **pulsar,** a rapidly rotating neutron star (see Figure 16.10), was discovered. If the collapsing star has a mass greater than 3 solar masses, it can collapse through the neutron star stage and form a **black hole.** These extremely interesting objects were discussed in Chapter 15.

If the mass of the star is less than about 1.4 solar masses (the *Chandrasekhar limit*), the star can support itself against gravitational collapse, because the free electrons exert a considerable outward pressure. Such hot stars are called **white dwarfs,** and for a solar mass, their radius is about 1% of the solar radius. Such is the future of our sun.

Our sun will become a white dwarf

In some ways, it is quite amazing that quantum physics, through the Pauli exclusion principle, plays such a major role in stellar evolution. Until now, we have tended to think of quantum physics as only playing a role inside the tiny confines of the atom, yet now we are talking about objects as massive as our sun!

16.4 Astronomical Objects

Galaxies

Collections of stars are called **galaxies;** the gravitational attraction between stars keeps a galaxy together. Our own galaxy is the **Milky Way,** and it is believed to be composed of some 10^{11} stars. The number of galaxies is extremely large, at least 10^{11}. In the early part of the 20th century there was considerable controversy whether in fact the Milky Way comprised the entire universe. Astronomers called objects that had a cloudy unresolved appearance *nebulae.* In 1924 Hubble showed that the Andromeda nebula is quite far away from us ($\sim 10^6$ lightyears), and in fact, is a separate galaxy. It is rightfully called the *Andromeda galaxy.* Although other, smaller galaxies are closer to our Milky Way, the Andromeda galaxy is the largest nearby one, and, on a clear night, can be observed with the naked eye.

Astronomers keep finding galaxies farther and farther away. For example, early in 1998 a galaxy, called 0140 + 326RD1 (or RD1 for short), was discovered by the 10-m-diameter Keck II telescope located atop Mauna Kea in Hawaii. It was the first object found with a redshift greater than 5. If the universe is 13 billion years old, we are seeing the light from this galaxy when it was only 820 million years old, practically a primeval galaxy in its infancy. Its recorded redshift of 5.34 means the galaxy is moving away from us at great speed, and its light traveled 12.2 billion years to reach us. This redshift record was broken a few weeks later, however, and a redshift over 6 was subsequently found. The formation of galaxies in the early universe is one of the great mysteries today, and the new data from the twin 10-m-diameter Keck telescopes and a dozen more 8-m-diameter telescopes to come on line in the next few years promise many revelations in cosmology.

Active Galactic Nuclei and Quasars

In the early 1960s astronomers discovered objects with tremendously strong radio signals that had optical spectra that could not be understood. In 1963 Maarten Schmidt showed that these objects had to be at least 3 billion light years away from Earth because of their large redshifts. He was able to decipher the shifted spectral lines of hydrogen. These objects were dubbed **quasars,** short for *quasi-stars* or *quasi-stellar objects,* and can outshine galaxies having hundreds of billions of stars.

Thousands of quasars have now been discovered, and today we know that most of their energy is in the infrared, but with considerable optical and x-ray components. Only about 10% or less of the quasars emit powerful radio signals. Scientists quickly surmised that only a supermassive black hole residing in the center of a host galaxy could produce such an intense energy. Quasars are among the most distant objects in the universe, and therefore are among the earliest objects formed in the universe, some more than 12 billion years ago. Part of the observational difficulty is that the enormous distance of the galaxy means that the host galaxy is faint against the much brighter light of the quasar. The fact that quasars can vary in brightness in just a few hours or days suggests their size is only a few light hours or light days, not much larger than our solar system.

Long ago when young galaxies were forming, the stars in their cores were closely packed. Collisions and mergers of these stars gave rise to the supermassive black holes. These black holes required huge amounts of mass fuel, which was provided by gas from the galaxy's interstellar medium, from a companion galaxy, or from a star that came too close. Because quasars are observed only from great distances, quasars are extremely rare today. Therefore, quasars must evolve into objects that are common today, most likely normal, quiescent galaxies. If this is the case, then normal galaxies should have massive black holes in their centers that are no longer quasars, because they are starved for fuel. We, of course, don't know what fraction of normal galaxies passed through the quasar phase.

Unraveling the mystery of quasars was so predominant and important in the 1970s that it was a major justification for the *Hubble Space Telescope* (HST). Spectacular images from the HST have indeed shown that quasars reside in galaxies. However, the images have revealed other surprising results. The variety of these galaxies is extensive; some are normal, while others are merging and colliding with surrounding galaxies, sucking their mass from them. With new telescopes on ground and a new detector installed in the HST by astronauts, it is expected that the supermassive black holes at the center of normal galaxies may be observed in the near future. Many more questions remain, but the study of quasars provides us with a window to study the conditions that prevailed early in the history of the universe.

An amazing array of exotic extragalactic objects is lumped into the category called **active galactic nuclei (AGN).** These include the extraordinarily luminous quasars, Seyfert galaxies, and blazars. The general behavior is of a supermassive black hole surrounded by an accretion disk providing the mass. As matter spirals into contact with the black hole, huge amounts of electromagnetic radiation and jets of plasma are spewed out into space, often in directions perpendicular to the accretion disk. We can see this effect in Figure 16.11 showing two views of the galaxy NGC 4261. In the photo on the left, taken with a ground-based telescope, we can see the mass spewing out in the jet. In the expanded view of the center on the right, we can see the central object, accretion disk, and lobes. Many of the apparent properties of different AGN are due to our having different perspectives with respect to the disk. When we look directly along the spewing jets, we see remarkable intensities. *Blazars* are active galactic nuclei having high energies with one of its relativistic jets pointed toward Earth. They are associated with strong radio emissions and were discovered in 1991 using the *Compton Gamma Ray Observatory.*

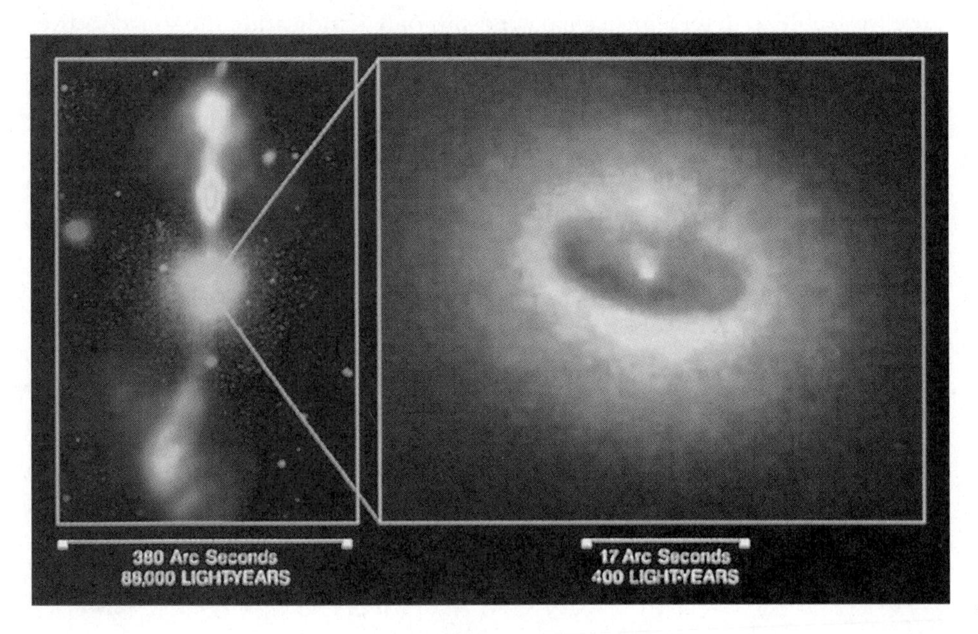

FIGURE 16.11 Photos of the active galaxy NGC 4261 believed to have a massive black hole: the image on the right is an enlarged view of the left image. Gas from the galaxy's interstellar medium, from a star that strays too close, or from another galaxy falls onto the massive black hole. An accretion disk forms, spewing light across the electromagnetic spectrum. In this photo both optical and radio images show the central object, accretion disk, and lobes. *Courtesy of W. Jaffe, H. Ford, and NASA.*

Gamma Ray Astrophysics

In the late 1960s the United States launched the *Vela* satellites to detect gamma-ray flashes from nuclear detonations for monitoring the Soviet Union's compliance with the Nuclear Test Ban Treaty. In a surprising discovery, *Vela* did detect gamma flashes, but they were from outer space, not from Earth. After declassification of these data in 1973, the scientific community set out to discover the source of these gamma rays. It is difficult to do gamma-ray astronomy on Earth, because the gamma rays are mostly absorbed in the Earth's atmosphere.

NASA's *Compton Gamma Ray Observatory* was launched in 1991 with a variety of detectors. The *Compton Observatory* is the grandest of new satellites that includes the Russian-French *Granat,* the Italian-Dutch *BeppoSAT,* as well as others that have some gamma-ray detection capability. Gamma-ray bursts are short flashes of electromagnetic radiation that appear about once a day at unpredictable times from random directions. The *Compton Observatory* has detected several thousand such bursts. Before 1991 it was thought that the source of the bursts were in our own Milky Way galaxy. Because the gamma-ray bursts have such a smooth distribution over the sky, it is now clear that the bursts have a cosmic origin rather than from our own galaxy.

The *Compton Observatory* has poor source directional determination, and distance determination is difficult for gamma ray detection. On the other hand, the *BeppoSAT* has capabilities for using x rays to determine the source direction of the bursts. Astronomers have been able to use information from the *BeppoSAT* to alert other astronomers to direct their optical telescopes in hopes of finding spectral data. They have learned that, although the gamma-ray bursts come and go quickly, there is an x-ray and optical afterglow that may last for days or months. Using the powerful Keck and Hubble telescopes, astronomers were able in 1997 to measure redshifts of the optical spectra to prove the bursts come from far away.

In late 1997 a gamma-ray burst was observed by both *Compton* and *BeppoSat* that apparently released several hundred times more energy than a supernova or about equal to the energy radiated by our entire galaxy over a period of 200 years. It is a stupendous amount of energy. Subsequent observations by several optical telescopes, including Kitt Peak, Keck II, and the *Hubble Space Telescope,* were able to determine a galaxy redshift of 3.4, indicating the source is over 11 billion light years away. To find such a brilliant source of energy is unprecedented. Speculation is that the amazing explosions may be due to rotating black holes, a black hole sucking in a neutron star, or something to do with early star formation. Of course, it is possible that the phenomenon is something completely unknown and more exotic. Future planned satellite missions will be able to locate gamma-ray bursts more precisely in real time and detect higher energy gammas.

Novae and Supernovae

Other interesting astronomical objects are **novae** and **supernovae.** Novae are simply stars that suddenly brighten, become visible (thus the word *nova* for "new"), and then fade over some period. There are believed to be two kinds of supernovae. The brightest is Type I. They arise in double-star systems when one star pulls matter from the other and becomes unstable resulting in an explosion. The other type, Type II, is believed to occur in very massive giant stars.

The Chinese may have spotted several supernovae during the past few thousand years. Good evidence exists that the explosion of the Crab supernova

(Type I) occurred in the year 1054. The Crab exploded only 6000 lightyears away, and it must have been an awesome sight. It was as bright as the planet Venus and could even be seen in the daytime. Both the Chinese and Japanese, and possibly American Indians, observed and noted the occurrence. The Crab nebula, shown in Figure 16.10, is a beautiful cloud of glowing gas easily seen with a telescope today. Other supernovae were observed by the Danish astronomer Tycho Brahe in 1572 and by Kepler in 1604. Curiously enough, there doesn't seem to have been a supernova visible to the naked eye after 1604 until 1987. The interest in supernovae increased dramatically with the unexpected observance of a supernova in 1987 (called the SN 1987A, see Figure 16.12).

Supernova Explosion. Astrophysicists now believe they have a good understanding of how the SN 1987A could have radiated more energy in 10 seconds than 100 of our suns could radiate in 10 billion years! We can describe a Type II supernova in very broad terms. The details are extremely interesting, and since SN 1987A, much has been learned because of the availability of a closely observed and measured occurrence. As hydrogen became exhausted by thermonuclear fusion and the helium content increased, gravity contracted the star to another phase. The helium nuclei began to fuse, and the central core of the star became denser and hotter. Eventually the helium, which had produced carbon and oxygen, was exhausted. Carbon began to fuse when the core temperature reached 8×10^8 K and the nuclei of neon, magnesium, and silicon were produced. Eventually most of the nuclei fused to form nuclei near iron.

Supernova 1987A

Before the explosion, the SN 1987A star had a mass of about 18 solar masses. At the center was a core of about 1.5 solar masses. The iron nuclei became so hot that they spewed out helium nuclei. The temperature and density skyrocketed, and neutrinos were radiated at an incredible rate. The gravitational force became so strong that a neutron star was formed. As the neutrons came closer together, the nuclear strong force eventually became effective and caused the neutrons to be even more attracted to each other. This attraction was only stopped when the neutrons came so close together that the nuclear force became repulsive to keep the nucleons from overlapping. This strongly repulsive nuclear force caused the material falling inward in the gravitational collapse to come to a screeching halt and to rebound. The reaction of the inner core stopping and the

FIGURE 16.12 Photos after and before the Supernova 1987A explosion taken by the Anglo-Australian Telescope in Australia. This is the first time that a progenitor star has been identified before the star actually exploded. © *Anglo-Australian Observatory, photography by David Malin.*

outside layers still collapsing under gravitation resulted in a shock wave moving outward. However, the shock wave was not quite able to blast its way back out through the mass of approximately 18 solar masses. The large flux of neutrinos, which normally just pass right through matter, was slowed down by the dense shock wave. The energy absorbed from the neutrinos was sufficient to give the shock wave the boost it needed to continue blasting its way through the outer layers of the star.

Neutrino astronomy is established

Because the explosion requires a high neutrino rate, neutrino production is a key indicator of a supernova explosion. Neutrinos were detected in Japan and the United States *3 hours* before the first visible light reached Earth. The neutrinos are able to pass more quickly through the star than visible light, which has a more circuitous path. The weak interaction has just enough strength to allow neutrinos to escape from the core. The travel time to Earth* and the energy spectrum of the neutrinos were consistent with predictions made from gravitational collapse and neutron-star formation. Our understanding of stellar evolution took a giant step forward, and the field of *neutrino astronomy* became well established with the singular event of supernova 1987A. Important information, including an upper mass limit of 16 eV/c^2 for the electron neutrino, was gleaned from the observation and associated calculations of SN 1987A. Computer simulations have been quite successful in understanding the ultraviolet burst some two hours after the explosion, the increase in luminosity for a week, and then the gradual decrease in brightness. The study of SN 1987A gave us confidence that observations of supernovae can be used as an indicator of cosmological distances. SN 1987A occurred in the Large Magellanic Cloud, our nearest neighbor galaxy, and estimates of the distance from this galaxy to Earth based solely on SN 1987A agreed with our best previous estimates (160,000 ly) to within 10%.

Supernova 1987A has been a veritable testing ground for theories and models on supernovae. The ionizing radiation has lit up a complex system of surrounding rings (see Figure 16.13). Astrophysicists would like to reconstruct how

*Note that SN 1987A occurred 160,000 years ago, and the light and neutrinos took this long to reach Earth.

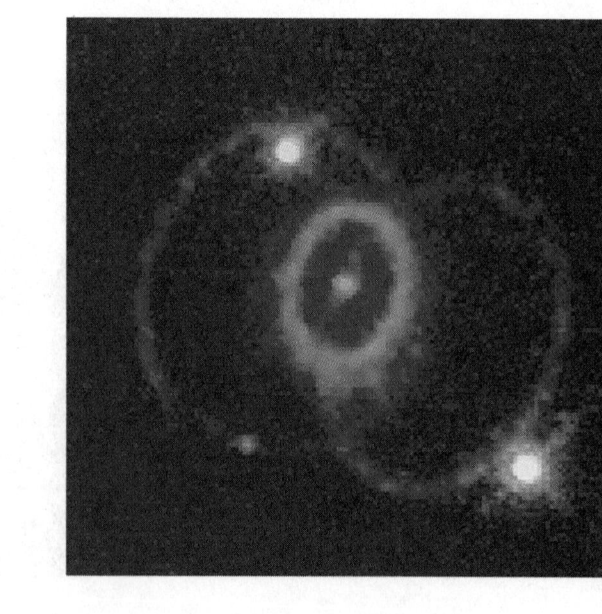

FIGURE 16.13 Supernova 1987A is the closest supernova since the telescope was invented. Its explosion sent a tremendous amount of gas, light, and neutrinos into interstellar space. In this *Hubble Space Telescope* image taken in 1994, large strange rings were discovered whose origins are still mysterious and unknown. It is thought that the rings may have been expelled before the main explosion. *Courtesy of Dr. Christopher Burrows, ESA/STScI and NASA.*

these rings were formed. One model is that the brightest inner ring represents gas thrown off when the progenitor star devoured a small binary companion some 20,000 years ago. The inner ring was lit up by x rays and ultraviolet light reaching the ring a few months after the supernova explosion, but it has been getting dimmer ever since. The *Hubble Space Telescope* first observed the two outer rings in 1994. All three rings were getting dimmer in the mid-1990s, but a gaseous shock wave triggered by the supernova explosion is working its way out to the brightest inner ring that is 160 billion km in diameter. By 1997 certain spots on the ring were already getting brighter, and by 2007 the ring will be ablaze with a fiery glow. It is hoped that a study of the inner ring will allow astrophysicists to learn more about the supernova's history, especially the origin of the outer rings and whether the original system was a single or binary system.

Example 16.6

Because galaxies have been observed with redshifts greater than 3 and quasars with redshifts more than 5, it is useful to have a plot of redshift versus recession velocity. Use Equation (2.32) for the relativistic Doppler shift to calculate such a plot up to redshifts of 4.

Solution: The term *redshift* used by astronomers and physicists refers to $\Delta\lambda/\lambda$. We can use Equation (2.32) to determine this quantity, where $\beta = v/c$.

$$\frac{\lambda}{\lambda_0} = \frac{\nu_0}{\nu} = \sqrt{\frac{1+\beta}{1-\beta}} \tag{16.9}$$

We determine $\Delta\lambda/\lambda_0$ from this equation.

$$\frac{\Delta\lambda}{\lambda_0} = \frac{\lambda - \lambda_0}{\lambda_0} = \sqrt{\frac{1+\beta}{1-\beta}} - 1 \tag{16.10}$$

We show the relationship between the redshift and recession velocity (β) in Figure 16.14. Note that as the redshift increases above 4, the velocity increases dramatically and approaches the speed of light.

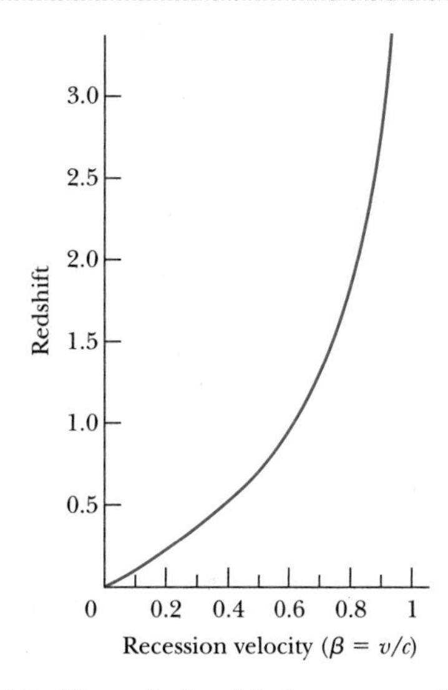

FIGURE 16.14 The relationship between redshift ($\Delta\lambda/\lambda_0$) and recession velocity ($\beta = v/c$) according to Equation (16.10).

Example 16.7

Astrophysicists were able to deduce a limit on the mass of electron neutrinos based on the spread of arrival times of neutrinos from supernova 1987A detected by two detectors in Japan and the United States. Find the difference in travel times between a massless neutrino and a neutrino having a mass of 16 eV/c^2. Assume the energy of the neutrino is 20 MeV.

Solution: Let t be the travel time for the massless neutrino, t' be the travel time for the neutrino having mass, and Δt be the time difference. We will have to use relativistic relations. The distance of SN 1987A from Earth is $d =$ 160,000 ly. If β is the value of v/c for the neutrino having mass, the time relations are

$$t = \frac{d}{c} \qquad t' = \frac{d}{\beta c}$$

$$\Delta t = t' - t = \frac{d}{c}\left(\frac{1}{\beta} - 1\right) = \frac{d}{c}\left(\frac{1-\beta}{\beta}\right)$$

From our earlier study of relativity, we have the total energy $E = \gamma mc^2$, $\gamma^2 = 1/(1-\beta^2)$, and $1 - \beta^2 = 1/\gamma^2$. The last equation can be written

$$(1-\beta)(1+\beta) = \frac{1}{\gamma^2}$$

or

$$(1-\beta) = \frac{1}{1+\beta}\frac{1}{\gamma^2} \approx \frac{1}{2\gamma^2}$$

because $\beta \approx 1$. The equation for Δt becomes, with $\beta \approx 1$,

$$\Delta t = \frac{d}{c}\frac{1}{2\gamma^2} = \frac{d}{2c}\left(\frac{mc^2}{E}\right)^2$$

We put in the appropriate numbers to obtain

$$\Delta t = \frac{1.6 \times 10^5 \text{ ly}}{2c}\left(\frac{16 \text{ eV}}{20 \text{ MeV}}\right)^2 = 5.1 \times 10^{-8} \text{ y} = 1.6 \text{ s}$$

This result is consistent with the actual spread in arrival times of a few seconds. Some assumptions had to be made in the actual calculation. For example, what if the slower neutrinos were emitted first and the faster neutrinos emitted last due to some effect in the supernova? Then the neutrinos might tend to arrive bunched together. A Monte Carlo simulation considering a wide class of possible neutrino emission models was used to assign an upper mass limit of 16 eV/c^2 for the electron neutrinos.

16.5 Problems with the Big Bang

By the early 1980s the Big Bang model was clearly the predominant theory for the origin of the universe. Nevertheless, there were at least three unexplained difficulties that the simple Big Bang model was not adequate to explain. They were

1. The *flatness* problem. Einstein's theory of relativity states that space must bend due to the gravitational attraction. Depending on the amount of matter per unit volume in the universe (its mass density), space curves in on itself such that parallel lines converge for a certain density called the *critical density.* For a mass density less than the critical density, parallel lines diverge, and the universe expands forever. This is called an *open universe.* For a mass density greater than the critical density, parallel lines converge, and the expansion of the universe will eventually be halted and then collapse. This is called a *closed universe.* In the 1980s it appeared that the universe may be just at that critical density where we have a *flat universe,* or very nearly so. Such an occurrence is quite an extraordinary circumstance.
2. The *homogeneity* problem. Why does the universe appear to be so homogeneous no matter in what direction we look? If the universe is 13 billion years old, then opposite sides of the universe are 26 billion light years apart. How can these regions have microwave radiation that is so similar? The temperature of the universe reflects the 3 K microwave background no matter in what direction we look.
3. The absence of magnetic monopoles. The occurrence of magnetic monopoles brings symmetry to Maxwell's equations of electromagnetism and also satisfies other physics theories. Why have we not yet detected magnetic monopoles?

The Inflationary Universe. These difficulties with the Big Bang model were relieved considerably in 1981 by a suggestion of Alan Guth, who proposed a variation of the standard Big Bang model. Guth proposed that at some time between roughly 10^{-35} s and 10^{-31} s after the Big Bang, the size of the universe suddenly expanded by a factor of 10^{50} due to the breaking up of the fundamental forces. We have to remember that the size of the universe was incredibly small at the time, and this remarkable expansion is unprecedented. It is like the electron of a hydrogen atom, which is normally only 10^{-10} m from a proton, suddenly finding itself 10^{24} lightyears away! After the inflationary period, the universe resumed its evolution according to the standard Big Bang model.

No matter how hard this may be to believe, inflation does mostly solve the three problems just listed. The inflationary theory *required* the mass density to be very close to the critical density. This solves the flatness problem. Guth argued that the universe was so tiny that it had already reached equilibrium before the inflation occurred; this explains the homogeneous universe. Subsequent to Guth's work, cosmologists have contributed many suggestions to his inflationary idea. Magnetic monopoles would have to occur along the boundaries or walls of different domains. These domains might be likened to different universes. The magnetic monopoles, if they exist, cannot be observed if they are at the edge of the universe. It is a neat, tidy package and much more complicated than we have been able to describe here. Inflationary theory is crucially connected to elementary particle theories.

Remaining Problems. Nevertheless, there are some remaining problems. Somehow cosmologists have to explain how the clumping of matter into galaxies, clusters of galaxies, and other strange objects like quasars and emitters of gamma-ray bursts have occurred. This is difficult to do in light of the homogeneous universe. Observations by the *Cosmic Background Explorer* satellite (COBE) launched in 1989 by NASA have shown that the homogeneity occurs over a much wider frequency range (see Figure 16.15) than could have been detected by ground-based observers. The COBE satellite searched for tiny *seeds* of inhomogeneity around which the first stars could have formed. Finally, in 1992 it was announced that scientists using the COBE data had succeeded in uncovering very tiny differences in temperature of radiation coming from matter at the edge of the universe. These tiny ripples in the temperature that were observed by COBE to be present only 300,000 years after the Big Bang are believed to account for the formation of stars and galaxies. The COBE results are consistent with the inflationary model of the Big Bang.

1992 COBE results

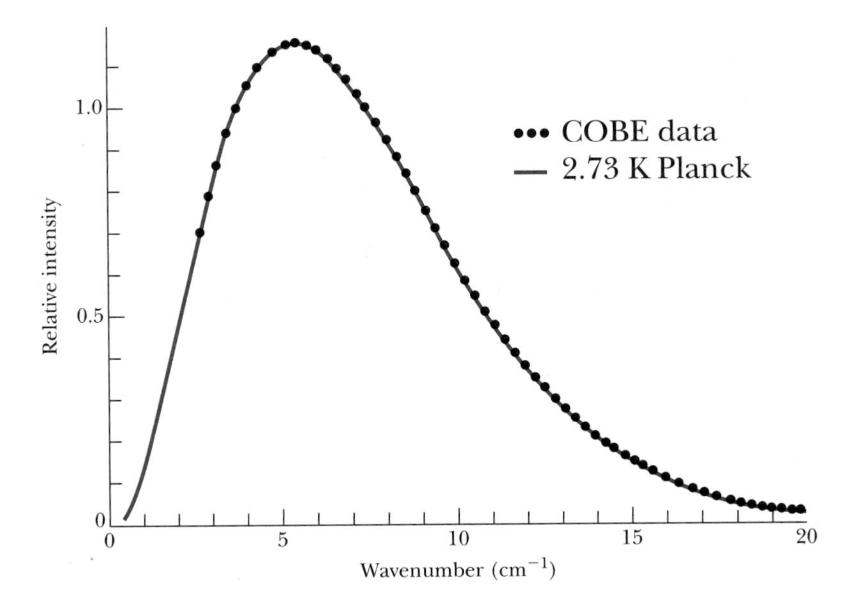

FIGURE 16.15 The spectrum of the intensity of the cosmic microwave background as measured with the COBE satellite. The data are smaller than the points shown, and the solid line is the blackbody radiation calculation for 2.73 K. The agreement is spectacular. The data are plotted as a function of wavenumber (inversely proportional to wavelength). *Courtesy of Nancy Burgess, NASA COBE science team.*

Missing mass problem

Another problem concerns the discrepancy between the mass of the universe required for the critical density (9×10^{-30} g/cm^3) and the apparent mass density (3×10^{-31} g/cm^3). This is known as the **missing mass problem.** The ratio of the actual mass density of the universe to the critical density is called Omega, Ω. Over 90% (maybe even 99%) of the mass of the universe may be in some form, called **cold dark matter,** that we have not yet observed. There are many theories concerning the source of this matter. *Dark matter* does not emit, scatter, or absorb light. It is not even made of atoms, and can only be seen by its gravitational effects. *Cold dark matter* moved much less slowly than the speed of light soon after the Big Bang. Dark matter might be due to neutrinos, WIMPs (Weakly Interacting Massive Particles), or primordial black holes. Astronomers and physicists have mounted many unsuccessful searches for the missing matter. One such search was on the observatory *Astro,* carried aloft by the space shuttle in 1990 to search for tau neutrinos. Some theorists thought that these mostly invisible particles could permeate the universe, and with their probable small mass, they could account for the cold dark matter. The *cold dark matter model* remains of interest to cosmologists and embraces the Big Bang, the inflationary theory, and particle physics theories. These searches for dark matter continue.

The Cosmological Constant. The cosmological constant was introduced by Einstein in 1917 because of certain problems in his original formulation of the general theory of relativity. Einstein introduced the constant, which accounts for the energy of perfect vacuum, in order to account for the homogeneous and isotropic universe. It was shown in 1922 that Einstein's theory does allow such a universe, and the constant is not needed. Einstein is reported to have later referred to the introduction of the cosmological constant Λ as his biggest blunder. It can be represented by

Cosmological constant

$$\Lambda = \frac{8\pi G}{c^4}\,\rho_V \tag{16.11}$$

where ρ_V is the vacuum energy density.

Now some physicists believe there indeed may be a justification for the energy in a vacuum due to some contributions that we now don't understand, for example by the field energy of some new interaction. Again, elementary particle theory becomes an indispensible part of a theory of the universe. A positive or negative cosmological constant has dramatic results for the future of the universe: If it is negative, the expansion of the universe will slow down, and if positive, the expansion of the universe will continue forever. A nonvanishing value for the cosmological constant solves certain problems, but some theorists believe it should be precisely zero. We shall discuss it again in Section 16.7.

Example 16.8

Determine the critical density of the universe.

Solution: One way to calculate the critical density ρ_c is by assuming the mass and radius of the universe are M and R, respectively. A galaxy of mass m and velocity v will be able to escape the universe with zero velocity if its total energy is precisely zero. We therefore determine the critical density by finding the condition for zero total energy. The sum of the kinetic and potential energy of the galaxy with respect to the universe is

$$\text{K.E.} + \text{P.E.} = \frac{1}{2}\,mv^2 - G\frac{mM}{R} = 0 \tag{16.12}$$

The mass M of the universe is related to the density ρ_c by $M = \frac{4}{3}\pi R^3\rho_c$. Equation (16.12) becomes

$$\frac{1}{2}mv^2 = G\frac{m}{R}\frac{4}{3}\pi R^3\rho_c$$

If we use $v = HR$ for the galaxy velocity, we rearrange the previous equation to give

$$\rho_c = \frac{3}{8\pi}\frac{H^2}{G} \qquad (16.13)$$

This is also the equation obtained from the general relativistic cosmological model of Einstein-de Sitter. If we insert the value of G and a (possible) value of $H = 21$ km/s per million lightyears, then we find $\rho_c = 9 \times 10^{-30}$ g/cm^3. This is the value reported in Section 16.5.

16.6 The Age of the Universe

In the last couple of decades the age of the universe has become perhaps the most sought after and the most controversial value in science. Until the mid-1990s, theory was well ahead of experiment, but with the breakthrough results of the *Hubble Space Telescope*, other productive satellite telescopes, and the advances in ground-based telescopes like those of the twin Keck together with adaptive optics systems, we are now in an observationally rich time. Today's theoretical ideas and models may be discarded tomorrow. It is an exciting time.

The age of the universe is intertwined with the eventual fate of the universe. At the present time, it is apparent that we can not determine how old the universe is without knowing its future. This is because we must interpret the observed electromagnetic radiation (light, ultraviolet, infrared, radio, x rays, and gammas) that comes from billions of years ago. Data must be consistent with models and theories, and when that is not so, new theories are required. Data from various sources are coming in at such a rapid rate that it is difficult for theories to keep up. Cosmologists can not determine the age of the universe without being able to understand how the universe was formed, especially in the early stages. The cosmic microwave background radiation is strong proof of the Big Bang, and inflation theory seems secure. However, the simple inflation theory does not explain all the data. The proof is in the details.

The Hubble constant H_0 determines an upper limit for the age of the universe. The Hubble time is $\tau = 1/H_0$, and because we believe the Hubble constant is 60–70 km/s per million parsec, the upper limit for the age becomes 14–16 billion years. There is still considerable controversy concerning the value of the Hubble constant, but the values are coalescing. In the mid-1990s cosmologists were in a quandary because of evidence that globular clusters were as old as 18 billion years. Globular clusters are aggregations containing up to millions of stars that are gravitationally bound; however, they are much smaller than galaxies. There are thousands of stars in each globular cluster that are about the same age. It was embarrassing that astronomers had found stars older than the universe itself! Finally, in 1998 it appeared that new analyses had found that the globular clusters were no more than 12 billion years old. The controversy had not been settled, but most astronomers expected that it would be as more data and analyses became available.

Let us now look into the ways that the age and the future of the universe are connected.

Redshift Measurements of Very Distant Objects. We use Einstein's theory of general relativity to relate the spacetime coordinates. We write those solutions, using a as our distance scale between galaxies, as

$$\left(\frac{da}{dt}\right)^2 = \frac{8\pi}{3} G\rho a^2 - kc^2 \tag{16.14}$$

$$-\frac{d^2a}{dt^2} = \frac{4\pi}{3} aG\left(\rho + 3\frac{P}{c^2}\right) \tag{16.15}$$

where ρ is the mass density of the universe, c is the speed of light, P is pressure (can be neglected), and k is a geometrical factor that determines the structure of the universe. A value of $k = +1$ indicates a closed universe; $k = -1$ indicates an open universe; $k = 0$ indicates a flat (Euclidean), but open universe. The case $k = 0$ represents the *Einstein-de Sitter* model. We can also now calculate the Hubble constant H using Equations (16.2) and (16.14). The result is

$$H^2 = \frac{8\pi\rho G}{3} - \frac{kc^2}{a^2} \tag{16.16}$$

Deceleration parameter

The **deceleration parameter q** is defined by

$$q = -a\frac{\dfrac{d^2a}{dt^2}}{\left(\dfrac{da}{dt}\right)^2} \tag{16.17}$$

and it tells us whether the universe is expanding more rapidly or slowing down (the expansion is slowing down if $q > 0$). Using the four previous equations, we determine q to be (neglecting the pressure P)

$$q = \frac{4\pi\rho G}{3H^2} \tag{16.18}$$

The deceleration parameter determines the deviation from Hubble's law for very distant objects. If $q > 1/2$, then k is positive, and the universe is closed. If $q \leq 1/2$, then k is negative, and the universe is open. If $q = 1/2$, $k = 0$, and the

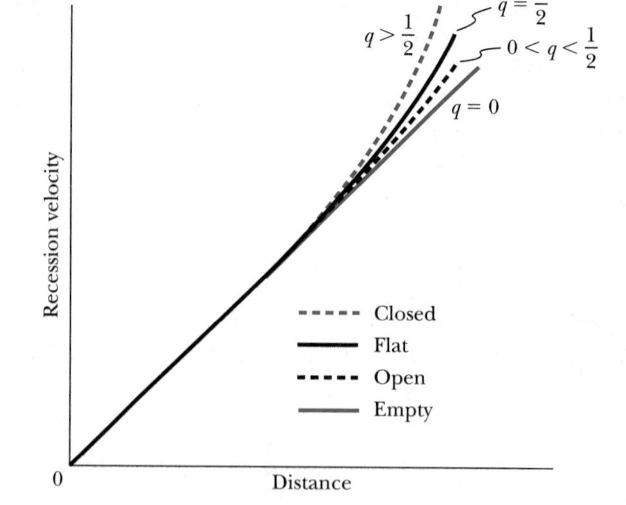

FIGURE 16.16 Recession velocities of galaxies are plotted as a function of their distance from us. Present data do not yet discriminate between various values of the deceleration parameter q to determine the evolution of the universe. More data for distances greater than 10^9 lightyears are needed.

universe is flat. Note that the critical value of the mass density ρ_c determined from Equation (16.18) for $q = 1/2$ is precisely the same as that determined in Equation (16.13) of Example 16.6. We show results for the relationship of the redshift vs. distance in Figure 16.16. It is not possible to determine the value of q from present data. With the new telescopes now operating and planned, it may be possible to determine q in the future.

Age of Universe Determination. Another method of determining the future of the universe has to do with the scale factor a (the approximate separation distance between galaxies) in Equation (16.2). As was discussed earlier, different models of the universe predict different relationships for a. The relationship of the scale distance a as a function of time is shown in Figure 16.17 for the three possibilities: Figure 16.17a—flat universe, $k = 0$, $q = 1/2$; Figure 16.17b—closed universe, $k > 0$, $q > 1/2$; Figure 16.17c—open universe, $k < 0$, $q < 1/2$.

Open or closed universe?

In Equation (16.2) we defined the Hubble constant to be $H = (1/a)(da/dt)$. We find the Hubble time τ to be

$$\tau = \frac{1}{H} = \frac{a}{\frac{da}{dt}} \tag{16.19}$$

In Figure 16.17a we draw a tangent to the a curve at the present time t_0. The intercept of this tangent with $a = 0$ determines the Hubble time τ as shown in

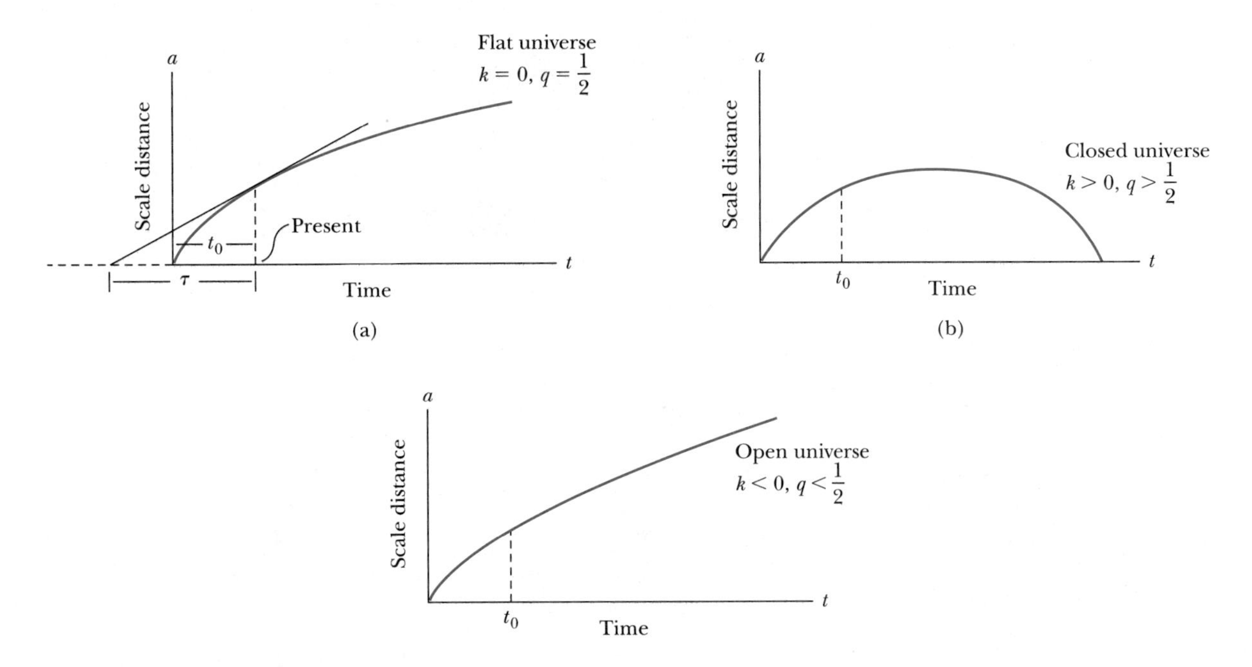

FIGURE 16.17 Representation of the distance scale parameter a as a function of time since the supposed Big Bang. The parameters k and q are indicated for the (a) flat, (b) closed, and (c) open universes. In (a) we draw a tangent to the curve at the present time t_0; the slope determines the Hubble constant, $H = 1/\tau$, and the intercept with $a = 0$ gives the Hubble time τ. The Hubble time can be determined in this manner for each graph. Note that the age of the universe must be less than the Hubble time.

FUTURE SPACE TELESCOPES

After decades of astrophysics and cosmology being dominated by theoretical models with little data to differentiate between sometimes wildly different projections, experimental astronomy has caught up. *The Hubble Space Telescope*, the *Compton Gamma Ray Observatory*, the twin Keck telescopes, and adaptive optics systems have all helped bring a horde of new experimental results. Many cosmologists are simply rubbing their hands with glee as they try to understand the new data. The future looks bright both for ground- and space-based telescopes.

But with a dozen additional 8-m-diameter ground-based telescopes coming into operation by the year 2010, why do we need telescopes in space? The answer is because the Earth's atmosphere absorbs the majority of the electromagnetic radiation coming from space (see Figure A). Only visible light, some radio waves, and a limited amount of infrared and ultraviolet light are not absorbed or distorted by the atmosphere. In order to see x rays, gammas, and most of the ultraviolet and infrared spectrum, we need to have telescopes above the Earth's atmosphere. The first observation of an x-ray source from space was made in 1962 on a rocket that stayed above the atmosphere for 6 minutes.

In the 1970s NASA scientists conceived of the *Great Observatories* program for space to cover most of the electromagnetic spectrum. The launch of the *Hubble Space Telescope* in 1990 (and its subsequent repair in 1993 and 1997) and the *Compton Gamma Ray Observatory* in 1991 were the first, and they have pro-

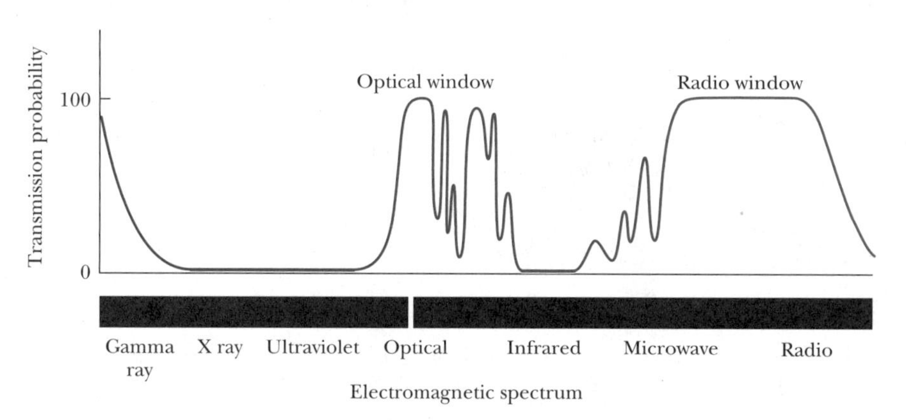

FIGURE A Much of the Earth's atmosphere is not transparent to the electromagnetic spectrum as shown by the transmission probabilities of the upper line. It is mostly optical and radio frequencies that are able to permeate the atmosphere.

Figure 16.17a. In the case of the flat universe, the relationship between a and t is $a^3 = Ct^2$, where C is a constant. Then if we take the derivative, we have

$$3a^2\,da = 2Ct\,dt$$

$$\frac{da}{dt} = \frac{2C}{3a^2}t$$

Equation (16.19) then gives

$$\tau = a\frac{3a^2}{2C}\frac{1}{t} = \frac{3a^3}{2C}\frac{1}{t} = \frac{3}{2}t \tag{16.20}$$

duced spectacular results. The third of the four observatories is the *Chandra X-ray Observatory* launched in 1999. The fourth will be the *Space Infrared Telescope Facility* (SIRTF). *Chandra* will be examining x rays from throughout the universe, like colliding galaxies and black holes. Work began in 1998 on SIRTF, which is planned for launch in 2001. It will be able to detect heat given off by objects in space. Conventional telescopes cannot see infrared light, but many objects like planets and unignited stars do not produce visible or ultraviolet light. Some of the most fascinating objects and processes in the universe may exist hidden behind vast clouds of dust and gas that populate the universe. These objects include black holes, quasars, and regions where stars are forming in galaxies and planets are forming around stars. SIRTF will be able to sense the heat of dark, faint or hidden objects at the edge of the universe.

The *Gamma-Ray Large Area Space Telescope* (GLAST) is under study by the United States, Japan, and Europe for possible flight before 2010. It will detect gamma rays from 10 MeV to more than 100 GeV energy that are emitted from active galactic nuclei, pulsars, stellar-mass black holes, supernova remnants, and gamma-ray bursts. Currently under development by the European Space Agency with help from Russia (Proton rocket launcher) and NASA (ground tracking stations) is the *International Gamma-Ray Astrophysics Laboratory* (INTEGRAL). It is a high-resolution device designed to detect gamma rays in the energy range 15 keV to 10 MeV. It will be able to image and accurately position celestial sources of gamma-ray emission. This energy range is particularly suitable for spectral line-forming processes such as nuclear excitation, radioactivity, positron annihilation, cyclotron emission and absorption. Unique astrophysical information will almost certainly lead to fundamental new discoveries.

The *Next Generation Space Telescope* (NGST) is presently under development by NASA and international partners such as the European Space Agency for possible launch around 2008. In some ways it is thought of as the replacement for the *Hubble Space Telescope*. Its main feature will be a large (6- to 8-m-diameter) mirror. It is expected to observe primarily infrared light to study high-redshift galaxies and protogalaxies. It would be placed in an Earth orbit beyond the moon.

There are many other missions under development or being planned. We can only mention a few. One smaller mission under development is the second *High Energy Transient Explorer* (HETE-2), which is intended to make gamma-ray and x-ray observations of gamma-ray bursts. It is hoped that HETE-2 will be launched before the year 2000. The original HETE was lost in 1996 due to a Pegasus launch failure. Another smaller mission is the Japanese-United States *Astro-E* that is expected to be launched in 2000. It has several x-ray instruments with large collecting area and high efficiency to study similar objects as the gamma-ray missions, including an x-ray telescope that covers a range of 10–700 keV and high resolution devices for lower energies. A successor to the remarkably successful COBE satellite is called the *Microwave Anisotropy Probe* (MAP). Scheduled for launch in 2000, MAP will measure anisotropy with much higher resolution than did COBE. The new information in these smaller size fluctuations should help clarify several cosmological questions.

In the case of the flat universe the age of the universe is $2\tau/3$ $(2/3H)$, and if we use a value of 65 from the midrange of Hubble's constant from 60–70 km/s per Mpc, we determine $\tau = 15$ billion years, and the age of the universe = 10 billion years. Detailed calculations show that for the closed universe $t_0 < 2\tau/3$, and for the open universe $\tau > t_0 > 2\tau/3$. These results are presented in Table 16.1. If the current suggestions of an open universe are correct, the age of the universe is between 10–15 billion years, which is consistent with other results. As Figure 16.17 shows, the universe cannot be older than the Hubble time τ. We are on the verge of pinning down the age of the universe and understanding much about the early stages of cosmology.

TABLE 16.1
Status of the Universe

Deceleration Parameter	Average Density	Age of Universe	State of Universe
$q = \frac{1}{2}$	ρ_c	$t_0 = \frac{2\tau}{3}$	Flat
$q > \frac{1}{2}$	Greater than ρ_c	$t_0 < \frac{2\tau}{3}$	Closed
$0 < q < \frac{1}{2}$	Less than ρ_c	$\frac{2\tau}{3} < t_0 < \tau$	Open
$q = 0$	0	$t_0 = \tau$	Open, empty

16.7 The Future

Long before we have to worry about what will happen to the universe, the sun will be a cold dark mass. In this section we first will discuss the demise of the sun, of which we are fairly certain. Less certain is the future of the universe.

The Demise of the Sun

The sun is about halfway through its life as a productive star. It became a star about 5 billion years ago, and it will be about another 5 billion years before it runs out of its thermonuclear fuel. As the hydrogen fuel is exhausted, the sun will contract under its gravitational attraction. The sun will heat up even more, with helium being at its core and a layer of hydrogen gas outside. The heat will cause the outside layers of the sun to expand. The surface will expand to such a size that it will engulf the planet Mercury, perhaps reaching as far as the Earth. We call such a star a **red giant.** The sun's surface temperature will cool from the present 5500 K to maybe 4000 K. However, the temperature of the Earth will dramatically increase because of the proximity to the sun's surface, and no life will then be able to exist on Earth.

Red giant

Eventually, the hot core of a star uses up its thermonuclear fuel, producing heavier and heavier nuclear masses so that mostly iron remains. The mass of the sun is not large enough for the gravitational attraction to form a neutron star or supernova. The lighter elements in the outer layers of the sun will boil off due to the reduced gravitational attraction at the surface, and the final mass will be about 0.5 to 0.6 times the mass of today's sun. The remaining sun will contract to about the size of the Earth. Slowly the sun will cool down, becoming a white dwarf, and then finally a cold black dwarf. The sun and its planets will be doomed to an eternally frozen death.

The Future of the Universe?

Type Ia supernovae serve as ideal candidates for measuring distances, because their brightness can be determined by measuring the rate at which they dim after exploding. For example, astrophysicists can tell whether they are relatively close dim objects or far away bright objects. Finding accurate distances is the most difficult problem in age of the universe determinations. In 1998 two groups using supernova data and another group using globular clusters determined that there was not enough mass in the universe to sustain a closed universe. These data caused renewed interest in Einstein's cosmological constant—a kind of "antigravity" effect that may exist in the vacuum of space and results in a *vacuum energy density*. The effect of the cosmological constant is to cause the universe to expand.

The cosmological constant Λ is related to the vacuum energy density ρ_V and the critical energy (mass) density ρ_c by

$$\Lambda = \frac{\rho_V}{\rho_c} \tag{16.21}$$

The supernova results suggest that $\Lambda = 0.75$. The values of Λ and Ω (the ratio of ordinary matter density to the critical density) determine the future of the universe. When $\Lambda + \Omega < 1$, the universe is open, when $\Lambda + \Omega > 1$, the universe is closed, and when $\Lambda + \Omega = 1$, the universe is flat. We show in Figure 16.18 the supernova data plotted with recession velocity vs. distance together with several of the predictions. The data favor a value of Ω between 0 and 1. There is still considerable controversy as to whether the universe is flat or open, but these data seem to indicate it is open. If this preliminary result is upheld by additional data, then one of the most fundamental questions in cosmology and astrophysics has been answered. The universe is doomed to a cold death eons in the future.

During the inflationary epoch, the vacuum energy density was very large, but later the simplest inflationary theory calls for it to be zero. A nonzero cosmological constant is a challenge for inflationary theory, but recent theoretical calculations suggest it can be fixed with a quantum correction. These new calculations have other predictions like multiple universes or too low mass density. Some theorists believe the universe will still turn out to be flat, and the question is not completely settled. Only additional experimental data in the next few years will be able to tell us.

Are There Other Earths Out There?

We are now in a position to question seriously whether there are other Earth-like planets in the universe. In the last few years, astronomers have identified many candidates for extrasolar planets, that is, planets revolving around stars other than our own sun. Initially, these candidates were identified through a wobble of

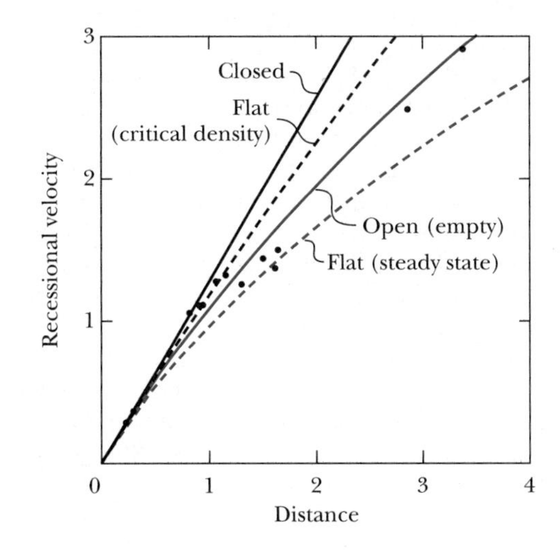

FIGURE 16.18 Recent data from supernova measurements are shown on a graph of recessional velocity versus distance. The curves are for various values of Ω and Λ: closed ($\Omega = 2$, $\Lambda = 0$); flat (critical density, $\Omega = 1$, $\Lambda = 0$); open (empty, $\Omega = 0$, $\Lambda = 0$); flat (steady state, ($\Omega = 0$, $\Lambda = 1$). The data indicate that Ω is between 0 and 1, and that the universe has less than the critical density needed for it to be closed. *Graph courtesy of Edward L. (Ned) Wright, UCLA.*

the star that was induced by an orbiting planet. The wobble's period and magnitude can be used to deduce the planet's orbit and minimum mass. Such observations use the Doppler effect and require extensive measurements.

One difficulty is distinguishing between planets and **brown dwarfs,** which are starlike objects with too little mass (<80 Jupiter masses) to create nuclear fusion. A brown dwarf forms the same way a star does, by gas clouds collapsing in on itself by gravitational attraction. Planets grow from dust and gas accreting in a circumstellar disk. Several observations of swirling dust around a star indicates a planet is forming. The first such observation of a dust disk was found around Beta Pictoris, a young star only 63 light years away. The data and information on extrasolar planets should explode in the coming years as more powerful telescopes are trained on likely spots.

Summary

Scientists believe the universe started with a primordial event called the Big Bang. The universe started as an extremely dense, hot fireball. After the initial 10^{-43} s, our current theories have been fairly successful in understanding the subsequent events. For example, by the time 10^{-13} s, quarks and leptons were formed, and by 10^{-3} s, neutrons and protons were also formed. Nuclei formed by 3 min, but atoms couldn't form for 700,000 years.

The three pieces of evidence that confirm the Big Bang include the redshift measurements of Edwin Hubble that indicate the universe is expanding, the observation of the cosmic microwave background radiation by Penzias and Wilson, and the agreement between the predictions of the primordial nucleosynthesis of the elements and known abundances. Hubble's law relates the recessional velocity and the distance from Earth R

$$v = HR \tag{16.1}$$

where H, Hubble's constant, is related to a scale factor a that is proportional to the distance between galaxies.

$$H = \frac{1}{a}\frac{da}{dt} \tag{16.2}$$

Stars were formed when matter collected together due to the gravitational force. Eventually nuclear fusion provides the energy radiated from stars. Neutrinos are emitted by stars due to the nuclear fusion process, but astrophysicists don't fully understand the observed rate of neutrinos detected from the sun.

After the thermonuclear processes are exhausted, massive stars undergo a gravitational contraction so strong that eventually the star consists mostly of neutrons. These neutron stars are incredibly dense. Stars of mass less than about 1.4 solar masses are able to exert enough pressure due to electrons to prevent a gravitational collapse. They eventually become white dwarfs.

There are many astronomical objects in the universe. Several kinds of galaxies, which are collections of many stars, have been observed. Quasars are tremendous sources of energy, and several have been found quite far away that must have been formed early in the universe.

Increased interest in supernovas occurred because of the SN 1987A event, which could be seen with the naked eye. This supernova occurred when a massive star underwent gravitational collapse after its thermonuclear fuel was exhausted. A tremendous explosion resulted that sent radiation streaming outward. On Earth we were able to see visible light and detect neutrinos, among other types of radiation.

Some problems with the Big Bang model were mostly solved in the early 1980s by a suggestion of Alan Guth that the universe had suddenly and dramatically expanded very early. This idea of inflation solved the flatness and homogeneity problems, but other problems remained. There is a discrepancy between the critical density and observed mass density of the universe. It appears that there is not enough mass in the universe to have a flat universe. Rather, it seems that the universe will expand forever.

In order to determine the future of the universe, it is first necessary to understand how the universe was formed. The Hubble constant is an important parameter in determining the age of the universe. It is believed to be 60–70 km/s per Mpc. For an open universe, the universe is then 10–15 billion years old.

Many new ground-based and space telescopes are now available, under construction, or being planned that will give new data and results that undoubtedly will be in conflict with present beliefs. This is the golden age of astrophysics and cosmology.

Questions

1. Explain why Hubble's constant is not actually a constant.
2. According to thermodynamic equilibrium, which should be the most abundant and least abundant quarks during the period from 10^{-13} s to 10^{-3} s?
3. If the gravitational attraction is important in a neutron star where the neutrons are close together, then why isn't the gravitational interaction important in a nucleus with many neutrons?
4. If all the distant galaxies are moving away from us, explain why we are not at the center of the universe.
5. How can you explain the fact that the Andromeda galaxy appears to be approaching us rather than receding?
6. Explain why the universe cannot be older than the Hubble time.
7. Explain why elements heavier than iron are not found in stars.
8. Why isn't it possible to know what is happening to our nearest neighbor stars today (in the next 24 hours)?
9. During which stage of the beginning of the universe would you expect deuterons to be formed? Explain.
10. What happened to the neutrons produced in the early stages of the universe that were not synthesized to deuterons or ^{4}He nuclei?
11. During what time period do free neutrons disappear? Explain.
12. Explain how it might be possible to confuse the redshifts from recession velocities with the gravitational redshifts. How can we distinguish the two?
13. Quasars are known to vary in brightness by just a few hours or days. What can we say about the size of these quasars?
14. Observations from the *Compton Gamma Ray Observatory* indicate that the gamma-ray bursts have an even distribution throughout the sky. How can we be sure that these bright phenomena are not coming from our own galaxy, the Milky Way?

Problems

16.1 Evidence of the Big Bang

1. Derive the conversion from parsecs to lightyears given the information in Example 16.1.

16.2 The Big Bang

2. Calculate the temperature for which the ratio of free protons to neutrons in the early stages of the universe was seven.
3. At what temperature was the universe hot enough for photons to produce π^0? Approximately what time was this? Use the mean value of the distribution.
4. Using the thermodynamic equilibrium factor $\exp\left(\frac{\Delta mc^2}{kT}\right)$, determine the relative abundances of the quarks during the time period from 10^{-13} s to 10^{-3} s. Assume the temperature is 10^{14} K and use mid-range quark masses from Table 14.5.
5. At what threshold temperature were the electron and muon first formed?
6. If the mass of the electron neutrino is 5 eV/c^2, at what time was it first formed in the universe? What if its mass is 10^{-4} eV/c^2?
7. Was π^0 or π^+ formed first in the early stages of the universe? What is the difference in mean temperature between when they were formed?
8. Calculate the temperature of the universe when photons can no longer disassociate deuterons. Use the mean value of the distribution.
9. Determine the temperature of the universe when it has cooled enough that photons no longer disassociate the hydrogen atom. Use the mean value of the distribution.

16.3 Stellar Evolution

10. Show that the result given in Equation (16.8) for the volume of a neutron star follows from the equation preceding it.
11. Calculate the density of a neutron star from the results given in Example 16.5 and compare that with the density of a nucleon and a nucleus.
12. Show that the radius of a neutron star decreases as the number of neutrons increases. Does this make sense? Shouldn't the radius increase with more neutrons?
13. Calculate the gravitational pressure for (a) the sun, and (b) the neutron star of Example 16.5.

16.4 Astronomical Objects

(*Note:* Use the intermediate value $H = 65 \text{ km} \cdot \text{s}^{-1} \cdot \text{Mpc}^{-1}$ for problems in this section.)

14. An object in *Hydra* is 4 Gly from us. What would we expect its recessional velocity to be?
15. An object in *Ursa Major* is determined to be receding from us with a velocity of 15,000 km/s. How far from us is it?
16. Using the redshift of 3.8 for 4C41.17, a powerful radio galaxy, determine the distance of the galaxy from us in Mpc.

16.5 Problems with the Big Bang

17. (a) Use the observed mass density of the universe to determine the average number of nucleons per cubic meter throughout the universe. (b) There are 60 stars within 16.6 ly of the sun. If each averages 1 solar mass, what is the mass density of nucleons in the neighborhood of the sun?

16.7 The Future

18. Show that Equation (16.16) for Hubble's constant follows from the preceding results.

19. In Example 16.8 show that the critical density ρ_c is 9×10^{-30} g/cm^3.

20. Calculate the deceleration parameter by assuming the Hubble constant to be midway in its range of values for (a) the critical density, and (b) the observed mass density.

General Problems

21. Assume a power law for the scale factor $a = Ct^n$, where C is a constant. (a) For what values of n is the universe accelerating and decelerating? (b) For deceleration, what is the dependence of H on time?

22. Use the blackbody spectrum to determine the peak wavelength for 2.7 K.

23. Calculate the critical density necessary for a closed universe for two extremes of the Hubble constant: 55 km · s^{-1} · Mpc^{-1} and 85 km · s^{-1} · Mpc^{-1}.

24. The time before which we don't know what happened in the universe (10^{-43} s) is called the *Planck time.* The theory needed is a quantum theory of gravity and concerns the three fundamental constants h, G, and c. (a) Use dimensional analysis to determine the exponents m, n, l if the Planck time $t_p = h^m\, G^n\, c^l$. (b) Calculate the Planck time using the expression you found in (a).

25. Let the wavelength of a photon produced during the early stages of the universe be λ, and denote the Doppler-shifted wavelength by λ_D we measure today. Show that

$$\lambda_D = \frac{\sqrt{1+\beta}}{\sqrt{1-\beta}}\,\lambda$$

where $\beta = v/c$.

26. On two occasions we have used the gravitational self-energy of a uniform sphere of mass M and radius R. Use integral calculus and start with a mass dm in the sphere. Calculate the work done to bring the remainder of the mass in from infinity. By this technique show that the self-potential energy of the mass is

$$\text{P.E.} = -\frac{3}{5}\frac{GM^2}{R}$$

27. Draw tangents on Figure 16.17b and 16.17c similar to that done in Figure 16.17a and determine the relationship between the Hubble time τ and the age of the universe.

28. Show that the extra time, Δt, that a neutrino with finite mass takes to reach the Earth from a supernova explosion compared to that taken for a zero mass particle is

$$\Delta t = 2.57\text{ s}\left(\frac{\text{distance}}{50\text{ kpc}}\right)\left(\frac{m_\nu c^2}{10\text{ eV}}\right)^2\left(\frac{10\text{ MeV}}{E}\right)^2$$

where $m_\nu c^2$ is the rest energy in eV and E is the energy in MeV of the neutrino.

29. Show that the mass density of radiation ρ_{rad} is given by

$$\rho_{\text{rad}} = \frac{4\sigma T^4}{c^3}$$

where σ is the Stefan-Boltzmann constant and T is the temperature. You might find the Stefan-Boltzmann law and $E = mc^2$ to be useful.

30. Use the mass density of radiation in the previous problem to determine the mass density of radiation when $T = 3$ K. How does this compare with the average density of matter in the universe? Does this mean we are in a radiation-dominated or matter-dominated universe?

31. Using the mass density of radiation from Problem 29, calculate the density for several temperatures between 10^{-2} K and 10^{30} K. Make a plot of ρ_{rad} vs. time using Figure 16.10. If the universe changed from being radiation dominated to matter dominated at 700,000 years, at what density for ρ_{rad} and ρ_{matter} did this occur?

32. The exponential fall off of the brightness of Supernova 1987A was due to the decay of ^{56}Ni ($t_{1/2}$ = 6.1 d) $\rightarrow$ ^{56}Co ($t_{1/2}$ = 77.1 d) $\rightarrow$ ^{56}Fe. If the energy were primarily due to the decay of ^{56}Ni, what fall off in brightness by the end of 300 days would you expect? What if it were due to the energy in the decay of ^{56}Co? The actual data showed a decrease in brightness about a factor of 100 after 300 days.

33. The Lyman alpha line (K_α) of hydrogen is measured in the laboratory to have a wavelength of 121.6 nm. In the quasar PKS 2000-330 the same line is determined to have a wavelength of 582.5 nm. What is its redshift and recession velocity?

34. The redshift parameter z is defined by $\Delta\lambda/\lambda$. Show that the Doppler redshift parameter is related to relative speed β by

$$1 + z = \sqrt{\left(\frac{1+\beta}{1-\beta}\right)}$$

35. In cases where the speed is small ($\beta \ll 1$), show that the Doppler redshift parameter is related to β by $z \approx \beta$.

36. In 1998 a galaxy named RD1 was discovered with a redshift of 5.34. (a) What is the speed of this galaxy with respect to us? (b) Use Hubble's law to determine how far away the galaxy is.

37. The first reaction in the proton–proton chain is $p + p \rightarrow d + \beta^+ + \nu$. Calculate the Q-value of the reaction and determine the maximum neutrino energy.

APPENDIX 1

Fundamental Constants

Quantity	Symbol	Value
Speed of light in vacuum	c	299 792 458 m/s (exact)
Permeability of vacuum	μ_0	$4\pi \times 10^{-7}$ N/A^2
Permittivity of vacuum	ϵ_0	$8.854\ 187\ 817 \times 10^{-12}$ F/m
Newtonian gravitational constant	G	$6.672\ 59(85) \times 10^{-11}$ m$^3 \cdot$ kg$^{-1} \cdot$ s^{-2}
Fine structure constant	$\alpha = \dfrac{e^2}{4\pi\epsilon_0\hbar c}$	$0.0072974 \approx \dfrac{1}{137}$
Planck constant	h	$6.626\ 0755(40) \times 10^{-34}$ J $\cdot$ s
Planck constant $h/2\pi$	$\hbar$	$1.054\ 572\ 66(63) \times 10^{-34}$ J $\cdot$ s
Elementary charge	e	$1.602\ 177\ 33(49) \times 10^{-19}$ C
Magnetic flux quantum	$\Phi_0 = \dfrac{h}{2e}$	$2.067\ 834\ 61(61) \times 10^{-15}$ T $\cdot$ m^2
Josephson frequency–voltage ratio	$2e/h$	$4.835\ 9767(14) \times 10^{14}$ Hz/V
Quantized Hall resistance	$R_H = h/e^2$	25 812.8056(12) Ω
Rydberg constant	$R_\infty = \dfrac{\alpha^2 m_e c}{2h}$	$1.097\ 373\ 1534(13) \times 10^7$ m^{-1}
Bohr radius	$a_0 = \dfrac{4\pi\epsilon_0\hbar^2}{m_e e^2}$	$5.291\ 772\ 49(24) \times 10^{-11}$ m
Compton wavelength	$\lambda_c = \dfrac{h}{m_e c}$	$2.426\ 310\ 58(22) \times 10^{-12}$ m
Classical electron radius	$r_e = \alpha^2 a_0$	$2.817\ 940\ 92(38) \times 10^{-15}$ m
Bohr magneton	$\mu_B = \dfrac{e\hbar}{2m_e}$	$9.274\ 0154(31) \times 10^{-24}$ J/T
Nuclear magneton	$\mu_N = \dfrac{e\hbar}{2m_p}$	$5.050\ 7866(17) \times 10^{-27}$ J/T
Avogadro constant	N_A	$6.022\ 1367(36) \times 10^{23}$ mol^{-1}
Molar gas constant	R	$8.314\ 510(70)$ J $\cdot$ mol$^{-1} \cdot$ K^{-1}
Boltzmann constant	k	$1.380\ 658(12) \times 10^{-23}$ J/K
Stefan-Boltzmann constant	σ	$5.670\ 51(19) \times 10^{-8}$ W $\cdot$ m$^{-2} \cdot$ K^{-4}
Wien displacement law constant	$\lambda_{max} T$	$2.897\ 756(24) \times 10^{-3}$ m $\cdot$ K
Atomic mass unit	u	$1.660\ 5402(10) \times 10^{-27}$ kg
Electron mass	m_e	$9.109\ 3897(54) \times 10^{-31}$ kg
Muon mass	m_μ	$1.883\ 5327(11) \times 10^{-28}$ kg
Proton mass	m_p	$1.672\ 6231(10) \times 10^{-27}$ kg
Neutron mass	m_n	$1.674\ 9286(10) \times 10^{-27}$ kg
Deuteron mass	m_d	$3.343\ 5860(20) \times 10^{-27}$ kg

Note: Digits in parentheses represent one standard deviation uncertainty in the final digits of the given value. Values from E. Richard Cohen and Barry N. Taylor, CODATA Task Group on Fundamental Constants, CODATA Bulletin No. 63, November 1986. The latest values of the fundamental constants can be found at the National Institute of Standards and Technology website at http://physics.nist.gov/cuu/Constants/index.html

APPENDIX 2

Conversion Factors

Length

	m	cm	km	ft	ly	pc
1 meter	1	10^2	10^{-3}	3.281	1.057×10^{-16}	3.241×10^{-17}
1 centimeter	10^{-2}	1	10^{-5}	3.281×10^{-2}	1.057×10^{-18}	3.241×10^{-19}
1 kilometer	10^3	10^5	1	3.281×10^3	1.057×10^{-18}	3.241×10^{-14}
1 foot	0.3048	30.48	3.048×10^{-4}	1	3.222×10^{-17}	9.878×10^{-18}
1 light year	9.461×10^{15}	9.461×10^{17}	9.461×10^{12}	3.104×10^{16}	1	0.3066
1 parsec	3.086×10^{16}	3.086×10^{18}	3.086×10^{13}	1.012×10^{17}	3.262	1

Mass

	kg	g	slug	u
1 kilogram	1	10^3	6.852×10^{-2}	6.022×10^{26}
1 gram	10^{-3}	1	6.852×10^{-5}	6.022×10^{23}
1 slug	14.59	1.459×10^4	1	8.789×10^{27}
1 atomic mass unit	1.661×10^{-27}	1.661×10^{-24}	1.138×10^{-28}	1

Time

	s	min	h	d	y
1 second	1	1.667×10^{-2}	2.778×10^{-4}	1.157×10^{-5}	3.169×10^{-8}
1 minute	60	1	1.667×10^{-2}	6.944×10^{-4}	1.901×10^{-6}
1 hour	3600	60	1	4.167×10^{-2}	1.141×10^{-4}
1 day	8.640×10^4	1440	24	1	2.738×10^{-3}
1 year	3.156×10^7	5.259×10^5	8.766×10^3	365.24	1

Speed

	m/s	cm/s	ft/s	mi/h	km/h
1 meter/second	1	10^2	3.281	2.237	3.600
1 centimeter/second	10^{-2}	1	3.281×10^{-2}	2.237×10^{-2}	0.036
1 foot/second	0.3048	30.48	1	0.6818	1.097
1 mile/hour	0.4470	44.70	1.467	1	1.609
1 kilometer/hour	0.2778	27.78	0.9113	0.6214	1

Note: 1 mi/min = 60 mi/h = 88 ft/s

Force

	N	dyn	lb
1 newton	1	10^5	0.2248
1 dyne	10^{-5}	1	2.248×10^{-6}
1 pound	4.448	4.448×10^5	1

Work, Energy, Heat

	J	erg	ft · lb
1 joule	1	10^7	0.7376
1 erg	10^{-7}	1	7.376×10^{-8}
1 ft · lb	1.356	1.356×10^7	1
1 eV	1.602×10^{-19}	1.602×10^{-12}	1.182×10^{-19}
1 cal	4.186	4.186×10^7	3.087
1 Btu	1.055×10^3	1.055×10^{10}	7.779×10^2
1 kWh	3.600×10^6	3.600×10^{13}	2.655×10^6

	eV	cal	Btu	kWh
1 joule	6.242×10^{18}	0.2388	9.478×10^{-4}	2.778×10^{-7}
1 erg	6.242×10^{11}	2.388×10^{-8}	9.478×10^{-11}	2.778×10^{-14}
1 ft · lb	8.464×10^{18}	0.3239	1.285×10^{-3}	3.766×10^{-7}
1 eV	1	3.827×10^{-20}	1.519×10^{-22}	4.450×10^{-26}
1 cal	2.613×10^{19}	1	3.968×10^{-3}	1.163×10^{-6}
1 Btu	6.585×10^{21}	2.520×10^2	1	2.930×10^{-4}
1 kWh	2.247×10^{25}	8.598×10^5	3.412×10^3	1

APPENDIX 3

Mathematical Relations

Expansions

$$(1 \pm x)^n = 1 \pm nx + \frac{n(n-1)}{2!}x^2 \pm \frac{n(n-1)(n-2)}{3!}x^3 + \cdots \qquad |x| < 1$$

$$(1 \pm x)^{\frac{1}{2}} = 1 \pm \tfrac{1}{2}x - \tfrac{1}{8}x^2 \pm \tfrac{1}{16}x^3 - \cdots \qquad |x| < 1$$

$$(1 \pm x)^{-\frac{1}{2}} = 1 \mp \tfrac{1}{2}x + \tfrac{3}{8}x^2 \mp \tfrac{5}{16}x^3 + \cdots \qquad |x| < 1$$

$$(1 \pm x)^{-1} = 1 \mp x + x^2 \mp x^3 + \cdots \qquad |x| < 1$$

$$\sin x = x - \frac{x^3}{3!} + \frac{x^5}{5!} - \frac{x^7}{7!} + \cdots$$

$$\cos x = 1 - \frac{x^2}{2!} + \frac{x^4}{4!} - \frac{x^6}{6!} + \cdots$$

$$\tan x = x + \frac{x^3}{3!} + \frac{2}{15}x^5 + \cdots \qquad |x| < \pi/2$$

$$e^x = 1 + x + \frac{x^2}{2!} + \frac{x^3}{3!} + \cdots$$

Functions and Relations

$$\sin(A \pm B) = \sin A \cos B \pm \cos A \sin B$$

$$\cos(A \pm B) = \cos A \cos B \mp \sin A \sin B$$

$$\sinh x = \frac{e^x - e^{-x}}{2}$$

$$\cosh x = \frac{e^x + e^{-x}}{2}$$

Cartesian form: $z = x + iy$

Complex conjugate: $z^* = x - iy$, $i = \sqrt{-1}$

Polar form: $z = |z|e^{i\theta}$

$$z^* = |z|e^{-i\theta}$$

$$zz^* = |z|^2 = x^2 + y^2$$

Real part of z: $\text{Re } z = \frac{1}{2}(z + z^*) = x$

Imaginary part of z: $\text{Im } z = -\frac{1}{2}(z - z^*) = y$

$$\sin x = \frac{e^{ix} - e^{-ix}}{2i}$$

$$\cos x = \frac{e^{ix} + e^{-ix}}{2}$$

$$e^{ix} = \cos x + i \sin x$$

Integrals

$$\int \sin^2 x\, dx = -\frac{1}{2}\cos x \sin x + \frac{1}{2}x = \frac{1}{2}x - \frac{1}{4}\sin 2x$$

$$\int x \sin^2 x\, dx = \frac{x^2}{4} - \frac{x \sin 2x}{4} - \frac{\cos 2x}{8}$$

$$\int x^2 \sin^2 x\, dx = \frac{x^3}{6} - \left(\frac{x^2}{4} - \frac{1}{8}\right)\sin 2x - \frac{x \cos 2x}{4}$$

$$\int_0^\infty e^{-x/\alpha}\, dx = \alpha$$

$$\int_0^\infty x e^{-x/\alpha}\, dx = \alpha^2$$

$$\int_0^\infty x^2 e^{-x/\alpha}\, dx = 2\alpha^3$$

$$\int_0^\infty x^n e^{-x/\alpha}\, dx = n!\alpha^{n+1}$$

APPENDIX 4

Periodic Table of the Elements

- Metals
- Metalloids
- Nonmetals

	1	2	3	4	5	6	7	8	9	10	11	12	13	14	15	16	17	18
1	Hydrogen 1 **H** 1.0079																Hydrogen 1 **H** 1.0079	Helium 2 **He** 4.0026
2	Lithium 3 **Li** 6.941	Beryllium 4 **Be** 9.0122											Boron 5 **B** 10.811	Carbon 6 **C** 12.0107	Nitrogen 7 **N** 14.0067	Oxygen 8 **O** 15.9994	Fluorine 9 **F** 18.9984	Neon 10 **Ne** 20.1797
3	Sodium 11 **Na** 22.9898	Magnesium 12 **Mg** 24.3050											Aluminum 13 **Al** 26.9815	Silicon 14 **Si** 28.0855	Phosphorus 15 **P** 30.9738	Sulfur 16 **S** 32.066	Chlorine 17 **Cl** 35.4527	Argon 18 **Ar** 39.948
4	Potassium 19 **K** 39.0983	Calcium 20 **Ca** 40.078	Scandium 21 **Sc** 44.9559	Titanium 22 **Ti** 47.867	Vanadium 23 **V** 50.9415	Chromium 24 **Cr** 51.9961	Manganese 25 **Mn** 54.9380	Iron 26 **Fe** 55.845	Cobalt 27 **Co** 58.9332	Nickel 28 **Ni** 58.6934	Copper 29 **Cu** 63.546	Zinc 30 **Zn** 65.39	Gallium 31 **Ga** 69.723	Germanium 32 **Ge** 72.61	Arsenic 33 **As** 74.9216	Selenium 34 **Se** 78.96	Bromine 35 **Br** 79.904	Krypton 36 **Kr** 83.80
5	Rubidium 37 **Rb** 85.4678	Strontium 38 **Sr** 87.62	Yttrium 39 **Y** 88.9059	Zirconium 40 **Zr** 91.224	Niobium 41 **Nb** 92.9064	Molybdenum 42 **Mo** 95.94	Technetium 43 **Tc** (98)	Ruthenium 44 **Ru** 101.07	Rhodium 45 **Rh** 102.9055	Palladium 46 **Pd** 106.42	Silver 47 **Ag** 107.8682	Cadmium 48 **Cd** 112.411	Indium 49 **In** 114.88	Tin 50 **Sn** 118.710	Antimony 51 **Sb** 121.76	Tellurium 52 **Te** 127.60	Iodine 53 **I** 126.9045	Xenon 54 **Xe** 131.29
6	Cesium 55 **Cs** 132.9054	Barium 56 **Ba** 137.327	Lanthanum 57 **La** 138.9055	Hafnium 72 **Hf** 178.49	Tantalum 73 **Ta** 180.9479	Tungsten 74 **W** 183.84	Rhenium 75 **Re** 186.207	Osmium 76 **Os** 190.23	Iridium 77 **Ir** 192.217	Platinum 78 **Pt** 195.078	Gold 79 **Au** 196.9666	Mercury 80 **Hg** 200.59	Thallium 81 **Tl** 204.3833	Lead 82 **Pb** 207.2	Bismuth 83 **Bi** 208.9804	Polonium 84 **Po** (209)	Astatine 85 **At** (210)	Radon 86 **Rn** (222)
7	Francium 87 **Fr** (223)	Radium 88 **Ra** (226)	Actinium 89 **Ac** (227)	Rutherfordium 104 **Rf** (261)	Hahnium 105 **Db** (262)	Seaborgium 106 **Sg** (266)	Bohrium 107 **Bh** (264)	Hassium 108 **Hs** (269)	Meitnerium 109 **Mt** (268)	Ununnilium 110 **Uun*** (269)	Unununium 111 **Uuu*** (272)	Ununbium 112 **Uub*** (277)						

Cerium 58 **Ce** 140.115	Praseodymium 59 **Pr** 140.9076	Neodymium 60 **Nd** 144.24	Promethium 61 **Pm** (145)	Samarium 62 **Sm** 150.36	Europium 63 **Eu** 151.965	Gadolinium 64 **Gd** 157.25	Terbium 65 **Tb** 158.9253	Dysprosium 66 **Dy** 162.50	Holmium 67 **Ho** 164.9303	Erbium 68 **Er** 167.26	Thulium 69 **Tm** 168.9342	Ytterbium 70 **Yb** 173.04	Lutetium 71 **Lu** 174.967
Thorium 90 **Th** 232.0381	Protactinium 91 **Pa** 231.0359	Uranium 92 **U** 238.0289	Neptunium 93 **Np** (237)	Plutonium 94 **Pu** (244)	Americium 95 **Am** (243)	Curium 96 **Cm** (247)	Berkelium 97 **Bk** (247)	Californium 98 **Cf** (251)	Einsteinium 99 **Es** (252)	Fermium 100 **Fm** (257)	Mendelevium 101 **Md** (258)	Nobelium 102 **No** (259)	Lawrencium 103 **Lr** (260)

Note: Averaged atomic masses are 1997 IUPAC values (up to four decimal places). An entry in parentheses indicates the mass number of the longest-lived isotope of an element that has no stable isotopes. The names of some elements above atomic number 104 remain in conflict.

*Official names have not been agreed upon for these elements.

APPENDIX 5

Mean Values and Distributions

Average values are encountered in several forms. A simple example is an arithmetic mean. If N students take an exam and receive scores S_1, S_2, S_3, and so on, the class average is

$$\text{Average score} = \frac{S_1 + S_2 + S_3 + \cdots + S_N}{N}$$

This calculation is more compact if we group the students receiving a given score S_i. If N_i students receive the score S_i, then the class average becomes

$$\text{Average score} = \frac{N_1 S_1 + N_2 S_2 + N_3 S_3 + \cdots}{N}$$

An example from physics of a *weighted* average brings us closer to the idea of mean value. In mechanics the center of mass of N particles of different mass (m_i) placed at various locations (x_i) along the x axis is

$$x_{\text{cm}} = \frac{m_1 x_1 + m_2 x_2 + \cdots}{m_1 + m_2 + \cdots} = \frac{\sum_i m_i x_i}{\sum_i m_i}$$

We define the mean value just like the weighted average, with the probabilities of the various possible events serving as the weights. Suppose some physical quantity x can have N possible values, which we shall call x_i, with $i = 1, 2, 3, \ldots, N$. Let P_i be the probability that x has the value x_i. Then the mean value of x is

Mean value

$$\bar{x} = \frac{P_1 x_1 + P_2 x_2 + \cdots}{P_1 + P_2 + \cdots} = \frac{\sum_i P_i x_i}{\sum_i P_i}$$

As usual, it is required that $\sum_i P_i = 1$ (the sum of all probabilities is one), so

$$\bar{x} = \sum P_i x_i \tag{A5.1}$$

The same definition of mean value holds if we are considering a function $f(x)$:

Mean value of a function

$$\overline{f(x)} = \sum P_i f(x_i) \tag{A5.2}$$

For any two functions $f(x)$ and $g(x)$ we have

$$\overline{f(x) + g(x)} = \overline{f(x)} + \overline{g(x)}$$

and

$$\overline{c\,[f(x) + g(x)]} = c\,[\overline{f(x) + g(x)}] = c\,\overline{f(x)} + c\,\overline{g(x)}$$

where c is a constant.

In physics we are often interested in how much a particular value differs from the mean value. The difference between a particular value and the mean value is known as the deviation (Δx_i):

Deviation

$$\Delta x_i = x_i - \overline{x}$$

The mean value of all deviations is easily computed:

$$\overline{\Delta x_i} = \overline{x_i - \overline{x}} = \overline{x_i} - \overline{x} = 0$$

This is just as one should expect because the positive and negative deviations will exactly cancel each other. However, we would like to get some idea of how large a typical deviation will be. To this end we define the **standard deviation**

Standard deviation

$$\sigma_x = \left[\overline{(\Delta x_i)^2}\right]^{1/2} = \left[\sum_i P_i(x_i - \overline{x})^2\right]^{1/2} \tag{A5.3}$$

Notice that in Equation (A5.3) the negative deviations have been eliminated by squaring. The standard deviation is an indication of the sharpness of the peak of the probability distribution curve. For a purely random (Gaussian) distribution it can be shown that the width of the probability curve at half of the peak value is slightly more than two standard deviations.

Equation (A5.3) can be rewritten

$$\sigma_x = \left[\sum_i P_i(x_i^2 - 2x_i\,\overline{x} + \overline{x}^2)\right]^{1/2}$$

$$= \left[\sum_i P_i x_i^2 - 2\overline{x}\sum_i P_i x_i + \overline{x}^2)\right]^{1/2}$$

The first sum is simply $\overline{x^2}$, and the second is $\overline{x}$. Thus

$$\sigma_x = \left[\overline{x^2} - \overline{x}^2\right]^{1/2} \tag{A5.4}$$

An important feature of many systems is that as N increases, the probability distribution begins to approach a smooth curve. Indeed, it will be to our advantage to think of such a curve as a function, which we shall call a **distribution function.** This will enable us to replace the sums used in this section by integrals, which is clearly a wise thing to do if we want to consider problems in which $N \approx 10^{23}$. The distribution function is not a probability per se, but rather a probability density. For example, if $f(x)$ is the distribution function describing the positions of a large number of particles lying along the x axis, then

$f(x)\ dx =$ The probability of finding a particle between x and $x + dx$

It is necessary to define the distribution function in this way so that we can say [by analogy with Equation (A5.1)]

$$\overline{x} = \int x\, f(x)\ dx$$

$$\overline{x^2} = \int x^2\, f(x)\ dx$$

and so on.

APPENDIX 6

Probability Integrals

$I_n = \int_0^\infty x^n \exp[-ax^2]\, dx$

Even and odd functions

Before proceeding to the calculation of these integrals, we introduce the concept of even and odd functions. An even function is one for which $f(x) = f(-x)$; similarly, for an odd function $f(x) = -f(-x)$. An example of an even function is the integrand of Equation (9.4)

$$\int_{-\infty}^{\infty} g(v_x)\, dv_x = \int_{-\infty}^{\infty} C' \exp[-\tfrac{1}{2}\beta m v_x^2]\, dv_x$$

This even integrand makes it possible to replace $\int_{-\infty}^{\infty}$ with $2\int_0^{\infty}$ because the curve is perfectly symmetric about $x = 0$ (see Figure 9.2).

On the other hand, consider the odd integrand of Equation (9.6):

$$\bar{v}_x = C' \int_{-\infty}^{\infty} v_x \exp[-\tfrac{1}{2}\, \beta m v_x^2]\, dv_x$$

The odd function is antisymmetric with respect to $x = 0$, and if we divide the integral into two parts ($-\infty$ to 0 and 0 to ∞), those two parts just cancel. The integral of an odd function from $-\infty$ to ∞ (or indeed over any symmetric limits) is therefore zero.

Based on these properties of even and odd functions, it makes sense to define

$$I_n = \int_0^{\infty} x^n \exp[-ax^2]\, dx \tag{A6.1}$$

and say that

$$\int_{-\infty}^{\infty} x^n \exp[-ax^2]\, dx = \begin{cases} 2I_n & \text{for even } n \\ 0 & \text{for odd } n \end{cases} \tag{A6.2}$$

Let us proceed to calculate the I_n, beginning with $n = 0$ [as in Equation (9.4)]. Now

$$I_0 = \int_0^{\infty} \exp[-ax^2]\, dx \tag{A6.3}$$

It is not possible to do this integral by straightforward substitution. Because this is a definite integral, x is a dummy variable. We would obtain the same result with any other variable, say y:

$$I_0 = \int_0^{\infty} \exp[-ay^2]\, dy \tag{A6.4}$$

Multiplying Equations (A6.3) and (A6.4) yields

$$I_0^2 = \int_0^{\infty}\int_0^{\infty} \exp[-a(x^2 + y^2)]\, dx\, dy \tag{A6.5}$$

We now switch to polar coordinates, with $r^2 = x^2 + y^2$ and $dx\,dy = r\,dr\,d\theta$. The limits of integration are 0 to ∞ for r but only 0 to $\pi/2$ for θ, because the limits on x and y restrict us to the first quadrant. Thus

$$I_0^2 = \int_0^\infty \exp[-ar^2]\; r\,dr \int_0^{\pi/2} d\theta \tag{A6.6}$$

The θ integral yields $\pi/2$, and the r integral is easily performed using a standard substitution (say $u = ar^2$):

$$\begin{aligned} I_0^2 &= \left(\frac{\pi}{2}\right)\left(\tfrac{1}{2}a\right) = \frac{\pi}{4a} \\ I_0 &= \sqrt{\frac{\pi}{4a}} = \frac{\sqrt{\pi}}{2}\,a^{-1/2} \end{aligned} \tag{A6.7}$$

Notice that Equation (A6.6) actually contains I_1, so we see that

$$I_1 = \int_0^\infty r \exp[-ar^2]\; dr = \frac{1}{2a} \tag{A6.8}$$

To calculate I_2 we return for a moment to I_0. Differentiating Equation (A6.3) with respect to a produces

$$\frac{dI_0}{da} = -\int_0^\infty x^2 \exp[-ax^2]\; dx = -I_2 \tag{A6.9}$$

But by Equation (A6.7)

$$\frac{dI_0}{da} = -\frac{\sqrt{\pi}}{4}\,a^{-3/2}$$

and so

$$I_2 = \frac{\sqrt{\pi}}{4}\,a^{-3/2} \tag{A6.10}$$

In a similar way one may use $\dfrac{dI_1}{da}$ to calculate I_3, $\dfrac{dI_2}{da}$ to calculate I_4, and so on, indefinitely. Some useful results are

$$I_3 = \tfrac{1}{2}\,a^{-2}$$

$$I_4 = \frac{3\sqrt{\pi}}{8}\,a^{-5/2}$$

$$I_5 = a^{-3}$$

It can be shown that in general

$$I_n = \frac{[(n-1)/2]!}{2a^{(n+1)/2}} \quad \text{for } odd\ n$$

$$I_n = \frac{1\cdot 3\cdot 5\cdots(n-1)}{2^{(n/2)+1}a^{(n/2)}}\sqrt{\frac{\pi}{a}} \quad \text{for } even\ n$$

APPENDIX 7

Integrals of the Type $\int_0^\infty \frac{x^{n-1}\,dx}{e^x - 1}$

Evaluation of these integrals introduces us to two special mathematical functions: the gamma function and the Riemann zeta function. A good handbook of integrals (see for example *Table of Integrals, Series, and Products* by Gradshteyn and Ryzhik) will give the result

$$\int_0^\infty \frac{x^{n-1}\,dx}{e^x - 1} = \Gamma(n)\ \zeta(n) \tag{A7.1}$$

The gamma function is in turn defined by the integral

The Gamma function

$$\Gamma(n) = \int_0^\infty e^{-x}x^{n-1}\,dx \tag{A7.2}$$

In many cases the gamma functions we encounter in physics have integer arguments. Then they are easily evaluated using the identity

$$\Gamma(n) = (n-1)! \tag{A7.3}$$

When n is not an integer one may use the recursion formula

$$\Gamma(n+1) = n\,\Gamma(n) \tag{A7.4}$$

with tabulated values of the gamma function [those tables generally range from $\Gamma(1)$ to $\Gamma(2)$] to find the desired value. Further, the fact that $\Gamma(\frac{3}{2}) = \sqrt{\pi}/2$ is useful, because half-integer powers occur in a number of applications. For other values one may resort to tables or computer packages such as Mathematica.

The Riemann zeta function

One may also use the appropriate tables or computer packages to obtain values for the Riemann zeta function. It is helpful to know that the zeta function can sometimes be expressed in closed form. For example,

$$\zeta(2) = \frac{\pi^2}{6} \quad \text{and} \quad \zeta(4) = \frac{\pi^4}{90} \tag{A7.5}$$

One integral of this form is

$$\int_0^\infty \frac{x^3}{e^x - 1}\,dx$$

which is found in Example 3.8. According to Equation (A7.1) the result is

$$\int_0^\infty \frac{x^3}{e^x - 1}\,dx = \Gamma(4)\ \zeta(4) = 3!\left[\frac{\pi^4}{90}\right] = \frac{\pi^4}{15}$$

This allows us to complete Example 3.8 and obtain the closed form expression for the total radiation emitted by a blackbody at a temperature T

$$R(T) = \frac{2\,\pi^5\,k^4}{15\,h^3\,c^2}T^4$$

In Section 9.7 we encounter an integral with a fractional value of n ($n = \frac{3}{2}$), with the result $\Gamma(\frac{3}{2})\ \zeta(\frac{3}{2})$. As was noted previously, $\Gamma(\frac{3}{2}) = \sqrt{\pi}/2$. A numerical evaluation yields $\zeta(\frac{3}{2}) \approx 2.61238$, so that $\Gamma(\frac{3}{2})\ \zeta(\frac{3}{2}) \approx 2.315$, as used in Equation (9.64).

The common thread running through these applications is that they concern collections of bosons. The Bose-Einstein distribution contains the factor $[B_2\exp(\beta E) - 1]^{-1}$, which can often be put into the form $e^x - 1$ for the purpose of doing this integral.

APPENDIX 8

Atomic Mass Table

Atomic masses are used in both atomic and nuclear physics calculations. Because of space limitations, we have not included all the known isotopes in this compilation, but we have listed the atomic masses of all known stable isotopes and all those with half lives greater than 10 s. In addition, for the light and heavy elements, we have included some additional isotopes.

We have listed the chemical symbol, atomic number Z, mass number A, atomic masses in atomic mass units (u), either the half life $t_{1/2}$ or natural abundance on Earth, and the decay modes. There are a few nuclides, for example, ^{40}K and ^{238}U, that are unstable, but have such long half life that they are still found naturally on Earth. In these few cases, we list both the half life and natural abundance. The most important decay modes are listed with the most likely ones given first. We present below the legend for the various decay modes.

The m listed after a few of the mass numbers means the data are for a metastable state of the particular nuclide, but we have omitted most metastable nuclides for brevity. An s listed at the end of the atomic mass value means that the mass is obtained from systematics, not measured. The time units for the half life are s (seconds), m (minutes), h (hours), and y (years). In some cases, for very short-lived isotopes of light masses, we have listed the width of the state Γ instead of the half life. The half life can be determined from the time τ after using the Heisenberg uncertainty principle.

The data for this appendix were obtained from the website (www.nndc.bnl.gov) of the National Nuclear Data Center at Brookhaven National Laboratory. Look at the web site for the latest values. The abundances, half lives, and decay modes are from J. K. Tuli, *Nuclear Wallet Cards,* 5th ed. (1995). The atomic mass values are from the "The 1995 Update to the Atomic Mass Evaluations", G. Audi and A. H. Wapstra, *Nuclear Physics* **A595**, pp. 409–480 (1995). We have rounded off the mass values to the nearest 0.000001 u.

The legend for the decay modes is clear except for a few cases. We list them all here.

Listed	Decay
β^-	β^-
$2\beta^-$	Double β^-
ϵ	Electron capture and/or β^+
IT	Isometric transition
α, 2α	α or 2α
SF	Spontaneous fission
$\beta^- n$, $\beta^- p$, $\beta^- \alpha$, $\beta^- 2\alpha$, $\beta^- n\alpha$, $\beta^- 3\alpha$	β^- decay followed by subsequent particles decaying
ϵn, ϵp, $\epsilon\alpha$, $\epsilon 2\alpha$, $\epsilon 3\alpha$, ϵSF	Electron capture and/or β^+ followed by other decays
n, p, $p2\alpha$	Neutron, proton, etc.
^{12}C, ^{14}C	^{12}C, ^{14}C

Z	A	Atomic Mass (u)	$t_{1/2}$ or Abundance	Decay Mode
Neutron (n)				
0	1	1.008665	10.4 m	β^-
Hydrogen (H)				
1	1	1.007825	99.985%	
1	2	2.014102	0.015%	
1	3	3.016049	12.33 y	β^-
Helium (He)				
2	3	3.016029	0.000137%	
2	4	4.002603	99.999863%	
2	5	5.012220	0.60 MeV	α, n
2	6	6.018888	806.7 ms	β^-
2	7	7.028030	160 keV	n
2	8	8.033922	119.0 ms	β^-, $\beta^- n$
Lithium (Li)				
3	5	5.012540	1.5MeV	α, p
3	6	6.015122	7.5%	
3	7	7.016004	92.5%	
3	8	8.022487	838 ms	β^-, $\beta^- 2\alpha$
3	9	9.026789	178.3 ms	β^-, $\beta^- n$
3	10	10.035481	1.2 MeV	n
3	11	11.043796	8.5 ms	β^-, $\beta^- n\alpha$
Beryllium (Be)				
4	6	6.019726	92 keV	$2p$
4	7	7.016929	53.29 d	ϵ
4	8	8.005305	6.8 eV	2α
4	9	9.012182	100%	
4	10	10.013534	1.51×10^6 y	β^-
4	11	11.021658	13.81 s	β^-, $\beta^- \alpha$
4	12	12.026921	23.6 ms	β^-, $\beta^- n$
Boron (B)				
5	8	8.024607	770 ms	$\epsilon\alpha$, ϵ, $\epsilon 2\alpha$
5	9	9.013329	0.54 keV	2α, p
5	10	10.012937	19.9%	
5	11	11.009305	80.1%	
5	12	12.014352	20.20 ms	β^-, $\beta^- 3\alpha$
5	13	13.017780	17.36 ms	β^-
5	14	14.025404	13.8 ms	β^-
5	15	15.031097	10.5 ms	β^-
Carbon (C)				
6	9	9.031040	126.5 ms	ϵ, ϵp, $\epsilon 2\alpha$
6	10	10.016853	19.255 s	ϵ
6	11	11.011434	20.39 m	ϵ
6	12	12.000000	98.89%	

Z	A	Atomic Mass (u)	$t_{1/2}$ or Abundance	Decay Mode
Carbon (C)				
6	13	13.003355	1.11%	
6	14	14.003242	5730 y	β^-
6	15	15.010599	2.449 s	β^-
6	16	16.014701	0.747 s	β^-
6	17	17.022584	193 ms	β^-, $\beta^- n$
6	18	18.026760	88 ms	β^-
Nitrogen (N)				
7	12	12.018613	11.000 ms	ϵ, $\epsilon 3\alpha$
7	13	13.005739	9.965 m	ϵ
7	14	14.003074	99.63%	
7	15	15.000109	0.37%	
7	16	16.006101	7.13 s	β^-
7	17	17.008450	4.173 s	β^-, $\beta^- n$
7	18	18.014082	624 ms	β^-, $\beta^- \alpha$, $\beta^- n$
7	19	19.017027	304 ms	β^-, $\beta^- n$
7	20	20.023370	100 ms	β^-, $\beta^- n$
7	21	21.027090	95 ms	$\beta^- n$, β^-
7	22	22.034440	24 ms	$\beta^- n$, β^-
Oxygen (O)				
8	13	13.024810	8.58 ms	ϵ
8	14	14.008595	70.606 s	ϵ
8	15	15.003065	122.24 s	ϵ
8	16	15.994915	99.76%	
8	17	16.999132	0.038%	
8	18	17.999160	0.20%	
8	19	19.003579	26.91 s	β^-
8	20	20.004076	13.51 s	β^-
8	21	21.008655	3.42 s	β^-
8	22	22.009970	2.25 s	β^-
Fluorine (F)				
9	16	16.011466	40 keV	p
9	17	17.002095	64.49 s	ϵ
9	18	18.000938	109.77 m	ϵ
9	19	18.998403	100%	
9	20	19.999981	11.00 s	β^-
9	21	20.999949	4.158 s	β^-
9	22	22.002999	4.23 s	β^-
9	23	23.003570	2.23 s	β^-
9	24	24.008100	340 ms	β^-
Neon (Ne)				
10	17	17.017700	109.2 ms	ϵ, ϵp
10	18	18.005697	1.672 s	ϵ
10	19	19.001880	17.22	ϵ

Z	A	Atomic Mass (u)	$t_{1/2}$ or Abundance	Decay Mode
Neon (Ne) *(continued)*				
10	20	19.992440	90.48%	
10	21	20.993847	0.27%	
10	22	21.991386	9.25%	
10	23	22.994467	37.24 s	β^-
10	24	23.993615	3.38 m	β^-
10	25	24.997790	602 ms	β^-
10	26	26.000460	230 ms	β^-
Sodium (Na)				
11	20	20.007348	447.9 ms	ϵ
11	21	20.997655	22.49 s	ϵ
11	22	21.994437	2.6019 y	ϵ
11	23	22.989770	100%	
11	24	23.990963	14.9590 h	β^-
11	25	24.989954	59.1 s	β^-
11	26	25.992590	1.072 s	β^-
11	27	26.994010	301 ms	$\beta^-, \beta^- n$
Magnesium (Mg)				
12	20	20.018863	95 ms	$\epsilon, \epsilon p$
12	21	21.011714	122 ms	$\epsilon, \epsilon p$
12	22	21.999574	3.857 s	ϵ
12	23	22.994125	11.317 s	ϵ
12	24	23.985042	78.99%	
12	25	24.985837	10.00%	
12	26	25.982593	11.01%	
12	27	26.984341	9.458 m	β^-
12	28	27.983877	20.91 h	β^-
12	29	28.988550	1.30 s	β^-
12	30	29.990460	335 ms	β^-
12	31	30.996550	230 ms	$\beta^-, \beta^- n$
12	32	31.999150	120 ms	$\beta^-, \beta^- n$
Aluminum (Al)				
13	23	23.007265	0.47 s	$\epsilon, \epsilon p$
13	24	23.999941	2.053 s	$\epsilon, \epsilon\alpha$
13	25	24.990429	7.183 s	ϵ
13	26	25.986892	7.4×10^5 y	ϵ
13	27	26.981538	100%	
13	28	27.981910	2.2414 m	β^-
13	29	28.980445	6.56 m	β^-
13	30	29.982960	3.60 s	β^-
13	31	30.983946	644 ms	β^-
Silicon (Si)				
14	24	24.011546	102 ms	$\epsilon, \epsilon p$
14	25	25.004107	220 ms	$\epsilon, \epsilon p$

Z	A	Atomic Mass (u)	$t_{1/2}$ or Abundance	Decay Mode
Silicon (Si)				
14	26	25.992330	2.234 s	ϵ
14	27	26.986705	4.16 s	ϵ
14	28	27.976927	92.23%	
14	29	28.976495	4.67%	
14	30	29.973770	3.10%	
14	31	30.975363	157.3 m	β^-
14	32	31.974148	172 y	β^-
14	33	32.978001	6.18 s	β^-
14	34	33.978576	2.77 s	β^-
14	35	34.984580	0.78 s	β^-
14	36	35.986690	0.45 s	$\beta^-, \beta^- n$
Phosphorus (P)				
15	27	26.999190	260 ms	$\epsilon, \epsilon p$
15	28	27.992312	270.3 ms	ϵ
15	29	28.981801	4.140 s	ϵ
15	30	29.978314	2.498 m	ϵ
15	31	30.973762	100%	
15	32	31.973907	14.262 d	β^-
15	33	32.971725	25.34 d	β^-
15	34	33.973636	12.43 s	β^-
15	35	34.973314	47.3 s	β^-
15	36	35.978260	5.6 s	β^-
15	37	36.979610	2.31 s	β^-
15	38	37.984470	0.64 s	$\beta^-, \beta^- n$
Sulfur (S)				
16	30	29.984903	1.178 s	ϵ
16	31	30.979554	2.572 s	ϵ
16	32	31.972071	95.02%	
16	33	32.971459	0.75%	
16	34	33.967867	4.21%	
16	35	34.969032	87.51 d	β^-
16	36	35.967081	0.02%	
16	37	36.971126	5.05 m	β^-
16	38	37.971163	170.3 m	β^-
16	39	38.975140	11.5 s	β^-
16	40	39.975470	8.8 s	β^-
Chlorine (Cl)				
17	33	32.977452	2.511 s	ϵ
17	34	33.973762	1.5264 s	ϵ
17	35	34.968853	75.77%	
17	36	35.968307	3.01×10^5 y	β^-, ϵ
17	37	36.965903	24.23%	
17	38	37.968011	37.24 m	β^-
17	39	38.968008	55.6 m	β^-

Z	A	Atomic Mass (u)	$t_{1/2}$ or Abundance	Decay Mode
Chlorine (Cl) *(continued)*				
17	40	39.970420	1.35 m	β^-
17	41	40.970650	38.4 s	β^-
17	42	41.973170	6.8 s	β^-
17	43	42.974200	3.3 s	β^-
Argon (Ar)				
18	35	34.975257	1.775 s	ϵ
18	36	35.967546	0.337%	
18	37	36.966776	35.04 d	ϵ
18	38	37.962732	0.063%	
18	39	38.964313	269 y	β^-
18	40	39.962383	99.600%	
18	41	40.964501	1.822 h	β^-
18	42	41.963050	32.9 y	β^-
18	43	42.965670	5.37 m	β^-
18	44	43.965365	11.87 m	β^-
18	45	44.968090	21.48 s	β^-
18	46	45.968090	8.4 s	β^-
Potassium (K)				
19	37	36.973377	1.226 s	ϵ
19	38	37.969080	7.636 m	ϵ
19	39	38.963707	93.2581%	
19	40	39.963999	1.277×10^9 y 0.0117%	β^-, ϵ
19	41	40.961826	6.7302%	
19	42	41.962403	12.360 h	β^-
19	43	42.960716	22.3 h	β^-
19	44	43.961560	22.13 m	β^-
19	45	44.960700	17.3 m	β^-
19	46	45.961976	105 s	β^-
19	47	46.961678	17.5 s	β^-
19	48	47.965513	6.8 s	β^-
19	49	48.967450	1.26 s	β^-, $\beta^- n$
Calcium (Ca)				
20	38	37.976319	440 ms	ϵ
20	39	38.970718	859.6 ms	ϵ
20	40	39.962591	96.941%	
20	41	40.962278	1.03×10^5 y	ϵ
20	42	41.958618	0.647%	
20	43	42.958767	0.135%	
20	44	43.955481	2.086%	
20	45	44.956186	162.6 d	β^-
20	46	45.953693	0.004%	
20	47	46.954546	4.536 d	β^-
20	48	47.952534	0.187%	

Z	A	Atomic Mass (u)	$t_{1/2}$ or Abundance	Decay Mode
Calcium (Ca)				
20	49	48.955673	8.718 m	β^-
20	50	49.957518	13.9 s	β^-
20	51	50.961470	10.0 s	β^-, $\beta^- n$
20	52	51.965100	4.6 s	β^-
Scandium (Sc)				
21	43	42.961151	3.891 h	ϵ
21	44	43.959403	3.927 h	ϵ
21	45	44.955910	100%	
21	46	45.955170	83.79 d	β^-
21	47	46.952408	3.349 d	β^-
21	48	47.952235	43.7 h	β^-
21	49	48.950024	57.2 m	β^-
21	50	49.952187	102.5 s	β^-
21	51	50.953603	12.4 s	β^-
Titanium (Ti)				
22	44	43.959690	49 y	ϵ
22	45	44.958124	184.8 m	ϵ
22	46	45.952630	8.25%	
22	47	46.951764	7.44%	
22	48	47.947947	73.72%	
22	49	48.947871	5.41%	
22	50	49.944792	5.18%	
22	51	50.946616	5.76 m	β^-
22	52	51.946898	1.7 m	β^-
22	53	52.949730	32.7 s	β^-
Vanadium (V)				
23	47	46.954907	32.6 m	ϵ
23	48	47.952255	15.974 d	ϵ
23	49	48.948517	330 d	ϵ
23	50	49.947163	1.4×10^{17} y 0.250%	ϵ, β^-
23	51	50.943964	99.750%	
23	52	51.944780	3.74 m	β^-
23	53	52.944343	1.61 m	β^-
23	54	53.946444	49.8 s	β^-
Chromium (Cr)				
24	48	47.954036	21.56 h	ϵ
24	49	48.951341	42.3 m	ϵ
24	50	49.946050	4.345%	
24	51	50.944772	27.702 d	ϵ
24	52	51.940512	83.79%	
24	53	52.940654	9.50%	
24	54	53.938885	2.365%	

Z	A	Atomic Mass (u)	$t_{1/2}$ or Abundance	Decay Mode
Chromium (Cr) *(continued)*				
24	55	54.940844	3.497 m	β^-
24	56	55.940645	5.94 m	β^-
24	57	56.943750	21.1 s	β^-
Manganese (Mn)				
25	51	50.948216	46.2 m	ϵ
25	52	51.945570	5.591 d	ϵ
25	53	52.941295	3.74×10^6 y	ϵ
25	54	53.940363	312.12 d	ϵ, β^-
25	55	54.938050	100%	
25	56	55.938909	2.5785 h	β^-
25	57	56.938287	85.4 s	β^-
25	58	57.939990	65.3 s	β^-
25	59	58.940450	4.6 s	β^-
25	60	59.943190	51 s	β^-
Iron (Fe)				
26	51	50.956825	305 ms	ϵ
26	52	51.948117	8.275 h	ϵ
26	53	52.945312	8.51 m	ϵ
26	54	53.939615	5.845%	
26	55	54.938298	2.73 y	ϵ
26	56	55.934942	91.754%	
26	57	56.935399	2.119%	
26	58	57.933280	0.282%	
26	59	58.934880	44.503 d	β^-
26	60	59.934077	1.5×10^6 y	β^-
26	61	60.936749	5.98 m	β^-
26	62	61.936770	68 s	β^-
Cobalt (Co)				
27	55	54.942003	17.53 h	ϵ
27	56	55.939844	77.27 d	ϵ
27	57	56.936296	271.79 d	ϵ
27	58	57.935758	70.82 d	ϵ
27	59	58.933200	100%	
27	60	59.933822	5.2714 y	β^-
27	61	60.932479	1.650 h	β^-
27	62	61.934054	1.50 m	β^-
27	63	62.933615	27.4 s	β^-
Nickel (Ni)				
28	56	55.942136	6.08 d	ϵ
28	57	56.939800	35.60 h	ϵ
28	58	57.935348	68.077%	
28	59	58.934352	7.6×10^4 y	ϵ
28	60	59.930791	26.223%	
28	61	60.931060	1.140%	
28	62	61.928349	3.634%	
28	63	62.929673	100.1 y	β^-
28	64	63.927970	0.926%	
28	65	64.930088	2.517 h	β^-
28	66	65.929115	54.6 h	β^-
28	67	66.931570	21 s	β^-
28	68	67.931845	19 s	β^-
28	69	68.935180	11.4 s	β^-
Copper (Cu)				
29	59	58.939504	81.5 s	ϵ
29	60	59.937368	23.7 s	ϵ
29	61	60.933462	3.333 h	ϵ
29	62	61.932587	9.74 m	ϵ
29	63	62.929601	69.17%	
29	64	63.929768	12.700 h	ϵ, β^-
29	65	64.927794	30.83%	
29	66	65.928873	5.088 m	β^-
29	67	66.927750	61.83 h	β^-
29	68	67.929640	31.1 s	β^-
29	69	68.929425	2.85 m	β^-
29	70	69.932409	4.5 s	β^-
29	71	70.932620	19.5 s	β^-
Zinc (Zn)				
30	60	59.941832	2.38 m	ϵ
30	61	60.939514	89.1 s	ϵ
30	62	61.934334	9.186 h	ϵ
30	63	62.933216	38.47 m	ϵ
30	64	63.929147	48.6%	
30	65	64.929245	244.26 d	ϵ
30	66	65.926037	27.9%	
30	67	66.927131	4.1%	
30	68	67.924848	18.8%	
30	69	68.926554	56.4 m	β^-
30	70	69.925325	5×10^{14} y	$2\beta^-$
			0.6%	
30	71	70.927727	2.45 m	β^-
30	72	71.926861	46.5 h	β^-
30	73	72.929780	23.5 s	β^-
30	74	73.929460	95.6 s	β^-
30	75	74.932940	10.2 s	β^-
Gallium (Ga)				
31	63	62.939140	32.4 s	ϵ
31	64	63.936838	2.630 m	ϵ

Z	A	Atomic Mass (u)	$t_{1/2}$ or Abundance	Decay Mode
Gallium (Ga) *(continued)*				
31	65	64.932739	15.2 m	ϵ
31	66	65.931592	9.49 h	ϵ
31	67	66.928205	3.2612 d	ϵ
31	68	67.927984	67.629 m	ϵ
31	69	68.925581	60.108%	
31	70	69.926028	21.14 m	β^-, ϵ
31	71	70.924705	39.892%	
31	72	71.926369	14.10 h	β^-
31	73	72.925170	4.86 h	β^-
31	74	73.926940	8.12 m	β^-
31	75	74.926501	126 s	β^-
31	76	75.928930	32.6 s	β^-
31	77	76.929280	13.2 s	β^-
Germanium (Ge)				
32	64	63.941570	63.7 s	ϵ
32	65	64.939440	30.9 s	ϵ
32	66	65.933850	2.26 h	ϵ
32	67	66.932738	18.9 m	ϵ
32	68	67.928097	270.82 d	ϵ
32	69	68.927972	39.05 h	ϵ
32	70	69.924250	21.23%	
32	71	70.924954	11.43 d	ϵ
32	72	71.922076	27.66%	
32	73	72.923459	7.73%	
32	74	73.921178	35.94%	
32	75	74.922859	82.78 m	β^-
32	76	75.921403	7.44%	
32	77	76.923548	11.30 h	β^-
32	78	77.922853	88.0 m	β^-
32	79	78.925400	18.98 s	β^-
32	80	79.925445	29.5 s	β^-
Arsenic (As)				
33	67	66.939190	42.5 s	ϵ
33	68	67.936790	151.6 s	ϵ
33	69	68.932280	15.2 m	ϵ
33	70	69.930930	52.6 m	ϵ
33	71	70.927115	65.28 h	ϵ
33	72	71.926753	26.0 h	ϵ
33	73	72.923825	80.30 d	ϵ
33	74	73.923929	17.77 d	ϵ, β^-
33	75	74.921596	100%	
33	76	75.922394	26.32 h	β^-
33	77	76.920648	38.83 h	β^-
33	78	77.921829	90.7 m	β^-
33	79	78.920948	9.01 m	β^-

Z	A	Atomic Mass (u)	$t_{1/2}$ or Abundance	Decay Mode
Arsenic (As)				
33	80	79.922578	15.2 s	β^-
33	81	80.922133	33.3 s	β^-
33	82	81.924500	19.1 s	β^-
33	83	82.924980	13.4 s	β^-
Selenium (Se)				
34	68	67.941870*s*	35.5 s	ϵ
34	69	68.939560	27.4 s	$\epsilon, \epsilon p$
34	70	69.933500*s*	41.1 m	ϵ
34	71	70.932270*s*	4.74 m	ϵ
34	72	71.927112	8.40 d	ϵ
34	73	72.926767	7.15 h	ϵ
34	74	73.922477	0.89%	
34	75	74.922524	119.779 d	ϵ
34	76	75.919214	9.36%	
34	77	76.919915	7.63%	
34	78	77.917310	23.78%	
34	79	78.918500	6.5×10^5 y	β^-
34	80	79.916522	49.61%	
34	81	80.917993	18.45 m	β^-
34	82	81.916700	1.1×10^{20} y 8.73%	$2\beta^-$
34	83	82.919119	22.3 m	β^-
34	84	83.918465	3.1 m	β^-
34	85	84.922240	31.7 s	β^-
34	86	85.924271	15.3 s	β^-
Bromine (Br)				
35	71	70.939250*s*	21.4 s	ϵ
35	72	71.936500	78.6 s	ϵ
35	73	72.931790	3.4 m	ϵ
35	74	73.929891	25.4 m	ϵ
35	75	74.925776	96.7 m	ϵ
35	76	75.924542	16.2 h	ϵ
35	77	76.921380	57.036 h	ϵ
35	78	77.921146	6.46 m	ϵ, β^-
35	79	78.918338	50.69%	
35	80	79.918530	17.68 m	β^-, ϵ
35	81	80.916291	49.31%	
35	82	81.916805	35.30 h	β^-
35	83	82.915180	2.40 h	β^-
35	84	83.916504	31.80 m	β^-
35	85	84.915608	2.90 m	β^-
35	86	85.918797	55.1 s	β^-
35	87	86.920711	55.60 s	$\beta^-, \beta^- n$
35	88	87.924070	16.34 s	$\beta^-, \beta^- n$

Z	A	Atomic Mass (u)	$t_{1/2}$ or Abundance	Decay Mode
Krypton (Kr)				
36	72	71.941910	17.2 s	ϵ
36	73	72.938930	27.0 s	ϵ, ϵp
36	74	73.933260	11.50 m	ϵ
36	75	74.931034	4.3 m	ϵ
36	76	75.925948	14.8 h	ϵ
36	77	76.924668	74.4 m	ϵ
36	78	77.920386	0.35%	
36	79	78.920083	35.04 h	ϵ
36	80	79.916378	2.25%	
36	81	80.916592	2.29×10^5 y	ϵ
36	82	81.913485	11.6%	
36	83	82.914136	11.5%	
36	84	83.911507	57.0%	
36	85	84.912527	10.756 y	β^-
36	86	85.910610	17.3%	
36	87	86.913354	76.3 m	β^-
36	88	87.914447	2.84 h	β^-
36	89	88.917630	3.15 m	β^-
36	90	89.919524	32.32 s	β^-
Rubidium (Rb)				
37	75	74.938569	19.0 s	ϵ
37	76	75.935071	36.5 s	ϵ
37	77	76.930407	3.78 m	ϵ
37	78	77.928141	17.66 m	ϵ
37	79	78.923997	22.9 m	ϵ
37	80	79.922519	33.4 s	ϵ
37	81	80.918994	4.576 h	ϵ
37	82	81.918208	1.273 m	ϵ
37	83	82.915112	86.2 d	ϵ
37	84	83.914385	32.77 d	ϵ, β^-
37	85	84.911789	72.17%	
37	86	85.911167	18.631 d	β^-, ϵ
37	87	86.909184	4.75×10^{10} y	β^-
			27.83%	
37	88	87.911319	17.78 m	β^-
37	89	88.912280	15.15 m	β^-
37	90	89.914809	158 s	β^-
37	91	90.916534	58.4 s	β^-
Strontium (Sr)				
38	78	77.932179	2.5 m	ϵ
38	79	78.929707	2.25 m	ϵ
38	80	79.924525	106.3 m	ϵ
38	81	80.923213	22.3 m	ϵ
38	82	81.918401	25.55 d	ϵ
38	83	82.917555	32.41 h	ϵ
38	84	83.913425	0.56%	

Z	A	Atomic Mass (u)	$t_{1/2}$ or Abundance	Decay Mode
Strontium (Sr)				
38	85	84.912933	64.84 d	ϵ
38	86	85.909262	9.86%	
38	87	86.908879	7.00%	
38	88	87.905614	82.58%	
38	89	88.907453	50.53 d	β^-
38	90	89.907738	28.78 y	β^-
38	91	90.910210	9.63 h	β^-
38	92	92.911030	2.71 h	β^-
38	93	92.914022	7.423 m	β^-
38	94	93.915360	75.3 s	β^-
38	95	94.919358	23.90 s	β^-
Yttrium (Y)				
39	80	79.934340*s*	35 s	ϵ
39	81	80.929130	72.4 s	ϵ
39	83	82.922350	7.08 m	ϵ
39	85	84.916427	2.68 h	ϵ
39	86	85.914888	14.74 h	ϵ
39	87	86.910878	79.8 h	ϵ
39	88	87.909503	106.65 d	ϵ
39	89	88.905848	100%	
39	90	89.907151	64.1 h	β^-
39	91	90.907303	58.51 d	β^-
39	92	91.908947	3.54 h	β^-
39	93	92.909582	10.18 h	β^-
39	94	93.911594	18.7 m	β^-
39	95	94.912824	10.3 m	β^-
Zirconium (Zr)				
40	81	80.936820	15 s	ϵ, ϵp
40	82	81.931090	32 s	ϵ
40	83	82.928650	44 s	ϵ, ϵp
40	84	83.923250*s*	25.9 m	ϵ
40	85	84.921470	7.86 m	ϵ
40	86	85.916470	16.5 h	ϵ
40	87	86.914817	1.68 h	ϵ
40	88	87.910226	83.4 d	ϵ
40	89	88.908889	78.41 h	ϵ
40	90	89.904704	51.45%	
40	91	90.905645	11.22%	
40	92	91.905040	17.15%	
40	93	92.906476	1.53×10^6 y	β^-
40	94	93.906316	17.38%	
40	95	94.908043	64.02 d	β^-
40	96	95.908276	3.9×10^{19} y	$2\beta^-$
			2.80%	
40	97	96.910951	16.90 h	β^-

Z	A	Atomic Mass (u)	$t_{1/2}$ or Abundance	Decay Mode
Zirconium (Zr) *(continued)*				
40	98	97.912746	30.7 s	β^-
40	99	98.916511	2.1 s	β^-
Niobium (Nb)				
41	84	83.933570*s*	12 s	ϵ, ϵp
41	85	84.927910	20.9 s	ϵ
41	86	85.925040	88 s	ϵ
41	87	86.920360	2.6 m	ϵ
41	88	87.917960*s*	14.5 m	ϵ
41	89	88.913500	1.18 h	ϵ
41	90	89.911264	14.60 h	ϵ
41	91	90.906991	6.8×10^2 y	ϵ
41	92	91.907193	3.47×10^7 y	ϵ, β^-
41	93	92.906378	100%	
41	93*m*	92.906412	16.13 y	IT
41	94	93.907284	2.03×10^4 y	β^-
41	95	94.906835	34.975 d	β^-
41	96	95.908100	23.35 h	β^-
41	97	96.908097	1.227 h	β^-
41	99	98.911618	15.0 s	β^-
Molybdenum (Mo)				
42	86	85.930700	19.6 s	ϵ
42	87	86.927330	14.5 s	ϵ, ϵp
42	88	87.921953	8.0 m	ϵ
42	89	88.919481	2.04 m	ϵ
42	90	89.913936	5.67 h	ϵ
42	91	90.911751	15.49 m	ϵ
42	92	91.906810	14.84%	
42	93	92.906812	4.0×10^3 y	ϵ
42	94	93.905088	9.25%	
42	95	94.905842	15.92%	
42	96	95.904679	16.68%	
42	97	96.906021	9.55%	
42	98	97.905408	24.13%	
42	99	98.907712	65.94 h	β^-
42	100	99.907477	1.2×10^{19} y 9.63%	$2\beta^-$
42	101	100.910347	14.61 m	β^-
42	102	101.910297	11.3 m	β^-
42	103	102.913200	67.5 s	β^-
42	104	103.913760	60 s	β^-
42	105	104.916970	35.6 s	β^-
Technetium (Tc)				
43	90	89.923560	8.7 s	ϵ
43	91	90.918430	3.14 m	ϵ
43	92	91.915260	4.23 m	ϵ

Z	A	Atomic Mass (u)	$t_{1/2}$ or Abundance	Decay Mode
Technetium (Tc)				
43	93	92.910248	2.75 h	ϵ
43	94	93.909656	293 m	ϵ
43	95	94.907656	20.0 h	ϵ
43	96	95.907871	4.28 d	ϵ
43	97	96.906365	2.6×10^6 y	ϵ
43	98	97.907216	4.2×10^6 y	β^-
43	99	98.906255	2.111×10^5 y	β^-
43	99*m*	98.906408	6.01 h	IT, β^-
43	100	99.907658	15.8 s	β^-
43	101	100.907314	14.22 m	β^-
43	103	102.909179	54.2 s	β^-
43	104	103.911440	18.3 m	β^-
43	105	104.911660	7.6 m	β^-
43	106	105.914355	35.6 s	β^-
43	107	106.915080	21.2 s	β^-
Ruthenium (Ru)				
44	92	91.920120*s*	3.65 m	ϵ
44	93	92.917050	59.7 s	ϵ
44	94	93.911360	51.8 m	ϵ
44	95	94.910413	1.643 h	ϵ
44	96	95.907598	5.52%	
44	97	96.907555	2.9 d	ϵ
44	98	97.905287	1.88%	
44	99	98.905939	12.7%	
44	100	99.904220	12.6%	
44	101	100.905582	17.0%	
44	102	101.904350	31.6%	
44	103	102.906324	39.26 d	β^-
44	104	103.905430	18.7%	
44	105	104.907750	4.44 h	β^-
44	106	105.907327	373.59 d	β^-
44	107	106.909910	3.75 m	β^-
44	108	107.910190	4.55 m	β^-
44	109	108.913200	34.5 s	β^-
44	110	109.913970	14.6 s	β^-
Rhodium (Rh)				
45	95	94.915900	5.02 m	ϵ
45	96	95.914518	9.90 m	ϵ
45	97	96.911340	30.7 m	ϵ
45	98	97.910716	8.7 m	ϵ
45	99	98.908132	16.1 d	ϵ
45	100	99.908117	20.8 h	ϵ
45	101	100.906164	3.3 y	ϵ
45	102	101.906843	207 d	ϵ, β^-
45	102*m*	101.906994	2.9 y	ϵ, IT
45	103	102.905504	100%	

Rhodium (Rh) *(continued)*

Z	A	Atomic Mass (u)	$t_{1/2}$ or Abundance	Decay Mode
45	104	103.906655	42.3 s	β^-, ϵ
45	105	104.905692	35.36 h	β^-
45	106	105.907285	29.80 s	β^-
45	107	106.906751	21.7 m	β^-
45	109	108.908736	80 s	β^-
45	110	109.910950	3.2 s	β^-
45	111	110.911660*s*	11 s	β^-

Palladium (Pd)

Z	A	Atomic Mass (u)	$t_{1/2}$ or Abundance	Decay Mode
46	96	95.918220	2.03 m	ϵ
46	97	96.916480	3.1 m	ϵ
46	98	97.912721	17.7 m	ϵ
46	99	98.911768	21.4 m	ϵ
46	100	99.908505	3.63 d	ϵ
46	101	100.908289	8.47 h	ϵ
46	102	101.905608	1.02%	
46	103	102.906087	16.991 d	ϵ
46	104	103.904035	11.14%	
46	105	104.905084	22.33%	
46	106	105.903483	27.33%	
46	107	106.905128	6.5×10^6 y	β^-
46	108	107.903894	26.46%	
46	109	108.905954	13.70 h	β^-
46	110	109.905152	11.72%	
46	111	110.907640	23.4 m	β^-
46	112	111.907313	21.03 h	β^-
46	113	112.910150	93 s	β^-
46	114	113.910365	2.42 m	β^-
46	115	114.913680	25 s	β^-
46	116	115.914160	11.8 s	β^-

Silver (Ag)

Z	A	Atomic Mass (u)	$t_{1/2}$ or Abundance	Decay Mode
47	97	96.924000*s*	19 s	ϵ
47	98	97.921760	46.7 s	ϵ, ϵp
47	99	98.917600	124 s	ϵ
47	100	99.916070	2.01 m	ϵ
47	101	100.912800	11.1 m	ϵ
47	102	101.912000	12.9 m	ϵ
47	103	102.908972	65.7 m	ϵ
47	104	103.908628	69.2 m	ϵ
47	105	104.906528	41.29 d	ϵ
47	106	105.906666	23.96 m	ϵ, β^-
47	107	106.905093	51.839%	
47	108	107.905954	2.37 m	β^-, ϵ
47	108*m*	107.906071	418 y	ϵ, IT
47	109	108.904756	48.161%	
47	110	109.906110	24.6 s	β^-, ϵ
47	110*m*	109.906237	249.79 d	β^-, IT

Silver (Ag)

Z	A	Atomic Mass (u)	$t_{1/2}$ or Abundance	Decay Mode
47	111	110.905295	7.45 d	β^-
47	112	111.907004	3.130 h	β^-
47	113	112.906566	5.37 h	β^-
47	115	114.908760	20.0 m	β^-
47	116	115.911360	2.68 m	β^-
47	117	116.911680	72.8 s	β^-

Cadmium (Cd)

Z	A	Atomic Mass (u)	$t_{1/2}$ or Abundance	Decay Mode
48	99	98.925010*s*	16 s	ϵ, ϵp, $\epsilon\alpha$
48	100	99.920230	49.1 s	ϵ
48	101	100.918680	1.2 m	ϵ
48	102	101.914780	5.5 m	ϵ
48	103	102.913419	7.3 m	ϵ
48	104	103.909848	57.7 m	ϵ
48	105	104.909468	55.5 m	ϵ
48	106	105.906458	1.25%	
48	107	106.906614	6.50 h	ϵ
48	108	107.904183	0.89%	
48	109	108.904986	462.6 d	ϵ
48	110	109.903006	12.49%	
48	111	110.904182	12.80%	
48	112	111.902757	24.13%	
48	113	112.904401	9.3×10^{15} y 12.22%	β^-
48	113*m*	112.904684	14.1 y	β^-, IT
48	114	113.903358	28.73%	
48	115	114.905431	53.46 h	β^-
48	116	115.904755	7.49%	
48	117	116.907218	2.49 h	β^-
48	118	117.906914	50.3 m	β^-
48	119	118.909920	2.69 m	β^-
48	120	119.909851	50.80 s	β^-
48	121	120.912980	13.5 s	β^-

Indium (In)

Z	A	Atomic Mass (u)	$t_{1/2}$ or Abundance	Decay Mode
49	101	100.926560*s*	16 s	ϵ, ϵp
49	102	101.924710	24 s	ϵ
49	103	102.919914	65 s	ϵ:
49	104	103.918340	1.8 m	ϵ
49	105	104.914673	5.07 m	ϵ
49	106	105.913461	6.2 m	ϵ
49	107	106.910292	32.4 m	ϵ
49	108	107.909720	58.0 m	ϵ
49	109	108.907154	4.2 h	ϵ
49	110	109.907169	69.1 m	ϵ
49	110	109.907169	4.9 h	ϵ
49	111	110.905111	2.8047 d	ϵ
49	112	111.905533	14.97 m	ϵ, β^-

Z	A	Atomic Mass (u)	$t_{1/2}$ or Abundance	Decay Mode
Indium (In) *(continued)*				
49	113	112.904061	4.29%	
49	114	113.904917	71.9 s	β^-, ϵ
49	115	114.903878	4.41×10^{14} y 95.71%	β^-
49	116	115.905260	14.10 s	β^-, ϵ
49	117	116.904516	43.2 m	β^-
49	119	118.905846	2.4 m	β^-
49	120	119.907960	3.08 s, 46.2 s, 47.3 s	β^-
49	121	120.907849	23.1 s	β^-
Tin (Sn)				
50	104	103.923190	20.8 s	ϵ
50	105	104.921390	31 s	ϵ, ϵp
50	106	105.916880	115 s	ϵ
50	107	106.915670	2.90 m	ϵ
50	108	107.911970	10.30 m	ϵ
50	109	108.911287	18.0 m	ϵ
50	110	109.907853	4.11 h	ϵ
50	111	110.907735	35.3 m	ϵ
50	112	111.904821	0.97%	
50	113	112.905173	115.09 d	ϵ
50	114	113.902782	0.65%	
50	115	114.903346	0.34%	
50	116	115.901744	14.54%	
50	117	116.902954	7.68%	
50	118	117.901606	24.22%	
50	119	118.903309	8.58%	
50	119*m*	118.903406	293.1 d	IT
50	120	119.902197	32.59%	
50	121	120.904237	27.06 h	β^-
50	121*m*	120.904243	55 y	IT, β^-
50	122	121.903440	4.63%	
50	123	122.905722	129.2 d	β^-
50	124	123.905275	5.79%	
50	125	124.907785	9.64 d	β^-
50	126	125.907654	1.0×10^5 y	β^-
50	127	126.910351	2.10 h	β^-
50	128	127.910535	59.07 m	β^-
50	129	128.913440	2.23 m	β^-
50	130	129.913850	3.72 m	β^-
50	131	130.916920	56.0 s	β^-
50	132	131.917744	39.7 s	β^-
Antimony (Sb)				
51	109	108.918136	17.0 s	ϵ
51	110	109.916760*s*	23.0 s	ϵ
51	111	110.913210*s*	75 s	ϵ
51	112	111.912395	51.4 s	ϵ

Z	A	Atomic Mass (u)	$t_{1/2}$ or Abundance	Decay Mode
Antimony (Sb)				
51	113	112.909378	6.67 m	ϵ
51	114	113.909100	3.49 m	ϵ
51	115	114.906599	32.1 m	ϵ
51	116	115.906797	15.8 m	ϵ
51	117	116.904840	2.80 h	ϵ
51	118	117.905532	3.6 m	ϵ
51	119	118.903946	38.19 h	ϵ
51	120	119.905074	15.89 m	ϵ
51	121	120.903818	57.21%	
51	122	121.905175	2.7238 d	β^-, ϵ
51	123	122.904216	42.79%	
51	124	123.905938	60.20 d	β^-
51	125	124.905248	2.7582 y	β^-
51	126	125.907250	12.46 d	β^-
51	127	126.906915	3.85 d	β^-
51	128	127.909167	9.01 h	β^-
51	129	128.909150	4.40 h	β^-
51	130	129.911546	39.5 m	β^-
51	131	130.911950	23.03 m	β^-
51	132	131.914413	2.79 m	β^-
51	133	132.915240	2.5 m	β^-
51	134	133.920550	10.22 s	β^-, $\beta^- n$
Tellurium (Te)				
52	110	109.922410	18.6 s	ϵ, α
52	111	110.921120	19.3 s	ϵ, ϵp
52	112	111.917060	2.0 m	ϵ
52	113	112.915930*s*	1.7 m	ϵ
52	114	113.912060*s*	15.2 m	ϵ
52	115	114.911580	5.8 m	ϵ
52	116	115.908420	2.49 h	ϵ
52	117	116.908634	62 m	ϵ
52	118	117.905825	6.00 d	ϵ
52	119	118.906408	16.03 h	ϵ
52	120	119.904020	0.096%	
52	121	120.904930	16.78 d	ϵ
52	121*m*	120.905245	154 d	IT, ϵ
52	122	121.903047	2.603%	
52	123	122.904273	1.3×10^{13} y 0.908%	ϵ
52	123*m*	122.904538	119.7 d	IT
52	124	123.902820	4.816%	
52	125	124.904425	7.139%	
52	126	125.903306	18.952%	
52	127	126.905217	9.35 h	β^-
52	127*m*	126.905312	109 d	IT, β^-
52	128	127.904461	7.7×10^{24} y 31.687%	$2\beta^-$

Z	A	Atomic Mass (u)	$t_{1/2}$ or Abundance	Decay Mode
Tellurium (Te) *(continued)*				
52	129	128.906596	69.6 m	β^-
52	130	129.906223	2.7×10^{21} y 33.799%	$2\beta^-$
52	131	130.908522	25.0 m	β^-
52	132	131.908524	3.204 d	β^-
52	133	132.910940	12.5 m	β^-
52	134	133.911540	41.8 m	β^-
52	135	134.916450	19.0 s	β^-
52	136	135.920100	17.5 s	$\beta^-, \beta^- n$
Iodine (I)				
53	115	114.917920*s*	1.3 m	ϵ
53	117	116.913650	2.22 m	ϵ
53	118	117.913380	13.7 m	ϵ
53	119	118.910180	19.1 m	ϵ
53	120	119.910048	81.0 m	ϵ
53	121	120.907366	2.12 h	ϵ
53	122	121.907592	3.63 m	ϵ
53	123	122.905598	13.27 h	ϵ
53	124	123.906211	4.1760 d	ϵ
53	125	124.904624	59.408 d	ϵ
53	126	125.905619	13.11 d	ϵ, β^-
53	127	126.904468	100%	
53	128	127.905805	24.99 m	β^-, ϵ
53	129	128.904987	1.57×10^7 y	β^-
53	130	129.906674	12.36 h	β^-
53	131	130.906124	8.02070 d	β^-
53	132	131.907995	2.295 h	β^-
53	133	132.907806	20.8 h	β^-
53	134	133.909877	52.5 m	β^-
53	135	134.910050	6.57 h	β^-
53	136	135.914660	83.4 s	β^-
53	137	136.917873	24.5 s	$\beta^-, \beta^- n$
Xenon (Xe)				
54	114	113.928150*s*	10.0 s	ϵ
54	115	114.926540*s*	18 s	$\epsilon\alpha, \epsilon, \epsilon p$
54	116	115.921740*s*	59 s	ϵ
54	117	116.920560	61 s	$\epsilon, \epsilon p$
54	118	117.916570	3.8 m	ϵ
54	119	118.915550	5.8 m	ϵ
54	120	119.912150	40 m	ϵ
54	121	120.911386	40.1 m	ϵ
54	122	121.908550	20.1 h	ϵ
54	123	122.908471	2.08 h	ϵ
54	124	123.905896	0.10%	
54	125	124.906398	16.9 h	ϵ
54	126	125.904269	0.09%	

Z	A	Atomic Mass (u)	$t_{1/2}$ or Abundance	Decay Mode
Xenon (Xe)				
54	127	126.905180	36.3446 d	ϵ
54	128	127.903530	1.91%	
54	129	128.904780	26.4%	
54	130	129.903508	4.1%	
54	131	130.905082	21.2%	
54	132	131.904154	26.9%	
54	133	132.905906	5.2475 d	β^-
54	134	133.905394	10.4%	
54	135	134.907207	9.14 h	β^-
54	136	135.907220	9.3×10^{19} y 8.9%	$2\beta^-$
54	137	136.911563	3.818 m	β^-
54	138	137.913990	14.08 m	$\beta^-, \beta^- n$
54	139	138.918787	39.68 s	β^-
54	140	139.921640	13.60 s	β^-
Cesium (Cs)				
55	118	117.926555	14 s	$\epsilon, \epsilon p$
55	119	118.922371	43.0 s	ϵ
55	120	119.920678	57 s	$\epsilon, \epsilon p$
55	120	119.920678	64 s	ϵ
55	121	120.917184	128 s	ϵ
55	122	121.916122	21.2 s	ϵ
55	123	122.912990	5.87 m	ϵ
55	124	123.912246	30.9 s	ϵ
55	125	124.909725	45 m	ϵ
55	126	125.909448	1.63 m	ϵ
55	127	126.907418	6.25 h	ϵ
55	128	127.907748	3.66 m	ϵ
55	129	128.906063	32.06 h	ϵ
55	130	129.906706	29.21 m	ϵ, β^-
55	131	130.905460	9.69 d	ϵ
55	132	131.906430	6.479 d	ϵ, β^-
55	133	132.905447	100%	
55	134	133.906713	2.0648 y	β^-, ϵ
55	135	134.905972	2.3×10^6 y	β^-
55	136	135.907306	19 s, 13.16 d	β^-
55	137	136.907084	30.1 y	β^-
55	138	137.911011	32.41 m	β^-
55	139	138.913358	9.27 m	β^-
55	140	139.917277	63.7 s	β^-
55	141	140.920044	24.94 s	$\beta^-, \beta^- n$
Barium (Ba)				
56	120	119.926050	32 s	ϵ
56	121	120.924490	29.5 s	$\epsilon, \epsilon p$
56	122	121.920260*s*	1.95 m	ϵ
56	123	122.918850*s*	2.7 m	ϵ

Z	A	Atomic Mass (u)	$t_{1/2}$ or Abundance	Decay Mode
Barium (Ba) *(continued)*				
56	124	123.915088	11.9 m	ϵ
56	125	124.914620	3.5 m	ϵ
56	126	125.911244	100 m	ϵ
56	127	126.911120	12.7 m	ϵ
56	128	127.908309	2.43 d	ϵ
56	129	128.908674	2.23 h	ϵ
56	130	129.906310	0.106%	
56	131	130.906931	11.50 d	ϵ
56	132	131.905056	0.101%	
56	133	132.906002	10.52 y	ϵ
56	134	133.904503	2.417%	
56	135	134.905683	6.592%	
56	136	135.904570	7.854%	
56	137	136.905821	11.23%	
56	138	137.905241	71.70%	
56	139	138.908835	83.06 m	β^-
56	140	139.910599	12.752 d	β^-
56	141	140.914406	18.27 m	β^-
56	142	141.916448	10.6 m	β^-
56	143	142.920617	14.33 s	β^-
56	144	143.922940	11.5 s	β^-, $\beta^- n$
Lanthanum (La)				
57	123	122.926240*s*	17 s	ϵ
57	124	123.924530*s*	29 s	ϵ
57	125	124.920670*s*	76 s	ϵ
57	126	125.919370*s*	54 s	ϵ
57	127	126.916160*s*	3.8 m	ϵ
57	128	127.915450	5.0 m	ϵ
57	129	128.912670	11.6 m	ϵ
57	130	129.912320*s*	8.7 m	ϵ
57	131	130.910110	59 m	ϵ
57	132	131.910110	4.8 h	ϵ
57	133	132.908400	3.912 h	ϵ
57	134	133.908490	6.45 m	ϵ
57	135	134.906971	19.5 h	ϵ
57	136	135.907650	9.87 m	ϵ
57	137	136.906470	6×10^4 y	ϵ
57	138	137.907107	1.05×10^{11} y	ϵ, β^-
			0.0902%	
57	139	138.906348	99.9098%	
57	140	139.909473	1.6781 d	β^-
57	141	140.910957	3.92 h	β^-
57	142	141.914074	91.1 m	β^-
57	143	142.916059	14.2 m	β^-
57	144	143.919590	40.8 s	β^-
57	145	144.921640	24.8 s	β^-

Z	A	Atomic Mass (u)	$t_{1/2}$ or Abundance	Decay Mode
Cerium (Ce)				
58	125	124.928540*s*	9.0 s	ϵ, ϵp
58	126	125.924100*s*	50 s	ϵ
58	127	126.922750*s*	32 s	ϵ
58	128	127.918870*s*	3 m	ϵ
58	129	128.918090*s*	3.5 m	ϵ
58	130	129.914690*s*	25 m	ϵ
58	131	130.914420	10.2 m	ϵ
58	131	130.914420	5.0 m	ϵ
58	132	131.911490*s*	3.51 h	ϵ
58	133	132.911550*s*	4.9 h	ϵ
58	134	133.909030	75.9 h	ϵ
58	135	134.909146	17.7 h	ϵ
58	136	135.907140	0.19%	
58	137	136.907780	9.0 h	ϵ
58	138	137.905986	0.25%	
58	139	138.906647	137.640 d	ϵ
58	140	139.905434	88.48%	
58	141	140.908271	32.501 d	β^-
58	142	141.909240	5×10^{16} y	$2\beta^-$
			11.08%	
58	143	142.912381	33.039 h	β^-
58	144	143.913643	284.893 d	β^-
58	146	144.917230	3.01 m	β^-
58	146	145.918690	13.52 m	β^-
58	147	146.922510	56.4 s	β^-
58	148	147.924390	56 s	β^-
Praseodymium (Pr)				
59	129	128.924860*s*	24 s	ϵ
59	130	129.923380*s*	40.0 s	ϵ
59	131	130.920060	1.53 m	ϵ
59	132	131.919120*s*	1.6 m	ϵ
59	133	132.916200*s*	6.5 m	ϵ
59	134	133.915670*s*	17 m	ϵ
59	135	134.913140	24 m	ϵ
59	136	135.912650	13.1 m	ϵ
59	137	136.910680	1.28 h	ϵ
59	138	137.910749	1.45 m	ϵ
59	139	138.908932	4.41 h	ϵ
59	140	139.909071	3.39 m	ϵ
59	141	140.907648	100%	
59	142	141.910040	19.12 h	β^-, ϵ
59	143	142.910812	13.57 d	β^-
59	144	143.913301	17.28 m	β^-
59	145	144.914507	5.984 h	β^-
59	146	145.917590	24.15 m	β^-
59	147	146.918980	13.4 m	β^-
59	148	147.922180	2.27 m	β^-

Praseodymium (Pr) *(continued)*

Z	A	Atomic Mass (u)	$t_{1/2}$ or Abundance	Decay Mode
59	149	148.923791	2.26 m	β^-
59	151	150.928230	18.90 s	β^-

Neodymium (Nd)

Z	A	Atomic Mass (u)	$t_{1/2}$ or Abundance	Decay Mode
60	130	129.928780*s*	28 s	ϵ
60	131	130.927100	27 s	ϵ, ϵp
60	132	131.923120*s*	1.75 m	ϵ
60	133	132.922210*s*	70 s	ϵ
60	134	133.918650*s*	8.5 m	ϵ
60	135	134.918240*s*	12.4 m	ϵ
60	136	135.915020	50.65 m	ϵ
60	137	136.914640	38.5 m	ϵ
60	138	137.911930*s*	5.04 h	ϵ
60	139	138.911920	29.7 m	ϵ
60	140	139.909310	3.37 d	ϵ
60	141	140.909605	2.49 h	ϵ
60	142	141.907719	27.13%	
60	143	142.909810	12.18%	
60	144	143.910083	2.29×10^{15} y 23.80%	α
60	145	144.912569	8.30%	
60	146	145.913112	17.19%	
60	147	146.916096	10.98 d	β^-
60	148	147.916889	5.76%	
60	149	148.920144	1.728 h	β^-
60	150	149.920887	1.1×10^{19} y 5.64%	$2\beta^-$
60	151	150.923825	12.44 m	β^-
60	152	151.924680	11.4 m	β^-
60	153	152.927695	28.9 s	β^-
60	154	153.929480	25.9 s	β^-

Promethium (Pm)

Z	A	Atomic Mass (u)	$t_{1/2}$ or Abundance	Decay Mode
61	133	132.929720*s*	12 s	ϵ
61	134	133.928490*s*	5 s	ϵ
61	135	134.924620*s*	40 s	ϵ
61	136	135.923450	47 s, 107 s	ϵ
61	137	136.920710*s*	2.4 m	ϵ
61	138	137.919450*s*	10 s	ϵ
61	139	138.916760	4.15 m	ϵ
61	141	140.913607	20.90 m	ϵ
61	142	141.912950	40.5 s	ϵ
61	143	142.910928	265 d	ϵ
61	144	143.912586	363 d	ϵ
61	145	144.912744	17.7 y	ϵ, α
61	146	145.914692	5.53 y	ϵ, β^-
61	147	146.915134	2.6234 y	β^-

Promethium (Pm)

Z	A	Atomic Mass (u)	$t_{1/2}$ or Abundance	Decay Mode
61	148	147.917468	5.370 d	β^-
61	149	148.918329	53.08 h	β^-
61	150	149.920979	2.68 h	β^-
61	151	150.921203	28.40 h	β^-
61	152	151.923490	4.12 m	β^-
61	153	152.924113	5.4 m	β^-
61	154	153.926550	1.73 m	β^-
61	155	154.928100	41.5 s	β^-
61	156	155.931060	26.70 s	β^-
61	157	156.933200*s*	10.56 s	β^-

Samarium (Sm)

Z	A	Atomic Mass (u)	$t_{1/2}$ or Abundance	Decay Mode
62	134	133.934020*s*	10 s	ϵ
62	135	134.932350*s*	10 s	ϵ, ϵp
62	136	135.928300*s*	47 s	ϵ
62	137	136.927050	45 s	ϵ
62	138	137.923540*s*	3.1 m	ϵ
62	139	138.922302	2.57 m	ϵ
62	140	139.918991	14.82 m	ϵ
62	141	140.918469	10.2 m	ϵ
62	142	141.915193	72.49 m	ϵ
62	143	142.914624	8.83 m	ϵ
62	144	143.911995	3.1%	
62	145	144.913406	340 d	ϵ
62	146	145.913037	10.3×10^{7} y	α
62	147	146.914893	1.06×10^{11} y 15.0%	α
62	148	147.914818	7×10^{15} y 11.3%	α
62	149	148.917180	2×10^{15} y 13.8%	α
62	150	149.917271	7.4%	
62	151	150.919928	90 y	β^-
62	152	151.919728	26.7%	
62	153	152.922094	46.27 h	β^-
62	154	153.922205	22.7%	
62	155	154.924636	22.3 m	β^-
62	156	155.925526	9.4 h	β^-
62	157	156.928350	482 s	β^-
62	158	157.929990	5.30 m	β^-
62	159	158.933200*s*	11.37 s	β^-

Europium (Eu)

Z	A	Atomic Mass (u)	$t_{1/2}$ or Abundance	Decay Mode
63	137	136.935210*s*	11 s	ϵ
63	138	137.933450*s*	12.1 s	ϵ
63	139	138.929840*s*	17.9 s	ϵ
63	141	140.924890	41.4 s	ϵ

Z	A	Atomic Mass (u)	$t_{1/2}$ or Abundance	Decay Mode
Europium (Eu) *(continued)*				
63	143	142.920287	2.57 m	ϵ
63	144	143.918774	10.2 s	ϵ
63	145	144.916261	5.93 d	ϵ
63	146	145.917200	4.59 d	ϵ
63	147	146.916741	24.1 d	ϵ, α
63	148	147.918154	54.5 d	ϵ, α
63	149	148.917926	93.1 d	ϵ
63	150	149.919698	36.9 y	ϵ
63	151	150.919846	47.8%	
63	152	151.921740	13.537 y	ϵ, β^-
63	153	152.921226	52.2%	
63	154	153.922975	8.593 y	β^-, ϵ
63	155	154.922889	4.7611 y	β^-
63	156	155.924751	15.19 d	β^-
63	157	156.925419	15.18 h	β^-
63	158	157.927840	45.9 m	β^-
63	159	158.929084	18.1 m	β^-
63	160	159.931970*s*	38 s	β^-
63	161	160.933680*s*	26 s	β^-
63	162	161.937040*s*	10.6 s	β^-
Gadolinium (Gd)				
64	140	139.933950*s*	15.8 s	ϵ
64	141	140.932210*s*	14 s	ϵ, ϵp
64	142	141.928230*s*	70.2 s	ϵ
64	143	142.926740	39 s	ϵ
64	144	143.922790*s*	4.5 m	ϵ
64	145	144.921690	23.0 m	ϵ
64	146	145.918305	48.27 d	ϵ
64	147	146.919089	38.06 h	ϵ
64	148	147.918110	74.6 y	α
64	149	148.919336	9.28 d	ϵ, α
64	150	149.918655	1.79×10^6 y	α
64	151	150.920344	124 d	ϵ, α
64	152	151.919788	1.08×10^{14} y	α
			0.20%	
64	153	152.921746	241.6 d	ϵ
64	154	153.920862	2.18%	
64	155	154.922619	14.80%	
64	156	155.922120	20.47%	
64	157	156.923957	15.65%	
64	158	157.924101	24.84%	
64	159	158.926385	18.479 h	β^-
64	160	159.927051	21.86%	
64	161	160.929666	3.66 m	β^-
64	162	161.930981	8.4 m	β^-
64	163	162.933990*s*	68 s	β^-
64	164	163.935860*s*	45 s	β^-
Terbium (Tb)				
65	143	142.934750*s*	12 s	ϵ
65	145	144.928880*s*	31.6 s	ϵ
65	147	146.924037	1.7 h	ϵ
65	148	147.924300	60 m	ϵ
65	149	148.923242	4.118 h	ϵ, α
65	150	149.923654	3.48 h	ϵ, α
65	151	150.923098	17.609 h	ϵ, α
65	152	151.924070	17.5 h	ϵ, α
65	153	152.923431	2.34 d	ϵ
65	154	153.924690	21.5 h	ϵ, β^-
65	155	154.923500	5.32 d	ϵ
65	156	155.924744	5.35 d	ϵ, β^-
65	157	156.924021	99 y	ϵ
65	158	157.925410	180 y	ϵ, β^-
65	159	158.925343	100%	
65	160	159.927164	72.3 d	β^-
65	161	160.927566	6.88 d	β^-
65	162	161.929480	7.60 m	β^-
65	163	162.930644	19.5 m	β^-
65	164	163.933350	3.0 m	β^-
65	165	164.934880*s*	2.11 m	β^-
Dysprosium (Dy)				
66	145	144.936950*s*	10.5 s	ϵ
66	146	145.932720	33.2 s	ϵ
66	147	146.930880	40 s	ϵ, ϵp
66	148	147.927180	3.1 m	ϵ
66	149	148.927334	4.20 m	ϵ
66	150	149.925580	7.17 m	ϵ, α
66	151	150.926180	17.9 m	ϵ, α
66	152	151.924714	2.38 h	ϵ, α
66	153	152.925761	6.4 h	ϵ, α
66	154	153.924422	3.0×10^6 y	α
66	155	154.925749	9.9 h	ϵ
66	156	155.924278	0.06%	
66	157	156.925461	8.14 h	ϵ
66	158	157.924405	0.10%	
66	159	158.925736	144.4 d	ϵ
66	160	159.925194	2.34%	
66	161	160.926930	18.9%	
66	162	161.926795	25.5%	
66	163	162.928728	24.9%	
66	164	163.929171	28.2%	
66	165	164.931700	2.334 h	β^-
66	166	165.932803	81.6 h	β^-
66	167	166.935650	6.20 m	β^-
66	168	167.937230*s*	8.7 m	β^-
66	169	168.940300	39 s	β^-

Z	A	Atomic Mass (u)	$t_{1/2}$ or Abundance	Decay Mode
Holmium (Ho)				
67	149	148.933790	21.1 s	ϵ
67	150	149.933350*s*	72 s	ϵ
67	151	150.931681	35.2 s	ϵ, α
67	152	151.931740	161.8 s	ϵ, α
67	153	152.930195	2.02 m	ϵ, α
67	154	153.930596	11.76 m	ϵ, α
67	155	154.929079	48 m	α, ϵ
67	156	155.929710*s*	56 m	ϵ
67	157	156.928190	12.6 m	ϵ
67	158	157.928950	11.3 m	ϵ
67	159	158.927709	33.05 m	ϵ
67	160	159.928726	25.6 m	ϵ
67	161	160.927852	2.48 h	ϵ
67	162	161.929092	150.0 m	ϵ
67	163	162.928730	4570 y	ϵ
67	164	163.930231	29 m	ϵ, β^-
67	165	164.930319	100%	
67	166	165.932281	26.763 h	β^-
67	166*m*	165.932287	1.20×10^3 y	β^-
67	167	166.933126	3.1 h	β^-
67	168	167.935500	2.99 m	β^-
67	169	168.936868	4.7 m	β^-
67	170	169.939610	2.76 m	β^-
67	171	170.941460	53 s	β^-
67	172	171.944820*s*	25 s	β^-
Erbium (Er)				
68	149	148.942170*s*	4 s	ϵ, ϵp
68	150	149.937760*s*	18.5 s	ϵ
68	151	150.937460*s*	23.5 s	ϵ
68	152	151.935080	10.3 s	α, ϵ
68	153	152.935093	37.1 s	α, ϵ
68	154	153.932777	3.73 m	ϵ, α
68	155	154.933200	5.3 m	ϵ, α
68	156	155.931020	19.5 m	ϵ
68	157	156.931950	18.65 m	ϵ, α
68	158	157.929910*s*	2.29 h	ϵ
68	159	158.930681	36 m	ϵ
68	160	159.929080	28.58 h	ϵ
68	161	160.930001	3.21 h	ϵ
68	162	161.928775	0.14%	
68	163	162.930029	75.0 m	ϵ
68	164	163.929197	1.61%	
68	165	164.930723	10.36 h	ϵ
68	166	165.930290	33.6%	
68	167	166.932045	22.95%	

Z	A	Atomic Mass (u)	$t_{1/2}$ or Abundance	Decay Mode
Erbium (Er)				
68	168	167.932368	26.8%	
68	169	168.934588	9.40 d	β^-
68	170	169.935460	14.9%	
68	171	170.938026	7.516 h	β^-
68	172	171.939352	49.3 h	β^-
68	173	172.942400*s*	1.4 m	β^-
68	174	173.944340*s*	3.3 m	β^-
Thulium (Tm)				
69	155	154.939192	21.6 s	ϵ, α
69	156	155.939010	83.8 s	ϵ, α
69	157	156.936760	3.63 m	ϵ
69	158	157.937000*s*	3.98 m	ϵ
69	159	158.934810	9.13 m	ϵ
69	160	159.935090	9.4 m	ϵ
69	161	160.933400	30.2 m	ϵ
69	162	161.933970	21.70 m	ϵ
69	163	162.932648	1.810 h	ϵ
69	164	163.933451	2.0 m	ϵ
69	164	163.933451	5.1 m	IT, ϵ
69	165	164.932432	30.06 h	ϵ
69	166	165.933553	7.70 h	ϵ
69	167	166.932849	9.25 d	ϵ
69	168	167.934170	93.1 d	ϵ, β^-
69	169	168.934211	100%	
69	170	169.935798	128.6 d	β^-, ϵ
69	171	170.936426	1.92 y	β^-
69	172	171.938396	63.6 h	β^-
69	173	172.939600	8.24 h	β^-
69	174	173.942160	5.4 m	β^-
69	175	174.943830	15.2 m	β^-
69	176	175.946990	1.9 m	β^-
69	177	176.949040*s*	85 s	β^-
Ytterbium (Yb)				
70	156	155.942850	26.1 s	ϵ, α
70	157	156.942660	38.6 s	ϵ, α
70	158	157.939858	1.49 m	ϵ, α
70	159	158.940150	1.58 m	ϵ
70	160	159.937560*s*	4.8 m	ϵ
70	161	160.937850*s*	4.2 m	ϵ
70	162	161.935750*s*	18.87 m	ϵ
70	163	162.936270	11.05 m	ϵ
70	164	163.934520*s*	75.8 m	ϵ
70	165	164.935398	9.9 m	ϵ
70	166	165.933880	56.7 h	ϵ

Ytterbium (Yb) *(continued)*

Z	A	Atomic Mass (u)	$t_{1/2}$ or Abundance	Decay Mode
70	167	166.934947	17.5 m	ϵ
70	168	167.933894	0.13%	
70	169	168.935187	32.026 d	ϵ
70	170	169.934759	3.05%	
70	171	170.936322	14.3%	
70	172	171.936378	21.9%	
70	173	172.938207	16.12%	
70	174	173.938858	31.8%	
70	175	174.941272	4.185 d	β^-
70	176	175.942568	12.7%	
70	177	176.945257	1.911 h	β^-
70	178	177.946643	74 m	β^-
70	179	178.950170*s*	8.0 m	β^-
70	180	179.952330*s*	2.4 m	β^-

Lutetium (Lu)

Z	A	Atomic Mass (u)	$t_{1/2}$ or Abundance	Decay Mode
71	158	157.949170*s*	10.4 s	ϵ, α
71	159	158.946620	12.1 s	ϵ, α
71	160	159.946020*s*	36.1 s	ϵ
71	161	160.943540*s*	72 s	ϵ
71	162	161.943220*s*	1.37 m	ϵ
71	163	162.941200	238 s	ϵ
71	164	163.941220*s*	3.14 m	ϵ
71	165	164.939610	12 m	ϵ
71	166	165.939760	2.65 m	ϵ
71	167	166.938310	51.5 m	ϵ
71	168	167.938700	5.5 m	ϵ
71	169	168.937649	34.06 h	ϵ
71	170	169.938472	2.00 d	ϵ
71	171	170.937910	8.24 d	ϵ
71	172	171.939082	6.70 d	ϵ
71	173	172.938927	1.37 y	ϵ
71	174	173.940334	3.31 y	ϵ
71	174*m*	173.940517	142 d	IT, ϵ
71	175	174.940768	97.41%	
71	176	175.942682	3.73×10^{10} y 2.59%	β^-
71	177	176.943755	6.734 d	β^-
71	177*m*	176.944796	160.4 d	β^-, IT
71	178	177.945951	28.4 m	β^-
71	179	178.947324	4.59 h	β^-
71	180	179.949880	5.7 m	β^-
71	181	180.951970*s*	3.5 m	β^-
71	182	181.955210*s*	2.0 m	β^-
71	183	182.957570*s*	58 s	β^-
71	184	183.961170*s*	20 s	β^-

Hafnium (Hf)

Z	A	Atomic Mass (u)	$t_{1/2}$ or Abundance	Decay Mode
72	160	159.950710	13.0 s	ϵ, α
72	161	160.950330	16.8 s	α, ϵ
72	162	161.947203	37.6 s	ϵ, α
72	163	162.947060*s*	40.0 s	ϵ
72	164	163.944420*s*	111 s	ϵ
72	165	164.944540*s*	76 s	ϵ
72	166	165.942250*s*	6.77 m	ϵ
72	167	166.942600*s*	2.05 m	ϵ
72	168	167.940630*s*	25.95 m	ϵ
72	169	168.941160	3.24 m	ϵ
72	170	169.939650*s*	16.01 h	ϵ
72	171	170.940490*s*	12.1 h	ϵ
72	172	171.939460	1.87 y	ϵ
72	173	172.940650*s*	23.6 h	ϵ
72	174	173.940040	2.0×10^{15} y 0.162%	α
72	175	174.941503	70 d	ϵ
72	176	175.941402	5.206%	
72	177	176.943220	18.606%	
72	178	177.943698	27.297%	
72	178*m*	177.946324	31 y	IT
72	179	178.945815	13.629%	
72	180	179.946549	35.100%	
72	181	180.949099	42.39 d	β^-
72	182	181.950553	9×10^6 y	β^-
72	183	182.953530	1.067 h	β^-
72	184	183.955450	4.12 h	β^-
72	185	184.958780*s*	3.5 m	

Tantalum (Ta)

Z	A	Atomic Mass (u)	$t_{1/2}$ or Abundance	Decay Mode
73	163	162.954320	11.0 s	ϵ, α
73	164	163.953570*s*	14.2 s	ϵ
73	165	164.950820*s*	31.0 s	ϵ
73	166	165.950470*s*	31.5 s	ϵ
73	167	166.947970*s*	1.33 m	ϵ
73	168	167.947790*s*	2.0 m	ϵ
73	169	168.945920*s*	4.9 m	ϵ
73	170	169.946090*s*	6.76 m	ϵ
73	171	170.944460*s*	23.3 m	ϵ
73	172	171.944740	36.8 m	ϵ
73	173	172.943540*s*	3.14 h	ϵ
73	174	173.944170	1.05 h	ϵ
73	175	174.943650*s*	10.5 h	ϵ
73	176	175.944740	8.09 h	ϵ
73	177	176.944472	56.6 h	ϵ
73	178	177.945750	9.31 m	ϵ
73	178	177.945750	2.36 h	ϵ

Z	A	Atomic Mass (u)	$t_{1/2}$ or Abundance	Decay Mode
Tantalum (Ta) *(continued)*				
73	179	178.945934	1.82 y	ϵ
73	180	179.947466	8.152 h	ϵ, β^-
73	180*m*	179.947546	1.2×10^{15} y 0.012%	β^-, ϵ
73	181	180.947996	99.988%	
73	182	181.950152	114.43 d	β^-
73	183	182.951373	5.1 d	β^-
73	184	183.954009	8.7 h	β^-
73	185	184.955559	49.4 m	β^-
73	186	185.958550	10.5 m	β^-
Tungsten (W)				
74	166	165.955020	18.8 s	ϵ, α
74	167	166.954670*s*	19.9 s	α, ϵ
74	168	167.951860*s*	51 s	ϵ,α
74	169	168.951760*s*	80 s	ϵ
74	170	169.949290*s*	2.42 m	ϵ
74	171	170.949460*s*	2.38 m	ϵ
74	172	171.947420*s*	6.6 m	ϵ
74	173	172.947830*s*	7.6 m	ϵ
74	174	173.946160*s*	31 m	ϵ
74	175	174.946770*s*	35.2 m	ϵ
74	176	175.945590*s*	2.5 h	ϵ
74	177	176.946620*s*	135 m	ϵ
74	178	177.945850	21.6 d	ϵ
74	179	178.947072	37.05 m	ϵ
74	180	179.946706	0.12%	
74	181	180.948198	121.2 d	ϵ
74	182	181.948206	26.498%	
74	183	182.950224	1.1×10^{17} y 14.314%	
74	184	183.950933	3×10^{17} y 30.642%	α?
74	185	184.953421	75.1 d	β^-
74	186	185.954362	28.426%	
74	187	186.957158	23.72 h	β^-
74	188	187.958487	69.4 d	β^-
74	189	188.961910	11.5 m	β^-
74	190	189.963180	30.0 m	β^-
Rhenium (Re)				
75	171	170.955550*s*	15.2 s	ϵ
75	172*m*	171.955290*s*	15 s	ϵ
75	173	172.953060*s*	1.98 m	ϵ
75	174	173.953110*s*	2.40 m	ϵ
75	175	174.951390*s*	5.89 m	ϵ
75	176	175.951570*s*	5.3 m	ϵ
75	177	176.950270*s*	14.0 m	ϵ
75	178	177.950850	13.2 m	ϵ
Rhenium (Re)				
75	179	178.949980	19.5 m	ϵ
75	180	179.950790	2.44 m	ϵ
75	181	180.950065	19.9 h	ϵ
75	182	181.951210	64.0 h	ϵ
75	183	182.950821	70.0 d	ϵ
75	184	183.952524	38.0 d	ϵ
75	184*m*	183.952726	169 d	IT, ϵ
75	185	184.952956	37.40%	
75	186	185.954987	90 h	β^-, ϵ
75	186*m*	185.955147	2.0×10^5 y	IT, β^-
75	187	186.955751	4.35×10^{10} y 62.60%	β^-, α
75	188	187.958112	17.021 h	β^-
75	189	188.959228	24.3 h	β^-
75	190	189.961820	3.1 m	β^-
75	191	190.963124	9.8 m	β^-
75	192	191.965960*s*	16 s	β^-
Osmium (Os)				
76	172	171.960080*s*	19.2 s	ϵ, α
76	173	172.959790*s*	16 s	ϵ, α
76	174	173.957120*s*	44 s	ϵ, α
76	175	174.957080*s*	1.4 m	ϵ
76	176	175.954950*s*	3.6 m	ϵ
76	177	176.955050*s*	2.8 m	ϵ
76	178	177.953350	5.0 m	ϵ
76	179	178.953950*s*	6.5 m	ϵ
76	180	179.952350*s*	21.5 m	ϵ
76	181	180.953270	105 m	ϵ
76	182	181.952186	22.10 h	ϵ
76	183	182.953110*s*	13.0 h	ϵ
76	184	183.952491	5.6×10^{13} y 0.020%	
76	185	184.954043	93.6 d	ϵ
76	186	185.953838	2.0×10^{15} y 1.58%	α
76	187	186.955748	1.6%	
76	188	187.955836	13.3%	
76	189	188.958145	16.1%	
76	190	189.958445	26.4%	
76	191	190.960928	15.4 d	β^-
76	192	191.961479	41.0%	
76	193	192.964148	30.11 h	β^-
76	194	193.965179	6.0 y	β^-
76	195	194.968120	6.5 m	β^-
76	196	195.969620	34.9 m	β^-
Iridium (Ir)				
77	177	176.961170*s*	30 s	ϵ, α
77	178	177.961080*s*	12 s	ϵ

Z	A	Atomic Mass (u)	$t_{1/2}$ or Abundance	Decay Mode
Iridium (Ir) *(continued)*				
77	179	178.959150*s*	79 s	ϵ
77	180	179.959250*s*	1.5 m	ϵ
77	181	180.957640	4.90 m	ϵ
77	182	181.958130	15 m	ϵ
77	183	182.956810*s*	57 m	ϵ
77	184	183.957390	3.09 h	ϵ
77	185	184.956590*s*	14.4 h	ϵ
77	186	185.957951	16.64 h	ϵ
77	187	186.957361	10.5 h	ϵ
77	188	187.958852	41.5 h	ϵ
77	189	188.958716	13.2 d	ϵ
77	190	189.960590	11.78 d	ϵ
77	191	190.960591	37.3%	
77	192	191.962602	73.830 d	β^-, ϵ
77	192*m*	191.962768	241 y	IT
77	193	192.962924	62.7%	
77	194	193.965076	19.15 h	β^-
77	194*m*	193.965280	171 d	β^-
77	195	194.965977	2.5 h	β^-
77	196	195.968380	52 s	β^-
77	197	196.969636	5.8 m	β^-
Platinum (Pt)				
78	177	176.968450*s*	11 s	ϵ, α
78	178	177.965710*s*	21.1 s	ϵ, α
78	179	178.965480*s*	21.2 s	ϵ, α
78	180	179.963220*s*	52 s	ϵ, α
78	181	180.963180*s*	51 s	ϵ, α
78	182	181.961270	2.2 m	ϵ, α
78	183	182.961730*s*	6.5 m	ϵ, α
78	184	183.959900*s*	17.3 m	ϵ, α
78	185	184.960750	70.9 m	ϵ
78	186	185.959430	2.0 h	ϵ, α
78	187	186.960560*s*	2.35 h	ϵ
78	188	187.959396	10.2 d	ϵ, α
78	189	188.960832	10.87h	ϵ
78	190	189.959930	6.5×10^{11} y	α
			0.01%	
78	191	190.961685	2.96 d	ϵ
78	192	191.961035	0.79%	
78	193	192.962985	50 y	ϵ
78	194	193.962664	32.9%	
78	195	194.964774	33.8%	
78	196	195.964935	25.3%	
78	197	196.967323	19.8915 h	β^-
78	198	197.967876	7.2%	
78	199	198.970576	30.8 m	β^-
78	200	199.971424	12.5 h	β^-
78	201	200.974500	2.5 m	β^-
78	202	201.975740*s*	44 h	β^-
Gold (Au)				
79	181	180.969950*s*	11.4 s	ϵ, α
79	182	181.969620*s*	15.6 s	ϵ, α
79	183	182.967620*s*	42.0 s	ϵ, α
79	184	183.967470*s*	12.0 s	ϵ
79	185	184.965810	4.25 m	ϵ, α
79	186	185.966000	10.7 m	ϵ
79	187	186.964560*s*	8.4 m	ϵ
79	188	187.965090*s*	8.84 m	ϵ
79	189	188.963890*s*	28.7 m	ϵ, α
79	190	189.964699	42.8 m	ϵ, α
79	191	190.963650	3.18 h	ϵ
79	192	191.964810	4.94 h	ϵ
79	193	192.964132	17.65 h	ϵ
79	194	193.965339	38.02 h	ϵ
79	195	194.965018	186.10 d	ϵ
79	196	195.966551	6.183 d	ϵ, β^-
79	197	196.966552	100%	
79	198	197.968225	2.6952 d	β^-
79	199	198.9687848	3.139 d	β^-
79	200	199.970720	48.4 m	β^-
79	201	200.971641	26 m	β^-
79	202	201.973790	28.8 s	β^-
79	203	202.975137	60 s	β^-
79	204	203.977710*s*	39.8 s	β^-
79	205	204.979610*s*	31 s	β^-
Mercury (Hg)				
80	182	181.974750*s*	10.83 s	ϵ, α
80	184	183.971900*s*	30.6 s	ϵ, α
80	185	184.971980*s*	49.1 s	ϵ, α
80	186	185.969460	1.38 m	ϵ, α
80	187	186.969790*s*	2.4 m	ϵ, α
80	188	187.967560*s*	3.25 m	ϵ, α
80	189	188.968130*s*	7.6 m	ϵ, α
80	190	189.966280*s*	20.0 m	ϵ, α
80	191	190.967060	49 m	ϵ
80	192	191.965570*s*	4.85 h	ϵ
80	193	192.966644	3.80 h	ϵ
80	194	193.965382	520 y	ϵ
80	195	194.966640	9.9 h	ϵ
80	196	195.965815	0.15%	
80	197	196.967195	64.14 h	ϵ
80	198	197.966752	9.97%	
80	199	198.968262	16.87%	

Z	A	Atomic Mass (u)	$t_{1/2}$ or Abundance	Decay Mode
Mercury (Hg) *(continued)*				
80	200	199.968309	23.10%	
80	201	200.970285	13.18%	
80	202	201.970626	29.86%	
80	203	202.972857	46.612 d	β^-
80	204	203.973476	6.87%	
80	205	204.976056	5.2 m	β^-
80	206	205.977499	8.15 m	β^-
80	207	206.982580	2.9 m	β^-
80	208	207.985940*s*	42 m	β^-

Z	A	Atomic Mass (u)	$t_{1/2}$ or Abundance	Decay Mode
Thallium (Tl)				
81	184	183.981760*s*	11 s	ϵ, α
81	185	184.979100*s*	19.5 s	ϵ
81	186	185.978550*s*	27.5 s	ϵ, α
81	187	186.976170*s*	51 s	ϵ, α
81	189	188.973690*s*	2.3 m	ϵ
81	193	192.970550*s*	21.6 m	ϵ
81	194	193.971050*s*	33.0 m	ϵ, α
81	195	194.969650*s*	1.16 h	ϵ
81	196	195.970520*s*	1.84 h	ϵ
81	197	196.969540	2.84 h	ϵ
81	198	197.970470	5.3 h	ϵ
81	199	198.969810	7.42 h	ϵ
81	200	199.970945	26.1 h	ϵ
81	201	200.970804	72.912 h	ϵ
81	202	201.972091	12.23 d	ϵ
81	203	202.972329	29.524%	
81	204	203.973849	3.78 y	β^-, ϵ
81	205	204.974412	70.476%	
81	206	205.976095	4.199 m	β^-
81	207	206.977408	4.77 m	β^-
81	208	207.982005	3.053 m	β^-
81	209	208.985349	2.20 m	β^-
81	210	209.990066	1.30 m	β^-, $\beta^- n$

Z	A	Atomic Mass (u)	$t_{1/2}$ or Abundance	Decay Mode
Lead (Pb)				
82	183	182.991930*s*	300 ms	α, ϵ
82	184	183.988200*s*	0.55 s	α
82	185	184.987580*s*	4.1 s	α
82	186	185.984300*s*	4.7 s	α
82	188	187.981060*s*	25.5 s	ϵ, α
82	189	188.980880*s*	51 s	ϵ, α
82	190	189.978180	1.2 m	ϵ, α
82	191	190.978200*s*	1.33 m	ϵ, α
82	192	191.975760*s*	3.5 m	ϵ, α
82	194	193.973970*s*	12.0 m	ϵ, α
82	195	194.974470*s*	15 m	ϵ
82	196	195.972710*s*	37 m	ϵ, α
82	197	196.973380*s*	8 m	ϵ
82	198	197.971980*s*	2.40 h	ϵ
82	199	198.972910	90 m	ϵ
82	200	199.971816	21.5 h	ϵ
82	201	200.972850	9.33 h	ϵ
82	202	201.972144	52.5×10^3 y	ϵ, α
82	203	202.973375	51.873 h	ϵ
82	204	203.973029	1.4×10^{17} y 1.4%	α
82	205	204.974467	1.52×10^7 y	ϵ
82	206	205.974449	24.1%	
82	207	206.975881	22.1%	
82	208	207.976636	52.4%	
82	209	208.981075	3.253 h	β^-
82	210	209.984173	22.3 y	β^-, α
82	211	210.988731	36.1 m	β^-
82	212	211.991888	10.64 h	β^-
82	213	212.996500*s*	10.2 m	β^-
82	214	213.999798	26.8 m	β^-

Z	A	Atomic Mass (u)	$t_{1/2}$ or Abundance	Decay Mode
Bismuth (Bi)				
83	191	190.986050*s*	12 s	α, ϵ
83	192	191.985370*s*	34.6 s	ϵ, α
83	193	192.983060*s*	67 s	ϵ, α
83	194	193.982750*s*	95 s	ϵ
83	195	194.980750*s*	183 s	ϵ, α
83	196	195.980610*s*	308 s	ϵ, α
83	197	196.978930	9.33 m	ϵ, α
83	198	197.979020	10.3 m	ϵ
83	199	198.977580	27 m	ϵ
83	200	199.978140	36.4 m	ϵ
83	201	200.976970	108 m	ϵ, α
83	202	201.977670	1.72 h	ϵ, α
83	203	202.976868	11.76 h	ϵ, α
83	204	203.977805	11.22 h	ϵ
83	205	204.977375	15.31 d	ϵ
83	206	205.978483	6.243 d	ϵ
83	207	206.978455	31.55 y	ϵ
83	208	207.979727	3.68×10^5 y	ϵ
83	209	208.980383	100%	
83	210	209.984105	5.013 d	β^-, α
83	210*m*	209.984396	3.04×10^6 y	α
83	211	210.987258	2.14 m	α, β^-
83	212	211.991272	60.55 m	β^-, α, $\beta^-\alpha$
83	213	212.994375	45.59 m	β^-, α
83	214	213.998699	19.9 m	β^-, α

Z	A	Atomic Mass (u)	$t_{1/2}$ or Abundance	Decay Mode
Bismuth (Bi) *(continued)*				
83	215	215.001830	7.6 m	β^-
83	216	216.006200*s*	3.6 m	β^-
Polonium (Po)				
84	194	193.988280	0.392 s	α
84	195	194.988050*s*	4.64 s	α, ϵ
84	196	195.985510*s*	5.8 s	α, ϵ
84	197	196.985570*s*	53.6 s	ϵ, α
84	198	197.983340*s*	1.76 m	α, ϵ
84	199	198.983600*s*	5.48 m	ϵ, α
84	200	199.981740*s*	11.5 m	ϵ, α
84	201	200.98210*s*	15.3 m	ϵ, α
84	202	201.980700*s*	44.7 m	ϵ, α
84	203	202.981410	36.7 m	ϵ, α
84	204	203.980307	3.53 h	ϵ, α
84	205	204.981170	1.66 h	ϵ, α
84	206	205.980465	8.8 d	ϵ, α
84	207	206.981578	5.80 h	ϵ, α
84	208	207.981231	2.898 y	α, ϵ
84	209	208.982416	102 y	α, ϵ
84	210	209.982857	138.376 d	α
84	211	210.986637	0.516 s	α
84	212	211.988852	0.299 μs	α
84	213	212.992843	4.2 μs	α
84	214	213.995186	164.3 μs	α
84	215	214.999415	1.781 ms	α, β^-
84	216	216.001905	0.145 s	α
84	217	217.006250*s*	10 s	α, β^-
84	218	218.008966	3.10 m	α, β^-
Astatine (At)				
85	196	195.995700*s*	0.3 s	α
85	197	196.993290*s*	0.35 s	α, ϵ
85	198	197.992750*s*	4.2 s	α, ϵ
85	199	198.990630*s*	7.2 s	α, ϵ
85	200	199.990290*s*	43 s	α, ϵ
85	201	200.988490	89 s	α, ϵ
85	202	201.988450	184 s	ϵ, α
85	203	202.986850	7.4 m	ϵ, α
85	204	203.987260	9.2 m	ϵ, α
85	205	204.986040	26.2 m	ϵ, α
85	206	205.986600	30.0 m	ϵ, α
85	207	206.985776	1.80 h	ϵ, α
85	208	207.986583	1.63 h	ϵ, α
85	209	208.986159	5.41 h	ϵ, α
85	210	209.987131	8.1 h	ϵ, α
85	211	210.987481	7.214 h	ϵ, α
85	212	211.990735	0.314 s	$\alpha, \epsilon, \beta^-$

Z	A	Atomic Mass (u)	$t_{1/2}$ or Abundance	Decay Mode
Astatine (At)				
85	213	212.992921	125 ns	α
85	214	213.996356	558 ns	α
85	215	214.998641	0.10 ms	α
85	216	216.002409	0.30 ms	$\alpha, \epsilon, \beta^-$
85	217	217.004710	32.3 ms	α, β^-
85	218	218.008681	1.6 s	α, β^-
85	219	219.011300	56 s	α, β^-
85	220	220.015300*s*	3.71 m	β^-
85	221	221.018140*s*	2.3 m	β^-
85	222	222.022330*s*	54 s	β^-
85	223	223.025340*s*	50 s	β^-
Radon (Rn)				
86	199	198.998310*s*	0.62 s	α, ϵ
86	200	199.995680*s*	1.06 s	α, ϵ
86	201	200.995540*s*	7.0 s	α, ϵ
86	202	201.993220*s*	9.85 s	ϵ, α
86	203	202.993320*s*	45 s	α, ϵ
86	204	203.991370*s*	1.24 m	α, ϵ
86	205	204.991670*s*	170 s	ϵ, α
86	206	205.990160*s*	5.67 m	α, ϵ
86	207	206.990730	9.25 m	ϵ, α
86	208	207.989631	24.35 m	α, ϵ
86	209	208.990380	28.5 m	ϵ, α
86	210	209.989680	2.4 h	α, ϵ
86	211	210.990585	14.6 h	ϵ, α
86	212	211.990689	23.9 m	α
86	213	212.993868	25.0 ms	α
86	214	213.995346	0.27 μs	α
86	215	214.998729	2.30 μs	α
86	216	216.000258	45 μs	α
86	217	217.003915	0.54 ms	α
86	218	218.005586	35 ms	α
86	219	219.009475	3.96 s	α
86	220	220.011384	55.6 s	α
86	221	221.015460*s*	25 m	β^-, α
86	222	222.017570	3.8235 d	α
86	223	223.021790*s*	23.2 m	β^-
86	224	224.024090*s*	107 m	β^-
86	225	225.028440*s*	4.5 m	β^-
86	226	226.030890*s*	6.0 m	β^-
86	227	227.035410*s*	22.5 s	β^-
86	228	228.038080*s*	65 s	β^-
Francium (Fr)				
87	201	201.003990*s*	48 ms	α, ϵ
87	202	202.003290*s*	0.34 s	α, ϵ
87	203	203.001050*s*	0.55 s	α, ϵ

Z	A	Atomic Mass (u)	$t_{1/2}$ or Abundance	Decay Mode
Francium (Fr) *(continued)*				
87	204	204.000590*s*	1.7 s	α, ϵ
87	205	204.998660	3.85 s	α, ϵ
87	206	205.998490	15.9 s	α, ϵ
87	207	206.996860	14.8 s	α, ϵ
87	208	207.997130	59.1 s	α, ϵ
87	209	208.995920	50.0 s	α, ϵ
87	210	209.996398	3.18 m	α, ϵ
87	211	210.995529	3.10 m	α, ϵ
87	212	211.996195	20.0 m	ϵ, α
87	213	212.996175	34.6 s	α, ϵ
87	214	213.998955	5.0 ms	α
87	215	215.000326	86 ns	α
87	216	216.003188	0.70 μs	α, ϵ
87	217	217.004616	16 μs	α
87	218	218.007563	1.0 ms	α
87	219	219.009241	20 ms	α
87	220	220.012313	27.4 s	α, β^-
87	221	221.014246	4.9 m	α, β^-
87	222	222.017544	14.2 m	β^-
87	223	223.019731	22.00 m	β^-, α
87	224	224.023240	3.30 m	β^-
87	225	225.025607	4.0 m	β^-
87	226	226.029340	48 s	β^-
87	227	227.031830	2.47 m	β^-
87	228	228.035720*s*	39 s	β^-
87	229	229.038430*s*	50 s	β^-
87	230	230.042510*s*	19.1 s	β^-
87	231	231.045410*s*	17.5 s	β^-
Radium (Ra)				
88	206	206.003780*s*	0.24 s	α
88	207	207.003730*s*	1.3 s	α, ϵ
88	208	208.001780*s*	1.7 s	α, ϵ
88	209	209.001940*s*	4.6 s	α, ϵ
88	210	210.000450*s*	3.7 s	α, ϵ
88	211	211.000890	13 s	α, ϵ
88	212	211.999783	13.0 s	α, ϵ
88	213	213.000350	2.74 m	α, ϵ
88	214	214.000091	2.46 s	α, ϵ
88	215	215.002704	1.59 ms	α
88	216	216.003518	182 ns	α, ϵ
88	217	217.006306	1.7 μs	α
88	218	218.007124	15.6 μs	α
88	219	219.010069	10 ms	α
88	220	220.011015	17 ms	α
88	221	221.013908	28 s	α, ^{14}C
88	222	222.015362	38.0 s	α, ^{14}C

Z	A	Atomic Mass (u)	$t_{1/2}$ or Abundance	Decay Mode
Radium (Ra)				
88	223	223.018497	11.435 d	α, ^{14}C
88	224	224.020202	3.66 d	α, ^{12}C
88	225	225.023604	14.9 d	β^-
88	226	226.025403	1600 y	α, ^{14}C
88	227	227.029171	42.2 m	β^-
88	228	228.031064	5.75 y	β^-
88	229	229.034820	4.0 m	β^-
88	230	230.037080	93 m	β^-
88	231	231.041220*s*	1.72 m	β^-
88	232	232.043690*s*	250 s	β^-
88	233	233.048000*s*	30 s	β^-
88	234	234.050550*s*	30 s	β^-
Actinium (Ac)				
89	209	209.009570	0.10 s	α, ϵ
89	210	210.009260	0.35 s	α, ϵ
89	211	211.007650	0.25 s	α
89	212	212.007810	0.93 s	α, ϵ
89	213	213.006570	0.80 s	α
89	214	214.006890	8.2 s	α, ϵ
89	215	215.006450	0.17 s	α, ϵ
89	216	216.008721	0.33 ms	α
89	217	217.009333	69 ns	α, ϵ
89	218	218.011630	1.06 μs	α
89	219	219.012400	11.8 μs	α
89	220	220.014750	26.1 ms	α, ϵ
89	221	221.015580	52 ms	α
89	222	222.017829	5.0 s	α, ϵ
89	223	223.019126	2.10 m	α, ϵ
89	224	224.021708	2.9 h	ϵ, α, β^-
89	225	225.023221	10.0 d	α
89	226	226.026090	29.4 h	β^-, ϵ, α
89	227	227.027747	21.773 y	β^-, α
89	228	228.031015	6.15 h	β^-, α
89	229	229.032930	62.7 m	β^-
89	230	230.036030	122 s	β^-
89	231	231.038550	7.5 m	β^-
89	232	232.042020	119 s	β^-
89	233	233.044550*s*	145 s	β^-
89	234	234.048420*s*	44 s	β^-
Thorium (Th)				
90	212	212.012920*s*	30 ms	α, ϵ
90	213	213.012960*s*	140 ms	α
90	214	214.011450*s*	100 ms	α
90	215	215.011730	1.2 s	α
90	216	216.011051	0.028 s	α, ϵ

Thorium (Th) *(continued)*

Z	A	Atomic Mass (u)	$t_{1/2}$ or Abundance	Decay Mode
90	217	217.013070	0.252 ms	α
90	218	218.013268	109 ns	α
90	219	219.015520	1.05 μs	α
90	220	220.015733	9.7 μs	α, ϵ
90	221	221.018171	1.68 ms	α
90	222	222.018454	2.2 ms	α
90	223	223.020795	0.60 s	α
90	224	224.021459	1.05 s	α
90	225	225.023941	8.72 m	α, ϵ
90	226	226.024891	30.6 m	α
90	227	227.027699	18.72 d	α
90	228	228.028731	1.9131 y	α
90	229	229.031755	7880 y	α
90	230	230.033127	7.538×10^4 y	α, SF
90	231	231.036297	25.52 h	β^-, α
90	232	232.038050	1.405×10^{10} y 100%	α, SF
90	233	233.041577	22.3 m	β^-
90	234	234.043595	24.10 d	β^-
90	235	235.047500	7.1 m	β^-
90	236	236.049710*s*	37.5 m	β^-
90	237	237.053890*s*	5.0 m	β^-

Protactinium (Pa)

Z	A	Atomic Mass (u)	$t_{1/2}$ or Abundance	Decay Mode
91	215	215.019100	15 ms	α
91	216	216.019110	105 ms	α, ϵ
91	217	217.018290	3.4 ms	α
91	218	218.020010	0.11 ms	α
91	219	219.019880	53 ns	α
91	220	220.021880	0.78 μs	α
91	221	221.021860	5.9 μs	α
91	222	222.023730*s*	3.3 ms	α
91	223	223.023960	5 ms	α
91	224	224.025610	0.95 s	α, ϵ
91	225	225.026120	1.7 s	α
91	226	226.027933	1.8 m	α, ϵ
91	227	227.028793	38.3 m	α, ϵ
91	228	228.031037	22 h	ϵ, α
91	229	229.032089	1.50 d	ϵ, α
91	230	230.034533	17.4 d	ϵ, β^-, α
91	231	231.035879	3.276×10^4 y	α, SF
91	232	232.038582	1.31 d	β^-, ϵ
91	233	233.040240	26.967 d	β^-
91	234	234.043302	6.70 h	β^-
91	235	235.045440	24.5 m	β^-
91	236	236.048680	9.1 m	β^-
91	237	237.051140	8.7 m	β^-
91	238	238.054500	2.3 m	β^-

Uranium (U)

Z	A	Atomic Mass (u)	$t_{1/2}$ or Abundance	Decay Mode
92	225	225.029380	95 ms	α
92	226	226.029340	0.20 s	α
92	227	227.031140	1.1 m	α
92	228	228.031366	9.1 m	α, ϵ
92	229	229.033496	58 m	ϵ, α
92	230	230.033927	20.8 d	α
92	231	231.036289	4.2 d	ϵ, α
92	232	232.037146	68.9 y	α
92	233	233.039628	1.592×10^5 y	α, SF
92	234	234.040946	2.455×10^5 y 0.0055%	α, SF
92	235	235.043923	703.8×10^6 y 0.720%	α, SF
92	236	236.045562	2.342×10^7 y	α, SF
92	237	237.048724	6.75 d	β^-
92	238	238.050783	4.468×10^9 y 99.2745%	α, SF
92	239	239.054288	23.45 m	β^-
92	240	240.056586	14.1 h	β^-, α
92	242	242.062930*s*	16.8 m	β^-

Neptunium (Np)

Z	A	Atomic Mass (u)	$t_{1/2}$ or Abundance	Decay Mode
93	227	227.034960	0.51 s	α
93	228	228.036180*s*	1.07 m	ϵ, ϵSF
93	229	229.036250	3.85 m	α, ϵ
93	230	230.037810	4.6 m	ϵ, α
93	231	231.038230	48.8 m	ϵ, α
93	232	232.040100*s*	14.7 m	ϵ
93	233	233.040730	36.2 m	ϵ, α
93	234	234.042889	4.4 d	ϵ
93	235	235.044056	396.1 d	ϵ, α
93	236	236.046560	154×10^3 y	ϵ, β^-, α
93	237	237.048167	2.144×10^6 y	α, SF
93	238	238.050940	2.117 d	β^-
93	239	239.052931	2.3565 d	β^-
93	240	240.056169	61.9 m	β^-
93	240*m*	240.056169	7.22 m	β^-, IT
93	241	241.058250	13.9 m	β^-
93	242	242.061640*s*	2.2 m	β^-
93	242	242.061640*s*	5.5 m	β^-
93	243	243.064270*s*	1.8 m	β^-
93	244	244.067850*s*	2.29 m	

Plutonium (Pu)

Z	A	Atomic Mass (u)	$t_{1/2}$ or Abundance	Decay Mode
94	230	230.039646	200 s	α
94	232	232.041179	34.1 m	ϵ, α
94	233	233.042990	20.9 m	ϵ, α
94	234	234.043305	8.8 h	ϵ, α

Plutonium (Pu) *(continued)*

Z	A	Atomic Mass (u)	$t_{1/2}$ or Abundance	Decay Mode
94	235	235.045282	25.3 m	ϵ, α
94	236	236.046048	2.858 y	α, SF
94	237	237.048404	45.2 d	ϵ, α
94	238	238.049553	87.7 y	α, SF
94	239	239.052156	24110 y	α, SF
94	240	240.953808	6564 y	α, SF
94	241	241.056845	14.35 y	β^-, α, SF
94	242	242.058737	3.733×10^5 y	α, SF
94	243	243.061997	4.956 h	β^-
94	244	244.064198	8.08×10^7 y	α, SF
94	245	245.067739	10.5 h	β^-
94	246	246.070198	10.84 d	β^-
94	247	247.074070s	2.27 d	β^-

Americium (Am)

Z	A	Atomic Mass (u)	$t_{1/2}$ or Abundance	Decay Mode
95	232	232.046590s	79 s	ϵ, α
95	234	234.047790s	2.32 m	α, ϵ
95	237	237.049970	73.0 m	ϵ, α
95	238	238.051980	98 m	ϵ, α
95	239	239.053018	11.9 h	ϵ, α
95	240	240.055288	50.8 h	ϵ, α
95	241	241.056823	432.7 y	α, SF
95	242	242.059543	16.02 h	β^-, ϵ
95	242m	242.059596	141 y	IT, α, SF
95	243	243.061373	7370 y	α, SF
95	244	244.064279	10.1 h	β^-
95	245	245.066445	2.05 h	β^-
95	246	246.069768	39 m	β^-
95	247	247.072090s	23.0 m	β^-

Curium (Cm)

Z	A	Atomic Mass (u)	$t_{1/2}$ or Abundance	Decay Mode
96	238	238.053020	2.4 h	ϵ, α
96	239	239.054950s	2.9 h	ϵ, α
96	240	240.055519	27 d	α, ϵ, SF
96	241	241.057647	32.8 d	ϵ, α
96	242	242.058829	162.79 d	α, SF
96	243	243.061382	29.1 y	α, ϵ, SF
96	244	244.062746	18.10 y	α, SF
96	245	245.065486	8500 y	α, SF
96	246	246.067218	4730 y	α, SF
96	247	247.070347	1.56×10^7 y	α
96	248	248.072342	3.40×0^5 y	α, SF
96	249	249.075947	64.15 m	β^-
96	250	250.078351	9700 y	SF, α, β^-
96	251	251.082278	16.8 m	β^-
96	252	252.084870s	2 d	β^-

Berkelium (Bk)

Z	A	Atomic Mass (u)	$t_{1/2}$ or Abundance	Decay Mode
97	240	240.059750s	4.8 m	ϵ, ϵSF
97	242	242.062050s	7.0 m	ϵ
97	243	243.063002	4.5 h	ϵ, α
97	244	244.065168	4.35 h	ϵ, α
97	245	245.066355	4.94 d	ϵ, α
97	246	246.068670	1.80 d	ϵ, α
97	247	247.070299	1380 y	α
97	248	248.073080s	23.7 h	β^-, ϵ, α
97	248	248.073080s	9 y	α
97	249	249.074980	320 d	β^-, α, SF
97	250	250.078311	3.217 h	β^-
97	251	251.080753	55.6 m	β^-, α

Californium (Cf)

Z	A	Atomic Mass (u)	$t_{1/2}$ or Abundance	Decay Mode
98	240	240.062300s	1.06 m	α
98	241	241.063720s	3.78 m	ϵ, α
98	242	242.063690	3.49 m	α
98	243	243.065420s	10.7 m	ϵ, α
98	244	244.065990	19.4 m	α
98	245	245.068040s	45.0 m	ϵ, α
98	246	246.068799	35.7 h	α, ϵ, SF
98	247	247.070992	3.11 h	ϵ, α
98	248	248.072178	333.5 d	α, SF
98	249	249.074847	351 y	α, SF
98	250	250.076400	13.08 y	α, SF
98	251	251.079580	898 y	α
98	252	252.081620	2.645 y	α, SF
98	253	253.085127	17.81 d	β^-, α
98	254	254.087316	60.5 d	SF, α
98	255	255.091040s	85 m	β^-
98	256	256.093440s	12.3 m	SF, β^-, α

Einsteinium (Es)

Z	A	Atomic Mass (u)	$t_{1/2}$ or Abundance	Decay Mode
99	243	243.069630s	21 s	ϵ, α
99	244	244.070970s	37 s	ϵ, α
99	245	245.071320s	1.1 m	ϵ, α
99	246	246.072970s	7.7 m	ϵ, α
99	247	247.073650s	4.55 m	ϵ, α
99	248	248.075460s	27 m	ϵ, α
99	249	249.076410s	102.2 m	ϵ, α
99	250	250.078650s	8.6 h	ϵ, α
99	251	251.079984	33 h	ϵ, α
99	252	252.082970	471.7 d	α, ϵ, β^-
99	253	253.084818	20.47 d	α, SF
99	254	254.088016	275.7 d	α, ϵ, SF, β^-

Einsteinium (Es) *(continued)*

Z	A	Atomic Mass (u)	$t_{1/2}$ or Abundance	Decay Mode
99	255	255.090266	39.8 d	β^-, α, SF
99	256	256.093590*s*	25.4 m	β^-

Fermium (Fm)

Z	A	Atomic Mass (u)	$t_{1/2}$ or Abundance	Decay Mode
100	242	242.073430*s*	0.8 ms	SF
100	243	243.074510*s*	0.18 s	α
100	244	244.074080*s*	3.3 ms	SF
100	245	245.075380*s*	4.2 s	α, SF
100	246	246.075280	1.1 s	α, SF, ϵ
100	248	248.077184	36 s	α, ϵ, SF
100	249	249.079020*s*	2.6 m	ϵ, α
100	250	250.079515	30 m	α, ϵ, SF
100	251	251.081566	5.30 h	ϵ, α
100	252	252.082460	25.39 h	α, SF
100	253	253.085176	3.00 d	ϵ, α
100	254	254.086848	3.240 h	α, SF
100	255	255.089955	20.07 h	α, SF
100	156	256.091767	157.6 m	SF, α
100	257	257.095099	100.5 d	α, SF
100	258	258.097070*s*	370 μs	SF
100	259	259.100590*s*	1.5 s	SF

Mendelevium (Md)

Z	A	Atomic Mass (u)	$t_{1/2}$ or Abundance	Decay Mode
101	247	247.081800*s*	2.9 s	α
101	248	248.082910*s*	7 s	ϵ, α, SF
101	249	249.083000*s*	24 s	α, ϵ
101	250	250.084490*s*	52 s	ϵ, α
101	251	251.084920*s*	4.0 m	ϵ, α
101	252	252.086630*s*	4.8 m	ϵ
101	253	253.087280*s*	6 m	ϵ
101	254	254.089730*s*	10 m	ϵ
101	254	254.089730*s*	28 m	ϵ
101	255	255.091075	27 m	ϵ, α, SF
101	256	256.094050	78.1 m	ϵ, α, SF
101	257	257.095535	5.52 h	ϵ, α, SF
101	258	258.098425	60 m	ϵ
101	258	258.098425	51.5 d	α, SF
101	259	259.100500*s*	1.60 h	SF, α
101	260	260.103650*s*	27.8 d	SF, α, ϵ, β^-

Nobelium (No)

Z	A	Atomic Mass (u)	$t_{1/2}$ or Abundance	Decay Mode
102	250	250.087490*s*	0.25 ms	SF, α
102	251	251.088960*s*	0.8 s	α, ϵ, SF
102	252	252.088966	2.30 s	α, SF
102	253	253.090650*s*	1.7 m	α, ϵ
102	254	254.090949	55 s	α, ϵ, SF

Nobelium (No)

Z	A	Atomic Mass (u)	$t_{1/2}$ or Abundance	Decay Mode
102	255	255.093232	3.1 m	α, ϵ
102	256	256.094276	2.91 s	α, SF
102	257	257.096850	25 s	α
102	258	258.098200*s*	1.2 ms	SF, α
102	259	259.101020*s*	58 m	α, ϵ, SF
102	260	260.102640*s*	106 ms	SF

Lawrencium (Lr)

Z	A	Atomic Mass (u)	$t_{1/2}$ or Abundance	Decay Mode
103	252	252.095330*s*	1 s	α, ϵ, SF
103	253	253.095260*s*	1.3 s	α, SF, ϵ
103	254	254.096590*s*	13 s	α, ϵ, SF
103	255	255.096770*s*	22 s	α, ϵ
103	256	256.098760*s*	28 s	α, ϵ, SF
103	257	257.099610*s*	0.646 s	α, SF
103	258	258.101880*s*	3.9 s	α, ϵ, SF
103	259	259.102990*s*	6.1 s	α, SF, ϵ
103	260	260.105570*s*	180 s	α, ϵ, SF
103	261	261.106940*s*	39 m	SF
103	262	262.109690*s*	3.6 h	ϵ, SF

Rutherfordium (Rf)

Z	A	Atomic Mass (u)	$t_{1/2}$ or Abundance	Decay Mode
104	253	253.100680*s*	1.8 s	α, SF
104	254	254.100170*s*	0.5 ms	SF, α
104	2555	255.101490*s*	1.5 s	SF, α
104	256	256.101180	6.7 ms	Sf, α
104	257	257.103070*s*	4.7 s	α, ϵ, SF
104	258	258.103570*s*	12 ms	SF, α
104	259	259.105630*s*	3.1 s	α, SF, ϵ
104	260	260.106430*s*	20.1 ms	SF, α
104	261	261.108750*s*	65 s	α, ϵ, SF
104	262	262.109920*s*	1.2 s	SF

Dubnium (Db)

Z	A	Atomic Mass (u)	$t_{1/2}$ or Abundance	Decay Mode
105	255	255.107400*s*	1.6 s	α, SF
105	256	256.108110*s*	2.6 s	α, SF, ϵ
105	257	257.107860*s*	1.3 s	α, SF, ϵ
105	258	258.109440*s*	4.4 s	α, ϵ, SF
105	258	258.109440*s*	20 s	ϵ
105	260	260.111430*s*	1.52 s	α, SF, ϵ
105	261	261.112110*s*	1.8 s	α, SF
105	262	262.114150*s*	34 s	α, SF, ϵ
105	263	263.115080*s*	27 s	SF, α

Seaborgium (Sg)

Z	A	Atomic Mass (u)	$t_{1/2}$ or Abundance	Decay Mode
106	259	259.114650*s*	0.9 s	α, SF
106	260	260.114440	3.6 ms	α, SF
106	261	261.116200*s*	0.23 s	α, SF

Z	A	Atomic Mass (u)	$t_{1/2}$ or Abundance	Decay Mode
Seaborgium (Sg) *(continued)*				
106	263	263.118310*s*	0.8 s	SF, α
106	265	265.121070*s*	16 s	α, SF
106	266	266.121930*s*	20 s	α, SF
Bohrium (Bh)				
107	261	261.121800*s*	11.8 ms	α, SF
107	262	262.123010*s*	102 ms	α, SF
Hassium (Hs)				
108	264	264.128410	0.08 ms	α
108	265	265.130000*s*	1.8 ms	α
108	267	267.131770*s*	60 ms	α
Meitnerium (Mt)				
109	266	266.137940*s*	3.4 ms	α, SF
109	268	268.138820*s*	70 ms	α
(Uun)				
110	269	269.145140*s*	0.17 ms	α
110	271	271.146080*s*	1.1 ms	α
110	272	272.146310*s*	8.6 ms	SF
(Uuu)				
111	272	272.153480*s*	1.5 ms	α

APPENDIX 9

Nobel Laureates in Physics

The following list gives the names and short descriptions of award citations* for all the physics laureates and a few chemistry laureates whose work was related to physics (denoted by C in front of their name).

Year	Nobel Laureate		Citation For
1901	Wilhelm Konrad Röntgen	1845–1923	Discovery of x rays
1902	Hendrik Antoon Lorentz	1853–1928	Their researches into the influence of magnetism upon radiation phenomena
	Pieter Zeeman	1865–1943	
1903	Antoine Henri Becquerel	1852–1908	His discovery of spontaneous radioactivity
	Pierre Curie	1859–1906	Their joint researches on the radiation phenomena discovered by Prof. Henri Becquerel
	Marie Sklowdowska-Curie	1867–1934	
1904	John William Strutt (Lord Rayleigh)	1842–1919	Investigations of the densities of the most important gases and his discovery of argon
	C Sir William Ramsay	1851–1939	His discovery of the inert gaseous elements in air and his determination of their place in the periodic system
1905	Philipp Eduard Anton von Lenard	1862–1947	His work on cathode rays
1906	Joseph John Thomson	1856–1940	His theoretical and experimental investigations on the conduction of electricity by gases
1907	Albert Abraham Michelson	1852–1931	His optical precision instruments and the spectroscopic and metrological investigations carried out with their aid
1908	Gabriel Lippman	1845–1921	His method of reproducing colors photographically based on the phenomena of interference
	C Ernest Rutherford	1871–1937	His investigations into the disintegration of the elements and the chemistry of radioactive substances
1909	Guglielmo Marconi	1874–1937	Their contributions to the development of wireless telegraphy
	Carl Ferdinand Braun	1850–1918	
1910	Johannes Diderik van der Waals	1837–1923	His work on the state of equations of gases and liquids
1911	Wilhelm Wien	1864–1928	His discoveries regarding the laws governing the radiation of heat
	C Marie Curie	1867–1934	Her services to the advancement of chemistry by the discovery of the elements radium and polonium, and by the isolation of radium and the study of its nature and compounds
1912	Nils Gustaf Dalén	1869–1937	His invention of automatic regulators for use in conjunction with gas accumulators for illuminating lighthouses and buoys

*From the Electronic Nobel Museum Project of the Nobel Foundation Website: www.nobel.se/prize/index.html

Year	Nobel Laureate		Citation For
1913	Heike Kamerlingh Onnes	1853–1926	His investigations of the properties of matter at low temperatures, which led, *inter alia,* to the production of liquid helium
1914	Max von Laue	1879–1960	His discovery of the diffraction of x rays by crystals
1915	William Henry Bragg William Lawrence Bragg	1862–1942 1890–1971	Their analysis of crystal structure by means of x rays
1917	Charles Glover Barkla	1877–1944	His discovery of the characteristic x rays of the elements
1918	Max Planck	1858–1947	His discovery of energy quanta
1919	Johannes Stark	1874–1957	His discovery of the Doppler effect in canal rays and of the splitting of spectral lines in electric fields
1920	Charles-Édouard Guillaume	1861–1938	The service he has rendered to precise measurement in physics by his discovery of anomalies in nickel steel alloys
1921	Albert Einstein	1879–1955	His services to Theoretical Physics, and especially for his discovery of the law of the photoelectric effect
	C Frederick Soddy	1877–1956	His contributions to our knowledge of the chemistry of radioactive substances, and his investigations into the origin and nature of isotopes
1922	Neils Bohr	1885–1962	His investigation of the structure of atoms and the radiation emanating from them
	C Francis W. Aston	1877–1945	His discovery, by means of his mass spectrograph, of isotopes in a large number of nonradioactive elements, and for his enunciation of the whole-number rule
1923	Robert Andrews Millikan	1868–1953	His work on the elementary charge of electricity and on the photoelectric effect
1924	Karl Manne Georg Siegbahn	1886–1978	His discoveries and researches in the field of x-ray spectroscopy
1925	James Franck Gustav Hertz	1882–1964 1887–1975	Their discovery of the laws governing the impact of an electron upon an atom
1926	Jean-Baptiste Perrin	1870–1942	His work on the discontinuous structure of matter, and especially for his discovery of sedimentation equilibrium
1927	Arthur Holly Compton	1892–1962	His discovery of the effect named after him
	Charles Thomson Rees Wilson	1869–1959	His method of making the paths of electrically charged particles visible by condensation of vapor
1928	Owen Willans Richardson	1879–1959	His work on the thermionic phenomenon, and especially for the discovery of the law named after him
1929	Prince Louis-Victor de Broglie	1892–1987	His discovery of the wave nature of electrons
1930	Sir Chandrasekhara Venkata Raman	1888–1970	His work on the scattering of light and for the discovery of the effect named after him
1932	Werner Heisenberg	1901–1976	The creation of quantum mechanics, the application of which has, *inter alia,* led to the discovery of the allotropic forms of hydrogen
1933	Erwin Schrödinger Paul Adrien Maurice Dirac	1887–1961 1902–1984	Their discovery of new productive forms of atomic theory
1934	C Harold C. Urey	1893–1981	His discovery of heavy hydrogen
1935	James Chadwick	1891–1974	His discovery of the neutron

Year	Nobel Laureate		Citation For
	C Frédéric Joliot	1900–1958	In recognition of their synthesis of new radioactive elements
	C Irène Joliot-Curie	1897–1956	
1936	Victor Franz Hess	1883–1964	His discovery of cosmic radiation
	Carl David Anderson	1905–1991	His discovery of the positron
	C Peter Debye	1884–1966	His contributions to our knowledge of molecular structure through his investigations on dipole moments and on the diffraction of x rays and electrons in gases
1937	Clinton Joseph Davisson	1881–1958	Their experimental discovery of the diffraction of electrons by crystals
	George Paget Thomson	1892–1975	
1938	Enrico Fermi	1901–1954	His demonstrations of the existence of new radioactive elements produced by neutron irradiation, and for his related discovery of nuclear reactions brought about by slow neutrons
1939	Ernest Orlando Lawrence	1901–1958	The invention and development of the cyclotron and for results obtained with it, especially with regard to artificial radioactive elements
1943	Otto Stern	1888–1969	His contributions to the development of the molecular ray method and his discovery of the magnetic moment of the proton
1944	Isidor Isaac Rabi	1898–1988	His resonance method for recording the magnetic properties of atomic nuclei
	C Otto Hahn	1879–1968	His discovery of the fission of heavy nuclei
1945	Wolfgang Pauli	1900–1958	His discovery of the Exclusion Principle, also called the Pauli Principle
1946	Percy Williams Bridgman	1882–1961	The invention of an apparatus to produce extremely high pressures and for the discoveries he made in the field of high-pressure physics.
1947	Sir Edward Victor Appleton	1892–1965	His investigations of the physics of the upper atmosphere, especially for the discovery of the Appleton layer
1948	Patrick Maynard Stuart Blackett	1897–1974	His development of the Wilson cloud chamber method and his discoveries therewith in nuclear physics and cosmic radiation
1949	Hideki Yukawa	1907–1981	His prediction of the existence of mesons on the basis of theoretical work on nuclear forces
1950	Cecil Frank Powell	1903–1969	His development of the photographic method of studying nuclear processes and his discoveries regarding mesons made with this method
1951	Sir John Douglas Cockcroft	1897–1967	Their pioneer work on the transmutation of atomic nuclei by artificially accelerated atomic particles
	Ernest Thomas Sinton Walton	1903–1995	
	C Edwin M. McMillan	1907–1991	Their discoveries in the chemistry of the transuranium elements
	C Glenn T. Seaborg	1912–1999	
1952	Felix Bloch	1905–1983	The development of new methods for nuclear magnetic precision measurements and discoveries in connection therewith
	Edward Mills Purcell	1912–1997	
1953	Frits Zernike	1888–1966	His demonstration of the phase contrast method, especially for his invention of the phase contrast microscope

Year	Nobel Laureate		Citation For
1954	Max Born	1882–1970	His fundamental research in quantum mechanics, especially his statistical interpretation of the wave function
	Walter Bothe	1891–1957	The coincidence method and his discoveries made therewith
1955	Willis Eugene Lamb, Jr.	1913–	His discoveries concerning the fine structure of the hydrogen spectrum
	Polykarp Kusch	1911–1993	His precision determination of the magnetic moment of the electron
1956	William Shockley	1910–1989	Their investigations on semiconductors and their discovery of the transistor effect
	John Bardeen	1908–1991	
	Walter Houser Brattain	1902–1987	
1957	Chen Ning Yang	1922–	Their penetrating investigation of the parity laws, which led to important discoveries regarding elementary particles
	Tsung Dao Lee	1926–	
1958	Pavel Alekseyevich Cherenkov	1904–1990	Their discovery and interpretation of the Cherenkov effect
	Ilya Mikhaylovich Frank	1908–1990	
	Igor Yevgenyevich Tamm	1895–1971	
1959	Emilio Gino Segrè	1905–1989	Their discovery of the antiproton
	Owen Chamberlain	1920–	
1960	Donald Arthur Glaser	1926–	The invention of the bubble chamber
	C Willard F. Libby	1908–1980	His method to use ^{14}C for age determination in several branches of science
1961	Robert Hofstadter	1915–1990	His pioneering studies of electron scattering in atomic nuclei and for his discoveries concerning the structure of the nucleons achieved thereby
	Rudolf Ludwig Mössbauer	1929–	His researches concerning the resonance absorption of γ rays and his discovery in this connection of the effect that bears his name
1962	Lev Davidovich Landau	1908–1968	His pioneering theories of condensed matter, especially liquid helium
1963	Eugene Paul Wigner	1902–1995	His contributions to the theory of the atomic nucleus and the elementary particles, particularly through the discovery and application of fundamental symmetry principles
	Maria Goeppert Mayer	1906–1972	Their discoveries concerning nuclear shell structure
	J. Hans D. Jensen	1907–1973	
1964	Charles H. Townes	1915–	Fundamental work in the field of quantum electronics, which has led to the construction of oscillators and amplifiers based on the maser-laser principle
	Nikolai G. Basov	1922–	
	Alexander M. Prokhorov	1916–	
1965	Shin'ichiro Tomonaga	1906–1979	Their fundamental work in quantum electrodynamics, with profound consequences for the physics of elementary particles
	Julian Schwinger	1918–1994	
	Richard P. Feynman	1918–1988	
1966	Alfred Kastler	1902–1984	The discovery and development of optical methods for studying Hertzian resonance in atoms
1967	Hans Albrecht Bethe	1906–	His contributions to the theory of nuclear reactions, especially his discoveries concerning the energy production in stars
1968	Luis W. Alvarez	1911–1988	His decisive contribution to elementary particle physics, in particular the discovery of a large number of resonance states, made possible through his development of the technique of using the hydrogen bubble chamber and data analysis

Year	Nobel Laureate		Citation For
1969	Murray Gell-Mann	1929–	His contributions and discoveries concerning the classification of elementary particles and their interactions
1970	Hannes Alfvén	1908–1995	Fundamental work and discoveries in magnetohydrodynamics with fruitful applications in different parts of plasma physics
	Louis-Eugène-Félix Néel	1904–	Fundamental work and discoveries concerning antiferromagnetism and ferrimagnetism, which have led to important applications in solid-state physics
1971	Dennis Gabor	1900–1979	His invention and development of the holographic method
1972	John Bardeen	1908–1991	Their theory of superconductivity, usually called the BCS theory
	Leon N. Cooper	1930–	
	J. Robert Schrieffer	1931–	
1973	Leo Esaki	1925–	His discovery of tunneling in semiconductors
	Ivar Giaever	1929–	His discovery of tunneling in superconductors
	Brian D. Josephson	1940–	His theoretical predictions of the properties of a supercurrent through a tunnel barrier
1974	Antony Hewish	1924–	The discovery of pulsars
	Sir Martin Ryle	1918–1984	His observations and inventions in radio astronomy
1975	Aage Bohr	1922–	The discovery of the connection between collective motion and particle motion in atomic nuclei and for the theory of the structure of the atomic nucleus based on this connection
	Ben R. Mottelson	1926–	
	L. James Rainwater	1917–1986	
1976	Burton Richter	1931–	Their pioneering work in the discovery of a heavy elementary particle of a new kind
	Samuel Chao Chung Ting	1936–	
1977	Philip Warren Anderson	1923–	Their fundamental theoretical investigations of the electronic structure of magnetic and disordered systems
	Nevill Francis Mott	1905–1996	
	John Hasbrouck Van Vleck	1899–1980	
1978	Pyotr L. Kapitza	1894–1984	His basic inventions and discoveries in the area of low-temperature physics
	Arno A. Penzias	1933–	Their discovery of cosmic microwave background radiation
	Robert Woodrow Wilson	1936–	
1979	Sheldon Lee Glashow	1932–	Their contributions to the theory of the unified weak and electromagnetic interaction between elementary particles, including, *inter alia,* the prediction of the weak neutral current
	Abdus Salam	1926–1996	
	Steven Weinberg	1933–	
1980	James W. Cronin	1931–	The discovery of violations of fundamental symmetry principles in the decay of neutral K-mesons
	Val L. Fitch	1923–	
1981	Nicolaas Bloembergen	1920–	Their contributions to the development of laser spectroscopy
	Arthur L. Schawlow	1921–1999	
	Kai M. Siegbahn	1918–	His contribution to the development of high-resolution electron spectroscopy
1982	Kenneth G. Wilson	1936–	His theory for critical phenomena in connection with phase transitions
1983	Subrahmanyan Chandrasekhar	1910–1995	His theoretical studies of the physical processes of importance to the structure and evolution of the stars
	William A. Fowler	1911–1995	His theoretical and experimental studies of the nuclear reactions of importance in the formation of the chemical elements in the universe

Year	Nobel Laureate		Citation For
1984	Carlo Rubbia Simon van der Meer	1934– 1925–	Their decisive contributions to the large project, which led to the discovery of the field particles W and Z, communicators of the weak interaction
1985	Klaus von Klitzing	1943–	The discovery of the quantized Hall effect
1986	Ernst Ruska	1906–1988	His fundamental work in electron optics and for the design of the first electron microscope
	Gerd Binnig Heinrich Rohrer	1947– 1933–	Their design of the scanning tunneling microscope
1987	J. Georg Bednorz Karl Alex Müller	1950– 1927–	Their important breakthrough in the discovery of superconductivity in ceramic materials
1988	Leon M. Lederman Melvin Schwartz Jack Steinberger	1922– 1932– 1921–	The neutrino beam method and the demonstration of the doublet structure of the leptons through the discovery of the muon neutrino
1989	Hans G. Dehmelt Wolfgang Paul	1922– 1913–1993	Their development of the ion-trap technique
	Norman F. Ramsey	1915–	The invention of the separated oscillatory fields method and its use in the hydrogen maser and other atomic clocks
1990	Jerome I. Friedman Henry W. Kendall Richard E. Taylor	1930– 1926– 1929–	Their pioneering investigations concerning deep inelastic scattering of electrons on protons and bound neutrons, which have been of essential importance for the development of the quark model in particle physics
1991	Pierre-Gilles de Gennes	1932–	His discovering that methods developed for studying order phenomena in simple systems can be generalized to more complex forms of matter, in particular to liquid crystals and polymers
	C Richard R. Ernst	1933–	His contributions to the development of the methodology of high-resolution nuclear magnetic resonance (NMR) spectroscopy
1992	Georges Charpak	1924–	His invention and development of particle detectors, particularly multi-wire proportional counters
1993	Russell A. Hulse Joseph H. Taylor Jr.	1950– 1941–	The discovery of a new type of pulsar
1994	Bertram N. Brockhouse	1918–	The development of neutron scattering
	Clifford G. Schull	1915–	The development of the neutron diffraction technique
1995	Martin L. Perl	1927–	The discovery of the tau lepton
	Frederick Reines	1918–1998	The detection of the neutrino
1996	David M. Lee Douglas D. Osheroff Robert C. Richardson	1931– 1945– 1937–	Their discovery of superfluidity in He-3
1997	Steven Chu Claude Cohen-Tannoudji William D. Phillips	1948– 1933– 1948–	Development of methods to cool and trap atoms with laser light
1998	Robert B. Laughlin Horst L. Störmer Daniel C. Tsui	1950– 1949– 1939–	Their discovery of a new form of quantum fluid with fractionally charged excitations

Answers to Odd-Numbered Problems

Chapter 2

3. $\theta = \sin^{-1}(v/c)$
13. K′ travels at a speed $c/2$ in the $-x$ direction
15. (a) 3.86×10^{-8} s (b) $x' = -10.4$ m, $y' = 5$ m, $z' = 10$m, $t' = 51.0$ ns
17. (a) $1 + 4.3 \times 10^{-15}$ (b) $1 + 4.7 \times 10^{-13}$ (c) $1 + 3.1 \times 10^{-12}$ (d) $1 + 3.1 \times 10^{-10}$ (e) 1.10 (f) 3.20
19. $0.14c$
21. $v = 1.4 \times 10^{-4}c$
23. $v = 4c/5$
25. $\Delta t = 16.7$ ns
27. $\Delta t' = 1.0 \times 10^{-5}$ s
29. 22.1 m
31. (a) $0.96c$ (b) $0.23c$
33. Each sees the other traveling at 62.1 m/s
35. $0.60c$ and $0.88c$
37. Classical: 15 muons; relativistic: 2710 muons
39. Mary receives signals at a rate ν' for t_1' and a rate ν'' for t_2'. Frank receives signals at a rate ν' for t_1 and a rate ν' for t_2.
45. (b) β (d) The lines are parallel.
47. Two events simultaneous in K are not simultaneous in K′.
51. 82 Hz
53. Van 2 receives signals from vans 1 and 3 at a rate $\nu = \nu_0 \sqrt{\frac{1-\beta}{1+\beta}}$ and vans 1 and 3 receive signals from van 2 at a rate $\nu = \nu_0 \sqrt{\frac{1-\beta}{1+\beta}}$. Vans 1 and 3 receive signals from each other at a rate $\nu = \nu_0/(1 + \sqrt{2}\beta)$.
59. 1.42×10^{-25} kg
61. 0.155 T
65. 0.999 999 986 c
67. (a) $p = 5.11$ keV/c, $K = 30$ eV, $E = 511.03$ keV (b) $p = 51.4$ keV/c, $K = 2.6$ keV, $E = 513.6$ keV (c) $p = 1055$ keV/c, $K = 661$ keV, $E = 1172$ keV
69. 2.25×10^{17} m
71. (a) $0.417c$ (b) $\sqrt{3}c/2$ (c) $0.996c$
75. 2.55×10^{14} J
79. 28.3 MeV
81. $v = 0.999\,999\,561c$, $p = 1.000938$ TeV/c, $E = 1$ TeV + 938 MeV
83. $v = 0.938c$, $p = 287.05$ MeV/c, $E = 306$ MeV
85. 2330 MeV
87. time $= L/\gamma v$, number $= \nu L(1 - \beta)/v$
89. (a) 52 (b) 9 years; 156 (c) Frank 312, Mary 520 (d) Frank 10 years, Mary 6 years (e) each one agrees with (d)
91. (a) $(1 - 3.97 \times 10^{-8})c$ (b) $(1 - 8.85 \times 10^{-8})c$

Chapter 3

3. Non-relativistically $V = 1137$ V; relativistically $V = 1141$ V
9. Lyman 91.2 nm; Balmer 364.7 nm
11. 93.6 cm between the first and second lines; 202.7 cm between the second and third lines
13. (a) 0.69 mm (b) 9.89 μm (c) 1.16 μm
15. 19.9
17. 966 nm
21. 9.35 μm
27. (a) 2.06×10^{29} (b) 6.05×10^{18} (c) 2.34×10^{14}
29. 264 nm; 4.7 eV
31. 1.79×10^{15} Hz
33. $h = 4.40 \times 10^{-15}$ eV · s; $\phi = 4.1$ eV
35. 0.0413 nm
37. 0.0620 nm
39. $K = 5420$ eV
41. yes; 1.32 fm; 938 MeV
45. 2.00243 nm, a change of 0.122%
47. (a) 2.04 MeV (b) 1.02 MeV
49. $r = 3.68$ km; 2×10^{13} nuclear arsenals

Chapter 4

5. (a) 1.69×10^{-12} m (b) 1.48×10^{-14} m
7. 36.2
9. (a) 2170 (b) 1347
11. (a) Al: 6.04 MeV, Au: 23.7 MeV (b) Al: 3.82 MeV, Au: 13.7 MeV
13. (a) 0.013° (b) The results are comparable.
15. (a) $1.75 \times 10^{-4}\,c$ (b) -14.4 eV
17. H: -13.6 eV; He^+: -54.4 eV; Li^+: -122.5 eV
21. $hc = 1239.8$ eV · nm; $e^2/4\pi\epsilon_0 = 1.4400$ eV · nm; $mc^2 = 511.00$ keV; $a_0 = 5.2918 \times 10^{-2}$ nm; $E_0 = 13.606$ eV
23. $n = 2$ and $n = 5$

23. $n = 2$ and $n = 5$
25. (a) 13.6 eV (b) 54.4 eV (c) 218 eV
27. 2.44×10^6
29. (a) 2.84×10^{-13} m (b) 2535 eV (c) 0.49 nm, 1.96 nm, 4.40 nm
31. (a) $2a_0$ (b) 243 nm
33. The lines nearly match for all even n_u in the He^+ series. Each "matching" pair actually differs by the ratio of the reduced masses $R_{He+}/R_H = 1.0004$ or a difference of 0.04%.
35. $R = 4.38889 \times 10^7\ m^{-1}$
37. No, because the K_α lines for Pt and Au are separated by less than 10^{-12} m.
39. Helium: 122 nm for K_α and 103 nm for K_β; Lithium: 30.4 nm for K_α and 25.6 nm for K_β
41. 4.33×10^{-4} eV
43. 4.13×10^{-15} eV · s
45. Magnesium: 1.00 nm for K_α and 31.0 nm for L_α; Iron: 0.194 nm for K_α and 1.90 nm for L_α
47. (a) 0.148 (b) 100.5 and 1.31×10^4

Chapter 5

1. 31.2° and 50.9°
3. 8.16 keV; we can observe up through $n = 4$
5. 3.68×10^{-35} m; no
7. 1.73×10^{-10} m
9. (a) $hc/\sqrt{K^2 + 2Kmc^2}$ (b) $h/\sqrt{2mK}$
11. (a) 8.27 keV (b) 66.9 eV (c) 0.036 eV (d) 9.17×10^{-3} eV
13. $K = 3$ keV, $E = 514$ keV, $p = 55.4$ keV/c, $\lambda = 22.4$ pm
15. (a) 2.60×10^{-11} m (b) 4.42×10^{-13} m
17. $d = 0.063$ nm, $\lambda = 0.122$ nm, $p = 10.2$ keV/c, $E = 511.102$ keV, $K = 102$ eV
19. 0.457 nm, 0.412 nm, 0.301 nm
21. (a) 0.571 Hz (b) 3.0 cm
23. (a) $\Psi = 0.006 \sin(6.5x - 275t)\cos(0.5x + 25t)$ (b) $v_{ph} = 42.3$ m/s, $u_{gr} = -50$ m/s (c) 2π m (d) 2π
25. v_{ph} is independent of wavelength.
27. $\Psi(x, 0) = -(2A_0/x)\sin(\Delta k x/2)\sin(k_0 x)$; width $\Delta x = \pi/\Delta k$; $\Delta k \Delta x = \pi$
29. The intensity is higher by a factor of four for the double slit.
31. $K = 2.67$ eV
33. $$\frac{E_2}{E_1} = \left[\frac{1 + h^2c^2/L^2E_0^2}{1 + h^2c^2/4L^2E_0^2}\right]^{1/2}$$
$$\frac{E_3}{E_1} = \left[\frac{1 + 9h^2c^2/4L^2E_0^2}{1 + h^2c^2/4L^2E_0^2}\right]^{1/2}$$
$$\frac{E_4}{E_1} = \left[\frac{1 + 4h^2c^2/L^2E_0^2}{1 + h^2c^2/4L^2E_0^2}\right]^{1/2}$$
35. 51.1 MeV; 1.30 MeV
37. $\hbar/4\pi$
39. 2.5×10^5 rad/s
41. (a) 3.29×10^{-3} eV (b) 0.18 nm
43. (a) 1.51 eV (b) 2.24×10^{-6} eV; yes
45. 3.34×10^{-33} J
49. $\Delta t = 4.0 \times 10^{-24}$ s; $\Delta E = 82$ MeV; $m = 82$ MeV/c^2, within a factor of 2 of the actual mass
51. $\hbar = 0.081$ J · s

Chapter 6

1. The function is not localized; it does not vanish at $\pm\infty$.
5. $A = 2\alpha^{-3/2}$
11. $A = \sqrt{2/\pi}$; probability = 0.091
13. $1/L$ (independent of n), in agreement with the classical result
15. 0.196; 0.609; 0.196
17. 11.3 GeV
19. 10.5 eV, 18.1 eV, 22.6 eV, 7.5 eV, 12.0 eV, 4.52 eV
21. (a) λ is longer for the infinite well (b) larger $\lambda \rightarrow$ lower E (c) when $E > V_0$ the particle is unbounded
23. $Ce^{ikL} + De^{ikL} = Be^{-\alpha L}$ $\quad \dfrac{C}{D} = \left(\dfrac{ik - 1}{ik + 1}\right)\dfrac{e^{2ikL}}{\alpha}$
25. In general we have
$$\psi(x) = A \sin\left(\frac{n_1 \pi x}{L}\right)\sin\left(\frac{n_2 \pi y}{L}\right)\sin\left(\frac{n_3 \pi z}{L}\right)$$
For $\psi_2(x)$ we can have $(n_1, n_2, n_3) = (1, 1, 2)$ or $(1, 2, 1)$ or $(2, 1, 1)$
For $\psi_3(x)$ we can have $(n_1, n_2, n_3) = (1, 2, 2)$ or $(2, 2, 1)$ or $(2, 1, 2)$
For $\psi_4(x)$ we can have $(n_1, n_2, n_3) = (1, 1, 3)$ or $(1, 3, 1)$ or $(3, 1, 1)$
For $\psi_5(x)$ we can have $(n_1, n_2, n_3) = (2, 2, 2)$
27. One quantum number is associated with each dimension.
29. $\Delta E_n = \hbar\omega$ for all n
31. $E = (4.136 \times 10^{-2}\ \text{eV})(n + \frac{1}{2})$; $k = 131$ N/m
33. $\langle p \rangle = 0$; $\langle p^2 \rangle = \frac{1}{2}\hbar\omega m$
35. (a) $E = (n + \frac{1}{2})\hbar\omega = (n + \frac{1}{2})(0.755\ \text{eV})$ (b) 1640 nm, 822 nm, 549 nm
37. (a) $p = \sqrt{2m(E - V_0)}$; $\lambda = h/\sqrt{2m(E - V_0)}$; $K = E - V_0$ (b) $p = \sqrt{2m(E + V_0)}$; $\lambda = h/\sqrt{2m(E + V_0)}$; $K = E + V_0$
41. (a) $L = 0.229$ nm or any integer multiple thereof (b) $L = 0.114$ nm or any odd integer multiple thereof
43. Let $E_0 = \hbar^2\pi^2/2mL^2$; $E_1 = 13E_0/4$; $E_2 = 4E_0$; $E_3 = 21E_0/4$; $E_4 = 25E_0/4$; $E_5 = 7E_0$; E_5 is degenerate
49. (b) $\langle x \rangle = 0$; $\langle x^2 \rangle = 5/2\alpha$
51. $V(r) = D a^2 (r - r_e)^2$
53. Let $E_0 = \pi^2\hbar^2/2mL^2$

$E_1 = E_0(1^2 + 1^2) = 2E_0$ $\quad n_1 = 1, n_2 = 1$

$E_2 = E_0(2^2 + 1^2) = 5E_0$ $\quad n_1 = 2, n_2 = 1$ or vice versa

$E_3 = E_0(2^2 + 2^2) = 8E_0$ $\quad n_1 = 2, n_2 = 2$

$E_5 = E_0(3^2 + 2^2) = 13E_0 \qquad n_1 = 3,\ n_2 = 2$ or vice versa

$E_6 = E_0(4^2 + 1^2) = 17E_0 \qquad n_1 = 4,\ n_2 = 1$ or vice versa

Chapter 7

5. $E = -E_0/4$, as predicted by the Bohr model
9. $\ell = 5$: $m_\ell = 0, \pm 1, \pm 2, \pm 3, \pm 4, \pm 5$ $\quad \ell = 4$: $m_\ell = 0, \pm 1, \pm 2, \pm 3, \pm 4$ $\quad \ell = 3$: $m_\ell = 0, \pm 1, \pm 2, \pm 3$ $\quad \ell = 2$: $m_\ell = 0, \pm 1, \pm 2$ $\quad \ell = 1$: $m_\ell = 0, \pm 1$ $\quad \ell = 0$: $m_\ell = 0$
11. $\psi_{310} = R_{31} Y_{10}$

$$= \frac{1}{81}\sqrt{\frac{2}{\pi}}\, a_0^{-3/2}\left(6 - \frac{r}{a_0}\right)\left(\frac{r}{a_0}\right) e^{-r/3a_0} \cos\theta$$

$\psi_{31\pm1} = R_{31} Y_{1\pm1}$

$$= \frac{1}{81\sqrt{\pi}}\, a_0^{-3/2}\left(6 - \frac{r}{a_0}\right)\left(\frac{r}{a_0}\right) e^{-r/3a_0} \sin\theta\, e^{\pm i\pi}$$

13. 36
15. 365
17. $0, \pm\hbar$
21. 30°
23. seven different states; with $B = 0$, $E = -E_0/25 = -0.544$ eV; with field on the levels are changed by $\Delta E = 0$ for $m_\ell = 0$, $\pm 1.74 \times 10^{-4}$ eV for $m_\ell = \pm 1$, $\pm 3.47 \times 10^{-4}$ eV for $m_\ell = \pm 2$, $\pm 5.21 \times 10^{-4}$ eV for $m_\ell = \pm 3$
25. The magnet should be designed so that the product of its length squared and its vertical magnetic field gradient be 57 T · m.
27. $n = 5$, $\ell = 3$, $m_\ell = 0, \pm 1, \pm 2, \pm 3$, and $m_s = \pm 1/2$, for a total degeneracy of 14
31. $r = (3 \pm \sqrt{5})\, a_0$
33. $r = 2a_0$
35. 0.054
37. $2s$: 1.1×10^{-15} $\quad 2p$: 2.0×10^{-26}
39. $0.24c$
41. $\psi_{100} = \dfrac{1}{\sqrt{\pi}}\left(\dfrac{Z}{a_0}\right)^{3/2} e^{-Zr/a_0}$
43. $E_0 = 2.53$ keV

Chapter 8

1. The first two electrons are in the $1s$ subshell and have $\ell = 0$, with $m_s = \pm 1/2$. The third electron (on the $2s$ level) has $\ell = 0$, with either $m_s = 1/2$ or $-1/2$. With four particles there are six possible interactions: the nucleus with electrons 1, 2, 3, electron 1 with electron 2, electron 1 with electron 3, or electron 2 with electron 3. In each case it is possible to have a coulomb interaction and a magnetic moment interaction.
3. 2, 4, 5
5. K: $4s^1$, V: $4s^2 3d^3$, Se: $4s^2 3d^{10} 4p^4$, Zr: $5s^2 4d^2$, Sm: $6s^2 4f^6$, U: $7s^2 6d^1 5f^3$
7. $1.36e$
9. (a) B (b) Na (c) Ar
11. $4^2P_{1/2}$
13. In the $3d$ state $\ell = 2$ and $s = 1/2$, so $j = 5/2$ or $3/2$. As usual $m_\ell = 0, \pm 1, \pm 2$. The value of m_j ranges from $-j$ to j, so its possible values are $\pm 1/2$, $\pm 3/2$, and $\pm 5/2$. As always $m_s = \pm 1/2$. The two possible term notations are $3D_{5/2}$ and $3D_{3/2}$.
15. $\pm\hbar/2, \pm 3\hbar/2, \pm 5\hbar/2, \pm 7\hbar/2$
19. 7.33×10^{-3} eV; 63.4 T
23. He, Ca, and Sr have singlet ground states and triplet excited states; Al has only a doublet.
27. (a) 2.4×10^{-5} eV (b) 4.53×10^{-5} eV
31. (a) 2 (b) 4/3 (c) 6/5
35. 3.86×10^{-5} eV

Chapter 9

3. (a) $\nu = \nu_0$ (b) standard deviation $= \dfrac{\nu_0}{c}\sqrt{\dfrac{kT}{m}}$

   (c) 3.66×10^{-6} for H_2 at 293 K; 2.25×10^{-5} for H at 5500 K
5. (a) $\displaystyle\int_c^\infty F(v)\, dv = 4\pi C \int_c^\infty v^2 \exp\left(-\frac{1}{2}\beta m v^2\right) dv$
7. (a) $\overline{v} = 3240$ m/s, $v^* = 2870$ m/s (b) $\overline{v} = 7240$ m/s, $v^* = 6420$ m/s
9. (a) 395 m/s (b) 428 m/s
13. (a) $\overline{E} = \frac{3}{2}kT$ (b) $\frac{1}{2}m\overline{v^2} = \overline{E} = \frac{3}{2}kT$; $\frac{1}{2}m\overline{v}^2 = 1.27kT$
15. 7790 K
21. (a) 5.86×10^{28} m^{-3} (b) 5.28×10^4 K (c) 5.28×10^6 K
23. (a) 5.50 eV (b) 1.39×10^6 m/s
25. 8.45×10^{28} m^{-3}; one conduction electron per atom
27. (a) 1.18×10^6 m/s (b) 1.83×10^6 m/s
33. 590 MeV

Chapter 10

1. (a) 1.4×10^{-3} eV (b) 1.4×10^{-2} eV
3. 1.1×10^{-11} m
5. $E_{rot} = n^2\hbar^2/2I$, which is similar to the quantum-mechanical result in the limit of large n
7. 24
9. (a) 1.46×10^{-46} kg · m^2 (b) 1.13×10^{-10} m
11. (a) 34.1 eV (b) no
15. (a) 2.4×10^{-47} kg · m^2 (b) 478 N/m
17. (a) 3.44 μm, infrared
19. (a) 7.37×10^{-13} m (b) 1.06×10^{-19} m
21. In a three-level system the population of the upper level must exceed the population of the ground state. This is not necessary in a four-level system.
23. 0.315 nm
25. $\alpha = 4 - \dfrac{4}{\sqrt{2}} - 2 + \dfrac{8}{\sqrt{5}} - \dfrac{4}{\sqrt{8}} + \ldots$

29. (c) 7.4×10^{-6} K^{-1}
31. 4×10^{14} N/m^2
35. $\overline{\mu} \cong \mu$
39. $0.95T_c$; $0.71T_c$; $0.32T_c$
41. ^{201}Hg: $T_c = 4.171$ K; ^{204}Hg: $T_c = 4.140$ K
43. 130 K
45. 1.07×10^{-5} m
47. (a) 307 A (b) The 307 A is much higher, more than a factor of 100
49. 3.33 m/s$^2 \cong g/3$
51. 17.4 TeV
53. 2.97×10^{-2} J/K

Chapter 11

1. (a) 236 Ω (b) 611 Ω (c) 5.48×10^7 Ω
3. The voltmeter reads positive.
5. 1.26×10^{-6} V/K
7. 5×10^{-7} V
9. (a) spring 0.839, winter 0.559, summer 0.985
(b) spring 0.643, winter 0.292, summer 0.891
(c) spring 0.500, winter 0.122, summer 0.799
11. (a) 195 mA (b) 41.6 mA (c) 1.87 mA
13. (a) 0.619 nm (b) about 2.6 times higher
17. (a) 55.5 mV (b) −40.6 mV
19. 2.70 eV
21. Si at $T = 0°$C: $F_{FD} = 4.93 \times 10^{-21}$
Si at $T = 75°$C: $F_{FD} = 1.17 \times 10^{-16}$
Ge at $T = 0°$C: $F_{FD} = 4.27 \times 10^{-13}$
Ge at $T = 75°$C: $F_{FD} = 1.98 \times 10^{-10}$
23. (a) At $T = 77$ K: $\dfrac{I_f}{I_r} = -1.5 \times 10^{98}$
At $T = 273$ K: $\dfrac{I_f}{I_r} = -4.9 \times 10^{27}$
At $T = 340$ K: $\dfrac{I_f}{I_r} = -1.7 \times 10^{22}$
At $T = 500$ K: $\dfrac{I_f}{I_r} = -1.3 \times 10^{15}$

Chapter 12

1. 54.9 MeV
3. In each case the atomic number equals the number of protons (Z), and the atomic charge is Ze.

$^3_2$He: $Z = 2$, $N = 1$, $A = 3$, $m = 3.02$ u

$^4_2$He: $Z = 2$, $N = 2$, $A = 4$, $m = 4.00$ u

$^{17}_8$O: $Z = 8$, $N = 9$, $A = 17$, $m = 17.0$ u

$^{42}_{20}$Ca: $Z = 20$, $N = 22$, $A = 42$, $m = 42.0$ u

$^{210}_{82}$Pb: $Z = 82$, $N = 128$, $A = 210$, $m = 210.0$ u

$^{235}_{92}$U: $Z = 92$, $N = 143$, $A = 235$, $m = 235.0$ u

5. Isotopes ^{36}Ca through ^{51}Ca; Isobars ^{40}Cl, ^{40}Ar, ^{40}K, ^{40}Sc; Isotones ^{34}Si, ^{35}P, ^{36}S, ^{37}Cl, ^{38}Ar, ^{39}K, ^{41}Sc, ^{42}Ti
7. -1.52×10^{-3}
9. 7.25×10^{-9}; yes
11. The electrostatic force is 50 times weaker than the strong force. The gravitational force is almost 10^{38} times weaker than the strong force.
13. (b) 5.67 MeV; 4.14 MeV; 6.74 MeV
15. 7.27 MeV
17. The electron binding energy is less by a factor of 227.
21. 0.57 μg
25. 6.67×10^3 Bq
27. 120 kg
29. Yes
33. 0.09 MeV
35. β^-, β^+, and EC
39. 0.337 atm
43. 3.65×10^9 y for $R = 0.76$ and 9.10×10^9 y for $R = 3.1$; no
45. 6.4×10^7 y
47. 7 alpha, 4 beta
49. (a) 51.9 g of Fm and 41.1 g of Cf (b) 0 g of Fm and between 94 g and 95 g Cf (d) ^{240}Pu (e) 2.34×10^7 y
57. (a) the chain that begins with ^{238}U (c) four days

Chapter 13

1. (a) $^4_2$He (b) $^1_1$H (c) $^{16}_8$O (d) $^{73}_{32}$Ge (e) $^{108}_{48}$Cd (f) $^3_2$He
3. 0.087
5. 0.199
7. (a) ^{16}O (n, α) ^{13}C (b) ^{16}O (d, n) ^{17}F
(c) ^{16}O (γ, p) ^{15}N (d) ^{16}O (α, p) ^{19}F
(e) ^{16}O $(d, {}^3\text{He})$ ^{15}N (f) ^{16}O $({}^7\text{Li}, p)$ ^{22}Ne
All the products listed above are stable except the one in part (b).
9. (b) 15.9 MeV (c) 2.34 MeV
11. (a) 2.0 MeV (b) 2.0 MeV
15. 10.8 MeV
17. (a) 4.94×10^{17} (b) 2.06×10^9 Bq (c) Put 1.94 kg of ^{60}Co into the reactor for one week.
19. 10.40 MeV
21. 6.98 MeV; 6.51 MeV; yes
23. No
25. $d + {}^{21}\text{Ne} \rightarrow {}^{22}\text{Ne} + p$ $d + {}^{22}\text{Ne} \rightarrow {}^{22}\text{Ne} + d$
$d + {}^{23}\text{Na} \rightarrow {}^{22}\text{Ne} + {}^3\text{He}$
$\alpha + {}^{18}\text{O} \rightarrow {}^{22}\text{Ne} + \gamma$ $\alpha + {}^{19}\text{F} \rightarrow {}^{22}\text{Ne} + p$
$\alpha + {}^{21}\text{Ne} \rightarrow {}^{22}\text{Ne} + {}^3\text{He}$ $\alpha + {}^{22}\text{Ne} \rightarrow {}^{22}\text{Ne} + {}^4\text{He}$
$d + {}^{25}\text{Mg} \rightarrow {}^{22}\text{Ne} + {}^5\text{Li}$ $d + {}^{26}\text{Mg} \rightarrow {}^{22}\text{Ne} + {}^6\text{Li}$
$d + {}^{27}\text{Al} \rightarrow {}^{22}\text{Ne} + {}^7\text{Be}$ (and so on)
$\alpha + {}^{23}\text{Ne} \rightarrow {}^{22}\text{Ne} + {}^5\text{Li}$ $\alpha + {}^{24}\text{Mg} \rightarrow {}^{22}\text{Ne} + {}^6\text{Be}$
$\alpha + {}^{27}\text{Al} \rightarrow {}^{22}\text{Ne} + {}^9\text{B}$ (and so on)

27. (a) 400 kg (b) 1.01×10^{27} (c) 276 Bq (d) 2.38×10^7
29. $^{236}\text{U} \longrightarrow {}^{95}\text{Y} + {}^{138}\text{I} + 3n$ $^{236}\text{U} \longrightarrow {}^{94}\text{Y} + {}^{140}\text{I} + 2n$ $^{236}\text{U} \longrightarrow {}^{97}\text{Y} + {}^{136}\text{I} + 3n$
31. Uranium: 2.30×10^7 kWh; Coal: 8.50 kWh
33. 4.16 MeV, 7.55 MeV, 10.05 MeV, 4.97 MeV, total 26.7 MeV
35. (b) $K = \frac{3}{2}kT$ (c) 4.64×10^7 K
37. 1.69×10^{10} K
39. (a)3.88×10^{-2} eV (b) 2720 m/s (c) 1.45×10^{-10} m
41. (b) 1.04×10^{10} Bq (c) 3.65×10^{10} Bq
43. (a) 1.92×10^{-14} A (b) 1.92×10^{-15} A
45. 8.95×10^{11} Bq
47. 11.6 W
49. (a) -13.37 MeV (b) 33.979 u
51. (a) 8620 eV (b) 0.607; 0.135; 0.0111
53. 6790 Bq

Chapter 14

1. 2.27×10^{23} Hz
3. The range of values is from 1.3×10^{-23} s to 5.0×10^{-23} s
5. 3.57×10^{51}
7. In both (a) and (b) baryon number is not conserved.
9. 74.5 MeV
11. (a) ν_μ (b) K^+
15. B^0: $\bar{b}d$; $\overline{\text{B}^0}$: $b\bar{d}$
17. 0.087
19. 2.84×10^{-7}
21. $c(1 - 9.0 \times 10^{-9})$
23. 5630 MeV
27. (a) $2mc^2$ (b) $\sqrt{2Kmc^2}$
29. (a) 1.0107 (b) 1.107 (c) 2.066
31. (a) 1585 MeV (b) 336 MeV
33. (a) allowed
    (b) ν_e should be $\bar{\nu}_e$ to conserve electron lepton number
    (c) strangeness is not conserved (changes by 2 units)
    (d) strangeness is not conserved in a strong interaction
    (e) allowed
35. (a) allowed (strangeness changes by one unit)
    (b) charge is not conserved
    (c) baryon number is not conserved
    (d) allowed (strangeness changes by one unit)
37. (a) 114.5 GeV (b) For colliding beams the available energy is the sum of the two beam energies, or 14 TeV. This is an improvement over the fixed-target result by a factor of 122.

Chapter 15

5. 1726.5 s; 1.16×10^{-5}%
7. 1.17×10^{-3} nm
9. 8541 Hz; 2.45×10^{-13}%
11. 2.82 m
13. 2.63×10^{21} kg; 1.32×10^{-9} solar masses; $r_s = 3.91 \times 10^{-6}$ m
15. 9.19×10^{-3} Hz
17. 2.18×10^{-8} kg; 1.22×10^{28} eV
19. In both cases $t = 1.35 \times 10^{-43}$ s

Chapter 16

3. 1.57×10^{12} K; 10^{-1} s
5. 5.93×10^9 K; 1.23×10^{12} K
7. π^+; 5.80×10^{10} K
9. 1.58×10^5 K
11. 7.14×10^{17} kg/m^3, about three times as dense as the nucleon and nucleus
13. (a) 5.00×10^{13} N/m^2 (b) 3.21×10^{33} N/m^2
15. 7.52 Mly
17. (a) 0.18 nucleons/m^3 (b) 4.41×10^6 nucleons/m^3
21. (a) deceleration for $0 < n < 1$ (b) $H = n/t$
23. 5.67×10^{-27} kg/m^3; 1.35×10^{-26} kg/m^3
27. $t_0 \approx \tau/2$
31. 0.84 kg/m^3
33. Redshift = 3.79; $v = 0.92c$
37. $Q = 0.43$ MeV; maximum neutrino energy $\cong 0.43$ MeV

Index

A

Absorption spectrum,136–137, 149, 315–316, 357–358
Accelerators, 497–502, 504–505
Acceptor levels, 369
Actinides, 256
Addition of velocities, 39–42, 45–46, 76
Adiabatic demagnetization, 347
Alkali metals, 255, 269–270
Alkaline earths, 255, 269–270
Alpha decay, 213, 216, 220, 412–415, 429
Alvarez, Luis, 498–499
Ampere, André Marie, 5
Ampere's law, 5
Anderson, Carl, 476–477
Angular equation, 225–228
Angular momentum, 229–235, 246–248
  fuzzy and sharp, 232
  intrinsic spin angular momentum, 238–239
  quantization of,133–134, 147, 161, 229–233, 312, 357
  total, 257–270
Antiferromagnetism, 336
Antimatter, 112–113, 117, 495
Antiparticle, 476
Associated Laguerre equation, 226
Associated Legendre equation, 225
Atomic clock, 44–45, 512
Atomic force microscopes, 214–215
Atomic mass units, 67
Atomic masses, 399–400
Atomic radii, 254
Atomic shells, 250–251
Avogadro, Amedeo, 6, 14
Azimuthal equation, 224–225, 227

B

Bahcall, John, 534
Balmer, Johann, 88
Balmer series, 89–91, 115, 136, 149
Band gap, 363–369, 377–378, 390–391
Band spectrum, 315
Band theory of solids, 362–369, 390
Bardeen, John, 341, 381, 383
Barkla, Charles, 152
Barriers, 208–220
Baryons, 481–482, 485
Beauty, 489
Becker, H., 394
Becquerel, Alexandre-Edmond, 378
Becquerel, Henri, 18
Becquerel, unit of, 409
Bednorz, J. Georg, 345
Bending of light (in a gravitational field), 510–511
Bernoulli, Daniel, 6
Beta decay, 415–421, 429
Bethe, Hans, 534
Big Bang, 423, 523–532, 544–547, 555–556
Binding energy, 68–70, 78, 136, 311, 399–400, 404–408. 428–429
Biot-Savart law, 270
Black holes, 514–520
Blackbody radiation, 17–18, 91–96, 115–117, 296–298, 308
  Planck law for, 18, 95–96, 116, 296–298, 308, 319
  spectral distribution, 92–93, 116
  Stefan-Boltzmann law, 93, 96
  ultraviolet catastrophe, 94–95
  Wien law, 93, 96, 117
Bohr atom, 132–149, 226, 390–391
Bohr, Niels, 118, 124, 125, 132–138, 179–180, 274, 405, 431, 438, 450
Bohr magneton, 234
Bohr radius, 134–137, 149, 243–245
Bohr-Einstein discussions, 179
Bohr's Correspondence Principle, *see* Correspondence Principle.
Bohr's Principle of Complementarity, *see* Principle of Complementarity.
Boltzmann, Ludwig, 6, 8, 15, 272, 285
Bonding, *see* Molecular bonding.
Bose condensation, 304–305
Bose, Satyendra Nath, 298
Bose-Einstein statistics, 286–287, 296–305, 308
Bosons, 286–287, 296–305
  gauge,480
  Higgs,480
Bothe, W., 394
Bottomness, 489

Boyle, Robert, 6, 14
Boyle's law, 6
Brackett series, 90
Bradley, James, 26
Bragg planes, 153–154
Bragg, William Henry, 153
Bragg, William Lawrence, 153–154, 156, 158, 474
Bragg's law, 154, 163, 182
Brattain, William, 381, 383
Breeder reactors, 451–452
Bremsstrahlung, 104–105, 117
Bridge rectifiers, 376–377
Brillouin zones, 365
Bronowski, J., 249
Brown, Robert, 15
Brownian motion, 15
Bubble chamber, 488

C

Carbon (CNO) cycle, 453
Carnot, Sadi, 5
Ceramics, 348
Chadwick, James, 124–125, 158, 394–395
Chain reaction, 445
Chamberlain, O., 112
Charge, conservation of, 3
Charles's law, 6
Charm, 489–490
Chu, Paul, 345, 348
Classical atomic model, 131–132, 149
Classical electron radius, 248
Classical statistics, *see* Maxwell-Boltzmann statistics.
Clausius, Rudolf, 5
Clement-Quinnell equation, 367–368
Closed universe, 544, 548–549, 552
COBE, 545
Cockroft, John D., 158, 392, 432
Coherent radiation, 320
Cold dark matter, 546
Color, 491
Compact disc/read only memory device (CDROM), 388–389
Compton, Arthur, 106–108
Compton effect, 106–110, 116–117, 171, 394, 428
Compton wavelength, 108
Conduction band, 366, 369, 377–378, 391
Conductivity, *see* Electrical conductivity and Thermal conductivity.
Confinement, 492–493
Conjugate variables, 179, 182, 184
Conservation laws, 11–13, 484–487
  baryon, 485
  hypercharge, 487
  lepton, 485–486
Conservation laws *(continued)*
  strangeness, 486
Cooper, Leon, 341
Cooper pairs, 341–344, 360
Correspondence Principle, 137–138, 197
Cosmic microwave background radiation, 524, 527–528
Cosmic rays, 476
Cosmological constant, 546
Cosmology, 113
Coulomb, Charles, 5
Coulomb energy, 403–405, 413–414, 429
Coulomb scattering, *see* Rutherford scattering.
Covalent bond, 312
Cowan, Clyde, 416
Critical density, 544–547, 556
Critical field, superconducting, 338–340
Cronin, James, 487
Cross sections, 127, 434–436
Crystal structures, 326–329, 350–351, 358–359
  amorphous solids, 326
  lattice, 326–327
  polycrystalline materials, 326
  sodium chloride structure, 326–329
Curie, Irene, 394, 409
Curie law, 334–335
Curie, Marie, 1, 409
Curie, Pierre, 1
Curie temperature, 336
Curie, unit of, 409
Cyclotron, 432–433

D

Dalton, John, 6, 14
Davis, Raymond Jr., 419, 534
Davisson, Clinton, 161–164, 183
de Broglie, Louis, 151, 159–160
de Broglie wavelength, 160–165, 182–183
de Broglie waves, *see* Particle waves.
Deaver, B. S. Jr., 342, 345
Debye-Scherrer pattern, 155, 172
Deceleration parameter, 548–549, 552
Degenerate states, 202, 220, 227, 229, 240, 247
Degrees of freedom, 7, 277–280
Density of states, 285–287, 293–294, 297
Deuteron, 393, 399–400
  binding energy, 399–400
  magnetic moment, 400
  mass, 399
  spin, 400
Dewar flask, 346
Dewar, James, 346
Diamagnetism, 333–334
  perfect, 338
Dicke, Robert, 527

Diffraction, *see* Light, diffraction of.
Diodes, 375–376, 391
Dirac, Paul Adrien Maurice, 288, 475–476
Discovery of the neutron, *see* Neutron, discovery of.
Disintegration energy, 412, 421
Dissociation energy, 70, 328–329
Distinguishable particles, 285–287
Donor levels, 369
Doppler effect, 51–56, 77, 306–307, 358
  applications of, 54–55
  transverse, 54–56
Doublet states, 239, 259–260
Drude, Paul, 289
Drude (classical) theory of electrical conduction, 289–290
Duane-Hunt rule, 106

E

Ehrenfest, Paul, 272
Eightfold way, 487
Einstein, Albert, 13, 16, 21–22, 27–28, 63, 80, 85, 100–102, 141, 274–275, 298, 506, 508, 510
Einstein's postulates (of relativity), 27–30
Electrical conductivity, 256, 269, 289–293, 361–363, 390
Electromagnetic waves, 10–11, 16–17
Electromagnetism, 5, 12–13, 16–17, 132, 149
  and relativity, 70–73, 78
Electron
  charge on, 84–87, 115
  charge-to-mass ratio, 82–83
  discovery of, 18, 82–84
Electron double-slit experiment, 172–174, 180–181, 183–184
Electron–phonon interaction, 341
Electron scattering, 161–165, 183
Electron volts, 66–67
Electroweak interaction, 11–13
Elementary particles, 91, 474–505
Elsewhere, 49–50
Emission spectrum, 136
Emissivity, 93
Endoergic (endothermic) reactions, 436
Energy, conservation of, 3, 11, 179
ENIAC computer, 386–387
Eötvös, R. V., 508
Equipartition theorem, 7, 277–280, 307
Equivalence of mass and energy, 63–65
Equivalence principle, *see* Principle of equivalence.
Ether, *see* Luminiferous ether.
Ether drag, 26–27, 42
Event horizon, 515–517
Exclusion principle, *see* Pauli exclusion principle.
Exoergic (exothermic) reactions, 436
Expectation values, 191–194, 197–198, 207–208, 218–219, 244, 248
Experimental verification (of special relativity), 42–46

F

Fairbank, W. M., 342
Families of matter, 943
Faraday, Michael, 5
Faraday's law, 5
Fermi energy, 288–289, 292–295, 308, 359
Fermi, Enrico, 1, 288, 397, 450
Fermi speed, 293, 295, 308
Fermi temperature, 293, 295, 308
Fermi, Unit of (femtometer), 449–450
Fermi-Dirac statistics, 286–296, 308, 362, 366–368, 390
Fermions, 250, 286, 306, 480
Ferrimagnetism, 336
Ferromagnetism, 335–336, 357
Feynman diagram, 475–476, 491
Feynman, Richard, 425, 523
Field emission (of electrons), 97
Fine structure, 140–141, 233
Fine structure constant,136
Finite square-well potential, 198–200, 212–213, 219
Fission, 75, 442–452
  fragments, 442–443
  induced, 443
  reactors, 446–452
  spontaneous, 442
  thermal neutron, 444
Fitch, Val, 487
Fitzgerald, George F., 17, 27, 124
Fizeau, H. L., 42, 76
Flat universe, 544, 549, 552
Flatness problem, 544
Flavor, 491
Flux jumps, 355
Forbidden transitions, 263
Fourier, Joseph, 273
Fourier series, 168
Fourier Transform Infrared Spectroscopy (FTIR), 316–317
Four-level system, 321–322
Frame dragging, 519–520
Franck, James, 144–146
Franck-Hertz experiment, 144–146, 150
Free electron number density, 289, 292, 372
Fresnel, Augustin, 9, 42
Freud, Sigmund, 2
Friedman, Jerome, 491
Friedrich, Walter, 152
Frisch, O. R., 431
Fundamental interactions, 477–479, 504
Fusion, 75, 323, 452–458, 472
Future of universe, 552–554

G

Galaxies, 538–539
Galilean invariance, 20–21
Galilean transformation, 21
Galileo, 3, 508
Gamma decay, 421–423, 429
Gamma rays, 112, 394, 428
Gamow, George, 185, 528
Gaussian wave function, *see* Wave function, Gaussian.
Gauss's law, 5
Gay-Lussac's law, 6, 14
Gedanken experiments, 34, 37, 41, 76, 173, 506
Geiger, Hans, 120, 125, 129–130, 147, 396
General relativity, 506–522
  tests of, 510–514, 521
Georgi, Howard, 495
Germer, L. H., 161–164, 183
Gibbs, J. Willard, 6
Glaser, Donald, 498
Glashow, Sheldon, 12, 478, 493, 495
Gluons, 478–479
Grand Unified Theories (GUTs), 13, 495–497
Gravitation, 11–13, 506–522
Gravitational mass, 508–510
Gravitational redshift, 510–513
Gravitational waves, 514, 516–517
Gravitons, 478–479, 516
Ground state, 196, 204–207, 219
Group velocity, 168–171, 183
Groups, 254
Guth, Alan, 544
Gyromagnetic radio, 239

H

Hadrons, 478, 481–484, 491
Hahn, Otto, 450
Hall effect, 292, 370–372, 390
Halogens, 255
Harmonic oscillator, 184, 202–208, 219, 280, 309, 313–314, 330, 359
Hawking, Stephen, 521
Heat capacity (molar), 7–8, 278–280, 290, 294–295, 307
  diatomic gas, 278–279, 307
  ideal gases, 278–279
  solid, 280, 290, 294–295
Heisenberg, Werner, 151, 177, 185, 222, 275, 475
Heisenberg uncertainty principle, 177–182, 205–206, 232, 275, 306, 318, 393, 430
Helium, discovery of, 88–89
Helium, liquid, 298–306, 308, 346–347
  creeping film, 299–300
  density of, 299
  dilution refrigerator, 347
  flow rate, 299–300
Helium, liquid *(continued)*
  specific heat of, 299
Helmholtz, Hermann, 1, 25
Henry, Joseph, 5
Hermite polynomials, 205–207
Hertz, Gustav, 144–146, 150
Hertz, Heinrich, 5, 10, 17, 82, 97
Hess, V. F., 476
Higgs particle, 480
Higgs, Peter, 480
Hofstadter, Robert, 397
Holes, 369–371, 375, 378
Holography, 324–325
  interferometric, 325
  reflection, 325
  transmission, 325
  white light, 325
Homogeneity (of the universe), 544
Hubble constant, 526, 548–551, 555
Hubble, Edwin, 524–527
Hubble space telescope, 518–519, 540, 542, 543, 550
Hubble time, 549–552
Hubble's law, 526, 548, 555
Hund's rules, 260–261, 265
Huygens, Christiaan, 9
Hydrogen atom, 222–248, 272
Hydrogen bond, 312
Hydrogenic atoms, 140, 223
Hyperfine structure, 248
Hyperons, 486

I

Ideal gases, 6–9
  internal energy of, 7, 278
  speed distribution of, 8, 280–283
Impact parameter, 123, 126–128, 148
Indistinguishable particles, 285–287
Inert gases, 255, 269–270
Inertial confinement, 323, 458
Inertial frame, 20, 28, 47, 508
Inertial mass, 508
Infinite square-well potential, 194–198, 219
Inflationary universe, 544–545
Integrated circuits, 387–388
Intrinsic spin, *see* Spin, intrinsic.
Inverse photoelectric effect, 106
Ionic bond, 311–312
Ionization energy, 254
Isobars, 395
Isomers, 422–423
Isospin, 488
Isotones, 395
Isotope effect, superconducting, 340, 357, 360

Isotopes, 395
  hydrogen, 395

J

jj coupling, 264–265
Joliot, F., 394
Jönsson, C., 172–173, 182–183
Josephson, Brian, 351
Josephson effect, 351
Josephson junction, 351–352
  SQUIDS, 352
Joule, James P., 273–274

K

Kelvin, Lord (William Thomson), 1, 5, 16
Kendall, Henry, 491
Kinetic theory of gases, 6–9, 275–283
Knipping, Paul, 152
Kronig, R. de L., 364
Kronig-Penney model, 364–365

L

Landé *g* factor, 266–268
Lanthanides, 256
Laplace, Pierre Simon de, 273, 306
Laplacian operator, 201
Laser cooling, 54–55
Lasers, 54–55, 116, 318–326, 358–359, 521
  free electron, 322–323
  helium-neon, 322
  in surgery, 325
  optical scanners, 326
  tunable, 322
Laughlin, Robert, 370
Law of Atmospheres, 308
Lawrence, Ernest O., 432, 497
Lawson criterion, 454, 457
Lee, Tsung Dao, 487
Lenard, Phillip, 80, 99
Length contraction, 36–39, 43–44, 72, 74–76
Leptons, 480–481, 493
Lie, Sophus, 2
Light
  diffraction of, 10, 87–90, 115–116, 152–153
  gravitational bending, *see* Bending of light,
  retardation, 513–514
  sources, 500
  speed of, 10, 22–56, 58–70, 73–79, 101–112, 323, 359
  Young's double-slit experiment with, 171–172
Light cone, 49–50, 74
Light emitting diode (LED), 377, 391
Line spectra, 87–91, 115–116, 135–136, 140, 149
Line spectra *(continued)*
  helium, 88–89,
  hydrogen, 87–91, 135–136, 140, 149
Liquid helium, *see* Helium, liquid.
Livingston, M. S., 432, 497
Lorentz force law, 5, 71–72, 334
Lorentz, Henrik A., 17, 27, 71
Lorentz transformations, 29–32, 48–49, 75
Lorentz-Fitzgerald contraction, 17, 27 , *see also* Length contraction.
Lorenz number, 332–333
Low-temperature methods, 346–348
LS coupling, *see* Russell-Saunders Coupling
Luminiferous ether, 17, 22–27, 74
Lyman series, 90, 115

M

Mach, Ernst, 15
Madelung constant, 327–328, 359
Maglev, 353
Magnetic moment
  deuteron, 400
  electron, 233–237, 398–399
  hydrogen atom, 233–237, 247
  neutron, 399
  proton, 399
  total, 265–267, 270
Magnetic monopoles, 495, 544
Magnetic properties of solids, 333–336, 359–360
Magnetic resonance imaging, 354–356
Magnetic susceptibility, 333–335, 338, 360
Magnetization, 333–335, 340, 347
Marconi, Guglielmo, 2
Marsden, Ernest, 120, 125, 129–130, 147, 396
Masers, 320
Mass excess, 399
Massless particles, 66
Matrix formulation of quantum theory,185
Matter waves, *see* Particle waves.
Matthias, Berndt, 344
Maxwell, James Clerk, 5, 8, 13, 17, 22, 71, 156–157, 273–275, 285
Maxwell speed distribution, 8, 15, 280–284, 307
  mean speed, 282–283
  most probable speed, 282
  root mean square speed, 283
Maxwell velocity distribution, 275–277, 306
  root mean square speed, 283
Maxwell-Boltzmann (classical) statistics, 8, 284–287, 307, 330, 334, 376
Maxwell's demon, 306
Maxwell's equations, 2, 5, 17, 22, 27–28, 71
Mean free path, 290
Mean values, 192

Meissner effect, 338–340, 357
Meissner, W., 338, 344
Meitner, Lise, 431
Mendeleev, Dmitri, 249–250
Mesons, 476–477, 481–482, 490
  B, 495
  K, 481–482, 491
  pion, 45, 476–477, 490–491
Metallic bond, 312
Metastable states, 263, 320–322
Michelson, Albert A., 17, 20, 22, 25–27, 74, 85
Michelson interferometer, 22–27, 317
Michelson-Morley experiment, 17, 22–27, 46, 71, 74–75, 80
Microkelvin laboratory, 348
Millikan oil drop experiment, 84–87, 114–115
Millikan, Robert A., 85–86, 102, 114
Minkowski diagrams, *see* Spacetime diagrams.
Minkowski, H. A., 48
Missing mass problem, 546
Molecular bonding and spectra, 311–318, 357–358
Momentum, conservation of, 3, 57–60, 77
Momentum–energy relation, in relativity, 65
Morley, Edward, 17, 25–27, 74
Moseley, H. G. J., 125, 142–144, 147, 149
Moseley plot, 142–143
Mössbauer effect, 511, 521
Mössbauer, R. L., 511
Müller, Karl Alex, 345
Multiplicity, 261
Muon, 477, 480–481
Muon decay, 43–44, 481
Muonic atom, 248

N

Ne'eman, Yuval, 487
Neutrino, 416–419, 480–481, 542–543, 546
Neutrino astronomy, 542–543
Neutron
  discovery of, 393–395, 428
  magnetic moment, 399
Neutron activation, 441, 460, 462
Neutron star, 514, 536–537
Newton, Isaac, 3–4, 9, 12, 74, 273, 521
Newton's laws of motion, 3–4, 20, 30, 32, 40, 58, 74–75, 273, 508, 521
Newton's *Opticks*, 9
Normalization, 176, 183–184, 188–189, 195–196, 207, 218–219, 227, 247
Nuclear
  binding energy, 399–408
  charge distribution, 397
  fusion, *see* Fusion.
  magneton, 398–399
  mass, 399–400
Nuclear *(continued)*
  mass density, 398
  models, 406–408
  photodisintegration, 400
  reactions, 431–436, 471
  stability, 402–408, 428–429
  volume, 398
Nuclear force, *see* Strong force.
Nuclear magnetic resonance,
Nuclear radius, 396–397
Nucleons, 395
Nucleosynthesis, 528–530
Nucleus
  designation, 395
  excited states, 421–423
  existence of, 122, 147
  magnetic moment, 398–399, 428
  size, 396
  spin, 398–400
Nuclides,
  even-even,404–405
  even-odd, 404–405
  odd-odd, 404–405

O

Occhialini, G. P., 477
Oersted, Christian, 5
Onnes, Heike Kamerlingh, 298–299, 337, 344, 347, 356
Open universe, 544, 549, 552–553
Operators, 193–194
  energy, 194
  momentum, 193–194, 201

P

Pair production and annihilation, 110–113, 117
Paramagnetism, 334–335, 360
Parsec, 527
Particle detectors, 418–419
Particle waves, 159–161, 182–184
Particles, 482–484
  composite, 484
  fundamental, 484
Paschen series, 90–91, 115
Pasteur, Louis, 1
Pauli exclusion principle, 250, 260, 265, 284, 286, 295, 301, 328, 357
Pauli, Wolfgang, 416, 494
Peebles, P. James E., 527
Peierls, Rudolf, 392
Penney, W. P., 364
Penzias, Arno, 524, 527–528
Perihelion shift of Mercury, 512
Periodic table, 249–256, 269–270

Periods, 254
Perrin, Jean, 16
PET scan, *see* Positron Emission Tomography.
Pfund series, 90, 115
Phase space, 275, 281, 291, 297
Phase velocity, 166, 168–171, 183
Phonons, 341
Photodisintegration, 400, 428
Photoelectric effect, 97–103, 110, 115–116, 171
Photons, 65–66, 101–117, 135, 151, 174, 242, 318–321, 358, 377–378, 394, 478
Photonuclear reaction, *see* Photodisintegration
Photovoltaic cells, 378–381, 390–391
Physical observables, 174, 192
Pi meson, *see* Meson, Pion.
Pierels, Rudolf, 392
Planck law, *see* Blackbody radiation, Planck law for.
Planck length, 522, 535
Planck mass, 522, 535
Planck, Max, 18, 80, 94–95, 100–101, 133
Planck time, 522, 535
Planck's constant, 95–96, 101–112, 117, 133–136, 150, 160–161, 173–175, 177–181, 184, 370–371, 511
Plasma, 287
Plum pudding model, 119, 131, 147–148
p-n–junction diode, 375–376
Population inversion, 321
Positron Emission Tomography (PET), 112–113, 459–460
Positronium, 112, 149
Positrons, 110–113, 117, 149, 475
Potential barriers, *see* Barriers.
Powell, C. P., 477
Principle of Complementarity, 174, 180
Principle of Equivalence, 507–510, 521
Principle of Superposition, 166, 187
Probability, 175–177, 183–184, 188–191, 200, 218, 228, 242–246, 248
  radial, 242–246, 248
Probability density, 175, 228, 242–246, 248
Proper length, 36–39
Proper time, 33–36
Proton decay, 495
Proton–proton chain, 453, 533
Pulsar,537

Q

Q value, 436–347
Quantized energy levels, 135–136, 138, 141–142, 149–150, 176, 196, 201–202, 205, 228–229, 312–318, 357–358
Quantum chromodynamics (QCD), 491
Quantum electrodynamics (QED), 46, 71
Quantum fluxoid, 342
Quantum Hall effect, 370–371
Quantum numbers, 134–135, 196, 202, 205, 219–220, 225–233, 241–242
  intrinsic spin quantum number, 238–239, 241–242
  magnetic quantum number, 231–237
  magnetic spin quantum number, 228, 238–239, 241–242, 245–248
  orbital angular momentum quantum number, 225, 228–229, 233–234, 241–242, 245–248
  principal quantum number, 134–135, 228–229, 241–242, 247–248
  total angular momentum, 257–265
Quantum statistics, 285–309
Quarks, 91, 115, 306, 396, 487–493, 504
  bottom, 489–490
  charmed, 498–490
  top, 489–490
Quasars, 538, 543

R

Radar, 53–54, 513
Radial equation, 225–227
Radioactive dating, 423–427, 429
Radioactivity, 18, 408–430
  activity, 409, 411
  artificial, 423
  branching, 424
  decay law, 409
  half-life, 410
  lifetime, 410
  natural, 423
Radioisotope, 458
Raman scattering, 317–318
Range parameter, *see* Repulsive range parameter.
Rayleigh, Lord (John William Strutt), 18, 156–158
Rayleigh scattering, 317
Rayleigh-Jeans law, 115
Rayleigh's criterion, 184
Reaction kinematics, 436–438, 471
Reaction mechanisms, 438–442, 471–472
  compound nucleus, 438
  direct reactions, 438, 442
Recession velocity, 524–527
Red giant, 552
Redshifts, 53–54, 525–526, 543
Reduced mass, 139, 150, 223
Reflection and transmission, 209–213, 220
Reines, Frederick, 416
Relativistic energy, 60–66
Relativistic force, 60
Relativistic kinetic energy, 61–64, 69, 78, 181
Relativistic mass, 59
Relativistic momentum, 57–60, 65, 78, 107–108, 181
Repulsive range parameter, 328–329
Resonances, 441

Rest energy, 63–65, 67–70, 78, 110–112
Rest mass, 59
Rigid rotator, 277, 312–313
Röntgen, Wilhelm, 81–82
Rotational (molecular) states and spectra, 312–318, 357–358
Rubbia, Carlo, 478, 499
Rumford, Count (Benjamin Thompson), 5, 273
Russell-Saunders coupling, 261–263
Rutherford, Ernest (Lord), 119–125, 129–131, 133, 142, 147, 156–158, 392–393, 431
Rutherford scattering, 122–131, 147–149
Rydberg atoms, 230–231
Rydberg constant, 90, 135, 139–140, 149
  for hydrogen, 90, 139
Rydberg equation, 90, 135, 139–140, 149

S

Salam, Abdus, 12, 478, 493
Scanning tunneling microscope, 214–215
Schawlow, Arthur, 325, 356
Schottky barrier, 384–385
Schrieffer, J. Robert, 341
Schrödinger, Erwin, 185–186, 475
Schrödinger wave equation
  in spherical coordinates, 223–225
  three-dimensional, 201–222
  time-dependent, 186–189
  time-independent, 190–191, 195, 198–199, 204, 209, 218
Schwarzschild, Karl, 515
Schwarzschild radius, 515, 521–522
Schwinger, Julian, 475
Secondary emission (of electrons), 97
Secular equilibrium, 430
Segrè, E.G., 112
Selection rules, 241–242
Semiconductor devices, 211, 218, 375–391
Semiconductors, 361–391
  electrical conductivity and resistivity, 362–363, 370–371
  impurity, 369–371
  intrinsic, 369
  n-type, 369–371, 375–377, 382, 390
  p-type, 370–371, 375–377, 382, 390
  theory of, 366–374, 389–390
Semi-empirical mass formula, 405
Semi-infinite potential well, 220
Semimetals, 366
Separation energy, 402
Separation of variables, 190, 224–226
Shockley, William, 362, 381, 383
Simple harmonic oscillator, *see* Harmonic oscillator.
Simultaneity, 28
Singlet states, 261–262, 270
Slipher, V. M., 525
Snow, C. P., 431
Solar cells, *see* Photovoltaic cells.
Solar constant, 378
Solar neutrinos, 534
Sommerfeld, Arnold, 140, 152, 295
Space Telescope, Future, 550–551
Space quantization, 232
Spacetime diagram, 48–51, 77
Spacetime intervals, 50–51
Spacetime invariant, 50
Speed of light, *see* Light, speed of.
Spherical harmonics, 227–228, 245–246
Spherical polar coordinates, 223–228, 242, 247
Spin, intrinsic, 238–240, 140, 248, 250–251, 286, 416, 480, 482–483
Spin–orbit interaction, 258
Spontaneous emission, 318
Square-well potential
  finite, 198–200, 219
  one-dimensional infinite, 194–198, 219
  three-dimensional infinite, 200–202, 219
Standard model, 494–495, 504
Standing waves, 161, 196, 201
Stark effect, 140, 230–231
Stationary states, 133, 191
Steady state theory, 524
Stefan-Boltzmann law, *see* Blackbody Radiation, Stefan-Boltzman law.
Stellar aberration, 26–27
Stellar evolution, 532–537, 555
Stern-Gerlach experiment, 237, 239, 248
Stimulated emission, 318–320, 358–359
Stokes's law, 85, 115
Stopping potential, 98–103, 116–117
Störmer, Horst, 370
Strangeness, 486, 490
Strassman, Fritz, 450
Strong (nuclear) force, 11–13, 399–402, 184, 428
  charge independence, 402
  short range, 401
Subshells, 250–256, 269–270
Sun
  temperature of, 93
Superconductivity, 336–357, 360
  BCS, theory of, 341–343
  energy gap, 342–344
  high temperature, 344–345, 348–350
  superconducting fullerenes, 350–351
  transition temperatures and critical fields, table, 338
  type I and type II, 340
Superfluidity, 298–304, 306
  Superfluid ^{3}He, 302–303
Supernovae, 541–542, 547
Superposition principle, *see* Principle of Superposition.

Superstrings, 496
Supersymmetry, 496
Symmetries, 487
  charge conjugation, 487
  parity, 487
  time reversal, 487
Synchronization of clocks, 29, 74
Synchrotron radiation, 500
Szilard, Leo, 450

T

Taylor, Richard, 491
Taylor series, 203, 330
Theory of everything, 530
Thermal conductivity, 269, 331–333
Thermal expansion, 329–331, 359
Thermionic emission (of electrons), 97, 144
Thermocouple, 374, 390
Thermodynamics, 5–6, 306
  First Law of, 5
  Second Law of, 5–6, 306, 357
  Third Law of, 6
  Zeroth Law of, 6
Thermoelectric effect, 373–374
Thomas factor, 270
Thomson, George P., 164
Thomson, J. J., 18, 81–84, 114, 119–120, 124, 131–132, 156–158, 164, 475
Thomson scattering, 106, 110
Thomson, William, *see* Kelvin.
Three-dimensional infinite potential well, 200–202, 219
Three-level system, 320–321, 357
Threshold energy, 437, 471
Threshold frequency, 98–99, 103
Tidal forces, 508
Time dilation, 33–36, 43–45, 47–76, 512
Tinkham, Michael, 343
Tokamak, 455–458
Tomography, 459–460
Tomonaga, Sin-itiro, 475
Topness, 489
Transistors, 381–384
  field effect (FET), 371, 384
  metal oxide semiconductor FET (MOSFET), 384
  metal semiconductor FET (MESFET), 385
  ohmic contact, 385
Transition metals, 256
Transition probabilities, 241
Transitions, allowed and forbidden, 241
Transmission, *see* Reflection and transmission.
Transuranic elements, 467–469
Triplet states, 261–262, 270
Tritium, 395
Truth, 489
Tsui, Daniel, 370
Tunnel diodes, 217
Tunneling, 210–217, 200, 220
Twin paradox, 46–48, 56, 77

U

Ultraviolet catastrophe, 18, 94
Uncertainty principle, *see* Heisenberg uncertainty principle.
Unified field theory, 13, 495

V

Valence, 119
Valence band, 369, 378, 386, 390
Van de Graaff accelerator, 432
Van de Graaff, Robert, 432
Van der Meer, Simon, 499
Van der Waals bond, 312
Velocity addition, *see* Addition of velocities.
Vibrational (molecular) states and spectra, 313–316, 357–358
von Klitzing, Klaus, 370
von Laue, Max, 152, 326
von Weizäcker, Carl, 405

W

Walton, E. T. S., 156, 158, 392, 432
Wave equation, Schrödinger, *see* Schrödinger wave equation.
Wave equations, 186
  classical, 166, 186
Wave function, 165–169, 175–177, 183–228, 242–248, 363–364
  Gaussian, 169–170, 177–178, 183
  radial, 226–228, 242–245, 247
Wave mechanics, 11, 185
Wave motion, 165–171, 183
Wave number, 166–171, 183, 187–188, 364–365
Wave packet, 166–171, 177–179, 187–189
Wave vector, 365
Wave–particle duality, 9–11, 171–174, 183, 186
Weak interaction, 12–13, 478–479
Weber, Joseph, 514, 516
Weinberg, Steven, 12, 478, 493
Wheeler, John, 450, 506
White dwarf, 537
Wiedemann-Franz law, 332–333
Wien law, *see* Blackbody Radiation, Wien law.

Wien, Wilhelm, 152
Wilson, C. T. R., 156, 158
Wilson, Robert, 524, 527–528
Work function, 97–103, 106, 116–117, 145–146
Worldlines, 48–49
Wu, C. S., 487

X

X rays, 18, 80–81, 104–106, 115–116, 141–144, 149–150, 152–155,159, 250
  Bragg method, 153–155, 159
  characteristic, 105, 141–144, 149–150
  Laue method, 152–153
  scattering, 152–155, 159, 182

Y

Yang, Chen Ning, 487
Young, Thomas, 5, 9, 22
Young's double-slit experiment, *see* Light, Young's double-slit experiment with.
Yukawa, Hideki, 476–477

Z

Zeeman effect, 18, 140, 233–236, 247, 265–271
  anomalous, 265–271
  normal, 140, 233–236, 247, 265–267, 270–271
Zeeman, Pieter, 18, 233
Zener diodes, 377
Zero-point energy, 205, 314–315, 329, 342
Zweig, George, 488